Image Principles, Neck, and the Brain

Magnetic Resonance Imaging Handbook

Image Principles, Neck, and the Brain
Imaging of the Cardiovascular System, Thorax, and Abdomen
Imaging of the Pelvis, Musculoskeletal System, and Special Applications to CAD

MAGNETIC RESONANCE IMAGING HANDBOOK

Image Principles, Neck, and the Brain

edited by
Luca Saba

CRC Press
Taylor & Francis Group
Boca Raton London New York

CRC Press is an imprint of the
Taylor & Francis Group, an **informa** business

CRC Press
Taylor & Francis Group
6000 Broken Sound Parkway NW, Suite 300
Boca Raton, FL 33487-2742

First issued in paperback 2019

© 2016 by Taylor & Francis Group, LLC
CRC Press is an imprint of Taylor & Francis Group, an Informa business

No claim to original U.S. Government works

ISBN-13: 978-0-4822-1613-4 (hbk)
ISBN-13: 978-0-367-86889-8 (pbk)

Library of Congress Cataloging-in-Publication Data

Names: Saba, Luca, editor.
Title: Magnetic resonance imaging handbook / [edited by] Luca Saba.
Description: Boca Raton : Taylor & Francis, 2016. | Includes bibliographical references and index.
Identifiers: LCCN 2015043923| ISBN 9781482216288 (set : hardcover : alk. paper) | ISBN 9781482216134 (v. 1 : hardcover : alk. paper) |
    ISBN 9781482216264 (v. 2 : hardcover : alk. paper) | ISBN 9781482216219 (v. 3 : hardcover : alk. paper)
Subjects: | MESH: Magnetic Resonance Imaging.
Classification: LCC RC386.6.M34 | NLM WN 185 | DDC 616.07/548--dc23
LC record available at http://lccn.loc.gov/2015043923

Visit the Taylor & Francis Web site at
http://www.taylorandfrancis.com

and the CRC Press Web site at
http://www.crcpress.com

This book is dedicated to Giovanni Saba.

Thank you.

Start by doing what's necessary; then do what's possible; and suddenly you are doing the impossible.

Francis of Assisi (1181–1226)

The whole is more than the sum of its parts.

Aristotle, Greek philosopher (ca. 384–322 BC)

Contents

Preface

Magnetic resonance imaging (MRI) is a medical imaging technique used in radiology to visualize internal structures of the body in detail. The introduction of MRI resulted in a fundamental and far-reaching improvement of the diagnostic process because this technique provides an excellent contrast between the different soft tissues of the body, which makes it especially useful in imaging the brain, muscles, heart, and cancers compared with other medical imaging techniques such as computed tomography or X-rays.

In the past 20 years, MRI technology has further improved with the introduction of systems up to 7 T and with the development of numerous postprocessing algorithms such as diffusion tensor imaging (DTI), functional MRI (fMRI), and spectroscopic imaging.

From these developments, the diagnostic potentialities of MRI have impressively improved with exceptional spatial resolution and the possibility of analyzing the morphology and function of several kinds of pathology.

The purpose of this book is to cover engineering and clinical benefits in the diagnosis of human pathologies using MRI. It will cover the protocols and potentialities of advanced MRI scanners with very high-quality MR images. Given these exciting developments in the MRI field, I hope that this book will be a timely and complete addition to the growing body of literature on this topic.

Luca Saba
University of Cagliari, Italy

Acknowledgments

It is not possible to overstate my gratitude to the many individuals who helped to produce this book; their enthusiasm and dedication were unbelievable.

I express my appreciation to CRC Press/Taylor & Francis Group for their professionalism in handling this project. Your help was wonderful and made producing this book an enjoyable and worthwhile experience.

Finally, I acknowledge Tiziana.

Editor

Professor Luca Saba earned his MD from the University of Cagliari, Italy, in 2002. Currently, he works at the Azienda Ospedaliero Universitaria of Cagliari. His research is focused on multidetector-row computed tomography, magnetic resonance, ultrasound, neuroradiology, and diagnostics in vascular sciences.

Professor Saba has published more than 180 papers in high-impact factor journals such as the *American Journal of Neuroradiology, Atherosclerosis, European Radiology, European Journal of Radiology, Acta Radiologica, Cardiovascular and Interventional Radiology, Journal of Computer Assisted Tomography, American Journal of Roentgenology, Neuroradiology, Clinical Radiology, Journal of Cardiovascular Surgery, Cerebrovascular Diseases, Brain Pathology, Medical Physics,* and *Atherosclerosis.* He is a well-known speaker and has spoken over 45 times at national and international conferences.

Dr. Saba has won 15 scientific and extracurricular awards during his career, and has presented more than 500 papers and posters at national and international congress events (Radiological Society of North America [RSNA], ESGAR, ECR, ISR, AOCR, AINR, JRS, Italian Society of Radiology [SIRM], and AINR). He has written 21 book chapters and is the editor of 10 books in the fields of computed tomography, cardiovascular surgery, plastic surgery, gynecological imaging, and neurodegenerative imaging.

He is a member of the SIRM, European Society of Radiology, RSNA, American Roentgen Ray Society, and European Society of Neuroradiology, and serves as the reviewer of more than 40 scientific journals.

Contributors

Fatih Alper
Department of Radiology
Ataturk University
Erzurum, Turkey

Jalal B. Andre
Harborview Medical Center
University of Washington School
   of Medicine
Seattle, Washington

Luigi Barberini
Department of Public Health
   Clinical and Molecular Medicine
University of Cagliari
Cagliari, Italy

and

Department of Radiology
Cittadella Universitaria di
   Monserrato
Monserrato, Italy

Roland Beisteiner
Department of Biomedical Imaging
   and Image-Guided Therapy
High Field MR Centre
and
Department of Neurology
Study Group Clinical fMRI
Medical University of Vienna
Vienna, Austria

Bavrina Bigjahan
Division of Neuroradiology
Department of Radiology
Keck School of Medicine
University of Southern California
Los Angeles, California

Wolfgang Bogner
Department of Biomedical Imaging
   and Image-Guided Therapy
High Field MR Centre
Medical University of Vienna
Vienna, Austria

Klaus Bohndorf
Department of Biomedical Imaging
   and Image-Guided Therapy
High Field MR Centre
Medical University of Vienna
Vienna, Austria

John Carr
Department of Radiological
   Sciences
University of California
Los Angeles, California

Antonia Ceccarelli
Department of Neurology
Icahn School of Medicine at
   Mount Sinai
New York, New York

Joan Cheng
Department of Radiology
Boston Medical Center
Boston University School of
   Medicine
Boston, Massachusetts

Marek Chmelík
Department of Biomedical Imaging
   and Image-Guided Therapy
High Field MR Centre
Medical University of Vienna
Vienna, Austria

Chris J. Conklin
Magnetic Resonance Imaging
   Physics Lab
Thomas Jefferson University
Philadelphia, Pennsylvania

Francesco D'Amore
Division of Neuroradiology
Department of Radiology
Keck School of Medicine
University of Southern California
Los Angeles, California

Ivana Delalle
Department of Pathology and
   Laboratory Medicine
Boston Medical Center
Boston University School of
   Medicine
Boston, Massachusetts

Irmak Durur-Subasi
Department of Radiology
Ataturk University
Erzurum, Turkey

Barbara Dymerska
Department of Biomedical Imaging
   and Image-Guided Therapy
High Field MR Centre
Medical University of Vienna
Vienna, Austria

Scott H. Faro
Departments of Radiology,
   Computer Engineering, and
   Bioengineering
Temple University
Philadelphia, Pennsylvania

Florian Fischmeister
Department of Biomedical Imaging
   and Image-Guided Therapy
High Field MR Centre
and
Department of Neurology
Study Group Clinical fMRI
Medical University of Vienna
Vienna, Austria

Samuel Frank
Department of Neurology
Boston Medical Center
Boston University School of
   Medicine
Boston, Massachusetts

Hiroyuki Fujii
Department of Radiology
Jichi Medical University
Tochigi, Japan

Akifumi Fujita
Department of Radiology
Boston Medical Center
Boston University School of
    Medicine
Boston, Massachusetts

and

Department of Radiology
Jichi Medical University School
    of Medicine
Shimotsuke, Japan

Günther Grabner
Department of Biomedical Imaging
    and Image-Guided Therapy
High Field MR Centre
Medical University of Vienna
Vienna, Austria

Stephan Gruber
Department of Biomedical Imaging
    and Image-Guided Therapy
High Field MR Centre
Medical University of Vienna
Vienna, Austria

Rehana Hafeez
Department of Surgery
Princess Royal University Hospital
Kent, United Kingdom

Sven Haller
Division of Interventional and
    Diagnostic Neuroradiology
Geneva University Hospitals
Geneva, Switzerland

Pradipta C. Hande
Department of Imaging
Breach Candy Hospital Trust
Mumbai, India

Gilbert Hangel
Department of Biomedical Imaging
    and Image-Guided Therapy
High Field MR Centre
Medical University of Vienna
Vienna, Austria

Valentina Hartwig
Italian National Research Council
    (CNR)
Institute of Clinical Physiology
    (IFC)
San Cataldo (Pisa), Italy

Michael E. Hayden
Physics Department
Simon Fraser University
Burnaby, British Columbia, Canada

Hongjian He
Center for Brain Imaging Science
    and Technology
Zhejiang University
Hangzhou, China

Ellen Hoeffner
Department of Radiology
University of Michigan Medical
    School
Ann Arbor, Michigan

Michael N. Hoff
Department of Radiology
University of Washington School
    of Medicine
Seattle, Washington

Matilde Inglese
Department of Neurology,
    Radiology and Neuroscience
Neurology Imaging Laboratory
Icahn School of Medicine at Mount
    Sinai
New York, New York

Willa Jin
Division of Neuroradiology
Department of Radiology
Keck School of Medicine
University of Southern California
Los Angeles, California

Vladimir Juras
Department of Biomedical Imaging
    and Image-Guided Therapy
High Field MR Centre
Medical University of Vienna
Vienna, Austria

Elzbieta Jurkiewicz
Department of Diagnostic Imaging
The Children's Memorial Health
    Institute
Warsaw, Poland

Shahmir Kamalian
Division of Neuroradiology
Department of Radiology
University of Massachusetts School
    of Medicine
Worcester, Massachusetts

Adem Karaman
Department of Radiology
Ataturk University
Erzurum, Turkey

Yukio Kimura
Department of Radiology
Jichi Medical University
Tochigi, Japan

Sara E. Kingston
Division of Neuroradiology
Department of Radiology
Keck School of Medicine
University of Southern California
Los Angeles, California

Gabriele A. Krombach
Department of Diagnostic and
    Interventional Radiology
University Hospital Giessen
Justus Liebig University Giessen
Giessen, Germany

Claudia Kronnerwetter
Department of Biomedical Imaging
    and Image-Guided Therapy
High Field MR Centre
Medical University of Vienna
Vienna, Austria

Martin Krššák
Department of Biomedical Imaging
  and Image-Guided Therapy
High Field MR Centre
Medical University of Vienna
Vienna, Austria

Alexander Lerner
Division of Neuroradiology
Department of Radiology
Keck School of Medicine
University of Southern California
Los Angeles, California

Ravi Kumar Lingam
Department of Radiology
Northwick Park and Central
  Middlesex Hospitals
London North West Hospitals NHS
  Trust
London, United Kingdom

Chia-Shang J. Liu
Division of Neuroradiology
Department of Radiology
Keck School of Medicine
University of Southern California
Los Angeles, California

Karl-Olof Lövblad
Division of Interventional and
  Diagnostic Neuroradiology
Geneva University Hospitals
Geneva, Switzerland

Jesica Makanyanga
Centre for Medical Imaging
University College London
London, United Kingdom

Eva Matt
Department of Biomedical Imaging
  and Image-Guided Therapy
High Field MR Centre
and
Department of Neurology
Study Group Clinical fMRI
Medical University of Vienna
Vienna, Austria

Devon M. Middleton
Department of Radiology
and
Department of Bioengineering
Temple University
Philadelphia, Pennsylvania

Lenka Minarikova
Department of Biomedical Imaging
  and Image-Guided Therapy
High Field MR Centre
Medical University of Vienna
Vienna, Austria

Feroze B. Mohamed
Magnetic Resonance Imaging
  Physics Lab
Thomas Jefferson University
Philadelphia, Pennsylvania

Pierre-Jean Nacher
Laboratoire Kastler Brossel
ENS-PSL Research University
CNRS, UPMC-Sorbonne
  Universités
Collège de France
Paris, France

Takashi Nakamura
Department of Radiology and
  Cancer Biology
Nagasaki University School of
  Dentistry
Nagasaki, Japan

Megha Nayyar
Division of Neuroradiology
Department of Radiology
Keck School of Medicine
University of Southern California
Los Angeles, California

Katarzyna Nowak
Department of Diagnostic Imaging
The Children's Memorial Health
  Institute
Warsaw, Poland

Douglas Pendse
Centre for Medical Imaging
University College London
London, United Kingdom

Vitor Mendes Pereira
Division of Interventional and
  Diagnostic Neuroradiology
Geneva University Hospitals
Geneva, Switzerland

Supada Prakkamakul
Department of Radiology
King Chulalongkorn Memorial
  Hospital
The Thai Red Cross Society
Bangkok, Thailand

Eytan Raz
Department of Radiology
NYU Langone Medical Center
New York, New York

Ahmed Abdel Khalek Abdel Razek
Department of Diagnostic
  Radiology
Mansoura University
Mansoura, Egypt

John H. Rees
Department of Radiology
Partners Imaging Center
Sarasota, Florida

and

Department of Radiology
Georgetown University
Washington, DC

Simon Robinson
Department of Biomedical Imaging
  and Image-Guided Therapy
High Field MR Centre
Medical University of Vienna
Vienna, Austria

Naoko Saito
Department of Radiology
Saitama International Medical
    Center
Saitama Medical University
Saitama, Japan

Osamu Sakai
Departments of Radiology,
    Otolaryngology, Head and Neck
    Surgery, and Radiation Oncology
Boston Medical Center
Boston University School of
    Medicine
Boston, Massachusetts

Benjamin Schmitt
Healthcare Sector
Siemens Australia
New South Wales, Australia

Mark S. Shiroishi
Division of Neuroradiology
Department of Radiology
Keck School of Medicine
University of Southern California
Los Angeles, California

James G. Smirniotopoulos
MedPix®
National Library of Medicine
National Institutes of Health
Bethesda, Maryland

Brent K. Stewart
Department of Radiology
University of Washington School
    of Medicine
Seattle, Washington

Bernhard Strasser
Department of Biomedical Imaging
    and Image-Guided Therapy
High Field MR Centre
Medical University of Vienna
Vienna, Austria

Misa Sumi
Department of Radiology
    and Cancer Biology
Nagasaki University School
    of Dentistry
Nagasaki, Japan

Kyunghyun Sung
Department of Radiological
    Sciences
University of California
Los Angeles, California

Benita Tamrazi
Division of Neuroradiology
Department of Radiology
Keck School of Medicine
University of Southern California
Los Angeles, California

Stuart A. Taylor
Centre for Medical Imaging
University College London
London, United Kingdom

Bruno A. Telles
Division of Neuroradiology
Department of Radiology
Keck School of Medicine
University of Southern California
Los Angeles, California

Joshua Thatcher
Department of Radiology
Boston Medical Center
Boston University School of
    Medicine
Boston, Massachusetts

Henrik S. Thomsen
Department of Diagnostic
    Radiology
Copenhagen University Hospital
    Herlev
Herlev, Denmark

Siegfried Trattnig
Department of Biomedical Imaging
    and Image-Guided Therapy
High Field MR Centre
Medical University of Vienna
Vienna, Austria

Daniel S. Treister
Division of Neuroradiology
Department of Radiology
Keck School of Medicine
University of Southern California
Los Angeles, California

Ioannis Tsougos
Department of Medicine
University of Thessaly
Larissa, Greece

Ram Vaidhyanath
Department of Radiology
University of Leicester
Leicester, United Kingdom

Maria Isabel Vargas
Division of Interventional and
    Diagnostic Neuroradiology
Geneva University Hospitals
Geneva, Switzerland

Memi Watanabe
Department of Radiology
Boston Medical Center
Boston University School of
    Medicine
Boston, Massachusetts

Albert J. Yoo
Division of Neurointervention
Texas Stroke Institute
Plano, Texas

Štefan Zbýň
Department of Biomedical Imaging
    and Image-Guided Therapy
High Field MR Centre
Medical University of Vienna
Vienna, Austria

Jianhui Zhong
Center for Brain Imaging Science
    and Technology
Zhejiang University
Hangzhou, China

and

Department of Imaging Sciences
University of Rochester
Rochester, New York

1

History and Physical Principles of Magnetic Resonance Imaging

Michael E. Hayden and Pierre-Jean Nacher

CONTENTS

Magnetic resonance imaging (MRI) was broadly introduced to the scientific community in 1973, when Paul C. Lauterbur published images representing the nuclear magnetic resonance (NMR) response of hydrogen nuclei in a pair of water-filled glass capillaries [1]. One-dimensional (1D) projections of this response were first obtained through a procedure that involved applying static magnetic field gradients to the sample, mapping NMR frequency onto the source position. A series of 1D projections, acquired along different gradient directions, were then combined to reconstruct a two-dimensional (2D) image, as illustrated in Figure 1.1.

separation in liquid ^{3}He–^{4}He mixtures at low temperatures [3]. However, it was Lauterbur who extended the method to two dimensions and recognized its potential for soft tissue imaging. Perhaps the most well-known 1D NMR images predating Lauterbur's 1973 paper were published just one year earlier, in connection with a Cornell University study of another low-temperature-phase transition, this time in pure liquid ^{3}He [4]. Three of the four authors of that report were awarded the 1996 Nobel Prize in Physics for the discovery of superfluidity in ^{3}He.

The term *Zeugmatography* remains obscure, but *MRI*—the field that emerged—is anything but.

A NEW WORD FOR A NEW WAY OF SEEING

Lauterbur coined the term *zeugmatogram* to describe his NMR images. This word is derived from the Greek ζευγμα ("zeugma"), meaning "that which is used for joining," in reference to the manner in which static magnetic field gradients were employed to localize the sample response to oscillating magnetic fields. Similar 1D NMR imaging methods had already been demonstrated as early as 1952 [2] and were employed in the 1956 discovery of phase

Lauterbur's simple but insightful demonstration launched a flurry of scientific, industrial, and clinical activity that has since profoundly influenced the practice and delivery of medicine in industrialized countries. Sophisticated extensions of his work are now fueling revolutions in neuroscience and our understanding of cognition.

This chapter starts by tracing the history of MRI from its roots in the field of NMR through to the present.

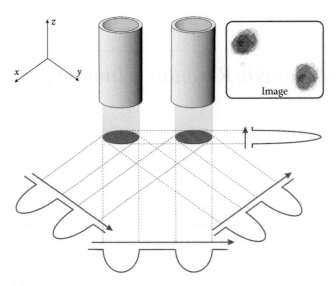

FIGURE 1.1
Principle underlying the first MRI experiment, performed by P.C. Lauterbur [1]. Two objects (water-filled capillaries) aligned with the z-axis are shown, along with their projection onto the x–y plane. Magnetic field gradients applied along various directions cause the NMR response to spread out in frequency, producing 1D projections reflecting the distribution of water (blue curves). Multiple projections, acquired along different gradient directions (indicated by red arrows), are then combined to reconstruct a 2D image. Inset: Lauterbur's NMR image of two 1 mm inner-diameter water-filled capillaries. (Data from Lauterbur, P., *Nature*, 242, 190–191, 1973, reprinted with permission from Nature Publishing Group.)

It is a story that is much richer and nuanced than can be adequately described in a few pages; our narrative certainly ignores many critical contributions to the field. The basic physical principles of NMR are then introduced; these form the basis for understanding the "NMR response" referred to above. Again, the treatment presented here is necessarily brief. The interested reader is encouraged to consult some of the excellent and extensive monographs that have been written on this topic [5–10]. Finally, the basic principles underlying magnetic resonance (MR) image generation itself are introduced; many of these topics are covered in greater detail in subsequent chapters, but again the interested reader will find valuable additional information about the underlying physics in more specialized references [11–15].

1.1 History

The foundations of NMR—and hence MRI—were laid during the 1940s, in experiments designed to directly detect the precession of nuclear magnetic moments in a magnetic field [16–18]. Those experiments, which

involved hydrogen atoms in liquids and solids, built on work carried out during the 1930s at Columbia University. There, a team led by Isidore I. Rabi showed that an oscillating magnetic field could be used to induce transitions between nuclear spin states of lithium and chlorine atoms in a molecular beam [19]. Rabi's pioneering experiments in turn employed spin-state selection and detection techniques similar to those developed in Frankfurt during the 1920s by Otto Stern and Walther Gerlach, in connection with their seminal discovery of spin quantization using a beam of silver atoms [20].

The extension of Rabi's 1938 observation of NMR in a beam of independent molecules to solid and liquid samples was successfully, independently, and essentially simultaneously accomplished in 1945 by Edward M. Purcell and Felix Bloch. A key feature of these experiments was the fact that both employed direct electromagnetic detection techniques to resolve the resonances. At the Massachusetts Institute of Technology (MIT) Radiation Laboratory, Purcell, Torrey, and Pound worked with 1 L of solid paraffin in a cavity tuned to resonate at 30 MHz. They observed a 0.4% change in radiofrequency (RF) signal amplitude across the cavity as the static magnetic field was swept through "an extremely sharp resonance"; this reduction in quality factor was attributed to energy dissipation associated with nuclear spin relaxation of H atoms [16]. Meanwhile at Stanford University, Bloch, Hansen, and Packard performed similar experiments on a 1.5 cm³ sample of water at 7.7 MHz. They used two orthogonal RF coils; the receive coil detected RF power when the nuclei of the water protons (H atoms) were resonantly excited by the transmit coil [18]. Although Rabi's work was crucial as the initial demonstration of NMR (he was awarded the Nobel Prize in Physics in 1944 "for his resonance method for recording the magnetic properties of atomic nuclei"), the conceptual and technical leap achieved by Bloch and Purcell really set the stage for the development of modern NMR and MRI.

EARLY ATTEMPTS AND FIRST SUCCESSES

The first reported attempt to observe nuclear spin transitions in solids was published in 1936 by Cornelius J. Gorter [21], who was based in Leiden. That experiment failed, as did a later attempt described in a 1942 paper [22]. Gorter's second paper contains the first published reference to *nuclear magnetic resonance*, a term that he attributed to Rabi. Meanwhile in Kazan, Yevgeny Zavoisky also failed to reliably detect NMR transitions in solids and liquids, but went on to discover electron spin resonance (ESR) in 1944.

I.I. Rabi   C.J. Gorter   Y.K. Zavoisky

Reprinted with permission from (l-r): The Nobel Foundation; Eddy de Jongh; World Scientific.

The first truly successful NMR experiments on solids and liquids were reported in early 1946 [16,17], by two independent teams. One of these teams was led by Felix Bloch at Stanford University. Bloch obtained a PhD from the University of Leipzig in 1928. He left Germany in 1933 and moved to Stanford University, where he spent most of his career. During the latter part of World War II, he spent time at the Harvard Radio Research Laboratory, where he worked on counter-radar measures and became acquainted with modern developments in electronics. The other team was led by Edward M. Purcell at the MIT. Purcell obtained a PhD from Harvard in 1938. He spent the war years at the MIT Radiation Laboratory, where he was influenced by Rabi and contributed to the development of radar and various microwave techniques. He returned to Harvard in 1945 and spent the rest of his career there. Bloch and Purcell were awarded the 1952 Nobel Prize in Physics "for their development of new methods for nuclear magnetic precision measurements and discoveries in connection therewith."

F. Bloch   E.M. Purcell

Reprinted with permission from The Nobel Foundation.

A final critical component of the modern NMR toolbox was contributed independently by Henry C. Torrey [23] and Erwin L. Hahn [24], who demonstrated the feasibility of pulsed NMR (an idea originally suggested by Bloch [25]) and observed free Larmor precession. Hahn further used pulsed NMR to generate and observe spin echoes [26].

The next 20 years saw the development of NMR as a powerful investigative tool in many areas of physics and even more so in chemistry. The sensitivity of the nucleus to its electronic environment in a molecule (the "chemical shift") and to spin–spin interactions were originally viewed by those in the nuclear physics community as annoying features of the technique. However, the enormous potential of NMR spectroscopy for analytical studies was soon revealed through the discovery of the three peaks of ethanol in Purcell's group [27]. Almost none of the early applications of NMR were medical, although a great deal of work was published on relaxation, diffusion, and exchange of water in cells and tissues, even in living human subjects [28] and whole animals [29].

As recounted above, MRI came into being in 1973 with Lauterbur's publication of true 2D NMR images (Figure 1.1), reconstructed from 1D projections acquired while magnetic field gradients were applied in various directions [1]. Soon thereafter, and quite independently, Peter Mansfield at the University of Nottingham introduced critical methods for efficient image generation, including slice selection [30] and fast "snapshot" acquisition schemes wherein entire 2D images could be obtained in a few tens of milliseconds [31].

RECOGNITION FOR KEY CONTRIBUTIONS

Richard R. Ernst developed Fourier transform methods that paved the way for modern MRI. He was awarded the 1991 Nobel Prize in Chemistry for "contributions to the development of the methodology of high-resolution nuclear magnetic resonance (NMR) spectroscopy." Paul C. Lauterbur and Sir Peter Mansfield were then jointly awarded the 2003 Nobel Prize in Physiology or Medicine "for their discoveries concerning magnetic resonance imaging."

R.R. Ernst   P.C. Lauterbur   P. Mansfield

Reprinted with permission from The Nobel Foundation.

Another early and essential contribution to MRI was made by Richard Ernst at the Swiss Federal Institute of Technology in Zurich. During the 1960s, he had introduced Fourier transform NMR spectroscopy [32]. In 1975, he realized that one should be able to generate 2D or three-dimensional (3D) NMR images by applying switched magnetic field gradients as NMR signals were acquired, and then employing the Fourier transform methods that are now a mainstay of modern MR image reconstruction [33].

During the 1970s, research in MRI was largely restricted to academic laboratories, most of them in the United Kingdom. This time period was marked by a series of important demonstrations: crude first *in vivo* images of a human finger (1975), hand (1976), thorax (1977), and head (1978). In 1980, William Edelstein, a postdoctoral fellow in John Mallard's group at the University of Aberdeen, implemented spin-warp (or Fourier) imaging and obtained the first clinically useful image of a human subject [34]. By this time, intense commercial investment in MRI had begun and clinical trials were being promoted. In 1983, Toshiba and Siemens brought the first commercial MRI scanners to market, equipped with 0.15 T (resistive) and 0.35 T (superconducting) magnets, respectively. Meanwhile, General Electric, one of the current leading manufacturers, recruited several of the pioneers in the field, including Edelstein. In 1985, it began to sell the first 1.5 T whole-body clinical MRI system.

Over the past three decades, MRI exams have become routine diagnostic procedures. In 2013, estimates place the number of operational scanners worldwide at more than 30 thousand and the number of exams performed every year at more than 100 million (Figure 1.2a) [35,36]. The ever-growing availability and performance of these systems has facilitated a remarkable and sustained growth in applications, as evidenced by measures such as the number of publications that make reference to MRI (Figure 1.2b).

One of the most obvious current trends in MRI technology is a concerted move toward a large installed base of 3 T systems, particularly for neurological imaging. Some of these scanners are now even being delivered as hybrid or dual-modality imaging systems, such as the promising combination of positron emission tomography and MRI

(see Chapter 19 in *Imaging of the Pelvis, Musculoskeletal System, and Special Applications to CAD*). There is also increasing interest in integrating the soft tissue imaging capability of MRI with interventional procedures, such as MR-guided focused ultrasound surgery (see Chapter 18 in *Imaging of the Pelvis, Musculoskeletal System, and Special Applications to CAD*). Another significant trend is in the area of image acquisition acceleration. The benefits of the latter include reduced motion or flow artifacts, the ability to capture anatomical motion (e.g., as desired in cardiac imaging), shorter scan times for patients, and more cost-efficient use of high demand resources. Improvements are being driven by concepts of sparse sampling (or compressed sensing) that exploit the spatial and/or temporal redundancies inherent in MRI data [37]. They are aided by parallel acquisition schemes built around the use of coil arrays, which provide direct access to spatial information and thus further enable under-sampling of image data [38,39].

Yet another promising initiative is in the area of hyperpolarization. The sensitivity of NMR as a probe is directly coupled to the orientation or alignment of nuclear spins in the applied magnetic field. At room temperature, the net equilibrium alignment (or polarization) of nuclei in any laboratory field is minuscule. In 1950, Alfred Kastler predicted that this polarization could be enhanced through "optical pumping" [40]; by the 1960s, this effect had been demonstrated in NMR experiments [41] and Kastler had been awarded the Nobel Prize in Physics (1966) "for the discovery and development of optical methods for studying Hertzian resonances in atoms." A number of such techniques are now capable of inducing up to a millionfold increase in NMR signal strength for specific nuclei. Examples include optical pumping [41,42], dynamic nuclear polarization [43], and para-hydrogen-induced polarization [44]. The enhancements provided by these methods are crucial when working with low-density or low-concentration nuclei, as encountered in MRI of gases

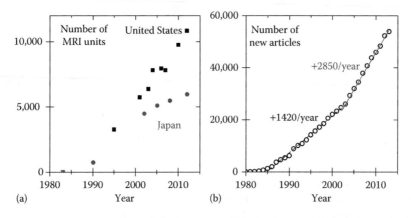

FIGURE 1.2
(a) Number of operational MRI units in two leading countries by year [35]. (b) Number of articles published in a given year making reference to "MRI" or "MR imaging" or "magnetic resonance imaging." (Data from Thomson Reuters *Web of Science* citation indexing service, 2014.)

in lung airspaces (see Chapters 10 and 13 in *Imaging of the Cardiovascular System, Thorax, and Abdomen*), ^{13}C nuclei in metabolites [45,46], injected Si nanoparticles [47,48], or "caged" ^{129}Xe [49,50].

In retrospect, the speed and extent to which the fields of NMR and MRI evolved is remarkable. Varian Associates played a key role in the rapid transition of NMR from the laboratory to a commercial product that revolutionized chemistry. The company was incorporated in 1948 and intentionally settled near Stanford; Martin Packard, part of Bloch's team, joined shortly thereafter. From that point in time onward, technical development of the field was primarily driven by industry. Similarly, in the case of MRI, as soon as the clinical potential of the technique was recognized, commercial interests drove the necessary technological developments. Throughout the 1980s, a number of companies, including General Electric, Picker, Toshiba, Siemens, and Hitachi, invested heavily in research and development, and promoted clinical evaluation of images. By the 1990s, the installed base had grown to the point where MRI exams were commonplace in industrialized countries. To this day, the number of facilities offering access to MRI continues to grow at an impressive rate, while scan times get shorter and scan quality and resolution continue to increase.

It has been argued that the remarkable evolution of MRI as a clinical imaging modality benefited enormously from the timing of its invention [51]. Had the idea been proposed two decades earlier, key components of the necessary technologies would simply not have been available. In particular, the need for rapid computation of Fourier transforms would have presented an enormous challenge. However, had the idea been proposed two decades later, the demand for new imaging modalities would almost certainly have been greatly reduced. By that point in time, imaging modalities based on well-controlled ionizing radiation (e.g., computed tomography scanners) had evolved to the point where significant hurdles would have been encountered in trying to convince radiologists and medical equipment manufacturers to invest in a new and entirely unproven technology. Even more important is the extent to which the regulatory environment has changed since the 1980s. The level of proof needed to obtain safety approval is now so high that if MRI was proposed today, few if any investors would likely be willing to fund its development.

1.2 Fundamentals of NMR

NMR is intrinsically a quantum mechanical phenomenon. It deals with the dynamics of microscopic objects (atomic nuclei) that behave according to the seemingly curious (but well-understood) laws of quantum mechanics. Fortunately, one does not need years of background study in quantum mechanics in order to appreciate and understand the essential elements of MRI. The reason is that MRI is invariably used to probe macroscopic objects, involving vast numbers of atomic nuclei. The collective behavior of these nuclei usually washes out the oddities of quantum mechanics, leaving something that bears resemblance to a familiar problem in classical mechanics: the precession of a spinning top in the earth's gravitational field. It leads to a simple but powerful mathematical description of nuclear dynamics that accurately predicts the outcome of many experiments. In this sense, it often provides a sufficient basis for developing intuition and interpreting experimental results.

Unfortunately, the tendency for many people—beginners and practitioners alike—is to lose sight of the fact that the classical picture of NMR dynamics is simply an analogy: it is not a correct description of dynamics at the microscopic scale, and it can lead to nonsensical explanations of the underlying physics. Examples of situations in which the analogy has been carried too far can be found on popular web sites purporting to explain NMR and MRI using pictures of toy tops or "spinning charged nuclei." As a rule of thumb, caution is advised whenever such props are encountered.

This section is organized into two parts. The first discusses the key factors that contribute to nuclear spin dynamics, leading to a set of phenomenological equations that encapsulate the essence of the classical description of the problem. These are the famous Bloch equations. The second part then outlines the various means by which the practitioner interacts with the atomic nuclei in a sample, both to induce collective motions and to detect the resulting response.

1.2.1 Bloch Equations and NMR Dynamics

In keeping with the spirit of this book, most of this chapter makes use of the classical picture of NMR. However, in order to motivate that picture, we start in Section 1.2.1.1 with a quick glimpse at a few quantum mechanical aspects of nuclear spin dynamics. The classical treatment of the problem is then presented in Section 1.2.1.2, leading to a statement of the Bloch equations in Section 1.2.1.3.

1.2.1.1 Spin: A Quantum Property

Particles such as the electron, the proton, and the neutron are characterized by their masses and electrical charges. They also possess "spin," an entirely quantum mechanical property that is associated with an intrinsic angular momentum. (This angular momentum has nothing at all to do with physical rotation.) Spin angular momentum **S** is a vector-like quantity; it has three spatial components and can be oriented in different ways. At the same time, it is different than an ordinary geometric vector. The total "amount" of spin (the length of the arrow) is fixed; it cannot be changed. Moreover, only a subset of all possible orientations is permitted. More precisely, when the component S_i of angular momentum is measured in any particular direction, it is only ever observed to have discrete or "quantized" values. For a spin 1/2 particle (such as the electron, the proton, or the neutron), only two values are possible: $S_i = \pm\hbar/2$, where $\hbar$ is Planck's constant h divided by 2π. Curiously, this is less than the total spin angular momentum of the particle ($S = \sqrt{3}\hbar/2$). Pictorial representations of spins and spin states relying on arrows and cones are commonplace in MRI, but they are best viewed with caution. None of them are entirely satisfactory when held up to careful scrutiny.

Particles with spin possess a magnetic moment $\mathbf{m} = \gamma\hbar\mathbf{S}$. Here, the constant of proportionality γ is known as the gyromagnetic ratio; each particle with a magnetic moment has a characteristic gyromagnetic ratio. Thus, even though the electron, the proton, and the neutron are all spin 1/2 particles, they have different magnetic moments (see Table 1.1). The same is true of strongly bound collections of particles, such as those that form the nucleus of an atom. The spins of the individual nucleons combine quantum mechanically to yield a well-defined total nuclear spin, usually denoted $\mathbf{I}$, that is characterized by its magnitude I and by a unique gyromagnetic ratio γ (Table 1.1). Because most applications of NMR and MRI involve nuclei with spin 1/2, this is the only case that is considered below. That being said, there are many important situations in which NMR is employed in connection with nuclei that have higher spin values; the features of the resulting spin dynamics are correspondingly more complex.

TABLE 1.1

Values of Reduced Gyromagnetic Ratios $\gamma/2\pi$ and Nuclear Polarizations for Selected Spin 1/2 Particles and Nuclei

Particle or Nucleus	$\gamma/2\pi$ (MHz/T)	Polarization ($\times 10^{-6}$/T)
Electron	28,025	2,295
Neutron	29.165	2.39
Proton, ^{1}H	42.577	3.49
^{3}He	−32.434	−2.66
^{13}C	10.705	0.877
^{15}N	−4.316	−0.353
^{19}F	40.052	3.28
^{31}P	17.235	1.41
^{129}Xe	−11.777	−0.964

Equilibrium polarizations are computed for room temperature ($T = 293$ K) using Equation 1.1 and are expressed on a per unit-magnetic field basis. Examples of nuclei with (a) spin 0 (not suitable for NMR): ^{4}He, ^{12}C, ^{14}C, ^{16}O; (b) spin 1: ^{2}H, ^{14}N; (c) spin 3/2: ^{23}Na, ^{31}K; and (d) spin 5/2: ^{17}O.

When an external magnetic field $\mathbf{B}$ is applied to a nucleus (or a particle) possessing a magnetic moment, an interaction takes place. The energy of the nucleus changes by an amount $-\mathbf{m} \cdot \mathbf{B}$, where the scalar (or dot) product "$\cdot$" accounts for the orientation of $\mathbf{m}$ relative to $\mathbf{B}$. The energy difference between the two states of the nucleus that have spin angular momentum components $\pm\hbar/2$ in the direction of $\mathbf{B}$ is $\Delta E = -\gamma\hbar B$. Peculiar features arise if the tools of quantum mechanics are brought to bear on this problem. These two particular states (often referred to as "spin-up" and "spin-down") are unique; they do not evolve in time. They are called stationary states. Other states of the nuclear spin $\mathbf{I}$ in a magnetic field are dynamic; they change as a function of time. If a weak magnetic field $\mathbf{B}_1$ is applied perpendicular to a static magnetic field $\mathbf{B}_0$ aligned with the z-axis, and B_1 is made to oscillate at an angular frequency $\omega_0 = \Delta E/\hbar = |\gamma B_0|$, the nuclear spin will execute a complex periodic oscillation back and forth between the spin-up and spin-down states via quantum superpositions of the two. If the field B_1 is eventually turned off when the nuclear spin happens to be "half-way" between the two stationary states, its transverse components I_x and I_y will continue to oscillate back and forth between their allowed values... at the angular frequency $\omega_0 = \Delta E/\hbar = |\gamma B_0|$. The phenomenon of magnetic resonance results from the time evolution of spin states in combined static and resonantly oscillating magnetic fields.

For a physical system containing several (or many) nuclei, a full quantum treatment of spin dynamics is only required in particular situations. It is important, for instance, when short-range quantum correlations between interacting spins of nuclei in a molecule are strong. This is usually not the case in problems relevant to MRI, and a semiclassical treatment of spin dynamics is thus sufficient.* Quantum statistical mechanics is used to evaluate the properties and time evolution of nuclear spins in a sample containing a large number of identical nuclei. Unlike individual spins that can be prepared in *pure* quantum states, a large quantum mechanical system is usually in a *mixed state*: a statistical sum of pure states in which many (or most) quantum correlations are washed out.† For instance, in equilibrium at a temperature T (a state known as "thermal equilibrium"), the probabilities p_{up} and p_{down} of observing the up and down states, respectively, are given by the Boltzmann factor:

---

* A discussion of the need for a fully quantum mechanical approach to the problem can be found in [52] and references therein.

† Consider the following analogy to experiments with polarized light. Pure states of polarization can be combined and transformed: Right and left circularly polarized light can be combined to form linearly or elliptically polarized light. Unpolarized light, however, is different; it cannot be converted to linearly polarized light or to any other polarized state.

$$\frac{p_{\text{down}}}{p_{\text{up}}} = \exp\left(-\frac{\Delta E}{k_B T}\right) \tag{1.1}$$

where k_B is Boltzmann's constant. This ratio is usually very close to 1. The energy difference ΔE set by most laboratory-scale magnetic fields is very small compared to the thermal energy $k_B T$ at room temperature, and thus the probability of observing a spin in its up or down states is very nearly the same. Examples of the very small difference $p_{\text{up}} - p_{\text{down}}$, which is known as the nuclear polarization, are listed in Table 1.1. At the same time, the probability of observing any other spin state (i.e., in a direction tilted away from $\mathbf{B}_0$) is equal to 0: The sum $p_{\text{up}} + p_{\text{down}} = 1$.

The nuclear magnetic properties of a sample can be determined from the quantum statistical description of its spin dynamics. Each nuclear magnetic moment **m** produces a magnetic field, similar to that produced by a tiny closed loop of electric current. Technically, this field is known as that of a magnetic dipole, or simply a "dipolar field." Its orientation depends on the orientation of **m** (and hence **I**) and decreases rapidly in strength as one moves away from the source. Adding up contributions from many nuclei leads naturally to the concept of a local magnetization density **M**, representing the magnetic moment per unit volume. The thermal equilibrium magnetization $\mathbf{M}_0$ is either parallel or antiparallel to the applied magnetic field $\mathbf{B}_0$ (depending on the sign of γ) and is proportional to both the local density of nuclear spins in the sample and the thermal equilibrium polarization. During experiments, the nuclear magnetization can be manipulated by applying static and/or time-varying magnetic fields, as discussed below and in Section 1.3. Similarly, the net magnetization of the sample can be inferred through monitoring the associated nuclear magnetic field. Normally, this involves detecting changes in magnetic flux passing through a coil of wire (or similar structure) as the nuclear magnetization evolves in time, as discussed in Section 1.2.2.2.

QUANTUM DYNAMICS OF A SPIN 1/2 PARTICLE

The spin-up and spin-down states of a spin 1/2 particle, which are often denoted $|+\rangle$ and $|-\rangle$, are called pure states. Repeated measurements of their spin angular momentum along a particular axis (as was done in the Stern–Gerlach experiment mentioned in Section 1.1) always yield $\hbar/2$ for the up state and $-\hbar/2$ for the down state. All other pure states, with a maximum spin projection

value $\hbar/2$ in a direction $\hat{u}$ other than $\pm z$, are "linear superpositions" of these states: $\alpha|+\rangle + \beta|-\rangle$, where α and β are complex coefficients such that $\alpha^2 + \beta^2 = 1$. For instance, $(|+\rangle + |-\rangle)/\sqrt{2}$ is the pure state in the $\hat{x}$ direction and $(|+\rangle + i|-\rangle)/\sqrt{2}$ is the pure state in the $\hat{y}$ direction. Unlike the up and down states, these quantum mechanical superpositions are not stationary; they evolve in time. In particular, the direction $\hat{u}$ evolves in exactly the manner predicted by the semiclassical treatment of the time evolution of **m** summarized in Section 1.2.1.2.

These concepts can be extended. For example, a two-level atomic system is formally identical to a spin 1/2 quantum system. Transitions between the two (quasi-) stationary states of the atom correspond to the emission or the absorption of a quantum of energy (a photon). This picture forms the basis for the popular—but incorrect—statement that NMR phenomena involve the emission or absorption of radio waves. To understand why this statement is wrong, one need only consider the fact that the electromagnetic wavelength associated with the Larmor frequency produced by a laboratory strength magnetic field is almost always large compared to the dimensions of typical samples and receive coils. In other words, NMR (and particularly MRI) is performed in the near-field electromagnetic regime and the photons that are involved are virtual [53,54].

1.2.1.2 Classical Magnetization Dynamics

A classical description of nuclear spin dynamics is obtained by considering a model system in which the macroscopic magnetic moment **m** and the resultant angular momentum **J** are coupled such that $\mathbf{m} = \gamma \mathbf{J}$. This seemingly innocuous relationship is the same as that which is obeyed by individual nuclei; only now **m** and **J** are purely classical quantities (not subject to the subtle restrictions imposed by quantum mechanics). This vector proportionality causes a gyroscopic response to an applied magnetic field, analogous to the dynamics of a spinning top in a gravitational field.

A MISLEADING ANALOGY: THE COMPASS NEEDLE

The relationship $\mathbf{m} = \gamma \mathbf{J}$ is not a general property of macroscopic objects. The magnetic moment of a compass, for example, is locked to the long axis of the needle, which is in turn free to rotate in a plane about its midpoint. (This is what makes the compass a useful device.) The angular momentum of the compass, however, is proportional to the angular rotation rate of the needle. Thus, $\mathbf{m} \neq \gamma \mathbf{J}$. A compass needle oscillates in a plane about its midpoint; it does not precess like a spinning top or like a collection of nuclear spins! Conversely, the nuclear magnetization in NMR is not locked to the physical orientation of the sample (or subject), as is the magnetization of a compass needle. Thus, in *magic angle spinning* (an NMR technique in which the sample is physically rotated at high speed), it is the lattice of the crystal structure that is spun, not the nuclear magnetization.

In a sample or subject, the local macroscopic magnetization density $\mathbf{M}$ associated with the magnetic moments of the nuclei obeys the classical equation of motion:

$$\frac{d\mathbf{M}}{dt} = \gamma \mathbf{M} \times \mathbf{B}. \tag{1.2}$$

An important feature of this equation is that the amplitude M of the local magnetization (the length of the vector $\mathbf{M}$) remains constant. At all times, the change in $\mathbf{M}$ is perpendicular to both $\mathbf{M}$ and $\mathbf{B}$. This behavior is encoded in the vector (or cross) product "$\times$" in Equation 1.2.

A common and convenient graphical tool for depicting the time evolution of $\mathbf{M}$ is the *Bloch sphere*: an imaginary sphere of radius M. With a coordinate system chosen such that the static magnetic field $\mathbf{B}_0$ is aligned with the z-axis, the corresponding thermal equilibrium magnetization $\mathbf{M}_0$ can be drawn as a vector pointing from the midpoint of the sphere to the "North Pole." In this case $\mathbf{M} = \mathbf{M}_0$ and $\mathbf{B} = \mathbf{B}_0$, and thus Equation 1.2 gives $dM/dt = 0$. That is, nothing happens; $\mathbf{M}$ remains aligned with $\mathbf{B}_0$. The same thing is true if $\mathbf{M}$ is somehow aligned to point to the "South Pole" of the Bloch sphere. In the language of Section 1.2.1.1, these situations correspond to the two stationary quantum spin states.

If instead $\mathbf{M}$ is somehow reoriented so that it is canted with respect to $\mathbf{B}_0$ by some angle other than 180°, as shown in Figure 1.3, Equation 1.2 gives a nonzero result for $d\mathbf{M}/dt$. In this case, the tip of the magnetization vector traces out a circular path at constant latitude, returning periodically to its starting point. This motion is referred to as *free precession*. Unless other processes intervene, it persists forever. It occurs at an angular velocity

$$\Omega_0 = -\gamma \mathbf{B}_0 \tag{1.3}$$

or an angular speed $\Omega_0 = -\gamma B_0$, both of which indicate the sense of the motion. The expression *angular Larmor*

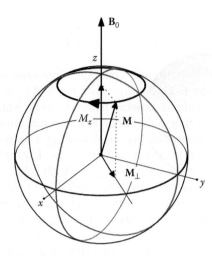

FIGURE 1.3
Graphical representation of solutions to Equation 1.2 for a constant field $\mathbf{B}_0$. The time-dependent magnetization vector $\mathbf{M}$ can be associated with a point on a sphere of radius M centered at the origin: the Bloch sphere. The longitudinal component M_z of $\mathbf{M}$ is static, whereas the transverse component $\mathbf{M}_\perp$ rotates in the transverse (x–y) plane at the angular Larmor frequency ω_0.

frequency is used by some authors to describe Ω_0, while others take it to mean $\omega_0 = |\gamma B_0|$. Care is thus required any time the absolute sense of rotation is needed. Importantly, the Larmor frequency is independent of the angle to which $\mathbf{M}$ is canted relative to $\mathbf{B}_0$.

The phenomenon of NMR enters when a time-varying magnetic field $\mathbf{B}_1$ is added. Imagine that a weak magnetic field $\mathbf{B}_1$ is added perpendicular to the static field $\mathbf{B}_0$ and that $\mathbf{B}_1$ rotates about $\mathbf{B}_0$ at the Larmor frequency. That is, the amplitude of $\mathbf{B}_1$ is constant while its direction changes. In this case $\mathbf{M}$ must precess about both $\mathbf{B}_0$ and $\mathbf{B}_1$, and so the tip of the vector traces out two simultaneous motions: a fast precession about $\mathbf{B}_0$ and a slow precession about the instantaneous orientation of $\mathbf{B}_1$. This produces tightly wound spiral trajectories such as the one shown in Figure 1.4a.

The examples shown in Figures 1.3 and 1.4a reveal that the longitudinal and transverse components of $\mathbf{M}$ exhibit very different dynamical behavior. The dynamics of the longitudinal component M_z (a scalar) correspond to a slow oscillation that involves variations in amplitude. The dynamics of the transverse component $\mathbf{M}_\perp$ (a 2D vector), however, involve a fast rotation about the z-axis at an angular frequency ω_0 combined with a slow oscillation that involves variations in amplitude. Here $\mathbf{M}_\perp$ can be decomposed into orthogonal components M_x and M_y. Equivalently, it can be represented as the quantity $M_x + iM_y$ in the complex plane. The latter approach enables one to use complex algebra rather than matrix algebra in the solution of Equation 1.2. With complex algebra, a rotation of the complex quantity $\mathbf{M}_\perp$ by an angle Φ is obtained by adding Φ to its phase, or multiplying by $e^{i\Phi}$.

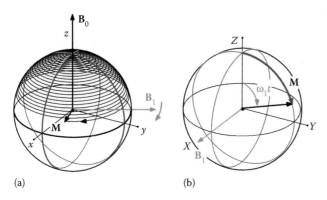

FIGURE 1.4
(a) Part of the spiral trajectory executed by the tip of the magnetization vector under resonance conditions. It results from the cumulative action of a weak magnetic field $\mathbf{B}_1$ in the x–y plane, rotating at the angular Larmor frequency ω_0. The amplitudes of the longitudinal (M_z) and transverse ($\mathbf{M}_\perp$) components of $\mathbf{M}$ oscillate periodically. (b) The same trajectory is shown in the rotating frame where $\mathbf{B}_1$ is fixed along the X-axis (see text).

Setting $\mathbf{B} = \mathbf{B}_0$ and using complex notation for $\mathbf{M}_\perp$ reduces Equation 1.2 to $d\mathbf{M}/dt = -i\gamma B_0\mathbf{M}_\perp$. The solution to this differential equation is the trajectory shown in Figure 1.3: that is, M_z is constant and $\mathbf{M}_\perp(t) = \mathbf{M}_\perp(0)e^{i\Omega_0 t}$, where $\Omega_0 = -\gamma B_0$. Note here that the sense in which the tip of the vector $\mathbf{M}$ traces out a circle depends on the sign of γ. When $\gamma > 0$ and the Bloch sphere is viewed from above, the sense of the free precession trajectory is clockwise. If $\gamma < 0$, as it is for some of the nuclei listed in Table 1.1, the sense of the free precession trajectory is counterclockwise. The sense of this rotation can be assessed experimentally by using two orthogonal detectors; the phase difference between the signals induced in the two detectors reveals the sense of the trajectory, and hence the sign of γ.

The motion of $\mathbf{M}$ is complicated by the rapid rotation of $\mathbf{M}_\perp$ about $\mathbf{B}_0$ at the Larmor frequency. It is technically simpler to display and compute trajectories of $\mathbf{M}$ if one works in a reference frame that is also rotating about the z-axis at the Larmor frequency. This is analogous to jumping onto a merry-go-round (or carousel) to better observe the wooden horses and their riders. More precisely, in a reference frame with axes X, Y, and Z rotating at an angular velocity Ω about the Z-axis (with $Z \equiv z$), Equation 1.2 becomes

$$\frac{d\mathbf{M}}{dt} = \gamma\mathbf{M} \times \mathbf{B}_0 + \mathbf{M} \times \Omega$$
$$= \gamma\mathbf{M} \times \left(\mathbf{B}_0 + \frac{\Omega}{\gamma}\right) \tag{1.4}$$

where $\mathbf{B}$ has been set to $\mathbf{B}_0$. Comparison of Equations 1.2 and 1.4 reveals that the magnetization now behaves as if it is responding to an apparent field

$$\mathbf{B}_{app} = \mathbf{B}_0 + \frac{\Omega}{\gamma} \tag{1.5}$$

and that in this frame its angular velocity is $\Omega_0 - \Omega$. If the rotation rate is chosen such that $\Omega = \Omega_0$, the apparent magnetic field $\mathbf{B}_{app}$ vanishes and the magnetization vector $\mathbf{M}$ appears stationary.

Likewise, when a transverse magnetic field $\mathbf{B}_1$ rotating about $\mathbf{B}_0$ is added, the time evolution of $\mathbf{M}$ is best described using a reference frame in which $\mathbf{B}_1$ appears stationary. That is, in a reference frame that is synchronous with $\mathbf{B}_1$. In any such frame, Equation 1.2 becomes

$$\frac{d\mathbf{M}}{dt} = \gamma\mathbf{M} \times (\mathbf{B}_{app} + \mathbf{B}_1). \tag{1.6}$$

If the field $\mathbf{B}_1$ is resonant with the free Larmor precession of the magnetization (i.e., $\Omega = \Omega_0$ so that $\mathbf{B}_{app} = 0$), then $\mathbf{M}$ simply rotates around $\mathbf{B}_1$ at an angular velocity $\Omega_1 = -\gamma\mathbf{B}_1$ and its tip traces out a great circle on the Bloch sphere. This scenario is sketched in Figure 1.4b, where the particular rotating frame that was chosen is the one in which $\mathbf{B}_1$ is aligned with the X-axis. Viewed in the laboratory frame, this motion produces the spiral trajectory shown in Figure 1.4a. The pitch of the spiral is given by the ratio of magnetic field amplitudes B_1/B_0.

This last example forms the basis for pulsed NMR. If the field $\mathbf{B}_1$ is applied on resonance for a finite period of time τ, $\mathbf{M}$ traces out an arc on the surface of the Bloch sphere that subtends an angle $\theta = \omega_1\tau$. Afterward, $\mathbf{M}$ undergoes free precession as shown in Figure 1.3. Here, $\omega_1 = |\Omega_1|$ denotes the angular nutation frequency and the finite-duration $\mathbf{B}_1$ field is referred to as a tipping pulse. The angle θ through which $\mathbf{M}$ is rotated is variously referred to as the tip, flip, rotation, or nutation angle. It can be controlled through the amplitude of B_1 or the time τ. Starting from thermal equilibrium where $\mathbf{M} = \mathbf{M}_0$, a 90° or $\pi/2$ tipping pulse will rotate the magnetization into the transverse plane, at which point it will undergo free precession. Alternately, a 180° or π pulse will invert the magnetization, transforming it from $\mathbf{M}_0$ to $-\mathbf{M}_0$.

If the field $\mathbf{B}_1$ is applied off resonance rather than precisely at the Larmor frequency, it is still convenient to work in a frame that is synchronous with $\mathbf{B}_1$ and employ Equation 1.6. The magnetization now rotates around an effective field $\mathbf{B}_{eff} = \mathbf{B}_{app} + \mathbf{B}_1$ as shown in Figure 1.5a, and the trajectories traced out by the tip of $\mathbf{M}$ are no longer great circles. If one starts from an initial magnetization $\mathbf{M}_0$ aligned with $\mathbf{B}_0$ it is no longer possible to reach the antipodal point on the Bloch sphere where $\mathbf{M} = -\mathbf{M}_0$; that is, perfect π-pulses leading to magnetization reversal are only possible at resonance. As the detuning $|\Omega - \Omega_0|$ increases, the circular

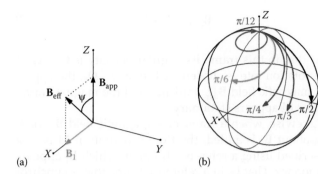

FIGURE 1.5
(a) Components $\mathbf{B}_1$ and $\mathbf{B}_{app}$ of the effective magnetic field $\mathbf{B}_{eff}$ in the rotating frame. (b) Examples of trajectories on the Bloch sphere for several detunings. In each case, the same evolution time $\tau = \pi/2\Omega_1$ is employed. Values of the angle ψ, which characterize the extent of the detuning (i.e., $\tan \psi = \Omega_1/|\Omega - \Omega_0|$) are indicated.

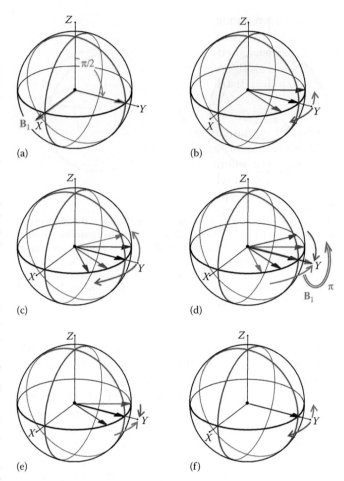

FIGURE 1.6
Spin dephasing and echo formation in the rotating frame: (a) A $\pi/2$ tipping pulse rotates the thermal equilibrium magnetization into the transverse plane. Over time (b,c) some spins precess faster than average while others precess slower. A distribution of azimuthal angles builds up, represented schematically by arrows indicating isochromats. (d) A π pulse applied along the Y-axis when the distribution is as shown in (c) inverts the magnetization. Over time (e,f), the magnetization is refocused, leading to a revival (spin echo) in panel (f). At times later than those shown here the spins continue to dephase.

trajectories traced by the tip of $\mathbf{M}$ on the Bloch sphere (in the rotating frame) become smaller and they are traveled at faster rates (Figure 1.5b). This illustrates the fact that NMR tipping pulses are frequency-selective. Only spins that precess at frequencies close to that of the applied $\mathbf{B}_1$ field are influenced to any significant degree.

As another example, consider situations where the Larmor precession frequency is not the same for all nuclei in a sample. This inhomogeneity could be the result of an intentionally applied magnetic field gradient, or it could be the result of an intrinsic property of the sample. In either case, there is no unique rotating reference frame in which $\mathbf{B}_{eff} = 0$ everywhere. Despite this, it is still advantageous to work in a rotating reference frame, a common choice being a frame that rotates at the average Larmor frequency. Figure 1.6 shows a sequence in which (a) a $\pi/2$ pulse is used to rotate the thermal equilibrium magnetization $\mathbf{M}_0$ into the transverse plane, at which point (b–c) the magnetization undergoes free precession. Unlike the situation pictured in Figure 1.3, the Larmor precession rate is not uniform; some spins precess faster than the average rate and some spins precess slower. This distribution is shown schematically as a series of arrows representing "isochromats": idealized collections of spins that precess at the same rate. Over time the distribution of angles subtended by these isochromats grows and the net transverse magnetization diminishes. If the sequence pictured in Figure 1.6b and c is allowed to continue indefinitely, the net transverse magnetization will average to zero.

Hahn recognized that the ordered dephasing of isochromats evident in Figure 1.6 can be "undone" by applying another tipping pulse, leading to the formation of a spin echo (or a Hahn echo) [26]. For example, if a π pulse at the average Larmor frequency is applied along the

Y-axis at the time pictured in Figure 1.6c, the isochromats undergo a 180° rotation. Thus in Figure 1.6d the isochromats that precess at the fastest rate are behind the average in terms of total accumulated phase. Conversely, the isochromats that precess at the slowest rate are ahead of the average. As the magnetization continues to execute free precession the angular width of the distribution narrows, forming a revival or spin echo in Figure 1.6f.

1.2.1.3 Irreversibility: The Bloch Equations

The classical theory of nuclear spin dynamics summarized above provides tools for manipulating nuclear spins, and for establishing states characterized by various forms of phase coherence (or correlation) between local magnetization vectors. These states do not persist forever. Given enough time, and the absence of further manipulations, we expect any interacting spin system to return to thermal equilibrium. That is, a state where **M** is aligned with $\mathbf{B}_0$ and has a magnitude M_0 that is set by the Boltzmann distribution (Equation 1.1).

The processes through which a spin system returns to thermal equilibrium can be complex. They typically involve an exchange of energy between the spins and their environment, and are usually mediated by random magnetic interactions. In many practical circumstances the rate at which this exchange proceeds can be characterized by a phenomenological timescale referred to as the "spin–lattice relaxation time" or the "longitudinal relaxation time." By convention, it is designated by the symbol T_1. If the longitudinal component of **M** is displaced from thermal equilibrium, the equation of motion governing its return is

$$\frac{dM_z}{dt} = -\frac{(M_z - M_0)}{T_1}.$$　　(1.7)

Thus, if $\mathbf{M}_0$ is inverted at time $t = 0$ through application of a π pulse, the return of M_z to equilibrium is given by

$$M_z(t) = M_0\left[1 - 2\exp\left(\frac{-t}{T_1}\right)\right].$$　　(1.8)

Similarly, coherence of the local transverse magnetization is degraded over time by random interactions or processes that destroy correlations. This degradation is distinct from the dephasing of spins in an inhomogeneous magnetic field described in connection with Figure 1.6, in the sense that it is irreversible. No subsequent manipulation of the nuclear spins can produce an echo or revival. Again, in many practical circumstances, the rate at which coherences of the transverse magnetization are attenuated can be characterized by a phenomenological timescale referred to as the "spin–spin relaxation time" or the "transverse relaxation time." By convention, it is designated by the symbol T_2. If the magnetization is manipulated so as to establish a transverse component, and then allowed to undergo free precession, the equation of motion describing the inevitable attenuation of $\mathbf{M}_\perp$ is

$$\frac{\mathbf{M}_\perp}{dt} = -\frac{\mathbf{M}_\perp}{T_2}.$$　　(1.9)

Thus, if a $\pi/2$ pulse is applied to a spin system in thermal equilibrium, producing $M_\perp = M_0$ at time $t = 0$, the transverse magnetization subsequently satisfies

$$M_\perp(t) = M_0\exp\left(\frac{-t}{T_2}\right).$$　　(1.10)

The full equations of motion for the local magnetization density **M** that give rise to these exponential relaxation functions are known as the Bloch equations. They can be written

$$\frac{d\mathbf{M}}{dt} = \gamma\mathbf{M}\times\mathbf{B} - \frac{\mathbf{M}_\perp}{T_2} - \frac{(M_z - M_0)}{T_1}$$　　(1.11)

where the first term on the right accounts for precession and the other terms account for relaxation. A more compact expression is obtained by introducing the relaxation matrix

$$[R] = \begin{pmatrix} \dfrac{1}{T_2} & 0 & 0 \\ 0 & \dfrac{1}{T_2} & 0 \\ 0 & 0 & \dfrac{1}{T_1} \end{pmatrix}$$　　(1.12)

and writing

$$\frac{d\mathbf{M}}{dt} = \gamma\mathbf{M}\times\mathbf{B} + [R](\mathbf{M}_0 - \mathbf{M}).$$　　(1.13)

Even though the original formulation of these equations was based on two phenomenological parameters, there are many important situations in which T_1 and T_2 can be derived from quantum mechanical principles.

ORIGIN AND REGIMES OF RELAXATION

In liquids and gases, rapid random fluctuations of intermolecular orientations and distances permit the use of perturbative methods to evaluate relaxation. This was originally described by Bloembergen, Purcell, and Pound, and has since often been called the BPP theory [55]. A key parameter in this theory is the correlation time, τ_c, which characterizes the relevant fluctuations in the relaxation process. An important result of the BPP theory is that it predicts exponential relaxation for both M_z and $M_\perp$, and provides values for the corresponding relaxation time constants T_1 and T_2. In the weak field–fast motion limit ($\omega_0\tau_c \ll 1$), T_1 and T_2 are equal and proportional to $1/(\omega_0\tau_c)$. In the opposite high field–slow motion regime, $T_2 \propto 1/(\omega_0\tau_c) \ll T_1 \propto \omega_0\tau_c$. A minimum in T_1 is obtained in the intermediate regime, for $\omega_0\tau_c = 1$. Over the years, more elaborate descriptions have been developed, encompassing different situations of interest in NMR spectroscopy [56] but seldom relevant for MRI of tissues.

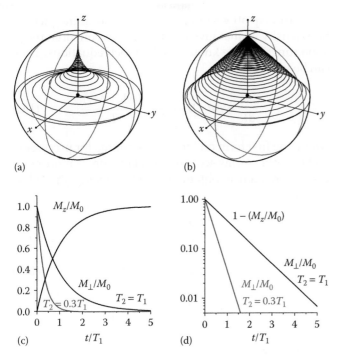

FIGURE 1.7
Influence of relaxation on the time evolution of **M**. Trajectories are shown for an initial transverse magnetization aligned with the y-axis for (a) $T_2 = 0.3T_1$ and (b) $T_2 = T_1$. Panel (c) shows the normalized magnitudes of M_z and $M_\perp$. Panel (d) shows the normalized magnitude of $M_\perp$ again, on a logarithmic scale.

Examples illustrating the combined influence of T_1 and T_2 on the time evolution of **M** are shown in Figure 1.7. One of the most striking differences relative to examples in Section 1.2.1.2 is that the trajectory of the tip of **M** is no longer restricted to the surface of the Bloch sphere. In Figure 1.7a, T_2 is substantially shorter than T_1. The resulting spiral collapses toward the z-axis much faster than it climbs toward the North Pole. In Figure 1.7b, the two relaxation times are equal. In this case the spiral trajectory is constrained to a cone-like surface. Figure 1.7c shows the recovery of the longitudinal magnetization M_z and the decay of the transverse magnetization $M_\perp$ for both scenarios. Note that the timescale has been normalized to T_1, and thus the recovery of M_z is the same for both cases. Finally, the decays of $M_\perp$ are shown again in Figure 1.7d, in order highlight the fact that they are exponential.

One often encounters situations where the Bloch equations alone are not sufficient to characterize the time evolution of **M**. Random atomic or molecular motions, for example, bring in the irreversible effects of diffusion. For liquids and gases these effects can often be characterized in terms of a diffusion coefficient D. More generally, when anisotropic media such as nerve tissue are involved, Equation 1.13 becomes

$$\frac{d\mathbf{M}}{dt} = \gamma \mathbf{M} \times \mathbf{B} + [R](\mathbf{M}_0 - \mathbf{M}) - \nabla \cdot [D]\nabla M \qquad (1.14)$$

where $[D]$ is the diffusion tensor. Equation 1.14 is known as the Bloch–Torry equation [57]. It provides the basis for extracting information about diffusion from NMR experiments and is central to understanding diffusion-weighted and diffusion tensor MRI—both of which are discussed in Chapter 4.

1.2.2 Electrodynamics of NMR

A myriad of MRI sequences exist, but they all start with the preparation of atomic nuclei in some well-defined state other than thermal equilibrium. They all also involve a mapping of the local Larmor precession frequency onto position. Information about the spatial distribution of some aspect of the nuclear magnetization **M** (such as the number of contributing nuclei, their local environment, or their displacement over time) is then inferred through monitoring the transverse component of **M** (i.e., $M_\perp$) as it evolves in time (and ultimately relaxes toward thermal equilibrium). A complete image is built up by repeating this measurement over and over again, as the mapping is systematically varied over an appropriate range of parameters.

The general procedures alluded to above are discussed further in Section 1.3 and in more detail in Chapter 2. Before getting to that point, however, it is worth examining the methods by which nuclear spin dynamics are initiated and manipulated. This is the focus of Section 1.2.2.1. It is also useful to understand the methods used to monitor nuclear spin precession. This is covered in Section 1.2.2.2. In both cases, the discussion is limited to the basic physics that is involved, as opposed to the instrumentation that is employed. Next, in order to set the stage for a discussion of specific imaging sequences, it is helpful to examine factors that influence the amplitude of signals that are detected in MRI and the extent to which these signals are influenced by unavoidable (intrinsic) sources of noise. This is the topic of Section 1.2.2.3. Finally, some of the same physics that is wrapped up in signal detection is responsible for undesirable effects: the deposition of RF energy into the subject and peripheral nerve stimulation. These issues are briefly summarized in Section 1.2.2.4 and are discussed further in Chapter 10.

1.2.2.1 Ampère's Law: Currents, Coils, and Fields

The dynamics of nuclear spin precession are controlled during MRI sequences by imposing magnetic fields. These magnetic fields are in turn produced by electric currents, and in some cases by magnetized (and/or magnetizable) materials. The precise manner in which magnetic fields are generated and influenced by these currents and magnetic materials is encapsulated in a mathematical expression known as Ampère's law.[*] The various fields that are typically required for MRI are summarized in Sections 1.2.2.1.1 through 1.2.2.1.3.

1.2.2.1.1 Static Field

A magnetic field is required to establish an energy difference between the spin-up and the spin-down states of the nucleus (Section 1.2.1.1). This is done by immersing the subject in a homogeneous magnetic field $\mathbf{B}_0$. This field is variously referred to as the static field, the main field, the homogeneous field, or simply "Bee-zero" or "Bee-naught". It is conventional to choose the direction of $\mathbf{B}_0$ as the direction that defines the z-axis of the coordinate system.

The simplest and most effective way to produce a strong, homogeneous, and accessible magnetic field is with a solenoid: a single wire wrapped many times around the circumference of a cylinder. When current flows through the wire, a homogeneous magnetic field is produced along the axis of the cylinder.[†] The central field produced by a thin uniformly wound cylindrical solenoid of length L and diameter D such as the one shown schematically in Figure 1.8 is given by

$$B_0 = \frac{\mu_0 n I}{\sqrt{1 + \left(\dfrac{D}{L}\right)^2}} \qquad (1.15)$$

where $\mu_0 = 4\pi \times 10^{-7}$ Tm/A is the permeability of free space and n represents the winding density. Choosing $L = 1.8$ m, $D = 0.9$ m, $I = 190$ A, and 7×10^3 turns/m (i.e., a wire wrapped around the cylinder approximately 12,600 times for a length totaling a few tens of kilometers) yields $B_0 = 1.5$ T, which is typical of the fields employed in the majority of clinical imagers in service today.[‡] In practice, superconducting solenoids are usually employed to generate this magnetic field. These "magnets" are operated in a "persistent" mode at cryogenic temperatures such that the current flows through the solenoid in a closed loop, without a power supply and effectively without dissipation. The main field magnet and the associated cryogenic vessel account for the bulk of the tubelike infrastructure that one normally sees when looking at an MRI scanner.

A key function of the main magnetic field is to set a uniform Larmor precession frequency for all of the

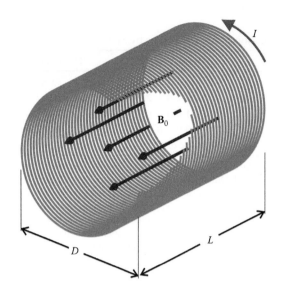

FIGURE 1.8
A solenoid of length L and diameter D carrying current I. The central magnetic field B_0 is given by Equation 1.15. The large number of turns required for an MRI magnet results in a coil that has many layers. Normally, the winding density is varied to improve the field homogeneity. A second (larger) solenoid wound outside the primary magnet, and in the opposite sense, provides "active shielding." The net field at the center of the combined magnet is somewhat weaker than that of primary magnet alone, but the external field is dramatically suppressed.

nuclei that are going to be imaged.[§] In practice, a simple finite length solenoid (or other winding pattern, or other magnet geometry) is not sufficient to produce the native magnetic field homogeneity that is required for high-resolution MRI. A number of shim (or correction) coils and judiciously placed ferromagnetic (magnetizable) materials are also used to create a central volume over which a very high degree of magnetic field uniformity (and hence nuclear spin precession rate) is achieved. The subject is then positioned such that the region to be imaged coincides with this "sweet spot." In typical whole-body clinical MRI systems, B_0 varies by at most a few parts-per-million over a 30 cm-diameter sphere at the center of the magnet (distortions equivalent to adding or subtracting a fraction of the Earth's magnetic field to B_0).

The nominal Larmor precession rate for ^{1}H nuclei (protons) in a 1.5 T imager is 63.87 MHz (see Table 1.1). In a 3 T imager, it is 127.73 MHz. Yet higher static magnetic fields can and have been employed for MRI, and yield substantial improvements in signal-to-noise ratio (SNR).[¶] Lower fields can also be employed and offer other advantages

---

[*] Additional considerations come into play when high frequencies are involved. In such cases, Maxwell's equations (which include the physics described by Ampère's law) are employed.

[†] The field is perfectly homogeneous only for an infinitely long solenoid.

[‡] To put this in perspective, the Earth's magnetic field is of order 10^{-4} T or 1 G.

[§] It is also usually responsible for establishing the thermal equilibrium polarization of the nuclei, and hence the maximum amplitude of the signals that can be detected.

[¶] They also present new challenges, such as limitations on the amplitude and duration of RF pulses that can be applied, as discussed in Sections 1.2.2.2 and 1.2.2.4.

TABLE 1.2

Classification of MRI Scanners by Static Field Strength

Range	Field Strength (T)	^{1}H Frequency (MHz)
High	$B_0 > 2$	$f > 85$
Conventional	$0.5 < B_0 < 2$	$20 < f < 85$
Low	$0.1 < B_0 < 0.5$	$4 < f < 20$
Very-low	$0.001 < B_0 < 0.1$	$0.04 < f < 4$
Ultra-low	$B_0 < 0.001$	$f < 0.04$

Over time the definition of what constitutes a "high-field" system has drifted upward. During the 1980s, a 1 T scanner would have been considered a high-field system. Most clinical systems in service today operate at 1.5 T, but the fastest growing segment of the market is for 3 T systems. The production of low-field scanners has dropped significantly in recent years. Scanners that operate in the very-low and ultra-low regimes are employed for niche applications.

such as the possibility of open geometries where ready access to the subject is possible and the imager environment is less likely to be claustrophobic. The magnets at the heart of these low-field systems can be wound from superconducting wire or ordinary copper wire, or they can be constructed from permanent magnets. Imagers that operate at yet lower magnetic fields have been developed in connection with a variety of different research initiatives. A crude classification of MR imagers by magnetic field strength is given in Table 1.2.

1.2.2.1.2 Gradient Fields

Spatial resolution in MRI is accomplished by inducing well-defined distortions of the main magnetic field, which produce the desired mapping between nuclear precession frequency and position. These distortions are created by running currents through sets of coils that are designed for this purpose.* This is precisely what Lauterbur did in his original demonstration (Figure 1.1). It is conventional to refer to these distortions as magnetic field gradients and to the coils that produce them as gradient coils.

In a solenoidal main field geometry, a longitudinal field gradient $G_z = dB_z/dz$ increases the field strength toward one end of the cylinder and decreases it toward the other. Transverse gradients ($G_x = dB_z/dx$ and $G_y = dB_z/dy$) increase the field strength on one side of the cylinder and decrease it on the other (to the left and the right of the subject, or above and below, or in fact in any transverse direction that is desired). Simple sets of coils generating such field gradients are sketched in

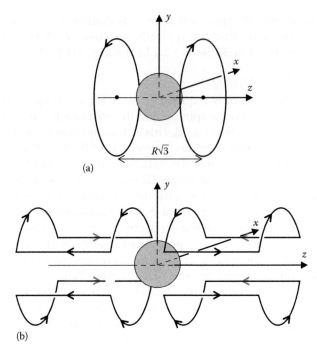

(a)

(b)

FIGURE 1.9
Elementary coil winding patterns that generate field gradients over a central region (indicated by the gray sphere). The main magnetic field B_0 is assumed to be in the $\hat{z}$-direction. (a) A Maxwell coil pair, with currents flowing in opposite directions, creates a z-odd field variation and hence produces a longitudinal gradient G_z. For the specified coil spacing, the z^3 term in the series expansion of the field vanishes, and G_z has a high uniformity. (b) A Golay coil consists of four saddle coils with currents flowing in the specified directions. It produces a transverse gradient G_y. Current flowing in the wires parallel to the z-axis does not produce a field in the $\hat{z}$-direction and does not affect G_y. The positions of the arcs and the angle that they subtend are chosen so as to maximize gradient uniformity. Note that the orientation of the x, y, and z axes differs from that chosen in Section 1.2.1: the z-axis is still aligned with B_0, but is now shown as being horizontal as it is for almost all clinical MRI systems.

Figure 1.9. Improved performance is obtained using analytical or numerical methods to obtain complex yet compact winding patterns for which gradient strengths and uniformity are maximized and the associated coil inductance is minimized [58,59].

Normally the field distortions induced by gradient coils are very small. For example, at 1.5 T they are typically less than 1% of B_0 over the Field of View (FOV). A careful examination of the magnetic fields produced by gradient coils reveals that they always do more than simply provide the nominal "desired effect." For example, a coil that produces a longitudinal field gradient dB_z/dz always produces a gradient in orthogonal components of the field (dB_x/dx and/or dB_y/dy).† In many

---

* Large magnetostatic forces are exerted on these coils whenever current flows through them. Imaging sequences normally require rapid switching or pulsing of the field distortions. The sudden changes in mechanical stress exerted on the coil support structure produce the characteristic patterns of acoustic noise that are generated by MRI scanners.

---

† The term *field gradient*, which is ubiquitous in MRI, is really a misnomer. It is technically a tensor quantity with nine components. In free space, constraints imposed by Maxwell's equations reduce the number of independent terms in the tensor to 5.

instances, the effects produced by these "concomitant gradients" (or Maxwell terms) are negligible, simply because the magnitude of the transverse components of the distorted field (B_x and B_y) are so small compared to the longitudinal field B_0. However, this is not always the case, particularly when very strong gradients or weak static fields are employed.

1.2.2.1.3 RF Field

Up to this point, all of the coils that have been discussed serve to control and manipulate the longitudinal magnetic field B_z, and hence the Larmor precession frequency. They do not produce fields that induce transition between nuclear energy levels. For this one usually needs a field directed orthogonal to B_z that oscillates at the Larmor frequency or close to it. That is, the field B_1 of Section 1.2.1.2 that causes the net nuclear magnetization **M** to rotate about an axis perpendicular to z, as long as it is applied.

In practice, the rotating field B_1 is often obtained as one of the two counter-rotating components of a linearly polarized oscillating field (see Figure 1.10). The other component, which rotates in the opposite sense, is detuned from the nuclear resonance by twice the Larmor frequency. It thus has almost no effect on nuclear spin dynamics. This linearly polarized oscillating field is produced by driving a time-varying current $i_0 \cos \Omega t$ through a coil with an appropriate geometry. The amplitude of this current (and hence the amplitude of the field B_1) and the time period over which it is applied control the tip (or flip) angle. Dozens of such coils have been developed and employed for MRI; they are generically known as transmit coils (or antennas), TX coils, RF coils, or even B_1 coils ("Bee-one coils"). Here, the reference to radiofrequencies is simply the fact that Larmor precession frequencies are typically in the radiofrequency range of the electromagnetic spectrum.

More often than not RF coils are "tuned"; inductive and capacitive elements in the circuit are balanced so that the net electrical impedance is purely resistive at the Larmor frequency. This facilitates efficient coupling between the transmitter and the coil, and hence efficient production of the largest possible B_1 field amplitudes.

Normally, RF coils are designed to produce a reasonably homogeneous oscillating field over the volume to be imaged, in order to generate reasonably uniform flip angles [60,61]. Some designs involve little more than a circular conducting loop, tuned to resonate at the Larmor frequency. A current flowing through this loop produces a magnetic field that is roughly aligned with its axis, and that is reasonably strong out to a distance of order its radius. This is an example of a "surface coil" (see Figure 1.11); it is convenient in situations where the tissues of interest are close to the surface and flip angle homogeneity is not terribly important. Other RF coils involve more sophisticated arrangements of current paths. The birdcage coil, for example, involves a series of long straight parallel conductors uniformly arranged around the periphery of a cylinder (Figure 1.11). These conductors act like inductors, and are carefully and individually tuned with capacitors to produce a resonance at the desired frequency; at resonance, the current flowing through the wires at any instant in time

FIGURE 1.11
A surface coil (lower) and a birdcage coil (upper), both of which can function as a transmit (TX) coil, a receive (RX) coil, or a combined TX/RX coil. A few of the capacitive and inductive elements in each coil are shown schematically.

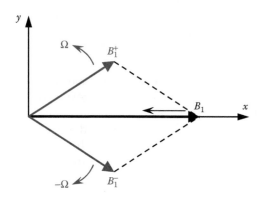

FIGURE 1.10
A linearly polarized oscillating magnetic field $B_1(t) = B_1^m \cos \Omega t$ aligned with the x-axis can be decomposed into two rotating components with equal and constant amplitudes but opposite angular velocities. That is, $B_1(t) = (B_1^m / 2)[\exp(i\Omega t) + \exp(-i\Omega t)]$.

is a sinusoidal function of the azimuthal angle. This arrangement is an example of a "volume coil"; it produces a very homogeneous B_1 field directed perpendicular to the axis of the cylinder. Importantly, the intensity of this field remains constant in time while its direction rotates. In the reference frame rotating at the Larmor frequency, this field is constant.

The use of coils like the birdcage that produce rotating or "circularly polarized" B_1 fields (as opposed to linearly polarized fields; see Figure 1.10) can be very important at high frequencies where essentially all of the energy delivered to the coil by the transmitter is ultimately dissipated in the subject. In effect, half of a linearly polarized oscillating field is wasted from an NMR perspective. It does, however, contribute to the rate at which energy is deposited in the subject. This point is discussed further in Section 1.2.2.4.

1.2.2.2 Faraday's Law: NMR Detection

A changing magnetic field creates an electric field **E**. The faster the magnetic field changes, the more intense is the resulting electric field. This is the essence of Faraday's law of induction,[*] which forms the basis for the detection of most NMR signals.

The net precessing nuclear magnetization that is established after a tipping pulse is applied has associated with it a small magnetic field whose orientation rotates about the z-axis at the Larmor frequency. This changing magnetic field produces an associated electric field, which also changes as a function of time. If an open loop of wire (or a coil with many turns) is placed near the region in which the precessing nuclei are situated, and arranged so that it intercepts some of the changing magnetic flux produced by those nuclei, an *electromotive force* (emf) or potential will be established between its two ends (see Figure 1.12). This emf is proportional to the amplitude of the transverse component of the precessing magnetization and to the precession frequency. It depends on the actual distribution of magnetization in an extended sample and on the geometry of the coil and sample. The principle of reciprocity [62] enables one to conveniently compute this elementary emf as a function of the coil shape and source position. The total emf, $e(t)$, induced around the loop is obtained by integration over the entire sample volume, taking into account variations in the phase of the nuclear magnetization across the sample and geometrical weighing factors. Typically, the detected emf is very small and needs to be amplified as soon as possible to avoid unnecessary degradation of signal quality due to interference from external noise sources. The

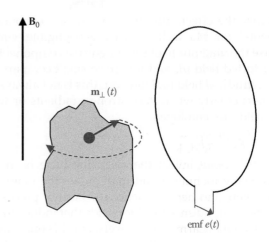

FIGURE 1.12
An elementary volume (the red sphere) within the magnetized sample (the gray volume) has a net magnetic moment $\mathbf{m}_\perp$. Its precession in the applied field $\mathbf{B}_0$ gives rise to an oscillating magnetic flux in the nearby detection loop, hence to an emf $e(t)$ oscillating at the corresponding Larmor frequency ω_0.

use of tuned detection coils conveniently provides a significant enhancement prior to amplification. For a tuned coil with quality factor Q, the resulting *signal* is $S(t) = Qe(t)$.

The coil used to detect nuclear precession signals is another RF coil, in the sense that it operates at the same frequency as the B_1 coil. It is typically referred to as a detection coil, a receive coil (or antenna), an RX coil, or a pickup coil. In some cases the same physical coil is used for both transmit and receive functions, but more often than not one wants to minimize cross talk or interference between the two. When two coils are used, efforts are made to ensure that they are orthogonal. If this is not done, the intense B_1 field produced by the transmit coil will induce an enormous emf across the terminals of the receive coil, complicating attempts to detect subsequent nuclear induction signals.

Just as is the case for B_1 coils, one is often interested in using volume detection coils designed to have reasonably uniform sensitivity to precessing magnetization over the entire volume of interest. In such cases, signals are approximately proportional to the total (integrated) magnetization of the sample. Alternatively, surface coils or arrays of surface coils can be employed. The recorded signal(s) from these arrays provide coarse information about the location of the magnetization within the sample [38,39].

1.2.2.3 Signal Amplitude Considerations

It is often stated that NMR is an inherently insensitive experimental probe. What is meant by this is that the conventional signature of NMR precession—the induced emf $e(t)$ discussed in Section 1.2.2.2—is

---

[*] As was the case with Ampère's law in Section 1.2.2.1, Faraday's law provides a precise mathematical connection between B and E.

invariably a weak signal that is readily obscured by noise. Weak or not, those signals still encode a wealth of information about the local magnetic environment in which the sample nuclei are immersed.

Qualitative appreciation for the relative strength or weakness of conventional NMR signals can be gleaned through considering alternative methods for monitoring nuclear precession that were developed early in the history of the field. Rabi's 1938 demonstration of NMR [19], for example, involved measuring the flux of molecules in a weak but fully polarized beam. A deflection of the beam, and hence a change in detected intensity, was observed when an oscillating magnetic field was applied at the Larmor frequency. Two decades later, Brossel and Cagnac [63] realized Kastler's proposal [64] for optical detection of nuclear magnetic resonance in optically polarized atomic vapors. A change in the polarization of fluorescent light emitted by low-density Hg vapor was observed in response to a change in the nuclear spin state, driven by an oscillating magnetic field applied at the Larmor frequency.

These pioneering experiments have two features in common that distinguish them from conventional NMR. First, in both cases the measurement was indirect, involving the detection of molecules or visible photons rather than the magnetic field produced by nuclei. The energy scale associated with each detected event (molecule or photon) was in the electron-volt range: 9 orders of magnitude more than the energy difference between the two nuclear spin states in the applied magnetic field (see Table 1.1 for characteristic values). Second, in both cases the nuclear polarization was very far from equilibrium ($|p_{up} - p_{down}| \approx 1$), dramatically enhancing the contrast between measurements performed on- and off-resonance. Recall here for scale that the equilibrium nuclear polarization at room temperature in a 1 T magnetic field is of order 10^{-6} or 1 ppm, and yet smaller in weaker fields. Combined, these features represent an astounding 15 order-of-magnitude advantage (or enhancement in signal amplitude) relative to direct detection of the nuclear transition under equivalent conditions. Of course, this discrepancy is compensated in part through the huge increase in density that is obtained when liquid or solid samples are employed rather than molecular beams or dilute atomic vapors. Nevertheless, one is left with the naïve impression that direct detection of NMR presents a daunting signal acquisition problem relative to the highly leveraged schemes described above.

A proper evaluation of this problem requires consideration of two parameters: the amplitude of the detected signal and the amplitude of the detected noise. The quality of the signal is then expressed in terms of a signal-to-noise ratio. For given sample and coil geometries the detected emf $e(t)$ in conventional NMR scales

as B_0^2; one factor of B_0 comes from the dependence of the equilibrium magnetization on field (Equation 1.1) and the other comes from the fact that the induced emf is proportional to the time derivative of the precessing magnetization (Faraday's law), and hence $\omega_0 = |\gamma B_0|$ (Equation 1.3). Estimating the field dependence of the detected noise is more involved and requires an understanding of its physical origin. Here it is useful to make a distinction between extrinsic and intrinsic sources. Noise from extrinsic sources, such as RF interference (often referred to as electromagnetic interference or EMI) or noise generated by amplifiers and recording electronics, can typically be suppressed or minimized through design: MRI systems are typically installed in a shielded room (or Faraday cage) for precisely this reason. Noise from intrinsic sources, however, is unavoidable: it typically arises from thermal agitation of electrical charges (Johnson noise) in the sample and the detection coil. If the sample is the dominant source of intrinsic noise, Faraday's law introduces the same factor of $\omega_0 = |\gamma B_0|$ to the corresponding induced emf as it does for the signal, and thus the SNR increases linearly with B_0. If the coil is the dominant source of intrinsic noise, however, SNR increases more rapidly as the operating field is increased, scaling as $B_0^{7/4}$ [65,66].

For most clinical imaging applications, the sample (i.e., tissues in the subject's body) acts as the dominant source of intrinsic noise, particularly as the operating field is increased. Conversely, in low or very-low magnetic fields, when small samples are employed (as is the case for MRI of small animals or in MR microscopy experiments) or when nonconducting samples are probed, the detection coil tends to dominate the intrinsic noise. In such cases it can be advantageous to use cold probes or even superconducting coils [67]. The spectral density of thermal noise appearing across a resistor R at temperature T is given by $\sqrt{4k_B RT}$, and thus gains are realized through reducing both R and T. More exotic options for enhancing SNR in MRI for niche applications are being explored, such as the use of SQUID-based detectors [68] and optical magnetometers [69] for ultra-low-field applications and force detection in magnetic resonance force microscopy for submicron resolution MRI [70].

1.2.2.4 Health Safety Considerations

The application of time-varying magnetic fields can lead to undesirable effects. Intense RF tipping pulses induce strong Faraday electric fields, which can in turn drive eddy currents and cause energy dissipation in the tissues of a subject. This energy dissipation rate has a strong frequency dependence, scaling as ω^2 over many decades in frequency [66]. In high-field MRI systems, where RF frequencies in the VHF band of the electromagnetic spectrum (30–300 MHz) and above are employed, situations can arise in which most of the RF power delivered to the B_1 coil is dissipated in the subject [71,72]. Likewise, fast switching of magnetic field gradients can also induce strong Faraday electric fields. In this case the characteristic frequencies are much lower and energy dissipation in the subject is usually not a concern. Instead, the induced electric fields E can cause peripheral nerve stimulation [73]. In both cases, restrictions and standards imposed by regulatory bodies and international commissions [73] limit the maximum permissible Faraday electric fields that can be induced. For RF fields, these limitations are normally expressed in term of specific absorption rates (SARs); for switched magnetic field gradients, limitations are variously expressed in terms of peak values of dB/dt and/or E, as well as direct volunteer-based observations of nerve stimulation thresholds. These issues are discussed further in Chapter 10.

1.3 Fundamentals of MRI

This chapter began with a brief and qualitative description of Paul Lauterbur's first published MRI experiment, as summarized in Figure 1.1. In this section, we revisit that experiment and examine somewhat more precisely the nature of the 1D projections of the "NMR response" that he obtained. We then survey a few key modern approaches to MR image acquisition, based on the use of pulsed NMR and Fourier transform methods. In effect, this sets the stage for the remainder of the book.

This section is organized into three parts: The first deals with methods for generating 1D projection images; the second deals with the acquisition of data for 2D and 3D images; and the third identifies the primary methods through which the "NMR response" is tuned or adapted to reflect different aspects of the nuclear environment. It is at this point in the process—the sensitization of NMR signals to different "contrast mechanisms"—that crucial connections are formed between the acquired data and the underlying structure and function of the tissues that are imaged.

1.3.1 Effect of a Field Gradient: 1D Imaging

Consider a uniform magnetic field $\mathbf{B}_0 = B_0 \hat{z}$ upon which is superimposed a uniform gradient $G \equiv dB_z/dk$, in the direction $\hat{k}$. The strength of the resulting magnetic field is a function of position $\mathbf{r}$. As long as concomitant gradients can be ignored, it can be written:

$$B_0(\mathbf{r}) = B_0(0) + G\hat{\mathbf{k}} \cdot \mathbf{r}. \qquad (1.16)$$

The Larmor precession frequency of nuclei subjected to this field similarly becomes a linear function of position, and takes on the same value in any given plane oriented perpendicular to $\hat{k}$. If a time-varying magnetic field $\mathbf{B}_1$ oriented perpendicular to $\hat{z}$ is then applied at angular frequency ω, resonance will only occur in the vicinity of one such plane. For a continuous wave (CW) NMR experiment, where the field B_1 is applied continuously, the width of this region is of order $\delta r = 1/(\gamma G T_2)$.

If the NMR signal is received using a coil that has uniform coupling to all parts of the sample (e.g., a long solenoid or a birdcage coil), its amplitude is proportional to the number of nuclei in the band that is excited. Sweeping or stepping the frequency ω or the field B_0 causes the resonant band to translate across the sample, generating a 1D map or projection image of the magnetization density. The frequency (or field) scale for this mapping is set by the strength G of the field gradient. More precisely, the projection represents the convolution of a Lorentzian line shape (whose width is set by the spin–spin relaxation rate $1/T_2$) with the net nuclear magnetization density. As long as $1/T_2$ is appropriately small, the signal reflects the spatial distribution of M_0. This is illustrated in Figure 1.13a, which shows the spectrum that is expected for a magnetic field gradient applied perpendicular to the axis of a cylindrical sample of radius a. It thus represents the signal that Lauterbur would have observed had he used only one tube in his experiments. By using two tubes instead of one, and by varying the direction $\hat{k}$ in which the gradient was applied relative to their axes, he was able to resolve both

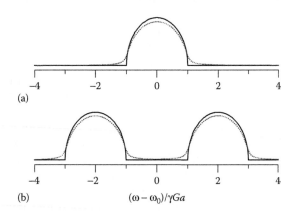

(a)

(b) $(\omega - \omega_0)/\gamma Ga$

FIGURE 1.13

(a) Anticipated CW NMR spectra for a cylindrical sample of radius a in a uniform magnetic field gradient G applied perpendicular to its axis. *Solid line*: Ignoring the effects of relaxation, the spectrum is proportional to $\sqrt{1 - (\omega - \omega_0)^2 / (\gamma Ga)^2}$, which is simply the projected width of the cylinder at a distance $(\omega - \omega_0)/(\gamma Ga)$ from its axis. *Dotted line*: With some relaxation added to the Bloch equations (but ignoring the effects of diffusion), sharp features are smoothed out; here $\gamma GaT_2 = 2$. (b) Anticipated spectra for two parallel cylindrical samples, a situation modeling Lauterbur's experiment summarized in Figure 1.1. Solid and dotted lines correspond to relaxation being ignored or included, as in (a).

their physical extent and their apparent separation, as illustrated in Figure 1.13b. In the remainder of this section we will assume that $1/(\gamma G a T_2) \ll 1$ and that the effects of diffusion can be ignored.

The procedure outlined above—sweeping the static magnetic field (or the frequency at which the field B_1 is applied) and collecting CW NMR spectra for various field gradient orientations—is time consuming. Most modern implementations of Lauterbur's experiment employ pulsed NMR, in which the field B_1 is only applied for finite periods of time. Subsequent to these tipping pulses, the nuclear magnetization **M** undergoes free precession and induces a time-varying emf in the detection coil:

$$e(t) \propto \int_{\text{sample}} M_\perp^0(\mathbf{r}) \cos[\gamma B_0(\mathbf{r})t + \varphi]\, e^{-(t/T_2)}. \qquad (1.17)$$

Here $M_\perp^0(\mathbf{r})$ is the local amplitude of the transverse magnetization immediately after the tipping pulse (at time $t = 0$), the local position-dependent Larmor frequency $\gamma B_0(\mathbf{r})$ depends on the local magnetic field strength (given by Equation 1.16) and the phase φ is a parameter that depends on the particular tipping pulse that is applied and on the position of the coil with respect to the sample.

Normally, the signal that is actually recorded during a pulsed NMR experiment is obtained by mixing the detected high-frequency emf $e(t)$ with a reference signal at a comparable frequency and fixed phase. The reference signal is often referred to as the local oscillator (LO), and also forms the basis for generating the field B_1. A common choice is thus to set $\omega_{\text{LO}} = \gamma B_0(0)$. The resulting "signal"[*] is complex:

$$S(t) = \int_{\text{sample}} M_\perp^0(\mathbf{r}) \exp[i(\gamma G \hat{k} \cdot \mathbf{r})t]\, e^{-(t/T_2)}. \qquad (1.18)$$

It has two components: one that is in-phase with the LO and another that is 90° out of phase (or "in-quadrature") with the LO. These are referred to as the real and imaginary parts of the signal, respectively. Equation 1.18 represents the sum of contributions to the detected emf arriving from all parts of the sample, as viewed in a frame of reference rotating at the local oscillator frequency ω_{LO}. The complex nature of $S(t)$ keeps track of the sense of rotation in the rotating frame, and thus discriminates between frequencies that are above or below ω_{LO}.

Figure 1.14a shows the anticipated signal $S(t)$ following a single tipping pulse applied to the same cylindrical sample used to generate Figure 1.13a. The rapid apparent attenuation of this "free induction decay"

---

[*] $S(t)$ is only called the signal for convenience. It is merely proportional to the voltages that are digitized and recorded.

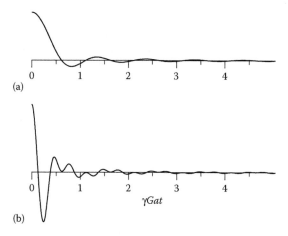

FIGURE 1.14
(a) Real part of the NMR free induction decay signal (Equation 1.18) anticipated for a cylindrical sample of radius a in a uniform magnetic field gradient of amplitude G directed perpendicular to its axis, as was the case in Figure 1.13a. The phase of the signal is constant, and its imaginary or quadrature component (not displayed) is zero at all times. The effects of relaxation and diffusion have been ignored. (b) Anticipated signal for two parallel cylinders, analogous to the situation in Figure 1.13b.

(FID) merely reflects the fact that a magnetic field gradient has been imposed. The nuclear precession frequency varies as a function of position across the sample or subject, and the net emf detected by the coil is subject to destructive interference between the contributions generated at different locations (see Figure 1.12). For a sample of size a, the characteristic timescale for the apparent decay of $S(t)$ is of order $1/(\gamma G a)$. Figure 1.14b shows the expected signal for the sample of Figure 1.13b: the high-frequency modulation within the same overall signal lifetime corresponds to interference, or "beats," between contributions from the two tubes. This illustrates how an applied frequency-encoding gradient links spatial characteristics of a sample to spectral features in the acquired data.

The information present in Figure 1.14 can be used to recover the same spectra as were obtained in the CW NMR experiment. All that is required is a calculation of the Fourier transform of $S(t)$ after setting $S(-t) = S^*(t)$, where the $*$ indicates complex conjugation. This Fourier transform represents the distribution of frequencies present in the FID, and hence the distribution of nuclear precession frequencies in the sample as the FID was recorded. Thus, apart from the necessary computations, the time required to obtain a 1D projection image of the sample via pulsed NMR is dramatically reduced relative to the CW approach outlined earlier.

The dephasing of the precessing magnetization that is responsible for the apparent decay in Figure 1.14 occurs on a timescale that is short compared to that set by relaxation (i.e., T_2). It can be represented pictorially with a series of isochromats, as was done in Figure 1.6.

Similarly, it can be refocused in order to generate spin- or gradient-echoes. Therefore, for example, if the direction of the field gradient is inverted at time τ, the recorded signal at subsequent times (i.e., $t > \tau$) becomes

$$S(t) = \int_{\text{sample}} M_{\perp}^0(\mathbf{r}) \exp[i(\gamma G \hat{k} \cdot \mathbf{r})(2\tau - t)] \, e^{-(t/T_2)}. \quad (1.19)$$

Spins that were initially precessing faster than average end up precessing slower after the inversion, and vice versa. By the time $t = 2\tau$, the net phase accumulated by all spins in the sample is the same and an echo is formed. The amplitude of the recorded signal at $t = 2\tau$ is the same as it was at $t = 0$, to the extent that relaxation and diffusion can be ignored. Moreover, the recorded signal at times $t > 2\tau$ evolves just as it did immediately after the tipping pulse (see Figure 1.14). And, just as was the case for the initial FID, calculating the Fourier transform of the recorded echo data yields a 1D projection image of the sample.

This gradient echo formation procedure can be generalized. For example, the sense in which G is applied can be periodically reversed at times τ, 3τ, 5τ, ... as shown in Figure 1.15a to form an echo train; that is, a periodic revival of phase coherence across the sample. The inevitable decay of the peak response every time an echo is formed provides information about irreversible processes such as relaxation and diffusion. More generally, the strength of the field gradient before and after the reversal can be changed; an echo is then formed every time the time integral of $G(t)$ vanishes.

HARD OR SOFT PULSES: WHICH SPINS ARE EXCITED?

When a magnetic field gradient G is present, the spatial uniformity of the flip angle induced by an RF pulse depends critically on the amplitude of the field B_1. This effect was alluded to in connection with Figure 1.5. When an intense RF pulse is applied, the effective magnetic field $\mathbf{B}_{\text{eff}}$ in the rotating frame is dominated by $\mathbf{B}_1$. The angle ψ shown in Figure 1.5 is thus always very close to 90° and the trajectory traced by the tip of the magnetization vector $\mathbf{M}$ on the Bloch sphere is essentially part of a great circle. The condition required for this to be true everywhere in the sample is that $B_1 \gg Ga$, where a is the size or extent of the object in the direction that G is applied. This condition defines what is known as a "hard pulse." The FID pictured in Figure 1.14 was implicitly launched using a hard RF pulse; even though a magnetic field gradient was present, the trajectory of $\mathbf{M}$ during the pulse and the initial phase of the transverse magnetization immediately afterward were effectively the same at all points in the sample.

A very different result is obtained when a "soft pulse" is applied. In this limit, $B_1 \ll Ga$, and the effective magnetic field $\mathbf{B}_{\text{eff}}$ in the rotating frame is dominated by the apparent field $\mathbf{B}_{\text{app}}$ (Equation 1.5), which—because of the applied gradient—is a function of position. To first approximation, only spins located near the plane $\hat{k} \cdot \mathbf{r} = (\omega - \omega_0)/(\gamma G)$ are strongly influenced by B_1. This forms the basis for "slice selection," which is discussed further in Section 1.3.2. The width of the region over which a soft pulse is effective in rotating the magnetization depends on the duration of the pulse, and the spatial profile of the flip angle can be controlled by the shape of the RF pulse envelope in time.

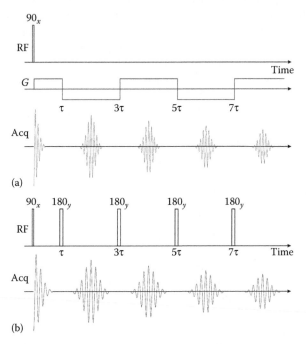

FIGURE 1.15
Schematic diagram showing the synchronization of applied B_1 fields (RF), switched gradient fields (G), and acquired nuclear induction signals (Acq) for two pulse sequences. Both begin with a 90° RF tipping pulse that rotates the thermal equilibrium magnetization into the transverse plane. In the first example (a), the sense of the applied magnetic field gradient G is periodically reversed, creating a train of gradient echoes. In the second example (b), the phase of the magnetization is periodically inverted by applying 180° tipping pulses, creating a train of spin echoes. The time between the initial RF pulse and the peak of the first echo (or the time between successive echoes) is referred to as the echo time and is conventionally denoted TE or T_E. Often, sequences such as those shown here are repeated, in which case the repetition time (the time between successive applications of the sequence) is conventionally denoted TR or T_R.

Alternatively, the spin-echo procedure outlined in connection with Figure 1.6 can be used to generate echoes and echo trains as shown in Figure 1.15b. In this case the gradient G remains the same, but the phase of the magnetization is inverted by applying 180° tipping pulses about any axis in the transverse plane. The phase advance of "fast" spins near one end of the sample suddenly becomes a phase lag, and vice versa for "slow" spins on the other end. If the inversion occurs at time τ, an echo is formed at time 2τ.

The signals that are obtained when spin- and gradient-echoes are formed are similar, but they are not identical. Experimental constraints often dictate the selection of one approach over the other. For example, a weakness of the gradient-echo technique is that the inversion of field gradients is usually imperfect. The current delivered to the gradient coils (and hence the field they produce) can certainly be inverted, but the static magnetic field itself is never perfectly uniform. Thus, inverting the applied magnetic field gradient G is not quite the same as inverting the total magnetic field gradient. Even worse, when strong gradients are applied (or weak magnetic fields are employed), the notion of a magnetic field gradient itself breaks down, and the influence of orthogonal components of the field on the time evolution of $\mathbf{M}$ needs to be considered. In either case, a progressive loss of phase coherence occurs on a timescale that is often shorter than the transverse relaxation time T_2. This particular limitation can be eliminated by generating spin echoes instead of gradient echoes. To the extent that perfect and uniform 180° tipping pulses can be generated, inverting the phase of the magnetization leads to coherent echo formation irrespective of imperfections in B_0. However, a limiting factor for spin-echo sequences when human subjects are involved is the need to employ short, high-amplitude B_1 fields in order to obtain uniform 180° rotations. In high-field systems, the energy associated with these pulses is invariably deposited in the subject and can pose a safety hazard (see Section 1.2.2.4 and Chapter 10).

Ultimately, the precision of the spectrum representing the 1D distribution of the precessing magnetization that is obtained from the Fourier transform of the recorded signal is limited. The time variation of $S(t)$ is sampled at discrete points in time (at a sampling frequency f_s) and over a finite period of time (T_{obs}) that is usually centered* on

the echo formation time 2τ or $2n\tau$. The spectrum obtained from these data (through a discrete Fourier transform procedure) has a frequency resolution $1/T_{obs}$ and extends over a frequency range $\pm f_s/2$. This frequency resolution limits the maximum spatial resolution of the 1D projection image described above to $1/(\gamma G T_{obs})$. Factors such as a finite transverse relaxation time T_2 (see Figure 1.13), diffusion, and noise (see Section 1.2.2.3) all serve to reduce this maximum resolution. The frequency range also plays an important role; it imposes a finite FOV given by $f_s/(\gamma G)$. The FOV must be larger than the physical extent of the sample or subject; otherwise folding artifacts associated with undersampling of high-frequency components of the signal can occur. Artifacts in MRI are discussed further in Chapter 9.

1.3.2 2D and 3D Imaging Methods

Paul Lauterbur's 2D NMR images of water-filled capillaries (such as the one shown in Figure 1.1) were obtained through a process of mathematical inference known as "back projection." This involves combining several 1D or "line" images, each acquired in a different direction in the same plane, to form a 2D image on a grid of points by iteratively modeling the unknown distribution of nuclear magnetization. Although this approach to image reconstruction played an important role in the early evolution of MRI, it is not often used today. Modern MR image reconstruction relies heavily on the phase and frequency encoding of the (precessing) nuclear magnetization in a sample, multidimensional Fourier transform techniques, and the selective excitation of nuclei in specific, well-defined planes or bands intersecting the sample or subject. The general principles by which MR image data are acquired are discussed next. Further examples are presented in Chapter 2. We begin with the process of "slice selection," which is ubiquitous in modern MRI.

A soft RF tipping pulse applied in the presence of a magnetic field gradient $G\hat{k}$ is selective. Maximum rotation of the magnetization vector is obtained in the plane defined by $\hat{k} \cdot \mathbf{r} = (\omega - \omega_0)/(\gamma G)$; elsewhere the effect is much smaller. For small tip angles,[†] the spatial width of the region that is influenced is proportional to the spectral width of the pulse, which is in turn inversely proportional to its duration in time τ. Rectangular RF pulse envelopes, where the field B_1 has a constant amplitude for a finite period of time, yield awkward sinc-shaped slice profiles in space; the plane on which the maximum rotation is obtained is symmetrically flanked by a series

---

* The observation window does not have to be centered on the echo time; a recording of half of the echo starting or ending with the signal at the echo time is sufficient. This is possible because of the symmetry of echoes, which result from the fact that the phase of the precessing magnetization was uniform at the start of the experiment. A full recording of a FID starting immediately after a tipping pulse (as implied by the example shown in Figure 1.14) is often technically difficult to acquire; a delay following the tipping pulse is normally required to avoid saturation of the detection electronics by the applied B_1 field. A symmetric echo is usually recorded to increase SNR, but asymmetric echoes are often used in ultrafast acquisition schemes.

---

† More sophisticated pulse shaping procedures are required when large tip angles are desired [74], because of the nonlinear dependence of flip angle on RF amplitude and detuning illustrated in Figure 1.5.

of side lobes. A sinc-shaped RF pulse envelope is much more useful. Modulating the amplitude of B_1 in time such that

$$B_1(t) = B_1(0) \frac{\sin\left(\frac{\gamma G \Delta z\, t}{2}\right)}{\left(\frac{\gamma G \Delta z\, t}{2}\right)} \equiv \mathrm{sinc}\left(\frac{\gamma G \Delta z\, t}{2}\right) \quad (1.20)$$

yields a rectangular (uniform) slice of width Δz in space. Starting from thermal equilibrium, a soft sinc-shaped, small-tip-angle RF pulse applied in the presence of a "slice selection gradient" conveniently produces a transverse magnetization that is uniform in amplitude over this slice and zero elsewhere. This would be ideal for imaging if not for the fact that, by the end of the pulse, the magnetization is strongly dephased by the gradient. The simplest and most convenient way to deal with this is by generating a gradient-recalled echo. Rather than turning off the slice selection gradient at the end of the RF pulse, its direction is momentarily reversed as shown in Figure 1.16. At the point in time where the echo is formed ($t = T_e$), the transverse magnetization in the sample or subject is uniform in a well-defined slice of thickness Δz and zero elsewhere. This slice can be positioned anywhere in the sample by choosing an appropriate frequency for B_1. As long as a 2D image can be generated from the selectively excited magnetization in each slice, a full 3D image can be recorded.

A strategy for generating a 2D image following slice selection is summarized in Figure 1.17. It proceeds as

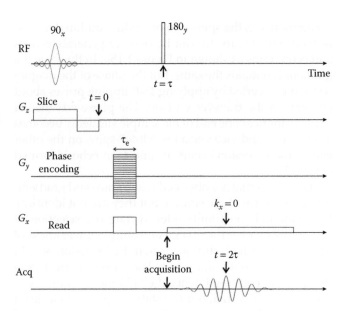

FIGURE 1.17
Outline of a sequence for 2D Fourier transform imaging. The first RF pulse in the presence of a gradient ($G_z \hat{z}$) excites one rectangular slice of the sample (see Figure 1.16). The second RF pulse at $t = \tau$ generates a spin echo that peaks at time $t = 2\tau$. A phase-encoding gradient ($\mathbf{G}_e \equiv G_y \hat{y}$) applied during the first free evolution period establishes a gradient in the phase of the transverse magnetization along the y-axis. Each time this sequence is repeated, a different phase encoding gradient is employed. An orthogonal read (or frequency-encoding) gradient ($\mathbf{G}_r \equiv G_x \hat{x}$) is then applied while the echo is acquired. The read gradient causes the transverse magnetization to dephase, and thus an extra gradient pulse is applied along $\hat{x}$ during the first free evolution period to compensate. The net phase shift caused by the read gradient at the echo maximum is thus zero. In practice, refinements to this basic scheme (such as the use of a selective 180° RF pulse) are often either necessary or desirable.

follows: A gradient $\mathbf{G}_e$ is first applied in the plane of the slice for a time period τ_e. This establishes a gradient in the phase of the transverse magnetization as it undergoes free precession. This gradient is consequently referred to as a phase-encoding gradient. Next, a 180° RF pulse is applied at time $t = \tau > \tau_e$ to invert the phase of the precessing magnetization (relative to the phase of the pulse). A second gradient $\mathbf{G}_r$ is then applied in the plane of the slice but perpendicular to $\mathbf{G}_e$, in anticipation of the spin echo generated by the 180° rotation. This gradient is known as a read gradient* because it is applied during signal acquisition.

As a concrete example, assume that a rectangular slice of thickness Δz has been selected perpendicular to the z-axis and that $\mathbf{G}_r$ and $\mathbf{G}_e$ happen to be applied along the x and y axes, respectively. The recorded signal (analogous to Equation 1.19) in the vicinity of the echo is of the form

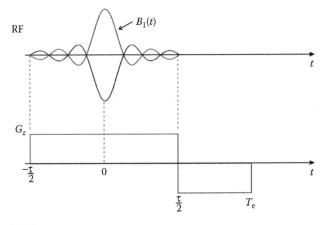

FIGURE 1.16
A soft sinc-shaped RF pulse for selective excitation of the transverse magnetization in a well-defined slice. The slice selection gradient G_z is reversed at the end of the RF pulse in order to unwind the accumulated phase of the magnetization. An echo is formed at the same instant that the gradient is turned off ($t = T_e$). The schematic depiction of the RF pulse only indicates the envelope or amplitude of $B_1(t)$ (see Equation 1.20). The high-frequency oscillation of B_1 at $\omega \sim \omega_0$ is not shown. Note that truncation of the sinc function distorts the rectangular slice profile. We have ignored such effects in this discussion.

---

* It is also referred to as a frequency-encoding gradient, in connection with the manner in which it is employed for Fourier transform spectroscopy. For imaging, signals are processed in terms of the spatial modulation of the magnetization in the sample (k-space) rather than frequency.

$$S(t > \tau) = \int_{\text{sample}} M_\perp^0(\mathbf{r}) \exp[i(k_x(t)x + k_y y)] e^{-(t/T_2)} \quad (1.21)$$

where $k_x(t) = \gamma(2\tau - t)G_r$ and $k_y = \gamma \tau_e G_e$ are interpreted as components of a wave vector, describing the spatial modulation of the magnetization in the sample. Equivalently, the recorded signal represents

$$
\begin{aligned}
S(k_x, k_y) &= \int_{\text{sample}} M_\perp^0(\mathbf{r}) \exp[i(k_x x + k_y y)] e^{-(t/T_2)} \\
&= \Delta z \int_{\text{slice}} M_\perp^0(x, y) \exp[i(k_x x + k_y y)] e^{-(t/T_2)}
\end{aligned} \quad (1.22)
$$

at a fixed value of k_y and over a range of k_x. More generally, ignoring attenuation and assuming $S(k_x, k_y)$ is known for all values of k_x and k_y, the 2D inverse Fourier transform of Equation 1.22 is $M_\perp^0(x, y)\Delta z \propto M_\perp^0(x, y)$, which is the desired 2D spatial distribution of nuclear magnetization in the slice.

In practice, the sequence shown in Figure 1.17 is repeated N_e times for evenly spaced values of k_y corresponding to phase-encoding gradients in the range $-G_{\max} < G_e < G_{\max}$. Each iteration probes $S(k_x, k_y)$ for evenly spaced values of k_x set by the sampling rate f_s and the acquisition time. The result is a 2D Cartesian array of data spanning a range of "k-space," as shown in Figure 1.18a. The (discrete) 2D inverse Fourier transform of these data corresponds to a 2D image of the transverse nuclear magnetization in the slice. The FOVs in the direction of the read gradient and the direction of the phase-encoding gradient are $f_s/(\gamma G_r)$ and $N_e/(2\gamma\tau_e G_{\max})$, respectively.

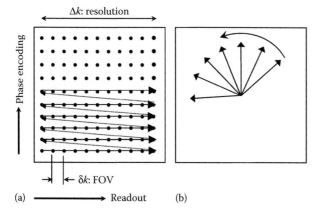

FIGURE 1.18
Examples of trajectories for 2D k-space sampling. (a) Cartesian mapping of the plane is obtained from a series of data acquisitions, each with different phase encoding in the y-direction. (b) Radial mapping is obtained from acquisitions along different azimuthal directions, each one of which samples the center of k-space.

k-SPACE IN MRI

The signals that are detected and recorded in MRI do not usually come from localized regions of the sample or subject. Rather, they represent spatially modulated depictions of the nuclear magnetization integrated over the entire volume to which the receive coils are sensitive. This spatial modulation is imposed and controlled by the linear magnetic field gradients that are applied and can be characterized by a wave vector **k**. A mathematical analysis of this problem shows that the detected signals are nothing more than Fourier transforms (or a spatial frequency representation) of the nuclear magnetization distribution that is being imaged.

The space in which the components k_x and k_y of the wave vector **k** are the natural parameters of the recorded signal $S(k_x, k_y)$ is conventionally referred to as k-space or *reciprocal space*. As long as enough data are acquired to characterize $S(k_x, k_y)$ in reciprocal space, one need only perform an inverse Fourier transform to reconstruct an image of the magnetization distribution in real space [75–78].

Imaging sequences, such as the one shown in Figure 1.18, are often thought of as being recipes or instructions for acquiring data that span k-space. Strictly speaking, standard receive coils only sense the average nuclear magnetization and hence only monitor the centre of k-space. It is actually the Fourier transform of the magnetization that traverses k-space as the imaging sequence is executed. Nevertheless, all of the information needed for image reconstruction is still acquired.

The concept of k-space or reciprocal space is commonplace in disciplines such as crystallography, solid-state physics, and optics. Often one is able to view or resolve key features of a complex system much more clearly in k-space than in real space. An important example is the phenomenon of Bragg diffraction, which occurs when coherent short-wavelength radiation is scattered from the lattice planes of a crystalline solid or other periodic structure. This is not the case in MRI. Generally, little or no useful information is evident when one looks at raw k-space MRI data. It is only when the real space image is reconstructed that useful information is revealed.

Many variants of this basic strategy exist. For example, the spin echoes generated by the sequence shown in Figure 1.17 are readily replaced with gradient-recalled echoes, which in some cases are useful for fast imaging. Even faster rates are possible when echo trains are formed after each excitation. In this way a different phase-encoding gradient can be used each time the magnetization is refocused, enabling rapid passage through k-space. This is the basic idea behind echo-planar imaging. Other approaches include radial acquisition schemes, as suggested in Figure 1.18b. These enable frequent resampling of the center of Fourier space, which can be used to help minimize movement artifacts to which other sequences are susceptible. It also permits fast time resolution using sliding window methods. Yet other strategies for acquiring data in k-space employ spiral (or interleaved spiral) trajectories or partial (e.g., half-plane) acquisition schemes [79].

1.3.3 Contrast

As described, the imaging strategies outlined in Sections 1.3.1 and 1.3.2 all have one thing in common: the acquired signal—and hence the image that is generated—nominally reflects the local magnetization density in the sample or subject. If proton NMR is employed, the signal strength scales with the density of H atoms, which are abundant in the water and lipids of all tissues. As a result, these strategies yield anatomical images with poor contrast and are of little use from a clinical perspective. It is only when they are "tuned" or modified to probe physical processes that are tissue-specific that the real potential of MRI is realized. Sensitization—or "weighting"—of the acquired signal to these processes can yield significantly enhanced contrast between organs with similar proton density, or between regions characterized by normal and pathological behavior. A brief summary of common image weighting schemes is given below. Much more information can be found in later chapters.

Sensitization of acquired NMR signals to nuclear relaxation—irreversible processes characterized by the phenomenological parameters T_1 and T_2—is the most obvious and most widely employed method of enhancing image contrast. In fact, Lauterbur's original demonstration of 2D MRI included a T_1-sensitized image [1]. And, in connection with this image, he noted that longer than normal values of T_1 had been observed in malignant tumors [80].

T_1 weighting of images is naturally obtained when the sequence repetition time T_R (or TR) is comparable to the longitudinal relaxation time. Endless repetition of the sequence under these conditions does not provide enough time for the magnetization to return to thermal equilibrium between iterations. As a result, the steady-state magnetization is suppressed relative to its thermal equilibrium value M_0 such that

$$M = M_0 \frac{1 - \exp\left[-\left(\dfrac{T_R}{T_1}\right)\right]}{1 - \cos\theta \exp\left[-\left(\dfrac{T_R}{T_1}\right)\right]} \qquad (1.23)$$

where θ is the tip angle of the RF pulse (see Figure 1.4).

For appropriate choices of θ and T_R, the magnetization in regions characterized by short values of T_1 will recover more than that in regions characterized by long values of T_1. They will thus provide larger amplitude signals and ultimately be rendered as more intense regions on an image. In normal tissues, fat (lipids) is characterized by shorter values of T_1 than water, and thus it appears white in T_1-weighted MR images. Thus, cerebral white matter appears white on a T_1-weighted MR image because it contains more lipids than does gray matter.

T_2 weighting of images is obtained when the echo formation time T_e (or TE) is comparable to the transverse relaxation time T_2. Under these conditions, significant irreversible dephasing of the transverse magnetization occurs between the initial RF excitation (tipping pulse) and the measurement of the signal amplitude. This effect is responsible for the factor $\exp(-t/T_2)$ appearing in the various expressions for $S(t)$ and $S(k_x, k_y)$ in Section 1.3.2; when $t = T_e \sim T_2$, the resulting signal attenuation becomes significant. This is precisely the opposite of the effect observed for T_1 weighting. That is, regions characterized by strong T_2 relaxation yield relatively weak NMR signals and are rendered as being dark on a T_2-weighted image. Regions characterized by strong T_1 relaxation yield relatively strong NMR signals and are rendered as being bright on a T_1-weighted image. Lipid-rich regions tend to be characterized by stronger relaxation (both T_1 and T_2) than water-rich regions. They thus tend to appear relatively brighter in T_1-weighted images and darker in T_2-weighted images.

The intrinsic T_1- or T_2-weighted contrast induced by tissue structure or pathology is not always strong enough to reveal features or to enable a sensitive and specific diagnosis. In such cases, it is sometimes possible to enhance contrast through the introduction of paramagnetic contrast agents. Gadolinium-based contrast media injected into the bloodstream, for example, enhance T_1 relaxation and yield a local increase in T_1-weighted signal intensity wherever blood perfusion is present. Small iron oxide and other superparamagnetic particles, on the other hand, enhance T_2 relaxation and lead to a corresponding decrease in T_2-weighted signal intensity wherever they are present. The use of contrast agents in MRI is discussed in Chapter 3.

A third physical process that can be used for image contrast is diffusion, as was discussed in Section 1.2.1.3.

Like relaxation, diffusion naturally causes an irreversible degradation of NMR signal coherence. One method for diffusion-weighting MR images involves adding a bipolar field gradient pulse between the initial RF excitation and the rest of the normal sequence. The purpose of this bipolar gradient is to imprint a helix-like pattern on the phase of the magnetization along the direction of the diffusion-sensitizing gradient, and then unwind it. If the nuclear spins contributing to the NMR signal are stationary, this manipulation has no effect. However, if diffusion occurs on the timescale of the pulse, the net phase accumulated by a particular spin depends on the (random) path it happened to follow in the interim. The net result is an attenuation of the net transverse magnetization by a factor $\exp(-D\gamma^2 G^2 \tau_d^3)$, where D is the relevant diffusion coefficient, G is the sensitizing gradient amplitude, and τ_d is a timescale associated with the duration of its application.* As with T_2 weighting, regions characterized by significant diffusion yield less signal than those characterized by little diffusion, and are thus rendered as being darker.

Diffusive motions are often *restricted*. Atoms or molecules might be relatively free to wander short distances ("free diffusion"), but then encounter barriers that impede longer range motion. In such cases, when pulsed gradient NMR techniques are used to measure D, an *apparent diffusion coefficient* (ADC) is observed. This ADC is invariably smaller than the free diffusion coefficient, but the factor by which it is reduced depends on the timescale over which the measurement is made. ADC imaging of the brain is routinely performed in cases of ischemic or hemorrhagic stroke (see Chapter 23).

An additional factor arises when the confining structures are anisotropic, as is the case with nerve fiber tracts. In this case, ADC mapping can be performed as a function of the direction in which the sensitizing gradient is applied. The resulting diffusion tensor images provide information about both the direction and the magnitude of the underlying diffusion processes. Pulsed-field diffusion tensor MRI of the brain enables visualization of white matter fiber tracts and can be used to map subtle changes associated with diseases such as multiple sclerosis or epilepsy.

Further discussion of diffusion-sensitized MRI can be found in Chapter 4. The same general strategies used to characterize diffusion can be adapted to probe displacements and velocities (flow imaging; Chapter 9 in *Imaging*

of the Cardiovascular System, Thorax, and Abdomen) and have many applications to angiography.

References

1. Lauterbur PC (1973) Image formation by induced local interactions: Examples employing nuclear magnetic resonance. *Nature* 242:190–191.
2. Carr HY (1952) Free precession techniques in nuclear magnetic resonance. PhD thesis, Harvard University, Cambridge, MA.
3. Walters GK, Fairbank WM (1956) Phase separation in He³-He⁴ solutions. *Phys. Rev.* 103:262–263.
4. Osheroff DD, Gully WJ, Richardson RC, Lee DM (1972) New magnetic phenomena in liquid He³ below 3 mK. *Phys. Rev. Lett.* 29:920.
5. Abragam A (1961) *Principles of Nuclear Magnetism.* Oxford University Press, Oxford.
6. Cowan B (1997) *Nuclear Magnetic Resonance and Relaxation.* Cambridge University Press, Cambridge.
7. Ernst RR, Bodenhausen G, Wokaun A (1987) *Principles of Nuclear Magnetic Resonance in One and Two Dimensions.* Oxford University Press, New York.
8. Slichter CP (1990) *Principles of Magnetic Resonance.* Springer-Verlag, Berlin, Germany.
9. Levitt MH (2001) *Spin Dynamics: Basics of Nuclear Magnetic Resonance.* John Wiley & Sons, West Sussex, England.
10. Callaghan PT (2011) *Translational Dynamics and Magnetic Resonance: Principles of Pulsed Gradient Spin Echo NMR.* Oxford University Press, Oxford.
11. Blümich B (2000) *NMR Imaging of Materials.* Clarendon Press, Oxford.
12. Brown MA, Semelka RC (2010) *MRI: Basic Principles and Applications.* John Wiley & Sons, Hoboken, NJ.
13. Callaghan PT (1991) *Principles of Nuclear Magnetic Resonance Microscopy.* Oxford University Press, New York.
14. Kuperman V (2000) *Magnetic Resonance Imaging: Physical Principles and Applications.* Academic Press, New York.
15. Kimmich R (1997) *NMR Tomography, Diffusometry, Relaxometry.* Springer-Verlag, Berlin, Germany.
16. Purcell EM, Torrey HC, Pound RV (1946) Resonance absorption by nuclear magnetic moments in a solid. *Phys. Rev.* 69:37–38.
17. Bloch F, Hansen WW, Packard M (1946) Nuclear induction. *Phys. Rev.* 69:127.
18. Bloch F, Hansen WW, Packard M (1946) The nuclear induction experiment. *Phys. Rev.* 70:474–485.
19. Rabi II, Zacharias JR, Millman S, Kusch P (1938) A new method of measuring nuclear magnetic moment. *Phys. Rev.* 53:318.
20. Gerlach W, Stern O (1922) Das magnetische moment des silberatoms. *Z. Physik* 9:353.

---

* An alternate and commonly employed variant of this measurement involves the insertion of a time delay Δ between two short oppositely directed gradient pulses. In this case, the degree to which the signal is attenuated by diffusion during this time period is characterized by a parameter that is conventionally referred to as the "*b*-value," as discussed further in Chapter 4.

21. Gorter CJ (1936) Negative result of an attempt to detect nuclear magnetic spins. *Physica* 3:995.

22. Gorter CJ, Broer LJF (1942) Negative result of an attempt to observe nuclear magnetic resonance in solids. *Physica* 9:591–596.

23. Torrey HC (1949) Transient nutations in nuclear magnetic resonance. *Phys. Rev.* 76:1059–1068.

24. Hahn EL (1950) Nuclear induction due to free Larmor precession. *Phys. Rev.* 77:297–299.

25. Bloch F (1946) Nuclear induction. *Phys. Rev.* 70:460–474.

26. Hahn EL (1950) Spin echoes. *Phys. Rev.* 80:580–594.

27. Arnold JT, Dharmatti SS, Packard ME (1951) Chemical effects on nuclear induction signals from organic compounds. *J. Chem. Phys.* 19:507.

28. Singer JR (1959) Blood flow rates by nuclear magnetic resonance measurement. *Science* 130:1652–1653.

29. Jackson JA, Langham WH (1968) Whole-body NMR spectrometer. *Rev. Sci. Instrum.* 39:510.

30. Garroway AN, Grannell PK, Mansfield P (1974) Image formation in NMR by a selective irradiative process. *J. Phys. C: Solid State Phys.* 7:L457–L462.

31. Mansfield P, Maudsley AA (1977) Planar spin imaging by NMR. *J. Magn. Reson.* 27:101–119.

32. Ernst RR, Anderson WA (1966) Application of Fourier transform spectroscopy to magnetic resonance. *Rev. Sci. Instrum.* 37:93.

33. Kumar A, Welti D, Ernst RR (1975) NMR Fourier zeugmatography. *J. Magn. Reson.* 18:69–83.

34. Edelstein WA, Hutchison JMS, Johnson G, Redpath T (1980) Spin warp NMR imaging and applications to human whole-body imaging. *Phys. Med. Biol.* 25:751–756.

35. OECD (2013) Magnetic resonance imaging units, total, Health: Key tables from OECD, No. 36. doi:10.1787/magresimaging-table-2013-2-en.

36. OECD (2013) Magnetic resonance imaging (MRI) exams, total, Health: Key tables from OECD, No. 46. doi:10.1787/mri-exam-total-table-2013-2-en.

37. Lustig M, Donoho D, Pauly JM (2007) Sparse MRI: The application of compressed sensing for rapid MR imaging. *Magn. Reson. Med.* 58:1182–1195.

38. Sodickson DK, Manning WJ (1997) Simultaneous acquisition of spatial harmonics (SMASH): Fast imaging with radiofrequency coil arrays. *Magn. Reson. Med.* 38:591–603.

39. Pruessmann KP, Weiger M, Scheidegger MB, Boesiger P (1999) SENSE: Sensitivity encoding for fast MRI. *Magn. Reson. Med.* 42:952–962.

40. Kastler A (1950) Quelques suggestions concernant la production optique et la détection optique d'une inégalite de population des niveaux de quantification spatiale des atomes—Application à l'experience de Stern et Gerlach et à la résonance magnétique. *J. Phys. Radium.* 11:255–265.

41. Colegrove FD, Schearer LD, Walters GK (1963) Polarization of He³ gas by optical pumping. *Phys. Rev.* 132:2561–2572.

42. Happer W (1972) Optical pumping. *Rev. Mod. Phys.* 44:169–249.

43. Abragam A, Goldman M (1978) Principles of dynamic nuclear polarisation. *Rep. Prog. Phys.* 41:395.

44. Natterer J, Bargon J (1997) Parahydrogen induced polarization. *Prog. Nucl. Magn. Reson. Spectrosc.* 31:293–315.

45. Ardenkjaer-Larsen JH, Fridlund B, Gram A, Hansson G et al. (2003) Increase in signal-to-noise ratio of >10,000 times in liquid-state NMR. *PNAS* 100:10158–10163.

46. Mansson S, Johansson E, Magnusson P, Chai CM et al. (2006) C-13 imaging—A new diagnostic platform. *Eur. Radiol.* 16:57–67.

47. Ackerman JJH (2013) Magnetic resonance imaging: Silicon for the future. *Nat. Nanotechnol.* 8:313–315.

48. Cassidy MC, Chan HR, Ross BD, Bhattacharya PK et al. (2013) In vivo magnetic resonance imaging of hyperpolarized silicon particles. *Nat. Nanotechnol.* 8:363–368.

49. Fogarty HA, Berthault P, Brotin T, Huber G et al. (2007) A cryptophane core optimized for xenon encapsulation. *J. Am. Chem. Soc.* 129:10332.

50. Klippel S, Doepfert J, Jayapaul J, Kunth M et al. (2014) Cell tracking with caged xenon: Using cryptophanes as MRI reporters upon cellular internalization. *Angew. Chem. Int. Ed.* 53:493–496.

51. Young IR (2004) Significant events in the development of MRI. *J. Magn. Reson. Imaging* 20:183–186.

52. Jeener J (2000) Equivalence between the "classical" and the "Warren" approaches for the effects of long range dipolar couplings in liquid nuclear magnetic resonance. *J. Chem. Phys.* 112:5091–5094.

53. Hoult DI, Ginsberg NS (2001) The quantum origins of the free induction decay signal and spin noise. *J. Magn. Reson.* 148:182–199.

54. Hoult DI (2009) The origins and present status of the radio wave controversy in NMR. *Concepts Magn. Reson. A* 34A:193–216.

55. Bloembergen N, Purcell EM, Pound RV (1948) Relaxation effects in nuclear magnetic resonance absorption. *Phys. Rev.* 73:679–712.

56. Murali N, Krishnan V (2003) A primer for nuclear magnetic relaxation in liquids. *Concepts Magn. Reson.* 17A:86–116.

57. Torrey HC (1956) Bloch equations with diffusion terms. *Phys. Rev.* 104:563–565.

58. Turner R (1993) Gradient coil design: A review of methods. *Magn. Reson. Imaging* 11:903–920.

59. Hidalgo-Tobon SS (2010) Theory of gradient coil design methods for magnetic resonance imaging. *Concepts Magn. Reson. A* 36A:223–242.

60. Mispelter J, Lupu M, Briguet A (2006) *NMR Probeheads for Biophysical and Biomedical Experiments.* Imperial College Press, London.

61. Vaughan JT, Griffiths JR (eds.) (2012) *RF Coils for MRI.* John Wiley & Sons, Chichester, England.

62. Hoult DI (2000) The principle of reciprocity in signal strength calculations: A mathematical guide. *Concepts Magn. Reson.* 12:173–187.

63. Cagnac B, Brossel J, Kastler A (1958) Résonance magnétique nucléaire du mercure Hg-201 aligné par pompage optique. *C.R. Acad. Sci.* 246:1827–1830.

64. Kastler A (1957) Optical methods of atomic orientation and of magnetic resonance. *J. Opt. Soc. Am.* 47:460–465.

65. Hoult DI, Lauterbur PC (1979) Sensitivity of the zeugmatographic experiment involving human samples. *J. Magn. Reson.* 34:425–433.

66. Hayden ME, Bidinosti CP, Chapple EM (2012) Specific absorption rates and signal-to-noise ratio limitations for MRI in very-low magnetic fields. *Concepts Magn. Reson. A* 40A:281–294.

67. Bittoun J, Querleux B, Darrasse L (2006) Advances in MR imaging of the skin. *NMR Biomed.* 19:723–730.

68. Mossle M, Myers WR, Lee SK, Kelso N et al. (2005) SQUID-detected in vivo MRI at microtesla magnetic fields. *IEEE Trans. Appl. Supercond.* 15:757–760. (2004 *Applied Super-conductivity Conference*, Jacksonville, FL, October 3–8, 2004.)

69. Budker D, Romalis M (2007) Optical magnetometry. *Nat. Phys.* 3:227–234.

70. Mamin HJ, Poggio M, Degen CL, Rugar D (2007) Nuclear magnetic resonance imaging with 90-nm resolution. *Nat. Nanotechnol.* 2:301–306.

71. Homann H, Boernert P, Eggers H, Nehrke K et al. (2011) Toward individualized SAR models and in vivo validation. *Magn. Reson. Med.* 66:1767–1776.

72. Wolf S, Diehl D, Gebhardt M, Mallow J et al. (2013) SAR Simulations for high-field MRI: How much detail, effort, and accuracy is needed? *Magn. Reson. Med.* 69:1157–1168.

73. International Electrotechnical Commission (2010) Medical electrical equipment part 2-33: Particular requirements for the basic safety and essential performance of magnetic resonance equipment for medical diagnosis, Geneva, International Standard IEC 60601-2-33.

74. Pauly J, Leroux P, Nishimura D, Macovski A (1991) Parameter relations for the Shinnar-Le Roux selective excitation pulse design algorithm. *IEEE Trans. Med. Imaging* 10:53–65.

75. Ljunggren S (1983) A simple graphical representation of Fourier-based imaging methods. *J. Magn. Reson.* 54:338–343.

76. Xiang QS, Henkelman RM (1993) K-space description for MR imaging of dynamic objects. *Magn. Reson. Med.* 29:422–428.

77. Hennig J (1999) K-space sampling strategies. *Eur. Radiol.* 9:1020–1031.

78. Paschal CB, Morris HD (2004) k-Space in the clinic. *J. Magn. Reson. Imaging* 19:145–159.

79. McGibney G, Smith MR, Nichols ST, Crawley A (1993) Quantitative-evaluation of several partial Fourier re-construction algorithms used in MRI. *Magn. Reson. Med.* 30:51–59.

80. Weisman ID, Bennett LH, Maxwell LR, Woods MW et al. (1972) Recognition of cancer in vivo by nuclear magnetic resonance. *Science* 178:1288–1290.

2

Introduction to the Basics of MRI to Introduce the Macroscopic Magnetization M

Luigi Barberini

CONTENTS

2.1 Introduction: The Basic of MRI to Introduce the Macroscopic M

MRI is one of the most important technical tools for the diagnostic process developed in the recent period. Modern MRI systems provide highly detailed images of tissue in the body with fast scan, reducing the examination time for the best comfort of the patient. With the MRI, it is possible to study the human body and the internal organs producing high-quality images without the use of ionizing radiation. This imaging technique is based on the resonance physics phenomenon that practically is the way to exchange energy between *physical systems* at particular frequencies, which is the characteristic, in some way, of these systems; these frequencies are called *natural frequencies* of the system. Under these conditions, energy flows from one system to the other with none or little dumping and can be efficiently stored by the living matter under investigation. The relaxation phenomena of the living matter can give much information to be used for the image construction.

The resonance phenomenon can occur between mechanical waves and vibrational physical systems like diapason, for example, and at the level of electrons and nuclear spin interacting with electromagnetic (EM) microwaves and radio waves.

In order to acquire information about the structural and functional properties of the molecules in biofluids, tissues, and organs in living systems, the nuclear magnetic resonance (NMR) technique can be used. Such information can be acquired from *in vitro* samples, for example, liquid mixtures of molecules contained in particular tubes, from samples in the solid state with appropriate coils (or probes); furthermore, molecular properties can also be acquired directly on the living tissues and organs (*in vivo*) in order to obtain morphological and functional images of tissues and organs. Several kinds of nuclei have the possibility to get coupled to the EM field: they are the ones without zero spin values. For these systems, characteristic frequencies of resonance are generated putting the matter into a static magnetic field. The most important nucleus in the NMR is the proton, with the nucleus of the hydrogen usually indicated by ^1H. ^1H has the spin value $S = 1/2$; hydrogen atoms have an inherent magnetic moment as a result of their nuclear spin. When placed in a strong magnetic field, the magnetic moments of these hydrogen nuclei tend to align along the B_0 direction with a motion of precession defined by the Larmor frequency; part of these spins align with a precession motion along the same direction of B_0 and part align in the opposite direction creating a two-level energy system (Figure 2.1).

This balance, maintained by the thermal agitation, generates a net magnetization because a little amount of spins prefers to stay in the lower energetic state; the net magnetization is the sum over all spins of the microscopic magnetic moments. We can represent the precession motion in a particular reference frame, a system rotating at the same precession of the spin and magnetization. In this reference frame, **M** is a vector aligned along B_0 (Figure 2.2).

For the physics of nuclear spins, we will return later on the references systems; for the moment, we can consider the one in which spins are aligned to B_0.

For the *bare* hydrogen, the characteristic frequency of precession defined as the Larmor angular velocity, at the 1 T magnetic static field is

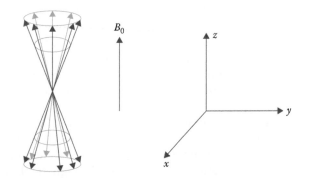

FIGURE 2.1
Spins motion and directions of reference.

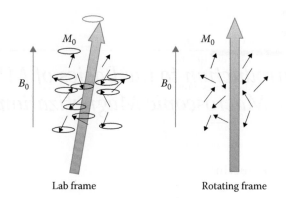

FIGURE 2.2
Spins behavior and reference frames.

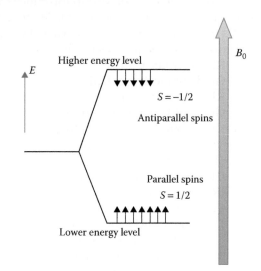

FIGURE 2.3
Zeeman levels and spins directions in space.

$$\omega = \frac{-eg}{2m} B_0 \qquad (2.1)$$

In this way, protons can exchange energy with EM wave through the coupling of the nuclear magnetic moment and the magnetic component of the EM wave; the exchanged energy corresponds to the transition of the spins from the lower energy level to the higher level (Figure 2.3).

This energy corresponds, in the EM spectrum region, to radiofrequency (RF) waves (Figure 2.4).

So, the EM field can be coupled to the biological matter protons at the *characteristic* frequency of the Larmor precession of the net magnetization M around the static magnetic field B_0. With the term *characteristic*, we indicate that it depends on the nucleus considered, proton, carbon, phosphorus (all nuclei with not null magnetic moment) and, most of all, on the applied magnetic field.

From the macroscopic point of view, the physical entity of interest in MRI is the magnetization vector **M** and its components M_P and M_l.

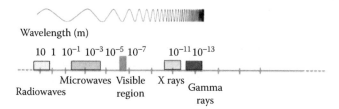

FIGURE 2.4
Electromagnetic spectrum.

The magnetization M_P (sometimes also indicated by the symbol M_0 to emphasize the relation with B_0) of the substance is the average of all the spin magnetic moments of each atom μ_i. We write

$$M_P = \sum \mu_i$$

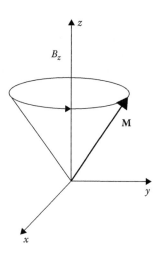

FIGURE 2.5
Precession of the **M** vector in the lab system.

which is pointed along of the z-axis.

The dynamic of the **M** vector is totally described by the phenomenological Block equations. In order to derive these equations, we have to consider the physical status of the spins. Thermal equilibrium defines, in the interaction spins-EM field, the distribution of the hydrogen spins on the two energetic levels as described the Maxwell–Boltzmann distribution:

$$\frac{N_1}{N_2} = \mu_0 B_0 e^{-(\Delta E/RT)} \tag{2.2}$$

The projection of this vector along the z direction defines the M_P or M_0. In order to produce a detectable time-varying signal to process by means of the MRI system, it is mandatory to rotate the M_0 vector in the plane orthogonal to z. Proper stimulation by means of a resonant magnetic, or an RF field at the resonant frequency of the hydrogen nuclei, can force the magnetic moments of the nuclei to partially, or completely, tip into a plane perpendicular to the applied field. The angle of this rotation is called the *flip angle* and indicated by FA; this rotation is strictly related to the temporal duration of the pulse. This is the excitation process of the biological and living matter at the resonance condition or phase concordance between radiation and spins. These RF pulses are indicated with the $B_1(t)$ expression after the B_0 symbol for the static magnetic field. The precession motion complicates the real motion of the **M** in the laboratory frame (Figure 2.5). But the dynamic of the **M** vector, when an RF pulse is acting on the spins (exchange of energy in resonant mode), can be decomposed as a rotation of **M** around the z-axis and a change in the angle between z and **M** (FA), called *nutation*. This composite motion can be represented by a conic helix trajectory of the **M** vector in the lab frame with parametric equations (Figure 2.6):

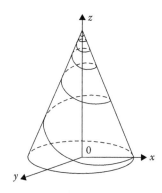

FIGURE 2.6
Conic helix trajectory of **M** vector in the laboratory frame.

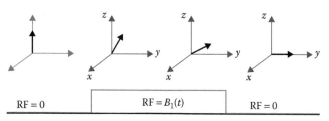

Nutation of M_0 induced by RF pulse $B_1(t)$ in the rotating frame

FIGURE 2.7
Effects of RF pulse P90 for the M orientation.

$$x(t) = (R_0(1-t) + R_1)\sin(N2\pi t)$$

$$y(t) = (R_0(1-t) + R_1)\cos(N2\pi t) \tag{2.3}$$

$$z(t) = Ht$$

The motion of **M** toward the x–y plane is the one of interest for the description of the transversal magnetization generation (Figure 2.7).

So, the hydrogen atoms, which are abundant in tissue, are placed in a strong magnetic field B_0, and then excited by a resonant magnetic excitation pulse $B_1(t)$ applied as long as needed to rotate magnetization; after the $B_1(t)$ field is switched off, the spin relaxation process starts, and the net magnetization laying in the x–y plane begins to decay due to the local spin–spin interaction. Meanwhile, due to the spin–lattice interaction, there is a loss of energy toward the bulk of matter related to the tissue nature and to the physical state of the matter in the tissue, and the longitudinal magnetization begins to rise. Spin–spin and spin–lattice interactions are different phenomena. Parameters of these phenomena are different: they contribute to the loss of magnetization along the x–y plane and to the recovery of the M_0 magnetization, and they depend on the tissue nature and on its physical status. The magnetization M_P (sometimes also indicated with the symbol M_0 to emphasize the relation with B_0) of the substance is the average of all the spin magnetic moments of each atom μ_i. We write

$$M_\| = \sum \mu_i$$

which is pointed along the z-axis.

The field $B_1(t)$ exerts a torque force on the vector M_0 leading it to the y-axis (Figure 2.8).

Then, the relaxation phenomena give back the energy absorbed from EM wave, transporting the RF pulse to the external environment with all the information about the location and the physical status of the nuclear spins. But the relaxation is a double process: spins lose the coherence, and the summation of the spins generating the $M_\perp$ magnetization decreases (Figure 2.9).

At the same time, but with different time constant, the M_P starts to recover toward the initial value M_0.

Meanwhile, the longitudinal magnetization begins the recovery process, but with a different temporal constant T_1, due to the interaction of the spins with the bulk of tissues (Figure 2.10).

In the frame laboratory, the magnetization vector **M** rotates in the x–y plane and in the z–y plane. Rotational motion in the x–y plane is not related to an energy exchange, so we can imagine a frame transformation to put in rotation the xyz frame around the z-axis to hide this behavior of $B_1(t)$. This is called the *rotating frame*. In the rotating frame, the magnetization vector **M** rotates only in the z–y plane without precession. The properties of interest of the living matter are resumed by the macroscopic magnetization $\overline{M}$ and its dynamic; the information used to realize images with structural and functional informative content is derived by the dynamic of the $\overline{M}$ vector equations. $\overline{M}$ is the net magnetic moment per volume unit (a vector quantity) of a sample in a given region, considered as the integrated effect of all the individual microscopic nuclear magnetic moments μ_i. $\overline{M}$ will be the actor of our play in MRI.

The microscopic magnetic moments and the macroscopic magnetization are *vectors*, entities endowed by intensity, line, and direction of action and phase. As usual, in order to work with these entities, it is better to define the reference work system used: the natural frame reference for MRI study is the *laboratory* system where a left-hand system is defined by the z direction along the B_0 field direction and x and y are oriented as shown in Figure 2.11.

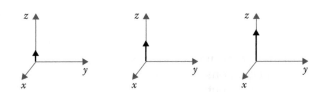

FIGURE 2.10
Longitudinal magnetization recovery. Please note that in this representation, transversal magnetization is not reported; in this way, we highlight that longitudinal and transversal magnetization vary with different physical phenomena and different time constants.

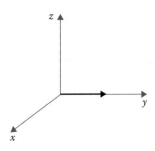

FIGURE 2.8
Transversal magnetization M after P90 RF pulse.

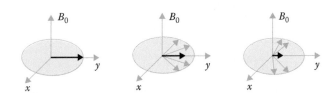

FIGURE 2.9
Dephasing of nuclear spins.

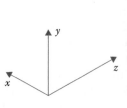
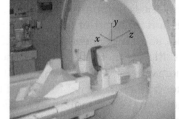

FIGURE 2.11
Spatial orientation in MRI.

In this frame, as mentioned before, the motion of **M** is complicated by the precession phenomenon. The same complication arises considering the excitation of the magnetization with the RF pulse. The equation for the **M** dynamic, in the laboratory reference frame, is

$$\left(\frac{d\overset{i}{M}}{dt}\right)_{lab} = \overset{r}{M} \times (\omega\overset{r}{z}) \tag{2.4}$$

The helical motion of the **M** vector in the lab frame can be simplified applying the transformation group of rotation. If we consider that it is possible to define a new reference frame rotating in the x–y plane with angular speed ω, we can define the previous equation in the rotating frame, and it is possible to demonstrate

$$\left(\frac{d\overset{i}{M}}{dt}\right)_{rotating} = (\omega - \omega_0)\overset{r}{M} \times \overset{r}{z} \tag{2.5}$$

If the rotation speed is equal to the Larmor frequency, then temporal derivative is null and **M** becomes a stationary vector. The same transformation can be applied to the RF pulse (in order to produce the alignment of the magnetization in the x–y plane), and simplifications of the same level can be introduced for the RF pulse description. The dynamic of the RF pulse $B_1(t)$ in the lab system is

$$\vec{B}_1(t) = \hat{x}B_1(t)\cos(\omega t) - \hat{y}B_1(t)\sin(\omega t) \tag{2.6}$$

and in the rotating frame is

$$\vec{B}_1(t) = \hat{x}B_1(t) \tag{2.7}$$

The rotating frame introduces a great amount of simplification in equations and formulas, and it is useful for the discussion about the relaxation process of the **M** net magnetization.

The most important phenomena for the use of MR in the investigation about the matter structure are the relaxation mechanisms of **M** that happens after the truncation of the EM (no-ionizing) fields exciting the living tissues. Generally speaking, living tissues and molecules should be transparent to the EM window of the RF and this part of the radiation spectra should not be able to exchange energy with the molecules. But using a static magnetic field, molecules are able to exchange energy with the EM field in the region of the RF and tissues become opaque to RF in different ways related to the status of the molecule bulk. In this way, it becomes possible to excite selectively the living matter and to record the relaxation MRI signal, which brings a lot of information about the molecules status.

The relaxation of the macroscopic magnetization is a dynamic phenomenon. This dynamic is different for each kind of tissue, and it is related to the bulk physical status, the molecular environment, in addition to the spin parameters. By using the RF pulses and a slow controlled linear variation of the static magnetic field, it is possible to *manipulate* the status of the single nuclear spin and the level of coherence between each other; from the macroscopic point of view, we can manipulate the **M** dynamic in the excitation and relaxation period. It is important to highlight that the possibility to *manipulate* the signals in imaging to produce images of interest has its maximum expression in the MRI; only the MRI, with the great amount of *free parameters* characterizing the phenomenon gives to researchers the possibility of acting on tissues and receiving different signals with different information.

2.2 Bloch Equations for the Macroscopic Magnetization (*M*) Dependences: From Protons to Echo Tissues Signals (Intrinsic MRI Parameters Fast Description: T_1 T_2 T_2^* PD)

Usually, in MRI we do not deal with microscopic entities as the spins: the vector used to describe the behavior of the biological matter in the MRI scanner is the net macroscopic magnetization **M**. As previously reported, at the thermal equilibrium, **M** is a vector with a precession motion around the B_0 field (Figure 2.12).

We introduce a perturbation of the thermodynamic equilibrium with the transmission of RF pulses for the excitation of tissue's nuclei. As previously represented,

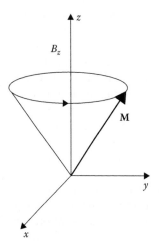

FIGURE 2.12
M vector in the lab system.

the RF pulses are labeled in the MRI theory as $B_1(t)$ in order to emphasize that it is a magnetic field time variant that is interacting with a nuclear magnetic spin. As previously reported, RF pulses are EM waves in the RF range of the spectrum.

During this energy transfer, part of the M_P is transformed in $M_\perp \cdot M_P$ is the result of the N_1/N_2 fraction of spin, orthogonally orientated to B_0. Spins are in coherent motion due to the $B_1(t)$ perturbation: Zeeman levels are coherent energetic states. After the end of the perturbation, that is, after the RF pulse is switched off, the energy transfer from the EM wave to spins is interrupted, thermal agitation of the spins starts to destroy the coherence between spins, and this introduces a decay of the orthogonal component of the **M**. At the same time, part of the $M_\perp$ can be recovered in the z direction and the longitudinal magnetization starts to arise in this direction. Energy and coherence loosing are relaxation phenomena. The most important parameters of these phenomena are PD, proton density; T_2, time constant for the molecule–molecule interaction describing the orthogonal magnetization recover; and T_1, time constant for the molecule–bulk interaction, describing the longitudinal magnetization recover. Furthermore, it is possible to calculate the effects of local decreasing of the T_2 values of tissues: in fact, the blood oxygenation level dependent (BOLD) mechanism realizes the explanation of a faster decrease of the local T_2 (called T_2^*) in tissue due to the activation of the neurons located in the brain area devoted to a task execution. Default brain activity leaves other areas in the default state of activation with a usual signal of $M_\perp$ decreasing with the usual T_2 time constant. These properties of interest can be all used in the construction of images with morphological and functional significance.

The molecule–molecule interaction is studied along the B_0 field direction; meanwhile, the molecule–bulk interaction is better studied in the x–y plane orthogonal to the direction of the B_0 field, that is, the z direction. So pulses must be used to selectively orientate the macroscopic magnetization. To produce an image with the MRI system, we must first stimulate the hydrogen nuclei in a specific region of the body. Excited nuclei, after the end of perturbation, radiate in the space a signal in the form of the EM wave until the energy absorbed during the excitation phase is completely released; this is the process called *relaxation*. Relaxation produces the NMR signal used to create images of differentiated tissue types and to produce images for MRI systems. So, initial MR signal amplitudes are directly related to hydrogen nuclei density in the tissue being imaged. Primarily, it will be notable for the concentration of mobile hydrogen atoms within a sample of tissue; the concentration of hydrogen atoms in water molecules or in some groups of fat molecules within tissues is defined by the term

proton density (PD). Then, the relaxation process of the magnetization determines the evolution of the MRI signal.

The passage from the microscopic to the macroscopic scale can be described by the summation:

$$M = \sum_i \mu_i \tag{2.8}$$

So, from the time variation of each single spin magnetic momentum expressed as

$$\frac{d\mu}{dt} = \mu \times \gamma \mathbf{B} \tag{2.9}$$

we arrive to the **M** vector dynamic after the summation of μ over the volume; this gives the following equation of macroscopic magnetization dynamic:

$$\frac{d\mathbf{M}}{dt} = \mathbf{M} \times \gamma \mathbf{B} \tag{2.10}$$

as previously reported as Equation 2.4, that is, **M** precession along **B** with a frequency equal to $\omega = \gamma B$.

Vector **M**, and its temporal derivatives, can be represented by the three vector components in the lab frame. In order to relate the time evolution of magnetization to the external magnetic fields and to the relaxation times, Bloch equations can be used:

$$\frac{dM_x(t)}{dt} = \gamma(M(t) \times B(t))_x - \frac{M_x(t)}{T_2}$$

$$\frac{dM_y(t)}{dt} = \gamma(M(t) \times B(t))_z - \frac{M_y(t)}{T_2} \tag{2.11}$$

$$\frac{dM_z(t)}{dt} = \gamma(M(t) \times B(t))_z - \frac{M_z(t) - M_0}{T_1}$$

The component along the z direction is called *longitudinal magnetization* $M_0 = M_{||}$. The other two components can be represented as a vectorial sum lying in the x–y plane orthogonal to the z direction: for this reason, it is represented as $M_\perp$.

The phenomenological Bloch equations explain the evolution of the magnetization moment during its precession. The magnetic field $B_1(t)$ is time dependent, the same as the magnetization moment $M(t)$. The important phenomenon related to the evolution of the magnetization vector is the relaxation following the precession induced by the excitation. As previously reported, two different relaxation processes occur: the longitudinal relaxation and the transverse relaxation.

As illustrated in Figures 2.9 and 2.10, interactions between spins result in the destruction of the coherence phase between spins, and in this way, the sum over the spin is incoherent and tends to decay toward zero. Meanwhile, the energy absorbed by excitation is released as a result of the energy exchanges between spins and bulk, or lattice: spins excited to the upper level come back to the lower restoring the $M_0 = M_{||}$ macroscopic magnetization. It is really interesting to note that time constants of these processes are quite different because of different physics processes underlying the phenomena.

Typically, the solution of the differential equation with the form of Equation 2.2 has an exponential form; the dynamics of the $M_{||}$, $M_\perp$ vectors are different and their solutions will have different temporal constants. The typical time evolution constant for the $M_{||}$ is labeled T_1 and the $M_\perp$ vector has a temporal constant labeled T_2. The graphical evolution of the two vectors is shown in Figures 2.13 and 2.14.

The pure T_2 time constant is the observed decay parameter of the FID because of the loss of phase coherence among spins due to the spin–spin interaction for the spin thermal agitation. But commonly, the coherence loss is due to a combination of the static magnetic field in homogeneity and spin–spin transverse relaxation, with the result of more rapid loss in the transverse magnetization and MRI signal decay:

$$\frac{1}{T_2{}^*} = \frac{1}{T_2} + \Delta\omega = \frac{1}{T_2} + \gamma\Delta B$$

$T_2{}^*$ depends on the local magnetic field nonuniformities ΔB; due to this phenomenon, the protons precess at slightly different frequencies. The $T_2{}^*$ effect causes a faster loss in spins coherence and transverse magnetization, and the $T_2{}^*$ time results lower than the T_2.

In order to control the magnetization and weighting the images in PD, T_1, T_2, and $T_2{}^*$, defined as the intrinsic MRI parameters, we need to submit the matter to a proper sequence of RF pulses and gradient waveforms. The parameters used to describe the temporal properties of these sequences are called *extrinsic parameters* for

the contrast control: the number of pulses, the intensity, the timing, the power of the pulses, and the gradient shape and length allow the sequence designers to realize a proper project of sequence to act on the magnetization in order to produce the proper contrast (Figure 2.15).

Before describing these parameters, we have to introduce and discuss the electronic chain components for the production and control of the RF pulses and gradients.

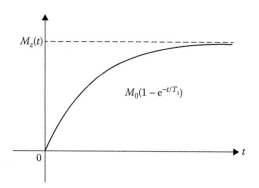

FIGURE 2.13
Longitudinal relaxation.

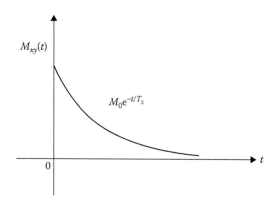

FIGURE 2.14
Transversal relaxation.

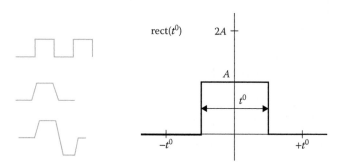

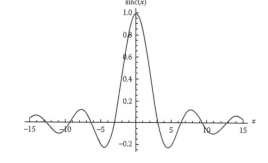

FIGURE 2.15
Principal gradient and RF pulses shapes.

2.3 MR Hardware for Sequences Production: Electronic Chain for the RF System and Gradient System

Among the main hardware elements in the MRI scanner systems that we can consider for the sequences description, there are static magnetic field and shimming coils, gradient magnetic field coils, RF receiver–transmitter coils, pulse waveform generator and timing, gradient waveform generator and timing, frequency synthesizer, amplificators and analog-to-digital—digital-to-analog (ADC–DAC) converters, power unit, and the image signal processing unit.

A schematic representation of a scanner is shown in Figure 2.16.

In order to have all the elements for the general comprehension of the MRI, it is mandatory to have a look at some important concepts related to the electronic components listed before and related to some important operations performed on the signals in MRI. Signals to and from scanner must be digitalized, processed, and stored. Electronic circuits at very high speeds perform part of these operations. This is the reason why companies involved in the MRI scanner market spend a lot of money in the electronic development for the new systems. The digital *treatment* of a signal is characterized by some important issues: the analogical signal must be measured with appropriate transducers (coils) functioning as *receivers* (or *transmitters* when they have to put in resonance the nuclear spin); in the receiving process, analogical signals must be sampled in order to allow the

storage on the electronic memory for the mathematical treatment. At this stage, we tackle the problem of the amplification of the signal: in some cases, the intensity of the electric signal emitted by transducers could be very low and it is necessary to provide *amplification* to this signal in order to acquire, store, and process the information in a proper way, avoiding storing *noise*. The electronic components devoted to this operation are the amplifiers. After amplification, the ADC provides the conversion of that signal in a digital form easily managed and stored by a computer. We also need to operate the inverse operation and convert a digital signal generated by the MRI system in the analogical format. For this operation, a DAC circuit is implemented in the electronic cabinets of the MRI system.

The analog–digital conversion is accomplished through two basic steps: sampling and quantization. Given a suitably amplified analogical signal, the sampling process consists in measuring that signal for a number of times per second, storing the measured values instead of the signal itself (Figure 2.17).

The signal is measured at regular intervals with a frequency-defined sampling frequency, which must be twice as much as the max frequency in the signal in order to avoid the lost of information about the signal (Nyquist theorem). In terms of time, the interval between two sampling is called *dwell time* Δt, and it represents the inverse of the full readout bandwidth. The readout bandwidth is the range of spin frequency in all the field of views (FOVs). Quantization is the process of mapping a large set of input values to a smaller and countable set, such as rounding values to some unit of

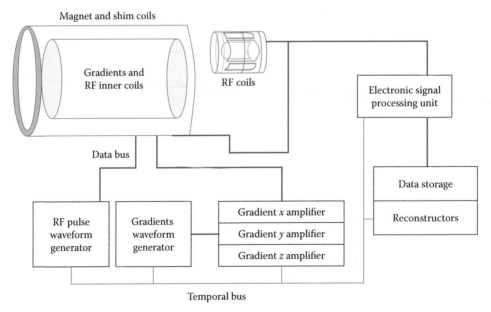

FIGURE 2.16
Flowchart of an MRI scanner.

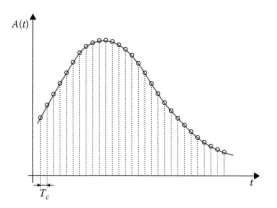

FIGURE 2.17
Sampling of the analogical signals.

precision. The difference between the actual analogical value and the quantized digital value is called the *quantization error*.

The signal coming out from transducers may contain variations related to the information of interest and some variations that are only *noise*. So, there is the need to filter the signal as better as possible in order to increase the part of the signal of interest and drastically reduce the noise. This is called the *digital signal processing*. Digital filters can be designed to operate the signal-to-noise ratio (SNR) and as much as possible to improve the quality of the image for the visual diagnostic process or for the computer-aided diagnosis (CAD).

2.3.1 RF Electronic Unit

The EM waves used in the NMR and MRI technologies are in the EM spectral region of the RFs. This wide range is named in such a way because some radio transmissions are in this region. The usual radio stations for the music and news operating in the so-called frequency modulation are in the MHz range, or thousands of MHz. So the electronic terminology for the radio transmissions is frequently used to describe the process and equipment in MR. In fact, the operative process to excite nuclei in the matter and to recoil the signals echoing from the matter is an RF transmission and receiving process (Figure 2.18).

So, there is a part of the scanner, named transmitter, used to generate the RF pulses necessary to exchange energy with the hydrogen nuclei. The range of frequencies in the transmit excitation pulse and the magnitude of the gradient field determine the width of the image slice. A typical transmit pulse will produce an output signal with a relatively narrow bandwidth, about ±1 kHz.

The shape of excitation wave in the time domain usually requires particular properties of the signal; for example, some RF pulses are used to prepare the magnetization for a selective excitation in order to saturate the energy level of some kind of biological matter to exclude it from the generation of the signal in the de-excitation of the matter. As reported in Section 2.3.3, these waveforms are usually digitally generated at the baseband and then up-converted by a mixer to the appropriate central frequency.

In electronics, a mixer or a frequency mixer is a non-linear electrical circuit that creates new frequencies from two signals applied to it (Figure 2.19).

Traditional transmitter circuitries require relatively low-speed DACs to generate the baseband waveform, as the bandwidth of this signal is relatively small. But with recent advances in DAC technology, other potential transmitter architectures are achievable. Very high speed, high-resolution DACs can be used for the direct RF

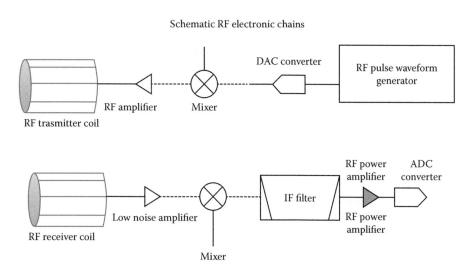

FIGURE 2.18
Electronic chains in MRI.

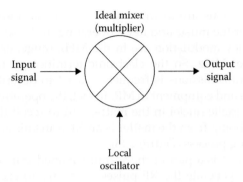

FIGURE 2.19
Ideal mixer of RF signals.

generation of transmit pulses up to 300 MHz. Therefore, waveform generation and up conversion over a broad band of frequencies can be entirely accomplished in the digital domain.

The biological matter excited by temporary RF pulses starts a rebalance process to emit the acquired energy, and this energy flows from the matter to the receiving circuit in the scanner.

An RF receiver is used to process the signals from the receiver coils. Recent MRI scanners have eight and more receiving channels to process the signals from multiple coils. The signals range from approximately 1 to 300 MHz, with the frequency range highly dependent on the applied-static magnetic field strength. The bandwidth of the received signal is small, typically less than 20 kHz, and is dependent on the magnitude of the gradient field. A traditional MRI receiver configuration has a low-noise amplifier (LNA) followed by a mixer. The mixer mixes the signal of interest to a low frequency for conversion by a high-resolution, low-speed, 12–32-bit ADC. In this receiver architecture, the ADCs used have relatively low sampling rates, below 1 MHz. Because of the low-bandwidth requirements, ADCs with higher sampling rates (1–5 MHz) can be used to convert multiple channels by means of the time multiplexing of the receiving channels through an analogical multiplexer into a single ADC.

Again, with the development of higher-performance ADCs, novel receiver architectures are now possible. High-resolution high-input bandwidth, 12–32-bit ADCs with samples rates up to 100 MHz can also be used for the direct sampling of the signals, hence eliminating the need for analogical mixers in the receive chain.

2.3.2 Gradient Electronic Unit

After the basic description of the electronics used to realize the RF pulses, we can now discuss the properties of the pulses. These properties are related to the magnetization modulation for the contrast control in the MRI. As previously mentioned, the MRI system stimulates hydrogen nuclei in a specific plane selected as slice in the body, and then determines the location of those nuclei within that plane as they relax to their ground state. These two tasks are realized using gradient coils with suitable shapes localized in the magnet. These coils, controlled by a chain of electronics devices, cause the magnetic field within a localized area to vary linearly as a function of spatial location (Figure 2.20).

As a result, the resonant frequencies of the hydrogen nuclei are spatially dependent within the gradient range variation, and consequently, by varying the frequency of the excitation pulses, it is possible to control the area to be stimulated. After the interruption of the excitation processes, the location of the stimulated nuclei precessing back to their ground state, can be determined by the emitted resonant RF and phase information.

An MRI system has gradient coils acting on the three axes *X*, *Y*, and *Z*. In a block diagram of the electronic chain for the gradient system, there is a control unit called the *waveform generator*, used to create a time-controlled waveform. This digital signal is sent to a DAC unit and then amplified, and directed to the proper segment of the gradient coil system.

To achieve adequate image quality and frame rates, the gradient coils in the MRI imaging system must rapidly change the strong static magnetic field in the

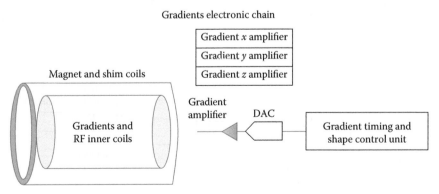

FIGURE 2.20
Schematic representation of the electronic chain for the gradients systems.

area of interest. A power electronic supply unit is used to drive the gradient coils. These electronic units operate at high voltage (up to few kilovolts) and high current (several amperes). In addition to the high power requirements, low noise and stability of the gradients are really important because any ripple in the coil current causes noise in the subsequent RF pickup. That noise directly affects the integrity of the images.

2.3.3 RF Pulses and Macroscopic Net Magnetization *M* Effects

We can start this section explaining that the RF pulses can primarily be used to excite the magnetization, to produce the inversion of this vector, and finally to refocus the spins in order to produce coherence and macroscopic magnetization. RF pulses can also be used for a saturation of undesired signals.

In order to better understand the use of the RF pulses, we have to highlight some important aspects: the pulse shape, the functionality of the pulse related to timing and duration, and the selectivity properties in space and in the spectrum. The duration of the RF pulse is also called *pulse width* and is typically measured in seconds or milliseconds. Regarding the pulse, it is also defined as an RF bandwidth Δf, measured in hertz, as a measure of the pulse frequency content and it represents the frequency profile for spin manipulation. Another parameter that is commonly used to describe the RF pulse effects is the FA θ. The FA is usually measured in degrees, and it describes the nutation angle produced by the pulse on the magnetization M. For example, an excitation pulse that tips the longitudinal magnetization completely into the transverse plane has a FA of 90°, and it is called *P90 pulse*. The FA is related to the duration of the pulse RF.

Another important parameter for the pulse property description is the shape. Principal shapes of RF pulses are described by the functions Rect and Sinc, with different mathematical properties that we do not treat in this context.

A Rect pulse is a pulse shaped like a RECT function in the time domain, which is zero for $|t| > T$; it is also called *hard pulse* because it is time independent. On the contrary, pulses that are time variant are defined as *soft pulses*. Hard pulses can be used when no spatial or spectral selection is required and are convenient because the pulse length can be very short. Usually, hard pulses are activated without a concurrent gradient. The bandwidth of a hard pulse, however, is broad enough to affect spins with a wide range of resonant frequencies.

A Sinc pulse is a pulse shaped like a SINC function in the time domain, that is a sine x over x.

The excitation of the magnetization vector entails the rotation of this vector from the magnetic field B_0 in the z direction to the orthogonal x–y plane. It is possible to acquire an MR signal only if the signal is produced by the x–y plane magnetization, and for this reason, all the pulse sequences have at least one excitation pulse. This kind of pulse is called P90 pulse because it moves the M_0 vector completely in the plane orthogonal to the primary z-axis, rotating M_0 to 90°. Because the duration of the pulse is related to the nutation angle, the doubling of the duration leads to the inversion of the direction of the $\mathbf{M}$ vector. This rotation can be applied to the magnetization $M_{||}$ and to the $M_\perp$ vector along each axis. Of particular interest is the effect of the inversion applied to $M_\perp$ after a short time since P90: $M_\perp$ begins to decrease due to the dephasing action of the thermal agitation. Loosing coherence in the motion, we will find the faster spins in the external trajectory, as shown in Figure 2.21.

Operating the inversion along the y-axis as shown in Figure 2.22, we have the $\mathbf{a}$ vector, which is faster than $\mathbf{b}$, still in the external position to $\mathbf{b}$ but now rotating in the opposite direction. Spins start to recover phase and $M_\perp$ has again the maximum intensity. This generates the echo signal that can be measured with the coils. So, by applying the RF pulses to invert magnetization, it is possible to generate an echo of the magnetization to be measured as needed.

RF pulses can be applied in different ways with specific functions related also to the shape of the RF pulse expressed by the *bandwidth*. This function can determine the selectivity on particular spectral regions. With this kind of a strategy, it is possible to control the magnetization transferred on the plane orthogonal to B_0. In this way, the contrast in the images will be dependent on the kind and the amount of magnetization transferred; in fact, the pulses used to transfer the proper magnetization for the contrast of interest can be spectrally selective and can be used to reduce the MR signal from some types of tissue leaving other types virtually unaltered. These kinds of pulses can only reduce the MR signal. Consequently, there is an increase in the image contrast, because the effect is tissue-specific. A widely used application of these kinds of pulses can be found in the

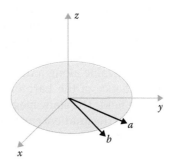

FIGURE 2.21
Rotating spins with different phases and velocity.

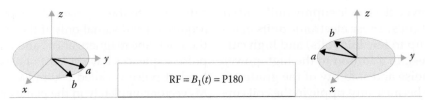

Inversion of the rotation plane for the spins induced by RF pulse $B_1(t) =$ P180

FIGURE 2.22
Spins after P180 pulse inversion.

magnetic resonance angiography (MRA), applied particularly to the cranial district, and performed with the so-called time-of-flight (TOF) sequence, reported in the following. Spectral selective pulses are used to attenuate the MR signal from brain parenchyma, while leaving the signal from blood unaltered.

This is just an example. All these functionalities and others not mentioned before but used in the modern MRI are performed by means of a sophisticated technology for the engineering of the shape, duration, and integration of the RF pulses with a gradient system. From the mathematical point of view, many of these arguments are very difficult to treat and we will have a qualitative approach to this argument.

Before moving further, let us have a look at the analog properties related to the gradient of the static magnetic field: the application of properly shaped and timed gradients allows us to further manipulate the spins and the magnetization.

Concluding, a proper mathematical description for these arguments is out of the topic of this chapter: interested readers can consult a proper treatment of the arguments in

1. *Principles of Magnetic Resonance imaging* by Yi Wang

2. *Handbook of MRI Pulse Sequences* by Bernstein et al.

2.3.4 Gradient Waveforms and Macroscopic Net Magnetization *M* Effects

A controlled variation of the static magnetic field in MR is a powerful tool for the manipulation of spins, at the microscopic level, and of the **M** at the macroscopic level. Introducing a linear variation of the magnetic field overlapped to the static B_0 value along a direction of action produces the dependence of the Larmor frequency on the local value of the field. The static magnetic field during the switching-on of the gradient can be expressed by the following equation:

$$B_0(x) = B_0 + \left| \overset{i}{G_x} \right| \times \left| \overset{r}{x} \right| \tag{2.12}$$

$$\omega(x) = \frac{-eg}{2m} B_0(x) \tag{2.13}$$

Introducing a linear gradient field for each direction, we add a spatial label to give a different Larmor frequency at each point of the space in the magnet bore. In this way, we introduce a code for the spatial position by means of the different Larmor frequency assigned to the protons located in that place. Easy, powerful, and extremely important for the imaging technology based on the NMR, the spatial encoding of the Larmor frequency is used in the image reconstruction process to realize the slice selection along the *z*-axis and in the *x–y* plane to have the 2D *image signal*.

The shape of the gradient can be designed to produce particular effects on the nuclei polarization and consequently on *M*. Gradients can also be used to compensate inhomogeneity in the static magnetic field. This is an important operation because the magnetic field is altered by the introduction of the sample, the patient body for MRI. The *shimming* compensation, usually performed automatically in MRI, is an important operation for the quality of image or for the *cleanness* of the signals from the tomograph acquired for the spectroscopic analysis, both *in vivo* and *in vitro*. Shimming is performed, in an active way, by coils with particular shape around the FOV and controlled by the computer.

It is important to have a visual representation of the gradients. Let us consider a gradient operating along the *x*-axis and a gradient operating along the *y*-axis. We will use the negative and positive variations of the field along each direction, the so-called positive lobe and negative lobe of the gradients.

MR pulse sequences usually contain gradient waveforms for the frequency and phase encoding, and one for the slice selection. Commonly, each gradient pulse has two lobes, positive and negative, with different shapes: trapezoidal, triangular, and sinusoidal, depending on the requested imaging parameters (Figure 2.23).

Along each axes, we have a field profile as reported in Figure 2.24.

It is important to understand that these variations can be applied simultaneously on both directions or with an opportune timing considering, for example,

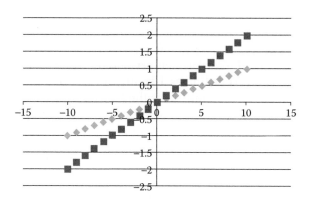

FIGURE 2.23
Strength and slope of gradient.

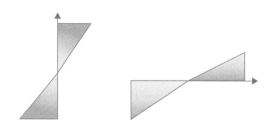

FIGURE 2.24
Linear gradient profiles for *y*- and *x*-axes.

the possibility of introducing the application of particular RF pulses. This is what happens in a particular kind of sequence *to prepare the magnetization* of some tissues.

An important consideration: gradients along the spatial axes allow for the labeling of the space with a different Larmor frequency for each point of the field of analysis. The receiver will capture the spectral frequency range depending on the excitation of RF pulses and on the gradient intensity and the gradient-related local magnetic field values.

If used properly in a combined way, both RF pulses and gradients give the researcher the possibility of manipulating spins and sampling signals in an extremely variable and plastic way. The primary use of gradients is the slice selections, that is, the selection of a slice of tissue to represent along the *z*-axes (the axial projection of the *x*–*y* plane) along the *x*-axis (the longitudinal projection of the *z*–*y* plane) and along the *y*-axis (the coronol projection of the *x*–*z* plane). After a plane selection, gradients act along the *x* direction for the so-called frequency encoding and along the *y* direction for the so-called phase encoding. Figure 2.24 gives the spatial representation of the gradient profile. In the sequence engineering, the time profile of the combinations of the characteristics and timing of the radio pulses and magnetic field gradients is important; they are used to manage the manipulation of the NMR signal and represent the so-called sequences. Different time lines are used for each gradient; one time line is used for the RF pulse representation

and one time line is used to represent the signal collection time. To keep pulse sequence diagrams as easy as possible, authors are used to simplify the representation of the gradient shapes in the drawings. One of the most common simplifications is to draw gradients as perfect rectangles, whereas they should be drawn as trapezoids (or some similar shape) having a more gradual rate of rise and fall; in fact, performances of the electronic gradient unit are expressed by their intensity, in mT/m, and their speed of variation, in mT/m/s. Another frequent simplification in the gradient drawing is that, in many references, the smaller downward projecting lobe in the slice-select gradient immediately after the RF pulse is not represented. This slice-rephasing lobe exists, and it helps to correct the phase dispersion of transverse magnetization that occurs with the application of the main slice-select gradient. Without the application of the rephasing lobe, we will tackle with an intravoxel phase dispersion, resulting in signal loss (Figure 2.25).

How gradients are used in MRI? This is a very important part of the *sequence* engineering. The *z* gradient is also called the gradient of selective excitation. It is applied along the *z* direction modifying in this way the Larmor frequency in this direction. Then the slice will be selected using a narrow-band RF pulse with the primary frequency centered on the value of interest.

The *y* gradient is called the *phase gradient* and it is applied for short time intervals. It used to de-phase the spins for a certain amount, and then it is switched off. After this procedure, spins return to rotate together, with the same angular velocity but with a constant phase shift dependent on the *y* coordinate. It is switched on and off with different intensities.

The last gradient is the *x* gradient, also called the readout gradient, is switched on during the *reading* of the signals from the P90 or P180 pulses.

Gradients can be activated several times in order to cover all the lines of the matrixes and all the slices selected. The process of timing of these pulses is very important. It is also energy consuming, and an apposite power unit is frequently devoted to the gradients supply. Finally, to describe the timing and the effects of RF pulses and gradients, we will use a diagram with these lines (Figure 2.26).

In each line, we will report the shape and the temporization for RF pulses and gradients. But now, before going further with the sequence description, it is better

FIGURE 2.25
Rephasing lobes.

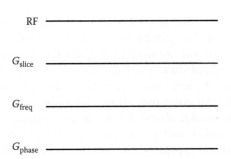

FIGURE 2.26
Representation for the sequence.

to give some more detail about the relation between the space position of the spin in the image FOV and the Larmor frequency and the energy of the RFs. We will introduce the important concept of the *k*-space in MRI.

2.4 *k*-Space and Complexification of the Space

Which is the best *workspace* for the MRI? We can start this discussion with some important considerations expressed in the previous paragraph: gradients along the spatial axes allow for the space labeling with a different Larmor frequency for each point of the field of analysis in view (FOV). The receiver, or coil, will capture frequencies in a spectral range depending on the excitation RF pulses and gradients activated during the excitation and the reading process.

So, there is a correspondence, through the Larmor frequency, between energy and spatial coordinates in the laboratory frame, expressed by the dependence of the Larmor frequency from the **r** vector. So, $\omega(\mathbf{r})$ has a spatial distribution in the volume, or region of interest. Further, in MRI, the nature of the signal image is discontinued, due to the sampling and quantization of the space in voxels and pixels, and consequently also the image has a discontinued structure. The voxel or pixel in the image is a quantization of the spatial coordinates. There is a sort of lattice in the space, and coordinates are the multiple of axial spatial resolution of the gradients used. This is the reason why in imaging we speak about the image as the spatial matrix and treat the image as a matrix of pixel in 2D or voxel in the 3D or 4D matrix in the case of the fMRI. Properties of this matrix or lattice are quite complicated. So, the matter is represented with a periodic structure, like the lattice model used in the solid-state physics. Variations in the position, in a space nestled in gradients of magnetic field along all the axes, means to consider different energy states of the spins with differences in terms of frequency and phase associated with different voxels.

Well, where is the problem? Obviously, in Mathematics, the proper manipulation of the properties of these two workspaces is required: one related to the spatial coordinates and the complementary and the other related to the energetic coordinates. Usually, at this point of time, we should need to introduce the Fourier transform, the mathematics tool to correlate the spaces of some functions. But I would like to propose an innovative pathway related to the possibility of making the workspace of the MRI *complex*. So, let us introduce the complex numbers in MRI.

Let me give a brief excursion about the linear algebra used in the most part of our university studies: the great limitations to the calculus are given by the restriction to work in the $\mathbb{R}$ ensemble! But $\mathbb{R}$ is not a *good ambient* to work in. The natural environment for algebra, and particularly for matrices algebra, is $\mathbb{C}$, the complex space, where the application of algebra is straightforward and easy to understand. In the MRI Physics, we are in the same situation. We have to use a space with properties useful to calculation and explication of spin properties: this space is called the *k-space* which is properly defined in the complex space as a complex, discontinuous and with *reticular properties* space. *k*-space actually simplifies the comprehension and the use of data acquisition and evaluation methods used in MRI and in NMR microscopy. Obviously, you need a strong mathematics background to properly handle the *k*-space for these purposes, and most of this mathematics is related to the Fourier transform. Students of physics meet the *k*-space concept during the studies about solid-state physics, based on the concepts of reciprocal lattice, Brillouin zones, and Fermi levels. Also Bloch, the author of the phenomenological magnetization equations, produced a great amount of study of the EM interaction in periodic structures in the complex space. Bloch defined the auto-solution of the photon solid matter interaction problem.

Applications of the *k*-space formalism regard many fields in medicine, such as X-ray scattering, ultrasound echography, positron emission tomography, and electron microscopy. Its introduction into NMR and MRI can be found in the pioneering works of Ugurbil and Twieg. However, considering its deep roots in physics, the concept has never been considered as particularly important in other disciplines; and it was just used at minimum due to the fact that the most part of our algebra is in $\mathbb{R}$ instead of $\mathbb{C}$. In fact, we need complex numbers to better introduce the *k*-space concept very useful in the MRI.

The first step toward this is the introduction of a generic vector $\mathbf{K} = (K_x, K_y, K_z)$ that we will later associate with the energy values of the spaces; then we can define the product of this vector with the spatial vector **r**:

$$\mathbf{K} \cdot \mathbf{r} = (k_x x, k_y y, k_z z)$$

Then, we can consider the complex exponential function of this product that can be indicated as

$$e^{(ik.r)}$$

term1

where i is the imaginary unit to define a complex number; this is the function that can be used for the complexation of a generic vector **P** represented in the space.

There are several representations of the complex number (Figure 2.27). Using the $\cos(\theta)$ e $\sin(\theta)$ components of a vector in the 2D space, it is possible to relate the Cartesian coordinates with the component of the complex number z as

$$z = \rho[\cos(\Phi) + i\sin(\Phi)] = \rho e^{i\Phi} \tag{2.14}$$

Now, we can define a new vector with dependence from the spatial coordinates r, $s(\mathbf{r})$. Starting by $s(r)$ we can use the definition of term 1 and the vector **K** to define the function $\tilde{s}(k \cdot r)$ defined in a infinitesimal volume dV:

$$\tilde{s}(k \cdot r) = s(\mathbf{r}) * e^{(ik \cdot r)} dV = s(\mathbf{r}) * e^{(ik \cdot r)} dx * dy * dz$$

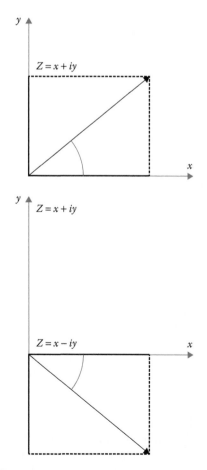

FIGURE 2.27
Complex number in the vectorial representation.

$$\int s(r)e^{ikr} \, dV = S(K)$$

FIGURE 2.28
Complex function $S(K)$.

In order to remove the dependence of this quantity from the spatial coordinates, we can operate an integration summing in this way all contributions from each voxel in the volume of interest V, and we can define a new entity, the distribution $S(\mathbf{K})$ with the direct dependence from the **k** vector alone and that has the property to be a complex function of a real variable (Figure 2.28).

In this way, we shift the functional dependence from the r coordinates to **k** vector. The new function $S(\mathbf{K})$ has the important property to be complex and to be related to the time evolution of the *coordinates* of k. Information in $S(k)$ is related to all the coordinates of the field of interest FOV; integration has the function to sum all the components from all the coordinates of the space. This will introduce an important characteristic of the k-space images: each pixel of the images will have energetic information from the spins of all the voxel of interest VOI!

Now, according to the standard NMR theory, and using the arguments previously introduced, we can do some other considerations:

1. The use of gradients means to change the frequency and the phase of the spins and magnetization in the voxels.
2. Generally speaking, gradients are time-dependent.
3. The k-space is specifically related to the frequency and phase of spins and macroscopic magnetization.

From all these considerations, we can say that protons contained into a volume $dxdydz$ give an NMR signal:

$$\int q(r)e^{2\pi ik(t)r} dV = Q(K) \tag{2.15}$$

Using the term-to-term identification method, we can see that formally $S(k) = Q(K)$. The time evolution of the vector **K** is related to the way to change the energy relation between the spatial label and the magnetic field. These physical entities are related to the way to combine RF pulses and gradient values, which means to manage a *trajectory* defined in the complex k-space and depending on the gradients and pulses applied.

We have seen that using RF pulses and gradients, we can move along any desired k-space path. In this path, we build up a record of the $Q(K(t))$ or $S(K(t))$ values for a subset of the *visited* k-space points. *Visited* means *measured*. These are the signals revealed by the coils. This signal contains, as a real function of the r real coordinates, information about the spatial labels of the signal, but, as a complex function of variable k, that is also a time-dependent function; this information is hidden in the $K(t)$ and in the image. So, by means of the RF and gradient pulses, you can select the k-space points to select and sampling the NMR signal! The trajectory in the k-space is related to the way the proper pulses and gradient sequence are managed for the image contrast of interest!

But what kind of image do we record in this way? Is it the one shown in Figure 2.29?

This is not an image that can be used for the representation, but it is the real representation of the signals measured by the scanner. This kind of image is sent by the *reconstructor* PC of the scanner to provide the representation for tissue and organ imaging. k-space images give a real representation of the signals acquired. Before treating how these images are used to realize the imaging of the inner organs, we have to make other important observations about the k-space. The function $Q(K(t))$ is complex. This is in accordance with the fact that the actual NMR signal has a phase, and it is completely detected by two signals acquisition along axes; these signals have a 90° phase difference and they are detected using the coils in *quadrature detection* mode, with orthogonal RF reference signals. The two channels, often denoted as u and v, provide two time-dependent output signals $u(t)$ and $v(t)$, which behave as the Cartesian components of a complex signal $S(t) = u(t) + iv(t)$. Again, a complex representation of the NMR signal gives us the possibility of getting all the information from the scanning of the magnetization vector also in the complex space of interest. This is the point: our information, the information in MRI, regards, at the same time, both the energy and the position of the *emitter*! We need complex numbers to treat this kind of information! But they cannot be related to the usual space images of the Imaging.

So, what has happened? The introduction of this formalism has several consequences, and the descriptions of some of which are beyond the scope of this chapter. However, there is an important point to consider: with the introduction of the complex number formalism, it is easier to discuss about the frequency and phase of a vector, and tissues magnetization is a vector, and not only about the module or intensity. The phase and frequency are two really important parameters in the MRI because they are related to different physical states of protons in different positions in the matter that can be revealed to exploit information to use in the imaging contrast. Different x, y, and z coordinates mean different k_x, k_y, and k_z *energy* point in the k-space.

With this complex vector transporting all the different energetic state of the spins in different positions, we can accurately differentiate the states of tissues magnetization in order to produce MRI signals dependence related to the intrinsic tissue parameters by means of the extrinsic parameters to settle up on the scanner: our contrast of image. Each energetic state of each tissue in different location, because of the great amount of intrinsic parameters in the MRI, can be labeled in different ways with these parameters, in order to distinguish among them and among different pathological states. But this sensibility and specificity of the imaging technique are related to the possibility of calibrating electronically pulses and gradient as the best we need. This is the power of the MRI. The possibility of increasing the sensibility and the specificity of the exams improving the technology of RF and gradients pulses! It is important to understand that in some situation it is better to have powerful gradients and RF systems instead of a high MR static field. Using the information stored by the timing and powering sections of the scanner during the production and control of the RF and gradients signals, we can map the k-space as requested by our exam: slice thickness, interslice gap, and tissues to excite or to saturate in order to change the magnetization contribution to the signal recorded. We can choose all these parameters by selecting the way to sample the k-space; then, the reconstruction will be all, and we need not convert the physiological image as shown in Figure 2.29.

Spatial localization for the slice selection and the y-axis position are encoded by the difference of the Larmor frequency induced by the z gradient and the y gradient; the second axis in the x–y plane can be encoded using the differences in the phases activated by a short and fast gradient pulse applied along the x-axis and related to the different position of the protons along the x-axis. A little dephasing is applied but it expires as the gradient pulse is switched off. After this controlled dephasing, spins move back in a coherent way, but with a different phase. In order to map all the position along x-axis the pulse gradient must be applied several times at different intensity, while the frequency encoding is obtained with a single longer gradient. So for each frequency gradient switched-on for a particular x position

FIGURE 2.29.
k-space representation of the image in MRI.

there are a sequence of faster and much more intense gradient along the y-axis.

In this way, the encoding for a single y position represents a sort of *offset* for the frequency encoding along the x-axis. The frequency can be the same but with different starting phase. The combined use of these two gradients allows us to elaborate a strategy of k-space covering that we can call *trajectory* in the k-space. We have to point out that k-space is a lattice, a reticular disposition of points along the x- and y-axes, and this is due to the discrete nature of the coordinates in the MRI space partitioned in pixel and voxel, due to the sampling process of the NMR signals and to the digitalization of the RF pulses. Motion on this lattice to cover points of the k-space is controlled by gradients and RF pulses application.

The other important property of the $Q(K(t))$ mentioned before is that, as represented in Figure 2.29, the image of this function in the plane usually appears extremely variable and without connection to the image of the organs and tissues in the body: it is difficult to relate the images in the k-space with the real images, but there are some important properties of these images: we can demonstrate *experimentally* that, since we always image spatially bounded objects, the area of the k-space where $Q(K(t))$ reaches important values is limited to a central region of the k-space image and it is related to all the structures and tissue of the real image. At large distances from the origin, $Q(K(t))$ becomes smaller than the experimental noise and it has no sense to continue mapping values along the spectral coordinates. With these concepts in mind, we can do another important consideration about the pathway to use in the covering of the k-space: it is possible to demonstrate that $S(-k) = S^*(k)$, with the star denoting complex conjugation operation on the complex number. This means that there is a inner symmetry in the k-space, related to the nature

of the quantity $S(k)$ or, in our case, $Q(k)$. Therefore, we could chart $Q(k)$ only in a suitable part of the k-space.

As mathematicians will have noticed, we have not yet introduced mathematically the important tools defined by the Fourier transform. All discussions introduced by the complexation of the k-space are related to this strategy of theoretical treatment of the argument. I'm really interested to the diffusion of the importance to do calculation in $\mathbb{C}$ instead of $\mathbb{R}$, but we will use the tool within the ImageJ software and for this reason we have to illustrate the properties of FFT, the algorithm to calculate the Fourier transform.

We can illustrate these steps in *imaging way*. Using the open source software ImageJ freely available from the Web, we can easily correlate dicom medical images with their k-space. The mathematical operation to do this is the Fourier transform. We will not discuss about the Operator Fourier Transform. We can avoid this using the concept of complexation of the space, much easier; using ImageJ we can charge a slice from an MRI structural study and we can move from the image to the related k-space and vice versa.

Using the area selection tool, we can capture portions of the k-space to delete or to replicate in other parts of the image, operating geometrical operations like reflection or axis inversion.

From the initial image of a brain slice, we can open the corresponding K-image space and select a square inner part to delete the external region of the k-space (Figure 2.30).

Which is the effect on the image? Visual comparison of the two images seems to reveal no differences, and only a mathematical subtraction of the second image from the first one can be evidence of a slight change exhibited in the following comparison and only in some details of the image, as reported before. So we can say

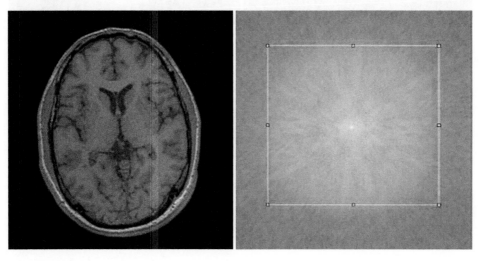

FIGURE 2.30
MRI image and the k-space representation.

that external part of the *k*-space is related to *details* of the image with low importance (Figure 2.31).

Selecting a smaller square we have these effects with a more significant effect of alteration. Other geometric property can be exhibited with this *k*-space operation we can remove the lower half part of the *k*-space and substitute it with a reflected upper part (Figures 2.32 and 2.33).

The effects of this operation on the image are shown in Figure 2.34.

One of the most important properties of the $\mathbb{C}$ space is the ability to make integral calculation easy. In the $\mathbb{C}$ space the mathematical tricks used to calculate the integrals are frequently transformed into a proper choice of the *integration path*. The functions of complex variables exhibit unbelievable simplification properties along particular paths. Using the $\mathbb{C}$ space properties and the c-number properties is easy to demonstrate that some important mathematics operators are *well-defined*

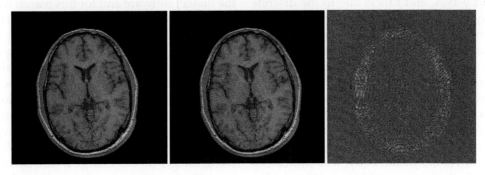

FIGURE 2.31
Effects of the elimination of the external part of the *k*-space in the image.

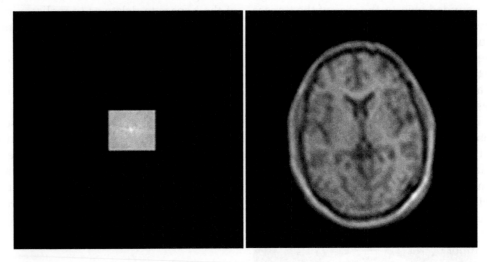

FIGURE 2.32
Image related to a smaller centered part of the *k*-space in Figure 2.30.

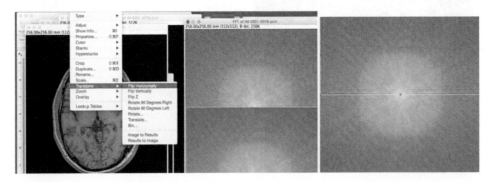

FIGURE 2.33
Manipulation of the *k*-space image: removal of the lower half with the mirror relative to the axis passing through the origin.

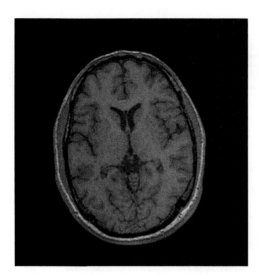

FIGURE 2.34
Geometric operations of parts of *k*-space duplication make the same starting image.

with the exponential representation of the complexness of the operator. Path and rotation are exhaustively explained in $\mathbb{C}$ space and the physical applications are better defined. This is the same situation for the MRI physics. As shown in Figures 2.32 and 2.33 and previous images the properties of the *k*-space can be used to describe some properties of the related images acquired. Also, using the interpretation of the K-vector introduced in the beginning of the paragraph as an energy vector, we can map the different energy values as different particulars of the tissues to describe MRI. We need complex numbers because in this way it is natural to introduce the phase parameter for the spins or magnetization vector. Phase became an important extrinsic parameter to control the image contrast in MRI. The combined choice of pulses and gradients is the path definition in the *k*-space to select part of the echoes signals to excite and acquire. This is important to understand. It is a key concept in MRI. We can select the elements of the *k*-space to better *construct* the image's characteristics we want to acquire. Energy selection in the *k*-space contains the spatial selection in the usual space but it is not only a spatial selection. This is the reason because the *k*-space has no direct representation of the image. The *k*-space contains the spatial selection but it is not only a spatial selection, there is also the association with the energy selected and also with the tissues selected. In other words, in the *k*-space we have all the instruments to properly select the contrast to visualize. Some other details to better understand the *k*-space.

We can magnify the structure of the raw image of the *k*-space related to an image of the brain: we can note that, obviously, also the *k*-space is *quantized* and has a pixel structure like the image representing the real inner head structures. Obviously, because there is a sampling and quantization process related to the acquisition of the MRI signals. But the pixel of *k*-space is not related to a single pixel of the real space image.

In Figure 2.35, you can see the central portion of *k*-space image of the slice reported in Figure 2.37 acquired with the use of ImageJ software.

We can select a squared area of 3×3 pixels around the center and then after the clearing of the image around the selected square we can reconstruct the image (Figures 2.36 and 2.37).

In the visual comparison between the starting image and the elaborated one it is clear that the small portion of central *k*-space has information about the entire pixels of the *spatial* image. This is due to the fact that in our formula for the $S(k)$ or, in our case, $Q(k)$, we have integrated the signal along the spatial coordinates *dxdydz* of the V volume of interest.

FIGURE 2.35
Pixel structure of the *k*-space image.

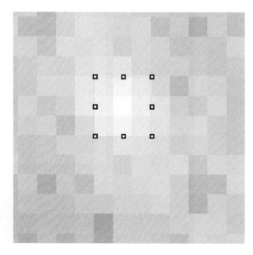

FIGURE 2.36
Pixel structure of the *k*-space image with higher magnification.

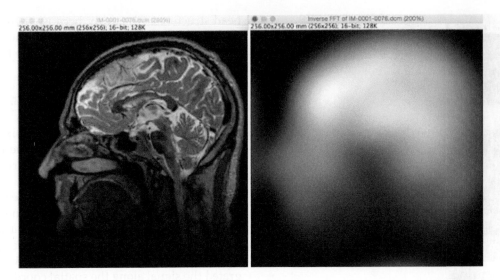

FIGURE 2.37
Details from a small portion of the *k*-space related to all the pixels in the real space.

Increasing the dimension of the square around the center, we can increment the detailed definition of all the *spatial* images. Selection of the *k*-space is shown in Figure 2.38.

Again, we have a greater amount of information and details for all the spatial pixels of the image. Let me show you some more interesting examples of the *k*-space properties. Let us select the portion of *k*-space around the center as shown in Figure 2.39.

The reconstructed image from Figure 2.39 is shown in Figure 2.40 with the selection (Figure 2.41) and the reconstructed image from Figure 2.41 is shown in Figure 2.42.

The geometry of the details is related to the paths of *k*-space covered: so a good strategy of *k*-space covering

can give to the researcher the possibility to acquire images faster and in order to increase as better as possible the SNR. External portion of the *k*-space can be related only to noise of image and it is a good strategy to acquire only in smaller part centered to the origin of the *k*-space.

At the end of this chapter and with all these instruments, we can resume an operative definition for the term *sequences*: in MRI, a sequence is a preselected set of defined RF and gradient pulses; some of these pulses can be repeated several times during the acquisition process defined as *scan*. The time interval between pulses, the amplitude and shape of the gradient waveforms, are used to manipulate spins and, at the macroscopic level, the net magnetization. After the first decay

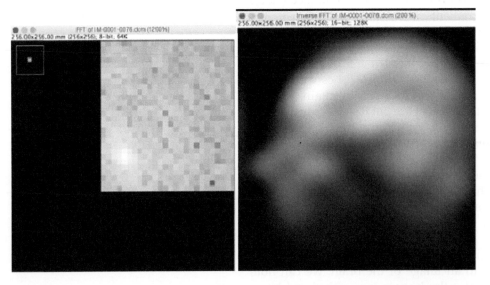

FIGURE 2.38
Portion of the *k*-space and related image.

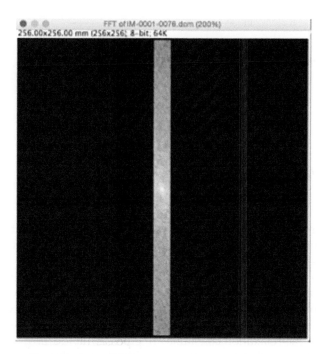

FIGURE 2.39
Horizontal slice of the *k*-space related to all the pixels in the real space.

elements of the electronic chain in the scanner for the MRI process. By means of different pulse sequences, the radiologist can image the same tissue in various ways, and combinations of sequences using different parameters can reveal important diagnostic information about all the tissue. Magnetic resonance is dynamic and flexible; this technology allows radiologists to tailor the imaging study to the anatomic part of interest, to the functions operated by these parts, and to the disease process being studied. In this chapter, I have omitted the discussion of many topics of interest for a complete *k*-space description of many MRI techniques. The discussion of such topics assumes aspects beyond an introductory exposition.

Further, many important aspects of *k*-space were just barely mentioned, avoiding a proper mathematics description. I have tried to highlight the importance of *k*-space workspace with its versatility, which makes MRI so attractive to physicists and mathematicians, engineers, and obviously physicians.

of magnetization spins, defined FID, the manipulation of spins with RF pulses or frequency and phase gradients can generate a sort of echo of the FID signal: this mechanism use the dephasing–rephasing technique to produce new signal as echoes of the first one. By operating a control over the NMR signal reception and sampling, it is possible to affect the characteristics of the MR images in order to produce the proper contrast between the tissues of interest. Pulse sequences are generated by computer programs that control all the hardware

2.5 Sequences and Contrast Control in the Imaging for MR

Gradients and pulses technology represent the most important and recent technical developments in the diagnostic imaging in MRI: gradients and pulses combinations realize the possibility to excite and detect, particularly selective modalities, signals to and from protons in the living matter due to a great amount of parameters that can be used to control these two phases of the images construction process in MRI: excitation

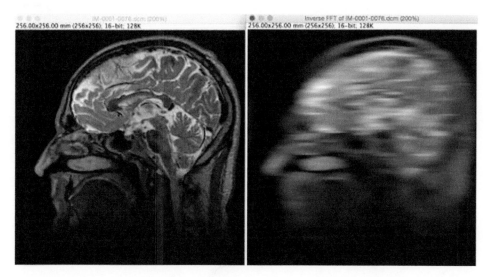

FIGURE 2.40
Comparison between the starting image and the reconstructed image for the portion of *k*-space in Figure 2.39.

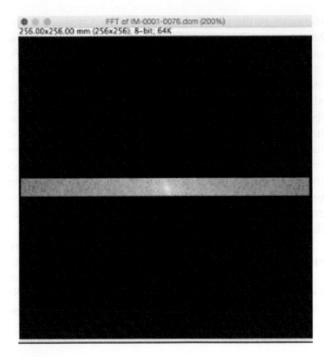

FIGURE 2.41
Vertical slice of the *k*-space.

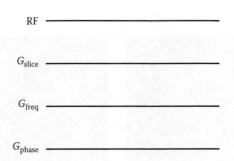

FIGURE 2.43
Representation lines for gradients and RF pulses timing in the sequence.

and acquisition. In order to proper illustrate the timing and technical characteristics of the sequences, several kinds of diagrams are used in the literature, where it is easy to represent graphically the effects of pulses and gradients waveforms on the magnetization. We will use the graph with four lines (Figure 2.43).

By combining shape, timing, intensity for the RF, and gradient pulses in this diagram, it is possible to explain the actions expressed on the spins and magnetization vectors; in other words, we can represent the most important properties of the sequences reading the data coming from the *k*-space. We will use subscripts x to denote the readout (or frequency) directions and the y to denote phase-encoded directions. In order to understand the behavior of the spins under the sequence solicitation we have to discuss the values of the relaxation parameters for the different tissues. Typical values for the relaxation constants are given in Table 2.1.

Paramagnetic contrast media, like the gadolinium, are also used to locally reduce the constants T_1 and T_2, as tissues with particular pathological state, as *fresh lesions* or *recent lesion* in multiple sclerosis, capture these media. In the clinical practice, proton MRI is used for the diagnosis: the most important signals in these conditions are due to the protons of the bulk water, defined as free, which is not linked and constrained to other molecules. Another important source of *proton signals* are the protons of the water linked to lipidic molecules with a structural geometry of droplets, a typical structure of the body fat. The semiliquid physical state of the lipidic droplets generates a strong spin–lattice interaction with a short relaxation time. Furthermore, fat can be related to some pathological

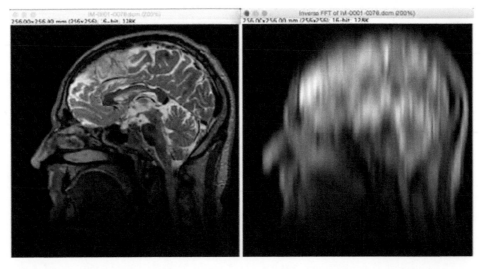

FIGURE 2.42
Comparison between the starting image and the reconstructed image for the portion of *k*-space in Figure 2.41.

TABLE 2.1

Relaxation Parameters of Some Tissues

	T_1 (ms)	T_2 (ms)
Water	3000	3000
Gray matter	810	100
White matter	680	90
Liver	420	45
Fat	240	85
CSF	800	110

FIGURE 2.44
FID and echo signal shapes in MRI.

conditions like the ones related to some kind of tumors, and in this case it is important to map differently this lipidic molecules by the natural lipids in order to build-up images with contrast power to differentiate, if possible, physiological fat by pathological one.

To try to solve this contrast problem usually the different component of the **M** vector are selectively excited with preparation pulses and dedicated sequences. Typical pulse sequence timing diagram uses the temporal lines to present the RF pulses, the three gradients applied, slice-phase frequency, and sometimes a fifth line to represent the signal recorded (see Figure 2.26).

Using this representation, it is possible to classify the sequences in different ways; from an historical point of view, the presentation of the classification based on the readout signal is interesting: sampling directly the FID signal or sampling a single echo of the FID signal or multiple echoes created with different strategies. The echo can be created from an FID with the use of a procedure of spin dephasing and rephasing using RF pulses or gradients waveform. Signal shapes are different: the FID has the typical damped sinusoid shape with T_2 temporal constant behavior. The echo signal shape has an increasing amplitude sinusoid with a dumping after the maximum amplitude. This shape is due to the fact that the recording of the signal usually begins when the spin is still in rephasing and proceed until dephasing again for the thermal agitation (Figure 2.44).

Techniques used for the echo creation are mainly differentiable about the pulse used: it is possible to create an echo with the use of a P180I RF pulse inverting the decreasing M plan of rotation (Figure 2.45).

It is possible to create a single or multiple echoes using the phase gradient applied several times in combination with the slice and frequency selector gradients. The FID sampling sequence is not used in MRI and we will not discuss about them. We will start describing the spin-echo (SE) sequence and the gradient-echo (GE) sequences.

Usually, in the description of the properties and peculiarities of the sequences we discuss about the T_e, T_R, and the FA, three parameters settable by the computer control. PC allows us to control the amount and the typology of the collected signal; by changing T_R, T_e, and FA, it is possible to characterize the timing in the sequences of pulses and rising and falling of gradients according to the requirements to read the signal of interest among the signals emitted by the prepared magnetization.

The echo time represents the time elapsing between the application of RF excitation pulse and the peak of the signal induced in the coil. It is usually measured in milliseconds. According to the T_e values, it is possible to control the amount of T_2 relaxation.

Another important extrinsic parameter is the repetition time or (T_R) is the time from the application of an excitation pulse to the application of the next pulse. If we leave all tissues to relax completely, we can recover the total amount of magnetization (long T_R); for T_R shorter than T_1's parameter of the tissues, a lower amount of magnetization will be recovered. T_R determines how much longitudinal magnetization recovers between each pulse. It is measured in milliseconds (Figure 2.46).

Another parameter can be used for the same reasons: the FA. It can be used to define the angle of excitation for a field echo-pulse sequence. It is the angle at which the net magnetization is rotated or tipped relative to the main

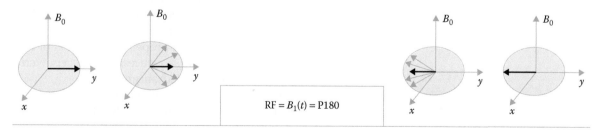

FIGURE 2.45
Echo from inversion.

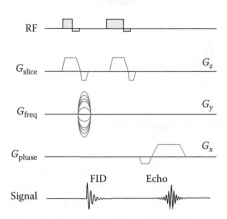

FIGURE 2.46
T_e and T_R representation.

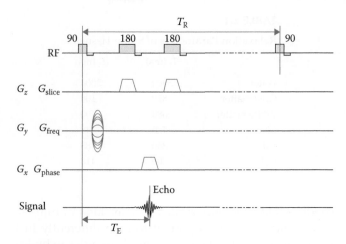

FIGURE 2.48
Spin-echo diagram.

magnetic field direction by applying an RF excitation pulse at the Larmor frequency. It is also referred to as the *tip angle*, *nutation angle*, or the *angle of nutation*.

The RF pulse power (which is proportional to the square of the pulse amplitude) is calibrated to the number of spins that are tilted. FAs between 0° and 90° are typically used in gradient-echo sequences, 90° and a series of 180° pulses in SE sequences, and an initial 180° pulse followed by a 90° and a 180° pulse in inversion recovery (IR) sequences (Figure 2.47).

These are the most important extrinsic or technical parameters used in the MRI set up. Using these parameters, we can now describe the properties of the most diffused family of sequences and the innovations introduced.

2.5.1 SE Sequences

The SE sequence has been the first type of sequences produced for the spectroscopic use and then has been adapted for use in imaging. The schema of this kind of sequences is based on the use of a 90° initial pulse to flip the magnetization in the *x–y* plane and then a 180° pulse to invert the rotation plane of the dephasing spins for the refocusing process. The 180° pulse is transmitted

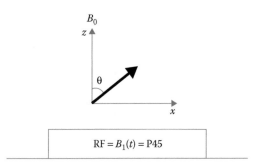

FIGURE 2.47
FA representation.

exactly at the halftime between the excitation pulse and the echo maximum signal, this time is defined $T_e/2$ (Figure 2.48).

Let us see the features of this kind of sequences. After the first pulse, spins are aligned in the *x–y* plane and the Fid signal produced during the recovery of the longitudinal magnetization is:

$$S = k\rho(1 - e^{-T_R/T_1})$$

In order to read the echo signal, we must apply the P180 pulse because the readable signal decay is only given by spin–spin relaxation. But, the application of the P180 pulse inverts the dephasing process (inverting, at the same time, *x* and *y* directions) and the echo is generated. The *S* signal becomes

$$S = k\rho(1 - e^{(-T_R/T_1)})e^{(-T_e/T_2)}$$

The use of the refocusing P180 pulse means that all the signal losses due to the B_0 inhomogeneity, to the susceptibility, and water/fat dephasing are recovered and neglected. Therefore, by means of the SE sequences it is possible to generate images with a weighting in the signal of the T_1, T_2, and PD components. Among the different advantages of this sequence, we can mention a high value of the SNR, a high spatial resolution and a low sensibility to different kind of artifacts like the inhomogeneity artifacts because of the refocusing P180 pulse.

It is clear from previous equation the MRI signal properties achieved with a proper manipulation of the T_R and T_e parameters in the sequences allow us a powerful multi parameter contrast mechanism in MRI.

By using a short repetition time and a short echo time, it is possible to approximate the signal with the T_1-weighted term; for a longer T_R compared to a short T_e, the T_2-weighted term is predominating. By considering finally in both terms of T_R long and T_e

long, we can achieve a weighting of the PD component in the equation.

So, the SE sequences can be used to produce a contrast control on the T_1, T_2, and PD values of tissues, it is possible to achieve for a single tissue a three-value label.

We have to consider some important disadvantages in the SE sequences: first, this family of sequences using a couple of pulses, the second one lasting twice as much as the former. In addition, there is a great amount of energy released into the tissues with a consequent increase in the specific absorption rate of RF power (SAR). The SAR is defined as the RF power absorbed per unit of mass of a tissue; it is measured in watt per kilogram (W/kg) and it describes the effects of heating of the patient's tissue. The organizations for the health quality assurance fixed precise limit to the SAR during the MRI examination in order to prevent tissues damages.

Another important point to consider about the SE sequences is the fact that for large FOV, the acquisition time is really long and this is particularly evident for the T_2 and DP sequences. Early attempts to reduce these negative effects in the SE sequences were undertaken using techniques to reduce the number of the k-space points to sample for the image production. The k-space has particular properties that can be used to reduce the energy release by RF pulses in SE sequences. As previously reported, k-space data is made up of complex values representing the M_x and M_y components of magnetization. The complex data in the right half of k-space is the complex conjugate of the data in the left half of k-space. Similarly, the data in the top half of k-space is the complex conjugates of the data in the bottom half of k-space.

Before we have to give some definition, the number of excitations (NEX) defines how many times each line of k-space data is acquired during the scan. The multiple acquisitions of the k-space lines is a technique adopted to increase the SNR by taking an average of the signals acquired and approximately SNR increase as the square root of the NEXs.

Now, the fractional NEX imaging technique takes advantage by the complex conjugate relationship between the top and bottom parts of k-space; these are symmetric and it is possible to reduce the number of phase-encoding steps, that is, the number of y gradients switching on. But due to the fact that fewer data points are collected in fractional NEX imaging, the SNR becomes poorer as NEX is decreased. So, it is faster but with a reduced SNR.

In *Half-NEX* imaging, phase-encoding steps +8 through −128 of +128 to −128 are recorded. Steps −128 through 0 are generated from the complex conjugate relationship between the halves of k-space. Phase-encoding steps −8 through 0 are recorded to assure the center of k-space is at 0 and there is a smooth transition between

the halves. Fractional NEX imaging sequences use NEX values between Nex = 1 and Nex = 1/2. The advantage of fractional NEX imaging is that an image can be recorded faster than with Nex = 1 but with the same contrast between the tissues as in the Nex = 1 case.

Another technique adopted to obtain a reduction in time acquisition and SAR is to reduce the FA, instead of the classic P90 pulse is used a shorter pulse; this reduce the time of acquisition and the energy released to tissue but still we also have a reduction in SNR, especially for the T_2-weighted sequences.

Another important technological solution to the SE problems was achieved by suppressing the second pulse in the sequence: the echo signal can be generated by the use of the gradients instead of RF pulses, with an important reduction in the acquisition time and energy released; this solution has generated a new family of sequences, the so-called gradient-echo sequences.

2.5.2 GE Sequences

The gradient-echo family of sequences has been studied expressly for the imaging purposes. In fact, they use the system gradients to operate on the spins as well as for the space labeling. The gradient-echo sequence consists of a series of excitation pulses, all P90 pulses, each separated by a time interval defined *repetition time* T_R; acquisition of data starting after a time interval defined T_e time. There is no P180 pulse and this results in a reduction of the T_R. In this kind of sequences, the sampling of the signal is carried out using a gradient starting with a FID dephasing along the x-axis and then protons are rephased by means of the inverted lobe of this gradient. The control contrast in the image can is obtained changing both T_R and T_e sequence parameters (Figure 2.49).

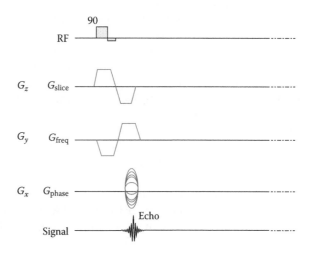

FIGURE 2.49

Representation for a typical gradient-echo sequence.

Also, the FA is used to produce a better T_1 weighing in the case of T_1 long tissues contrast, in order to avoid a long busy time of electronic for long TRs. But with this kind of sequences, the most important parameter is the T_e: there is a great amount of sequences used with the schema of the GE, namely the FLASH and FISP sequences, the FAST, and the SPGR sequences. The FLASH sequence is a GE fast sequence, producing signals of echo at low FAs; FISP sequence is a modification, with the exception that the spoiler pulse is eliminated. In this way, any transverse magnetization still present at the time of the next RF pulse is incorporated into the steady state. Using a small FA selection, a little amount of longitudinal magnetization is lost and the image contrast results quite independent of T_1. Using a combination of very short T_e (T_R [10:50] ms) and a range of FAs in [30°:45°], it is possible to reduce the T_2^* effects, so that the images become PD weighted. Increasing the FA, the contrast becomes increasingly dependent on T_1 and T_2^*: in this *region of work*, FISP exhibits very different contrast in comparison to FLASH sequences. FISP and FLASH can be used for orthopedic imaging, 3D MPR, cardiography, and angiography.

Generally speaking, gradient-echo sequences are used to control the image contrast weighting images in T_1 and T_2^*; particularly, a T_2^* decay occurs between the dephasing and rephasing gradients and generates image contrast. In order to obtain good values for the SNR usually short T_e are used in this kind of sequences.

These sequences are generally sensible to B_0 inhomogeneity, magnetic susceptibility, and water-fat incoherence unrecoverable, especially using T_R shorter than 200–150 ms; for these T_R there is a residual magnetization along the x-axis for water and fat (see Table 2.1) disturbing the signal. In fact, for the T_R shorter than these ranges, the contrast became a function of the ratio T_2/T_1 and tissue differentiation it is undetermined in T_1 and T_2. To resolve this problem and to be able to use T_R shorter than 150 ms, introduction of a series of preparative pulses or gradient waveform is considered. The use of preparation RF pulses has generated the family of the IR sequence, which we will describe in the follow. The use of different gradient shapes along the z, x, and y directions has generated several releases of the gradient-echo sequences: flow compensation sequences, rewinder sequences, diffusion sequences, spoiled sequences, and bipolar lobes sequences, particularly interesting for the use in the phase contrast PC MRA.

A general advantage for the gradient-echo sequences is fast imaging sequence because they can use shorter T_R and T_e than SE; furthermore, by using a low FA, it is possible to deposit less energy than SE, with an important decrease in SAR. But there is an important disadvantage in these families of sequence: the great difficulty in generating good T_2-weighted images.

As mentioned before there are many types of sequences based on the gradient-echo diagram:

- The so-called conventional gradient-echo sequence, like the gradient recalled acquisition in the steady state: It is possible to generate T_1 weighting, with larger FA value, and T_2^* weighting images, with longer T_e.

- The spoiled gradient-echo SPGR: Spoiling destroys accumulated transverse coherence and maximizes T_1 contrast.

- Contrast-enhanced gradient-echo steady-state free precession: Those are sequences with poor SNR, and, for this reason, they are rarely used; but these sequence can generate heavily T_2^*-weighted images.

The idea of the magnetization preparation in order to control the contrast in the MRI has an interesting result in the production of sequences defined as IR sequence. This family of sequence is similar to the SE and pickup the idea to prepare tissues to the requested contrast. By using a proper initial pulse, it is possible to neglect the unwanted signals and to highlight the contrast between the other tissues. Those are the family of the IR sequences.

2.5.3 IR Sequences

IR can be considered as a variant of an SE sequence as it begins with an inverting pulse, P180I, and then the usual single 90° excitation pulse is applied, after a time defined TI time of inversion, followed, as in SE, by the usual refocusing P180 pulse. In this way, the contrast mostly depends on the magnitude of the longitudinal magnetization, as in SE sequence, defined this time by the proper choice of the delay time TI time of inversion.

In this sequence, the time between the middle of initial (inverting) 180° RF pulse and middle of the subsequent exciting 90° pulse to detect the amount of longitudinal magnetization defines the extrinsic parameter TI.

There is the possibility to calibrate the inversion pulse P180I in order to suppress a particular tissue signal: during the inversion process, all the spins are altered by the pulse, and magnetization is the macroscopic expression of the local spin magnetic momentum. But by using gradients to differentiate frequency and phase of nuclear spins, it is possible to calibrate the P180I for a specific tissue by reducing and centering the pulse band of frequency. So, frequency, phase, and duration of the pulse P180I can be calibrated in a different manner, so that a particular tissue is suppressed or a particular contrast is enhanced.

Using broadband inversion pulses and TI selection, signal from voxel excited with this sequence of pulses and repetition time T_R is

$$S = k\rho(1 - 2e^{(-TI/T_1)} + e^{(-T_R/T_1)})$$

So, the important property of IR sequences is related to the choice of TI: if a TI is chosen such that the longitudinal magnetization of a tissue is null, the latter cannot emit a signal because the tissue will not have a transverse magnetization component because it will have no longitudinal magnetization to excite with the following P90 pulse of the IR sequence. The IR technique thus allows the signal of a given tissue to be suppressed by selecting a TI adapted to the T_1 of this tissue. Two different sequences are particularly used in TI: short-tau IR (STIR) also called short T_1 IR is the elective fat suppression technique with an inversion time TI $= T_1 \ln 2$ where the signal of fat is zero. This equates to approximately 140 ms at 1.5 T. Tissues with different T_1 values can be distinguished by use of this technique. But most tissues recover more slowly than fat, and so STIR images have intrinsically lower SNR. Particular care has to be taken during the interpretation of contrast between tissues because of the incomplete relaxation of the water signal of tissues when the image is acquired. IR imaging allows homogeneous and global fat suppression, and it can be used with low-field-strength magnets for large volume of interest. However, this technique is not specific for fat, and the signal intensities of the tissue with a long T_1 and the tissue with a short T_1 may be difficult to interpret.

Another IR-like technique used for the attenuation of unwanted signals is the fluid-attenuated IR (FLAIR) used to suppress water. This technique has a long TI to remove the effects of fluid from the images. For fluid suppression, the inversion time (long TI) is set to the zero-crossing point of fluid, resulting in signal erasing. Using conventional T_2 contrast some lesion in the parenchyma can be fuzzed by bright fluid signals; lesions result more visible, with a better contrast, using the FLAIR, which is an important technique for the differentiation of brain and spine lesions.

Another important use of the *inversion* techniques is for cardiac MRI: in this case, the sequence is used to null the signal from normal myocardium during delayed enhanced imaging. The normal myocardium will be dark in contrast to the enhanced abnormal myocardium. The appropriate TI at which the normal myocardium is dark occurs about 330 ms after the RF pulse, but varies from person to person. To determine the appropriate TI for an individual, a TI scout series is obtained where each image in the series has a progressively larger TI. Alternatively, a newer automated sequence known as *phase-sensitive IR* can be used, which does not require a

TI scout. IR pulses that are used to null the signal from a desired tissue allow us to accentuate surrounding pathology.

2.6 Sequence Engineering as a Strategy for Developing Innovative Diagnostic Applications

MRI is in constant development in several fields of the medical diagnosis. It is impossible to report all advancements in the several disciplines of medicine. We can report just a few examples for the modern brain functional and angiography, and we will discuss about the peculiarities of these sequences.

2.6.1 Functional MRI Sequences

2.6.1.1 Echo-Planar Imaging (Functional MRI)

Fast imaging in brain function studies is required in many field of diagnosis research; as in the case of functional connectivity brain imaging, we want to be able to follow the oscillation of the time series of the T_2^* contrast related to the tasks operated by the patients or to the default mode network of the resting state activity. The lines of k-space can be collected in a variety of different view orders to be fast; important application are achieved through echo-planar trajectories, employing an oscillating spatial selection gradient in the x direction and a unipolar gradient in the y direction. The speed of acquisition is greater than other techniques and this allows us the possibility to catch a *photo* of the brain functionalities related to the local contrast T_2^* in order to capture the functionality of the several brain areas. Using the BOLD mechanism to reveal the neuronal population activated in some areas of the brain generate a local T_2^* contrast related to the differential hemoglobin/deoxyhemoglobin ratio alteration related to hyperactivation of the neurons. The neurovascular coupling enables researchers to follow the neuron activity by means of an indirect mechanism of the BOLD effect.

Innovative sequence for the functional brain imaging is the EPI; EPI is really fast and it is possible to relate the oscillation of the brain T_2^* signal to the variation of the neurons activity. The technique records an entire image in a T_R period. To understand the EPI mechanism, the concept of k-space as previously discussed is of great help. The k-space is equivalent to the space defined by the frequency- and phase-encoding directions. Conventional imaging sequences record one line of k-space for each phase-encoding step. Since one phase-encoding step occurs for each T_R seconds, the time required to produce an image is determined by the product of T_R and the number of phase-encoding steps. EPI measures all lines of k-space in a single T_R period (single shot).

Basically, the EPI schema can be based on both SE and GE mechanism, while the GE schema gradients can be used for the readout of the signal.

In this schema, we use the inversion of the y gradient to produce a single line of dephased echoes; inversion in the lobe gradient allows the reading of the single x-lines of dephased spins moving in opposite direction in the k-space (Figure 2.50).

The inversion of the k_x reading direction is determined by the gradient lobe value; alternating polarity of a series of readout gradient lobes allows us to sample each line of the k-space very fast. From the technical point of view, EPI sequence can use a double mechanism of phase-encoding waveforms: with constant gradient Gy, and with the *blip* gradients (Figure 2.51).

In this way, the diagram of the EPI is represented in Figures 2.52 and 2.53.

FIGURE 2.50
Bipolar symmetric gradient shape.

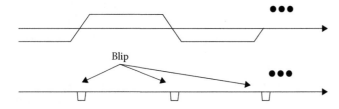

FIGURE 2.51
Continues and blip gradient shape.

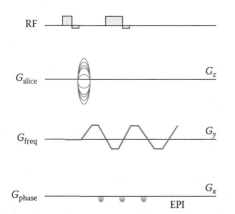

FIGURE 2.52
EPI diagram.

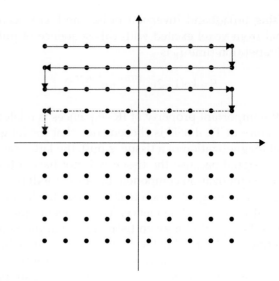

FIGURE 2.53
Oscillation of the gradients in the k-space covering.

The acquisition time of a single slice can be of the order of tens mS. In this way, it is possible to acquire a complete 3D volume of the brain within a single T_R. The EPI sequence can be realized with both SE and gradient-echo schemes. By observing the timing diagram for an EPI sequence in the SE format, we can do some consideration. There is a 90° slice selective RF pulse, which is applied in conjunction with a slice-selection gradient. There is an initial phase-encoding gradient pulse and an initial frequency-encoding gradient pulse to position the spins at the corner of k-space. Next, there is a 180° pulse. Since the echo-planar sequence is typically a single slice sequence, the 180° pulse need not be a slice selective pulse. The phase- and frequency-encoding directions are next cycled so as to traverse k-space. This is equivalent to putting 128 or 256 phase- and frequency-encoding gradients in the usual period when the echo is recorded. If we zoom into this region of the timing diagram, it will be clearer. You can see that there is a phase-encoding gradient, followed by a frequency-encoding gradient, during which time a signal is recorded. Next, there is another phase-encoding gradient followed by the reverse polarity frequency-encoding gradient during which time a signal is recorded. Looking at the k-space trajectory map at the same time as we are zoomed into the phase- and frequency-encoding gradient area, we can see how the gradients trace out k-space. The rate at which k-space is traversed is so rapid that it is possible, depending on the image matrix, to obtain 15–30 images per second.

2.6.2 Magnetic Resonance Angiography

MRA is a powerful technique used to visualize alterations inside blood vessels, with particular interest in the arteries and veins; MRA is appreciated as alternative to the techniques like X-rays in computed tomography angiography and fluoroscopy. MRA to image blood vessels is less invasive compared to other techniques and no dose of ionizing radiation is released to the patients. MRA techniques can generate images of arteries and veins with the spatial resolution to reveal and evaluate the presence of stenosis, occlusions, aneurysms, and other abnormalities present in the lumen of the vessels.

Generally speaking, improvements in technology have produced a great amount of MRA techniques; they can be classified into two general families: *flow-dependent* methods and *flow-independent* methods. The most diffused techniques based on the *flow effects* in MRA are the phase-contrast MRA (PC-MRA) and the TOF-MRA.

MRA can be operated using techniques of contrast enhancement using pharmacological solutions with paramagnetic agents or using the effects of inherent contrast agents; the most frequently applied MRA exams involve the use of intravenous contrast agents, particularly those containing gadolinium to shorten the T_1 of blood to values usually shorter than the T_1 of all other tissues (see Table 2.1). Sequences weighed with a short T_R produce bright images of the blood. MRA is frequently used to evaluate the arteries of the neck and brain, the thoracic and abdominal aorta, the renal arteries, and the leg vascular system, and for this reason, it is an important application of the MRI.

2.6.3 Flow-Dependent Angiography

A class of methods for MRA is based on blood flowing in the vascular system. Those methods are referred to as *flow-dependent MRA*. They take advantage of the fact that the blood within vessels is flowing and it represents a paramagnetic agent moving along pathways of the vascular system; in this way, it is possible to distinguish these vessels from other static tissue and imaging of the vasculature can be produced. Flow-dependent MRA can be divided into different categories: there is phase-contrast MRA (PC-MRA) techniques, which utilizes phase differences to distinguish blood from static tissue, and TOF-MRA techniques; the spins that are moving for the blood circulation experience fewer excitation pulses than static tissue. Both these techniques do not use contrast agents and are suitable for investigation of the intracranial circulatory system alteration.

2.6.3.1 *Phase-Contrast*

The use of bipolar and symmetric gradient allows the creation of images of blood flow; in this way, it is possible to encode the velocity of moving blood with the phase of the NMR signal. The application of a bipolar gradient occurs between the excitation pulse and the readout of the signal. Two symmetric lobes of equal area form a bipolar gradient. By definition, the total area of a bipolar symmetric gradient G is null. The bipolar gradient can be applied along any axis or combination of axes, depending on the direction along which flow is to be measured (e.g., x). The phase accumulated during the application of the gradient is 0 for stationary spins: this phase does not vary during the application of the bipolar gradient. On the contrary, for spins moving with a constant velocity v_x, along the direction of the applied bipolar gradient, the accumulated phase is proportional to both v_x and the first moment of the bipolar gradient; in this way, it is possible to provide an estimation of v_x. Note that to measure phase variation of interest, the MRI signal is manipulated by bipolar gradients (varying magnetic fields) that are preset to a maximum expected flow velocity. An image acquisition that is reverse of the bipolar gradient is then acquired and the difference of the two images is calculated. Static tissues such as muscle or bone will subtract out; however, moving tissues such as blood will acquire a different phase since it moves constantly through the gradient, thus also giving its speed of the flow. Because of the possibility to acquire only images of the blood flow in just one direction at a time with the phase-contrast technique, we need three separate image acquisitions in all directions to give a complete image of flow. This is a slow technique but it is still used because, in addition to imaging flowing blood, it is possible to perform quantitative measurements of blood flow that are useful for the diagnostic process.

2.6.3.2 *Time-of-Flight*

By using the gradient-echo technique, it is also possible to realize the imaging of vessels.

The source of diverse flow effects is the difference between the unsaturated and pre-saturated spins, and it is possible to create a bright image of the vessel without the invasive use of contrast media. The TOF technique uses a short echo time T_e and flow compensation to make the blood flowing in the vessels much brighter than stationary tissue. As the flowing blood enters the area being imaged, it has seen a limited NEX pulses so it is not saturated; this gives it a much higher signal than the saturated stationary tissue. Due to the fact that this method is strictly dependent by the flowing blood, areas with slow flow (such as large aneurysms) or blood flow in the plane of the image may not be well visualized. This technique is most commonly used in the head and neck districts, and it can give detailed and high-resolution images.

2.6.4 Contrast-Enhanced Angiography

Injection of MRI contrast agents is currently the most common method of performing MRA due to high resolution images of the vascular system that can be achieved. The contrast medium is injected into a vein, and images are acquired both in precontrast and during the first pass of the agent through the arteries. By subtracting these two acquisitions in the postprocessing stage, an image is obtained, which in principle only shows blood vessels, and not the surrounding tissue (digital subtraction angiography—DSA). Providing a correct timing of the pulses in the sequence may result in good-quality images. Timing is important due to the short "half-life" of the most used contrast agents. An alternative is to use contrast agents that do not leave the vascular system within a few minutes, but can remain in the circulation up to an hour (this kind of contrast agents are called *blood-pool agent*: the higher the molecular weight, the higher is the relativities of the media). With a longer time of contrast enhancement available for image acquisition, higher resolution imaging is achievable. Recent developments in MRA technology have made it possible to create high-quality contrast-enhanced MRA images without the digital subtraction of a noncontrast-enhanced mask image. This approach has been shown to improve diagnostic quality, because it prevents motion subtraction artifacts, as well as increases image background noise, both of which as direct results of the image subtraction. An important condition for this approach is to have excellent body fat suppression over large image areas, which is possible by using Dixon acquisition methods. Traditional MRA suppresses signals originating from body fat during the actual image acquisition, which is a method that is sensitive to the inhomogeneities of the magnetic fields, with the consequence of insufficient fat signal suppression. Dixon proposed a method to better distinguish and separate image signals created by fat or water; in this method, the difference in magnetic resonance frequencies between fat- and water-bound protons can be used for the separation of water and fat signals; these differences are related to the chemical shift effect. Difference in Larmor frequency generated by the shield on the B_0 by the two different molecular environments (fat and water) is important for the tissues discrimination (Figure 2.54).

The pioneering work of Dixon, published in 1984, presented this imaging technique based on the use of the phase differences to calculate water and fat components in the *post-processing* stage. Dixon's method relies on the acquisition of the images in the condition with fat and water *in phase*, and then in *phase opposition*. These images are then added together to get *water-only* images, and subtracted to get *fat-only* images. Therefore, this

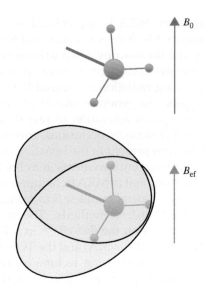

FIGURE 2.54
Chemical shift effects: schematic representation of the shielding action of the different fields induced by the molecule structure.

sequence type can deliver up to four contrasts images in just one measurement: in phase, opposed phase, water, and fat images.

Originally, the technique proposed by Dixon was based on the so-called two-point sequence, with the acquisition of two images at different T_e; however, Dixon-based fat suppression method can be very effective in regions of human body characterized by high magnetic susceptibility, where other techniques fail in the production of images without artifacts. Both the insensitivity to magnetic field in homogeneity and the possibility of direct image-based water and fat quantification have currently generated high interest in the researches for improvements to the basic method. So, the combination of Dixon methods with gradient-echo sequences allows, for example, the liver imaging with four image types in one breath hold. But, both GE and SE sequences can be mixed with the Dixon mechanism. Using the combination of the Dixon methods with SE sequences, it is possible to achieve an excellent fat suppression with high-resolution imaging. Both SE and GE families have the fast version with the multi-echo acquisition at different phase-encoding gradient values. They have a different commercial identification as FSE and TSE. Fast SE (FSE) imaging and turbo SE (TSE) imaging are different commercial implementations of the RF refocusing echoes technique described in the pioneer work of Hennig et al. in 1986. Practically, the FSE–TSE pulse sequence insert, in a conventional SE sequence, a series of P180 refocusing pulse to generate a train of echoes; however, unlike the usual multi-echo sequence that collects all echoes in a train with the same phase encoding, in the FSE–TSE technique, the phase-encoding gradient is changed at each echo of the train. In this way, by changing the phase-encoding gradient between each echo, it is possible to acquire several k-space lines in a single repetition time T_R.

With this technique and using the *water-only images*, the body fat can be neglected with high efficiency, so there is no need of images for subtraction in the post–processing, and it is possible to achieve high-quality MR image for angiography. These are just some examples of possible applications: technology makes available a great amount of solutions for the imaging problem.

2.7 Conclusion

The *sequence* theme in MRI is continuously in evolution due to the technological improvement of the electronic technology of the *gradients* and *radiofrequency* pulses. Being faster and more powerful, gradients are able to explore innovative solutions for the proper k-space *measuring* for the MRI of multiparametric contrast, and RF is becoming highly selective and fast, to manipulate as better as possible all the magnetization states. All the companies are investing in R&D in this field. In this way, with the collaboration of expert physicians it is possible to test and realize new diagnostic sequences with the capability to get a greater amount of information with a shorter time of exam reducing the exam stress for the patients. Angiography and brain functional MRI are only two of the possible examples of the successful application of MRI technology. The functional connectivity in MRI leads to important understanding of network activity for the brain functions, and this kind of measurement derives from a stable and fast technology to follow the oscillations in the MRI signal. The discussions in this chapter can only be an introduction to the methods of sequence description and design. This chapter was realized with the idea to give the basic knowledge for interpretation of a sequence and to operate in the scanner control to optimize the improvements effects on the image. It is not exhaustive of the argument, but it is only to give a basic knowledge. There is a great amount of

materials published for a deep discussion of the topics; some important papers and books on the arguments are reported in the bibliography at the end of this chapter. I hope I was able to give you the importance of the appropriate formalism to use based on the complex functions: Mathematics is important for better exploitation of the innovation in MRI and the reader can find many books and also a number of web sites of help for this exploration. This is only the first step.

Bibliography

Brown T.R., Kincaid B.M., Ugurbil K., NMR chemical shift imaging in three dimension, *Proc. Natl. Acad. Sci. USA* 79, 3523–3526 (1982).

Carr James C., Carroll Timothy J., Editors, *Magnetic Resonance Angiography: Principles and Applications*, Springer, New York (2011).

Hennig J., Nauerth A., Friedburg H., RARE imaging: A fast imaging method for clinical MR, *Magn. Reson. Med.* 3, 823–833 (1986).

Holodny Andrei I., *Clinical fMRI and Diffusion Tomography: Paradigm Selection, Neurological Assessment, and Case-Based Analysis*, Springer, New York (January 2015).

Ljunggren S., A simple graphical representation of Fourier-based imaging methods, *J. Magn. Reson.* 54, 338–343 (1983).

Poldrack R.A., Mumford J.A., Nichols T.E., *Handbook of Functional MRI Data Analysis*, Cambridge University Press, New York (2011).

Prof. StamSykora web page, *K-space formulation of MRI*, http://www.ebyte.it/library/educards/mri/K-SpaceMRI.html.

Twieg D.B., The k-trajectory formulation of the NMR imaging process with applications in analysis and synthesis of imaging methods, *Med. Phys.* 10, 610–621 (1983); Building on earlier work in the field of optics, *Proc. Soc. Photo-Opt. Instrum. Eng.* 347, 354 (1982).

Wang Y., *Principles of Magnetic Resonance Imaging.* CreateSpace Independent Publishing, USA.

Weadock W., Chenevert T., Emerging concepts in MR angiography, *Magn. Reson. Imaging Clin. N. Am.* 17, doi: 10.1016/j.mric.2009.02.005 (Saunders 2009).

3

Contrast Agents for Magnetic Resonance Imaging

Henrik S. Thomsen

CONTENTS

FOCUS POINT

Generally, gadolinium-based contrast media for magnetic resonance imaging (MRI) are safe in the amount approved by the medicines agencies. However, significant adverse reactions may occur. Particularly, identifying patients at risk prior to administration of contrast agent can lead to a risk reduction. In case of a significant adverse reaction to a patient, it is very important to be ready for a possible instant treatment. In many countries, neither manganese-based nor iron-based contrast media are commercially accessible for the time being.

3.1 Contrast Media

All commercially accessible MRI contrast media at present are based on the atom gadolinium. Gadolinium is

1. Part of the lanthanide group in the periodic table.

2. The atom that provides the highest relaxivity per atom.

3. Toxic for the body even in small doses.

Before using gadolinium diagnostically in human beings, the atom has to be detoxified, which is done by binding it to a chelate (De Häen 2001). Administration of contrast media is used in one-third of the MRI examinations because unenhanced MRI is providing additional information regarding the soft tissue compared to computed tomography (CT), even if this examination is performed with iodine-based contrast media. MRI most often is used in relation to brain imaging and chest imaging, including breast and abdomen (e.g., liver, bowel, kidney, and prostate). In the past, contrast media were mandatory to perform MR angiography, but this is no longer the case.

Iron-based and manganese-based contrast agents have also been available previously. However, these are no longer obtainable with the exception of a few countries (perhaps none at the time of printing). All the contrast agents were organ-specific: hepatocytes, macrophages, bowel lumen, and lymph nodes.

3.2 Gadolinium-Based Contrast Media

MRI contrast agents are diagnostic pharmaceutical compounds. They primarily contain *paramagnetic ions* (e.g., gadolinium and manganese) affecting the MR signal properties of the surrounding tissue. Paramagnetic agents are positive enhancers that reduce the T_1 and T_2 relaxation times and increase tissue signal intensity on T_1-weighed MR images and have just about no effect on T_2-weighed images. *Superparamagnetic ions* (e.g., iron-based agents) have a signal-enhancing effect on T_1-weighted images at low concentrations; however, they have a negative effect on the tissue signal in higher concentrations, which is seen on the T_2/T_2^*-weighted images. Conversely, it has nearly no effect on T_1-weighted images (Thomsen et al. 2016).

In the 1980s, copper (Cu^{2+}), manganese (Mn^{2+}), and gadolinium (Gd^{3+}) were contemplated for use as paramagnetic ions for MRI (De Häen 2001). However, gadolinium (atomic number 64 and atomic weight of 157) proved to be the most powerful and unfortunately also the most toxic of the ions with its seven unpaired electrons. Today, gadolinium is the active constituent in all commercially available MR contrast media due to its high magnetic moment and a relatively slow electronic relaxation time (Table 3.1).

3.2.1 Chemistry

Gadolinium administration cannot be undertaken as a simple inorganic solution, such as chlorides and sulfates, since the simple salts of gadolinium hydrolyze instantaneously, forming insoluble oxides and hydroxides at pH 7, which are retained in macrophage organs such as liver and bone for a very long period of time (Thomsen et al. 2016). In regard to mice, their LD_{50} values for injected raw gadolinium ions are approximately 0.1 to 0.3 mmol/kg, similar to the standard human dose of a current gadolinium-based contrast media. If gadolinium is to be used in a contrast medium, it requires that is bound firmly and stably to a carrier molecule, which can solubilize it and furthermore prevent hydrolysis. Yet, it must still allow for catalysis of water proton

TABLE 3.1

Gadolinium-Based Agents: Brand Names and Characteristics

Name	Brand Name	Organ Specific	Extra-Cellular	Chelate	Ionicity	Hepatobiliary Excretion	Protein-Binding	Risk of NSF[a]
Gadodiamide	Omniscan	No	Yes	Linear	Nonionic	No	No	High
Gadoversetamide	Optimark	No	Yes	Linear	Nonionic	No	No	High
Gadopentetate dimeglumine	Magnevist	No	Yes	Linear	Ionic	No	No	High
Gadobenate dimeglumine	Multihance	Yes (Liver)	Mainly	Linear	Ionic	Yes (1%–4%)	Yes (4%)	Intermediate
Gadoxetate disodium	Primovist, Eovist	Yes (Liver)	No	Linear	Ionic	Yes (42%–51%)	Yes (10%)	Intermediate
Gadofosveset trisodium	Vasovist, Ablavar	Yes (Blood)	No	Linear	Ionic	Yes (5%)	Yes (90%)	Intermediate
Gadobutrol	Gadovist, Gadavist	No	Yes	Macrocyclic	Nonionic	No	No	Low
Gadoteridol	Prohance	No	Yes	Macrocyclic	Nonionic	No	No	Low
Gadoterate meglumine	Dotarem, Magnescope	No	Yes	Macrocyclic	Ionic	No	No	Low

[a] According to the classification provided by the European Medicines Agency.

relaxation and excrete rapidly, thus carrying the heavy metal ion out of the body.

Since the 1950s, certain polycarboxylic acids (e.g., diethylene tetraaminepenta-acetate [DTPA]) have been recognized as a great coordinator of metal ions (Dawson et al. 1999). Gadolinium(III) is a 3+ charged ion with nominally nine sites at which another chemical can bind it. When using the DTPA ligand, eight out of the nine sites are occupied by three N and five O⁻ in the carboxyl groups. The additional ninth gadolinium site is used for a fast exchange of water molecules between the *inner sphere* (bonded directly to gadolinium) and the bulk (nonbonded) water (Morcos 2007). This feature is a necessity in regard to strong water proton relaxation. With the exception of protein binders, all the commercial gadolinium chelates achieve a similar effect on the proton relaxation and visibility in MRIs (Figures 3.1 and 3.2).

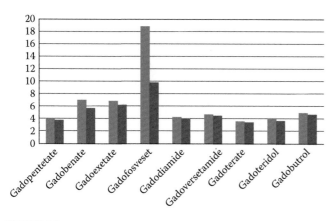

FIGURE 3.1
r_1-Relaxivity in plasma of the nine commercially available gadolinium-based contrast media at 1.5 (blue) and 3.0 (red) T (mM^{-1}s^{-1}).

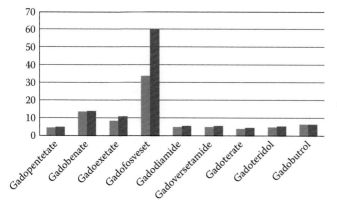

FIGURE 3.2
r_2-Relaxivity in plasma of the nine commercially available gadolinium-based contrast media at 1.5 (blue) and 3.0 (red) T (mM^{-1}s^{-1}).

In regard to gadolinium-based contrast media, the choice of ligand structure influences the chemical and biological stability of the compound as well as the colligative properties of its formulations (Thomsen et al. 1999, 2016). These in turn then influence tolerance, pharmacokinetics, clinical uses, and contraindications. Two fundamental structural types are used (Figures 3.3 and 3.4):

1. Linear seven-ring structures derived from DTPA
2. Macrocyclic eight-ring structures derived from dodecane tetra-acetic acid (DOTA)

The fundamental difference in these structures is that the chain of nitrogen and carbon atoms that constitute the backbone of the chelating agent is open on both ends in linear agents and closed into a single loop in macrocycles.

3.2.2 Stability

There are significant differences in the liability to release free gadolinium ions between the various agents (Morcos 2014b). The instability of gadolinium-based contrast agents is an important factor in the pathogenesis of the serious complication of nephrogenic systemic fibrosis (NSF). Coordination sites of gadolinium represent the number of atoms or ligands directly bonded to the metal center such as Gd^{3+}. The bonding between the metal center (Gd^{3+}) and the ligands is through valent bonds in which shared electron pairs are donated to the metal ion by the ligand. In an ionic linear molecule such as Gd-DTPA, Gd^{3+} is coordinated with five carboxyl groups and three amino nitrogen atoms. The three negatively charged carboxyl groups neutralize the three positive charges of the Gd ion, and the remaining two carboxyl groups are neutralized by two meglumine cations. In a nonionic linear molecule such as gadodiamide or gadoversetamide, the number of carboxyl groups is reduced to three because each of the other two carboxyl groups has been replaced by a nonionic methyl amide. Although both amide carbonyl atoms are directly coordinated to Gd^{3+}, the binding is weaker compared to that with carboxyl groups. This weakens the grip of the nonionic chelate on the Gd^{3+} and decreases the stability of the complex.

The other feature influencing the binding between the Gd^{3+} and the chelate is whether the configuration of the molecule is cyclic or linear. The macrocyclic molecule offers better protection and binding of Gd^{3+} because it is a preorganized rigid ring of almost optimal size to cage the Gd ion. In contrast, the linear structure, which is a flexible open chain, provides weaker protection of the gadolinium ion.

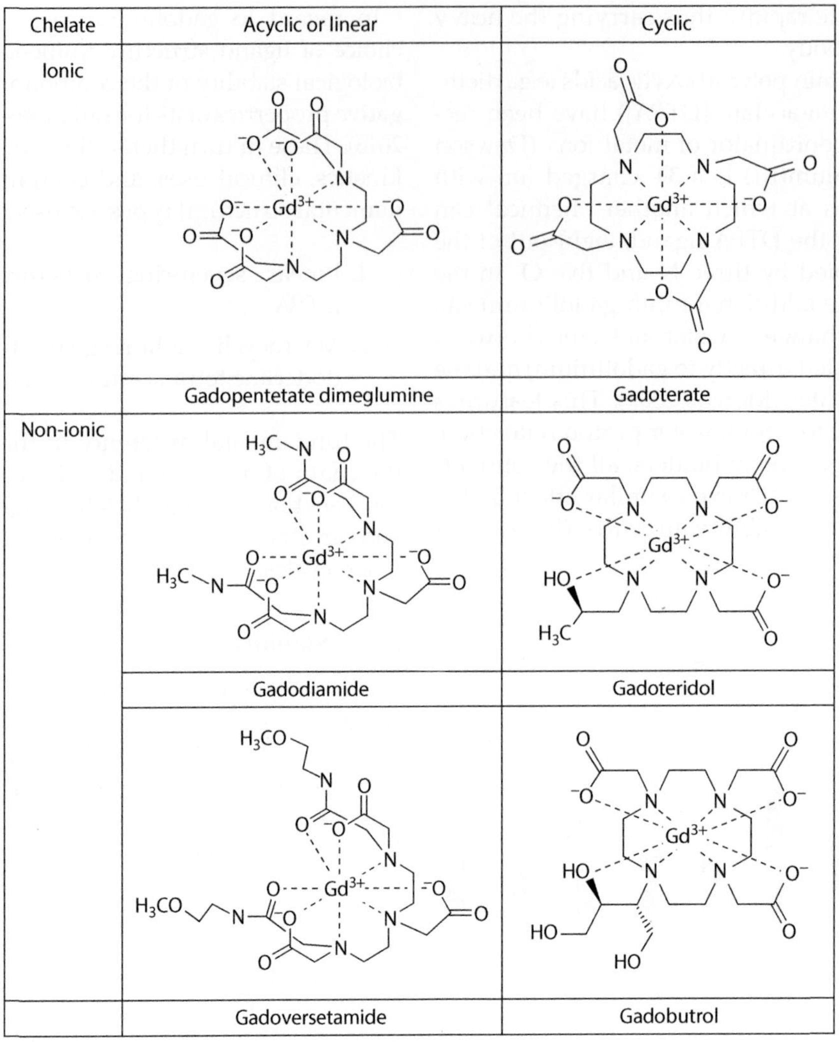

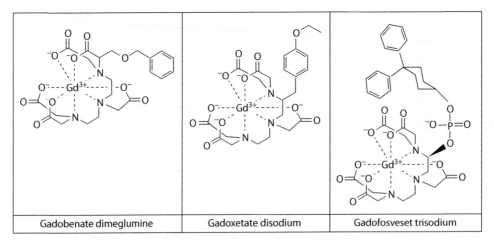

Chelate	Acyclic or linear	Cyclic
Ionic	Gadopentetate dimeglumine	Gadoterate
Non-ionic	Gadodiamide	Gadoteridol
	Gadoversetamide	Gadobutrol

FIGURE 3.3
Gadolinium-based contrast agents: Chemical structure of extracellular agents.

Gadobenate dimeglumine	Gadoxetate disodium	Gadofosveset trisodium

FIGURE 3.4
Gadolinium-based contrast agents: Chemical structure of the protein binders.

The in-vitro measurements used to assess the stability of the chelate molecules are as follows (Thomsen et al. 2014):

1. The *thermodynamic stability constant* (measured under very alkaline conditions (pH ~ 11); at this pH, there are no competing hydrogen ions for the chelate and a theoretical maximum stability for the chelate is obtained) (Figure 3.5)
2. The *conditional stability constant* (measured at physiological pH of 7.4) (Figure 3.5)
3. The *kinetic stability* (dissociation half-life under very acidic conditions—pH 1) (Figure 3.6)

The higher the value of these measurements, the higher is the stability of the molecule.

The amount of excess chelate in the gadolinium-based contrast agent preparations is another marker of the stability of these agents. A large amount of excess chelate

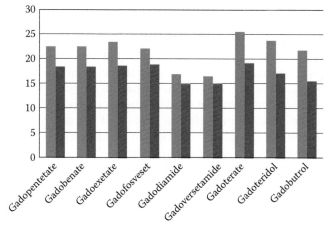

FIGURE 3.5
Thermodynamic (blue) and conditional stability (red) constants of the currently available gadolinium-based contrast media.

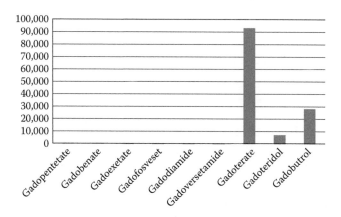

FIGURE 3.6
$T\frac{1}{2}$ for the dissociation of the gadolinium-based contrast media at pH 1 and 25°C (seconds).

is present in agents of low stability. The excess chelate is included in the preparation to ensure the absence of free Gd^{3+} in solution. The addition of excess chelate to gadodiamide (nonionic linear chelate) dramatically reduces the acute toxicity of nonformulated preparations (no excess chelate) by a factor of 2.5 as shown by acute toxicity studies (intravenous LD_{50}).

The dissociation of gadolinium-based contrast agents incubated in human serum at 37°C is another suitable method for predicting the stability of these agents in vivo. The examination has been known for many years; the dissociation of gadolinium from gadodiamide with time was already shown in 1995 (Maton et al. 1995). A recent study of dissociation with time showed that the highest release of gadolinium (Gd^{3+}) is observed with the nonionic linear chelates gadodiamide and gadoversitamide (Frenzel et al. 2008). The addition of phosphate to serum markedly increased the release of Gd^{3+} and the total amount of Gd^{3+} released at day 15 increased from 20% to around 35% of the total dose of these agents. With the ionic linear chelate gadopentetate dimeglumine, phosphate did not increase the total amount of Gd^{3+} released, which remained at 2%, but the speed of the release was increased in day 1 to 2% and remained at this level up to day 15. The smaller release of free Gd^{3+} from the ionic linear chelates compared to the nonionic chelates is due to a higher thermodynamic stability of the former (Figure 3.5). No release of Gd^{3+} was observed with the macrocyclic gadolinium-based contrast agents even after phosphate was added to the serum.

As expected from the chemical structure, all the above-mentioned stability measurements confirm the superior stability of the macrocyclic chelates and the low stability of the nonionic linear chelates.

3.2.3 Transmetallation

Transmetallation was initially reported in the literature in 1988 (Tweedle et al. 1988). Transmetallation of gadolinium-based contrast media leads to emission of free gadolinium by substitution of the Gd^{3+} within the chelate molecule by body cations such as iron, copper, zinc, and calcium. Zinc mainly displaces a significant amount of Gd^{3+} due to its high concentration in the blood (55–125 µmol l^{-1}) (Morcos 2014b). Copper is present in very small amounts (1–10 µmol l^{-1}); calcium ions have low-affinity to organic ligands. Iron ions are tightly bound by the storage proteins ferritin and hemosiderin and are not available for transmetallation with Gd^{3+}. Transmetallation between Gd^{3+} and zinc results in the formation of zinc chelate, which is excreted in urine. The released Gd^{3+} becomes attached to endogenous anions such as phosphate, citrate, hydroxide, or carbonate, which deposit in the tissues as insoluble compounds. In-vivo, in-vitro, and human studies have shown that linear chelates, particularly the nonionic ones, cause a large increase in zinc excretion in urine. The nonionic linear chelate gadodiamide induced a 32% decrease in plasma zinc after a single injection in healthy volunteers (Kimura et al. 2005). This is thought to be secondary to transmetallation and the presence of excess chelate in the gadodiamide preparation. In patients undergoing contrast-enhanced MRI examination, gadodiamide caused a large increase in zinc excretion in the urine, which was almost three times greater than that induced by the ionic linear molecule Gd-DTPA. In comparison, the ionic macrocyclic Gd-DOTA had no effect on zinc excretion. Ex-vivo studies have also confirmed that all macrocyclic gadolinium-based contrast agents are insensitive to transmetallation by zinc ions compared to the open-chain complexes.

3.2.4 Osmolality

The osmolality of the various gadolinium-based contrast media is shown in Figure 3.7. As with iodine-based contrast agents, one way to reduce osmolality, without a reduction in concentration of the active agent, is to make a nonionic (gadolinium-chelate) molecule (Thomsen et al. 2016). Although nonionic contrast agents for radiography showed vast improvements in tolerance over the ionic agents, the same has not been seen for gadolinium-based contrast agents. The primary reason for this is that the dose of gadolinium-based contrast media for MRI is only about 10% of the dose of iodine-based contrast media for radiography including CT. The only major difference between ionic and nonionic gadolinium-based contrast agents is the introduction of necroses, at least in animals, when the vein is missed and extravasation occurs (see Section 3.4.1).

In order to reduce the osmolality of the linear chelate, the carboxyl groups are reduced to three in the nonionic linear chelate compared to the DPTA chelate (ionic linear). The other two carboxyl groups have simply been

replaced by nonionic methyl amide. The amide has a weaker binding to gadolinium(III) in comparison to the negatively charged carboxyl groups. These changes decrease the grip of the chelate on the gadolinium atom. A similar decrease in the grip of the macrocyclic chelate of the gadolinium atom is not the case when the chelate becomes nonionic.

3.2.5 Viscosity

Viscosity (Figure 3.8) is not an issue with gadolinium-based contrast agents like iodine-based contrast media (Thomsen et al. 2016). All are low and present no practical difficulties.

3.2.6 Pharmacokinetics

3.2.6.1 *Extracellular Agents*

The pharmacokinetics depends on whether the contrast agent is an extracellular agent or an organ-specific agent (Table 3.1). Six of the nine commercially available agents are almost pure extracellular and behave in blood just like the well-known iodine-based agents (Thomsen et al. 2016). The only exceptions are the protein binders, which can cross hepatic cell membranes for hepatobiliary excretion. Their pharmacokinetics follow the two-compartment model, where one compartment is the intravascular plasma space and the other is the interstitial space; entering cells in or outside circulation is insignificant, other than for hepatobiliary excretion in the protein binder gadolinium-based contrast media. All agents can visualize the arteries (Figure 3.9) and they can be used for perfusion studies (Figure 3.10).

On injection into circulating blood, they are simultaneously diluted in the circulating plasma volume and immediately begin to pass more or less freely out of the circulation into the interstitium, with the intact blood–brain barrier as the only exception (Thomsen et al. 2016).

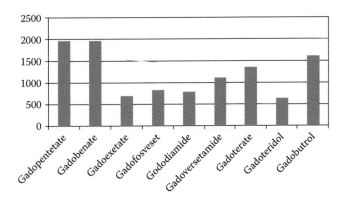

FIGURE 3.7
Osmolality of the gadolinium-based contrast media (mOsm/kg).

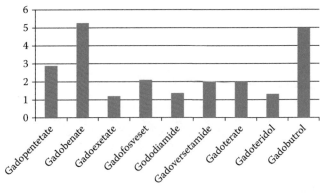

FIGURE 3.8
Viscosity of the gadolinium-based contrast media (mPa/s).

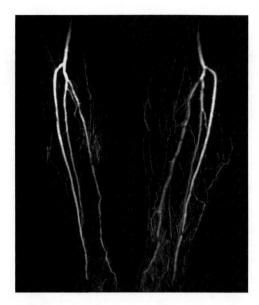

FIGURE 3.9
MR angiography of the cural arteries.

A normal brain enhances only slightly, whereas a process damaging the blood–brain barrier (e.g., tumor or abscess) will enhance (Figure 3.11a and b).

When the equilibrium point is reached, decreasing plasma levels and increasing interstitial levels are transiently equal. Subsequently, the extracellular contrast media are continuously removed from the blood via renal filtration and an opposite movement of contrast molecules from the interstitial space into the plasma (Figure 3.12a–c).

Tiny amounts are excreted in saliva, sweat, and tears, and less than 1% is taken up by the liver and excreted in the bile, except for the protein-binding gadolinium-based contrast media. The hepatobiliary contrast media probably use the organic anion transport mechanism that is designed to transport, metabolize, and excrete organic chemicals that bear a negative charge. If the renal function is severely impaired, then excretion by other routes becomes important. It is also in these circumstances of a prolonged residence time and/or pH that the stability of contrast medium becomes more important.

None of the current nine gadolinium-based contrast media is approved for other administrations (e.g., intra-arterial, oral intake, and injection into body cavities) than the intravenous one.

3.2.6.2 Organ-Specific Agents (Protein Binders)

Another class of gadolinium-based contrast media is the protein binders (Bellin and Leander 2014, Thomsen et al. 2016). Protein binding, predominantly serum albumin binding, can produce the following:

1. Variable hepatobiliary versus renal excretion
2. Greater relativity where albumin is present
3. Extended plasma half-life versus both excretion and leakage into extracellular space

Three protein-binding gadolinium-based contrast agents are commercially available. All are based on the DTPA-base structure and are ionic.

Under similar conditions, gadobenate binds only slightly, ~5% to albumin; gadoxetate binds ~10%; and gadofosveset binds ~90%. These values are approximates for comparison and highly dependent on conditions of the measurement, including gadolinium concentration, which changes with time after injection, and also is complicated by binding of multiple gadolinium-based contrast agents to each albumin molecule.

The biodistribution and pharmacokinetics of the three agents differ, but not proportional to albumin binding alone. Gadobenate is excreted mostly into the urine at an indistinguishable rate from the nonprotein-binding agents, with only ~4% hepatobility excretion in men; in rats the rate is much higher. The consequence of the added lipophilicity in gadoxetate is ~10% albumin binding along with very rapid hepatic uptake and then hepatobiliary excretion, ~50%, accompanied by relatively rapid renal excretion. In gadofosveset, the hepatobiliary excretion is 5%; due to the protein binding its $T\frac{1}{2}$ is 18 h or 12 times longer than that of the extracellular gadolinium-based contrast agents.

3.3 Adverse Reactions

The totally inert contrast medium does not exist. The ideal agent should enhance the appropriate lesion(s) and structure(s), cause no reaction, and leave the body rapidly and unchanged. All contrast media cause adverse reactions. They are divided into acute adverse reactions, which occur within 60 min after the administration of the agent, and which may be either renal or nonrenal, late adverse reactions, which occur between 1 h after the administration and within 7 days, and the very late reactions, which occur more than 7 days after the administration of the agent.

3.3.1 Acute Nonrenal Adverse Reactions

The acute nonrenal adverse reactions are the same for all types of extracellular contrast media (Clement and Webb 2014). They are seen more frequently after exposure to iodine-based contrast media than after gadolinium-based contrast media, after which reactions

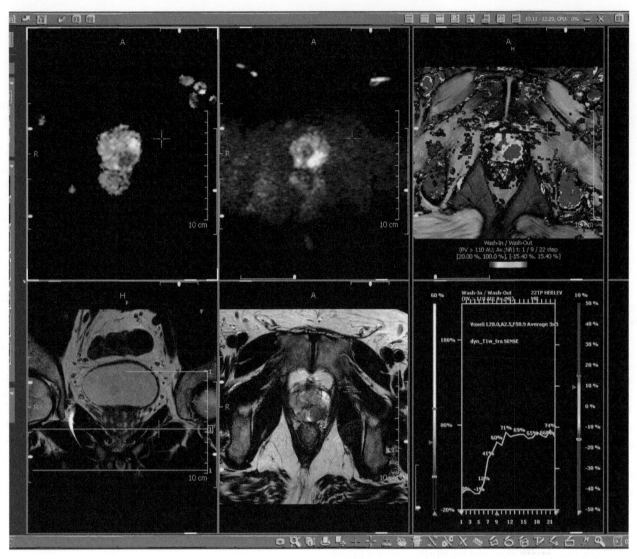

FIGURE 3.10
Multiparametric MRI of the prostate. Upper panel: diffusion-weighed imaging (b = 1400, ADC map) and perfusion images. Lower panel: T_2 coronal and axial, perfusion curve. There is a malignant tumor in the left posterior part of the peripheral zone. The diffusion is decreased and the perfusion is increased, decreased signal intensity of the tumor on the T_2-weighed images compared to the rest of the gland.

are seen more frequently than after contrast agents for ultrasonography. In general, these reactions are mild and self-limiting, but they are also the most dreaded, most serious, and fatal complications of contrast medium injection. The reason is that they occur without warning cannot be reliably predicted and are not preventable. These reactions usually begin either during or immediately after the injection of the contrast medium. Adverse drug reactions are more frequent in patients who have had a previous adverse reaction to a contrast medium, asthmatics, allergic and atopic patients, and patients on β-adrenergic blockers. They may be mild (requiring no treatment) and include nausea, mild vomiting, urticaria, and itching. Moderately severe adverse reactions, which may require treatment, include severe vomiting, marked urticaria, bronchospasm, facial/laryngeal edema, and vasovagal attack. Severe adverse reactions always need prompt intervention; they include hypotensive shock, respiratory arrest, and convulsions. A crash cart with appropriate drugs, instructions, tubes, fluids, and so on should always be available in the room where contrast media are given. Detailed treatment protocols are beyond the scope of this chapter and the reader is referred to www.esur.org for more details.

The incidence for gadolinium-based contrast media with regard to mild reactions is below 10%, for moderate reactions below 0.2%, and for severe reactions <0.01%; there are no difference between the high- and low-osmolar agents according to prospective studies. In this area, retrospective trials, for example, based on MedWatch and

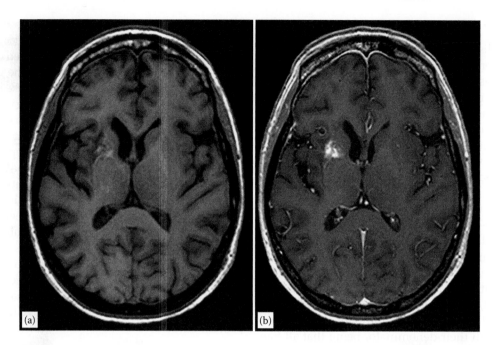

FIGURE 3.11
Contrast-enhanced MRI of the brain in a patient with metastases. T_1-weighed images: (a) without gadolinium-based contrast medium and (b) with gadolinium-based contrast medium.

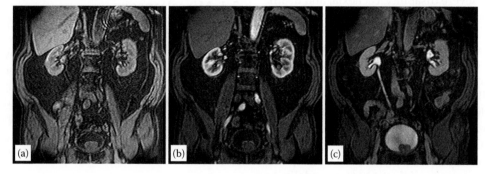

FIGURE 3.12
MR-urography: (a) unenhanced T_1-weighed images, (b) cortical phase, and (c) excretion phase no abnormalities.

local registries, have no value as the registration is not systematic and may be influenced by many factors. Furthermore, the denominator is unknown. How data is collected also affects the incidence, particularly the mild reactions. The lowest incidence is always found in retrospective studies, whereas the highest incidence is seen when a structured interview with a long list of events is performed. Patients undergoing unenhanced MRI report the same events as those undergoing enhanced MRI.

3.3.2 Acute Renal Adverse Reactions

An acute renal adverse reaction (contrast-induced nephropathy [CIN]) is well known from iodine-based contrast media, whereas CIN is rarely seen after gadolinium-based contrast media, especially at the doses approved for contrast-enhanced MRI. At higher doses, gadolinium-based contrast media may be at least as nephrotoxic as the iodine-based contrast media, if not more according to animal studies (Thomsen and Leander 2014). Since CIN occurs so seldom after gadolinium-based contrast media, there is no reason for determining the glomerular filtration rate and hydrate the patient as part of preventing CIN.

3.3.3 Late Adverse Reactions

Late adverse reactions, for example, skin rashes, which are seen after administration of iodine-based contrast media—especially the dimeric agents—have not been reported after administration of gadolinium-based contrast media.

3.3.4 Very Late Adverse Reactions

NSF is a new disease, which was seen for the first time in 1997 (Thomsen 2014). This happened at approximately the same time as radiologists began to transfer patients from enhanced radiographic examinations to MRI and furthermore started performing magnetic resonance arteriography with double or triple dose of the allegedly *nonnephrotoxic* contrast media. The gadolinium-based contrast agents were also used for some radiographic examinations such as conventional arteriography and CT.

In 2006, two European groups found a correlation between the occurrence of NSF and exposure to one of the gadolinium-based contrast media (Grobner 2006, Marckmann et al. 2006). Afterward, it very rapidly became clear that unconfounded cases (i.e., only one agent ever used in the patient) were only seen after exposure to the least stable agents (gadodiamide, gadoversetamide, and, to a lesser extent, gadopentetate dimeglumine; Table 3.1). NSF occurred all over the world, hit both younger and older people, with their denominator being that all had reduced renal function or was on dialysis (Thomsen 2014). The severity of the disease varied from a single small plaque on a lower extremity to nearly universal disease, causing severe disability, cachexia, and death.

It can begin hours to several years after the last exposure to the gadolinium-based contrast medium; however, not all patients with renal impairment develop NSF after exposure to one of the high-risk agents. Thus, another yet unknown factor must be present. It is mostly believed that transmetallation (see Section 3.2.3) is the course. When the gadolinium is free (dissociated), it localizes in the skin, the liver, and especially in the bone probably as a complex of calcium and phosphate. This may trigger the circulating fibrocytes and fibrosis may develop.

After each intravenous injection of a gadolinium-based contrast medium, a tiny amount of unchelated gadolinium is left in the body. More is left after the nonionic linear chelates than after the ionic linear chelates, which again leave more than the macrocyclic agents and more in patients with reduced renal function than in patients with normal function. The longer the agents stay in the body, the poorer the renal function is.

Whether the accumulation has any long-term effects even in patients with normal renal function is still unknown. Hence, it is advisable in all patients—independent of renal function—to use an agent that leaves the smallest amount of gadolinium, and to use the lowest dose for sufficient diagnostic work-up of the patient. Nonionic linear gadolinium contrast media (gadoversetamide and gadodiamide) and the ionic linear gadolinium contrast media (gadopentetate dimeglumine) are now contraindicated to use in patients with GFR < 30 mL/min × 1.73 m² and should only be used with caution in patients with GFR between 30 and 60 mL/min × 1.73 m².

Therefore, it is mandatory to determine the renal function before use of these agents (ESUR 2013).

At least, it looks like NSF has been wiped out in Europe, Japan, and the United States owing to the guidelines for use of gadolinium-based contrast agents issued by the European Society of Urogenital Radiology and American College of Radiology (Thomsen 2010). It is recommended to keep the least stable agents away from the more stable agents should departments decide to keep using the least stable agents in patients with normal renal function. This way mistakes in terms of giving patients with reduced renal function or on dialysis the contraindicated agents can be avoided—sadly, mistakes have been made.

Recently it has been shown that gadolinium released from the chelate may accumulate in the skin (erythromatous plaques) and in the globus pallidus as well as the dentate nuclei after multiple injections, even in patients with normal renal function. These changes have not been seen after exposure to macrocyclic agents or the protein binders.

3.4 Miscellaneous

3.4.1 Contrast Media Extravasation

Contrast media may enter the extravascular space (extravasation) during injection (Jakobsen 2014). These injuries are mostly of a minor character, but in rare cases skin ulceration, soft tissue necrosis, and compartment syndrome are seen. Ice packs, elevation of the limbs, and careful monitoring of the lesions can be performed. The risk factors may be either technique- or patient-related:

1. Technique-related—use of a power injector, less optimal injection sites, including lower limb and small distal veins, large volume of contrast medium, and high osmolar contrast medium
2. Patient-related—inability to communicate, fragile or damaged veins, arterial insufficiency, compromised lymphatics, and/or venous drainage and obesity

In order to decrease the risk, nonionic gadolinium-based agents may be used and an appropriately sized plastic cannula should be placed in a suitable vein and tested with saline before use with contrast media.

3.4.2 Interaction

The lack of systematic examinations makes it unknown, with minor exceptions, whether MR contrast media interfere with the various clinical biochemical analyses. It has been documented that the nonionic linear chelates gadodiamide and gadovertisamide may cause falsely low levels of calcium (spurious hypercalcemia) when the calcium level is examined by assays using the ortho-cresolphthalein complex one method (Morcos 2014a). However, the same is not true with assays using the Arsenazo III method. Caution should be exercised when using colorimetric assays for angiotensin-converting enzyme, calcium, iron, magnesium, total iron-binding capacity, and zinc in serum samples collected shortly after injection of gadolinium-based contrast media. Hence, it is recommended not to do biochemical assays within 24 h after administration of contrast media (and longer if the patient has reduced renal function or is on dialysis) (Morcos 2014a).

3.4.3 Pregnancy and Lactation

Gadolinium-based contrast agents should only be administered when the clinical symptoms and/or the unenhanced images are strongly indicative of a need for contrast enhancement. In that case, only the most stable agents (i.e., macrocyclic agents) should be used (Webb and Thomsen 2013). In Europe, the use of the least stable agents (gadopentetate dimeglumine, gadodiamide, and gadovertisamide) is contraindicated in pregnant women. The issue of concern is that gadolinium may cross the placenta and eventually enter into the fetus and amniotic fluid and here it is very slowly excreted.

For lactating women, the situation is dissimilar, as the extracellular contrast media are not protein bound or lipophilic. Thus, it is only in tiny amounts that are transferred via the milk to the newborn, considerable less than has been approved for the examination of a child. Lactation can continue if one of the most stable agents (i.e., macrocyclic agents) is used. Yet, it must stop for 24 h if one of the least stable agents is used. No information is available in regard to the protein-binding gadolinium-based contrast media (ESUR 2013).

3.5 Contrast Media Administration

3.5.1 Routes of Administration

The gadolinium-based contrast agents that are commercially available are presently approved for intravenous injection only; all other administrations will be off-label. Currently, it is tested whether the extracellular agents can be used for lymphography.

3.5.2 Dose

The standard dose for extracellular agents is 0.1 mmol/kg body weight for all examinations. In some cases, double or triple dose is used for visualization of, for example,

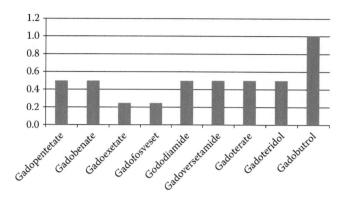

FIGURE 3.13
Commercially available concentrations of the nine gadolinium-based contrast media in the vials, syringes, and so on.

small brain metastases. With the newest scanners, triple dose is no longer a prerequisite for an adequate peripheral angiography. The protein binders can be used in lower doses than the extracellular agents due to their higher relaxivity caused by the protein binding (Figures 3.1 and 3.2). For extracellular agents, the relaxivity is almost identical for 1.5 and 3 T. Thus, no adjustment of dose is necessary when applying a higher field strength. For the protein binders, the relaxivity decreases when the field strength increases, especially in regard to gadofosveset.

With exception of gadofosveset, gadoxetate, and gadobutrol, gadolinium-based contrast agents are delivered with a concentration at 0.5 mmol/mL (Figure 3.13). Gadobutrol has a higher concentration (1 mmol/mL) and the protein binders gadofosveset and gadoxetate are delivered in a lower concentration (0.25 mmol/mL).

3.6 Other Contrast Media

3.6.1 Manganese-Based Contrast Media

Manganese is another paramagnetic ion. It has the same relaxivity properties as gadolinium, but it occurs naturally in the body in contrast to gadolinium (Chabanova et al. 2011). Manganese is one of the least toxic metal ions and is excreted by the hepatobiliary system. Agents containing manganese are no longer commercially accessible; however, a new oral intake agent is in phase 3 testing at present. As a result of an addition of promotors increasing the uptake of manganese over the bowelwall, this agent can be used as a hepatobiliary agent.

3.6.2 Iron-Based Contrast Media

Superparamagnetic agents are exceptionally effective T_2 relaxation agents producing signal loss on

T_2- and T_2*-weighted images. Furthermore, they have a T_1 effect, which is markedly lower than their T_2 effect. However, superparamagnetic agents are no longer commercially available. In many of the agents, iron was the active ion. These iron agents were metabolized by macrophages, and they entered the iron pool of the body (Bellin 2009).

In patients with iron deficiency, ferumoxytol is approved for intravenous treatment and by some institutions it is administered during MRI (Qju et al. 2012). Studies of whether ferumoxytol can be applied for angiography and perfusion studies are being undertaken at present. Nevertheless, it is off-label in regard to MR use.

3.6.3 Water

Water may be used as a contrast agent in radiography. It does not necessitate approval from the authorities.

References

Bellin MF. Non-gadolinium-based contrast agents. In: Thomsen HS, Webb JAW, eds. *Contrast Media: Safety Issues and ESUR Guidelines*. 2nd ed. Berlin, Germany: Springer Verlag; 2009; 205–211.

Bellin MF, Leander P. Organ-specific gadolinium-based contrast media. Gadolinium chelates and stability. In: Thomsen HS, Webb JAW, eds. *Contrast Media: Safety Issues and ESUR Guidelines*. 3rd ed. Berlin, Germany: Springer Verlag; 2014; 219–225.

Chabanova E, Logager VB, Moller JM, Thomsen HS. Manganese based MR contrast agents: Formulation and clinical applications. *Open Drug Safety J* 2011; 2: 29–38.

Clement O, Webb JAW. Acute adverse reactions to contrast media: Mechanisms and prevention. Gadolinium chelates and stability. In: Thomsen HS, Webb JAW, eds. *Contrast Media: Safety Issues and ESUR Guidelines*. 3rd ed. Berlin, Germany: Springer Verlag; 2014; 51–60.

Dawson P, Cosgrove DO, Grainger RG (eds). *Textbook of Contrast Media*. Oxford: Isis Medical Media; 1999.

De Haën C. Conception of the first magnetic resonance imaging contrast agents: A brief story. *Top Magn Reson Imaging* 2001; 12: 221–230.

European Society of Urogenital Radiology (ESUR). Contrast media guidelines version 8.1; 2013. Available at: www.esur.org.

Frenzel T, Lengsfeld P, Schirmer H et al. Stability of gadolinium based magnetic resonance imaging contrast agents in human serum at 37°C. *Invest Radiol* 2008; 43: 817–828.

Grobner T. Gadolinium: A specific trigger for the development of nephrogenic fibrosing dermopathy and nephrogenic systemic fibrosis? *Nephrol Dial Transplant* 2006; 21: 1104–1108.

Jakobsen JÅ. Contrast media extravation injury. In: Thomsen HS, Webb JAW, eds. *Contrast Media: Safety Issues and ESUR Guidelines*. 3rd ed. Berlin, Germany: Springer Verlag; 2014; 131–137.

Kimura J, Ishguchi T, Matsuda J et al. Human comparative study of zinc and copper excretion via urine after administration of magnetic resonance imaging contrast agents. *Radiation Med* 2005; 23: 322–326.

Marckmann P, Skov L, Rossen K et al. Nephrogenic systemic fibrosis: Suspected causative role of gadodiamide used for contrast-enhanced magnetic resonance imaging. *J Am Soc Nephrol* 2006; 17: 2359–2362.

Maton F, van der Elst L, Muller RN. Influence of human proteins on the relaxivity of Gd(III) complexes; 1995. Available at: http://www.b.dk/upload/webred/bmsandbox/uploads/2012/03/bb81a28663015b7a1d828ee723de7a26.pdf.

Morcos SK. Nephrogenic systemic fibrosis following administration of extracellular gadolinium based contrast agents: Is stability of the contrast agent molecule an important factor in pathogenesis of this condition? *Br J Radiol* 2007; 80: 73–76.

Morcos SK. Contrast media and interactions with other drugs and clinical tests. In: Thomsen HS, Webb JAW, eds. *Contrast Media: Safety Issues and ESUR Guidelines*. 3rd ed. Berlin, Germany: Springer Verlag; 2014a; 125–130.

Morcos SK. Gadolinium chelates and stability. In: Thomsen HS, Webb JAW, eds. *Contrast Media: Safety Issues and ESUR Guidelines*. 3rd ed. Berlin, Germany: Springer Verlag; 2014b; 175–180.

Qju D, Zaharchuk G, Christen T, Ni WW, Moseley ME. Contrast-enhanced functional blood volume imaging (CE-fBVI): Enhanced sensitivity for brain activation in humans using the ultrasmall superparamagnetic iron oxide agent ferumoxytol. *Neuroimage* 2012; 62: 1726–1731.

Thomsen HS. Contrast-enhanced MRI in patients with impaired renal function: Recent recommendations to minimize the risk of nephrogenic systemic fibrosis. *Solutions Contrast Imaging* 2010; 1(2): 1–8.

Thomsen HS. Nephrogenic systemic fibrosis and gadolinium-based contrast media. In: Thomsen HS, Webb JAW, eds. *Contrast Media: Safety Issues and ESUR Guidelines*. 3rd ed. Berlin, Germany: Springer Verlag; 2014; 207–218.

Thomsen HS, Dawson P, Tweedle M. MR and CT contrast agents for perfusion imaging and regulatory issues. In: Bammer R, ed. *MR & CT Perfusion Imaging: Clinical Applications and Theoretical Principles*. Philadelphia, PA: Lippincott Williams & Wilkins; 2016: Chap. 11.

Thomsen HS, Leander P. Radiography with gadolinium-based contrast media. In: Thomsen HS, Webb JAW, eds. *Contrast Media: Safety Issues and ESUR Guidelines*. 3rd ed. Berlin, Germany: Springer Verlag; 2014; 193–200.

Thomsen HS, Müller RN, Mattrey RR, eds. *Trends in Contrast Media*. Berlin, Germany: Springer Verlag; 1999.

Tweedle MF, Gaughan GT, Hagan J et al. Considerations involving paramagnetic coordination compounds as useful NMR contrast agents. *Nucl Med Biol. Int J Radiat Appl Instrum* 1988; Part B 15: 31–36.

Webb JAW, Thomsen HS. Gadolinium contrast media during pregnancy and lactation. *Acta Radiol* 2013; 54: 599–600.

4

Diffusion Imaging: Basic Principles

Ioannis Tsougos

CONTENTS

4.1 Introduction

Magnetic resonance imaging (MRI) with its excellent soft tissue visualization and variety of imaging sequences has evolved to one of the most important noninvasive diagnostic tools for the detection and evaluation of treatment response of cerebral tumors. Nevertheless, conventional MRI presents limitations regarding certain tumor properties, such as infiltration and grading [1]. It is evident that a more accurate detection of infiltrating cells beyond the tumoral margin and a more precise tumor grading would strongly facilitate surgical resection as well as postsurgical treatment procedures. Hence, biopsy still remains the gold standard, although it provides histopathological information about a limited portion of the lesion. Therefore, advanced MRI techniques have been incorporated into the clinical routine in order to aid tumor diagnosis. Diffusion-weighted

imaging (DWI) and diffusion tensor imaging (DTI) provide noninvasively significant structural and functional information in a cellular level, highlighting aspects of the underlying brain pathophysiology.

In theory, DWI is based on the freedom of motion of water molecules, which can reflect tissue microstructure, while DTI allows the estimation of water motion anisotropy. Hence, the possibility to characterize tumoral and peritumoral tissue microstructure, based on water diffusion findings, provided clinicians a whole new perspective on improving the management of brain tumors. Although initially DWI was established as an important method in the assessment of stroke, a large number of studies has been conducted in order to assess whether the quantitative information derived by DWI and DTI may aid differential diagnosis and tumor grading, especially in cases of ambiguous cerebral neoplasms. Many researchers have reported increased diagnostic value while using DWI and/or DTI [2–6]; but on the other hand, there is a significant number of studies reporting the opposite [7–10]. Nevertheless, the most probable explanation for these controversial observations may be the complexity of the underlying pathophysiology, which may eventually result in similar diffusion and perfusion patterns.

This chapter aims to evaluate and review the principles and recent results that have been obtained using the diffusion properties of tissues in the form of DWI and DTI, regarding tumor characterization and grading, and discusses how the available quantitative data information can be exploited through advanced methods of analysis, in order to optimize clinical decision making of the most common cerebral neoplasms.

4.2 Diffusion Imaging: Basic Principles

4.2.1 Diffusion-Weighted Imaging

FOCUS POINTS

- Diffusion is considered as the result of the random walk of water molecules.
- Molecular diffusion in tissues is not free, but reflects interactions with many obstacles, such as macromolecules, fibers, and membranes.
- DWI represents the microscopic motion of water molecules hence probes local tissue microstructure.

Diffusion is considered as the result of the random walk of water molecules inside a medium due to their thermal energy and is described by the *Brownian* law by a diffusion constant D. Water makes up to 60%–80% of the human body weight. For pure water at ~37°C, D is approximately 3.4×10^{-3} mm^2/s [11]. In an isotropic medium, diffusion is equally distributed toward all directions and is usually described as a drop of ink in a glass of water. The most important thing however is to consider that within an anisotropic medium such as human tissue, water motion is restricted. Therefore, inside an even more complex environment, such as the human brain, cell membranes, neuronal axons, and other macromolecules act as biological barriers to free water motion; hence, water mobility is considered to be anisotropic. More specifically, the highly organized white-matter bundles, due to their myelin sheaths, force water to move along their axes, rather than perpendicular to them, as an apt analogy of a bundle of spaghetti as seen in Figure 4.1.

DWI is an advanced MRI technique, which is based on the aforementioned Brownian motion of molecules to acquire images. When the patient is inserted into the

FIGURE 4.1
Freeze-fractured section through a bundle of myelinated nerve fibers and the apt analogy of a bundle of spaghetti. Water molecules move easier along the axes, rather than perpendicular to them.

homogeneous magnetic field of the MR scanner, the nuclear spins are lined up along the direction of the static magnetic field. If a radiofrequency pulse is applied, the protons will spin at different rates depending on the strength, duration, and direction of the gradient. If an equal and opposite gradient is applied, the protons will be refocused. Thus, it is evident that stationary protons will provide a null signal after this counter process, while mobile protons, which have changed positions between the two gradients, will present a signal loss. This signal loss is dependent on the degree of diffusion weighting (DW) and is referred to as the *b*-value [12]. Therefore, by ignoring the stationary protons and measuring the signal of the mobile protons, the amount of diffusion that has occurred in a specific direction can be determined. The *b*-value is described by the following mathematical equation:

$$b = (\gamma G \delta)^2 \left(\Delta - \frac{\delta}{3} \right) \qquad (4.1)$$

In Equation 4.1, Δ is the temporal separation of the gradient pulses, δ is their duration, G is the gradient amplitude, and γ is the gyromagnetic ratio of protons [13]. The diffusion time is assigned as $(\Delta - \delta/3)$, where the second term in the expression accounts for the finite duration of the pulsed field gradients. The units for the *b*-value are s/mm^2, and the range of values typically used in clinical DW is 800–1500 s/mm^2. The formula for the *b*-factor implies that we can increase DW by increasing either gradient timing δ or Δ, or gradient strength, G.

It can be shown that for a fixed DW, the signal in a diffusion-weighted experiment is given by the following equation:

$$S = S_0 e^{-T_e/T_2} e^{-bD}$$

or

$$S = S_0 \exp(-bADC) \qquad (4.2)$$

S_0 is the signal intensity in the absence of any T_2 or DW, T_e is the echo time, and D is the apparent diffusivity, usually called the apparent diffusion coefficient (ADC). The term *apparent* reveals that it is often an average measure of a number of complicated processes inside the tissues and does not necessarily reflect the magnitude of intrinsic self-diffusivity of water [14,15].

The first exponential term in Equation 4.2 is the weighting due to transverse (T_2) relaxation and the second term shows that diffusion induces an exponential attenuation to the signal [12]. As the diffusing spins are moving inside the field gradient, each spin is affected differently by the field; thus, the alignment of the spins with each other is destroyed. Since the measured signal is a summation of tiny signals from all individual spins, the misalignment, or *dephasing*, caused by the gradient pulses results in a drop in signal intensity; the longer the diffusion distance, the more dephasing, the lower the signal [16]. Figure 4.2 depicts a simplified version

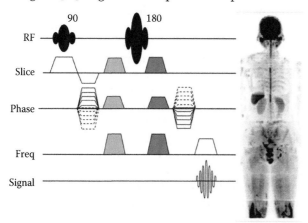

FIGURE 4.2
A pulse-gradient spin-echo pulse sequence for diffusion imaging.

of the most commonly used pulse-gradient spin-echo pulse sequence for diffusion imaging.

The goal of DWI is to estimate the magnitude of diffusion within each voxel, that is, the tissue microstructure, and this can be measured by the term ADC. By collecting a series of DW images with different *b*-values, a parametric map of ADC values can be obtained in order to facilitate qualitative measurements. The intensity of each image pixel on the ADC map reflects the strength of diffusion in the pixel. Therefore, a low value of ADC (dark signal) indicates restricted water movement, whereas a high value (bright signal) of ADC represents free diffusion in the sampled tissue [17]. Figure 4.3 shows a typical DWI obtained from a healthy human subject.

For example, in cerebral regions where water diffuses freely, such as cerebrospinal fluid (CSF) inside the ventricles, there is a drop in signal on the acquired DW images, whereas in areas that contain many more cellular structures and constituents (gray matter or white matter), water motion is relatively restricted and the signal on DW images is increased. Consequently, regions of CSF will present higher ADC values on the parametric maps than other brain tissues.

The most widely used diffusion-weighted acquisition technique is the single-shot echo-planar imaging (EPI). This is because in a clinical environment, certain requirements are imposed for diffusion studies—(a) reasonable imaging time should be maintained (i.e., short time), (b) multiple slices (15–20) are required to cover most of the brain, (c) good spatial resolution (~5–8 mm thick, 1–3 mm in-plane) is required, and (d) a reasonably short T_e (120 ms) to reduce T_2 decay and an adequate diffusion sensitivity (ADC ~0.2–1 × 10^{-3} mm^2/s for brain tissues) are also needed. This sequence is fast and insensitive to small motion, which is essential and readily available on most clinical MRI scanners. However, EPI is sensitive to magnetic field inhomogeneities, which cause distortions in the image data. Alternative DWI techniques include multishot EPI with navigator echo correction or diffusion-weighted propeller and parallel imaging

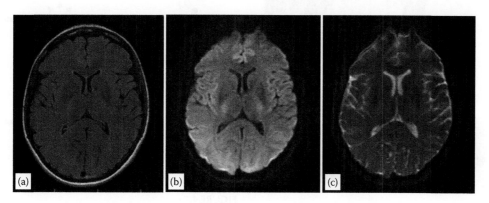

FIGURE 4.3
(a) FLAIR image, (b) diffusion-weighted image, and (c) ADC image collected from a healthy male subject.

methods, such as sensitivity encoding (SENSE) [18]. The application of such techniques increases the bandwidth per voxel in the phase-encoded direction, thus reducing artifacts arising from field inhomogeneities, like those induced by eddy currents and local susceptibility gradients.

DWI is undoubtedly a very useful clinical tool; however, it encounters a limitation. The DWI sequence is sensitive, but not specific for the detection of restricted diffusion, and one should not use only signal changes to quantify diffusion properties, as the signal from DWI is prone to the underlying T_2-weighted signal, referred to as the T_2 *shine-through* effect. That is, the increased signal in areas of cytotoxic edema on T_2-weighted images may be present on the DWI images as well [18]. To determine whether this signal hyperintensity on DWI images truly represent decreased diffusion, an ADC map should be used. The ADC sequence is not as sensitive as the DWI sequence for restricted diffusion, but it is more specific, as the ADC images are not susceptible to the T_2 *shine-through* effect [17]. Figure 4.4 shows a typical ADC parametric color map.

So inevitably, at this point, the question arises: What is the optimal *b*-value for clinical DWI?

It has been shown that for typical imaging experiments, the optimal *b*-factor is about 1257 s/mm^2 [18,19]. However, most studies are limited to DWI using *b*-factors of 0 and 1000. An upper *b*-factor around 1000 has been available for most clinical scanners until now and DWI using these standard values has been shown to be effective in detecting and delineating restricted diffusion, for example, in acute ischemic lesions of the brain.

Nevertheless, it may be more important to consider that in a clinical setting, it is advisable to maintain the same *b*-value for all examinations, making it easier to learn to interpret these images and become aware of the appearance of findings in various disease processes.

4.2.2 Diffusion Tensor Imaging

FOCUS POINTS

- ADC is directionally dependent.
- A single ADC is inadequate for characterizing diffusion in vivo.
- DTI represents a further development of DWI.
- The diffusion tensor describes an ellipsoid that represents the directional movement of water molecules inside a voxel.

DTI was developed to remedy the main limitation of DWI (see Section 4.2.1), taking advantage of the preferential water diffusion inside the brain tissue [20,21]. The water diffusion in the brain is not an isotropic process due to the natural intracellular (neurofilaments and organelles) and extracellular (glial cells and myelin sheaths) barriers that restrict diffusion toward certain directions. Hence, water molecules diffuse mainly along the direction of white-matter axons, rather than perpendicular to them (Figure 4.1). Under these circumstances, diffusion can become highly directional along the length of the tract, and is called *anisotropic* [12] (Figure 4.5). This means that we talk about media that have different diffusion properties in different directions. In other words, in certain regions of the brain,

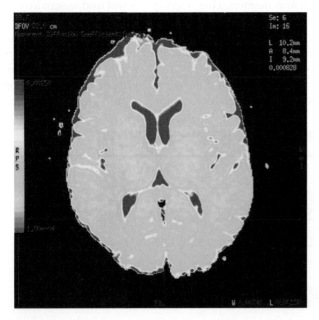

FIGURE 4.4
A typical ADC parametric color map.

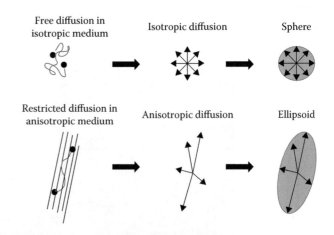

FIGURE 4.5
Free diffusion in isotropic medium results in a sphere, while restricted diffusion in an anisotropic medium results in an ellipsoid.

ADC is directionally dependent; it is therefore also clear that a single ADC would be inadequate for characterizing diffusion and a more compound mathematical description is required.

In that sense, DTI measures both the magnitude and the direction of proton movement within a voxel, for multiple dimensions of movement, using a mathematical model to represent this information, called the *diffusion tensor* (DT) [17]. Under the assumption that the probability of molecular displacements follow a multivariate Gaussian distribution over the observation diffusion time, the diffusion process can be described by a 3×3 tensor matrix, proportional to the variance of the Gaussian distribution. Thus, the diffusion tensor, D, is characterized by nine elements:

$$D = \begin{bmatrix} Dxx & Dxy & Dxz \\ Dyx & Dyy & Dyx \\ Dzx & Dzy & Dzz \end{bmatrix}$$

Now, consider that the directional movement of water molecules inside a voxel can be represented by an ellipsoid (Figure 4.5), which in turn can be described by the tensor in that specific voxel. This tensor consists of the 3×3 matrix derived from diffusivity measurements in at least six different directions. The tensor matrix is diagonally symmetric ($D_{ij} = D_{ji}$) meaning that the matrix is fully determined by six parameters. If the tensor is completely aligned with the anisotropic medium then the off-diagonal elements become zero and the tensor is diagonalized. This diagonalization provides three eigenvectors that describe the orientation of the three axes of the ellipsoid, and three eigenvalues that represent the magnitude of the axes (apparent diffusivities) in the corresponding directions (Figure 4.6). The major axis is considered to be oriented

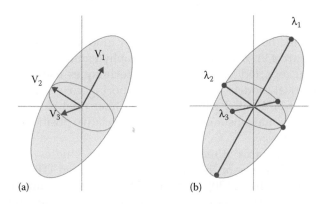

(a)   (b)

FIGURE 4.6
(a) Three eigenvectors describe the orientation of the three axes of the ellipsoid and (b) three eigenvalues represent the magnitude of the axes (apparent diffusivities) of the ellipsoid.

in the direction of maximum diffusivity, which has been shown to coincide with tract orientation [12,19]. Therefore, there is a transition through the diffusion tensor from the x, y, z coordinate system defined by the scanner's geometry to a new independent coordinate system, in which axes are dictated by the directional diffusivity information.

Depending on the local diffusion, the ellipsoid may be *prolate*, *oblate*, or *spherical*. Prolate shapes are expected in highly organized tracts where the fiber bundles all have similar orientations; oblate shapes are expected when fiber orientations are more variable but remain limited to a single plane, whereas spherical shapes are expected in areas that allow isotropic diffusion [22].

The local diffusion anisotropy can be quantified by the calculation of *rotationally invariant* parameters. The most commonly reported indices that can be calculated are the mean diffusivity (MD) and fractional anisotropy (FA), although there are several others that can be derived from DTI. MD is the mean of the eigenvalues, and represents a directionally measured average of water diffusivity, whereas FA derives from the standard deviation of the three eigenvalues. Figure 4.7 depicts the comparison of a T_2-weighted, average DC, FA map, and color-coded orientation map.

On an FA map, the signal brightness of a voxel describes the degree of anisotropy in the given voxel. FA ranges from 0 to 1, depending on the underlying tissue architecture. A value closer to 0 indicates that the diffusion in the voxel is isotropic (unrestricted water movement), such as in areas of CSF, whereas a value closer to 1 describes a highly anisotropic medium, where in areas such as in the corpus callosum, water molecules diffuse along a single axis [12].

However, the interpretation of such grayscale images is rather difficult since the information of the three components of the vector cannot be displayed in one single. Hence as it is shown in Figure 4.7c, diffusion directionality in various regions of interest can be further represented by a directionally encoded color (DEC) FA map. More specifically, the eigenvector with the largest eigenvalue defines the orientation of the ellipsoid in each voxel, which can then be color coded to evaluate and display information about the direction of white-matter tracts. Hence, ellipsoids describing diffusion from left to right are colored red (x-axis), ellipsoids describing anterioposterior (y-axis) diffusion are colored green, and diffusion in the craniocaudal direction is colored blue (z-axis) [23]. This procedure provides a user-friendly and a convenient summary map from which one can determine the degree of anisotropy (in terms of signal brightness) and the fiber orientation in the voxel (in terms of color hue). A neuroradiologist can then combine and correlate this information with normal brain anatomy,

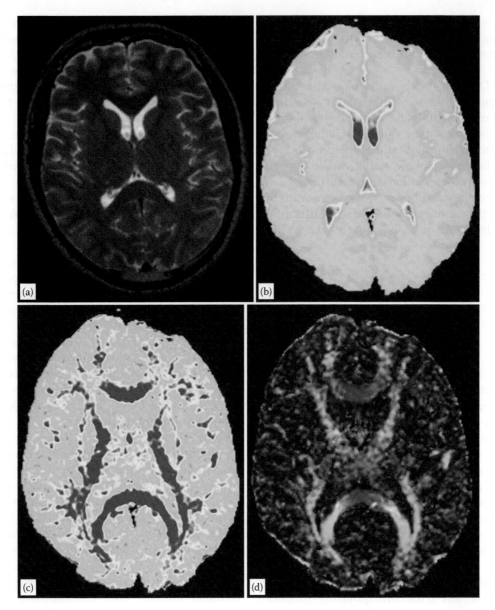

FIGURE 4.7
Comparison of (a) T_2-weighted, (b) average DC (c) FA map, and (d) color-coded orientation map. Images were acquired using a 3.0 T scanner. The colors represent the orientations of fibers; red: right–left, green: anterior–posterior, and blue: superior–inferior.

identify specific white-matter tracts, and assess the impact of a lesion on neighboring white-matter fibers.

By now, it must be evident that the underlying tissue structure dictates diffusion anisotropy, and in the human brain, this is mainly the white-matter architecture. Hence, by combining FA values with directionality, it would be possible to obtain estimates of fiber orientation. This idea has led to the development of fiber tractography enabling the mapping of white-matter tracts noninvasively [24] as seen in Figure 4.8.

Different algorithms have been developed for fiber tractography, but the main idea is that following the tensor's orientation on a voxel-by-voxel basis, it is possible to identify intravoxel connections and display

specific fiber tracts using computer graphic techniques (Figure 4.9). A variety of tractography techniques have been reported [25–28]. All these techniques use mathematical models to identify neighboring voxels, which might be located within the same fiber tract based on the regional tensor orientations and relative positions of the voxels.

Toward this direction, a number of studies have created atlases of the human brain based on DTI and tractography [29,30]. According to this, a very important differential diagnostic parameter regarding the displacement or disruption of a specific fiber tract by a pathology may be assessed by 3D tractograms [31,32], as this is displayed in Figure 4.10.

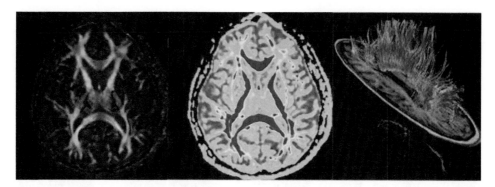

FIGURE 4.8
MR fiber tractography.

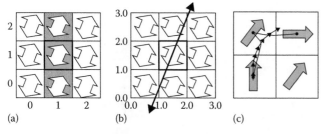

FIGURE 4.9
Double-headed arrows indicate fiber orientations at each pixel. Tracking is initiated from the center pixel. In the discrete number field (a), the coordinate of the seed pixels is {1,1}. If it is judged that the vector is pointing to {1,2} and {1,0}, shaded pixels are connected. In the continuous number field (b), the seed point is {1.50, 1.50} and a line, instead of a series of pixels is propagated. (c) shows an example of the interpolation approach to perform nonlinear line propagation. (From S. Mori and P.C.M. van Zijl: Fiber tracking: Principles and strategies: A technical review. *NMR Biomed.*, 2002. Vol. 15. pp. 468–480. Copyright Wiley-VCH Verlag GmbH & Co. KGaA. Reproduced with permission.)

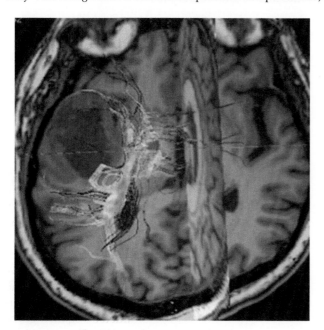

FIGURE 4.10
Displacement or disruption of a specific fiber tract by a pathology may be assessed by 3D tractograms.

In order to produce tracts, the user needs to define a *seed* region of interest (ROI) on the color orientation map that is very useful in visualizing the white-matter tract orientation. In most software applications, this is defined as *structural view*. Depositing a seed ROI results in a white-matter track oriented through the ROI. To display white-matter tracks oriented from one ROI to another ROI, one needs to position a second ROI on the image and define a *target* ROI.

These techniques also provide useful information in terms of presurgical planning; nonetheless, they present limitations such as in the cases of complex tracts (crossing or branching fibers), which should be taken into consideration when these methods are used for preoperative guidance.

In DTI, the diffusion gradients are applied in multiple directions, and based on previous reports, the number of noncolinear gradients applied varies (ranging from 6 to 55). There is much debate in the literature; however, an optimal number has not yet been defined [33–35]. As one can imagine, the main drawback of an increased number of gradients in DTI is the imaging time, which increases simultaneously, and may not be applicable in clinical practice [36]. Therefore, as always, there is a trade-off between the imaging time and the number of gradients applied in order to obtain sufficient diffusion information.

DTI has also been applied in the spinal cord for the evaluation of acute and chronic trauma, tumors of the spinal canal, degenerative myelopathy, demyelinating and infectious diseases, and so on, and there are strong indications that it can be a sensitive and specific method.

Figure 4.11 illustrates a case of spinal cord diffusion tensor tractography.

It has to be noted that there are still many technical limitations in the application of spinal cord DTI, especially in thoracic and lumbar segments. Nevertheless, the wider use of higher field scanners (3 T or more), and the further development of acquisition and postprocessing techniques should result in the increased role of this promising advanced technique in both research and clinical practice.

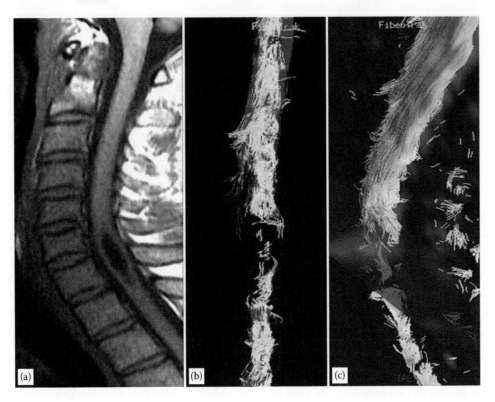

FIGURE 4.11
Spinal cord diffusion tensor tractography: (a) FA map, (b) ROI placement on colored orientation map, and (c) fiber tracts. (Courtesy of Mt Sinai Hospital.)

4.3 DWI and DTI in the Assessment of Cerebral Tumors

Precise tumor differentiation and diagnosis play an essential role in the selection of the optimum clinical care, as the nature of the tumor and the grade critically define the therapeutic approach. Despite the utilization of advanced MRI techniques, such as DWI/DTI, tumor characterization, and grading, is in some cases a challenging process, as there are imaging characteristics and contrast-enhancement patterns that may be similar in many cases. The parameters extracted from these techniques provide useful information in a microscopic level; however, their accurate interpretation is not always straightforward as microstructure similarities exist between pathologies, and one should be very careful in correctly combining and evaluating all the available MR data.

Brain tumors are a diverse group of neoplasms arising from different cells within the central nervous system (CNS) or from systemic tumors that have metastasized to the CNS. Brain tumors include a number of histologic types with markedly different tumor growth rates. In this chapter, we will discuss the main types of tumors

and the potential ability of DWI/DTI techniques to contribute to an accurate diagnosis.

4.3.1 Gliomas

FOCUS POINTS: DWI/DTI IN GLIOMAS

DWI

- Ambiguous results in the differentiation of lower and higher glioma grades.
- Large variations of ADC values exist between the two groups.

DTI

- FA can distinguish HGG from LGG.
- High anisotropy implies that the tissue is symmetrically organized.

Glioma is a general term used to describe any tumor that arises from the supportive tissue of the brain. The name of this tissue originates from the Greek word

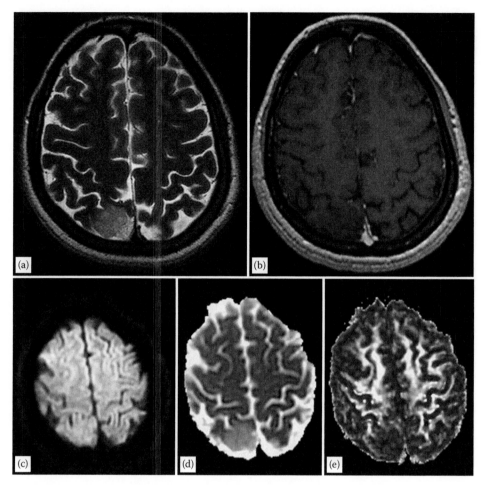

FIGURE 4.12
Low-grade glioma, presenting (a) high signal intensity on a T_2-weighted image, (b) no contrast enhancement on a 3D-SPGR image, and (c) an isointense signal on a diffusion-weighted image. The lesion shows (d) increased ADC and (e) lower FA.

glia (gluey), meaning that it helps to keep the neurons in place and functioning well. Gliomas represent the most common cerebral neoplasms and the preoperative assessment of their grade is important for therapeutic decision making. As previously mentioned, gliomas arise from supporting glial cells in the brain and the predominant cell type determines the pathological classification. Gliomas are categorized according to their grade, which is determined by pathologic evaluation of the tumor.

- Low-grade gliomas (LGGs) are well differentiated (not anaplastic); these are benign and indicate a better prognosis for the patient.
- High-grade gliomas (HGGs) are undifferentiated or anaplastic; these are malignant and portend a worse prognosis.

LGGs consist of grade I and II glial tumors. Grade I tumors are usually considered benign because they progress very slowly over time. Grade II tumors present nuclear atypia;

however, cellularity and vascularity is low and normal brain is usually mixed in with the tumor [37]. Depending on their cell origin, they may be termed as oligodendroglioma, astrocytoma, or mixed type. On conventional MR images, LGGs present a homogenous structure whereas contrast enhancement and peritumoral edema are usually uncommon (Figure 4.12, LGG) [38].

HGGs consist of grade III and grade IV glial tumors. Grade III present mitoses and anaplasia, and their most common subtype is anaplastic astrocytoma (AA) (Figure 4.13), whereas grade IV gliomas are characterized by increased cellularity and vascularity with extended necrosis, and are usually termed as *glioblastoma multiforme* (GBM) (Figure 4.14). HGGs present heterogeneous contrast-enhancement patterns, necrotic or cystic areas, hemorrhage, and infiltrative edema. Nonetheless, the imaging characteristic of these two main glioma categories are not always grade specific, as in some cases LGGs may show similar morphological features to HGGs and the latter may present relatively benign imaging findings [3,39]. Hence, these imaging

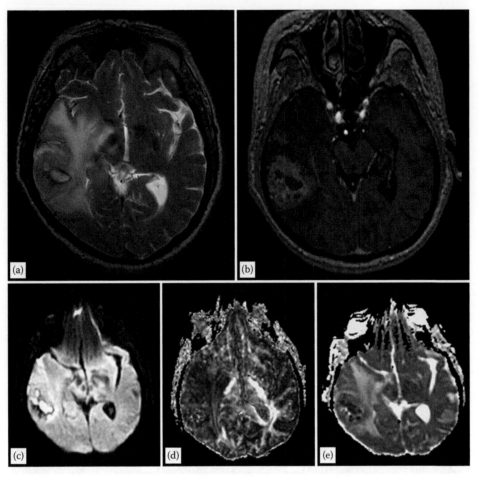

FIGURE 4.13
Anaplastic astrocytoma. (a) T_2-weighted image shows increased signal intensity with peritumoral edema, (b) heterogeneous contrast enhancement on a postcontrast 3D-SPGR image, and (c) restricted diffusion in the solid portion of the tumor. The lesion is hypointense on the ADC map (d) and presents low FA (e).

similarities may potentially lead to inaccurate tumor staging based on conventional MRI alone.

4.3.2 DWI Contribution

Several groups have studied the contribution of DWI metrics in the differentiation of lower and higher glioma grades, but their results have been ambiguous. Regarding LGG, it is probable that due to their cellular structure, these tumors usually present higher ADC values compared to HGG [38,40]; however, in many cases, there is an overlapping between the ADC values of both groups. Zonari et al. reported that even though diffusion was higher in LGGs, large variations of ADC values existed between the two groups, thus no significant differences were observed [41]. A significant issue that always plays a major role in the evaluation of diffusion results is the exact placement of the ROI area of measurement. Nevertheless, several studies have also concluded that DWI metrics, either from the solid part of the tumor

or from the peritumoral edema, are inadequate to provide information about the degree of differentiation of glial tumors [42–44]. On the other hand, in the study of Kono et al., the difference of ADC values between glioblastomas and grade II astrocytomas reached statistical significance; however, the authors reported that peritumoral neoplastic cell infiltration cannot be revealed using individual ADC values or even by evaluating ADC maps [45]. In the same study, an inverse relationship was observed between diffusion and tumor cellularity, where lower ADC values suggested malignant gliomas, whereas higher ADC values suggested low-grade astrocytomas. Nevertheless, the authors concluded that even though ADC values cannot be reliably used in individual cases to differentiate tumor type, a combination of routine image interpretation and ADC results in a higher diagnostic value. On the other hand, it has been shown that DWI metrics may be useful in the differentiation of nonenhancing gliomas, as ADC values in AAs can be significantly lower in the solid portions of

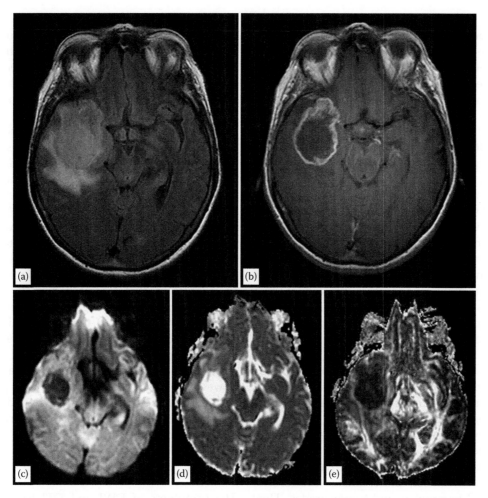

FIGURE 4.14
Glioblastoma multiforme. Axial T_2-weighted (a) and T_1-weighted postcontrast (b) images demonstrate a right temporal lesion with surrounding edema and ring-shaped enhancement. On the DW image, the lesion presents low signal intensity (c), resulting in higher intratumoral ADC (d) and lower intratumoral FA (e).

the tumors compared to LGG [39]. However, it is probable that no differences will be observed for ADC in the peritumoral area of these tumor groups.

4.3.3 DTI Contribution

Conflicting results are also reported in the literature regarding the ability of DTI parameters to discriminate between LGGs and HGGs. Studies have shown that MD measured in the intratumoral area [9,46,47] or in the peritumoral area [9,57] cannot be used as a predictor of lower and higher glioma grades. Contrary to these findings, it has also been reported that MD within grade I gliomas was significantly higher compared to grade III and grade IV gliomas, respectively, but no differences were observed in MD values between grade II and grade III gliomas [48]. However, Lee et al. showed that MD is significantly lower in the nonenhancing regions of HGG compared to LGGs,

although HGG may present relatively benign imaging findings, such as the absence of contrast enhancement [49]. Similar tendencies were observed in other studies as well, although these differences did not reach statistical significance [3].

FA has also been investigated regarding glioma grading and tumor infiltration. Studies of the relationship between DTI and histological malignancy of gliomas showed that FA can distinguish HGGs from LGGs; thus, it can be useful in deciding the surgical strategy or the selected site of stereotactic biopsy [46–50]. Inoue et al. reported that the FA value could be used to distinguish HGGs from LGGs, because FA is significantly higher in HGGs than LGGs and that high anisotropy implies that the tissue is symmetrically organized. A cutoff value of 0.188 was determined between the two groups [48]. In the same study, a positive correlation of FA with cell density of gliomas was observed in agreement with the results from Beppu et al. [50].

Nevertheless, the correlation between FA and glioma cellularity has been disputed. Stadlbauer et al. reported a negative correlation; however, the authors concluded that FA is a better indicator than MD for the assessment and delineation of different degrees of pathologic changes in gliomas [46]. In agreement with their noninfiltrating nature, the presence of well-preserved fibers in the periphery of LGG would suggest higher FA values, in contrast to HGG where peritumoral tracts are disarranged or disrupted [51,52]. Even in the case of nonenhancing HGGs, the evaluation of peritumoral FA has been reported to provide useful information and differentiation from LGGs [3]. As always, the determination of a certain cutoff value is difficult; nevertheless, Liu et al. observed that the mean and maximal FA values were significantly lower in LGG, and proposed a cutoff value of 0.129 between the two groups. Moreover, the authors proposed that the diagnostic accuracy improves if these two parameters are combined rather than evaluated separately. Hence, this combination may be useful in the preoperative grading of nonenhancing gliomas [3]. Similarly, Ferda et al. reported that although the evaluation of the FA maps might not be sufficient for glioma grading, the combination of the contrast-enhancement pattern and FA maps evaluation improves the possibility of distinguishing low- and high-grade glial tumors [53].

One of the very latest studies by Server et al. evaluated the diagnostic accuracy of axial diffusivity (AD), radial diffusivity (RD), ADC, and FA values derived from DTI for grading of glial tumors, and estimated the correlation between DTI parameters and tumor grades. They concluded that these are useful DTI parameters for differentiation between LGGs and HGGs and reported a diagnostic accuracy of more than 90% [54].

Nonetheless, controversies regarding the contribution of FA still exist in the literature, as a number of studies have concluded that the utility of DTI metrics in the grading of enhancing and nonenhancing gliomas, regardless of the area of measurement, is still limited [9,47,49,51,52].

4.3.4 Meningiomas

FOCUS POINTS: DWI/DTI IN MENINGIOMAS

- Inverse relationship between water diffusion and malignancy.
- Lower ADC and higher FA values for atypical/malignant meningiomas compared to benign.
- FA is significantly different between subtypes of benign meningiomas, whereas ADC cannot contribute.

Meningiomas are a diverse set of tumors arising from the meninges, the membranous layers surrounding the CNS. Meningiomas are the most common extra-axial cerebral tumors, and their characteristic location enables their relatively straightforward diagnosis. Though the majority of meningiomas are benign, they can have malignant presentations. Classification of meningiomas is based upon the WHO classification system as follows. The benign meningiomas or grade I can be up to 90% and usually full recovery is achieved with surgical resection. Grade II (atypical) and grade III (malignant) meningiomas are less common but more aggressive than grade I; thus, they are more likely to recur even after complete resection. According to the WHO classification, the differences between benign and atypical/malignant meningiomas relate to the number of mitoses, cellularity, and nucleus-to-cytoplasm ratio as well as their histologic patterns [55]. Meningiomas are highly vascular lesions that derive blood mostly from meningeal arteries, regardless of the grade. Their tumor capillaries present complete lack of the blood–brain barrier); thus, increased contrast leakage and permeability is observed on perfusion images. Conventional MRI provides useful information regarding their localization and morphology; however, in cases where meningiomas present atypical imaging findings, mimicking high-grade tumors, their histologic grading is of significant importance for beneficial treatment planning (Figure 4.15).

4.3.5 DWI and DTI Contribution

The usefulness of DWI and DTI techniques either in meningioma grading or in the differentiation of benign and atypical/malignant subtypes has been previously systematically investigated. Studies in the literature have shown that DWI and DTI metrics from the intratumoral region are useful in meningioma grading [5,56–58]. An inverse relationship between water diffusion and malignancy has been reported, since lower ADC and higher FA values have been indicated for atypical/malignant meningiomas compared to benign. The increased mitotic activity, necrosis, the high nucleus-to-cytoplasm ratio as well as the uninterrupted patternless cell growth present in high-grade meningiomas [59] lead to restricted water diffusion, which is reflected in the related diffusion parameters. On the contrary, benign meningiomas show the histological lack of coherent organization, as they consist of oval- or spindle-shaped neoplastic cells that form whorls, fascicles, cords, or nodules, forcing water molecules to move in a relatively isotropic way [60]. Regarding the differentiation between subtypes of the same grade, FA was significantly different between subtypes of benign meningiomas [62], whereas ADC did not contribute either in benign or in malignant subtype discrimination [62,63]. Nevertheless, as in the case of gliomas, controversies still exist in the literature regarding meningiomas, as a number of previous

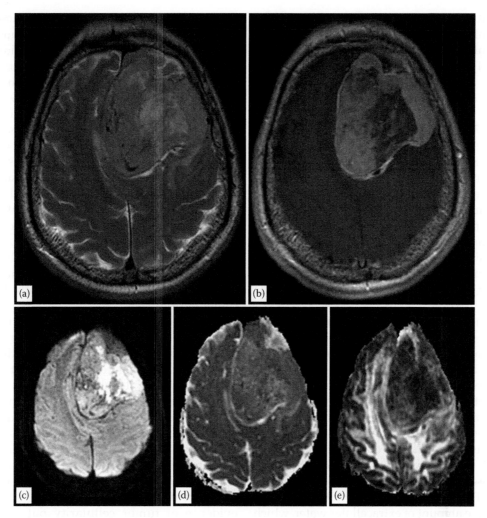

FIGURE 4.15
Atypical meningioma. Axial T_2-weighted (a) and postcontrast T_1-weighted (b) images demonstrate a large heterogeneous enhanced left frontal mass with an intense mass effect. The lesion presents areas of restricted diffusion (c), isointensity on the ADC map (d), and hypointensity on the FA map (e).

studies conclude that diffusion quantification, derived either from the tumor [40,61,64,65], or from the peritumoral edema [5,57], cannot provide a significant information about meningioma grading.

DWI/DTI metrics can also contribute in the differentiation between atypical meningiomas and gliomas. Similarities in diffusion properties have been reported between meningiomas and LGGs [9,45]; however, in the study of Tropine et al., significant differences in MD and FA were observed between the two groups [47]. LGGs had higher MD and lower FA values in the intratumoral region compared to meningiomas, which might be attributed to their lower tumor cellularity. No differences were observed in the related parameters of the peritumoral edema, most probably due to the noninfiltrating nature of both tumor types [47].

Meningiomas that exhibit atypical MRI features such as cystic and necrotic areas, ring-like enhancement, and parenchymal invasion may resemble malignant brain lesions such as gliomas or metastatic brain tumors, leading to false radiological reports and misinterpreted treatment decisions. For such cases that can rise up to 15% of meningiomas, the results in the literature are mixed. A number of studies suggest that diffusion and perfusion quantification can be helpful in correctly characterizing these lesions and thus aid treatment planning. Lower ADC and higher FA values have been observed in the solid portion of meningiomas as compared to HGGs [4,5,7,47]. The differences in diffusion properties between these two tumor types indicate a higher level of fibrous organization in meningiomas compared to HGGs, which present a more incoherent cellular structure [4]. Contrary to these observations, the contribution of DWI/DTI metrics in the preoperative differentiation of these two groups has been reported insignificant [8,9]. Furthermore, research has also been conducted to identify differences related to the characteristic nature of these lesions, which is infiltration versus noninfiltration; however, the results obtained were also insignificant [4,8,9,40,42,66,67].

Apart from high-grade gliomas, meningiomas that present atypical imaging findings may resemble solitary metastatic tumors (MTs) as well. The diffusion profiles in meningiomas and metastases present great similarities that have also been reported by several groups [8,9,45,67]. Based on these studies, diffusion and anisotropy changes, either from the solid region or from the periphery of the tumor, are inadequate to distinguish meningiomas from metastases. These findings may be explained by the fact that atypical/malignant meningiomas often present a heterogeneous cellular structure, with necrotic and cystic portions, thus inducing unhindered water diffusion comparable to that of MT. Furthermore, as noninfiltrating lesions, their surrounding edema is purely vasogenic, and cannot provide distinct information in terms of DWI/DTI measurements.

Once again, the key of differentiation might be in the nature of their surrounding edema, since a previous study demonstrated that ADC and FA values are significantly different between meningiomas and metastases [68]. Based on this study, as the mechanisms of edema formation in metastatic brain tumors and meningiomas may derive from different factors, the classification of peritumoral edema in purely vasogenic and infiltrating might not be sufficient. Therefore, the authors suggest that DTI could potentially identify subtle differences in the *purely vasogenic* edema associated with different tumor groups. The hypothesis is that MD is primarily determined by the amount of extracellular water, which is probably greater in metastases than in meningiomas. On the other hand, the formation of peritumoral edema in meningioma has been suggested to be related to many factors, such as tumor size, histological subtype, vascularity, secretory activity, tumor-related venous obstruction, and expression of sex hormones and receptors. However, none of these factors has been identified as the principle cause of edema formation [68].

4.3.6 Cerebral Metastases

FOCUS POINTS: DWI/DTI IN METASTASES

- Direct differentiation of secondary tumors from LGGs using ADC.
- Conflicting results in the literature regarding differentiation ability between metastases and HGGs.
- HGGs present elevated FA in the intratumoral and peritumoral region compared to metastases.

Cerebral metastases account for approximately 25%–50% of intracranial tumors in hospitalized patients and unfortunately their incidence is rapidly increasing. The most common primary cancers that metastasize to the brain are lung, breast, colon, malignant melanoma, and gastrointestinal cancers [69,70]. The differentiation of metastases from other malignant tumors on the conventional MRI is usually straightforward due to the clinical history of the patient or the existence of multiple well-circumscribed lesions. However, the occurrence of a solitary-enhancing lesion without the knowledge of a primary tumor complicates differential diagnosis, because it may present similar imaging characteristics and contrast-enhancement patterns like those of HGGs (Figure 4.16). Hence, the accurate characterization of these lesions is clinically important as medical staging, surgical planning, and therapeutic approach differ significantly between these tumor entities [71,72].

The differential diagnosis between metastases and primary HGGs represents one of the most common problems in the clinical routine and thus it has been extensively investigated in the literature. HGGs are characterized by the ability to recruit and synthesize vascular networks for further growth and proliferation. Hence, tumor cells are expected to be present in their periphery along with increased edema concentration (infiltrating edema). On the other hand, metastatic brain tumors arise within the brain parenchyma and usually grow by expansion, displacing the surrounding brain tissue, and with no histologic evidence of tumor cellularity outside the contrast-enhanced margin of the tumor (pure vasogenic edema) [73].

Due to their heterogeneous cellular structure, increased ADC and lower FA have been reported in both the tumoral part of metastases and HGGs. Moreover, the presence of increased edema in the periphery of both lesions, even though of different nature (vasogenic vs. infiltrating), does not provide any significant information that may allow a distinct differentiation between the two tumor groups. Hence, because of the aforementioned similarities, a large number of studies in the literature have concluded that the contribution of DWI and DTI metrics, either in the tumor or the peritumoral area, is still limited [6,10,11,40,71,74,75].

Once again, in the literature, conflicting results are presented. There are a number of studies that, contrary to these observations, report that the diffusion profiles of HGGs and metastases, either within or around the tumors, differ and that DWI/DTI measurements may be indicative for tumor discrimination [8,9,72,74–76]. Based on these studies, HGGs present elevated FA in the intratumoral and peritumoral region compared to metastases, whereas the latter present increased

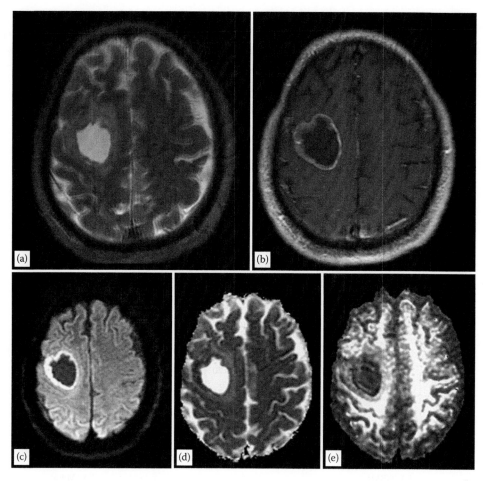

FIGURE 4.16
Intracranial lung metastasis. (a) T_2-weighted image, (b) ring-shaped enhancement on a T_1-weighted postcontrast edema, (c) restricted diffusion in the periphery of the tumor, increased intratumoral ADC (d), and decreased FA (e).

ADC values in their periphery. Restricted diffusion in HGGs might be explained by higher cellularity in their solid part compared to metastases, as well as by the presence of tumor cells in their periphery due to their infiltrating nature [11,77]. On the other hand, metastases have been associated with increased edema concentration, as a result of their leaky tumor capillaries, leading to higher ADC values in the peritumoral parenchyma [9,72]. Nevertheless, it is evident that the utility of DWI and DTI in preoperative differentiation of solitary MT from HGGs remains controversial. Our studies have shown that based solely on the mean values and the corresponding SDs of FA and ADC measurements (Figure 4.17), the mean value of FA in glioblastomas is higher than in metastases, indicating roughly that FA might be a more appropriate index to quantify the diffusion properties in the intratumoral region of these two groups [10].

The higher FA of glioblastomas in this patient cohort (35 GBM and 14 metastases) may be attributed to the fact that glioblastomas present higher cellularity in the solid part of their lesion than do brain metastases [77]. Whether FA is positively or negatively correlated with tumor cell density and vascularity, and whether it can be used to assess tumor grading, is still under wide dispute.

Differentiation of secondary tumors from LGGs is usually straightforward due to absence of contrast enhancement in LGGs, and the presence of minimal or no peritumoral edema around them. However, if low-grade glial tumors do not present typical imaging and contrast-enhancement findings, ADC in their intratumoral region has been reported significantly higher than in metastases, allowing their distinct differentiation [45]. In the same study, no differences were observed for peritumoral ADC values.

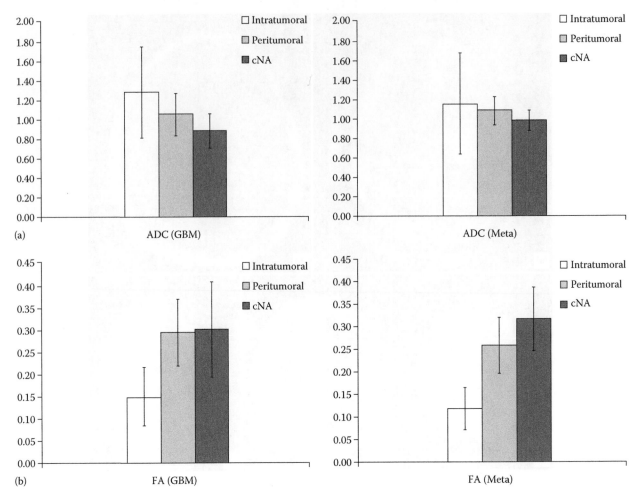

FIGURE 4.17

A comparison of (a) ADC and (b) FA measurements for glioblastomas (GBMs) and metastases (Meta), in terms of the mean values with their corresponding standard deviation in the intratumoral region (upper row), peritumoral region (lower row), and the contralateral normal area (cNA).

Discrimination between different types of metastases in terms of diffusion measurements has also been investigated [42,78]. ADC values between all types showed a wide variability with considerable overlapping and no statistical differences, with the greatest differences seen in metastases from primary lymphoma [42]. Nevertheless, it would be of great interest to examine a larger number of metastases in order to correctly classify them. The common signal intensity of necrotic/cystic components of cerebral metastases may relate to an increase in free water, showing hypointensity on DW images and increased ADC. However, it must be noted that in the presence of extracellular met-hemoglobin and/or increase viscosity, DW images can show hyperintensity with decreased ADC [17].

4.3.7 Primary Cerebral Lymphoma

> **FOCUS POINTS: DWI/DTI IN PCL**
>
> - PCLs have been reported with significantly lower ADC values than HGGs and metastases.
> - Higher cellularity and the absence of neo-angiogenesis give PCLs characteristic diffusion features.

Primary cerebral lymphomas (PCLs) are aggressive neoplasms with increased incidence in immunocompetent as well as immunocompromised patients. The incidence

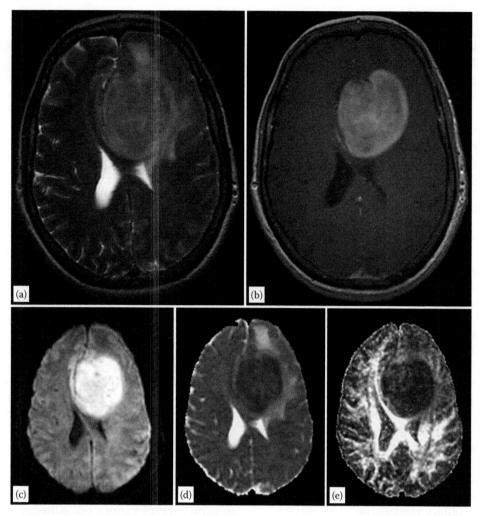

FIGURE 4.18
Primary cerebral lymphoma. High signal intensity with peritumoral edema on a T_2-weighted image (a) and intense contrast enhancement on a T_1-weighted postcontrast image (b). The lesion is hyperintense on the DW image (c), which results in a hypointense appearance on the ADC (d) and FA (e) maps.

of PCL has been substantially increased over the last three decades and currently accounts for about 6% of all cerebral tumors [79]. The morphology of lymphomas tends to be round or oval lesions in appearance and peritumoral edema is typically identified around them. Because these tumors are usually infiltrative in nature and not encapsulated, the borders of the MR signal change may not necessarily reflect the true tumor margin [80]. One of the most significant histopathologic characteristics of a PCL is the angiocentric growth with neoplastic cells forming multiple, thick layers around blood vessels. Tumor invasion of endothelial cells in the perivascular spaces and within the vessel walls can be often observed; however, neoangiogenesis is not a prominent feature [10,39,81]. More importantly, PCLs present a remarkable contrast enhancement on conventional MR

images due to the complete absence of blood–brain barrier (Figure 4.18). However, based on conventional MRI findings alone, it is often difficult to distinguish them from HGGs and solitary metastases, because of their diffuse infiltrative growth [78]. It is evident that differentiation of PCLs from other high-grade malignancies is very important since there are significant differences in the presurgical staging, intraoperative management, and postoperative treatment among these tumors.

Under this perspective, many studies have been performed using advanced MRI techniques in order to differentiate PCLs from glioblastomas and metastases. DWI and DTI have been found very helpful in distinguishing these tumor groups [6,10,45,82–85]. One of the main aspects is that as highly cellular tumors, PCLs have a relatively decreased amount of extracellular space, causing a restriction to free-water diffusibility. Hence, PCLs have been reported with significantly lower ADC values than HGGs and metastases. Similar to ADC, lower FA values have been also observed in PCLs compared to high-grade malignancies [10,83]. Lower FA values are not to be expected in PCLs because of their high cellularity; however, the relationship of FA and tumor cellularity still remains controversial, as both positive and negative correlations have been reported [46,48,50]. Despite the encouraging results regarding diffusion quantification in the discrimination of PCLs from both MT and HGGs, there have been studies in the literature concluding that the contribution of DWI and DTI is insignificant in both the intratumoral [6,8,47,84,86] and peritumoral area [8,10] of these lesions.

It is true that the differentiation of PCLs from LGGs is usually direct, because of the intense contrast enhancement and peritumoral edema of PCLs compared to LGGs. However, if LGGs lack their conventional imaging findings, ADC in their intratumoral region has been reported higher than PCLs, whereas FA values are lower [84]. Bendini et al. observed significant differences in perfusion properties as well between the two groups; however, these findings are in disagreement with the ones reported by Kremer et al. [7,87].

Overall, regarding the differentiation of PCL from gliomas and metastases, it is the higher cellularity, the absence of neoangiogenesis, and the different patterns of contrast leakage that give PCLs characteristic diffusion features, which enable their differentiation from glial tumors and solitary metastases.

4.3.8 Intracranial Abscesses

Brain abscesses are focal lesions caused by an infectious process of microorganism or pathogens that produce an area of focal cerebritis leading to the accumulation of purulent exudates in the brain tissue. They usually consist of a capsule of collagenous substance that begins to grow and encapsulate the purulent focus [88,89]. In some cases, radiologic diagnosis of cerebral abscesses may be challenging due to the variable appearance of these lesions secondary to different offending microbes and different stages of manifestation. On conventional MRI, abscesses present increased signal intensity on T_2-weighted images with associated peritumoral edema, increased signal intensity on DW images, and ring-shaped contrast enhancement. These features are nonspecific, and cystic or necrotic tumors (glioblastomas and solitary metastases) may contain pus, a highly viscous, thick, mucoid fluid consisting of inflammatory cells, bacteria, proteinaceous exudate, and fibrinogen, thus complicating their direct differentiation from cerebral abscesses

[78,89]. The pathophysiology and the imaging findings vary greatly depending on the organism causing the infection. DWI is useful for differentiating vasogenic edema from cytotoxic edema and also separates abscesses from cystic and necrotic tumors.

DWI and DTI metrics have proven to be beneficial in differentiating between abscesses and other cystic lesions [2,89–95]. Based on these studies, lower ADC values are observed in the central cavity of abscesses compared to glioblastomas and metastases. This is attributed to the high viscosity and cellularity of pus, which results in substantially restricted diffusion, contrary to the low viscosity in the cystic or necrotic areas of tumors that facilitates free diffusion and results in higher ADC values [91,92,95]. Significant differences in ADC values have been also observed between the capsular wall of abscesses and the peripheral tumor wall. Chan et al. reported that the capsular wall was hypointense on DW images and higher ADC values were measured in the area, compared to the hyperintese tumor wall, associated with lower ADC values [89]. The authors suggested that inflammation induced increased extracellular fluid accumulation in the abscess wall, thus water diffusion was unhindered. On the contrary, the higher cellularity in the peripheral wall of the tumors, due to closely packed malignant cells, resulted in restricted diffusion [89]. Furthermore, these intracavity histological differences are reported to be responsible for the higher FA values measured in cerebral abscesses compared to cystic or necrotic tumors. This may be attributed to the more organized structure of inflammatory cells, owing to cell adhesion secondary to expression of various cell adhesion molecules on the surface of inflammatory cells [94,96].

Overall, the high viscosity and cellularity of intracranial abscesses are the main biological factors that define the diffusion and perfusion characteristics of these lesions. DWI and DTI metrics can identify these distinct characteristics and provide a direct differentiation from cystic and necrotic tumors (Figure 4.19).

4.3.9 Stroke

FOCUS POINTS: DWI/DTI IN STROKE

- DWI is the most sensitive MRI technique to diagnose hyperacute cerebral infarction.

- An ischemic event results in restricted diffusion of the affected tissue, which can be seen as early as 30 min after ictus.

- DTI can provide unique and detailed information about white-matter anatomy following stroke.

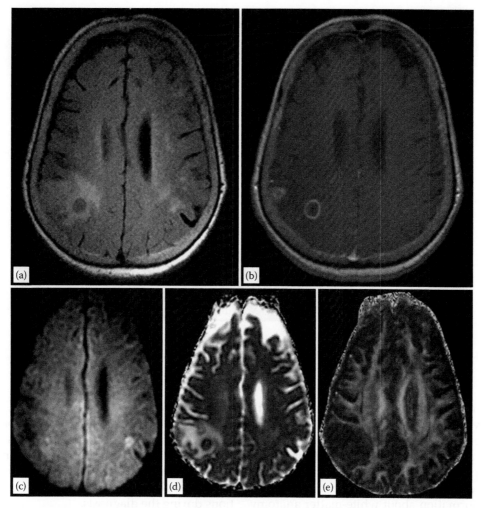

FIGURE 4.19
Intracranial abscesses. (a) T_2 FLAIR image of the lesion with associated peritumoral edema, (b) ring-shaped contrast enhancement on a T_1-weighted postcontrast image. The lesion is hyperintense on the DW image (c), which results in a hypointense appearance on the ADC (d) and FA (e) maps.

An ischemic stroke resulting from a blockage in the blood vessels that supply blood to the brain can lead to a cerebral infarction (Figure 4.20). Stroke caused by cerebral infarction should be distinguished from two other kinds of stroke: cerebral hemorrhage and subarachnoid hemorrhage. A cerebral infarction occurs when a blood vessel that supplies a part of the brain becomes blocked or leakage occurs outside the vessel walls.

DWI has proven to be the most sensitive MRI technique to diagnose hyperacute cerebral infarction, as it is based on alterations in the motion of water molecules. The technique is very sensitive and is not significantly affected by patient motion, since it can usually be accomplished in less than 1–2 min. The ischemic event results in restricted diffusion of the affected tissue, which can be seen as early as 30 min after ictus.

But when is diffusion imaging particularly useful? It has been shown that the brain water diffusion constant drops rapidly after the induction of ischemia, leading to hyperintensity in DWI. The evolution of the diffusion constant with time is a complex phenomenon, but it appears that most chronic strokes show elevated diffusion constants. Therefore, the diffusion characteristics of a lesion may help establish its age.

Studies that have used DWI in patients with symptoms of large hyperacute middle cerebral artery (MCA) territory infarction have highlighted that many ischemic lesions become visible very quickly, hence differentiating stroke from other forms of brain injury [97,98].

DTI can also be used to detect reduced diffusion in acute cerebral ischemia. Although FA is variable in acute stroke, it decreases with time after stroke. DTI may be especially helpful in distinguishing stroke mimics, which might appear with increased average diffusivity, from acute cerebral ischemia, which appears with reduced average diffusivity.

Hence, DTI is also beginning to provide new insights into the pathophysiology of acute stroke. The time

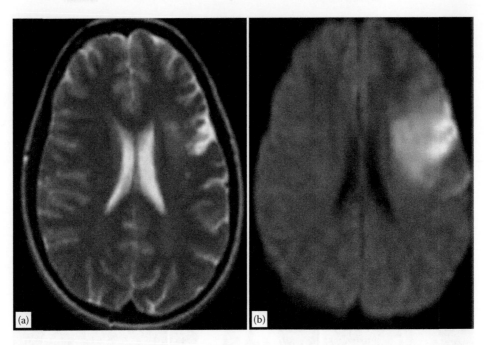

FIGURE 4.20
T_2-weighted image (a) and diffusion-weighted image (b) of an ischemic stroke leading to a cerebral infarction.

course of FA and MD changes in combination with ADC and T_2 have been delineated and this seems to be important in determining stroke onset time. FA ratios correlate with clinical neurological scales and may prove to be important in determining tissue viability and patient outcome. Moreover, DTI can provide unique and detailed information about white-matter anatomy following stroke. Fiber tract reconstruction using tract-based techniques and cross-sectional ROI delineation have been used to quantify white-matter integrity as it can be seen in Figure 4.21 [99,100]. Nevertheless, it is evident that further investigation is needed in order to reveal newly available information that DTI may provide.

4.3.10 Demyelinating Disease/Multiple Sclerosis

Multiple sclerosis (MS), also known as disseminated sclerosis or encephalomyelitis disseminata, is an inflammatory disease in which the insulating covers of nerve cells in the brain and spinal cord are damaged. T_2-weighted and fluid-attenuated inversion-recovery (FLAIR) images are quite sensitive for depicting focal lesions in patients with MS, but lack histopathologic specificity.

Moreover, tumefactive MS is a condition in which the CNS of a person has multiple demyelinating lesions with atypical characteristics for those of standard MS. It is called *tumefactive* as the lesions are *tumor-like* and they

mimic tumors clinically, radiologically, and sometimes pathologically [101]. These atypical lesion characteristics may include large intracranial lesions of size greater than 2 cm with a mass effect, edema, and an open-ring enhancement. Thus, a tumefactive lesion may mimic a malignant glioma or cerebral abscess causing complications during the diagnosis.

The evaluation of DWI in MS patients has shown that there can be increased values of ADC and decreased FA in normal-appearing white matter, which would clearly differentiate them from healthy control subjects, where these abnormalities are not seen [102]. MS plaques usually show hyper- or isointensity and increased ADC on DW images, in both active plaques (contrast enhancing) as well as chronic plaques.

On the other hand, MS plaques are reported to have decreased anisotropy [103]. This is thought to be related to an increase in the extracellular space due to demyelination, perivascular inflammation with vasogenic edema, and gliosis. An enhancing portion of MS plaques has slightly increased ADC, histologically representing prominent inflammation with mild demyelination, while the nonenhancing portions tend to have more increased ADC, representing scarring with mild inflammation and myelin loss [104]. It has been proposed that ADC values of MS plaques may be related to the severity of MS, since the ADC values in secondary-progressive MS are higher than those in relapsing-remitting MS [105].

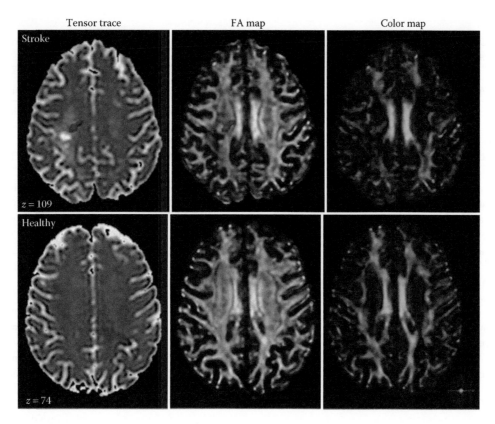

FIGURE 4.21

Axial images generated from tensor calculation. The lesion location is noted with a red arrow in each panel. (Reprinted from *NeuroImage*, 59, M.R. Borich, K.P. Wadden, L.A. Boyd, Establishing the reproducibility of two approaches to quantify white matter tract integrity in stroke, pp. 2393–2400, 2012, with permission from Elsevier.)

4.4 Summary

Tables 4.1 through 4.3 have been created in an attempt to facilitate the reader to better understand the aforementioned mechanisms as well as to summarize the correlation and the potential diagnostic outcome of T_2/DWI signals and ADC/FA measurements regarding different types of cerebral pathology.

Table 4.1 represents a summary of the correlation of T_2/DWI signals and ADC/FA measurements regarding different types of cerebral pathology. It must be noted that the relationship is indicative, which can be altered under certain conditions.

Table 4.2 depicts a summary of the diagnostic outcome of recent published studies regarding glioma grading and the differentiation of MT from HGGs, using DWI and DTI techniques.

Finally, Table 4.3 depicts a summary of the diagnostic outcome of recent published studies regarding different tumor comparisons, using DWI and DTI techniques.

TABLE 4.1

Summary of the Correlation of T_2/DWI Signals and ADC/FA Measurements Regarding Different Types of Cerebral Pathology

Pathology	T_2	DWI	ADC	FA
Ischemia	–	↑	↓	↓
Demyelination (MS)	↑	↑	˅	↓
Anaplastic astrocytoma	↑	↑	˄	↓
Low-grade astrocytoma	↑	↑	˄	↓
Abscess	↑	↑	↓	↑[a]
GBM	↑	˄	↑	↓
Necrotic tumor	↑	↓	↑	↓
Meningioma (classic)	↑	↑	↑	˅
Meningioma (atypical)	↑	↑	↓	˄[b]
Metastasis	↑	↑	↓	↓
Vasogenic edema	↑	–	↑	[b]

Note: It must be noted that the relationship is indicative and that it can be altered under certain conditions.

[a] Compared to cystic or necrotic tumors.

[b] Contradicting results.

↓, decrease; ↑, increase; ˄, moderate increase; ˅, moderate decrease; –, isointense/no change.

TABLE 4.2

Summary of the Diagnostic Outcome of Recent Published Studies Regarding Glioma Grading and the Differentiation of Metastatic Tumors (MT) from High-Grade Gliomas Using DWI and DTI Techniques

Type of Pathology	Authors	No. of Patients	Area of Measurement	Technique	Diagnostic Outcome
Glioma grading	Kono et al. [45]	17	Intra/peritumoral	DWI	Intratumoral ADC higher in LGGs than GBM
	Beppu et al. [50]	31	Intratumoral	DTI	FA lower in LGGs than HGGs
	Inoue et al. [48]	41	Intratumoral	DTI	MD higher in LGGs than HGGs
					FA lower in LGGs than HGGs
	Fan et al. [39]	22	Intra/peritumoral	DWI	Intratumoral ADC higher in LGGs than AA
	Stadlbauer et al. [46]	20	Intratumoral	DTI	FA higher in LGGs than HGGs
	Lee et al. [49]	27	Intratumoral	DTI	MD higher in LGGs than HGGs
	Chen et al. [52]	31	Intra/peritumoral	DTI	Peritumoral FA higher in LGGs than HGGs
	Liu et al. [3]	52	Intratumoral	DTI	FA lower in LGGs than HGGs
	Svolos et al. [107]	73	Intra/peritumoral	DWI	Intratumoral ADC higher in LGGs than HGGs
MT versus HGGs	Lu et al. [9]	20	Intra/peritumoral	DTI	Peritumoral MD higher in MT than GBM
	Server et al. [7]	82	Intra/peritumoral	DWI	Intratumoral ADC lower in MT than GBM
	Pavlisa et al. [74]	40	Intra/peritumoral	DWI	Peritumoral ADC higher in MT than GBM
	Wang et al. [75]	63	Intra/peritumoral	DTI	Intra/peritumoral FA lower in MT than GBM
	Lee et al. [76]	73	Intra/peritumoral	DWI	Peritumoral ADC higher in MT than GBM
	Wang et al. [9]	51	Intra/peritumoral	DTI	Intra/peritumoral FA lower in MT than GBM
	Svolos et al. [107]	71	Intra/peritumoral	DTI	Peritumoral FA lower in MT than HGG

4.5 Pitfalls and Artifacts

DW images are basically T_2 weighted and the interpretation of signal intensity on DW images requires a correlation between $b0$ images, ADC maps, and exponential images to uncover the underlying pathophysiologic condition. In that sense, it is crucial to understand the possible artifacts in order to avoid misinterpretation.

Understanding inherent artifacts and the way to reduce them is the way to improve the quality and accuracy of DWI. Artifacts may be related to the MR equipment itself, such as eddy currents and nonlinear gradients, or may as well be due to the properties of the subject introduced into the magnetic field, such as T_2 effects and anisotropy and susceptibility issues.

4.5.1 Eddy Current Artifacts

Electrical currents that may be induced in a conductor by a changing magnetic field are called *eddy currents*. Eddy currents can occur both in patients as well as in the MR scanner cables or wires, gradient coils, and radiofrequency shields. The stronger the applied magnetic field, the greater the electrical conductivity of the conductor, and the faster the field changes, the greater the currents that are developed and the greater the fields produced. Hence, eddy currents can become particularly severe when gradients are turned on and off quickly, as in EPI pulse sequences used in diffusion imaging. The problem is that diffusion measurements can be particularly vulnerable to eddy currents produced by the diffusion gradient pulses. The effect of these gradients on the image is approximately the same to that of susceptibility gradients. In ADC maps, the miscalculated diffusion coefficient may be neglected but in DTI, any miscalculation can remarkably alter the direction corresponding to the largest principal diffusivity. The effect of this miscalculation primarily affects tractography by altering the fiber direction in the affected voxels. Nevertheless, measurements of the tensor trace can also be affected but usually to a lower degree.

4.5.2 Nonlinearity of Gradient Fields Artifacts

The gradient coils are designed to add a contribution to the main magnetic field, which increases linearly along one of three orthogonal directions. However, the region where this linearity assumption holds is limited is due to the fact that the coil is not infinitely large [106]. It is because when moving away from the magnet's isocenter, the field gradient decreases, hence producing distortions in the outer regions of the anatomical image.

4.5.3 T_2 Effect

The signal intensity in a DW image depends on the T_2 time, since in order to have significant DW, the time required for diffusion sensitization entails long echo times (T_e). Hence, tissues with long T_2 values, for example, due to an old ischemic insult can appear as bright tissue with

TABLE 4.3

Summary of the Diagnostic Outcome of Recently Published Studies Regarding Different Tumor Comparisons, Using DWI and DTI Techniques

Comparative Cases	Authors	No. of Patients	Area of Measurement	Technique	Diagnostic Outcome
MNG versus HGGs	Tropine et al. [47]	22	Intra/peritumoral	DTI	Intratumoral MD lower in MNG than HGGs
					Intratumoral FA higher in MNG than HGGs
	De Belder et al. [4]	35	Intra/peritumoral	DTI	Intratumoral ADC lower in MNG than HGGs
					Intratumoral FA higher in MNG than HGGs
					Peritumoral FA higher in MNG than HGGs
	Svolos et al. [107]	77	Intra/peritumoral	DWI	Intra/peritumoral ADC lower in MNG than HGGs
				DTI	Intra/peritumoral FA higher in MNG than HGGs
MNG versus MT	Toh et al. [68]	26	Peritumoral	DTI	MD lower in MNG than MT
					FA higher in MNG than MT
	Svolos et al. [107]	42	Intra/peritumoral	DWI	Intra/peritumoral ADC lower in MNG than MT
				DTI	Intra/peritumoral FA higher in MNG than MT
PCL versus HGG	Guo et al. [82]	28	Intratumoral	DWI	ADC lower in PCL than HGG
	Yamasaki et al. [40]	44	Intratumoral	DWI	ADC lower in PCL than GBM
	Calli et al. [6]	25	Intratumoral	DWI	ADC lower in PCL than GBM
	Toh et al. [83]	20	Intratumoral	DTI	MD and FA lower in PCL than GBM
	Kinoshita et al. [84]	14	Intratumoral	DTI	MD lower in PCL than HGG
	Server et al. [7]	64	Intra/peritumoral	DWI	Intratumoral ADC lower in PCL than HGG
	Wang et al. [9]	42	Intra/peritumoral	DTI	Intratumoral MD and FA lower in PCL than GBM
					Peritumoral FA lower in PCL than GBM
PCL versus MT	Yamasaki et al. [46]	37	Intratumoral	DWI	ADC lower in PCL than MT
	Server et al. [7]	28	Intra/peritumoral	DWI	Intratumoral ADC lower in PCL than MT
	Wang et al. [9]	41	Intra/peritumoral	DTI	Intratumoral MD lower in PCL than MT
Abscess versus cystic/necrotic tumor	Hartmann et al. [91]	17	Intratumoral	DWI	ADC lower in abscess than other tumors
	Chan et al. [89]	12	Intra/peritumoral		Intratumoral ADC lower in abscess than other tumors
				DWI	Peritumoral ADC higher in abscess than other tumors
	Chang et al. [2]	26	Intratumoral	DWI	ADC lower in abscess than other tumors
	Lai et al. [92]	14	Intratumoral	DWI	ADC lower in abscess than other tumors
	Nadal-Desbarats et al. [93]	26	Intratumoral	DWI	ADC lower in abscess than other tumors
	Nath et al. [94]	53	Intratumoral	DTI	MD lower and FA higher in abscess than other tumors
	Reiche et al. [95]	17	Intratumoral	DWI	ADC lower in abscess than other tumors

a low-diffusion coefficient, for example, due to a recent stroke. This is a phenomenon often called T_2 *shinethrough* and complicates the delineation of an infarct. However, the T_2 effect is diminished by using the maps of diffusion coefficients or the trace of the diffusion tensor since these are independent of T_2. Moreover, at large *b*-values, the image contrast is dominated by diffusion. Also, with the development of clinical scanners with stronger gradient values, it is becoming possible to record high *b*-value diffusion images with shorter T_e and hence less T_2 weighting.

4.5.4 Susceptibility Artifacts

Susceptibility artifacts occur as the result of microscopic gradients or variations in the magnetic field strength that occur near the interfaces of substance of different magnetic susceptibility. Single-shot EPI is sensitive to susceptibility artifacts near the skull base, especially near the air in the sinus and mastoid, and has to be taken into account. A good approach is to always have phase encoding along the anterior–posterior direction for axial DW images or use coronal and sagittal images in the hippocampus and brain stem to identify possible artifacts.

4.5.5 Conclusions and Future Perspectives

The usefulness of MRI in the detection of cerebral pathology has been well established. Nevertheless, conventional MRI's main advantages, like the excellent soft tissue visualization and the great variety of imaging sequences, can be in many cases nonspecific. Hence, advanced techniques, like DWI and DTI, have been used in the clinical routine to improve diagnostic accuracy.

At this point, it should be realized that over the years, MR systems have evolved from imaging modalities to advanced systems that produce a variety of numerical parameters. This variety of quantitative information derived from these techniques provides significant structural and functional information in a cellular level, highlighting aspects of the underlying brain pathophysiology. Exploiting these advanced technological and imaging capabilities of MR systems is of great importance to optimize the diagnosis and treatment of cerebral pathology.

Despite the indisputable, contribution of advanced techniques to the preoperative assessment of cerebral pathology [2–6,107], no single technique can provide a robust tumor characterization, while the reported results in the literature can be conflicting and complicate even further clinical decision making [7–11]. It is precisely these controversies that reflect the complex underlying pathophysiologic mechanisms that are present in cerebral lesions and may prevent the clear discrimination between pathologies.

Usually, the most critical elements in the determination of tumor grade and prognosis are the tumor cellularity and vascularity. These two factors can be quantified through diffusion and perfusion metrics; however, as they are closely correlated, their evaluation and interpretation is difficult based on individual numeric parameters.

Conventional methods of data analysis, such as searching for statistical significances of the related parameters between different tumor groups may be efficient in some cases. However, in more demanding diagnostic problems, like tumors that have similar pathophysiological profiles, their efficiency might be limited. It is interesting to note that in the last few years, diagnostic interest has been focused on the combination of different parameters provided by advanced MRI techniques, and the incremental diagnostic and predictive value that multiparametric analysis may yield. Different methods of data analysis have been evaluated, such as logistic regression (LR) and receiver operating characteristic (ROC) analysis [10,41,108], as well as more sophisticated techniques like machine learning algorithms [109–112], using various parametric combinations.

It has been reported by Server et al. that the combination of DWI and magnetic resonance spectroscopic imaging (MRSI) increased the accuracy of preoperative differentiation of LGGs versus HGGs, compared to DWI or MRSI alone. In this study, the four-factor model, consisting of intratumoral mean ADC, maximum ADC, peritumoral Cho/Cr, and Cho/N-acetyl aspartate (NAA) ratios, resulted in 92.5% accuracy, 91.5% sensitivity, 100% specificity, 100% positive predictive value (PPV), and 60% negative predictive value (NPV). Wang et al. investigated the differentiation of glioblastomas, solitary brain metastases,

and PCLs [10]. The authors showed that the best model to discriminate glioblastomas from nonglioblastomas consisted of ADC and FA from the enhancing region of the tumors. The accuracy, sensitivity, and specificity scores were 93.8%, 89%, and 93%, respectively. Additionally, the best model to differentiate PCLs from metastases consisted of ADC from the enhancing regions and the planar anisotropy coefficient (CP) from the immediate peritumoral area. The accuracy, sensitivity, and specificity scores were 90.9%, 77%, and 94%, respectively [10].

Similarly Zonari et al. showed that the differentiation of LGGs and HGGs is more efficient if DWI, dynamic susceptibility contrast imaging, and MRSI data are combined than evaluated independently [46].

Therefore, it seems that multiparametric analysis may substantially improve diagnostic accuracies over conventional MRI alone, and highlight the underlying pathophysiology. However, this process is quite demanding and time consuming, due to the numeric nature of the acquired MR data. Recent studies have reported that machine learning techniques may be used as an automated computer analysis tool in order to aid tumor diagnosis [109,112–114]. The use of such techniques allows the manipulation and evaluation of a large amount of quantitative data during clinical practice. A variety of features, such as morphological (e.g., tumor shape and texture) and conventional (e.g., signal intensity) extracted from different MR sequences have been evaluated with very interesting results [115,116]. However, the most important aspect of machine learning techniques is their additional ability to provide predictive outcomes in contrast to conventional statistical methods, which are limited to producing diagnostic results retrospectively.

In conclusion, the characterization of tissue microstructure both in the tumoral and peritumoral areas, based on diffusion findings, results in increased diagnostic value. Nevertheless, further research is still required to determine more precisely how diffusion relates to tumor grade within tumor types, as well as to understand the relationship between the different types of tissue edema and ADC. On the other hand, it is evident that DTI shows great promise in its ability to show visually the effect of tumors on the surrounding WM tracts, but there still exist great controversies, and the conclusions are yet to be fully confirmed by histopathology. Although each MRI technique provides important information regarding certain tumor properties, it is clear that none of them separately can allow a direct tumor characterization. Tumor cellularity and vascularity, which can be quantified through DWI and DTI metrics, are two factors closely correlated in a nonlinear way and thus difficult to evaluate and interpret through conventional methods of data analysis. Hence, it is evident that the combination of diffusion parameters, in either a statistical model or a classification scheme, should further improve the diagnostic outcome. LR and ROC analysis may be useful in the characterization and grading of brain tumors using parametric combinations; however, the discrimination accuracy and specificity may be further improved, especially for tumors that present similar histopathological profiles, if sophisticated machine learning algorithms are used.

References

1. B. Hakyemez, C. Erdogan, G. Gokalp et al., "Solitary metastases and high-grade gliomas: Radiological differentiation by morphometric analysis and perfusion-weighted MRI," *Clin Radiol*, vol. 65, no. 1, pp. 15–20, 2010.
2. S.C. Chang, P.H. Lai, W.L. Chen et al., "Diffusion-weighted MRI features of brain abscess and cystic or necrotic brain tumors: Comparison with conventional MRI," *Clin Imaging,* vol. 26, no. 4, pp. 227–236, 2002.
3. X. Liu, W. Tian, B. Kolar et al., "MR diffusion tensor and perfusion-weighted imaging in preoperative grading of supratentorial nonenhancing gliomas," *Neuro Oncol*, vol. 13, no. 4, pp. 447–455, 2011.
4. F.E. De Belder, A.R. Oot, W. Van Hecke et al., "Diffusion tensor imaging provides an insight into the microstructure of meningiomas, high-grade gliomas, and peritumoral edema," *J Comput Assist Tomogr*, vol. 36, no. 5, pp. 577–582, 2012.
5. B. Hakyemez, N. Yildirim, C. Erdogan et al., "Meningiomas with conventional MRI findings resembling intraaxial tumors: Can perfusion-weighted MRI be helpful in differentiation?," *Neuroradiology*, vol. 48, no. 10, pp. 695–702, 2006.
6. C. Calli, O. Kitis, N. Yunten et al., "Perfusion and diffusion MR imaging in enhancing malignant cerebral tumors," *Eur J Radiol*, vol. 58, no. 3, pp. 394–403, 2006.
7. A. Server, B. Kulle, J. Maehlen et al., "Quantitative apparent diffusion coefficients in the characterization of brain tumors and associated peritumoral edema," *Acta Radiol*, vol. 50, no. 6, pp. 682–689, 2009.
8. S. Lu, D. Ahn, G. Johnson et al., "Diffusion-tensor MR imaging of intracranial neoplasia and associated peritumoral edema: Introduction of the tumor infiltration index," *Radiology*, vol. 232, no. 1, pp. 221–228, 2004.
9. S. Wang, S. Kim, S. Chawla et al., "Differentiation between glioblastomas, solitary brain metastases, and primary cerebral lymphomas using diffusion tensor and dynamic susceptibility contrast-enhanced MR imaging," *AJNR Am J Neuroradiol*, vol. 32, no.3, pp. 507–514, 2011.
10. I. Tsougos, P. Svolos, E. Kousi et al., "Differentiation of glioblastoma multiforme from metastatic brain tumor using proton magnetic resonance spectroscopy, diffusion and perfusion metrics at 3 T," *Cancer Imaging*, vol. 12, pp. 423–436, 2012.
11. J. Gillard, A. Waldman, P. Barker, *Clinical MR Neuroimaging: Diffusion, Perfusion, Spectroscopy*, Cambridge University Press, 2005.
12. S.J. Price, "The role of advanced MR imaging in understanding brain tumour pathology," *Br J Neurosurg*, vol. 21, no. 6, pp. 562–575, 2007.
13. E.O. Stejskal, J.E. Tanner, "Spin diffusion measurements: Spin echoes in the presence of a time-dependent field gradient," *J Chem Phys*, vol. 42, pp. 288–292, 1965.
14. J.E. Tanner, "Transient diffusion in a system partitioned by permeable barriers. Application to NMR measurements with a pulsed field gradient," *J Chem Physiol*, vol. 69, pp. 1748–1754, 1978.
15. D. Le Bihan, E. Breton, D. Lallemand et al., "MR imaging of intravoxel incoherent motions: Application to diffusion and perfusion in neurologic disorders," *Radiology*, vol. 161, pp. 401–407, 1986.
16. T. Moritani, S. Ekholm, P.L. Westesson, *Diffusion-Weighted MR Imaging of the Brain*, 2nd ed., Springer, 2009.
17. J.M. Debnam, D. Schellingerhout, "Diffusion MR imaging of the brain in patients with cancer," *Int J Mol Imaging*, vol. 2011, pp. 714021, 2011.
18. D.K. Jones, M.A. Horsfield, A. Simmons, "Optimal strategies for measuring diffusion in anisotropic systems by magnetic resonance imaging," *Magn Reson Med*, vol. 42, pp. 515–525, 1999.
19. A.S. Field, A.L. Alexander, "Diffusion tensor imaging in cerebral tumor diagnosis and therapy," *Top Magn Reson Imaging,* vol. 15, no. 5, pp. 315–324, 2004.
20. P. Mukherjee, J.I. Berman, S.W. Chung et al., "Diffusion tensor MR imaging and fiber tractography: Theoretic underpinnings," *AJNR Am J Neuroradiol*, vol. 29, no. 4, pp. 632–641, 2008.
21. D. Le Bihan, "Looking into the functional architecture of the brain with diffusion MRI," *Nat Rev Neurosc*, vol. 4, no. 6, pp. 469–480, 2003.
22. A.L. Alexander, K. Hasan, G. Kindlmann et al., "A geometric analysis of diffusion tensor measurements of the human brain," *Magn Reson Med*, vol. 44, no. 2, pp. 283–291, 2000.
23. S. Pajevic, C. Pierpaoli, "Color schemes to represent the orientation of anisotropic tissues from diffusion tensor data: Application to white matter fiber tract mapping of the human brain," *Magn Reson Med*, vol. 42, no. 3, pp. 526–540, 1999.
24. C.F. Westin, S.E. Maier, H. Mamata et al., "Processing and visualization for diffusion tensor MRI," *Medical Image Anal*, vol. 6, no. 2, pp. 93–108, 2002.
25. S. Mori, B.J. Crain, V.P. Chacko et al., "Three-dimensional tracking of axonal projections in the brain by magnetic resonance imaging," *Ann Neurol*, vol. 45, no. 2, pp. 265–269, 1999.
26. D.K. Jones, A. Simmons, S.C. Williams et al., "Non-invasive assessment of axonal fiber connectivity in the human brain via diffusion tensor MRI," *Magn Reson Med*, vol. 42, no.1, pp. 37–41, 1999.
27. S. Mori, P.C.M. van Zijl, "Fiber tracking: Principles and strategies. A technical review," *NMR Biomed*, vol. 15, pp. 468–480, 2002.
28. G.J. Parker, K.E. Stephan, G.J. Barker et al., "Initial demonstration of in vivo tracing of axonal projections in the macaque brain and comparison with the human brain using diffusion tensor imaging and fast marching tractography," *Neuroimaging*, vol. 15, no. 4, pp. 797–809, 2002.
29. S. Wakana, H. Jiang, L.M. Nagae-Poetscher et al., "Fiber tract-based atlas of human white matter anatomy," *Radiology*, vol. 230, no. 1, pp. 77–87, 2004.

30. B.J. Jellison, A.S. Field, J. Medow et al., "Diffusion tensor imaging of cerebral white matter: A pictorial review of physics, fiber tract anatomy, and tumor imaging patterns," *AJNR Am J Neuroradiol*, vol. 25, no. 3, pp. 356–369, 2004.

31. S. Mori, K. Fredericksen, P.C. van Zijl et al., "Brain white matter anatomy of tumor patients using diffusion tensor imaging," *Ann Neurol*, vol. 51, no. 3, pp. 377–380, 2002.

32. L. Bello, A. Castellano, E. Fava et al., "Intraoperative use of diffusion tensor imaging fiber tractography and subcortical mapping for resection of gliomas: Technical considerations," *Neurosurg. Focus*, vol. 28, no. 2, pp. E6, 2010.

33. K.M. Hasan, D.L. Parker, A.L. Alexander, "Comparison of gradient encoding schemes for diffusion-tensor MRI," *J Magn Reson Imaging*, vol. 13, no. 5, pp. 769–780, 2001.

34. D.K. Jones, "The effect of gradient sampling schemes on measures derived from diffusion tensor MRI: A Monte Carlo study," *Magn Reson Med*, vol. 51, no. 4, pp. 807–815, 2004.

35. P.G. Nucifora, R. Verma, S.K. Lee et al., "Diffusion tensor MR imaging and tractography: Exploring brain microstructure and connectivity," *Radiology*, vol. 245, no. 2, pp. 367–384, 2007.

36. A. Gupta, A. Shah, R.J. Young et al., "Imaging of brain tumors: Functional magnetic resonance imaging and diffusion tensor imaging," *Neuroimaging Clin N Am*, vol. 20, no. 3, pp. 379–400, 2010.

37. H.H. Batjer, C.M. Loftus, *Textbook of Neurological Surgery. Principles and Practice: Volume Two*, Lippincott, Williams & Wilkins, Chapter 102, pp. 1257–1270, 2003.

38. S.J. Price, "Advances in imaging low grade gliomas," *Advances and Technical Standards in Neurosurgery*, Springer, vol. 35, pp. 1–34, 2010.

39. G.G. Fan, Q.L. Deng, Z.H. Wu et al., "Usefulness of diffusion/perfusion-weighted MRI in patients with non-enhancing supratentorial brain gliomas: A valuable tool to predict tumor grading?," *Br J Radiol*, vol. 79, no. 944, pp. 652–658, 2006.

40. F. Yamasaki, K. Kurisu, K. Satoh et al., "Apparent diffusion coefficient of human brain tumors at MR imaging," *Radiology*, vol. 235, no. 3, pp. 985–991, 2005.

41. P. Zonari, P. Baraldi, G. Crisi, "Multimodal MRI in the characterization of glial neoplasms: The combined role of single-voxel MR spectroscopy, diffusion imaging and echo-planar perfusion imaging," *Neuroradiology*, vol. 49, no. 10, pp. 795–803, 2007.

42. L. Rizzo, S.G. Crasto, P.G. Moruno et al., "Role of diffusion- and perfusion-weighted MR imaging for brain tumor characterization," *Radiol Med*, vol. 114, no. 4, pp. 645–659, 2009.

43. D. Pauleit, K.J. Langen, F. Floeth et al., "Can the apparent diffusion coefficient be used as a noninvasive parameter to distinguish tumor tissue from peritumoral tissue in cerebral gliomas?," *J Magn Reson Imaging*, vol. 20, no. 5, pp. 758–764, 2004.

44. W.W. Lam, W.S. Poon, C. Metreweli, "Diffusion MR imaging in glioma: Does it have any role in the pre-operation determination of grading of glioma?," *Clin Radiol*, vol. 57, no. 3, pp. 219–225, 2002.

45. K. Kono, Y. Inoue, K. Nakayama et al., "The role of diffusion-weighted imaging in patients with brain tumors," *AJNR Am J Neuroradiol*, vol. 22, no. 6, pp. 1081–1088, 2001.

46. A. Stadlbauer, O. Ganslandt, R. Buslei et al., "Gliomas: Histopathologic evaluation of changes in directionality and magnitude of water diffusion at diffusion-tensor MR imaging," *Radiology*, vol. 240, no. 3, pp. 803–810, 2006.

47. A. Tropine, G. Vucurevic, P. Delani et al., "Contribution of diffusion tensor imaging to delineation of gliomas and glioblastomas," *J Magn Reson Imaging*, vol. 20, no. 6, pp. 905–912, 2004.

48. T. Inoue, K. Ogasawara, T. Beppu et al., "Diffusion tensor imaging for preoperative evaluation of tumor grade in gliomas," *Clin Neurol Neurosurg*, vol. 107, no. 3, pp. 174–180, 2005.

49. H.Y. Lee, D.G. Na, I.C. Song et al., "Diffusion-tensor imaging for glioma grading at 3-T magnetic resonance imaging: Analysis of fractional anisotropy and mean diffusivity," *J Comput Assist Tomogr*, vol. 32, no. 2, pp. 298–303, 2008.

50. T. Beppu, T. Inoue, Y. Shibata et al., "Measurement of fractional anisotropy using diffusion tensor MRI in supratentorial astrocytic tumors," *J Neurooncol*, vol. 63, no. 2, pp. 109–116, 2003.

51. E. Goebell, S. Paustenbach, O. Vaeterlein et al., "Low-grade and anaplastic gliomas: Differences in architecture evaluated with diffusion-tensor MR imaging," *Radiology*, vol. 239, no. 1, pp. 217–222, 2006.

52. Y. Chen, Y. Shi, Z. Song, "Differences in the architecture of low-grade and high-grade gliomas evaluated using fiber density index and fractional anisotropy," *J Clin Neurosci*, vol. 17, no. 7, pp. 824–829, 2010.

53. J. Ferda, J. Kastner, P. Mukensnabl et al., "Diffusion tensor magnetic resonance imaging of glial brain tumors," *Eur J Radiol*, vol. 74, no. 3, pp. 428–436, 2010.

54. A. Server, B. Graff, R. Josefsen, T. Orheim, T. Schellhorn, W. Nordhoy, P. Nakstad. "Analysis of diffusion tensor imaging metrics for gliomas grading at 3 T," *Eur J Radiol*, vol. 83, pp. e156–e165, 2014.

55. A. Perry, D.N. Louis, B.W. Scheithauer et al., "Meningiomas." In: D. N. Louis, H. Ohgaki, O. D. Wiestler et al., eds. *WHO Classification of Tumours of the Central Nervous System*, Lyon, France: IARC Press; pp. 164–72, 2007.

56. V.A. Nagar, J.R. Ye, W.H. Ng et al., "Diffusion-weighted MR imaging: Diagnosing atypical or malignant meningiomas and detecting tumor dedifferentiation," *AJNR Am J Neuroradiol*, vol. 29, no. 6, pp. 1147–1152, 2008.

57. C.H. Toh, M. Castillo, A.M. Wong et al., "Differentiation between classic and atypical meningiomas with use of diffusion tensor imaging," *AJNR Am J Neuroradiol*, vol. 29, no. 9, pp. 1630–1635, 2008.

58. C.G. Filippi, M.A. Edgar, A.M. Ulug et al., "Appearance of meningiomas on diffusion-weighted images: Correlating diffusion constants with histopathologic findings," *AJNR Am J Neuroradiol*, vol. 22, no. 1, pp. 65–72, 2001.

59. M.P. Buetow, P.C. Buetow, J.G. Smirniotopoulos, "Typical, atypical, and misleading features in meningioma," *RadioGraphics*, vol. 11, no. 6, pp. 1087–1106, 1991.

60. D. Ellison, S. Love, L. Chimelli et al., "Meningiomas." In: *Neuropathology: A Reference Text of CNS Pathology*, Edinburgh: Mosby, pp. 703–16, 2004.

61. H. Zhang, L.A. Rodiger, T. Shen et al., "Perfusion MR imaging for differentiation of benign and malignant meningiomas," *Neuroradiology*, vol. 50, no. 6, pp. 525–530, 2008.

62. A. Tropine, P.D. Dellani, M. Glaser et al., "Differentiation of fibroblastic meningiomas from other benign subtypes using diffusion tensor imaging," *J Magn Reson Imaging*, vol. 25, no. 4, pp. 703–708, 2007.

63. D.T. Ginat, R. Mangla, G. Yeaney et al., "Correlation of diffusion and perfusion MRI with Ki-67 in high-grade meningiomas," *AJR Am J Roentgenol*, vol. 195, no. 6, pp. 1391–1395, 2010.

64. L. Santelli, G. Ramondo, A. Della Puppa et al., "Diffusion-weighted imaging does not predict histological grading in meningiomas," *Acta Neurochir*, vol. 152, no. 8, pp. 1315–1319, 2010.

65. G. Pavlisa, M. Rados, L. Pazanin et al., "Characteristics of typical and atypical meningiomas on ADC maps with respect to schwannomas," *Clin Imaging*, vol. 32, no. 1, pp. 22–27, 2008.

66. J.M. Provenzale, P. McGraw, P. Mhatre et al., "Peritumoral brain regions in gliomas and meningiomas: Investigation with isotropic diffusion-weighted MR imaging and diffusion-tensor MR imaging," *Radiology*, vol. 232, no. 2, pp. 451–460, 2004.

67. D. van Westen, J. Latt, E. Englund et al., "Tumor extension in high-grade gliomas assessed with diffusion magnetic resonance imaging: Values and lesion-to-brain ratios of apparent diffusion coefficient and fractional anisotropy," *Acta Radiol*, vol. 47, no. 3, pp. 311–319, 2006.

68. C.H. Toh, A.M. Wong, K.C. Wei et al., "Peritumoral edema of meningiomas and metastatic brain tumors: Differences in diffusion characteristics evaluated with diffusion-tensor MR imaging," *Neuroradiology*, vol. 49, no. 6, pp. 489–494, 2007.

69. R. Sawaya, "Considerations in the diagnosis and management of brain metastases," *Oncology*, vol. 15, no. 9, pp. 1144–1154, 2001.

70. R.A. Patchell, "Brain metastases," *Neurol Clin.*, vol. 9, no. 4, pp. 817–827, 1991.

71. S.K. Lee, "Diffusion tensor and perfusion imaging of brain tumors in high-field MR imaging," *Neuroimaging Clin N Am*, vol. 22, no. 2, pp. 123–134, 2012.

72. I.C. Chiang, Y.T. Kuo, C.Y. Lu et al., "Distinction between high-grade gliomas and solitary metastases using peritumoral 3-T magnetic resonance spectroscopy, diffusion, and perfusion imagings," *Neuroradiology*, vol. 46, no. 8, pp. 619–627, 2004.

73. M. Bertossi, D. Virgintino, E. Maiorano et al., "Ultrastructural and morphometric investigation of human brain capillaries in normal and peritumoral tissues," *Ultrastruct Pathol*, vol. 21, no. 1, pp. 41–49, 1997.

74. G. Pavlisa, M. Rados, L. Pavic et al., "The differences of water diffusion between brain tissue infiltrated by tumor and peritumoral vasogenic edema," *Clin Imaging*, vol. 33, no. 2, pp. 96–101, 2009.

75. S. Wang, S. Kim, S. Chawla et al., "Differentiation between glioblastomas and solitary brain metastases using diffusion tensor imaging," *NeuroImage*, vol. 44, no. 3, pp. 653–660, 2009.

76. E.J. Lee, K. terBrugge, D. Mikulis et al., "Diagnostic value of peritumoral minimum apparent diffusion coefficient for differentiation of glioblastoma multiforme from solitary metastatic lesions," *AJR Am J Roentgenol*, vol. 196, no. 1, pp. 71–76, 2011.

77. D.A. Altman, D.S. Atkinson Jr., D.J. Brat, "Best cases from the AFIP: Glioblastoma multiforme," *RadioGraphics*, vol. 27, no. 3, pp. 883–888, 2007.

78. B. Hakyemez, C. Erdogan, N. Bolca et al., "Evaluation of different cerebral mass lesions by perfusion-weighted MR imaging," *J Magn Reson Imaging*, vol. 24, no. 4, pp. 817–824, 2006.

79. T.S. Surawicz, B.J. McCarthy, V. Kupelian et al., "Descriptive epidemiology of primary brain and CNS tumors: Results from the Central Brain Tumor Registry of the United States, 1990–1994," *Neuro Oncol*, vol. 1, no.1, pp. 14– 25, 1999.

80. J.L. Go, S.C. Lee, P.E Kim, "Imaging of primary central nervous system lymphoma," *Neurosurg Focus*, vol. 21, no. 5, pp. E4, 2006.

81. B. Bataille, V. Delwail, E. Menet et al., "Primary intracerebral malignant lymphoma: Report of 248 cases," *J Neurosurg*, vol. 92, no. 2, pp. 261–266, 2000.

82. A.C. Guo, T.J. Cummings, R.C Dash et al., "Lymphomas and high-grade astrocytomas: Comparison of water diffusibility and histologic characteristics," *Radiology*, vol. 224, no. 1, pp. 177–183, 2002.

83. C.H. Toh, M. Castillo, A.M. Wong et al., "Primary cerebral lymphoma and glioblastoma multiforme: Differences in diffusion characteristics evaluated with diffusion tensor imaging," *AJNR Am J Neuroradiol*, vol. 29, no. 3, pp. 471–475, 2008.

84. M. Kinoshita, N. Hashimoto, T. Goto et al., "Fractional anisotropy and tumor cell density of the tumor core show positive correlation in diffusion tensor magnetic resonance imaging of malignant brain tumors," *NeuroImage*, vol. 43, no. 1, pp. 29–35, 2008.

85. T.W. Stadnik, C. Chaskis, A. Michotte et al., "Diffusion-weighted MR imaging of intracerebral masses: Comparison with conventional MR imaging and histologic findings," *AJNR Am J Neuroradiol*, vol. 22, no. 5, pp. 969–976, 2001.

86. N. Rollin, J. Guyotat, N. Streichenberger et al., "Clinical relevance of diffusion and perfusion magnetic resonance imaging in assessing intra-axial brain tumors," *Neuroradiology*, vol. 48, no. 3, pp. 150–159, 2006.

87. M. Bendini, E. Marton, A. Feletti et al., "Primary and metastatic intraaxial brain tumors: prospective comparison of multivoxel 2D chemical-shift imaging (CSI) proton MR spectroscopy, perfusion MRI, and histopathological findings in a group of 159 patients," *Acta Neurochir*, vol. 153, no. 2, pp. 403–412, 2011.

88. E. Grigoriadis, W.L. Gold, "Pyogenic brain abscess caused by *Streptococcus pneumoniae*: Case report and review," *Clin Infect Dis*, vol. 25, no. 5, pp. 1108–12, 1997.

89. J.H. Chan, E.Y. Tsui, L.F. Chau et al., "Discrimination of an infected brain tumor from a cerebral abscess by combined MR perfusion and diffusion imaging," *Comput Med Imaging Graph*, vol. 26, no. 1, pp. 19–23, 2002.

90. A.C. Guo, J.M. Provenzale, L.C. Cruz Jr. et al., "Cerebral abscesses: Investigation using apparent diffusion coefficient maps," *Neuroradiology*, vol. 43, no. 5, pp. 370–374, 2001.

91. M. Hartmann, O. Jansen, S. Heiland et al., "Restricted diffusion within ring enhancement is not pathognomonic for brain abscess," *AJNR Am J Neuroradiol*, vol. 22, no. 9, pp. 1738–1742, 2001.

92. P.H. Lai, J.T. Ho, W.L. Chen et al., "Brain abscess and necrotic brain tumor: Discrimination with proton MR spectroscopy and diffusion-weighted imaging," *AJNR Am J Neuroradiol*, vol. 23, no. 8, pp. 1369–1377, 2002.

93. L. Nadal-Desbarats, S. Herlidou, G. de Marco et al., "Differential MRI diagnosis between brain abscesses and necrotic or cystic brain tumors using the apparent diffusion coefficient and normalized diffusion-weighted images," *Magn Reson Imaging*, vol. 21, no. 6, pp. 645–650, 2003.

94. K. Nath, M. Agarwal, M. Ramola et al., "Role of diffusion tensor imaging metrics and in vivo proton magnetic resonance spectroscopy in the differential diagnosis of cystic intracranial mass lesions," *Magn Reson Imaging*, vol. 27, no. 2, pp. 198–206, 2009.

95. W. Reiche, V. Schuchardt, T. Hagen et al., "Differential diagnosis of intracranial ring enhancing cystic mass lesions: Role of diffusion-weighted imaging (DWI) and diffusion-tensor imaging (DTI)," *Clin Neurol Neurosurg*, vol. 112, no. 3, pp. 218–225.

96. R.K. Gupta, K.M. Hasan, A.M. Mishra et al., "High fractional anisotropy in brain abscesses versus other cystic intracranial lesions," *AJNR Am J Neuroradiol*, vol. 26, no. 5, pp. 1107–14, 2005.

97. P.D. Schellinger, J.B. Fiebach, O. Jansen et al., "Stroke magnetic resonance imaging within 6 hours after onset of hyperacute cerebral ischemia," *Ann Neurol*, vol. 49, no. 4, pp. 460–469, 2001.

98. M. Hermier, N. Nighoghossian, P. Adeleine et al., "Early magnetic resonance imaging prediction of arterial recanalization and late infarct volume in acute carotid artery stroke," *J Cereb Blood Flow Metab*, vol. 23, no. 2, pp. 240–248, 2003.

99. M.R. Borich, K.P. Wadden, L.A. Boyd, "Establishing the reproducibility of two approaches to quantify white matter tract integrity in stroke," *NeuroImage*, vol. 59, no. 3, pp. 2393–2400, 2012.

100. C.H. Park, N. Kou, M.H. Boudrias, E.D. Playford, N.S. Ward, "Assessing a standardised approach to measuring corticospinal integrity after stroke with DTI," *NeuroImage Clin*, vol. 2, pp. 521–533, 2013.

101. L. Xia, S. Lin, Z. Wang, S. Li, L. Xu, J. Wu, S. Hao, C. Gao, "Tumefactive demyelinating lesions: Nine cases and a review of the literature," *Neurosurg Rev*, vol. 32, pp. 171–179, 2009.

102. A.C. Guo, J.R. MacFall, J.M. Provenzale, "Multiple sclerosis: Diffusion tensor MR imaging for evaluation of normal appearing white matter," *Radiology*, vol. 222, pp. 729–736, 2002.

103. A.L. Tievsky, T. Ptak, J. Farkas, "Investigation of apparent diffusion coefficient and diffusion tensor anisotropy in acute and chronic multiple sclerosis lesion," *AJNR Am J Neuroradiol*, vol. 20, pp. 1491–1499, 1999.

104. S. Roychowdhury, J.A. Maldjian, R.I. Grossman, "Multiple sclerosis: Comparison of trace apparent diffusion coefficients with MR enhancement pattern of lesions," *AJNR Am J Neuroradiol*, vol. 21, pp. 869–874, 2000.

105. A. Castriota Scanderbeg, F. Tomaiuolo, U. Sabatini, U. Nocentini, M.G. Grasso, C. Caltagirone, "Demyelinating plaques in relapsing-remitting and secondary-progressive multiple sclerosis: Assessment with diffusion MR imaging," *AJNR Am J Neuroradiol*, vol. 21, pp. 862–868, 2000.

106. L. Wald, F. Schmitt, A. Dale, "Systematic spatial distortion in MRI due to gradient non-linearities," *NeuroImage*, vol. 13, p. S50, 2001.

107. P. Svolos, E. Tsolaki, K. Theodorou et al., "Investigating brain tumor differentiation with diffusion and perfusion metrics at 3T MRI using pattern recognition techniques," *Magn Reson Imaging*, vol. 31, no. 9, pp. 1567–1577, 2013.

108. P. Svolos, E. Kousi, E. Kapsalaki, K. Theodorou, I. Fezoulidis, C. Kappas, I. Tsougos, "The role of diffusion and perfusion weighted imaging in the differential diagnosis of cerebral tumors: A review and future perspectives," *Cancer Imaging*, vol. 14, pp. 20, 2014.

109. M. Law, S. Yang, H. Wang et al., "Glioma grading: Sensitivity, specificity, and predictive values of perfusion MR imaging and proton MR spectroscopic imaging compared with conventional MR imaging," *AJNR Am J Neuroradiol*, vol. 24, no. 10, pp. 1989–1998, 2003.

110. P. Georgiadis, S. Kostopoulos, D. Cavouras et al., "Quantitative combination of volumetric MR imaging and MR spectroscopy data for the discrimination of meningiomas from metastatic brain tumors by means of pattern recognition," *Magn Reson Imaging*, vol. 29, no. 4, pp. 525–535, 2011.

111. E.I. Zacharaki, V.G. Kanas, C. Davatzikos, "Investigating machine learning techniques for MRI-based classification of brain neoplasms," *Int J Comput Assist Radiol Surg*, vol. 6, no. 6, pp. 821–828, 2011.

112. A. Devos, A.W. Simonetti, M. van der Graaf et al., "The use of multivariate MR imaging intensities versus metabolic data from MR spectroscopic imaging for brain tumour classification," *J Magn Reson*, vol. 173, no. 2, pp. 218–228, 2005.

113. E. Tsolaki, P. Svolos, E. Kousi et al, "Automated differentiation of glioblastomas from intracranial metastases using 3 T MR spectroscopic and perfusion data," *Int J Comput Assist Radiol Surg*, vol. 8, no. 5, pp. 751–761, 2013.

114. X. Hu, K.K. Wong, G.S. Young et al., "Support vector machine multiparametric MRI identification of pseudoprogression from tumor recurrence in patients with resected glioblastoma," *J Magn Reson Imaging*, vol. 33, no. 2, pp. 296–305, 2011.

115. R. Verma, E.I. Zacharaki, Y. Ou et al., "Multiparametric tissue characterization of brain neoplasms and their recurrence using pattern classification of MR images," *Acad Radiol*, vol. 15, no. 8, pp. 966–977, 2008.

116. E.I. Zacharaki, S. Wang, S. Chawla et al, "Classification of brain tumor type and grade using MRI texture and shape in a machine learning scheme," *Magn Reson Med*, vol. 62, no. 6, pp. 1609–1618, 2009.

117. L. Blanchet, P.W. Krooshof, G.J. Postma et al., "Discrimination between metastasis and glioblastoma multiforme based on morphometric analysis of MR images," *AJNR Am J Neuroradiol*, vol. 32, no. 1, pp. 67–73, 2011.

5

Arterial Spin-Labeled Perfusion Imaging

Hongjian He and Jianhui Zhong

CONTENTS

In physiology, the term *perfusion* refers to the process of blood delivery into a capillary bed in a biological tissue. A healthy adult with a body weight of 60 kg has approximately 4.2–4.8 L of blood on average, which is approximately 6%–8% of the human body weight. Blood circulates within the body through blood vessels and delivers necessary substances such as nutrients and oxygen to the cells and also washes metabolic waste product away from tissue. Differentiated from the function and components, arterial blood carries oxygen and venous blood carries carbon dioxides. The human brain gets 20% of the total blood flow, even though it is only 2% of the body weight. The amount of blood supply guarantees the energy consumption needed by this most important organ of the body.

In magnetic resonance imaging (MRI), perfusion is usually termed as *cerebral blood flow* (CBF) and is expressed as the amount of arterial blood delivered to a local volume of tissue per unit time, in units of milliliter of blood per 100 g of tissue per minute (mL/100 g/min).

Perfusion measurements provide information about tissue viability and function and are therefore of fundamental significance in neuroscience, medical research, and clinical diagnostics. In particular, local perfusion changes reflect regional cerebral activity and metabolism and thus can be used as an index for mapping functional neuroanatomy (Yang, 2002). Abnormal state of perfusion could be related with brain disorders, such as stroke, ischemia, and tumor. It is therefore important to measure brain perfusion.

There have been various clinical methods available for investigating perfusion in disease, such as computed tomography (CT), nuclear medicine with positron-emission tomography (PET), and single-photon emission CT (SPECT). These methods have been validated in a variety of neurologic disorders, including stroke, tumor, and seizure (Wintermark et al., 2005). Their disadvantages are invasive, high cost, with radiation, and not suitable for repeated measurements. Therefore, MRI-based perfusion imaging techniques potentially provide attractive alternatives because of their noninvasive nature, repeatability, availability of complementary anatomic and functional MRI (fMRI) scans, and superior temporal and spatial resolution compared to PET and SPECT (Wang et al., 2011). Currently, two classes of techniques have been widely used in clinical MRI to measure resting perfusion state: dynamic susceptibility contrast (DSC) and arterial spin labeling (ASL).

DSC relies on the measurement of T_2- or T_2^*-related signal decrease during the first pass of an exogenous endovascular tracer through the capillary bed. The signal contrast comes from magnetic susceptibility change of blood altered by exogenous endovascular tracer. Using commercially available software, various perfusion-related parameters can be derived in each voxel, including CBF, cerebral blood volume (CBV), and vascular mean transit time (MTT). These indexes do not afford quantification assessment of brain hemodynamics but provide indicators of hemodynamic disturbances, which are useful in clinical setting (Wintermark et al., 2005).

The second class is an ASL technique. The goal of this chapter is to review this technique, draw a brief picture of its basis, and give examples of its applications.

5.1 Introduction to ASL

Unlike the DSC method, the ASL technique (Detre et al., 1992; Williams et al., 1992) uses magnetization of water protons in the arterial blood stream as an endogenous, freely diffusible tracer for perfusion measurements. It allows for measuring regional CBF without injecting any exogenous contrast agent, and is completely noninvasive and repeatable. This is very suitable for perfusion studies of healthy participants and in patient groups requiring longitudinal investigations (Wang et al., 2011). It also bypasses the concerns regarding nephrogenic systemic fibrosis (NSF) in patients with significant renal insufficiency, because it is performed without gadolinium (Deibler et al., 2008), or in pediatric populations, where the use of radioactive tracers or exogenous contrast agents may be restricted (Petersen et al., 2006).

The principle of ASL relies on magnetically labeled blood in the brain-feeding arteries as endogenous tracer. In an ASL sequence, the longitudinal magnetization of the water in arterial blood is first flipped by a 180° radiofrequency (RF) inversion pulse, and a volume of blood is tagged. After a given sufficiently delayed inversion time (TI), the tagged blood reaches at the issue plane of interest, known as the *image slice*. Those labeled water molecules, in proportion to the local CBF, flow into each tissue element in this slice. These water molecules exchange (both in physical position and in magnetization) with the

water molecules in the tissue capillary bed, which causes a change in longitudinal magnetization, resulting in a change in the nuclear magnetic resonance signal. At this point, an acquisition pulse sequence is played out and a tag image is obtained, which contains signal from both labeled water and static tissue water (Bernstein et al., 2004). Another control image is necessary to calculate perfusion parameters (Figure 5.1). This is simply done by repeating the above experiment, but without prior labeling of the arterial blood. The subtraction difference between control (M_C) and label (M_L) image is perfusion weighted. If the tag and control images are carefully done, the signal from static spins subtracts out in the difference image, leaving just the signal difference of arterial blood. On the contrary, the arterial blood signal was fully relaxed in the control image, but inverted in the tag image. Therefore, the blood signal does not subtract out, and the resulting ASL difference image $|M_C - M_L|$ can be viewed as a qualitative perfusion-weighted image. The difference is directly proportional to how much arterial blood was delivered during the interval of TI (Figure 5.2). This labeled blood contributes to the image signal with a very small ratio, approximately 1%–2% of the image magnitude. It is hard to directly differentiate this signal from image noise with a single pair of tag/control images. In a conventional ASL scan, 20–40 pairs of control and label maps are acquired and averaged to achieve sufficient signal-to-noise ratio (SNR). In this way, one obtains a perfusion-weighted image with high spatial and temporal resolution, and can also quantify the CBF with a priori knowledge of the tissue relaxation times and the set of scan parameters.

A basic ASL pulse sequence usually has two different elements: the labeling or tagging and the image

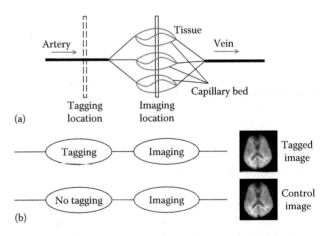

FIGURE 5.1
The principal of arterial spin tagging (a) and an associated conceptual pulse sequence. (a) Arterial blood is labeled (dashed box), flows into the imaging plan (solid box) in tissue site, and then drained into the vein. (b) Both tagged image and untagged control image are required to calculate the ASL signal. (Modified from Bernstein, M.A., King, K.F., Zhou, X.J., 2004. *Handbook of MRI Pulse Sequences*. Amsterdam, the Netherlands: Elsevier Academic Press. With permission.)

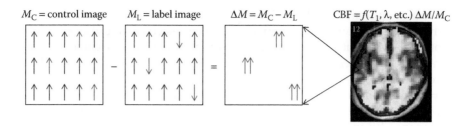

M_C = control image M_L = label image $\Delta M = M_C - M_L$ CBF = $f(T_1, \lambda,$ etc.$)$ $\Delta M/M_C$

FIGURE 5.2

Basic principle of ASL difference signal. The difference between a control image and labeled image with inverted blood water is proportional to CBF. This difference signal can be converted to CBF value with a function of physiological and MR parameters. (Modified from Borogovac, A. and Asllani, I., *Int. J. Biomed. Imaging*, 2012, 1–13, 2012.)

readout (Figure 5.1b). They are separated by a time interval to allow the labeled blood to enter the imaged slices. Because ASL uses a subtractive scheme, the tagging pulse is a principal module to obtain tag and control images with high quality. Till now, there have been four major alternative strategies for arterial spin tagging: continuous labeling, pseudocontinuous labeling, pulsed labeling, and velocity-selective labeling. The first two labeling approaches, continuous and pseudocontinuous, constitute long label scenarios and are usually grouped together as appeared in the literature (Alsop et al., 2013; Bernstein et al., 2004; Wong, 2014). The pulsed labeling method is fundamentally different from continuous labeling in both the spatial extent and the duration of the tagging (Figure 5.3), and these differences give rise to the strengths and weaknesses of each approach (Alsop et al., 2013).

5.1.1 Continuous ASL

Historically, continuous ASL (CASL) was developed before other techniques (Detre et al., 1992; Williams et al., 1992). In this approach, the inversion occurs at a thin plane referred to as the *labeling plane*, generally positioned in the carotids. The inversion labeling pulse is applied over the labeling period, which usually lasts for as long as 1–3 s. As blood flows through the labeling plane, water spins are *continuously* inversed by an effective continuous and constant RF pulse (Figure 5.3). This process is called *adiabatic fast passage* (AFP) or *flow-driven adiabatic inversion* (Alsop et al., 2013). For adiabatic inversion to occur, both *fast* and *adiabatic* are two main conditions that must be met: first, the entire labeling process needs to be faster than the relaxation times; second, the orientation of the effective magnetic field (B_{eff}) needs to change at sufficiently slow rate so that the angle between B_{eff} and the net magnetization remains constant. The average flow velocity in tagging plane in the carotids ensures that both adiabatic conditions are met. Theoretically, the labeling pulse must be long enough to reach a steady state. However, taking into consideration both hardware and experimental restrictions, the labeling pulse is typically about 2 s (Borogovac and Asllani, 2012). CASL was originally implemented for human use as a single-slice technique but was later extended to multislice imaging (Alsop et al., 2013; Kimura et al., 2005).

One of the major drawbacks of CASL is the requirement for a long labeling pulse to bring about the

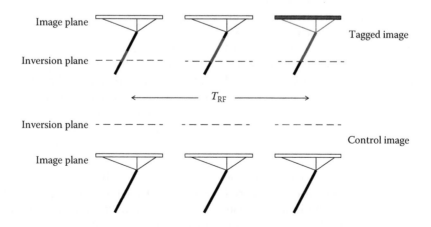

Image plane

Inversion plane

Tagged image

T_{RF}

Inversion plane

Control image

Image plane

FIGURE 5.3

Continuous arterial spin labeling. The magnetization of blood is labeled at the inversion plane via a continuous RF pulse (above). However, in the control procedure, the labeling pulse is not applied (or applied above the image slice as shown) so that no arterial blood is tagged. (From Buxton, R.B., *Introduction to Functional Magnetic Resonance Imaging: Principles and Techniques*, Cambridge University Press, New York, 2009. With permission.)

adiabatic inversion. This requirement allows brain magnetization to reach a steady state, which maximizes the signal difference of $|M_C - M_L|$, but at the expense of both magnetization transfer (MT) effects and increased specific absorption rate (SAR).

Protons associated with macromolecules would have a broad-frequency spectrum, and can also become saturated by a long off-resonance RF pulses, that is, up to several kilohertz. These spins transfer saturation to MRI-visible water protons, thereby causing a reduction in MR signal, which is commonly referred as the *MT effect*. This effect would not affect control image, as tagging pulse is not necessary there. Consequently, the ASL difference signal would reflect not only the blood flow but also the loss in signal due to this MT effect. In CASL, a relatively weak gradient is used during labeling, and the RF pulse, which is on-resonance at the labeling plane, is a few kilohertz off-resonance in the imaging volume, leading to MT effects that could be much larger than the ASL signal. The use of control pulses has been recommended to minimize MT effects, such as an amplitude-modulated RF pulse with the same duration as the labeling pulse during the acquisition of the control image (Figure 5.3) (Alsop and Detre, 1998). However, these strategies are imperfect and lead to a significant decrease in labeling efficiency (Wong, 2014).

In CASL, the extremely long pulse width (e.g., several seconds) needed to achieve a continuous arterial spin inversion often exceeds the RF amplifier capability or FDA regulatory limits on SAR (Bernstein et al., 2004). One way to reduce global SAR is to use a second transmit coil placed over the neck to label the carotid and vertebral arteries, and specifically dedicated for labeling. Since labeling is achieved independently from imaging in this way, the MT effects are completely avoided (Talagala et al., 2004). However, this approach does not avoid local SAR, and the requirement of an additional hardware makes it difficult for routine applications. The approach is still of particular interest at ultrahigh static field strengths such as 7 T. On clinical 3- and 1.5-T scanners, pseudo-CASL (pCASL) uses a series of short and shaped pulses separated by a time delay, which is mostly preferred (Alsop et al., 2013; Borogovac and Asllani, 2012; Wong, 2014).

5.1.2 Pseudocontinuous ASL

As mentioned in Section 5.1.1, pCASL is a modified form of ASL to address some of the weaknesses of the CASL approach. The main difference when compared to CASL is that instead of a long continuous pulse, pCASL employs a long series of short RF and large gradient pulses as shown in Figure 5.4. When the gradient and RF pulses are adjusted such that the mean values over time are similar to those of conventional CASL, the

pulse train results in a flow-driven adiabatic inversion. A slice-selective RF pulse, typically Hamming shaped, is applied to provide excitation at the labeling plane, and the gradient amplitude during the RF pulse is approximately 10 times that of a CASL experiment (Wong, 2014). Larger gradients can increase the resonance offset of the pulses relative to brain tissue, and thereby decrease MT effects and increase labeling efficiency (Alsop et al., 2013).

The first immediate advantage of pCASL is that it requires no additional hardware to generate continuous low-level RF demanded by CASL, and therefore can be implemented on unmodified clinical scanners with the standard RF coils. Second, MT effects are nearly eliminated in pCASL. This is ascribed to the much larger gradient amplitude during tagging period.

Several variants of pCASL have been proposed. Some of them provide images with better SNR, while others may provide more vascular information. For instance, one approach takes advantage of the time gaps between RF pulses, and applies gradient pulses during the labeling pulse train to modulate the labeling across vessels within the labeling plane. It could obtain the mapping of vascular territories in this manner (Helle et al., 2010; 2013; Wong, 2007).

One disadvantage of pCASL is about its reduced labeling efficiency. pCASL does not use an adiabatic inversion strictly in the tagging process, and is more sensitive to resonance offsets in the labeling plane than CASL. This could affect the quality of perfusion images and further perfusion quantification. Estimation and correction approaches have been introduced using modified pCASL labeling and postprocessing methods (Alsop et al., 2013; Aslan et al., 2010; Wong, 2014; Wu et al., 2007). A detailed recommendation for the implementation of pCASL can be found in a recent well-documented review (Alsop et al., 2013).

5.1.3 Pulsed ASL

In contrast to CASL, pulsed ASL (PASL) uses a single RF pulse or short train of pulses to rapidly invert the magnetization of a thick slab of tissue, referred to as the *inversion slab*, which contains feeding arteries. The inversion happens quickly with a total duration of approximately 10–20 ms, and is usually implemented using a slab-selective adiabatic inversion pulse (Figure 5.5) (Edelman et al., 1994).

PASL is a family of pulse sequences differing in the strategies by which the tagged and control images are acquired. But overall, the methods share the same basic principles (Bernstein et al., 2004). The differences involve implementation of the label and control RF pulses, the manner in which spins distal to the imaging region are labeled, and the image contrast in the static tissue signal

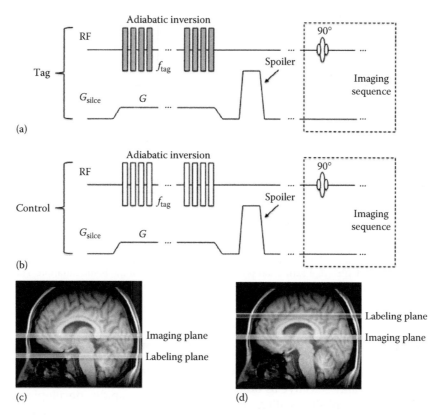

FIGURE 5.4
Pulse sequence for pCASL. Very similar to CASL, a pair of tag (c) and control (d) images is acquired using tag (a) and control pulses (b), respectively. The inversion process is consisted of a long series of short RF and large gradient pulse. (From Bernstein, M.A. et al., *Handbook of MRI Pulse Sequences*, Elsevier Academic Press, Amsterdam, the Netherlands, 2004. With permission.)

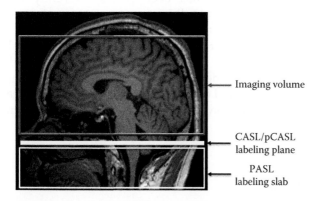

FIGURE 5.5
Comparison of the labeling region between PASL, CASL, and pCASL. PASL inverts the magnetization of a thick slab of tissue that contains feeding arteries. (From Alsop, D.C. et al., *Magn. Reson. Med.*, 73, 102–116, 2013. With permission.)

(Alsop et al., 2013; Wong, 2014). As a brief comparison with these variants, inflow from vessels above the imaging slab may produce ASL signal differently: positive signal for flow-sensitive alternating inversion recovery (FAIR), negative signal for echo-planar imaging (EPI) and signal targeting with alternating RF (EPISTAR), and no signal for proximal inversion with a control for off-resonance effects (PICORE) (Alsop et al., 2013) [4]. In order to achieve efficient inversion, RF with adiabatic inversion pulses are used to generate an effective field that rotates from +Z to –Z. Distinct from CASL, the frequency sweep is implemented by the pulse itself in this case, rather than being dependent upon flow along a gradient.

The inversion efficiency for PASL is high and consistent, while its SAR is low relative to other ASL techniques. However, the SNR of the PASL approach is fundamentally lower. In addition, the size of labeling slab is limited by both the RF transmit coil for its spatial coverage and B1 homogeneity, and the transition zone between inverted and inverted blood. For brain PASL, approximately 15–20 cm in length has been recommended as a good compromise between these factors (Alsop et al., 2013; Wong, 2014).

Another drawback of PASL is that it creates a tagged bolus with unknown and relatively short temporal width (Alsop et al., 2013). It is possible and necessary to control and remove the tail portion of the labeled bolus in both label and control conditions by means of the QUIPSS II modification (Figure 5.6) (Wong et al., 1998a).

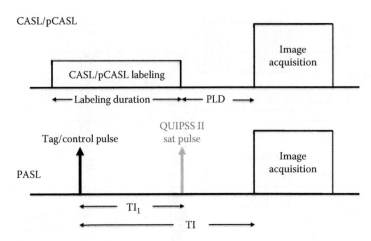

FIGURE 5.6

Timing diagram for the three ASL techniques. The delay time between the end of RF pulse and image acquisition is referred to as PLD in CASL/pCASL, but TI in PASL. For QUIPSS II, a saturation pulse is applied to control the bolus duration. The PLD in CASL/pCASL is analogous to the quantity (TI–TI_1). (From Alsop, D.C. et al., *Magn. Reson. Med.*, 73, 102–116, 2013. With permission.)

5.1.4 Velocity-Selective ASL

In velocity-selective ASL (VSASL), the tagging pulse is velocity selective, instead of spatially selective as other ASL tagging manners. This is accomplished with an RF and a gradient-pulse train that effectively dephases spins flowing faster than a specified cutoff velocity (V_c) while rephasing the signal from slower inflow (Figure 5.7). For a typical V_c of 1 cm/s, the sequence dephases spins in arterioles that are approximately 50 μm or more in diameter (Liu and Brown, 2007). This technique is introduced to address the issue in organs with very slow blood flow or in the presence of vascular disease, where transit delay time may be significantly longer than T_1 relaxation, and the SNR of image would be poor due to T_1 decay (Wong, 2014). VSASL is still relatively new and requires additional validation for routine clinical care (Alsop et al., 2013).

5.2 General Comparison

ASL is still a rapidly developing field, both in terms of technical innovation and applications. Given by the many variants of ASL, it is complex to define the best strategy among them. Studies have suggested that all above-mentioned tagging methods are capable of measuring CBF within a few minutes of scanning, while general consensus has been reached recently that pCASL gives the overall best performance and reproducibility (Aslan et al., 2010; Chen et al., 2011a; Gevers et al., 2011; Jahng et al., 2005; Wang et al., 2011; Wu et al., 2014). However, a realistic choice of study obviously involves the considerations of hardware, brain coverage, image quality, and disease. It is often based on practical requirement and availability in clinic rather than scientific consideration.

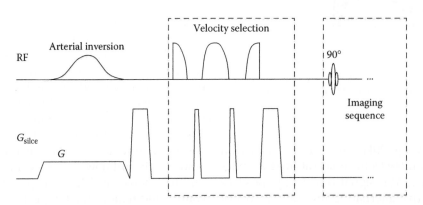

FIGURE 5.7

Basic principle of the velocity-selective technique.

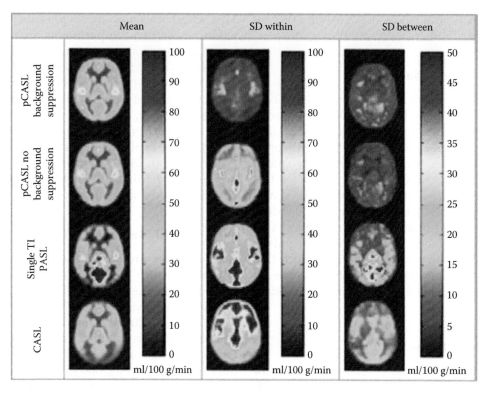

FIGURE 5.8

Mean perfusion images obtained using pCASL (with or without background suppression), PASL, and CASL. Their within session and between session SD maps are also presented to show regional perfusion variability. (From Gevers, S. et al., *J. Cerebr. Blood Flow Metab.*, 31, 1706–1715, 2011. With permission.)

In general, PASL has been more widely used because it is easier to implement and conceptually more straightforward. Due to its shorter labeling pulses, PASL is less affected by MT effect than the standard CASL. However, PASL has drawbacks, such as low SNR, high sensitivity to transit times, and slice profile artifacts, which can limit brain coverage. These factors should always draw great cautious approach (Borogovac and Asllani, 2012). CASL is more difficult to implement practically with a clinical hardware, but can achieve higher SNR for whole-brain imaging. pCASL takes advantages of CASL's superior SNR and PASL's high labeling efficiency. It is claimed that pCASL provides 50% improvement in SNR than PASL, and an 18% increase in labeling efficiency than CASL (Wu et al., 2007). A general comparison of perfusion images obtained with these techniques is shown in Figure 5.8, and a more detailed comparison between pulsed and continuous ASL techniques can be found in (Wong et al., 1998b). In addition, the potential of pCASL to selectively label vessels varying in size and orientation without compromising the SNR may prove invaluable in studies for disease diagnosis, progression, and treatment (Borogovac and Asllani, 2012). Due to these merits, pCASL labeling approach with background suppression and 3D RARE-based readout has been recommended as the optimal protocol set for a broad range of applications in both clinic and brain research in a recent article by a group of active researchers and clinicians worldwide in this field (Alsop et al., 2013).

5.3 Optimization and Quantification

5.3.1 Labeling Efficiency

Inversion pulse for labeling is an essential part for all ASL sequences. Flow quantification could be very sensitive to tagging efficiency, as pointed out by a theoretical analysis (Wu et al., 2010). In a condition of perfect inversion labeling, the difference in blood magnetization between the label and control states should be $2M_{0B}$, if measured immediately after the inversion pulse. The labeling efficiency is defined as the ratio between the actual magnetization difference and $2M_{0B}$, with a maximum value of 1. However, inversion pulses are imperfect, and other factors such as flow velocity can affect this parameter as well. An experimental study found the labeling efficiency value to be 0.86±0.06 in vivo for pCASL, and it is tightly correlated with flow velocity (Aslan et al., 2010). The labeling efficiency should be estimated on a subject-specific basis, but in practice, most studies uses typical values of 0.7, 0.9, and 0.98 for CASL, pCASL, and PASL, respectively (Wong, 2014).

5.3.2 Signal-to-Noise Ratio

In ASL, the difference in signal between label and control conditions is inherently low, typically only approximately 1%–2% of the static tissue signals. It is crucial to improve SNR with noise reduction methods wherever possible. In some cases, a combination of low spatial resolution and signal averaging is employed for this purpose. Moreover, background suppression is also beneficial to suppress the static tissue signal and reduce noise from motion and other system instabilities. This is accomplished using one or more inversion pulses, applied after the labeling pulses, and timed such that over the range of T_1 values present in the tissues, the M_z of the tissues is near zero at the time of image acquisition (Wong, 2014).

5.3.3 Arterial Transit Time

Transit delays are a significant confounding factor for the interpretation of the ASL signal in terms of CBF. Essentially, the transit time of the blood from the tagging region to the imaged voxel includes two parts: (1) arterial transit time (ATT), which is the average time for the blood to reach the microvasculature in the region of interest, and (2) tissue transit time, which represents the time for labeled blood to exchange with local tissue (Alsop and Detre, 1996; Buxton, 2009). ATT is considered as main contribution in most ASL models. Because the accurate estimation of this parameter requires long scanning time (Chen et al., 2012), ATT is not measured routinely, rather generally assumed to be linearly increasing with the ascending image slices or even uniformly distributed throughout the brain. However, the transit delay can vary by several tenths of a second across a single image slice, and is likely to decrease with the increased blood flow caused by activation in an fMRI study. A pediatric study also finds ATT in an increasing trend with age (Jain et al., 2012). In particular, when the upstream arterial occlusion or stenosis happens in the case of the acute ischemic, blood flow is going to be much slower and transit time could go even longer than T_1. Consequently, CBF cannot be estimated properly using this sequence.

5.3.4 Water Exchange

As mentioned above, the labeled bloods will exchange with local tissue in tissue transit time (St Lawrence et al., 2012). The exchange causes the T_1 relaxation of labeled water shifts from that of blood to that of tissue.

This effect becomes more prominent when the T_1 values of the two compartments are significantly different and is small in the gray matter, where T_1 values of blood and tissue are not dramatically different. However, the T_1 value of white matter is significantly lower than that of blood, so the sensitivity to exchange is higher (Wong, 2014). Therefore, it is important to have estimates of the exchange times and tissue T_1 values when applying ASL in the white matter (van Gelderen et al., 2008; van Osch et al., 2009) or other organs with significantly different T_1 relaxation times between blood and tissue.

5.3.5 Blood T_1 Relaxation Effect

Labeled water protons in blood will experience a T_1 relaxation, which means the ASL signal has a lifetime related to the blood longitudinal relaxation time. This lifetime is approximately 1350 ms at 1.5 T and 1650 ms at 3 T (Lu et al., 2004). Many implementation choices of ASL are influenced by the fact that this lifetime is similar to the transport time from the labeling position to the tissue ATT. Ideally, one should choose a delay time just longer than the longest value of ATT present in the subject, thus the entire labeled bolus is delivered into the tissue. In practice, a delay too short will not allow for complete delivery of the labeled blood water to the tissue, whereas a delay too long will result in strong T_1 decay and therefore a reduced SNR (Alsop et al., 2013). A delay time of 2000 ms is recommended for the clinical adult population given the potential for a wide variety of pathologies (Alsop et al., 2013).

5.3.6 Quantification

In most routine clinical practice, most disorders of perfusion can be easily visualized via changes of ASL difference $|M_c - M_L|$. More importantly, ASL perfusion provides quantitative CBF values with proper postprocessing model and calculation (Buxton et al., 1998a; Liu and Wong, 2005). It is recommend that quantitative CBF maps would provide more useful information, and the quantification can be performed with many selections of postprocessing software (i.e., Wang et al., 2008) for ASL data. This is very necessary and useful in many cases, even in clinical evaluations, such as comparison before and after a cerebrovascular dilator, or before and after a neurointerventional procedure. The ability of quantification CBF in observing functional brain activation also makes it useful in neurosurgical planning (Wintermark et al., 2005) and neuroscience studies. In Section 5.4, we will discuss some applications in clinical and basic research by using quantitative ASL.

5.4 Applications

Blood flow delivers glucose and oxygen to tissue to maintain basal ATP production and to replenish it during increased neuronal activity. In cerebrovascular disorders, there is obvious utility for measurement of CBF, as it is tightly regulated to meet the demands of brain's metabolic processes and provides useful information about cerebrovascular condition and regional metabolism (Borogovac and Asllani, 2012; Buxton et al., 1998a; 1998b). While measurements of perfusion are of direct diagnostic value in vascular disorders, perfusion parameters also serve as biomarkers for a broader range of physiological and pathophysiological functions. For instance, a close coupling between CBF and metabolism allows regional brain function to be assessed through measurements of cerebral perfusion and increased vascularity of neoplasms allows tumor perfusion to be used as a measure of tumor grade and to monitor the response to tumor therapy. Therefore, it is of great importance to have the ASL technique with the advantage of using endogenous contrast to measure brain CBF in vivo and noninvasively. So far, this technique has been widely used in basic and clinical neuroscience, as examples shown below.

5.4.1 Cerebrovascular Disease and Stroke

In 1997, Siewer et al. reported a qualitative and quantitative comparison between PASL and DSC perfusion MRI results for the evaluation of acute cerebrovascular disease. They found agreement between techniques in assessment of perfusion (hypoperfused or delayed, normal, or hyperperfused) in 17 of 21 imaging studies, and disagreement in four. The disagreement was attributed to significant delayed perfusion in those cases (Siewert et al., 1997). CASL perfusion was also compared with

PET in patients with chronic occlusive cerebrovascular disease (Figure 5.9). Although CBF is found underestimated in affected regions, significant correlation between regional CBF values from these two techniques is reported (Kimura et al., 2005). Similar consistence evidence is also found with other modalities (Figure 5.10) (Uchihashi et al., 2011). This proves the practical usefulness of ASL in this disease.

Stroke is a heterogeneous syndrome caused by multiple disease mechanisms, but all result in a disruption of CBF with subsequent tissue damage (Markus, 2004). The main role of perfusion neuroimaging in the management of acute ischemic stroke is to confirm the presence of reduced regional blood flow and contribute to identification of the ischemic penumbra, regions of hypoperfusion that may be salvaged by thrombolytic and/ or endovascular recanalization therapy (Figure 5.11). In addition to that, hyperperfusion also exists in stroke (Figure 5.12). It may indicate metabolic failures such as low oxygen extraction fraction, and reflect vasoparalysis and greater regional vulnerability to hemorrhagic transformation. ASL has clinical uses in detecting both hypo- and hyperperfusion lesions (Wang et al., 2012). It is reported that a fast 2.5-min ASL perfusion scan may be adequate for screening patients with acute stroke with contraindications to gadolinium-based contrast agents (Bokkers et al., 2012).

5.4.2 Vascular Malformations

Moyamoya disease is implicated in approximately 6% of childhood strokes. Children and young adults in this disease have progressive bilateral stenosis of the supraclinoid ICA, anterior cerebral artery, MCA, and to a much lesser extent, the posterior cerebral artery, with the formation of netlike collateral vessel networks, termed *Moyamoya* vessels (Goetti et al., 2014). In the young population, use of ionizing radiation and

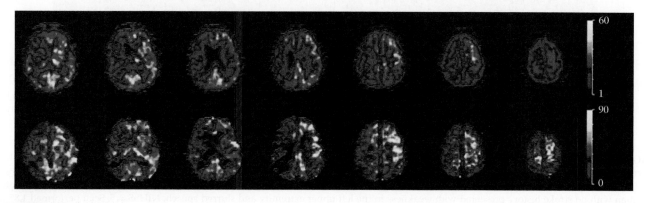

FIGURE 5.9
Large hypoperfused region is shown both in PET (upper) and CASL (lower) CBF maps of a subject with right internal carotid artery occlusion. The color bar is expressed in mL/minute/100 g. (From Kimura, H. et al., *J. Magn. Reson. Imaging*, 22, 189–198, 2005. With permission.)

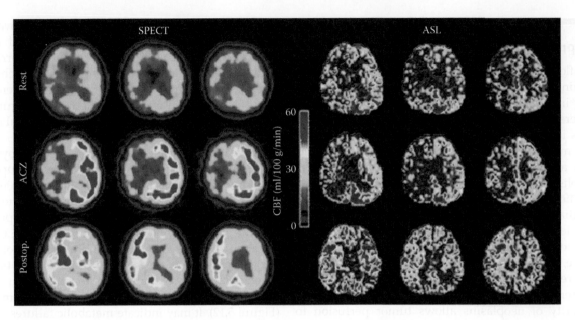

FIGURE 5.10
CBF maps obtained by ASL (right) in patients with severe right carotid stenosis, compared with that by SPECT (left). Preoperative resting images (top row). Preoperative hypoperfusion (top), poor vasoreactivity revealed by acetazolamide (ACZ) (middle), and postoperative hyperperfusion (bottom) in the right internal carotid artery region can be seen in both SPECT and ASL. (From Uchihashi, Y. et al., *Am. J. Neuroradiol.*, 32, 1545–1551, 2011. With permission.)

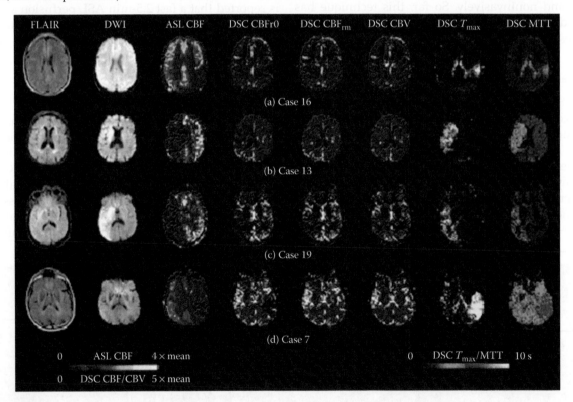

FIGURE 5.11
Acute ischemic stroke cases showing hypoperfused lesions in baseline scans. (a) Case 16: a 39-year-old man with no stroke history presented with slurred speech, National Institutes of Health Stroke Scale (NIHSS) = 2. Scan performed 1.37 h after onset. (b) Case 13: a 93-year-old woman with no stroke history presented with weakness in the left upper extremity and slurred speech, NIHSS = 5. Scan performed 1.85 h after onset. (c) Case 19: a 70-year-old man with a history of hypertension presented with left-sided weakness and slurred speech, NIHSS = 20. Scan performed 8.18 h after onset. (d) Case 7: a 68-year-old man with a history of atrial fibrillation, hypertension, and dyslipidemia presented with slurred speech, right-sided weakness, and gait disturbance, NIHSS = 18. Scan performed 1.38 h after onset. The patient received postscan intravenous tissue plasminogen activator (IV tPA). (From Wang, D.J.J. et al., *Stroke*, 43, 1018–1024, 2012. With permission.)

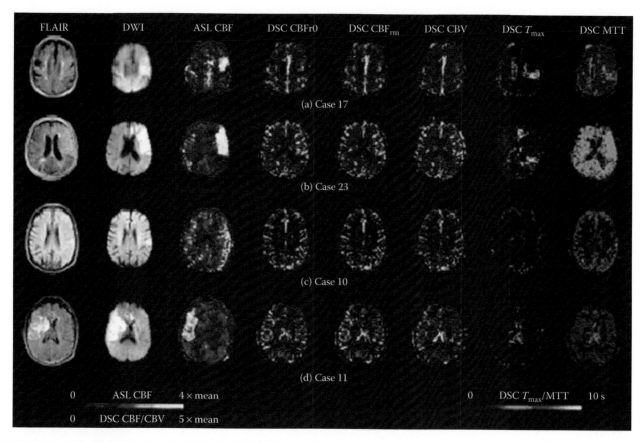

FIGURE 5.12

Acute ischemic stroke cases showing hyperperfused lesions on baseline scans. (a) Case 17: a 90-year-old woman with a history of atrial fibrillation and atrial flutter presented with slurred speech and right-sided weakness, NIHSS = 16. IV tPA was administered before the scan. (b) Case 23: a 93-year-old woman with a history of hypertension and atrial fibrillation presented with right-sided numbness and weakness and slurred speech, NIHSS = 23. IV tPA was administered before the scan. (c) Case 10: a 86-year-old man with a history of hypertension presented with right-sided weakness, slurred speech, and right lower facial droop, NIHSS = 11. The patient received postscan IV tPA. (d) Case 11: a 75-year-old man with no stroke history presented with left-sided weakness, NIHSS = 18. IV tPA was administered before the scan. AIS indicates acute ischemic stroke; NIHSS, National Institutes of Health Stroke Scale; IV tPA, intravenous tissue-type plasminogen activator. (From Wang, D.J.J. et al., *Stroke*, 43, 1018–1024, 2012. With permission.)

exogenous agents should be minimized or even avoided whenever possible. ASL perfusion thus could be a better alternation. Quantitative CBF with pCASL has been compared with both $H_2[^{15}O]$-PET (Goetti et al., 2014) and FDG-PET (Cha et al., 2013), and good agreement and significant correlations were found in both qualitative perfusion scoring and quantitative perfusion assessment of relative CBF with cerebellar normalization. Similar conclusion was also drawn from a comparison between multidelay pCASL and CT perfusions (Figure 5.13) (Wang et al., 2014).

5.4.3 Brain Tumor

Angiogenesis has a key role both in the development and in the malignant transformation of tumors, as malignant tumors require new blood vessels to maintain growth above a few millimeters in diameter. Quantitative measurements of tumor blood flow are thus important for

understanding tumor physiology and can be valuable in selecting and evaluating therapies (Silva et al., 2000). ASL has been used to evaluate meningioma (Perini et al., 2008) and gliomas (Garzón et al., 2011; Wolf et al., 2005), and the glioma grade can be differentiated by perfusion measurement (Figure 5.14) (Garzón et al., 2011; Noguchi et al., 2008; Warmuth et al., 2003; Wolf et al., 2005).

5.4.4 Epilepsy

Epilepsy is a functional disorder that may not be accompanied by gross abnormalities on structural imaging in its early stage. In temporal lobe epilepsy (TLE), an important application is the determination of laterality by the measurement of interictal hypometabolism. In two studies of presurgical lateralization, ASL was found in good agreement with FDG-PET, making it a surrogate in assessment and treatment planning (Alsop et al., 2002; Wolf et al., 2001). A multimodality study

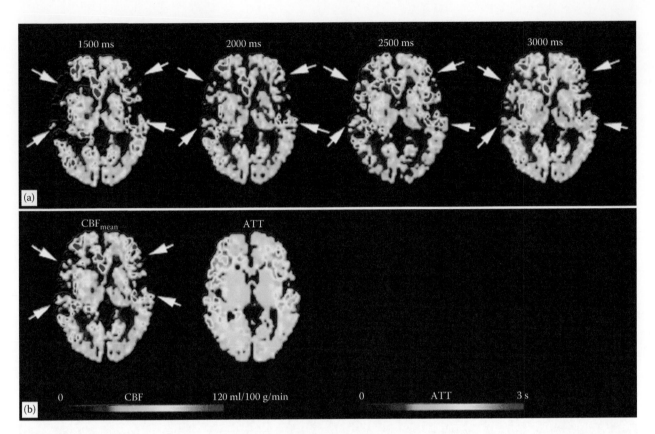

FIGURE 5.13
Perfusion maps in a 32-year-old female patient with Moyamoya disease. (a) pCASL CBF images for PLD of 1500, 2000, 2500, and 3000 ms. (b) CBF$_{mean}$ and ATT images of the same slice. pCASL with short PLD CBF maps showed enlarged abnormal perfusion territories (arrows) with prolonged ATT, whereas the problem is mitigated in perfusion images with long PLDs and in CBF$_{mean}$ image. (From Wang, R. et al., *Eur. Radiol.*, 24, 1135–1144, 2014. With permission.)

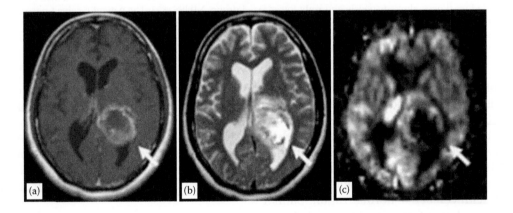

FIGURE 5.14
Postcontrast T_1-weighted image (a), T_2-weighted image (b), and ASL perfusion image (c) in a 44-year-old woman with glioblastoma. The tumor was depicted as a high-intensity area with a central low perfusion area on the perfusion image. (From Noguchi, T. et al., *Am. J. Neuroradiol.*, 29, 688–693, 2008. With permission.)

combining electroencephalography (EEG), ASL, and FDG-PET found in most of the patients hypoperfusion and hypometabolism in interictal phase, but hyperperfusion in early postictal phase (Storti et al., 2013).

One issue of perfusion imaging in TLE is the structure of mesial temporal lobe. The baseline perfusion in this region is relatively small, and thus measuring error of CBF could be severe since ASL signal change is a few percent as well. Moreover, this temporal region suffers more from susceptibility artifacts. Much attention to these problems should be paid in clinical applications of ASL (Wolf and Detre, 2007).

5.4.5 Alzheimer Disease

Different courses of Alzheimer disease (AD) are observed in clinical practice. The rapidly progressive form could be associated with the presence of a major microcirculatory involvement and hemodynamic insufficiency (Diomedi and Misaggi, 2013). One study had quantitatively compared CBF, in 71 patients with this disease, 35 patients with mild cognitive impairment (MCI), and 73 subjects who visited a memory clinic with subjective complaints, by using a whole-brain 3D pCASL and suggested that CBF helps in detecting functional changes in the prodromal and more advanced stages of AD and is a marker for disease severity (Binnewijzend et al., 2013). A study involving concurrent ASL and FDG-PET scans in 17 patients with AD and 19 healthy controls confirms that ASL could detect functional deficits in specific regions of interest in the brains of patients with AD with a similar sensitivity to FDG-PET imaging (Figure 5.15) (Chen et al., 2011b; Musiek et al., 2012). These studies shed light on a possible faster screening of patients who present with cognitive deficits, a procedure that could reduce both the cost of imaging and the time to diagnosis in patients with AD (Malpass, 2012).

5.4.6 Chronic Pain

The underlying mechanism of chronic pain is not clearly understood yet. It is believed to be associated with structural, functional, and neurochemical alterations distributed across multiple brain networks. In addition, it is important to measure the pain objectively and quantitatively, if possible. One recent study uses ASL to evaluate clinical pain, and it found that patients' baseline clinical pain correlated positively with connectivity strength between brain default mode network (DMN) and right insula. This result supports the use of resting ASL to define a neuroimaging biomarker for chronic pain perception (Loggia et al., 2013).

5.4.7 Functional MRI

The mechanism underlying detection of brain activation is that increased cerebral activity is accompanied by local changes in CBF and concentration of deoxygenated hemoglobin. The changes in CBF and blood-oxygenation-dependent (BOLD) signal induced by a task in one scenario can be described by physiological models (Blockley et al., 2013; Buxton et al., 2004). BOLD-fMRI has been popular and successful since it is the first development (Ogawa et al., 1990; Ogawa and Sung, 2007), but it cannot provide a quantitative measurement. Recently, ASL perfusion imaging techniques have been successfully used for cerebral activation studies (Borogovac and Asllani, 2012; Wang et al., 2003). As opposed to BOLD, ASL is not based on local magnetic susceptibility, so a T_2^*-weighted imaging sequence is not required or desirable. The use of spin-echo-based sequences allows ASL measurements to be performed in regions of high static field in homogeneity. More importantly, because the ASL technique is capable to quantify CBF, it could be more promising for patient cohort applications (Yang, 2002).

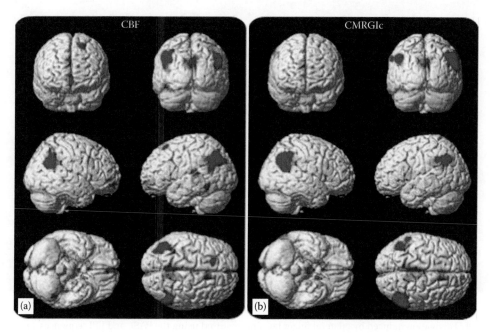

FIGURE 5.15
Excellent agreement between the ASL hypoperfusion (a) and FDG-PET hypometabolism (b) maps related with AD disease in the bilateral angular gyri and posterior cingulate. (From Chen, Y. et al., *Neurol.*, 77, 1977–1985, 2011b. With permission.)

As it is measuring the arterial blood signal, perfusion fMRI has the potential to provide better localization of the functional signal to sites of neural activity than BOLD fMRI. For example, a recent comparison study was performed with a hand movement task. In this study, subjects were asked to perform active and passive hand task, while both ASL and BOLD fMRI data were acquired separately. As the study claims, there is a better colocalization of activation volume with ASL than that with BOLD (Galazzo et al., 2013). Moreover, because ASL uses pairwise subtraction and subsequent calibration process to produce an absolute measure of CBF, the slow drifts present in BOLD contrast images are eliminated in ASL. This makes ASL as a potential solution to the fundamental limitations of BOLD methods at low task frequencies (Figure 5.16), permitting the measurement of changes in neural activity that span longer periods of time—minutes, hours, and even days (Wang et al., 2003).

It is possible to derive both CBF and BOLD measurements simultaneously using the ASL technique. For example, dual echo with EPI or spiral readouts are

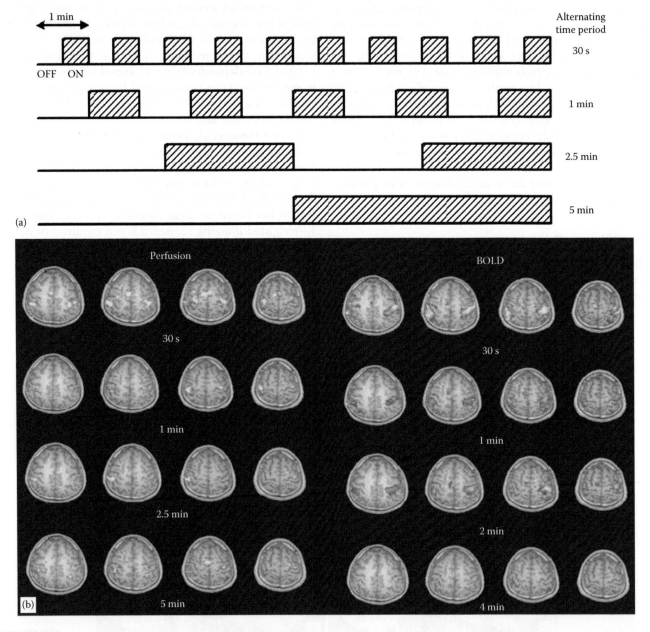

FIGURE 5.16
The perfusion and BOLD activation induced by bilateral finger tapping acquired using the PASL and EPI sequences. (a) The perfusion experimental designs. BOLD experiment uses very similar design except for slight differences in last two alternating time periods (2.5/5 min vs. 2/4 min). (b) Group-level motor activation maps overlaid on anatomical images. Perfusion fMRI is better in detecting slow-task (>2 min) activity. (From Wang, J. et al., *Magn. Reson. Med.*, 49, 796–802, 2003. With permission.)

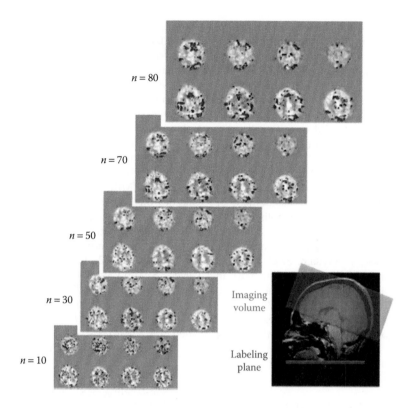

FIGURE 5.17
Real-time functional MRI using pseudocontinuous arterial spin labeling. The evolution of the statistical map during visual-motor stimulation task shows that the statistical maps became more detailed with a lower number of false positives and more significant scores in the active regions as more data were available. (From Hernandez-Garcia, L. et al., *Magn. Reson. Med.*, 65, 1570–1577, 2011. With permission.)

particularly useful for simultaneous perfusion and BOLD imaging. Images acquired at a short echo time (e.g., 3 ms) can be used to form a perfusion time series with relatively little BOLD weighting, whereas images acquired at a later echo time (e.g., 30 ms) can be used to form the BOLD time series. This is because EPI or spiral images can exhibit BOLD weighting with a sufficiently long echo time (Liu and Brown, 2007). A BOLD time series can be formed from the control and tag images through running average approaches that are analogous to the subtraction approaches used to form the perfusion time series (Liu and Wong, 2005).

In the very recent years, real-time fMRI becomes a hot topic as an exciting extension to conventional fMRI. A real-time experiment based on ASL was first implemented by Hernandez-Garcia et al. (2011). In real-time fMRI, the results are immediately available as the subject is being scanned, and the results can be used to reveal and guide the subject's cognitive processes. It can also facilitate the experimenter's parameter selections or a clinician's interventions (Figure 5.17).

In summary, ASL has shown its prominence as tools for quantitatively measuring brain perfusion. pCASL holds the best promise at the moment, and even more new ASL-based implementations are under development to take full potentials. These techniques

are gaining increasing use for both clinic and brain researches in cerebrovascular disease, disorders, neuroscience, and so on.

References

Alsop, D.C., Connelly, A., Duncan, J.S., Hufnagel, A., Pierpaoli, C., Rugg-Gunn, F.J., 2002. Diffusion and perfusion MRI in epilepsy. *Epilepsia* 43, 69–77.

Alsop, D.C., Detre, J.A., 1996. Reduced transit-time sensitivity in noninvasive magnetic resonance imaging of human cerebral blood flow. *J. Cereb. Blood Flow Metab.* 16, 1236–1249.

Alsop, D.C., Detre, J.A., 1998. Multisection cerebral blood flow MR imaging with continuous arterial spin labeling. *Radiology* 208, 410–416.

Alsop, D.C., Detre, J.A., Golay, X., Günther, M., Hendrikse, J., Hernandez-Garcia, L., Lu, H. et al., 2013. Recommended implementation of arterial spin-labeled perfusion MRI for clinical applications: A consensus of the ISMRM perfusion study group and the European consortium for ASL in dementia. *Magn. Reson. Med.* 73, 102–116.

Aslan, S., Xu, F., Wang, P.L., Uh, J., Yezhuvath, U.S., van Osch, M., Lu, H., 2010. Estimation of labeling efficiency in pseudocontinuous arterial spin labeling. *Magn. Reson. Med.* 63, 765–771.

Bernstein, M.A., King, K.F., Zhou, X.J., 2004. *Handbook of MRI Pulse Sequences*. Amsterdam, the Netherlands: Elsevier Academic Press.

Binnewijzend, M.A.A., Kuijer, J.P.A., Benedictus, M.R., van der Flier, W.M., Wink, A.M., Wattjes, M.P., van Berckel, B.N.M. et al., 2013. Cerebral blood flow measured with 3D pseudocontinuous arterial spin-labeling MR imaging in Alzheimer disease and mild cognitive impairment: A marker for disease severity. *Radiology* 267, 221–230.

Blockley, N.P., Griffeth, V.E.M., Simon, A.B., Buxton, R.B., 2013. A review of calibrated blood oxygenation level-dependent (BOLD) methods for the measurement of task-induced changes in brain oxygen metabolism. *NMR Biomed.* 26, 987–1003.

Bokkers, R.P.H., Hernandez, D.A., Merino, J.G., Mirasol, R.V., van Osch, M.J., Hendrikse, J., Warach, S. et al., 2012. Whole-brain arterial spin labeling perfusion MRI in patients with acute stroke. *Stroke* 43, 1290–1294.

Borogovac, A., Asllani, I., 2012. Arterial spin labeling (ASL) fMRI: Advantages, theoretical constrains and experimental challenges in neurosciences. *Int. J. Biomed. Imaging* 2012, 1–13.

Buxton, R.B., 2009. *Introduction to Functional Magnetic Resonance Imaging: Principles and Techniques*. New York: Cambridge University Press.

Buxton, R.B., Frank, L.R., Wong, E.C., Siewert, B., Warach, S., Edelman, R.R., 1998a. A general kinetic model for quantitative perfusion imaging with arterial spin labeling. *Magn. Reson. Med.* 40, 383–396.

Buxton, R.B., Uludağ, K., Dubowitz, D.J., Liu, T.T., 2004. Modeling the hemodynamic response to brain activation. *NeuroImage* 23(Suppl 1), S220–233.

Buxton, R.B., Wong, E.C., Frank, L.R., 1998b. Dynamics of blood flow and oxygenation changes during brain activation: The balloon model. *Magn. Reson. Med.* 39, 855–864.

Cha, Y.-H.K., Jog, M.A., Kim, Y.-C., Chakrapani, S., Kraman, S.M., Wang, D.J., 2013. Regional correlation between resting state FDG PET and pCASL perfusion MRI. *J. Cerebr. Blood Flow Metab.* 33, 1909–1914.

Chen, Y., Wang, D.J.J., Detre, J.A., 2011a. Test–retest reliability of arterial spin labeling with common labeling strategies. *J. Magn. Reson. Imaging* 33, 940–949.

Chen, Y., Wang, D.J.J., Detre, J.A., 2012. Comparison of arterial transit times estimated using arterial spin labeling. *Magn. Reson. Mater Phys.* 25, 135–144.

Chen, Y., Wolk, D.A., Eddin, J.S.R., Korczykowski, M., Martinez, P.M., Pvlusiek, E.S., Newberg, A.B. et al., 2011b. Voxel-level comparison of arterial spin-labeled perfusion MRI and FDG-PET in Alzheimer disease. *Neurology* 77, 1977–1985.

Deibler, A.R., Pollock, J.M., Kraft, R.A., Tan, H., Burdette, J.H., Maldjian, J.A., 2008. Arterial spin-labeling in routine clinical practice, part 1: Technique and artifacts. *Am. J. Neuroradiol.* 29, 1228–1234.

Detre, J.A., Leigh, J.S., Williams, D.S., Koretsky, A.P., 1992. Perfusion imaging. *Magn. Reson. Med.* 23, 37–45.

Diomedi, M., Misaggi, G., 2013. Vascular contribution to Alzheimer disease: Predictors of rapid progression. *CNS Neurol. Disord. Drug Target.* 12, 532–537.

Edelman, R.R., Siewert, B., Adamis, M., Gaa, J., Laub, G., Wielopolski, P., 1994. Signal targeting with alternating radiofrequency (STAR) sequences: Application to MR angiography. *Magn. Reson. Med.* 31, 233–238.

Galazzo, I., Storti, S.F., Formaggio, E., Pizzini, F.B., Fiaschi, A., Beltramello, A., Bertoldo, A. et al., 2014. Investigation of brain hemodynamic changes induced by active and passive movements: A combined arterial spin labeling: BOLD fMRI study. *J. Magn. Reson. Imaging* 40, 937–48.

Garzón, B., Emblem, K.E., Mouridsen, K., Nedregaard, B., Due-Tønnessen, P., Nome, T., Hald, J.K. et al., 2011. Multiparametric analysis of magnetic resonance images for glioma grading and patient survival time prediction. *Acta Radiol.* 52, 1052–1060.

Gevers, S., van Osch, M.J., Bokkers, R., 2011. Intra- and multi-center reproducibility of pulsed, continuous and pseudo-continuous arterial spin labeling methods for measuring cerebral perfusion. *J. Cerebr. Blood Flow Metab.* 31, 1706–1715.

Goetti, R., Warnock, G., Kuhn, F.P., Guggenberger, R., O'Gorman, R., Buck, A., Khan, N. et al., 2014. Quantitative cerebral perfusion imaging in children and young adults with Moyamoya disease: Comparison of arterial spin-labeling–MRI and $H_2[^{15}O]$-PET. *Am. J. Neuroradiol.* 35, 1022–1028.

Helle, M., Norris, D.G., Rüfer, S., Alfke, K., 2010. Superselective pseudocontinuous arterial spin labeling. *Magn. Reson. Med.* 64, 777–786.

Helle, M., Rüfer, S., Osch, M., Nabavi, A., 2013. Superselective arterial spin labeling applied for flow territory mapping in various cerebrovascular diseases. *J. Magn. Reson. Imaging* 38, 496–503.

Hernandez-Garcia, L., Jahanian, H., Greenwald, M.K., Zubieta, J.K., Peltier, S.J., 2011. Real-time functional MRI using pseudo-continuous arterial spin labeling. *Magn. Reson. Med.* 65, 1570–1577.

Jahng, G.-H., Song, E., Zhu, X.-P., Matson, G.B., Weiner, M.W., Schuff, N., 2005. Human brain: Reliability and reproducibility of pulsed arterial spin-labeling perfusion MR imaging. *Radiology* 234, 909–916.

Jain, V., Duda, J., Avants, B., Giannetta, M., Xie, S.X., Roberts, T., Detre, J.A. et al., 2012. Longitudinal reproducibility and accuracy of pseudo-continuous arterial spin-labeled perfusion MR imaging in typically developing children. *Radiology* 263, 527–536.

Kimura, H., Kado, H., Koshimoto, Y., Tsuchida, T., Yonekura, Y., Itoh, H., 2005. Multislice continuous arterial spin-labeled perfusion MRI in patients with chronic occlusive cerebrovascular disease: A correlative study with CO_2 PET validation. *J. Magn. Reson. Imaging* 22, 189–198.

Liu, T.T., Brown, G.G., 2007. Measurement of cerebral perfusion with arterial spin labeling: Part 1. Methods. *J. Int. Neuropsychol. Soc.* 13, 517–525.

Liu, T.T., Wong, E.C., 2005. A signal processing model for arterial spin labeling functional MRI. *NeuroImage* 24, 207–215.

Loggia, M.L., Kim, J., Gollub, R.L., Vangel, M.G., Kirsch, I., Kong, J., Wasan, A.D. et al. 2013. Default mode network connectivity encodes clinical pain: An arterial spin labeling study. *PAIN* 154, 24–33.

Lu, H., Clingman, C., Golay, X., 2004. Determining the longitudinal relaxation time (T_1) of blood at 3.0 Tesla. *Magn. Reson. Med.* 52, 679–682.

Malpass, K., 2011. Alzheimer disease: Arterial spin-labeled MRI for diagnosis and monitoring of AD. *Nat. Rev. Neurol.* 8, 3.

Markus, H.S., 2004. Cerebral perfusion and stroke. *J. Neurol. Neurosurg. Psychiatr.* 75, 353–361.

Musiek, E.S., Chen, Y., Korczykowski, M., Saboury, B., Martinez, P.M., Reddin, J.S., Alavi, A. et al. 2012. Direct comparison of fluorodeoxyglucose positron emission tomography and arterial spin labeling magnetic resonance imaging in Alzheimer's disease. *Alzheimer Dement.* 8, 51–59.

Noguchi, T., Yoshiura, T., Hiwatashi, A., Togao, O., Yamashita, K., Nagao, E., Shono, T. et al. 2008. Perfusion imaging of brain tumors using arterial spin-labeling: Correlation with histopathologic vascular density. *Am. J. Neuroradiol.* 29, 688–693.

Ogawa, S., Lee, T.M., Kay, A.R., 1990. Brain magnetic resonance imaging with contrast dependent on blood oxygenation. *Proc. Natl. Acad. Sci. U. S. A.* 87, 9868–9872.

Ogawa, S., Sung, Y.-W., 2007. Functional magnetic resonance imaging. *Scholarpedia* 2, 3105.

Perini, R., Choe, R., Yodh, A.G., Sehgal, C., Divgi, C.R., Rosen, M.A., 2008. Non-invasive assessment of tumor neovasculature: Techniques and clinical applications. *Cancer Metastasis Rev.* 27, 615–630.

Petersen, E.T., Zimine, I., Ho, Y.-C.L., Golay, X., 2006. Non-invasive measurement of perfusion: A critical review of arterial spin labelling techniques. *Br. J. Radiol.* 79, 688–701.

Siewert, B., Schlaug, G., Edelman, R.R., Warach, S., 1997. Comparison of EPISTAR and T_2*-weighted gadolinium-enhanced perfusion imaging in patients with acute cerebral ischemia. *Neurology* 48, 673–679.

Silva, A.C., Kim, S.G., Garwood, M., 2000. Imaging blood flow in brain tumors using arterial spin labeling. *Magn. Reson. Med.* 44, 169–173.

St. Lawrence, K.S., Owen, D., Wang, D.J.J., 2012. A two-stage approach for measuring vascular water exchange and arterial transit time by diffusion-weighted perfusion MRI. *Magn. Reson. Med.* 67, 1275–1284.

Storti, S.F., Boscolo Galazzo, I., Del Felice, A., Pizzini, F.B., Arcaro, C., Formaggio, E., Mai, R. et al. 2014. Combining ESI, ASL and PET for quantitative assessment of drug-resistant focal epilepsy. *NeuroImage* 102, 49–59.

Talagala, S.L., Ye, F.Q., Ledden, P.J., 2004. Whole-brain 3D perfusion MRI at 3.0 T using CASL with a separate labeling coil. *Magn. Reson. Med.* 52, 131–140.

Uchihashi, Y., Hosoda, K., Zimine, I., Fujita, A., Fujii, M., Sugimura, K., Kohmura, E., 2011. Clinical application of arterial spin-labeling MR imaging in patients with carotid stenosis: Quantitative comparative study with single-photon emission CT. *Am. J. Neuroradiol.* 32, 1545–1551.

van Gelderen, P., de Zwart, J.A., Duyn, J.H., 2008. Pitfalls of MRI measurement of white matter perfusion based on arterial spin labeling. *Magn. Reson. Med.* 59, 788–795.

van Osch, M.J.P., Teeuwisse, W.M., van Walderveen, M.A.A., Hendrikse, J., Kies, D.A., van Buchem, M.A., 2009. Can arterial spin labeling detect white matter perfusion signal? *Magn. Reson. Med.* 62, 165–173.

Wang, D.J.J., Alger, J.R., Qiao, J.X., Hao, Q., Hou, S., Fiaz, R., Günther, M. et al., 2012. The value of arterial spin-labeled perfusion imaging in acute ischemic stroke comparison with dynamic susceptibility contrast-enhanced MRI. *Stroke* 43, 1018–1024.

Wang, J., Aguirre, G.K., Kimberg, D.Y., Roc, A.C., Li, L., Detre, J.A., 2003. Arterial spin labeling perfusion fMRI with very low task frequency. *Magn. Reson. Med.* 49, 796–802.

Wang, R., Yu, S., Alger, J.R., Zuo, Z., Chen, J., Wang, R., Wang, R. et al. 2014. Multi-delay arterial spin labeling perfusion MRI in Moyamoya disease—Comparison with CT perfusion imaging. *Eur. Radiol.* 24, 1135–1144.

Wang, Y., Saykin, A.J., Pfeuffer, J., Lin, C., Mosier, K.M., Shen, L., Kim, S. et al. 2011. Regional reproducibility of pulsed arterial spin labeling perfusion imaging at 3T. *NeuroImage* 54, 1188–1195.

Wang, Z., Aguirre, G.K., Rao, H., Wang, J., Fernandez-Seara, M.A., Childress, A.R., Detre, J.A., 2008. Empirical optimization of ASL data analysis using an ASL data processing toolbox: ASLtbx. *Magn. Reson. Imaging* 26, 261–269.

Warmuth, C., Günther, M., Zimmer, C., 2003. Quantification of blood flow in brain tumors: Comparison of arterial spin labeling and dynamic susceptibility-weighted contrast-enhanced MR imaging. *Radiology* 228, 523–532.

Williams, D.S., Detre, J.A., Leigh, J.S., Koretsky, A.P., 1992. Magnetic resonance imaging of perfusion using spin inversion of arterial water. *Proc. Natl. Acad. Sci. U.S.A.* 89, 212–216.

Wintermark, M., Sesay, M., Barbier, E., Borbély, K., Dillon, W.P., 2005. Comparative overview of brain perfusion imaging techniques. *Stroke* 36, e83–99.

Wolf, R.L., Alsop, D.C., Levy-Reis, I., Meyer, P.T., Maldjian, J.A., Gonzalez-Atavales, J., French, J.A. et al. 2001. Detection of mesial temporal lobe hypoperfusion in patients with temporal lobe epilepsy by use of arterial spin labeled perfusion MR imaging. *Am. J. Neuroradiol.* 22, 1334–1341.

Wolf, R.L., Detre, J.A., 2007. Clinical neuroimaging using arterial spin-labeled perfusion magnetic resonance imaging. *Neurotherapeutics* 4, 346–359.

Wolf, R.L., Wang, J., Wang, S., Melhem, E.R., O'Rourke, D.M., Judy, K.D., Detre, J.A., 2005. Grading of CNS neoplasms using continuous arterial spin labeled perfusion MR imaging at 3 Tesla. *J. Magn. Reson. Imaging* 22, 475–482.

Wong, E.C., 2007. Vessel-encoded arterial spin-labeling using pseudocontinuous tagging. *Magn. Reson. Med.* 58, 1086–1091.

Wong, E.C., 2014. An introduction to ASL labeling techniques. *J. Magn. Reson. Imaging* 40, 1–10.

Wong, E.C., Buxton, R.B., Frank, L.R., 1998a. Quantitative imaging of perfusion using a single subtraction (QUIPSS and QUIPSS II). *Magn. Reson. Med.* 39, 702–708.

Wong, E.C., Buxton, R.B., Frank, L.R., 1998b. A theoretical and experimental comparison of continuous and pulsed arterial spin labeling techniques for quantitative perfusion imaging. *Magn. Reson. Med.* 40, 348–355.

Wu, B., Lou, X., Wu, X., Ma, L., 2014. Intra- and interscanner reliability and reproducibility of 3D whole-brain pseudo-continuous arterial spin-labeling MR perfusion at 3T. *J. Magn. Reson. Imaging* 39, 402–409.

Wu, W.C., Fernández-Seara, M., Detre, J.A., Wehrli, F.W., Wang, J., 2007. A theoretical and experimental investigation of the tagging efficiency of pseudo continuous arterial spin labeling. *Magn. Reson. Med.* 58, 1020–1027.

Wu, W.C., St Lawrence, K.S., Licht, D.J., Wang, D.J.J., 2010. Quantification issues in arterial spin labeling perfusion magnetic resonance imaging. *Topics Magn. Reson. Imaging* 21, 65–73.

Yang, Y., 2002. Perfusion MR imaging with pulsed arterial spin-labeling: Basic principles and applications in functional brain imaging. *Concept Magn. Reson.* 14, 347–357.

6

Kinetic Modeling for T_1-Weighted Dynamic Contrast-Enhanced Magnetic Resonance Imaging

Kyunghyun Sung and John Carr

CONTENTS

6.1 Introduction

Small molecular weight paramagnetic contrast agents that diffuse from the vascular to extravascular space are routinely used in magnetic resonance imaging (MRI) to improve contrast between healthy and diseased tissue. Since attaining FDA approval in 1988, gadolinium-based contrast agents, in particular, have facilitated major improvements in the sensitivity of detection, delineation, and characterization of cancer using MRI [1,2]. Paramagnetic contrast agents increase the relaxivity of proximal water protons, shortening their apparent T_1 and T_2. Accumulation of contrast agent in a tissue will reduce the observed signal in a T_2- or T_2*-weighted image while increasing or *enhancing* the observed signal on a T_1-weighted image.

A common data acquisition approach for dynamic contrast-enhanced MRI (DCE-MRI) is to acquire T_1-weighted images, inject a gadolinium-based contrast agent, and continuously acquire a time series of T_1-weighted images as the contrast agent circulates through the tissue microvasculature (Figure 6.1). Volumetric or multislice images may be acquired, and rapid imaging approaches applied to enhance temporal and spatial resolution. Dynamic MR images are typically obtained every few seconds for 5–6 min, and the contrast medium is typically injected intravenously after 4–10 baseline images using a power injector. Before the contrast agent injection, precontrast T_1 measurements are necessary to obtain to convert the dynamic MRI data into the contrast agent concentration, where the contrast agent concentration changes over time can be used to extract microvascular functional properties by fitting the uptake of the gadolinium concentration to either a model-free approach [3–7] or a pharmacokinetic model [8,9].

Gadolinium-diethylenetriaminepenta-acetic acid (Gd-DTPA) is the most commonly used contrast agent for DCE-MRI. Due to its relatively small size (<1000 Da), the contrast agent is able to diffuse from the vascular

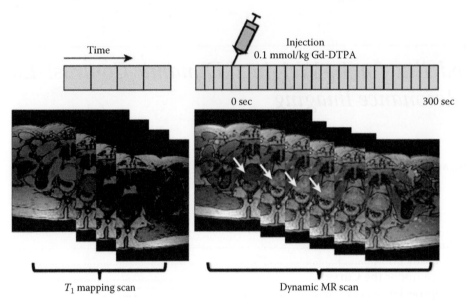

FIGURE 6.1
Schematic representation of the DCE-MRI data acquisition. Twenty slices are acquired at each data point, and the data acquisition continues for 5–6 min. Numbers indicate seconds after contrast enhancement (assuming 4 s per data point). Precontrast T_1 map is obtained by acquiring an identical 3D acquisition with different flip angles (e.g., 2°, 5°, 10°, and 15°).

space into the extravascular space. The observed signal intensity change on a contrast-enhanced, T_1-weighted image results predominantly from this process, the dynamics of which can be evaluated to reveal functional information about the tissue.

Diseased tissue is often characterized by changes in capillary permeability surface area product, perfusion, and blood volume, or extravascular volume. DCE-MRI is sensitive to such changes and may reveal useful functional information in pathologies that exhibit hypervascularity or compromised vascular integrity. Pathologies where DCE-MRI has been utilized include multiple sclerosis [10], osteoarthritis [11], Sjören's syndrome [12], and kidney function [13]. The predominant application of DCE-MRI has been in cancer imaging, specifically imaging the angiogenic processes associated with cancer development. Angiogenesis is a well-established marker of cancer growth and aggressiveness [14]. DCE-MRI is sensitive to angiogenic-induced microvascular changes and can provide an essential tool for assessing cancer therapies, many of which target the angiogenic process.

Cancer treatments are often cytostatic, especially when applied in isolation. A therapy may disrupt tumor biology, killing part of a tumor but not shrinking its size. Conventional radiological measures of treatment response, such as the RECIST criterion [15], based on parameters such as tumor size may not be helpful for therapeutic assessment, especially in the case of antiangiogenic therapies. DCE-MRI can provide a method

that is sensitive to functional changes in tumor biology associated with angiogenesis, making it an ideal tool for the assessment of antiangiogenic therapies.

Modeling the tracer kinetics in DCE-MRI has become the method of choice for the assessment of new oncology drugs, investigation of pathophysiology, as well as monitoring and more recently predicting treatment response [16]. In contrast to heuristic semiquantitative measures of tracer uptake, kinetic model parameters can reflect underlying functional physiological properties within the tissue. Pharmacokinetic-model-derived parameters such as K^{trans} and v_e have been recommended as a primary end-point in the assessment of antiangiogenic and antivascular therapies [17].

In this chapter, DCE-MRI quantitative tracer kinetic analyses methods will be reviewed, ranging from simple model-free approaches to state-of-the-art pharmacokinetic modeling.

6.2 Model-Free Quantification

Model-free quantification methods aim to provide semiquantitative parameters closely related to contrast agent accumulation without assuming any pharmacokinetic modeling. Many simple and straightforward approaches have been proposed to provide relatively reliable information of the kinetics of contrast agent accumulation,

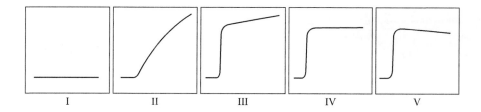

FIGURE 6.2
Classification system for qualitative assessment of time-intensity curves: (I) type 1, no enhancement; (II) type 2, slow and sustained enhancement; (III) type 3, rapid initial and sustained late enhancement; (IV) type 4, rapid initial and stable late enhancement; and (V) type 5, rapid initial and decreasing late enhancement. (From Daniel, B.L. et al., *Radiology*, 209, 499–509, 1998. With permission.)

and here we describe three most common methods such as simple curve classification, initial slope of the curve, and area under the curve [3–7].

6.2.1 Curve Classification

One of the most straightforward approaches is to categorize types of the time–intensity curve. Each type of the curves can be either systemically classified or visually inspected by radiologists without knowledge of the morphologic appearance, size, or pathologic diagnosis of the lesion. An example of the classification system, proposed by Daniel et al. [18], is shown in Figure 6.2: type 1—no enhancement; type 2—slow sustained enhancement; type 3–rapid initial and sustained late enhancement; type 4—rapid initial and stable late enhancement; type 5—rapid initial and decreasing late enhancement. A change in the curve type to a higher number (e.g., from type 2 to type 3 or type 4) is considered a transformation to a more aggressive type. A change in the curve type to a lower number (e.g., from type 4 to type 3 or type 2) is considered a transformation to a less aggressive type [19].

6.2.2 Initial Area under the Curve

Another simple yet effective method is an integration of the relative enhancement curve in the tissue of interest over time. The relative enhancement at time t, $E(t)$, is defined by

$$E(t) = \frac{S(t) - S_{\text{baseline}}}{S_{\text{baseline}}} \quad (6.1)$$

where:
  $S(t)$ is the signal intensity t seconds after contrast injection
  S_{baseline} is the average signal intensity before contrast injection

The value of initial area under the curve (IAUC) is dependent on the period over which integration is performed and can be computed by

$$\text{IAUC}_t = \int_0^t E(t')dt' \quad (6.2)$$

where t is the time after the contrast injection. Figure 6.3 shows an example of the IAUC calculation over the tumor region of interest (ROI) (Figure 6.3a and b), as well as pixel-by-pixel IAUC maps for two different breast cancer patients (Figure 6.3c and d). One simple variation is to use the gadolinium concentration curve, $Gd(t)$, instead of the relative enhancement, $E(t)$, to avoid any potential bias due to various baseline T_1 values, and the value of IAUGC (initial area under the gadolinium curve) can be computed as

$$\text{IAUC}_t = \int_0^t Gd(t')dt' \quad (6.3)$$

where t is typically chosen to be 60–120 s after contrast injection

The values of both IAUC and IAUGC are proportional to the contrast agent delivery to the tissue of interest, and therefore are dependent on the arterial input function (AIF). To overcome potential variation due to subject-dependent AIFs, IAUGC can be further improved by normalizing IAUGC by the blood signal, defined as the area under the AIF curve. The blood-normalized IAUGC is commonly computed up to 90 s after contrast injection, and it is shown that the blood-normalized IAUGC has great potential to be a useful semiquantitative imaging biomarker in the development of antiangiogenic and antivascular therapies [20].

In practice, AIF is difficult to measure, and an alternative approach is to normalize IAUC against reference measurements obtained from normal tissues. For example, IAUC measurements in tumors have been made using a reference IAUC obtained from muscle [3], but extra care must be taken to ensure that the normalization of tissue is not affected by any interventional procedures. An alternative method to minimize the influence of variable AIFs is to use a power injector or other highly reproducible contrast agent administration protocol.

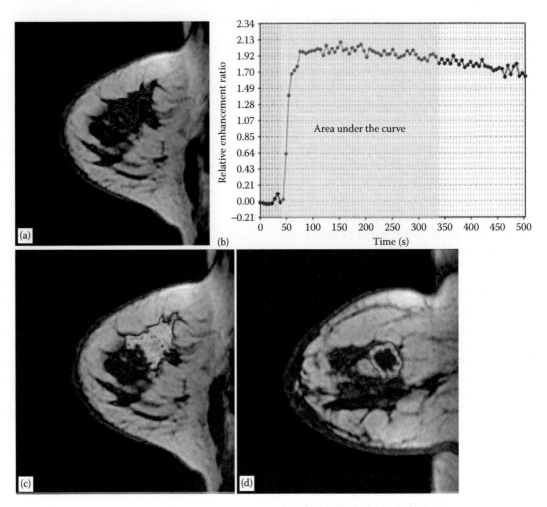

FIGURE 6.3
An example of the IAUC calculation over the tumor ROI (a), and the integration is performed over 300 s after contrast injection (b). Pixel-by-pixel IAUC maps on two different breast cancer patients (c and d).

6.2.3 Other Approaches

Other approaches to measure semiquantitative parameters include measurement of the peak concentration, initial slope of the curve, and time to maximum enhancement. These parameters are typically simpler to estimate than IAUC, but are known to be more prone to data noise and motion and therefore less reproducible than IAUC. An example of the initial slope calculation is shown in Figure 6.4.

6.2.4 Interpretation

The main drawback of using semiquantitative parameters is the difficulty in their interpretation. IAUC is proportional to the contrast agent delivery to the tissue of interest and can be further normalized by the blood signal to improve consistency, but it is still hard to directly correlate with physiological understanding, such as blood flow and microvascular endothelial

permeability. IAUC generally provides information of contrast agent kinetics determined by a combination of many factors such as blood flow, endothelial permeability, and extracellular extravascular space volume. Other semiquantitative parameters including measurement of the peak concentration, initial slope of the curve, and time to maximum enhancement have similar limitations in their interpretation, as they are similarly dependent on the combinations of the same factors without knowing the exact proportions.

6.3 Tissue Compartmentalization and Functional Parameters

Each imaging voxel contains vast, complex, vasculature, and cellular structures observable on the microscopic scale. Capillaries, for example, have a diameter

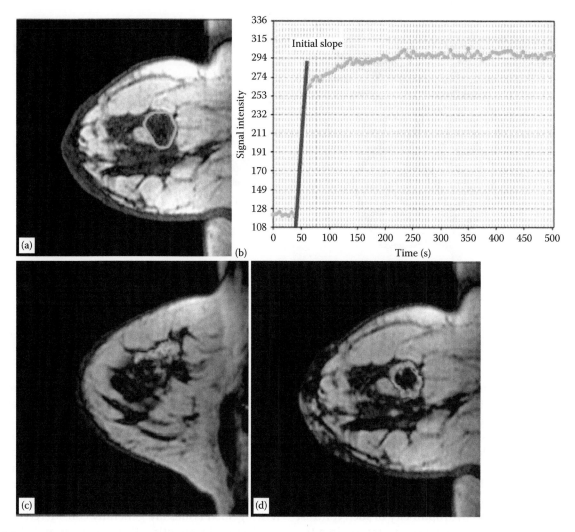

FIGURE 6.4
An example of the initial slope calculation over the tumor ROI (a and b) and pixel-by-pixel initial slope maps on two different breast cancer patients (c and d).

of 5–10 μm, while imaging voxels acquired in DCE-MRI typically cover tens of thousands of cubic microns. (In a typical DCE-MRI acquisition, the spatial resolution is limited by the requirement for high temporal resolution. Temporal resolution of 5–10 s is desirable. Voxel sizes of 10–20 mm³ are typical at the time of writing.)

During a DCE-MRI experiment, the aggregated effect of contrast agent accumulation and washout, within the vast microscopic structures contained within each acquired imaging voxel, is observed on a series of T_1-weighted images. In order to model contrast agent kinetics, the tissue is simplified by compartmentalization (Figure 6.5). This step allows the microscopic characteristics of the tissue to be aggregated, and the key mechanisms involved in contrast agent distribution to be modeled [21,22]. Bulk measurements that reflect the aggregated tissue characteristics over the whole imaging voxel may be extracted by fitting a compartmentalized model to DCE-MRI data. A model that links

contrast agent kinetics to tissue physiology allows quantification of meaningful physiological parameters using DCE-MRI. Compartmental modeling has been used extensively in DCE-MRI since the work of Tofts et al. [8,9].

The imaging voxel is most commonly split into three compartments. The intravascular or blood plasma space, the intracellular space, and the extravascular extracellular space (EES) are often referred to as the *interstitial space*. The total volume of all three compartments in a unit volume of tissue can be characterized as follows:

$$v_e + v_p + v_i = 1 \tag{6.4}$$

where:
v_i is the volume of the intracellular space
v_p is the intravascular or blood plasma volume
v_e is the volume of EES

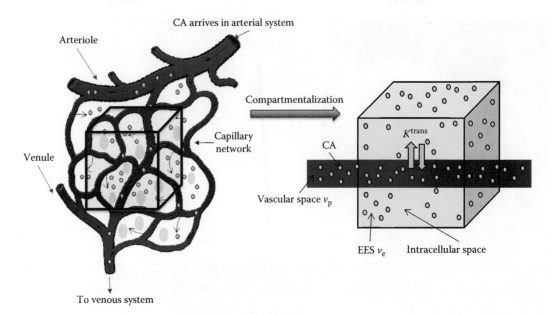

FIGURE 6.5
The vast microvasculature and cellular structures contained within the imaging voxel (black cube) are simplified by compartmentalization allowing the major mechanisms involved in contrast agent distribution to be modeled.

The volumes are commonly expressed as a fraction of total tissue volume or in their absolute units of mL per mL tissue, or, mL per gram tissue when tissue density is known. DCE-MRI is typically performed using low-molecular-weight gadolinium-based contrast agents. Such contrast agents diffuse from the intravascular space into the extracellular extravascular space, but do not cross cell membranes, and enter the intracellular space due to the size and hydrophilicity of such contrast agents [23,24]. The contrast agent concentration in a tissue or voxel can therefore be defined as

$$C_t = v_p C_p + v_e C_e \qquad (6.5)$$

where:

C_t is the total contrast agent concentration within the tissue
C_p is the contrast agent concentration in the blood plasma space
C_e is the contrast agent concentration in the EES

The major mechanisms of contrast agent distribution from which parameters related to tissue function are extracted from DCE-MRI data are perfusion (or blood flow) F, most commonly reported in units of milliliter of blood (g tissue)$^{-1}$ min^{-1}, capillary permeability surface area product PS, most commonly reported in units of ml g^{-1} min^{-1}, EES volume v_e and blood volume v_b both of which are usually reported in mL per mL of tissue.

Compartmentalized models used to fit DCE-MRI data provide a simplified description of the mechanisms responsible for contrast agent distribution, containing parameters measureable with DCE-MRI. In fact, there are other mechanisms and functional parameters that may impact contrast agent distribution. Such parameters include lymphatic drainage and contrast agent diffusion between voxels [25]. As improved MRI hardware facilitates better data quality, there has been interest in attempting to include some of these factors within compartmentalized models. However, at the time of writing such issues are generally considered confounding factors in quantification accuracy and the current accepted models are assumed to be adequate for accurate quantification.

6.4. Pharmacokinetic Modeling

6.4.1 History of Physiological Modeling

Tissue compartmentalization and tracer kinetic modeling have been used extensively to model functional physiological parameters since the work of Fick over 100 years ago. Fick used a single compartment model with oxygen as tracer to estimate blood flow in the brain. The model assumed oxygen uptake to be proportional to blood flow. The equation, which is the genesis of compartmental modeling, stated that the oxygen uptake of a tissue, Q (mol per unit time), is equal to blood flow, F (mL min^{-1} g^{-1}), multiplied by the difference in arterial and venous oxygen concentration, C_a and C_v, respectively such that

$$\frac{dQ}{dt} = F(C_a(t) - C_v(t)) \qquad (6.6)$$

Although utilizing oxygen as a tracer led to inaccurate quantification (due to consumption of the tracer), the approach taken formed the basis of pharmacokinetic modeling. Kety expanded upon the work of Fick and accurately quantified brain perfusion in 1944 utilizing nitrous oxide as a more suitable tracer [26]. Pharmacokinetic modeling in DCE-MRI adapts the early work of Fick and Kety to MRI where an injectable magnetic contrast agent is used as a tracer rather than an inhaled gas.

6.4.2 Conversion from Signal Intensity to Contrast Agent Concentration

During DCE-MRI, an image volume is dynamically acquired at a specified temporal resolution following injection of a contrast agent bolus over a specified time period (Figure 6.1). The contrast agent accumulation and washout are recorded as a change in signal intensity on a dynamic series of T_1-weighted images. In the presence of contrast agent of relaxivity, r_1, and concentration of C, T_1 is reduced from its baseline, precontrast value T_{10} as

$$\frac{1}{T_1} = \frac{1}{T_{10}} + r_1 C \qquad (6.7)$$

Pharmacokinetic modeling requires the concentration of contrast agent in each image to be calculated.

Contrast agent concentration can be calculated using the well-established relationship between T_1-weighted signal intensity change in the presence of contrast agent and the spoiled gradient-echo signal equation, assuming acquisition parameters flip angle and repetition time of the imaging sequence are known [8,10].

6.4.3 Pharmacokinetic Modeling DCE-MRI

The etymology of the term *pharmacokinetic* stems from the ancient Greek language. *Pharmakon* (*drug*) and *kinetikos* (*moving*) are combined to give us a term that describes the study of drugs or exogenously administered agents moving around the body.

In DCE-MRI, the agent of interest is an intravenously administered bolus of low-molecular-weight paramagnetic contrast agent. Following the conversion of signal intensity to contrast agent concentration, pharmacokinetic analysis can be carried out on the dynamic time series. Contrast agent concentration is plotted as a function of time in each pixel or over a ROI to generate curves to which pharmacokinetic models can be fitted (Figure 6.6). Least squares minimization fitting algorithms are utilized for this purpose; parameters of interest can be fitted as free parameters or, more commonly, be fitted with appropriate physiological constraints.

Pharmacokinetic analysis has significant advantages when compared to some of the model-free approaches discussed earlier in Section 6.2. Perhaps most importantly, no subjective interpretation of data is required, absolute quantification of physiological parameters

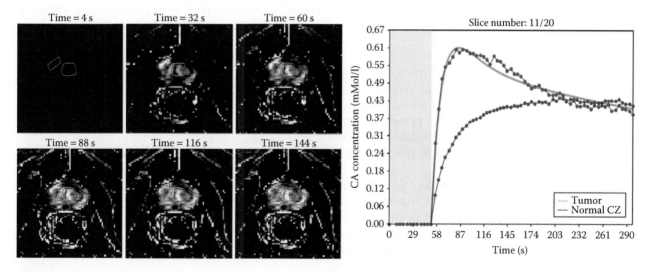

FIGURE 6.6

Coarsely sampled time series showing CA concentration images acquired during prostrate DCE-MRI with the corresponding time course plotted at the fully acquired temporal resolution shown in two regions of interest (ROIs) marked in red and blue on the images. The pathologically confirmed tumor region (red ROI on the images) shows rapid wash-in and washout (red line on plot) compared to healthy central zone tissue (blue ROI on the images and blue line on plot). An example of a pharmacokinetic model fitting is shown for the tumor, fitted red line on plot.

that are independent of data acquisition (field strength, readout parameters, etc.) is achievable. Absolute quantification allows statistical analysis of DCE-MRI data acquired at different sites, on different scanners and even with different read out parameters (e.g., T_e and T_R). Quantitative pharmacokinetic-modeled parameters are used extensively in drug trails and multicenter MRI studies [14,27]. Theoretically, even DCE-MRI data, acquired at lower SNR, temporal and spatial resolution should provide equivalent quantitative results to data acquired with state-of-the-art technology. In fact, only the levels of precision would be expected to show differences.

Second, the complexity of the pharmacokinetic model that is fitted to DCE-MRI data can be selected to reflect the quality of the data acquired [28]. Multiple pharmacokinetic models ranging from the simplest, to most complex may be fitted to acquire data.

The simplest pharmacokinetic model used in DCE-MRI quantifies the size of the EES v_e, and the bulk transfer coefficient of contrast agent K^{trans} [29]. However, more complex models that quantify absolute physiological parameters such as perfusion, F, and capillary permeability surface area product, PS, can also be fitted when the data quality is sufficient to estimate such parameters with acceptable levels of accuracy and precision. In a large multicenter study, simple pharmacokinetic-modeled parameters may be extracted from all the data while more complex parameters are only extracted from higher quality data. (High-quality data in DCE-MRI refers to data that has high SNR, low temporal resolution, and is optimized to minimize confounding factors such as motion and B1 inhomogeneity that are discussed in Chapter 7.)

6.4.4 Tofts Models

The most commonly used pharmacokinetic model in DCE-MRI is the simple Tofts model, also referred to as the *simple Kety model* due to the mathematical equivalence of approaches. Tofts applied the principles proposed by Kety where nitrous oxide is used as a tracer to DCE-MRI where low-molecular-weight paramagnetic contrast agent is the tracer used [8].

The simple Tofts model is a single compartment model. The injected bolus of contrast agent travels through the vasculature and due to its small molecular size (less than 1000 Da) diffuses from the capillaries into the EES. As the leakage from the capillaries is diffusive, it is also reversible. The diffusive leakage is proportional to concentration difference between the vascular or blood plasma space C_p and EES C_e. This relationship is described by the general rate equation proposed by Kety modified for DCE-MRI by Tofts as

$$v_e \frac{dC_e(t)}{dt} = K^{trans}(C_p(t) - C_e(t)) \tag{6.8}$$

where:

$C_e(t)$ is the contrast agent concentration in the EES

K^{trans} is the volume transfer coefficient between v_p and v_e

$C_p(t)$ is concentration of the contrast agent in the vascular or blood plasma space

Pharmacokinetic modeling of DCE-MRI data requires an estimate of the contrast agent concentration in the vascular or blood plasma space $C_p(t)$ entering the tissue or voxel. This is commonly referred to as the AIF and can be assumed or measured. The effect of contrast agent extravasation from the measured input $C_p(t)$, to the imaging voxel, is observed as a signal change on the series of T_1-weighted images. Pharmacokinetic models are fitted to this data to estimate model parameters. Many methods have been proposed for acquiring an accurate AIF [30,31] and for the form an assumed AIF should take [32]. An example of the commonly used population averaged AIF proposed by Parker and the modified Fritz–Hansen AIF are shown in Figure 6.7. Achieving an accurate AIF is a significant confounding factor for accurate quantification of physiological parameters using DCE-MRI. This and other sources of error are discussed in further detail in Section 6.5.

The rate of transfer of contrast agent from the vascular space to EES, K^{trans}, is dependent on blood flow, F, permeability of the microvessel walls, P, and the surface area of the perfusing vessels, S. The latter two terms are combined and referred to as permeability surface area product, PS. The transfer coefficient of contrast agent, K^{trans}, from the plasma space v_p, to the EES is defined as

$$K^{trans} = EFp(1 - Hct) \tag{6.9}$$

where:

F is the blood flow

Hct is the hematocrit fraction

p is the tissue density

E is the extraction fraction defined as

$$E = 1 - e - \frac{PS}{F(1 - Hct)} \tag{6.10}$$

where:

PS is the permeability surface area product

F is the blood flow

p is the tissue density

Hct is the blood hematocrit

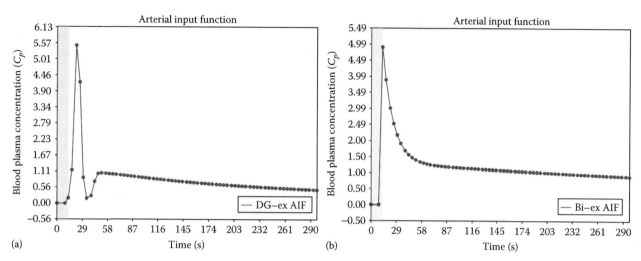

FIGURE 6.7
(a) The Parker population averaged AIF and (b) the modified Fritz–Hansen AIF.

As described in the Section 6.3, the contrast agent concentration in a tissue or voxel is the sum of the intravascular and extravascular concentrations in their respective intravascular and extravascular volumes. In many tissues, the vascular volume is a small fraction of total volume within the imaging voxel (Equation 6.5). When this is the case, it is often assumed that the vascular contribution to the observed signal is negligible and as such the total contrast agent concentration within the voxel $C(t) \sim v_e C_e(t)$ [10]. Concentration of contrast agent observed in a tissue during a DCE-MRI may then be defined as

$$C_t(t) = K^{trans} \int_0^t C_p(t') e^{-K^{trans}(t-t')/v_e} \, dt' \quad (6.11)$$

which may also be expressed as

$$C_t(t) = C_p(t) \otimes H(t) \quad (6.12)$$

where $H(t)$ is the impulse residue function:

$$H(t) = K^{trans} e^{-\left(K^{trans} t/v_e\right)} \quad (6.13)$$

This is commonly referred to as the Tofts, standard Tofts, or simple Kety model.

The standard Tofts model provides an approximation that is appropriate for tissues with relatively low blood volume. This approximation in not appropriate in tissues where blood volume is high as is the case in many pathologies as well as in healthy tissues such as lung and kidney. The model was therefore extended to account for contrast agent in the blood plasma volume. This is commonly referred to as the extended Tofts model defined as

$$C_t(t) = v_p C_p(t) + K^{trans} \int_0^t C_p(t') e^{-\left(K^{trans}(t-t')/v_e\right)} dt' \quad (6.14)$$

which may also be expressed as

$$C_t(t) = v_p C_p(t) + C_p(t) \otimes H(t) \quad (6.15)$$

where $H(t)$ is the impulse residue function:

$$H(t) = K^{trans} e^{-\left(K^{trans} t/v_e\right)} \quad (6.16)$$

To quantify pharmacokinetic model parameters such as v_p, v_e, and K^{trans} from DCE-MRI data, the model of choice is fitted to the time concentration curves in a voxel or ROI with the parameters of interest fitted as free parameters using a nonlinear least squares minimization [33].

6.4.5 Moving toward Absolute Quantification of *F* and PS

The Tofts and extended Tofts models assume that contrast agent is instantaneous and well mixed in each of the compartments it occupies. However, this is a physiologically inaccurate assumption as contrast agent concentration would be expected to vary with distance along the capillary [34]. The tissue homogeneity model offers a more complex, but physiologically accurate pharmacokinetic model where the concentration gradient along the capillary is included (Figure 6.8). An adiabatic approximation that $dC_e/dt \ll dC_p/dt$ provides a solution to the tissue homogeneity model such that the model can be fitted to DCE-MRI data [35].

As the adiabatic approximation to the tissue homogeneity (AATH) model accounts for the concentration

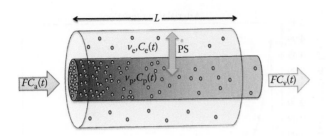

FIGURE 6.8
AATH model accounts for the gradient of contrast agent concentration $C_p(t)$ between the arterial $C_a(t)$ and venous $C_v(t)$ ends of the capillary. Contrast agent concentration in the extravascular space $C_e(t)$ is assumed to be well mixed. In the schematic, F represents perfusion, PS is capillary permeability surface area product, v_p is the volume of the blood plasma space, and v_e is the volume of the extravascular extracellular space.

gradient of contrast agent along the capillary an additional model parameter τ, which is the mean time taken for a molecule of contrast agent to traverse the capillary bed, is fitted. The model is defined as

$$C_t(t) = F\rho(1 - Hct)\int_0^\tau C_p(t - t')dt'$$

$$+ EF\rho(1 - Hct)\int_\tau^t C_p(t - t')e^{-(EF\rho(1-Hct)(t'-\tau)/v_e)}dt' \tag{6.17}$$

This may be expressed as

$$C_t(t) = C_p(t) \otimes H(t) \tag{6.18}$$

where:

$$H(t) = F\rho(1 - Hct), (0 \le t \le \tau)$$

$$H(t) = EF\rho(1 - Hct)e^{-(EF\rho(1-Hct)/v_e)}(t \ge \tau) \tag{6.19}$$

where:
  τ is the mean capillary transit time
  F is the perfusion (plasma blood flow)
  Hct is the hematocrit fraction
  p is tissue density
  E is the extraction fraction
  v_e is the volume of the EES

Values of K^{trans} can be generated from previous relationship between K^{trans}, F, and E (Equation 6.9), while v_p may be generated via the relationships $v_p = \tau Fp(1 - Hct)$.

At times shorter than $\tau (0 \le t \le \tau)$ the residue function $H(t)$ remains constant, as the contrast agent has not yet had time to traverse and leave the capillary. Contrast agent concentration within the tissue is therefore equal

to the inflow of the contrast agent before τ. After τ, contrast agent not extravasated into the EES is washed out, at this time the residue function takes the same form as the Tofts model.

More complex pharmacokinetic models such as the described AATH model allow K^{trans} to be *decoded* into its constituent parts of extraction fraction, E, and perfusion F. When data quality is sufficient, these parameters can be quantified from DCE-MRI data to acceptable levels of accuracy and precision [36].

With improving MR technology, the importance of pharmacokinetic modeling to facilitate quantification of absolute physiological parameters can be expected to significantly increase.

6.5 Sources of Error

The main sources of error associated with kinetic modeling include measuring an accurate AIF and establishing an accurate relationship between signal change and contrast agent concentration. All these issues need to be resolved to increase confidence in quantified estimates of physiological parameters, and we describe three major sources of error such as AIF, the blood hematocrit fraction, and the T_1 measurement.

6.5.1 Arterial Input Function

The pharmacokinetic modeling requires knowledge of AIF, but measuring AIF is very difficult in practice. AIF can be either measured for each subject if fast MRI techniques are available or assumed by a population average AIF [10,30,32]. If the DCE-MRI data acquisition is not properly implemented, the subject-based AIF can introduce extra variation, contaminating the final quantitative measures, while the population-based AIF includes potential discrepancy between actual and assumed AIFs. Figure 6.9 shows different pharmacokinetic model parameters (K^{trans} and k_{ep}) when various AIFs are either assumed (Weimann AIF [10] and Parker AIF [32]) or measured on the same contrast agent concentration curve. The appropriate choice for AIF will be dependent on the tissue being studied, available pulse sequences, and the desired temporal resolution.

6.5.2 Blood Hematocrit Fraction

The blood hematocrit fraction (Hct) plays an important role in the quantitative DCE-MRI analysis since the contrast agent only occupies the plasma compartment of the blood. Any AIF measurements are required to include the following relationship to covert the measured blood concentration (C_b) into the blood plasma concentration (C_p); $C_p = C_b/(1 - Hct)$. In practice, a value

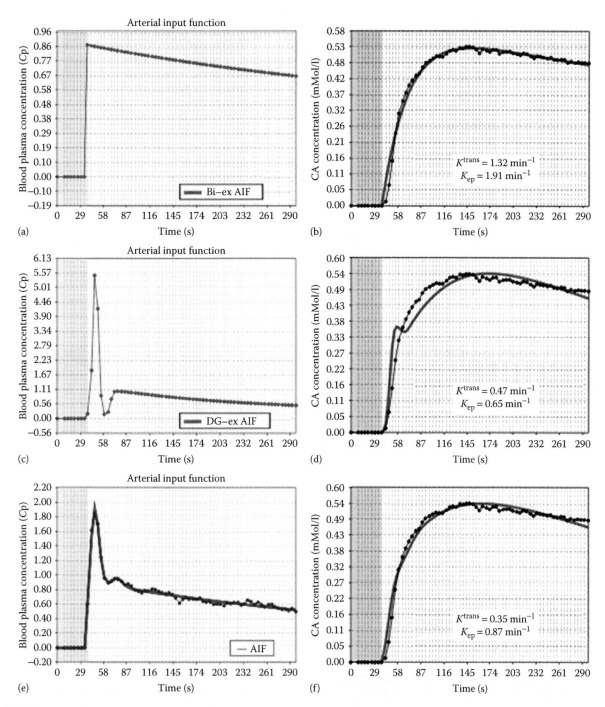

FIGURE 6.9
Pharmacokinetic model parameters (K^{trans} and k_{ep}) when various AIFs are used for quantitative DCE-MRI analysis: Weimann AIF and its corresponding contrast agent concentration curve (a and b), Parker AIF and its corresponding contrast agent concentration curve (c and d), and subject-based AIF and its corresponding contrast agent concentration curve (e and f).

of *Hct* is not explicitly measured but generally assumed to be 0.42 [37], which may lead to errors since *Hct* may vary in patients with cancer.

Another consideration is that *Hct* may vary between large vessels and the microvasculature since the packing of red blood cells in the capillary bed is less dense, leading to a smaller value of *Hct*. Additionally, these variations can be more pronounced for the poorly formed and variably perfused vessels within a tumor.

6.5.3 *T$_1$* Measurement

One common method to measure *T$_1$* values is variable flip angle (VFA) imaging, also known as driven

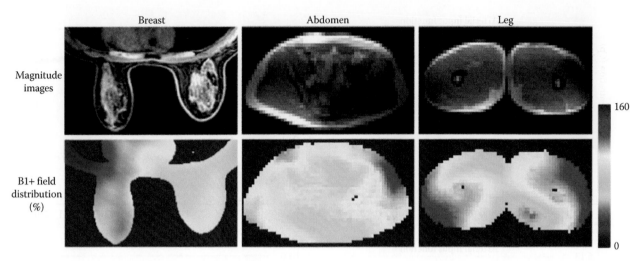

FIGURE 6.10
Relative B1+ field distribution (%) in the breast, abdomen and leg at 3 T.

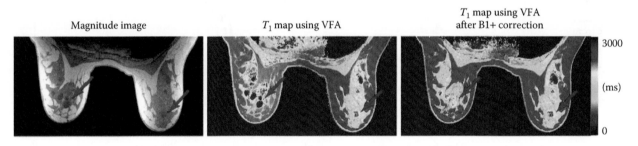

FIGURE 6.11
T_1 maps using VFA before and after B1+ correction. The magnitude image for an anatomical reference (left), the VFA T_1 map (middle), and the VFA T_1 map after B1+ correction (right). The arrow shows heterogeneous T_1 of several breast masses. (From Sung, K. et al., *Magn. Reson. Med.*, 70, 1306–1318, 2013. With permission.)

equilibrium single-pulse observation of T_1 (DESPOT1), which uses several short T_R RF-spoiled gradient-echo (SPGR) acquisitions with VFAs [38–40]. Using the SPGR signal equation and linear fitting, T_1 maps can be estimated. VFA imaging is widely used in DCE-MRI since it is highly time-efficient and allows rapid 3D volumetric T_1 mapping with the same pulse sequence used to measure contrast uptake [38,41]. Even though many efforts have been made to improve the accuracy of VFA imaging, VFA methods seem to be less accurate in vivo due to their high sensitivity to any flip angle variations [42–44].

Nonuniformity of the transmit RF (B1+) field can lead to flip angle variations, known to be significant, especially with current high-field MRI systems. At 3 T, which offers a higher SNR and improved temporal and spatial resolution, noticeable B1+ variations over the body have been observed by many studies [45–47], causing huge errors in T_1 calculation [48–50]. Figure 6.10 shows the relative flip angle distribution in various body areas at 3 T. Therefore, the B1+ variation should be carefully addressed for quantitative DCE-MRI analysis, and the

accuracy of VFA can be improved by compensating for the B1+ variation [49,51], as shown in Figure 6.11.

Inaccurate measurement of T_1 directly impacts on establishing an accurate relationship between signal change and contrast agent concentration. Any kinetic modeling based on the contrast agent concentration will be significantly impacted. Figure 6.12 shows an example of the *chain reaction* of errors in quantitative DCE-MRI analysis. T_1 map has been overestimated by 50% due to the B1+ inhomogeneity at 3 T, and in turn the contrast agent concentration is underestimated, which causes underestimation of K^{trans} by 55%.

6.6 Summary

Both semiquantitative and quantitative kinetic modeling are possible for DCE-MRI analysis to provide useful information of microvascular functional parameters. The IAUC is commonly used for model-free approaches,

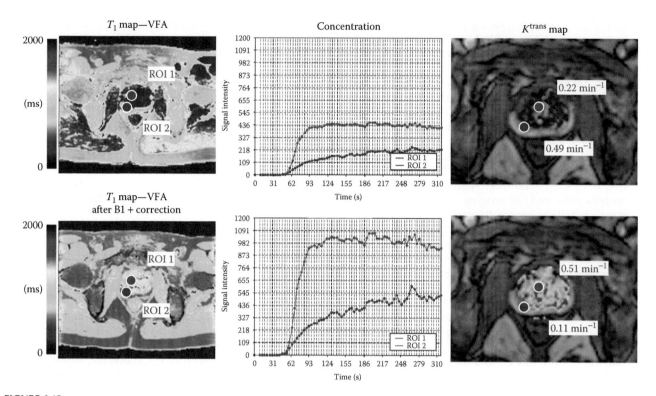

FIGURE 6.12
Overall impact of the inaccurate T_1 maps due to B1+ inhomogeneity at 3 T. The overestimated T_1 map results in underestimation of contrast agent concentration, which finally causes the underestimation of the K^{trans} map (55% underestimation).

but its interpretation is somewhat limited due to the compounding factors contributing the value of IAUC. The primary physiological parameters that can be measured with quantitative DCE-MRI are K^{trans} (related to capillary permeability, surface area, and perfusion) and v_e (related to the size of EES). This requires good control of pulse sequence parameters and an accurate measurement of tissue T_1 before injection of contrast agent. The possible and optimum acquisition protocols and models will depend on tissue that is being imaged.

References

1. Jackson A, O'Connor JPB, Parker GJM, Jayson GC: Imaging tumor vascular heterogeneity and angiogenesis using dynamic contrast-enhanced magnetic resonance imaging. *Clin Cancer Res* 2007, **13**:3449–59.
2. Caravan P, Ellison JJ, McMurry TJ, Lauffer RB: Gadolinium(III) chelates as MRI contrast agents: Structure, dynamics, and applications. *Chem Rev* 1999, **99**:2293–352.
3. Evelhoch JL: Key factors in the acquisition of contrast kinetic data for oncology. *J Magn Reson Imaging* 1999, **10**:254–9.
4. Erlemann R, Reiser MF, Peters PE, Vasallo P, Nommensen B, Kusnierz-Glaz CR, Ritter J, Roessner A: Musculoskeletal neoplasms: static and dynamic Gd-DTPA–enhanced MR imaging. *Radiology* 1989, **171**:767–73.
5. Hawighorst H, Libicher M, Knopp MV, Moehler T, Kauffmann GW, Kaick GV: Evaluation of angiogenesis and perfusion of bone marrow lesions: Role of semiquantitative and quantitative dynamic MRI. *J Magn Reson Imaging* 1999, **10**:286–94.
6. Galbraith SM, Lodge MA, Taylor NJ, Rustin GJS, Bentzen S, Stirling JJ, Padhani AR: Reproducibility of dynamic contrast-enhanced MRI in human muscle and tumours: Comparison of quantitative and semi-quantitative analysis. *NMR Biomed* 2002, **15**:132–42.
7. Østergaard M, Stoltenberg M, Løvgreen-Nielsen P, Volck B, Sonne-Holm S, Lorenzen I: Quantification of synovistis by MRI: Correlation between dynamic and static gadolinium-enhanced magnetic resonance imaging and microscopic and macroscopic signs of synovial inflammation. *Magn Reson Imaging* 1998, **16**:743–54.
8. Tofts PS, Brix G, Buckley DL, Evelhoch JL, Henderson E, Knopp MV, Larsson HB et al.: Estimating kinetic parameters from dynamic contrast-enhanced T(1)-weighted MRI of a diffusable tracer: Standardized quantities and symbols. *J Magn Reson Imaging* 1999, **10**:223–32.
9. Sourbron SP, Buckley DL: Tracer kinetic modelling in MRI: Estimating perfusion and capillary permeability. *Phys Med Biol* 2011:R1–R33.

10. Tofts PS, Kermode AG: Measurement of the blood-brain barrier permeability and leakage space using dynamic MR imaging. 1. Fundamental concepts. *Magn Reson Med* 1991, **17**:357–67.

11. Jans L, De Coninck T, Wittoek R, Lambrecht V, Huysse W, Verbruggen G, Verstraete K: 3 T DCE-MRI assessment of synovitis of the interphalangeal joints in patients with erosive osteoarthritis for treatment response monitoring. *Skeletal Radiol* 2013, **42**:255–60.

12. Roberts C, Parker GJM, Rose CJ, Watson Y, O'Connor JP, Stivaros SM, Jackson A, Rushton VE: Glandular function in Sjögren syndrome: Assessment with dynamic contrast-enhanced MR imaging and tracer kinetic modeling—Initial experience. *Radiology* 2008, **246**:845–53.

13. Buckley DL, Shurrab AE, Cheung CM, Jones AP, Mamtora H, Kalra PA: Measurement of single kidney function using dynamic contrast-enhanced MRI: Comparison of two models in human subjects. *J Magn Reson Imaging* 2006, **24**:1117–23.

14. O'Connor JPB, Jackson A, Parker GJM, Jayson GC: DCE-MRI biomarkers in the clinical evaluation of antiangiogenic and vascular disrupting agents. *Br J Cancer* 2007, **96**:189–95.

15. Cervera Deval J: RECIST and the radiologist. *Radiologia* 2012.

16. George ML, Padhani AR, Dzik-jurasz AS, Padhani AR, Brown G, Tait DM, Eccles SA, Eccles SA, Swift RI: Non-invasive methods of assessing angiogenesis and their value in predicting response to treatment in colorectal cancer. *Br J Surg* 2001, **88**:1628–36.

17. Leach MO, Brindle KM, Evelhoch JL, Griffiths JR, Horsman MR, Jackson A, Jayson GC et al.: The assessment of antiangiogenic and antivascular therapies in early-stage clinical trials using magnetic resonance imaging: Issues and recommendations. *Br J Cancer* 2005, **92**:1599–610.

18. Daniel BL, Yen YF, Glover GH, Ikeda DM, Birdwell RL, Sawyer-Glover AM, Black JW, Plevritis SK, Jeffrey SS, Herfkens RJ: Breast disease: Dynamic spiral MR imaging. *Radiology* 1998, **209**:499–509.

19. Hayes C, Padhani AR, Leach MO: Assessing changes in tumour vascular function using dynamic contrast-enhanced magnetic resonance imaging. *NMR Biomed* 2002, **15**:154–63.

20. Leach MO, Brindle KM, Evelhoch JL, Griffiths JR, Horsman MR, Jackson A, Jayson G, et al.: Assessment of antiangiogenic and antivascular therapeutics using MRI: Recommendations for appropriate methodology for clinical trials. *Br J Radiol* 2003, **76 Spec No**:S87–91.

21. Tofts PS, Kermode AG: Measurement of the blood-brain barrier permeability and leakage space using dynamic MR imaging. 1. Fundamental concepts. *Magn Reson Med* 1991, **17**:357–67.

22. Larsson HB, Stubgaard M, Frederiksen JL, Jensen M, Henriksen O, Paulson OB: Quantitation of blood-brain barrier defect by magnetic resonance imaging and gadolinium-DTPA in patients with multiple sclerosis and brain tumors. *Magn Reson Med* 1990, **16**:117–31.

23. Parker GJ, Tofts PS: Pharmacokinetic analysis of neoplasms using contrast-enhanced dynamic magnetic resonance imaging. *Top Magn Reson Imaging* 1999, **10**:130–42.

24. Weinmann HJ, Brasch RC, Press WR, Wesbey GE: Characteristics of gadolinium-DTPA complex: A potential NMR contrast agent. *AJR Am J Roentgenol* 1984, **142**:619–24.

25. Roberts C, Issa B, Stone A, Jackson A, Waterton JC, Parker GJM: Comparative study into the robustness of compartmental modeling and model-free analysis in DCE-MRI studies. *J Magn Reson Imaging* 2006, **23**:554–63.

26. Kety SS: The theory and applications of the exchange of inert gas at the lungs and tissues. *Pharmacol Rev* 1951, **3**:1–41.

27. Zee Y-K, O'Connor JPB, Parker GJM, Jackson A, Clamp AR, Taylor M Ben, Clarke NW, Jayson GC: Imaging angiogenesis of genitourinary tumors. *Nat Rev Urol* 2010, **7**:69–82.

28. Buckley DL: Uncertainty in the analysis of tracer kinetics using dynamic contrast-enhanced T1-weighted MRI. *Magn Reson Med* 2002, **47**:601–6.

29. Sourbron SP, Buckley DL: On the scope and interpretation of the Tofts models for DCE-MRI. *Magn Reson Med* 2011, **66**:735–45.

30. Fritz-Hansen T, Rostrup E, Larsson HB, Søndergaard L, Ring P, Henriksen O: Measurement of the arterial concentration of Gd-DTPA using MRI: A step toward quantitative perfusion imaging. *Magn Reson Med* 1996, **36**:225–31.

31. Roberts C, Little R, Watson Y, Zhao S, Buckley DL, Parker GJM: The effect of blood inflow and B(1)-field inhomogeneity on measurement of the arterial input function in axial 3D spoiled gradient echo dynamic contrast-enhanced MRI. *Magn Reson Med* 2011, **65**:108–19.

32. Parker GJM, Roberts C, Macdonald A, Buonaccorsi GA, Cheung S, Buckley DL, Jackson A, Watson Y, Davies K, Jayson GC: Experimentally-derived functional form for a population-averaged high-temporal-resolution arterial input function for dynamic contrast-enhanced MRI. *Magn Reson Med* 2006, **56**:993–1000.

33. Ahearn TS, Staff RT, Redpath TW, Semple SIK: The use of the Levenberg-Marquardt curve-fitting algorithm in pharmacokinetic modelling of DCE-MRI data. *Phys Med Biol* 2005, **50**:N85–92.

34. St Lawrence KS, Lee TY: An adiabatic approximation to the tissue homogeneity model for water exchange in the brain: II. Experimental validation. *J Cereb Blood Flow Metab* 1998, **18**:1378–85.

35. Naish JH, Kershaw LE, Buckley DL, Jackson A, Waterton JC, Parker GJM: Modeling of contrast agent kinetics in the lung using T1-weighted dynamic contrast-enhanced MRI. *Magn Reson Med* 2009, **61**:1507–14.

36. Kershaw LE, Cheng H-LM: Temporal resolution and SNR requirements for accurate DCE-MRI data analysis using the AATH model. *Magn Reson Med* 2010, **64**:1772–80.

37. Tofts PS, Brix G, Buckley DL, Evelhoch JL, Henderson E, Knopp MV, Larsson HBW, Lee T-Y, Mayr NA, Parker GJM: Estimating kinetic parameters from dynamic contrast-enhanced T1-weighted MRI of a diffusable tracer: Standardized quantities and symbols. *J Magn Reson Imaging* 1999, **10**:223–32.

38. Deoni SCL, Rutt BK, Peters TM: Rapid combined T1 and T_2 mapping using gradient recalled acquisition in the steady state. *Magn Reson Med* 2003, **49**:515–26.

39. Brookes JA, Redpath TW, Gilbert FJ, Murray AD, Staff RT: Accuracy of T_1 measurement in dynamic contrast-enhanced breast MRI using two- and three-dimensional variable flip angle fast low-angle shot. *J Magn Reson Imaging* 1999, **9**:163–71.

40. Zhu XP, Li KL, Kamaly-Asl ID, Checkley DR, Tessier JJ, Waterton JC, Jackson A: Quantification of endothelial permeability, leakage space, and blood volume in brain tumors using combined T_1 and T_2^* contrast-enhanced dynamic MR imaging. *J Magn Reson Imaging* 2000, **11**:575–85.

41. Wang HZ, Riederer SJ, Lee JN: Optimizing the precision in T1 relaxation estimation using limited flip angles. *Magn Reson Med* 1987, **5**:399–416.

42. Treier R, Steingoetter A, Fried M, Schwizer W, Boesiger P: Optimized and combined T1 and B1 mapping technique for fast and accurate T1 quantification in contrast-enhanced abdominal MRI. *Magn Reson Med* 2007, **57**:568–76.

43. Deoni SCL: High-resolution T1 mapping of the brain at 3T with driven equilibrium single pulse observation of T1 with high-speed incorporation of RF field inhomogeneities (DESPOT1-HIFI). *J Magn Reson Imaging* 2007, **26**:1106–11.

44. Sung K, Daniel BL, Hargreaves BA: Transmit B1+ field inhomogeneity and T1 estimation errors in breast DCE-MRI at 3 tesla. *J Magn Reson Imaging* 2013, **38**:454–9.

45. Greenman RL, Shirosky JE, Mulkern RV, Rofsky NM: Double inversion black-blood fast spin-echo imaging of the human heart: A comparison between 1.5T and 3.0T. *J Magn Reson Imaging* 2003, **17**:648–55.

46. Sung K, Nayak KS: Measurement and characterization of RF nonuniformity over the heart at 3T using body coil transmission. *J Magn Reson Imaging* 2008, **27**:643–8.

47. Kuhl CK, Kooijman H, Gieseke J, Schild HH: Effect of B1 inhomogeneity on breast MR imaging at 3.0 T. *Radiology* 2007, **244**:929–30.

48. Azlan CA, Di Giovanni P, Ahearn TS, Semple SIK, Gilbert FJ, Redpath TW: B1 transmission-field inhomogeneity and enhancement ratio errors in dynamic contrast-enhanced MRI (DCE-MRI) of the breast at 3T. *J Magn Reson Imaging* 2010, **31**:234–9.

49. Di Giovanni P, Azlan CA, Ahearn TS, Semple SI, Gilbert FJ, Redpath TW: The accuracy of pharmacokinetic parameter measurement in DCE-MRI of the breast at 3 T. *Phys Med Biol* 2010, **55**:121–32.

50. Sung K, Daniel BL, Hargreaves BA: Transmit B1+ field inhomogeneity and T1 estimation errors in breast DCE-MRI at 3 Tesla. *J Magn Reson Imaging* 2013, 38: 454–459.

51. Sung K, Saranathan M, Daniel BL, Hargreaves BA: Simultaneous T1 and B1+ mapping using reference region variable flip angle imaging. *Magn Reson Med* 2013, **70**:1306–18.

7

BOLD Functional Magnetic Resonance Imaging

Chris J. Conklin, Devon M. Middleton, Scott H. Faro, and Feroze B. Mohamed

CONTENTS

7.1 Introduction

In the last few decades, there has been a constant need to develop new imaging techniques that aid in the investigation of biological function of the brain in conjunction with the associated anatomical information. Blood oxygenation level-dependent imaging (BOLD) is one such method that has been researched and developed to measure and map the location of activation in the brain related to neuronal activity. Although measured indirectly, this method has enabled the neuroscience community at large to noninvasively study brain functions in a reliable and fairly reproducible manner in humans. In the last two decades, this ability has enabled the mapping of the functions of various regions of the brain, and has helped to understand the complex connections in brain circuitry; it continues to be an exciting method to study brain functions. In recent years, this technique has moved from bench-top research to clinical bedside applications, particularly in the area of presurgical localization of cortical structures. This chapter discusses the principles underlying this technique, and the application of this method in clinical medicine.

7.2 Magnetic Resonance Physics

In order to generate magnetic resonance (MR) signals, it is necessary to manipulate the magnetization of protons within the human body. Given that our bodies are composed mostly of water (two-third by weight), the hydrogen atom is the ideal candidate. When placed in a large, static magnetic field ($B0$), the hydrogen nucleus acts like a compass needle as it becomes magnetized. Each nucleus precesses about $B0$, referred to as a *spin*. These spins can align themselves in both parallel and antiparallel manner with respect to $B0$, with the parallel spins being in a lower energy state and the antiparallel spins in a high-energy state. While the distribution of parallel and antiparallel spins is completely random, a small but significant number of unpaired nuclei exist aligned with the main field $B0$. Summing these parallel magnetic moments yields a net magnetization vector ($M0$), which forms the basis for MR signal synthesis.

Manipulation of $M0$ is what generates the stimulated echo, or signal, containing the relevant tissue-specific information. The unpaired nuclei that make up the net magnetization vector can absorb electromagnetic photons in the radiofrequency (RF) spectrum. This frequency for absorption, generally referred to as *Larmor frequency*, is a linear function of the gyromagnetic ratio of the hydrogen nucleus and the static magnetic field strength. When this resonance condition is met, the unpaired nuclei undergo a series of energy transitions through the absorption and emission of electromagnetic photons. During equilibrium, there is insufficient energy to precipitate the required energy level transitions. These transitions are achieved by the introduction of transient, time-varying magnetic fields, where the necessary condition for resonance is that the frequency of the oscillating magnetic field matches with the frequency of spin precession. The oscillating field is called a *radiofrequency* (RF) *pulse*.

As previously stated, when the parallel spins that make up the net magnetization are exposed to an RF pulse, they absorb photons and jump to a higher energy state. The spins undergoing this energy transition are *excited* and the specific RF pulse that provokes this response is typically known as an *excitation pulse*. A favorable consequence resultant from this excitation is that transitioning to a higher energy states means that the collection of unpaired spins become *tilted* such that they are no longer parallel with the static field $B0$. As the number of unpaired nuclei absorbing photons increases, the net magnetization vector will continue to reorient itself further from the parallel state. This degree of deflection of the net magnetization relative to $B0$ is called the *flip angle* and is representative of the number of spins present in the higher energy state. The power of the excitation pulse is ultimately what dictates this quantity. Power is a function of the duration, strength, and the frequency of the applied pulse. Tipping the net magnetization through interaction with an RF excitation pulse creates a transverse magnetization component that plays a key role in signal generation.

Through the application of transient excitation pulses, it becomes possible to track and evaluate the dynamic changes the net magnetization vector experiences. When the pulse is applied, the magnetization vector rotates away from the longitudinally aligned $B0$ field. After the excitation pulse has played out, two processes dominate to restore equilibrium, each independent of the other. The transverse magnetization vector decays as a function of spin–spin interactions. As each spin is in essence a single magnetic moment, perturbations between neighboring nuclei will induce inhomogeneities in the magnetic field. These inhomogeneities create a local field gradient that alters the precessional frequency of spatially dependent nuclei. With spins precessing at different rates, a loss in coherence occurs, known as *dephasing*. Given that the transverse magnetization vector is the sum of all field contributions, this phase incoherence gradually reduces the magnitude of the transverse component as the phases of spins become uniformly distributed. While the transverse magnetization decays over time, the longitudinal magnetization begins to regrow. As interactions between nuclei and the surrounding microenvironment occur, the spins return to an equilibrium state as they drop to a lower energy level, which is parallel to the static field $B0$. The combination of both of these relaxation phenomena enables the measurement of an MR signal for a region of local excitation. In short, the MR signal arises from an induced current resulting from the net magnetic moment created by a collection of coherent (in-phase) spins precessing about the longitudinal axis of the static field $B0$. This precession is effected by applying an RF pulse at the Larmor frequency of the nucleus of interest, which results in a tipping of the net magnetization $M0$ such that it is no longer parallel with the $B0$ field, with stronger RF pulses yielding greater change in $M0$. The rates at which transverse magnetization decays and longitudinal magnetization recovers are affected by interactions in the local microenvironment and thus can be exploited to achieve tissue contrast as local microenvironments vary between differing biological tissues.

The notion of field inhomogeneities and their effect on spin coherence also play a critical role in the spatial localization of a received MR signal. By inducing precise and well-defined inhomogeneities along three orthogonal axes, spin coherence can be predictably anticipated and exploited. As mentioned previously, the strength of the field dictates the precessional frequency. Subsequently, the rate at which the spins precess can be varied through the application of linearly changing magnetic fields, called *gradients*. Three gradients are present that can spatially distribute the precessional frequencies and phases in a specific region of a magnetized sample. When these gradients are applied, the frequency content of the received MR signal can be analyzed. This is known as *frequency encoding* and is used to fill in one axis of a three-dimensional k-space. The k-space is the frequency analog of spatial domain signals, that is, in two dimensions, k-space data is aligned on frequency and phase axes, rather than the familiar x–y of the spatial domain. By continuing to apply gradients at specific time points during acquisition, it becomes possible to construct a grid of MR signals, each with a unique frequency, phase, and slice coordinate. Upon transformation back to the spatial domain, through the use of a Fourier transform, reconstruction is achieved that enables clinicians and researchers to visualize the inherent water characteristics of the imaging volume. Extension of this concept allows for the imaging of brain regions under heightened activation by tracking the oxygen content of blood.

7.3 Neuro-Physiological Basis of BOLD Signal

Functional MR imaging (fMRI) has become ubiquitous in modern imaging communities. This branch of imaging focuses on probing physiological changes during certain events or behaviors. BOLD imaging is the most popular and well-studied method of measuring activation in the brain using MR. BOLD imaging relies on the differing magnetic characteristics of oxyhemoglobin and deoxyhemoglobin (dHb), which result in changes in a measured MR signal. This phenomenon was initially demonstrated ex vivo by imaging blood with varying levels of oxygen [1]. Deoxygenated blood creates magnetic susceptibility differences in the regional microenvironment, thereby inducing local field inhomogeneities. However, the knowledge that changes in the concentrations of oxyhemoglobin and dHb can be measured by MR was not in and of itself sufficient to measure neuronal activation without evidence that those changes were representative of changes in local brain activity. It was shown that there is an increase in $R2$ values, the T_2 relaxation rate defined as the inverse of the transverse magnetization decay time, as relative dHb concentration increases due to diffusion of water through these local field variations. These results postulated the feasibility of using blood oxygen as an endogenous contrast agent for rapid in-vivo imaging of functional regions. Clearly, certain physiologic phenomena influence the demand for oxygen in regions of heightened neuronal activity, as stressed tissues have a higher metabolic demand than those at rest. Early findings suggested the local uncoupling of cerebral blood flow (CBF) and cerebral metabolic rate of oxygen ($CMRO_2$) during induced neuronal excitation. Dynamic regulation of blood flow due to neuronal activity is independent of $CMRO_2$ when studied in comparison between rest (somatosensory) and active conditions. Physiological changes induced by regional stimulation showed a significant increase in CBF relative to $CMRO_2$, while also raising the partial pressure of oxygen (PO_2) and decreasing oxygen extraction fraction. This was suggestive of the extent to which increased CBF, rather than actual oxygen uptake, was indicative of local neuronal activity [2]. In addition, regional cerebral metabolic rate of glucose (CMR_{glu}) and CBF are significantly increased by localized neuronal excitation as was first demonstrated through visual cortex stimulation. Through visual stimulation, it was shown that despite a local increase in both CBF and CMR_{glu}, a minimal amount of the increased glucose uptake was oxidized in the visual cortex. This seemed to suggest that an increase in CBF due to brain excitation satisfies biochemical or neural physiologic needs beyond oxidative metabolism [3]. This early work established the concept that neuronal activation should be expected to exhibit local increases

in the relative concentration of oxyhemoglobin versus dHb due to increased CBF, but decreased oxygen extraction fraction. Knowledge of this change in concentration allowed for the known differences in magnetic character between oxy/deoxyhemoglobin to be exploited via MR.

By 1990, in-vivo functional brain mappings of rats was made possible through the presence of dHb in venous blood [4]. dHb serves as an endogenous contrast agent in the presence of gradient-echo (GRE) acquisition schemes that enables the quantification of blood flow during heightened states of neuronal activity. This was an attractive alternative and/or complement to positron emission tomography (PET) mappings for tracking similar measurements. BOLD imaging may have even more utility in comparison to PET as the presence of an intrinsic paramagnetic agent eliminates the need for invasive tracer methods. As previously detailed, BOLD contrast is predominantly influenced by blood flow and oxygen extraction by surrounding microstructure. Arterial blood flow is highly oxygenated in contradistinction to venous flow, which has an abundance of deoxygenated blood. This venous flow contributes greatly to the enhanced (dark) contrast noticed in BOLD [4]. Due to the abundance of water concentration in tissue, physiologic changes are difficult to access. This is predominantly a function of the insensitivity that metabolic reactions have on the generated echo. BOLD imaging offers an indirect method of evaluating physiologic response to neuronal stimulation [5,6]. Figure 7.1 summarizes the physiologic processes behind the BOLD response.

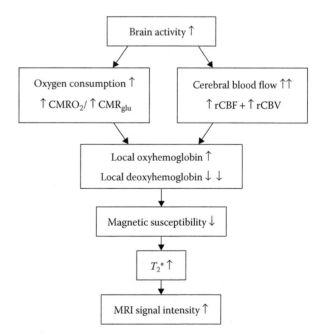

FIGURE 7.1
Physiologic processes involved in BOLD response.

Spatial specificity in BOLD imaging is a function of the local hemodynamic response as well as the vascular structure of the imaging region. Capillaries are closest in proximity to neurons and therefore the site of the activation. After neural excitation, the dHb drain from the capillaries to intracortical veins and then into pial veins. The farther the deoxygenated blood travels downstream, the larger the vessel diameter. This has the confounding effect of combining with blood from other areas and adversely influencing BOLD specificity. Mitigation of large draining vessel contamination can be addressed from an acquisition standpoint (SE-EPI or early dip BOLD) [7]. It is also important to understand that due to the multicompartmental nature of signal generation, any changes to the microenvironment, through biochemical/physical response, will influence the received BOLD signal [8].

signal when inflowing blood replaces the saturated spins from the excited imaging region. However, while increases in blood flow due to neuronal activation enhance the functional response, large vessel contributions reduce spatial specificity. BOLD contrast parameter T_2^* is largely influenced by phase incoherence through paramagnetic dHb content [1]. Water quickly exchanges between the red blood cells and plasma while also diffusing due to susceptibility-induced field inhomogeneities. These processes cause phase incoherence that reduce T_2^* values. Another phenomenon that reduces T_2^* is the angle between the static field **B0** and vessel orientation. Given that multiple vessels are contained within a single imaging voxel, the subsequent phase dispersion can reduce T_2^* values. All of these factors are important to consider when analyzing the BOLD effect [8].

7.4 Evolution of BOLD MRI Signal

Given the nature of the BOLD effect, it is important to discuss the mechanisms that contribute to the overall signal. Quantitatively, the MR signal intensity can be shown through the following relation:

$$S = \sum_i \rho_i \times V_i \times M_{ss,i} \times e^{-T_e/T_{2,i}^*}$$

where:

$$M_{ss} = \frac{\left(1 - e^{-T_e/T_{2,i}^*}\right)\sin\alpha}{\left(1 - \cos\alpha \times e^{-T_e/T_{2,i}^*}\right)}$$

and T_R is the repetition time, T_1^* is the apparent longitudinal relaxation rate in the presence of inflow, and α is the flip angle of the excitation pulse. Also, ρ is the water proton spin density, V is the volume fraction, T_e is the echo time, T_2^* is the apparent transverse relaxation rate that accounts for the inhomogeneities induced by BOLD contrast, and i is the compartment representing either blood, extravascular tissue, and/or cerebrospinal fluid contributions in the imaging voxel [8]. This equation shows the dependencies that influence the overall MR signal. While some imaging parameters will influence the MR signal in general, there are specific physiologic processes that can alter the above values, particularly the BOLD effect. Both volume fraction V and spin density ρ are closely correlated with the water content of the tissue and blood. The apparent longitudinal relaxation T_1^* is inversely proportional to increases in inflow and CBF. There is an increase in relative MR

7.5 Acquisition Schemes

With an understanding of the basics behind MR signal generation and the neural physiologic consequence of BOLD, a deeper investigation of signal acquisition techniques is required. Two common flexible imaging schemes exist that allow for the probing of different tissue contrasts. These are the spin-echo (SE) and GRE sequences, both of which are sensitive to blood oxygen levels.

7.5.1 SE Imaging

SE sequences require the use of two types of RF pulses: an excitation and a refocusing pulse(s). The excitation pulse tips the net magnetization vector, initially aligned with the static field **B0**, into the transverse plane. As previously discussed, once the net magnetization is in the transverse plane, spins begin to lose phase coherence and the *excited* MR signal disappears and returns to a lower energy state. The loss of coherence is known as a *free induction decay* (FID) signal and does contain relevant information about the microstructure of the imaging volume particularly T_2 and T_2^* characteristics. This will be elaborated upon in the sections to follow. While the spins begin to interact with one another and precess at different rates, a refocusing pulse, typically 180° for conventional SE sequences, is applied, thereby allowing spins to come back into phase and generate a signal echo. An echo contains different information than a FID as it probes the T_1- and T_2-contrast mechanisms of the targeted tissue. Figure 7.2 shows a typical SE pulse sequence diagram.

The first line representing the RF pulses was detailed in the above discussion. A typical SE sequence contains

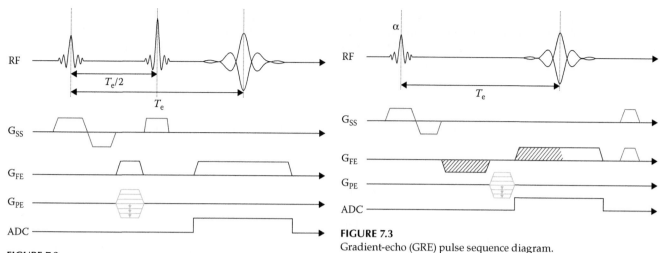

FIGURE 7.2
Spin-echo (SE) pulse sequence diagram.

FIGURE 7.3
Gradient-echo (GRE) pulse sequence diagram.

a 90° excitation pulse followed by a 180° refocusing pulse at time $T_e/2$. T_e refers to the echo time and, as the name suggests, is the time it takes for the induced signal from coherent refocused spins, or echo, to be generated. This echo contains the encoded spatial information corresponding to a single acquired slice. Spatial selectivity is achieved through the use of three independent gradients aligned along the orthogonal coordinate axes. Slabs of spins are excited by the application of a slice selection gradient that linearly varies the resonant frequency of hydrogen nuclei along through the targeted slice. By adjusting the bandwidth of the excitation pulse, it is possible to selectively influence a well-defined region of spins. As this region tips its net magnetization onto the transverse plane, further signal localization is required to resolve the information contained in each imaging voxel. Two other gradients are played out along the phase and frequency directions. Phase encoding is achieved by pulsing a gradient, while the net magnetization is in the transverse plane but prior to the readout. The degree of linear phase variation is influenced by the area of the gradient lobe, thus allowing the location of spins to be determined based on their phase during precession. The final encoding step is along the frequency axis where both a prephasing gradient lobe and a readout gradient lobe are required to offset phase variations and ensure that an echo is generated. This sequence of events can be repeated multiple times to acquire the necessary amount of slices for full coverage of the image region as long as the phase encoding gradient is stepped up appropriately.

7.5.2 GRE Imaging

Another technique that can be used to probe different contrast mechanisms is the GRE. It also holds great utility as a fast scanning acquisition scheme ideally

suited for vascular imaging, which includes BOLD. Unlike the SE sequence, no refocusing RF pulse is required to generate an echo; rather, gradient reversal along the frequency encode axis is sufficient. Applying a prephasing gradient lobe induces phase incoherence that is reversed by pulsing another readout gradient lobe of opposite polarity. The echo is formed when the areas of both the rephasing and the readout gradients are equal as illustrated by the blue-shaded regions in Figure 7.3. It is also important to mention that the RF excitation pulse of a GRE sequence is typically not 90°, but some other angle α. A smaller flip angle does not project the longitudinal magnetization completely onto the transverse plane leading to a shorter repetition time for faster acquisitions. GRE sequences also employ spoiler gradients (shown in purple in Figure 7.3) to spoil residual transverse magnetization prior to the delivery on the next excitation pulse. This is important as any residual signal can create undesirable image artifacts. Concurrent spoilers played out on multiple axes also help to reduce overall imaging time. GRE acquisitions are also more sensitive to magnetic susceptibility differences compared to the SE counterpart, which is advantageous when investigating BOLD effects.

7.5.3 Echo-Planar Imaging

Echo-planar imaging (EPI) sequences are used for rapid acquisition. This is important for BOLD imaging as the transient physiological changes induced by neural stimulation must be captured in the shortest allowable time while still allowing tracking the BOLD effect on MR signal. The sampling rate of EPI provides the temporal resolution needed to accurately interpret the observed changes. While built upon the same principles of SE and GRE in terms of echo generation, EPI is different from simple SE/GRE sequences in the manner in which its spatial encoding gradients are applied. Since EPI is a fast scanning technique, it requires very high performance gradients that allow for rapid on–off switching. This enables the acquisition of an image (i.e., filling the *k*-space) after the application of a single RF pulse, called *single-shot* (SS) *EPI*, commonly used for BOLD acquisition. In SS EPI, all *k*-space lines are filled by constant gradient oscillations (blips), thereby producing multiple GREs in a single acquisition. The blipped nature of the phase-encoded gradient also ensure that each echo is placed on its own unique line in *k*-space. A multishot EPI sequence is also used where the filling of *k*-space can be broken up into segments. This technique places less stress on the gradients while also reducing phase error accumulations. However, it takes longer than its SS counterpart and is more susceptible to motion-related artifacts.

The EPI sequence shown in Figure 7.4 is the GRE variant where the MR signal is obtained through the FID.

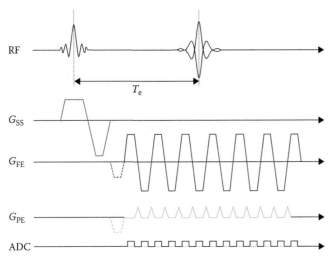

FIGURE 7.4
Echo-planar imaging (EPI) pulse sequence diagram.

This is the preferred choice of acquisition scheme in BOLD imaging studies due to the inherent T_2^* weighting that is central to the GRE sequence. It is important to mention that although T_2 contrast will be present, as spin dephasing in the transverse plane is unavoidable, the inherent sensitivity that GRE possess to local field inhomogeneities ensures that T_2^* processes prevail. It is precisely these inhomogeneities that are critical to observation of the BOLD effect due to the susceptibility differences created by the presence of dHb.

Recent advances in pulse sequence development have enabled the imaging of high susceptibility regions in the brain, such as the orbital frontal cortex, and regions of temporal lobe closer to the temporal bone. These areas are typically characterized by the loss of signal in the interface regions of two different biological structures (e.g., air–tissue interface). This loss of signal can be overcome by new acquisition methods such as *z*-shim sequences. *z*-shim allows for recovering lost signal from these regions of high susceptibility, thus enabling investigators to obtain reliable BOLD signal. Typical multishot *z*-shim modes are implemented as follows. *z*-shimming is implemented by applying oscillation gradients along the *z*, or slice select, axis in conjunction with phase-encoded blips. The net effect of this gradient scheme is to alter spin coherence such that isochromats in high susceptibility regions end up in phase to generate a complete echo. However, this has the undesirable effect of dephasing spins in all other regions. Reconstruction is performed to create a maximum intensity projection image from the in-phase and out-of-phase acquisitions, but at the cost of temporal resolution. This temporal resolution can be recovered through the use of a SS *z*-shim variant. The SS version has increased temporal resolution while minimizing geometric distortions, making it the more ideal option. The *z*-shim sequence shown below in Figure 7.5 is a SS version, meaning

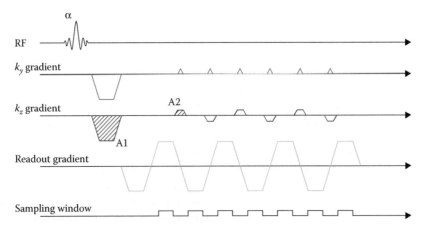

FIGURE 7.5
Single-shot (SS) *z*-shim pulse sequence diagram.

that an entire imaging volume is acquired within one T_R. This is accomplished by acquiring one image for all EPI lines corresponding to the positive readout gradient (RO+), while the second image is acquired on the remaining lines associated with negative readout gradients (RO−). Each of these *images* is linked to a z-shim parameter that is calculated through calibration scans that seek to optimize regional signal intensities across all slices. The oscillating nature of the z-gradient that is played out during the EPI train effectively controls the settings for the RO+ and RO− lines. This has the effect of quickly adjusting the sampled slice for each line of the EPI readout [9].

7.6 Paradigm Design

In order to probe-specific regions with BOLD imaging, it is necessary to design a stimulus delivery method for inducing a desired neuronal response. Given that the temporal progression of the BOLD signal is an order of magnitude less than the growth of the neural activity of interest, effective paradigm design is crucial for good experimental methodology. Stimulus delivery is achieved through a combination of hardware and software systems. Real-time feedback and user interaction is typically desired to isolate the locus of complex neuronal processes. This is contingent upon the nature of the hemodynamic response, which can vary not only between subjects but also between brain regions within the same subject. This necessitates proper paradigm design that can account for potential variances between hemodynamic responses. Broadly, paradigm design can be broken into three distinct models: block, event-related, and mixed.

The most common clinical paradigm used for BOLD functional magnetic resonance imaging (fMRI) studies is the block design. Block designs rely on grouping trials of similar focus into blocks of finite duration. Each group is referred to as a *condition*. Multiple conditions can exist in a single paradigm, which can stimulate different regions of the brain. Information received from each of these distinct blocks can then be compared among one another to determine active regions unique to a given condition. While the design of such paradigms is still debated and is a subject of its own, several key points exist that can motivate further investigation. Based on the nature of BOLD contrast, fMRI acquisition and analysis seeks to track and exploit the relative signal differences caused by neural changes associated with task-specific events. These differences between blocks require sufficient localized signal

changes to make statistically significant conclusions about the nature and extent of activation. As trials are presented in block fashion, the duration of stimulus delivery is critical to properly average the signal of interest. This duration is directly proportional to the magnetic field strength. Higher field strength requires less time to acquire the necessary amount of data and the converse is also true. An understanding of your MR system is needed to begin designing experiments.

Also related to block designs is the issue of how the comparisons between groups are going to maximize the extraction of differences that will be used to isolate a locus of task relevant activation. Comparisons between blocks can be classified as either loose or tight. Loose task comparisons center around stimuli that are not closely coupled, such as using a crosshair for a resting condition to control for visual activation. Comparisons between loose designs allow for a wider spectrum of activation patterns that may reveal more about the underlying network. In contradistinction to loose designs are tight designs. Tight designs attempt to control for as many variables as possible to ensure that only those regions that differ appear in the activation mapping. A tight design is more popular in cognitive studies where precise regions responsible for specific tasks are desired, such as in parsing semantic networks or memory.

While block designs are the paradigm of choice for clinical applications, event-related paradigms have gained traction in the cognitive community and are worth mentioning (Figure 7.6). In contrast to block designs, the event-related variant is capable of correlating specific, isolated trials with an increase in neural activity. The signal averaging across a block paradigm consists of activation that may be a mixture

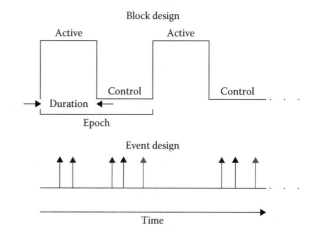

FIGURE 7.6
Paradigm designs: standard block design (top) and an event related design (bottom).

of multiple trials. However, this may confound the cognitive function of interest. Event-related analysis can explore a variety of conditions by the delivery of stimulus at discrete time points. These trials can be repeated at well-defined instances and intermixed with other trials for comparison. By adopting this strategy, the hemodynamic response induces a transient BOLD signal associated with a specific stimulus that can be tracked and subsequently averaged. One downside is that activity could be missed upon the application of successive trials given the nature of the hemodynamic response. This can be resolved by using a mixed model that both averages the sustained changes seen in block designs and is sensitive to transient signal fluctuations [10].

7.7 BOLD Data Processing

With a proper paradigm and acquisition scheme, the next critical step is processing the functional data for clinical interpretation. The typical processing pipeline can be broken into two distinct sections: preprocessing and statistical inference. Preprocessing includes strategies for rendering the data in preparation for statistical testing. Generally speaking, steps typically include motion correction, coregistration, segmentation, normalization, and smoothing. Each of these steps has a specific purpose to ensure that data is in the proper coordinate space with sufficient signal to make comparisons across tasks.

The BOLD signal is analyzed at a voxel level and as such, any motion between scans or within a scan can have an adverse effect on the BOLD signal, which is only about a 3%–5% change relative to baseline. Even despite choosing an acquisition sequence that is less sensitive to motion, both continuous and intermittent spikes in movement will reduce detection sensitivity. This is because motion can manifest itself as the same signal change induced by the hemodynamic response responsible for BOLD contrast due to spin history effects. Realignment algorithms can help ensure that false-positive activation does not appear in the final activation mappings by applying a linear transformation to the acquired volumes. A rigid body transformation is the common choice and when applied to six affine parameters (x, y, z, pitch, roll, and yaw), slices within a given volume are aligned with one another and to each volume in the total functional dataset.

For the clinical application, there is a benefit to overlaying the generated activation map onto a high-resolution image given the low spatial resolution of EPI data for interpretation by clinicians. This requires coregistration of the BOLD map to an anatomic map of the same subject. Unlike realignment, coregistration is a technique to map scans of different contrasts. In BOLD studies, the functional EPI volumes are registered to an anatomic image that best highlights the pathology of the lesion. These anatomic images could include T_1, T_1-Gad, T_2, or FLAIR contrasts. A different algorithm is needed to perform this operation as no linear relationship exists between signal intensities received from different weighted images. Coregistration can also aid in a more accurate spatial normalization as higher signal-to-noise ratio and in-plane resolution make individual voxels more apparent. This process requires maximizing the mutual information between two images [11].

7.7.1 Segmentation

Segmentation is a method for parsing MR images by tissue type. Classification can be achieved through a variety of algorithms, with one of the most popular being Gaussian mixture models, which predict the tissue by using spatial prior probabilities.

7.7.2 Normalization

Unlike previous preprocessing steps normalization is highly nonlinear. In order to compare results across studies and between subjects, the data needs to be transformed into a common coordinate space. This mapping occurs by applying a deformation field to warp a reference image to a standard template. This may not be an essential step in clinical imaging but is used very heavily for population studies in cognitive fMRI studies.

7.7.3 Smoothing

The final preprocessing step relates to smoothing the BOLD data. Smoothing increases signal-to-noise ratio as well as spatial correlations. This increase in spatial correlations has the added benefit of reducing the number of multiple comparisons during statistical inference effectively reducing type I (false-positive) error rates. Simply stated, smoothing is implemented by averaging the signal intensities of neighboring voxels. Averaging occurs by convolving the image of interest with a Gaussian kernel. Gaussian kernels are characterized by their full width half maximum (FWHM) and the convention is to use a kernel size of twice the voxel size. Other techniques exist that define the FWHM as anticipated width of the hemodynamic response or expected spatial extent of the BOLD contrast mechanism. It behooves both researchers and clinicians to choose an appropriate kernel width as too much smoothing (large filter width) can degrade the signal-to-noise ratio, while causing extinction of smaller, yet potentially significant, activated regions. On the other hand, merging of close yet distinct activation regions can occur if insufficient smoothing is performed. The application of a smoothing kernel to the functional dataset is synonymous to passing the data through a low-pass filter as edges and fine details, which are the high-frequency components of the received echo, are lost. However, smoothing is a necessary step that will increase confidence in the statistical conclusions regarding task-related activations.

With the functional dataset appropriately preprocessed, the final step is determining whether the relative signal changes between tasks are statistically significant. This analysis will provide clinicians with the activation maps characterizing task-induced brain function. Statistical inference can be broadly classified into two groups. Classical statistics employ the general linear model and standard *t*-tests to calculate confidence metrics representative of the probability of the observed activation given no effect. Bayesian inference determines the probability of observing the activation given the effect.

The general linear model used for statistical analysis is a regression model that relates the received signal with the expected. In essence, the model weights the basis functions that serve as predictors in an effort to minimize differences with the observed signal. For example, the design matrix for a blocked paradigm would contain information pertaining to two conditions that would explain the observed effect. This type of analysis is classified as a mass-univariate problem and is applied at a voxel level. The general linear model is ideally suited for BOLD studies, as the experimental design, confounds, and data that characterize the study can be integrated into a linear framework [12]. With the model, created and estimated statistically relevant information can be inferred.

Finally, the interpretation of BOLD results is widely debated, and is only superseded by the larger question of statistical significance versus clinical relevance. Statistical significance quantifies the likelihood that a difference exists between two conditions but ignores the importance of that difference. On the other hand, clinical relevance speaks to the magnitude of the observed difference and is assessed by the clinician. These two work in tandem as statistical significance is a necessary precondition for clinical relevance. With this distinction in place, the next issue is thresholding. Type I and type II errors should be balanced through adaptive methods as opposed to standard *p*-value inspection. Once confidence with the calculated statistical map is gained, the extent of activation and laterality metrics can be obtained that will help in neurosurgical planning [11].

7.8 Clinical Applications

Current clinical applications of functional BOLD include the identification of eloquent cortex, which includes primary language cortex and sensorimotor cortex regions of activation. The location of the eloquent cortex in relation to an intraaxial/parenchymal brain lesion is a critical part of the presurgical planning for brain surgery. The primary language cortices include Broca's region within the inferior lateral dominant frontal lobe and Wernicke's region within the dominant posterior superior temporal lobe. The sensorimotor cortex lies within the posterior frontal and anterior parietal regions.

In relation to language evaluation, there are many cognitive tasks that can be performed. However, for illustrative purpose, we are going to discuss two of the most commonly used tasks for clinical applications: sentence completion and word generation. For sentence completion, a simple block design is implemented alternating between trials of sentences and nonsense. These blocks will typically contain several fill in the blank sentences that the patient will covertly complete. An example of a sentence displayed during active blocks would be the following: "Crying babies make a lot of _____." The patient would then think about the word that best completes the sentence, in this case *noise* or a comparable variant, until the next sentence is displayed. These trials are typically repeated 3–4 times during a 15–30 s block, though the patient's level of lucidness will ultimately dictate the duration of the sentence stimulus. Offset between these active sentence-completion blocks are blocks of nonsense or gibberish conditions of the same duration as the active trial [13].

The second task that probes the eloquent cortex is word generation. The structure and duration is similar to that of the sentence completion task described above but with different stimulus delivered during the active and control blocks. Active conditions consist of displaying letters and having the patient subvocalize as many words as possible that start with the letter shown during the trial. For example, if the letter *D* appears, the patient will think of words like *dog, donut, driveway,* and so on. The control task consists of pictures that do not resemble any recognizable shape. This controls for visual activation while not evoking a language response. Both the sentence completion and word generation tasks attempt to isolate brain regions responsible for language comprehension and expression such that presurgical strategies can be developed to improve quality of life [13]. Figure 7.7 shows an example where a patient with a low-signal lesion in the left anterior temporal lobe (low-grade glioma) performs the sentence completion task. Heightened activation in primary

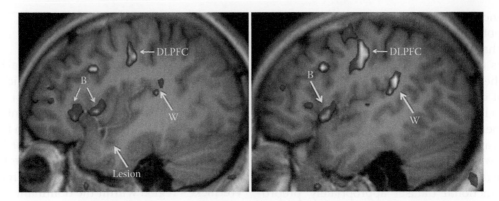

FIGURE 7.7
Two contiguous sagittal T_1 slices with overlaid sentence completion (SC) BOLD maps (red) of the left hemisphere. B, Broca inferior left frontal lobe; W, Wernicke superior posterior left temporal lobe; DLPFC, dorsal lateral pre frontal cortex.

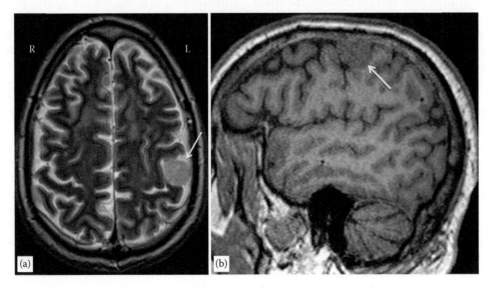

FIGURE 7.8
(a) (axial FSE T_2) A left fronto-parietal extra axial mass (a slight increased signal on T_2—denoted by arrow) representing a meningioma. (b) A sagittal FSE T_1 demonstrating that the lesion is isointense on T_1 (denoted by arrow).

language centers provides information that could aid in presurgical planning.

The next essential task for presurgical mapping is investigation of the motor strip and supplementary motor cortex. This paradigm is the simplest to both implement and comply with. Active conditions consist of watching an animated video of both left- and right-hand (or foot) flexions. The patient is to follow along and either makes a fist for hand stimuli, or extend the toes parallel with the body's axis for feet studies. Both movements are repeated for the entire duration of the block. Rest of the conditions shows just a picture of the hands or feet with no movement. The patient will simply lie on the table and stay still while focusing on the stimulus. Again, both active and control blocks will oscillate in a boxcar fashion through the acquisition. Figure 7.8 is

another example of a patient with a left fronto-parietal meningioma. Please note that all axial MR images are presented in radiologic coordinates (right side of the image represents the left side of the brain).

The same patient from Figure 7.8 performed sentence completion and finger tapping paradigms that were then used to determine whether eloquent and/or motor cortex activation were in close proximity to the lesion. Once again, this helps both neurosurgeons and neuroradiologists develop a strategy to minimize the loss of functionality during resection. Figure 7.9 illustrates the activation mappings from the aforementioned tasks. Figure 7.9a demonstrates left motor mouth activation (short arrow), during a language sentence completion task, adjacent to the anterior margin of the meningioma (long arrow) and the supplemental

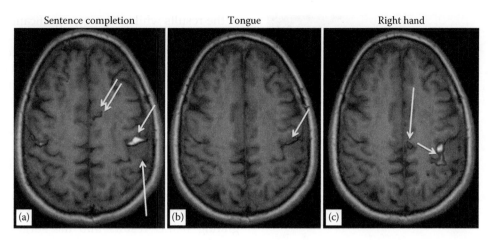

FIGURE 7.9
Activation maps for sensory motor cortex. (a) The sentence completion task, (b) the tongue paradigm, and (c) the right-hand squeeze.

motor area for language (SMA-L, double arrows). Figure 7.9b shows a tongue BOLD paradigm with activation in the left sensory motor cortex in relation to the meningioma (arrow). This proves the activation seen in the SC paradigm represented motor tongue activation. Right-hand squeeze results are displayed in Figure 7.9c. The results show the more medial hand region within the sensory motor cortex in relation to the lesion. Also note the left supplemental motor area (SMA-M), large arrow, is posterior to the SMA-L shown in Figure 7.9a.

In order to decrease false-negative BOLD activation, arising from neurovascular uncoupling (NVU) phenomena, it is important to perform breath-hold BOLD experiments on the patients, which can provide the status of the cerebral vascular reserve (CVR) in the brain. If there is no BOLD activation observed in an expected location, it is important to be confident that it represents a true negative. A breath hold can be used to show normal patterns of CVR; therefore, if eloquent cortex was present, a BOLD fMRI study should be able to show this. If no CVR is demonstrated in a region of interest, then there may be eloquent cortex activation present, but BOLD fMRI may not be able to detect the activation due to this lack of CVR (false negative). This phenomenon revolves around NVU. The physiologic underpinning of BOLD functional imaging relates to CBF. It has been demonstrated that in response to stimulus, an increase in CBF manifests locally to regions of neuronal activation [11,14]. Typical brain functioning relies on cerebrovascular autoregulation, the process by which normal blood flow is maintained despite dynamic physiological stressors. This autoregulation modulates blood flow by varying vascular resistance, which is a function of the cerebrovascular reserve. If blood flow augmentation is altered due to a tumor, diseased/occluded vascular networks, or a steal phenomenon from an arteriovascular

malformation, then neuronal activation will be adversely affected. This has the undesirable consequence of yielding false-negative results with the potential for gross underestimation of cortical activation. In a clinical setting, this could be detrimental to neurosurgical patients as proper surgical planning is contingent on accurate localization of the eloquent cortex containing language and sensorimotor regions. It has been shown that NVU is highly prevalent for patients with low-grade gliomas and can provide a more comprehensive evaluation of presurgical mappings [11,15].

The final example shows a case where a patient with an arteriovenous malformation (AVM) in the left temporal lobe underwent BOLD testing. In addition to the standard language paradigms appropriate for this type of lesion, a breath-hold (CVR) task was also performed. By imaging while having the patient alternate between normal breathing and breath hold conditions, it is possible to test for NVU. As detailed before, this is important because it can help decide whether false negatives are present in the activation mapping used for presurgical consult. In this particular case, the vascular steal effect from the AVM caused a decrease in CVR in the inferior and middle portions of the left temporal lobe, but had no impact on local regions of primary language activations in the frontal and superior temporal lobes. Figure 7.10 shows an arteriogram of the AVM in the inferior left temporal lobe. Arrows show the left middle cerebral feeding artery (small arrow), the serpiginous AVM nidus (double arrows), and the prominent early draining vein (big arrow).

Figure 7.11 is an axial map with 5-mm slice thickness through the whole brain showing the AVM in the inferior left temporal lobe.

As previously stated, the CVR breath-hold task was performed to check for NVU and steal effects associated with AVMs. Figure 7.12 shows the CVR map overlaid on

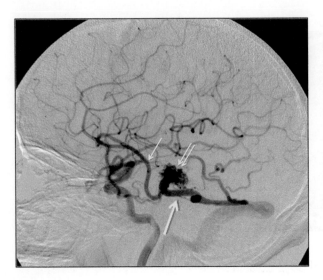

FIGURE 7.10
Lateral digital subtraction cerebral arteriogram (left internal carotid injection) of a congenital arteriovenous malformation (AVM) in the inferior left temporal lobe.

the axial FSE T_2 map from Figure 7.11. The results show a decrease in CVR in the region of AVM nidus (short arrows). Please note that there is a bilaterally symmetric decrease in CVR in the adjacent inferior temporal lobes (long arrows) that is due to loss of BOLD signal related to susceptibility artifact from the inferior petrous portion of the temporal bone (double arrows). The remainder of the brain shows bilaterally symmetric normal CVR in the gray matter cortex.

With the CVR map, false negatives can be more easily identified. Figures 7.13 and 7.14 are activation mappings of sentence completion and word generation, respectively, for the AVM patient under investigation. It is clear upon inspection that both figures consistently show bilateral inferior lateral frontal lobe, $(L > R)$; Broca's regions (arrows); and bilateral posterior superior lateral temporal lobe Wernike's region (double arrows) activation.

The BOLD testing performed on the AVM patient helped to determine that eloquent cortex and language centers were remote from the lesion.

While these techniques enhance our understanding and progress the nature of presurgical mapping, there is much advancement to be made. Most BOLD applications are largely qualitative. Currently, the common clinical practice is to use a stringent p value with a minimum value of .05 but typically around $p < .01$. If multiple corrections are used then the p value can be comfortably set to .05. As no consensus on thresholding exists, it is a subjective decision made by a trained neuroradiologist. Once a threshold is found that yields confidence in the results while showing significant areas of activation, the process stops. This necessitates a need for more quantitative methods that are specific to an individual patient as opposed to global, population norms. Further research would also include further intraoperative correlation of the location of the eloquent cortex in relation to an intracranial lesion.

BOLD imaging is currently a great tool to aid in presurgical planning and will continue to evolve to the point of being an integral component in the direction of surgical approach. These surgical considerations will have definite clinical impact, not necessarily clinical outcome. From a clinical impact perspective, integrating BOLD results into the surgical environment could aid in guiding the operative procedure with the patient asleep, resulting in an overall reduction in surgery time. This is in comparison to performing a more involved awake craniotomy with intraoperative cortical stimulation for brain mapping and monitoring of cognitive functioning, which could have the profound impact of lower morbidity rates due to less potential for infection. Aggressive procedures and success rates are largely time based. If eloquent cortex is intimately associated with the lesion, this would allow for only a partial resection which would clearly affect patient outcome.

7.9 Conclusion

BOLD fMRI has continued potential for significant impact in both clinical and research settings as imaging acquisition schemes and analysis methods improve. As capabilities of modern MRI scanners increase resulting in improved spatial and temporal resolution, it becomes possible to more accurately isolate the locus of brain-related activations in response to neural stimulation. The hemodynamic mechanisms that govern BOLD signal generation may enable clinicians and researchers to fully map connectivity networks as blood propagates through the task-related system. In addition, research into resting state BOLD activation may reveal new information regarding brain function. Increased use of these task-free techniques in presurgical planning has the potential to greatly improve surgeons' abilities to preserve function when operating. Continuous refinement of paradigm design, as well as data processing and analysis, can be paired with state-of-the-art hardware to greatly improve results for both researchers and clinicians.

FSE T_2

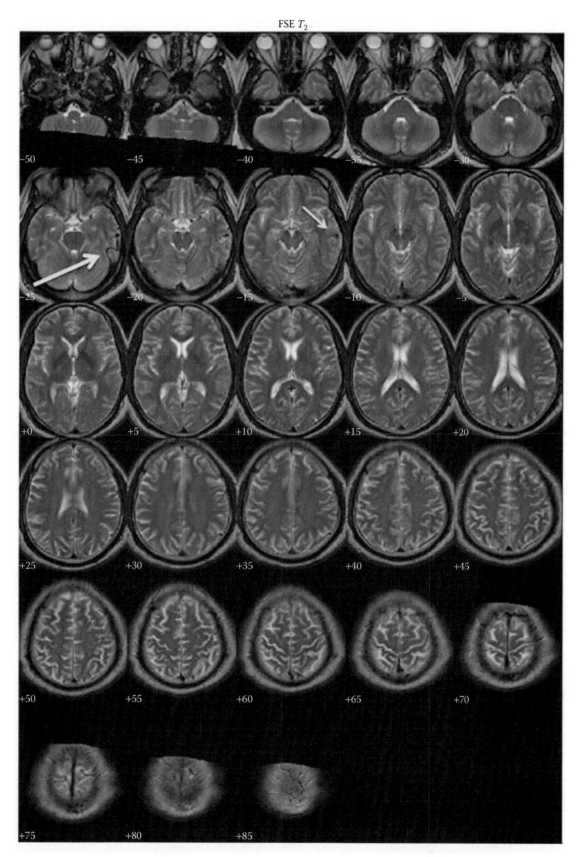

FIGURE 7.11
Axial FSE T_2 showing the left temporal lobe AVM with the AVM nidus (short arrow) and the draining vein (long arrow).

CVR breath hold

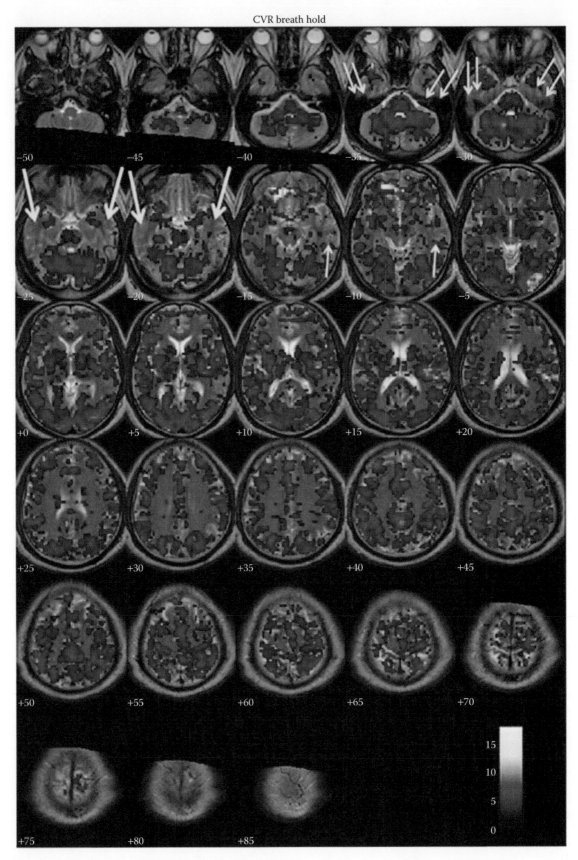

FIGURE 7.12
Axial breath-hold BOLD task at $p < .001$ overlaid on an axial FSE T_2.

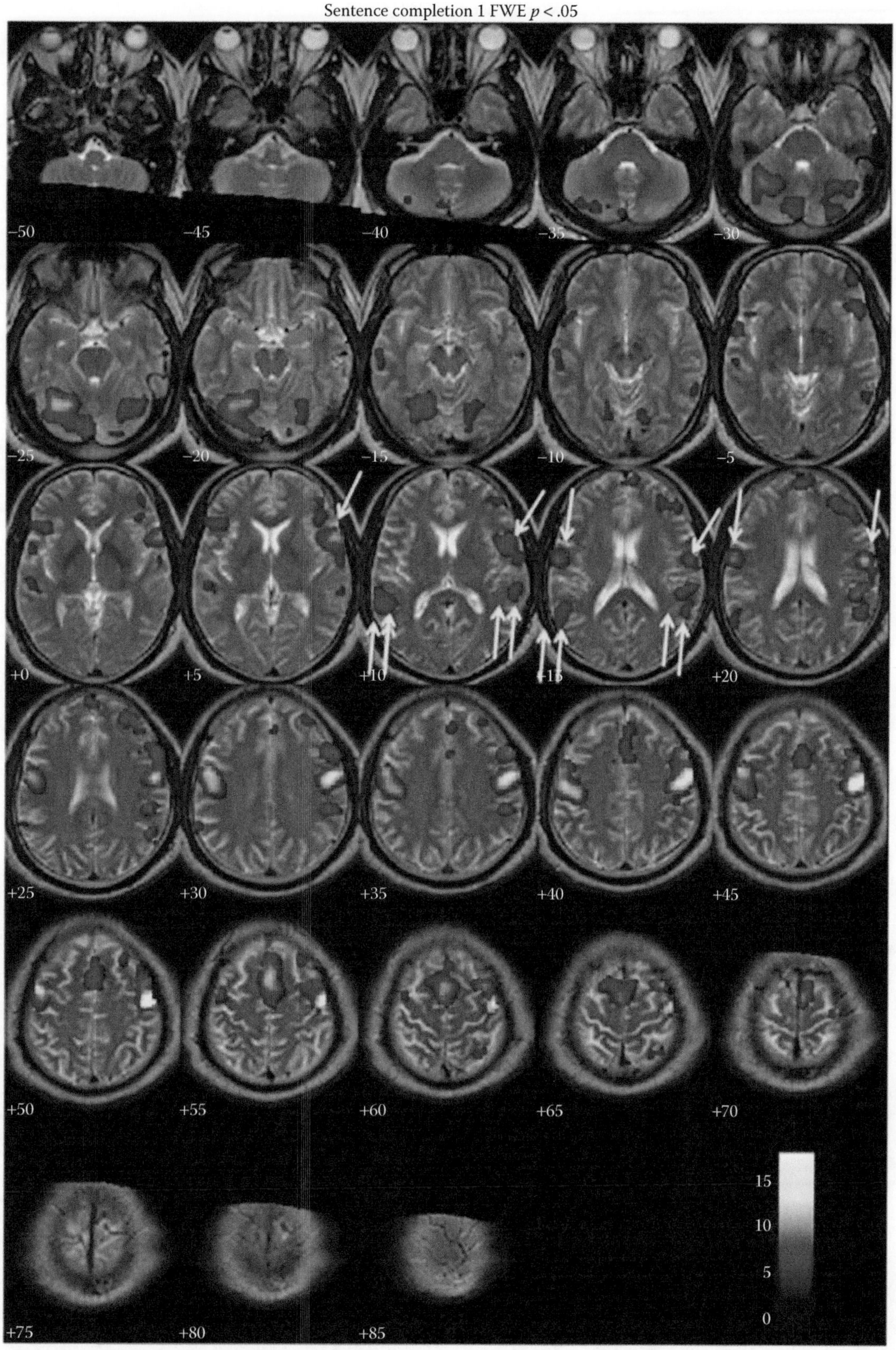

FIGURE 7.13
Sentence completion BOLD map overlay on axial FSE T_2 corrected with family wise error (FWE) $p < .05$. The arrows point to Broca's regions while the double arrows indicate Wernicke's regions.

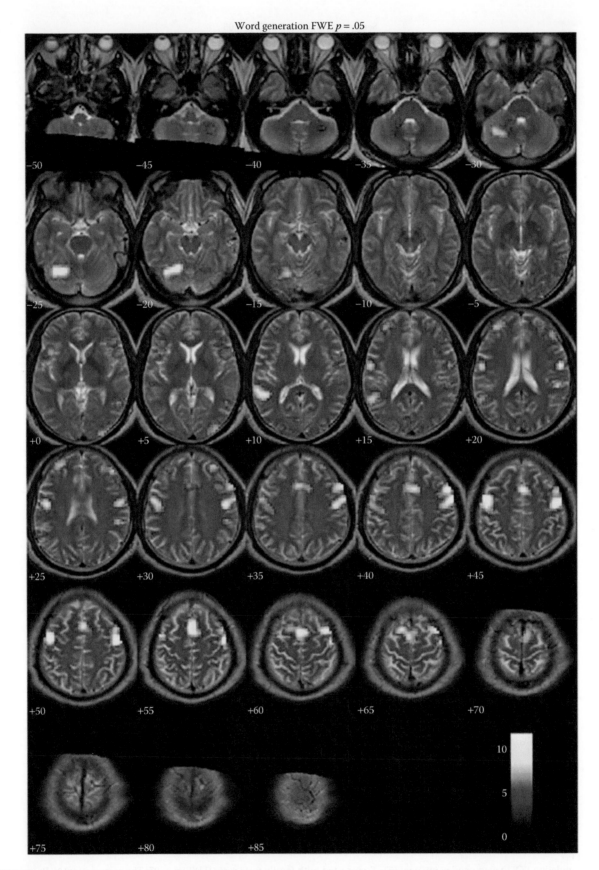

FIGURE 7.14

Word generation BOLD task showing similar activation patterns to the sentence completion task shown in Figure 7.13 at the same threshold.

References

1. Thulborn KR et al. Oxygenation dependence of the transverse relaxation time of water protons in whole blood at high field. *Biochemica.* 1982; 714: 265–70.

2. Fox PT, Raichle ME. Focal physiological uncoupling of cerebral blood flow and oxidative metabolism during somatosensory stimulation in human subjects. *PNAS.* 1986; 83: 1140–44.

3. Fox PT et al. Nonoxidative glucose consumption during focal physiologic neural activity. *Science.* 1988; 241: 462–64.

4. Ogawa S et al. Brain magnetic resonance imaging with contrast dependent on blood oxygenation. *PNAS.* 1990; 87: 9868–72.

5. Ogawa S et al. Oxygenation-sensitive contrast in magnetic resonance image of rodent brain at high magnetic fields. *MRM.* 1990; 14: 68–78.

6. Ogawa S et al. Magnetic resonance imaging of blood vessels at high fields: In vivo and in vitro measurements and image simulation. *MRM.* 1990; 16: 9–18.

7. Duong, TQ et al. Localized cerebral blood flow response at submillimeter columnar resolution. *PNAS.* 2001; 98: 10904–909.

8. Kim SG, Ogawa S. Biophysical and physiological origins of blood oxygenation level-dependent fMRI signals. *JCBFM.* 2012; 32: 1188–1206.

9. Hoge SW et al. Efficient single-shot z-shim EPI via spatial and temporal encoding. *IEEE ISBI.* 2011; 1565–68.

10. Jezzard P et al (Eds). *Functional MRI: An Introduction to Methods.* Oxford, England: Oxford University Press, 2001.

11. Conklin CJ et al. Technical considerations for functional magnetic resonance imaging analysis. *Neuroimaging Clinics.* 2014; 24: 695–704.

12. Friston KJ et al. Statistical parametric maps in functional imaging: A general linear approach. *HBM.* 1995; 2: 189–210.

13. Faro SH et al. (Eds). *Functional Neuroradiology: Principles and Clinical Applications.* New York: Springer, 2011.

14. Ye FQ et al. Quantitation of regional cerebral blood flow increases during motor activation: A multislice, steady-state, arterial spin tagging study. *MRM.* 1999; 42: 404–07.

15. Pillai JJ, Zaca D. Clinical utility of cerebrovascular reactivity mapping in patients with low grade gliomas. *WJCO.* 2011; 12: 397–403.

References

8

Magnetic Resonance Enterography and Colonography: Technical Considerations

Jesica Makanyanga, Douglas Pendse, Rehana Hafeez, and Stuart A. Taylor

CONTENTS

8.1 Introduction

The use of magnetic resonance imaging (MRI) has revolutionized imaging of the gastrointestinal tract. Not only does MRI provide impressive tissue contrast to help detect and characterize intestinal pathology, but it also does not impart ionizing radiation, which is a major advantage over computed tomography and traditional fluoroscopic techniques.

The clinical dissemination of the MRI of the small and large bowel is increasing rapidly and particularly in the field of IBD; consensus recommendations now advocate first-line use [1] over alternate imaging techniques.

Both MR enterography (MRE) and MR colonography (MRC) are dependent on careful attention to technique. Poor-quality studies are easily produced if considerations regarding patient preparation, bowel distension, and sequence selection are ignored. For example, inappropriate MRI sequence selection can result in artifacts that distort or destroy the image, and suboptimal bowel distension lowers accuracy for detecting enteric disease.

This chapter reviews the basic requirements for high-quality MRI of the bowel and focuses on its major clinical use in inflammatory bowel disease (IBD). Both MRE and MRC will be considered. Facets common to both (such as MRI scanner considerations and sequence collection) will be considered together, but where appropriate, descriptions of the two techniques will be separated to reflect the difference in approach and indications.

8.2 MRI Considerations

The field strength of an MRI scanner relates to the static magnetic field ($B0$) generated by the scanner. It is this magnetic field that polarizes atomic nuclei, thereby creating magnetization in the tissue. In most modern MRI scanners, this is achieved by using superconducting magnets. Clinical MRI systems in use worldwide are predominately 1.5 tesla (1.5 T) systems, although the availability and use of 3.0 tesla (3 T) systems has been increasing. The advent of 3 T field strengths has brought about a significant advantage in terms of increased signal-to-noise ratio (SNR) compared to 1.5 T systems. The theoretical SNR gain from 3 to 1.5 T systems is that of a factor of 2, although real-life increases are more modest, being limited by many factors, including field inhomogeneity and hardware issues. The increase in SNR can be exploited for gains in spatial resolution, decrease in scanning time, or a combination of both.

The contrast-to-noise ratio (CNR) is also increased at 3 T as compared to 1.5 T. The reasons for this are multifactorial, but include the overall decrease in noise in the image, the changes in longitudinal (T_1) relaxation time of tissue as magnetic field strength increases, and the improved T_1 shortening effect of gadolinium at 3 T [2]. The net effect is one of improved postcontrast imaging at 3 T in terms of both spatial resolution and contrast [2,3].

A number of MR artifacts are increased with increasing field strength at 3 T. The two clinically important artifacts that are more prevalent with 3 T abdominal imaging are susceptibility-dependent artifacts and chemical-shift artifacts [4]. These artifacts are important to recognize and can be mitigated against by a careful choice of image protocol.

Clinical studies of MRE in small bowel disease have largely been conducted at 1.5 T field-strength systems, although a number of studies evaluating the technique at 3.0 T have been reported [3,5–12]. A direct comparison of MRE at both 1.5 and 3 T has been performed, with patients undergoing both examinations and ileocolonoscopy as a reference standard [7]. This study in 26 patients with Crohn's disease demonstrated equivalent accuracy of both 1.5 and 3 T enterography for detecting features such as bowel wall thickening, enhancement, fistulae, and strictures, although 3 T MRI was superior at detecting ulcerations.

8.3 Motion Artifacts and Motion Suppression

MRI is a technique that is particularly sensitive to various *artifacts*, which degrade the image. These MR artifacts are features within the MR-generated image that do not represent real features of the subject being scanned. One of the most common MR artifacts is the one caused by the effect of motion within the scanned volume. At one extreme, severe motion artifact will result in a total *blur* of the image, rendering it to no diagnostic use. More subtle motion artifact may cause *ghosting* of an image, which is otherwise acceptable. It is important for the radiologist to understand the cause of the MR motion artifacts, so that (1) steps can be made in imaging protocols to reduce their effect and (2) features in the final image caused by motion artifact are not misinterpreted as pathology.

In abdominal imaging, there are four main sources of movement that need to be considered: (1) respiratory motion, (2) peristalsis of bowel, (3) vascular motion, and (4) patient movement. Although the extent of artifact from these various types of motion may differ, the underlying mechanism causing the artifact is the same. Motion effects on MRI result in artifacts that occur in the direction of the phase-encoding gradient [13]. Typically, in the axial abdominal imaging, this is in the anterior–posterior direction. Patient movement and peristalsis occur sporadically and produce blurring of the image in the phase-encoding direction. Motions that occur with regularity such as vascular motion (Figure 8.1) and breathing (Figure 8.2) produce more well-defined ghost images in the phase-encoding direction.

The extension of the field of view outside of the abdomen can help to differentiate movement artifacts from true imaging features. This can be particularly helpful in the case of vascular movement artifact, which in the case of the aorta may mimic hyperintense lesions within other organs (Figure 8.1). The review of the anterior abdominal wall shows the same ghosting artifact anterior to the patient, therefore, clearly demonstrating that the pseudolesion seen at first glance is indeed an artifact.

In small bowel imaging, two key approaches are routinely used to reduce the effect of motion artifact on imaging. First, breath-hold imaging techniques are employed whenever possible. Breath-hold imaging reduces much of the movement artifact, but limits

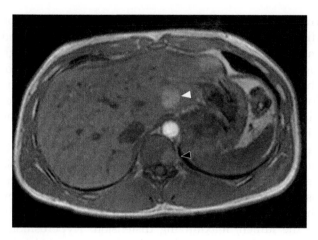

FIGURE 8.1
Axial in-phase T_1-weighted image (modified Dixon technique) of the upper abdomen showing artifact from vascular motion. Aortic pulsation results in ghost artifacts in the phase-encoding direction, seen in this image as bright circles overlying the left lobe of the liver (white arrow) and the vertebral body (black arrow). These pseudolesions could be mistaken for pathology, but as equidistant replicas of the pulsating structure should be identified as artifact.

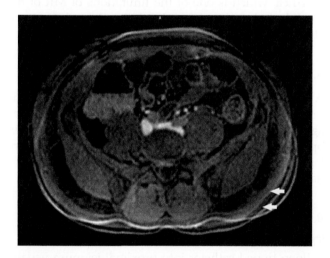

FIGURE 8.2
Axial T_2-weighted single-shot fast spin-echo sequence of the abdomen showing ghost artifact due to patient movement. Ghost lines can be seen adjacent to the abdominal wall due to patient movement (white arrows).

acquisition times to less than 20 s or so, although breath holds can be batched together for longer imaging protocols. Even very short breath holds may cause problems for some patients such as those with respiratory disease, the elderly, or children. Successful breath-hold imaging also relies on good communication between the radiographer/technician and the patient. The second technique employs spasmolytic agents to reduce peristaltic activity of the bowel to combat motion artifact.

In the United Kingdom and Europe, the use of hyoscine *N*-butylbromide (*Buscopan®*, Boehringer Ingelheim) as a spasmolytic agent is common. *Buscopan* is

TABLE 8.1

Focus Points: Reducing Motion Artifact

1	Breath-hold imaging
2	Spasmolytic—*Buscopan* or Glucagon
3	Extending field of view outside the abdomen to allow artifacts to be readily identified

an anticholinergic agent, which relaxes enteric smooth muscle via its effect on the parasympathetic ganglia. Administration during MRE is typically by intravenous administration of 20–40 mg *Buscopan*, although both intramuscular administration [14,15] and split-dose regimens have also been evaluated [15].

Buscopan is associated with side effects, including cardiac and ocular events, as well as antimuscarinic effects on the urinary bladder. The potential for cardiac side effects is often of concern to radiologists, although cardiac complications during radiological procedures involving antimuscarinics are rare and there is little evidence to directly implicate *Buscopan* [16]. The common ocular side effect of *Buscopan* is that of blurred vision. In most cases, this is mild, although patients should be warned not to drive until effects have subsided. More rarely, pupillary dilatation associated caused by *Buscopan* may precipitate acute angle-closure glaucoma. Open-angle glaucoma is not a contraindication to *Buscopan*. Recommendations for the use of *Buscopan* in radiology suggest that *Buscopan* can be given to most patients. Use of *Buscopan* is contraindicated in those with unstable cardiac disease and all patients should be advised to seek medical help if they develop painful blurred vision within 12 h of injection [16]. Male patients should also be alerted to the worsening of voiding difficulties including urinary retention, after administration.

Glucagon (*GlucaGen®*, Novo Nordisk) is often administered as a spasmolytic, particularly in the United States, where intravenous hyoscine butylbromide not licensed for use. Glucagon has no significant cardiovascular side effects but may produce nausea and emesis. Studies evaluating the use of glucagon in MRE show efficacy in reducing motion-related artifact in both the adult [17] and pediatric population [18,19] (Table 8.1).

8.4 Distending the Bowel

Enteric MRI relies on adequate bowel distension. Distension allows more accurate assessment of the bowel wall for pathological processes such as IBD or neoplasia. Collapsed bowel can both hide and mimic pathology and can be a major limitation to the technique if appropriate efforts are not made to achieve luminal distension.

Approaches to distending the bowel depend on the main indication for the examination. For example, if the focus is the colon, enema techniques are usually required, whereas the small bowel can be distended per os. The proximal small bowel is best distended via enteroclysis, whereas ileal distension is adequate using oral administration of contrast agents.

It is possible to combine both small bowel and colonic distension into one examination so as to provide information on the whole bowel.

8.4.1 Small Bowel MRI-Luminal Contrast

There is a range of possible oral contrast agents that have been employed to distend the small bowel. In general, they have hyperosmolar properties and/or are nonabsorbable such that they remain in the bowel lumen, affecting good distension throughout the length of the small bowel. Water, for example, is a very poor distension agent as it is readily absorbed by the gut and therefore provides very poor ileal distension [20]. Conventionally, oral luminal contrast agents are classified by the signal characteristics on MRI.

8.4.1.1 Positive Agents

These agents demonstrate high signal intensity on T_1- and T_2-weighted images. Their main advantage is that they help delineate the bowel wall on most sequences. Examples are blueberry and pineapple juice, diluted gadolinium, and milk with high fat content. In general, positive agents are rarely used in clinical practice given their expense and limited availability compared to other contrast agents. Furthermore, high luminal signal on T_1-weighted images is a disadvantage when assessing mural enhancement after intravenous gadolinium administration.

8.4.1.2 Negative Agents

Negative agents have low signal intensity on both T_1- and T_2-weighted images, and include oral superparamagantic iron oxide particles. They provide better visualization of bowel wall edema on T_2-weighted images, and mucosal enhancement on T_1-weighted images after IV gadolinium administration. They also help to discriminate between intraluminal and extraluminal fluid (such as abscess). Advocated by several European groups, negative contrast agents tend to have a less pleasant taste compared to other agents and are more expensive [21]. Their use is mainly limited to a few centers.

8.4.1.3 Biphasic Agents

Biphasic agents have low signal intensity on T_1-weighted images and high intensity on T_2-weighted sequences. They are the most commonly used agents for MR of the bowel. Examples include water, polyethylene glycol, diluted barium with sorbitol (2%), mannitol (2.5%), and mannitol (2.5%) with locust bean gum (0.2%). The high signal on T_2-weighted images facilitate delineation of the bowel wall against the luminal content, while the low signal on T_1-weighted images allows better appreciation of contrast enhancement. Although water has biphasic properties, as noted above, in reality, it is a very poor oral contrast agent prior to enteric MRI as it is quickly absorbed and bowel distension is poor. Biphasic agents used in clinical practice therefore have hyperosmolar properties such that they resist absorption by the bowel (and may draw in luminal fluid) and remain in the lumen, maximizing distension. Hyperosmolar agents, such as polyethylene glycol and mannitol (with or without locust bean gum), are the most widely used agents in clinical practice, and are among this group [22]. There is some evidence in volunteers that a combination of mannitol (2.5%) and locust bean gum (0.2%) provides the best bowel distention [22]. Because of their limited absorption, such agents can cause hyperosmolar diarrhea, which is one of the limitations of MR of the small bowel, and patients should be warned appropriately (Table 8.2).

8.4.2 Route of Administration

After the choice of oral contrast agent to be used, the practitioner must then decide how to administer the agent to distend the small bowel. There are two main options: either the patient drinks the contrast agent (enterography) or the contrast agent is infused via a naso-jejunal tube (enteroclysis).

8.4.2.1 MR Enteroclysis

This method involves transnasal or oral placement of a balloon tipped catheter into proximal jejunum usually under fluoroscopic guidance.

TABLE 8.2

Focus Points: Bowel Distension—Types of Contrast

Type of Contrast Agent	Advantages	Disadvantages
Positive agents (e.g., diluted gadolinium)	Delineate bowel wall well	Expensive Impair assessment of contrast-enhanced images
Negative agents (e.g., superparamagnetic iron oxide)	Better assessment of bowel wall edema and enhancement Differentiate intra and extraluminal fluid	Expensive
Biphasic agents (e.g., mannitol 2.5%)	Cheap Maximize distension	Hyperosmolar diarrhea

The 1.5–2 L of enteric contrast is then infused in a controlled fashion to distend the bowel. The tube balloon can be inflated to prevent the reflux of fluid into stomach. Real-time filling of the bowel can be monitored by heavily T_2-weighted sequences [23]. The high filling volume leads to secondary paralysis of the small bowel and helps avoid motion artifacts, although additional antiperistaltic agents are usually used (Figure 8.3).

8.4.2.2 MR Enterography

MRE is the more commonly performed alternative to enteroclysis and has better patient compliance. The patient drinks up to 1.5 L of the oral contrast agent over the 40–45 min before procedure. Although good quality images can be acquired with ingested volumes below 1 L, active encouragement of the patient is important to ensure compliance with the drink regimen. In particular, the patient should be encouraged to drink small volume continuously over the 45 min, rather than consuming a large volume at the beginning or end of the time period.

8.4.2.3 Choice of Technique

The advantage of enteroclysis lies mainly in the superior distention of the proximal small bowel (jejunum) in comparison to oral administration [24,25]. If proximal small bowel pathology is suspected (e.g., small bowel tumors), consideration should be made to performing enteroclysis. However, patients find enteroclysis more unpleasant than enterography [23], which may affect compliance. This is important in the context of chronic disease such as IBD where patients often undergo multiple imaging examinations over time. Furthermore, enteroclysis exposes patients to radiation during tube placement, thus negating one of the major advantages of MRI. It also requires logistical organization such that the fluoroscopy suite, radiologists, and MRI scanner are available at the same time.

Enterography achieves similar ileal distension to enteroclysis [25] and is better tolerated by patients (Figures 8.4 and 8.5). It is cheaper and simpler to perform than enteroclysis and in most centers, the MRE service is run by MRI radiographers and nurses without the need for the presence of a radiologist. Small comparative studies have shown equivalent diagnostic accuracy of enterography with enteroclysis in the context of IBD [25,26]. For these reasons, enterography is the workhorse of the MRI of the small bowel and although some centers used MR enteroclysis as their first-line

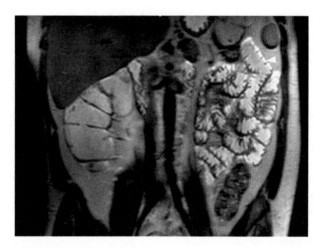

FIGURE 8 4
Coronal T_2-weighted single-shot fast spin-echo sequence MRE showing well-distended jejunal loops at MR enterography (white arrow) with oral mannitol 2.5% contrast 1600 ml taken 40 min prior to MR.

FIGURE 8.3
2.5% mannitol 1600 ml and feeding tube for MR enteroclysis.

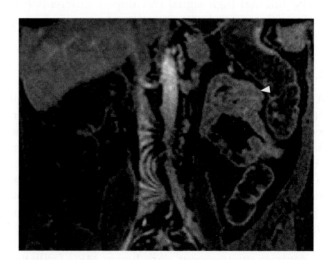

FIGURE 8.5
Coronal T_1-weighted volume interpolated breath-hold examination MR enterography showing collapsed jejunal loops (white arrow) with oral mannitol 2.5% contrast 1600 ml taken 40 min prior to MR. The water-based enteric contrast is black on this T_1-weighted sequence.

test, many reserve the more invasive investigation for specific clinical indications where proximal distension is just as important as ileal.

8.4.3 MRC-Luminal Distention

MRC requires the colon to be distended, akin to the small bowel in MRE. Typical MRC protocols require bowel preparation to cleanse the bowel before the imaging. There are two methods to perform MRC depending on the nature of fluid used to fill the colon: the bright lumen or the dark lumen technique.

8.4.3.1 Bright Lumen MRC

Distention of bowel with fluid labeled with an agent with strong T_1-weighted signal is referred to as *bright lumen*, and the method first described by Lubolt in 1997 [27]. Typically, a gadolinium/water mixture (5 mmol/L) is applied rectally after conventional bowel cleansing resulting in a high signal on both T_1 and T_2 sequences. Bowel distension can be monitored in real time using a fast T_2-weighted TrueFisp sequences [28]. 1.5–2.5 L of fluid is usually used to fill the whole colon. When the target is colorectal neoplasia, two datasets are collected (one in prone and the other in supine position) to move residual air pockets, which can reduce accuracy. Colorectal lesions appear as solid filling defects, which require differentiation from air bubbles and stool residue.

Changing the patient position during the procedure however does prolong the procedure and may precipitate the escape of contrast into the small bowel, reducing colonic distension and diagnostic quality [29].

The luminal high signal can impair differentiation between lumen and colonic wall and mural enhancement is less well appreciated.

8.4.3.2 Dark Lumen MRC

Although the dark lumen technique is a more recent development, it has largely replaced the bright lumen technique as the method of choice for MRC [29–31]. The dark lumen is achieved by filling the colon with water, air, or carbon dioxide. One advantage of dark lumen is that mural contrast enhancement is readily appreciated. As water itself is biphasic, it gives low signal on T_1-weighted images (Figure 8.6) and high signal intensity of T_2-weighted images.

Moreover, the advantage of clearer appreciation of IV contrast enhancement, dark lumen MRC facilitates appreciation of inflammatory changes such as submucosal edema, mesenteric fat stranding, mesenteric hypervascularity, and fibro-fatty proliferation [32]. It is also helpful in diagnosing polyps and bowel masses.

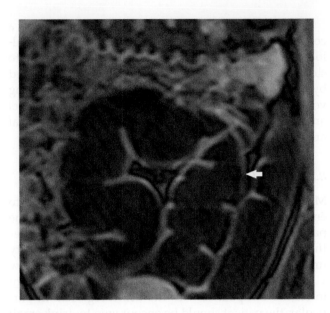

FIGURE 8.6
Coronal T_1-weighted volume interpolated breath-hold examination. Dark lumen MRC performed with rectal water enema preparation showing good distension of the sigmoid colon.

Because residual air has a similar T_1 signal to water, the risk of false-positive diagnosis is reduced compared to the bright lumen technique and there is usually no need to change patient position during the scan [33].

8.4.3.3 Distension Protocol

If a liquid distension agent is used, it is usually administered via an enema bag attached to a soft-tipped rectal catheter using gravity to slowly instill the fluid. Colonic distension using gas is achieved via gentle insufflation by using a handheld balloon (room air) or enema bag (carbon dioxide). The patient is usually examined in prone position and two surface array coils are used to obtain homogenous and complete MR signal reception. As for MRE, the administration of antiperistaltic agents helps to reduce motion artifacts.

8.4.3.4 Reduced Bowel Preparation Protocols

Although MRC is often performed after full laxation, it is known that bowel preparation is often poorly tolerated: 75% of patients undergoing bowel preparation express symptoms ranging from *feeling unwell* to *inability to sleep* [34].

Fecal tagging is a concept introduced to avoid bowel cleansing prior to radiological colonic imaging. It involves ingestion of contrast compounds such as barium sulfate, ferumoxil solution containing 5% gastrograffin, 1% barium, and 0.2% locust bean gum added to regular meals 36 h prior to examination in order to modify the signal intensity of colonic content to match that of the rectal enema. In effect, the fecal material becomes *virtually invisible* [35].

An alternative approach is to reduce stool signal by increasing its water content (so-called fecal cracking). Ajaj et al. [36] used the oral stool softener lactulose and rectal stool softener docusate sodium for this purpose.

The idea of reduced laxative protocols is to improve the patient experience of the examination and there is evidence that this is the case. Florie and colleagues [37] used limited bowel preparation with lactulose and fecal tagging with gadolinium and compared patient experience and preference with conventional colonoscopy. Patients found MRC with limited bowel preparation less burdensome and less painful. Overall, patient preference was higher for MRC immediately and even after 5 weeks. Similar higher acceptance for MRC with fecal tagging is found in other studies [32,38].

Despite the advantages to the patient, clinical uptake of reduced laxative protocols is limited. This is due to reported reduced diagnostic accuracy. For example, when using reduced laxative MRC for the detection of colonic neoplasia, Goehde et al. found poorly tagged stool signal in 18% of patients, although the overall sensitivity was 100% for detection of polyps larger than 2 cm, while it was 40% for polyps between 10 and 19 mm in comparison to conventional colonoscopy [39].

The role of MRC with fecal tagging in the detection of IBD activity has also yielded disappointing results. One study reported just 32% sensitivity and 88% specificity for detection of inflamed segment in comparison to conventional colonoscopy [40]. The difficulty in identification of inflammatory changes, particularly in establishing precise measurements of wall thickness and obscuration of the presence of ulceration, perhaps makes the technique less suitable for adequate assessment of less advanced IBD. It maybe that use of novel imaging sequences such as diffusion weighted may overcome some of the problems of reduced sensitivity with reduced laxative MRC protocols [41].

As noted above, it is possible to combine both MRE and MRC into one examination by administering oral contrast and a rectal water enema to the cleansed colon. Such an approach gives optimum visualization of the whole small and large bowel [11], but is relatively arduous for patients and thus used when it is necessary to accurately examine both the small and large bowel, for example, in the staging of complex IBD. In general, most enteric MR examinations are targeted to the particular organ of interest.

An alternative to performing colonic distension via rectal enema is to prolong the oral contrast phase in order to fill the colon as well as the small bowel. For example, good-quality colonic distension has been achieved using

TABLE 8.3

Focus Points: Small Bowel and Colonic Distension—Methods of Contrast Administration

1	Small bowel
	Enterography—contrast-ingested orally
	Enteroclysis—contrast administered via a jejunal tube
2	Colon
	Enterography—contrast-ingested orally in large volume up to 3 h before imaging for colonic distension to fill the colon
	Rectal tube—administration of water, air, or CO_2

oral polyethylene glycol solution ingested 1–2 h prior to MRI using colonic ultrasound to monitor the adequacy of colonic distension [42] (Table 8.3).

8.5 MRI Sequence Protocol

8.5.1 Positioning and Field of View

It is essential for successful MRE and MRC that both patient and coil positioning be optimized to allow for coverage of the entire small bowel and colon. Both prone and supine positions have been used, but patient comfort is paramount in order to keep movement to a minimum, and therefore supine positioning is often used. However, prone imaging reduces the volume to be scanned and may improve image quality (Figure 8.7).

Coronal imaging allows for the largest volume of bowel to be visualized in any given image and all MRE protocols should include coronal images. This imaging plane is particularly useful for cine imaging and post-contrast imaging.

In general, both MRE and MRC utilize similar sequence protocols, although as described above, the use of IV contrast is paramount to dark lumen MRC (Tables 8.4 and 8.5).

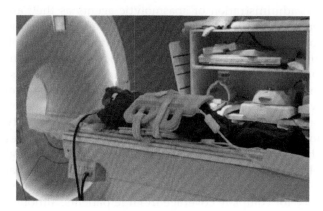

FIGURE 8.7
Patient positioned in prone position for MRE.

TABLE 8.4

Focus Points: Typical Sequences Used in a Clinical MRE Protocol

Sequence	Uses
Balanced steady-state free precession	Anatomy, particularly mesentery Cine imaging
Single-shot T_2-weighted spin-echo with and without fat saturation	Anatomy, determining increased signal intensity in the bowel wall or mesentery
Diffusion-weighted imaging[a]	Detecting inflammation in absence of contrast
T_1-weighted precontrast	Baseline for assessment of contrast enhancement
Dynamic contrast-enhanced image[a]	Quantifying contrast enhancement
T_1-weighted postcontrast	Assessment of contrast enhancement

[a] Optional.

TABLE 8.5

Typical MRE Sequence Parameters at 1.5 T

Parameters	Coronal/Axial SSTSE	Coronal/Axial TrueFisp with and without Fat Saturation	Baseline Volume Interpolated Gradient-Echo	Dynamic Contrast-Enhanced MRI
Field of view (mm)	Variable	Variable	Variable	Variable
Number of slices	20/26	25/34	40	40
Stacks	1/3	1/3	1	1
Repetition time (ms)	1200/800	3.98/4.25	3.07	2.73
Echo time (ms)	86/86	1.72/2.13	1.08	0.9
Image matrix	256/256	256/256	256	256
Slice thickness (mm)	4/4	4/4	3.5	3.5
Averages	1	1	1	1
Flip angle			15°	15°

8.5.2 Spin-Echo Sequences

Arguably, the most important sequences in any MRE protocol are the single-shot T_2-weighted spin-echo techniques. The sequences used vary between vendors, but may include single-shot turbo spin-echo (SS-TSE) [*Philips*], fast spin-echo (SS-FSE) [*GE, Hitachi*], or half-Fourier acquisition single-shot turbo spin-echo (HASTE) [*Siemens*]. The techniques are rapid to acquire, minimizing the respiratory and peristaltic artifacts. They should however ideally be performed following the administration of spasmolytic agents as they are prone to luminal flow artifacts. The sequences allow for detailed morphological examination of the bowel, and when there is sufficient enteric contrast distension, allows for a detailed luminal analysis and assessment of mural thickness. The T_2 signal of the bowel wall should be assessed on these sequences; however, fat saturation should be used to increase sensitivity for mural edema. Ideally, these rapid T_2w techniques should be performed in both the axial and the coronal planes supplemented by fat-saturated T_2 imaging in at least one plane. Multiple breath holds are often necessary for complete coverage in large patients, although variations in breath holds may degrade the volume coverage (Figures 8.8 and 8.9).

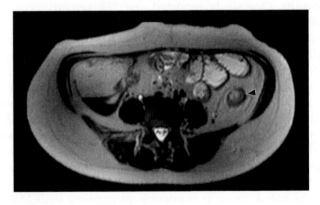

FIGURE 8.8

Axial single-shot T_2-weighted spin-echo sequence in a patient with Crohn's disease showing an abnormal strictured descending colon (black arrow).

8.5.3 Gradient-Echo Sequences

Balanced steady-state free precession (bSSFP) sequences such as TruFISP [*Siemens*], FIESTA [*GE*], and bFFE [*Philips*] are gradient-echo techniques that use a short repetition time and are very rapid to acquire, eliminating motion artifacts. The resulting images are very crisp with chemical shift artifacts, seen as *Indian ink* outlining

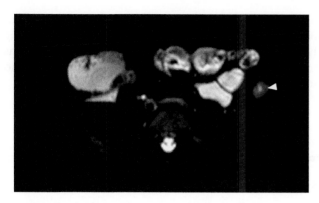

FIGURE 8.9
Axial single-shot T_2-weighted spin-echo weighted image with fat saturation of the same patient with Crohn's disease showing signal intensity of strictured descending colon (white arrow).

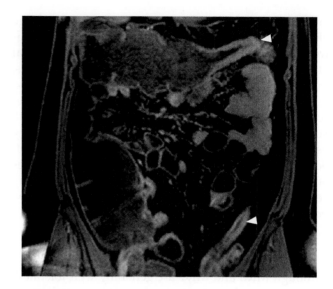

FIGURE 8.10
Coronal T_1-weighted sequence the same patient with Crohn's disease showing mild enhancement of strictured colon (arrowheads).

of bowel and vascular structures. These sequences are useful in assessment of the mesentery and for giving an overview of the bowel wall. As the sequences are very quick to acquire, they add very little time to the overall imaging protocol. They are however on their own insufficient for complete evaluation of the bowel wall.

8.5.4 Cine Imaging

Repeated acquisition of rapid gradient-echo imaging in the coronal plane allows for cine imaging of bowel motility. It has been shown that such sequences increase the diagnostic accuracy of MRE by alerting the radiologist to areas of potential abnormality [43]. A quantified reduction in bowel motility has also been linked to disease activity in IBD [44–46].

The temporal sampling rate and slice thickness will be limited by field strength but ideally imaging should be performed at least at 1- to 2-frames per second. Cine imaging coverage should include the entire small bowel, necessitating multiple slices to be imaged in every patient. Cine imaging is usually performed during multiple breath holds and prior to administration of spasmolytics.

8.5.5 Post IV Contrast Imaging

Another important part of the MR protocols is the post-contrast imaging. The sequences are technically challenging, as rapid acquisition of a large imaging volume is necessary. Precontrast and at least one postcontrast fat-saturated imaging volume should be acquired in the axial and/or coronal plane, each within a single breath hold. Standard intravenous MRI contrast agents at usual clinical doses are sufficient with either hand or power injection.

The 3D spoiled gradient-echo fat-saturated techniques are normally employed. The timing of postcontrast-enhanced image acquisition is variable but many acquire

images between 30 and 40 s (enteric phase) and then again at 70 s. Delayed imaging (up to 7 min) may help characterize fibrosis in IBD strictures [47], but is not practical in most clinical protocols.

Dynamic contrast-enhanced MRI (DCE-MRI) has been advocated, particularly in the context of assessing disease activity in IBD [48]. DCE allows for the assessment of relative enhancement between different parts of the bowel at multiple time points following contrast administration. It may also be used for the quantification of perfusion parameters such as the contrast transfer coefficient (K^{trans}), although this has yet to be of proven benefit in the assessment disease. In reality, for most clinical protocols, a simple pre- and post-IV contrast sequences at one or two time points is sufficient (Figure 8.10).

8.5.6 Diffusion-Weighted Imaging

Of increasing interest in MRE imaging is the use of diffusion-weighted imaging (DWI). DWI aims to quantify the diffusion of water (and other small molecules) within tissue, and its use is widespread in both oncological imaging and neuroimaging. In the small bowel, diffusion is of interest as a tool to assess inflammation, for both diagnosis and assessment of treatment response in IBD [49–51]. Data suggests disease activity generally leads to restricted diffusion [50–56].

Diffusion imaging is of particular challenge in the bowel as the technique is particularly vulnerable to movement and susceptibility artifacts, notably from intraluminal gas. Imaging may be done with breath holds (limiting the acquisition time and therefore SNR) or free breathing (potentially increasing the image blur). Most centers currently acquire enteric diffusion-weighted images in free breathing, usually in the axial plane at *B* values of 0 and 600 or 800. At high *B* values, the normal bowel wall is usually poorly visualized, and areas of abnormal restricted diffusion, for example, in active Crohn's disease are seen as areas of high signal. Subjective assessment of the image is therefore very useful, although calculation of apparent diffusion coefficient (ADC) using a monoexponential fit with *B* values of 0 and either 600 s/mm² [49,50] or 800 s/mm² [51] may be performed and the use of multiple *B* values and biexponential fitting has been investigated [57].

8.5.6.1 Post Examination

Although MRE and MRC are generally well tolerated, the use of oral contrast agents and/or rectal water enema will usually lead to post procedure diarrhea. Patients should be warned before the examination so they can plan their travel home afterward. Easy access to toilet facilities near the MRI suite is mandatory. If *Buscopan* is administered, patients should be warned about the effect on eyesight in the context of driving, and should be advised to seek urgent medical attention should they develop a painful red eye. Allergic reactions to either oral or IV contrast agents are rare, but patients should be advised to seek medical attention if they develop allergic symptoms such as a rash or bronchospasm.

8.6 Normal Anatomy

Normal bowel wall thickness at MRE is between 1 and 2 mm. Diameter of the small bowel decreases from duodenum through to ileum as does the number and thickness of small bowel folds [58]. Signal intensity on T_2 imaging is usually uniform throughout the bowel; however, T_1 enhancement after contrast maybe increased in the proximal small bowel. Enhancement should however be less than that of adjacent blood vessels.

8.7 Conclusion

MR of the small bowel and colon are increasingly implemented as first-line imaging techniques for suspected enteric pathology, particularly in the context of IBD. Diagnostic accuracy is reliant on careful attention being paid to technique. In particular, consideration must be made to minimize artifacts inherent in MRI of the abdomen, achieving good bowel distension and ensuring the correct sequences are performed. This chapter has presented the options available to the practitioner and provided guidance on achieving high quality studies.

References

1. Panes J, Bouhnik Y, Reinisch W et al. (2013) Imaging techniques for assessment of inflammatory bowel disease: Joint ECCO and ESGAR evidence-based consensus guidelines. *J Crohns Colitis* 7(7):556–585. doi: 10.1016/j.crohns.2013.02.020
2. Barth MM, Smith MP, Pedrosa I et al. (2007) Body MR imaging at 3.0 T: Understanding the opportunities and challenges. *RadioGraphics* 27:1445–1462. doi: 10.1148/rg.275065204
3. Chang KJ, Kamel IR, Macura KJ, Bluemke DA. (2008) 3.0-T MR imaging of the abdomen: Comparison with 1.5 T. *RadioGraphics* 28:1983–1998. doi: 10.1148/rg.287075154
4. Schick F. (2005) Whole-body MRI at high field: Technical limits and clinical potential. *Eur Radiol* 15:946–959. doi: 10.1007/s00330-005-2678-0
5. Adamek HE, Schantzen W, Rinas U et al. (2012) Ultra-high-field magnetic resonance enterography in the diagnosis of ileitis (neo-)terminalis: A prospective study. *J Clin Gastroenterol* 46:311–316. doi: 10.1097/MCG.0b013e31822fec0c
6. Makanyanga JC, Pendsé D, Dikaios N et al. (2014) Evaluation of Crohn's disease activity: Initial validation of a magnetic resonance enterography global score (MEGS) against faecal calprotectin. *Eur Radiol* 24:277–287. doi: 10.1007/s00330-013-3010-z
7. Fiorino G, Bonifacio C, Padrenostro M et al. (2013) Comparison between 1.5 and 3.0 Tesla magnetic resonance enterography for the assessment of disease activity and complications in ileo-colonic Crohn's disease. *Dig Dis Sci* 58:3246–3255. doi: 10.1007/s10620-013-2781-z
8. Van Gemert-Horsthuis K, Florie J, Hommes DW et al. (2006) Feasibility of evaluating Crohn's disease activity at 3.0 Tesla. *J Magn Reson Imaging* 24:340–348. doi: 10.1002/jmri.20650
9. Ziech MLW, Lavini C, Caan MW A et al. (2012) Dynamic contrast-enhanced MRI in patients with luminal Crohn's disease. *Eur J Radiol* 81:3019–3027. doi: 10.1016/j.ejrad.2012.03.028
10. Ordás I, Rimola J, García-bosch O et al. (2012) Diagnostic accuracy of magnetic resonance colonography for the evaluation of disease activity and severity in ulcerative colitis: A prospective study. *Gut* 62(11): 1566–1572. doi: 10.1136/gutjnl-2012-303240
11. Rimola J, Rodriguez S, García-Bosch O et al. (2009) Magnetic resonance for assessment of disease activity and severity in ileocolonic Crohn's disease. *Gut* 58:1113–1120. doi: 10.1136/gut.2008.167957

12. Rimola J, Ordás I, Rodriguez S et al. (2011) Magnetic resonance imaging for evaluation of Crohn's disease: Validation of parameters of severity and quantitative index of activity. *Inflamm Bowel Dis* 17:1759–1768. doi: 10.1002/ibd.21551

13. Wood ML, Runge VM, Henkelman RM. (1988) Overcoming motion in abdominal MR imaging. *AJR Am J Roentgenol* 150:513–522. doi: 10.2214/ajr.150.3.513

14. Wagner M, Klessen C, Rief M et al. (2008) High-resolution T_2-weighted abdominal magnetic resonance imaging using respiratory triggering: Impact of butylscopolamine on image quality. *Acta Radiol* 49:376–382. doi: 10.1080/02841850801894806

15. Gutzeit A, Binkert CA, Koh D-M et al. (2012) Evaluation of the anti-peristaltic effect of glucagon and hyoscine on the small bowel: Comparison of intravenous and intramuscular drug administration. *Eur Radiol* 22:1186–1194. doi: 10.1007/s00330-011-2366-1

16. Dyde R, Chapman AH, Gale R et al. (2008) Precautions to be taken by radiologists and radiographers when prescribing hyoscine-N-butylbromide. *Clin Radiol* 63:739–743. doi: 10.1016/j.crad.2008.02.008

17. Froehlich JM, Daenzer M, von Weymarn C et al. (2009) Aperistaltic effect of hyoscine N-butylbromide versus glucagon on the small bowel assessed by magnetic resonance imaging. *Eur Radiol* 19:1387–1393. doi: 10.1007/s00330-008-1293-2

18. Absah I, Bruining DH, Matsumoto JM et al. (2012) MR enterography in pediatric inflammatory bowel disease: Retrospective assessment of patient tolerance, image quality, and initial performance estimates. *AJR Am J Roentgenol* 199:W367–375. doi: 10.2214/AJR.11.8363

19. Dillman JR, Smith EA, Khalatbari S, Strouse PJ. (2013) I.v. glucagon use in pediatric MR enterography: Effect on image quality, length of examination, and patient tolerance. *AJR Am J Roentgenol* 201:185–189. doi: 10.2214/AJR.12.9787

20. Lomas DJ, Graves MJ. (1999) Small bowel MRI using water as a contrast medium. *Br J Radiol* 72:994–997.

21. Rieber A, Aschoff A, Nüssle K et al. (2000) MRI in the diagnosis of small bowel disease: Use of positive and negative oral contrast media in combination with enteroclysis. *Eur Radiol* 10:1377–1382.

22. Lauenstein T, Schneemann H. (2003) Optimization of oral contrast agents for MR imaging of the small bowel. *Radiology* 228:279–283.

23. Lawrance IC, Welman CJ, Shipman P, Murray K. (2009) Small bowel MRI enteroclysis or follow through: Which is optimal? *World J Gastroenterol* 15:5300–5306. doi: 10.3748/wjg.15.5300

24. Ajaj WM, Lauenstein TC, Pelster G et al. (2005) Magnetic resonance colonography for the detection of inflammatory diseases of the large bowel: Quantifying the inflammatory activity. *Gut* 54:257–263. doi: 10.1136/gut.2003.037085

25. Negaard A, Paulsen V, Sandvik L et al. (2007) A prospective randomized comparison between two MRI studies of the small bowel in Crohn's disease, the oral contrast method and MR enteroclysis. *Eur Radiol* 17:2294–2301. doi: 10.1007/s00330-007-0648-4

26. Masselli G, Casciani E, Polettini E, Gualdi G. (2008) Comparison of MR enteroclysis with MR enterography and conventional enteroclysis in patients with Crohn's disease. *Eur Radiol* 18:438–447. doi: 10.1007/s00330-007-0763-2

27. Luboldt W, Bauerfeind P, Steiner P et al. (1997) Preliminary assessment of three-dimensional magnetic resonance imaging for various colonic disorders. *Lancet* 349:1288–1291. doi: 10.1016/S0140-6736(96)11332-5

28. Luboldt W, Debatin JF. (2000) Abdominal imaging invited update virtual endoscopic colonography based on 3D MRI. *Abdom Imaging* 572:568–572.

29. Lauenstein TC, Herborn CU, Vogt FM et al. (2001) Dark lumen MR-colonography: Initial experience. *Rofo* 173:785–789. doi: 10.1055/s-2001-16987

30. Ajaj W, Pelster G, Treichel U, Vogt F. (2003) Dark lumen magnetic resonance colonography: Comparison with conventional colonoscopy for the detection of colorectal pathology. *Gut* 52:1738–1744.

31. Schreyer AG, Scheibl K, Heiss P et al. (2006) MR colonography in inflammatory bowel disease. *Abdom Imaging* 31:302–307. doi: 10.1007/s00261-005-0377-6

32. Achiam MP, Chabanova E, Logager V et al. (2007) Implementation of MR colonography. *Abdom Imaging* 32:457–462. doi: 10.1007/s00261-006-9143-7

33. Debatin JF, Lauenstein TC. (2003) Virtual magnetic resonance colonography. *Gut* 52(Suppl 4):iv17–22.

34. Elwood JM, Ali G, Schlup MM et al. (1995) Flexible sigmoidoscopy or colonoscopy for colorectal screening: A randomized trial of performance and acceptability. *Cancer Detect Prev* 19:337–347.

35. Papanikolaou N, Grammatikakis J, Maris T et al. (2003) MR colonography with fecal tagging: Comparison between 2D turbo FLASH and 3D FLASH sequences. *Eur Radiol* 13:448–452. doi: 10.1007/s00330-002-1808-1

36. Ajaj W, Lauenstein TC, Schneemann H et al. (2005) Magnetic resonance colonography without bowel cleansing using oral and rectal stool softeners (fecal cracking)—A feasibility study. *Eur Radiol* 15:2079–2087. doi: 10.1007/s00330-005-2838-2

37. Florie J, Birnie E, van Gelder RE et al. (2007) MR colonography with limited bowel preparation: Patient acceptance compared with that of full-preparation colonoscopy. *Radiology* 245:150–159. doi: 245/1/150 [pii] 10.1148/radiol.2451061244

38. Achiam MP, Logager V, Chabanova E et al. (2010) Patient acceptance of MR colonography with improved fecal tagging versus conventional colonoscopy. *Eur J Radiol* 73:143–147. doi: S0720-048X(08)00551-2 [pii] 10.1016/j.ejrad.2008.10.003

39. Goehde SC, Descher E, Boekstegers A et al. (2005) Dark lumen MR colonography based on fecal tagging for detection of colorectal masses: Accuracy and patient acceptance. *Abdom Imaging* 30:576–83. doi: 10.1007/s00261-004-0290-4

40. Langhorst J, Kuhle CA, Ajaj W et al. (2007) MR colonography without bowel purgation for the assessment of inflammatory bowel diseases: Diagnostic accuracy and patient acceptance. *Inflamm Bowel Dis* 13:1001–1008. doi: 10.1002/ibd.20140

41. Oussalah A, Laurent V, Bruot O et al. (2010) Diffusion-weighted magnetic resonance without bowel preparation for detecting colonic inflammation in inflammatory bowel disease. *Gut* 59:1056–1065. doi: gut.2009.197665 [pii] 10.1136/gut.2009.197665

42. Bakir B, Acunas B, Bugra D et al. (2009) MR colonography after oral administration of polyethylene glycol-electrolyte solution. *Radiology* 251:901–909.

43. Froehlich JM, Waldherr C, Stoupis C et al. (2010) MR motility imaging in Crohn's disease improves lesion detection compared with standard MR imaging. *Eur Radiol* 20:1945–1951. doi: 10.1007/s00330-010-1759-x

44. Bickelhaupt S, Pazahr S, Chuck N et al. (2013) Crohn's disease: Small bowel motility impairment correlates with inflammatory-related markers C-reactive protein and calprotectin. *Neurogastroenterol Motil* 25:467–473. doi: 10.1111/nmo.12088

45. Bickelhaupt S, Froehlich JM, Cattin R et al. (2013) Differentiation between active and chronic Crohn's disease using MRI small-bowel motility examinations—Initial experience. *Clin Radiol* 68(12):1247–53. doi: 10.1016/j.crad.2013.06.024

46. Menys A, Atkinson D, Odille F et al. (2012) Quantified terminal ileal motility during MR enterography as a potential biomarker of Crohn's disease activity: A preliminary study. *Eur Radiol* 22:2494–2501. doi: 10.1007/s00330-012-2514-2

47. Zappa M, Stefanescu C, Cazals-Hatem D et al. (2011) Which magnetic resonance imaging findings accurately evaluate inflammation in small bowel Crohn's disease? A retrospective comparison with surgical pathologic analysis. *Inflamm Bowel Dis* 17:984–993. doi: 10.1002/ibd.21414

48. Makanyanga J, Punwani S, Taylor SA. (2012) Assessment of wall inflammation and fibrosis in Crohn's disease: Value of T_1-weighted gadolinium-enhanced MR imaging. *Abdom Imaging* 37:933–943. doi: 10.1007/s00261-011-9821-y

49. Oto A, Zhu F, Kulkarni K et al. (2009) Evaluation of diffusion-weighted MR imaging for detection of bowel inflammation in patients with Crohn's disease. *Acad Radiol* 16:597–603. doi: 10.1016/j.acra.2008.11.009

50. Oussalah A, Laurent V, Bruot O et al. (2010) Diffusion-weighted magnetic resonance without bowel preparation for detecting colonic inflammation in inflammatory bowel disease. *Gut* 59:1056–1065. doi: 10.1136/gut.2009.197665

51. Buisson A, Joubert A, Montoriol P-F et al. (2013) Diffusion-weighted magnetic resonance imaging for detecting and assessing ileal inflammation in Crohn's disease. *Aliment Pharmacol Ther* 37:537–545. doi: 10.1111/apt.12201

52. Oto A, Zhu F, Kulkarni F et al. (2009) Evaluation of diffusion-weighted MR imaging for detection of bowel inflammation in patients with Crohn's disease. *Acad Radiol* 16:597–603. doi: 10.1016/j.acra

53. Kiryu S, Dodanuki K, Takao H et al. (2009) Free-breathing diffusion-weighted imaging for the assessment of inflammatory activity in Crohn's disease. *J Magn Reson Imaging* 29:880–886. doi: 10.1002/jmri.21725

54. Tielbeek J A W, Ziech MLW, Li Z et al. (2014) Evaluation of conventional, dynamic contrast enhanced and diffusion weighted MRI for quantitative Crohn's disease assessment with histopathology of surgical specimens. *Eur Radiol* 24(3):619–629. doi: 10.1007/s00330-013-3015-7

55. Bittman M, Freiman M, Callahan MJ, Rossello JP, Warfield S. (2012) Diffusion-weighted imaging (DWI) biomarkers for the evaluation of Crohn's ileitis. *Pediatr Radiol* S261–S262. doi: 10.1007/s00247

56. Barber I, Cadavid L, Castellote A, Alvarez M, Enriquez G. (2011) Intestinal MRI in inflammatory bowel disease in paediatric patients: The added value of diffusion-weighted imaging (DWI). *Pediatr Radiol* S262. doi: 10.1007/s00247

57. Freiman M, Perez-Rossello JM, Callahan MJ et al. (2013) Characterization of fast and slow diffusion from diffusion-weighted MRI of pediatric Crohn's disease. *J Magn Reson Imaging* 37:156–163. doi: 10.1002/jmri.23781

58. Cronin CG, Delappe E, Lohan DG et al. (2014) Normal small bowel wall characteristics on MR enterography. *Eur J Radiol* 75:207–211. doi: 10.1016/j.ejrad.2009.04.066

9

Artifacts in Magnetic Resonance Imaging

Michael N. Hoff, Jalal B. Andre, and Brent K. Stewart

CONTENTS

9.1 Introduction

Magnetic resonance imaging (MRI) is a powerful diagnostic imaging modality, but artifacts may compromise MR images. An artifact is any image feature that is not inherently present in the original object. Artifacts are undesirable, especially in instances when medical diagnosis and therapy depend on the accurate representation of patient anatomy/morphology; accurate delineation of true object features from artifacts in images is thus essential. This chapter provides (a) a comprehensive description of MRI artifacts with examples [1–10] to facilitate their radiological differentiation and (b) proposed remedies for MRI artifacts to enable clinical courses of action.

Here artifacts are described according to the phenomena causing them, although emphasis is placed on subdividing them further according to artifact appearance. They are divided into three groups: (1) *sample* artifacts caused by the patient or objects placed in the scanner; (2) *sequence* artifacts caused by the imaging sequence; and (3) *system* artifacts caused by the hardware, equipment, and environment. Methods for artifact avoidance are listed following the artifact descriptions, unless service repair is the only remedy.

9.2 Sample Artifacts

Sample artifacts are defined as any artifact that originates from the patient. Section 9.2.1 details *motion* artifacts, a common artifact caused by *periodic* (respiration, arterial pulsation) and *aperiodic* (patient motion, cerebrospinal fluid [CSF] pulsation) motion frequency. Section 9.2.2

details *field inhomogeneity* artifacts, which are divided into *susceptibility effects* (typically seen as signal overlap, loss, and distortion), *chemical shift*, and *static field variations* of the equipment (the latter is left for later description in Section 9.4). Finally, Section 9.2.3 discusses *the magic angle effect*.

9.2.1 Motion

MRI data are not acquired in a spatially contiguous manner; rather, each digital sample contributes to the entire image. Each sample voxel contains magnetization with both magnitude and phase. Motion modulates this magnitude and/or phase, disrupting the one-to-one correspondence of the full dataset with the final image. This yields image artifacts [11,12] such as *ghosting, blurring,* and/or *signal loss* [13–16]. Most motion occurs *inter-view,* or on a timescale longer than the repetition time (T_R). This contributes to artifact along the phase-encoding axis. Faster *intra-view* motion that occurs during data sampling between the radiofrequency (RF) excitation and the center of the following echo propagates artifact along the frequency-encoding (FE) axis [11].

If motion is periodic, repeating itself over regular intervals in time (such as with cardiac motion, arterial pulsation, and respiration), there are various techniques available to mitigate the subsequent ghosting artifacts. If, on the other hand, the motion is aperiodic and relatively random as a function of time (as is commonly seen in bulk patient motion, eye movement, swallowing, or gastrointestinal peristalsis), it is difficult to fully correct subsequent blurring artifacts. However, recent developments indicate that reduction of artifacts stemming from random bulk motion may be on the horizon. Here, we evaluate both periodic and aperiodic artifacts in terms of their appearance, causation, and methods of avoidance.

9.2.1.1 Ghosting

Ghosts are replicas of image features overlapping true anatomical structures. They can resemble dissections and thrombi [16], and generally degrade diagnostic image quality. Periodic motion such as cardiac motion, arterial pulsation, and respiration generate ghosts across the field of view (FOV) by coherently phase-shifting acquired data. Typically, this *inter-view* motion yields ghosts with intensity dependent on the image intensity of the moving structure [17] and the degree of data modulation [18]. Ghost spacing depends on the motion frequency and the repetition time. Similar to aliasing artifacts (see Section 9.3.1), ghosts will fold over to the opposite side of an image if they extend beyond the FOV. Figure 9.1 depicts examples of ghosting artifacts.

9.2.1.2 Blurring

General blurring or smearing across the FOV of an image may be caused by aperiodic motion such as peristalsis, swallowing, eye movement, and bulk translational motion, as seen in Figure 9.2. Unpredictable motion introduces random phase shifts to the sampled data, yielding blurred incoherent ghosts smeared across the FOV. Occasionally, motion such as respiration may be semiperiodic, blending, ghosting, and blurring.

9.2.1.3 Intravoxel Dephasing

Darkened regions may occur when *intra-view* vascular or CSF flow causes intravoxel magnetization cancellation, as depicted in Figure 9.3a. Dark contours at organ boundaries may also arise from shear that causes similar magnetization cancellation [19]. Dark flow artifacts also arise in balanced steady-state free precession (bSSFP) imaging. If spins pass through the banded signal loss regions characteristic to bSSFP (see Section 9.2.2.1.2), signal loss can spread [20,21].

9.2.1.4 Misregistration

Misregistration of flowing blood in a vessel [22,23] and false contours [24] can occur if blood flows obliquely

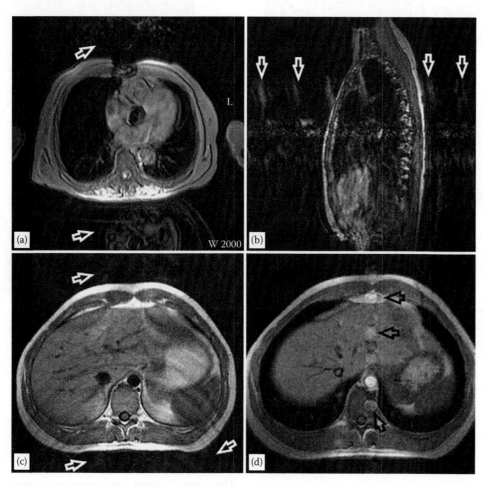

FIGURE 9.1
Arrows indicate periodic motion ghosting caused by (a) cardiac motion, (b and c) respiration, and (d) arterial pulsation. (© Koninklijke Philips N.V. All rights are reserved; Wijnen P: Philips Medical Systems, 2013; Advanced TIQ Artifacts in MRI, nly171172013; L1-TIQ Achievea more complex artifacts-PT.)

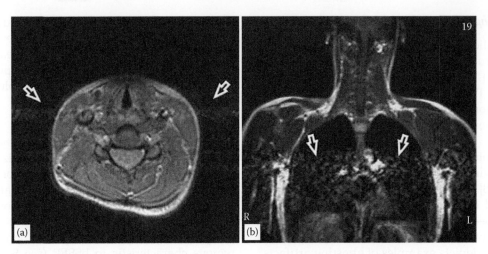

FIGURE 9.2
Arrows indicate incoherent ghosting or blurring caused by aperiodic motion: (a) bulk motion and (b) peristalsis. (© Koninklijke Philips N.V. All rights are reserved; L1-TIQ Achieva common artifacts-PT, Wijnen P: Philips Medical Systems, 2013; Advanced TIQ Artifacts in MRI, nly171172013.)

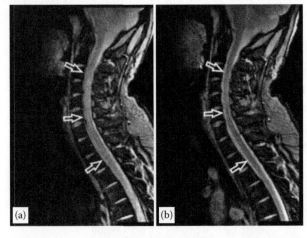

FIGURE 9.3
Arrows show (a) darker regions of CSF indicative of flow-induced intravoxel dephasing. (b) Flow compensation shows artifact mitigation. (© Koninklijke Philips N.V. All rights are reserved; L1-TIQ Achievea more complex artifacts-PT, Wijnen P2013.)

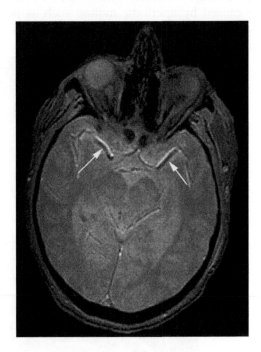

FIGURE 9.4
Arrows indicate flow misregistration when oblique-flow artifact occurs as vessels are at an oblique angle relative to the imaging plane. (From Storey, P., Artifacts and solutions, in: Edelman, R.R. et al., eds., *Clinical Magnetic Resonance Imaging*, 3rd ed., Saunders, Philadelphia, PA, 2006, pp. 577–629. With permission.)

with respect to the image plane during the time delay between phase and FE gradients. The degree of signal displacement depends on the flow velocity and the time delay in this *oblique-flow artifact* depicted in Figure 9.4. Modulated signal intensity in the center of a vessel may be caused by through-plane misregistration if laminar flow occurs during the time delay between slice-selection (SS) and *in-plane* (in the image plane) encoding. The degree of intensity modulation depends upon the direction of flow relative to the in-plane direction.

9.2.1.5 Motion-Related Banding

Image intensity oscillations near the peripheral regions of a uniformly moving anatomical region may manifest

as *motion-related bands*. Single-shot techniques acquire an entire image in one T_R; while they mitigate ghost artifacts, they generate a smoothly varying modulation to the sampled data manifesting as image oscillations [25]. These oscillations have amplitude dependent on tissue boundary contrast, and become progressively narrower with increasing distance from boundaries. Motion-related banding occurs *inter-view* and thus occurs

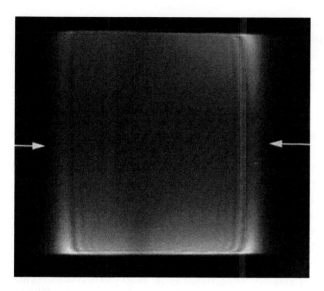

FIGURE 9.5
Arrows indicate motion-related banding in a phantom moving in the left/right phase-encoding direction. (From Storey, P., Artifacts and solutions, in: Edelman, R.R. et al., eds., *Clinical Magnetic Resonance Imaging*, 3rd ed., Saunders, Philadelphia, PA, 2006, pp. 577–629. With permission.)

along the phase-encoding direction (horizontal in the Figure 9.5 phantom image), and can mimic perfusion deficits in images.

9.2.1.6 Avoiding Motion Artifacts

Simple steps can be employed to mitigate some motion artifacts. Cushioning, vacuum devices [26], and sedation can reduce motion via support or immobilization. Swapping the phase-encoding and frequency-encoding directions can reduce artifacts caused by *inter-view* motion; for example, respiration occurs anterior–posterior, so spine imaging benefits from superior–inferior phase encoding. *Fast imaging* can acquire signal before motion and is able to interfere with the encoding process [27], which can be useful for aperiodic motion, motion in inhomogeneous fields, and motion-related banding.

Sequence modifications are common in motion artifact mitigation. PROPELLER (periodically rotated overlapping parallel lines with enhanced reconstruction) oversamples the central data region in k-space such that inconsistent data may be discarded, while providing reference information for improved image coregistration [28]. Saturating moving tissue or selective excitation is an option if imaging in moving regions is unnecessary. *Spatial saturation* is effective for flow artifact suppression [29]. *Spectral saturation* selectively saturates fat or water using their precession frequencies, but is sensitive to field inhomogeneity. Finally, *short-tau inversion recovery* (STIR) sequences can selectively null fat if an echo time is chosen such

that T_1-based relaxation causes fat to have no signal [30,31]. *Gradient moment nulling, flow compensation, gradient motion rephasing,* and *gradient compensation* [32–34] reduce phase offsets by adding a gradient along the FE or SS directions to mitigate *intra-view* blurring and intravoxel dephasing [16], or along the phase-encoding direction to suppress oblique-flow misregistration [35].

Gating, or control of sequence timing, is a prominent technique for dealing with cardiac-motion-related artifacts. Data acquisition is *triggered* by an electrocardiogram signal of the cardiac cycle [36], and data corrupted by excessive motion is discarded [37]. *Prospective gating* acquires the data at the same moment of each cardiac cycle, and does not acquire at the end of diastole; in general, it requires extra scan time, preparation, and processing [38–40]. Typically, prospective gating is combined with breath holding and *navigator echoes* [41] that monitor and adjust the acquired images to the patient's position within the scanner. *Retrospective gating* acquires data throughout the cardiac cycle, and reorganizes data during image reconstruction. This permits imaging of the entire cardiac cycle while minimizing the effects of cardiac motion [42,43].

There are other noteworthy techniques for motion-related artifact reduction. *Pseudogating* purposefully overlays ghosts onto the true image [44], while *adaptive correction* shifts the data reference position to track motion [45], typically using navigator echoes. *Averaging* techniques cancel some of the random, motion-induced magnetization variation in the acquired data [44]; a coherent application of this concept is the *ghost-interference* technique [46]. *Ordered phase-encoding* reorders data acquisition such that motion-induced magnetization modulation increases incrementally across k-space. While this requires *a priori* motion information, data translation can then reduce ghosts and blurring [18]. Recently, position-tracking using small RF coils [47–49] has shown promise for correcting motion-related artifacts. This technique prospectively tracks patient motion to permit the correction of associated artifacts, which may prove valuable for aperiodic, unpredictable motion.

9.2.2 Field Inhomogeneity

Magnetic field inhomogeneity can cause a variety of MRI artifacts. The magnetic field experienced by a patient within an MRI scanner is a superposition of gradient, static B_0, and RF-excitation/B_1 magnetic fields, and accurate MR imaging presupposes that only the linear field gradients are responsible for spatial field changes. However, the system and sample-related sources can also cause variations in the magnetic field. System or hardware-related sources of static field inhomogeneity such as poor shimming (shimming is the correction of field inhomogeneity by metals of coils in the magnet), gradient nonlinearity, and B_1 inhomogeneity are discussed in Section 9.4. Here, two sample-related sources of inhomogeneous magnetic fields are considered: the susceptibility effect (a substances' tendency to become magnetized in the magnet and generate subsidiary magnetic fields), and chemical shift (the deviation of nuclei's precession frequency caused by their local environment).

9.2.2.1 Susceptibility Effect

Field inhomogeneity artifacts may be caused by a tissue's or implanted device's tendency to be magnetized in the presence of an externally applied magnetic field. The resultant subsidiary magnetic fields combine with the B_0 and gradient fields to yield a spatially nonlinear field and subsequently a nonlinear variation in nuclear precession frequencies. Subsequent image artifacts are exacerbated by increasing B_0 [47] field strength and for greater susceptibility values [48], as are typical near metals (see Figure 9.6) and air cavities. Artifacts may be classified according to two contributing phenomena: signal cancellation stemming from intravoxel phase cancellation, and *in-plane* (within the image plane) and *through-plane* (propagating from image slice to slice) signal displacement. These phenomena yield multiple mechanisms that generate the distortion, signal loss, and signal overlap artifacts visualized in Figure 9.8 from Kaur et al. [1] and described in turn below.

9.2.2.1.1 Signal Distortion

Skewed and distorted image structure is a common MR artifact indicative of susceptibility artifact. Typically, associated frequency shifts generate spatial signal displacement along the FE axis; however, in echo-planar imaging (EPI), phase offsets accumulate over multiple phase encoding gradient applications, yielding distortion along the phase-encoding direction. SS suffers from similar through-plane frequency shifts causing *slice offset* (Figure 9.7a).

9.2.2.1.2 Signal Loss

Signal loss refers to the darkened image intensity that is common near metals (see Figures 9.6b, d, and 9.8); it may be caused by signal displacement and/or cancellation. Signal-displacement-induced signal loss occurs via various mechanisms, including (1) *in-plane pixel gap*, when no signal is mapped to a particular target pixel (Figure 9.7h); (2) *slice thinning*, due to through-plane frequency shifts (Figure 9.7f, and the dark gray region of Figure 9.9 from Lu et al. [3]); and (3) *slice skipping* (Figure 9.7g), when large field perturbations cause precession frequencies to extend beyond the RF excitation bandwidth [10]. *Signal cancellation-induced signal loss*, or *intravoxel dephasing*, occurs when inhomogeneous magnetic fields cause magnetization vectors in a voxel to dephase and cancel out. *General dephasing* (Figure 9.7i) is prevalent in spoiled gradient-echo (GRE) imaging (Figure 9.10 from Muir et al. [9]), and occurs due to phase incoherence as magnetization vectors point in essentially all directions and cancel. Band artifacts in bSSFP imaging stem from a more coherent form of intravoxel dephasing, here called *biphasic cancellation* (Figure 9.7j) since magnetization vectors take only two opposed phase values and cancel. Water diffusion in microscopically inhomogeneous fields can also yield dephasing [50].

9.2.2.1.3 Signal Pile-Up

Signal pile-up, or overlap, occurs when displaced signal generates spuriously bright regions in images. In-plane pixel shifts can yield *in-plane signal overlap* (Figure 9.7e) if signal from disparate locations are mapped to the

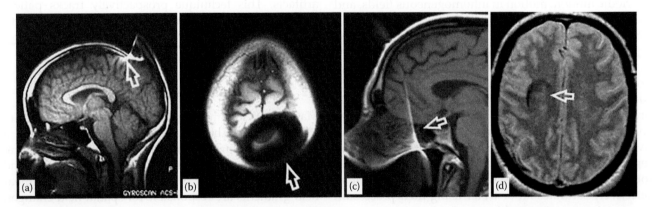

FIGURE 9.6
Arrows indicate susceptibility artifacts caused by metals in MR brain images, specifically near (a) and (b) a hairpin and (c) and (d) dental restorations. (© Koninklijke Philips N.V. All rights are reserved; L1-TIQ Achieva common artifacts-PT, Wijnen P: Philips Medical Systems, 2013.)

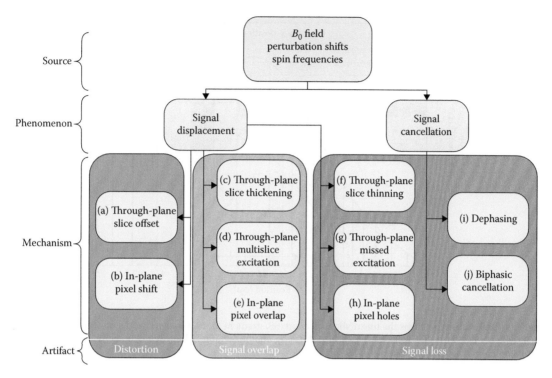

FIGURE 9.7
B_0 Inhomogeneity artifact categorization map. The effects of a source B_0 field perturbation on MR images may be broken down into two phenomena responsible for multiple artifact-generating mechanisms. Subsequent artifacts are seen as distortion, signal overlap, and signal loss.

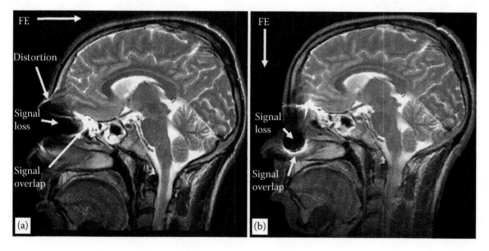

FIGURE 9.8
Metal flake in the sinus causes artifacts in turbo spin-echo brain images. Frequency-encoding (FE) direction is (a) horizontal and (b) vertical. Distortion, signal loss, and signal overlap are indicated. (© Radiography, 2007, with permission from Kaur, P. et al., *Radiography*, 13, 291–306, 2007; L1-TIQ Achieva common artifacts-PT. Wijnen P: Philips Medical Systems, 2013.)

same target pixel [51]. Through-plane frequency shifts can cause *slice thickening* (Figure 9.7c) as a broadened range of spins along the slice direction is excited and piled up in an image. The light-gray region of Figure 9.9 demonstrates *through-plane slice splitting* (Figure 9.7d), in which a field perturbation causes multiple slice locations along the slice select axis to have similar precession

frequencies, and thus all are excited and overlapped in a 2D image.

9.2.2.1.4 Failed Fat Suppression

Failed fat suppression is indicated by signal from fatty tissue and/or suppressed signal in nonfatty tissue following fat suppression methods, as shown

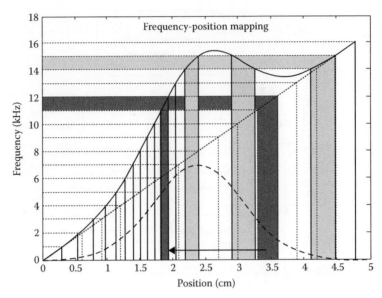

FIGURE 9.9
Field inhomogeneity (dashed line) perturbs the linear slice selection (SS) gradient (dotted line) to yield a nonlinear SS range of spin frequencies (solid line). Dark gray region at ~11.5 kHz RF shows slice thinning and offset, and light gray region at ~14.5kHz RF shows slice splitting, thinning, and thickening. (© Magnetic Resonance in Medicine, 2009, with permission from Lu, W.M. et al., *Magn. Reson. Med.*, 62, 66–76, 2009.)

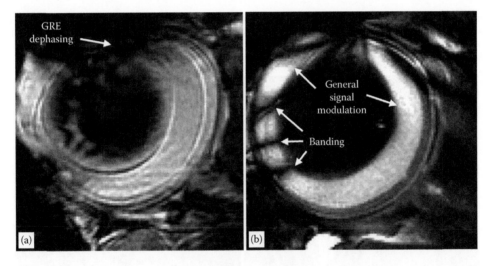

FIGURE 9.10
(a) General dephasing in a gradient-echo image of the eye. (b) bSSFP image of an eye shows banding/biphasic signal cancellation. (© Magnetic Resonance in Medicine, 2011, with permission from Muir, E.R. and Duong, T.Q., *Magn. Reson. Med.*, 66, 1416–1421, 2011.)

in Figure 9.11. Frequency-selective fat suppression techniques may fail when precession frequencies become unpredictable, as may occur near metals [52]. Failed fat suppression is exacerbated in asymmetric anatomic structures, in regions far from the isocenter, and at lower magnetic fields when the water-fat shift shrinks and becomes more error prone.

9.2.2.1.5 Avoiding Susceptibility Artifacts

Many methods of susceptibility artifact reduction are available clinically. Improved shimming, increased imaging bandwidths via stronger gradients [53,54], reduced voxel size [55,56], lower B_0 field strength [57,58], fast spin-echo imaging, minimization of echo time (T_e) [59–61], three-dimensional imaging using two dimensions of phase encoding, and orienting the long axis of a metal object parallel to the FE direction [62] and/or the B_0 field [63,64] are all ways to mitigate susceptibility artifact. Failed fat suppression can be avoided by shimming over the region of interest, or by employing alternate methods of fat suppression like STIR that do not inherently depend on frequency [31]. Removal of any metals from the scanner and calibration of the scanner center frequency may also reduce artifacts.

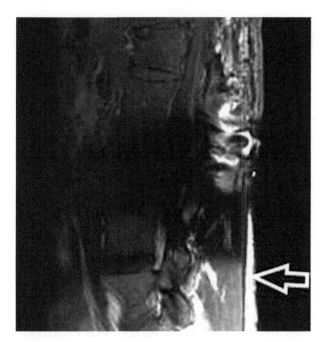

FIGURE 9.11
Arrows indicate failed fat suppression caused by poor shimming near metals. (© Koninklijke Philips N.V. All rights are reserved; Advanced TIQ Artifacts in MRI, nly171172013.)

Many advanced techniques exist for more comprehensive susceptibility artifact correction. *Ultra-short TE (UTE)* applications [65,66] minimize artifacts by lessening the time during which off-resonant phase can accumulate [67,68], but suffer from residual distortion, blurring, banding, and eddy current artifacts [69]. *Field mapping* techniques need prior knowledge to remap displaced signal using magnetic field estimates [51,70–76]. *Susceptibility mapping* [77,78] shows promise among these field mapping techniques, but requires a three-dimensional map of the susceptibility variation and experience difficulties near susceptibility interfaces [79]. *View angle tilting* (VAT) mitigates in-plane distortion with compensation gradients that tilt the slice angle [80], but suffers from residual blurring and through-plane artifacts [81]. *SEMAC, or slice encoding for metal artifact correction* [3,82,83], addresses VAT's through-plane artifacts via added through-plane phase encoding. An RF pulse of shifted central excitation frequency can excite spins that are beyond the original RF excitation bandwidth, as depicted in Figure 9.12 [10]; *MAVRIC, or multiacquisition variable-resonance image combination* [84], sums several such images acquired at variable central RFs. Both SEMAC and MAVRIC are being developed to further reduce residual artifact [69] and minimize the original ~15–20 min scan time requirement [85]. *Biphasic cancellation* in bSSFP imaging is commonly mitigated using techniques that combine images with modified excitation similar to MAVRIC [86–90], but recently analytical solutions have been developed which calculate susceptibility-independent signal [91,92].

9.2.2.2 Chemical Shift

Bright or dark contours at tissue interfaces in MR images are often due to *chemical shift*. It stems from a shift in a given molecule's resonant frequency due to magnetic field variations induced by its molecular environment (since each molecule has unique binding partners and bond lengths/angles). Here, artifacts are split into two groups: type I and type II, which manifest as a result of susceptibility signal displacement and signal cancellation, respectively.

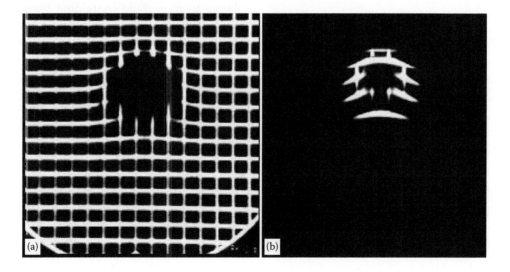

FIGURE 9.12
Through-plane slice skipping/missed excitation. (a) Surgical instrument metal particles cause signal loss in a grid phantom. (b) Shifting the RF pulse frequency permits signal loss recovery. (© Adapted with permission from Heindel, W. et al., *J. Comput. Assist. Tomogr.*, 10, 596–599, 1986.)

9.2.2.2.1 Chemical Shift Type I: Signal Displacement

Bright and dark contours in images at opposite interfaces between adipose and watery tissues are typically type I chemical shift displacement artifacts, as depicted in Figure 9.13. Fat and water have different resonant frequencies that disrupt the linearity of the FE process, yielding a spatial shift of fat relative to watery tissue in images along the FE direction [93]. SS direction shifts yield a nonplanar slice with water and fat from different through-plane regions. Chemical shift type I artifacts can result in erroneous determination of subchondral bone thickness [94], but can be useful for the determination of ambiguous tissues and identification of fatty lesions, such as intracranial lipomas and dermoid tumors [95].

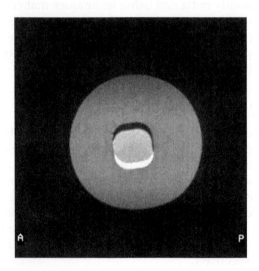

FIGURE 9.13
MR phantom image shows water and oil relatively displaced. (© Koninklijke Philips N.V. All rights are reserved; Advanced TIQ Artifacts in MRI, nly171172013.)

9.2.2.2.2 Chemical Shift Type II: Black Boundary

Chemical shift type II artifacts are dark borders between adipose and watery tissues. The different precession frequencies of fat and water yield generally different phase orientations at any given time; if signal is acquired at time T_e when the fat and water magnetization vectors are *out-of-phase* (directed 180° relative to one another), the magnetization in any voxels containing both fat and water will destructively interfere to cancel signal. The resultant *black boundaries* at interfaces [96] are demonstrated in Figure 9.14b, where such out-of-phase images can provide assistance in delineating soft-tissue lesions and organs (such as the kidneys, liver, and adrenal glands) from surrounding fat [95].

9.2.2.2.3 Avoiding Chemical Shift Artifacts

Spin-echo imaging can alleviate most chemical shift artifacts. Type I artifacts may be mitigated with stronger imaging gradients, since the subsequently larger bandwidth of encoded frequencies yields displacement across fewer image pixels (as with susceptibility displacements). Spectral presaturation can suppress type 1 contours by saturating either fat or water. Type II artifacts can be avoided by setting the T_e to occur when fat and water are in-phase (as shown in Figure 9.14a). *Slice-direction chemical shift* can be alleviated by reversing the slice gradient polarity between spin-echo pulses such that only part of the fat component will be focused by the 180° pulse [24].

9.2.3 Magic Angle Effect

Unexpected high intensity signal in collagenous tissue can stem from the *magic angle effect*. Due to structural rigidity, collagenous tissue has a short T_2 relaxation time relative to more fluid-containing tissue such as blood or fat. Solid tissue thus appears darker due to faster relaxation. However, the strength of the dipole–dipole

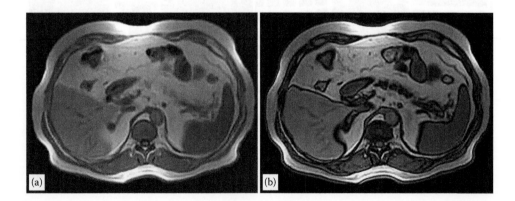

FIGURE 9.14
(a) In-phase and (b) out-of-phase gradient-echo images of the abdomen. The out-of-phase image demonstrates type II chemical shift black boundary artifact. (© Koninklijke Philips N.V. All rights are reserved; Advanced TIQ Artifacts in MRI, nly171172013.)

interaction responsible for T_2 relaxation is dependent on the direction of fibers relative to the main magnetic field. At a magic angle of ~55° for the fiber/field separation angle, the strength of the dipole-dipole reaction is zero; this lengthens the T_2 relaxation time and causes bright signal. Subsequent signal intensity increases that occur in tendons and hyaline cartilage [97] can mimic frank tear, tendonitis, meniscal tears, and general degeneration.

9.2.3.1 Avoiding the Magic Angle Effect

Increasing the echo time will ultimately dim the brightened signal, although this simultaneously introduces T_2 weighting [98].

9.3 Sequence Artifacts

Sequence artifacts are defined as any artifact that originates from the imaging method. While many artifacts such as *Gibb's/Truncation* artifact, *partial volume effect*, and *zero-fill artifact* arise due to the acquisition of insufficient data, *aliasing* is the most common and restrictive. SS-specific artifacts such as *crosstalk* and *staircase* are also discussed, followed by technique-specific artifacts including MR *angiography* (MRA) artifacts, the *spin-echo free induction decay* (FID) artifact, *stimulated echo* artifacts, *non-Cartesian imaging* artifacts, diffusion T_2 *shine-through*, and *parallel imaging* artifacts.

9.3.1 Aliasing

Aliasing (a.k.a. *wraparound* or *folding*) is a ghost-like *alias* of a peripheral area of an image that appears wrapped around to the opposite edge of the image. This is exacerbated if any signal intensity variation exists in that aliased, peripheral area, since its superposition with the true anatomical magnetization can yield spatially periodic intensity variation, or *Moiré fringes* (a.k.a. *banding interference* or *zebra artifacts*). Aliasing arises when the discrete digital sampling of MRI is not fine enough to uniquely identify signal [99].

9.3.1.1 FE Direction Aliasing

Although rare, aliases may occur along the frequency-encoded direction on older MR systems. Signal uniqueness requires that signal is sampled at twice the highest signal frequency (*Nyquist rate*) or above. Signals sampled at a frequency below the Nyquist rate will be aliased and appear to originate from a lower-frequency signal. An analogy to this is demonstrated in Figure 9.15: a red sine wave that represents signal encoded with some frequency is insufficiently sampled by the blue dots, since these samples could also represent the lower-frequency, blue wave. Detection of this red signal would be misinterpreted by the FE process, aliasing the signal to the incorrect, blue-wave frequency-encoded image location.

9.3.1.2 Phase-Encoding-Direction Aliasing

Aliasing occurs along the phase-encoding direction if excited signal exists beyond the prescribed phase-encoding FOV. This prescription sets 2π of phase variation to the FOV; Figure 9.16 shows that any magnetization beyond this FOV will have redundant phase values. A Fourier transformation of sampled data will confuse the redundant values and map them to the same location in the final image. Figure 9.17 demonstrates phase-encoding aliasing of the body and head.

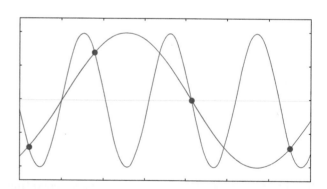

FIGURE 9.15
Aliasing will occur if the sampling (blue dots) is not frequent enough to sample the red-signal sine wave, since the lower frequency, blue-signal sine wave can be assumed.

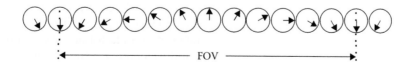

FIGURE 9.16
Twelve pixels in the field of view (FOV) have magnetization with slowly changing phase encoding from 0 to 2π. Redundant phase encoded values will be superimposed in the final image via phase-encoding-direction aliasing. (From Xiang, Q.-S., *Introduction to Magnetic Resonance Imaging*, UBC-PHYS 542 Class Notes, 2004. With permission.)

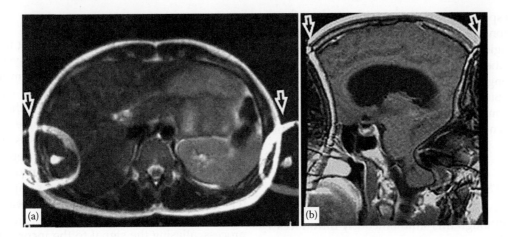

FIGURE 9.17
Arrows indicate peripheral anatomical regions beyond the field of view (FOV) that were aliased into MR images of (a) the body and (b) head along the phase-encoding direction. (© Koninklijke Philips N.V. All rights are reserved; L1-TIQ Achieva common artifacts-PT, Wijnen P: Philips Medical Systems, 2013.)

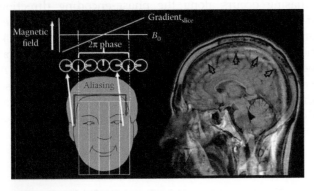

FIGURE 9.18
Slice-direction aliasing. (a) An excited sagittal slab of the head that has been binned into slices with phase encoding. A peripheral region to the left of the slab is encoded with the same phase as a slice on the right, and will be aliased onto it in image reconstruction. (b) Arrows in brain image demonstrate slice-direction aliasing. (© Koninklijke Philips N.V. All rights are reserved; Advanced TIQ Artifacts in MRI, nly171172013.)

9.3.1.3 SS Direction Aliasing

The 3D MRI employs slice direction phase encoding following slab excitation. Slices with redundant phase encoding at the slab periphery will thus be aliased (see Figure 9.18). Since viewers are not always aware of the anatomical features in these peripheral slices, when their features are aliased into the image of concern, they may not be recognized.

9.3.1.4 Avoiding Aliasing

FE aliasing is corrected with a low-pass filter that restricts high-frequency signals beyond the FOV from aliasing into the image. Since this can cause shading near image edges, a larger FOV is typically acquired

and low-pass filtered, followed by image truncation to the desired FOV.

Simple ways to avoid phase-encoding aliasing are to adjust the patient and image to the center of the FOV, and to set the phase-encoding direction to the narrower dimension. Aliases are eliminated if the FOV is large enough to cover all excited signal, but this can take extra scan time if equivalent spatial resolution is desired. Thus, it may be preferable to avoid exciting tissue outside the FOV; this may be achieved with a smaller surface coil or via spatial presaturation.

Slice-direction aliasing may be prevented using the remedies prescribed above for the phase-encoding direction. Additionally, phase encoding the entire slab and throwing away peripheral aliased slices manually or via oversampling can limit slice aliasing. Figure 9.19 demonstrates this with images acquired with and without oversampling.

9.3.2 Gibb's Ringing

Truncation artifacts (a.k.a. *Gibb's ringing* or *spectral leakage*) are image ripples that are parallel to and decreasing in intensity with distance from signal edges; Figure 9.20 demonstrates these artifacts in the brain. They arise due to insufficient acquisition of high spatial frequency data, typically along the phase-encoding direction where undersampling is frequently employed to reduce scan time. This effective truncation of the sampled data yields ringing artifacts near sharper edges that need more data for accurate representation. The ripple width depends on the image resolution, and its intensity depends on the interface contrast. The danger of these artifacts is that they can mimic the appearance of narrow structures such as a syrinx, the spinal cord, and intervertebral disks [100].

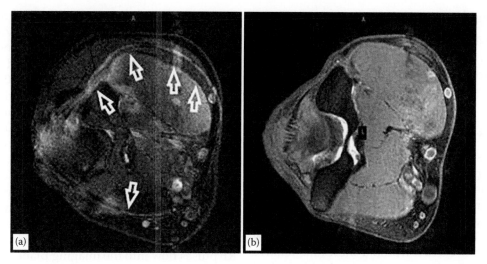

FIGURE 9.19
(a) Arrows in MR image indicate slice aliasing. (b) Oversampling peripheral slab regions eliminates aliasing. (© Koninklijke Philips N.V. All rights are reserved; Advanced TIQ Artifacts in MRI, nly171172013.)

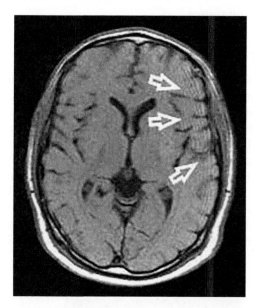

FIGURE 9.20
Arrows in brain image indicate Gibb's ringing artifact. (© Koninklijke Philips N.V. All rights are reserved; L1-TIQ Achieva common artifacts-PT, Wijnen P: Philips Medical Systems, 2013.)

9.3.2.1 Avoiding Gibb's Ringing

Gibb's ringing may be avoided by reducing the FOV or by increasing the matrix size, but these changes can cause aliasing and scan time increases, respectively. Parallel imaging may be employed to achieve higher resolution, lower signal-to-noise ratio (SNR) images without increasing scan time [101]. Typically, high spatial frequency data is progressively attenuated using filters to reduce ringing, but this can yield some image blurring. Alternatively, there are various

computationally intensive schemes that approximate missing high-spatial frequencies using the acquired data [102].

9.3.3 Partial Volume Effect

Partial volume effects appear as inaccurate pixel intensities. When voxels are larger than regional object features (e.g., more than one tissue lies in a voxel), image reconstruction forms an average of those features. This occurs at tissue boundaries, and chemical shift type II black boundary artifacts are an example of how this can yield phase cancellation and darkened signal. A similar partial volume effect occurs in inversion recovery imaging. Excitation is executed when one tissue has zero magnetization from T_1 recovery; if other tissues with higher and lower T_1 values exist in a voxel, the excitation will cause their magnetizations to have opposed phase and cancel [24].

9.3.3.1 Avoiding Partial Volume Effects

Choosing smaller voxels will mitigate partial volume effects, at the expense of SNR. As noted earlier, black boundary artifacts may be corrected by choosing a T_e at which fat and water are in-phase.

9.3.4 Zero-Fill Artifact

Zero-fill or *zebra artifact* manifest as alternating darkened image regions, often in an oblique direction. This artifact is caused by missing data prior to image reconstruction, potentially due to the data being set to zero.

9.3.4.1 Avoiding Zero-Fill Artifacts

Full data acquisition, variation of sequence parameters, or employing a surface coil can all help alleviate this artifact. If the artifact persists, it is a hardware issue and should be addressed by a service engineer.

9.3.5 Crosstalk

Crosstalk (a.k.a. *slice overlap* or *multistack artifact*) appears as darkened signal stripes in images that often mimic saturation bands, as demonstrated in Figure 9.21. Crosstalk is a direct result of imperfect slice profiles, where regions just beyond slice boundaries (side lobes) are unintentionally excited. If a slice containing a recently excited side lobe is excited within a short time relative to the side lobe's T_1 relaxation time, the doubly excited side lobe can suffer from magnetization cancellation and signal loss. [103].

9.3.5.1 Avoiding Crosstalk

Crosstalk is avoided by not imaging regions recently excited. *Slice interleaving* sequentially acquires slices that are not contiguous, and using *slice gaps* spaces slices out to ensure recently excited side lobes are not imaged. Alternatively, 3D imaging may be employed to avoid contiguous slice excitation altogether.

9.3.6 MRA Artifacts

MRA suffers from several artifacts inherent to the process of imaging blood vessels and associated organs. The *staircase artifact* appears as discrete signal dots along the direction of vessel flow in 3D imaging, due to postprocessing errors of thick slices. *Venetian blind artifacts* are a similar step-like artifact that can mimic luminal disease, caused by tissue moving in and out of regions of suppressed signal. Anomalous intensity oscillations in contrast-enhanced MRA images often stem from the *Maki* artifact [104,105], caused by premature acquisition of data prior to contrast arrival. *Laminar flow saturation* artifacts are partial volume effects manifesting as speciously low signal near vessel walls, which can result in overestimation of stenosis. *Flow-related aliasing* (a.k.a. *phase aliasing*) artifacts appear as dark regions of vessels when blood flows faster than the chosen velocity-encoding threshold. The *entry-slice phenomena* appear as intensity anomalies due to flow. *Inflow enhancement* occurs in GRE imaging when protons unsaturated by previous RF pulses flow into the imaging plane to yield brighter signal than stationary, partially saturated protons, while spin-echo imaging's inflowing spins may suffer from incomplete refocusing and darker signal as they may not be subjected to the full train of excitation and refocusing pulses.

9.3.6.1 Avoiding Angiography Artifacts

Staircase and venetian blind artifacts may be avoided by reducing slice thickness and using multiple overlapping thin slabs or slices. Maki artifacts are avoided by using a test bolus, specialized timing algorithms, *fluoroscopic triggering* [106–108], parallel imaging [101], and/or undersampling [109–111]. Phase aliasing may be mitigated using techniques such as the antialiasing acquisition (AAA) shown in Figure 9.22 [7], which applies bipolar velocity sensitization to compute flow-sensitized images without aliasing. Entry-slice phenomena in GRE imaging may be avoided if inflowing spins beyond the FOV are presaturated.

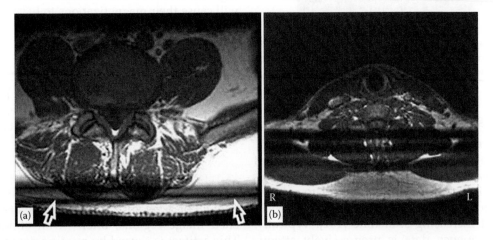

FIGURE 9.21
Arrows indicate darkened signal bands caused by crosstalk artifact.

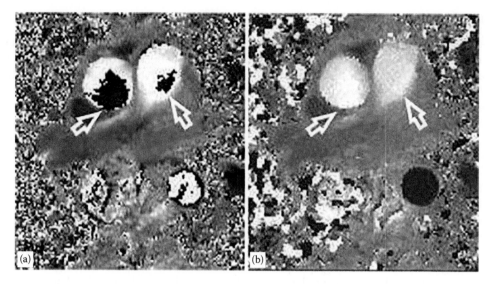

FIGURE 9.22
(a) Arrows in phase-velocity map indicates phase aliasing. (b) Antialiasing technique generates a velocity map free of phase aliasing. (From Xiang, Q.-S., Anti-aliasing acquisition for velocity encoded CINE. *Proceedings of the International Society for Magnetic Resonance in Medicine*, Glasgow, Scotland, 2001, p.1978. With permission.)

9.3.7 Spin-Echo FID Artifact

A line along the FE direction at the center of a spin-echo image may be caused by incomplete spoiling of the FID signal, and is thus denoted a *FID* artifact. It manifests as a line at zero-phase-encoding, since the FID occurs prior to application of the phase-encoding gradient; B_1 inhomogeneity and a poor slice selective excitation profile exacerbate its occurrence.

9.3.7.1 Avoiding Spin-Echo FID Artifact

The standard way to address the FID artifact is to manipulate the phase of the RF excitation pulse via RF *phase cycling*. Figure 9.23 demonstrates how the line arti-fact can be shifted to the edge of the FOV using phase cycling.

9.3.8 Stimulated Echoes

Stimulated echo artifacts can appear as a *ghost* inverted about the middle of an image along the phase-encoding direction, or a *zipper*-like line with alternating bright and dark pixels along the FE direction at the center of the image like an FID artifact. Stimulated echoes are tertiary echoes in sequences of at least three RF pulses. The mirrored ghosts arise from inaccuracies in the 90° and/or 180° pulses, and can yield band-like intensity modulations due to interference when over-laid over the true signal. The zipper artifact occurs when the stimulated echo magnetization is not phase encoded, assigning the signal to a line at zero-phase

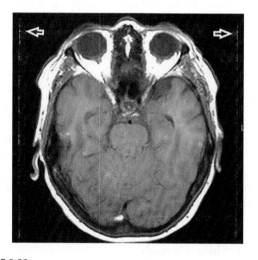

FIGURE 9.23
Arrows in the MR brain image indicate that a spin-echo FID artifact is shifted the image periphery using phase cycling. (© Koninklijke Philips N.V. All rights are reserved; L1-TIQ Achieva more complex artifacts-PT, Wijnen P2013.)

encoding [24]. Zipper artifacts may be exacerbated by imperfect slice profiles.

9.3.8.1 Avoiding Stimulated Echo Artifacts

Crusher gradients can be employed to dephase any unwanted magnetization that could contribute to stimulated echo coherences [112]. Mirrored ghosts are often alleviated via hardware repair if they arise from a transmitter problem. Similar to FID artifacts, RF phase-cycling can push a stimulated echo zipper artifact to the periphery of the image.

9.3.9 Non-Cartesian Sampling Artifacts

If data is not sampled in the standard Cartesian (rectangular or square) fashion, blurring, streaking, aliasing, and/or noise can arise. Data sampled by *radial* (a.k.a. *projection reconstruction*), *spiral*, and periodically rotated overlapping parallel lines with enhanced reconstruction (*PROPELLER*) trajectories must be *gridded* prior to image processing, which can yield some image blurring when there are trajectory errors. Patient motion in radial and PROPELLER imaging may manifest some blurring and streaking near edges and perpendicular to the direction of motion [113,114]. Radial imaging can suffer from aliasing, radial streaking, and diffuse noise due to the sparser sampling at higher spatial frequencies. If there is excited signal beyond the FOV, radial imaging shows a region of diffuse brightness near the image periphery, while spiral images show circular smearing. The long readout time of spiral imaging also makes it vulnerable to susceptibility effects and field inhomogeneity.

9.3.9.1 Avoiding Non-Cartesian Sampling Artifacts

The easiest way to avoid non-Cartesian sampling artifacts is to use Cartesian sampling. Radial acquisition motion artifacts may be mitigated by choosing a view order that spreads the error over all data [115], or by correcting with phase information [116]. PROPELLER imaging can discard or exploit the oversampled low-spatial frequency data to correct artifacts [28]. Spiral imaging motion artifacts may be corrected using navigators [117]. All techniques benefit from the standard corrections for aliasing (e.g., using a proximal lower sensitivity surface coil) and field inhomogeneity (fat suppression [118–120] and field mapping [121–123]).

9.3.10 T_2 Shine-Through

T_2 *shine-through* refers to bright signal in diffusion-weighted imaging (DWI) that reflects slow T_2 relaxation of tissue as opposed to restricted diffusion; high signal from a T_2 image without diffusion sensitization is *shining through* to the diffusion image. This artifact can be caused by various pathological abnormalities such as subacute infarctions or epidermoid cysts.

9.3.10.1 Avoiding T_2 Shine-Through

T_2 shine-through may be avoided by acquiring diffusion images with varying degrees of diffusion sensitization and computing an apparent diffusion coefficient (ADC) map. The ADC map is decoupled from the T_2 characteristics of tissue, displaying signal primarily weighted by diffusion.

9.3.11 Parallel Imaging Artifacts

Parallel imaging [124–127] is the process of detecting MR signal using multiple coil elements, where each coil detects unique yet complementary spatial information. One method called sensitivity encoding for fast MRI (SENSE) performs a weighted combination of multiple reduced-FOV, aliased coil images. The weights are chosen to yield an alias-free image using each coil's spatial sensitivity. Parallel imaging permits SNR increases or scan time reduction, but may suffer from inhomogeneous noise and/or residual aliasing.

9.3.11.1 Parallel Imaging Inhomogeneous Noise

Figure 9.24 demonstrates spatially varying image noise in a reconstructed parallel image. Since multiple structures may be overlapped in aliased component images, the calculation of weights can lack sufficient source information and yield amplified noise in the reconstruction. Since weights and the intensity of overlapping structures vary from pixel to pixel, the noise also varies regionally.

9.3.11.2 Parallel Imaging Aliasing

Residual aliases such as those demonstrated in the reconstructed parallel image in Figure 9.25 may arise due to corruption of the coil sensitivity maps. This can occur if there is anatomical displacement due to motion between calibration and regular image acquisition, or if the base coil images have RF overflow, excessive aliasing, or ghosting. The weights calculated for component image combination may then be incorrect, yielding residual aliasing.

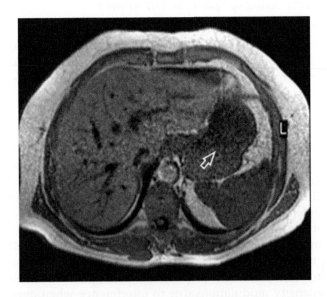

FIGURE 9.24
Arrow indicates inhomogeneous noise in an abdomen MR image attributed to parallel imaging. (© Koninklijke Philips N.V. All rights are reserved; L1-TIQ Achieva more complex artifacts-PT, Wijnen P2013.)

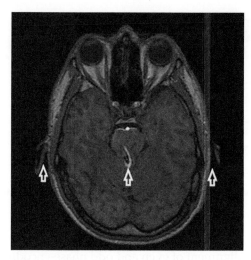

FIGURE 9.25
Arrows indicate aliasing from parallel imaging in an MR brain image. (© Koninklijke Philips N.V. All rights are reserved; L1-TIQ Achieva more complex artifacts-PT, Wijnen P2013.)

9.3.11.3 Avoiding Parallel Imaging Artifacts

Placing firm limitations on the degree of SENSE undersampling can mitigate both noise and aliasing artifacts. Careful coil placement and an FOV prescription with adequate coverage can reduce inhomogeneous noise. Repeating the calibration scan can prevent residual aliasing.

9.4 System Artifacts

The system artifact is defined as any artifact that is caused by the MRI equipment. Accordingly artifacts are grouped based on whether they affect the B_0 field and gradient systems, or the B_1 field and RF systems. If an *avoidance* section is omitted for a particular artifact, it is because avoidance of the artifact is best achieved via adjustments and/or repairs performed by a service engineer. Many of these artifacts are historic and automatically corrected on modern scanners.

9.4.1 B_0 Field and Gradient Systems Artifacts

Artifacts affecting the B_0 field and gradients include *static field variations* stemming from hardware issues associated with the main B_0 field and the gradients, as well as *temporal field variations* caused by gradient hardware problems.

9.4.1.1 Static Field Variations

Static-field-variation artifacts due to hardware issues appear similar to artifacts caused by the susceptibility effect (see Section 9.2.2.1), but are instead caused by poor magnetic field shimming, faulty hardware, and field drop-off at the bore periphery. Additionally, gradient nonlinearity and eddy currents can perturb the static B_0 field and generate similar artifacts.

9.4.1.1.1 Gradient Nonlinearity

Gradient nonlinearity causes images to appear distorted at the periphery relative to the center of the FOV (see Figure 9.26). While gradients are linear near the magnet isocenter, they drop off toward the periphery of the magnet bore, yielding distortions similar to those described in Section 9.2.2.1.1. When excited nuclei are completely beyond the gradient coil range, a bright, star-shaped artifact near the isocenter of the image can arise. When peripheral gradient-induced distortions are aliased, cone-shaped artifacts may arise.

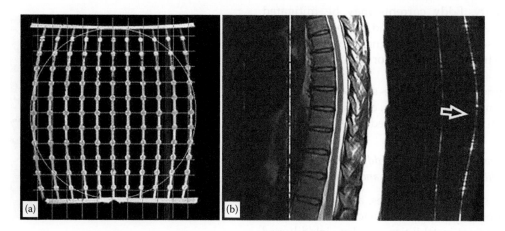

FIGURE 9.26
(a) MR phantom image with nonlinear gradient, weaker near the top and bottom. (b) MR spine image with RF interference line artifacts caused by a flickering light bulb. Arrow indicates the effects of gradient nonlinearity on lines near the image periphery. (© Koninklijke Philips N.V. All rights are reserved; Advanced TIQ Artifacts in MRI, nly171172013.)

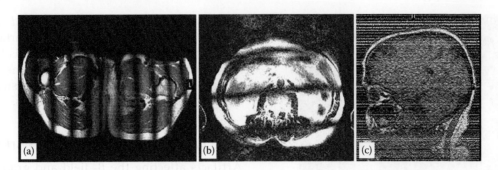

FIGURE 9.27
MR images suffer from (a) poor eddy current compensation and (b and c) gradient instability. (© Koninklijke Philips N.V. All rights are reserved; L1-TIQ Achieva more complex artifacts-PT, Wijnen P2013.)

9.4.1.1.2 Eddy Currents

Eddy current artifacts manifest as distortion, signal loss, and heightened sensitivity to motion. Rapidly changing magnetic field gradients typical of diffusion-weighted sequences can induce electrical currents in conductive materials such as the cryostat and RF coils, and these eddy currents induce secondary magnetic fields that contaminate the overall B_0 field in a manner similar to susceptibility effects. Eddy currents can also cause distortion of the gradient pulse profile, which can degrade magnet performance.

9.4.1.1.3 Avoiding Static-Field-Variation Artifacts

Field variation in an image may be minimized by positioning the patient's imaged region near the magnet isocenter, shimming on the region of interest, increasing the imaging bandwidth, and employing spin-echo imaging. Current MRI scanners automatically compensate for gradient nonlinearity. Residual cone-shaped artifacts are corrected with the dealiasing techniques described in Section 9.3.1.4, while star-shaped artifacts may be avoided by using local RF coils with limited spatial sensitivity. Eddy currents may be mitigated with self-shielded gradient coils and compensated via a dual spin-echo sequence [128], magnetic field map [129], empirical registration technique, or diffusion gradient reversal.

Service engineers can mitigate static-field-variation artifacts with improved shimming, calibration, and gradient waveform modifications to reduce eddy currents.

9.4.1.2 Temporal Field Variation: Eddy Currents, Instabilities, Timing Errors, and Failure

Eddy currents, gradient instabilities, and timing errors can yield temporally variable artifacts, as shown in Figure 9.27. Gradient instabilities can also yield blurred ghosts that appear similar to some RF artifacts discussed in Section 9.4.2. The EPI $N/2$ *ghost*, an alias that is shifted by half of the phase-encoding FOV, is caused by the misalignment of even and odd phase-encoding lines due to delays between gradients and the RF receiver, eddy currents, gradient imperfections, and filter asymmetries. Gradient failure can yield entirely unexpected images; for example, if the gradients are not amplified correctly, or fine gain calibration is not performed, an image like the one in Figure 9.28 may arise.

9.4.1.2.1 Avoiding Temporal Field Variation Artifacts

Eddy currents may be reduced as described in Section 9.4.1.1.3. Various techniques exist to mitigate $N/2$ ghosting, including those that employ added reference scans for echo realignment, phased-array processing [130], and additional images with added phase [76].

9.4.2 B_1 Field and RF Systems Artifacts

Artifacts affecting the B_1 field and RF systems include B_1 inhomogeneity, *RF overflow*, detector artifacts, *RF interference*, and temporal field variations (receiver variation, instabilities, poor connections, and failure).

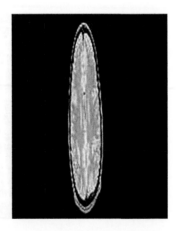

FIGURE 9.28
Brain MRI generated with a defective gradient amplifier. (© Koninklijke Philips N.V. All rights are reserved; L1-TIQ Achieva common artifacts-PT, Wijnen P: Philips Medical Systems, 2013.)

9.4.2.1 B₁ *Inhomogeneity*

Signal intensity variations throughout an image are typical manifestations of RF/B_1 inhomogeneity artifacts. Causes of RF inhomogeneity include eddy currents, the *quadrupole effect*, and *standing waves*; all effects are exacerbated by stronger B_0 field strength.

9.4.2.1.1 *Eddy Currents and the Quadrupole Effect*

In addition to the B_0 effects discussed above, eddy currents can generate signal loss in images due to B_1 effects. Signal loss near metal implants can arise when eddy currents generated in the metal shield the RF field, yielding a lack of RF exposure to tissue nearby [131–134]. Diagonal intensity variations from left to right and anterior to posterior in an axial image may be due to an eddy-current-induced B_1 disturbance called the *quadrupole effect* (see Figure 9.29). It commonly occurs in fat-suppressed spectral presaturation with inversion recovery (SPIR) images of the abdomen and pelvis, and is dependent on patient size and tissue composition.

9.4.2.1.2 *Standing Wave Artifact*

Standing wave (a.k.a. *dielectric, field-focusing effect*, or B_1 *doming*) artifacts leave bright spots, signal attenuation, and general signal modulation throughout an image. RF waves from multiple coils/elements/boundaries interfere with one another, yielding spurious bright regions in the center of the image, and a higher propensity for eddy currents to shield RF and suppress signal [135,136]. It is more likely at increasing field strength as the shrinking RF wavelength approaches the dimensions of the tissue itself.

9.4.2.1.3 *Avoiding RF Inhomogeneity Artifacts*

The quadrupole effect may be mitigated by suppressing eddy currents as described previously, or by reducing the flip angle. On the other hand, high flip angle compensation techniques [137] and "active" implanted materials that amplify the local RF via coupling also reduce

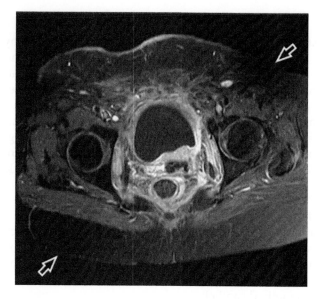

FIGURE 9.29
An abdominal spectral presaturation with inversion recovery MR image with arrows indicating the quadrupole effect. (© Koninklijke Philips N.V. All rights are reserved; L1-TIQ Achieva more complex artifacts-PT, Wijnen P2013.)

RF shielding effects [138]. Figure 9.30 demonstrates the ability of spectral attenuated inversion recovery (SPAIR) or STIR sequences to circumvent quadrupole-effect-induced fat suppression difficulties.

Standing waves are often corrected via RF shimming of independently controlled RF transmit elements to achieve B_1 uniformity. Postprocessing compensation techniques such as *surface coil intensity correction* (SCIC) algorithms may also be employed, and RF flip angle correction maps may be generated using empirical knowledge, although accuracy may not be sufficient for inversion recovery or magnetization transfer techniques. Improved coil design [139,140] and conductive dielectric patient pads assist in the mitigation of standing wave artifacts.

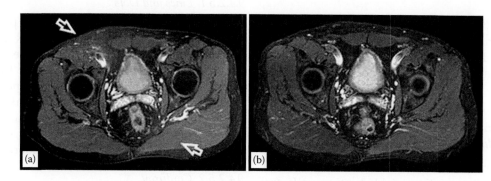

FIGURE 9.30
(a) Spectral presaturation with inversion recovery MR image with arrows indicating quadrupole-effect-induced failed fat suppression. (b) Spectral attenuated inversion recovery MR image has predictable signal in the same abdominal region. (© Koninklijke Philips N.V. All rights are reserved; Advanced TIQ Artifacts in MRI, nly171172013.)

9.4.2.2 RF Overflow

RF overflow appears as image contrast contaminated by a slowly varying shading pattern across the FOV, and a *halo*-like brightness in background noise regions near the patient. During prescanning, the RF receiver makes signal estimates to prescribe an adequate dynamic signal range. If the MRI signal in subsequent scans exceeds this estimated range, there will be an overflow in the analog-to-digital converter. Since calibration is executed on central slices, this is often seen in peripheral slices of a stack.

9.4.2.2.1 Avoiding RF Overflow

The receiver gain should be set to amplify the signal enough to avoid quantization error, but not so much as to yield RF overflow artifacts. Automatic correction methods are available if reacquisition is not an option [141], and if high fat signal causes overflow, a fat saturation method can help.

9.4.2.3 RF Detector Artifacts

Artifacts such as *DC offset* and *quadrature ghost* can arise due to RF detector problems.

9.4.2.3.1 Quadrature Ghosts

A quadrature ghost artifact occurs when a ghost of the true image appears rotated by 180° about the image center point. The artifact arises when the two receiver channels are not exactly 90° out of phase, or if channel outputs are variably amplified. If transmission asynchronicity exists, SS can form an image of the wrong slice.

9.4.2.3.2 DC Offset Artifact

The *DC offset* (a.k.a. *DC artifact*, *DC line*, or *central-point artifact*) appears as a dot or line at the center of an image. A DC (direct current) line runs along the phase-encoding direction where the FE gradient is zero, as seen in Figure 9.31. This artifact is generated when a DC offset voltage exists in the signal amplifiers.

9.4.2.3.3 Avoiding RF Detector Artifacts

The digital transceivers and software corrections standard in newer MRI systems prevent RF detector artifacts. Residual artifacts should be repaired by a service engineer.

9.4.2.4 Temporal RF Variation: Receiver Sensitivity, Instabilities, Poor Connections, and Failure

Various temporal RF variation artifacts exist in MRI. Receiver sensitivity variations yield ghosting artifacts if there is a temporal variation in the receiver coil's electrical characteristics; this may be caused by coil-loading variations stemming from patient respiration.

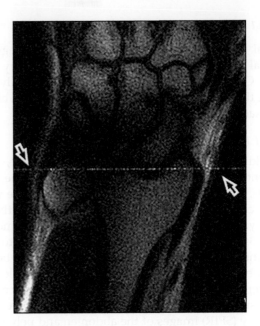

FIGURE 9.31
Arrows indicate a DC offset artifact in an MR wrist image. (© Koninklijke Philips N.V. All rights are reserved; Advanced TIQ Artifacts in MRI, nly171172013.)

Figure 9.32 depicts artifacts associated with poor connections, RF instabilities, and failure. Blurred ghosting and a variety of noise-like artifacts can stem from instability in the RF transmitter and loose connections. Shading and inhomogeneous brightness across MRI images can be caused by incorrect RF transceiver tuning, unbalanced RF energy deposition, and localized RF attenuation.

9.4.2.5 RF Interference

RF interference artifacts can arise due to faulty RF equipment and stray RF signal that lead to *lines* or *dots*, or data corruption leading to *corduroy* artifact.

9.4.2.5.1 Lines and Dots

An RF interference artifact manifests as parallel straight lines or dots running along the MR image phase-encoding direction at the FE value of the contaminating RF signal. Figure 9.33 shows some of these artifacts. Interference can originate from RF signals outside the magnet room (due to an open door or leaking RF shield), RF signals inside the magnet room (due to monitoring devices, cables, or flickering light bulbs), or from system malfunctions (bad diodes or poor soldering).

9.4.2.5.2 Corduroy

Figure 9.34 displays patterns of angled adjacent parallel lines overlaid in MR images that are attributed to corduroy (a.k.a. *zebra*, *herringbone*, *criss-cross*, or *data*

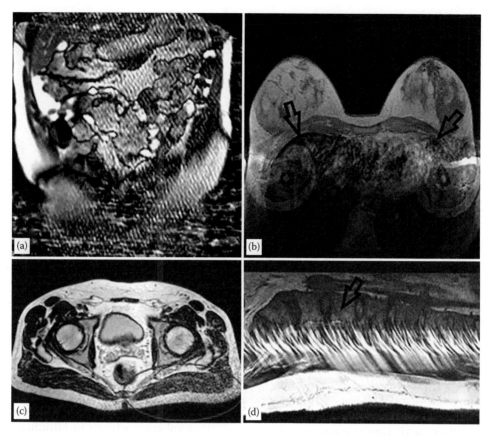

FIGURE 9.32
MR images of temporal RF variation artifacts such as (a) a loose RF connection, (c) RF coil channel defects, and (b and d) RF instabilities. (© Koninklijke Philips N.V. All rights are reserved; L1-TIQ Achieva more complex artifacts-PT, Wijnen P2013.)

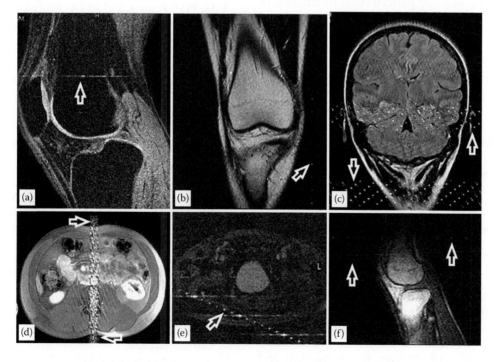

FIGURE 9.33
MR images display dot and line RF interference artifacts caused by (a, b, and d) RF leakage, (c and e) unfiltered cables, and (f) system malfunction. (© Koninklijke Philips N.V. All rights are reserved; L1-TIQ Achieva common artifacts-PT, Wijnen P: Philips Medical Systems, 2013; Advanced TIQ Artifacts in MRI, nly171172013; L1-TIQ Achieve a more complex artifacts-PT, Wijnen P2013.)

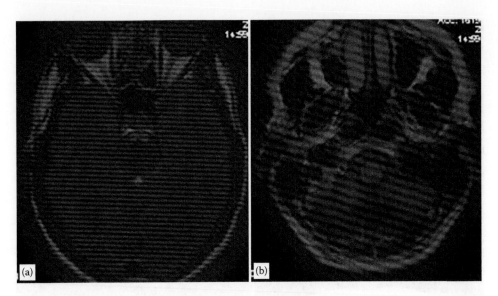

FIGURE 9.34
Corduroy artifact in MR brain images caused by data spikes due to ceiling incorrectly grounded metal ceiling tiles. (© Koninklijke Philips N.V. All rights are reserved; Advanced TIQ Artifacts in MRI, nly171172013.)

corruption error) artifact. Electrostatic discharges from metal-on-metal contact, patient blankets, clothing, or faulty lighting can cause a spike in *k*-space, especially when the humidity is low. The data spike causes an image corduroy pattern with line density and angle depending on the frequency and phase of the corrupted data point. Several *k*-space spikes can result in complicated overlapping corduroy patterns that resemble random noise.

9.4.2.5.3 *Avoiding RF Interference Artifacts*

Line and dot artifacts are avoided by closing the magnet room door, checking all third-party devices in the magnet room, validating the integrity of the RF shielding, and making repairs to any faulty equipment. Corduroy artifacts may be corrected by removing or repairing any materials that might generate electrostatic discharges.

9.5 Summary

The goal of this chapter is to phenomenologically describe various MRI artifacts, and to offer correction remedies. Armed with this information, technologists, scientists, and engineers have tools to understand and reduce artifacts. It is the responsibility of the radiologist to differentiate any residual artifacts from the anatomical and morphological content of an image.

References

1. Kaur P, Senthil KS, Tripathi RP, Khushu S, Kaushik S. Protocol error artifacts in MRI: Sources and remedies revisited. *Radiography* 2007;13(4):291–306.
2. L1-TIQ Achieva common artifacts-PT. Wijnen P: Philips Medical Systems; 2013.
3. Lu WM, Pauly KB, Gold GE, Pauly JM, Hargreaves BA. SEMAC: Slice encoding for metal artifact correction in MRI. *Magn Reson Med* 2009;62(1):66–76.
4. Xiang Q-S. Introduction to Magnetic Resonance Imaging, UBC-PHYS 542 Class Notes. 2004.
5. Advanced TIQ Artifacts in MRI. nly171172013.
6. L1-TIQ Achiev a more complex artifacts-PT. Wijnen P2013.
7. Xiang Q-S. Anti-aliasing acquisition for velocity encoded CINE. 2001. *Proceedings of the International Society for Magnetic Resonance in Medicine*, Glasgow, Scotland. p. 1978.
8. Storey P. Artifacts and solutions. In: Edelman RR, Hesselink J, Zlatkin M, editors. *Clinical Magnetic Resonance Imaging*. 3rd ed: Saunders; 2006. pp. 577–629.
9. Muir ER, Duong TQ. Layer-specific functional and anatomical MRI of the retina with passband balanced SSFP. *Magn Reson Med* 2011;66:1416–1421.
10. Heindel W, Friedmann G, Bunke J, Thomas B, Firsching R, Ernestus RI. Artifacts in MR imaging after surgical intervention. *J Comput Assist Tomogr* 1986;10(4):596–599.
11. Xiang Q-S, Henkelman RM. K-space description for MR imaging of dynamic objects. *Magn Reson Med* 1993;29:422–428.
12. Lauzon ML, Rutt BK. Generalized K-space analysis and correction of motion effects in MR-imaging. *Magn Reson Med* 1993;30(4):438–446.

13. Haacke EM, Patrick JL. Reducing motion artifacts in two-dimensional Fourier transform imaging. *Magn Reson Imaging* 1986;4(4):359–376.

14. Schultz CL, Alfidi RJ, Nelson AD, Kopiwoda SY, Clampitt ME. The effect of motion on two-dimensional fourier transformation magnetic-resonance images. *Radiology* 1984;152(1):117–121.

15. Wood ML, Henkelman RM. MR image artifacts from periodic motion. *Med Phys* 1985;12(2):143–151.

16. Ehman RL, Felmlee JP. Flow artifact reduction in MRI—A review of the roles of gradient moment nulling and spatial presaturation. *Magn Reson Med* 1990;14(2):293–307.

17. Wood ML, Mark HR. The magnetic field dependence of the breathing artifact. *Magn Reson Imaging* 1986;4(5):387–392.

18. Macgowan CK, Wood ML. Phase-encode reordering to minimize errors caused by motion. *Magn Reson Med* 1996;35(3):391–398.

19. Wedeen VJ, Weisskoff RM, Poncelet BP. MRI signal void due to in-plane motion is all-or-none. *Magn Reson Med* 1994;32(1):116–120.

20. Patz S. Some factors that influence the steady state in steady-state free precession. *Magn Reson Imaging* 1988;6:405–413.

21. Storey P, Li W, Chen Q, Edelman RR. Flow artifacts in steady-state free precession cine imaging. *Magn Reson Med* 2004;51:115–122.

22. Larson TC, Kelly WM, Ehman RL, Wehrli FW. Spatial misregistration of vascular flow during MR imaging of the CNS—Cause and clinical-significance. *Am J Neuroradiol* 1990;11(5):1041–1048.

23. Nishimura DG, Jackson JI, Pauly JM. On the nature and reduction of the displacement artifact in flow images. *Magn Reson Med* 1991;22(2):481–492.

24. Wood ML, Henkelman RM. Artifacts. In: Stark DD, Bradley W, editors. *Magnetic Resonance Imaging*. 3rd ed. Volume 1. St. Louis, MO: Mosby; 1999. pp. 215–230.

25. Storey P, Chen Q, Li W, Edelman RR, Prasad PV. Band artifacts due to bulk motion. *Magn Reson Med* 2002;48(6):1028–1036.

26. Bale RJ, Lottersberger C, Vogele M, Prassl A, Czermak B, Dessl A, Sweeney RA, Waldenberger P, Jaschke W. A novel vacuum device for extremity immobilisation during digital angiography: Preliminary clinical experiences. *Eur Radiol* 2002;12(12):2890–2894.

27. Coakley FV, Glenn OA, Qayyum A, Barkovich AJ, Goldstein R, Filly RA. Fetal MRI: A developing technique for the developing patient. *AJR Am J Roentgenol* 2004;182(1):243–252.

28. Pipe JG. Motion correction with PROPELLER MRI: Application to head motion and free-breathing cardiac imaging. *Magn Reson Med* 1999;42(5):963–969.

29. Felmlee JP, Ehman RL. Spatial presaturation—A method for suppressing flow artifacts and improving depiction of vascular anatomy in MR imaging. *Radiology* 1987;164(2):559–564.

30. Bydder GM, Pennock JM, Steiner RE, Khenia S, Payne JA, Young IR. The short T_1 inversion recovery sequence: An approach to MR imaging of the abdomen. *Magn Reson Imaging* 1985;3:251–254.

31. Bydder GM, Young IR. MR imaging: Clinical use of the inversion recovery sequence. *J Comput Assist Tomogr* 1985;9(4):659–675.

32. Haacke EM, Lenz GW. Improving MR image quality in the presence of motion by using rephasing gradients. *AJR Am J Roentgenol* 1987;148(6):1251–1258.

33. Pattany PM, Phillips JJ, Chiu LC, Lipcamon JD, Duerk JL, McNally JM, Mohapatra SN. Motion artifact suppression technique (MAST) for MR imaging. *J Comput Assist Tomogr* 1987;11(3):369–377.

34. Wood ML, Bronskill MJ, Mulkern RV, Santyr GE. Physical MR desktop data. *J Magn Reson Imaging* 1993;3 Suppl:19–24.

35. Bernstein MA, King KF, Zhou XJ. *Handbook of MRI Pulse Sequences*. Burlington, VT: Elsevier Academic Press; 2004.

36. Lanzer P, Barta C, Botvinick EH, Wiesendanger HUD, Modin G, Higgins CB. ECG-synchronized cardiac MR imaging—Method and evaluation. *Radiology* 1985;155(3):681–686.

37. Runge VM, Clanton JA, Partain CL, James AE. Respiratory gating in magnetic-resonance imaging at 0.5 tesla. *Radiology* 1984;151(2):521–523.

38. Liu YL, Riederer SJ, Rossman PJ, Grimm RC, Debbins JP, Ehman RL. A monitoring, feedback, and triggering system for reproducible breath-hold MR-imaging. *Magn Reson Med* 1993;30(4):507–511.

39. Fu ZW, Wang Y, Grimm RC, Rossman PJ, Felmlee JP, Riederer SJ, Ehman RL. Orbital navigator echoes for motion measurements in magnetic-resonance-imaging. *Magn Reson Med* 1995;34(5):746–753.

40. Sachs TS, Meyer CH, Hu BS, Kohli J, Nishimura DG, Macovski A. Real-time motion detection in spiral MRI using navigators. *Magn Reson Med* 1994;32(5):639–645.

41. Wang Y, Rossman PJ, Grimm RC, Riederer SJ, Ehman RL. Navigator-echo-base real-time respiratory gating and triggering for reduction of respiration effects in three-dimensional coronary MR angiography. *Radiology* 1996;198(1):55–60.

42. Glover GH, Pelc NJ. A rapid-gated cine MRI technique. In Kressel HY, ed. *Magnetic Resonance Annual*. New York: Raven Press; 1988, pp. 299–333.

43. Lenz GW, Haacke EM, White RD. Retrospective cardiac gating—A review of technical aspects and future-directions. *Magn Reson Imaging* 1989;7(5):445–455.

44. Wood ML, Henkelman RM. Suppression of respiratory motion artifacts in magnetic resonance imaging. *Med Phys* 1986;13(6):794–805.

45. Ehman RL, Felmlee JP. Adaptive technique for high-definition MR imaging of moving structures. *Radiology* 1989;173(1):255–263.

46. Xiang Q-S, Bronskill MJ, Henkelman RM. Two-point interference method for suppression of ghost artifacts due to motion. *J Magn Reson Imaging: JMRI* 1993;3:900–906.

47. Dumoulin CL, Souza SP, Darrow RD. Real-time position monitoring of invasive devices using magnetic resonance. *Magn Reson Med* 1993;29(3):411–415.

48. Derbyshire JA, Wright GA, Henkelman RM, Hinks RS. Dynamic scan-plane tracking using MR position monitoring. *J Magn Reson Imaging* 1998;8(4):924–932.

49. Ooi MB, Krueger S, Thomas WJ, Swaminathan SV, Brown TR. Prospective real-time correction for arbitrary head motion using active markers. *Magn Reson Med* 2009;62(4):943–954.

50. Schenck JF. The role of magnetic susceptibility in magnetic resonance imaging: MRI magnetic compatibility of the first and second kinds. *Med Phys* 1996;23(6):815–850.

51. Sekihara K, Matsui S, Kohno H. NMR imaging for magnets with large nonuniformities. *IEEE Trans Med Imaging* 1985;4:193–199.

52. Hargreaves BA, Worters PW, Pauly KB, Pauly JM, Koch KM, Gold GE. Metal-induced artifacts in MRI. *AJR Am J Roentgenol* 2011;197:547–555.

53. Lüdeke KM, Röschmann P, Tischler R. Susceptibility artefacts in NMR imaging. *Magn Reson Imaging* 1985;3(4):329–343.

54. O'Donnell M, Edelstein WA. NMR imaging in the presence of magnetic field inhomogeneities and gradient field nonlinearities. *Med Phys* 1985;12:20.

55. Petersilge CA, Lewin JS, Duerk JL, Yoo JU, Ghaneyem AJ. Optimizing imaging parameters for MR evaluation of the spine with titanium pedicle screws. *AJR Am J Roentgenol* 1996;166(5):1213–1218.

56. Young IR, Cox IJ, Bryant DJ, Bydder GM. The benefits of increasing spatial resolution as a means of reducing artifacts due to field inhomogeneities. *Magn Reson Imaging* 1988;6(5):585–590.

57. Laakman R, Kaufman B, Han J, Nelson A, Clampitt M, O'Block A, Haaga J, Alfidi R. MR imaging in patients with metallic implants. *Radiology* 1985;157:711–714.

58. Farahani K, Sinha U, Sinha S, Chiu LCL, Lufkin RB. Effect of field-strength on susceptibility artifacts in magnetic resonance imaging. *Comput Med Imaging Graph* 1990;14(6):409–413.

59. Czervionke LF, Daniels DL, Wehrli FW, Mark LP, Hendrix LE, Strandt JA, Williams AL, Haughton VM. Magnetic susceptibility artifacts in gradient-recalled echo MR imaging. *Am J Neuroradiol* 1988;9(6):1149–1155.

60. Posse S, Aue.WP. Susceptibility artifacts in spin-echo and gradient-echo imaging. *J Magn Reson* 1990;88(3):473–492.

61. Tartaglino LM, Flanders AE, Vinitski S, Friedman DP. Metallic artifacts on MR-images on the postoperative spine-reduction with fast spin echo techniques. *Radiology* 1994;190(2):565–569.

62. Eustace S, Goldberg R, Williamson D, Melhem ER, Oladipo O, Yucel EK, Jara H. MR imaging of soft tissues adjacent to orthopaedic hardware: Techniques to minimize susceptibility artefact. *Clin Radiol* 1997;52:589–594.

63. Guermazi A. Metallic artefacts in MR imaging: Effects of main field orientation and strength. *Clin Radiol* 2003;58:322–328.

64. Ladd ME, Erhart P, Debatin JF, Romanowski BJ, Boesiger P, McKinnon GC. Biopsy needle susceptibility artifacts. *Magn Reson Med* 1996;36(4):646–651.

65. Robson MD, Gatehouse PD, Bydder M, Bydder GM. Magnetic resonance: An introduction to ultrashort TE (UTE) imaging. *J Comput Assist Tomogr* 2003;27(6):825–846.

66. Idiyatullin D, Corum C, Park JY, Garwood M. Fast and quiet MRI using a swept radiofrequency. *J Magn Reson* 2006;181(2):342–349.

67. Rahmer J, Bornert P, Dries SPM. Assessment of anterior cruciate ligament reconstruction using 3D ultrashort echo-time MR imaging. *J Magn Reson Imaging* 2009;29(2):443–448.

68. Gold GE, Thedens D, Pauly JM, Fechner KP, Bergman G, Beaulieu CF, Macovski A. MR imaging of articular cartilage of the knee: New methods using ultrashort TEs. *AJR Am J Roentgenol* 1998;170(5):1223–1226.

69. Koch KM, Hargreaves BA, Pauly KB, Chen W, Gold GE, King KF. Magnetic resonance imaging near metal implants. *J Magn Reson Imaging: JMRI* 2010;32:773–787.

70. Kim JK, Plewes DB, Henkelman RM. Phase constrained encoding (PACE): A technique for MRI in large static field inhomogeneities. *Magn Reson Med* 1995;33:497–505.

71. Maudsley AA, Oppelt A, Ganssen A. Rapid measurement of magnetic field distributions using nuclear magnetic resonance. *Siemens Forschungs-Und Entwicklungsberichte-Siemens Research and Development Reports* 1979;8(6):326–331.

72. Willcott MR, Mee GL, Chesick JP. Magnetic field mapping in NMR imaging. *Magn Reson Imaging* 1987;5(4):301–306.

73. Yamamoto E, Kohno H. *Corrections of Field Errors Caused by Field Errors in MR Imaging.* 1987; New York. p. 402.

74. Chang H, Fitzpatrick JM. A technique for accurate magnetic resonance imaging in the presence of field inhomogeneities. *IEEE Trans Med Imaging* 1992;11(3):319–329.

75. Hoff MN, Xiang Q-S. *Eliminating Metal Artifact Distortion Using 3D-PLACE.* 2009; Honolulu, HI. p. 570.

76. Xiang Q-S, Ye FQ. Correction for geometric distortion and N/2 ghosting in EPI by phase labeling for additional coordinate encoding (PLACE). *Magn Reson Med* 2007;57:731–741.

77. Marques JP, Bowtell R. Application of a Fourier-based method for rapid calculation of field inhomogeneity due to spatial variation of magnetic susceptibility. *Concepts Magn Res Part B: Mag Res Eng* 2005;25B:65–78.

78. Salomir R, de Senneville BD, Moonen CT. A fast calculation method for magnetic field inhomogeneity due to an arbitrary distribution of bulk susceptibility. *Concepts Magn Resonance* 2003;19B:26–34.

79. Koch KM, Papademetris X, Rothman DL, de Graaf RA. Rapid calculations of susceptibility-induced magnetostatic field perturbations for in vivo magnetic resonance. *Phys Med Biol* 2006;51:6381–6402.

80. Cho ZH, Kim DJ, Kim YK. Total inhomogeneity correction including chemical shifts and susceptibility by view angle tilting. *Med Phys* 1988;15:7.

81. Butts K, Pauly JM, Gold GE. Reduction of blurring in view angle tilting MRI. *Magn Reson Med* 2005;53:418–424.

82. Yang QX, Williams GD, Demeure RJ, Mosher TJ, Smith MB. Removal of local field gradient artifacts in T-2*-weighted images at high fields by gradient-echo slice excitation profile imaging. *Magn Reson Med* 1998;39(3):402–409.

83. Lu W, Pauly KB, Gold GE, Pauly JM, Hargreaves BA. *Towards Artifact-Free MRI Near Metallic Implants.* 2008; Toronto, ON, Canada. p. 838.

84. Koch KM, Lorbiecki JE, Hinks RS, King KF. A multispectral three-dimensional acquisition technique for imaging near metal implants. *Magn Reson Med* 2009;61:381–390.

85. Hargreaves BA, Chen W, Lu W, Alley MT, Gold GE, Brau ACS, Pauly JM, Pauly KB. Accelerated slice encoding for metal artifact correction. *J Magn Res Imaging: JMRI* 2010;31:987–996.

86. Schwenk A. NMR pulse technique with high sensitivity for slowly relaxing systems. *J Magn Reson* 1971;5:376–389.

87. Ernst RR; Varian Associates, assignee. Fourier Transform NMR Spectroscopy Employing a Phase Modulated RF Carrier. US patent 3,968,424. July 6, 1976.

88. Bangerter NK, Hargreaves BA, Vasanawala SS, Pauly JM, Gold GE, Nishimura DG. Analysis of multiple-acquisition SSFP. *Magn Reson Med* 2004;51:1038–1047.

89. Elliott AM, Bernstein MA, Ward HA, Lane J, Witte RJ. Nonlinear averaging reconstruction method for phase-cycle SSFP. *Magn Reson Imaging* 2007;25:359–364.

90. Casselman JW, Kuhweide R, Deimling M, Ampe W, Dehaene I, Meeus L. Constructive interference in steady state-3DFT MR imaging of the inner ear and cerebellopontine angle. *Am J Neuroradiol* 1993;14(1):47–57.

91. Xiang Q-S, Hoff MN. Banding artifact removal for bSSFP imaging with an elliptical signal model. *Magn Reson Med* 2014;71(3):927–933.

92. Hoff MN, Xiang Q-S. *An Algebraic Solution for Banding Artifact Removal in bSSFP Imaging.* 2011; Montreal, QC, Canada. p. 2824.

93. Soila KP, Viamonte M, Starewicz PM. Chemical-shift misregistration effect in magnetic-resonance imaging. *Radiology* 1984;153(3):819–820.

94. McGibbon CA, Bencardino J, Palmer WE. Subchondral bone and cartilage thickness from MRI: Effects of chemical-shift artifact. *MAGMA* 2003;16(1):1–9.

95. Hood MN, Ho VB, Smirniotopoulos JG, Szumowski J. Chemical shift: The artifact and clinical tool revisited. *RadioGraphics* 1999;19(2):357–371.

96. Wehrli FW, Perkins TG, Shimakawa A, Roberts F. Chemical shift-induced amplitude modulations in images obtained with gradient refocusing. *Magn Reson Imaging* 1987;5(2):157–158.

97. Xia Y. Magic-angle effect in magnetic resonance imaging of articular cartilage: A review. *Invest Radiol* 2000;35(10):602–621.

98. Li T, Mirowitz SA. Manifestation of magic angle phenomenon: Comparative study on effects of varying echo time and tendon orientation among various MR sequences. *Magn Reson Imaging* 2003;21(7):741–744.

99. Bracewell RN. *The Fourier Transform and Its Applications.* New York: McGraw-Hill; 1986.

100. Bronskill MJ, McVeigh ER, Kucharczyk W, Henkelman RM. Syrinx-like artifacts on MR images of the spinal-cord. *Radiology* 1988;166(2):485–488.

101. Maki JH, Wilson GJ, Eubank WB, Hoogeveen RM. Utilizing SENSE to achieve lower station sub-millimeter isotropic resolution and minimal venous enhancement in peripheral MR angiography. *J Magn Reson Imaging* 2002;15(4):484–491.

102. Liang ZP, Boada FE, Constable RT, Haacke EM, Lauterbur PC, Smith MR. Constrained reconstruction methods in MR imaging. *Rev Magn Reson Med* 1992;4:67–185.

103. Kucharczyk W, Crawley AP, Kelly WM, Henkelman RM. Effect of multislice interference on image-contrast in T_2-weighted and T_1-weighted MR images. *Am J Neuroradiol* 1988;9(3):443–451.

104. Lee VS, Martin DJ, Krinsky GA, Rofsky NM. Gadolinium-enhanced MR angiography: Artifacts and pitfalls. *AJR Am J Roentgenol* 2000;175(1):197–205.

105. Maki JH, Prince MR, Londy FJ, Chenevert TL. The effects of time varying intravascular signal intensity and k-space acquisition order on three-dimensional MR angiography image quality. *J Magn Reson Imaging* 1996;6(4):642–651.

106. Ho VB, Foo TK. Optimization of gadolinium-enhanced magnetic resonance angiography using an automated bolus-detection algorithm (MR SmartPrep). Original investigation. *Invest Radiol* 1998;33(9):515–523.

107. Riederer SJ, Bernstein MA, Breen JF, Busse RF, Ehman RL, Fain SB, Hulshizer TC et al. Three-dimensional contrast-enhanced MR angiography with real-time fluoroscopic triggering: Design specifications and technical reliability in 330 patient studies. *Radiology* 2000;215(2):584–593.

108. Tatli S, Lipton MJ, Davison BD, Skorstad RB, Yucel EK. From the RSNA refresher courses—MR imaging of aortic and peripheral vascular disease. *RadioGraphics* 2003;23:S59–S78.

109. Korosec FR, Frayne R, Grist TM, Mistretta CA. Time-resolved contrast-enhanced 3D MR angiography. *Magn Reson Med* 1996;36(3):345–351.

110. Du J, Carroll TJ, Wagner HJ, Vigen K, Fain SB, Block WF, Korosec FR, Grist TM, Mistretta CA. Time-resolved, undersampled projection reconstruction imaging for high-resolution CE-MRA of the distal runoff vessels. *Magn Reson Med* 2002;48(3):516–522.

111. Barger AV, Block WF, Toropov Y, Grist TM, Mistretta CA. Time-resolved contrast-enhanced imaging with isotropic resolution and broad coverage using an undersampled 3D projection trajectory. *Magn Reson Med* 2002;48(2):297–305.

112. Crawley AP, Henkelman RM. A stimulated echo artifact from slice interference in magnetic resonance imaging. *Med Phys* 1987;14(5):842–848.

113. Schaeffter T, Weiss S, Eggers H, Rasche V. Projection reconstruction balanced fast field echo for interactive real-time cardiac imaging. *Magn Reson Med* 2001;46(6):1238–1241.

114. Glover GH, Pauly JM. Projection reconstruction techniques for reduction of motion effects in MRI. *Magn Reson Med* 1992;28(2):275–289.

115. Theilmann RJ, Gmitro AF, Altbach MI, Trouard TP. View-ordering in radial fast spin-echo imaging. *Magn Reson Med* 2004;51(4):768–774.

116. Shankaranarayanan A, Wendt M, Lewin JS, Duerk JL. Two-step navigatorless correction algorithm for radial k-space MRI acquisitions. *Magn Reson Med* 2001;45(2): 277–288.

117. Moriguchi H, Lewin JS, Duerk JL. Novel interleaved spiral imaging motion correction technique using orbital navigators. *Magn Reson Med* 2003;50(2):423–428.

118. Bornert P, Stuber M, Botnar RM, Kissinger KV, Manning WJ. Comparison of fat suppression strategies in 3D spiral coronary magnetic resonance angiography. *J Magn Reson Imaging* 2002;15(4):462–466.

119. Nayak KS, Cunningham CH, Santos JM, Pauly JM. Real-time cardiac MRI at 3 tesla. *Magn Reson Med* 2004;51(4):655–660.

120. Moriguchi H, Lewin JS, Duerk JL. Dixon techniques in spiral trajectories with off-resonance correction: A new approach for fat signal suppression without spatial-spectral RF pulses. *Magn Reson Med* 2003;50(5):915–924.

121. Moriguchi H, Dale BM, Lewin JS, Duerk JL. Block regional off-resonance correction (BRORC): A fast and effective deblurring method for spiral imaging. *Magn Reson Med* 2003;50:643–648.

122. Ahunbay E, Pipe JG. Rapid method for deblurring spiral MR images. *Magn Reson Med* 2000;44(3):491–494.

123. Nayak KS, Tsai CM, Meyer CH, Nishimura DG. Efficient off-resonance correction for spiral imaging. *Magn Reson Med* 2001;45(3):521–524.

124. Sodickson DK, Manning WJ. Simultaneous acquisition of spatial harmonics (SMASH): Fast imaging with radiofrequency coil arrays. *Magn Reson Med* 1997;38:591–603.

125. Griswold MA, Jakob PM, Heidemann RM, Nittka M, Jellus V, Wang J, Kiefer B, Haase A. Generalized autocalibrating partially parallel acquisitions (GRAPPA). *Magn Reson Med* 2002;47(6):1202–1210.

126. Roemer PB, Edelstein WA, Hayes CE, Souza SP, Mueller OM. The NMR phased-array. *Magn Reson Med* 1990;16(2):192–225.

127. Pruessmann KP, Weiger M, Scheidegger MB, Boesiger P. SENSE: Sensitivity encoding for fast MRI. *Magn Reson Med* 1999;42(5):952–962.

128. Reese TG, Heid O, Weisskoff RM, Wedeen VJ. Reduction of eddy-current-induced distortion in diffusion MRI using a twice-refocused spin echo. *Magn Reson Med* 2003;49(1):177–182.

129. Jezzard P, Barnett AS, Pierpaoli C. Characterization of and correction for eddy current artifacts in echo planar diffusion imaging. *Magn Reson Med* 1998;39(5):801–812.

130. Kellman P, McVeigh ER. Ghost artifact cancellation using phased array processing. *Magn Reson Med* 2001;46(2):335–343.

131. Bartels LW, Smits HFM, Bakker CJG, Viergever MA. MR imaging of vascular stents: Effects of susceptibility, flow, and radiofrequency eddy currents. *J Vasc Interv Radiol* 2001;12(3):365–371.

132. Teitelbaum G, Bradley W, Klein B. MR imaging artifacts, ferromagnetism, and magnetic torque of intravascular filters, stents, and coils. *Radiology* 1988;166:657–664.

133. Augustiny N, Vonschulthess GK, Meier D, Bosiger P. MR imaging of large nonferromagnetic metallic implants at 1.5 T. *J Comput Assist Tomogr* 1987;11(4):678–683.

134. New P, Rosen B, Brady TJ, Buonanno F, Kistler J, Burt C, Hinshaw W, Newhouse J, Pohost G, Taveras J. Potential hazards and artifacts of ferromagnetic and nonferromagnetic surgical and dental materials and devices in nuclear magnetic resonance imaging. *Radiology* 1983;147:139.

135. Alecci M, Collins CM, Smith MB, Jezzard P. Radio frequency magnetic field mapping of a 3 Tesla birdcage coil: Experimental and theoretical dependence on sample properties. *Magn Reson Med* 2001;46(2):379–385.

136. Yang QX, Wang J, Zhang X, Collins CM, Smith MB, Liu H, Zhu XH, Vaughan JT, Ugurbil K, Chen W. Analysis of wave behavior in lossy dielectric samples at high field. *Magn Reson Med* 2002;47(5):982–989.

137. Meyer JM, Buecker A, Spuentrup E, Schuermann K, Huetten M, Hilgers RD, van Vaals JJ, Guenther RW. Improved in-stent magnetic resonance angiography with high flip angle excitation. *Invest Radiol* 2001;36(11):677–681.

138. Quick HH, Kuehl H, Kaiser G, Bosk S, Debatin JF, Ladd ME. Inductively coupled stent antennas in MRI. *Magn Reson Med* 2002;48(5):781–790.

139. Alecci M, Collins CM, Wilson J, Liu W, Smith MB, Jezzard P. Theoretical and experimental evaluation of detached endcaps for 3 T birdcage coils. *Magn Reson Med* 2003;49(2):363–370.

140. Liu W, Collins CM, Delp PJ, Smith MB. Effects of endring/shield configuration on homogeneity and signal-to-noise ratio in a birdcage-type coil loaded with a human head. *Magn Reson Med* 2004;51(1):217–221.

141. Jackson J, Macovski A, Nishimura D. Low-frequency restoration. *Magn Reson Med* 1989;11(2):248–257.

10

Risk of Magnetic Resonance: The Safety-Biological Effects

Valentina Hartwig

CONTENTS

10.1 Introduction

Magnetic resonance (MR) is a well-known diagnostic technique today widely used especially for cardiac and neurological applications. MR has an excellent spatial resolution and a good temporal resolution, and allows obtaining high-quality clinical images often essential to the diagnosis or monitoring of several diseases. Moreover, MR is considered a safe technology because it has just the ability to change the position of atoms, but not to alter their structure, composition, and properties, as the ionizing radiations do. In fact, the electromagnetic radiations involved in an MR procedure (see Figure 10.1) have no enough energy to detach electrons from atoms or molecules, such as other higher energetic radiations can do (X-rays, radiations used in nuclear medicine, etc.). However, as in any sanitary intervention, even in an MR diagnostic procedure there are intrinsic hazards that must be understood, acknowledged, and taken into consideration; for this reason, the analysis of the interaction between the magnetic fields and the biological tissues that undergo an MR procedure is essential. These interactions are caused from different physical phenomena which, on the one hand, are responsible for the signal generation that contribute to the final image but, on the other hand, can cause dangerous biological effects for the patient or signal artifacts. The study of these interactions has become more important in the past years due to growing interest for high-static magnetic

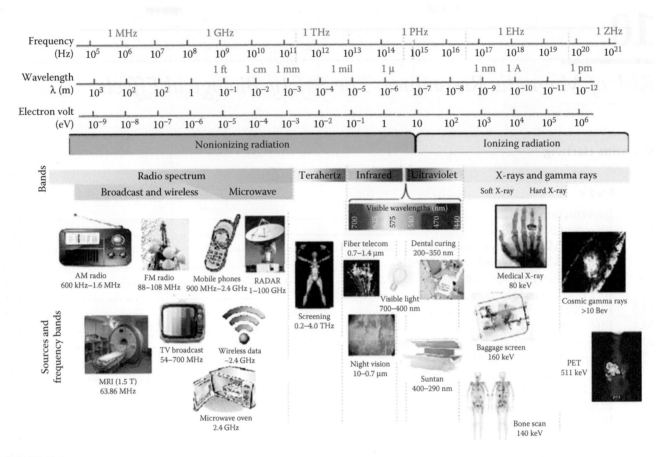

FIGURE 10.1
Electromagnetic spectrum and some sources of radiation.

field (MF) MR scanners, which assure a higher signal-to-noise ratio (SNR), and hence a better quality for the final images, but imply heavier risks for patients and occupational workers. Although the radiation used is not ionizing, there are several effects to be considered for safety assurance and engineering aspects relative to MR signal and image generation. The knowledge of these phenomena is important not only for the design of transmission/reception coils and acquisition sequences but also for the choice of acquisition parameters for each diagnostic exam.

During an magnetic resonance imaging (MRI) examination, three types of MFs are employed to produce three-dimensional images: (i) a high static MF (B_0), which generates a net magnetization vector in the human body, that is, a measure of the proton density; (ii) three gradient MFs (100–1000 Hz one for each direction) used to localize aligned protons inside the body, thus allowing spatial reconstruction of tissue sections into images; (iii) an RF MF (B_1 @ 10–400 MHz), which energizes the magnetization vector allowing its detection by the MRI scanner, converting tissue properties into MR images. Different levels of contrast are based on the different magnetic properties

TABLE 10.1

MRI Magnetic Field Components and Recognized Hazards

Static Magnetic field	Projectile effects
	Magnetic induction effects
	Magnetomechanical effects
	Electron spin interaction effects
Field gradients	Peripheral nerve and muscle stimulation
	Acoustic noise
Radiofrequency field	Thermal heating
	Nonthermal effects
	Burns
Other hazards	Quench hazards
	Contrast agents
	Claustrophobia
	Pregnancy

and physical structures of the biological tissues (i.e., density of hydrogen atoms).

This chapter deals with the biological effects and safety problems in MRI, first describing the biological effects of each MF involved in an MRI procedure, summarized in Table 10.1.

There are also dedicated sections for other MR-related safety issues (such as quench hazards in superconducting magnets, contrast agents, claustrophobia, and risks for pregnancy), risks for MR occupational workers, and subjects with biomedical implants. Section 10.9 is dedicated relatively to MR safety guidelines and regulations.

Finally, Section 10.10 describes known methods to quantify the interactions between MR fields and biological tissues: the dosimetric parameters are defined and the different kinds of radiofrequency (RF) dosimetry, experimental, and theoretical dosimetry are described in detail.

10.2 Effects of Static MF

Safety of static MF has been discussed for more than a century; in 1921, Drinker and Thompson [1] carried out numerous experiments to investigate possible effects on workers exposed to MF in industrial applications. They concluded that the static MF had no significant hazard effects on human health.

More than 400 papers were published on the biological effects of static MF, but the results were contradictory and confusing [2].

With the advent of MRI at the beginning of the 1980s, the interest in understanding the potential hazards associated with static MF exposure has increased. A recent review concluded that it was very difficult to prove the existence of significant biological effects of static MF [3], with the exception of some sensory effects such as nausea, vertigo, and metallic taste [4].

The strength of the static MF B_0 of a scanner is expressed in tesla (T). One tesla equals 10,000 Gauss (G). The Earth's MF is approximately 0.05 mT (0.5 G). There are three principal types of MR magnets: superconducting, permanent, and resistive magnets. The most widely used are superconducting magnets, which rely on liquid helium to cool the specially constructed coil windings to extremely low temperatures close to absolute zero. Permanent magnets are bar magnets and do not require cooling or an electrical power source to operate. Resistive electromagnets require a permanent electrical source to operate. Typically, most MR magnets (superconducting and permanent magnets) are permanently on, even in the event of a power failure.

The major recognized mechanical risk associated with MR static MF is the presence of ferromagnetic devices and equipment, including biomedical implants. These pieces of equipment will be subject to the attractive (projectile effect) and rotational forces, caused by the static field, whose magnitude depends on their mass and distance from the bore entrance [5].

The inadvertent introduction or the presence of ferromagnetic objects (e.g., oxygen tanks and wheelchairs) into the MR environment caused the documented serious injuries and few fatalities in MR system magnets.

The projectile effect caused the most serious accident reported to date: a 6-year-old boy died after an MRI exam, when the machine's powerful MF jerked a metal oxygen tank across the room, crushing the child's head [6].

Although there is no evidence of health hazards associated with the exposure of patients to strong static MFs, three physical mechanisms of interaction between tissues and static MFs that could lead to potential pathological effects are known: magnetic induction, magneto-mechanical, and electronic interactions.

10.2.1 Magnetic Induction

The current density flowing in biological tissues exposed to an external electric field is determined by the electric field $\vec{E}$ and the electric conductivity of the tissues. If the tissue moves with a velocity $\vec{v}$ and is exposed to a static MF $\vec{B}$, there is an additional term $(\vec{v} \times \vec{B})$ in the expression of the electric current density flowing in the tissue, known as a motion-induced electric field:

$$\vec{J} = \sigma \cdot (\vec{E} + \vec{v} \times \vec{B}) \quad \left(A/m^2\right) \tag{10.1}$$

This additional electric current can produce biological effects by disrupting physiological electrical signals of the human body, such as neuronal conduction and biopotentials. Moreover, the blood flow in a static MF induces electromotive force (emf) which could produce a biological effect on the electric activity of the heart. However, for the static MF intensity values used today for diagnostic imaging, no significant vital sign changes, for example, electrocardiogram (ECG) recordings, have been reported up to now [7].

Another possible magnetic induction effect is the induction of magnetohydrodynamic forces and pressure: when a static MF is applied to a biological tissue, and ionic currents are present, a net force applied to the moving ions is present. Hence, these forces principally act on flowing liquids, such as blood and on the endolymphatic tissues of the inner ear, inducing the sensations of nausea and vertigo sometimes reported at higher field strengths [8,9].

10.2.2 Magneto-Mechanical Effects

Materials may be classified into three large groups according to the values of their susceptibility: diamagnetic materials ($-1 < \chi < 0$), paramagnetic materials

$(0 < \chi < 0.01)$, and ferromagnetic materials ($\chi > 0.01$). The majority of human tissues is diamagnetic or weakly paramagnetic [3]. Static fields induce a torque on paramagnetic molecules orienting in a configuration that minimizes their energy. This effect is also present on diamagnetic macromolecules with differing magnetic susceptibilities along the principal axes of symmetry. However, due to the very small values of magnetic susceptibility, these forces are too small to affect biological material in vivo [3].

If the static field is spatially inhomogeneous, it forms a gradient that produces a net translational force on both diamagnetic and paramagnetic materials that are proportional to the product of magnetic flux density (B) and its gradient (dB/dx). For ferromagnetic objects, such as metal of high-magnetic susceptibility, this force causes dangerous acceleration.

10.2.3 Electron Spin Interaction

A static MF affects the kinetics and recombination of radical pairs because it splits them into two energy levels increasing the number of radicals escaping the recombination reaction [10,12]. This radical pair mechanism with the Earth's MF is used by birds as a source of navigational information during migration [13].

10.2.4 Other Effects

Movements of the head in a high-MF cause time-varying MFs, which could directly stimulate the retina and/or optic nerve and produce a phenomenon known as *magnetophosphenes* that are flashes of light. These phenomena are experienced frequently with MF above 4 T [7]. Movement in high-MF could also generate a metallic taste [14].

According to the World Health Organization (WHO) [15], there has been no conclusive evidence for irreversible or hazardous bioeffects of static MFs, so the projectile effect for ferromagnetic materials remains the main safety concern for static MFs [4,16].

10.3 Effects of Gradient MF

During an MRI examination, the gradient MF, which serves for the spatial localization in the image reconstruction process, is often switched on and off rapidly, for varying durations and with varying maximum strength, depending on the pulse sequence and scanner under consideration.

The time variation of the MF induces in the patient, undergoing an MR scan, an electric field according to Maxwell's third equation [17]:

$$\vec{\nabla} \times \vec{E} = -\frac{\partial \vec{B}}{\partial t} \tag{10.2}$$

This electric field could stimulate nerves and muscles and generate cardiac stimulation or even ventricular fibrillation. While the latter is a primary concern, being a life-threatening condition, possible peripheral nerve stimulation (PNS) may cause discomfort and could not be tolerated by the subjects thus interfering with the examination (e.g., due to patient movements) or would result in a request to stop the examination [18]. In 1989, in his review Reilly presented a rational methodology for evaluating excitation thresholds for peripheral nerves stimulated by high-gradient-induced electric fields. Since cardiac stimulation threshold is very high than PNS threshold (PNST) for gradient ramp duration less than 1000 µs, patients could be protected from gradient-induced ventricular fibrillation by not exceeding these thresholds.

According to Faraday's law, the induced electric field leads to an induced current: the threshold current density for ventricular fibrillation is about 1.2 A/m², so it can be avoided by keeping the current densities below 0.4 A/m² [20]. Clinical scanners are designed with restrictions to ensure that only peripheral nerve excitation, if at all, is possible. The most effective in producing excitation is generally the y-gradient, but the shape of the RF pulses used is important [21].

New generation of MR systems is characterized by higher static MF and faster gradient fields; however, on modern MR systems, dB/dt values are calculated prior to scan execution and thus monitored to prevent the initiation of a scan if safety standards are exceeded.

Figure 10.2 shows the mean thresholds for peripheral nerve and cardiac stimulation versus the time variation length.

During image acquisition, the gradient coils are switched on and off using rapid alterations of currents within them. These currents produce significant forces that act upon gradient coils and produce noise in the bore of the magnet, which can reach up to 140 dB for some sequences [22]. The noise is manifested as loud tapping, knocking, or chirping sounds. Levels greater than 100 dB may result in disturbance in microcirculation in the brain and produce mechanical damage [23].

Another problem with gradient MF is the induction of electric currents (so-called eddy currents) on conductivity materials such as metallic biomedical implants or electrode cables for vital parameter measurements (e.g., ECG) (see Section 10.8).

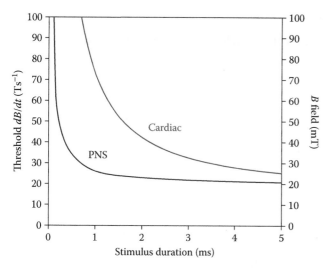

FIGURE 10.2
Mean thresholds for peripheral nerve and cardiac stimulation.

10.4 Effects of RF MF

During an MRI exam, the patient is exposed to a MF B_1 in the range of 7.5–300 MHz, that is, the RF range of the electromagnetic radiation spectrum (0–3000 GHz), which includes radar, ultrahigh frequency (UHF), very high frequency (VHF) television, AM and FM radio, and microwave communication frequencies.

The primary bioeffects associated with the RF radiation used for MR procedures are directly related to the thermogenic qualities of this electromagnetic field [24,25]. However, RF radiation also causes athermal effects in biological systems that are produced without increased elevation in temperature [26–28].

Hence, the RF biological effects can be classified into two categories [28,29]:

1. *Thermal effects*: They are caused due to tissue heating by direct absorption of energy from the electric fields [30].

2. *Nonthermal effects*: They are caused due to direct energy transfer from the field to the living system, which might be strongly nonlinear, and are dependent on the field frequency. The nonthermal effects have not been studied in association with the use of MR systems and are not well understood [29].

Since 1985, many investigations have been conducted to characterize the thermal effects of MR procedure-related heating. RF energy interactions with biological materials (in a physical sense) are complex and the results obtained from animal and in-vitro experiments

are not always directly applicable to human beings since the coupling of radiation to biological tissues is primarily dependent on the organism's size, anatomical features, duration of exposure, and sensitivity of the involved tissues (i.e., thermal sensibility) [31,32]. For example, certain organs, such as the eyes and testes, are particularly sensitive to heating due to lack of perfusion, so the presence of *hot spots* in their correspondence can be very dangerous for patient safety [33].

The temperature increase of the tissues due to the RF energy absorption depends on parameters such as the electrical and geometrical tissue properties, the type of RF pulse used, its repetition time, and the frequency of the radiation. The dosimetric parameter used to describe the absorption of RF radiation is the specific absorption rate (SAR) [34], that is, the mass normalized rate at which RF power is coupled to biological tissue. SAR is indicated in units of watts per kilogram [26,32]. Usually, SAR is averaged in 1 g of tissue to indicate the relative amount of RF radiation that an individual encounters during an MR with respect to the whole body.

During an MR scan, the patient temperature is not easy to be detected so SAR represents a convenient parameter to control the possible temperature increases. However, SAR is a complex function of numerous variables including the frequency, the type of RF pulse used, the repetition time, the type of RF coil used, the volume of tissue contained within the coil, the configuration of the anatomical region exposed, the orientation of the body to the field vectors, and other factors [29,35]. Generally, the MRI scanner software permits to monitor the SAR for the whole body: these values have to be always below the limits values set by International Electrotechnical Commission (IEC) standard [35,36] and must be recognizable by the software, so that if the SAR value exceeds the standard limits, the software stops the scanning process. The admitted SAR is usually 4 W/kg for a whole-body scanner, calculated for a body temperature increase up to 0.6°C and a scanning period of 20–30 min [37].

Most of the reported accidents are burns due to excessive local heating in the presence of conducting materials close to the patient such as the leads of equipment for monitoring physiological parameters (heart rate, blood pressure, oxygen saturation, and temperature) [38]. These metallic devices can act as an *antenna* and concentrate the RF energy, especially at the tip of the devices. A high risk of thermal injury can be for electrically conductive implants such as wires or leads, particularly when such wires or leads form large loops.

This kind of risk can be more serious in the case of internal biomedical implants (aneurism clips, stents, etc.), especially for implants that have elongated configurations and/or are electronically activated (neurostimulation systems and cardiac pacemakers) [39,40] (see Section 10.8).

Moreover, tattoos and permanent cosmetics, realized with iron oxide or other metal-based pigments, can cause reactions or adverse events (including first- and second-degree burns) [41].

10.5 Effects of Combination of Static, Gradient, and RF MFs

During an MRI scan, patients are exposed to combination of static, gradient, and RF fields. Besides minor adverse events, such as nausea and rare allergic reactions or tissue necrosis, associated with gadolinium-based contrast agents used routinely for MR examinations [42], more relevant for human health are the effects on biological parameters.

Recently, some biophysical properties of erythrocytes were analyzed in 25 patients during an MRI scan [43]. The results showed a significant decrease in red blood cells membrane permeability, membrane elasticity, and erythrocytes sedimentation rate during MRI, but the removal of the MF resulted in a rapid return to the normal conditions.

In a recent work, [28] lymphocyte cultures from healthy subjects had been exposed into magnetic resonance device in order to build dose–effect curves, using micronuclei induction as biological marker. Furthermore, micronuclei induction has also been evaluated in circulating lymphocytes of individuals after cardiac scan. A dose-dependent increase of micronuclei frequency was observed in vitro. Moreover, after in-vivo scan, a significant increase in micronuclei is found till 24 h, after the frequencies slowly return to control value.

Recently, the genotoxic potential of 3 T clinical MRI scans in cultured human lymphocytes in vitro was investigated [44]. Human lymphocytes were exposed to electromagnetic fields generated during a clinical routine MRI brain examination protocols. The results of this study suggest that exposure to 3 T MRI induces genotoxic effects in human lymphocytes.

Unluckily, very few works deal with the biological effects due to simultaneous exposure to the three types of MF.

Until a wider knowledge of the potential risk related to diagnostic MRI is available, a prudent attention should be adopted in order to avoid unnecessary examinations; according to the precautionary principle, it is being understood that MRI procedures are relatively safer than any other clinical tests using ionizing radiations.

10.6 Other MR-Related Safety Issues

10.6.1 Quench Hazards in Superconducting Magnets

Most of the clinical MR whole-body scanners have magnets made with superconducting materials that need liquid helium or liquid nitrogen to maintain the magnet at a temperature enabling superconductivity (close to absolute zero –273°C). The helium or nitrogen and magnet coils are maintained in a vacuum. The temperature of cryogens is approximately –269°C or 4.17 K. In the event of a so-called quench [20], the liquefied gases expand and boil off to the outside. This could happen for a disruption to the temperature or loss of the vacuum and causes an immediate loss of the MF. During a quench, the gas rushes out of the magnet and, under normal conditions, can be vented outside the scan room and building. In this case, the MR staff should remove the patient from the scan room as quickly as possible to avoid breathing gases and frostbite.

For this reason, appropriate local emergency procedures should be in place and included in the training program for all authorized personnel.

Moreover, it is also possible that the high pressure inside MR rooms makes it impossible to open the scanner door.

Typically, MR clinical scanners are provided with an emergency quench button in order to turn off the MF in the case of an emergency such as a person trapped against the side of the magnet by a ferromagnetic item. Moreover, the MR room is provided by some oxygen monitors at critical locations to detect any increase in cryogens caused by a quench.

10.6.2 Contrast Agents

Contrast agents are often used for clinical MRI and injected into patients as a part of the MRI procedure. There are a variety of contrast agents, such as liver contrast agents, water in the gut, and a variety of chelates. Gadolinium-based contrast agents are the most common contrast agents used in MR, in particular, for brain and spine imaging, as well as for contrast-enhanced MR angiography.

Side effects from MR contrast agents occur infrequently and generally are minor effects such as nausea, headache, and taste perversion.

In a patient with kidney disease, some gadolinium-based contrast may cause nephrogenic systemic fibrosis (NSF); so in this kind of patients, glomerular filtration rate should be measured prior to deciding on the use of gadolinium contrast agents [45].

Particular attention should be made before administering contrast agents to pregnant women and lactating mothers since some agents could cross the placenta readily and be excreted in breast milk.

10.6.3 Claustrophobia

Claustrophobia and anxiety-related symptoms are frequent causes of failure in completing MRI procedures; in fact, approximately 2,000,000 MR procedures worldwide cannot be completed due to this situation. This kind of patients experience a feeling of confinement or being closed in hence conscious sedation may be necessary to complete the examinations [46].

To reduce claustrophobia episodes, a detailed explanation of the MRI procedure should be given to the patients and special equipment such as mirrors to look outside the machine or emergency bells should be adopted. Recently, open scanner configurations have helped to reduce claustrophobic reactions while maintaining image quality and diagnostic accuracy.

10.6.4 Pregnancy

There are only few studies directed toward determining the relative safety of using MR procedures in pregnant patients and no effects on the fetus have been demonstrated [47]. MR environment related risks are difficult to assess for pregnant patients due to the great variability of the factors that are present in this setting (e.g., differences in field strengths, pulse sequences, exposure times, etc.).

However, pregnant patients in the first trimester are generally considered at risk from MR procedures so current guidelines recommended that pregnant women undergo MR only when essential [48]. Nevertheless, MR procedures are still preferable to any imaging involving ionizing radiation.

10.7 Risks for MR Occupational Workers

MR personnel (radiographers, anesthetists, technicians, etc.) prepare and assist patients before and after each clinical exam, so they are repetitively and lengthily exposed to great static MFs during their daily work shift. Moreover, MR operators move inside the MR scanner room and these movements in significant spatial heterogeneous static MFs cause exposure to low-frequency (<1 Hz) time-varying MFs [49], inducing electrical current in the worker's body. This may result in some short-term sensory effects such as dizziness, headaches, metallic taste, nausea, vertigo, magnet phosphenes, and some changes in cognitive functions [50]. These physiological effects have been occasionally observed in some people exposed to MF above 2 T; these effects are transient but may result in difficulties in the work.

No evidence has been found for any irreversible or serious adverse health effects up to 8 T (i.e., clinically significant cardiovascular or neurological effects).

A recent review summarizes studies on health effects of occupational exposure to static MFs [4] and concludes that there is insufficient scientific data for establishing health risks of work exposure to static MF. This conclusion is also drawn in a recent WHO monograph [15]. However, to protect workers, several guidelines are present in literature reporting exposure limits (see Section 10.9).

10.8 Biomedical Implants

Patients with biomedical implants may be at risk in an MR environment due to MF interactions with some kind of implants such as metallic implants [51,52]. Most of the problems with certain biomedical implants are primarily due to movement or dislodgment of ferromagnetic objects caused by MF-related translational attraction and torque, which is proportional to the strength of the static MF and depends on the mass of the object, its shape, and its susceptibility [51].

Others possible hazards related to the presence of biomedical implants regard induction of currents in conductive materials, local heating, malfunction of active implants, and imaging artifacts that could be misinterpreted.

Excessive heating and the induction of electrical currents are typically associated with implants that have elongated configuration [53] (such as gating leads, guide wires, infusion pumps, and physiologic monitors) and/ or active devices (such as cardiac pacemakers and neurostimulation systems) [54,55].

Artifacts and image distortions, due to the presence of metallic implants, are generally caused by a disruption of the local MF and they are dependent on the magnetic susceptibility of the object in the body and the specific pulse sequence parameters used for imaging. Such a kind of artifact is seen typically as a local distortion of the image and/or a signal loss or high-signal intensity area along the edges of the signal void [56,57].

Generally, each biomedical implant should be evaluated regarding safety in the MR environment using standard technique. First of all, the relative magnetic susceptibility for an object may be determined in order to make a decision for possible risks associated with the exposure to the MR system.

The US Food and Drug Administration (FDA) recognized the need for standards that address MR-related safety issues for implants and other medical devices; also to respond to the growing use of MR technology, the FDA requested that a new standard activity be organized within the American Society for Testing and Materials (ASTM) International.

Hence, ASTM has drawn a series of standard test procedures for the evaluation of medical device safety in the MR environment. In particular, ASTM gives a standard test method for the measurement of magnetically induced displacement force on passive implants, measurement of RF-induced heating near passive implants during MR imaging, and evaluation of MR image artifacts from passive implants [58–61].

Moreover, there are, in the literature, numerous studies to assess interactions associated with MR MFs for implants and devices [52,62–64].

The MR task group of ASTM developed a set of terms regarding the safety of medical devices in the MR environment associated with corresponding icons (see Table 10.2) [61].

In addition, ASTM defined an *MR-compatible* implant as a device that is "MR safe and has been demonstrated to neither significantly affect the quality of the diagnostic information, nor have its operations affected by the MR device" [58, pp. 1–2]. The definition of MR safe and MR compatible should be accompanied by the description of MR conditions in which the device was tested since these definitions can be respected only under one set of conditions and not under more extreme MR conditions.

Patients with implanted pacemaker or cardioverter defibrillator (ICD) are ~5 million worldwide and many of these patients had an indication for an MRI scan. Current guidelines for MR safety strongly discourage MR examination of patient and ICD except in cases of urgent need, since the majority of current cardiac implantable is considered MR conditional or unsafe. Various mechanisms could cause problems to patients undergoing MR procedures including movements of the pulse generator or leads, modification of the device function, excessive heating of the leads, and induced currents in the leads [65].

However, some recent studies have demonstrated that new generation pacemaker and ICD-reduced concerns regarding the short- and long-term effects of MRI. These devices may increase the number of centers that are able to safely perform MRI and thus expand access to scans for patients with these devices [66–68]. Most of these studies have been conducted with 1.5 and 3.0 T MR scanners so there is a need to evaluate the safety of scanning these devices at higher field strengths.

Reference Manual for Magnetic Resonance Safety, Implants and Devices [52] is an annually revised textbook that includes guidelines and recommendations for MR safety and patient management and contains a list of thousands of implants and devices tested in the MR environment, including information for objects tested at 3 T or higher.

TABLE 10.2

Terms and Symbols for Safety of Medical Devices in the MR Environment

MR Safe	Used for items that are nonconducting, nonmetallic, and nonmagnetic, such as a plastic Petri dish, and pose no known hazards in all MR environments
MR Conditional	Used for an item that has been demonstrated to pose no known hazards in a specified MR environment with specified conditions of use. Conditions that define the MR environment include static magnetic field strength, radiofrequency fields, specific absorption rate, and other factors. For MR conditional items, the item labeling includes results of testing sufficient to characterize the behavior of the item in the MR environment
MR Unsafe	Defines an item that is known to pose hazards in all MRI environments, such as a pair of ferromagnetic scissors

10.9 MR Safety Guidelines and Regulations

There are many organizations involved in MRI safety legislation, guidelines, and best practice, which are all continually evolving. Most of these organizations are part of the WHO's framework (Figure 10.3).

At the international level, MR safety is regulated by the International Commission of Non-Ionizing Radiation Protection (ICNIRP) that produces and implements MRI safety standards and guidelines [69–73]. In the European Union, the IEC published the *International Standard* 60601-2-33 (Edition 2.1) in 2006 [74], in which advices on exposure limits for patients and volunteers are reported. Proposals and guidelines are incorporated in EU Directives and standards produced by the European Committee for Electrotechnical Standardization (CENELEC) that supports the preparation and adaptation of the national legislation.

In the United Kingdom, the UK Health Protection Agency's (HPA) published advice on "Protection of

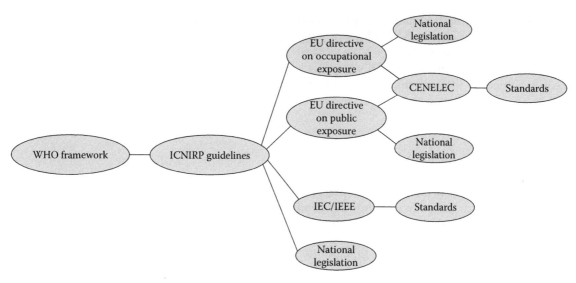

FIGURE 10.3
MRI safety legislation organizations.

Patients and Volunteers Undergoing MRI Procedures" in 2008 [75], based on ICNIRP's recommendations. Also, the Medicines and Healthcare products Regulatory Agency (MHRA) produced guidelines based on the UK and international MR safety advice [76].

In the United States, the FDA gives publication entitled *Criteria for Significant Risk Investigations of Magnetic Resonance Diagnostic Devices* that contains advice on exposure limits [77]. Moreover, the American College of Radiology produces guidance documents [78].

In 2004, the European Union adopted the Directive 2004/40/EU on the minimum health and safety requirements for exposure to electromagnetic fields in workplaces [79]. This Directive contains a set of limits and action values affecting different frequencies of electromagnetic fields that may set significant restrictions on the use of clinical MRI. Initially, all EU member states would have to bring into force laws and regulations to comply with the directive by April 30, 2008. In 2005, the European medical community alerted the Commission to concerns regarding the potential negative impact on the use and development of certain MRI activities.

For this reason, in 2008, the transposition deadline of the original Directive was delayed until April 2012. Afterward, in April 2012 the implementation of Directive 2004/40/EC was further delayed from April 30, 2012 to October 31, 2013. Finally, on June 29, 2013, the European Commission published the new Directive 2013/35/EU on the minimum health and safety requirements regarding the exposure of workers to the risks arising from electromagnetic fields [80] that should be receipt by member states by July 1, 2016.

The new Directive addresses all known direct biophysical effects and indirect effects caused by electromagnetic fields, while it does not address suggested long-term effects of exposure to electromagnetic fields, since there is currently no well-established scientific evidence of a causal relationship. Article 10 of the Directive states that the exposure limits may be exceeded during "the installation, testing, use, development, maintenance of or research related to magnetic resonance imaging (MRI) equipment for patients in the health sector," provided that certain conditions are met [80]. Unfortunately, these conditions are very difficult to interpret and so ambiguous as to allow many possibilities for interpretation. Before the deadline for implementation of the new directive, the European Commission will issue a practical guide, which should include MRI as well as all other applications of electromagnetic fields (EMFs).

10.10 Methods to Quantify Interaction between MR Field and Biological Tissue: Dosimetry

As stated previously, all three fields used in MRI interact with the human body in different ways and these interactions can have implications for patient safety and comfort. Over recent years, engineering is concerned to study the behavior of the MF fields within the human body to assure patient's and workers' safety. For this reason, several methods to quantify and monitor the patient and operator exposure to MF in MRI have recently been designed and the term *dosimetry* has begun to be used also for nonionizing techniques as MRI. However, due to the simultaneous presence of all three MFs, is very complex to assess the exposure from MRI equipment. For this reason, generally the contribution of static MF, RF MF, and gradient is evaluated separately.

10.10.1 Static MF

The static MF B_0 can be measured using a Hall-effect gaussmeter with three-axis probes. To monitor directly the exposure of MRI staff to static MF, a personal dosimeter should be used: there are only few personal MF dosimeter in the market that use a combination of Hall-effect sensors and induction coils and integrators to measure both static and time-varying MFs [81–85]. The use of such dosimeters is recommended in order to evaluate the regulatory guidelines and determine any underlying exposure risks.

No standardized assessment procedures are yet available in the literature dealing with static and time-varying MFs due to movement in static MF.

Some basic restrictions are specified in terms of induced electric field or current density [72,79]: experimental measurement of exposure levels in terms of induced current density is very complex and thus cannot be used in daily routine assessment.

Numerical calculations can be used to evaluate male and female healthcare workers exposure to MFs: by using numerical simulations of tissue-equivalent body models, it is possible to accurately estimate the induced current density [86]. Unfortunately, these methods require long computational times and complex body models generally obtained from the segmentation of MRI or computed tomography dataset. However, the more recent methods are very fast and accurate [86].

The induced electric field can be calculated by using analytical method based on classical electromagnetic theory of Maxwell's equations: the third equation expresses the concept that an electric conductor, such as the human body, which moves in spatial heterogeneous static MFs B, induces an electric field E. This equation in its integral form is known as *Faraday's law*:

$$\oint_\Gamma \bar{E} \cdot d\bar{l} = -\frac{d\Phi_B}{dt} \tag{10.3}$$

where:
Φ_B is the magnetic flux and the integral of electric field $\bar{E}$ is calculated over a path $\bar{l}$ on the surface Γ

ICNIRP guidelines provide a model for calculating induced current density [72], according to which

Equation 10.2 can be solved for a circular loop in the human body with a radius r:

$$\oint_\Gamma \bar{E} \cdot d\bar{l} \rightarrow \oint_\Gamma E \cdot dl \cdot \cos(0)$$

$$= E\oint_\Gamma dl = E2\pi r = -\frac{\partial(\pi r^2 B)}{\partial t} \rightarrow E \tag{10.4}$$

$$= -\frac{r}{2}\frac{dB}{dt} = k\frac{dB}{dt}$$

where k is a geometry factor for a given subject. Generally, a radius of 0.64 m is assumed for a typical current loop in the body [72,87], while for the calculation of induced current density in the head, a radius of 0.07 m can be chosen [87].

Finally, Equation 10.3 can be simplified as follows:

$$E = k\frac{dB}{dt} = k\left(\frac{\partial B}{\partial x} \cdot v_x + \frac{\partial B}{\partial y} \cdot v_y + \frac{\partial B}{\partial z} \cdot v_z\right) \text{(V/m)} \tag{10.5}$$

where v_x, v_y, and v_z are the worker walking speed components. From the induced electric field in the operator's body, the electrical current density can be calculated as follows:

$$J = \sigma \cdot E \text{ (A/m}^2) \tag{10.6}$$

where σ is the mean electrical conductivity of human tissues (@ 0.2 S/m [72]).

Using this analytical model, it is possible to create software tools for estimating the induced current density due to worker movement in the static MF of an MR scanner; by means of a simple graphical user interface, the tool can easily simulate typical operator movements during clinical MR exams and estimate the maximum current density relative to a specific path for comparing the limits indicated by the regulations [88] (Figure 10.4).

10.10.2 Gradient MF

The switched MF can be induced in the human body electric currents strong enough to cause PNS, so it is necessary to calculate gradient-induced fields, for example, to achieve a higher rate of switching without inducing PNS [89–92].

According to early measurements [93], significant gradient fields also exist beyond the bore of the scanner; hence, they interact with both patient and MRI staff.

The electric fields induced in the body by gradient fields have a high frequency as opposed to those caused by the movement of the operator that have a low frequency.

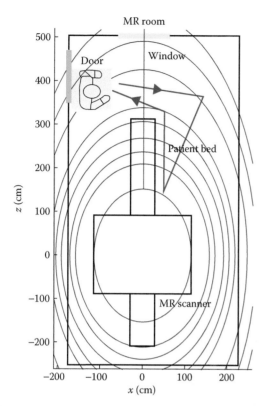

FIGURE 10.4
Simulation of typical operator movements during a clinical MR exam. (From Hartwig, V. et al., *MAGMA*, 24, 323–330, 2011. With permission.)

The imaging gradient fields inside and outside the magnet bore can be measured using device based on three axis induction coil that detect dB/dt: the measured value can be compared to reference levels of the regulations.

Otherwise, it is possible to calculate the induced electric current using the analytical model described in Equations 10.4–10.6, considering the patient as a set of uniform cylinders and using empirical model for the nerve and cardiac stimulation threshold [94].

Recently, a number of authors have used numerical methods to calculate gradient-induced fields in human body [90,91,94–96]. Brand and Heid [95] performed a numerical analysis of the electric fields induced by a complete whole-body gradient system using three human body models of different complexities. The results of their analysis correlated to the experimental observations with a body model that resembles the human body but applying a model with inhomogeneous conductivity, numerical stability was not reached.

Liu and Crozier [96] used a finite difference time domain (FDTD) method to calculate gradient-induced electric fields in a heterogeneous human body model and found that the largest electric fields always occur at the periphery of the body.

Generally, the numerical methods have shown success in comparison with analytical field calculations but require a further investigation and better interpretation.

10.10.3 RF Magnetic Field

The vast majority of works about MRI safety are related to RF fields: biological effects due to RF energy absorption have been the most analyzed in literature regarding MR safety issues.

The presence of human body can cause perturbations in B_1 field of 10% and hence can degrade the image quality. Moreover, the RF energy absorption causes a temperature increase of tissues: this RF heating must be monitored to avoid excess heating of the subject. These kinds of effects depend on several factors such as electrical and geometric tissue properties, applied pulse sequence, its repetition time, and frequency. Furthermore, heat-sensitive organs, such as the eyes and testes, are of particular concern, although there is no evidence of any deleterious effect of MRI on either.

The quantification of the magnitude and distribution of RF energy absorbed in the biological objects is known as RF dosimetry, which is based on the calculation of the SAR. There are different kinds of methods to evaluate this parameter known as theoretical and experimental dosimetry. Theoretical dosimetry regards the SAR estimation on the body using calculations: it includes analytical and numerical methods.

Experimental dosimetry is based on the direct measurements of parameters relative to the SAR, such as the temperature increase or the induced electrical field on the subject. These kinds of measurement are primarily performed on animal tissues or human models, for example, in the presence of biomedical implants.

SAR is defined as the total absorbed power P in watts (W) per kilogram of tissue. If is the electric conductivity of a uniform conducting medium, the power deposition can be calculated as

$$P = J \cdot E = \sigma \cdot E^2 \ (\text{W}) \qquad (10.7)$$

where J is the induced current density and E is the induced electric field.

SAR is then calculated as

$$\text{SAR} = \frac{\sigma \cdot E^2}{2 \cdot \rho} \ (\text{W/kg}) \qquad (10.8)$$

where ρ is the tissue density or as:

$$\text{SAR} \propto \frac{\sigma \cdot B_1^2}{2 \cdot \rho} \ (\text{W/kg}) \qquad (10.9)$$

if a rectangular RF pulse with a sinusoidal B_1 is assumed.

All commercial MRI scanners have a SAR calculation tool that is based on the knowledge of the energy deposited by a *standard RF* pulse derived empirically. This parameter is a function of the patient's weight, inserted by the operator before starting the MRI exam, the strength of the B_0 field and the coil that is used. The standard RF pulse is defined as a rectangular pulse with a flip angle of 180° and a duration of 1 ms.

Then, the average SAR is calculated as follows:

$$\text{SAR}_{\text{avg}} = \frac{N_{\text{stdRF}} \cdot J_{\text{stdRF}}}{T_R \cdot M_{\text{pat}}} \qquad (10.10)$$

where:

N_{stdRF} is the number of standard RF pulses in sequence
J_{stdRF} is the energy deposited by standard RF pulse
M_{pat} is the patient weight

If the estimated SAR value exceeds the regulation safety limit, the scanner does not permit to perform the exam until the operator changes some sequence parameters, such as T_R, to reduce the power deposition on the subject.

10.10.3.1 Analytical Dosimetry

The well-known analytical model for SAR calculation dates back to 1985 when Bottomley [34] estimated RF power absorption using a simple geometrical representation of the exposed subject. Thereafter, other authors evaluated the whole-body or partial-body average SAR for a specific MRI sequence using Bottomley's model [97]. In this model, the exposed human body is modelled by a set of four homogeneous cylinders (representing the head, trunk, and legs) (Figure 10.5) in each of which the time-averaged power deposition for mass unit can be estimated as

$$\text{SAR}_{\text{avg}} = \frac{k\sigma R^2}{\rho} \qquad (10.11)$$

where:

R is the cylinder radius
k is a constant that include the MRI sequence parameters (f: RF pulse frequency, ϑ: RF pulse flip angle, γ: gyromagnetic ratio, τ: RF pulse length, T_R: MR sequence repetition time):

$$k = \frac{\pi f^2 \vartheta^2}{\gamma^2 \tau T_R} \qquad (10.12)$$

According to Bottomley's model, the maximum SAR, that is, the peak SAR, is equal to

$$\text{SAR}_{\text{peak}} = 2 \cdot \text{SAR}_{\text{avg}} \qquad (10.13)$$

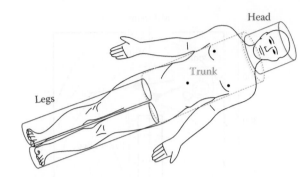

FIGURE 10.5
Model of exposed human body for average SAR estimation.

This analytical model can be used to estimate the average whole-body SAR for a specific MRI sequence, but it does not consider the dependence from the complex geometry of the body. By this method, it is not possible to estimate the local SAR distribution on the entire body volume to evaluate the presence of unexpected hot spots.

10.10.3.2 Numerical Dosimetry

RF fields involved in MRI imaging have a frequency for which the size of the human body becomes comparable to or even larger than the effective wavelength. For frequencies at which the wavelength is considerably larger than the dimensions of the body, the so-called quasistatic approximations are valid so the analytical method, based on a simplified solution of the Maxwell's equations, can be used. But when the wavelength is comparable to body size, the full-time dependence of Maxwell's equations must be considered, hence numerical methods must be used. For problems involving the human body, the crossover frequency between these approaches is approximately 10 MHz [98].

In the last decade, commercial high-frequency MRI systems with a higher field were developed so that this crossover frequency is easily reached and exceeded. Moreover, analytical calculation is very useful in simple shapes but they have limitations in application to the human body [99,100]. For these reasons, recent works have concentrated on numerical methods to quantify the interactions between RF fields and dielectrically inhomogeneous human body in terms of spatial distributions of internal electric fields, currents, and SAR [99, 101–105].

Numerical methods concern the use of mathematical models of the body exposed to RF field and of the RF transmission coil and provide a numerical solution of Maxwell's equations with specific boundary conditions. Powerful personal computers or other dedicated computing platforms have to be used to implement these methods. A classification of numerical methods can be made according to whether the computation is made in the time domain or frequency domain, and if the method is based on differential or integral equations. The more largely used numerical methods for local SAR estimation are as follows:

- Finite element method (FEM)
- Methods of moments (MoM)
- FDTD

FEM [106] is a differential equation approach based on the discretization in mesh elements, usually tetrahedral elements, of the entire computational domain including the surrounding environment. In FEM, the field equations are determined in terms of polynomial with unknown coefficients defined on the mesh nodes or along element edges, which are then determined from solving a matrix equation. The equations are resolved in frequency domain.

FEM has been used in the literature for the evaluation of SAR and field inhomogeneity of an anatomically accurate human head model [107–111].

However, the matrix equation dimension can be very large when it is used on an accurate human model, so FEM is not the most used method for MRI applications.

The most widely used computer code based of FEM is HFSS from ANSYS [112].

MoM is a frequency domain method based on the resolution of integral equations reducing them to a system of linear equations [113]. This method is highly efficient when treating problems with perfectly conducting sources (i.e., RF coils) and homogeneous media. The most advantage feature of MoM is that only the structure in question, and not the surrounding environment, is discretized. Even so, computational memory requirements scale in proportion to the size of the problem and the required frequency, so for realistic dosimetric problems, the computation can be significantly expensive. Accordingly, the analysis of an unloaded coil can be done with high accuracy and efficiency [114].

As FEM, MoM solve Maxwell's equations in the frequency domain so to compute the solution over a frequency band, the computations have to be repeated at each frequency.

Recently, some works have been proposed to use hybrid techniques for the analysis of RF coils loaded with a dielectric object known as MoM/FDTD or MoM/FEM methods. In these approaches, MoM is used together with FEM or FDTD in order to exploit its strength for modeling an RF coil and FDTD or FEM for calculating the field inside a complicated dielectric object [115–117].

A commercial software based on MoM is FEKO by EM Software & Systems [118], which also implements a hybrid MoM/FEM algorithm.

Currently, FDTD and related methods are the most used in numerical field calculations that consider realistic human body models, though some recent works have proposed the use of hybrid approaches using one numerical method in region containing the human body and another in regions containing the RF coils. FDTD was introduced for the first time by Yee in 1996 [119]; it solves Maxwell's equations in the time domain using a discretization of the temporal (time discretization) and spatial (space discretization) parameters with the finite difference approximation. Typically, minimum spatial sampling is at intervals of 10–20 per wavelength and temporal sampling is sufficiently small to maintain stability of the algorithm. To solve the equation for the electrical and MFs on a 3D object, this object must be enclosed in a box; the computational domain is then divided into cubic cells (Yee's cells). To discretize the fields, the electric field components are assigned at the center of each edge of the cells and the MF components are assigned at the center of each face of the cells. Starting from initial values for the electric and magnetic fields and given a specific excitation, the process iteratively calculates the field values of each cell using a central finite-difference approximation to resolve spatial and temporal derivates in Maxwell's equations.

The condition of the cell size can be expressed as

$$\max(\Delta x, \Delta y, \Delta z) < \frac{\lambda}{20\sqrt{\varepsilon_{r\max}}} \qquad (10.14)$$

where:

$\Delta x, \Delta y$, and Δz are the dimensions of the cell, respectively
γ is the free-space wavelength
$\varepsilon_{r\max}$ is the maximum relative permittivity in the computational domain

The time axis is divided uniformly into many small intervals Δt, which must satisfy the following condition (Courant–Friedrichs–Lewy stability criterion) [119]:

$$\Delta t \leq \frac{\sqrt{\mu\varepsilon_{r\max}}}{\sqrt{\frac{1}{\Delta x^2} + \frac{1}{\Delta y^2} + \frac{1}{\Delta z^2}}} \qquad (10.15)$$

where μ is the medium permeability. This condition ensures that the wave cannot propagate faster than the speed of the light in the medium.

The electric field is then calculated at time instants $t = n\Delta t$ and MF is calculated at time instants $t = (n + 1/2)$ Δt with a temporal shift equals to $\Delta t/2$ starting from $n = 1$ and proceeding until the field reaches steady state (leapfrog time stepping).

The box that contains the volume of interest will reflect the field back into the object space, not permitting to simulate the propagation of the electromagnetic wave in the free space. To overcome this problem, the most popular approach is to place a layer of absorbing material at the exterior surface to absorb the power incident from the inside of the computational domain. The best artificial absorbing material is the so-called perfectly matched layer (PML) [120], which is a nonphysical material with an excellent absorption capability.

A well-known problem with FDTD simulations on uniform Cartesian grid is the handling of the curved surfaces or highly irregular geometrically detailed boundaries (staircase approximation). For this reason, a nonuniform grid or a grid with high-spatial resolution (and large memory requirement) must be used. Recent progress in this area has been reported in [121].

Since FDTD method is simple to implement, highly efficient, and does not require solving any matrix equations; it has been widely used for numerical simulation of electromagnetic problems, including MRI applications [122–128]. Today, thanks to the available realistic human (with a maximum resolution of 1 mm for each dimension) [129,130], it is possible to obtain SAR and electromagnetic field distribution maps with high resolution [101,102,124] (Figure 10.6), which are also useful to evaluate the possible presence of hot spot at metallic implant locations [131], optimize new acquisition sequence for high-field MRI scanner, and design new RF coils [127,128] such as phasedarray coils [132,133].

The three most used commercial software based on FDTD method are XFDTD by Remcom [134], SEMCAD X by SPEAG [135], and Microwave Studio (MWS) by CST Computer Simulation Technology [136]. There are also many other in-house developed codes used for specific applications of modeling and simulation of RF coils and interactions between RF fields in human body.

10.10.3.2.1 Electromagnetic Numerical Models

For a typical electromagnetic numerical simulation problem, numerical models of RF coil and imaging subject (usually human body or human head) are needed. The electromagnetic model of RF coil (and RF shield) can be constructed in a straightforward manner because its dimensions and electromagnetic properties (permeability, permittivity, and conductivity) are known. The electromagnetic models of human body are more complex since such an object consists of many irregularly shaped tissues having very different electromagnetic properties. To construct an anatomically accurate electromagnetic model of human body, generally it is possible to first collect a set of tomographic images using MRI or X-ray computed tomography and then segment each image and interpolate all of them in the third dimension to create a 3D geometric model. This procedure is

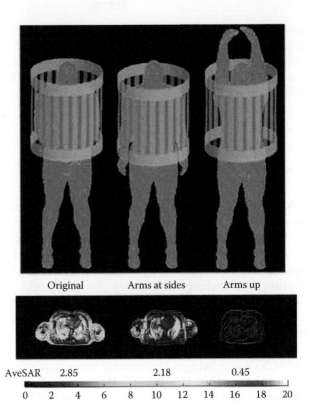

FIGURE 10.6
SAR distribution on a human body model in various postures. (From Collins, C.M. and Zhangwei, W.: Calculation of radiofrequency electromagnetic fields and their effects in MRI of human subjects. *Magn. Reson. Med.* 2011. 65. 1470–1482. Copyright Wiley-VCH Verlag GmbH & Co. KGaA. Reproduced with permission.)

very difficult and time consuming so it is not possible to construct a specific numerical model for each individual study. The most used approach is to use one of the available models and then scale it to appropriate size in order to obtain a good approximation of the specific individual. Another important issue is relative to the electromagnetic properties for each tissue that compose the model: permeability is approximately the same as that of free space, permittivity and conductivity are different for different tissues, and their values change with the frequency. Moreover, the permittivities and conductivities of human tissues are very difficult to measure accurately so there are some discrepancies among the published data [137].

The best known numerical model available in the literature is derived from the National Library of Medicine (NLM) Visible Human Project, which represents a large male with arms placed along his side: the model is commercially available in the form of a 3D mesh of cells with a resolution of 10, 5, 3, 2, or 1 mm, each cell represents a tissue type with appropriate dielectric properties [134].

Generally, commercial software for electromagnetic numerical simulation are available together with

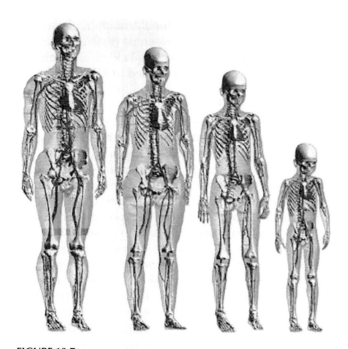

FIGURE 10.7
The Virtual Family. (From Christ, A. et al., The Virtual Family—Development of surface-based anatomical models of two adults and two children for dosimetric simulations, *Phys. Med. Biol.*, 55, N23–N38, 2010, Copyright 2010, Institute of Physics and Engineering in Medicine. Reproduced by permission of IOP Publishing.)

several numerical models of human body. For example, SPEAG provides a set of numerical model including possible human phantoms (the virtual family: whole-body adult and child, male and female, developed by IT'IS Foundation [138]) (Figure 10.7) animal phantoms (rat and mice), and other generic phantoms [135]. Moreover, Remcom developed a special purpose software tool (VariPose) [134] that permits to reposition the voxels of the male visible human mesh including internal anatomical structures. Using this tool, the body can be posed in realistic positions depending on the specific user application.

10.10.3.3 Experimental Dosimetry

10.10.3.3.1 Calorimetric Method

The measurement of temperature increases due to RF power deposition is generally the most experimental method for the SAR estimation, especially in the presence of metallic biomedical implants which could create hot spots [139]. The temperature rise is measured in a subject, or in an appropriate phantom, over a period sufficient to obtain good signal respect to noise using temperature probes at one or more locations; this kind of measurement is known as calorimetric method [140]. SAR is then estimated from the temperature changes

using the tissue bioheat equation first proposed by Pennes [141]:

$$c\frac{dT}{dt} = W_M + W_B + W_C + \text{SAR} \tag{10.16}$$

where:
c is the specific heat capacity of the tissue in J/kg °C
dT/dt is the temperature increment due to the exposure
W_M is the heat generated from the metabolic activity
W_B is the heat loss due to the blood perfusion
W_C is the heat loss due to the thermal conduction in tissue

For in-vivo measurement, it is very difficult to estimate, the parameters relative to metabolic activity, blood perfusion, and thermal conduction heat exchange, whereas for in-vitro SAR evaluations, that is, measurements on phantom simulating biological tissue properties, the terms W_M and W_B are not present. Moreover, using a phantom with a sufficient viscosity, the term W_C is negligible. In this case, the bioheat equation can be simplified as

$$c\frac{dT}{dt} = \text{SAR} \tag{10.17}$$

where the specific heat capacity of the phantom is generally assumed equal to the specific heat of water (4184 J/kg °C). The temperature measured by the probe is usually plotted and the time rate of irradiation-induced temperature rise ($\Delta T/\Delta t$) is graphically determined using a slope-determining algorithm: SAR is then determined from the initial linear slope of $\Delta T/\Delta t$ [142]:

$$\text{SAR} = \frac{\Delta T \cdot c}{\Delta t} \tag{10.18}$$

When SAR is measured using a single-point probe in an object with one or more hot spots not coincident with the probe tip, significant errors can occur. In this case, the temperature measured by the probe may not rise immediately after the RF irradiation but after several seconds, it begins to rise more quickly as heat is conducted from a nearby hot spot. When RF ceases, the temperature continues to rise as heat is conducted away from the hot spot to cooler regions where the probe is located (Figure 10.8). The apparent slope could be falsely interpreted as the local SAR at the location of the probe tip; hence, to avoid errors in the SAR estimation, the initial linear portion of the temperature rise should be selected. This method is the standard method to measure the RF-induced heating near medical implants and its surroundings [142], but it is not used for the estimation of in-vivo local SAR also because it is not possible to obtain a map of the SAR value on the entire subject surface or volume since it is only a point-by-point measure.

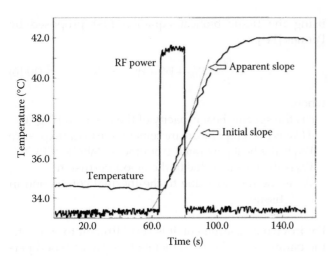

FIGURE 10.8
RF-induced heating near medical implant.

10.10.3.3.2 Method based on B_1 mapping via MRI

Recent studies have shown that starting from the knowledge of the B_1 field distribution on the subject under RF exposure, it is possible to estimate tissue electrical parameters and SAR in vivo [143–146]. B_1 mapping is a well-known MR application regarding the measurement of the active magnetic component of the applied RF field [147–149]. There are some methods in the literature to estimate the transmission RF field B_1^+ (or the MF strength H_1^+), which is the MF produced by the RF transmission coil, and the receive sensitivity of a receive coil to pick up MR signal, which is related to the MF B_1^- (or H_1^-) received from the sample [150]. For regular RF pulses at Larmor frequency on resonance, the flip angle of the magnetization from B_0 direction is proportional to the magnitude of the locally transmit B_1 field. For these reasons, some B_1 mapping methods are based on the local flip angle distribution [145]; among these, the double-angle method (DAM) [148] is considered the simplest and requires the acquisition of two images with different nominal flip angle α and 2α. Another well-known method for B_1^+ mapping involves a voxel-by-voxel fitting of the signal from a series of images with different nominal flip angles [151]; this method is based on the approximation of the Bloch equation for the signal intensity and is the most accurate but requires long acquisition and computational time and high-power deposition on the subject.

Other known B_1^+/actual flip angle mapping methods are actual flip angle imaging (AFI) [152] based on pulse steady-state sequence, echo-planar imaging-based method [153], the Cunningham et al. [154] modification of the DAM, and Morrel's [155] phase-sensitive technique.

Recently, a novel method for B_1^+ mapping based on the Bloch–Siegert shift was presented [156]: this method permits to acquire B_1 maps very quickly but requires a specific sequence to be implemented in MRI scanner.

10.10.3.3.3 MR Thermometry

To directly visualize the spatial distribution of SAR, it is possible to use methods for MR thermal imaging that involve the noninvasive temperature monitoring with MRI and are based on temperature-sensitive MR parameters such as the proton resonance frequency (PRF), the diffusion coefficient (D), T_1- and T_2-relaxation times, magnetization transfer, the proton density, as well as temperature-sensitive contrast agents [157].

MRI has been shown to be an excellent modality to noninvasively monitor thermal therapy in vivo to ensure efficacy and safety of the treatment and is regularly used in a number of clinical applications [158–161]. However, MR thermometry is rarely used to predict SAR distribution [162,163]: generally, phase-difference imaging, based on PRF shift, is used to measure 3D temperature distribution in phantoms before and after RF heating to map the power absorbed. Once the temperature change is calculated on the entire sample, the local SAR distribution can be obtained by the model of Equation 10.18.

10.10.3.3.4 Other Experimental Methods

Recently, a novel prototype RF dosimeter for measuring average head or body SAR has been presented [164] and patented in 2011. This device is entirely independent of the MRI scanner; it is based on a transducer for the time-averaged RF voltage from which the equivalent SAR can be determined after calibration. The transducer is constituted by some copper loops oriented orthogonal to each other. The SAR transducer must be placed in the scanner and connected to a true rms RF voltage-measuring device outside the magnet, then the scanner is started using the MRI pulse sequence for which the SAR measurement is desired. The average and peak SAR is finally calculated using the measured rms power and the subject mass.

10.11 Conclusion

MR is considered a safe technology since it uses no ionizing electromagnetic radiations that have no enough energy to detach electrons from atoms or molecules, such as other higher energetic radiations can do (X-rays, radiations used in nuclear medicine, etc.). However, there are several effects to consider for safety assurance and engineering aspects relative to MR signal and image generation. This chapter summarizes the biological effects of the three types of fields that are employed in an MRI exam. Given the increasingly high number of clinical MRI exams and the rapid development of MR technology, which has made available a large number of MR scanner with high static MF (>3 T), the consideration of possible risks and health effects associated with the MRI procedures is gaining increasing importance.

Today, the effect recognized as the greatest potential hazard is the so-called projectile effect due to the static MF; this effect can only be minimized by a strict and careful screening of all individuals entering the MR environment. Transient phenomena such as slight nausea and vertigo, headache, tingling/numbness, and visual disturbances reported in association with patients moving in high fields are transient, ceasing after leaving the magnet, and are usually reduced or avoided by making sure the patient moves slowly while in the main field.

For scanner with high-field strengths, high-performance gradients are mandatory in order to take advantage of the potential for faster acquisitions or higher resolution scans available via the increased SNR from the main field. But, these gradients become available, possibility of cardiac stimulations increases. New strategies for overcoming this problem have been developed such as the design of high-performance gradients linear over a smaller volume for neurological or cardiac imaging applications.

Regarding RF fields, safety SAR limits create some of the most difficult challenges for high-field imaging by forcing tradeoffs between image acquisition rates resolution and slice coverage. A technical challenge for high-field imaging will be pulse sequence design, RF pulse design, new acquisition techniques, and hardware designs that help overcome the limitations imposed by these tradeoffs.

The unresolved problems relative to potential for hazardous situations to occur in the MR environment are still a lot, although several world research groups are working for solving them: the significance of these problems is also highlighted to the presence of *MR safety and compatibility group* inside the major world MR research groups.

Until a wider knowledge of the potential risk related to diagnostic MRI is available, a prudent attention should be adopted; to prevent incidents and accidents, it is necessary to be aware of the latest information pertaining to MR biologic effects, use current evidence-based guidelines to ensure safety for patients and staff members, and follow proper recommendations pertaining to biomedical implants and devices.

References

1. Drinker CK, Thomson RM. (1921) Does the magnetic field constitute an industrial hazard? *J Ind Hyg* 3:117–129.
2. Davis LD, Pappajohn K, Plavnieks IM. (1962) Bibliography of the biological effects of magnetic fields. *Fed Proc* 21:1–38.
3. Schenck JF. (2000) Safety of strong, static magnetic fields. *J Magn Reson Imaging* 12:2–19.
4. Franco G, Perduri R, Murolo A. (2008) Effetti biologici da esposizione occupazionale a campi magnetostatici utilizzati in imaging a risonanza magnetica nucleare: una rassegna. *Med Lav* 99:16–28.
5. McRobbie DW, Moore EA, Graves MJ et al. (2006) *MRI— From Picture to Proton.* Cambridge University Press: Cambridge, New York.
6. McNeil DG. (August 19, 2005) M.R.I's Strong Magnets Cited in Accidents. *New York Times.*
7. Chakeres DW, Kangarlu A, Boudoulas H et al. (2003) Effect of static magnetic field exposure of up to 8 Tesla on sequential human vital sign measurements. *J Magn Reson Imaging* 18:346–352.
8. Schenck JF, Dumoulin CL, Redington RW et al. (1992) Human exposure to 4.0 T magnetic fields in a whole-body scanner. *Med Phys* 19:1089–1098.
9. Schenck JF. (1992) Health and physiological effects of human exposure to whole-body 4 Tesla magnetic fields during MRI. *Ann NY Acad Sci* 649:285–301.
10. Kleinman MH, Shevchenko T, Bohne C. (1998) Magnetic field effects on the dynamics of radical pairs: The partition effects in vescicles. *Photochem Photobiol* 68:710–718.
11. Scaiano JC, Cozens FL, Mohtat N. (1995) Influence of combined AC-DC magnetic fields on free radical in organized and biological systems. Development of a model and application of the radical pair mechanism to radicals in micelles. *Photochem Photobiol* 62:818–829.
12. Everson RW, Timmel CR, Brocklehurst B et al. (2000) The effects of weak magnetic fields on radical recombination reactions in micelles. *Int J Radiat Biol* 76:1509–1522.
13. Ritz T, Thalau P, Phillips JB et al. (2004) Resonance effects indicate a radical-pair mechanism for avian magnetic compass. *Nature* 429:177–180.
14. Kangarlu A, Burgess RE, Zhu H et al. (1999) Cognitive, cardiac, and physiological safety studies in ultra high field magnetic resonance imaging. *Magn Reson Imaging* 17:1407–1416.
15. World Health Organization. (2006) Environmental Health Criteria 232. Static fields. World Health Organization: Geneva, Switzerland.
16. International Commission on Non-Ionizing Radiation Protection (ICNIRP). (2009) Guidelines on limits of exposure to static magnetic field. *Health Phys* 96:504–514.
17. Formica D, Silvestri S. (2004) Biological effects of exposure to magnetic resonance imaging: An overview. *BioMed Eng OnLine* 3:11.

18. Vogt FM, Ladd ME, Hunold P et al. (2004) Increased time rate of change of gradient fields: Effect on peripheral nerve stimulation at clinical MR imaging. *Radiology* 233:548–554.

19. Reilly JP. (1989) Peripheral nerve stimulation by induced electric currents: exposure to time-varying magnetic fields. *Med Biol Eng Comput* 27:101–110.

20. Simmons A, Hakansson K. (2011) Magnetic resonance safety. *Methods Mol Biol* 711:17–28.

21. Abart J et al. (1997) Peripheral nerve stimulation by time-varying magnetic fields. *J Comput Assist Tomogr* 21(4):532–538.

22. Shellock FG, Ziarati M, Atkinson D et al. (1998) Determination of gradient magnetic field-induced acoustic noise associated with the use of echo planar and three-dimensional, fast spin echo techniques. *J Magn Reson Imaging* 8(5):1154–1157.

23. McJury M, Shellock FG. (2000) Auditory noise associated with MR procedures: A review. *J Magn Res Imag* 12:37–45.

24. Shellock FG, Kanal E. (1996) *Magnetic Resonance: Bioeffects, Safety, and Patient Management*, 2nd ed. Lippincott-Raven: New York.

25. Persson BRR, Stahlberg F. (1989) *Health and Safety of Clinical NMR Examinations*: CRC Press: Boca Raton, FL.

26. Michaelson SM, Lin JC. (1987) *Biological Effects and Health Implications of Radiofrequency Radiation*. Plenum: New York.

27. Beers J. (1989) Biological effects of weak electromagnetic fields from 0 Hz to 200 MHz: A survey of the literature with special emphasis on possible magnetic resonance effects. *Magn Reson Imaging* 7:309–331.

28. Simi S, Ballardin M, Casella M et al. (2008) Is the genotoxic effect of magnetic resonance negligible? Low persistence of micronucleus frequency in lymphocytes of individuals after cardiac scan. *Mut Res* 645:39–43.

29. Polk C. (1995) Biological effects of nonionizing electromagnetic fields. In: Bronzino JD (ed) *Handbook of Biomedical Engineering*. CRC Press: Boca Raton, FL.

30. Shellock FG. (2000) Radiofrequency energy-induced heating during MR procedures: A review. *J Magn Reson Imaging* 12:30–36.

31. Shellock FG. (1990) MRI bioeffects and safety. In: Atlas S (ed) *Magnetic Resonance Imaging of the Brain and Spine*. Raven Press: New York.

32. Gordon CJ. (1987) Normalizing the thermal effects of radiofrequency radiation: Body mass versus total body surface area. *Bioelectromagnetics* 8:111–118.

33. Shellock FG, Crues JV. (1988) Corneal temperature changes associated with high-field MR imaging using a head coil. *Radiology* 167:809–811.

34. Bottomley PA, Redington RW, Edelstein WA, Schenck JF. (1985) Estimating radiofrequency power deposition in body NMR imaging. *Magn Reson Med* 2:336–349.

35. Bottomley PA, Edelstein WA. (1981) Power deposition in whole body NMR imaging. *Med Phys* 8:510–512.

36. International Electrotechnical Commission. (2002) *IEC 60601-2-33 Particular Requirements for Basic Safety and Essential Performance of Magnetic Resonance Equipment for Medical Diagnosis*, 2nd ed., International Electrotechnical Commission: Geneva, Switzerland.

37. Shellock FG, Schaefer DJ, Crues JV. (1989) Alterations in body and skin temperatures caused by MR imaging: Is the recommended exposure for radiofrequency radiation too conservatice? *Brit J Radiol* 62:904–909.

38. Manufacturer and User Facility Device Experience Database—(MAUDE) accessed September 2013.

39. Dempsey MF, Condon B, Hadley DM. (2001) Investigation of the factors responsible for burns during MRI. *J Magn Reson Imaging* 13:627–631.

40. Kanal E, Shellock FG. (1990) Burns associated with clinical MR examinations. *Radiology* 175:585.

41. Kreidstein ML, Giguere D, Freiberg A. (1997) MRI interaction with tattoo pigments: Case report, pathophysiology, and management. *Plast Reconstr Surg* 99:1717–1720.

42. Shellock FG, Spinazzi A. (2008) MRI safety update 2008: Part 1, MRI contrast agents and nephrogenic systemic fibrosis. *Am J Roentgenol* 191:1129–1139.

43. Ali MA. (2007) Magnetic resonance imaging and associated alteration in some biophysical properties of blood. *Rom J Biophys* 17:277–286.

44. Joong WL, Myeong SK, Kim JY et al. (2011) Genotoxic effects of 3 T magnetic resonance imaging in cultured human lymphocytes. *Bioelectromagnetics* 32:535–542.

45. Stikova E. (2012) Magnetic resonance imaging safety: Principles and guidelines. *Prilozi* 33(1):441–72.

46. Eshed I, Althoff CE, Hamm B et al. (2007) Claustrophobia and premature dermination of magnetic resonance imaging examinations. *J Magn Reson Imaging* 26(2):401–404.

47. Alorainy IA, Albadr FB, Abujamea AH. (2006) Attitude towards MRI safety during pregnancy. *Ann Saudi Med* 26(4):306–309.

48. Eskandar OS, Eckford SD, Watkinson T. (2010) Safety of diagnostic imaging in pregnancy. Part 2: Magnetic resonance imaging, ultrasound scanning and Doppler assessment. *Obstetrician Gynaecol* 12:171–177.

49. Kannala S, Toivo T, Alanko T et al. (2009) Occupational exposure measurements of static and pulsed gradient magnetic fields in the vicinity of MRI scanners. *Phys Med Biol* 54:2243–2257.

50. Glover PM. (2009) Interaction of MRI field gradients with the human body. *Phys Med Biol* 54:R99–115.

51. Shellock FG. (2002) Biomedical implants and devices: Assessment of magnetic field interactions with a 3.0-Tesla MR system. *J Magn Reson Imaging* 16:721–732.

52. Shellock FG. *Reference Manual for Magnetic Resonance Safety, Implants and Devices*: 2013 edition. Biomedical Research Publishing Group.

53. Kainz W. (2007) MR heating tests of MR critical implants. *J Magn Reson Imaging* 26:450–451.

54. Mattei E, Calcagnini G, Censi F et al. (2010) Numerical model for estimating RF-induced heating on a pacemaker implant during MRI: Experimental validation. *IEEE Trans Biomed Eng* 57(8):2045–2052.

55. Angelone LM, Potthast A, Segonne F et al. (2004) Metallic electrodes and leads in simultaneous EEG-MRI: Specific absorption rate (SAR) simulation studies. *Bioelectromagnetics* 25(4):285–295.

56. Bennett LH, Wang PS. Donahue MJ. (1996) Artifacts in magnetic resonance imaging from metals. *Appl Phys* 79:4712.

57. Shellock FG. (2001) Metallic neurosurgical implants: Evaluation of magnetic field interactions, heating, and artifacts at 1.5 Tesla. *J Magn Reson Imaging* 14:295–299.

58. ASTM F2182—11a Standard Test Method for Measurement of Radio Frequency Induced Heating on or Near Passive Implants during Magnetic Resonance Imaging.

59. ASTM F2119—07. (2013) Standard Test Method for Evaluation of MR Image Artifacts from Passive Implants.

60. ASTM F2052—06e1 Standard Test Method for Measurement of Magnetically Induced Displacement Force on Medical Devices in the Magnetic Resonance Environment.

61. ASTM F2503—13 Standard Practice for Marking Medical Devices and Other Items for Safety in the Magnetic Resonance Environment.

62. Schaefers G, Melzer A. (2006) Testing methods for MR safety and compatibility of medical devices. *Minimally Invasive Therapy* 15:71–75.

63. Vanello N, Hartwig V, Tesconi M et al. (2008) Sensing glove for brain studies: Design and assessment of its compatibility for fMRI with a robust test. *IEEE/ASME Trans Mechatron* 13:345–354.

64. Bassen HI, Angelone LM. (2012) Evaluation of unintended electrical stimulation from MR gradient fields. *Frontiers Biosci* 4:1731–1742.

65. Mattei E, Triventi M, Calcagnini G et al. (2008) Complexity of MRI induced heating on metallic leads: Experimental measurements of 374 configurations. *BioMed Eng OnLine* 7:11.

66. Martin ET, Coman JA, Shellock FG et al. (2004) Magnetic resonance imaging and cardiac pacemaker safety at 1.5-Tesla. *J Am Coll Cardiol* 43:1315–1324.

67. Sommer T, Vahlhaus C, Lauck G et al. (2000) MR imaging and cardiac pacemakers: In-vitro evaluation and in-vivo studies in 51 patients at 0.5 T. *Radiology* 215:869–879.

68. Nazarian S, Roguin A, Zviman MM et al. (2006) Clinical utility and safety of a protocol for noncardiac and cardiac magnetic resonance imaging of patients with permanent pacemakers and implantable-cardioverter defibrillators at 1.5 tesla. *Circulation* 114:1277–1284.

69. ICNIRP. (2004) Statement on medical magnetic resonance (MR) procedures: Protection of patients. *Health Phys* 87(2):197–216.

70. Amendment to the ICNIRP. (2009) Statement on medical magnetic resonance (MR) procedures: Protection of patients. *Health Phys* 97(3):259–261.

71. ICNIRP. (2009) Guidelines on limits of exposure to static magnetic fields. *Health Phys* 96(4):504–514.

72. ICNIRP. (1998) Guidelines for limiting exposure to time-varying electric, magnetic, and electromagnetic fields (up to 300 GHz). *Health Phys* 74(4):494–522.

73. ICNIRP. (2009) Statement on the "guidelines for limiting exposure to time-varying electric, magnetic and electromagnetic fields (up to 300 GHz)." *Health Phys* 97(3):257–259.

74. IEC 60601-2-33 Consol. ed2.1 (incl. am1): Medical electrical equipment—Part 2: Particular requirements for the safety of magnetic resonance equipment for medical diagnosis, 2006.

75. HPA Protection of patients and volunteers undergoing MRI procedures: Advice from the Health Protection Agency, 2008.

76. MHRA Device Bulletin: Safety Guidelines for Magnetic Resonance Imaging Equipment in Clinical Use (DB2007(03)), 2007.

77. FDA Guidance for Industry and FDA Staff: Criteria for Significant Risk Investigations of Magnetic Resonance Diagnostic Devices, 2003.

78. ACR. (2007) Guidance document for MR safe practices. *AJR Am J Roentgenol* 188:1–27.

79. Directive 2004/40/EC of the European Parliament and of the Council of April 29, 2004, on the minimum health and safety requirements regarding the exposure of workers to the risks arising from physical agents (electromagnetic fields). Official Journal of the European Union L 159 of April 30, 2004 (and corrigenda L 184 of May 24, 2004).

80. Directive 2013/35/EU of the European Parliament and of the Council. Official Journal of the European Union 2004 L179/1 [cited October 15, 2013]. Available from: http://new.eur-lex.europa.eu/legal-content/EN/TXT/PDF/?uri5CELEX:32013L0035&qid51381829385026&from5EN.

81. Cavagnetto F, Prati P, Ariola V et al. (1993) A personal dosimeter prototype for static magnetic fields. *Health Physics* 65:172–177.

82. Cavin I, Gowland P, Glover P et al. (2005) Static B0 field monitoring at 3 T and 7 T using an MRI portable dosimeter. In: *Proceedings of ISMRM Workshop on MRI Safety: Update, Practical Information and Future Implications.* McLean, VA, November 5–6. Berkeley, CA: International Society for Magnetic Resonance in Medicine.

83. Bradley JK, Nyekiova M, Price DL et al. (2007) Occupational exposure to static and time-varying gradient magnetic fields in MR units. *J Magn Reson Imag* 26:1204–1209.

84. Fuentes MA, Trakic A, Wilson SJ, Crozier S. (2008) Analysis and measurements of magnetic field exposures for healthcare workers in selected MR environments. *IEEE Trans Biomed Eng* 55:1355–1364.

85. de Vocht F, Muller F, Engels H, Kromhout H. (2009) Personal exposure to static and time-varying magnetic fields during MRI system test procedure. *J Magn Reson Imag* 30:1223–1228.

86. Crozier S, Trakic A, Wang H, Liu F. (2007) Numerical study of currents in workers induced by body-motion around high-ultrahigh field MRI magnets. *J Magn Reson Imag* 26:1261–1277.

87. Riches SF, Collins DJ, Charles-Edwards GD et al. (2007) Measurements of occupational exposure to switched gradient and spatially-varying magnetic fields in areas adjacent to 1.5 T clinical MRI systems. *J Magn Reson Imaging* 26:1346–1352.

88. Hartwig V, Vanello N et al. (2011) A novel tool for estimation of magnetic resonance occupational exposure to spatially varying magnetic fields. *MAGMA* 24(6):323–330.

89. Forbes LK, Crozier S. (2001) On a possible mechanism for peripheral nerve stimulation during magnetic resonance imaging scans. *Phys Med Biol* 46:591–608.

90. Liu F, Zhao W, Crozier S. (2003) On the induced electric field gradients in the human body for magnetic stimulation by gradient coils in MRI. *IEEE Trans Biomed Eng* 50:804–811.

91. McKinnon G. (2003) Simplifying gradient coil modeling in FDTD calculations. In *Proceedings of the 11th Annual Meeting of SMRM*, Toronto, Ontario, Canada, p. 2437.

92. Mao W, Chronik BA, Feldman RE et al. (2006) Calculations of the complete electric field within a loaded gradient coil. *Magn Reson Med* 55:1424–1432.

93. McRobbie DW, Cross T. (2005) Occupational exposure to time varying magnetic gradient fields (dB/dt) in MRI and European limits. In: *Proceedings of ISMRM Workshop on MRI Safety: Update, Practical Information and Future Implications*. McLean, VA, November 5–6. Berkeley, CA: International Society for Magnetic Resonance in Medicine.

94. Crozier S, Trakic A, Wang H, Liu F. (2007) Numerical study of currents in occupational workers induced by body-motion around high ultrahigh field MRI magnets. In: *Proceedings of the Joint Annual Meeting ISMRM-ESMRMB*, May 19–25 Berlin, Germany, p. 1098.

95. Brand M, Heid O. (2002) Induction of electric fields due to gradient switching: A numerical approach. *Magn Reson Med* 48:731–734.

96. Liu F, Crozier S. (2004) A distributed equivalent magnetic current based FDTD method for the calculation of the E-fields induced by gradient coils. *J Magn Reson* 169:323–327.

97. Brix G, Reinl M, Brinker G. (2001) Sampling and evaluation of specific absorption rates during patient examinations performed on 1.5-Tesla MR systems. *Magn Reson Imaging* 9:769–779.

98. Hand JW. (2008) Modelling the interaction of electromagnetic fields (10 MHz–10 GHz) with the human body: Methods and applications. *Phys Med Biol* 53:R243–R286.

99. Collins CM, Li S, Smith MB. (1998) SAR and B1 field distributions in a heterogeneous human head model within a birdcage coil. *Magn Reson Med* 40:847–856.

100. Ibrahim TS. (2004) Optimization of RF coils for high field imaging: Why the head is different than symmetrical phantoms. In: *ISMRM 12th Scientific Meeting*, May 15–21, Kyoto, Japan, p. 1643.

101. Collins CM, Smith MB. (2001) Calculations of B1 distribution, SNR, and SAR for a surface coil against an anatomically-accurate human body model. *Magn Reson Med* 45:692–699.

102. Collins CM, Zhangwei W. (2011) Calculation of radiofrequency electromagnetic fields and their effects in MRI of human subjects. *Magn Reson Med* 65:1470–1482.

103. Liu W, Collins CM, Smith MB. (2005) Calculations of B1 distribution, specific energy absorption rate, and intrinsic signal-to-noise ratio for a body-size birdcage coil loaded with different human subjects at 64 and 128 MHz. *Appl Magn Reson* 29:5–18.

104. Collins CM, Liu W, Wang JH et al. (2004) Temperature and SAR calculations for a human head within volume and surface coils at 64 and 300 MHz. *J Magn Reson Imaging* 19:650–656.

105. Hand JW, Lagendijk JJW, Hajnal JV et al. (2000) SAR and temperature changes in the leg due to an RF decoupling coil at frequencies between 64 and 213 MHz. *J Magn Reson Imag* 12:68–74.

106. Jin JM. (1993) *The Finite Element Method in Electromagnetics*. Wiley: New York.

107. Jin JM, Chen J. (1997) On the SAR and field inhomogeneity of birdcage coils loaded with human head. *Magn Reson Med* 38:953–963.

108. Yang QX, Maramis H, Li CS, Smith MB. (1994) Three-dimensional full wave solution of MRI radio frequency resonator. In: *Proceedings of the SMR, 2nd Meeting*, p. 1110.

109. Harrison JG, Vaughan JT. (1996) Finite element modeling of head coils for high-frequency magnetic resonance imaging applications. In: *12th Annual Review of Progress in Applied Computational Electromagnetics*, pp. 1220–1226.

110. Simunic D, Wach P, Renhart W, Stollberger R. (1996) Spatial distribution of high-frequency electromagnetic energy in human head during MRI: Numerical results and measurements. *IEEE Trans Biomed Eng* 43:88–94.

111. Jin JM, Chen J, Gan H et al. (1996) Computation of electromagnetic fields for high-frequency magnetic resonance imaging applications. *Phys Med Biol* 41:2719–2738.

112. ANSYS HFSS, http://www.ansys.com/Products/Simulation+Technology/Electromagnetics/Signal+Integrity/ANSYS+HFSS.

113. Harrington RF. (1967) Matrix methods for field problems. *Proc. IEEE* 55:136–149.

114. Rogovich A, Monorchio A, Nepa P et al. (2004) Design of magnetic resonance imaging (MRI) RF coils by using the method of moments. In: *Proceedings of the IEEE International Symposium on Antennas and Propagation and USNC/URSI National Radio Science Meeting*, vol. 1, Monterey, CA, June 20–26, pp. 950–953.

115. Liu F, Beck BL, Xu B et al. (2005) Numerical modeling of 11.1 T MRI of a human head using a MoM/FDTD method. *Concepts Magn Reson Part B Magn Reson Eng* 24B:28–38.

116. Li BK, Liu F, Crozier S. (2006) High-field magnetic resonance imaging with reduced field/tissue RF aΛrtifacts—A modeling study using hybrid MoM/FEM and FDTD technique. *IEEE Trans Electromag Compat* 48(4):628–633.

117. Li BK, Liu F, Weber E, Crozier S. (2009) Hybrid numerical techniques for the modelling of radiofrequency coils in MRI. *NMR Biomed* 22(9):937–951.

118. FEKO, EM Software & Systems-S.A. (Pty) Ltd. (EMSS-SA), http://www.feko.info/.

119. Yee KS. (1996) Numerical solution of initial boundary value problems involving Maxwell's equations in isotropic media. *IEEE Trans Ant Propag* 14:302–307.

120. Berenger JP. (1994) A perfectly matched layer for the absorption of electromagnetic waves. *J Computational Phys* 114:185–200.

121. Schild S, Chavannes N, Kuster N. (2007) A robust method to accurately treat arbitrarily curved 3-D thin conductive sheets in FDTD. *IEEE Trans Antennas Propag* 55:3587–3594.

122. Liu W, Collins CM, Smith MB. (2005) Calculations of B1 distribution, specific absorption rate, and intrinsic signal-to-noise ratio for a body-size birdcage coil loaded with different human subjects at 64 and 128 MHz. *Appl Magn Reson* 29:5–18.

123. Van den Berg CAT, van den Bergen B, van de Kamer JB et al. (2007) Simultaneous B+1 homogenization and specific absorption rate hotspot suppression using a magnetic resonance phased array transmit coil. *Mag Res Med* 57:577–586.

124. Hartwig V, Giovannetti G, Vanello N et al. (2010) Numerical calculation of peak to average specific absorption rate on different human thorax models for magnetic resonance safety considerations. *Appl Magn Reson* 38:337–348.

125. Giovannetti G, Frijia F, Hartwig V et al. (2010) A novel magnetic resonance phased array coil designed with FDTD algorithm. *Appl Magn Reson* 39(3):225–231.

126. Giovannetti G, Hartwig V, Landini L et al. (2011) Sample-induced resistance estimation in magnetic resonance experiments: Simulation and comparison of two methods. *Appl Magn Reson* 40:351–361.

127. Hartwig V, Tassano S, Mattii A et al. (2013) Computational analysis of a radiofrequency knee coil for low-field MRI using FDTD. *Appl Magn Reson* 44(3):389–400.

128. Morelli MS, Hartwig V, Tassano S et al. (2013) FDTD analysis of a radiofrequency knee coil for low-field MRI: Sample-induced resistance and decoupling evaluation. *Appl Magn Reson* 44:1393–1403.

129. Mazzurana M, Sandrini L, Vaccari A et al. (2004) Development of numerical phantoms by MRI for RF electromagnetic dosimetry: A female model. *Radiat Prot Dosim* 111:445–451.

130. Makris N, Angelone L, Tulloch S et al. (2008) MRI-based anatomical model of the human head for specific absorption rate mapping. *Med Biol Eng Comput* 46:1239–1251.

131. Ho HS. (2001) Safety of metallic implants in magnetic resonance imaging. *J. Magn. Reson. Imaging* 14:472–477.

132. Giovannetti G, Hartwig V, Viti V et al. (2008) Low field elliptical MR coil array designed by FDTD. *Concepts Magn Reson Part B* 33B:32–38.

133. Giovannetti G, Frijia F, Hartwig V et al. (2010) A novel magnetic resonance phased array coil designed with FDTD algorithm. *Appl Magn Reson* 39(3):225–231.

134. XFdtd® EM Simulation Software, Remcom, http://www.remcom.com/xf7.

135. SEMCAD X, SPEAG, http://www.speag.com/products/semcad/overview/.

136. CST MICROWAVE STUDIO, CST, https://www.cst.com/Products/CSTMWS.

137. Gabriel C. (1996) Compilation of the dielectric properties of body tissues at RF and microwave frequencies, Occupational and Environmental Health Directorate, Radio-frequency Radiation Division, Brooks Air Force Base, TX, Tech. Rep. ALOE-TR-1996-0037.

138. Christ A, Kainz W, Hahn EG. et al. (2010) The Virtual Family—Development of surface-based anatomical models of two adults and two children for dosimetric simulations. *Phys Med Biol* 55(2):N23–N38.

139. Mattei E, Triventi M, Calcagnini G et al. (2008) Complexity of MRI induced heating on metallic leads: Experimental measurements of 374 configurations. *BioMed Eng OnLine* 7:11.

140. Yeung CJ, Atalar E. (2001) A Green's function approach to local RF heating in interventional MRI. *Med Phys* 28(5):826–832.

141. Pennes HH. (1948) Analysis of tissue and arterial temperatures in the resting human forearm. *J Appl Physiol* 1:93–122.

142. ASTM F2182-02a. (2002) *Standard test method for measurement of Radio Frequency induced heating near passive implants during Magnetic Resonance Imaging.* ASTM International: West Conshohocken, PA.

143. Bulumulla SB, Yeo TB, Zhu Y. (2009) Direct calculation of tissue electrical parameters from B1 maps. *Proc Intl Soc Mag Reson Med* 17:3043.

144. Cloos MA, Bonmassar G. (2009) Towards direct B1 based local SAR estimation. *Proc Intl Soc Mag Reson Med* 17:3037.

145. Hartwig V, Vanello N, Giovannetti G et al. (2011) B1+/actual flip angle and reception sensitivity mapping methods: Simulation and comparison. *Mag Reson Imag* 29:717–722.

146. Katscher U, Voigt T, Findeklee C. (2009) Determination of electric conductivity and local SAR via B1 mapping. *IEEE Trans Med Imag* 28(9):1365–1374.

147. Akoka S, Franconi F, Seguin F, le Pape A. (1993) Radiofrequency map of an NMR coil by imaging. *Magn Reson Imag* 11:437–441.

148. Stollberger R, Wach P. (1996) Imaging of the active B1 field in vivo. *Magn Reson Med* 35:246–251.

149. Collins CM, Yang QX, Wang JH et al. (2005) Different excitation and reception distributions with a single-loop transmit-receive surface coil near a head-sized spherical phantom at 300 MHz. *Magn Reson Med* 47:1026–1028.

150. Helluy X, Webb AG. (2005) Characterization of B1 field focusing effects in magnetic resonance spectroscopy and imaging at 17.6 tesla. *Concepts Magn Reson Part B Magn Reson Eng* 27B(1):8–16.

151. Hornak JP, Szumowski J, Bryant RG. (1988) Magnetic field mapping. *Magn Reson Med* 6:158–163.

152. Yarnykh VL. (2007) Actual flip-angle imaging in the pulse steady state: A method for rapid three dimensional mapping of the transmitted radiofrequency field. *Magn Reson Med* 57:192–200.

153. Schmitt F, Turner R, Stehling MK. (1998) *Echo-Planar Imaging: Theory, Technique and Application.* Berlin, Germany: Springer.

154. Cunningham CH, Pauly JM, Nayak KS. (2006) Saturated double angle method for rapid B1+ mapping. *Magn Reson Med* 55(6):1326–1333.

155. Morrell GR. (2008) A phase-sensitive method of flip angle mapping. *Magn Reson Med* 60(4):889–894.

156. Sacolick LI, Wiesinger F, Hancu I, Vogel MW. (2010) B1 mapping by Bloch–Siegert shift. *Magn Reson Med* 63:1315–1322.

157. Rieke V, Pauly KB. (2008) MR Thermometry. *J Mag Res Imag* 27:376–390.

158. Diederich CJ, Nau WH, Ross AB et al. (2004) Catheter-based ultrasound applicators for selective thermal ablation: Progress towards MRI guided applications in prostate. *Int J Hyperthermia* 20:739–756.

159. Botnar RM, Steiner P, Dubno B et al. (2001) Temperature quantification using the proton frequency shift technique: In vitro and in vivo validation in an open 0.5 Tesla interventional MR scanner during RF ablation. *J Magn Reson Imaging* 13:437–444.

160. Quesson B, de Zwart JA, Moonen CT. (2000) Magnetic resonance temperature imaging for guidance of thermotherapy. *J Magn Reson Imaging* 12:525–533.

161. Parker DL. (1984) Applications of NMR imaging in hyperthermia: an evaluation of the potential for localized tissue heating and noninvasive temperature monitoring. *IEEE Trans Biomed Eng* 31:161–167.

162. Cline H, Mallozzi R, Li Z et al. (2004) Radiofrequency power deposition utilizing thermal imaging. *Magn Reson Med* 51:1129–1137.

163. Shapiro EM, Borthakur A, Shapiro MJ et al. (2002) Fast MRI of RF heating via phase difference mapping. *Magn Reson Med* 47:492–498.

164. Stralka JP, Bottomley PA. (2007) A prototype RF dosimeter for independent measurement of the average specific absorption rate (SAR) during MRI. *J Magn Reson Imaging* 26:1296–1302.

11

7 T Magnetic Resonance Imaging and Spectroscopy: Methods and Applications

Simon Robinson, Roland Beisteiner, Wolfgang Bogner, Klaus Bohndorf, Marek Chmelík,
Barbara Dymerska, Florian Fischmeister, Stephan Gruber, Gilbert Hangel, Vladimir Juras,
Claudia Kronnerwetter, Eva Matt, Günther Grabner, Martin Krššák, Lenka Minarikova,
Benjamin Schmitt, Bernhard Strasser, Štefan Zbýň, and Siegfried Trattnig

CONTENTS

11.1 Introduction

Ultra-high-field magnetic resonance (UHF MR) systems (7 Tesla [7 T] and higher field strength) are expected to yield a twofold-to-threefold improvement in image signal-to-noise ratio (SNR) over 3 Tesla (3 T) MR scanners. Advances in multichannel radiofrequency (RF) technology provide an additional twofold-to-sixfold improvement in SNR over single-channel RF coils (Wiggins et al., 2005). Many fields require new methods to deal with increased B_0 and B_1 inhomogeneity, however, and sequences need to be modified to adapt to modified relaxation times of ^{1}H and X nuclei at 7 T. In the following sections, we review the state of the art of 7 T methods and nascent applications in neuroimaging, musculoskeletal (MSK) imaging, studies of metabolism, and oncology.

11.2 Neuroimaging

11.2.1 T_1-Weighted Imaging

The interest in accurate visualization and/or segmentation of cortical and deep gray matter structures sets a trend toward T_1-weighted imaging at UHFs. One advantage at 7 T is improved T_1-contrast with respect to lower field systems. Additionally, high SNR allows measurements to be carried out with higher resolution or/and shorter measurement time. Most commonly, T_1-weighted

imaging is performed with the magnetization-prepared rapid gradient-echo (MPRAGE) sequences (Mugler and Brookeman, 1990) and, more recently, MP2RAGE (Marques et al., 2010). MP2RAGE differs from its predecessor in the inclusion of two rather than one image readouts after each adiabatic inversion pulse. Those two full readouts are acquired with identical parameters other than the effective inversion time (TI_1, TI_2), which yield different T_1-weighting. Otherwise, the two images are affected to the same extent by the receive bias field (B_1^-), proton density (M_0), and T_2^*. In a final step, those two gradient-echo images (GRE_{TL1} and GRE_{TL2}) are combined as follows:

$$MP2RAGE = \frac{GRE_{TL1} \times GRE_{TL2}}{GRE_{TL1}^2 + GRE_{TL2}^2}$$

This operation removes M_0, T_2^*, and the (B_1^-) dependencies, which is especially important at 7 T, where bias field inhomogeneities are prominent.

The image characteristics of MPRAGE and MP2RAGE are demonstrated in a 7 T measurement (Magnetom, Siemens Healthcare, Erlangen, Germany) of a healthy volunteer acquired with a 32-channel head coil (Nova Medical, Wilmington, MA). The geometry and resolution of the MPRAGE and MP2RAGE acquisitions were matched, but the sequence parameters were adjusted to obtain optimal T_1-contrast and high SNR for each sequence. The results are presented in Figure 11.1. Bias field inhomogeneities are present in the MPRAGE image (Figure 11.1, left), as an

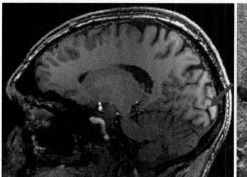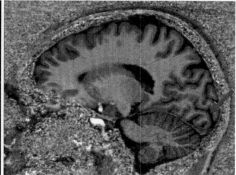

FIGURE 11.1
Comparison of image features in MPRAGE (left) and MP2RAGE images (right) of a healthy volunteer. The geometry and resolution of the images are the same. The red arrows point to a region where the white/gray matter boundary is not visible in MPRAGE, but is clear in MP2RAGE.

increasing white matter intensity from anterior to posterior part of the brain. This is accompanied by reduced T_1-contrast, such that there is almost complete loss of white/gray-matter-contrast border (Figure 11.1, left, red arrow). These unwanted effects are effectively removed when the two echo images acquired in the MP2RAGE sequence are combined (Figure right); white matter intensity and T_1-contrast are spatially uniform in the whole brain.

After the sequence was first described in 2010, the suitability of MP2RAGE for high-field applications has been quickly acknowledged, with several studies in which it was used being published in the subsequent four years. One example concerns high-resolution (HR) visualization of small, deep brain structures, such as the thalamic nuclei (Keuken et al., 2013; Marques and Gruetter, 2013) and brain stem (Marques and Gruetter, 2013). In a paper by Marques and Gruetter (2013), MP2RAGE sequence parameters were adjusted to yield optimal contrast for the range of T_1-values expected to be found in the structures investigated. This optimization allowed better delineation of medio dorsal, ventral lateral, and pulvinar nuclei in thalamus or white matter bundles in the brain stem. MP2RAGE, optimized for deep brain structure imaging at 7 T, opens the possibility of studying age-related changes in subthalamic nuclei (Keuken et al., 2013) and characterizing early stages of Parkinson's disease (Jubault et al., 2009).

Another study presents an automated hypothalamus segmentation, which was achievable thanks to MP2RAGE measurement with submillimeter resolution at 7 T (Schindler et al., 2013). This type of computer-assisted segmentation has a potential application in the analysis of the correlation between hypothalamus structure and psychiatric disorders, such as schizophrenia (Tognin et al., 2012), narcolepsy

(Kim et al., 2009), and frontotemporal dementia (Piguet et al., 2011). Recent studies have also employed the MP2RAGE sequence at 7 T for whole-brain automated tissue segmentation (Bazin et al., 2013; Fujimoto et al., 2014). It could also be useful for cortical thickness determination, providing insight into normal brain development and neurodegenerative disorders (Lusebrink et al., 2013). Certain segmentation procedures can provide less reliable results when applied to MP2RAGE data, as they do not deal well with the high levels of background noise (see Figure 11.1) (Fujimoto et al., 2014). Since the two gradient-echo images are acquired at different inversion times, MP2RAGE has the attractive feature that the sequence can be used for quantitative T_1-mapping (Weiss et al., 2013). Finally, MP2RAGE is increasingly being used as an anatomical reference at 7 T in functional magnetic resonance imaging (fMRI) (Da Costa et al., 2011; Koopmans et al., 2012; Van der Zwaag et al., 2012) and spectroscopy (Xin et al., 2013).

11.2.2 T_2-Weighted Imaging

Fast spin-echo (FSE)-based imaging techniques are essential tools for clinical MRI because the most common applications and contrasts are based on that readout scheme. In current routine practice, FSE contrasts are predominantly used in a two-dimensional (2D) fashion.

If the commonly known FSE contrasts from 1.5 to 3 T are to be transitioned to higher magnetic fields, such as 7 T, restraints of the specific absorption rate (SAR) limit the applicability of standard implementations; especially the large flip angles in the train of refocusing RF pulses, which are usually between 150° and 180° if executed as standard sinc-type pulses, exhibit high B_1 amplitudes and, thus, contribute significantly to the

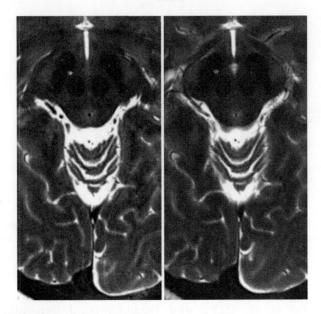

FIGURE 11.2

T_2 FSE image of the brain comparing 0.4 mm (left) and 0.3 mm (right) in-plane resolution at 1 mm slice thickness. The level of detail is visibly increased by the higher resolution and can be well observed, and SNR is sufficient despite using Gaussian-type RF pulses instead of sinc-shaped pulses.

generally high SAR depositions of FSE-based imaging techniques (Figure 11.2).

Since similar patterns of exceeding SAR values were noticed when transitioning FSE sequence implementations from 1.5 to 3 T, several methods for mitigating SAR load were already developed. These include application of variable-rate selective excitation (VERSE) modifications (Hargreaves et al., 2004) to the refocusing pulses and the concept of using hyperechoes (Hennig and Scheffler, 2001). However, modifying RF pulses with VERSE can impair the accuracy of their slice profile by making the pulse profile more susceptible to inhomogeneities of the static magnetic field. Since such inhomogeneities are commonly stronger at 7 T, the applicability of VERSE at 7 T remains limited. Hyperechoes are most effective means to reduce SAR for FSE sequences with high numbers of RF pulses in the echo train, which is the case for, for example, T_2-weighted FSE contrasts. However, this concept will only have limited applicability for proton density or T_1-weighted contrasts. Another approach would then be to use refocusing pulses that are still fast enough to allow common sequence timings, but exhibit lower intrinsic B_1 amplitudes than sinc-type pulses. Such pulses, for example, Gaussian-shape pulses, however, are likely to exhibit less accurate slice profiles compared with their sinc-type counterparts. Having less uniform slice profiles, especially with lower signal

strengths in parts of the slice profile, may pose significant problems with SNR at clinical field strengths and may impair diagnostic accuracy. This problem, however, is considerably mitigated at field strengths of 7 T or higher, thus making these pulse alternatives viable options (Figure 11.3).

Another concern for SAR in FSE contrasts is inversion recovery (IR)-type contrasts, such as fluid-attenuated inversion recovery (FLAIR) or short-tau IR (STIR), which use high-B_1 inversion pulses in addition to readouts with long echo trains. Hence, it will be mandatory for IR contrasts at 7 T to replace the standard high-power inversion pulses, which are usually adiabatic, with SAR-reduced equivalents. Since IR pulses for 2D FSE readouts need to be slice specific, approaches for high-field adiabatic inversion pulses must ensure that the adiabatic condition will be met over a broad range of offset frequencies that can accumulate in a slice profile due to field inhomogeneities. Concepts for SAR-reduced, slice-selective adiabatic inversion that can potentially cover large frequency ranges at reasonable pulse durations include gradient offset-independent adiabaticity (GOIA) pulses using wideband uniform rate smooth truncation (WURST) modulation for RF and gradient waveforms (Andronesi et al., 2010) or B_1-reduced frequency offset corrected inversion (FOCI) pulses (O'Brien et al., 2013; Ordidge et al., 1996).

During the setup of FSE sequences at 7 T, one remaining major concern will then be the mitigations of chemical shift displacement (CSD) artifacts that can occur due to different RF bandwidths of the excitation pulse and the refocusing pulses. Such a difference will induce visually shifted locations of voxels according to their fat and water fractions on the one hand. On the other hand, it may also lead to partial suppression of fat signals if lipid signals are excited at a particular frequency, but the bandwidth of the refocusing pulse does not allow for these signals to be refocused entirely (Figure 11.3). Due to the increased chemical shift dispersion at 7 versus 3 T and due to the tendency of RF pulses to exhibit generally lower bandwidths for SAR reduction, CSD artifacts are likely to be a concern at 7 T and require careful setup of RF pulse parameters to mitigate these effects (Figure 11.4).

Due to the changes in relaxation times compared with 3 T- that is, decreased T_2 and increased T_2- the FSE T_2 contrasts provide excellent options for discriminating tissue types at 7 T. Suitable parameters for T_2 imaging at 7 T revealing a high level of detail in a clinically relevant scan time can be: 768 × 1024 matrix size, field of view = 172 × 230 × 3 mm³, giving voxel sizes of 0.2 × 0.2 mm in-plane at a slice thickness of 3.0 mm with a 0.3 mm gap, (time to echo/time of repetition) T_e/T_R = 55/3700 ms, full-brain coverage with 32 slices in an acquisition time of 3 min 48 s (Figure 11.5).

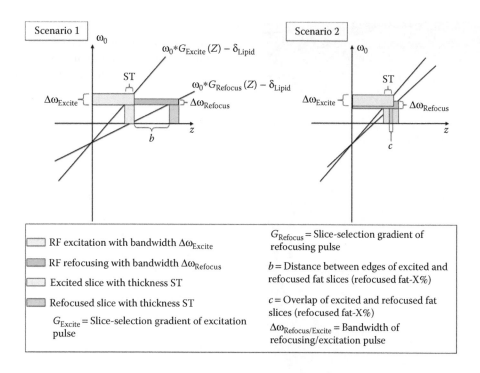

FIGURE 11.3
Schematic illustration of volume selection for lipid resonance signal during excitation and refocusing. The two scenarios shall clarify the dependency of the relative amount of refocused fat on bandwidth of refocusing pulses (Scenario 1: Small refocusing bandwidth; Scenario 2: Large refocusing bandwidth). The RF bandwidth is inversely proportional to the duration of the RF pulse.

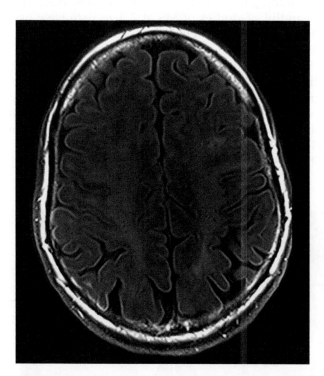

FIGURE 11.4
FLAIR FSE image of a brain acquired at 7 T demonstrating the occurrence of chemical shift displacement of lipid signals. The lipid signals from the skull originate from a more apical position than the water signals of the current slice position. Due to curvature of the skull, these lipid signals from a more apical position a visible as bright overlay on the outer edges of the brain.

11.2.3 Fluid-Attenuated Inversion Recovery

FLAIR is an IR sequence in which the inversion time is chosen close to the crossing point of cerebrospinal fluid (CSF), so that fluid contributes little or no signal (Hajnal et al., 1992). High SAR is associated with the use of multiple 180° refocusing pulses and adiabatic inversion pulses, which are needed both to reduce variations in the transmit field at 7 T, which would cause image heterogeneity, and to avoid inadequate nulling of the CSF (Garwood and DelaBarre, 2001). SAR is dependent on coil design and sequence parameters but it is generally the case that at 7 T, 2D turbo spin echo (TSE) and 2D FLAIR are SAR limited so that only thin slab (rather than whole-brain coverage) is practicable (Zwanenburg et al., 2010).

HR whole-brain FLAIR imaging is feasible within SAR limitations using a three-dimensional (3D) FLAIR-TSE sequence, which deploys 3D "sampling perfection with application-optimized contrasts using different flip angle evolutions" (SPACE) (Grinstead et al., 2010). SPACE denotes 3D FLAIR-TSE with the addition of a restore pulse, which allows a very large turbo factor. Variable flip angles entail lower RF power and allow longer echo trains and higher turbo factors (Hodel et al., 2011). CSF signal can be suppressed with the optimum inversion time and T_2 weighting achieved (for hyperintense lesions) and low

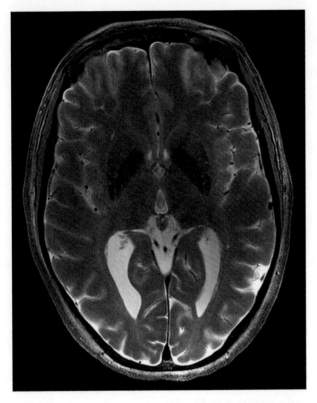

FIGURE 11.5
T_2 FSE image of a brain acquired at 7 T demonstrating high SNR and excellent tissue discrimination properties at very high resolution of 0.2 × 0.2 mm in-plane, which can be achieved for full brain coverage in 3:48 min scan time.

gray–white matter contrast via the optimum selection of T_e and T_R. This allows a FLAIR contrast similar to that of standard 2D FLAIR. Echo time and repetition time have to be modified in the light of longer T_1 and shorter T_2 proton relaxation times. In Figure 11.6, gray–white matter contrast was reduced and lesion–white

matter contrast was maximized to yield optimum FLAIR contrast using T_e/T_R of 279/8000 ms. CSF signal was minimized using an inversion time of 2180 ms (Figure 11.6). The characteristics of 3D FLAIR-SPACE at 7 T are illustrated and compared with those of a 2D FLAIR-TSE sequence at 3 T in Figures 11.7 and 11.8. Both 3 T 2D FLAIR-TSE sequence and 7 T 3D FLAIR-SPACE show minimal CSF signal and low gray–white matter contrast.

11.2.3.1 Application to MS

Diagnosing multiple sclerosis (MS) depends on the identification of multifocal demyelinating lesions, which appear hyperintense on T_2-weighted MRI. By nulling the CSF signal, FLAIR sequences generate higher tissue separability than standard TSE (Rossi et al., 2010). FLAIR has become the reference method in the evaluation of patients with MS at 1.5 and 3 T. Increased SNR and tissue contrast at UHF (7 T and above) suggest that the study of MS may benefit from the use of FLAIR at 7 T.

There are very few studies of MS to date at 7 T (Ge et al., 2008; Kollia et al., 2009; Tallantyre et al., 2010). Although FLAIR is the clinical standard sequence for MS investigations at 1.5–3 T, to our knowledge, only a handful of studies of MS at 7 T to date have used 3D FLAIR (De Graaf et al., 2011; Kilsdonk et al., 2012, 2014).

In these pilot data of patients with an established diagnosis of MS, lesions located on the border between gray and white matter could also be visualized with excellent contrast to the background (see the example of a lesion in the intersecting planes a, b, and c in Figure 11.9: The appearance of a large lesion in 3D 7 T FLAIR [white arrow in the intersecting places a, b, and c]).

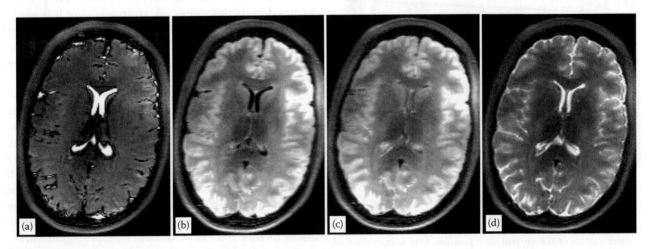

FIGURE 11.6
Tissue contrast with different IR-times: (a) 1000, (b) 2180, (c) 2500, and (d) 3500 ms, indicating different levels of CSF nulling.

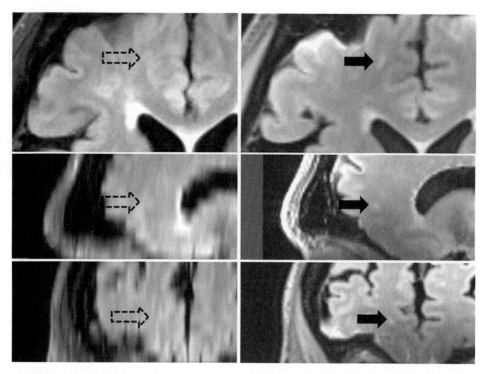

FIGURE 11.7
Comparison of 2D 3 T FLAIR (left-hand column) and 3D 7 T FLAIR data (right-hand column) and in axial (top row), sagittal (middle row), and coronal orientations (bottom row). Full arrows indicate a small lesion visible in 3D, 7 T FLAIR only. Dashed arrows indicate the position of the same small lesion in 2D FLAIR.

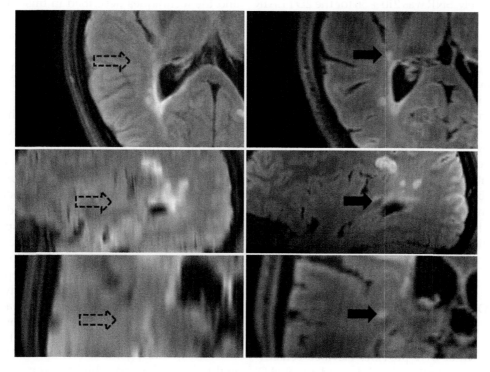

FIGURE 11.8
Comparison between 2D 3 T FLAIR (left-hand column) and 3D 7 T FLAIR (right-hand column). A lesion which is clearly visible in 3D 7 T FLAIR at the position of solid arrows shows nearly no contrast in 2D 3 T FLAIR (dashed arrow positions).

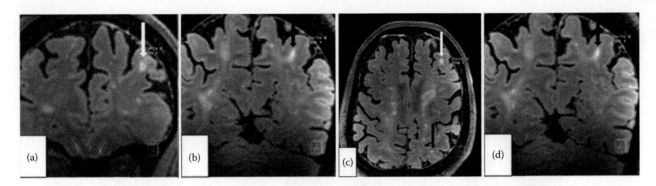

FIGURE 11.9
The appearance of a large lesion in 3D 7 T FLAIR (white arrow in the intersecting places a, b, c). The black arrow in (d) demonstrates the degree of contrast achieved in some small lesions.

The black arrow in d demonstrates the degree of contrast achieved in some small lesions. The ability to visualize lesions in the coronal plane in addition to the acquisition plane (which was oblique axial) allowed hyperintensities to be precisely categorized as gray or white matter lesions. For instance, the coronal section (d) through the lesion indicated with a black arrow in Figure 11.9c is constrained to white matter. This was not clear in 2D FLAIR at 3 T.

Image SNR and contrast-to-noise ratio (CNR) in the 2D sequence at 3 T and the 3D sequence at 7 T were calculated according to Erhardt (2008) and Bogner et al. (2009b), respectively. SNR was 210 ± 56 for the 2D FLAIR sequence at 3 T and 433 ± 81 for the 3D FLAIR-SPACE at 7 T. CNR was 8.0 ± 1.7 for the 2D FLAIR at 3 T and 17.5 ± 6.7 for the 3D FLAIR SPACE at 7 T.

These pilot data demonstrate that an optimized 3D FLAIR sequence at 7 T based on SPACE allows measurement of the whole brain within SAR limits. Isotropic 3D measurements could be acquired with eightfold higher resolution than standard 2D FLAIR at 3 T, reducing partial volume effects and allowing brain lesions to be visualized in any plane without loss of information. A number of small lesions in MS patients could be visualized compared to 2D FLAIR at 3 T. This concurs with the findings of a recent 7 T MS study (Kilsdonk et al., 2014) showing that the highest number of cortical gray matter lesion is detected with 3D FLAIR technique.

11.2.4 Blood Oxygen Level-Dependent fMRI

11.2.4.1 BOLD Sensitivity and Specificity at UHF

In fMRI, high SNR at UHF provides elevated sensitivity to the blood oxygenation level-dependent (BOLD) signal changes that take place close to active neurons. This has been demonstrated both in healthy subjects (Olman et al., 2010; Van der Zwaag et al., 2009) and, more recently, patients (Beisteiner et al., 2011b). In the context of presurgical planning—the most established clinical application of fMRI—increased BOLD sensitivity increases the proportion of successful localizations for a given experiment duration or allows the number of repetitions of the fMRI experiment (*runs*) to be decreased. It also offers the possibility to increase resolution and to detect very weak activation in previously undiagnosable patients with large and rapidly evolving tumors and complex pathologies. Improving the reliability of clinical fMRI through the use of UHF and the techniques described in this section offers the possibility to increasingly supplant invasive diagnostic procedures such as intraoperative cortical stimulation. The latter method is the historic gold standard mapping method for functional localization, but one which prolongs surgery and carries the risk of inducing intraoperative seizures.

Early studies on the field strength characteristics of the BOLD signal reported an increase in sensitivity to signal changes of tissue (Gati et al., 1997; Yacoub et al., 2001) rather than *draining vein* origin (Frahm et al., 1994), indicating that the BOLD response at UHF is better localized to the source of activation. More recent fMRI studies have not confirmed an increase in the tissue specificity of the BOLD response (Geissler et al., 2013), however, largely because of the enduringly large contribution of physiological artifacts. The early methodological studies reported a more than linear increase in BOLD sensitivity with field strength (Gati et al., 1997; Yacoub et al., 2001). More recently, Triantafyllou et al. have shown that BOLD sensitivity gain is strongly dependent on resolution, with the greatest gains to be made at HR, where thermal noise makes up a higher proportion of the total (Triantafyllou et al., 2005). Applied fMRI activation studies have shown increases in BOLD sensitivity which are significant but more modest than expected (Beisteiner et al., 2011b; Van der Zwaag et al., 2009), a fact which is attributable, at least in part, to increased Nyquist ghost and parallel imaging reconstruction artifacts at UHF. In the light of these physiological and technical artifacts, 7 T fMRI has only partially realized its full promise. In the following sections, we briefly describe the most recent and most promising approaches to solving the remaining problems affecting UHF fMRI and indicate the data quality that can be achieved when these methods are applied.

11.2.4.2 Challenges and Solutions

In contrast to the spin-echo methods discussed at the beginning of this chapter, gradient-echo echo planar imaging (EPI) is not very sensitive to the flip angle variation that arises from inhomogeneous B_1^+ transmission. Long echo train length and low resolution (typically above 1 mm isotropic voxels) can lead to severe signal dropout in regions of high B_0 inhomogeneities, however, and the low bandwidth per pixel in the phase-encoded direction can yield significant distortions.

Signal dropout in 7 T EPI can be reduced using the same approaches used in the high field (3–4 T) regime; by increasing resolution (Merboldt et al., 2000; Robinson et al., 2008), optimizing slice angle (Deichmann et al., 2003), using a compensation gradient in the z direction combined with a phase-encoded polarity selected for the brain regions of primary interest (De Panfilis and Schwarzbauer, 2005) and parallel imaging acceleration (Griswold et al., 2002; Pruessmann et al., 1999). Deploying parallel imaging allows phase-encoding steps to be skipped, enabling the echo time to be reduced to a value equal to or lower than T_2^* (even for HR acquisitions), and reduces field gradient-related shifts in the position of contributing echoes, which can lead to Type 2 signal loss (Deichmann et al., 2002). It is important to remember, however, that parallel imaging reduces SNR by a factor $g^* R^{1/2}$, where R is the acceleration factor and g is the coil geometry factor, which is greater than 1: the use of excessive acceleration can negate the SNR advantage of high field. A recent radical development in fast imaging is *multiband* or *simultaneous multislice* imaging, in which multiple slices are excited and also read out simultaneously. The sensitivities of measurements with coils of different sensitivity (Feinberg et al., 2010), assisted by slice-dependent shifts in the image position (blipped CAIPIRINHA) (Setsompop et al., 2012), are used to separate the overlapping signal in images. Where conventional (*in-plane*) parallel imaging allows a reduction in T_e, at a cost to SNR of at least the square root of the acceleration factor, simultaneous multislice imaging allows T_R to be reduced (by a factor equal to

the number of slices simultaneously acquired) but is not associated with the square root of the acceleration factor loss in SNR.

Distortion in 7 T EPI can be minimized by effective shimming, using improved algorithms (Fillmer et al., 2014b) and higher-order shimming with a shim insert (e.g., that from Resonance Research Inc., Billerica, MA) or dynamic shimming (Juchem et al., 2010). Unlike signal loss, which arises from gradients in ΔB_0, distortion, which arises from ΔB_0 itself, can be corrected. Point spread function mapping (Zaitsev et al., 2004; Zeng and Constable, 2002) or maps of ΔB_0 acquired with multi-echo reference scans (Jezzard and Balaban, 1995) can be used to correct for distortion as long as the signal does not pile up or swap position (Robinson and Jovicich, 2011). In clinical fMRI at 7 T, this has been shown to be an effective remedy for distortions in primary motor regions (Dymerska et al., 2014).

Parallel imaging reconstruction artifacts can be reduced by improving the quality of the reference lines, either by acquiring these with fast low angle shot (FLASH) (Talagala et al., 2013) or by acquiring all the segments required for each slice in rapid succession, with low flip angle, in a slice-by-slice as opposed to a slice-by-segment fashion, which acquires the reference scans over a time which is short compared to the respiratory cycle (Polimeni, 2013). Nyquist ghost artifacts, which are more pronounced at higher field, particularly in tasks that involve head motion, can be reduced by performing a local rather than a global phase correction with the navigators. An alternative to trying to make the Nyquist ghost correction more robust is offered by a novel sequence called IDEA EPI (Poser et al., 2013) in which, compared to a regular EPI scheme, every second phase blip is omitted and the remaining blip size doubled. The data are assigned to two undersampled k-spaces, one for the positive readouts and one for the negative readouts. The fact that all lines in each k-space are acquired in the same direction means there is no need to reverse alternate lines, the source of the Nyquist ghost artifact in EPI. The undersampled k-spaces are reconstructed using SENSE (Pruessmann et al., 1999) or generalized autocalibrating partially parallel acquisition (Griswold et al., 2002) and the images combined.

We end this section with an example of the data quality that is achievable using some of the techniques described. Gradient-echo EPI data were acquired from a single healthy subject with a 7 T MR whole-body system (Magnetom, Siemens Healthcare, Erlangen, Germany) using a 32-channel head coil (Nova Medical, Wilmington, NC), with a 128 × 128 matrix with a square 192 × 192 mm field of view and 50 slices of 1.5 mm thickness (cubic voxels of 1.5 mm side length), with in-plane generalized autocalibrating partially parallel acquisition factor 2 and multiband factor 2, $T_e/T_R = 22/1500$ ms. The subject was instructed to do nothing for an acquisition duration of 5 min (200 volumes). Data were corrected for distortions using a multichannel field map (Robinson and Jovicich, 2011), motion corrected and smoothed with a Gaussian kernel with a full width half maximum of 2 mm. Independent component analysis (ICA) was performed using FSL's MELODIC (Beckmann and Smith, 2004), yielding 60 components, including a component reflecting the motor resting state network (Biswal et al., 1995) (number 38, ranked

by the percentage of variance explained) illustrated in Figure 11.10: EPI data quality at 7 T with a 32-channel RF coil. A resting state network in motor regions, identified using ICA (Beckmann et al., 2005) from a single 5-min run without stimulus, is overlaid (see white arrows) on a single EPI time point. A number of recent studies have explored the possibility of performing presurgical identification of the precentral sulcus via functional connectivity in primary motor regions rather than via a motor task

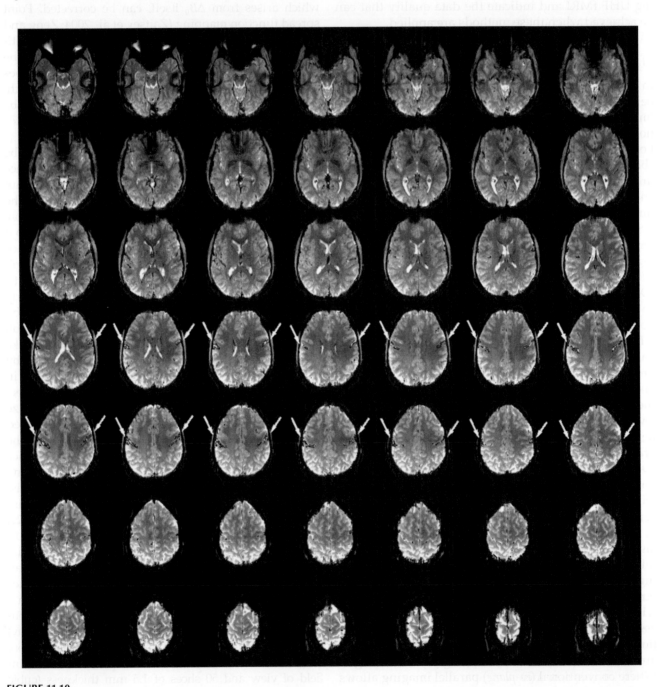

FIGURE 11.10

EPI data quality at 7 T with a 32-channel RF coil. A resting-state network in motor regions, identified using independent component analysis from a single 5-min run without stimulus, is overlaid (see white arrows) on a single EPI time point (Beckmann, C.F. et al., *Philos. Trans. R Soc. Lond. B Biol. Sci.*, 360, 1001–1013, 2005.)

(see, e.g., Shimony et al., 2009; Zhang et al., 2009). These results show that reliable, HR mapping of primary motor regions can be achieved in under 5 min, even in subjects who are not able to perform a task.

11.2.4.3 Clinical Applications

UHF fMRI offers increased sensitivity and specificity of BOLD signal and the possibility to acquire high spatial resolution images as has been discussed in previous sections of this chapter. These advantages of UHF are particularly important when investigating subtle activations and/or when measuring small brain structures. As functional activation might be weaker in pathologically compromised brain areas, these benefits encourage clinical applications of 7 T fMRI such as presurgical functional localization, assessing and monitoring neuroplasticity in various neuropathologies, and identifying noninvasive biomarkers for brain diseases (Beisteiner, 2013). However, UHF fMRI studies in patients are still rare due to methodological challenges such as image distortions or increased effects of physiological noise, as well as due to the limited availability of 7 T MR systems in clinical settings. In the following, the main benefits of UHF for functional patient investigations, increased sensitivity, and increased spatial resolution are discussed.

11.2.4.4 Clinical Relevance of Improved BOLD Sensitivity

Higher magnetic field strengths significantly improve SNR (Triantafyllou et al., 2005; Vaughan et al., 2001) and sensitivity of BOLD signal changes (Van der Zwaag et al., 2009; Yacoub et al., 2001). In healthy participants, the resulting stronger task-related activation at 7 T compared to lower field strengths has been demonstrated for visual areas (Yacoub et al., 2001) for the motor cortex (Van der Zwaag et al., 2009), for memory-related BOLD responses in the hippocampus (Theysohn et al., 2013), for pain elicited activation in the periaqueductal gray (Hahn et al., 2013), and for amygdala activation (Sladky et al., 2013). The latter study reported that signal change in the amygdala during a facial emotion discrimination task was twice as high at 7 T as at 3 T.

Pathological alterations of neuronal tissue such as tumors, lesions, or neurodegeneration may weaken functional activation if the vascular response is reduced (Ulmer et al., 2004). This circumstance underlines the need for increasing sensitivity of BOLD signals in patient investigations. Thus, using higher field strengths could lead to a more reliable localization of essential cortical motor, language, or memory areas, which are important tasks during presurgical patient evaluation. Moreover, higher sensitivity improves the detection of subtle differences between normal and pathological conditions,

which is important for classification and prognosis of various neurological diseases (e.g., Alzheimer's disease).

In the first clinical fMRI study using 7 T, Beisteiner et al. (2011b) compared the BOLD response at 3 and 7 T during a simple motor task in 17 patients referred for presurgical diagnosis. They found increased functional activation within the primary motor hand area at 7 T compared to 3 T despite increased artifacts such as enhanced ghosting and head motion artifacts at 7 T. Increased patient artifacts at 7 T are mainly driven by two factors: a generally higher sensitivity of 7 T for physiological artifacts and the currently less comfortable examination conditions (less space inside the magnet) that reduce patient compliance. While motor tasks typically elicit robust activations in well-defined cortical regions, cognitive tasks such as those probing language evoke more complex activation patterns often involving more artifact-prone inferior frontal regions. A subsequent study by the same authors (Geißler et al., 2014) addressed language function in presurgical patients. Functional activation measured during an overt speech task was analyzed within two eloquent language areas: the posterior superior temporal gyrus (Wernicke's area) and the inferior frontal cortex (Broca's area). While for Wernicke's area 7 T showed a significantly improved functional sensitivity compared to 3 T (Figure 11.11), this higher field strength benefit was not observable in Broca's area related to increased inferior frontal artifacts. These results indicate that higher 7 T sensitivity for BOLD signals strongly depends on the local artifact situation.

The example above demonstrates that methods for controlling artifacts (e.g., improved shimming and field map corrections, see Section 11.1) are crucial when conducting UHF measurements in clinical applications. Such an approach of motion artifact correction was recently presented by our group (Robinson et al., 2013). fMRI data of neurological patients performing hand and chin motor tasks were analyzed using the classic general linear model (GLM) approach and ICA. While the GLM results were strongly contaminated by motion artifacts, ICA revealed activations more concordant

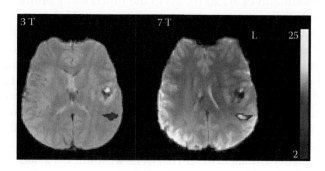

FIGURE 11.11
Activation in Wernicke's area in a representative patient at 3 and 7 T.

with known primary motor areas that were nearly completely free of such artifacts.

The clinical importance of high sensitivity for weak functional activations also arises in the assessment and monitoring of neuroplastic changes as a result of cortical (e.g., tumor and stroke), subcortical (e.g., Parkinson's disease, brain stem or cerebellar lesions), or peripheral nerve damage. The latter was investigated by Beisteiner et al. (2011a) in patients suffering from a complete brachial plexus lesion who underwent a nerve reconstruction of the lesioned musculocutaneous nerve (innervating the biceps) via new inputs from the intact phrenic nerve (end to side connection). Cortical diaphragm activations are weak, and in this study, the increased functional 7 T sensitivity was central to demonstrating that the cortical diaphragm area was independently active for movement of the affected arm and forced respiration after surgery. Thus, the healthy diaphragm area was able to add a new—phylogenetically not designed—function (moving the arm) to its existing dedication (moving the diaphragm for breathing). Detection of such new types of neuroplasticity depends on the ability to monitor small brain activation changes. Thus, high functional sensitivity provided by 7 T is of particular relevance for improved and early detection of neuronal reorganization.

11.2.4.5 High Spatial Resolution fMRI and Possible Clinical Applications

A higher SNR achieved with UHF strength allows spatial resolution to be increased. HR images (voxel side length ≤1.5 mm) have the benefit that the proportion of genuine gray matter voxels increases. This increase in the number of tissue-specific voxels is due to a reduced partial volume effect, which is of particular importance when investigating small brain structures such as subcortical nuclei or intracortical structures.

Studies of healthy participants impressively demonstrate the possibilities of HR fMRI in the primary cortical areas and in subcortical regions. For the visual cortex, HR 7 T fMRI improved retinotopic mapping (Hoffmann et al., 2009), led to the functional identification of orientation columns (Swisher et al., 2010; Yacoub et al., 2008) and of ocular dominance columns (Yacoub et al., 2007), and has enabled functional mapping of the cortical layer structure (De Martino et al., 2013b; Koopmans et al., 2011; Olman et al., 2012; Polimeni et al., 2010; Siero et al., 2011). Within the relative small cortical auditory system, HR functional imaging revealed tonotopic maps (Da Costa et al., 2011) and led to the identification of regions sensitive to the spatial origin of sound (Van der Zwaag et al., 2011). HR fMRI studies also led to a better understanding of the functional organization of primary sensorimotor cortices confirming early reports

(Beisteiner et al., 2001) of somatotopy of the finger representation within the primary sensorimotor cortex (Besle et al., 2013; Kuehn et al., 2014; Martuzzi et al., 2014; Sanchez-Panchuelo et al., 2010, 2013; Stringer et al., 2011).

In subcortical areas, HR fMRI revealed tonotopic maps in the inferior colliculus (De Martino et al., 2013a) and a functional digit somatotopy in the cerebellum (Van der Zwaag et al., 2013). Furthermore, the activation of small subcortical structures such as the mediodorsal and intralaminar thalamic nuclei (Metzger et al., 2010) and subregions of the periaqueductal gray (Satpute et al., 2013) was demonstrated using HR functional imaging at 7 T.

Morphological studies have also shown the clinical benefit of HR 7 T imaging, for example, for quantifying hippocampus atrophy in patients with Alzheimer's disease (Kerchner et al., 2010) or healthy subjects with high risk for developing Alzheimer's disease (Kerchner et al., 2014). UHF has also been used for detecting cerebral microbleeds (Theysohn et al., 2011) and cavernomas (Schlamann et al., 2010).

Despite promising results of HR fMRI in healthy subjects and morphological HR imaging in neurological patients, HR functional imaging has not been applied for clinical populations yet. Evidently, HR fMRI possesses considerable potential to characterize abnormal functioning of very small brain structures. Highly focal pathology often occurs at the very beginning of a disease (Kerchner et al., 2014), and since disturbed function typically precedes disturbed morphology, the capability for small structure fMRI may soon open new diagnostic possibilities.

11.2.5 Susceptibility-Weighted Imaging

11.2.5.1 Methodological Aspects: Image Acquisition, Reconstruction, Processing, and Generation of SWI

Susceptibility-weighted imaging (SWI) is an imaging technique first introduced by Reichenbach and Haacke as *MR BOLD venography* (1997). It was subsequently renamed SWI as it became clear that the potential of the technique reached beyond the depiction of veins. SWI are generated from both magnitude and phase data. The MR signal phase, the inherent counterpart of the signal magnitude in all MR acquisitions, reflects the underlying magnetic susceptibility, making it sensitive to deoxyhemoglobin in venous vessels, as well as other sources of iron.

Higher SNR and enhanced susceptibility effects at higher field strength (Barth et al., 2003) have led to wide clinical application of phase imaging (Rauscher et al., 2005), SWI, and burgeoning-related imaging methods such as quantitative susceptibility mapping (QSM) (Liu et al., 2009; Schweser et al., 2011; Shmueli et al., 2009; Wharton et al., 2010) and the estimation of tissue

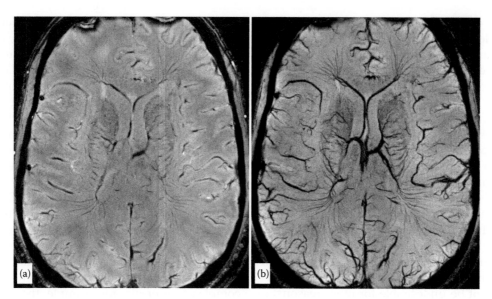

FIGURE 11.12

SWI example: (a) the original SWI image (resolution 0.23 × 0.23 × 1.2 mm) and (b) an mIP over 7 slices (8.4 mm).

conductivity (Voigt et al., 2011). Figure 11.12 shows an example of HR SWI at 7 T and a minimum intensity projection (mIP or minIP) SWI, sometimes referred to as HRBV (HR BOLD venography) image (Lee et al., 1999).

This section reviews suitable acquisition parameters for SWI at 7 T and assesses the many new and evolving approaches to generating high-quality SWI, particularly how to combine phase images from a number of RF coils (in the absence of a body coil) and how to remove phase wraps and create HR SWI of optimum quality. We conclude by looking in more detail at the sources of contrast in SWI and a number of clinical applications.

SWI is based on a transverse, 3D, fully flow-compensated T_2*-weighted gradient-echo acquisition. A relatively long echo time (T_e) is used to generate T_2*-weighted contrast. For 7 T systems, it has been shown that a T_e of 14 ms gives an optimized contrast between veins and the surrounding tissue and that vessel visibility can be optimized by using a slice thickness that is at least twice the in-plane voxel size (Deistung et al., 2008). In our clinical research, we use a resolution of 0.3 × 0.3 × 1.2 mm; other parameters are listed in full in Grabner et al. (2012).

11.2.5.2 SWI Calculation Step 1: Combining Phase Data from a Number of RF Coils

Complex data from multiple coils are optimally combined by weighting coil values at each voxel by the complex sensitivity of the relevant coil at that voxel (Bydder et al., 2002; Roemer et al., 1990). Accurate maps of coil sensitivities are required and are ideally generated by

division of complex data acquired with each coil element by complex data acquired with a volume reference coil (Pruessmann et al., 1999). Some high field systems, and most very high field systems, do not have integrated body RF coils or alternative volume coils. A simple solution in such cases is to take advantage of the fact that *nuisance* sources of phase (macroscopic deviations from B_0 and receiver-dependent contributions) vary slowly over the object, whereas *interesting* sources of phase (veins and iron-rich structures such as the basal ganglia) are comparatively small. Homodyne filtering of the complex data (Noll et al., 1991) therefore provides a means to remove nuisance sources, allowing data from multiple RF coils to be combined. It also removed many wraps in the process. At very high field, however, the RF wavelength is shorter (circa 10 cm at 7 T compared to circa 30 cm at 3 T) and macroscopic deviations from B_0 are also more severe, making it hard to choose a filter that both removes these sources and preserves phase information in relatively large structures such as the putamen, which is several centimeters in extent. To calculate SWI of the best possible quality, the problems of combining and unwrapping phase data are currently best tackled separately, and phase data are processed independently of the magnitude.

The magnitude signals from a number of RF coils can be combined in a weighted sum (where the magnitude itself is used for the weighting), giving rise to the sum of squares reconstruction (Larsson et al., 2003). The MR signal phase is more challenging. The phase from each RF coil is subject to a different, time-independent *phase offset*—the phase distribution over the object at time zero. This contribution needs to be identified and removed in order to be able to combine the signals. We briefly present a number of approaches to achieving this both for the case where there is reference data available from a volume coil and for the case where no such data are available.

The phase offset is dominated by the path length of the MR signal from a particular location in the object to the receiver coil in question given a finite MR wavelength (Brunner et al., 2009; Robinson et al., 2011). If measurements are made at two echo times, the phase offset of each channel can be determined explicitly (Robinson et al., 2011). Removing this contribution *matches* the phase images throughout the object, allowing them to be combined. While this leads to a near-perfect solution, it requires either a HR multi-echo SWI (which has SNR advantages over single-echo SWI [Denk and Rauscher, 2009] but may not be undesirable for other reasons) or a separate low resolution dual-echo scan, and entails phase unwrapping.

A recently proposed method has demonstrated that phase data can be reconstructed by setting the phase to zero at a point in the image (e.g., the center) in all channels to provide approximate matching (Hammond et al., 2008), sufficient to calculate a combined *virtual reference coil* (VRC) image, then calculating the difference between each RF coil of the VRC image before finally subtracting this difference, matching the phase images throughout the image and allowing the data to be combined without loss or interference (Parker et al., 2013). This approach has the advantage that it can be applied to single-echo data and no reference scan required.

11.2.5.3 SWI Calculation Step 2: Phase Unwrapping

The fact that the measured phase of the MR signal can only be defined over a full circle (360° or 2π radians) means that values of the true phase, which are outside this range, become aliased. For instance, all of the phase values −330°, 30°, and 390° all yield the measured value 30°. Unwrapping is the process of identifying the number of wraps that have occurred so that the true phase may be restored.

Unwrapping methods can be classified as either *spatial* or *temporal*. Spatial methods are image based and use some regional characteristics of phase or complex values at a single-echo time—in most cases abrupt change—to identify wraps. Temporal methods, on the other hand, are based on identifying wraps via the evolution of the phase in each voxel at two or more echo times. Spatial methods usually employ region-growing algorithms that begin in a region of stable phase (as defined, e.g., by local phase coherence [Xu and Cumming, 1999]) and proceed either in 2D or 3D, identifying a wrap when the phase difference between adjacent voxels exceeds

π. Spatial methods, such as PRELUDE (Jenkinson, 2003) and PHUN (Witoszynskyj et al., 2009), are capable of resolving highly wrapped images from a single-echo time. The main disadvantages of spatial approaches are that they tend to be computationally intensive and, in most implementations, need to distinguish the object (which should be unwrapped) from the background (which should not be unwrapped). Alternative fast, deterministic Fourier-based spatial unwrapping methods (Bagher-Ebadian et al., 2008) have recently been adopted from the field of optics (Volkov and Zhu, 2003) and are particularly being applied in QSM.

Temporal methods require phase images acquired at two or more echo times. Where the phase evolution between two measurements is between $+\pi$ and $-\pi$, a wrap-free image can be created by calculating the difference between the two images from the angle of the complex sum or via, for example, the Hermitian inner product method (Scharnhorst, 2001). This is a simple and robust approach. At 7 T, however, echo spacings would have to be unrealizably short (circa 1 ms) in order to avoid wraps developing in the inter-echo period. While using such a short-echo spacing has been demonstrated to be feasible for SWI at 3 T (Feng et al., 2012), dB/dt limitations mean that this cannot be achieved in a HR multi-echo acquisition at 7 T or higher field strengths. This limitation has been overcome by using a triple-echo acquisition with unevenly spaced echoes in two methods presented very recently by Dagher et al. (2013) and Robinson et al. (2013).

Once phase data have been combined and unwrapped in this way, it is high-pass filtered and then rescaled so that negative phase values all have a value of 1, and positive phase values decrease linearly from 1 to 0, raised to a power (typically between 3 and 6, where a higher power provides more vessel weighting) then multiplied by the magnitude. It is then common to calculate an mIP over a number of slices. Both the mIP and the original SWI will generally be of interest to the radiologist.

11.2.5.4 Information in SWI

SWI is sensitive to paramagnetic substances, such as deoxygenated blood, blood products, and iron deposits, and has therefore found application in many different fields (Hingwala et al., 2010; Ong and Stuckey, 2011; Robinson and Bhuta, 2010), such as MR venographic methods (Reichenbach and Haacke, 2001), arteriovenous malformations (Essig et al., 1999), occult venous disease (Lee et al., 1999), MS (Grabner et al., 2011; Tan et al., 2000), tumors (Barth et al., 2003; Schad, 2001), stroke (Hermier et al., 2003, 2005), and functional brain imaging (Essig et al., 1999).

SWI and phase imaging have been proposed as methods to image iron stores in the human brain *in vivo*

(Haacke et al., 2005). Ogg et al. (1999) demonstrated a positive correlation between phase values and age-related changes in regional brain iron stores. Haacke et al. (2005) showed that the SWI phase is proportional to the iron concentration in brain tissue and presented average phase behavior over 75 subjects for regions known to accumulate iron (Haacke et al., 2007). Xu et al. (2008b) showed age-related iron accumulation in the putamen and the red nucleus also using phase information.

Both the transverse relaxation rate R2 and the effective transverse relaxation rate R2* have been shown to correlate with brain iron concentration (Langkammer et al., 2010). A technique referred to as field-dependent correlation imaging (FDRI) analyzes the enhancement of R2 due to iron at two different magnetic field strengths to quantify tissue iron stores (Bartzokis et al., 1993). FDRI is of limited use in clinical studies as the subject needs to be scanned at two different field strengths.

SWI phase reflects the magnetic susceptibility of tissues only indirectly as it is the convolution of the underlying susceptibility with the typical pattern of a (orientation dependent) magnetic dipole that determines the phase (Pathak et al., 2008; Schäfer et al., 2008, 2009). Intricate geometrical shapes, for instance, lead to very complex spatial phase patterns (Moser et al., 2010; Schäfer et al., 2009). To overcome the non-local relationship between SWI phase and the underlying magnetic susceptibility distribution, a relatively new method known as QSM has been introduced (Schweser et al., 2011; Shmueli et al., 2009; Wharton and Bowtell, 2010). The computation of QSM maps of the brain is currently based on a complex post-processing framework that includes post-processing of phase images (Schofield and Zhu, 2003; Schweser et al., 2011), dipole inversion (Schweser et al., 2012), and the computation of quantitative susceptibility information from relative values (Schweser et al., 2011) with white matter as reference tissue (Deistung et al., 2013). QSM provides quantitative analysis (Langkammer et al., 2012) as well as excellent anatomical delineation of cortical and deep gray matter structures (Deistung et al., 2013; Schweser et al., 2011, 2012).

11.2.5.5 Clinical Applications

SWI is sensitive to paramagnetic structures such as iron deposits, bleeding, and veins, making its potential applications far reaching. The characterization of cerebral tumors is an area which benefits from the SWI contrast mechanism. Tumors' characterization is to a certain degree reliant on understanding angiogenesis and micro-hemorrhage, which is particularly important for infiltrating gliomas in order to optimize postoperative treatment. Additionally, low-grade gliomas (WHO grade I and II) typically show a progression to high-grade gliomas (WHO grade III and IV), where the formation of pathological microvessels is a marker for such a malignant transformation. Another example of the use of SWI in tumor imaging is the monitoring of therapy response in gliomas (Grabner et al., 2012). Figure 11.13 gives a follow-up example of a male patient during therapy over four weeks with a two-week interval.

In addition to tumor imaging, UHF SWI has been used to visualize microbleeds in vascular dementia (Theysohn et al., 2011) and venous vessels (Grabner et al., 2011) as well as possible pathological iron content increases (Hammond et al., 2008) in MS lesions. Further clinical UHF SWI research fields are stroke (Chakeres et al., 2002), cerebral venous diseases, breast imaging, and neurodegenerative diseases, which tend to accumulate iron such as Alzheimer's.

11.2.6 Sodium MRI

Sodium is involved in many essential cellular processes such as ion homeostasis and osmotic balance (Somjen, 2004). A high concentration gradient maintained in healthy tissue between intracellular (ISC: 10–15 mmol/L) and extracellular sodium concentration (ESC: ~145 mmol/L) (Robinson and Flashner, 1979) is used to control other important processes such as cellular proliferation (Koch and Leffert, 1979) and cellular energy metabolism (Skou, 1957). Owing to the importance of sodium in these critical biochemical processes, noninvasive monitoring of sodium homeostasis via sodium MRI has been used for the evaluation of tissue viability and treatment response in many studies.

Sodium MRI is a promising noninvasive method but a challenging one. Due to the significantly lower tissue concentration and MR sensitivity, the *in vivo* sodium MR signal is ~22,000 times smaller than that of conventional proton MRI. In order to achieve adequate SNR, sodium MRI requires longer measurement times (15–40 min) and results in low-resolution images (2–10 mm). This disadvantage of sodium MRI is reduced by using an UHF 7 T MR systems, which can provide 2.2–2.7 times higher SNR compared to 3 T systems (Fleysher et al., 2009; Qian et al., 2012b) (Figure 11.14). The challenges of proton MRI at 7 T MR systems (such as RF power deposition, susceptibility artifacts, or RF coils homogeneity) are much less pronounced in sodium MRI, which only benefits from 7 T (or higher) field strength. Additional SNR increase can be achieved by employing dedicated sodium coils. By using an array head coil, Qian et al. obtained uniform image sensitivity and two-times higher SNR compared to that from a volume coil (Qian et al., 2012a). Sodium SNR can be further increased by using sequences with sampling density-weighted apodization (Konstandin and Nagel, 2013). However, this

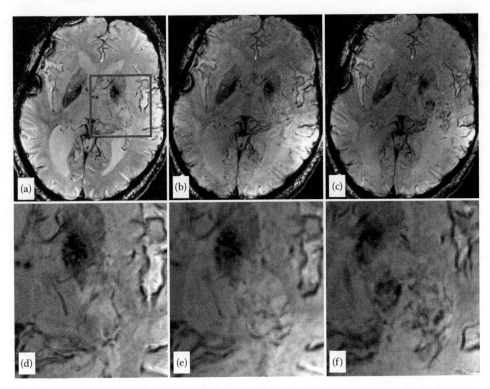

FIGURE 11.13
Follow-up study of a tumor patient under therapy: (a–c) show the SWI image at baseline, after two and after four weeks, respectively. (d–f) represent the area marked with the red rectangle in (a) for the individual time points. Note the continuous increase in irregular intratumoral SWI hypointensities, most likely corresponding to growing tumor microvasculature.

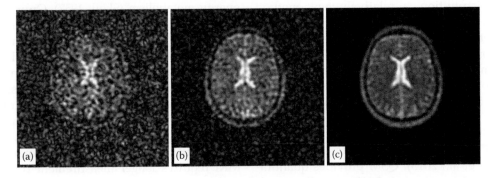

FIGURE 11.14
Transverse sodium MR images of a human brain from (a) 1.5, (b) 3, and (c) 7 T demonstrate the increase of SNR with increasing magnetic field strength. All images were acquired using the same 3D density-adapted projection reconstruction sequence with the same parameters: a nominal resolution of $4 \times 4 \times 4$ mm^3, 13,000 projections, T_e of 0.2 ms (@ 1.5 and 3 T) and 0.5 ms (@ 7 T), T_R of 50 ms, flip angle of 77°, acquisition time of 10:50 min.

method is beneficial only with short readout times (below 7 ms).

Sodium in brain parenchyma is characterized by biexponential relaxation with very short T_1 (~37 ms) and T_2 (fast ~3 ms, slow component ~42 ms) relaxation times (Fleysher et al., 2013; Nagel et al., 2011). While short T_1 is beneficial (allowing faster averaging and improving SNR), short T_2 leads to rapid MR signal loss. To minimize this loss, many sodium neuroimaging studies at 7 T have used non-Cartesian MR techniques such as

acquisition-weighted stack of spirals (Qian et al., 2012b), twisted projection imaging (TPI) (Qian et al., 2012a), or 3D radial (Nagel et al., 2011), which enable acquisition with ultra-short echo time (UTE).

Evaluation of tissue sodium concentration in brain with MRI has shown very promising results. When compared to healthy tissue, increased tissue sodium concentration was found in brain tumors (Fiege et al., 2013), in areas completely infarcted after stroke (Tsang et al., 2011), in demyelinating MS lesions (Zaaraoui et al., 2012), and in patients with Huntington's disease (Reetz et al., 2012). However, most of these studies were performed at lower field (4.7 T or below). Tissue sodium concentration represents a volume-weighted average of sodium in the intra- and extracellular space. Since the evaluation of tissue sodium concentration cannot differentiate between early cellular changes reflected by increased ISC and later changes in cellular volume fraction, a number of methods have been developed to selectively evaluate ISC. Fleysher et al. demonstrated noninvasive *in vivo* quantification of ISC and intracellular sodium volume fraction in healthy human brain by measuring triple-quantum filtered (TQF) and conventional sodium imaging at 7 T (Fleysher et al., 2013). Unfortunately, the signal in TQF MRI is approximately 10 times lower than that of conventional sodium MRI. Nagel et al. employed the IR and double-readout methods for suppression of sodium signal from predominantly extracellular compartments in the brain (Nagel et al., 2011). In contrast to conventional sodium MRI, these methods were more sensitive to ISC, enabled further differentiation between WHO grade I–III and WHO grade IV gliomas, and showed correlation with proliferation rate of tumor cells. Benkhedah et al. modified the quantum-filtered approach and introduced a sequence for biexponentially weighted sodium imaging with up to 200% higher SNR compared to the TQF method (Benkhedah et al., 2012). The same group has recently improved the biexponentially weighted approach by using only two instead of three RF pulses, reducing the minimum repetition time and increasing SNR by ~20% compared to the three-pulse method (Benkhedah et al., 2014). Novel sodium imaging methods at 7 T MR system are opening many possibilities for functional metabolic studies in healthy and diseased brains at 7 T.

11.2.7 Neuro Magnetic Resonance Spectroscopy

11.2.7.1 Introduction

While magnetic resonance spectroscopy (MRS) of animals such as rats at 7 T can be dated back to the 1990s, the feasibility of *in vivo* MRS in the human brain at 7 T was first shown in 2001 (Tkáč et al., 2001). Since then, despite considerable research, only few clinical applications were published until 2010, with numbers only increasing in the last few years. To understand this, it is necessary to discuss the benefits and especially limitations of spectroscopy at UHF strengths (Moser et al., 2012; Posse et al., 2013).

The two main advantages of increased field strength for MR spectroscopy are increases in signal as well as in chemical shift dispersion proportional to B_0. In theory, this allows more metabolites to be detected due to better separation between nearby resonances, lowered

j-coupling, and increased SNR. Additionally, the higher spectral shift between lipids and metabolites allows easier fat saturation, but also increases chemical shift displacement errors (CSDE) and therefore metabolite misregistration and lipid contamination.

As the SAR is proportional to B_0^2, pulse design is constrained, especially for fat and water suppression. Lower T_2 values lead to signal loss, especially for long T_es. Increased B_1 inhomogeneities lead to inconsistent flip angles that also hinder inversion and suppression techniques. Stronger susceptibility effects increase the importance of B_0 homogeneity, but obtaining good shim values for larger volumes requires better shim techniques or hardware.

Non-proton MRS, principally ^{31}P-spectroscopy, at lower B_0 suffers from low sensitivity and benefits greatly from the SNR increase at 7 T, based partly on the inherent SNR increase and partly on the lowered T_1 values, while not being affected by water and lipid contamination. Additionally, the increased excitation wavelengths in comparison to proton MRS reduce coil design constraints.

Coping with all these limitations requires sophisticated protocols that may include additional scans and pulses or extra hardware to facilitate improved shimming and suppression schemes. Extensive post-processing may be necessary while general MRS problems like localization and the balancing of SNR and measurement times persist. Nonetheless, comparison of ^{1}H MRS in the brain between 3 and 7 T has confirmed improved SNR, quantification, and spectral resolution at 7 T (Stephenson et al., 2011).

The next section will give an overview of these approaches to improved 7 T MRS in the last few years.

11.2.7.2 MRS Methods

A variety of acquisition and excitation strategies for 7 T MRS have been proposed. These include the optimization of the spin-echo full-intensity acquired localized (SPECIAL) spectroscopy with TEs less than 6 ms for 7 T (Mekle et al., 2009); time of echo-optimized point-resolved spectroscopy (PRESS) to detect glutamate (Glu), glutamine (Gln), and glutathione (GSH) (Choi et al., 2010a); and a semi-localized by adiabatic selective refocusing (semi-LASER) localization at short TEs (Boer et al., 2011b). Further, B_1^+-insensitive composite pulses to improve flip angle homogeneity for thin slices (Moore et al., 2012) and a combination of SPECIAL and semi-LASER MRS (Fuchs et al., 2013) have been proposed. Optimal PRESS timings for the simultaneous detection of Glu and Gln at 7 T were presented by Snyder and Wilman (2010).

For the measurement of GABA, Mescher–Garwood (MEGA)-edited semi-LASER MRS (Andreychenko et al., 2012; Arteaga de Castro et al., 2013) have been adapted.

PRESS sequences have been proposed for the measurement of glycine (Choi et al., 2009b) and serine (Choi et al., 2009a) as well as MEGA editing to detect NAAG in the frontal brain (Choi et al., 2010b).

The use of the generalized least squares algorithm for the combination of multichannel MRS data to reduce control values has been proposed (An et al., 2013). Unconventional approaches at MRS have also been introduced with the use of fast Padé transforms (Belkić and Belkić, 2006), the implementation of 2D localized correlated spectroscopy (L-COSY) (Verma et al., 2013), and a single-shot Carr–Purcell–Meiboom–Gill sequence to map T_2 values of singlet resonances (Ronen et al., 2013b).

11.2.7.3 MRSI Methods

Magnetic resonance spectroscopic imaging (MRSI) combines MRS and MRI approaches to obtain spectral information of multiple volumes with a single measurement, with the most common drawbacks being long measurement times and contamination due to field inhomogeneities, CSDE, and incomplete outer volume suppression (OVS). Improvements in sequence and pulse design are required to enable 7 T brain MRSI.

The use of semi-LASER localization (Scheenen et al., 2008) and spectroscopic missing pulse steady-state free precession (Schuster et al., 2008) showed high-quality spectra in small voxels for 7 T MRSI. MRSI PRESS sequences using adiabatic spatial-spectral pulses for B_1 insensitivity were proposed to achieve a larger region of interest (Balchandani et al., 2008; Xu et al., 2008a). Adiabatic fast passage refocusing pulses improve sensitivity and are insensitive to transmit B_1 inhomogeneity (Zhu et al., 2013).

Double-echo J-refocused coherence transfer was proposed to minimize macromolecule signals and J modulation at long TEs and used for the detection of NAA, Glu, and Gln (Pan et al., 2010). As further developments, selective homonuclear polarization transfer spectroscopic imaging was used for the imaging of GABA (Pan et al., 2013b).

Due to the shortened T_2 relaxation times at 7 T, the shortest possible TEs are desired to minimize SNR losses (Avdievich et al., 2009). Broadband or frequency modulated pulses are necessary to overcome the increased CSDE problems. In both cases, the pulse durations are prolonged, leading to unfavorably long TEs. To resolve this contradiction, the direct acquisition of the free induction decay (FID) was proposed, eliminating the CSDE-problematic refocusing pulses and enabling very short acquisition delays (Henning et al., 2009). Lipid suppression was performed by extensive amounts of OVS slabs, causing a very long T_R of 5 s (Henning et al., 2009). Instead, Boer et al. used an optimized B_0-shimming routine, enabling them to use frequency-selective lipid suppression. With that, they could reduce the T_R to 1 s (Boer et al., 2011a). Bogner et al. showed that using a HR matrix size of 64 × 64 in combination with spatial Hamming filtering also leads to negligible lipid artifacts (Bogner et al., 2012).

New B_1-insensitive fat suppression pulses that do not degrade relevant metabolite signals have been proposed, such as spectrally selective adiabatic pulses for volumetric MRSI and spatial-spectral adiabatic pulses for multi-slice MRSI (Balchandani and Spielman, 2008). Using a ring-shaped B_1^+-excitation, OVS of extracranial resonances could be facilitated with negligible effect on the intracranial metabolites (Hetherington, 2011). A combination of three dual-band pulses and eight OVS pulses was proposed for water and lipid suppression, with high-bandwidth frequency-modulated slice-selective pulses to minimize CSDE (Zhu et al., 2013).

The problem with increased B_0-inhomogeneities has been addressed in several approaches. Snaar et al. showed in 2011 that the use of pads with a high dielectric constant can also reduce the B_0-inhomogeneities of the temporal lobe (Snaar et al., 2011). Pan et al. compared the effects of using first and second versus first–third and first–fourth-order shim terms and found a substantial decrease in the global and local B_0-inhomogeneities when using higher shim terms (Pan et al., 2012). Boer et al. addressed the problem by dynamically shimming each slice separately in a multi-slice MRSI measurement, resulting in B_0-inhomogeneities comparable to a first–fourth shim system (Boer et al., 2012a). Hetherington et al. used a first–third conventional shim system in combination with an RF shim coil and could reliably quantify NAA, Cho, and Cr in the hippocampus (Hetherington et al., 2014). Duerst et al. showed the high potential of dynamic shimming based on the results of a B_0-field camera (Duerst et al., 2014). The only software-based shim advance was proposed by Fillmer et al., using a shim region of less interest adjacent to the region of interest to reduce artifacts, for example, lipid-contaminated voxels by the neighboring skull (Fillmer et al., 2014a).

With the recent advent of parallel transmit hardware, B_1^+ shimming has become feasible within the last few years. Boer et al. performed B_1^+ shimming for each slice of the multi-slice measurement separately, as well as a global B_1^+ shim for the water suppression and a ring-like B_1^+-shim for suppressing the lipids using eight transmit channels (Boer et al., 2012a). Emir et al. used 16 transmit channels for B_1^+-shimming, enabling quantification of GABA, Glu, and GSH in single voxels in deep brain regions such as putamen, substantia nigra, and pons (Emir et al., 2012a).

One problem of MRS in general is its low sensitivity. Using strong B_0 fields inherently increases sensitivity.

Nevertheless, SNR can be improved further with the use of array coils with optimized signal combination of the individual channels. In contrary to conventional MR imaging, the phase of the signal has to be taken into account in order to achieve the best results possible. Strasser et al. proposed measuring a short imaging sequence with similar parameters to the MRSI sequence to allowing phase matching and weighting of spectra from the individual RF coils without the need for a reference coil (Strasser et al., 2013). In comparison to the standard method proposed by Brown et al. (2004), their proposed method showed an SNR increase of 29%. Hall et al. proposed weighting the signal of the individual channels with the signal/noise^2, rather than with the signal/noise or the signal itself, resulting in an SNR increase of 1.9% and 8%, respectively (Hall et al., 2014).

The SNR gain and line width decrease for proton echo-planar spectroscopic imaging (PEPSI) at 7 T have matched predictions (Otazo et al., 2006).

The increased SNR provided by 7 T can be traded for shorter measurement times: Zhu et al. (2013) used a SENSE acceleration of 2×2 to reduce the measurement time from ~50 to 12 min while still maintaining a good SNR and spectral quality. Kirchner et al. (2014) showed the feasibility of a SENSE acceleration up to 3×3 with a matrix size of 20×16 while optimizing the spatial response function to reduce voxel bleeding.

11.2.7.4 P MRS, diffMRS, fMRS, Quantification

Initial experiences with ^{31}P MRS at 7 T were first presented in 2003 (Lei et al., 2003), and later increased sensitivity and spectral resolution were shown (Qiao et al., 2006). To study reaction fluxes of the PCr-ATP-Pi network in the occipital lobe, a ^{31}P MRS sequence using multiple single-site magnetization transfer was proposed (Du et al., 2007). An improved image-selected-*in vivo*-spectroscopy (ISIS) sequence was adapted to 7 T and the SNR increase over 3 T verified (Bogner et al., 2011). The application of gradient modulated GOIA-WURST pulses to ^{31}P MRSI reduced CSDE while maintaining clinically acceptable measurement times (Chmelik et al., 2013a). Two interesting, but very specialized improvements were shown in the field of diffusion-weighted MRS and functional MRS. A combination of MRSI and diffusion weighting uses fiber tracking to differentiate voxels of the cingulum bundles from the surrounding tissue (Mandl et al., 2012). Measurement of T_2^* BOLD signal changes from the unsuppressed water peak was proposed as an application of functional MRS (Koush et al., 2011, 2013).

In general, quantification precision in MRS is expected to improve substantially with B_0 up to 11.7 T, with higher gains for J-coupled metabolites (Deelchand et al., 2013). Using an internal water reference, it was shown that

quantification for 7 T MRS needs half as much SNR as at 4 T to obtain the same precision (Tkáč et al., 2009). Spatial artifacts related to the limited RF pulse bandwidth may be accounted for in quantification by prior knowledge gained from density matrix simulations (Kaiser et al., 2008).

11.2.7.5 Research Applications

Methods to overcome MRS limitations at 7 T have resulted in some preclinical studies offering insights into human brain metabolism. The mapping of regional differences in metabolite distribution is the focus of several publications. Measurements in the frontal and parietal white matter and the insular, thalamic, and occipital gray matter showed regional differences for NAA, Cr, and Cho (Grams et al., 2011). Gly concentrations in white matter and gray matter in the frontal and occipital lobe were found to be 0.1 and 1.1 mM using segmented PRESS-MRS (Banerjee et al., 2012). Using B_1^+-shimming, concentrations of 16 metabolites and T_2 relaxation times were quantified within the occipital lobe, motor cortex, basal ganglia, and cerebellum (Marjańska et al., 2012). Macromolecule signals in the occipital lobe were found to be lower for white matter than for gray matter (Schaller et al., 2013b). An age-related change of Glu in the medial frontal cortex by 0.33 mM/year was found by MRS in healthy volunteers aged 18–31 (Marsman et al., 2013). A reproducibility study of STEAM and MEGA-PRESS found mean coefficients of variation in the range 10%–20% (Wijtenburg et al., 2013).

Spectroscopy has been utilized to show metabolic differences caused by diseases (mostly with single-voxel methods at 7 T), with the main research focus on GABA in different brain regions: MRS in the gray matter of manifest Huntington's disease showed lower NAA and Cr concentrations in comparison to controls (van den Bogaard et al., 2011). Lower NAA/Cr ratios in the cerebellum were detected in migraine patients (Zielman et al., 2014). An increase in GABA was found in the pons and putamen in Parkinson's disease (Emir et al., 2012b). GABA increase and Glu in the visual cortex could be successfully measured during gabapentin treatment (Cai et al., 2012). In the measurement of GABA by MRS, the *in vivo* T_2 relaxation time at 7 T was estimated to be 63 ± 19 ms (Intrapiromkul et al., 2013). Further, the thalamic GABA/NAA ratio was found to increase for well-controlled epilepsy and decrease for poorly controlled epilepsy (Pan et al., 2013b). Investigation of GABA, Glu, and Gln in different subregions of the anterior cingulate cortex showed different relative concentration patterns (Dou et al., 2013). GABA measurements in the hypothalamus and occipital cortex showed a trend of GABA decrease in the hypothalamus during hypoglycemia (Moheet et al., 2014).

Applications of multi-voxel MRS (MRSI) have been limited so far at 7 T, but show high potential. For example, MRSI in adult X-linked adrenoleukodystrophy has shown changes in metabolic ratios such as lower NAA/Cr for different phenotypes compared to controls (Ratai et al., 2008). Spectral edited MRSI has reported significantly higher GSH in gray matter relative to white matter in controls and a significant decrease in GSH in gray matter and white matter lesions in MS patients (Srinivasan et al., 2010). Using third-order B_0 shimming, RF-shimming, and OVS in MRSI, a decrease in NAA/Cho and NAA/Cr in the hippocampi of mild traumatic brain injury victims could be detected (Hetherington et al., 2014). Abnormalities in NAA/Cr ratios have been successfully correlated with the outcome of surgical resection in epilepsy (Pan et al., 2013a). Preliminary MRSI results show also great promise for the investigations of brain tumors via HR metabolic mapping (e.g., Cho and Cho/NAA maps of a low-grade glioma obtained at 7 T—Figure 11.15).

Due to significantly improved sensitivity, MRS sequences at 7 T have also been applied to observe functional metabolic changes or metabolic diffusion weighting that is inherently limited to lower B_0: Functional MRS in the visual cortex has been shown to detect concentration changes of ± 0.2 μmol/g for most metabolites—the same order of magnitude as the variations due to stimulations (Mangia et al., 2006). Metabolic changes in the visual and motor cortex—predominantly increases in lactate and Glu—have been shown by several groups (Lin et al., 2012; Mangia et al., 2007; Schaller et al., 2013a, 2014). Functional diffusion-weighted MRS in the visual cortex has shown significant changes under stimulation in the apparent diffusion coefficients of NAA, Cr, and Cho (Branzoli et al., 2013). Furthermore, the application of diffusion-weighted PRESS MRS has been used to measure axonal and glial microstructures (Ronen et al., 2013a) and apparent diffusion coefficients for several metabolites, showing significant differences between gray matter and white matter (Kan et al., 2012).

In summary, MRS and MRSI of the brain at 7 T are used successfully in research applications but require further methodological development to unlock their full potential. In particular, the improvements in spatial resolution or detection of low-concentration metabolites that are otherwise difficult to quantify will open the window for research applications beyond those currently being investigated at lower field strengths. While preliminary clinical applications are scarce, as few developed methods overcome all technical limitations at once, we can expect to see an increase in the number of patient studies in the near future. MRS and MRSI at 7 T have great potential in the characterization of multiple sklerosis, tumors, Alzheimer's disease, and psychiatric disorders.

11.3 MSK Imaging

11.3.1 Methodology

11.3.1.1 Imaging Sequences

UHF imaging at 7 T allows significant increase in SNR and resolution (Ugurbil et al., 2003). In principle, all clinically relevant T_1, proton density, and T_2-weighted spin-echo sequences can be applied at 7 T. However, in order to obtain maximum SNR and CNR and to address potential safety problems, several problems have to be taken into account.

11.3.1.1.1 Pros and Cons of Chemical Shift

The chemical shift difference between water and fat resonance is 1040 Hz at 7 T compared to 440 Hz at 3 T and 220 Hz at 1.5 T, respectively. Therefore, the chemical shift artifact in the frequency-encoded direction is substantially higher at 7 T. This artifact can be reduced by increasing the RF bandwidth. The increase in the

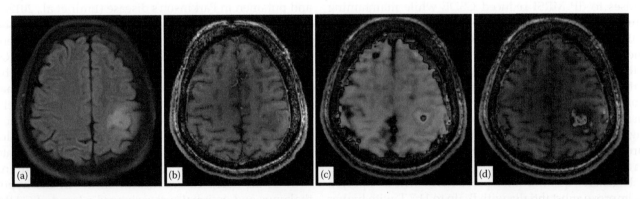

FIGURE 11.15
Glioma grade II of a 52-year-old patient showing T_2-weighted (a) and T_1-weighted (b) MRI, as well as choline maps (c) and ratio maps of Cho/NAA (d). Metabolic maps illustrate the increased cell proliferation and loss of neurons in infiltrated tumor areas.

chemical shift has advantages, however. Fat suppression at 7 T is much easier to achieve and is more robust than at 3 T.

11.3.1.1.2 Longer Repetition Times

At 7 T, the relaxation time T_1 increases in the range of 20%–35% depending on the tissues examined (Regatte and Schweitzer, 2007). Accordingly, T_R has to be increased in the same range to avoid saturation effects. Longer repetition times address—in part at least—the problem of the SAR, which increases with higher field strength and RF power. Longer repetition times spread the applied energy over a longer time.

11.3.1.1.3 Scan Time Acceleration

A spin-echo turbo factor can be used to reduce acquisition time. To compensate the higher energy of the 180° pulses which may easily exceed the exposure limits, an increase in the repetition time and a reduction in the number of slices obtained are necessary. In principle, parallel imaging can be utilized with higher speed reduction factors at higher field strengths, significantly reducing the acquisition time (Wiesinger et al., 2004). However, substantial development is needed to develop safe and robust parallel transmit technology (Moser et al., 2012).

Gradient-echo sequences work at 7 T. As magnetic susceptibility is field strength dependent, differences of precession of different protons in one voxel become more evident resulting in lower T_2^* values. Echo times have to be adapted, that is, shortened accordingly.

11.3.1.2 Susceptibility-Weighted Imaging

In the late 1990s, Reichenbach and Haacke developed a method for depicting the venous vessels of the brain using the influence of deoxygenated hemoglobin on both the phase and the magnitude of T_2^*-weighted gradient-echo scans as the source of contrast (Reichenbach et al., 1997). This imaging of the static BOLD effect evolved to be called *SWI* (Haacke et al., 2004). Higher SNR and enhanced susceptibility effects at higher field strength (Barth et al., 2003) have led to clinical application of the method in the field of neuroimaging, particularly at high static magnetic field strength (3 T) and UHF (7 T).

To our knowledge, the sole example of the use of SWI in MSK at 7 T is the work by Nissi et al. 2013, studying piglets *in situ* and *in vivo* (Nissi et al., 2013). They could demonstrate blood vessels in the growing chondroepiphysis. However, the methodological challenges in creating high-quality SWI images in joints are not solved yet.

In addition to the magnitude, SWI uses the phase of the MR signal. Phase is an inherent property of the MR signal, but one which is omitted in most MR applications (such as conventional anatomical imaging). The processing of phase signals from a number of RFs is problematic. Novel techniques have already been developed and published to overcome these problems in the near future and to generate artifact-free SWI of the growing skeleton (Bagher-Ebadian et al., 2008; Parker et al., 2013; Robinson et al., 2011).

11.3.1.3 UTE/vTE

Imaging and quantitative analysis of rapidly relaxing connective tissues are of great interest also at UHF. Since tissues such as tendon, bone, ligament, or calcified cartilage may have T_2 of 1 ms or even lower, special sequences with UTEs are required. A 3D-UTE pulse sequence using a 3D projection reconstruction acquisition trajectory and spoiling gradients with hard pulse excitation has been used for cortical bone imaging (Krug et al., 2011) and for tendon T_2^* regional variation analysis (Juras et al., 2012b) (Figure 11.16a). This sequence provides isotropic resolution at very short echo times—down to 0.064 ms—albeit with some drawbacks such as stripe and blurring artifacts. This approach benefits substantially from UHF; the higher SNR allows for much more accurate curve fitting despite the multicomponent nature of the decay. Another option is provided by the 3D multi-echo Cartesian Spoiled Gradient Echo (SPGR) technique adapted to use a variable echo time (vTE) in combination with an asymmetric readout only for the first echo (Deligianni et al., 2013) (Figure 11.16b). This technique was successfully used in quantitative analysis of tendons, ligaments, and menisci (Juras et al., 2013a, 2014). The advantage over the radial UTE technique is the higher resolution and reduced artifacts. This Fourier-encoded method also benefits from higher robustness and allows HR imaging within clinically feasible scan times of about 2–3 min.

11.3.1.3.1 Relaxation Rates T_2, T_2^*, T_1

In general, UHF MR provides an increased SNR compared to lower field strengths; however, the translation of this advantage to greater temporal and spatial resolution requires some technical difficulties to be overcome (Chang et al., 2010). Proton T_1 increases with increasing field strength, whereas T_2, on the contrary, decreases (T_2^* even more due to larger susceptibility effects in larger static magnetic fields) (Schmitt et al., 2006). Another technical issue to surmount is the increased chemical shift artifact that results in greater pixel misregistration (Regatte and Schweitzer, 2007). The construction of RF coils for UHF MR is also challenging as B_1 also becomes more inhomogeneous (Vaughan et al., 2001). To acquire images or relaxation constant maps suitable for diagnosis or further evaluation, it is important to adjust sequence parameters from lower field strength protocols

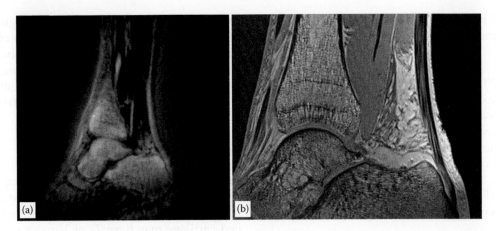

FIGURE 11.16
(a) An example of UTE image calculated as a subtraction of images measured at TE of 0.07 and 9.0 ms; the Achilles tendon is rich in signal due to substantial presence of short T_2 components. (b) The example of vTE image acquired at TE of 0.89 ms with delineated tendon bundles.

appropriately. In the field of MSK, T_1 mapping is most often used for delayed gadolinium-enhanced MRI of cartilage, since it is able to quantify the Gd-DTPA^{2-} concentration which is indirectly proportional to proteoglycan content. IR is widely used (Welsch et al., 2008) as a reference method for T_1 mapping, although there is a tendency to replace this technique with more time effective techniques, such as volumetric interpolated breath-hold examination (VIBE) (Bittersohl et al., 2009) or triple-echo steady state (Heule et al., 2014). Transversal relaxation constant (T_2) can be measured at UHF with various methods, typically multi-echo spin echo (Welsch et al., 2008), fully balanced steady-state free precession (Krug et al., 2007) or double-echo steady state (Welsch et al., 2009), and triple-echo steady state (Heule et al., 2014) with shorter acquisition times. To map T_2^*, especially for rapidly relaxation tissues (tendon, ligaments, and bones), sequences with UTEs must be employed (3D-UTE [Juras et al., 2012b], vTE [Deligianni et al., 2013]).

11.3.1.3.2 Sodium MR Imaging

Glycosaminoglycans (GAG) play a central role in homeostasis of articular cartilage and other connective tissues. Since negative groups of GAGs are in equilibrium with positively charged sodium ions, sodium imaging is a very sensitive method for the detection of changes in GAG content. Sodium MR imaging therefore offers direct and noninvasive evaluation of GAG content in various connective tissues.

Sodium imaging is a challenging method. Low MR sensitivity, short T_2 relaxation times (Madelin et al., 2012), and low sodium concentrations result in a sodium MR signal that is 4,000–5,000 times smaller than the proton MR signal from cartilage. Therefore, sodium imaging requires longer measurement times (10–30 min) and generates images with lower resolution compared to morphological proton MR imaging. Although proton MR imaging with

UHF (7 T and above) MR systems poses various challenges (e.g., RF power deposition, susceptibility artifacts, and homogeneity of RF coils), these problems are much less pronounced in sodium MR imaging. In fact, sodium imaging at UHF MR systems only benefits from higher SNR and, consequently, from higher spatial/temporal resolution in the images (Regatte and Schweitzer, 2007; Staroswiecki et al., 2010). Another SNR gain in sodium images is provided by the dedicated phase-array RF coils (Brown et al., 2013; Moon et al., 2013) and MR sequences with UTE acquisition (Boada et al., 1997; Nagel et al., 2009; Nielles-Vallespin et al., 2007; Qian and Boada, 2008).

Several 7 T studies have used noninvasive sodium MR imaging for the evaluation of pathological changes in GAG content of cartilage, cartilage repair tissue, Achilles tendon, or intervertebral discs.

11.3.1.3.3 gagCEST Imaging

Chemical exchange saturation transfer (CEST) is a method that generates contrast in MR images based on chemical exchange between free water protons and protons of solute molecules. Off-resonant spins are selectively pre-saturated by RF irradiation and then undergo chemical exchange with bulk water protons. This exchange then reduces MR signal in bulk water in CEST imaging (Guivel-Scharen et al., 1998). The most common CEST measurement is the acquisition of multiple sets of images with pre-saturation at different offset frequencies around the water resonance and one reference data set without saturation or with saturation at a very large offset frequency (Van Zijl and Yadav, 2011). The amide and hydroxyl protons of GAGs have exchange properties that can be used for CEST experiments (Ling et al., 2008c). Experiments in bovine cartilage samples have demonstrated that CEST contrast mediated by GAGs (gagCEST) can be used as a biomarker for cartilage GAG content (Ling et al., 2008a, b). The gagCEST imaging benefits from increasing frequency differences at UHF strengths (7 T and above), where the RF spillover is reduced. On the other hand, this method requires good shimming, correction for B_0 homogeneity, and compensation for the patient movement during the scans.

gagCEST has been proposed for the noninvasive investigation of changes in GAG content of various connective tissues. Initial 7 T experiments have shown that gagCEST contrast is sensitive to the GAG content in articular cartilage and in tissue after cartilage repair surgery.

11.3.2 Clinical Applications

11.3.2.1 Cartilage Repair

Cartilage imaging at higher magnetic field strengths is expected to improve the visualization of cartilage pathology, the segmentation of cartilage, and the results of cartilage repair procedures. With decreasing voxel size, partial volume effects can be reduced, which may enable the measurement of cartilage volume and thickness more reliably than with 1.5 and 3 T systems. Krug et al. (2011) employed a voxel size of $312 \times 312 \times 1000$ μ^3 in their 7 T images of the knee joint used for cartilage segmentation, and Regatte and Schweitzer used a voxel size of $254 \times 254 \times 1000$ μ^3 (Regatte and Schweitzer, 2007). Using specially designed coils in smaller joint like the wrist, a resolution of $190 \times 190 \times 500$ μ^3 can be achieved (Raghuraman et al., 2013)

In 2012, Chang et al. discussed the advantages of a new birdcage-transmit, 28-channel receive coil, and a quadrature coil for morphologic MRI using a fat-suppressed 3D-FLASH sequence. The SNR gain was approximately 300%–400% at the periphery of the field

of view and 17% for both femoral and tibial cartilage within the center of the field of view (Chang et al., 2012b).

Juras et al. compared SNR and CNR of optimized clinical sequences for imaging of the ankles with dedicated coils at 3 and 7 T. The mean SNR increase at 7 T compared to 3 T for 3D GRE and 2D TSE was 60.9% and 86.7%, respectively. In contrast, an average SNR decrease of almost 25% was observed in the 2D TSE sequence (Juras et al., 2012a).

The potential advantages of morphological imaging of cartilage repair at 7 T in comparison to lower field strength have not been demonstrated, however, and there is an absence of comparative studies. This is especially true for the studies that use UHF MRI to predict clinical outcome.

11.3.2.2 Osteoarthritis

Based on the general considerations above (morphologic imaging of cartilage repair), an improved visualization of osteoarthritic changes at 7 T is to be expected. However, clinically orientated studies comparing the accuracy in the delineation of osteoarthritic changes at different field strength including 7 T do not exist in the literature.

11.3.2.3 $T_2/T_2{}^*$

Transversal relaxation constants (T_2) and $T_2{}^*$ $(T_2$ plus local field inhomogeneities) are valuable clinical parameters reflecting the interplay between collagen matrix and water molecules. Due to altered collagen content and orientation as well as non-physiological zonal stratification, $T_2/T_2{}^*$ can be used as a quantitative and qualitative marker for cartilage repair maturation (Welsch et al., 2008) (Figure 11.17). At UHF, the absolute $T_2/T_2{}^*$ values in

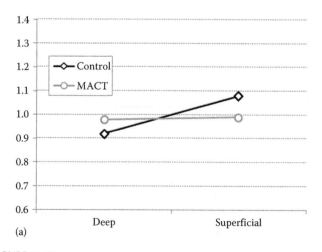

(a)

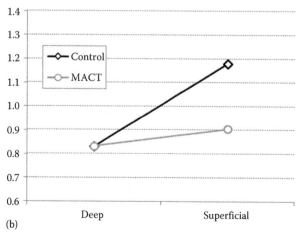

(b)

FIGURE 11.17

Relative zonal T_2 (a) and $T_2{}^*$ (b) assessment for healthy cartilage and cartilage repair tissue after MACT. Whereas the mean value of healthy cartilage (T_2: 56.6 ms; $T_2{}^*$: 18.6 ms) was set to 100%, relative zonal (deep and superficial) T_2 and $T_2{}^*$ values mirror the zonal variation in control cartilage and cartilage repair tissue. (From Welsch, G.H. et al., *Invest. Radiol.*, 43, 619–626, 2008. With permission.)

cartilage are expected to be lower, although it is quite difficult to validate this experimentally for a number of reasons (including the partial volume effect and the influence of multiple T_2/T_2^* components) (Welsch et al., 2011). Nevertheless, HR T_2/T_2^* mapping at UHF allows various cartilage repair procedures to be evaluated in more detail, for example, in the tissue adjacent to a repair site. It was shown that the absolute values of T_2 were significantly higher in repair cartilage (51.6 ± 7.6 ms, on average) and in both healthy cartilage (45.9 ± 4.7 ms) and adjacent cartilage (40.0 ± 3.7 ms) (Chang et al., 2013b).

11.3.2.4 Tendons

Using spectroscopic methods, it has been shown that the T_2^* decay in the Achilles tendon is a multicomponent process [33]. However, with clinical sequences, it is difficult to acquire signal from the second, third, and fourth components, since these have a small component ratio. The first component has the largest ratio, but it is in the submillisecond range, and can thus barely be acquired with conventional echo times. The 3D-UTE sequence provides the ability to detect MR signal from a large variety of rapidly relaxing tissues and materials, including tendons. Juras et al. used a 3D-UTE sequence at 7 T to estimate T_2^* in tendons in order to investigate the potential feasibility of using this parameter as a marker for Achilles tendinopathy (Juras et al., 2013b). It seems that ultra-short bicomponent T_2^* measurements in the human Achilles tendon *in vivo* are feasible using a 3D-UTE sequence. Higher SNR at 7 T allows the calculation of both T_2^* components very accurately. Moreover, HR T_2^* maps are suitable for regional T_2^* differences description, which provides insight into the collagen content and orientation in the Achilles tendon. Recently, a vTE, based on a gradient-echo sequence, has been used for T_2^* mapping in tendons with the advantage of low blurring and highly resolved image acquisition. Although the shortest echo time was only slightly below 1 ms, it was possible to show that the short component of T_2^* provides a robust marker of Achilles tendinopathy, suggesting that it reflects changes in water content and collagen orientation, and moreover, more accurately than a simple mono-exponentially calculated T_2^* (Juras et al., 2013b).

11.3.2.5 Menisci

Higher spatial resolution in UHF may be beneficial for imaging small structures, such as the meniscus. UHF 7 T MRI can detect the higher frequency and severity of medial meniscal lesions in asymptomatic dancers compared to healthy controls (Chang et al., 2013a). In the future, one can expect an increase in the number of quantitative studies of mono- and multi-component T_2/T_2^* analysis thanks to higher SNR at UHF, as this allows more robust calculation of short T_2/T_2^* component and hence possible utilization as a clinical marker for tissue degeneration.

11.3.2.6 Applications of Sodium MRI and gagCEST—Cartilage Repair

Several 7 T studies have demonstrated that noninvasive sodium MR imaging can provide information on the GAG content in native cartilage and in tissue after cartilage repair. Trattnig et al. have shown that sodium imaging was able to differentiate between repair tissue with lower GAG content and native cartilage of patients after matrix-associated autologous chondrocyte transplantation (MACT) without the application of contrast agent (Trattnig et al., 2010). Zbyn et al. reported significantly higher sodium values in repair tissue of patients after MACT than in bone marrow stimulation (BMS) patients, suggesting higher quality of MACT than BMS repair tissue (Zbyn et al., 2012). The fact that MACT and BMS patients showed similar morphological results underlines the added value of sodium imaging and shows that sodium imaging can be useful for noninvasive evaluation of the performance of new cartilage repair techniques. To assess contamination from synovial fluid, Chang et al. evaluated cartilage repair and native tissue using sodium imaging with and without fluid suppression (Chang et al., 2012a). The results presented suggest that sodium imaging with fluid suppression using IR (Madelin et al., 2010) seems to be more accurate in the assessment of sodium concentration in repair tissue. Sodium imaging may enable the noninvasive *in vivo* evaluation of methods for cartilage repair and follow-up of patients after cartilage repair surgery.

The usefulness of gagCEST imaging for the evaluation of GAG content in articular cartilage and repair tissue at 7 T has also been demonstrated. Schmitt et al. compared gagCEST and sodium values in patients after microfracturing and after MACT (Schmitt et al., 2011). The observed correlation between the two techniques validated the sensitivity of gagCEST to changes in GAG content of cartilage. Moreover, gagCEST was able to differentiate between native cartilage and cartilage repair tissue of patients. In the following study, significantly lower gagCEST values were found in repair tissue of patients after autologous osteochondral transplantation (Krusche-Mandl et al., 2012). The clinical outcome of autologous osteochondral transplantation patients did not correlate with morphological evaluations or with the sodium and gagCEST results. These results suggest that biochemical MR imaging provides unique insight into the composition of native cartilage and tissue after cartilage repair. The gagCEST technique seems to be especially useful for detecting early pathological changes that coincide with a primary loss of GAG content.

11.3.2.7 Applications of Sodium MRI and gagCEST—Osteoarthritis

The first 7 T paper reporting sodium MR imaging for the evaluation of patients with osteoarthritis (OA) was published by Wang et al. (2009b). The authors showed 30%–60% lower sodium concentration in the cartilage of OA patients compared to volunteers (~260 mmol/L). Madelin et al. calculated sodium concentration in the cartilage of volunteers and OA patients using sodium imaging with and without fluid suppression at 7 T (Madelin et al., 2013). Fluid suppression resulted in greater difference in cartilage sodium concentrations between volunteers and OA patients. The authors found that the sodium concentration in cartilage from IR-prepared fluid-suppressed measurements (Madelin et al., 2010) was a significant predictor of OA. Thus, evaluation of cartilage sodium concentration with fluid-suppressed MR imaging at 7 T might be a potential biomarker for OA. Sodium MR imaging might help to detect early signs of cartilage degeneration that precede morphological changes visible on proton MR images. Sodium imaging may also be useful for the noninvasive *in vivo* evaluation of disease-modifying treatments for OA.

11.3.2.8 Applications for Sodium MRI—Intervertebral Discs

Although several authors have imaged intervertebral disks (IVDs) using sodium MR imaging, only one application of 7 T MRI has been published to date. Noebauer-Huhmann et al. demonstrated the feasibility of sodium MR imaging of the IVDs in asymptomatic volunteers *in vivo* (Noebauer-Huhmann et al., 2012). Results of this study showed decreased sodium values in more degenerated discs (higher Pfirrmann score) than in the healthy discs. Therefore, sodium imaging seems to be a promising tool for the noninvasive evaluation of changes in proteoglycan content of discs and for the detection of early degenerative changes in IVDs.

11.4 Metabolism

11.4.1 Introduction

MRS is a noninvasive state-of-the-art methods for the assessment of human tissue biochemistry and metabolism. General advantages and challenges of MRS at 7 T and related metabolic techniques and applications in the brain have been addressed in Section 11.2. In this section, we provide an overview of the human adipose tissue, muscle, liver, and heart metabolic studies at 7 T. The applications are grouped according to the nuclei of interest involved in metabolic processes and detectable by MRS at 7 T—proton (^{1}H), phosphorus (^{31}P), and carbon (^{13}C).

Subsections include methodological information (relaxation times, localization techniques, benefits over lower fields, and available static and dynamic metabolic information) and related 7 T studies. Studies focused on the assessment of the metabolism and staging of neoplastic lesions will be discussed in the following section.

11.4.2 Proton MRS

11.4.2.1 Lipid Profiling

Adipose tissue and its anatomic distribution strongly influence the risk of a number of diseases. The spectral resolution of the ^{1}H MR signal at higher fields is improved, and spectra can provide information about lipid composition including the degree of fatty acid chain saturation, both mono- and polyunsaturation. Well-resolved single-voxel 7 T spectra were demonstrated in adipose tissue and bone marrow (Ren et al., 2008), muscle (Ramadan et al., 2010; Ren et al., 2010; Wang et al., 2009a), liver (Gajdosik et al., 2013), and breast (Dimitrov et al., 2012). An example of a single-voxel ^{1}H MRS spectrum acquired from subcutaneous adipose tissue at 7 T is shown in Figure 11.18.

11.4.2.2 Relaxation Times

The basic requirement for the appropriate adjustment of sequence timing and accurate quantification of metabolite concentrations is knowledge of relaxation parameters. In general, both relaxation times (spin–lattice and spin–spin) are tissue and B_0-dependent and should be assessed or for every specific experimental setup including magnetic field applied, tissue, and pathophysiological conditions studied. Several studies have assessed ^{1}H relaxation times of water and metabolite resonance lines in adipose tissue, bone marrow (Ren et al., 2008), muscle (Ren et al., 2010; Wang et al., 2009a), and liver (Gajdosik et al., 2013). Tables 11.1 and 11.2 summarize assessed relaxation rates (times) and provide information about the method used and the number of subjects included in the experiments. In general, ^{1}H T_1 relaxation times at 7 T are longer and T_2 times are shorter than at lower magnetic fields.

11.4.2.3 Localization

Although an alternative localization scheme using the effect of longitudinal traveling waves has been demonstrated (Webb et al., 2010), most published 7 T metabolic studies use conventional stimulated echo-based localization techniques. Shorter 90° RF-pulses included in the STEAM localization scheme can be achieved with

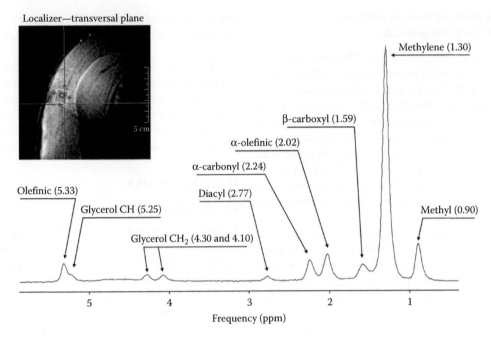

FIGURE 11.18

^{1}H NMR spectrum of subcutaneous fat from a young healthy male acquired at 7 T. Ten resonances can be resolved. A localizer image shows the position of the voxel in the subcutaneous fat tissue (12 × 12 × 12 mm³, STEAM, T_e 20 ms, T_R 2 s, 1 avg) from which the spectrum was acquired.

TABLE 11.1

Summary of Published *In Vivo* T_1 (ms) Relaxation Times (Mean ± SD) of [1]H Metabolites at 7 T Assessed in Various Human Tissues

Tissue Type	Adipose (Ren et al. 2008)	Marrow (Ren et al. 2008)	Muscle			Liver (Gajdosik et al. 2013)
			Soleus (Ren et al. 2010)	Tibialis Anterior (Wang et al. 2009a)	Soleus (Ramadan et al. 2010)	
Method	IR STEAM	IR STEAM	IR STEAM	PS STEAM	PS STEAM	IR STEAM
Number of Subjects	*n* = 7	*n* = 7	*n* = 25	*n* = 4	*n* = 6	*n* = 5
Methyl IMCL[a]	1080 ± 50	1160 ± 40	1380 ± 100	1946 ± 606	1735 ± 132	1026 ± 162
(0.9 ppm) EMCL[a]			1200 ± 110	1427 ± 338	1470 ± 102	
Methylene IMCL[a]	530 ± 40	550 ± 30	580 ± 40	1494 ± 158	1350 ± 193	514 ± 25
(1.3 ppm) EMCL[a]			570 ± 40	2187 ± 1001	1031 ± 166	
β-Carboxyl (1.59 ppm)	320 ± 50	390 ± 40	–	–	1343 ± 173	–
α-Olefinic (2.02 ppm)	390 ± 30	420 ± 20	–	1372 ± 239	972 ± 170	488 ± 220
α-Carboxyl (2.24 ppm)	400 ± 20	440 ± 20	–	1971 ± 235	1420 ± 130	476 ± 89
Diacyl (2.77 ppm)	580 ± 30	600 ± 30	–	–	–	479 ± 260
tCr (3.02 ppm)	–	–	950	1632 ± 164	1320 ± 121	–
CCC (3.22 ppm)	–	–	–	1339 ± 40	1216 ± 85	1084 ± 52
Water (4.70 ppm)	–	–	1140	–	1514 ± 10	1362 ± 83
Olefinic (5.3 ppm)	–	–	–	2036 ± 404	1195 ± 122	–
Carnosine (7 ppm)	–	–	–	–	908 ± 165	–
(8 ppm)					1464 ± 210	

[a] In case of muscle, IMCL and EMCL are splitted. EMCLs are shifted by ~0.2 ppm downfield; IR, inversion recovery; PS, progressive saturation.

surface coils, are less SAR demanding, and lead to lower CSDE (Ren et al., 2008, 2010). Higher B_1 inhomogeneities at 7 T complicate pulse calibration even with volume coils. The acquisition voxel should be used for pulse calibration (Versluis et al., 2010). A fully adiabatic LASER sequence may be a good option to overcome the problems mentioned (B_1 inhomogeneity, CSDE) but it faces SAR limits at 7 T. The application of sequences with longer echo times (e.g., 280 ms) without water suppression has been proposed to simplify spectral quantification

TABLE 11.2

Summary of Published *In Vivo* T_2 (ms) Relaxation Times (Mean ± SD) of [1]H Metabolites at 7 T Assessed in Various Human Tissues

Tissue Type	Adipose (Ren et al. 2008)	Marrow (Ren et al. 2008)	Muscle			Liver (Gajdosik et al. 2013)
			Soleus (Ren et al. 2010)	Tibialis Anterior (Wang et al. 2009a),	Soleus (Ramadan et al. 2010)	
Number of Subjects	$n = 7$	$n = 7$	$n = 25$	$n = 4$	$n = 6$	$n = 5$
Methyl IMCL[a]	67 ± 8	74 ± 6	97 ± 11	42 ± 12	85 ± 23	34 ± 10
(0.9 ppm) EMCL[a]			74 ± 10	53 ± 1	117 ± 28	
Methylene IMCL[a]	63 ± 5	69 ± 4	66 ± 5	56 ± 4	78 ± 12	41 ± 8
(1.3 ppm) EMCL[a]			51 ± 4	64 ± 7	64 ± 10	
β-Carboxyl (1.59 ppm)	30 ± 6	33 ± 6	–	–	65 ± 9	–
α-Olefinic (2.02 ppm)	39 ± 3	42 ± 2	–	28 ± 7	36 ± 5	44 ± 19
α-Carboxyl (2.24 ppm)	55 ± 4	60 ± 3	–	29 ± 8	49 ± 10	39 ± 15
Diacyl (2.77 ppm)	58 ± 3	59 ± 3	–	–	–	44 ± 5
tCr (3.02 ppm)	–	–	–	53 ± 6	101 ± 29	–
CCC (3.22 ppm)	–	–	–	39 ± 3	93 ± 18	32 ± 9
Water (4.70 ppm)	–	–	–	–	22 ± 2	15 ± 2
Olefinic (5.3 ppm)	–	–	–	56 ± 21	72 ± 19	–
Carnosine (7 ppm) (8 ppm)	–	–	–	59 ± 18 / 86 ± 19	–	–

[a] In case of muscle, IMCL and EMCL signals are splitted.

of intramyocellular lipids (IMCL) in the soleus muscle (Ren et al., 2010). Multi-voxel MRSI localization with combined FID (water signal) and echo (fat signals) acquisition has been shown to provide high-quality ^{1}H MRS spectra from various muscle types, bone marrow, and subcutaneous tissue within one experiment (Just Kukurova et al., 2014).

11.4.2.4 Applications

The main application area of skeletal muscle ^{1}H MRS was found in the interrelating fields of exercise physiology and pathophysiology of type 2 diabetes. As has been established at lower fields, the IMCL pool can play a dual role in this respect. On the one hand, it has been found to be increased in populations with insulin resistance and type 2 diabetes mellitus (Anderwald et al., 2002; Krssak et al., 1999), where its catabolites interact with skeletal muscle glucose transport and utilization. A recent UHF study has also found lower IMCL accumulation in the skeletal muscle of middle-aged subjects predisposed to familial longevity (Wijsman et al., 2010). On the other hand, IMCL content was found to be increased in the population of highly trained population of endurance runners (Thamer et al., 2003; Zehnder et al., 2006) who can effectively use this energy pool during prolonged submaximal exercise (Krssak et al., 2000; Zehnder et al., 2006). As has been observed at lower magnetic fields, resonance position of extramyocellular

lipid (EMCL) compartment is dependent on the muscle fiber orientation within the magnetic field and partly overlaps with that of IMCL (Khuu et al., 2009). At 7 T, increased spectral resolution has enhanced knowledge about the magnetic field inhomogeneity distribution, which influences the orientation-dependent frequency position. Thus, new insights were gained about the frequency distribution of resonance line of lipids methylene group of EMCL compartment (Khuu et al., 2009), and refined spectral deconvolution of IMCL resonance can improve the quantification of this important metabolic pool. SNR values (measured for the water peak) at 7 T were 90% higher than the values measured at 3 T, and test–retest measurement of IMCL levels showed improved reproducibility at 7 T compared with 3 T (2% vs. 6%) (Stephenson et al., 2011).

Other ^{1}H MRS visible metabolites—carnosine, carnitine, lactate, and creatine—are also interesting from the point of view of exercise physiology. Again, the presence of their resonance lines in the MR spectrum of skeletal muscle has been shown at lower fields (Meyerspeer et al., 2007; Ozdemir et al., 2007; Wachter et al., 2002), but enhanced sensitivity at 7 T allows the dynamic assessment of acetylcarnitine production and its decay during exercise and following recovery (Ren et al., 2013) as well as real-time monitoring of lactate accumulation and disposal in human skeletal muscle following exhaustive exercise (Ren et al., 2012). The absolute quantification of skeletal muscle carnosine concentration can also be improved (Kukurova et al., 2014), allowing this measurement to easily be implemented in more complex experiments in the field of integrative physiology. Acute exercise-related changes in skeletal muscle can also be assessed by monitoring phosphocreatine–creatine exchange. Even though ^{31}P MRS is the method of choice here (see below), the ^{1}H-based creatine-specific chemical exchange saturation transfer (Cr-CEST) method, as demonstrated by Kogan et al. (2014), can complement these studies.

11.4.3 Phosphorous MRS

The main focus of phosphorus (^{31}P) MRS lies in the observation of tissue bioenergetics under various physiological and pathological conditions. Detection of high-energy ^{31}P metabolites (nucleoside triphosphate—NTP, phosphocreatine—PCr, and inorganic phosphate—Pi), intracellular pH, and the intracellular free magnesium concentration ([Mg_{2+}]) are the analytical basis for this method (Gupta et al., 1978; Kemp et al., 2007; Szendroedi et al., 2009). In addition, information about phosphomonoesters (PME; i.e., phosphocholine) and phosphodiesters (PDE; i.e., glycerophosphocholine) tissue concentration have been proposed as possible diagnostic markers for cancer (Ackerstaff et al., 2003; Arias-Mendoza et al., 2006; Klomp et al., 2011b), inflammatory liver disease (Dezortova et al., 2005), and neurodegenerative diseases (Forlenza et al., 2005).

The major benefit of ^{31}P-MRS at 7 T is increased SNR (Bogner et al., 2009a, 2011; Qiao et al., 2006; Rodgers et al., 2013; Valkovic et al., 2013a). Due to the physical properties of ^{31}P nuclei, ^{31}P-MRS at high magnetic fields does not face the problems posed by ^{1}H-MRS. Water and fat suppression are not necessary. Higher chemical shift dispersion of the ^{31}P MR spectrum improves the spectral resolution from 3 to 7 T significantly (Bogner et al., 2009a), separating the signals of ^{31}P metabolites and making ^{31}P MRS less sensitive to B_0 field inhomogeneities. Increased spectral resolution enabled the detection of an alkaline inorganic phosphate pool with short T_1 in human muscle (Kan et al., 2010). A further benefit of ^{31}P-MRS at UHFs (7 T) are the shorter T_1 relaxation times, allowing for faster signal averaging with lower saturation effects (Bogner et al., 2009a). This leads to a more than linear increase in SNR per time unit (Bogner et al., 2011; Rodgers et al., 2013), which can be transferred to higher spatial (Chmelik et al., 2014; Parasoglou et al., 2013b; Steinseifer et al., 2013) and/or temporal resolution (Parasoglou et al., 2013c; Valkovic et al., 2013a).

^{31}P-MRS is primarily applied with sensitive surface coils that provide inhomogeneous B_1^+ excitation and approximate localization due to their B_1^+/B_1^- profile. However, many clinical studies require further localization of the MRS signal. The very short T_2 relaxation times of some phosphorous metabolites, which decrease with increasing B_0 (Bogner et al., 2009a), make FID acquisition immediately after the excitation pulse (e.g., FID-CSI [Brown et al., 1982; Maudsley et al., 1983] or ISIS [Ordidge et al., 1986]) the preferred choice. A short T_2 leads to a significantly decreased signal with echo-based methods. ISIS (Ordidge et al., 1986) localization is an option when an MR spectrum from a single volume is desired. It has recently been shown that technical limitations of 7 T, such as CSDE due to gradient slice selection, can be overcome by the implementation of broadband inversion GOIA pulses—E-ISIS (Bogner et al., 2011), goISICS (Chmelik et al., 2013a)—or by the implementation of adiabatic refocusing semi-LASER pulses (Meyerspeer et al., 2011). Representative ^{31}P hepatic spectra obtained with optimized localization techniques at 7 T are displayed in Figure 11.19: representative hepatic ^{31}P spectra obtained at 7 T by 1D-ISIS slab selection (a—white solid lines, b), 3D-ISIS voxel (a—white dashed box, c), 2D-CSI (a—yellow dashed lines, d), and 3D-CSI (a—red lines, e). Problems of B_1^+ inhomogeneities may be addressed either by direct measurement of the flip angle distribution (Chmelik et al., 2013b) or by the implementation of adiabatic pulses (Tannus and Garwood, 1997).

11.4.3.1 Relaxation Times

Table 11.3 summarizes assessed relaxation rates (times) of ^{31}P-containing metabolites acquired at 7 T in muscle (Bogner et al., 2009a; Rodgers et al., 2013), liver (Chmelik et al., 2014), and heart (Rodgers et al., 2013) and provides information about the method used and the number of subjects included in the experiments. In general, ^{31}P T_1 and T_2 relaxation times at 7 T are shorter than at lower magnetic field.

11.4.3.2 Applications

Direct quantification of the dynamics of PCr to ATP exchange, as described above, is the main field for the applications of ^{31}P MRS. SNR gain through 7 T can be transferred into elimination of partial volume effects by localizing the signal from specific muscle group (Meyerspeer et al., 2011, 2012; Parasoglou et al., 2012, 2013c) or into dynamic 3D-imaging of PCr recovery (Meyerspeer et al., 2011, 2012; Parasoglou et al., 2012, 2013c). The results obtained point to the methodological advantage, yielding increased dynamic window of dynamic PCr changes and better correspondence of experimental data and theoretical model (Meyerspeer et al., 2011, 2012; Parasoglou et al., 2012, 2013c). Additionally, based on the experiments combining accurately localized ^{31}P MRS and T_2^*-weighted MRI, strong correlations of the skeletal muscle T_2^*-weighted signal changes, tissue pH time courses, and parameters

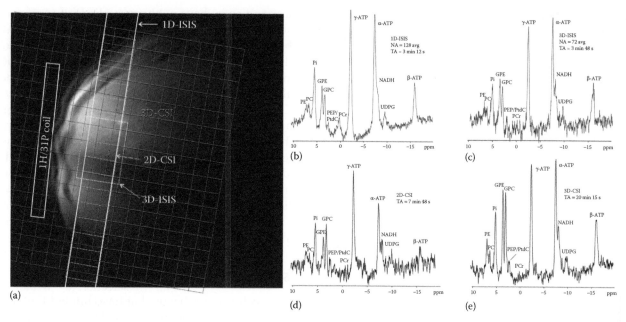

FIGURE 11.19

Representative hepatic ^{31}P spectra obtained at 7 T by 1D-ISIS slab selection (a—white solid lines, b), 3D-ISIS voxel (a—white dashed box, c), 2D-CSI (a—yellow dashed lines, d), and 3D-CSI (a—red lines, e).

TABLE 11.3

Summary of Published *In Vivo* T_1 and T_2 (ms) Relaxation Times (Mean $\pm$ SD) of [31]P Metabolites at 7 T Assessed in Various Human Tissues

Tissue Type	T_1				T_2
	Muscle (Bogner et al. 2009a)	Muscle (Rodgers et al. 2013)	Liver (Chmelik et al. 2014)	Heart (Rodgers et al. 2013)	Muscle (Bogner et al. 2009a)
Method	IR	LL-CSI	IR 1D-ISIS	LL-CSI	Sel. SE
Number of Subjects	$n = 8$	$n = 2$	$n = 8$	$n = 6$	$n = 8$
PE (6.78 ppm)	3.1 ± 0.9	–	4.41 ± 1.55	–	–
PC (5.88 ppm)		–	3.74 ± 1.31	–	–
2.3-DPG (5.4 and 6.4 ppm)	–	–	–	3.05 ± 0.41	–
Pi (5.02 ppm)	6.3 ± 1.0	6.65 ± 0.23	0.70 ± 0.33	–	109 ± 17
GPE (3.2 ppm)	5.7 ± 1.5	–	6.19 ± 0.91	–	314 ± 35
GPC (2.76 ppm)		–	5.94 ± 0.73	–	
PCr (0 ppm)	4.0 ± 0.2	3.96 ± 0.07	–	3.09 ± 0.32	217 ± 14
-ATP (−2.48 ppm)	3.3 ± 0.2	4.12 ± 0.15	0.50 ± 0.08	1.82 ± 0.09	29 ± 3.3
α-ATP (−7.52 ppm)	1.8 ± 0.1	1.70 ± 0.02	0.46 ± 0.07	1.39 ± 0.09	–
β-ATP (−16.26 ppm)	1.8 ± 0.1	1.42 ± 0.12	0.56 ± 0.07	1.02 ± 0.17	–

Notes: IR, inversion recovery; sel SE, frequency-selective spin echo; LL-CSI, look locker chemical shift imaging.

of energy metabolism during- and post-exercise were observed (Schmid et al., 2014).

Similarly, measurements of Pi to ATP flux, applying the observation of magnetization transfer between these two entities, yield identical values to those measured at lower field. Again, SNR gain can be transferred into improved temporal resolution (Valkovic et al., 2013a, b, 2014a) or tissue-specific localization (Parasoglou et al.,

2013a; Valkovic et al., 2013a, b, 2014a). Application of this method confirmed the previously observed (Schmid et al., 2012) interrelation between measures of basal and excited ATP metabolism in skeletal muscle (Valkovic et al., 2013a, b, 2014a; Van Oorschot et al., 2011) and points to a link between decreased Pi to ATP flux and increased hepatic lipid accumulation in patients with non-alcholic fatty liver disease (NAFLD) (Valkovic et al., 2014b).

11.4.3.3 ^{13}C MRS

The performance of ^{13}C MRS benefits from increased SNR and spectral resolution at 7 T. On the other hand, proton decoupling usually applied for further SNR improvement and simplification of spectral pattern is very SAR demanding. The majority of 7 T studies focus on the optimization of ^{13}C coil design and in particular on improving the decoupling efficiency in combined ^{1}H and ^{13}C channel experiments (McDougall et al., 2014; Meyerspeer et al., 2013; Roig et al., 2014). Recently, it was confirmed that ^{13}C MRS at 7 T is capable of sequentially measuring dynamic changes in glycogen levels (Stephenson et al., 2011), again with increased time resolution and tissue specificity resolution (Krssak et al., 2014). In addition, the ^{13}C spectrum from adipose tissue reveals the detailed chemical profile of fatty acid chains, which, for example, allows omega-6/omega-3 lipid ratio (Cheshkov et al., 2012; Dimitrov et al., 2010a) to be calculated and trans-fatty acids to be detected (Dimitrov et al., 2010b).

In summary, multinuclear *in vivo* MRS and its application benefits from ultra-high magnetic field through increased sensitivity and signal specificity. These gains are usually transferred into improved localization of spectroscopic volume of interest and/or temporal resolution. Shorter acquisition times can further lead into more sophisticated combinations of various measurements.

11.5 Oncology

11.5.1 Magnetic Resonance Spectroscopy

While there is currently no literature showing results on MRS at 7 T in brain tumors, there are a number of papers that present initial experience applying either ^{1}H-MRS or ^{31}P-MRS to investigations of prostate or breast tumors (De Graaf et al., 2013; Klomp et al., 2009, 2011a, b; Korteweg et al., 2011; Lagemaat et al., 2014b). While the number of patients measured is still very limited for breast MRS, with only a few patients measured per study (De Graaf et al., 2013; Klomp et al., 2011b; Korteweg et al., 2011), there are at least two larger patient studies including 12 and 15 patients (with overlap) studied at 7 T via ^{31}P-MRS in the prostate (Lagemaat et al., 2014a, b). A first ^{1}H-MRS study investigating the feasibility of measuring polyamine resonances in prostate cancer was limited to only a single patient (Klomp et al., 2011a). However, a number of very recent publications exist that show significantly improved MRS techniques specifically tailored for the investigation of breast and prostate

cancer and some have not been applied yet (Arteaga de Castro et al., 2012; Boer et al., 2012b; De Graaf et al., 2013; Kobus et al., 2012; Luttje et al., 2013; Van den Bergen et al., 2011; Van der Kemp et al., 2013; Wijnen et al., 2012). These studies had different approaches to improving data quality, including coil hardware improvements (Arteaga de Castro et al., 2012; Kobus et al., 2012; Van den Bergen et al., 2011), the use of signal enhancement techniques such as decoupling or nuclear overhauser enhancement (Lagemaat et al., 2014a), or solving problems related to inadequate B_0 shimming (Boer et al., 2012b; Van der Kemp et al., 2013). Although first results are promising, indicating improvement in the specificity of ^{31}P-MRS, the low spatial resolution may be a significant problem for clinical use (Lagemaat et al., 2014b).

11.5.2 Diffusion-Weighted Imaging

Due to technical problems that have limited the image quality obtained by diffusion-weighted imaging (DWI) at 7 T, there are currently only preliminary unilateral breast DWI results of three patients (Korteweg et al., 2011) reported at 7 T. However, first conference presentations show the high potential of 7 T DWI for tumor imaging as shown recently for breast cancer (Minarikova et al., 2014; Zaric et al.) (Figure 11.20). This was possible due to recent improvements in DWI and diffusion tensor imaging sequences that substantially reduced signal loss due to long echo times and off-resonance artifacts such as distortions and blurring and which will lead to further application of DWI in a clinical context (Eichner et al., 2014; Heidemann et al., 2010, 2012; Jeong et al., 2013).

11.5.3 T_1/Dynamic Contrast-Enhanced Imaging

Contrast enhancement using gadolinium T_1-weighted imaging was shown to be efficient in a study on 5 patients with astrocytomas, displaying similar contrast enhancement at 7 T to that seen at 1.5 T (Moenninghoff et al., 2010).

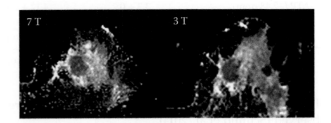

FIGURE 11.20
Example of ADC maps constructed from DWI MRI of breast with malignant lesion (invasive ductal carcinoma grade III [DCI G3], diameter: 25 mm, BIRADS 5) at 7 (left) and 3 T (right) of a 51-year-old female patient.

Furthermore, dynamic contrast-enhanced (DCE) imaging was tried with healthy volunteers in liver and renal studies. In this study, 16 healthy volunteers underwent DCE MRI using coronal T_1-weighted spoiled gradient-echo sequence (3D-FLASH) imaging, yielding homogeneous enhancement of the liver with a statistically significant increase in SNR from non-enhanced MRI. In one healthy volunteer, an incidental hemangioma was detected (Umutlu et al., 2013).

A renal study was performed on 10 healthy volunteers using 3D-FLASH, where the images were acquired during precontrast and at 20, 70, and 120 s delay. The DCE MRI provided homogeneous enhancement of the renal parenchyma in conjunction with imaging at high spatiotemporal resolution within a breath hold (Umutlu et al., 2011).

DCE imaging has also been successfully applied in breast cancer research. The feasibility of such a study at 7 T was demonstrated using a single-loop coil and 15 subjects, including 5 patients with histologically confirmed breast cancer (Umutlu et al., 2010).

A more recent study included 27 patients with breast cancer, using a bilateral coil and a transversal T_1-weighted time-resolved angiography with stochastic trajectories sequence and showed that DCE-MRI of the breast at 7 T is clinically applicable (Pinker et al., 2014).

Furthermore, a comparison of 3 and 7 T breast DCE MRI using the T_1-weighted time-resolved angiography with stochastic trajectories sequence showed an improved temporal and spatial resolution at higher field strength (Gruber et al., 2014).

Figure 11.21 gives an example of a patient with breast cancer measured at 3 and 7 T. As published previously in 3 T imaging trials, the magnetic field strength increase comes with the potential for a decrease in dosage of contrast media. However, only limited data are available on r1 relaxivity changes of gadolinium-based contrast agents at ultra-high magnetic field strength (Noebauer-Huhmann et al., 2010).

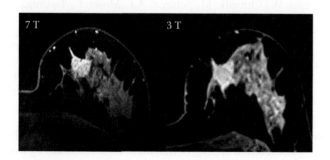

FIGURE 11.21
Example of DCE MRI of breast with malignant lesion (DCI G3) on 7 and 3 T of the same patient shown in Figure 11.1.

References

Ackerstaff, E., Glunde, K., Bhujwalla, Z.M., 2003. Choline phospholipid metabolism: A target in cancer cells? *Journal of Cellular Biochemistry* 90, 525–533.

An, L., Willem van der Veen, J., Li, S., Thomasson, D.M., Shen, J., 2013. Combination of multichannel single-voxel MRS signals using generalized least squares. *Journal of Magnetic Resonance Imaging* 37, 1445–1450.

Anderwald, C., Bernroider, E., Krssak, M., Stingl, H., Brehm, A., Bischof, M.G., Nowotny, P., Roden, M., Waldhausl, W., 2002. Effects of insulin treatment in type 2 diabetic patients on intracellular lipid content in liver and skeletal muscle. *Diabetes* 51, 3025–3032.

Andreychenko, A., Boer, V.O., Arteaga de Castro, C.S., Luijten, P.R., Klomp, D.W.J., 2012. Efficient spectral editing at 7 T: GABA detection with MEGA-sLASER. *Magnetic Resonance in Medicine: Official Journal of the Society of Magnetic Resonance in Medicine/Society of Magnetic Resonance in Medicine* 68, 1018–1025.

Andronesi, O.C., Ramadan, S., Ratai, E.M., Jennings, D., Mountford, C.E., Sorensen, A.G., 2010. Spectroscopic imaging with improved gradient modulated constant adiabaticity pulses on high-field clinical scanners. *Journal of Magnetic Resonance* 203, 283–293.

Arias-Mendoza, F., Payne, G.S., Zakian, K.L., Schwarz, A.J., Stubbs, M., Stoyanova, R., Ballon, D. et al., 2006. In vivo 31P MR spectral patterns and reproducibility in cancer patients studied in a multi-institutional trial. *NMR in Biomedicine* 19, 504–512.

Arteaga de Castro, C.S., Boer, V.O., Andreychenko, A., Wijnen, J.P., van der Heide, U.A., Luijten, P.R., Klomp, D.W.J., 2013. Improved efficiency on editing MRS of lactate and γ-aminobutyric acid by inclusion of frequency offset corrected inversion pulses at high fields. *NMR in Biomedicine* 26, 1213–1219.

Arteaga de Castro, C.S., van den Bergen, B., Luijten, P.R., van der Heide, U.A., van Vulpen, M., Klomp, D.W., 2012. Improving SNR and B1 transmit field for an endorectal coil in 7 T MRI and MRS of prostate cancer. *Magnetic Resonance in Medicine* 68, 311–318.

Avdievich, N.I., Pan, J.W., Baehring, J.M., Spencer, D.D., Hetherington, H.P., 2009. Short echo spectroscopic imaging of the human brain at 7T using transceiver arrays. *Magnetic Resonance in Medicine: Official Journal of the Society of Magnetic Resonance in Medicine/Society of Magnetic Resonance in Medicine* 62, 17–25.

Bagher-Ebadian, H., Jiang, Q., Ewing, J.R., 2008. A modified Fourier-based phase unwrapping algorithm with an application to MRI venography. *Journal of Magnetic Resonance Imaging* 27, 649–652.

Balchandani, P., Pauly, J., Spielman, D., 2008. Interleaved narrow-band PRESS sequence with adiabatic spatial-spectral refocusing pulses for 1H MRSI at 7T. *Magnetic Resonance in Medicine: Official Journal of the Society of Magnetic Resonance in Medicine/Society of Magnetic Resonance in Medicine* 59, 973–979.

Balchandani, P., Spielman, D., 2008. Fat suppression for 1H MRSI at 7T using spectrally selective adiabatic inversion recovery. *Magnetic Resonance in Medicine: Official Journal of the Society of Magnetic Resonance in Medicine/Society of Magnetic Resonance in Medicine* 59, 980–988.

Banerjee, A., Ganji, S., Hulsey, K., Dimitrov, I., Maher, E., Ghose, S., Tamminga, C., Choi, C., 2012. Measurement of glycine in gray and white matter in the human brain in vivo by 1H MRS at 7.0 T. *Magnetic Resonance in Medicine: Official Journal of the Society of Magnetic Resonance in Medicine/Society of Magnetic Resonance in Medicine* 68, 325–331.

Barth, M., Nobauer-Huhmann, I.M., Reichenbach, J.R., Mlynarik, V., Schoggl, A., Matula, C., Trattnig, S., 2003. High-resolution three-dimensional contrast-enhanced blood oxygenation level-dependent magnetic resonance venography of brain tumors at 3 Tesla: First clinical experience and comparison with 1.5 Tesla. *Investigative Radiology* 38, 409–414.

Bartzokis, G., Aravagiri, M., Oldendorf, W.H., Mintz, J., Marder, S.R., 1993. Field dependent transverse relaxation rate increase may be a specific measure of tissue iron stores. *Magnetic Resonance in Medicine* 29, 459–464.

Bazin, P.L., Weiss, M., Dinse, J., Schafer, A., Trampel, R., Turner, R., 2013. A computational framework for ultra-high resolution cortical segmentation at 7 Tesla. *NeuroImage* 93, 201–209.

Beckmann, C.F., DeLuca, M., Devlin, J.T., Smith, S.M., 2005. Investigations into resting-state connectivity using independent component analysis. *Philosophical Transactions of the Royal Society of London B Biological Sciences* 360, 1001–1013.

Beckmann, C.F., Smith, S.M., 2004. Probabilistic independent component analysis for functional magnetic resonance imaging. *IEEE Transactions on Medical Imaging* 23, 137–152.

Beisteiner, R., 2013. Improving clinical fMRI: Better paradigms or higher field strength? *American Journal of Neuroradiology* 34, 1972–1973.

Beisteiner, R., Hollinger, I., Rath, J., Wurnig, M., Hilbert, M., Klinger, N., Geissler, A. et al., 2011a. New type of cortical neuroplasticity after nerve repair in brachial plexus lesions. *Archives of Neurology* 68, 1467–1470.

Beisteiner, R., Robinson, S., Wurnig, M., Hilbert, M., Merksa, K., Rath, J., Hollinger, I. et al., 2011b. Clinical fMRI: Evidence for a 7T benefit over 3T. *NeuroImage* 57, 1015–1021.

Beisteiner, R., Windischberger, C., Lanzenberger, R., Edward, V., Cunnington, R., Erdler, M., Gartus, A., Streibl, B., Moser, E., Deecke, L., 2001. Finger somatotopy in human motor cortex. *NeuroImage* 13, 1016–1026.

Belkić, D., Belkić, K., 2006. In vivo magnetic resonance spectroscopy by the fast Padé transform. *Physics in Medicine and Biology* 51, 1049–1075.

Benkhedah, N., Bachert, P., Nagel, A.M., 2014. Two-pulse biexponential-weighted (23)Na imaging. *Journal of Magnetic Resonance* 240, 67–76.

Benkhedah, N., Bachert, P., Semmler, W., Nagel, A.M., 2012. Three-dimensional biexponential weighted (23) Na imaging of the human brain with higher SNR and shorter acquisition time. *Magnetic Resonance in Medicine* 70, 754–765.

Besle, J., Sanchez-Panchuelo, R.M., Bowtell, R., Francis, S., Schluppeck, D., 2013. Event-related fMRI at 7T reveals overlapping cortical representations for adjacent fingertips in S1 of individual subjects. *Human Brain Mapping* 35, 2027–2043.

Biswal, B., Yetkin, F.Z., Haughton, V.M., Hyde, J.S., 1995. Functional connectivity in the motor cortex of resting human brain using echo-planar MRI. *Magnetic Resonance in Medicine* 34, 537–541.

Bittersohl, B., Steppacher, S., Haamberg, T., Kim, Y.J., Werlen, S., Beck, M., Siebenrock, K.A., Mamisch, T.C., 2009. Cartilage damage in femoroacetabular impingement (FAI): Preliminary results on comparison of standard diagnostic vs delayed gadolinium-enhanced magnetic resonance imaging of cartilage (dGEMRIC). *Osteoarthritis and Cartilage* 17, 1297–1306.

Boada, F.E., Christensen, J.D., Gillen, J.S., Thulborn, K.R., 1997. Three-dimensional projection imaging with half the number of projections. *Magnetic Resonance in Medicine* 37, 470–477.

Boer, V.O., Klomp, D.W.J., Juchem, C., Luijten, P.R., de Graaf, R.A., 2012a. Multislice 1H MRSI of the human brain at 7 T using dynamic B0 and B1 shimming. *Magnetic Resonance in Medicine: Official Journal of the Society of Magnetic Resonance in Medicine/Society of Magnetic Resonance in Medicine* 68, 662–670.

Boer, V.O., Siero, J.C.W., Hoogduin, H., van Gorp, J.S., Luijten, P.R., Klomp, D.W.J., 2011a. High-field MRS of the human brain at short TE and TR. *NMR in Biomedicine* 24, 1081–1088.

Boer, V.O., van de Bank, B.L., van Vliet, G., Luijten, P.R., Klomp, D.W., 2012b. Direct B0 field monitoring and real-time B0 field updating in the human breast at 7 Tesla. *Magnetic Resonance in Medicine* 67, 586–591.

Boer, V.O., van Lier, A.L.H.M.W., Hoogduin, J.M., Wijnen, J.P., Luijten, P.R., Klomp, D.W.J., 2011b. 7-T (1) H MRS with adiabatic refocusing at short TE using radiofrequency focusing with a dual-channel volume transmit coil. *NMR in Biomedicine* 24, 1038–1046.

Bogner, W., Chmelik, M., Andronesi, O.C., Sorensen, A.G., Trattnig, S., Gruber, S., 2011. In vivo 31P spectroscopy by fully adiabatic extended image selected in vivo spectroscopy: A comparison between 3 T and 7 T. *Magnetic Resonance in Medicine* 66, 923–930.

Bogner, W., Chmelik, M., Schmid, A.I., Moser, E., Trattnig, S., Gruber, S., 2009a. Assessment of (31)P relaxation times in the human calf muscle: A comparison between 3 T and 7 T in vivo. *Magnetic Resonance in Medicine* 62(3), 574–582.

Bogner, W., Gruber, S., Pinker, K., Grabner, G., Stadlbauer, A., Weber, M., Moser, E., Helbich, T.H., Trattnig, S., 2009b. Diffusion-weighted MR for differentiation of breast lesions at 3.0 T: How does selection of diffusion protocols affect diagnosis? *Radiology* 253, 341–351.

Bogner, W., Gruber, S., Trattnig, S., Chmelik, M., 2012. High-resolution mapping of human brain metabolites by free induction decay (1)H MRSI at 7T. *NMR in Biomedicine* 25, 873–882.

Branzoli, F., Techawiboonwong, A., Kan, H., Webb, A., Ronen, I., 2013. Functional diffusion-weighted magnetic resonance spectroscopy of the human primary visual cortex at 7 T. *Magnetic Resonance in Medicine: Official Journal of the Society of Magnetic Resonance in Medicine/Society of Magnetic Resonance in Medicine* 69, 303–309.

Brown, M.A., 2004. Time-domain combination of MR spectroscopy data acquired using phased-array coils. *Magnetic Resonance in Medicine: Official Journal of the Society of Magnetic Resonance in Medicine/Society of Magnetic Resonance in Medicine* 52, 1207–1213.

Brown, R., Madelin, G., Lattanzi, R., Chang, G., Regatte, R.R., Sodickson, D.K., Wiggins, G.C., 2013. Design of a nested eight-channel sodium and four-channel proton coil for 7T knee imaging. *Magnetic Resonance in Medicine* 70, 259–268.

Brown, T.R., Kincaid, B.M., Ugurbil, K., 1982. NMR chemical shift imaging in three dimensions. *Proceedings of the National Academy of Sciences of the USA* 79, 3523–3526.

Brunner, D.O., De Zanche, N., Frohlich, J., Paska, J., Pruessmann, K.P., 2009. Travelling-wave nuclear magnetic resonance. *Nature* 457, 994–998.

Bydder, M., Larkman, D.J., Hajnal, J.V., 2002. Combination of signals from array coils using image-based estimation of coil sensitivity profiles. *Magnetic Resonance in Medicine* 47, 539–548.

Cai, K., Nanga, R.P., Lamprou, L., Schinstine, C., Elliott, M., Hariharan, H., Reddy, R., Epperson, C.N., 2012. The impact of gabapentin administration on brain GABA and glutamate concentrations: A 7T ^{1}H-MRS study. *Neuropsychopharmacology: Official Publication of the American College of Neuropsychopharmacology* 37, 2764–2771.

Chakeres, D.W., Abduljalil, A.M., Novak, P., Novak, V., 2002. Comparison of 1.5 and 8 tesla high-resolution magnetic resonance imaging of lacunar infarcts. *Journal of Computer Assisted Tomography* 26, 628–632.

Chang, G., Diamond, M., Nevsky, G., Regatte, R.R., Weiss, D.S., 2013a. Early knee changes in dancers identified by ultra-high-field 7 T MRI. *Scandinavian Journal of Medicine & Science in Sports* 24, 678–682.

Chang, G., Madelin, G., Sherman, O.H., Strauss, E.J., Xia, D., Recht, M.P., Jerschow, A., Regatte, R.R., 2012a. Improved assessment of cartilage repair tissue using fluid-suppressed Na-23 inversion recovery MRI at 7 Tesla: Preliminary results. *European Radiology* 22, 1341–1349.

Chang, G., Wang, L.G., Cardenas-Blanco, A., Schweitzer, M.E., Recht, M.P., Regatte, R.R., 2010. Biochemical and physiological MR imaging of skeletal muscle at 7 Tesla and above. *Seminars in Musculoskeletal Radiology* 14, 269–278.

Chang, G., Wiggins, G.C., Xia, D., Lattanzi, R., Madelin, G., Raya, J.G., Finnerty, M., Fujita, H., Recht, M.P., Regatte, R.R., 2012b. Comparison of a 28-channel receive array coil and quadrature volume coil for morphologic imaging and T2 mapping of knee cartilage at 7T. *Journal of Magnetic Resonance Imaging* 35, 441–448.

Chang, G., Xia, D., Sherman, O., Strauss, E., Jazrawi, L., Recht, M.P., Regatte, R., 2013b. High resolution morphologic imaging and T2 mapping of cartilage at 7 Tesla: Comparison of cartilage repair patients and healthy controls. *Magnetic Resonance Materials in Physics, Biology, and Medicine* 26, 539–548.

Cheshkov, S., Dimitrov, I.E., Rispoli, J.V., 2012. Proton decoupled 13C MRS of the breast at 7T. *Proceedings of the 21st Annual ISMRM Meeting*, Melbourne, Australia.

Chmelik, M., Kukurova, I.J., Gruber, S., Krssak, M., Valkovic, L., Trattnig, S., Bogner, W., 2013a. Fully adiabatic 31P 2D-CSI with reduced chemical shift displacement error at 7 T--GOIA-1D-ISIS/2D-CSI. *Magnetic Resonance in Medicine* 69, 1233–1244.

Chmelik, M., Povazan, M., Jiru, F., Just-Kukurova, I., Dezortova, M., Krssak, M., Bogner, W., Hajek, M., Trattnig, S., Valkovic, L., 2013b. Flip-angle mapping of 31P coils by steady-state MR spectroscopic imaging. *Journal of Magnetic Resonance Imaging* doi:10.1002/jmri.24401.

Chmelik, M., Povazan, M., Krssak, M., Gruber, S., Tkacov, M., Trattnig, S., Bogner, W., 2014. In vivo (31) P magnetic resonance spectroscopy of the human liver at 7 T: An initial experience. *NMR in Biomedicine* 27, 478–485.

Choi, C., Dimitrov, I., Douglas, D., Zhao, C., Hawesa, H., Ghose, S., Tamminga, C.A., 2009a. In vivo detection of serine in the human brain by proton magnetic resonance spectroscopy (1H-MRS) at 7 Tesla. *Magnetic Resonance in Medicine: Official Journal of the Society of Magnetic Resonance in Medicine/Society of Magnetic Resonance in Medicine* 62, 1042–1046.

Choi, C., Dimitrov, I.E., Douglas, D., Patel, A., Kaiser, L.G., Amezcua, C.A., Maher, E.a., 2010a. Improvement of resolution for brain coupled metabolites by optimized (1)H MRS at 7T. *NMR in Biomedicine* 23, 1044–1052.

Choi, C., Douglas, D., Hawesa, H., Jindal, A., Storey, C., Dimitrov, I., 2009b. Measurement of glycine in human prefrontal brain by point-resolved spectroscopy at 7.0 tesla in vivo. *Magnetic Resonance in Medicine: Official Journal of the Society of Magnetic Resonance in Medicine/Society of Magnetic Resonance in Medicine* 62, 1305–1310.

Choi, C., Ghose, S., Uh, J., Patel, A., Dimitrov, I.E., Lu, H., Douglas, D., Ganji, S., 2010b. Measurement of N-acetylaspartylglutamate in the human frontal brain by 1H-MRS at 7 T. *Magnetic Resonance in Medicine: Official Journal of the Society of Magnetic Resonance in Medicine/Society of Magnetic Resonance in Medicine* 64, 1247–1251.

Da Costa, S., van der Zwaag, W., Marques, J.P., Frackowiak, R.S., Clarke, S., Saenz, M., 2011. Human primary auditory cortex follows the shape of Heschl's gyrus. *The Journal of Neuroscience* 31, 14067–14075.

Dagher, J., Reese, T., Bilgin, A., 2013. High-resolution, large dynamic range field map estimation. *Magnetic Resonance in Medicine* 71, 105–117.

De Graaf, R.A., Klomp, D.W., Luijten, P.R., Boer, V.O., 2013. Intramolecular zero-quantum-coherence 2D NMR spectroscopy of lipids in the human breast at 7 T. *Magnetic Resonance in Medicine* 71, 451–457.

De Graaf, W.L., Zwanenburg, J.J., Visser, F., Wattjes, M.P., Pouwels, P.J., Geurts, J.J., Polman, C.H., Barkhof, F., Luijten, P.R., Castelijns, J.A., 2011. Lesion detection at seven Tesla in multiple sclerosis using magnetisation prepared 3D-FLAIR and 3D-DIR. *European Radiology* 22, 221–231.

De Martino, F., Moerel, M., van de Moortele, P.F., Ugurbil, K., Goebel, R., Yacoub, E., Formisano, E., 2013a. Spatial organization of frequency preference and selectivity in the human inferior colliculus. *Nature Communications* 4, 1386.

De Martino, F., Zimmermann, J., Muckli, L., Ugurbil, K., Yacoub, E., Goebel, R., 2013b. Cortical depth dependent functional responses in humans at 7T: Improved specificity with 3D GRASE. *PLoS One* 8, e60514.

De Panfilis, C., Schwarzbauer, C., 2005. Positive or negative blips? The effect of phase encoding scheme on susceptibility-induced signal losses in EPI. *NeuroImage* 25, 112–121.

Deelchand, D.K., Iltis, I., Henry, P.-G., 2013. Improved quantification precision of human brain short echo-time (1) H magnetic resonance spectroscopy at high magnetic field: A simulation study. *Magnetic Resonance in Medicine: Official Journal of the Society of Magnetic Resonance in Medicine/Society of Magnetic Resonance in Medicine* 72, 20–25.

Deichmann, R., Gottfried, J.A., Hutton, C., Turner, R., 2003. Optimized EPI for fMRI studies of the orbitofrontal cortex. *NeuroImage* 19, 430–441.

Deichmann, R., Josephs, O., Hutton, C., Corfield, D.R., Turner, R., 2002. Compensation of susceptibility-induced BOLD sensitivity losses in echo-planar fMRI imaging. *NeuroImage* 15, 120–135.

Deistung, A., Rauscher, A., Sedlacik, J., Stadler, J., Witoszynskyj, S., Reichenbach, J.R., 2008. Susceptibility weighted imaging at ultra high magnetic field strengths: Theoretical considerations and experimental results. *Magnetic Resonance in Medicine* 60, 1155–1168.

Deistung, A., Schafer, A., Schweser, F., Biedermann, U., Turner, R., Reichenbach, J.R., 2013. Toward in vivo histology: A comparison of quantitative susceptibility mapping (QSM) with magnitude-, phase-, and R2*-imaging at ultra-high magnetic field strength. *NeuroImage* 65, 299–314.

Deligianni, X., Bar, P., Scheffler, K., Trattnig, S., Bieri, O., 2013. High-resolution Fourier-encoded sub-millisecond echo time musculoskeletal imaging at 3 Tesla and 7 Tesla. *Magnetic Resonance in Medicine* 70, 1434–1439.

Denk, C., Rauscher, A., 2009. Susceptibility weighted imaging with multiple echoes. *Journal of Magnetic Resonance Imaging* 31, 185–191.

Dezortova, M., Taimr, P., Skoch, A., Spicak, J., Hajek, M., 2005. Etiology and functional status of liver cirrhosis by 31P MR spectroscopy. *World Journal of Gastroenterology* 11, 6926–6931.

Dimitrov, I., Ren, J., Douglas, D., Davis, J., Sherry, A.D., Malloy, C.R., 2010a. Composition of fatty acids in adipose tissue by in vivo 13C MRS at 7T. *Proceedings of the 18th Annual ISMRM Meeting*, Stockholm, Sweden, p. 320.

Dimitrov, I., Ren, J., Douglas, D., Sherry, A.D., Malloy, C.R., 2010b. In vivo detection of trans-fatty acids by 13C MRS at 7T. *Proceedings of the 18th Annual ISMRM Meeting*, Stockholm, Sweden, p. 374.

Dimitrov, I.E., Douglas, D., Ren, J., Smith, N.B., Webb, A.G., Sherry, A.D., Malloy, C.R., 2012. In vivo determination of human breast fat composition by (1)H magnetic resonance spectroscopy at 7 T. *Magnetic Resonance in Medicine* 67, 20–26.

Dou, W., Palomero-Gallagher, N., van Tol, M.-J., Kaufmann, J., Zhong, K., Bernstein, H.-G., Heinze, H.-J., Speck, O., Walter, M., 2013. Systematic regional variations of GABA, glutamine, and glutamate concentrations follow receptor fingerprints of human cingulate cortex. *The Journal of Neuroscience: The Official Journal of the Society for Neuroscience* 33, 12698–12704.

Du, F., Zhu, X.-H., Qiao, H., Zhang, X., Chen, W., 2007. Efficient in vivo 31P magnetization transfer approach for noninvasively determining multiple kinetic parameters and metabolic fluxes of ATP metabolism in the human brain. *Magnetic Resonance in Medicine: Official Journal of the Society of Magnetic Resonance in Medicine/Society of Magnetic Resonance in Medicine* 57, 103–114.

Duerst, Y., Wilm, B.J., Dietrich, B.E., Vannesjo, S.J., Barmet, C., Schmid, T., Brunner, D.O., Pruessmann, K.P., 2014. Real-time feedback for spatiotemporal field stabilization in MR systems. *Magnetic Resonance in Medicine: Official Journal of the Society of Magnetic Resonance in Medicine/Society of Magnetic Resonance in Medicine* 73, 1–10.

Dymerska, B., Fischmeister, F., Geissler, A., Matt, E., Trattnig, S., Beisteiner, R., Robinson, S.D., 2014. Clinical relevance of EPI distortion correction in presurgical fMRI at 7 Tesla. *Proceedings of the 23rd Annual Meeting of the ISMRM*, Milan, Italy.

Eichner, C., Setsompop, K., Koopmans, P.J., Lutzkendorf, R., Norris, D.G., Turner, R., Wald, L.L., Heidemann, R.M., 2014. Slice accelerated diffusion-weighted imaging at ultra-high field strength. *Magnetic Resonance in Medicine* 71, 1518–1525.

Emir, U.E., Auerbach, E.J., Van De Moortele, P.-F., Marjańska, M., Uğurbil, K., Terpstra, M., Tkáč, I., Oz, G., 2012a. Regional neurochemical profiles in the human brain measured by ¹H MRS at 7 T using local B1 shimming. *NMR in Biomedicine* 25, 152–160.

Emir, U.E., Tuite, P.J., Öz, G., 2012b. Elevated pontine and putamenal GABA levels in mild-moderate Parkinson disease detected by 7 tesla proton MRS. *PLoS One* 7, e30918.

Erhardt, A., 2008. *Einführung in die Digitale Bildverarbeitung: Grundlagen, Systeme und Anwendungen.* Vieweg+Teubner, Wiesbaden, Germany, 1, 2.

Essig, M., Reichenbach, J.R., Schad, L.R., Schoenberg, S.O., Debus, J., Kaiser, W.A., 1999. High-resolution MR venography of cerebral arteriovenous malformations. *Magnetic Resonance Imaging* 17, 1417–1425.

Feinberg, D.A., Moeller, S., Smith, S.M., Auerbach, E., Ramanna, S., Gunther, M., Glasser, M.F., Miller, K.L., Ugurbil, K., Yacoub, E., 2010. Multiplexed echo planar imaging for sub-second whole brain FMRI and fast diffusion imaging. *PLoS One* 5, e15710.

Feng, W., Neelavalli, J., Haacke, E.M., 2012. Catalytic multi-echo phase unwrapping scheme (CAMPUS) in multi-echo gradient echo imaging: Removing phase wraps on a voxel-by-voxel basis. *Magnetic Resonance in Medicine* 70, 117–126.

Fiege, D.P., Romanzetti, S., Mirkes, C.C., Brenner, D., Shah, N.J., 2013. Simultaneous single-quantum and triple-quantum-filtered MRI of 23Na (SISTINA). *Magnetic Resonance in Medicine* 69, 1691–1696.

Fillmer, A., Kirchner, T., Cameron, D., Henning, A., 2014a. Constrained image-based B0 shimming accounting for "local minimum traps" in the optimization and field inhomogeneities outside the region of interest. *Magnetic*

Resonance in Medicine: Official Journal of the Society of Magnetic Resonance in Medicine/Society of Magnetic Resonance in Medicine 73, 1370–1380.

Fillmer, A., Kirchner, T., Cameron, D., Henning, A., 2014b. Constrained image-based B shimming accounting for "local minimum traps" in the optimization and field inhomogeneities outside the region of interest. Magnetic Resonance in Medicine 73, 1370–1380.

Fleysher, L., Oesingmann, N., Brown, R., Sodickson, D.K., Wiggins, G.C., Inglese, M., 2013. Noninvasive quantification of intracellular sodium in human brain using ultra-high-field MRI. NMR in Biomedicine 26, 9–19.

Fleysher, L., Oesingmann, N., Stoeckel, B., Grossman, R.I., Inglese, M., 2009. Sodium long-component T(2)(*) mapping in human brain at 7 Tesla. Magnetic Resonance in Medicine 62, 1338–1341.

Forlenza, O.V., Wacker, P., Nunes, P.V., Yacubian, J., Castro, C.C., Otaduy, M.C., Gattaz, W.F., 2005. Reduced phospholipid breakdown in Alzheimer's brains: A 31P spectroscopy study. Psychopharmacology (Berlin) 180, 359–365.

Frahm, J., Merboldt, K., Hänicke, W., Kleinschmidt, A., Boecker, H., 1994. Brain or vein-oxygenation or flow? On signal physiology in functional MRI of human brain activation. NMR in Biomedicine 7, 45–53.

Fuchs, A., Luttje, M., Boesiger, P., Henning, A., 2013. SPECIAL semi-LASER with lipid artifact compensation for 1H MRS at 7 T. Magnetic Resonance in Medicine: Official Journal of the Society of Magnetic Resonance in Medicine/Society of Magnetic Resonance in Medicine 69, 603–612.

Fujimoto, K., Polimeni, J.R., van der Kouwe, A.J., Reuter, M., Kober, T., Benner, T., Fischl, B., Wald, L.L., 2014. Quantitative comparison of cortical surface reconstructions from MP2RAGE and multi-echo MPRAGE data at 3 and 7T. NeuroImage 90, 60–73.

Gajdosik, M., Chmelik, M., Just-Kukurova, I., Bogner, W., Valkovic, L., Trattnig, S., Krssak, M., 2013. In vivo relaxation behavior of liver compounds at 7 tesla, measured by single-voxel proton MR spectroscopy. Journal of Magnetic Resonance Imaging 40, 1365–1374.

Garwood, M., DelaBarre, L., 2001. The return of the frequency sweep: Designing adiabatic pulses for contemporary NMR. Journal of Magnetic Resonance 153, 155–177.

Gati, J.S., Menon, R.S., Ugurbil, K., Rutt, B.K., 1997. Experimental determination of the BOLD field strength dependence in vessels and tissue. Magnetic Resonance in Medicine 38, 296–302.

Ge, Y., Zohrabian, V.M., Grossman, R.I., 2008. Seven-Tesla magnetic resonance imaging: New vision of microvascular abnormalities in multiple sclerosis. Archives of Neurology 65, 812–816.

Geissler, A., Fischmeister, F.P., Grabner, G., Wurnig, M., Rath, J., Foki, T., Matt, E., Trattnig, S., Beisteiner, R., Robinson, S.D., 2013. Comparing the microvascular specificity of the 3 T and 7 T BOLD response using ICA and susceptibility-weighted imaging. Frontiers in Human Neuroscience 7, 474.

Geißler, A., Matt, E., Fischmeister, F., Wurnig, M., Dymerska, B., Knosp, E., Feucht, M. et al., 2014. Differential functional benefits of ultra highfield MR systems within the language network 103, 163–170.

Grabner, G., Dal-Bianco, A., Schernthaner, M., Vass, K., Lassmann, H., Trattnig, S., 2011. Analysis of multiple sclerosis lesions using a fusion of 3.0 T FLAIR and 7.0 T SWI phase: FLAIR SWI. Journal of Magnetic Resonance Imaging 33, 543–549.

Grabner, G., Nobauer, I., Elandt, K., Kronnerwetter, C., Woehrer, A., Marosi, C., Prayer, D., Trattnig, S., Preusser, M., 2012. Longitudinal brain imaging of five malignant glioma patients treated with bevacizumab using susceptibility-weighted magnetic resonance imaging at 7 T. Magnetic Resonance Imaging 30, 139–147.

Grams, A.E., Brote, I., Maderwald, S., Kollia, K., Ladd, M.E., Forsting, M., Gizewski, E.R., 2011. Cerebral magnetic resonance spectroscopy at 7 Tesla: Standard values and regional differences. Academic Radiology 18, 584–587.

Grinstead, J.W., Speck, O., Paul, D., Silbert, L., Perkins, L., Rooney, W., 2010. Whole-brain FLAIR using 3D TSE with variable flip angle readouts optimized for 7 Tesla #3034. Joint Annual Meeting of the ISMRM-ESMRMB. ISMRM, Stockholm, Sweden.

Griswold, M.A., Jakob, P.M., Heidemann, R.M., Nittka, M., Jellus, V., Wang, J., Kiefer, B., Haase, A., 2002. Generalized autocalibrating partially parallel acquisitions (GRAPPA). Magnetic Resonance in Medicine 47, 1202–1210.

Gruber, S., Pinker, K., Zaric, O., Minarikova, L., Chmelik, M., Baltzer, P., Boubela, R.N., Helbich, T., Bogner, W., Trattnig, S., 2014. Dynamic contrast-enhanced magnetic resonance imaging of breast tumors at 3 and 7 T: A comparison. Investigative Radiology 49, 354–362.

Guivel-Scharen, V., Sinnwell, T., Wolff, S.D., Balaban, R.S., 1998. Detection of proton chemical exchange between metabolites and water in biological tissues. Journal of Magnetic Resonance 133, 36–45.

Gupta, R.K., Benovic, J.L., Rose, Z.B., 1978. The determination of the free magnesium level in the human red blood cell by 31P NMR. The Journal of Biological Chemistry 253, 6172–6176.

Haacke, E.M., Ayaz, M., Khan, A., Manova, E.S., Krishnamurthy, B., Gollapalli, L., Ciulla, C., Kim, I., Petersen, F., Kirsch, W., 2007. Establishing a baseline phase behavior in magnetic resonance imaging to determine normal vs. abnormal iron content in the brain. Journal of Magnetic Resonance Imaging 26, 256–264.

Haacke, E.M., Cheng, N.Y., House, M.J., Liu, Q., Neelavalli, J., Ogg, R.J., Khan, A., Ayaz, M., Kirsch, W., Obenaus, A., 2005. Imaging iron stores in the brain using magnetic resonance imaging. Magnetic Resonance Imaging 23, 1–25.

Haacke, E.M., Xu, Y.B., Cheng, Y.C.N., Reichenbach, J.R., 2004. Susceptibility weighted imaging (SWI). Magnetic Resonance in Medicine 52, 612–618.

Hahn, A., Kranz, G.S., Seidel, E.M., Sladky, R., Kraus, C., Kublbock, M., Pfabigan, D.M. et al., 2013. Comparing neural response to painful electrical stimulation with functional MRI at 3 and 7 T. NeuroImage 82, 336–343.

Hajnal, J.V., Bryant, D.J., Kasuboski, L., Pattany, P.M., De Coene, B., Lewis, P.D., Pennock, J.M., Oatridge, A., Young, I.R., Bydder, G.M., 1992. Use of fluid attenuated inversion recovery (FLAIR) pulse sequences in MRI of the brain. Journal of Computer Assisted Tomography 16, 841–844.

Hall, E.L., Stephenson, M.C., Price, D., Morris, P.G., 2014. Methodology for improved detection of low concentration metabolites in MRS: Optimised combination of signals from multi-element coil arrays. *NeuroImage* 86, 35–42.

Hammond, K.E., Metcalf, M., Carvajal, L., Okuda, D.T., Srinivasan, R., Vigneron, D., Nelson, S.J., Pelletier, D., 2008. Quantitative in vivo magnetic resonance imaging of multiple sclerosis at 7 Tesla with sensitivity to iron. *Annals of Neurology* 64, 707–713.

Hargreaves, B.A., Cunningham, C.H., Nishimura, D.G., Conolly, S.M., 2004. Variable-rate selective excitation for rapid MRI sequences. *Magnetic Resonance in Medicine: Official Journal of the Society of Magnetic Resonance in Medicine/Society of Magnetic Resonance in Medicine* 52, 590–597.

Heidemann, R.M., Anwander, A., Feiweier, T., Knosche, T.R., Turner, R., 2012. k-space and q-space: Combining ultra-high spatial and angular resolution in diffusion imaging using ZOOPPA at 7 T. *NeuroImage* 60, 967–978.

Heidemann, R.M., Porter, D.A., Anwander, A., Feiweier, T., Heberlein, K., Knosche, T.R., Turner, R., 2010. Diffusion imaging in humans at 7T using readout-segmented EPI and GRAPPA. *Magnetic Resonance in Medicine* 64, 9–14.

Hennig, J., Scheffler, K., 2001. Hyperechoes. *Magnetic Resonance in Medicine: Official Journal of the Society of Magnetic Resonance in Medicine/Society of Magnetic Resonance in Medicine* 46, 6–12.

Henning, A., Fuchs, A., Murdoch, J.B., Boesiger, P., 2009. Slice-selective FID acquisition, localized by outer volume suppression (FIDLOVS) for (1)H-MRSI of the human brain at 7 T with minimal signal loss. *NMR in Biomedicine* 22, 683–696.

Hermier, M., Nighoghossian, N., Derex, L., Adeleine, P., Wiart, M., Berthezene, Y., Cotton, F. et al., 2003. Hypointense transcerebral veins at T2*-weighted MRI: A marker of hemorrhagic transformation risk in patients treated with intravenous tissue plasminogen activator. *Journal of Cerebral Blood Flow & Metabolism* 23, 1362–1370.

Hermier, M., Nighoghossian, N., Derex, L., Wiart, M., Nemoz, C., Berthezene, Y., Froment, J.C., 2005. Hypointense leptomeningeal vessels at T2*-weighted MRI in acute ischemic stroke. *Neurology* 65, 652–653.

Hetherington, H.P., 2011. RF Shimming for spectroscopic localization in the human brain at 7T. *Magnetic Resonance in Medicine: Official Journal of the Society of Magnetic Resonance in Medicine/Society of Magnetic Resonance in Medicine* 63, 9–19.

Hetherington, H.P., Hamid, H., Kulas, J., Ling, G., Bandak, F., de Lanerolle, N.C., Pan, J.W., 2014. MRSI of the medial temporal lobe at 7 T in explosive b mild traumatic brain injury. *Magnetic Resonance in Medicine: Official Journal of the Society of Magnetic Resonance in Medicine/Society of Magnetic Resonance in Medicine* 71, 1358–1367.

Heule, R., Ganter, C., Bieri, O., 2014. Triple Echo Steady-State (TESS) relaxometry. *Magnetic Resonance in Medicine* 71, 230–237.

Hingwala, D., Kesavadas, C., Thomas, B., Kapilamoorthy, T.R., 2010. Clinical utility of susceptibility-weighted imaging in vascular diseases of the brain. *Neurology India* 58, 602–607.

Hodel, J., Silvera, J., Bekaert, O., Rahmouni, A., Bastuji-Garin, S., Vignaud, A., Petit, E., Durning, B., Decq, P., 2011. Intracranial cerebrospinal fluid spaces imaging using a pulse-triggered three-dimensional turbo spin echo MR sequence with variable flip-angle distribution. *European Radiology* 21, 402–410.

Hoffmann, M.B., Stadler, J., Kanowski, M., Speck, O., 2009. Retinotopic mapping of the human visual cortex at a magnetic field strength of 7T. *Clinical Neurophysiology* 120, 108–116.

Intrapiromkul, J., Zhu, H., Cheng, Y., Barker, P.B., Edden, R.a. E., 2013. Determining the in vivo transverse relaxation time of GABA in the human brain at 7T. *Journal of Magnetic Resonance Imaging* 38, 1224–1229.

Jenkinson, M., 2003. Fast, automated, N-dimensional phase-unwrapping algorithm. *Magnetic Resonance in Medicine* 49, 193–197.

Jeong, H.K., Gore, J.C., Anderson, A.W., 2013. High-resolution human diffusion tensor imaging using 2-D navigated multishot SENSE EPI at 7 T. *Magnetic Resonance in Medicine* 69, 793–802.

Jezzard, P., Balaban, R.S., 1995. Correction for geometric distortion in echo planar images from B0 field variations. *Magnetic Resonance in Medicine* 34, 65–73.

Jubault, T., Brambati, S.M., Degroot, C., Kullmann, B., Strafella, A.P., Lafontaine, A.L., Chouinard, S., Monchi, O., 2009. Regional brain stem atrophy in idiopathic Parkinson's disease detected by anatomical MRI. *PLoS One* 4, e8247.

Juchem, C., Nixon, T.W., Diduch, P., Rothman, D.L., Starewicz, P., de Graaf, R.A., 2010. Dynamic shimming of the human brain at 7 Tesla. *Concepts in Magnetic Resonance Part B Magnetic Resonance Engineering* 37B, 116–128.

Juras, V., Apprich, S., Szomolanyi, P., Bieri, O., Deligianni, X., Trattnig, S., 2013a. Bi-exponential T2 analysis of healthy and diseased Achilles tendons: An in vivo preliminary magnetic resonance study and correlation with clinical score. *European Radiology* 23, 2814–2822.

Juras, V., Apprich, S., Szomolanyi, P., Bieri, O., Deligianni, X., Trattnig, S., 2013b. Bi-exponential T2* analysis of healthy and diseased Achilles tendons: An in vivo preliminary magnetic resonance study and correlation with clinical score. *European Radiology* 23, 2814–2822.

Juras, V., Apprich, S., Zbyn, S., Zak, L., Deligianni, X., Szomolanyi, P., Bieri, O., Trattnig, S., 2014. Quantitative MRI analysis of menisci using biexponential T-2* fitting with a variable echo time sequence. *Magnetic Resonance in Medicine* 71, 1015–1023.

Juras, V., Welsch, G., Bar, P., Kronnerwetter, C., Fujita, H., Trattnig, S., 2012a. Comparison of 3 T and 7 T MRI clinical sequences for ankle imaging. *European Journal of Radiology* 81, 1846–1850.

Juras, V., Zbyn, S., Pressl, C., Valkovic, L., Szomolanyi, P., Frollo, I., Trattnig, S., 2012b. Regional variations of T-2* in healthy and pathologic achilles tendon in vivo at 7 Tesla: Preliminary results. *Magnetic Resonance in Medicine* 68, 1607–1613.

Just Kukurova, I., Valkovic, L., Bogner, W., Gajdosik, M., Krssak, M., Gruber, S., Trattnig, S., Chmelik, M., 2014. Two-dimensional spectroscopic imaging with combined free induction decay and long-TE acquisition (FID echo spectroscopic imaging, FIDESI) for the detection of intramyocellular lipids in calf muscle at 7 T. *NMR in Biomedicine* 27, 980–987.

Kaiser, L.G., Young, K., Matson, G.B., 2008. Numerical simulations of localized high field 1H MR spectroscopy. *Journal of Magnetic Resonance* (San Diego, CA, 1997) 195, 67–75.

Kan, H.E., Klomp, D.W., Wong, C.S., Boer, V.O., Webb, A.G., Luijten, P.R., Jeneson, J.A., 2010. In vivo 31P MRS detection of an alkaline inorganic phosphate pool with short T1 in human resting skeletal muscle. *NMR in Biomedicine* 23, 995–1000.

Kan, H.E., Techawiboonwong, A., van Osch, M.J.P., Versluis, M.J., Deelchand, D.K., Henry, P.-G., Marjańska, M., van Buchem, M.A., Webb, A.G., Ronen, I., 2012. Differences in apparent diffusion coefficients of brain metabolites between grey and white matter in the human brain measured at 7 T. *Magnetic Resonance in Medicine: Official Journal of the Society of Magnetic Resonance in Medicine/Society of Magnetic Resonance in Medicine* 67, 1203–1209.

Kemp, G.J., Meyerspeer, M., Moser, E., 2007. Absolute quantification of phosphorus metabolite concentrations in human muscle in vivo by 31P MRS: A quantitative review. *NMR in Biomedicine* 20, 555–565.

Kerchner, G.A., Berdnik, D., Shen, J.C., Bernstein, J.D., Fenesy, M.C., Deutsch, G.K., Wyss-Coray, T., Rutt, B.K., 2014. APOE epsilon 4 worsens hippocampal CA1 apical neuropil atrophy and episodic memory. *Neurology* 82, 691–697.

Kerchner, G.A., Hess, C.P., Hammond-Rosenbluth, K.E., Xu, D., Rabinovici, G.D., Kelley, D.A., Vigneron, D.B., Nelson, S.J., Miller, B.L., 2010. Hippocampal CA1 apical neuropil atrophy in mild Alzheimer disease visualized with 7-T MRI. *Neurology* 75, 1381–1387.

Keuken, M.C., Bazin, P.L., Schafer, A., Neumann, J., Turner, R., Forstmann, B.U., 2013. Ultra-high 7T MRI of structural age-related changes of the subthalamic nucleus. *Journal of Neuroscience* 33, 4896–4900.

Khuu, A., Ren, J., Dimitrov, I., Woessner, D., Murdoch, J., Sherry, A.D., Malloy, C.R., 2009. Orientation of lipid strands in the extracellular compartment of muscle: Effect on quantitation of intramyocellular lipids. *Magnetic Resonance in Medicine* 61, 16–21.

Kilsdonk, I.D., de Graaf, W.L., Soriano, A.L., Zwanenburg, J.J., Visser, F., Kuijer, J.P., Geurts, J.J. et al., 2012. Multicontrast MR imaging at 7T in multiple sclerosis: Highest lesion detection in cortical gray matter with 3D-FLAIR. *American Journal of Neuroradiology* 34, 791–796.

Kilsdonk, I.D., Wattjes, M.P., Lopez-Soriano, A., Kuijer, J.P., de Jong, M.C., de Graaf, W.L., Conijn, M.M. et al., 2014. Improved differentiation between MS and vascular brain lesions using FLAIR* at 7 Tesla. *European Radiology* 24, 841–849.

Kim, S.J., Lyoo, I.K., Lee, Y.S., Lee, J.Y., Yoon, S.J., Kim, J.E., Kim, J.H., Hong, S.J., Jeong, D.U., 2009. Gray matter deficits in young adults with narcolepsy. *Acta Neurologica Scandinavica* 119, 61–67.

Kirchner, T., Fillmer, A., Tsao, J., Pruessmann, K.P., Henning, A., 2014. Reduction of voxel bleeding in highly accelerated parallel (1) H MRSI by direct control of the spatial response function. *Magnetic Resonance in Medicine: Official Journal of the Society of Magnetic Resonance in Medicine/Society of Magnetic Resonance in Medicine* 73, 1–12.

Klomp, D.W., Bitz, A.K., Heerschap, A., Scheenen, T.W., 2009. Proton spectroscopic imaging of the human prostate at 7 T. *NMR in Biomedicine* 22, 495–501.

Klomp, D.W., Scheenen, T.W., Arteaga, C.S., van Asten, J., Boer, V.O., Luijten, P.R., 2011a. Detection of fully refocused polyamine spins in prostate cancer at 7 T. *NMR in Biomedicine* 24, 299–306.

Klomp, D.W., van de Bank, B.L., Raaijmakers, A., Korteweg, M.A., Possanzini, C., Boer, V.O., van de Berg, C.A., van de Bosch, M.A., Luijten, P.R., 2011b. 31P MRSI and 1H MRS at 7 T: Initial results in human breast cancer. *NMR in Biomedicine* 24, 1337–1342.

Kobus, T., Bitz, A.K., van Uden, M.J., Lagemaat, M.W., Rothgang, E., Orzada, S., Heerschap, A., Scheenen, T.W., 2012. In vivo 31P MR spectroscopic imaging of the human prostate at 7 T: Safety and feasibility. *Magnetic Resonance in Medicine* 68, 1683–1695.

Koch, K.S., Leffert, H.L., 1979. Increased sodium ion influx is necessary to initiate rat hepatocyte proliferation. *Cell* 18, 153–163.

Kogan, F., Haris, M., Singh, A., Cai, K., Debrosse, C., Nanga, R.P., Hariharan, H., Reddy, R., 2014. Method for high-resolution imaging of creatine in vivo using chemical exchange saturation transfer. *Magnetic Resonance in Medicine* 71, 164–172.

Kollia, K., Maderwald, S., Putzki, N., Schlamann, M., Theysohn, J.M., Kraff, O., Ladd, M.E., Forsting, M., Wanke, I., 2009. First clinical study on ultra-high-field MR imaging in patients with multiple sclerosis: Comparison of 1.5T and 7T. *American Journal of Neuroradiology* 30, 699–702.

Konstandin, S., Nagel, A.M., 2013. Performance of sampling density-weighted and postfiltered density-adapted projection reconstruction in sodium magnetic resonance imaging. *Magnetic Resonance in Medicine* 69, 495–502.

Koopmans, P.J., Barth, M., Orzada, S., Norris, D.G., 2011. Multi-echo fMRI of the cortical laminae in humans at 7 T. *NeuroImage* 56, 1276–1285.

Koopmans, P.J., Boyacioglu, R., Barth, M., Norris, D.G., 2012. Whole brain, high resolution spin-echo resting state fMRI using PINS multiplexing at 7 T. *NeuroImage* 62, 1939–1946.

Korteweg, M.A., Veldhuis, W.B., Visser, F., Luijten, P.R., Mali, W.P., van Diest, P.J., van den Bosch, M.A., Klomp, D.J., 2011. Feasibility of 7 Tesla breast magnetic resonance imaging determination of intrinsic sensitivity and high-resolution magnetic resonance imaging, diffusion-weighted imaging, and (1)H-magnetic resonance spectroscopy of breast cancer patients receiving neoadjuvant therapy. *Investigative Radiology* 46, 370–376.

Koush, Y., Elliott, M.A., Mathiak, K., 2011. Single voxel proton spectroscopy for neurofeedback at 7 Tesla. *Materials* 4, 1548–1563.

Koush, Y., Elliott, M.A., Scharnowski, F., Mathiak, K., 2013. Real-time automated spectral assessment of the BOLD response for neurofeedback at 3 and 7T. *Journal of Neuroscience Methods* 218, 148–160.

Krssak, M., Falk Petersen, K., Dresner, A., DiPietro, L., Vogel, S.M., Rothman, D.L., Roden, M., Shulman, G.I., 1999. Intramyocellular lipid concentrations are correlated with insulin sensitivity in humans: A 1H NMR spectroscopy study. *Diabetologia* 42, 113–116.

Krssak, M., Gajdosik, M., Valkovic, L., Bogner, W., Krebs, M., Luger, A., Trattnig, S., Chmelik, M., 2014. Detection of hepatic glycogen by 1D ISIS localized 13C MRS at 7T. *Proceedings of the 22nd Annual ISMRM Meeting*, Milan, Italy, p. 1437.

Krssak, M., Petersen, K.F., Bergeron, R., Price, T., Laurent, D., Rothman, D.L., Roden, M., Shulman, G.I., 2000. Intramuscular glycogen and intramyocellular lipid utilization during prolonged exercise and recovery in man: A 13C and 1H nuclear magnetic resonance spectroscopy study. *The Journal of Clinical Endocrinology & Metabolism* 85, 748–754.

Krug, R., Carballido-Gamio, J., Banerjee, S., Stahl, R., Carvajal, L., Xu, D., Vigneron, D., Kelley, D.A.C., Link, T.M., Majumdar, S., 2007. In vivo bone and cartilage MRI using fully-balanced steady-state free-precession at 7 Tesla. *Magnetic Resonance in Medicine* 58, 1294–1298.

Krug, R., Larson, P.E.Z., Wang, C.S., Burghardt, A.J., Kelley, D.A.C., Link, T.M., Zhang, X.L., Vigneron, D.B., Majumdar, S., 2011. Ultrashort echo time MRI of cortical bone at 7 Tesla field strength: A feasibility study. *Journal of Magnetic Resonance Imaging* 34, 691–695.

Krusche-Mandl, I., Schmitt, B., Zak, L., Apprich, S., Aldrian, S., Juras, V., Friedrich, K.M., Marlovits, S., Weber, M., Trattnig, S., 2012. Long-term results 8 years after autologous osteochondral transplantation: 7 T gagCEST and sodium magnetic resonance imaging with morphological and clinical correlation. *Osteoarthritis Cartilage* 20, 357–363.

Kuehn, E., Mueller, K., Turner, R., Schutz-Bosbach, S., 2014. The functional architecture of S1 during touch observation described with 7 T fMRI. *Brain Structure and Function* 219, 119–140.

Kukurova, I.J., Krssak, M., Chmelik, M., Gajdosik, M., Trattnig, S., Valkovic, L., 2014. Carnosine at 7T: Quantification and relaxation times in m. gastrocnemius. *Proceedings of the 23rd Annual ISMRM Meeting*, Milan, Italy.

Lagemaat, M.W., Maas, M.C., Vos, E.K., Bitz, A.K., Orzada, S., Weiland, E., van Uden, M.J., Kobus, T., Heerschap, A., Scheenen, T.W., 2014a. 31P MR spectroscopic imaging of the human prostate at 7 T: T1 relaxation times, Nuclear Overhauser Effect, and spectral characterization. *Magnetic Resonance in Medicine* 73, 909–920.

Lagemaat, M.W., Vos, E.K., Maas, M.C., Bitz, A.K., Orzada, S., van Uden, M.J., Kobus, T., Heerschap, A., Scheenen, T.W.J., 2014b. Phosphorus magnetic resonance spectroscopic imaging at 7 T in patients with prostate cancer. *Investigative Radiology* 49, 363–372.

Langkammer, C., Krebs, N., Goessler, W., Scheurer, E., Ebner, F., Yen, K., Fazekas, F., Ropele, S., 2010. Quantitative MR imaging of brain iron: A postmortem validation study. *Radiology* 257, 455–462.

Langkammer, C., Schweser, F., Krebs, N., Deistung, A., Goessler, W., Scheurer, E., Sommer, K. et al., 2012. Quantitative susceptibility mapping (QSM) as a means to measure brain iron? A post mortem validation study. *NeuroImage* 62, 1593–1599.

Larsson, E.G., Erdogmus, D., Yan, R., Principe, J.C., Fitzsimmons, J.R., 2003. SNR-optimality of sum-of-squares reconstruction for phased-array magnetic resonance imaging. *Journal of Magnetic Resonance* 163, 121–123.

Lee, B.C., Vo, K.D., Kido, D.K., Mukherjee, P., Reichenbach, J., Lin, W., Yoon, M.S., Haacke, M., 1999. MR high-resolution blood oxygenation level-dependent venography of occult (low-flow) vascular lesions. *American Journal of Neuroradiology* 20, 1239–1242.

Lei, H., Zhu, X.-H., Zhang, X.-L., Ugurbil, K., Chen, W., 2003. In vivo 31P magnetic resonance spectroscopy of human brain at 7 T: An initial experience. *Magnetic Resonance in Medicine: Official Journal of the Society of Magnetic Resonance in Medicine/Society of Magnetic Resonance in Medicine* 49, 199–205.

Lin, Y., Stephenson, M.C., Xin, L., Napolitano, A., Morris, P.G., 2012. Investigating the metabolic changes due to visual stimulation using functional proton magnetic resonance spectroscopy at 7T. *Journal of Cerebral Blood Flow & Metabolism* 32, 1484–1495.

Ling, W., Eliav, U., Navon, G., Jerschow, A., 2008a. Chemical exchange saturation transfer by intermolecular double-quantum coherence. *Journal of Magnetic Resonance* 194, 29–32.

Ling, W., Regatte, R.R., Navon, G., Jerschow, A., 2008b. Assessment of glycosaminoglycan concentration in vivo by chemical exchange-dependent saturation transfer (gagCEST). *Proceedings of the National Academy of Sciences of the USA* 105, 2266–2270.

Ling, W., Regatte, R.R., Schweitzer, M.E., Jerschow, A., 2008c. Characterization of bovine patellar cartilage by NMR. *NMR in Biomedicine* 21, 289–295.

Liu, T., Spincemaille, P., de Rochefort, L., Kressler, B., Wang, Y., 2009. Calculation of susceptibility through multiple orientation sampling (COSMOS): A method for conditioning the inverse problem from measured magnetic field map to susceptibility source image in MRI. *Magnetic Resonance in Medicine* 61, 196–204.

Lusebrink, F., Wollrab, A., Speck, O., 2013. Cortical thickness determination of the human brain using high resolution 3T and 7T MRI data. *NeuroImage* 70, 122–131.

Luttje, M.P., Italiaander, M.G., Arteaga de Castro, C.S., van der Kemp, W.J., Luijten, P.R., van Vulpen, M., van der Heide, U.A., Klomp, D.W., 2013. 31P MR spectroscopic imaging combined with H MR spectroscopic imaging in the human prostate using a double tuned endorectal coil at 7T. *Magnetic Resonance in Medicine* 72, 1516–1521.

Madelin, G., Babb, J., Xia, D., Chang, G., Krasnokutsky, S., Abramson, S.B., Jerschow, A., Regatte, R.R., 2013. Articular cartilage: Evaluation with fluid-suppressed 7.0-T sodium MR imaging in subjects with and subjects without osteoarthritis. *Radiology* 268, 481–491.

Madelin, G., Jerschow, A., Regatte, R.R., 2012. Sodium relaxation times in the knee joint in vivo at 7T. *NMR in Biomedicine* 25, 530–537.

Madelin, G., Lee, J.S., Inati, S., Jerschow, A., Regatte, R.R., 2010. Sodium inversion recovery MRI of the knee joint in vivo at 7T. *Journal of Magnetic Resonance* 207, 42–52.

Mandl, R.C.W., van den Heuvel, M.P., Klomp, D.W.J., Boer, V.O., Siero, J.C.W., Luijten, P.R., Hulshoff Pol, H.E., 2012. Tract-based magnetic resonance spectroscopy of the cingulum bundles at 7 T. *Human Brain Mapping* 33, 1503–1511.

Mangia, S., Tkác, I., Gruetter, R., Van De Moortele, P.-F., Giove, F., Maraviglia, B., Uğurbil, K., 2006. Sensitivity of single-voxel 1H-MRS in investigating the metabolism of the activated human visual cortex at 7 T. *Magnetic Resonance Imaging* 24, 343–348.

Mangia, S., Tkác, I., Gruetter, R., Van de Moortele, P.-F., Maraviglia, B., Uğurbil, K., 2007. Sustained neuronal activation raises oxidative metabolism to a new steady-state level: Evidence from 1H NMR spectroscopy in the human visual cortex. *Journal of Cerebral Blood Flow & Metabolism* 27, 1055–1063.

Marjańska, M., Auerbach, E.J., Valabrègue, R., Van de Moortele, P.-F., Adriany, G., Garwood, M., 2012. Localized 1H NMR spectroscopy in different regions of human brain in vivo at 7T: T2 relaxation times and concentrations of cerebral metabolites, *NMR in Biomedicine* 25, 332–339.

Marques, J.P., Gruetter, R., 2013. New developments and applications of the MP2RAGE sequence—Focusing the contrast and high spatial resolution R1 mapping. *PLoS One* 8, e69294.

Marques, J.P., Kober, T., Krueger, G., van der Zwaag, W., Van de Moortele, P.F., Gruetter, R., 2010. MP2RAGE, a self bias-field corrected sequence for improved segmentation and T1-mapping at high field. *NeuroImage* 49, 1271–1281.

Marsman, A., Mandl, R.C.W., van den Heuvel, M.P., Boer, V.O., Wijnen, J.P., Klomp, D.W.J., Luijten, P.R., Hilleke E, H.P., 2013. Glutamate changes in healthy young adulthood. *European Neuropsychopharmacology: The Journal of the European College of Neuropsychopharmacology* 23, 1484–1490.

Martuzzi, R., van der Zwaag, W., Farthouat, J., Gruetter, R., Blanke, O., 2014. Human finger somatotopy in areas 3b, 1, and 2: A 7T fMRI study using a natural stimulus. *Human Brain Mapping* 35, 213–226.

Maudsley, A.A., Hilal, S.K., Perman, W.H., Simon, H.E., 1983. Spatially resolved high-resolution spectroscopy by 4-dimensional NMR. *Journal of Magnetic Resonance* 51, 147–152.

McDougall, M.P., Cheshkov, S., Rispoli, J., Malloy, C., Dimitrov, I., Wright, S.M., 2014. Quadrature transmit coil for breast imaging at 7 tesla using forced current excitation for improved homogeneity. *Journal of Magnetic Resonance Imaging*40, 1165–1173.

Mekle, R., Mlynárik, V., Gambarota, G., Hergt, M., Krueger, G., Gruetter, R., 2009. MR spectroscopy of the human brain with enhanced signal intensity at ultrashort echo times on a clinical platform at 3T and 7T. *Magnetic Resonance in Medicine: Official Journal of the Society of Magnetic Resonance in Medicine/Society of Magnetic Resonance in Medicine* 61, 1279–1285.

Merboldt, K.D., Finsterbusch, J., Frahm, J., 2000. Reducing inhomogeneity artifacts in functional MRI of human brain activation-thin sections vs gradient compensation. *Journal of Magnetic Resonance* 145, 184–191.

Metzger, C.D., Eckert, U., Steiner, J., Sartorius, A., Buchmann, J.E., Stadler, J., Tempelmann, C. et al., 2010. High field FMRI reveals thalamocortical integration of segregated cognitive and emotional processing in mediodorsal and intralaminar thalamic nuclei. *Frontiers in Neuroanatomy* 4, 138.

Meyerspeer, M., Kemp, G.J., Mlynarik, V., Krssak, M., Szendroedi, J., Nowotny, P., Roden, M., Moser, E., 2007. Direct noninvasive quantification of lactate and high energy phosphates simultaneously in exercising human skeletal muscle by localized magnetic resonance spectroscopy. *Magnetic Resonance in Medicine* 57, 654–660.

Meyerspeer, M., Robinson, S., Nabuurs, C.I., Scheenen, T., Schoisengeier, A., Unger, E., Kemp, G.J., Moser, E., 2012. Comparing localized and nonlocalized dynamic 31P magnetic resonance spectroscopy in exercising muscle at 7 T. *Magnetic Resonance in Medicine* 68, 1713–1723.

Meyerspeer, M., Roig, E.S., Gruetter, R., Magill, A.W., 2013. An improved trap design for decoupling multinuclear RF coils. *Magnetic Resonance in Medicine* doi:10.1002/mrm.24931.

Meyerspeer, M., Scheenen, T., Schmid, A.I., Mandl, T., Unger, E., Moser, E., 2011. Semi-LASER localized dynamic 31P magnetic resonance spectroscopy in exercising muscle at ultra-high magnetic field. *Magnetic Resonance in Medicine* 65, 1207–1215.

Minarikova, L., Wolfgang, B., Zaric, O., Pinker-Domenig, K., Helbich, T., Trattnig, S., Gruber, S., 2014. *Breast Diffusion-Weighted Imaging at 3 and 7 Tesla: Comparison Study.* ISMRM-ESMRMB, Milano, Italy, p. 0704.

Moenninghoff, C., Maderwald, S., Theysohn, J.M., Kraff, O., Ladd, M.E., El Hindy, N., van de Nes, J., Forsting, M., Wanke, I., 2010. Imaging of adult astrocytic brain tumours with 7 T MRI: Preliminary results. *European Radiology* 20, 704–713.

Moheet, A., Emir, U.E., Terpstra, M., Kumar, A., Eberly, L.E., Seaquist, E.R., Öz, G., 2014. Initial experience with seven tesla magnetic resonance spectroscopy of hypothalamic GABA during hyperinsulinemic euglycemia and hypoglycemia in healthy humans. *Magnetic Resonance in Medicine: Official Journal of the Society of Magnetic Resonance in Medicine/Society of Magnetic Resonance in Medicine* 71, 12–18.

Moon, C.H., Kim, J.H., Zhao, T., Bae, K.T., 2013. Quantitative (23) Na MRI of human knee cartilage using dual-tuned (1) H/(23) Na transceiver array radiofrequency coil at 7 tesla. *Journal of Magnetic Resonance Imaging* 38, 1063–1072.

Moore, J., Jankiewicz, M., Anderson, A.W., Gore, J.C., 2012. Slice-selective excitation with B1+-insensitive composite pulses. *Journal of Magnetic Resonance* (San Diego, CA, 1997) 214, 200–211.

Moser, E., Meyerspeer, M., Fischmeister, F.P.S., Grabner, G., Bauer, H., Trattnig, S., 2010. Windows on the human body—In vivo high-field magnetic resonance research and applications in medicine and psychology. *Sensors* (Basel, Switzerland) 10, 5724–5757.

Moser, E., Stahlberg, F., Ladd, M.E., Trattnig, S., 2012. 7-T MR—From research to clinical applications? *NMR in Biomedicine* 25, 695–716.

Mugler, J.P., 3rd, Brookeman, J.R., 1990. Three-dimensional magnetization-prepared rapid gradient-echo imaging (3D MP RAGE). *Magnetic Resonance in Medicine* 15, 152–157.

Nagel, A.M., Bock, M., Hartmann, C., Gerigk, L., Neumann, J.O., Weber, M.A., Bendszus, M. et al., 2011. The potential of relaxation-weighted sodium magnetic resonance imaging as demonstrated on brain tumors. *Investigative Radiology* 46, 539–547.

Nagel, A.M., Laun, F.B., Weber, M.A., Matthies, C., Semmler, W., Schad, L.R., 2009. Sodium MRI using a density-adapted 3D radial acquisition technique. *Magnetic Resonance in Medicine* 62, 1565–1573.

Nielles-Vallespin, S., Weber, M.A., Bock, M., Bongers, A., Speier, P., Combs, S.E., Wohrle, J., Lehmann-Horn, F., Essig, M., Schad, L.R., 2007. 3D radial projection technique with ultrashort echo times for sodium MRI: Clinical applications in human brain and skeletal muscle. *Magnetic Resonance in Medicine* 57, 74–81.

Nissi, J., Toth, F., Zang, J., Schmitter, S., Benson, M., Carlson, C., Ellermann, J., 2013. Susceptibility weighted imaging of cartilage canals in porcine epiphyseal growth cartilage ex vivo and in vivo. *Magnetic Resonance in Medicine* 71, 2197–2205, doi:10.1002/mrm.24863.

Noebauer-Huhmann, I.M., Juras, V., Pfirrmann, C.W., Szomolanyi, P., Zbyn, S., Messner, A., Wimmer, J. et al., 2012. Sodium MR imaging of the lumbar intervertebral disk at 7 T: Correlation with T2 mapping and modified Pfirrmann score at 3 T—Preliminary results. *Radiology* 265, 555–564.

Noebauer-Huhmann, I.M., Szomolanyi, P., Juras, V., Kraff, O., Ladd, M.E., Trattnig, S., 2010. Gadolinium-based magnetic resonance contrast agents at 7 Tesla: In vitro T1 relaxivities in human blood plasma. *Investigative Radiology* 45, 554–558.

Noll, D.D., Nishimura, D.G., Makovski, A., 1991. Homodyne detection in magnetic resonance imaging. *IEEE Transactions on Medical Imaging* 10, 154–163.

O'Brien, K.R., Magill, A.W., Delacoste, J., Marques, J.P., Kober, T., Fautz, H.P., Lazeyras, F., Krueger, G., 2013. Dielectric pads and low-B1+ adiabatic pulses: Complementary techniques to optimize structural T1 w whole-brain MP2RAGE scans at 7 tesla. *Journal of Magnetic Resonance Imaging* 40, 804–812.

Ogg, R.J., Langston, J.W., Haacke, E.M., Steen, R.G., Taylor, J.S., 1999. The correlation between phase shifts in gradient-echo MR images and regional brain iron concentration. *Magnetic Resonance Imaging* 17, 1141–1148.

Olman, C.A., Harel, N., Feinberg, D.A., He, S., Zhang, P., Ugurbil, K., Yacoub, E., 2012. Layer-specific fMRI reflects different neuronal computations at different depths in human V1. *PLoS One* 7, e32536.

Olman, C.A., Van de Moortele, P.F., Schumacher, J.F., Guy, J.R., Ugurbil, K., Yacoub, E., 2010. Retinotopic mapping with spin echo BOLD at 7T. *Magnetic Resonance Imaging* 28, 1258–1269.

Ong, B.C., Stuckey, S.L., 2011. Susceptibility weighted imaging: A pictorial review. *Journal of Medical Imaging and Radiation Oncology* 54, 435–449.

Ordidge, R.J, Connelly, A., Lohman, J.A.B., 1986. Image-selected in vivo spectroscopy (ISIS)—A new technique for spatially selective NMR-spectroscopy. *Journal of Magnetic Resonance* 66, 283–294.

Ordidge, R.J., Wylezinska, M., Hugg, J.W., Butterworth, E., Franconi, F., 1996. Frequency offset corrected inversion (FOCI) pulses for use in localized spectroscopy. *Magnetic Resonance in Medicine: Official Journal of the Society of Magnetic Resonance in Medicine/Society of Magnetic Resonance in Medicine* 36, 562–566.

Otazo, R., Mueller, B., Ugurbil, K., Wald, L., Posse, S., 2006. Signal-to-noise ratio and spectral linewidth improvements between 1.5 and 7 Tesla in proton echo-planar spectroscopic imaging. *Magnetic Resonance in Medicine: Official Journal of the Society of Magnetic Resonance in Medicine/Society of Magnetic Resonance in Medicine* 56, 1200–1210.

Ozdemir, M.S., Reyngoudt, H., De Deene, Y., Sazak, H.S., Fieremans, E., Delputte, S., D'Asseler, Y., Derave, W., Lemahieu, I., Achten, E., 2007. Absolute quantification of carnosine in human calf muscle by proton magnetic resonance spectroscopy. *Physics in Medicine and Biology* 52, 6781–6794.

Pan, J.W., Avdievich, N., Hetherington, H.P., 2010. J-refocused coherence transfer spectroscopic imaging at 7 T in human brain. *Magnetic Resonance in Medicine: Official Journal of the Society of Magnetic Resonance in Medicine/Society of Magnetic Resonance in Medicine* 1246, 1237–1246.

Pan, J.W., Duckrow, R.B., Gerrard, J., Ong, C., Hirsch, L.J., Resor, S.R., Zhang, Y. et al., 2013a. 7T MR spectroscopic imaging in the localization of surgical epilepsy. *Epilepsia* 54, 1668–1678.

Pan, J.W., Duckrow, R.B., Spencer, D.D., Avdievich, N.I., Hetherington, H.P., 2013b. Selective homonuclear polarization transfer for spectroscopic imaging of GABA at 7T. *Magnetic Resonance in Medicine: Official Journal of the Society of Magnetic Resonance in Medicine/Society of Magnetic Resonance in Medicine* 69, 310–316.

Pan, J.W., Lo, K.-M., Hetherington, H.P., 2012. Role of very high order and degree B0 shimming for spectroscopic imaging of the human brain at 7 tesla. *Magnetic Resonance in Medicine: Official Journal of the Society of Magnetic Resonance in Medicine/Society of Magnetic Resonance in Medicine* 68, 1007–1017.

Parasoglou, P., Feng, L., Xia, D., Otazo, R., Regatte, R.R., 2012. Rapid 3D-imaging of phosphocreatine recovery kinetics in the human lower leg muscles with compressed sensing. *Magnetic Resonance in Medicine* 68, 1738–1746.

Parasoglou, P., Xia, D., Chang, G., Convit, A., Regatte, R.R., 2013a. Three-dimensional mapping of the creatine kinase enzyme reaction rate in muscles of the lower leg. *NMR in Biomedicine* 26, 1142–1151.

Parasoglou, P., Xia, D., Chang, G., Regatte, R.R., 2013b. 3D-mapping of phosphocreatine concentration in the human calf muscle at 7 T: Comparison to 3 T. *Magnetic Resonance in Medicine* 70, 1619–1625.

Parasoglou, P., Xia, D., Chang, G., Regatte, R.R., 2013c. Dynamic three-dimensional imaging of phosphocreatine recovery kinetics in the human lower leg muscles at 3T and 7T: A preliminary study. *NMR in Biomedicine* 26, 348–356.

Parker, D.L., Payne, A., Todd, N., Hadley, J.R., 2013. Phase reconstruction from multiple coil data using a virtual reference coil. *Magnetic Resonance in Medicine* 72, 563–569.

Pathak, A.P., Ward, B.D., Schmainda, K.M., 2008. A novel technique for modeling susceptibility-based contrast mechanisms for arbitrary microvascular geometries: The finite perturber method. *NeuroImage* 40, 1130–1143.

Piguet, O., Petersen, A., Yin Ka Lam, B., Gabery, S., Murphy, K., Hodges, J.R., Halliday, G.M., 2011. Eating and hypothalamus changes in behavioral-variant frontotemporal dementia. *Annals of Neurology* 69, 312–319.

Pinker, K., Bogner, W., Baltzer, P., Trattnig, S., Gruber, S., Abeyakoon, O., Bernathova, M., Zaric, O., Dubsky, P., Bago-Horvath, Z., Weber, M., Leithner, D., Helbich, T.H., 2014. Clinical application of bilateral high temporal and spatial resolution dynamic contrast-enhanced magnetic resonance imaging of the breast at 7 T. *European Radiology* 24, 913–920.

Polimeni, J., Bhat, H., Benner, T., Feiweier, T., Inati, S., Thomas, W., Heberlein, K., and Wald, L., 2013. Sequential-segment multi-shot auto-calibration for GRAPPA EPI: Maximizing temporal SNR and reducing motion sensitivity. *Proceedings of the 23 Annual Meeting of the ISMRM*, Salt Lake City, UT, #2646.

Polimeni, J.R., Fischl, B., Greve, D.N., Wald, L.L., 2010. Laminar analysis of 7T BOLD using an imposed spatial activation pattern in human V1. *NeuroImage* 52, 1334–1346.

Poser, B.A., Barth, M., Goa, P.E., Deng, W., Stenger, V.A., 2013. Single-shot echo-planar imaging with Nyquist ghost compensation: Interleaved dual echo with acceleration (IDEA) echo-planar imaging (EPI). *Magnetic Resonance in Medicine* 69, 37–47.

Posse, S., Otazo, R., Dager, S.R., Alger, J., 2013. MR spectroscopic imaging: Principles and recent advances. *Journal of Magnetic Resonance Imaging* 37, 1301–1325.

Pruessmann, K., Weiger, M., Scheidegger, M., Boesiger, P., 1999. SENSE: Sensitivity encoding for fast MRI. *Magnetic Resonance in Medicine* 42, 952–962.

Qian, Y., Boada, F.E., 2008. Acquisition-weighted stack of spirals for fast high-resolution three-dimensional ultra-short echo time MR imaging. *Magnetic Resonance in Medicine* 60, 135–145.

Qian, Y., Zhao, T., Wiggins, G.C., Wald, L.L., Zheng, H., Weimer, J., Boada, F.E., 2012a. Sodium imaging of human brain at 7 T with 15-channel array coil. *Magnetic Resonance in Medicine* 68, 1807–1814.

Qian, Y., Zhao, T., Zheng, H., Weimer, J., Boada, F.E., 2012b. High-resolution sodium imaging of human brain at 7 T. *Magnetic Resonance in Medicine* 68, 227–233.

Qiao, H., Zhang, X., Zhu, X.-H., Du, F., Chen, W., 2006. In vivo 31P MRS of human brain at high/ultrahigh fields: A quantitative comparison of NMR detection sensitivity and spectral resolution between 4 T and 7 T. *Magnetic Resonance Imaging* 24, 1281–1286.

Raghuraman, S., Mueller, M.F., Zbyn, S., Baer, P., Breuer, F.A., Friedrich, K.M., Trattnig, S., Lanz, T., Jakob, P.M., 2013. 12-channel receive array with a volume transmit coil for hand/wrist imaging at 7 T. *Journal of Magnetic Resonance Imaging* 38, 238–244.

Ramadan, S., Ratai, E.M., Wald, L.L., Mountford, C.E., 2010. In vivo 1D and 2D correlation MR spectroscopy of the soleus muscle at 7T. *Journal of Magnetic Resonance* 204, 91–98.

Ratai, E., Kok, T., Wiggins, C., Wiggins, G., Grant, E., Gagoski, B., O'Neill, G., Adalsteinsson, E., Eichler, F., 2008. Seven-Tesla proton magnetic resonance spectroscopic imaging in adult X-linked adrenoleukodystrophy. *Archives of Neurology* 65, 1488–1494.

Rauscher, A., Sedlacik, J., Barth, M., Mentzel, H.J., Reichenbach, J.R., 2005. Magnetic susceptibility-weighted MR phase imaging of the human brain. *American Journal of Neuroradiology* 26, 736–742.

Reetz, K., Romanzetti, S., Dogan, I., Sass, C., Werner, C.J., Schiefer, J., Schulz, J.B., Shah, N.J., 2012. Increased brain tissue sodium concentration in Huntington's Disease—A sodium imaging study at 4 T. *NeuroImage* 63, 517–524.

Regatte, R.R., Schweitzer, M.E., 2007. Ultra-high-field MRI of the musculoskeletal system at 7.0T. *Journal of Magnetic Resonance Imaging* 25, 262–269.

Reichenbach, J.R., Haacke, E.M., 2001. High-resolution BOLD venographic imaging: A window into brain function. *NMR in Biomedicine* 14, 453–467.

Reichenbach, J.R., Venkatesan, R., Schillinger, D.J., Kido, D.K., Haacke, E.M., 1997. Small vessels in the human brain: MR venography with deoxyhemoglobin as an intrinsic contrast agent. *Radiology* 204, 272–277.

Ren, J., Dean Sherry, A., Malloy, C.R., 2012. Noninvasive monitoring of lactate dynamics in human forearm muscle after exhaustive exercise by (1) H-magnetic resonance spectroscopy at 7 tesla. *Magnetic Resonance in Medicine* 70, 610–619.

Ren, J., Dimitrov, I., Sherry, A.D., Malloy, C.R., 2008. Composition of adipose tissue and marrow fat in humans by 1H NMR at 7 Tesla. *The Journal of Lipid Research* 49, 2055–2062.

Ren, J., Lakoski, S., Haller, R.G., Sherry, A.D., Malloy, C.R., 2013. Dynamic monitoring of carnitine and acetylcarnitine in the trimethylamine signal after exercise in human skeletal muscle by 7T 1H-MRS. *Magnetic Resonance in Medicine* 69, 7–17.

Ren, J., Sherry, A.D., Malloy, C.R., 2010. 1H MRS of intramyocellular lipids in soleus muscle at 7 T: Spectral simplification by using long echo times without water suppression. *Magnetic Resonance in Medicine* 64, 662–671.

Robinson, J.D., Flashner, M.S., 1979. The (Na+ + K+)-activated ATPase. Enzymatic and transport properties. *Biochimica et Biophysica Acta* 549, 145–176.

Robinson, R.J., Bhuta, S., 2010. Susceptibility-weighted imaging of the brain: Current utility and potential applications. *Journal of Neuroimaging* 21, e189–e204.

Robinson, S., Grabner, G., Witoszynskyj, S., Trattnig, S., 2011. Combining phase images from multi-channel RF coils using 3D phase offset maps derived from a dual-echo scan. *Magnetic Resonance in Medicine* 65, 1638–1648.

Robinson, S., Jovicich, J., 2011. B0 mapping with multi-channel RF coils at high field. *Magnetic Resonance in Medicine* 66, 976–988.

Robinson, S., Pripfl, J., Bauer, H., Moser, E., 2008. The impact of EPI voxel size on SNR and BOLD sensitivity in the anterior medio-temporal lobe: A comparative group study of deactivation of the Default Mode. *Magnetic Resonance Materials in Physics* 21, 279–290.

Robinson, S.D., Schopf, V., Cardoso, P., Geissler, A., Fischmeister, F.P., Wurnig, M., Trattnig, S., Beisteiner, R., 2013. Applying independent component analysis to clinical fMRI at 7 T. *Frontiers in Human Neuroscience* 7, 496.

Rodgers, C.T., Clarke, W.T., Snyder, C., Vaughan, J.T., Neubauer, S., Robson, M.D., 2013. Human cardiac P magnetic resonance spectroscopy at 7 tesla. *Magnetic Resonance in Medicine* 72, 304–315.

Roemer, P.B., Edelstein, W.A., Hayes, C.E., Souza, S.P., Mueller, O.M., 1990. The NMR phased array. *Magnetic Resonance in Medicine* 16, 192–225.

Roig, E.S., Magill, A.W., Donati, G., Meyerspeer, M., Xin, L., Ipek, O., Gruetter, R., 2014. A double-quadrature radio-frequency coil design for proton-decoupled carbon-13 magnetic resonance spectroscopy in humans at 7T. *Magnetic Resonance in Medicine* doi:10.1002/mrm.25171.

Ronen, I., Ercan, E., Webb, A., 2013a. Axonal and glial micro-structural information obtained with diffusion-weighted magnetic resonance spectroscopy at 7T. *Frontiers in Integrative Neuroscience* 7, 13.

Ronen, I., Ercan, E., Webb, A., 2013b. Rapid multi-echo measurement of brain metabolite T2 values at 7T using a single-shot spectroscopic Carr-Purcell-Meiboom-Gill sequence and prior information. *NMR in Biomedicine* 26, 1291–1298.

Rossi, M., Ruottinen, H., Elovaara, I., Ryymin, P., Soimakallio, S., Eskola, H., Dastidar, P., 2010. Brain iron deposition and sequence characteristics in Parkinsonism: Comparison of SWI, T2* maps, T2-weighted-, and FLAIR-SPACE. *Investigative Radiology* 45, 795–802.

Sanchez-Panchuelo, R.M., Besle, J., Mougin, O., Gowland, P., Bowtell, R., Schluppeck, D., Francis, S., 2013. Regional structural differences across functionally parcellated Brodmann areas of human primary somatosensory cortex. *NeuroImage* 93, 221–230.

Sanchez-Panchuelo, R.M., Francis, S., Bowtell, R., Schluppeck, D., 2010. Mapping human somatosensory cortex in individual subjects with 7T functional MRI. *Journal of Neurophysiology* 103, 2544–2556.

Satpute, A.B., Wager, T.D., Cohen-Adad, J., Bianciardi, M., Choi, J.K., Buhle, J.T., Wald, L.L., Barrett, L.F., 2013. Identification of discrete functional subregions of the human periaqueductal gray. *Proceedings of the National Academy of Sciences of the USA* 110, 17101–17106.

Schad, L.R., 2001. Improved target volume characterization in stereotactic treatment planning of brain lesions by using high-resolution BOLD MR-venography. *NMR in Biomedicine* 14, 478–483.

Schäfer, A., Wharton, S., Gowland, P., R., B., 2009. Using magnetic field simulation to study susceptibility-related phase contrast in gradient echo MRI. *NeuroImage* doi:10.1016/j.neuroimage.2009.05.093.

Schäfer, A., Wharton, S., R., B., 2008. Calculation of susceptibility maps from phase image data. *Proceedings of the 16th Annual Meeting of the ISMRM*, Toronto, Canada, p. 641.

Schaller, B., Mekle, R., Xin, L., Kunz, N., Gruetter, R., 2013a. Net increase of lactate and glutamate concentration in activated human visual cortex detected with magnetic resonance spectroscopy at 7 tesla. *Journal of Neuroscience Research* 91, 1076–1083.

Schaller, B., Xin, L., Gruetter, R., 2013b. Is the macromolecule signal tissue-specific in healthy human brain? A (1) H MRS study at 7 tesla in the occipital lobe. *Magnetic Resonance in Medicine: Official Journal of the Society of Magnetic Resonance in Medicine/Society of Magnetic Resonance in Medicine* 72, 1–7.

Schaller, B., Xin, L., O'Brien, K., Magill, A.W., Gruetter, R., 2014. Are glutamate and lactate increases ubiquitous to physiological activation? A (1)H functional MR spectros-copy study during motor activation in human brain at 7 Tesla. *NeuroImage* 93 Pt 1, 138–145.

Scharnhorst, K., 2001. Angles in complex vector spaces. *Acta Applicandae Mathematicae* 69, 95–103.

Scheenen, T.W.J., Heerschap, A., Klomp, D.W.J., 2008. Towards 1H-MRSI of the human brain at 7T with slice-selective adia-batic refocusing pulses. *Magma* (New York, NY) 21, 95–101.

Schindler, S., Schonknecht, P., Schmidt, L., Anwander, A., Strauss, M., Trampel, R., Bazin, P.L. et al., 2013. Devel-opment and evaluation of an algorithm for the computer-assisted segmentation of the human hypothalamus on 7-Tesla magnetic resonance images. *PLoS One* 8, e66394.

Schlamann, M., Maderwald, S., Becker, W., Kraff, O., Theysohn, J.M., Mueller, O., Sure, U. et al., 2010. Cerebral cavern-ous hemangiomas at 7 Tesla: Initial experience. *Academic Radiology* 17, 3–6.

Schmid, A.I., Schewzow, K., Fiedler, G.B., Goluch, S., Laistler, E., Wolzt, M., Moser, E., Meyerspeer, M., 2014. Exercising calf muscle T2 * changes correlate with pH, PCr recov-ery and maximum oxidative phosphorylation. *NMR in Biomedicine* 27, 553–560.

Schmid, A.I., Schrauwen-Hinderling, V.B., Andreas, M., Wolzt, M., Moser, E., Roden, M., 2012. Comparison of measur-ing energy metabolism by different (31) P-magnetic reso-nance spectroscopy techniques in resting, ischemic, and exercising muscle. *Magnetic Resonance in Medicine* 67, 898–905.

Schmitt, B., Zbyn, S., Stelzeneder, D., Jellus, V., Paul, D., Lauer, L., Bachert, P., Trattnig, S., 2011. Cartilage quality assess-ment by using glycosaminoglycan chemical exchange saturation transfer and (23)Na MR imaging at 7 T. *Radiology* 260, 257–264.

Schmitt, F., Potthast, A., Stoeckel, B., Triantafyllou, C., Wiggins, C.J., Wiggins, G., Wald, L.L., 2006. Aspects of clini-cal imaging at 7 T. *Ultra High Field Magnetic Resonance Imaging* 26, 59–103.

Schofield, M.A., Zhu, Y., 2003. Fast phase unwrapping algo-rithm for interferometric applications. *Optics Letters* 28, 1194–1196.

Schuster, C., Dreher, W., Stadler, J., Bernarding, J., Leibfritz, D., 2008. Fast three-dimensional 1H MR spectroscopic imaging at 7 Tesla using "spectroscopic missing pulse—SSFP." *Magnetic Resonance in Medicine: Official Journal of the Society of Magnetic Resonance in Medicine/Society of Magnetic Resonance in Medicine* 60, 1243–1249.

Schweser, F., Deistung, A., Lehr, B.W., Reichenbach, J.R., 2011. Quantitative imaging of intrinsic magnetic tissue prop-erties using MRI signal phase: An approach to in vivo brain iron metabolism? *NeuroImage* 54, 2789–2807.

Schweser, F., Sommer, K., Deistung, A., Reichenbach, J.R., 2012. Quantitative susceptibility mapping for investigat-ing subtle susceptibility variations in the human brain. *NeuroImage* 62, 2083–2100.

Setsompop, K., Gagoski, B.A., Polimeni, J.R., Witzel, T., Wedeen, V.J., Wald, L.L., 2012. Blipped-controlled aliasing in parallel imaging for simultaneous multislice echo planar imaging with reduced g-factor penalty. *Magnetic Resonance in Medicine* 67, 1210–1224.

Shimony, J.S., Zhang, D., Johnston, J.M., Fox, M.D., Roy, A., Leuthardt, E.C., 2009. Resting-state spontaneous fluctuations in brain activity: A new paradigm for presurgical planning using fMRI. *Academic Radiology* 16, 578–583.

Shmueli, K., de Zwart, J.A., van Gelderen, P., Li, T.Q., Dodd, S.J., Duyn, J.H., 2009. Magnetic susceptibility mapping of brain tissue in vivo using MRI phase data. *Magnetic Resonance in Medicine* 62, 1510–1522.

Siero, J.C., Petridou, N., Hoogduin, H., Luijten, P.R., Ramsey, N.F., 2011. Cortical depth-dependent temporal dynamics of the BOLD response in the human brain. *Journal of Cerebral Blood Flow & Metabolism* 31, 1999–2008.

Skou, J.C., 1957. The influence of some cations on an adenosine triphosphatase from peripheral nerves. *Biochimica et Biophysica Acta* 23, 394–401.

Sladky, R., Baldinger, P., Kranz, G.S., Trostl, J., Hoflich, A., Lanzenberger, R., Moser, E., Windischberger, C., 2013. High-resolution functional MRI of the human amygdala at 7 T. *European Journal of Radiology* 82, 728–733.

Snaar, J.E.M., Teeuwisse, W.M., Versluis, M.J., van Buchem, M.A., Kan, H.E., Smith, N.B., Webb, A. G., 2011. Improvements in high-field localized MRS of the medial temporal lobe in humans using new deformable high-dielectric materials. *NMR in Biomedicine* 24, 873–879.

Snyder, J., Wilman, A., 2010. Field strength dependence of PRESS timings for simultaneous detection of glutamate and glutamine from 1.5 to 7T. *Journal of Magnetic Resonance* (San Diego, CA, 1997) 203, 66–72.

Somjen, G.G., 2004. *Ions in the Brain: Normal Function, Seizures, and Stroke.* Oxford University Press, New York.

Srinivasan, R., Ratiney, H., Hammond-Rosenbluth, K.E., Pelletier, D., Nelson, S.J., 2010. MR spectroscopic imaging of glutathione in the white and gray matter at 7 T with an application to multiple sclerosis. *Magnetic Resonance Imaging* 28, 163–170.

Staroswiecki, E., Bangerter, N.K., Gurney, P.T., Grafendorfer, T., Gold, G.E., Hargreaves, B.A., 2010. In vivo sodium imaging of human patellar cartilage with a 3D cones sequence at 3 T and 7 T. *Journal of Magnetic Resonance Imaging* 32, 446–451.

Steinseifer, I.K., Wijnen, J.P., Hamans, B.C., Heerschap, A., Scheenen, T.W., 2013. Metabolic imaging of multiple x-nucleus resonances. *Magnetic Resonance in Medicine* 70, 169–175.

Stephenson, M.C., Gunner, F., Napolitano, A., Greenhaff, P.L., Macdonald, I.A., Saeed, N., Vennart, W., Francis, S.T., Morris, P.G., 2011. Applications of multi-nuclear magnetic resonance spectroscopy at 7T. *World Journal of Radiology* 3, 105–113.

Strasser, B., Chmelik, M., Robinson, S.D., Hangel, G., Gruber, S., Trattnig, S., Bogner, W., 2013. Coil combination of multichannel MRSI data at 7 T: MUSICAL. *NMR in Biomedicine* 26, 1796–1805.

Stringer, E.A., Chen, L.M., Friedman, R.M., Gatenby, C., Gore, J.C., 2011. Differentiation of somatosensory cortices by high-resolution fMRI at 7 T. *NeuroImage* 54, 1012–1020.

Swisher, J.D., Gatenby, J.C., Gore, J.C., Wolfe, B.A., Moon, C.H., Kim, S.G., Tong, F., 2010. Multiscale pattern analysis of orientation-selective activity in the primary visual cortex. *Journal of Neuroscience* 30, 325–330.

Szendroedi, J., Chmelik, M., Schmid, A.I., Nowotny, P., Brehm, A., Krssak, M., Moser, E., Roden, M., 2009. Abnormal hepatic energy homeostasis in type 2 diabetes. *Hepatology* 50, 1079–1086.

Talagala, S.L., Sarlls, J.E., Inati, S.J., 2013. Improved temporal SNR of accelerated EPI using a FLASH based GRAPPA reference scan. *Proceedings of the 23rd Annual Meeting of the ISMRM*, Utah, U.S.A #2658.

Tallantyre, E.C., Morgan, P.S., Dixon, J.E., Al-Radaideh, A., Brookes, M.J., Morris, P.G., Evangelou, N., 2010. 3 Tesla and 7 Tesla MRI of multiple sclerosis cortical lesions. *Journal of Magnetic Resonance Imaging* 32, 971–977.

Tan, I.L., van Schijndel, R.A., Pouwels, P.J., van Walderveen, M.A., Reichenbach, J.R., Manoliu, R.A., Barkhof, F., 2000. MR venography of multiple sclerosis. *American Journal of Neuroradiology* 21, 1039–1042.

Tannus, A., Garwood, M., 1997. Adiabatic pulses. *NMR in Biomedicine* 10, 423–434.

Thamer, C., Machann, J., Bachmann, O., Haap, M., Dahl, D., Wietek, B., Tschritter, O. et al., 2003. Intramyocellular lipids: anthropometric determinants and relationships with maximal aerobic capacity and insulin sensitivity. *Journal of Clinical Endocrinology & Metabolism* 88, 1785–1791.

Theysohn, J.M., Kraff, O., Maderwald, S., Barth, M., Ladd, S.C., Forsting, M., Ladd, M.E., Gizewski, E.R., 2011. 7 tesla MRI of microbleeds and white matter lesions as seen in vascular dementia. *Journal of Magnetic Resonance Imaging* 33, 782–791.

Theysohn, N., Qin, S., Maderwald, S., Poser, B.A., Theysohn, J.M., Ladd, M.E., Norris, D.G., Gizewski, E.R., Fernandez, G., Tendolkar, I., 2013. Memory-related hippocampal activity can be measured robustly using FMRI at 7 tesla. *Journal of Neuroimaging* 23, 445–451.

Tkáč, I., Andersen, P., Adriany, G., 2001. In vivo 1H NMR spectroscopy of the human brain at 7 T. *Magnetic Resonance in Medicine: Official Journal of the Society of Magnetic Resonance in Medicine/Society of Magnetic Resonance in Medicine* 456, 451–456.

Tkáč, I., Oz, G., Adriany, G., Uğurbil, K., Gruetter, R., 2009. In vivo 1H NMR spectroscopy of the human brain at high magnetic fields: metabolite quantification at 4T vs. 7T. *Magnetic Resonance in Medicine: Official Journal of the Society of Magnetic Resonance in Medicine/Society of Magnetic Resonance in Medicine* 62, 868–879.

Tognin, S., Rambaldelli, G., Perlini, C., Bellani, M., Marinelli, V., Zoccatelli, G., Alessandrini, F. et al., 2012. Enlarged hypothalamic volumes in schizophrenia. *Psychiatry Research* 204, 75–81.

Trattnig, S., Welsch, G.H., Juras, V., Szomolanyi, P., Mayerhoefer, M.E., Stelzeneder, D., Mamisch, T.C., Bieri, O., Scheffler, K., Zbyn, S., 2010. 23Na MR imaging at 7 T after knee matrix-associated autologous chondrocyte transplantation preliminary results. *Radiology* 257, 175–184.

Triantafyllou, C., Hoge, R.D., Krueger, G., Wiggins, C.J., Potthast, A., Wiggins, G.C., Wald, L.L., 2005. Comparison of physiological noise at 1.5 T, 3 T and 7 T and optimization of fMRI acquisition parameters. *NeuroImage* 26, 243–250.

Tsang, A., Stobbe, R.W., Asdaghi, N., Hussain, M.S., Bhagat, Y.A., Beaulieu, C., Emery, D., Butcher, K.S., 2011. Relationship between sodium intensity and perfusion deficits in acute ischemic stroke. *Journal of Magnetic Resonance Imaging* 33, 41–47.

Ugurbil, K., Adriany, G., Andersen, P., Chen, W., Garwood, M., Gruetter, R., Henry, P.G. et al., 2003. Ultrahigh field magnetic resonance imaging and spectroscopy. *Magnetic Resonance Imaging* 21, 1263–1281.

Ulmer, J.L., Hacein-Bey, L., Mathews, V.P., Mueller, W.M., DeYoe, E.A., Prost, R.W., Meyer, G.A., Krouwer, H.G., Schmainda, K.M., 2004. Lesion-induced Pseudo-dominance at functional magnetic resonance imaging: Implications for preoperative assessments. *Neurosurgery* 55, 569–581.

Umutlu, L., Bitz, A.K., Maderwald, S., Orzada, S., Kinner, S., Kraff, O., Brote, I. et al., 2013. Contrast-enhanced ultra-high-field liver MRI: A feasibility trial. *European Journal of Radiology* 82, 760–767.

Umutlu, L., Kraff, O., Orzada, S., Fischer, A., Kinner, S., Maderwald, S., Antoch, G. et al., 2011. Dynamic contrast-enhanced renal MRI at 7 Tesla: Preliminary results. *Investigative Radiology* 46, 425–433.

Umutlu, L., Maderwald, S., Kraff, O., Theysohn, J.M., Kuemmel, S., Hauth, E.A., Forsting, M. et al., 2010. Dynamic contrast-enhanced breast MRI at 7 Tesla utilizing a single-loop coil: A feasibility trial. *Academic Radiology* 17, 1050–1056.

Valkovic, L., Bogner, W., Gajdosik, M., Povazan, M., Kukurova, I.J., Krssak, M., Gruber, S., Frollo, I., Trattnig, S., Chmelik, M., 2014a. One-dimensional image-selected in vivo spectroscopy localized phosphorus saturation transfer at 7T. *Magnetic Resonance in Medicine* 72, 1509–1515.

Valkovic, L., Chmelik, M., Just Kukurova, I., Krssak, M., Gruber, S., Frollo, I., Trattnig, S., Bogner, W., 2013a. Time-resolved phosphorous magnetization transfer of the human calf muscle at 3 T and 7 T: A feasibility study. *European Journal of Radiology* 82, 745–751.

Valkovic, L., Gajdosik, M., Traussnigg, S., Wolf, P., Chmelik, M., Kienbacher, C., Bogner, W. et al., 2014b. Application of localized P MRS saturation transfer at 7 T for measurement of ATP metabolism in the liver: reproducibility and initial clinical application in patients with non-alcoholic fatty liver disease. *European Radiology* 24, 1602–1609.

Valkovic, L., Ukropcova, B., Chmelik, M., Balaz, M., Bogner, W., Schmid, A.I., Frollo, I. et al., 2013b. Interrelation of 31P-MRS metabolism measurements in resting and exercised quadriceps muscle of overweight-to-obese sedentary individuals. *NMR in Biomedicine* 26, 1714–1722.

Van den Bergen, B., Klomp, D.W., Raaijmakers, A.J., de Castro, C.A., Boer, V.O., Kroeze, H., Luijten, P.R., Lagendijk, J.J., van den Berg, C.A., 2011. Uniform prostate imaging and spectroscopy at 7 T: Comparison between a microstrip array and an endorectal coil. *NMR Biomedicine* 24, 358–365.

Van den Bogaard, S.J.A., Dumas, E.M., Teeuwisse, W.M., Kan, H.E., Webb, A., Roos, R.A. C., van der Grond, J., 2011. Exploratory 7-Tesla magnetic resonance spectroscopy in Huntington's disease provides in vivo evidence for impaired energy metabolism. *Journal of Neurology* 258, 2230–2239.

Van der Kemp, W.J., Boer, V.O., Luijten, P.R., Stehouwer, B.L., Veldhuis, W.B., Klomp, D.W., 2013. Adiabatic multi-echo (3)(1)P spectroscopic imaging (AMESING) at 7 T for the measurement of transverse relaxation times and regaining of sensitivity in tissues with short T(2) values. *NMR in Biomedicine* 26, 1299–1307.

Van der Zwaag, W., Francis, S., Head, K., Peters, A., Gowland, P., Morris, P., Bowtell, R., 2009. fMRI at 1.5, 3 and 7 T: Characterising BOLD signal changes. *NeuroImage* 47, 1425–1434.

Van der Zwaag, W., Gentile, G., Gruetter, R., Spierer, L., Clarke, S., 2011. Where sound position influences sound object representations: A 7-T fMRI study. *NeuroImage* 54, 1803–1811.

Van der Zwaag, W., Kusters, R., Magill, A., Gruetter, R., Martuzzi, R., Blanke, O., Marques, J.P., 2012. Digit somatotopy in the human cerebellum: A 7T fMRI study. *NeuroImage* 67, 354–362.

Van der Zwaag, W., Kusters, R., Magill, A., Gruetter, R., Martuzzi, R., Blanke, O., Marques, J.P., 2013. Digit somatotopy in the human cerebellum: A 7T fMRI study. *NeuroImage* 67, 354–362.

Van Oorschot, J.W., Schmitz, J.P., Webb, A., Nicolay, K., Jeneson, J.A., Kan, H.E., 2011. 31P MR spectroscopy and computational modeling identify a direct relation between Pi content of an alkaline compartment in resting muscle and phosphocreatine resynthesis kinetics in active muscle in humans. *PLoS One* 8, e76628.

Van Zijl, P.C., Yadav, N.N., 2011. Chemical exchange saturation transfer (CEST): What is in a name and what isn't? *Magnetic Resonance in Medicine* 65, 927–948.

Vaughan, J.T., Garwood, M., Collins, C.M., Liu, W., DelaBarre, L., Adriany, G., Andersen, P. et al., 2001. 7T vs. 4T: RF power, homogeneity, and signal-to-noise comparison in head images. *Magnetic Resonance in Medicine* 46, 24–30.

Verma, G., Hariharan, H., Nagarajan, R., Nanga, R.P.R., Delikatny, E.J., Albert Thomas, M., Poptani, H., 2013. Implementation of two-dimensional L-COSY at 7 tesla: An investigation of reproducibility in human brain. *Journal of Magnetic Resonance Imaging* 40, 1319–1327.

Versluis, M.J., Kan, H.E., van Buchem, M.A., Webb, A.G., 2010. Improved signal to noise in proton spectroscopy of the human calf muscle at 7 T using localized B1 calibration. *Magnetic Resonance in Medicine* 63, 207–211.

Voigt, T., Katscher, U., Doessel, O., 2011. Quantitative conductivity and permittivity imaging of the human brain using electric properties tomography. *Magnetic Resonance in Medicine* 66, 456–466.

Volkov, V.V., Zhu, Y., 2003. Deterministic phase unwrapping in the presence of noise. *Optics Letters* 28, 2156–2158.

Wachter, S., Vogt, M., Kreis, R., Boesch, C., Bigler, P., Hoppeler, H., Krahenbuhl, S., 2002. Long-term administration of L-carnitine to humans: effect on skeletal muscle carnitine content and physical performance. *Clinica Chimica Acta* 318, 51–61.

Wang, L., Salibi, N., Wu, Y., Schweitzer, M.E., Regatte, R.R., 2009a. Relaxation times of skeletal muscle metabolites at 7T. *Journal of Magnetic Resonance Imaging* 29, 1457–1464.

Wang, L., Wu, Y., Chang, G., Oesingmann, N., Schweitzer, M.E., Jerschow, A., Regatte, R.R., 2009b. Rapid isotropic 3D-sodium MRI of the knee joint in vivo at 7T. *Journal of Magnetic Resonance Imaging* 30, 606–614.

Webb, A.G., Collins, C.M., Versluis, M.J., Kan, H.E., Smith, N.B., 2010. MRI and localized proton spectroscopy in human leg muscle at 7 Tesla using longitudinal traveling waves. *Magnetic Resonance in Medicine* 63, 297–302.

Weiss, M., Geyer, S., Lohmann, G., Trampel, R., Turner, R., 2013. Quantitative T1 mapping at 7 Tesla identifies primary functional areas in the living human brain. *Proceedings of the 19th Annual Meeting of the Organisation for Human Brain Mapping*, Seattle, WA, p. 2360.

Welsch, G.H., Apprich, S., Zbyn, S., Mamisch, T.C., Mlynarik, V., Scheffler, K., Bieri, O., Trattnig, S., 2011. Biochemical (T2, T2* and magnetisation transfer ratio) MRI of knee cartilage: Feasibility at ultra-high field (7T) compared with high field (3T) strength. *European Radiology* 21, 1136–1143.

Welsch, G.H., Mamisch, T.C., Hughes, T., Zilkens, C., Quirbach, S., Scheffler, K., Kraff, O., Schweitzer, M.E., Szomolanyi, P., Trattnig, S., 2008. In vivo biochemical 7.0 Tesla magnetic resonance—Preliminary results of dGEMRIC, zonal T2, and T2* mapping of articular cartilage. *Investigative Radiology* 43, 619–626.

Welsch, G.H., Scheffler, K., Mamisch, T.C., Hughes, T., Millington, S., Deimling, M., Trattnig, S., 2009. Rapid estimation of cartilage T2 based on Double Echo at Steady State (DESS) with 3 Tesla. *Magnetic Resonance in Medicine* 62, 544–549.

Wharton, S., Bowtell, R., 2010. Whole-brain susceptibility mapping at high field: A comparison of multiple- and single-orientation methods. *NeuroImage* 53, 515–525.

Wharton, S., Schafer, A., Bowtell, R., 2010. Susceptibility mapping in the human brain using threshold-based k-space division. *Magnetic Resonance in Medicine* 63, 1292–1304.

Wiesinger, F., Boesiger, P., Pruessmann, K.P., 2004. Electrodynamics and ultimate SNR in parallel MR imaging. *Magnetic Resonance in Medicine* 52, 376–390.

Wiggins, G.C., Potthast, A., Triantafyllou, C., Wiggins, C.J., Wald, L.L., 2005. Eight-channel phased array coil and detunable TEM volume coil for 7 T brain imaging. *Magnetic Resonance in Medicine* 54, 235–240.

Wijnen, J.P., van der Kemp, W.J., Luttje, M.P., Korteweg, M.A., Luijten, P.R., Klomp, D.W., 2012. Quantitative 31P magnetic resonance spectroscopy of the human breast at 7 T. *Magnetic Resonance in Medicine* 68, 339–348.

Wijsman, C.A., van Opstal, A.M., Kan, H.E., Maier, A.B., Westendorp, R.G., Slagboom, P.E., Webb, A.G., Mooijaart, S.P., van Heemst, D., 2010. Proton magnetic resonance spectroscopy shows lower intramyocellular lipid accumulation in middle-aged subjects predisposed to familial longevity. *American Journal of Physiology Endocrinology and Metabolism* 302, E344–E348.

Wijtenburg, S.A., Rowland, L.M., Edden, R.a. E., Barker, P.B., 2013. Reproducibility of brain spectroscopy at 7T using conventional localization and spectral editing techniques. *Journal of Magnetic Resonance Imaging* 38, 460–467.

Witoszynskyj, S., Rauscher, A., Reichenbach, J.R., Barth, M., 2009. Phase unwrapping of MR images using Phi UN—A fast and robust region growing algorithm. *Medical Image Analysis* 13, 257–268.

Xin, L., Schaller, B., Mlynarik, V., Lu, H., Gruetter, R., 2013. Proton T1 relaxation times of metabolites in human occipital white and gray matter at 7 T. *Magnetic Resonance in Medicine* 69, 931–936.

Xu, D., Cunningham, C.H., Chen, A.P., Li, Y., Kelley, D.A. C., Mukherjee, P., Pauly, J.M., Nelson, S.J., Vigneron, D.B., 2008a. Phased array 3D MR spectroscopic imaging of the brain at 7 T. *Magnetic Resonance Imaging* 26, 1201–1206.

Xu, W., Cumming, I., 1999. A region-growing algorithm for InSAR phase unwrapping. *IEEE Transactions on Geoscience and Remote Sensing* 37, 124–134.

Xu, X., Wang, Q., Zhang, M., 2008b. Age, gender, and hemispheric differences in iron deposition in the human brain: An in vivo MRI study. *NeuroImage* 40, 35–42.

Yacoub, E., Harel, N., Ugurbil, K., 2008. High-field fMRI unveils orientation columns in humans. *Proceedings of the National Academy of Sciences of the USA* 105, 10607–10612.

Yacoub, E., Shmuel, A., Logothetis, N., Ugurbil, K., 2007. Robust detection of ocular dominance columns in humans using Hahn Spin Echo BOLD functional MRI at 7 Tesla. *NeuroImage* 37, 1161–1177.

Yacoub, E., Shmuel, A., Pfeuffer, J., Van De Moortele, P.F., Adriany, G., Andersen, P., Vaughan, J.T., Merkle, H., Ugurbil, K., Hu, X., 2001. Imaging brain function in humans at 7 Tesla. *Magnetic Resonance in Medicine* 45, 588–594.

Zaaraoui, W., Konstandin, S., Audoin, B., Nagel, A.M., Rico, A., Malikova, I., Soulier, E. et al., 2012. Distribution of brain sodium accumulation correlates with disability in multiple sclerosis: A cross-sectional 23Na MR imaging study. *Radiology* 264, 859–867.

Zaitsev, M., Hennig, J., Speck, O., 2004. Point spread function mapping with parallel imaging techniques and high acceleration factors: Fast, robust, and flexible method for echo-planar imaging distortion correction. *Magnetic Resonance in Medicine* 52, 1156–1166.

Zaric, O., Pinker-Domenig, K., Gruber, S., Porter, D., Helbich, T., Trattnig, S., Bogner, W., 2013. *Diffusion Weighted Imaging of the Breast at 7T—Ready for Clinical Application?* ISMRM, Salt Lake City, UT, p. 6203.

Zbyn, S., Stelzeneder, D., Welsch, G.H., Negrin, L.L., Juras, V., Mayerhoefer, M.E., Szomolanyi, P. et al., 2012. Evaluation of native hyaline cartilage and repair tissue after two cartilage repair surgery techniques with 23Na MR imaging at 7 T: Initial experience. *Osteoarthritis Cartilage* 20, 837–845.

Zehnder, M., Christ, E.R., Ith, M., Acheson, K.J., Pouteau, E., Kreis, R., Trepp, R., Diem, P., Boesch, C., Decombaz, J., 2006. Intramyocellular lipid stores increase markedly in athletes after 1.5 days lipid supplementation and are utilized during exercise in proportion to their content. *European Journal of Applied Physiology* 98, 341–354.

Zeng, H., Constable, R.T., 2002. Image distortion correction in EPI: Comparison of field mapping with point spread function mapping. *Magnetic Resonance in Medicine* 48, 137–146.

Zhang, D., Johnston, J.M., Fox, M.D., Leuthardt, E.C., Grubb, R.L., Chicoine, M.R., Smyth, M.D., Snyder, A.Z., Raichle, M.E., Shimony, J.S., 2009. Preoperative sensorimotor mapping in brain tumor patients using spontaneous fluctuations in neuronal activity imaged with functional magnetic resonance imaging: Initial experience. *Neurosurgery* 65, 226–236.

Zhu, H., Soher, B.J., Ouwerkerk, R., Schär, M., Barker, P.B., 2013. Spin-echo magnetic resonance spectroscopic imaging at 7 T with frequency-modulated refocusing pulses. *Magnetic Resonance in Medicine: Official Journal of the Society of Magnetic Resonance in Medicine/Society of Magnetic Resonance in Medicine* 69, 1217–1225.

Zielman, R., Teeuwisse, W., Bakels, F., Van der Grond, J., Webb, A., van Buchem, M., Ferrari, M., Kruit, M., Terwindt, G., 2014. Biochemical changes in the brain of hemiplegic migraine patients measured with 7 tesla 1H-MRS. *Cephalalgia: An International Journal of Headache* 34, 1–9.

Zwanenburg, J.J., Hendrikse, J., Visser, F., Takahara, T., Luijten, P.R., 2010. Fluid attenuated inversion recovery (FLAIR) MRI at 7.0 Tesla: Comparison with 1.5 and 3.0 Tesla. *European Radiology* 20, 915–922.

12

Pathology of the Paranasal Sinuses (Infective–Neoplastic)

Ellen Hoeffner

CONTENTS

12.1 Introduction

Computed tomography (CT) is the main imaging modality used to assess paranasal sinus disease as this provides the highest resolution and, with the use of bone algorithm, the best definition of bony structures [1]. The role of magnetic resonance imaging (MRI) in paranasal sinus disease consists primarily of assessing intracranial and orbital complications of sinus inflammatory disease, helping to differentiate aggressive infectious or inflammatory disease from neoplasm, and assessing the extent of neoplasm [1,2].

Most patients with sinus disease require no imaging. Patients who do not respond to medical therapy, have unilateral or recurrent symptoms concerning for neoplasm or for surgical planning will generally be imaged initially by CT [2]. Immunocompromised patients should be imaged more acutely due to the concern over aggressive inflammatory processes in this patient population [3]. Findings on CT that are related to an aggressive inflammatory or infectious process or a neoplasm include bone erosion, remodeling or new bone formation, complete unilateral sinus opacification, extensive soft-tissue mass, necrosis, and lymphadenopathy [4–6]. Further imaging of the sinuses with MRI should be considered in these situations and should be viewed as a complimentary imaging study to CT (Figure 12.1). CT provides an excellent assessment of the osseous structures, the internal architecture of some fibro-osseous lesions, and orbital fat. MRI is better at separating tumor from inflammatory mucosal disease and secretions, defining the extent of disease spread into adjacent structures and detecting perineural tumor spread (PNTS) [3,7,8].

An MRI protocol of the sinuses should include thin-slice technique (3 mm or less) through the sinuses, palate, orbits, skull base, and adjacent intracranial contents in axial, sagittal, and coronal planes. T_1- and T_2-weighted sequences should both be obtained as sinonasal secretions can have variable signal intensities depending on their protein concentration. Multiplanar, thin-slice post-contrast images should be obtained with fat suppression [3,7,8]. Inclusion of the adjacent structures is mandatory to fully map disease extension, including affected areas within each sinus and any extension into the periantral and facial soft tissues, orbits, palate, pterygopalatine fossa (PPF), and the anterior and middle cranial fossae [7,8].

12.2 Malignant Sinonasal Tumors

Tumors of the nasal cavity and paranasal sinuses are rare, comprising 3.6% of upper aerodigestive tract malignancies, approximately 0.2% of all cancers in males and 0.1% in females (excluding nonmelanoma skin cancer) [9,10]. They often present at a locally advanced stage as they can grow silently in the hollow sinuses and have nonspecific presenting symptoms that overlap with those of infection or inflammation, and can coexist with infection [9,11,12]. The most common symptoms are nasal obstruction, usually unilateral, epistaxis, swelling, and facial or dental pain [13–15]. Pain and numbness often indicate perineural spread and an advanced stage [8]. Most patients with sinonasal malignancies are diagnosed in the sixth decade or later and a majority are male [16]. Risk factors for the development of sinonasal malignancies are primarily occupational exposures including wood dust, leather dust, nickel compounds, radium-226 and radium-228 and their decay compounds, and acids used in the production of isopropyl alcohol. Other possible associations include exposure to formaldehyde, textile dust, and chromium compounds. Tobacco smoking has been associated with an increased

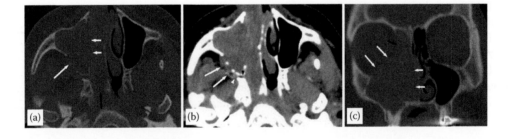

FIGURE 12.1
Features of aggressive sinonasal lesion on CT. (a) Axial non-contrast CT image in bone algorithm shows a unilateral process involving the right maxillary sinus and nasal cavity with osseous destruction of posterior (long arrow) and medial (short arrows) walls. (b) Corresponding CT image in soft tissue algorithm shows soft tissue spread into the retro maxillary fat (arrows), pterygomaxillary fissure (white arrowhead), and pterygopalatine fossa (black arrowhead). (c) Coronal non-contrast CT image shows destruction of medial maxillary sinus wall with soft tissue spread into nasal cavity (short white arrows), destruction of orbital floor with soft tissue extension into orbit (long white arrows), and destruction of lamina papyracea (short black arrow) with adjacent opacification of the ethmoid sinus. Such aggressive features warrant further assessment with MRI. Biopsy revealed squamous cell carcinoma.

risk, being stronger for squamous cell carcinoma than adenocarcinoma. Human papillomavirus (HPV) has been identified in inverted papilloma (IP) and associated squamous cell carcinoma (IP) (SSC) [10,13]. Treatment usually consists of the combination of surgery and radiation [17–19]. Chemotherapy may be used in an effort to further reduce the risk of local recurrence. For patients with unresectable disease or for those unfit for surgery, radiation or radiation plus chemotherapy is typically used. The 5-year survival has increased only slightly over the past 30–40 years from 49.7% for patients diagnosed in 1973 to 56.4% for those diagnosed in 2001 [16].

Multiple tissue types are normally found in the sinonasal cavity (epithelial, mesenchymal, muscle, nervous, and vascular) resulting in a variety of primary tumors that can originate in the sinonasal cavity. Tumors of epithelial origin are most common and include SSC (52%), adenocarcinoma (12%), adenoid cystic carcinoma (ACC) (6%), and esthesioneuroblastoma (6%) [16]. Other common tissue types include sinonasal undifferentiated carcinoma (SNUC), melanoma, lymphoma, and sarcomas [1,7,16]. The most common areas of involvement are the maxillary sinus (36%–80%) and nasal cavity (24%–44%) followed by the ethmoid sinuses (10%). Involvement of the sphenoid and frontal sinuses is rare, approximately, 3% and 1%, respectively [3,16,19].

12.3 Imaging

There is a significant overlap in the imaging features of different histologic types of sinonasal malignancies and definitive radiological diagnosis by imaging alone is difficult [7]. The major contribution of the radiologist is the precise mapping of tumor extent, which impacts decision-making regarding treatment options and prognosis [7,8,20]. With the exception of assessing for osseous involvement, MRI is the best imaging modality to

precisely describe the areas within each sinus involved by neoplasm as well as describe tumor extension into critical areas that affect treatment planning [7]. These critical areas include extension intracranially into the anterior and middle cranial fossae, the PPF, the orbits, the palate, the skull base, and PNTS [8,20,21].

MRI is the best imaging modality to distinguish neoplasm from often coexisting acute and subacute inflammatory secretions and tissue. Such secretions and inflamed mucosa generally have high water content with high signal on T_2-weighted images while most neoplasms are of intermediate to low signal on T_2-weighted images due to their high cellularity [8,20,22,23]. On postcontrast T_1-weighted images, inflamed mucosa typically shows peripheral enhancement, while tumors have more solid enhancement [20] (Figure 12.2). There are some pitfalls with this imaging approach. Small sinonasal tumors may be obscured by the high T_2-weighted signal of adjacent inflammation [7]. A minority of sinonasal tumors may have high signal on T_2-weighted images, including pleomorphic adenomas of minor salivary gland origin, nerve sheath tumors, hemangiomas, IPs, and chondrosarcomas [3,8,21] (Figure 12.3). Sinonasal melanomas often have increased signal on precontrast T_1-weighted images, which has been attributed to the presence of paramagnetic melanin and sometimes coexistent hemorrhage [20,24] (Figure 12.4). Additionally, the signal intensity of sinonasal secretions can vary on both T_1- and T_2-weighted images as the secretions become chronically obstructed, with increased protein concentration and viscosity and decreased free water. Sinonasal secretions are typically 95% water, with the remaining 5% consisting mainly of proteins, resulting in the aforementioned high signal on T_2-weighted images and low signal on T_1-weighted images. With a protein concentration of approximately 20%–25%, secretions remain of high signal on T_2-weighted images and become hyperintense on T_1-weighted images. As the protein concentration further increases, secretions become hypointense on

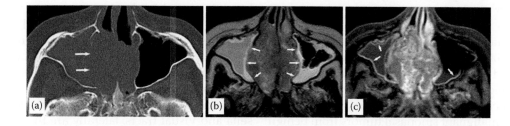

FIGURE 12.2
(a) Axial CT image in a patient with squamous cell carcinoma centered in the nasal cavity. (a) There is a nasal cavity mass destroying the medial wall of right maxillary sinus (arrows) with complete maxillary sinus opacification. Margins of tumor and extent into maxillary sinus are poorly delineated. (b) Axial T_2-weighted MR image shows mass of intermediate signal (arrows). The mass is clearly delineated from high signal inflammatory mucosal thickening and secretions in the maxillary sinuses. (c) On axial postcontrast, fat-saturated T_1-weighted image the mass has solid enhancement. Inflammatory mucosal thickening shows peripheral enhancement (arrows) with intermediate to low signal non-enhancing secretions.

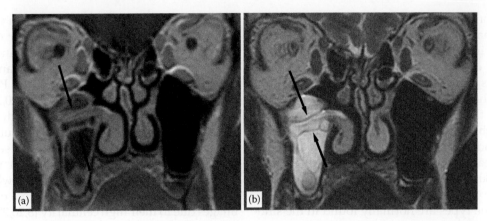

FIGURE 12.3

(a) A heterogeneously enhancing mass (arrows) is seen in the superior portion of the right maxillary sinus on T_1-weighted postcontrast MR image. (b) The mass (arrows) is of high signal on T_2-weighted images making it difficult to distinguish from surrounding secretions. Biopsy revealed inverted papilloma.

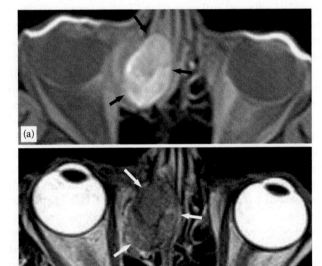

FIGURE 12.4

Sinonasal melanoma. (a) Axial precontrast, fat-saturated T_1-weighted MR image shows a bright mass (arrows) centered in the right ethmoid region. (b) On axial T_2-weighted image the mass (arrows) is of intermediate to low signal.

T_2-weighted images (Figure 12.5). Above a protein concentration of 35%–40%, secretions are markedly hypointense on T_1- and T_2-weighted images [20,25].

Diffusion-weighted imaging (DWI) with apparent diffusion coefficient (ADC) mapping may be helpful in distinguishing malignant sinonasal tumors from benign tumors and inflammatory lesions. A study in a small cohort of patients concluded that areas with low and or extremely low ADCs in malignant tumors were significantly greater than those in benign and inflammatory processes (Figure 12.6). The same authors showed that multiparametric MR imaging using the ADC maps

and time-signal intensity curves from dynamic contrast-enhanced MRI exams may provide additional benefit in distinguishing malignant form nonmalignant sinonasal disease [26,27].

12.4 Patterns of Spread

Sinonasal cancers typically spread by direct or perineural extension. For the maxillary sinus, the most significant structures with regard to treatment and prognosis are superior and posterior to the sinus. Superiorly are the ethmoid air cells and orbit, while the pterygoid plates, PPF, and masticator space are posterior. Inferior and medial spread to the alveolus and the nasal cavity, respectively, is more amenable to en bloc resection (Figure 12.7). Superior and lateral spread in the ethmoid sinuses is of greater concern as the superiorly located, thin cribriform plate, and fovea ethmoidalis provide little resistance to intracranial spread. The laterally located, also thin, lamina papyracea when violated, leads to intraorbital extension. Medial spread to the nasal cavity and inferior spread to the maxillary sinus are less worrisome (Figure 12.8). Although cancer arising in the sphenoid sinus is rare, this sinus located in the central skull base is surrounded by vital structures making resection difficult. Superiorly located are the sella and visual tracts; laterally are the internal carotid arteries and cavernous sinuses; inferiorly are the vidian nerve, clivus, and nasopharynx; and anteriorly are the posterior ethmoid air cells (Figure 12.9). The frontal sinus, although also a rare site for malignant tumors, is bordered inferiorly by the orbit and posteriorly by the intracranial compartment [13,20].

FIGURE 12.5
Patient with squamous cell carcinoma of right maxillary and ethmoid sinuses and right nasal cavity. (a) Coronal CT image shows an aggressive process completely opacifying the right maxillary and ethmoid sinuses and right nasal cavity with dehiscence of bone at the right cribriform plate (short arrow) and fovea ethmoidalis (long arrow), concerning for possible intracranial extension. (b) Secretions with high protein content are seen in the right maxillary sinus with high signal on T_1-weighted MR image (short arrows) extending to the fovea ethmoidalis (long arrow). The tumor can be distinguished by its intermediate signal. (c) On T_2-weighted MR image secretions are of intermediate to low signal (arrows) and are difficult to distinguish from the mass. Cribriform biopsy was negative for tumor involvement.

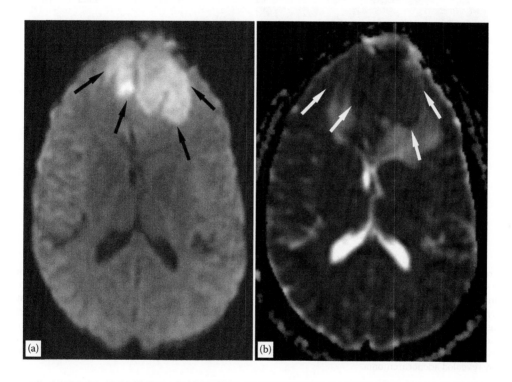

FIGURE 12.6
Restricted diffusion in a sinonasal undifferentiated tumor with intracranial extension. (a) Axial diffusion-weighted image (DWI) shows high signal within the intracranial portion of the mass (arrows). (b) Corresponding apparent diffusion coefficient (ADC) map shows low signal in the mass (arrows) consistent with restricted diffusion. There is surrounding high signal vasogenic edema.

12.4.1 Marrow Invasion

Although CT is superior at detecting bone remodeling or erosion, MRI can detect bone marrow invasion particularly at the skull base. The combination of CT and MRI may afford greater accuracy in assessing the status of osseous structures [28]. Marrow invasion in adults is seen as the replacement of the normal high signal fatty marrow on precontrast T_1-weighted images. Normal appearing fatty marrow at the skull base is generally an indicator of the absence of skull base invasion. However, hypointense signal on precontrast T_1-weighted images may be related to marrow invasion

versus reactive change, hematologic disease, or edema [20,21,28]. Associated enhancement on fat-suppressed postcontrast T_1-weighted images may be helpful in distinguishing tumor from other causes of abnormal marrow signal [29,30] (Figure 12.10).

12.4.2 Orbital Invasion

Given the close proximity of the paranasal sinuses to the orbits, with multiple foramina and perforating vessels and nerves and thin surrounding bony structures, including the lacrimal fossa and lamina papyracea,

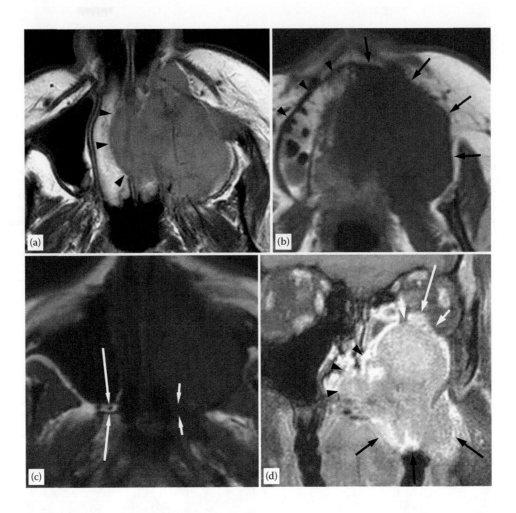

FIGURE 12.7
Patterns of local spread in a maxillary sinus neoplasm. (a) An intermediate signal mass is present in left maxillary sinus on axial T_2-weighted image with medial spread into nasal cavity (arrowheads). (b) There is inferior spread into the maxilla, seen well on precontrast T_1-weighted image, with mass infiltrating and replacing normal high signal fatty marrow of left maxilla (arrow). Normal high signal fatty marrow present in right maxilla (arrowheads). (c) There is posterior spread into the pterygopalatine fossa with intermediate signal (short arrows) replacing the normal fatty signal (long arrows) seen in the right pterygopalatine fossa, on this axial precontrast T_1-weighted image. (d) Coronal postcontrast, fat-saturated T_1-weighted image shows medial spread into nasal cavity (black arrowheads), inferior spread into maxilla and buccal space (black arrows), and superior spread into the orbit (short white arrows) with tumor infiltration abutting the inferior rectus muscle (long white arrow). Biopsy revealed adenocarcinoma.

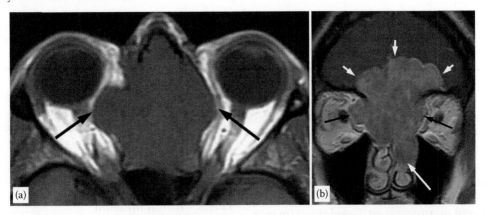

FIGURE 12.8
Patterns of local spread in an ethmoid sinus neoplasm. (a) An intermediate signal mass centered in the ethmoid sinuses is extending laterally into both orbits (arrow), with orbital extension well seen on axial precontrast T_1-weighted image. (b) Coronal postcontrast T_1-weighted image shows superior intracranial extension (short white arrows) through the cribriform plate and fovea ethmoidalis bilaterally. Lateral orbital extension (black arrows) and inferior nasal cavity extension (long white arrow) are also present. Biopsy revealed a sinonasal undifferentiated carcinoma.

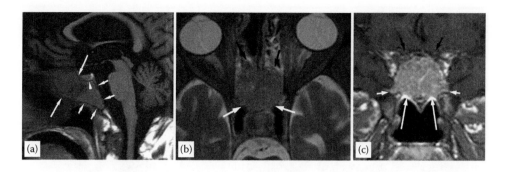

FIGURE 12.9

Patterns of local spread in a sphenoid sinus neoplasm. (a) Sagittal precontrast T_1-weighted image shows an intermediate signal sphenoid sinus mass (long white arrows) with extension into the sphenoid bone and clivus (short white arrows) with replacement of normal fatty marrow. There is no definite extension into immediately adjacent sella (white arrowheads). (b) Axial T_2-weighted image shows the mass abutting the cavernous portions of both internal carotid arteries (white arrows) while anteriorly there is extension into the ethmoid sinuses (black arrows). (c) Coronal postcontrast T_1-weighted image shows extension of the mass into the foramen rotundum (short white arrows) and the vidian canal (long white arrows) bilaterally. Superiorly the mass is in close proximity to the optic nerves (black arrows).

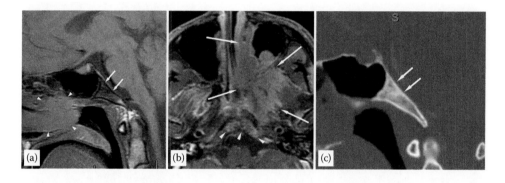

FIGURE 12.10

(a) Sagittal T_1-weighted precontrast image shows abnormal hypointense marrow signal (arrows) replacing the normal fatty marrow of the clivus in a patient with adenocarcinoma of the left nasal cavity (white arrowheads). (b) Axial postcontrast, fat-saturated T_1-weighted image shows enhancement in the clivus (arrowheads). The left nasal cavity mass (arrows) with posterior extension is also seen. Findings are concerning for tumor infiltration into the clivus. (c) On a sagittal reformatted CT image only subtle sclerosis (arrows) is present in the clivus.

there are many routes by which a sinonasal tumor can gain access to the orbit [21]. Orbital invasion is most common with ethmoid tumors, with invasion of orbital wall in 66%–82% and involvement of orbital periosteum in 30%–50% of patients with ethmoid malignancy. Bone erosion or invasion occurs in 60%–80% of maxillary sinus malignancies [31–34]. Distinction needs to be made between bony orbital wall erosion, involvement of the periorbita (the orbital periosteum), and deeper invasion of the orbital soft tissues [28]. The periorbita is a resilient structure that is thought to provide a barrier to deeper invasion of the orbital contents [21,28,34]. If tumor has not extended through the periorbita, the eye can generally be preserved while maintaining oncologic safety and a functional eye [34,35]. Precontrast T_1-weighted images provide an excellent contrast between the usually intermediate signal tumor and the hyperintense orbital fat. The periorbita is seen as a hypointense structure on T_1- and T_2-weighted sequences and enhances less than the extraocular muscles (Figure 12.11). Various imaging criteria have been proposed on CT and MRI to predict tumor invasion of the orbit. These criteria included the tumor's relationship to the periorbita (abutting, displacing, or bowing laterally), the interface between the tumor and the periorbita (nodular or smooth), the extraocular muscles (displaced, enlarged, or abnormal signal intensity/density), and orbital bone integrity. Extension of tumor beyond a thickened or disrupted periorbita on MRI has also been suggested as a sign of extension through the periorbita [36,37] (Figure 12.12). Definitive assessment of the periorbita, however, must be done intraoperatively by the frozen section [28,37].

12.4.3 Dural and Intracranial Invasion

Due to their close relationship to the anterior and middle cranial fossae, sinonasal tumors often extend to these structures and once through the bone can invade the dura and brain parenchyma [8]. Such an extension is important with regard to treatment planning and

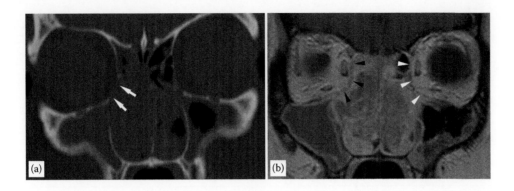

FIGURE 12.11

(a) Coronal reformatted CT image shows erosion of bony orbital wall (arrows) in a patient with squamous cell carcinoma of the nasal cavity. (b) Coronal postcontrast T_1-weighted image in same patient shows a linear hypointense structure (white arrowheads) between the orbits and the sinuses, including the area of osseous erosion on CT (black arrowheads). Findings suggest the periorbita is intact. Biopsy of the periorbita at time of surgery was negative for tumor invasion. A portion of the lamina was resected but remaining orbital structures were preserved.

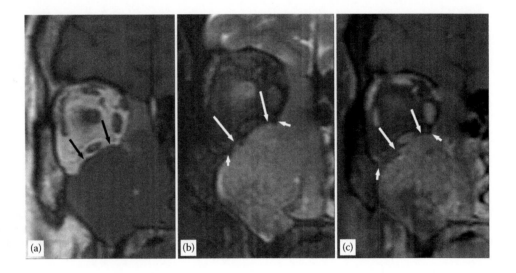

FIGURE 12.12

(a) On a coronal precontrast T_1-weighted image in a patient with right maxillary sinus squamous cell carcinoma, there is a nodular interface between the tumor and the periorbita with extension into orbital fat (arrows). (b) Coronal T_2-weighted image shows nodular interface with periorbita and orbital fat (long arrows) and more normal appearing periorbita at margins of mass (short arrows). (c) Coronal post-contrast, fat-saturated T_1-weighted image shows nodular interface (long arrows) with loss of normal low signal of periorbita (short arrows). At time of surgery biopsy confirmed invasion through the periorbita and right orbital exenteration was performed.

prognosis [28]. In general, tumors involving the skull base most commonly affect the anterior cranial fossa with much less frequent involvement of the middle cranial fossa or the combined anterior and middle cranial fossae. If surgical resection is considered for tumors that extend into the anterior cranial fossa, extensive surgery with a craniofacial approach is typically required [38]. Dural and intracranial invasion is associated with significantly worse recurrence-free survival, disease-specific survival, and overall survival [39]. MRI is superior to CT for detecting such intracranial extension, particularly dural, pial, and parenchymal invasion [20]. Various imaging findings have been described as being indicative of dural invasion. Linear, thin dural enhancement adjacent to a tumor may reflect reactive dural enhancement caused by increased vascularity, inflammation, or dural hypertrophy incited by the adjacent neoplasm; particularly, if the dural enhancement is separated from the tumor by a thin hypointense zone on MRI (Figure 12.13). This hypointense zone has been attributed to cortical bone and the epidural space [40,41]. Findings more specific for the dural invasion include loss of the hypointense zone between the tumor and the enhancing dura and nodular dural thickening. The thickness of the enhancing dura is also an important factor. A 2012 study suggests dural thickening ≥2 mm is predictive of dural invasion while an older study proposed dural thickening >5 mm as suggestive of invasion

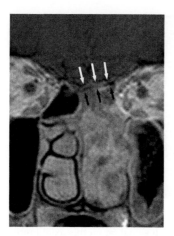

FIGURE 12.13
Linear dural enhancement (white arrows) separated from adjacent left nasal cavity and ethmoid sinus teratocarcinoma (black arrows) by a thin hypointense line composed of cortical bone and the adjacent epidural space. Dural biopsy was negative for tumor involvement and dural enhancement on MRI was reactive.

[40–42]. However, lesser degrees of dural enhancement can be associated with invasion (Figure 12.14). These factors and potential false-negative results with frozen section dural biopsies have led to many surgeons resecting all potentially involved dura rather than risk leaving residual tumor [28,40–42]. Dural invasion is also suggested by the presence of pial enhancement and hyperintense signal in adjacent brain parenchyma on T_2-weighted images [28].

12.4.4 Perineural Tumor Spread

Perineural invasion (PNI) is a histologic finding at the primary tumor site that refers to tumor cells invading in, around, and through small nerves, tumor cells within any layers of the nerve sheath or at least 33% of nerve circumference surrounded by tumor cells. The most recent theory of pathogenesis is that PNI involves reciprocal signaling interactions between tumor cells and nerves. Nerve cells express signaling proteins that could initiate and sustain cancer invasion if the cancer cells have the appropriate receptors [28,43,44]. A retrospective 2009 single center study demonstrated PNI in 20% of paranasal sinus cancers with the highest propensity in sinonasal undifferentiated tumor, ACC, and SCC [44]. Although there are conflicting results on the prognostic significance of PNI, it is generally thought to be associated with increased risk of locoregional recurrence and many oncologists consider it an indication for adjuvant radiotherapy, especially if extensive [45].

In contrast, PNTS refers to the ability of a tumor to use a nerve or nerve sheath as a conduit to spread away from the primary focus. Although starting microscopically, this term reflects radiographically visible large nerve involvement by tumor and is a mode of metastasis rather than a microscopic finding [45,46]. PNTS is asymptomatic in 40% of patients; thus, imaging and careful assessment of the images is critical in diagnosing such spread, which can alter treatment and prognosis [47]. Symptomatic patients may have pain, numbness, burning, and dysesthesias along the course of the

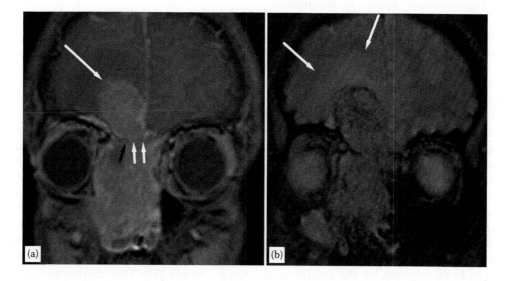

FIGURE 12.14
(a) Coronal postcontrast, fat-saturated T_1-weighted image in a patient with nasal adenocarcinoma shows a large nodular mass extending intracranially with a focal area of leptomeningeal enhancement extending into a sulcus (long arrow). There is loss of the hypointense zone between the tumor and the intracranial disease (white arrows). Normal hypointense line is seen more laterally (black arrow). (b) Coronal T_2-weighted image shows hyperintense edema in adjacent brain (arrows). Findings concerning for dural invasion which was confirmed on dural biopsy.

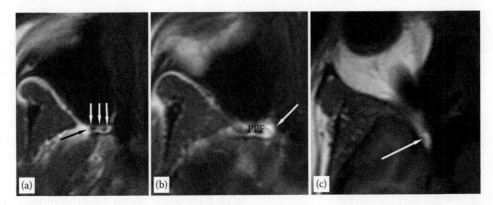

FIGURE 12.15
Normal PPF and adjacent structures. (a) Normal PPF (white arrows) contains mainly high signal fat on this axial precontrast T_1-weighted image. Lateral to the PPF is the pterygomaxillary fissure (black arrow) leading into the masticator space. (b) More superior axial precontrast T_1-weighted image shows sphenopalatine foramen (arrow) at the medial margin of the PPF. (c) Axial precontrast T_1-weighted image more superiorly shows normal fat in the inferior orbital fissure (arrow).

affected nerve and motor denervation changes [46]. Gross perineural spread can be resected to the level of the skull base foramina by experienced surgeons, but generally requires extensive, radical resection. Extension into the cavernous sinus and more proximally generally is a contraindication to radical curative surgery and in these instances any surgery is usually for palliation [20,21,28,48]. Patients with PNTS often have recurrence, which adversely affects long-term survival. Treatment failures may be related to initially undetected perineural spread [48].

PPF, located between the maxilla and the pterygoid process of the sphenoid bone, is an important anatomic structure for the detection of perineural spread of sinonasal malignancies. From the PPF, tumor can spread through the pterygomaxillary fissure to the masticator space, through the sphenopalatine foramen into the nasal cavity, into the inferior orbital fissure and subsequently the orbit, and along the foramen rotundum or vidian canal to the cavernous sinus with further spread intracranially [20,21]. The PPF is often best assessed on T_1-weighted precontrast images where high signal fat in the fossa should normally be seen and tumor is identified by loss of the normal fat (Figure 12.15). Both retrograde and antegrade spread can occur along a nerve. Typically, the tumor spreads retrograde toward the brain stem. Once reaching a junction or coming into close proximity to another nerve, PNTS may involve other branches of the nerve and spread antegrade [47].

MRI detection of perineural spread requires thin section (≤3 mm) postcontrast T_1-weighted images along the entire course of the nerve, from its most peripheral branches to its brain-stem nuclei. Fat saturation is usually used as it is thought to improve the detection of subtle areas of contrast enhancement. Fat saturation, however, can increase the potential for susceptibility artifact at the skull base; particularly, adjacent to

the sphenoid sinus, which may obscure the foramen rotundum, foramen ovale, vidian canal, and cavernous sinus. To prevent this possible pitfall, some radiologists have proposed using pre- and postcontrast T_1-wieghted images without fat saturation [28,48–50] (Figure 12.16). MRI findings of PNTS include obliteration of fat in the PPF, obliteration of fat around a nerve below the neural foramen, enlargement of a foramen, thickening of a nerve, and nerve enhancement due to disruption of the blood-nerve barrier (Figures 12.17. and 12.18). Findings concerning for extension to the cavernous sinus include loss of normal fluid signal in Meckel cave, enhancement and thickening of the cavernous sinus, and lateral bulging or convexity of the cavernous sinus [21,28,45,48–51] (Figure 12.18).

12.5 Staging

12.5.1 Epithelial Malignancies

Staging systems are used to estimate the extent of disease and provide some basis to determine prognosis. Staging of sinonasal neoplasms is most commonly done using a tumor, node, metastasis (TNM) system developed by the American Joint Committee on Cancer (AJCC) staging manual, most recently updated in 2010. There is a staging system for maxillary sinus tumors and tumors of the nasal cavity and ethmoid sinuses. Because of their rarity, there is no TNM staging for frontal or sphenoid sinus cancers [52]. This staging mainly applies to epithelial sinonasal malignancies, which are the most common. There are separate staging systems for olfactory neuroblastoma (ONB) and sinonasal melanoma, which will be discussed later.

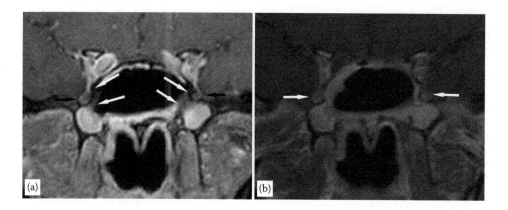

FIGURE 12.16

(a) On a coronal postcontrast, fat-saturated T_1-weighted image, susceptibility artifact (white arrows) partially obscures the bilateral foramen rotundum (black arrows). (b) The bilateral foramen rotundum (white arrows) are not obscured on this coronal postcontrast T_1-weighted image without fat saturation.

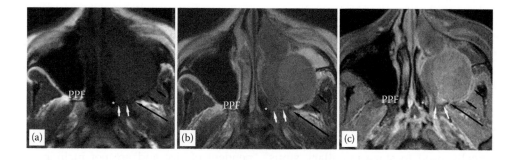

FIGURE 12.17

Involvement of the PPF in a patient with adenoid cystic carcinoma of the left nasal cavity and maxillary sinus. (a) On an axial precontrast T_1-weighted image there is intermediate signal tissue replacing normal fat in the left pterygopalatine fossa (PPF) (white arrows) with extension to the sphenopalatine foramen (*) and through the pterygomaxillary fissure (long black arrow) into the retromaxillary fat (short black arrows). Normal right PPF is seen (PPF). Similar involvement of left PPF (white arrows), sphenopalatine foramen (*), pterygomaxillary fissure (long black arrow), and retromaxillary fat (short black arrows) is seen, but less conspicuous on (b) axial T_2-weighted image and (c) axial postcontrast, fat-saturated T_1-weighted image.

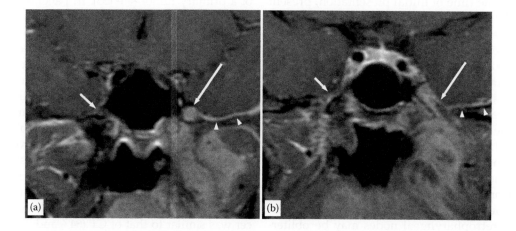

FIGURE 12.18

(a) Coronal postcontrast, fat-saturated T_1-weighted image demonstrates an enlarged and enhancing left maxillary branch (V2) of cranial nerve V in an enlarged foramen rotundum (long arrow) compatible with PNTS in a patient with ACC of left maxillary sinus. Compare to normal right side (short arrow). Abnormal dural enhancement is also seen along floor of left middle cranial fossa (arrowheads). (b) In the same patient, an enlarged and enhancing mandibular branch (V3) of cranial nerve V is seen at the level of the foramen ovale (long arrow) compatible with PNTS. Compare to normal right side (short arrow). Abnormal dural enhancement is also seen (arrowheads).

TABLE 12.1

Primary Tumor (T) Staging for Maxillary Sinus Cancers

TX	Primary tumor cannot be assessed
T0	No evidence of primary tumor
Tis	Carcinoma *in situ*
T1	Tumor limited to maxillary sinus mucosa with no erosion or destruction of bone
T2	Tumor causing bone erosion or destruction including extension to hard palate and/or middle nasal meatus, except extension to posterior wall of maxillary sinus and pterygoid plates
T3	Tumor invades any of the following: bone of the posterior wall of maxillary sinus, subcutaneous tissues, floor or medial wall of orbit, pterygoid fossa, or ethmoid sinuses
T4a	Moderately advanced local disease Tumor invades anterior orbital contents, skin of cheek, pterygoid plates, infratemporal fossa, cribriform plate, or sphenoid or frontal sinuses
T4b	Very advanced local disease Tumor invades any of the following: orbital apex, dura, brain, middle cranial fossa, cranial nerves other than maxillary division of trigeminal nerve (V2), nasopharynx, or clivus.

Source: Edge, S.B. et al. (eds.), *American Joint Committee on Cancer Staging Manual*, 7th ed., Springer, New York, 2010.

TABLE 12.2

Primary Tumor (T) Staging for Nasal Cavity and Ethmoid Sinus Cancers

TX	Primary tumor cannot be assessed
T0	No evidence of primary tumor
Tis	Carcinoma *in situ*
T1	Tumor restricted to any one subsite, with or without bony invasion
T2	Tumor invading two subsites in a single region or extending to involve an adjacent region within the nasoethmoidal complex, with or without bony invasion
T3	Tumor extends to invade the medial wall or floor of the orbit, maxillary sinus, palate, or cribriform plate
T4a	Moderately advanced local disease Tumor invades any of the following: anterior orbital contents, skin of nose or cheek, minimal extension to anterior cranial fossa, pterygoid plates, or sphenoid or frontal sinuses
T4b	Very advanced local disease Tumor invades any of the following: orbital apex, dura, brain, middle cranial fossa, cranial nerves other than (V2), nasopharynx, or clivus

Source: Edge, S.B. et al. (eds.), *American Joint Committee on Cancer Staging Manual*, 7th ed., Springer, New York, 2010.

Ohngren's line, extending from the medial canthus of the orbit to the angle of the mandible on a lateral radiograph, was historically used to stage maxillary sinus cancers. It divided the maxillary sinus into an anteroominferior and posterosuperior half, with posterosuperior tumors having a poorer prognosis due to closer proximity to PPF, orbit, and skull base [8]. The current staging for maxillary sinus tumors, nasal cavity, and ethmoid sinus tumors is shown in Tables 12.1 and 12.2 [52].

Lymph node metastases from sinonasal malignancies are rare at the time of initial presentation, present in less than 15% of patients, and more common with maxillary sinus primaries than those in the ethmoid sinus. Nodal metastases, either at presentation or developing later, are associated with a poor prognosis and a decreased overall survival compared to patients with no nodal metastases [53,54]. Lymph from the anterior half of nasal cavity drains into level 1b nodes while lymph from the posterior half drains into retropharyngeal nodes and into level 2–5 nodes. Frontal and ethmoid sinus lymph drains into level 1 nodes. Sphenoid sinus lymphatics drain into retropharyngeal nodes. Lymphatics from the maxillary sinus drain into level 1b and 2–4 nodes and retropharyngeal nodes. As lymphatics to the retropharyngeal nodes may be obliterated by recurrent childhood infections, levels 1 and 2 may represent the primary nodal stations for sinonasal malignancies [21].

The AJCC staging for lymph node involvement is based on the size, number, and location (ipsilateral, contralateral,

or bilateral) of the involved lymph node or nodes. In addition to size, there are other imaging features of lymph nodes concerning for metastatic spread, which are independent of size and are not included in the N-staging system. These features consist of focal internal inhomogeneity (commonly attributed to necrosis), rounded shape with loss of typical elongated configuration, thickened, irregular enhancing rim with infiltration of adjacent fat or soft-tissue structures (suggestive of extracapsular spread [ECS]), and nodal clustering with ≥3 borderline lymph nodes in the first or second lymph node drainage region of a primary tumor site [55]. In patients with suspected or known head and neck cancer, internal heterogeneity >3 mm is the most reliable finding of nodal metastasis [56]. A recent retrospective study demonstrated poor accuracy for detection of ECS by CT using the criteria of capsular contour irregularity, poorly defined margins, and infiltration of adjacent fat [57]. Prior to gadolinium-based IV contrast administration, normal nodes are isointense to muscle on T_1-weighted images and variably hyperintense to muscle on T_2-weighted and short T1 inversion recovery (STIR) images and enhance homogeneously after contrast is given [55]. The sensitivity and specificity of MRI (67% and 79%, respectively) for detecting lymph node metastases in patients with squamous cell head and neck cancer was similar to that of CT (64% and 75%, respectively) in a 2012 metaanalysis. Combined PET-CT, however, has greater sensitivity and specificity than either CT or MRI [58,59]. PET-CT is also useful for the detection of distant metastases in high-risk patients [3]. The current N and M staging is shown in Table 12.3 [52].

TABLE 12.3

Regional Lymph Node (N) and Distant Metastasis (M) Spread for Paranasal Sinus and Nasal Cavity Cancers

Regional Lymph Nodes (N)

NX	Regional lymph nodes cannot be assessed
N0	No regional lymph node metastasis
N1	Metastasis in a single ipsilateral lymph node, ≤3 cm in greatest dimension
N2	Metastasis in a single ipsilateral lymph node, >3 cm but ≤6 cm in greatest dimension, or metastases in multiple ipsilateral lymph nodes, ≤6 cm in greatest dimension, or in bilateral or contralateral lymph nodes, ≤6 cm in greatest dimension
N2a	Metastasis in a single ipsilateral lymph node, >3 cm but ≤6 cm in greatest dimension
N2b	Metastases in multiple ipsilateral lymph nodes, ≤6 cm in greatest dimension
N2c	Metastases in bilateral or contralateral lymph nodes, ≤6 cm in greatest dimension
N3	Metastasis in a lymph node, >6 cm in greatest dimension

Distant Metastasis (M)

M0	No distant metastasis
M1	Distant metastasis

Source: Edge, S.B. et al. (eds.), *American Joint Committee on Cancer Staging Manual*, 7th ed., Springer, New York, 2010.

12.5.2 ONB (Esthesioneuroblastoma)

The Kadish staging system was introduced for the staging of ONB in 1976 [60]. A modified version was introduced by Morita in 1993 in which stage A tumors are confined to the nasal cavity, stage B tumors have extension to the paranasal sinuses, stage C tumors have extension beyond the nasal cavity and paranasal sinuses, and stage D tumors have regional lymph node or distant metastasis [61]. The modified Kadish staging is often used, although other staging classifications have been proposed, namely the Dulguerov classification [62] (Table 12.4). The Kadish staging system may also be employed for SNUC [63].

12.5.3 Sinonasal Melanoma

The AJCC staging system for mucosal melanomas arising from the head and neck (Table 12.5) is generally used to stage sinonasal melanomas. The staging system begins with T3 and stage III disease, reflecting the poor prognosis with this malignancy. T3 is disease limited to the mucosa, T4a is moderately advanced disease with tumor involving deep soft tissues, cartilage, bone, or overlying skin, and T4b is very advanced disease with tumor involving the skull base, dura, brain, lower cranial nerves (IX–XII), masticator space, carotid artery, prevertebral space, or mediastinal structures [64,65].

TABLE 12.4

Olfactory Neuroblastoma Staging Systems

Modified Kadish Staging System[a]

A	Tumor confined to the nasal cavity
B	Tumor confined to the nasal cavity and paranasal sinuses
C	Tumor extent beyond the nasal cavity and paranasal sinuses, including involvement of the cribriform plate, base of skull, orbit, or intracranial cavity
D	Tumor with metastasis to cervical lymph nodes or distant sites

Dluguerov Classification[b]

T1	Tumor involving the nasal cavity and/or paranasal sinuses (excluding sphenoid), sparing the most superior ethmoids
T2	Tumor involving the nasal cavity and/or paranasal sinuses (including sphenoid), with extension to or erosion of the cribriform plate
T3	Tumor extending into the orbit or protruding into the anterior cranial fossa, without dural invasion
T4	Tumor involving the brain
N1	Any form of cervical lymph node metastasis
M0	No metastases
M1	Distant metastasis

[a] Morita, A. et al., *Neurosurgery*, 32, 706–714, 1993.
[b] Dulguerov, P. and Calcaterra, T., *Laryngoscope*, 102, 843–849, 1992.

TABLE 12.5

Staging for Mucosal Melanoma of Head and Neck

Primary Tumor (T)

T3	Mucosal disease
T4a	Moderately advanced disease. Tumor involving deep soft tissue, cartilage, bone or overlying skin
T4b	Very advanced disease. Tumor involving skull base, dura, brain, lower cranial nerves (IX–XII), masticator space, carotid artery, prevertebral space, or mediastinal structures

Regional lymph nodes (N)

NX	Regional lymph nodes cannot be assessed
N0	No regional lymph node metastases
N1	Regional lymph node metastases present

Distant metastasis (M)

M0	No distant metastasis
M1	Distant metastasis present

Source: Edge, S.B. et al. (eds.), *American Joint Committee on Cancer Staging Manual*, 7th ed., Springer, New York, 2010.

12.6 Neoplasms

12.6.1 Squamous Cell Carcinoma

SSC is the most common histologic type affecting the sinonasal structures, although its reported incidence is widely variable, ranging from 38% to 80% of sinonasal malignancies [8,10,16]. Part of this variation is related to the time period during which the data was collected as well as the geographical distribution from which the data was collected; however, overall there has been a decline in squamous cell histology over time [10,16]. Males have a greater proportion of SCC than females [10]. Occupational exposure to wood dust increases the risk of developing sinonasal SCC 20-times. Smoking increases the risk 2–3 times. HPV has also been implicated in the malignant transformation of IPs [10,13,66]. SCC occurs most commonly in the maxillary sinus then the nasal cavity. Lymph node metastases are present in 10%–20% of patients at diagnosis, especially if the tumor erodes through the maxillary gingiva [66].

12.6.2 Adenocarcinoma Nonsalivary Gland Type

Combined, adenocarcinomas and salivary-type carcinomas comprise approximately 10%–20% of sinonasal malignancies. Nonsalivary gland-type sinonasal adenocarcinomas have been classified by the World Health Organization (WHO) into intestinal-type adenocarcinoma (ITAC) and non-ITAC, with the latter being subdivided into low-grade and high-grade carcinomas [67]. They are much more common in males, with a male-to-female ratio of approximately 6:1 for ITAC, probably as a result of greater occupational hazards in men. Occupational exposure to wood dust leads to a 500–990-times increased risk of developing ITAC. Forty percent of ITAC are found in the ethmoid sinuses, with 27% and 20% in the nasal cavity and maxillary sinus, respectively. Of the non-ITAC, the low-grade tumors have a predilection for the ethmoid sinues while the high grade tumors have a predilection for the maxillary sinuses [66,67]. Unlike other histological types of sinonasal cancer, survival has increased from approximately 50%–66% over the past 40 years [16].

12.6.3 Salivary Gland-Type Carcinomas

Several hundred salivary glands line the sinonasal tract; thus, a variety of salivary gland tumors can arise in this region, most being malignant. ACC is the most common, accounting for approximately 6% of all sinonasal malignancies [16]. These most frequently occur in the maxillary sinus (approximately 60%) followed by the nasal cavity (approximately 25%). ACC have a predilection for perineural spread, found in approximately 60% of these tumors. Recurrence rates over 60% have been reported, and they may occur up to 15 years after treatment. Approximately, 50% of patients have distant metastases, most commonly to lung, brain, and bones. Long-term prognosis is poor with a 10-year survival rate of 7%. Most patients die of local spread rather than metastatic disease [67–69]. Other salivary gland-type carcinomas arising in the sinonasal region are rarer and include acinic cell carcinomas and mucoepidermoid carcinoma among other cell types [67].

12.6.4 Melanoma

The incidence of sinonasal melanoma has been increasing over the past 40 years, although it remains a rare malignancy accounting for approximately 6% of all sinonasal malignancies [16]. Melanomas more commonly arise in the nasal cavity than the paranasal sinuses, especially from the nasal septum [68]. Lymph node metastases are present in 30%–40% of patients and have little impact on survival [20,70]. Overall survival is poor regardless of treatment, with 5-year survival rates of approximately 30% [68,70]. Surgery is the treatment of choice. Radiation may help locoregional control but does not improve overall survival [70].

12.6.5 Sinonasal Undifferentiated Carcinoma

SNUC is an undifferentiated epithelial neoplasm of uncertain histogenesis that is highly aggressive, often presenting with locally extensive disease. SNUC accounts for approximately 3% of sinonasal malignancies and is 2–3 times more common in males [16,67,71]. The nasal cavity, maxillary antrum, and ethmoid sinus are typically involved, alone or in combination, although in some cases the site of origin is difficult to determine due to the size and breach of anatomical boundaries associated with these tumors. Compared to other sinonasal neoplasms, symptoms are of relatively short duration (weeks to months) [67,71]. Lymph node metastases are present in 10%–30% of patients, while distant metastases are unusual. Both the AJCC and Kadish staging systems have been used for SNUC [62]. Treatment generally consists of surgery with adjuvant chemoradiation; however, prognosis is poor with 5-year survival ranging between 18% and 35% [16,67].

12.6.6 Olfactory Neuroblastoma

ONB or esthesioneuroblastoma arises from the olfactory membrane of the sinonasal tract. The commonest sites of origin for ONB are the epithelium overlying the cribriform plate, the superior nasal septum, and the superior turbinate. They comprise approximately 3%–6% of all sinonasal tumors with the incidence increasing over the past 40 years [16,67]. Both genders are affected equally and most tumors arise in the fifth and sixth decades, although historically a bimodal peak has been described, in the second and sixth decades [67,72,73]. Lymph node metastases are rare, occurring in approximately 20% of patients [72]. The best survival results are obtained with surgery and radiation therapy [74]. Prognosis is relatively good for sinonasal malignancies with 5-year survival of greater than 70% [16,67]. Although imaging features are generally nonspecific, a dumbbell-shaped mass crossing the cribriform plate with peritumoral cyst along superior margin of the intracranial component is suggestive of ONB [75].

12.6.7 Others

A variety of other less common malignant neoplasms can involve the sinonasal region. Among those are lymphoma, plasmacytoma, sarcomas (including chondrosarcoma, osteosarcoma, fibrosarcoma, and rhabomyosarcoma), and neuroendocrine tumors [67] (Figure 12.19).

12.6.8 Imaging Follow-Up

The goal of follow-up imaging after treatment is to detect recurrent tumor, especially in the first 2 years after the treatment. Long-term follow up is often performed for ACC, ONB, melanoma, and chondrosarcoma, all of which have a tendency for delayed recurrence [11,20]. Early detection of recurrence allows for salvage therapy with curative intent and associated survival benefit in some patients [76,77]. Local recurrence generally occurs at the resection margin in patients treated surgically and at the primary site in those treated with radiation; thus, comparison with prior imaging and knowledge of prior treatment is imperative when interpreting post-treatment images. Enlarging neck lymph nodes or the appearance of a new abnormal node are also concerning for recurrence [11,78]. A baseline posttreatment MRI scan 3–6 months after the completion of therapy has been proposed by some authors, with increasing size of an enhancing mass of intermediate to high signal on T_2-weighted images in surgical bed on subsequent follow-up exams concerning for recurrence. Stability or retraction of soft tissue suggests posttreatment change. A mass of decreased T_2-weighted signal in the operative bed is suggestive of scar or fibrotic tissue [78–80] (Figures 12.20 and 12.21). However, there is overlap in the imaging appearance of recurrence and post-therapeutic change in patients with head and neck cancer. There is poor sensitivity and specificity for detecting recurrence using conventional anatomic imaging with MRI and CT, with reported sensitivities less than 80% and specificities less than 60% [81–83]. DWI has been shown to improve sensitivity and specificity for detecting recurrent head and neck cancers,

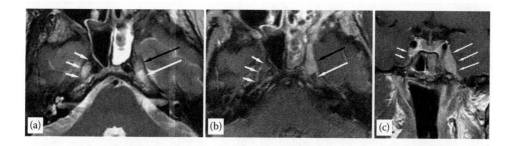

FIGURE 12.19
Cavernous sinus invasion in a patient with adenoid cystic carcinoma. (a) Axial T_2-weighted image shows intermediate signal tumor (long black arrow) extending into the anterior part of Meckel cave on the left. Normal fluid signal is seen more posteriorly on the left (long white arrow) and in right Meckel cave (short white arrows). (b) Axial postcontrast, fat-saturated T_1-weighted image shows enhancing tumor anteriorly in Meckel cave on the left (long black arrow). The remainder of Meckel cave on the left (long white arrow) and the entirety on the right (short white arrows) has normal fluid signal. (c) Coronal postcontrast T_1-weighted image shows thickening, lateral bowing and enhancement of abnormal left cavernous sinus (long white arrows) with normal cavernous sinus seen on right (short white arrows).

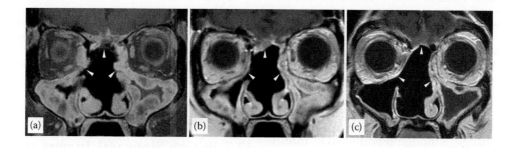

FIGURE 12.20
Resolving posttreatment changes in a patient treated with subcranial resection of squamous cell carcinoma centered in the nasal cavity and post-op radiation therapy. (a) Coronal postcontrast T_1-weighted image obtained 5 months following surgery and 1.5 months following completion of radiation therapy shows enhancing soft tissue along the margins of the resection cavity (arrowheads). Subsequent imaging obtained 1 year after surgery (b) and 2.5 years after surgery (c) show a progressive decrease in the enhancing soft tissue along margins of the surgical cavity (arrowheads).

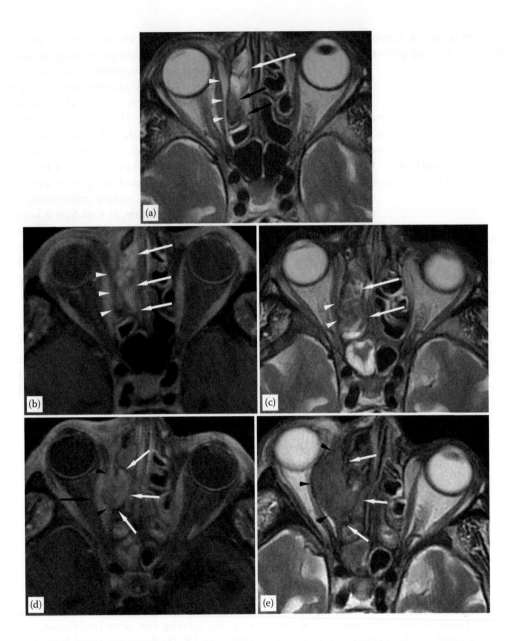

FIGURE 12.21

Recurrent squamous cell cancer in a patient treated with radiation and chemotherapy for right maxillary sinus primary. (a) Opacification of right ethmoid sinus with high signal anteriorly (long white arrow) and intermediate signal posteriorly (short black arrows) on axial T_2-weighted image. No extension into orbital fat (white arrowheads) is seen. (b) Following contrast there is heterogeneous enhancement of right ethmoid sinus (white arrows) on axial postcontrast, fat-saturated T_1-weighted image. No evidence of extension of process into orbit (arrowheads). (c) 3 month follow up axial T_2-weighted MRI image shows progression of low signal in right ethmoid sinus (white arrows) with infiltration into orbit adjacent to the medial rectus muscle (white arrowheads). (d) Corresponding axial postcontrast, fat-saturated T_1-weighted image shows a focal enhancing mass in right ethmoid sinus (white arrows) with extension into medial orbit (black arrowheads) and displacement of medial rectus muscle (black arrow). (e) Further follow-up shows enlarging mass (white arrows) with greater orbital extension (black arrowheads) on axial T_2-weighted image despite the patient receiving chemotherapy.

with recurrent tumor having lower ADC values than post-treatment change, although reported ADC threshold values are variable and there is overlap in ADC values for recurrence and posttreatment change [83,84]. PET/CT has also shown promise in detecting recurrent head and neck cancer compared to conventional MRI and CT [59,81,85,86].

12.7 Benign Sinonasal Tumors and Tumor-Like Conditions

A large variety of benign tumors can occur in the sinonasal region, the most common of which are papillomas, osteomas, fibrous dysplasia (FD), and neurogenic tumors.

Only those in which MRI has a role in their diagnosis and treatment planning or those that can mimic an aggressive lesion on MRI will be discussed in this chapter [13,67,68,87].

12.7.1 Inverted Papilloma

Papillomas arising from the Schneiderian mucosa of the sinonasal cavity represent 0.4% to 4.7% of all sinonasal tumors, the IP being the most common, comprising nearly half of Schneiderian papillomas. Its name is derived from its pathologic appearance in which its epithelium invaginates into and proliferates in the underlying stroma. Although their exact role in the etiology of IP is not certain, human papillomas and Epstein–Barr viruses have been detected in IP. These tumors most commonly occur between the ages of 40 and 70 years and are 2–5 times more common in men. They characteristically arise from the lateral wall of the nasal cavity near the middle turbinate and commonly extend into the maxillary and ethmoid sinuses [3,4,67].

These can behave aggressively with reported recurrence rates of 13%–67% [3]. Additionally, the incidence of carcinoma-ex-IP ranges from 3% to 27% and may be concurrent with or develop subsequent to the IP. SSCs are the most common, but other types may occur [8,67].

The imaging appearance is nonspecific and variable, ranging from a small polypoid lesion to an expansile nasal mass with osseous remodeling and erosion. IP are typically isointense on T_1-weighted images and of low-to-intermediate signal on T_2-weighted images with solid enhancement [1,7,8]. A septated striated appearance on T_2-weighted images or convoluted cerebriform pattern on postcontrast T_1-weighted images, with alternating areas of hypo- and hyperintense signal, have been described as being suggestive of IP (Figure 12.22). However, this pattern is not seen in all IPs and has also been described in other tumors including adenocarcinoma, SCC, ONB, melanoma, and undifferentiated carcinoma [88–90] (Figure 12.23). Areas of necrosis or adjacent infiltrative bone invasion in a mass thought to be an IP should raise the possibility to coexistent carcinoma [8]. CT may play

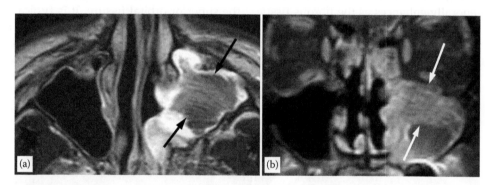

FIGURE 12.22
Inverted papilloma. (a) Axial T_2-weighted image shows a left maxillary sinus mass with a striated appearance (arrows) with alternating areas of high signal and lower signal. (b) Coronal postcontrast, fat-saturated T_1-weighted image shows mass in left maxillary sinus (arrows) with a cerebriform appearance with alternating areas of higher and lower signal.

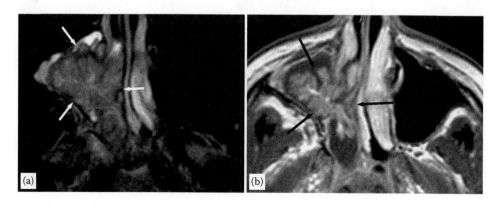

FIGURE 12.23
(a) Axial T_2-weighted image shows a striated appearance to a right maxillary sinus mass (arrows). (b) A cerebriform appearance is seen on an axial postcontrast T_1-weighted image (arrows). Biopsy revealed a squamous cell carcinoma. Striated, cerebriform appearance not specific for inverted papilloma.

a complimentary role in the imaging assessment of IP, with a cone-shaped area of hyperostosis correlating to the site of origin in 90% of cases [91].

12.7.2 Fibrous Dysplasia

FD is a benign, idiopathic condition in which medullary bone is replaced by poorly organized fibro-osseous tissue and immature woven bone. Monostotic (MFD) and polyostotic (PFD) forms exist with the craniofacial bones involved in 25% of cases of MFD and 40%–50% in PFD. New lesions usually do not develop after the fusion of growth plates, although some lesions will continue to grow. Lesion stability tends to occur by early adulthood [3,8]. FD can encroach on the paranasal sinuses, nasal cavity, orbit, and neurovascular foramina to produce symptoms. Involvement of frontal and sphenoid sinuses can result in mucocele formation. Malignant transformation occurs in less than 1% of lesions [9]. On MRI FD can have an aggressive appearance, mimicking a neoplasm. It is of variable signal on T_1- and T_2-weighted images depending on the amount of fibrous tissue and cystic components. It generally enhances as the fibrous tissue is very vascular (Figure 12.24) [3,8,67]. A CT can generally confirm the diagnosis of FD. The imaging appearance of FD can overlap with other benign fibro-osseous lesions [3].

12.7.3 Juvenile Nasopharyngeal Angiofibroma

A juvenile nasopharyngeal angiofibroma (JNA) is a lesion composed of vascular and fibrous tissue that represents less than 1% of all head and neck neoplasms. Some consider it a vascular malformation or hamartoma rather than a neoplasm. JNA is thought to be hormonally responsive as it occurs almost exclusively in males between 10 and 18 years of age [8,87]. It has generally been thought to arise from posterior choanal tissue near the PPF and sphenopalatine foramen; however, more recently some have suggested it arises from the pterygoid (vidian) canal [8,92]. There is commonly a extension into the sphenopalatine foramen, nasopharynx, and nasal cavity. In nearly 90% of cases, there is a spread to the PPF resulting in widening of the fossa and bowing of posterior wall of maxillary sinus. It can extend further laterally through the pterygomaxillary fissure. There is sphenoid sinus involvement in approximately 60% of cases. The maxillary and ethmoid sinuses are involved in 43% and 35% of cases, respectively. From the PPF, it can spread into the orbit via the inferior orbital fissure with subsequent intracranial spread via the superior orbital fissure. It is typically of intermediate signal on T_1-weighted images and intermediate to bright signal on T_2-weighted images. Flow voids and intense enhancement on MRI reflect the high vascularity of a JNA [8,68,87,92]. The primary goal of imaging is

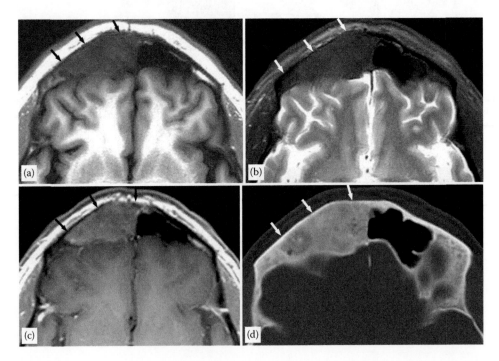

FIGURE 12.24
Fibrous dysplasia. Expansile lesion (arrows) involving the right frontal bone and frontal sinus is of intermediate signal on T_1-weighted images (a), low signal on T_2-weighted images (b), and homogeneously enhances following contrast (c). On CT the lesion has the typical appearance of fibrous dysplasia (d).

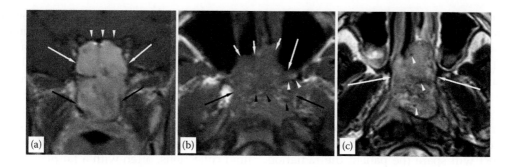

FIGURE 12.25

Juvenile nasal angiofibroma. (a) Coronal postcontrast T_1-weighted image shows an intensely enhancing mass in the sphenoid sinus (white arrows) and nasal cavity (black arrows). There is no intracranial extension (white arrowheads). (b) On axial precontrast T_1-weighted image the mass (black arrows) is of slightly heterogeneous intermediate signal and fills the nasopharynx with extension into the posterior nasal cavity (short white arrows) and the left PPF (white arrowheads). There is bowing of posterior wall of left maxillary sinus (long white arrow). Flow voids (black arrowheads) are present. (c) On axial T_2-weighted image the mass (white arrows) is of heterogeneously bright signal with flow voids (white arrowheads).

to map the tumor for the surgeon, particularly assessing for intracranial spread (Figure 12.25). The primary treatment is surgery while unresectable intracranial disease is treated with radiation therapy [8]. Recurrence rates ranging from 10% to 61% have been reported, often associated with incomplete surgical resection [93].

12.8 Sinonasal Infectious and Inflammatory Disease

12.8.1 Acute Rhinosinusitis Complications

MRI has no role in the imaging assessment of patients with uncomplicated acute rhinosinusitis; although when acute and subacute inflamed mucosa and secretions are imaged, they have high T_2-weighted signal intensities with peripheral enhancement of the mucosa [2,25]. Although CT is often obtained first, as it is faster and more widely available, MRI does play a role in evaluating the complications of acute rhinosinusitis, particularly the intracranial complications [2].

Complications from sinusitis arise from the close proximity of the sinuses to the orbit, face, and intracranial compartment with spread via retrograde thrombophlebitis of valveless veins that drain both the sinuses and the intracranial compartment or direct extension through areas of bony discontinuity related to normal sutures and foramina, congenital or acquired dehiscence, or infectious osteitis [94–96]. The exact prevalence of complications is difficult to determine but is presumed to be low and has decreased with the introduction of antibiotic treatment and earlier diagnosis of complications with CT and MRI. Factors that predispose to the development of complications include delay in treatment, bacterial resistance to antibiotics, incomplete treatment,

and immunosuppression [25]. Infection involving the ethmoid and frontal sinuses accounts for the majority of complications [94]. Acute sinusitis affects children more frequently than adults and children account for a greater percentage of complicated cases. Orbital complications typically involve younger children (3–8 years) while intracranial complications typically affect older children and adolescents (12–15 years). Complications are more common in otherwise healthy males, accounting for up to 60%–70% of complicated cases. This predisposition for young males may be related to the maximum vascularity of the diploic space and the continued development of the frontal sinuses, which occurs in this age group [95–97]. The causative organisms are the ones most commonly implicated in acute bacterial sinusitis, including *S. pneumonia*, *H. influenza*, and *Moraxella catarrhalis* [97]. Complications can be divided into extracranial and intracranial. Extracranial complications include orbital cellulitis, orbital abscess, and subperiosteal abscess. Intracranial complications include subdural empyema, epidural abscess, meningitis, brain abscess, and dural venous sinus thrombosis. Orbital and intracranial complications coexist in up to 45% of complicated cases [7].

12.8.2 Orbital Complications

Ethmoid sinusitis is the most common cause of orbital complications due to the thin lamina papyracea separating the ethmoid air cells from the orbit and the valveless anterior and posterior ethmoid veins, allowing for rapid access to the orbit [25]. The orbital septum, a connective tissue extension from the periosteum of the orbital rim to the tarsal plates of the eyelid, divides orbital complications into those that are preseptal (periorbital) or postseptal (orbital) and acts as a barrier for the spread of infection from the preseptal to the postseptal orbit [97].

Clinically, patients with orbital infections present with periorbital edema, chemosis, visual loss, restricted eye movement, and proptosis, with the latter three findings usually seen with postseptal infections [95].

Contrast-enhanced CT of the orbits is typically the initial imaging modality when an orbital complication of sinusitis is suspected [96]. However, there have been case reports suggesting the benefit of MRI over CT in detecting orbital complications of cellulitis [98,99]. Additionally, MRI will spare the patient the radiation exposure of CT to orbital structures.

With preseptal cellulitis, MRI shows edematous infiltration of the eyelid and periorbital soft tissues with high signal on T_2-weighted images and enhancement following contrast. A rim-enhancing fluid collection can be seen with preseptal abscess formation. Postseptal complications can consist of cellulitis, subperiosteal abscess, and orbital abscess. Like its preseptal counterpart, postseptal cellulitis is seen as edematous infiltration and enhancement of the orbital fat. Both intraconal and extraconal fat can be involved and the process may be diffuse or localized, often adjacent to the affected sinus [96]. When a sinus infection spreads into the orbit,

the tough periorbita may serve as a barrier to the spread of infection. If the infection develops into an abscess, it strips the periorbita from the bone and displaces it inward. This is termed a subperiosteal abscess and typically develops when infection spreads from the ethmoid sinuses into the medial orbit. Less commonly, infection may spread across the orbital roof from the frontal sinus or the orbital floor from the maxillary sinus. They are seen as elongated or lenticular rim-enhancing extraconal fluid collections, often with a fat plane visible between the collection and the extraocular muscles. If a collection of pus is not contained by the periorbita, an orbital abscess forms that may be intra- or extraconal [96]. In addition to conventional sequences, DWI may improve confidence in diagnosing an orbital abscess, with an abscess showing restricted diffusion [100] (Figure 12.26).

Except for mild preseptal cellulitis, which may be treated with oral antibiotics, most other cases of orbital cellulitis require parenteral antibiotics. Small abscess may be treated initially with a trial of parenteral antibiotics, but if no improvement occurs within 24–48 h, the abscess should be drained surgically. Larger abscess should be drained surgically [94,97].

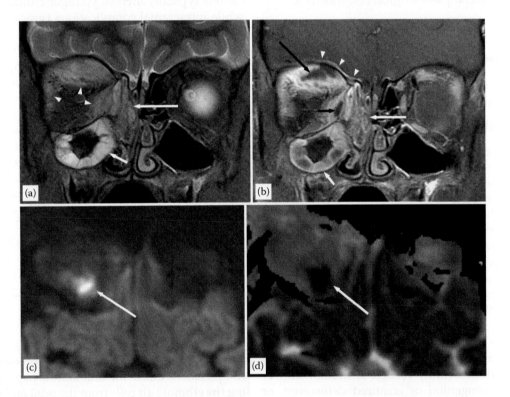

FIGURE 12.26
Sinusitis complicated by orbital cellulitis and subperiosteal abscesses. (a) Coronal T_2-weighted image shows edematous infiltration of postseptal soft tissues, mainly extraconal (white arrowheads). There are inflammatory mucosal changes in ethmoid (long white arrow) and maxillary (short white arrow) sinuses. (b) Coronal postcontrast, fat-saturated T_1-weighted image shows infiltrative enhancement in orbit with non-enhancing foci superiorly (long black arrow) and medially (short black arrow) compatible with subperiosteal abscesses, which were confirmed at surgery. Inflammatory mucosal thickening and enhancement is present in ethmoid (long white arrow) and maxillary (short white arrow) sinuses. Dural enhancement (arrowheads) along floor of anterior cranial fossa is likely reactive as patient had no evidence of intracranial infection. Restricted diffusion in superior abscess is seen as high signal (arrow) on DWI (c) and low signal (arrow) on ADC map (d). Restricted diffusion was also seen in medial abscess, not shown.

12.8.3 Intracranial Complications

Intracranial complications of sinusitis generally result from frontal sinusitis, with the frontal sinuses being particularly susceptible to intracranial spread of infection due to valveless diploic veins in the frontal bone, which drain the sinuses. Thrombophlebitis can spread through these veins to the dural plexus resulting in epidural abscess, subdural empyema, meningitis, cerebral abscess, and dural sinus thrombosis [96].

Epidural abscess is now the most common type of intracranial complication, replacing subdural empyema. These develop between the inner table of the skull and the outer periosteal dura. These enlarge slowly because of tight adherence of the dura to the skull, resulting in a more indolent course with headache and no central neurologic signs [96,101]. Epidural abscess have a lenticular shape and do not cross sutures. They are hyperintense on T_2-weighted images, and are associated with restricted diffusion and rim enhancement. They have variable signal on T_1-weighted images depending on the presence of protein and hemorrhage [96] (Figure 12.27).

Subdural empyemas are the second commonest intracranial complication. Once infection gains access to the subdural space, it can spread rapidly over the convexities, so patients may present more acutely with earlier development of focal neurologic symptoms [96,101]. These are crescentic-shaped collections that can cross sutures. Signal characteristics are similar to epidural abscesses. Restricted diffusion is an important finding of a subdural empyema, distinguishing it from a subdural effusion. Meningitis usually occurs in association with a subdural empyema. Associated meningeal enhancement is more readily seen on MRI than CT. Lack of nulling of the signal from cerebrospinal fluid (CSF) on fluid-attenuated inversion recovery sequences due to elevated protein and cellular content may also be seen, as can hydrocephalus [96] (Figure 12.28).

Spread of infection to the brain results in cerebritis and abscess formation. The classic clinical triad of fever, headache, and focal neurologic findings may not be present and abscesses are equally likely to present with or without focal signs [96,101,102]. Cerebritis is predominantly hypointense to normal brain on T_1-weithed

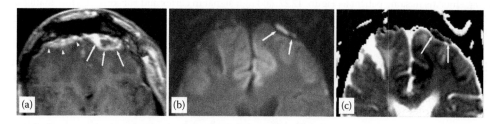

FIGURE 12.27
Epidural abscess and phlegmon complicating chronic frontal and sphenoid sinusitis in a patient with history of sinonasal cancer treated with subcranial resection and radiation. (a) Axial T_1-weighted postcontrast image shows thick walled, rim-enhancing, lenticular fluid collection over left frontal convexity (long white arrows) compatible with epidural abscess. Lenticular epidural enhancement (white arrowheads) over right frontal convexity concerning for phlegmon. Findings confirmed at surgery. Restricted diffusion in left frontal epidural abscess is seen as high signal (arrow) on DWI (b) and low signal (arrow) on ADC map (c).

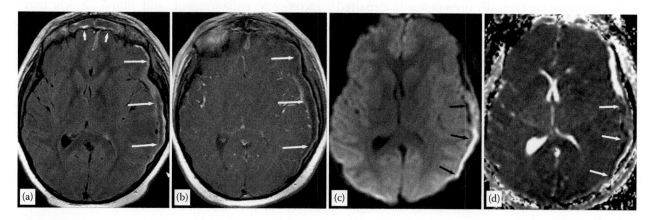

FIGURE 12.28
Subdural empyema complicating acute sinusitis. (a) Axial fluid-attenuated inversion recovery (FLAIR) image shows high signal crescentic extra-axial collection over left fronto-temporal convexity crossing coronal suture (long white arrows) compatible with subdural empyema. Inflammatory changes present in frontal sinuses (short white arrows). (b) Axial postcontrast T_1-weighted image shows enhancement surrounding subdural collection (arrows). Restricted diffusion in much of subdural empyema is seen as high signal (arrows) on DWI (c) and low signal (arrows) on ADC map (d).

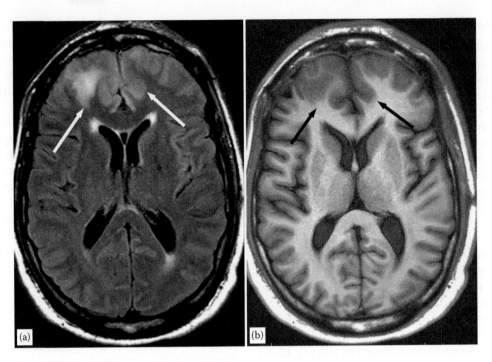

FIGURE 12.29
Cerebritis complicating sinusitis in same patient as Figure 12.27. (a) High signal in cortex and subcortical white matter of both frontal lobes (arrows) is seen on axial FLAIR image. (b) Corresponding hypointense signal (arrows) with mild mass effect, evidenced by sulcal effacement, is seen on T_1-weighted precontrast image.

images and hyperintense on T_2-weighted images with mass effect (Figure 12.29). As an abscess develops and becomes organized, a capsule of low signal on T_2-weighted images develops with surrounding edema. Centrally, the abscess shows prominent restricted diffusion with peripheral enhancement [96,103].

Cavernous sinus thrombosis can arise as an intracranial complication with or without orbital cellulitis. Along with fever and headache, orbital signs and symptoms (including periorbital swelling, diplopia, chemosis, and proptosis) are typically present due to impaired orbital venous drainage and involvement of the cranial nerves within the cavernous sinus. Palsies involving the third through sixth cranial nerves may be present, with sixth nerve palsy most commonly seen. On MRI, the cavernous sinus enhances less than usual, with focal poorly enhancing areas, often larger than 7 mm in size. Other findings include abnormal signal, isointense to gray matter on T_1-weitghted images, abnormal convex contour along the lateral margin of the cavernous sinus, and increased dural enhancement along the lateral wall of the cavernous sinus [96,104] (Figure 12.30). Distension, incomplete enhancement, and filling defects may be seen in the superior and inferior ophthalmic veins

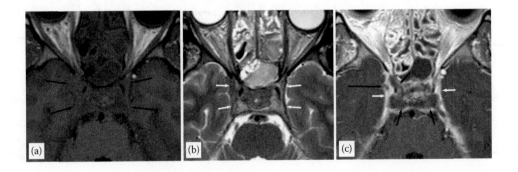

FIGURE 12.30
Acute sinusitis complicated by cavernous sinus thrombosis. (a) Abnormal intermediate signal soft tissue expanding the cavernous sinuses (black arrows) is seen on axial T_1-weighted, precontrast image. (b) Corresponding axial T_2-weighted images shows abnormal tissue is heterogeneously hyperintense (white arrows). (c) Axial postcontrast T_1-weighted image shows confluent non-enhancing foci (short black arrows), abnormal convex contour on right (long black arrow) and thickened, enhancing dura (short white arrows) along lateral margins of cavernous sinus.

and the sphenoparietal and inferior petrosal sinuses due to thrombosis. Restricted diffusion has also been described in the cavernous sinus and superior ophthalmic vein [104,105].

Except for isolated meningitis, the treatment of intracranial complications is surgical with concomitant intravenous antimicrobial therapy. Cavernous sinus thrombosis is treated with surgical debridement and IV antibiotics. The role of anticoagulation is debatable, although the use of anticoagulants may decrease morbidity [96,106].

12.9 Fungal Rhinosinusitis

Fungal rhinosinusitis is an infection of the sinonasal cavity in which a fungus is the primary pathogen or incites an inflammatiory reaction in the sinonasal cavity. Fungal rhinosinusitis is classified as noninvasive or invasive based on histology. Noninvasive disease includes allergic fungal sinusitis and mycotic colonization resulting in a fungus ball. The invasive disease consists of acute invasive fungal rhinosinusitis, chronic invasive fungal rhinosinusitis and granulomatous rhinosinusitis [25,87,107,108]. Infection can be caused by any fungus, although the most common organism is *Aspergillus*.

12.9.1 Fungus Ball

A fungus ball consists of a tightly packed, tangled mass of fungal hyphae usually involving a single sinus, most commonly the maxillary sinus [7,25,107]. These develop in immunocompetent patients, although often in response to changes in the local microenvironment such as after surgery, endodontic procedures, or radiation therapy. Patients may have symptoms of chronic sinusitis or be asymptomatic [25,107,109]. Affected patients may have nasal polyps and bacterial sinusitis [110]. Treatment involves surgical removal although recolonization of the sinus can occur [25,107]. On MRI fungus balls are iso- to hypointense on T_1-weighted images and markedly hypointense on T_2-weighted images due to calcification, paramagnetic metals, and absence of free water. There is no enhancement of the fungus ball. Surrounding inflamed mucosa with high signal on T_2-weighted images and enhancement may be seen. There is no infiltration of soft tissues beyond the sinus [7,111–113].

12.9.2 Allergic Fungal Rhinosinusitis

Allergic fugal rhinosinusitis (AFS) develops when a person, genetically predisposed to fungal allergy, inhales mold spores that elude clearance and expulsion from sinonasal cavity long enough for germination to occur. Germination increases the antigenicity of the fungus resulting in immune-mediated mucosal inflammation and greater production of viscous allergic mucin in which the fungus continues to grow. The inflammation and mucin result in the obstruction of normal sinonasal drainage pathways [107,114]. There is typically involvement of multiple sinuses, often bilateral, and the nasal cavity [115]. Patients are immunocompetent but with a history of allergic rhinitis, sinonasal polyposis, asthma, and eosinophilia. Common presenting symptoms are nasal congestion and headache [25,110,112]. Treatment consists of removal of polyps and allergic mucin and a 2–4 week course of systemic steroids following surgery. Long-term use of intranasal topical corticosteroids and nasal lavage are helpful and immunotherapy may be helpful in the treatment of AFR [107,110]. On MRI, there is sinus opacification of variable signal on T_1-weighted images, ranging from low to high signal and often of mixed signal. There is typically no enhancement in the center of the sinus. On T_2-weighted images, there is usually markedly hypointense signal. Peripherally inflamed mucosa can be seen. Expansion of the involved sinuses with bony thinning and smooth pressure-related erosion can lead to orbital and intracranial, usually extradural, extension [112–114].

12.9.3 Invasive Fungal Rhinosinusitis

Invasion of the mucosa, submucosa, bone, and vessels by fungal elements is what characterizes invasive fungal rhinosinusitis and distinguishes it from noninvasive fungal disease. The acute invasive and chronic invasive forms of fungal rhinosinusitis share similar imaging features with the main difference between these two invasive forms being the clinical course [7,87,110].

Acute invasive disease occurs in immunocompromised patients with severe neutropenia and in diabetic patients; particularly, patients with poorly controlled diabetes and diabetic ketoacidosis. In the former group, *Aspergillus* species accounts for 80% of infections while in the diabetic patients 80% of infections are caused by fungi in the order Zygomycetes, which includes mucor [107,110,113,116]. There is a fulminant progression of disease over a course of less than 4 weeks in which fungal organisms spread form the paranasal sinuses to the mucosa, bone, and blood vessels. Spread from the paranasal sinuses may be by direct invasion or hematogenous spread, with extension into the orbit and intracranially. Vessel wall invasion can result in thrombosis and arterial and/or venous infarction. There is a frequent involvement of the nasal cavity, which may be the primary site of infection, particularly the middle turbinate. There is no specific site predilection for any of the paranasal sinuses [7,110,112–114,116]. Initial symptoms are similar to those of acute sinusitis. Additional symptoms consist of purulent hemorrhagic nasal discharge, visual disturbances, headache, mental status change, seizure, and neurologic deficits. On exam, patients may have a painless, necrotic ulcer, or eschar of the palate or nasal septum [25,110,113]. Treatment consists of surgical debridement, systemic antifungal therapy, and most importantly the reversal of underlying causes of immunosuppression. Historically mortality of 50%–80% was reported, although more recent studies have shown decreased mortality of 7%–18%, attributed to earlier diagnosis and intervention [116,117].

Chronic invasive disease usually affects immunocompetent patients, although those with diabetes or a low level of immunocompromise are also at risk [7,113,114]. The most common causative organism is *Aspergillus fumigatus*. There is a dense accumulation of hyphae, occasional vascular invasion, and minimal inflammatory reaction [7,108]. This is a slowly progressive disease with a duration of greater than 12 weeks. Over several months to years, inhaled fugal organisms that are deposited in the sinonasal cavity invade the mucosa, bony walls, and blood vessels of the sinonasal region. The ethmoid and sphenoid sinuses are most commonly involved. Patients often have a history and symptoms of chronic rhinosinusitis. Symptoms related to invasive disease, including neurologic symptoms, local soft tissue swelling, and visual symptoms, may take months to years to develop. Treatment consists of surgery and systemic antifungal therapy [7,108,113,114].

With both acute invasive and chronic invasive rhinosinusitis on CT, there is often aggressive osseous destruction of the sinus walls, mucosal thickening, and sinus and nasal cavity opacification. Hyperdense attenuation in the affected sinuses is more common with chronic invasive disease. As with neoplastic diseases of the sinuses, osseous changes are best detected by CT and soft tissue extent by MRI. On MRI, mildly hypointense mucosal thickening on T_2-weighted images may be one of the earliest imaging findings. More typical is more extensive mucosal thickening and opacification of the sinuses, often of low signal on T_2-weighted images. Infiltration of the periantral fat is a sign of extra-sinus, invasive disease [7,113,114]. One study has shown soft-tissue abnormalities outside the sinuses on MRI is the most sensitive and specific single imaging parameter in evaluating for acute invasive fungal sinusitis [118]. Loss of normal contrast enhancement of the sinonasal mucosa has been described, possibly related to mucosal ischemia and necrosis [87,118] (Figure 12.31). Invasion into adjacent structures including the orbits, cavernous sinuses, and anterior cranial fossa can occur (Figure 12.32). Intracranial extension can result in cavernous sinus thrombosis, carotid artery invasion, thrombosis, or pseudoaneurysm formation resulting in infarction and hemorrhage, meningitis, extraaxial and parenchymal abscesses, and hematogenous dissemination [113].

Granulomatous invasive rhinosinusitis is a chronic process of greater than 12 weeks duration that affects immunocompetent patients. The disease primarily occurs in Sudan, Saudi Arabia, Pakistan, and India. The causative organism is generally *Aspergillus flavus*. Pathologically, there is a granulomatous response with noncaseating granulomas, prominent fibrosis, vasculitis, vascular proliferation, and perivascular fibrosis. Invasion beyond the sinuses to the orbit and brain occurs over time. Clinically, patients have symptoms of chronic sinusitis often with an enlarging mass in the cheek, orbit, nose, or sinuses. Proptosis is a common reported clinical feature. Treatment is surgery and systemic antifungal therapy. Limited reports of imaging

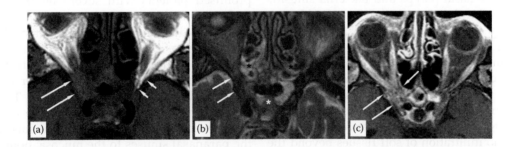

FIGURE 12.31
Patient with acute lymphocytic leukemia and acute right visual loss found to have invasive *Aspergillus* infection of sinonasal region. (a) There is abnormal intermediate signal soft tissue in right orbital apex and anterior cavernous sinus on axial precontrast T_1-weighted image (long white arrows) with adjacent sinus opacification. Compare to normal side (short white arrows). (b) Axial T_2-weighted images show corresponding abnormal low signal at right orbital apex and in anterior cavernous sinus (white arrows) with contiguous opacification of ethmoid (black asterisks) and sphenoid (white asterisk) sinuses. (c) Enhancement is present in orbital apex, anterior cavernous sinus, and adjacent sphenoid sinus on axial postcontrast, fat-saturated T_1-weighted image (white arrows).

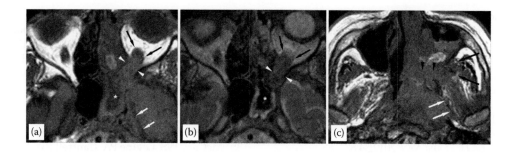

FIGURE 12.32
Chronic invasive *Aspergillus* rhinosinusitis in a diabetic patient. Contrast could not be given due to renal failure. (a) Axial T_1-weighted precontrast image shows heterogeneous opacification of left ethmoid and sphenoid sinuses with low signal in central sphenoid sinus (asterisk). There is infiltration into left inferior orbital fissure (white arrowheads), orbital apex (black arrows), and Meckel cave (white arrows). (b) Abnormal soft tissue in inferior orbital fissure (white arrowheads) and orbital apex (black arrows) is hypointense on axial T_2-weighted image. Opacified left sphenoid sinus (asterisk) is markedly hypointense centrally, compatible with fungal disease. (c) Axial T_1-weighted image more inferiorly shows extra-sinus involvement of PPF (black arrowheads) with extension through pterygomaxillary fissure into retro maxillary fat (black arrows) and parapharyngeal fat (white arrows).

findings describe soft tissue masses eroding the orbital walls with the involvement of nasal cavity and PPF [7,107,108,110,113,114,119].

12.10 Infectious and Noninfectious Granulomatous Disease

There are a variety of infectious and noninfectious granulomatoses that can involve the sinonasal structures. Actinomycosis is the most common infectious agent, often arising from a periodical dental abscess. Other infectious agents include nocardia, tuberculosis, leprosy, and syphilis. Noninfectious causes include granulomatous with polyangiitis (GPA, formerly Wegener's granulomatosis), sarcoidosis, and cocaine abuse. Symptoms are often nonspecific and vary with the underlying etiology of the granulomatous process. These diseases, however, share common imaging features. Nasal cavity involvement, particularly the nasal septum, typically precedes that of the paranasal sinuses. Sinus disease most frequently involves the maxillary and ethmoid sinuses, with rare involvement of the sphenoid sinuses and sparing of frontal sinuses. Bone findings can vary from sclerosis, to remodeling with expansion, especially in the nasal cavity, to erosion. On MRI, sinus opacification in the granulomatous processes typically has low-to-intermediate signal on T_1-weighted images and intermediate-to-high signal on T_2-weighted images with enhancement. Extension into the orbit, PPF, and inferior orbital fissure has been reported in GPA, as has thickening and enhancement of cranial nerves five and seven. Involvement may appear similar to neoplastic processes and biopsy is often required for diagnosis [2,25,87,120–122].

References

1. Dym RJ, Masri D, Shifteh K. Imaging of the paranasal sinuses. *Oral Maxillofacial Sug Clin N Am* 2012;24:175–89.
2. Madani G, Beale TJ. Sinonasal inflammatory disease. *Semin Ultrasound CT MRI* 2009;30:17–24.
3. Hartman MJ, Gentry LR. Aggressive inflammatory and neoplastic processes of the paranasal sinuses. *Magn Reson Imaging Clin N Am* 2012;20:447–71.
4. Madani G, Beale TJ. Differential diagnosis in sinonasal disease. *Semin Ultrasound CT MRI* 2009;30:39–45.
5. Yong Lee J. Unilateral paranasal sinus disease: Analysis of the clinical characteristics, diagnosis, pathology and computed tomography findings. *Acta Otolaryngol* 2008;128:621–26.
6. Ahsan F, El-Hakim H, Ah-See KW. Unilateral opacification of paranasal sinus CT scans. *Otolaryngol Head Neck Surg* 2005;133:178–80.
7. Mossa-Basha M, Blitz AM. Imaging of the paranasal sinuses. *Seminars in Roentgenology* 2013;48:14–34.
8. Som PM, Brandwein-Gensler MS, Kassel EE et al. Tumors and Tumor-like Conditions of the Sinonasal Cavities. In: Som PM, Curtain HD, editors. *Head and Neck Imaging*, 5th edition. St. Louis, MO: Elsevier Mosby; 2011. pp. 253–410.
9. Muir C, Weiland L. Upper aerodigestive tract cancers. *Cancer* 1995;75(1 Suppl):147–53.
10. Youlden DR, Cramb SM, Peters S et al. International comparisons of the incidence and mortality of sinonasal cancer. *Cancer Epidemiology* 2013;37:770–9.
11. Madani G, Beale TJ, Lund VJ. Imaging of sinonasal tumors. *Semin Ultrasound CT and MRI* 2009;30:25–36.
12. Harrison D. The management of malignant tumors of the nasal sinuses. *Otolaryngol Clin North Am* 1971;4:159–77.
13. Weymuller EA, Davis GE. Malignancies of the Paranasal Sinuses. In: Flint PW, Haughey BH, Lund VJ et al., editors. *Cummings Otolaryngology Head and Neck Surgery*, 5th edition. Philadelphia, PA: Elsevier Mosby; 2010. pp. 1121–32.

14. Guntinas-Lichius O, Kreppel MP, Stuetzer H et al. Single modality and multimodality treatment of nasal and paranasal sinuses cancer: A single institution experience of 229 patients. *Eur J Surg Oncol* 2007;33:222–8 (Up to date).

15. Blanco AI, Chao KS, Ozyigit G et al. Carcinoma of paranasal sinuses: Long-term outcomes with radiotherapy. *Int J Radiat Oncol Biol Phys* 2004;59:41–8.

16. Turner JH, Reh DD. Incidence and survival in patients with sinonasal cancer: A historical analysis of population-based data. *Head and Neck* 2012;34:877–85.

17. Katz TS, Mendenhall WM, Morris CG et al. Malignant tumors of the nasal cavity and paranasal sinuses. *Head Neck* 2002;24:821–9.

18. Michel J, Fakhry N, Mancini J et al. Sinonasal squamous cell carcinomas: Clinical outcomes and predictive factors. *Int J Oral Maxillofacc Surg* 2014;43:1–6.

19. Myers LL, Nussenbaum B, Bradford CR et al. Paranasal sinus malignancies: An 18-year single institution experience. *Laryngoscope* 2002;112:1964–69.

20. Loevner LA, Sonners AI. Imaging of neoplasms of the paranasal sinuses. *Magn Reson Imaging Clin N Am* 2002;10:467–93.

21. Raghavan P, Phillips CD. Magnetic resonance imaging of sinonasal malignancies. *Top Magn Reson Imaging* 2007;18:259–67.

22. Som PM, Shapiro MD, Biller HF et al. Sinonasal tumors and inflammatory tissues: Differentiation with MR imaging. *Radiology* 1988;167:803–8.

23. Kraus DH, Lanzieri CF, Wanamaker JR et al. Complementary use of computed tomography and magnetic resonance imaging in assessing skull base lesions. *Laryngoscope* 1992;102:623–9.

24. Yousem DM, Cheng L, Montone KT et al. Primary malignant melanoma of the sinonasal cavity: MR imaging evaluation. *RadioGraphics* 1996;16:1101–10.

25. Som PM, Brandwein MS, Wang BY. Inflammatory Diseases of the Sinonasal Cavities. In: Som PM, Curtain HD, editors. *Head and Neck Imagining*, 5th edition. St. Louis, MO: Elsevier Mosby; 2011. pp. 167–251.

26. Sasaki M, Eida S, Sumi M et al. Apparent diffusion coefficient mapping for sinonasal diseases: Differentiation of benign and malignant lesions. *AJNR Am J Neuroradiol* 2011;32:1100–6.

27. Sasaki M, Sumi M, Eida S et al. Multiparametric MR imaging of sinonasal diseases: Time-signal intensity curve—and apparent diffusion coefficient-based differentiation between benign and malignant lesions. *AJNR Am J Neuroradiol* 2011;32:2154–9.

28. Singh N, Eskander A, Huang SH et al. Imaging and resectability issues of sinonasal tumors. *Expert Rev Anticancer Ther* 2013;13:297–312.

29. Chong VFH, Fan YF. Skull base erosion in nasopharyngeal carcinoma: Detection by CT and MRI. *Clinical Radiology* 1996;51:625–31.

30. Barakos JA, Dillon WP, Chew WM. Orbit, skull base and pharynx: Contrast-enhanced fat suppression imaging. *Radiology* 1991;179:191–8.

31. Ganly I, Patel SG, Singh B et al. Craniofacial resection for malignant paranasal sinus tumors: Report of an international collaborative study. *Head Neck* 2005;27:575–584.

32. Suárez C, Llorente JL, Fernández de León R, Maseda E. Prognostic factors in sinonasal tumors involving the anterior skull base. *Head Neck* 2004;26:136–144.

33. Carrau RL, Segas J, Nuss DW et al. Squamous cell carcinoma of the sinonasal tract invading the orbit. *Laryngoscope* 1999;109:230–235.

34. Suarez C, Ferlito A, Lund VJ et al. Management of the orbit in malignant sinonasal tumors. *Head Neck* 2008;30:242–50.

35. McCary WS, Levine PA, Cantrell RW. Preservation of the eye in the treatment of sinonasal malignant neoplasms with orbital involvement. A confirmation of the original treatise. *Arch Otolaryngol Head Neck Surg* 1996;122:657–59.

36. Eisen MD, Yousen DM, Loevner LA et al. Preoperative imaging to predict orbital invasion by tumor. *Head Neck* 2000;22:456–62.

37. Kim HJ, Lee TH, Lee HS et al. Periorbita: Computed tomography and magnetic resonance imaging findings. *Am J Rhinol* 2006;20:371–4.

38. Taghi A, Ali A, Clarke P. Craniofacial resection and its role in the management of sinonasal malignancies. *Expert Rev Anticancer Ther* 2012;12:1169–76.

39. Patel SG, Singh B, Polluri A et al. Craniofacial surgery for malignant skull base tumors. Report of an international collaborative study. *Cancer* 2003;98:1179–87.

40. Eisen MD, Yousem DM, Montone KT et al. Use of preoperative MR to predict dural, perineural and venous sinus invasion of skull base tumors. *AJNR Am J Neuroradiol* 1996;17:19337–45.

41. Ahmadi J, Hinton DR, Segall HD et al. Dural invasion by craniofacial and calvarial neoplasms: MR imaging and histopathologic evaluation. *Radiology* 1993;188:747–9.

42. McIntyre JB, Perez C, Penta M et al. Patterns of dural involvement in sinonasal tumors: Prospective correlation of magnetic resonance imaging and histopathologic findings. *Int Forum Allergy Rhinol* 2012;2:336–41.

43. Liebig C, Ayala G, Wilks JA et al. Perineural invasion in cancer: A review of the literature. *Cancer* 2009;115:3379–91.

44. Gil Z, Carlson DL, Gupta A et al. Patterns and incidence of neural invasion in patients with cancers of the paransal sinuses. *Arch Otolaryngol Head Neck Surg* 2009;135:173–9.

45. Johnston M, Yu E, Kim J. Perineureal invasion and spread in head and neck cancer. *Expert Rev Anticancer Ther* 2012;12:359–71.

46. Moonis G, Cunnane MB, Emerick K et al. Patterns of perineural tumor spread in head and neck cancer. *Magn Reson Imagin Clin N Am* 2012;20:435–46.

47. Catalano PJ, Sen C, Biller HF. Cranial neuropathy secondary to perineural spread of cutaneous malignancies. *Am J Otol* 1995;16:772–7.

48. Nemzek WR, Hecht S, Gandour-Edwwards R et al. Perineural spread of head and neck tumors: How accurate is MR imaging. *AJNR Am J Neuroradiol* 1998;19:701–6.

49. Parker GD, Harnsberger HR. Clinical-radiologic issues in perineural spread of malignant diseases of the extracranial head and neck. *RadioGraphics* 1991;11:383–99.

50. Curtin HD. Detection of perineural spread: Fat suppression versus no fat suppression. *AJNR Am J Neuraodiol* 2004;25:1–3.

51. Laine FJ, Braun IF, Jensen ME et al. Perineural tumor extension through the foramen ovale: Evaluation with MR imaging. *Radiology* 1990;174:65–71.

52. Edge SB, Byrd DR, Compton CC et al., editors, *American Joint Committee on Cancer Staging Manual*, 7th ed, New York: Springer, 2010.

53. Dulguerov P, Jacobsen MS, Allal AS. Nasal and parana-sal sinus carcinoma: Are we making progress? *Cancer* 2001;92(12)3012–29.

54. Cantu G, Bimbi G, Miceli R, et al. Lymph node metas-tases in malignant tumors of the paranasal sinuses. Prognostic value and treatment. *Arch Otolaryngol Head Neck Surg* 2008;134:170–7.

55. Forghani R, Yu E, Levental M et al. Imaging evalua-tion of lymphadenopathy and patterns of lymph node spread in head and neck cancer. *Expert Rev Anticancer Ther* 2014;15:207–224.

56. van den Breckel MW, Stel HV, Castelijns JA et al. Cervical lymph node matastasis: Assessment of radio-logic findings. 1990;177:379–84.

57. Chai RL, Rath TJ, Johnson JT et al. Accuracy of com-puted tomography in the prediction of extracapsular spread of lymph node metastases in squamous cell car-cinoma of the head and neck. *JAMA Otolaryngol Head Neck Surg* 2013;139:1187–94.

58. Wu LM, Xu JR, Liu MJ et al. Value of magnetic reso-nance imaging for nodal staging in patients with head and neck squamous cell carcinoma: A meta-analysis. *Acad Radiol* 2012;19:331–40.

59. Johnson JT, Branstetter BF. PET/CT in head and neck oncology: State-of-the-art 2013. *Laryngoscope* 2014;124:913–5.

60. Kadish S, Goodman W, Wang CC. Olfactory neuroblas-toma: A clinical analysis of 17 cases. *Cancer* 1976;37:1571–6.

61. Morita A, Ebersold MJ, Olsen KD et al. Esthesioneuroblastoma: Prognosis and management. *Neurosurgery* 1993;32:706–14.

62. Dulguerov P, Calcaterra T. Esthesioneuroblastoma: The UCLA experience 1970–1990. *Laryngoscope* 1992;102:843–9.

63. Mendenhall WM, Mendenhall CM, Riggs CE et al. Sinonasal undifferentiated carcinoma. *Am J Clin Oncol* 2006;29:27–31.

64. Edge SB, Byrd DR, Compton CC et al. (eds.) Mucosal Melanoma of the Head and Neck. In: *American Joint Committee on Cancer Staging Manual*, 7th ed. New York: Springer; 2010. p. 97.

65. Kiovunen P, Back L, Pukkila M et al. Accuracy of the current TMN classification in predicting survival in patients with sinonasal mucosa melanoma. *Laryngoscope* 2013;122:1734–8.

66. Llorente JL, Lopez F, Suarez C et al. Sinonasal carcinoma: Clinical, pathological, genetic and therapeutic advances. *Nat Rev Clin Oncol* 2014;11:460–72.

67. Franchi A, Santucci M, Wenig BM. Adenocarcinoma. WHO histological classification of tumors of the nasal cavity and paranasal sinuses. In: Barnes L, Eveson JW, Reichardt P et al., editors. *Pathology and Genetics, Head and Neck Tumors*. Lyon, France: IARC Press; 2005. pp. 20–3.

68. Eggesbo HB. Imaging of sinonasal tumors. *Cancer Imaging* 2012;12:136–52.

69. Godivkar SM, Gadbail AR, Chole R et al. Adenoid cystic carcinoma: A rare clinical entity and literature review. *Oral Oncol* 2011;47:231–6.

70. Gal TJ, Silver N, Huang B. Demographics and trends in sino-nasal mucosal melanoma. *Laryngoscope* 2011;121;2026–33.

71. Bell D, Hanna EY. Sinonasal undifferentiated carci-noma: Morphological heterogeneity, diagnosis, manage-ment and biological markers. *Expert Rev Anticancer Ther* 2013;13:285–9.

72. Malouf GG, Casiraghi O, Deutssch E et al. Low- and high-grade esthesioneuroblastomas display a distinct natural history and outcome. *Eur J Cancer* 2013;49:1324–34.

73. Elkon D, Hightower SI, Lim ML et al. Esthesioneu-roblastoma. *Cancer* 1979;44:1087–94.

74. Platek ME, Merzianu M, Mashtare TL et al. Improved survival following surgery and radiation therapy for olfactory neuroblastoma: An analysis of SEER database. *Radiat Oncol* 2011;6:41.

75. Schuster JJ, Phillips CD, Levine PA. MRI of olfactory neuroblastoma and appearance after craniofacial resec-tion. *AJNR Am J Neuroradiol* 1994;15:1169–77.

76. Goodwin WJ Jr. Salvage surgery for patients with recur-rent squamous cell carcinoma of the upper aerodigestive tract: When do the ends justify the means? *Laryngoscope* 2000;110(Suppl 93):1–18.

77. Jeong WJ, Jung YH, Kwon SK et al. Role of surgical salvage for regional recurrence in laryngeal cancer. *Laryngoscope* 2007;117:74–77.

78. Hudgins PA, Burson JG, Gussack GS et al. CT and MR appearance of recurrent malignant head and neck neo-plasms after resection and flap reconstruction. *AJNR Am J Neuroradiol* 1994;15:1689–94.

79. Hermans R. Post-treatment imaging of head and neck cancer. *Cancer Imaging* 2004;4:S6–S15.

80. Lell M, Baum U, Greess H et al. Head and neck tumors: Imaging recurrent tumor and post-therapeutic changes with CT and MRI. *Eur J Radiol* 2000;33:239–47.

81. Gandhi D, Falen S, McCartney W et al. Value of 2-[18F]-fluoro-2-deoxy-D-glucose imaging with dual-head gamma camera in coincidence mode. Comparison with computed tomography/magnetic resonance imag-ing in patients with suspected recurrent head and neck cancers. *J Comput Assit Tomogr* 2005;29:513–19.

82. Klabbers BM, Lammertsma AA, Slotman BJ. The value of positron emission tomography for monitoring response to radiotherapy in head and neck cancer. *Mol Imaging Biol* 2003;5:257–70.

83. Tshering Vogel DW, Zbaeren P, Geretschlaeger A et al. Diffusion-weighted MR imaging including bi-exponential fitting for the detection of recurrent or residual tumour after (chemo)radiotherapy for laryngeal and hypopharyngeal cancers. *Eur Radiol* 2013;23:562–9.

84. Abdel Razak AAK, Kandeel AY, Soliman N et al. Role of diffusion-weighted echo-planar MR imaging in differ-entiation of residual or recurrent head and neck tumors and posttreatment changes. *AJNR Am J Neuroradiol* 2007;28:1146–52.

85. Al-Ibraheem A, Buck A, Krause BJ et al. Clinical applica-tions of FDG PET and PET/CT in head and neck cancer. *J Oncol* 2009;20:872–5.

86. Rangaswamy B, Fardanesh MR, Genden EM et al. Improvement in the detection of locoregional recurrence in head and neck malignancies: F-18 fluorodeoxyglusoce-poitron emission tomography/computed tomography compared to high resolution contrast-enhanced computed tomography and endoscopic examination. *Laryngoscope* 2013;123:2664–9.

87. Maroldi R, Ravanelli M, Borghesi A et al. Paranasal sinus imaging. *Eur J Radiol* 2008;66:372–86.

88. Yousem DM, Fellows DW, Kennedy DW et al. Inverted papilloma: Evaluation with MR imaging. *Radiology* 1992;185:501–5.

89. Ojiri H, Ujita M, Tada S, Fukuda K. Potentially distinctive features of sinonasal inverted papilloma on MR imaging. *Am J Roentgenol* 2000;175:465–8.

90. Jeon TY, Kim HJ, Chung SK et al. Sinonasal inveted papilloma: Value of convoluted cerbriform pattern on MR imaging. *AJNR Am J Neuroradiol* 2008;29:1556–60.

91. Lee DK, Chung SK, Dhong HJ et al. Focal hyperostosis on CT of sinonasal inverted papilloma as a predictor of tumor origin. *AJNR Am J Neuroradiol* 2007;28:618–21.

92. Liu ZF, Wang DH, Sun XC et al. The site of origin and expansive routes of juvenile nasopharyngeal angiofibroma (JNA). *Int J Pediatr Otorhinolaryngol* 2011;75:1088–92.

93. Lloyd G, Howard D, Phelps P et al. Juvenile angiofibroma: The lessons of 20 years of modern imaging. *J Laryngol Otol* 1999;113:127–34.

94. DeMuri GP, Wald ER. Sinusitis. In: Bennett JE, Dolin R, Blaser MJ, editors. *Mandell, Douglas and Bennett's Principles and Practice of Infectious Disease*, 8th ed. Philadelphia, PA: Elsevier Saunders; 2015. pp. 774–784.e2.

95. Osborn MK, Steinberg JP. Subdural empyema and other supperative complications of paranasal sinusitis. *Lancet Infect Dis* 2007;7:62–7.

96. Hoxworth JM, Glastonbury CM. Orbital and intracranial complications of acute sinusitis. *Neuroimg Clin N Am* 2010;20:511–26.

97. DeMuri GP, Wald ER. Complications of acute bacterial sinusitis in children. *Pediatr Infect Dis J* 2011;30:701–2.

98. McIntosh D, Mahadevan M. Acute orbital complications of sinusitis: The benefits of magnetic resonance imaging. *J Laryngol Otol* 2008;122:324–6.

99. McIntosh D, Mahadevan M. Failure of contrast enhanced computed tomography scans to identify an orbital abscess. The benefit of magnetic resonance imaging. *J Laryngol Otol* 2008;122:639–40.

100. Sepahdari AR, Aakalu VK, Kapur R et al. MRI of orbital cellultis and orbital abscess: The role of diffusion-weighted imaging. *AJR Am J Roentgenol* 2009;193:W244–W250.

101. Germiller JA, Monin DL, Sparano AM et al. Intracranial complications of sinusitis in children and adolescents and their outcome. *Arch Otolaryngol Head Neck Surg* 2006;132:969–76.

102. Adame N, Hedlund G, Byington CL. Sinogenic intracranial empyema in children. *Pediatrics* 2005;116:e461–7.

103. Chiang IC, Hsieh TJ, Chiu ML et al. Distinction between pyogenic brain abscess and necrotic brain tumour using 3-tesla MR spectroscopy, diffusion and perfusion imaging. *Br J Radiol* 2009;82:813–20.

104. Schuknecht B, Simmen D, Yuksel C et al. Tributary venous occlusion and septic cavernous sinus thrombosis: CT and MR findings. *AJNR Am J Neuroradiol* 1998;19:617–26.

105. Parmar H, Gandhi D, Mukherji SK et al. Restricted diffusion in the superior ophthalmic vein and cavernous sinus in a case of cavernous sinus thrombosis. *J Neuroophthalmol* 2009;29:16–20.

106. Desa V, Green R. Cavernous sinus thrombosis: Current therapy. *J Oral Maxillofac Surg* 2012;70:2085–91.

107. Ferguson BJ. Fungal rhinosinusitis. In: Flint PW, Haughey BH, Lund VJ et al., editors. *Cummings Otolaryngology Head and Neck Surgery*, 5th ed. Philadelphia, PA: Elsevier Mosby; 2010. pp. 709–16.

108. Chakrabarti A, Denning DW, Ferguson BJ et al. Fungal rhinosinusitis: A categorization and definitional schema addressing current controversies. *Laryngoscope* 2009;119:1809–18.

109. Grosjean P, Weber R. Fungus balls of the paranasal sinuses: A review. *Eur Arch Otorhinolaryngol* 2007;264:461–470.

110. deShazo RD, Chapin K, Swain RE. Fungal sinusitis. *N Engl J Med* 1997;337:254–9.

111. Seo YJ, Kim J, Kim K et al. Radiologic characteristics of sinonasal fungus ball: An analysis of 119 cases. *Acta Radilociga* 2011;52:790–5.

112. Ilica AT, Mossa-Basha M, Maluf F et al. Clinical and radiologic features of fungal disease of the paranasal sinuses. *J Comput Assist Tomogr* 2012;36:570–6.

113. Aribandi M, McCoy VA, Bazan C. Imaging features of invasive and noninvasive fungal sinusitis: A review. *RadioGraphics* 2007;27:1283–96.

114. Mossa-Basha M, Ilica AT, Maluf F et al. The many faces of fungal disease of the paranasal sinuses: CT and MRI findings. *Diagn Interv Radiol* 2013;19:195–200.

115. Mukherji SK, Figueroa RE, Ginsberg LE et al. Allergic fungal sinusitis: CT findings. *Radiology* 1998;207:417–22.

116. Gillespie MB, O'Malley BW Jr, Francis HW. An approach to fulminant invasive fungal rhinosinusitis in the immunocompromised host. *Arch Otolaryngol Head Neck Surg* 1998;124:520–6.

117. DelGaudio JM, Clemson LA. An early detection protocol for invasive fungal sinusitis in neutropenic patients successfully reduces extent of disease at presentation and long term morbidity. *Laryngoscope* 2009;119:180–3.

118. Groppo ER, El-Sayed IH, Aiken AH et al. Computed tomography and magnetic resonance imaging characteristics of acute invasive fungal sinusitis. *Arch Otolaryngol Head Neck Surg* 2011;37:1005–10.

119. Stringer SP, Ryan MW. Chronic invasive fungal rhinosinusitis. *Otolaryngol Clin North Am* 2000;33:375–87.

120. Muhle C, Reinhold-Keller E, Richter C et al. MRI of the nasal cavity, the paranasal sinuses and orbits in Wegener's granulomatosis. *Eur Radiol* 1997;7:566–70.

121. Keni SP, Wiley EL, Dutra JC, Mellott AL, Barr WG, Altman KW. Skull base Wegener's granulomatosis resulting in multiple cranial neuropathies. *Am J Otolaryngol* 2005;26:146–9.

122. Marsot-Dupuch K, De Givry SC, Ouayoun M. Wegener granulomatosis involving the pterygopalatine fossa: An unusual case of trigeminal neuropathy. *AJNR Am J Neuroradiol* 2002;23:312–5.

13

Magnetic Resonance Imaging of the Ear

Ravi Kumar Lingam and Ram Vaidhyanath

CONTENTS

13.1 Anatomy

13.1.1 External Ear

The pinna (auricle), external auditory canal (EAC), and the tympanic membrane form the external ear apparatus (Figure 13.1). The surface anatomy of all these structures can be seen easily on direct clinical and otoscopic examination and hence the diagnosis of many external ear diseases is a clinical one. The pinna is the visible part of the ear that resides outside the head and is essentially a skin-covered flap of cartilage. It amplifies and funnels sound into the EAC.

The EAC comprises a lateral fibrocartilaginous one-third and a medial bony two-thirds. The bony canal is approximately 16 mm in length and is directed medially, inferiorly, and anteriorly [1]. The isthmus is the narrowest portion of the external canal and lies just medial to the junction of the bony and cartilaginous part of the canal. The function of the external canal is to protect the tympanic membrane and act as a resonator to provide 10 dB of gain to the tympanic membrane. The skin of the lateral cartilaginous part of the canal has a substantial subcutaneous layer with hair follicles, sebaceous glands, and cerumen glands. There are many tiny foramina here, the foramina of Santorini, which can allow passage of canal inflammation into the adjacent deep neck spaces and temporomandibular joint. In contrast, the skin of the bony canal is very thin and bereft of adnexal structures. The bony external canal wall is best demonstrated on high-resolution computed tomography (CT), which is therefore the imaging modality of choice in the assessment of bony canal integrity in various canal disease processes. Not uncommonly, there is a defect in the anteroinferior bony canal, the foramen tympanicum (Foramen of Hushcke), just posteromedial to the temporomandibular joint, which can potentially allow spread of inflammation from the ear canal to the temporomandibular joint [2]. The tympanic membrane (eardrum) separates the inner bony ear canal from the middle ear cavity. The tympanic membrane is innervated by four cranial nerves: the auriculotemporal nerve (a branch of the trigeminal nerve); the posterior auricular nerve (a branch of the facial nerve); fibers of the Arnolds nerve from the vagus nerve; and Jacobsen's nerve, a branch of the glossopharyngeal nerve. Pain from these cranial nerve innervation sites can therefore be referred to the ear (referred otalgia).

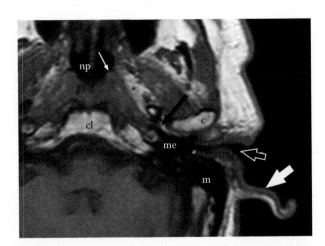

FIGURE 13.1
External ear anatomy and relations. Axial T_1W image shows salient external ear anatomy to include pinna (large white arrow), cartilaginous ear canal (open arrow) and bony canal (white *). Important anatomical relationships include middle ear cavity (me), mastoid (m), mandibular condyle (c) in temporomandibular joint, medial condylar fat space (black arrow), clivus (cl) of basiocciput with high signal bone marrow, parapharyngeal fat space (black *), and torus tubaris (small white arrow) at nasopharynx (np).

13.1.2 Middle Ear

The middle ear cleft is situated within each temporal bone and comprises the middle ear cavity, mastoid antrum, and mastoid air cells. The middle ear cavity can be divided into three parts: the epitympanum (attic), mesotympanum (middle ear cavity proper), and hypotympanum (below the level of the tympanic membrane) [3,4]. The epitympanum lies superior to the level of the tympanic membrane and communicates posteriorly with the mastoid antrum via a narrow tunnel, the aditus ad antrum. As the anterior and posterior boundaries of the aditus cannot be visualized well on magnetic resonance imaging (MRI), some authors have used the better visualized adjacent anterior and posterior limbs of the lateral semicircular canal, respectively, as indirect landmarks [3] (Figure 13.22). The roof of the middle ear clefts, the tegmen, forms part of the skull base and together with its periosteum and adjacent meninges appears as a low-signal line best appreciated on the coronal and sagittal images (Figure 13.34). The intracranial posterior wall of the mastoid, the sigmoid plate, is closely applied to the sigmoid venous sinus. Anteriorly, the middle ear cavity connects with the nasopharynx via the Eustachian tube. The anterior cartilaginous end of the tube projects into the posterolateral aspect of the nasopharynx as the torus tubaris (Figure 13.1).

The ossicular chain is situated within the middle ear cavity and comprises the malleus, incus, and stapes. The manubrium of the malleus attaches to the tympanic membrane. The head of the malleus articulates with the body of the incus in the epitympanum. The incus articulates with the stapes at the mesotympanum. The stapedial footplate articulates with the oval window of the labyrinth of the inner, thereby completing the conduit of sound transmission from the tympanic membrane to the inner ear. The ossicular chain is best imaged with high-resolution CT. On MRI, the ossicles are not visualized unless the middle ear is filled with fluid and hence

appear as low-signal structures within high signal fluid on the T_2-weighted images.

13.1.3 Inner Ear

The inner ears are situated in the otic capsule at the petrous temporal bones and receive sound transmission from the middle ear ossicular chains. Each inner ear comprises the bony labyrinth (cochlea, vestibule, semicircular canals, and vestibular aqueduct), which encloses a closed fluid filled system, the membranous labyrinth, in a bed of fluid, the perilymph [5] (Figure 13.2). The membranous labyrinth contains interconnecting fluid filled spaces, which constitutes the endolymphatic cavity and comprises the cochlea duct (in the cochlea), the utricle and saccule (in the vestibule), the semicircular ducts (in the semicircular canals), and the endolymphatic duct (in the vestibular aqueduct) [6,7]. The perilymphatic space of the bony labyrinth communicates with the subarachnoid space via the cochlear aqueduct (Figure 13.2a), which extends from the basal turn of the cochlea to the lateral margin of the jugular foramen. As the inner ear is fluid filled, it returns high signal on T_2-weighted images and hence high-resolution (thin section: typically less than 1 mm slice thickness) heavily T_2-weighted images will depict its anatomy well and the detail is even better when a scanner with a higher magnetic field is used [8].

The cochlea (Figure 13.2), the anterior part of the perilymph-filled bony labyrinth, is a tubular structure

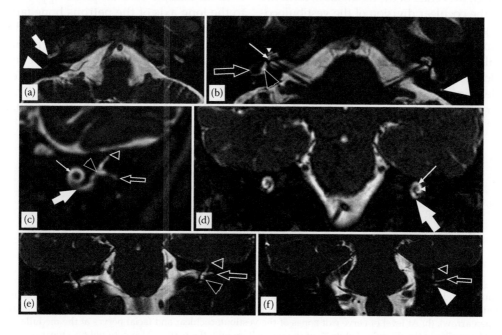

FIGURE 13.2
Inner ear anatomy. High-resolution heavily T_2W images through the inner ears. (a,b) axial planes where (a) is caudal to (b). (c) Sagittal image through inner ear. Coronal planes through inner ear from (d) anterior to (e) middle, at level of internal auditory meati, to (f) posterior. Cochlea apical turn (small white arrowhead), middle turn (small white arrow), basal turn (large white arrow), and modiolus (white *). Vestibule (black arrowhead). Semicircular canals: superior (open arrowhead), lateral (open white arrow), and posterior (large white arrow).

FIGURE 13.3
Inner ear (finer detail). High-resolution heavily T_2W images (a) sagittal, (b) coronal, and (c) axial images through the inner ear. Cochlea: modiolus (white*), scala tympani (large white arrowhead), and scala vestibuli (large white arrow). Ampulla of superior semicircular canal (large open arrow). Vestibule: utricle (small white arrowhead), saccule (small open arrow), macula of utricle (long white arrow), macula of saccule (short small white arrow).

wound in a spiral with approximately two-and-a-half turns around the modiolus, a conical central bony pillar with nervous tissue of the spiral ganglia. The turns of the cochlea, the basal, middle, and apical turns, are separated from each other by the interscalar septum which arises from the modiolus. From the modiolus, projects the thin bony spiral lamina which provides support to the membranous cochlear duct. Together, they cleave the turns of the cochlea into two approximately equal perilymphatic chambers, the scala vestibuli and scala tympani, which are both visible well on thin-slice T_2-weighted MRI [1,8,9] (Figure 13.3). The cochlear duct contains the organ of Corti where the sensory hair cells that mediate hearing are sited. As yet, the organ of Corti is not visible on imaging. From the spiral organ of Corti and the spiral ganglia, axons pass through the center of the modiolus to form the cochlear nerve, the cochlear division of the vestibulocochlear nerve (Figures 13.4 and 13.5). The round window is a small opening at the posterolateral aspect of the basal turn of the cochlea and covered by a membrane that separates the cochlea from the middle ear.

The vestibule (Figure 13.3) is an ovoid chamber of the bony labyrinth located posterosuperiorly to the cochlea [10]. It contains the membranous vestibule made of two endolymphatic subunits: saccule and utricle. The vestibule has two openings, the oval window, where the stapedial footplate is attached, and the vestibular aqueduct. The three semicircular canals arise from the vestibule and are oriented at right angles to each other. The semicircular canals house the semicircular ducts, which arise from the utricle. Each duct and canal has a slight nodular prominence, the ampulla, at one end. The superior semicircular canal is in an oblique sagittal plane and approximately perpendicular to the long axis of the petrous bone (apex). The bony roof of the superior semicircular canal forms the arcuate eminence, a landmark on the superior surface of the petrous apex. The posterior semicircular canal is in an oblique coronal plane approximately parallel to the long axis of the petrous bone (apex). The maculae of both the utricle and saccule and the ampullary cristae of the semicircular canals (Figure 13.3) contain the sensory organs (hair cells) that mediate balance and are better visualized on a higher field strength 3 T MRI scanner than on a 1.5 T scanner [8]. Afferent axons from the maculae and ampullary cristae terminate at

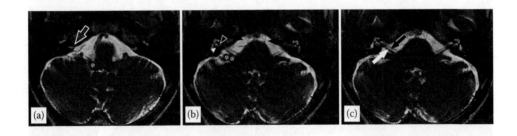

FIGURE 13.4
Cerebellopontine angles and internal auditory meati. High-resolution heavily T_2W axial images through the posterior cranial fossa and internal meati from caudal to cranial positions. (a) At level of upper medulla, the position of the cochlear nuclei (white *) is marked. Note cochlear aqueduct (white open arrow). (b) At level of nerve exit zones of the vestibulocochlear and facial nerves at the medullary–pontine junction. Note the intimate relationship of the flocculus (white **) of the cerebellum with the nerves. Also note right cochlear nerve (black arrowhead) and inferior vestibular nerve (white arrowhead) at the inferior aspect of the right internal meatus. The basilar artery (black *) and the abducen nerves (black open arrow) lie in the prepontine cistern which communicates freely with the cerebellopontine angles. (c) At the level of the internal auditory meati. The vestibulocochlear nerve (white arrow) lies posterior to the thinner facial nerve (black arrow) and they both can be seen traversing the cerebellopontine angles into the internal auditory meati toward the inner ears.

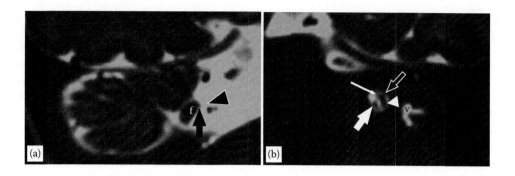

FIGURE 13.5

Vestibulocochlear nerve. High-resolution heavily T_2W sagittal image of vestibulocochlear nerve and adjacent facial nerve. (a) At the cerebellopontine angle, the vestibulocohlear nerve (black arrow) appears thicker and is posterior to the facial nerve (black arrowhead). (b) In the internal auditory meatus, the vestibulocochlear nerve divides into the cochlear nerve anteriorly (big white arrow), the superior vestibular nerve (open white arrow), and the inferior vestibular nerve (white arrowhead) posteriorly. The facial nerve (small white arrow) lies anterior to the superior vestibular nerve. Note flocculus (f) of the cerebellum and vestibule (v) of the inner ear.

the vestibular (Scarpa's) ganglion at the bottom of the internal auditory meatus, from which the vestibular division of the vestibulocochlear nerve arises. The vestibular ganglion has two parts, the superior and inferior vestibular ganglion; the former receives axons via the superior vestibular nerve from the superior and lateral canal ampullary cristae and the utricular macula, while the latter receives the inferior vestibular nerve from the posterior canal ampullary cristae and saccular macula [11,12].

The endolymphatic duct arises from the membranous vestibule and passes through the vestibular aqueduct to end into a flat endolymphatic sac. The endolymphatic sac fills most of the aqueduct and also protrudes through the inferior aperture of the aqueduct to lie between the periosteum of the petrous bone and the dura mater of the posterior cranial fossa. Usually, the vestibular aqueducts are not or barely visible on high-resolution MRI but can become prominent and larger than the adjacent posterior semicircular canal when they are pathologically enlarged [1].

13.1.4 Vestibulocochlear Nerve

In the internal auditory meatus, the vestibular and cochlear divisions form the eighth cranial nerve: the vestibulocochlear nerve [11]. The internal auditory meatus has a diameter of 2–8 mm [1]. The nerve runs with the facial nerve through the internal auditory meatus to exit at the porus acusticus into the cerebellopontine angle toward the brain stem. On the thin section T_2-weighted MRI through the internal auditory canal, the vestibular nerve and its divisions, the superior and inferior vestibular nerves, and the cochlear nerve can all be visualized well; particularly, on the sagittal view, with the facial nerve and their caliber can be assessed (Figures 13.4 and 13.5). Within the internal auditory meatus, the facial nerve lies in the anterosuperior quadrant, the cochlear nerve anteroinferiorly and the superior and inferior vestibular nerves posterosuperiorly and posteroinferiorly, respectively. It is not uncommon to visualize incidental small vascular loops, usually from the anterior inferior cerebellar artery, within the internal auditory meatus (Figure 13.6).

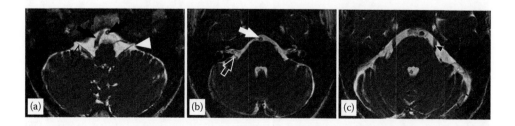

FIGURE 13.6

Cerebellopontine angles. High-resolution heavily T_2W axial images through the cerebellopontine angles from caudal to cranial positions. (a) At the inferior cerebellopontine angle on each side, the vagus–glossopharyngeal nerve complex (open black arrow) can be seen coursing toward the pars nervosa (white arrowhead) of the jugular fossa. The vertebral arteries (black arrows) are noted anteriorly in the cisternal space. (b) At the level of the internal auditory meati the vestibulocochlear nerves can be seen. A small vascular loop (normal variant) is seen at the proximal internal auditory meati in the region of the porus acusticus. The basilar artey (white arrow) is formed from the union of the two vertebral arteries. (c) At the level of the trigeminal nerves. Each trigeminal nerve (black arrowhead) can be seen entering the Meckel's cave (white *). Note fourth ventricle (black *).

The cerebellopontine angle is a triangular space located between the anterolateral surface of the pons and cerebellum and the posterior surface of the petrous bone [13] (Figures 13.4 and 13.6). It also contains other cranial nerves such as the trigeminal nerve and the abducent nerve superiorly, the superior cerebellar artery, and anterior inferior cerebellar artery, a variable number of draining veins and the flocculus of the cerebellum (Figures 13.4 and 13.5). It may also contain choroid plexus protruding through the lateral foramen of Luschka of the fourth ventricle. The lower cranial nerves, the glossopharyngeal, vagus, and the accessory nerves are located inferiorly.

13.1.5 Central Acoustic and Vestibular Pathways

The vestibulocochlear nerve enters the brain stem at the medullo–pontine junction. The cochlear portion of the nerve then divides into two branches, which extend to the dorsal and ventral cochlear nuclei at the upper medulla (Figure 13.7). From the dorsal nuclei, the axons cross over along the rhomboid fossa just beneath the medullary striae of the fourth ventricle to join the contralateral lateral lemniscus. The axons from the ventral cochlear nuclei extend in the trapezoid body to the superior olivary nucleus and nuclei of the trapezoid body and eventually ascend the pons in the contralateral lateral lemniscus (Figure 13.7). A second portion of the ventral auditory pathway does not cross over and ascends in the ipsilateral lateral lemniscus. The lateral lemnisci connect with the inferior colliculi (caudal bulges at the quadrigeminal plate [Figure 13.8]). From each inferior colliculus, the axons connect to the medial geniculate body and eventually via the auditory radiation to the primary auditory cortex in the anterior transverse temporal gyrus (of Herschl) [14] (Figure 13.8).

The axons in the vestibular portion of the eighth cranial nerve enter the four vestibular nuclei (superior, inferior, medial, and lateral) at the medullo–pontine junction and the flocculonodular lobe of the cerebellum (Figure 13.4). The vestibular nuclei also receive afferent signals from the spinal cord, reticular formation, and cerebellum. Efferent fibers run in the vestibulospinal tract and medial longitudinal fasciculus with numerous connections to the motor neurons of the eye, neck, trunk, and limbs. There are also superior connections to the cerebrum (thalamus and parietal cortical area around the intraparietal sulcus) [15].

Lesions affecting the central acoustic pathways (Figure 13.9) characteristically result in bilateral sensorineural hearing loss (SNHL), worse in the contralateral ear. Involvement of higher centers in the auditory cortex can result in complex auditory processing disorders such as auditory agnosia.

13.1.6 Facial Nerve

The facial nerve is composed of motor, sensory, and parasympathetic fibers from three brain stem nuclei. The largest is the motor nucleus situated in the ventrolateral pontine tegmentum and supplies efferent motor fibers to the muscles of facial expression, stapedius, stylohyoid, and the posterior belly of diagastric muscles [15]. The motor fibers loop dorsally around the abducent nerve nucleus creating a bulge in the floor of the fourth ventricle called the facial colliculus. The superior salivatory nucleus is located dorsal to the motor nucleus and contributes parasympathetic secretomotor fibers via the

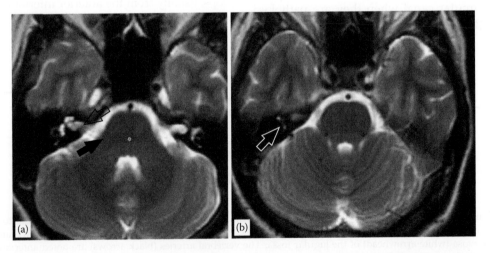

FIGURE 13.7
Central acoustic pathway anatomy. (a,b) Axial T_2W images through the brain stem show positions of the internal auditory meatus (open black arrow), cochlear and vestibular nuclei at medullary–pontine junction (black arrow), trapezoid body (white *), and lateral lemnisci (black *). (b) is cranial to (a) and shows postion of right superior semicircular canal (open white arrow).

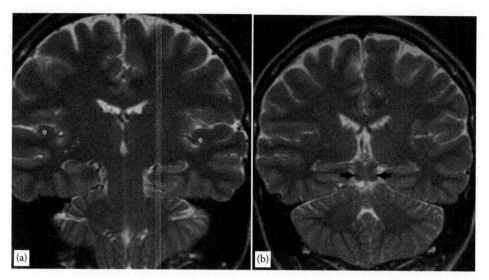

FIGURE 13.8
Central acoustic pathways. (a) Coronal T_2W image of the temporal lobes show positions of medial geniculate bodies (black *) and auditory cortex at the anterior transverse temporal gyrus (of Herschl) (white *). (b) Coronal T_2W image further posteriorly shows position of the paired inferior colliculi (black arrows) at the qudrigeminal plate of the midbrain, inferior to the paired superior colliculi.

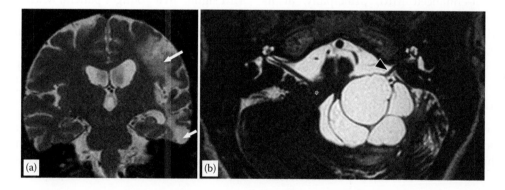

FIGURE 13.9
Central acoustic pathway lesions. (a) Coronal T_2W image shows extensive infarction (white arrows) at the left temporoparietal lobe in a patient presenting with asymmetric right sensorineural hearing loss following a stroke. The auditory cortex (black *) at the anterior transverse gyrus of the superior temporal lobe and its connections lie within the zone of infarction. (b) A recurrent choroid plexus papilloma at the fourth ventricle indents against the left vestibulocochlear nerve (black arrowhead), the nerve exit zone, and the left cochlear nuclei at the upper medulla in a patient presenting with left sensorineural hearing loss. Compare with region of the cochlear nuclei at right side of upper medulla (white *) and adjacent right nerve exit zone.

greater superficial petrosal nerve to the lacrimal glands and via the chorda tympani to the sublingual, submandibular, and minor salivary glands in the oral cavity. The nucleus solitarius located in the upper medulla receives afferent fibers of taste from the anterior two-thirds of the tongue via its chorda tympani branch. The parasympathetic secretomotor fibers combine with the special sensory fibers of the nucleus solitarius to form the nervus intermedius of Wrisberg. Both the facial nerve and the nervus intermedius exit the brain stem at the pontomedullary junction to enter the cerebellopontine angle [16] (Figure 13.4).

The facial nerve courses anterior to the vestibulocochlear nerve along its cisternal course in the cerebellopontine angle [10,11]. At the cerebellopontine angle, the facial nerve can be seen on high-resolution T_2-weighted MR images to course anterior to the thicker vestibulocochlear nerve (Figures 13.4 and 13.5). It continues into the internal auditory meatus where it lies anterior to the superior vestibular nerve and superior to both the cochlear nerve and inferior vestibular nerve (Figure 13.5).

The facial nerve exits the fundus of the internal auditory meatus to enter the bony petrous facial nerve canal and loses its visibility on noncontrast-enhanced

MR images. Within the facial nerve canal, the nerve is best appreciated on high-resolution CT. It is subdivided into the labyrinthine, tympanic, and mastoid segments, separated by the first and second genu. The geniculate ganglion is situated in the first genu. The first major branch of the facial nerve, the greater superficial petrosal nerve, exits the geniculate ganglion via the facial hiatus. The tympanic segment courses along the medial wall of the tympanic cavity and is consistently related inferiorly to the lateral semicircular canal. The mastoid segment extends from the second or posterior genu vertically to the stylomastoid foramen. The chorda tympani and the nerve to stapedius muscle emerge from the mastoid segment of the facial nerve.

The extracranial course of the facial nerve begins as the nerve exits the stylomastoid foramen and enters the parotid gland substance forming the parotid plexus. Within the parotid gland, the termination of the nerve is variable, mostly bifurcating into five motor branches namely the temporal, zygomatic, buccal, marginal mandibular, and cervical branches [16].

The facial nerve displays normal but variable enhancement on contrast-enhanced MRI; principally, at its mastoid and tympanic segments and at the geniculate ganglion [17] (Figures 13.10 and 13.34). The cisternal, intracanalicular, labyrinthine, and parotid segments of the facial nerve do not normally enhance.

13.1.7 Petrous Apex

The petrous apex is the pyramidal-shaped medial portion of the temporal bone [18] (Figure 13.11). It is obliquely positioned within the skull base, with its apex pointing anteromedially and its base located posterolaterally.

The petrous apex is bounded by the petrooccipital fissure medially, the petrosphenoidal fissure anteriorly, the inner ear and petrosquamous suture laterally, and the posterior cranial fossa behind [19]. It forms part of the floor of the middle cranial fossa and there is a marked depression at the floor for Meckel's cave anteromedially. Anteriorly, it contains a groove for the greater superficial petrosal nerve and forms the posterior border of the carotid canal. The abducens nerve runs in Dorello's canal along the medial portion of the petrous apex. Inferiorly, it is continuous with the inner surface of the mastoid portion, articulates with the basiocciput, and is related to the jugular bulb and inferior petrosal sinus. The inferior surface forms part of the exterior of the base of the skull and serves for the attachment of the levator veli palatini muscle and the cartilaginous portion of the auditory tube [20].

The petrous apex is divided by the internal auditory canal into an anterior and a posterior portion by the internal auditory meatus. The smaller posterior portion derived from the otic capsule is denser. The larger anterior portion of the petrous apex mostly contains bone marrow and, depending on patient's age, can show variable MRI signal. In a younger patient, when red marrow predominates, it returns intermediate signal on conventional sequences but in older patients, it can return high T_1 signal in the presence of fatty bone marrow [21] (Figure 13.11). In a third of patients, the petrous apex is pneumatized by tracts that directly communicate with middle ear cleft and return low signal, if aerated, and high T_2 signal, if associated with trapped fluid. It is therefore possible for the normal asymmetric signal at the petrous apices to be mistaken for a mass (pseudolesion). Pneumatization of the petrous apex also provides direct access for diseases to spread from the middle ear cleft [22].

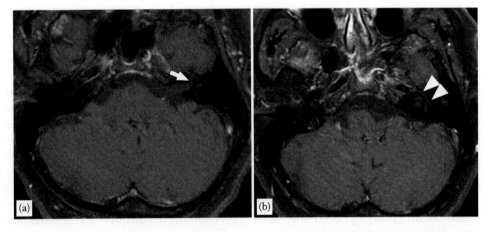

FIGURE 13.10

Normal enhancement pattern of the facial nerve. (a) The labyrinthine segment (white arrow) does not enhance normally but (b) the tympanic segment (white arrowheads) and the mastoid will show minimal enhancement due to vascularity in the epineurium and perineurium.

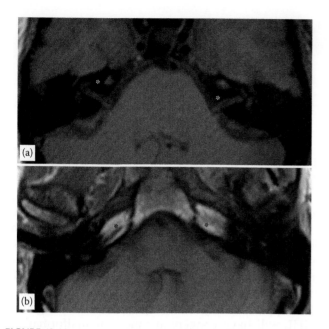

FIGURE 13.11
T_1-weighted axial images (a) show low signal petrous apex bilaterally in keeping with sclerosis/pneumatized apex marrow. Compare this with image (b) where the petrous apices show high signal in keeping with fatty marrow.

13.2 Clinical Applications and Imaging Strategy

Because of the complex anatomy and relations of the ear components, the temporal bone and the intracranial auditory pathways, the choice of cross-sectional imaging is crucial in demonstrating the relevant anatomy and associated pathology. The choice is usually between using MRI, CT, and using them both.

In investigating hearing loss, MRI is the modality of choice for SNHL as CT is for conductive hearing loss. For sensorineural loss, vertigo, and nonpulsatile tinnitus, high-resolution (thin section) volume-acquired heavily T_2-weighted spin echo (e.g., driven equilibrium [DRIVE] and sampling perfection with application optimized contrasts using different flip angle evolution [SPACE]) or gradient-echo (constructive interference in steady state [CISS] and fast imaging with steady-state precession [FISP]) sequences through the inner ears, internal auditory meati, and cerebellopontine angles are indicated. These sequences excellently delineate the high signal fluid filled inner ears within the low-signal bone and low-signal intracranial vestibulocochlear nerves within the high signal cerebrospinal fluid (CSF). They can be supplemented with contrast-enhanced T_1-weighted images to demonstrate inflammatory changes or small tumors at the inner ear or the nerves. For completeness, an MRI sequence of the brain is suggested to assess the central auditory pathways and other relevant intracranial structures. For pulsatile tinnitus with normal otoscopic examination, imaging evaluation can be performed with CT- or MR-arteriography and venography [22,23]. High-resolution CT or cone beam CT can complement the thin section MRI of the inner ears when assessing for congenital inner ear anomalies and suitability for cochlear implant.

CT is also the imaging modality of choice for evaluating congenital external and middle ear anomalies. While CT is the mainstay for imaging external and middle ear inflammatory diseases, MRI roles have been established for malignant external otitis and postoperative middle ear cholesteatoma. The latter requires non-echoplanar diffusion-weighted imaging complete with apparent diffusion coefficient (ADC) map. With regard to imaging for referred otalgia, MRI with multiplanar images is ideally suited to help detect potential causes in the temporomandibular joints, deep neck spaces, oral cavity, pharynx, and skull base.

The superior soft tissue detail and enhancement of MRI is often used to complement the bony detail provided by CT for staging of tumors at the external canal, middle ear cleft, and the petrous apex. Similarly, MRI with diffusion-weighted and contrast-enhanced sequences can complement CT in characterizing and accurately diagnosing the various pathologies in the petrous apex, obviating the need for invasive biopsy procedures in many cases [19].

The complex course of the facial nerve could warrant imaging evaluation with either CT or MRI or perhaps both, depending where along the nerve the likely cause is. In imaging for facial nerve palsy or related symptoms, MRI sequences, with intravenous gadolinium contrast agent, to cover the brain, internal auditory meatus, and skull base are indicated if the cause is suspected to be proximal or if it is associated with other cranial nerve palsies. For evaluating a cause at the temporal bones such as fracture or middle ear pathology, high-resolution CT temporal bone is the modality of choice. For extracranial causes such as a malignant parotid mass, ultrasound with fine-needle aspiration or biopsy and MRI of the extracranial head and neck with contrast are indicated.

13.3 Congenital Ear Anomalies

13.3.1 Congenital Middle and External Ear Anomalies

Congenital anomalies of the external and middle ear are best evaluated with CT and may be part of a syndrome. Occasionally, MRI can be of value and complement CT as in the case of perilabyrinthine CSF fistula [1] (Figure 13.12). In this condition, there is a congenital absence of a portion of the tegmen tympani. This can

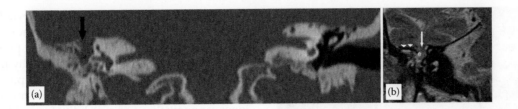

FIGURE 13.12
Perilabyrinthine fistula causing CSF otorrhoea. (a) Coronal CT image shows bony defect (black arrow) at right tegmen mastoid associated with soft tissue opacification (black *) of the mastoid. (b) Coronal T_2W MR image shows herniation of adjacent brain tissue through the tegmen defect (white arrow) into the mastoid (white *) in keeping with meningo-encephalocele. Note intact tegmen (white arrowheads) appear as low signal line on MRI.

uncommonly result in meningoceles, meningoencephaloceles, and fistula, which are usually present as CSF otorrhea in adulthood. They are treated by neurosurgical repair of the bony and meningeal defects.

13.3.2 Congenital Inner Ear Anomalies

As thin section volume acquired heavily T_2-weighted MRI sequences depict the inner ear anatomy well; especially, on scanners with a high field strength, it is ideally suited for the evaluation of congenital inner ear anomalies. MRI can perform better than CT in depicting finer anatomic detail of the inner ear such as the scala tympani and scala vestibuli (Figure 13.3). Even finer details such as the macula utriculi, macula sacculi, and crista ampullaris can be better appreciated with a 3 T MRI scanner. To date, the microscopic anatomy of the organ of Corti is not resolvable on imaging and hence associated anomalies such as the Ging–Siebenmann anomaly and the most common genetic anomaly, the Scheibe anomaly cannot be detected on imaging [24]. Approximately 20% of patients with congenital SNHL will have congenital inner ear malformations detected on imaging [24]. Absence of salient anatomy on MRI may not necessarily represent congenital aplasia but could be a result of acquired fibrous or bony replacement of the fluid in a normally developed inner ear (labyrinthine sclerosis or ossificans) (Figure 13.26). In these cases, clarification with high-resolution CT is key.

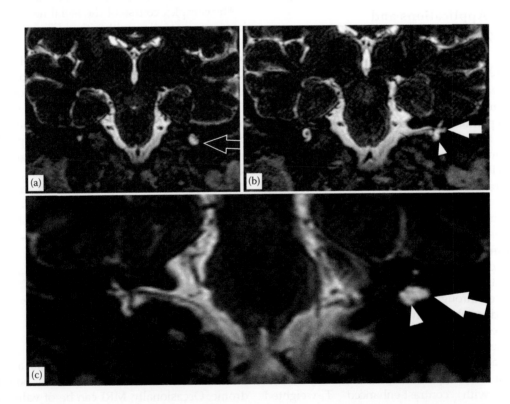

FIGURE 13.13
Left inner ear congenital dysplasia. (a–c) Coronal heavily T_2W images of the inner ear show gross dysplasia of the vestibule (arrowhead) and semicircular canals (white arrows). There is also gross dysplasia of the left cochlea with no separate turns seen (open arrow). Compare with normal right inner ear.

Congenital malformations can be either genetic or nongenetic. Genetic causes can be part of systemic syndrome such as Klippel–Feil syndrome or may occur alone [24]. Nongenetic caused include maternal infections, like toxoplasmosis and rubella, and ototoxic drugs. Over the years, various classifications have been proposed and modified for inner ear malformations; however, there is none that is all inclusive [24–28].

The inner ear develops during the fourth to eighth weeks of life; the earlier the insult and developmental arrest, the more severe the anomaly. The rare complete inner ear aplasia (Michel's deformity) results from development arrest in the third gestational week and the common cavity deformity with no differentiation in the fourth week. Following this, less severe aplasia, hypoplasia, and differentiation and partition anomalies can occur. Late developmental anomalies are isolated and involve structures that develop later such as the lateral semicircular canal and vestibular aqueduct. Incorporation of vascular channels via the fissula ante fenestram helps the growth phase and ossification that occurs via 14 ossification centers is complete by the 24th gestational week [29–31].

Isolated malformation of the lateral semicircular canal is one of the most common inner ear anomalies [23] and the canal can appear short and wide or less commonly narrowed. Less frequently, it can be involved with dysplasia of the vestibule or other semicircular canals (Figure 13.13). The rarer aplasia of all semicircular canals and ducts is associated with the CHARGE syndrome. Vestibular aqueduct anomalies are associated with other inner ear anomalies in the majority of cases, typically with cochlear partitioning abnormality such as the Mondini malformation. The vestibular aqueduct is typically enlarged (enlarged vestibular aqueduct syndrome) with a mid-transverse diameter more than 1.5 mm [24]. On imaging, it appears larger than the adjacent posterior semicircular canal (Figure 13.14).

Cochlear aplasia and hypoplasia (Figure 13.13) result from developmental arrest in the fifth and sixth gestational week, respectively. The commoner Mondini malformation, a partition anomaly, is characterized by a small cochlea with normal basal turns but with reduced number of upper turns secondary to modiolar hypoplasia and absence of interscalar septum (Figure 13.15). It results from an arrest in maturation in the seventh week and is frequently seen in patients with branchio-otorenal syndrome. More subtle partition defects have also been described, such as the modiolar deficiency, defects in the interscalar septum and cochlear scalar asymmetry, and may be appreciated on MRI [24]. With cochlear scalar asymmetry, the anterior scala vestibuli is typically larger than the posterior scala tympani. The *dwarf cochlea* has also been described as a small cochlear but with normal number of turns.

Anomalies of the internal auditory meatus can be associated with inner ear or vestibulocochlear nerve anomalies. The widened internal meatus is a feature of the X-linked mixed deafness and associated with modiolar deficiency, a hypoplastic cochlea with normal turns, enlarged vestibular aqueduct, and stapes fixation [24]. An atretic or stenosed canal, with a diameter less than 2 mm, can be associated with hypoplastic vestibulocochlear and facial nerves (Figure 13.16).

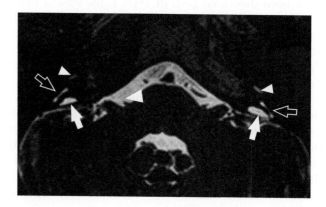

FIGURE 13.14
Enlarged vestibular aqueduct syndrome. High-resolution axial heavily T_2W image through the inner ears show bilateral enlarged vestibular aqueducts (white arrows) just posteromedial to the normal posterior semicircular canals (open arrows). Note basal turns of cochlea (arrowhead).

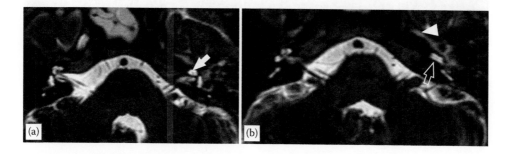

FIGURE 13.15
Left Mondini malformation. Axial high-resolution T_2W images of the inner ear show (a) poor differentiation of the apical and middle turns of the small left cochlea (white arrow) associated with modiolar hypoplasia (compare with normal right inner ear), but with (b) normal appearance of left basal turn (open arrow). Note fluid within left Eustachian tube (arrowhead).

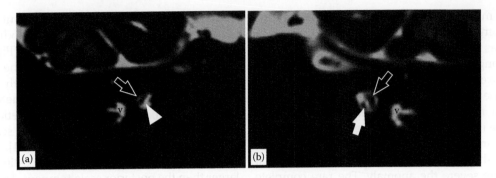

FIGURE 13.16
Right cochlear nerve aplasia. Coronal high-resolution heavily T_2W images through the distal internal meatus showing vestibular nerves (open arrows). (a) On the right the cochlear nerve is absent (arrowhead). (b) The normal left cochlear nerve is shown for comparison. Note vestibule (v) of inner ears.

13.3.3 Congenital Vestibulocochlear Nerve and Facial Nerve Anomalies

MRI has the advantage over CT in that it can also assess for congenital hypoplasia or aplasia of the vestibulo-cochlear nerve and its divisions (Figure 13.17). Three types of vestibulocochlear nerve aplasia and hypoplasia have been described [24]. In Type 1, the absent nerve is associated with a stenotic internal auditory canal. In Type 2, the common vestibulocochlear nerve is associated with aplasia or hypoplasia of the cochlear nerve: Type 2A is associated with inner ear anomaly whereas Type 2B is not. They may be associated with an abnormal course of the facial nerve where the nerve can leave the internal meatus in the middle third, run in a separate canal parallel to the internal meatus, or pass between the temporal bone and temporal lobe of the brain [24]. Congenital facial nerve anomalies involving the labyrinthine, tympanic, and mastoid segments can also be associated with inner ear anomalies and are best depicted with high-resolution CT.

13.4 Inflammatory Diseases

13.4.1 External Ear Inflammatory Diseases

13.4.1.1 Simple External Otitis

The EAC is also easily well examined clinically and imaging is seldom required for simple external otitis. When imaging is used, CT is the preferred modality as it can reliably demonstrate the extent of disease, bony anatomy, and involvement by disease and involvement of the adjacent middle ear cleft. MRI has a role in evaluating otitis externa when disease spills into the deep neck spaces or intracranially as in malignant external otitis.

13.4.1.2 Malignant (Necrotizing) External Otitis

Malignant external otitis (necrotizing external otitis) is an uncommon but severe inflammatory process of the external canal seen typically in elderly patients with diabetes. They present with severe unilateral earache

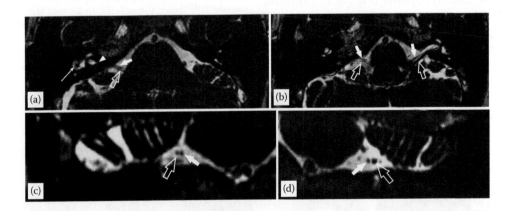

FIGURE 13.17
Right internal auditory meatus stenosis associated with hypoplastic vestibulocochlear nerve. High-resolution heavily T_2W (a,b) axial and (c,d) sagitttal oblique images through the cerebellopontine angles. The facial nerves (thick white arrows) and vestibulocochlear nerves (open arrows) are demonstrated. The right vestibulocochlear nerve is thinner compared to the left and is of similar calibre to the adjacent right facial nerve. The right internal auditory meatus is stenosed (white arrowhead) and the right lateral semicircular canal is dysplastic (thin white arrow).

and discharge, which has been unresponsive to topical agents. The incriminating organism is commonly *Pseudomonas aeruginosa*. Though uncommon, malignant external otitis has been reported in patients with AIDS, in whom *Aspergillus fumigatus* has also been implicated [32]. These patients are younger and not diabetics.

The disease usually starts at the bony–cartilaginous junction of the canal and spreads rapidly beyond the confines of the canal through tiny clefts in cartilaginous floor (the fissures of Santorini). It has a tendency to infiltrate the adjacent bones, spill into the deep neck spaces and temporomandibular joint (Figure 13.18), and involves the adjacent cranial nerves (typically VII, IX, X, XII, IX) at or around their exit foramina at the skull base [32,33] (Figure 13.19). Occlusion of the internal carotid artery can occur in extensive disease [34]. Intracranial extension, typically through the petroclival synchondrosis, may lead to complications such as meningeal disease and venous sinus thrombosis. Treatment is primarily with long-term antibiotics, although aggressive debridement and hyperbaric oxygen may be required for resistant cases.

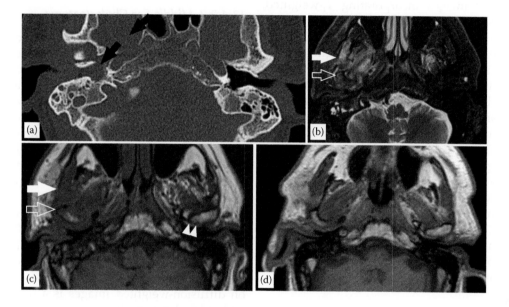

FIGURE 13.18
Malignant otitis externa with anterior extension. (a) CT shows right ear canal soft issue (*) with focal anterior bony wall destruction (black arrow). Axial (b) STIR image shows abnormal high inflammatory signal soft tissue at the adjacent masticator space (white arrow) which corresponds to intermediate signal on (c) T_1W image. Note preserved retrocondylar fat on the unaffected left side (white arrowheads). Bone marrow edema is also seen at the mandibular condyle as demostrated by high STIR and low T_1W signal (open arrow). Following treatment, the axial T_1W image (d) shows resolution of abnormal soft tissue at the mastictor space and mandibular condyle.

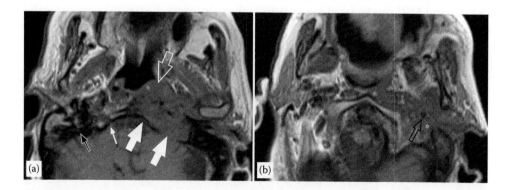

FIGURE 13.19
Left malignant otitis externa with medial extension and ipsilateral facial nerve and hypoglossal nerve palsies. (a and b) Axial T_1 weighted shows abnormal intermediate T_1W soft tissue at the left external canal extending medially into the post-stylod parapharyngeal and retropharyngeal spaces (black *) and invading the skull base and its foramina (large white arrows). The left facial nerve is involved at the stylomastoid foramen with abnormal soft tissue seen in its vicinity (white asterisk). The soft tissue asymmetry at the nasopharynx (open white arrow) can mimic a nasopharyngeal malignancy. Note styloid process (open black arrow), right jugular fossa (small black arrow), and right hypoglossal nerve foramen (small white arrow).

Initial imaging with CT can demonstrate associated bony canal and/or adjacent skull base erosion and destruction and hence confirm diagnosis [35]. MRI due to its superior soft-tissue resolution can demonstrate spread into deep neck spaces, bone marrow, and cranial cavity, and hence extent of disease better than CT, technetium-99 (Tc-99) bone scanning, and gallium-67 citrate (Ga-67 citrate) scanning [35,36]. Retrocondylar fat infiltration has been proposed as one of the most frequent diagnostic findings in patients with malignant external otitis (Figure 13.18) [33]. To demonstrate disease extent best, multiplanar images incorporating T_1-weighted, short-tau inversion recovery (STIR) and fat-suppressed contrast-enhanced T_1-weighted sequences should be acquired. MR is preferred to CT as the imaging modality for monitoring treatment response because it does not involve radiation and is able to detect soft tissue, medullary, and intracranial changes better. To optimize its use for this purpose, a pretreatment baseline MR is indicated, with which posttreatment scans can be compared (Figure 13.18). Abnormal soft tissue in the ear canal, though not always present, resolves early [34]. Infratemporal fossa soft tissue and skull base marrow changes usually improve with treatment but may persist for a long time (more than a year) and not disappear completely [34,35]. Hence, clinical and microbiological evaluation with erythrocyte sedimentation rate (ESR) is essential to supplement imaging findings when deciding to change or stop treatment [32,34].

13.4.1.3 Relapsing Polychondritis

At the pinna of the ear, inflammation of the cartilage can be associated with relapsing polychondritis, which typically spares the EAC but may involve the cartilages of the nose, upper respiratory tract, and peripheral joints. Involvement of the pinna is a clinically apparent as a red and tender ear and MRI can be used to confirm this, detect other involved sites in the head and neck, and monitor treatment response (Figure 13.20).

13.4.2 Middle Ear Inflammatory Diseases

13.4.2.1 Otitis Media

The use of conventional MRI in evaluating inflammatory middle ear disease is limited by its poor demonstration of the salient middle ear and temporal bone anatomy. Its use is primarily in uncommon cases where intracranial complications such as meningeal disease, intracranial abscess, and venous sinus thrombosis are suspected.

13.4.2.2 Middle Ear Cholesteatoma

Cholesteatoma, a sac lined with stratified squamous epithelium and filled with keratin debris, is usually sequelae of repeated middle ear infection. It can erode local body structures, cause otorrhea, hearing loss, and vertigo, and is treated with surgery. On MRI, they can be characterized by contrast-enhanced MRI acquired with a delay of 45 min postintravenous gadolinium contrast administration or by diffusion-weighted imaging (DWI). With the former, cholesteatoma appears as a nonenhancing area or a rim-enhancing soft tissue (Figure 13.58) while granulation tissue enhances. On DWI, by virtue of its keratin content, cholesteatoma appears as a high signal on images obtained with *b*-values of 800 or 1000 s/mm^2 and as a low signal on the corresponding postacquisition reconstructed apparent diffusion coefficient (ADC) map (Figures 13.21). The high signal of cholesteatoma on diffusion-weighted images is a result of restricted molecular diffusion but also due to a T_2 *shine-through effect* that occurs in tissues with high T_2 signal intensity and ADC values less than brain [37,38].

As primary middle ear cholesteatoma is essentially a clinical/otoscopic diagnosis and its complications (bony and ossicular chain erosion) and extent can be well assessed with CT; the use of DWI in primary middle ear cholesteatoma has not been extensively investigated. However, in a subgroup of patients in whom the entire middle ear and mastoid is opacified on CT, DWI can help

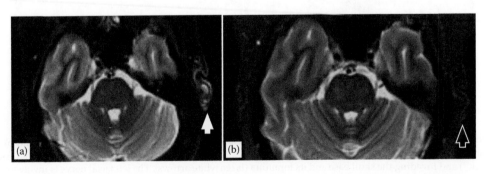

FIGURE 13.20
Polychondritis affecting the left pinna. Axial STIR images show (a) high signal at the left pinna (white arrow) which has resolved completely (open arrow) following treatment (b).

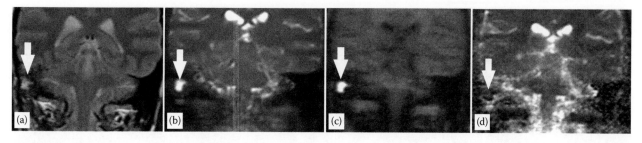

FIGURE 13.21

Residual postoperative right mastoid cholesteatoma. Coronal (a) T_2W shows abnormal right masotid soft tissue (arrow) which returns persistent high signal on the (b) HASTE DWI $b0$ and (c) HASTE DWI $b1000$ images, and low signal on (d) ADC map, in keeping with restricted diffusion typical of cholesteatoma. The left mastoid is clear.

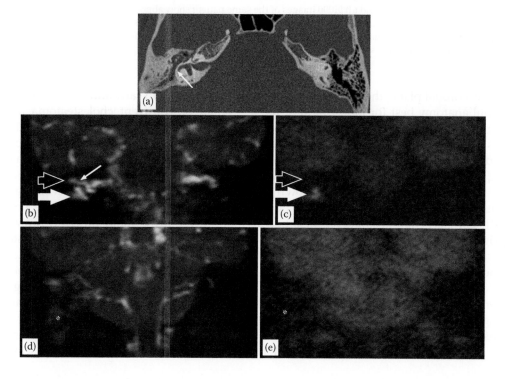

FIGURE 13.22

Staging extent of disease of primary cholesteatoma on non-echoplanar DWI for preoperative planning. (a) Axial CT shows complete soft tissue opacification (black *) of right middle ear cleft. The lateral semicircular canal (small white arrow) can be used as a landmark for locating the mastoid aditus on DWI (b). (b) Coronal HASTE DWI $b0$ image shows abnormal soft tissue in the mesotympanum (large white arrow) and mastoid aditus (open arrow). The mesotympanic soft tissue (large white arrow) retains high signal on (c) coronal HASTE $b1000$ image in keeping with cholesteatoma but the mastoid aditus soft tissue (open arrow) loses signal in keeping with non-cholesteatomatous soft tissue. At the mastoid, the abnormal high signal soft tissue (white *) on the (d) coronal HASTE DWI $b0$ image loses signal on the corresponding (e) $b1000$ image in keeping with non-cholesteatomatous soft tissue. DWI has therefore localized the cholesteatoma to the middle ear cavity with no mastoid extension.

to characterize and define the extent of the coexisting cholesteatoma and subsequently plan surgical approach [39] (Figure 13.22). Another small observational study has demonstrated that though primary middle ear cholesteatoma had low ADC values, other entities such as middle ear abscesses can have lower ADC values [40].

In contrast, DWI has now become established in the detection and management of postoperative middle ear cleft cholesteatoma, primarily after canal wall-up mastoidectomy, and offers a noninvasive alternative to traditional *second-look* or *relook* surgeries. In particular, the single-shot nonechoplanar type of DWI has been

shown to have a high diagnostic performance in detecting cholesteatoma and its use has now superseded that of CT, delayed contrast-enhanced MRI, and echoplanar type of DWI [37,38,41]. With single-shot nonechoplanar DWI, 2 mm thick slices can be obtained, the acquired $b0$ and $b1000$ images are copy referenced and the ADC map reconstructed postacquisition. It can be supplemented with the corresponding T_1- and T_2-weighted images for anatomic detail and further characterization. It performs better than its echoplanar counterpart as it has a higher spatial resolution without air–bone interface artifact and distortion [37,38] (Figure 13.23). When the images

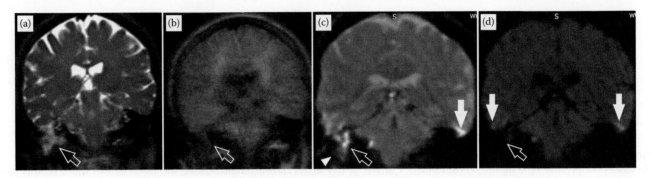

FIGURE 13.23

Detection of postoperative cholesteatoma: Non-echoplanar (HASTE) DWI versus echoplanar DWI. Coronal non-echoplanar DWI (a) *b*0 and (b) *b*1000 images and echoplanar DWI (c) *b*0 and (d) *b*1000 images of the same patient show abnormal high *b*0 signal soft tissue (open arrow) in the mastoid which loses signal on the *b*1000 images in keeping with facilitated diffusion demonstrated by non–cholesteatomatous soft tissue. Note inferior image quality of the echoplanar images with distortion at the mastoid (arrowhead) and artifacts (white arrow) at the tegmen mastoid which could masquerade as a cholesteatoma (and hence false-positive finding).

are acquired in the coronal plane, they demonstrate the middle ear spatial anatomy best for localization of the disease [39] (Figures 13.21 and 13.22).

Many studies on the use of single-shot nonechoplanar DWI have demonstrated sensitivities and specificities in detecting cholesteatoma in the range of 80%–100% and the ability to detect small cholesteatoma down to 2 to 3 mm in size [37,38,42–44]. The ADC value of postoperative middle ear cleft cholesteatoma has also been shown to be significantly lower than that of noncholesteatomatous tissue and can aid in the qualitative diagnosis of cholesteatoma [44]. False-positive findings that have been described include bone powder, silastic sheets, cholesterol granulomas, and cerumen in the ear canal [37,38] (Figure 13.24). Its main limitation however is its lack of sensitivity to detect cholesteatoma less than 2–3 mm or an evacuated cholesteatoma sac [37,38,43].

A negative DWI scan cannot therefore exclude a small residual or recurrent cholesteatoma and for DWI to replace second-look surgery, it would require serial scanning over a period of time prior to discharge [38]. Further studies are required to evaluate the duration of follow-up, cost-effectiveness, and risk of not carrying out *second-look* procedures. Avoiding surgery would confer financial benefits and reduce potential surgical morbidity but the main practical drawback would be the risk of a patient with residual cholesteatoma becoming lost in the follow-up pathway. In addition to detecting postoperative cholesteatoma, nonechoplanar DWI can also be used for surgical planning and approach by estimating the size of the choleasteatoma pearl and localizing disease [45]. It can also offer a detailed consenting process for the patient who can also visualize and be better informed of his/her disease.

13.4.3 Inner Ear Inflammation

13.4.3.1 Labyrinthithis

Inflammation of the inner ear, labyrinthitis, can be caused by infection, trauma, or autoimmune diseases. Infection of the inner ear can be secondary to middle ear infection, meningitis, or hematogenous transmission [46]. Meningitis is the most common cause of acquired SNHL in children with up to 35% of children with bacterial meningitis developing SNHL [47–49]. Autoimmune diseases that can affect the inner ear include rheumatoid arthritis, systemic lupus erythematosus, and polyarteritis nodosa. Clinically, labyrinthitis is associated with a new onset of SNHL and vertigo. Thin section contrast-enhanced MRI can demonstrate abnormal enhancement of the normally nonenhancing inner ear structures (Figure 13.25). Cochlear enhancement can be seen up till 2 months after start of the meningitis and is highly associated with the occurrence of SNHL [50].

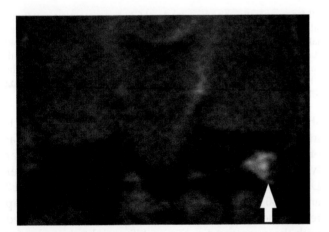

FIGURE 13.24

False-positive case for postoperative cholesteatoma. Coronal HASTE DWI *b*1000 image shows high signal cerumen in the left external canal, mimicking a cholesteatoma. Accurate location of abnormal signal is therefore important to reduce false postive rate.

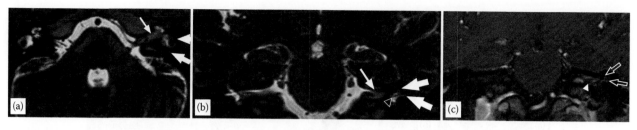

FIGURE 13.25

Acute labyrinthitis in a patient with HIV and cerebral toxoplasmosis with sudden onset left sensorineural hearing loss. (a) Axial and (b) coronal high-resolution heavily T_2W images show loss of normal T_2W signal of the left-sided semicircular canals (large white arrows) and slight loss of T_2W signal at the left internal auditory meatus (small white arrows). There is also low signal thickening of the utricular macula (black arrowhead). Compare with normal right inner ear and internal auditory meatus. (c) Coronal post-gadolinium-enhanced image shows abnormal enhancement of the left-sided macula, semicircular canals (open arrows), and internal meatus (white arrowhead) in keeping with abnormal enhancing inflammatory tissue.

A noncontrast-enhanced T_1W scan is additionally required to demonstrate any high signal intralabyrinthine blood (methemoglobin) or proteinacoues fluid that may be associated [46].

13.4.3.2 Labyrinthine Sclerosis and Ossification

Following the inflammation, repair takes place that include fibrosis and ossification, typically with suppurative bacterial labyrinthitis. If obliteration occurs, it usually starts within the first weeks after onset of the meningitis and ossification of the cochlear lumen can be complete within months after meningitis [49,51]. Both labyrinthine fibrosis and labyrinthine ossification result in the replacement of normal inner ear fluid and hence loss of the normal T_2W fluid signal, best appreciated on thin section 3D heavily T_2W sequences with a reported sensitivity of 92%–100% [49] (Figure 13.26). High-resolution CT has a lower sensitivity but can differentiate these entities, by only detecting labyrinthine

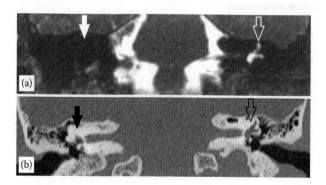

FIGURE 13.26

Labyrinthine ossification and sclerosis. Presentation of right sensorineural hearing loss following trauma. (a) Thin slice coronal T_2W image shows normal appearance of left vestibule and semicircular canals (open arrow) but nonvisualisation of the corresponding structures on the right (white arrow). (b) CT shows normal left vestibule and semicircular canals on the left (open black arrow) but substantial ossification of the corresponding structures on the right (black arrow). Low density sclerosis is also noted at the right vestibule (black *).

ossification characterized by bone density within or *white out* of the inner ear (Figure 13.26). Cochlear fibrosis or ossification can block the cochlear lumen and hamper cochlear implant electrode insertion, which is the treatment of choice in patients with associated bilateral sensorineural deafness. Obliterative cochlear ossification is an absolute contraindication for cochlear implantation [46]. Patients should preferably be imaged and implanted before obliteration occurs and the timing is crucial [49]. The combination of MRI and HRCT is required to optimize the surgeon's preoperative knowledge regarding cochlear patency.

13.4.4 Inflammation of the Facial Nerve

Like labyrinthitis, facial neuritis can be a result of infection, vasculitis, or immunologic irritation. The facial nerve contains efferent fibers to the facial muscles and afferent fibers from the tongue, lacrimal gland, and salivary glands and hence symptoms of facial neuritis include facial palsy abnormal tearing, reduced taste, hyperacusis, and postauricular pain. The most common cause, idiopathic facial nerve or Bell's palsy, although described as idiopathic has a strong association with herpes simplex virus infection [52,53]. Most patients with Bell's palsy do not require imaging in the acute phase and is only indicated in atypical progression, and another cause for nerve palsy need to be excluded [54]. MRI therefore has a dual purpose: to exclude other causes of facial nerve palsy and to positively depict facial nerve inflammation. On MRI, parts of the facial nerve can enhance normally and variably but pathologic enhancement is diagnosed when the intracanalicular and labyrinthine segments are involved (Figure 13.27) [55,56]. Asymmetric linear enhancement with or without thickening of the tympanic and mastoid segments relative to the contralateral side should also be considered abnormal [56]. In the less common Ramsay Hunt syndrome secondary to herpes zoster infection (Figure 13.28), abnormal linear facial nerve

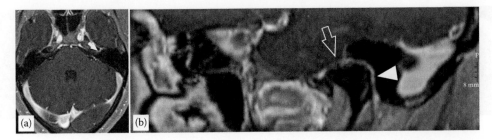

FIGURE 13.27
Patient with a non resolving left-sided Bell's palsy. Postcontrast MRI shows enhancement of the (a) labyrinthine (white arrow), (b) tympanic (white open arrow), and the mastoid (white arrowhead) segments of the facial nerve.

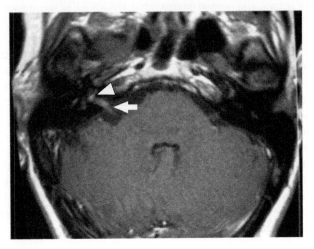

FIGURE 13.28
Ramsay Hunt syndrome. There is enhancement of the right vestibulocochlear nerve (white arrow) as well as the labyrinthine segment of the right facial nerve.

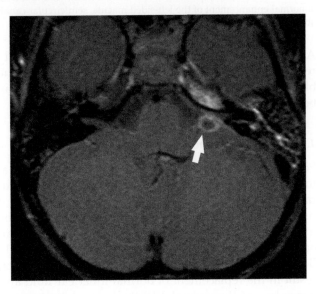

FIGURE 13.29
A small abscess (white arrow) in the left CPA seen in a young patient secondary to bacterial meningitis. The patient recovered without surgical intervention.

enhancement is accompanied by enhancement of the vestibular and cochlear nerves as a result of extension of inflammation from the facial to the vestibulocochlear nerves [57].

13.4.5 Cerebellopontine Angle Inflammation

Meningeal inflammatory diseases can affect the cerebellopontine angle and internal auditory meatus and therefore can cause facial nerve palsy or SNHL [58]. Both infective and noninfective meningitis can occur and the former can be complicated by abscess formation (Figure 13.29) [59]. Typically, meningitis is a clinical and microbiological diagnosis but imaging can be of value in atypical and complicated cases. Inflammatory meningeal disease is characteristically depicted on contrast-enhanced MRI as diffuse meningeal thickening and confluent enhancement. Meningeal disease secondary to tuberculosis or neurosarcoidosis (Figure 13.30) can be depicted this way or appear more focally with nodularity or as a discrete mass mimicking meningioma, lymphoma, or metastatic disease [60].

13.5 Tumors

13.5.1 External Ear Tumors

External ear tumors are easily visualized and typically diagnosed clinically. Biopsies can also be obtained easily for histological diagnosis. Imaging with CT and MRI can complement each other in defining the deep extent of the tumor, complications, and staging (Figure 13.31).

The most common benign tumors of the EAC are exostoses and osteomas: these bony tumors are best imaged with CT [61]. Less common benign soft tissue tumors include ceruminoma and schwannomas (Figure 13.32). Squamous cell carcinoma of the ear canal, though uncommon, is the commonest malignancy here and may be preceded by the long history of chronic ear infection. They can extend into the middle ear with consequent poorer prognosis. This disease

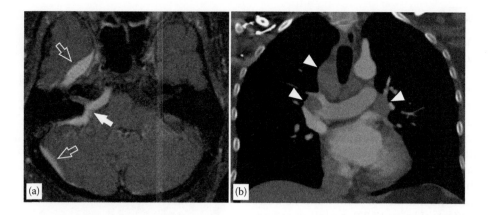

FIGURE 13.30
A 42-year-old male with sudden right-sided sensorineural hearing loss. MRI (a) following contrast administration shows a right CPA-based dural lesion (white arrow). Note further areas of dural enhancement (arrows). Coronal contrast-enhanced CT (b) shows mediastinal and hilar adenopathy (arrowheads). Endoscopic biopsy of the mediastinal node confirmed diagnosis of sarcoidosis.

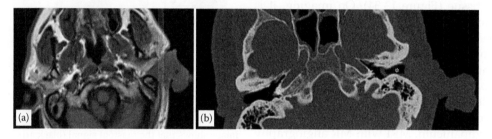

FIGURE 13.31
Squamous cell carcinoma left pinna. (a) Axial T_1W image shows abnormal soft tissue at left pinna (black *) extending into post-auricular region. (b) CT does not show adjacent bony invasion. The left bony canal (white *) is not involved.

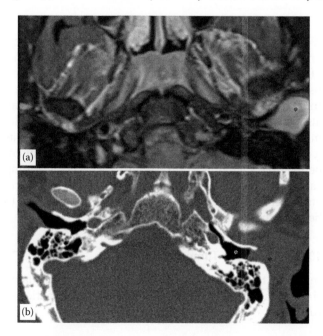

FIGURE 13.32
External canal schwannoma. A mass lesion (black *) is seen within the outer cartilaginous part of the left external auditory canal. (a) On the fat–saturated contrast-enhanced T_1W image, the mass demonstrates uniform enhancement. (b) On CT, it is seen abutting against the lateral aspect of the aerated bony canal (white *) and is not associated with bony erosion.

often erodes the adjacent bony canal wall and may extend into the surrounding soft tissue [62]. Certain infections such as necrotizing external otitis can mimic the imaging appearance of canal malignancy (Figure 13.19). CT is the imaging modality of choice as it defines canal disease and bony destruction well and contrast-enhanced MR can supplement it if there is extensive invasion beyond the confines of the canal or middle ear.

13.5.2 Middle Ear Tumors

Middle ear cavity tumors are uncommon. The most common intratympanic tumor is the paraganglioma or glomus tumor. Less common benign middle ear tumors include cholesterol granuloma, facial nerve schwannomas, middle ear adenomas and hemagiomas. Cholesterol granuloma demonstrates characteristic high T_1-weighted signal on MRI and also restricted diffusion on DWI. The rarer malignant primary middle ear tumors include squamous cell carcinoma and rhabdomyosarcoma. In all these entities, the soft tissue detail provided by contrast-enhanced MRI can supplement the bony detail of CT for complete imaging evaluation [61].

13.5.2.1 Middle Ear Glomus Tumors

Middle ear glomus tumors typically present clinically with pulsatile tinnitus associated with a retrotympanic mass on otoscopy. If confined within the tympanic cavity, it arises along the course of the tympanic branch of the glossopharyngeal nerve (Jacobson's nerve) and is classified as a glomus tympanicum tumor. Small glomus tympanicum tumors are usually sited at the cochlear promontory while larger lesions extend within the middle ear cavity without destroying the ossicular chain [61,62]. While high-resolution CT may be sufficient to demonstrate the middle ear tumor and its bony relationships and effects (Figure 13.33), gadolinium contrast-enhanced MRI may be required in a largely opacified middle ear cleft to delineate the avidly enhancing glomus tumor among coexisting fluid or other lesser enhancing soft tissue (Figure 13.34). On imaging, it is also necessary to differentiate it from vascular anomalies such as the aberrant carotid artery or dehiscent jugular bulb, or the glomus jugulare-tympanicum type of glomus tumor, which invades the middle ear cavity (typically hypotympanum) from the jugular bulb. While CT can demonstrate bony erosion and permeative changes at the jugular foramen typically associated with a glomus jugulare-tympanicum, MRI with MR angiography (MRA) sequences can help depict the jugular fossa mass and distinguish it from the normal or high jugular bulb [63]. MRA can also help confirm the diagnosis of an aberrant carotid artery and other rarer carotid anomalies

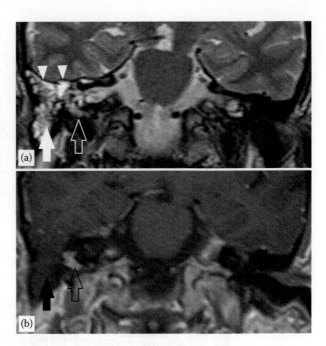

FIGURE 13.34

Glomus tympanicum. (a) Coronal T_2W image shows abnormal high T_2W signal at the rigt mastoid (white arrow) and meso-hypo-tympanum (white open arrow) of the right middle ear cavity. Note intact tegmen (white arrowhead). (b) The corresponding coronal T_1W image shows that the mastoid signal (black arrow) does not enhance in keeping with fluid whereas the meso-hypotympanic soft tissue enhances avidly in keeping with co-existing glomus tympanicum tumor (black open arrow).

(Figure 13.35) [61]. On MRI, large glomus tumors demonstrate a unique salt and pepper appearance with foci of high and low signals interspersed within the soft tissue on T_1- and T_2-weighted images (Figure 13.36). Intratumoral serpentine vessels with flow voids may also be seen to support the diagnosis. MRI can supplement the bony detail and relationship provided by CT as it delineates the infralabyrinthine soft tissue component better [61]. Dynamic contrast-enhanced MRI scanning can demonstrate the specific dropout effect in the early enhancement pattern that allows the glomus tumor to be distinguished from other tumors [64].

13.5.3 Inner Ear Tumors

Inner ear tumors are uncommon and include intralabyrinthine schwannoma and the rare endolymphatic sac tumor.

13.5.3.1 Intralabyrinthine Schwannoma

Intralabyrinthine schwannomas arise primarily from the terminal vestibulocochlear nerve within the cochlea, vestibule, or semicircular canals and are classified according to their location [65,66]. An intralabyrinthine

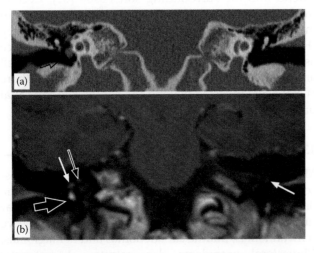

FIGURE 13.33

Glomus tympanicum. (a) Coronal CT shows a small soft tissue nodule (open black arrow) just lateral to the cochlea (black *), in the otherwise well-aerated right middle ear cavity. (b) Coronal contrast-enhanced T_1W image shows enhancement of this middle ear soft tissue (open white arrow) in keeping with glomus tympanicum. Note non- enhancement of labyrinthine part (black arrow) and mild normal enhancement of tympanic part (white arrow) of the facial nerve.

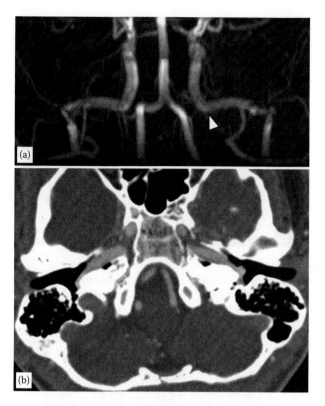

FIGURE 13.35
A 17-year-old patient with left-sided pulsatile tinnitus. Clinical examination revealed a red retro tympanic mass and a diagnosis of glomus tumor was made. MR angiography (a) (white arrowhead) and CT angiography (b) (black arrowhead) shows the aberrant ICA within the left middle ear cavity. The right petrous ICA is also aberrant (white *).

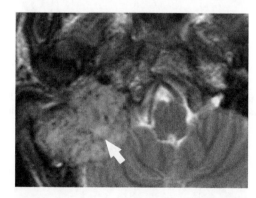

FIGURE 13.36
Glomus jugulare-tympanicum. Axial T_2W image shows large mass lesion at the right jugular fossa eroding into the adjacent middle ear cleft. The tumor shows typical *salt and pepper* appearance.

schwannoma appears as a focal low-signal lesion surrounded by normal high-signal intralabyrinthine fluid on thin section heavily T_2-weighted images [65] and shows intense homogeneous enhancement following intravenous gadolinium contrast administration (Figure 13.37). The most important differential diagnosis is labyrinthitis, which on MRI demonstrates diffuse, rather than focal, enhancement with blurred edges and no signal loss on the

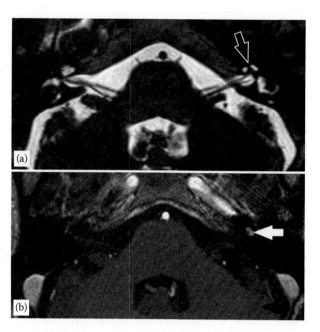

FIGURE 13.37
Intralabyrinthine schwannoma neuroma. (a) High-resolution (thin section) T_2W MR image through the inner ears and internal auditory meati show abnormal soft tissue at the modiolus and middle turn of the left cochlea (open arrow) which enhances (white arrow) on the axial contrast-enhanced T_1W image (b).

T_2-weighted images. The enhancement will also eventually disappear on follow-up imaging. Labyrinthitis ossificans, during its fibro-osseous phase, can also show T_2 hyperintensity and contrast enhancement but can be differentiated on CT, where bony encroachment on the membranous labyrinth would be a feature [65,66].

13.5.3.2 Endolymphatic Sac Tumor

Endolymphatic sac tumors are rare, locally aggressive papillary cystadenomatous tumors at the retrolabyrinthine petrous bone. They may extend into the medial mastoid to invade facial nerve or into posterior cranial fossa. They are associated with von Hippel–Lindau disease in which they can be bilateral [61]. They are treated by wide local resection.

13.5.4 Cerebellopontine Angle and Internal Auditory Canal Tumors

About 10% of all intracranial tumors arise in or involve the cerebellopontine angle.

The vast majority of these tumors are vestibular schwannomas. Meningiomas, epidermoid cysts, and arachnoid cysts account for approximately 10% of the remaining lesions, followed by a more uncommon and extremely heterogeneous group of tumors to include schwannomas of other cranial nerves, lipoma, hemangioma, chondroid tumors, chordoma, and metastases [61].

13.5.4.1 Vestibular Schwannoma

Vestibular schwannomas represent about 8%–10% of all intracranial tumors. The vestibular division of the vestibulocochlear nerve is far more commonly involved than the cochlear division and hence its name [61]. Its incidence is around 13 cases per million per year [67,68] but may be underestimated [69]. Most occur in patients between 30 and 70 years of age. Ninety-five percent of cases are sporadic and in 5%, it is associated with neurofibromatosis type 2 (NF-2). NF-2 is characterized by earlier onset and bilateral vestibular schwannomas, and is commonly associated with other nerve schwannomas, multiple meningiomas, neurofibromas, and glial tumors [61].

Vestibular schwannomas are composed almost entirely of Schwann cells, myelinating cells of the peripheral nervous system [70], and typically grow within a capsule that remains peripherally attached to the parent nerve. Histologically, the tumors are formed of interweaving bundles of spindle-shaped cells with cigar-shaped nuclei, with distinct *Antoni A and B* areas: *Antoni A* areas are densely cellular and interwoven and the predominant tissue in most schwannommas; whereas, *Antoni B* areas are less cellular and structured, and often contain areas of microcyst formation and hemorrhage [71]. Most tumors grow at a slow-to-medium rate of 0.02–0.2 cm per year but a minority of them grow at an alarming rate of 1 cm or more per year [61].

Vestibular schwannomas commonly present with unilateral SNHL, tinnitus, and vertigo, though occasionally they can present with just any one of these symptoms [61]. Decreased speech discrimination is also frequently associated. Facial nerve manifestations are relatively uncommon but trigeminal manifestations are more likely with larger tumors. Noncontrast-MRI consisting of thin section axial and coronal heavily T_2-weighted sequences is the imaging modality of choice for screening for vestibular schwannomas (Figure 13.38). It can detect small internal auditory canal lesions with 100% sensitivity and with excellent interobserver agreement [72]. Intravenous contrast-enhanced scans can supplement and depict the tumor well for the accurate measurement of its size (Figure 13.39).

As the glial–Schwann junction of the nerve is near the porus acusticus, the tumor can arise within the internal auditory meatus, porus acusticus, or occasionally in the cerebellopontine angle. They are small when they lie within the internal meatus (intracannalicular) and enhance uniformly, appearing as ovoid or cylindrical shaped with a convex medial margin (Figure 13.40). As the tumor grows, it often extends into the cerebellopontine angle cistern where it can expand without being impeded by the osseous confines of the internal auditory canal and hence giving an *ice cream on a cone* appearance,

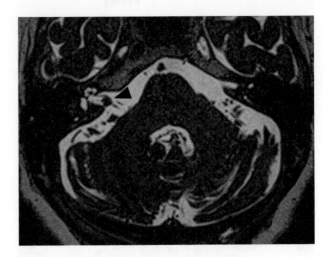

FIGURE 13.38
Axial image obtained using *Constructive Interference in the Steady State* (CISS) as part of screening MRI for right asymmetrical hearing loss. The small vestibular schwannoma (black arrowhead) is well demonstrated against the bright background CSF signal in the IAC.

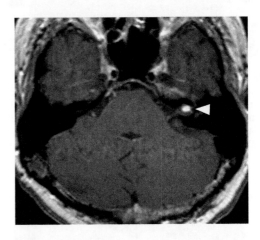

FIGURE 13.39
Postcontrast T_1-weighted fat-suppressed image shows a small enhancing intracanalicular vestibular schwannoma (white arrowhead).

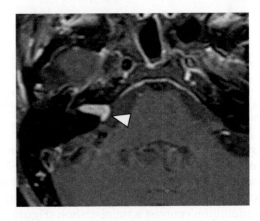

FIGURE 13.40
Convex medial margin in a patient with right intracanalicular vestibular schwannoma (white arrowhead).

with the cone representing the intracanalicular component (Figure 13.40). The round cisternal component typically remains centered at the porus acusticus and forms acute angle with the petrous bone (Figures 13.41 and 13.42). Because the tumor grows slowly, the bony canal may expand and be remodeled, particularly near the porus acusticus. As the tumor enlarges, resulting areas of internal necrosis and hemorrhage can cause the formation of cysts, which appear as central regions of nonenhancement (Figure 13.43). Large cerebellopontine components can cause significant mass effect on the brain stem and cerebellum to cause midline shift and hydrocephalus (Figure 13.44).

Screening for vestibular schwannoma is a significant burden in the modern otology and radiology

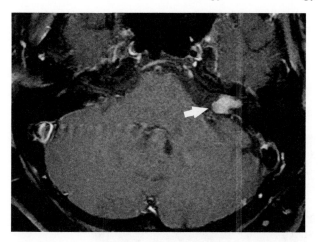

FIGURE 13.41
A small extra meatal component in another patient with vestibular schwannoma which just projects into the cerebellopontine angle cistern (white arrow).

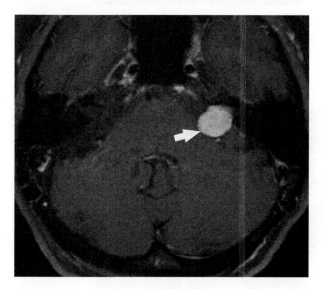

FIGURE 13.42
Left vestibular schwannoma reaches up to the cerebellopontine angle but does not cause any mass effect.

departments. However, coupled with treatment expertise, they allow early detection to be followed, where appropriate, by treatment with low morbidity and mortality [61,73,74]. There are two types of treatments for vestibular schwannoma: microsurgery stereotactic radiosurgery (gamma knife). The aim of the surgery is to completely remove tumor with functional preservation of the facial nerve and where possible, with preservation of hearing. Preservation of labyrinthine signal without associated abnormal enhancement correlates well with hearing preservation following surgery [61]. A variety of factors influence the choice of surgical approach and include preoperative hearing levels in both ears, tumor size and location, patient's age, and individual preference. The main surgical approaches in general use are the retrosigmoid, middle fossa, and translabyrinthine approaches [75]. Stereotactic radiosurgery has emerged as an effective minimally invasive alternative to the surgical removal of small-to-moderate sized vestibular schwannoma [76] (Figure 13.45). Advanced dose planning and optimizing software and introduction of robotics (automated-positioning systems) have improved ability to collimate and focus beam on the target and reduce collateral damage. In some cases, usually elderly or medically infirm patients or in patients with very small tumors, it may be reasonable to observe the tumor for potential growth with annual clinical and MRI assessments instead [61].

Differentiation of residual tumor from scar tissue in the internal auditory canal after vestibular schwannoma resection requires close, long-term follow-up (Figure 13.46). Nodular and progressive enhancements in the internal auditory canal indicate residual tumor. Linear enhancement in the internal auditory canal has been found to be a common finding after vestibular schwannoma resection not associated with residual tumor [77,78] (Figure 13.47). Factors including completeness of resection and baseline postoperative MRI findings provide valuable information regarding risk for recurrence, which may assist the clinician in determining an appropriate postoperative MRI surveillance schedule [79].

13.5.4.2 Meningioma

Meningiomas are the second most frequent tumors at the cerebellopontine angle and internal auditory meatus after vestibular schwanommas. Ten percent of all intracranial meningiomas arise in the posterior fossa. On MRI, meningiomas are isointense or mildly hypointense to gray matter on T_1 images. On T_2 sequences, they are mostly isointense to brain but can be hyperintense or hypointense with mottled or speckled heterogeneity. Most enhance avidly following contrast administration, with a characteristic dural tail (Figure 13.48). Intracannalicular meningiomas arise from the arachnoid

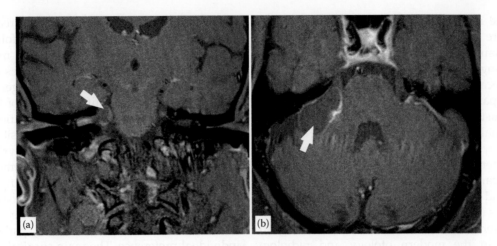

FIGURE 13.43
Coronal (a) and axial (b) T_1 fat-suppressed postcontrast images. Cystic changes within vestibular schwannoma is common and in this example the cystic component is larger than the solid part of the tumor. Cystic changes are also seen in Figure 13.4b.

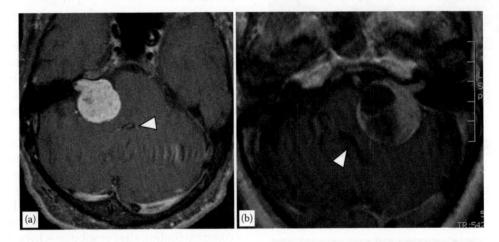

FIGURE 13.44
(a,b) Two different patients with vestibular schwannoma causing significant mass effect on the brain. Note the compression of the fourth ventricle in both patients (white arrowhead).

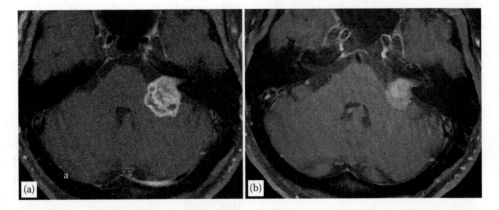

FIGURE 13.45
Patient with a large left vestibular schwannoma (a) treated with stereotactic radiosurgery which show reduction in size (b) of tumor mass.

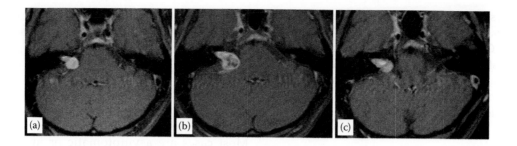

FIGURE 13.46
Cystic changes can occur within vestibular schwannoma following treatment with stereotactic radiosurgery and should not be interpreted as increase in size of the tumor. Treatment details are crucial while reporting posttreatment MRI examinations. In this patient the tumor (a) showed cystic changes following stereotactic radiosurgery (b) which resolved subsequently (c).

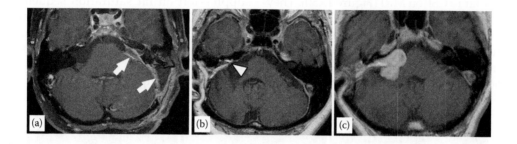

FIGURE 13.47
Post-surgical changes in three different patients. MR imaging immediately following surgery serves as a baseline scans (a). Note post-surgical changes (white arrows). On follow-up imaging linear enhancement (b) should not be misinterpreted as residual tumor which usually has nodular contrast enhancement (c).

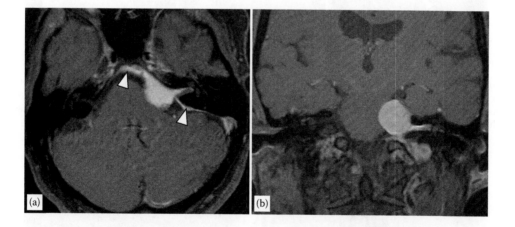

FIGURE 13.48
Cerebellopontine angle mengioma. (a) Axial and (b) coronal postcontrast T_1W images shown an enhancing lesion in the left CPA in a patient presenting with mild ataxia but with no hearing impairment. Appearances are in keeping with a meningioma. Note the dural enhancement (white arrowhead) also referred to as *dural tail* (a).

granulations within the canal and are almost impossible to differentiate from a vestibular schwannoma and the diagnosis is usually made at surgery [80].

13.5.4.3 Epidermoid Cyst

Epidermoid cysts are rare slow growing benign lesions that arise from ectopic inclusion of ectodermal cells

during closure of the neural tube between the third and the fifth weeks of embryonic life [81]. They account for 1% of all intracranial tumors and about 40% of these cysts are located in the cerebellopontine angle, representing the third most frequent lesion in this region [82]. They spread along paths of least resistance and envelop neural and vascular structures, and may cause trigeminal neuralgia, hemifacial spasm, and deafness [83].

Epidermoid cysts appear hypointense relative to brain on T_1W images and show high signal on T_2W images (Figure 13.49). On diffusion-weighted imaging, an epidermoid cyst shows characteristic restricted diffusion (Figure 13.49) and this differentiates it from an arachnoid cyst. It does not enhance following intravenous contrast administration.

13.5.4.4 Arachnoid Cyst

Arachnoid cysts are congenital benign lesions that occur in the intraarachnoid space secondary to the splitting of embryonic meninges. They contain CSF and hence appear similar to CSF on MRI (Figure 13.50). They account for approximately 1% of all intracranial masses mostly in the middle cranial fossa and only about 10% are found in the posterior fossa [84]. They are usually asymptomatic and are incidental findings. Symptoms from an arachnoid cyst are caused by an increase in the osmotic gradient of the liquid content of the cyst and subsequent enlargement of the cyst [85].

13.5.4.5 Lipoma

Intracranial lipomas account for only about 0.3% of all intracranial tumors and rarely occur within the cerebellopontine angle or internal meatus [86,87]. Most cases are asymptomatic or incidental findings but can mimic symptoms associated with a vestibular schwannoma [88]. Lipomas typically appear hyperintense compared to brain tissue on T_1W images and hypointense on T_2W images. They show uniform suppression on fat saturation sequences and do not enhance after the administration of intravenous contrast agent (Figure 13.51) [89]. Surgical treatment is reserved for a small subset of patients with intractable vestibular symptoms and most cases will only need follow-up imaging [90].

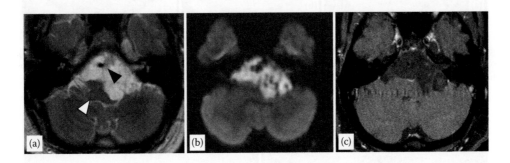

FIGURE 13.49
Cerebellopontine angle epidermoid cyst. MRI in a patient presenting with left trigeminal neuralgia. T_2 axial (a) image shows a large lobulated extra-axial mass in the left CPA and extending into the prepontine cistern. The basilar artery is encased by the lesion (black arrowhead) and there is mass effect on the brain stem (white arrow). There is restricted diffusion (b) and no contrast enhancement of the mass (c). Appearances are in keeping with a large epidermoid cyst.

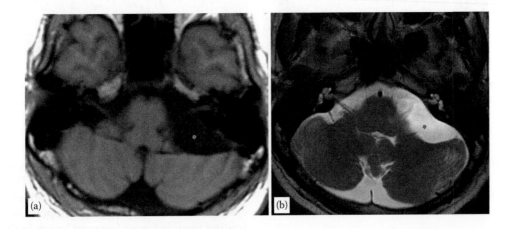

FIGURE 13.50
Cerebellopontine angle arachnoid cyst. Large uniform CSF signal intensity lesion in the left CPA confirms a diagnosis of arachnoid cyst. It is hypointense relative to brain in T_1-weighted sequence (a) and hyperintense on T_2-weighted sequence (b). There was no restricted diffusion or contrast enhancement (images not shown).

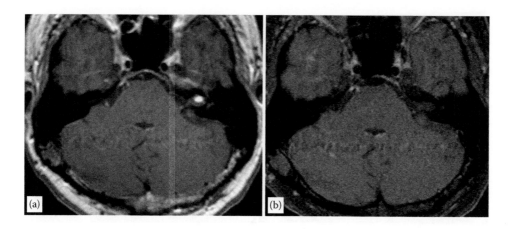

FIGURE 13.51

Lipoma within internal auditory meatus. (a) A small nodular hyperintense lesion on T_1-weighted sequence with the left IAC. (b) It shows uniform fat suppression and no contrast enhancement. This lesion can be erroneously called as a small vestibular schwannoma, if precontrast T_1 and postcontrast fat-suppressed images are not included in the scanning protocol.

13.5.4.6 Metastasis

Metastatic lesions in the cerebellopontine angle and internal auditory meatus are rare and associated with breast or lung carcinomas [91,92]. Leptomeningeal carcinomatosis secondary to adenocarcinoma, leukemia, or lymphoma can also occur (Figure 13.52).

13.5.4.7 Facial Nerve Tumors

Facial nerve schwannoma are uncommon benign tumors of Schwann cells and compose only 0.8% of all petrous bone mass lesions [93]. The clinical presentation varies with its anatomical location and involvement of adjacent structures. Facial nerve dysfunction results from compression rather than invasion of the facial nerve. Hearing loss can be sensorineural, if the lesion

is at the cerebellopontine angle or internal auditory meatus, or conductive, if within the tympanic segment. Other symptoms include hemifacial spasm, vertigo, and otalgia [93–95]. Clinical observation is preferred management strategy until facial function and surgery are considered when facial function deteriorates to House–Brackmann grade IV [96,97].

Facial nerve schwannomas can appear as focal or multisegmental enhancing mass lesions. At the cerebellopontine angle or internal auditory meatus, a facial nerve schwannoma appears very similar on MRI to a vestibular schwannoma, presenting as an enhancing mass. However, additional findings such as extension into and expansion of the adjacent labyrinthine facial nerve canal *tumor tail* and eccentricity of the mass in relation to the porus acusticus are pointers to a diagnosis of the former (Figure 13.53) [94–96]. A geniculate ganglion

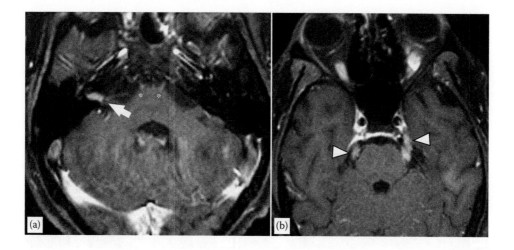

FIGURE 13.52

Meningeal leukaemia. Patient on treatment for acute leukaemia developed bilateral hearing loss which was initially thought to be drug related. Postcontrast MRI images (a and b) shows enhancement of the nerves in the right IAC. Further enhancement of the sixth cranial nerves (asterisks) and the trigeminal nerves (arrowhead) suggests leukemic infiltration.

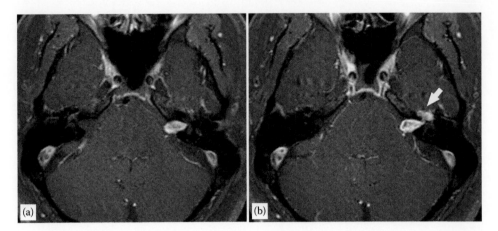

FIGURE 13.53
Right facial nerve schwannoma (a and b). Involvement of the labyrinthine segment (b) of the facial canal (white arrow) helps differentiate facial from VS.

schwannoma can present as a large mass mimicking a meningioma or a cerebral temporal lobe mass [61]. Within the facial nerve canal, they appear as enhancing sausage-shaped expanding long segments.

Other causes of a facial nerve mass include an hemangioma/vascular malformation, an epidermoid cyst/ congenital cholesteatoma and rarely a choristoma or a paraganglioma [61]. Hemangiomas and epidermoid cysts are usually small. The former usually arise at the geniculate ganglion and may contain bone spicules and hence appear hypointense on MRI while the latter demonstrates restricted diffusion and does not enhance with intravenous gadolinium contrast administration. Multisegment facial schwannomas (Figure 13.54)

should be differentiated from perineural spread of a malignant tumor.

13.6 Petrous Apex Lesions

A myriad of entities can involve the petrous apex. Many are incidentally depicted on imaging and are asymptomatic. Because of the deep location of the petrous apex at the skull base, it is crucial to be able to confidently characterize lesions there on imaging, thereby obviating the need of invasive biopsy or surgery for

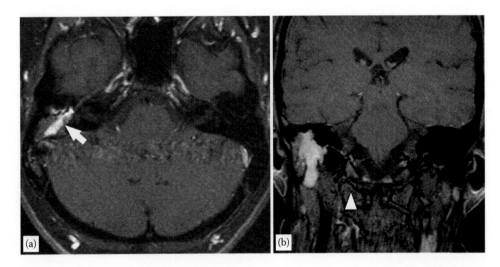

FIGURE 13.54
Facial nerve schwannoma can not only involve the (a) petrous segment (white arrow) but can also involve into (b) the infratemporal segment of the nerve (white arrowhead).

benign lesions and normal variants, the *leave me alone* lesions (Tables 13.1 and 13.2). MRI, supported by diffusion-weighted and contrast-enhanced sequences, and CT can complement each other in evaluating the petrous apex.

13.6.1 Common Normal Variants at the Petrous Apex

Asymmetric pneumatization and asymmetric marrow signal of the petrous apex are common and can be easily misconstrued as pathology when noted in one of the petrous apices by the unsuspecting radiologist (Figure 13.55). The former is a normal anatomical variant that can occur in nearly 10%–30% of the population and there is a positive correlation between the degree of mastoid segment pneumatization and aeration of the

TABLE 13.1

MRI Features of *Leave Me Alone Lesions*

Diagnosis	T_1-MRI	T_2-MRI
Asymmetric marrow	High	Intermediate/high
Trapped fluid	Low	High
Trapped fluid with high proteinaceous content	High	High

TABLE 13.2

MRI Features of Primary Petrous Lesions

Lesion	MRI-T_1	MRI-T_2	Contrast	Expansion
Cholesterol granuloma	High	High	No	Yes
Cholesteatoma	Low	High	No	Yes
Mucocoele	Low	High	No	Yes
Petrous apicitis	Low	High	Yes	No

petrous apex [98]. Where pneumatized, low MR signal is returned on all sequences. Conversely, asymmetric fatty marrow within the petrous apex returns typical high-signal intensity of fat on both T_1- and T_2-weighted images. It can be easily confused as T_1 high signal seen in cholesterol granuloma if fat suppression techniques are not used as part of the diagnostic work up. The asymmetrical high signal in the petrous apex can also be mistaken for a pathological lesion on postcontrast imaging if appropriate precontrast T_1 images are not available for comparison [99].

13.6.2 Trapped Fluid/Effusion of Petrous Apex

Trapped fluid within petrous apex air cells is thought to be the sequela of previous otitis media, which fails to drain due to obstructed communicating channels. It often presents as an incidental-imaging finding in patients with no otological complaints [100]. The fluid tends to follow high T_2 CSF signal on MRI (Figure 13.56) and demonstrates facilitated diffusion, but the T_1-weighted appearances may vary depending on the protein content of the trapped fluid. The T_1 shortening effect leads to an iso- or hyperintense signal on T_1-weighted images and may confound the radiological diagnosis. Hence, close attention to the signal characteristics is critical. The typically seen increased T_1-weighted signal intensity of fatty bone marrow within the nonpneumatized petrous apex can be ruled out due to lack of mass effect and suppression on fat-suppressed sequences, leaving cholesterol granuloma as the only differential diagnostic option. T_1-weighted signal intensity of a cholesterol granuloma is often even higher in signal intensity than the surrounding fatty bone marrow. Diagnostic support for trapped fluid can

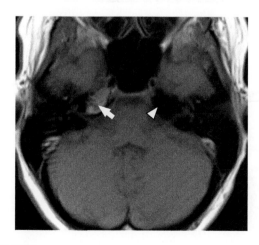

FIGURE 13.55
Asymmetrical pneumatization of the petrous apex makes the fatty right apex (white arrow) conspicuous. Compare this with the pneumatized/sclerotic left apex (arrowhead).

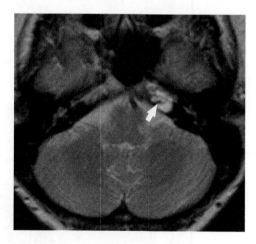

FIGURE 13.56
Petrous apex trapped fluid. High signal in the left petrous apex on T_2-weighted image of the brain (white arrow). CT temporal bone (not shown) did not show any bony changes in the left apex confirming the diagnosis of trapped fluid.

be provided by high-resolution CT, which confirms lack of bone remodeling, cortical disruption, or trabecular erosion [101,102]

13.6.3 Petrous Apex Cephalocele

Petrous apex cephaloceles are uncommon and represent cystic expansion and herniation of the Meckel's cave into the petrous apex. They are thought to represent developmental dehiscence occurring gradually due to a combination of a congenitally thin bone along the roof of the petrous apex and chronic CSF pulsation leading to herniation of the Meckel cave content. A coronal high-resolution T_2-weighted sequence is the best sequence to show the superomedial communication with Meckel cave [103–105] (Figure 13.57).

13.6.4 Congenital Cholesteatoma of Petrous Apex

Congenital cholesteatoma are believed to arise from squamous rests, which forms a cyst lined by stratified squamous epithelium. Internal desquamation of the lining results in the accumulation of keratinized debris,

forming a highly organized structure. Enlargement of the cholesteatoma occurs gradually as the advancing epithelium combined with host inflammatory response results in surrounding bone resorption [100]. In this context, they are analogous to intracranial epidermoids. In keeping with their slow growing nature, congenital cholesteatoma of the petrous apex tends to be asymptomatic. However, progressive expansion and consequent mass effect can cause clinical symptoms such as headaches, SNHL, and cranial nerve palsy, necessitating surgery [101]. In MRI, they appear hypointense on T_1-weighted images, hyperintense on T_2-weighted images, and show characteristic restricted diffusion and subtle peripheral rim enhancement (Figure 13.58) [102].

13.6.5 Petrous Apicitis

Petrous apicitis refers to overt infection of the petrous apex, usually as a complication of otomastoiditis. This is thought to develop only in a pneumatized petrous apex, which communicates with the middle ear cleft [106]. Typically, there is debris and fluid in the petrous apex with destruction of bony septa (analogous to

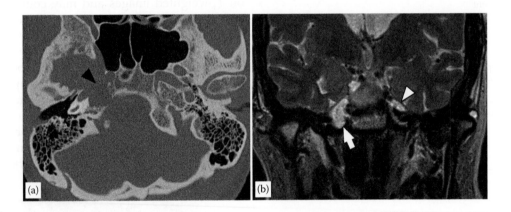

FIGURE 13.57
Petrous apex cephalocele. (a) CT performed for non-otological indication shows a rather aggressive looking lesion in the right petrous apex (black arrowhead). (b) Coronal T_2-weighted MRI confirms herniation of the right Meckel cave (white arrow). White arrowhead shows the contralateral normal left Meckel cave.

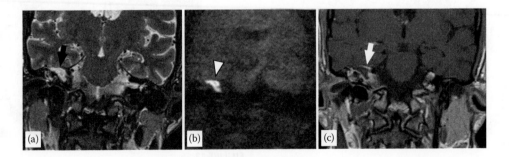

FIGURE 13.58
Petrous apex cholesteatoma. Coronal T_2-weighted (a) MR image in a patient investigated for headache. There is a well-defined focal hyperintense lesion in the right petrous apex (white arrow). This shows restricted diffusion (b) (white arrowhead) and thin peripheral rim of enhancement (c) (arrow).

coalescent mastoiditis), giving the appearance of low T_1W signal and high T_2W signal. Interspersed bony and dural enhancement is also usually seen, and is associated with the inflammation (Figure 13.59). Diffusion imaging can depict a petrous apex abscess [107,108]. Destruction of the bony cortex can result in intracranial complications, which can be better evaluated by MRI. This can cause the classic Gradenigo triad of otomastoiditis, sixth nerve palsy, and pain in the distribution of the fifth nerve, although patients rarely have all three of these symptoms [109,110]. It is important to differentiate this disease process from incidental trapped fluid or effusion of the petrous apex; unlike the former, the latter does not demonstrate enhancement in MRI and bony erosions on CT.

13.6.6 Mucocoele

Petrous apex air cells can become obstructed resulting in the accumulation of mucus within the obstructed

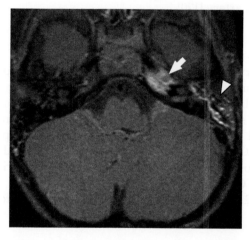

FIGURE 13.59
Petrous apicitis. A young patient presented with headache, retrobulbar pain, and diplopia. Postcontrast MRI shows enhancement of the left petrous apex with a small focal nonenhancing area (white arrow) in keeping with pus. Note the left mastoid disease (arrowhead).

air cell. This results in a well-circumscribed mass with smooth contour changes [111]. With expansion of the bony boundaries, mucocoeles may extend and compress adjacent structures. Patients usually present with localized pain and/or cranial nerve involvement. MRI shows a well-defined mass that is hypointense on T_1-weighted images, hyperintense on T_2-weighted enhance, and do not enhance after administration of contrast although there may be peripheral enhancement because of the inflammatory response (Figure 13.60). However, MRI appearance of petrous apex mucocoele can vary depending on the degree of hydration or inspissation of the contents [112].

13.6.7 Cholesterol Granuloma

Petrous apex cholesterol granulomas are expansile, cystic lesions containing cholesterol crystals. They are thought to be foreign body giant cell reaction to the deposition of cholesterol crystals. If symptomatic, hearing loss, tinnitus, and hemifacial spasm are the most common associated clinical findings. On MRI, cholesterol granulomas are typically hyperintense on both T_1- and T_2-weighted sequences and can demonstrated restricted diffusion (Figure 13.61). This is a reflection of accumulated blood products and proteinaceous debris. Small lesions may be relatively homogeneous but larger lesions can show heterogeneity [113]. They often have a distinct hypointense peripheral rim on T_2-weighted images due to hemosiderin deposition. As a rule, cholesterol granulomas do not enhance following contrast administration but subtle enhancement of the periphery of the lesion has been reported supposedly secondary to inflammation [114,115].

13.6.8 Petrous Apex Vascular Lesions

Petrous carotid aneurysms are rare and usually asymptomatic. Patients can present with hearing loss, pulsatile tinnitus, headaches, diplopia, and possibly life-threatening hemorrhage [116,117]. They are thought

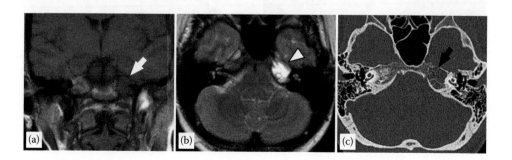

FIGURE 13.60
Petrous apex mucocele. Low signal lesion on T_1-weighted image (a) that is of high signal on T_2-weighted image (b) in a patient presenting with headache and left sixth cranial nerve palsy. Corresponding CT shows an expansile smoothly margined lesion (c).

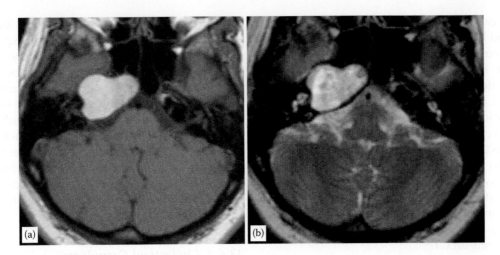

FIGURE 13.61
Petrous apex cholesterol granuloma. (a) axial T_1W and (b) T_2W images show a hyperintense lesion of the right petrous apex has the characteristic appearance of a cholesterol granuloma.

to arise because of developmental weakness of the arterial wall. However, trauma to the internal carotid artery is another major cause. The MR imaging appearance of pseudoaneurysm may mimic mucocoele or cholesterol granulomas [118]. MR imaging may show a mass of mixed signal intensity and flow voids with irregular enhancement following intravenous contrast administration (Figure 13.62). CT can show bony expansion of the petrous carotid canal. CT angiography is the best study to diagnose petrous internal carotid artery aneurysms.

Dural arteriovenous fistulas are abnormal shunts between meningeal vessels and the dural sinuses [119]. In the petrous apex, they are predominantly supplied by meningeal branches of the external and internal carotid artery [120].

13.6.9 Tumors of the Petrous Apex

Primary bony tumors of the petrous apex are uncommon and include chondroscarcomas and chordomas. Petroclival chondrosarcoma tend to occur off midline along synchondroses, centered along the petrosphenoidal and petrooccipital fissures within the petrous apex. On MR imaging, they are characteristically iso- to hypointense on T_1-weighted images, hyperintense on T_2-weighted images, and usually enhance avidly [121,122]. Chordomas, which arise from notochord

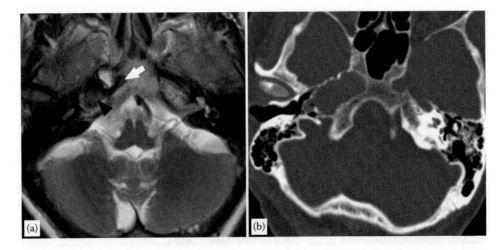

FIGURE 13.62
Right petrous internal carotid artery aneurysm shown here as a large flow void on (a) T_2 axial image (black arrowhead) and is continuous with the ICA (white arrow). (b) CT in the same patient shows the right carotid canal (black asterisk) continuous with the expansile aneurysm.

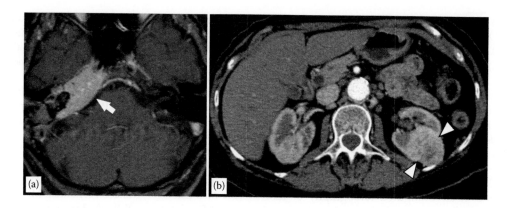

FIGURE 13.63
Metastatic renal carcinoma at the petrous apex. Clinical presentation of right hearing loss and tinnitus with a destructive lesion at the right petrous apex on CT. (a) MRI with contrast shows an enhancing petrous apex lesion (white arrow). (b) CT abdomen shows primary tumor at the left kidney (white arrowhead).

remnants, typically occur in midline, and may contain calcifications. Petrous apex involvement would then occur secondarily as the midline tumor invades superiorly and laterally [123,124]. The petrous apex can also be invaded by inner ear and middle ear malignancies such as glomus tumors and endolymphatic sac tumors.

Petrous apex is the most common site for metastases in the temporal bone and common tumors to metastasize to the petrous apex are breast, lung, prostate, and renal cancers [125,126] (Figure 13.63). Metastatic tumor involvement of the petrous apex may be due to hematogenous spread from distant tumors, direct extension of an extra- or intracranial tumor, or leptomeningeal extension of a distant or intracranial tumor [127]. Susceptibility of the petrous apex to hematogenous metastases is thought to be due to slow blood flow through the petrous apex marrow, which allows filtering and deposition of tumor cells [128].

References

1. Curtin HD, Sanelli P, and Som PM. Temporal bone: Embryology and anatomy. In: Som PM, Curtin HD, eds. *Head and Neck Imaging*. St. Louis, MO: Mosby; 2003: pp. 1057–1091.
2. Lacout A, Marsot-Dupuch K, Smoker WR, and Lasjaunias P. Foramen tympanicum, or foramen of Huschke: Pathologic cases and anatomic CT study. *American Journal of Neuroradiology* 2005; 26(6):1317–23.
3. Majithia A, Lingam RK, Nash R, Khemani S, Kalan A, and Singh A. Staging primary middle ear cholesteatoma with non-echoplanar (half-Fourier-acquisition single-shot turbo-spin-echo) diffusion-weighted magnetic resonance imaging helps plan surgery in 22 patients: Our experience. *Clinical Otolaryngology* 2012; 37(4):325–30.
4. Valvassori GE. Imaging of the temporal bone. In: Mafee MF, Valvassori, GE, and Becker, M, eds. *Imaging of the Head and Neck*. Stuttgart, Germany; New York: Thieme; 2005: pp. 3–133.
5. Swartz JD, and Loevner LA, eds. *Imaging of the Temporal Bone*, Thieme, 2009: pp. 299–302.
6. Harnsberger HR et al. Diagnostic and surgical imaging anatomy: Brain, head & neck, spine. *American Journal of Neuroradiology* 2007; 28(4):795.
7. Shah LM, and Wiggins RH III. Imaging of hearing loss. *Neuroimaging Clinics of North America* 2009; 19(3):287–289.
8. Pyykkö I, Zou J, Poe D, Nakashima T, and Naganawa S. Magnetic resonance imaging of the inner ear in Meniere's disease. *Otolaryngologic Clinics of North America*. 2010; 43(5):1059–80.
9. Abele TA, and Wiggins RH, Imaging of the temporal bone. *Radiologic Clinics of North America* 2015; 53(1):15–18.
10. Davidson HC. Imaging of the temporal bone. *Neuroimaging Clinics of North America* 2004; 14(4):721–725.
11. Mitsuoka H et al. Microanatomy of the cerebellopontine angle and internal auditory canal: Study with new magnetic resonance imaging technique using three-dimensional fast spin echo. *Neurosurgery* 1999; 44(3):561–566.
12. Arnold B, Jäger L, and Grevers G. Visualisation of inner ear structures by three-dimensional high-resolution magnetic resonance imaging. *Otology & Neurotology* 1996; 17(3):480–485.
13. Held, P et al. MRI of inner ear anatomy using 3D MP-RAGE and 3D CISS sequences. *The British Journal of Radiology* 1997; 70(833):465–472.
14. Kretschmann HJ, and Weinrich W. *Cranial Neuroimaging and Clinical Neuroanatomy* (2nd edition). Stuttgart, Germany; New York: Thieme; 1992.
15. Phillips CD, and Bubash LA. The facial nerve: Anatomy and common pathology. *Seminars in Ultrasound, CT and MRI* 2002; 23:202–205.
16. Jäger L, and Reiser M. CT and MR imaging of the normal and pathologic conditions of the facial nerve. *European Journal of Radiology* 2001; 40(2):133–135.

17. Hong HS, Yi B-H, Cha J-G, Park S-J, Kim DH, Lee HK, and Lee J-D. Enhancement pattern of the normal facial nerve at 3.0 T temporal MRI. *British Journal of Radiology.* 2010; 83(986):118–121.

18. Connor SEJ, Leung R, and Natas S. Imaging of the petrous apex: A pictorial review. *Imaging* 2008; 81(965):427–435.

19. Razek AA, and Huang BY. Lesions of the petrous apex: Classification and findings at CT and MR imaging. *RadioGraphics* 2011; 32(1):151–173.

20. Sennaroglu L, and Slattery WH. Petrous anatomy for middle fossa approach. *The Laryngoscope* 2003; 113(2):332–342.

21. Virapongse C et al. Computed tomography of temporal bone pneumatization: 1. Normal pattern and morphology. *American Journal of Roentgenology* 1985; 145:473–48.

22. Mundada P, Singh A, and Lingam RK. CT arteriography and venography in the evaluation of pulsatile tinnitus with normal otoscopic examination. *Laryngoscope* 2015 Apr;125(4):979–984.

23. Shweel M, and Hamdy B. Diagnostic utility of magnetic resonance imaging and magnetic resonance angiography in the radiological evaluation of pulsatile tinnitus. *American Journal of Otolaryngology.* 2013; 34: 710–717.

24. Romo LV, Casselman JW, and Robson CD. Temporal bone: Congenital anomalies. In: Som PM, Curtin HD, eds. *Head and Neck Imaging.* 4th ed. Vol 2. St Louis, MO: Mosby, 2003: pp. 1119–1140.

25. Peck JE. Development of hearing. Part I: Phylogeny. *Journal of the American Academy of Audiology* 1994; 5(5):291–299.

26. Jackler RK, and Dillon WP. Computed tomography and magnetic resonance imaging of the inner ear. *Otolaryngology—Head and Neck Surgery* 1988; 99(5): 494–504.

27. Marangos N, and Aschendorff A. Congenital deformities of the inner ear: Classification and aspects regarding cochlear implant surgery. 1997; 52:52–56.

28. Sennaroglu L, and Saatci I. A new classification for cochleovestibular malformations. *Laryngoscope* 2002; 112:2230–2241.

29. Peck JE. Development of hearing. Part II. Embryology. *Journal of the American Academy of Audiology* 1994; 5(6):359–365.

30. Joshi VM et al. CT and MR imaging of the inner ear and brain in children with congenital sensorineural hearing loss. *RadioGraphics* 2012; 32(3):683–685.

31. Jackler RK, Luxfor WM, and House WF. Congenital malformations of the inner ear: A classification based on embryogenesis. *The Laryngoscope* 1987; 97(S40):2–14.

32. Grandis RJ, Branstetter BF, and Yu VL. The changing face of malignant (necrotizing) external otitis: Clinical, radiological and anatomic correlation. *Lancet Infectious Disease* 2004; 4:34–39.

33. Mehrotra P, Elbadawey MR, and Zammit-Maempel I. Spectrum of radiological appearances of necrotising external otitis: A pictorial review. *The Journal of Laryngology and Otology* 2011; 125:1109–1115.

34. Al-Noury K, and Lotfy A. Computed tomography and magnetic resonance imaging findings before and after treatment of patients with malignant external otitis. *European Archives of Otorhinolaryngology* 2011; 268:1727–1734.

35. Grandis JR, Curtin HD, and Yu VL. Necrotizing(malignant) external otitis: Prospective comparison of CT and MR imaging in diagnosis and follow-up. *Radiology* 1995; 196:499–504.

36. Gherini SG, Brackmann DE, and Bradley WG. Magnetic resonance imaging and computerized tomography in malignant external otitis. *Laryngoscope* 1986; 96:542–48.

37. Mas-Estelles F, Mateos-Fernandez M, Carrascosa-Bisquet B, Facal de Casro F, Puchades-Roman I, and Morera-Perez C. Contemporary non-echoplanar diffusion-weighted imaging of middle ear cholesteatomas. *RadioGraphics* 2012; 32:1197–1213.

38. Khemani S, Singh A, Lingam RK, and Kalan A. Imaging of postoperative middle ear cholesteatoma. *Clinical Radiology* 2011; 66(8):760–767.

39. Majithia A, Lingam RK, Nash R, Khemani S, Kalan A, and Singh A. Staging primary middle ear cholesteatoma with non-echoplanar (half-Fourier-acquisition single-shot turbo-spin-echo) diffusion-weighted magnetic resonance imaging helps plan surgery in 22 patients: Our experience. *Clinical Otolaryngology* 2012; 37(4): 325–30.

40. Thiriat S, Riehm S, Kremer S, Martin E, and Veillon F. Apparent diffusion coefficient values of middle ear cholesteatoma differ from abscess and cholesteatoma admixed infection. *American Journal of Neuroradiology* 2009; 30(6):1123–6.

41. De Foer B et al. Middle ear cholesteatoma: Non-echoplanar diffusion-weighted MR imaging versus delayed gadolinium-enhanced T1-weighted MR imaging-value in detection. *Radiology* 2010; 255(3):866–872.

42. Li PM, Linos E, Gurgel RK, Fischbein NJ, and Blevins NH. Evaluating the utility of non-echoplanar diffusion-weighted imaging in the preoperative evaluation of cholesteatoma: A meta-analysis. *Laryngoscope* 2013; 123(5):1247–1250.

43. Huins CT, Singh A, Lingam RK, and Kalan A. Detecting cholesteatoma with non-echo planar (HASTE) diffusion-weighted magnetic resonance imaging. *Otolaryngology— Head and Neck Surgery* 2010; 143:141–6.

44. Lingam RK, Khatri P, Hughes J, and Singh A. Apparent diffusion coefficients for detection of postoperative middle ear cholesteatoma on non-echoplanar diffusion weighted images. *Radiology* 2013(Nov); 269(2):504–510.

45. Khemani S, Lingam RK, Kalan A, and Singh A. The value of non-echo planar HASTE diffusion-weighted MR imaging in the detection, localisation and prediction of extent of postoperative cholesteatoma. *Clinical Otolaryngology* 2011; 36(4):306–12.

46. Nemzek WR, and Swartz JD. Temporal Bone: Inflammatory Disease. In: Som PM, Curtin HD, eds. *Head and Neck Imaging.* St. Louis, MO: Mosby; 2003: pp. 1205–1229.

47. Koomen I, Grobbee DE, Roord JJ, Donders R, Jennekens-Schinkel A, and van Furth AM. Hearing loss at school age in survivors of bacterial meningitis: Assessment, incidence, and prediction. *Pediatrics* 2003; 112: 1049–53.

48. Fortnum HM. Hearing impairment after bacterial meningitis: A review. *Archive of Disease in Childhood* 1992; 67(9):1128–33.

49. Isaacson B, Booth T, Kutz JW Jr, Lee KH, and Roland PS. Labyrinthitis ossificans: How accurate is MRI in predicting cochlear obstruction? *Otolaryngology—Head Neck and Surgery* 2009; 140:692–6.

50. Van Loon MC, Hensen EF, de Foer B, Smit CF, Witte B, and Merkus P. Magnetic resonance imaging in the evaluation of patients with sensorineural hearing loss caused by meningitis: Implications for cochlear implantation. *Otology & Neurotology* 2013; 34(5):845–54.

51. Durisin M et al. Cochlear osteoneogenesis after meningitis in cochlear implant patients: A retrospective analysis. *Otology & Neurotology* 2010; 31:1072–8.

52. Gilden DH. Bell's palsy. *New England Journal of Medicine* 2004; 351(13):1323–1331.

53. Peitersen E. Bell's palsy: The spontaneous course of 2,500 peripheral facial nerve palsies of different etiologies. *Acta Oto-Laryngologica* 2002; 122(7):4–30.

54. Saatci I et al. MRI of the facial nerve in idiopathic facial palsy. *European Radiology* 1996; 6(5):631–636.

55. Saremi F et al. MRI of cranial nerve enhancement. *American Journal of Roentgenology* 2005; 185(6):1487–1497.

56. Martin-Duverneuil N et al. Contrast enhancement of the facial nerve on MRI: Normal or pathological? *Neuroradiology* 1999; 39:207–212

57. Iwasaki H et al. Vestibular and cochlear neuritis in patients with Ramsay Hunt syndrome: A Gd-enhanced MRI study. *Acta Oto-Laryngologica* 2013; 133(4):373–377.

58. Lutz J, and L Jäger. Inflammatory lesions of the brainstem and the cerebellopontine angle. *Der Radiologe* 2006; 46(3):205–215.

59. Bonneville F et al. Unusual lesions of the cerebellopontine angle: A segmental approach 1. *RadioGraphics* 2001; 21(2):419–438.

60. Smith JK, Matheus MG, and Castillo M. Imaging manifestations of neurosarcoidosis. *American Journal of Roentgenology* 2004; 182(2):289–295.

61. Maya MM, Lo WWM, and Kovanlikaya, I. Temporal bone tumors and cerebellopontine angle lesions. In: Som PM, and Curtin HD, eds. *Head and Neck Imaging*. St. Louis, MO: Mosby; 2003: pp. 1275–1360.

62. De Foer B, Kenis C, Vercruysse, JP, Somers T, Pouillon M, Offeciers E, and Casselman JW. Imaging of temporal bone tumors. *Neuroimaging Cliniccs of North America* 2009; 19:339–366.

63. Vogl TJ, Juergens M, Balzer JO, Mack MG, Bergman C, Grevers G, Lissner J, and Felix R. Glomus tumors of the skull base: Combined use of MR angiography and spin-echo imaging. *Radiology* 1994; 192(1):103–10.

64. Vogl TJ, Mack MG, Juergens M, Bergman C, Grevers G, Jacobsen TF, Lissner J, and Felix R. Skull base tumors: Gadodiamide injection-enhanced MR imaging-drop-out effect in the early enhancement pattern of paragangliomas versus different tumors. *Radiology* 1993; 188(2):339–46.

65. Salzman KL et al. Intralabyrinthine schwannoma: Imaging diagnosis and classification. *American Journal of Neuroradiology* 2012; 33(1):104–109.

66. Zhu AF, and McKinnon BJ. Transcanal surgical excision of an intracochlear schwannoma. *American Journal of Otolaryngology* 2012; 33(6):779–781.

67. Moffat DA, Hardy DG, and Baguley DM. Strategy and benefits of acoustic neuroma searching. *The Journal of Laryngology and Otology* 1989; 103(01):51–59.

68. Gal TJ, Shinn J, and Huang B. Current epidemiology and management trends in acoustic neuroma. *Otolaryngology—Head and Neck Surgery* 2010; 142(5):677–81.

69. Tos M et al. What is the real incidence of vestibular schwannoma? *Archives of Otolaryngology—Head and Neck Surgery* 2004; 130(2):216–220.

70. Pilch BZ, ed. *Head and Neck Surgical Pathology*. Lippincott Williams & Wilkins, 2001; pp. 72–73

71. Thompson LDR. Head and neck pathology: A volume in the series: Foundations in diagnostic pathology. *Elsevier Health Sciences* 2012; 17(2)

72. Abele TA et al. Diagnostic accuracy of screening MR imaging using unenhanced axial CISS and coronal T2WI for detection of small internal auditory canal lesions. *American Journal of Neuroradiology* 2014; 35(12):2366–2370.

73. Murphy MR, and Selesnick SH. Cost-effective diagnosis of acoustic neuromas: A philosophical, macroeconomic, and technological decision. *Otolaryngology—Head and Neck Surgery*; 2002; 127:253–259.

74. Hoistad DL et al. Update on conservative management of acoustic neuroma. *Otology & Neurotology* 2001; 22(5):682–685.

75. Silk PS, Lane JI, and Driscoll CL. Surgical approaches to vestibular schwannomas: What the radiologist needs to know 1. *RadioGraphics* 2009; 29(7):1955–1970.

76. Leksell L. *Stereotaxis and Radiosurgery: An Operative System*. Springfield, IL: Charles C Thomas; 1971.

77. Brors D et al. Postoperative magnetic resonance imaging findings after transtemporal and translabyrinthine vestibular schwannoma resection. *The Laryngoscope* 2003; 113(3):420–426.

78. Carlson ML et al. Magnetic resonance imaging surveillance following vestibular schwannoma resection. *The Laryngoscope* 2012; 122(2):378–388.

79. Carlson ML et al. Nodular enhancement within the internal auditory canal following retrosigmoid vestibular schwannoma resection: A unique radiological pattern: Clinical article. *Journal of Neurosurgery* 2011; 115(4):835–841.

80. Samii M and Gerganov V. *Surgery of Cerebellopontine Lesions*. Springer; 2013: pp. 417–418.

81. David EA, and Chen JM. Posterior fossa epidermoid cyst. *Otology & Neurotology* 2003; 24(4):699–700.

82. Di Giustino F et al. Cerebellopontine angle epidermoid cyst: Case Report. *International Journal of Otolaryngology—Head and Neck Surgery* 2013; 2:5.

83. Son DW, Choi CH, and Cha SH. Epidermoid tumors in the cerebellopontine angle presenting with trigeminal neuralgia. *Journal of Korean Neurosurgical Society* 2010; 47(4):271–277.

84. Helland CA, Morten Lund-Johansen, and Knut Wester. Location, sidedness, and sex distribution of intracranial arachnoid cysts in a population-based sample: Clinical article. *Journal of Neurosurgery* 2010; 113(5):934–939.

85. Samii M et al. Arachnoid cysts of the posterior fossa. *Surgical Neurology* 1999; 51(4):376–382.

86. Bacciu A et al. Lipomas of the internal auditory canal and cerebellopontine angle. *Annals of Otology, Rhinology and Laryngology* 2014; 123(1):58–64.

87. White JR et al. Lipomas of the cerebellopontine angle and internal auditory canal. *The Laryngoscope* 2013; 123(6):1531–1536.

88. Mukherjee P, Street I, and Irving RM. Intracranial lipomas affecting the cerebellopontine angle and internal auditory canal: A case series. *Otology & Neurotology* 2011; 32(4):670–675.

89. Méndez J et al. Lipoma of the internal auditory canal: MR findings. *European Radiology* 2002; 12(3):703–704.

90. Maiuri S, Cirillo S, Simonetti L et al. Intracranial lipomas: Diagnostic and therapeutic considerations. *Journal of Neurosurgical Science* 1998; 32:161–167.

91. Chiong Y et al. Isolated metastasis to the cerebellopontine angle secondary to breast cancer. *Canadian Journal of Surgery* 2009; 52(5):E213.

92. Huang T-W, and Young Y-H. Differentiation between cerebellopontine angle tumors in cancer patients. *Otology & Neurotology* 2002; 23(6):975–979.

93. Kirazli T et al. Facial nerve neuroma: Clinical, diagnostic, and surgical features. *Skull Base* 2004; 14(2):115.
    Wiggins RH et al. The many faces of facial nerve schwannoma. *American Journal of Neuroradiology* 2006; 27(3):694–699.

94. Thompson AL et al. Magnetic resonance imaging of facial nerve schwannoma. *The Laryngoscope* 2009; 119(12):2428–2436.

95. Martin-Duverneuil N, Behin A, and Chiras J. Imaging of facial nerve pathology. *Radiology of the Petrous Bone*. Springer, Berlin, Heidelberg; 2004: pp. 181–190.

96. Günther M et al. Surgical treatment of patients with facial neuromas—A report of 26 consecutive operations. *Otology & Neurotology* 2010; 31(9):1493–1497.

97. Andrea B et al. Intraoperatively diagnosed cerebellopontine angle facial nerve schwannoma: How to deal with it. *Annals of Otology, Rhinology and Laryngology* 2014; 123:647–653.

98. Isaacson B, Kutz JW, and Roland PS. Lesions of the petrous apex: Diagnosis and management. *Otolaryngologic Clinics of North America* 2007; 40(3):479–519.

99. Moore KR et al. 'Leave me alone' lesions of the petrous apex. *American Journal of Neuroradiology* 1998; 19(4):733–738.

100. Profant M, and Steno J. Petrous apex cholesteatoma. *Acta Oto-Laryngologica* 2000; 120(2):164–167.

101. Atlas MD, Moffat David A, and Hardy David G. Petrous apex cholesteatoma: Diagnostic and treatment dilemmas. *The Laryngoscope* 1992; 102(12):1363–1368.

102. Magliulo G. Petrous bone cholesteatoma: Clinical longitudinal study. *European Archives of Oto-Rhino-Laryngology* 2007; 264(2):115–120.

103. Moore KR et al. Petrous apex cephaloceles. *American Journal of Neuroradiology* 2001; 22(10):1867–1871.

104. Isaacson B et al. Invasive cerebrospinal fluid cysts and cephaloceles of the petrous apex. *Otology & Neurotology* 2006; 27(8):1131–1141.

105. Lin BM, Aygun N, and Agrawal Y. Cystic lesions of the petrous apex: Identification based on magnetic resonance imaging characteristics. *Otology & Neurotology* 2012; 33(9):75–76.

106. Gadre AK et al. Venous channels of the petrous apex: Their presence and clinical. *Otolaryngology—Head and Neck Surgery* 1997; 116(2):168–174.

107. Ibrahim M, Shah G, and Parmar H. Diffusion-weighted MRI identifies petrous apex abscess in Gradenigo syndrome. *Journal of Neuro-Ophthalmology* 2010; 30(1):34–36.

108. Lee YH et al. CT, MRI and gallium SPECT in the diagnosis and treatment of petrous apicitis presenting as multiple cranial neuropathies. *British Journal of Radiology* 2005; 78:948–951.

109. Chole RA, and Donald PJ. Petrous apicitis. Clinical considerations. *The Annals of Otology, Rhinology, and Laryngology* 1982; 92(6 Pt 1):544–551.

110. Kohan D, Heman-Ackah S, and Chandrasekhar S. *Neurotology*. Oxford University Press; 2014.

111. Memis A et al. Petrous apex mucocele: High resolution CT. *Neuroradiology* 1994; 36(8):632–633.

112. Larson TL, and Wong ML. Primary mucocele of the petrous apex: MR appearance. *American Journal of Neuroradiology* 1992; 13(1):203–204.

113. Jackler RK, and Cho M. A new theory to explain the genesis of petrous apex cholesterol granuloma. *Otology & Neurotology* 2003; 24(1):96–106.

114. Royer MC, and Pensak ML. Cholesterol granulomas. *Current Opinion in Otolaryngology and Head and Neck Surgery* 2007; 15(5):319–322.

115. Connor, SEJ, Leung R, and Natas S. Imaging of the petrous apex: A pictorial review. *The British Journal of Radiology* 2008; 81(965):427–435.

116. Liu JK et al. Aneurysms of the petrous internal carotid artery: Anatomy, origins, and treatment. *Neurosurgical Focus* 2004; 17(5):1–9.

117. Moonis G et al. Otologic manifestations of petrous carotid aneurysms. *American Journal of Neuroradiology* 2005; 26(6):1324–1327.

118. Jackler RK, and Parker DA. Radiographic differential diagnosis of petrous apex lesions. *American Journal of Otology* 1992; 13(6): 561–74.

119. McDougall CG, Halbach VV, Dowd CF, Higashida RT, Larsen DW, and Hieshima GB. Dural arteriovenous fistulas of the marginal sinus. *American Journal of Neuroradiology* 1997; 18(8):1565–1572.

120. Jung C et al. Intraosseous cranial dural arteriovenous fistula treated with transvenous embolisation. *American Journal of Neuroradiology* 2009; 30(6):1173–1177.

121. Andrew ER et al. Chondrosarcoma of the base of the skull: A clinicopathologic study of 200 cases with emphasis on its distinction from chordoma. *The American Journal of Surgical Pathology* 1999; 23(11):1370.

122. Neff B et al. Chondrosarcoma of the skull base. *The Laryngoscope* 2002; 112(1):134–139.

123. Erdem E et al. Comprehensive review of intracranial chordoma 1. *RadioGraphics* 2003; 23(4): 995–1009.

124. Gehanne C et al. Skull base chordoma: CT and MRI features. *Journal Belge De Radiologie* 2005; 88(6):325.

125. Isaacson B, Kutz JW, and Roland PS. Lesions of the petrous apex: Diagnosis and management. *Otolaryngologic Clinics of North America* 2007; 40(3):479–519.

126. Pontious MB, Kim YS, and Backous DD. Metastasis to the petrous apex: A report of an uncommon case. *Otolaryngology—Head and Neck Surgery* 2003; 129(6):751–753.

127. Cureoglu S et al. Otologic manifestations of metastatic tumours to the temporal bone. *Acta Oto-Laryngologica* 2004; 124(10):1117–1123.

128. Razek AA, and Huang BY. Lesions of the petrous apex: Classification and findings at CT and MR imaging. *RadioGraphics* 2011; 32(1):151–173.

14

Differential Diagnosis in Magnetic Resonance Imaging of the Ear

Gabriele A. Krombach

CONTENTS

14.1 Introduction

Magnetic resonance imaging (MRI) plays a major role for diagnosis in otolaryngology. Technical advances in MRI in the last two decades allow for submillimeter resolution, which enables delineation of the inner ear, diagnosis of vestibular schwannomas measuring few millimeters in diameter, and assessment of enhancement pattern of the labyrinth in patients with suspected labyrinthitis [1]. Since computed tomography (CT) does

not allow differentiation between fluid and soft tissue within the labyrinth, due to the small differences in density compared to the surrounding dense bone, MRI is of major advantage for detailed assessment of fibrosis or small tumors within the labyrinth.

Currently, many indications for imaging patients with pathologies of the middle ear are accepted in clinical routine, such as tumors, otitis media with complications, or vascular variants [2]. The external auditory canal is easily accessible for clinical examination. In patients with malignant external otitis or squamous cell carcinoma, MRI might be necessary for delineation of the extent [3,4]. Planning of therapy including the surgical access path requires the presentation of the positional relationship between pathology and anatomical structures [5–7]. The radiologist requires disclosure of results obtained from the history of the patient and the clinical examination. From several possible differential diagnoses for diseases of the ear, the correct diagnosis can often confidently be provided after considering clinical presentation, location of a pathology, signal intensity on different weightings, and contrast enhancement pattern. This chapter systematically presents differential diagnosis for the different entities and provides the reader with typical imaging characteristics that allow selecting the most probable diagnosis from a list of possibilities.

bone. MRI might be applied in patients with deep infiltration, if extent of the lesion into the intracranial space is suspected from CT. In such cases, a protocol, similar to that of imaging the middle ear and posterior fossa, is applied.

MRI is the modality of choice if the fluid-filled inner ear or the temporal bone needs to be imaged. The scan protocol must be adapted to the region and clinical question. Recommended imaging protocols are listed in Table 14.1.

With regard to clinical symptoms, MRI is the modality of choice for the evaluation of patients with vertigo, and objective tinnitus, if the tympanon is normal. If tympanoscopy reveals pathology, CT is the modality of choice and MRI might be performed in those patients, in whom CT is not conclusive and additional information is required.

14.2 Indications

The outer auditory canal is rarely a target for MRI. The normal external auditory canal is not sufficiently visible on MR images, due to the thin lining with skin, which is in direct contact with the periosteum without underpinning of connective tissue or subcutaneous fat and the lack of protons in the compact bone. The outer auditory canal can be assessed by otoscopy and in case of stenosis or surfers ear, CT is the imaging modality of choice. Moreover, tumor spread in cases of small squamous cell carcinoma or extent of tissue inflammation in patients with malignant external otitis is usually indications for CT, which readily delineates erosion of the

14.3 MRI Anatomy of the Ear

The outer third of the external auditory canal consists of fibrocartilaginous tissue and is visible on T_1- and T_2-weighted images (Figure 14.1). The external auditory canal courses slightly backward until the beginning of the middle third, where it curves anteriorly. The middle and inner third are constituted from compact bone, which does not provide signal on MRI, due to lack of free protons. The tympanic membrane is the border toward the middle ear and stretches from superior to inferior forming an angle of approximately 140° with respect to the superior wall of the external auditory canal.

TABLE 14.1

Sequence Recommendation Adapted from Guidelines of the German Society of Radiology [8] and French Society of Radiology [9]

Inner Ear and Cerebellopontine Angle

High-resolution 3D T_2-weighted sequence for delineation of anatomical structures slice orientation axial, slice thickness ≤0.7 mm, matrix 512

T_1-weighted spin-echo (SE) sequence without fat saturation slice orientation axial, slice thickness ≤3 mm, matrix 512 (oversampling, reduced field of view)

Repetition of T_1-weighted sequence after intravenous injection of contrast medium, slice orientation axial and coronal, fat saturation might be applied

Middle Ear/Posterior Scull Base

Fast T_2-weighted SE (turbo SE, TSE) slice orientation coronal and transversal, slice thickness 2–3 mm, matrix 512

T_1-weighted SE slice orientation transversal, slice thickness ≤2 mm, matrix 512

T_1-weighted SE after IV injection of contrast medium slice orientation transversal and coronal, fat saturation might be applied

Additional Sequences

Suspected cholesteatoma or epidermoid cyst

DWI (diffusion-weighted imaging)

Tumor in the jugular foramen, aberrant ICA, persistent stapedial artery

MRA (contrast-enhanced MR angiography)

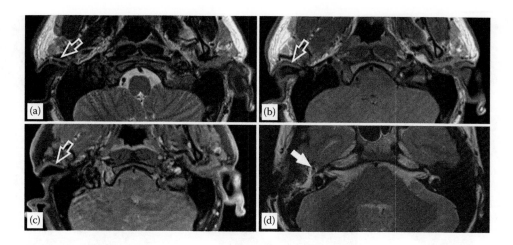

FIGURE 14.1

Outer auditory canal and middle ear (a), T_2-TSE (b), T_1-weighted TSE (c) T_1-weighted after IV injection of contrast medium (Gd-DTPA): The distal third of the outer auditory canal appears as a thin lined canal (open arrow), while the proximal part and the middle ear cannot be assessed in heathy stage. (d) (T_2-weighted TSE) Fluid filling of the middle ear due to malfunction of the Eustachian tube allows for visualization of the middle ear. The ossicles are visible as signal voids within the bright signal arising from the fluid (arrow).

The middle ear is air filled and contains the ossicular chain. Both remain invisible to MRI in its physiological state. The mastoid is situated posterior and inferior to the external auditory canal. Pneumatization of the mastoid is physiologic within a wide range between individuals. The mastoidal cells are lined with a thin layer of mucous membrane and filled with air. Similar to the structures, mentioned so far, the mastoid in its physiologic state cannot be visualized on MRI, since the mucous membrane is too thin for contributing sufficient signal. Fluid within the mastoidal cells in most cases contains only a small percentage of protein and thus usually has high signal intensity on T_2-weighted images. Due to the lack of signal in the immediate adjacent bone, contrast is high and fluid collections within the mastoidal cells and the middle ear will be conspicuous (Figure 14.1d). The presence of fluid in the mastoidal cells on one or both sides is a hint for a non- or malfunctioning Eustachian tube [10], and if accounted for, the middle ear should be assessed for the presence of fluid, too.

The inner ear consists of the bony labyrinth in which the membranous labyrinth is tightly fitted comparable to a hand in a glove. The bony labyrinth is made up of dense bone and the otic capsule is the hardest bone of the human body. The bony labyrinth itself is not visible on MRI in its physiologic state, since it lacks free protons. Only in pathologic states such as otosclerosis, which represents an autoimmune disease in which parts of the bony labyrinth become interspersed with inflammatory tissue, this inflammatory tissue can be assessed on T_1-weighted images and enhances after intravenous injection of contrast medium, if the inflammatory process is still ongoing [11–13]. On CT images, the bony labyrinth is readily assessable due to its dense bone structures, while fluid and solid structures such

as tumors or fibrotic tissue within the membranous labyrinth cannot be differentiated from each other due to the small difference in density, compared to the extremely high density of the ottic capsule [14]. MRI readily differentiates between fluid and soft tissue in the inner ear so that in this regard both modalities are truly complementary.

Since the membranous labyrinth is filled with peri- and endolymph, it provides ideal features for delineation on high-resolution heavily T_2-weighted images (Figures 14.2 and 14.3). The endo- and perilymph-filled systems are closed tubes, so that fluid mixture is not possible. Perilymph is similar to cerebrospinal fluid (CSF), regarding chemical composition [15]. It contains high levels of sodium and low concentrations of potassium. Endolymph resembles intracellular fluid and contains high levels of potassium and low levels of sodium [16]. The endolymph-filled structures, namely the cochlear duct, which contains the organ of Corti, saccule, and utricle within the vestibule, and the semicircular ducts, are positioned within the perilymph-filled system. The later prevents direct transmission of shock waves to the endolymphatic system and plays a major role in conducting sound waves to the sensory epithelium, the organ of Corti. In the cochlea, sound waves are transmitted via the perilymphatic system. If a cross section through a cochlear winding is exposed, three parallel tubes, arranged in a triangular shape, are visible (Figure 14.4). On the bottom, the scala vestibuli, filled with perilymph, in the middle, the endolymph-filled cochlear duct (scala media), and on top, the scala tympani are located. Sound waves are conducted from the tympanic membrane via the ossicular chain to the oval window (Figure 14.5). Here, they are transmitted to the scale vestibule, which is filled with perilymph. Sound

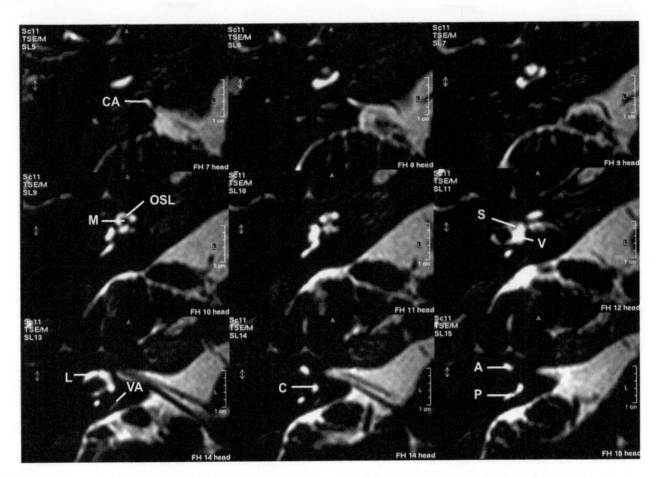

FIGURE 14.2
T_2-weighted TSE, consecutive slices. CA, cochlear aqueduct; OSL, osseous spiral lamina; M, modiolus; V, vestibule; S, sacculus; L, lateral semicircular canal; VA, vestibular aqueduct; C, common crus; A, anterior semicircular canal; P, posterior semicircular canal.

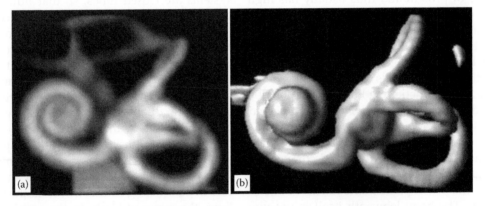

FIGURE 14.3
(a) Maximum intensity projection (MIP) and (b) 3D volume rendered reconstruction of a T_2-weighted TSE dataset. The 2.5 windings are visible on the MIP, while the orientation of the semicircular canals can be appreciated on the 3D reconstruction.

waves are conducted via the scala vestibuli through the windings of the cochlea. According to their frequency, sound waves cause deflection of the Reissner membrane, are thereby transmitted to the cochlear duct, and cause bending of the tectorial membrane, which causes excitation of the cells of the organ of Corti. The location where the sound waves are transmitted depends exactly on the frequency (Figure 14.5). They travel through all windings of the cochlea up to the apex. At the apex of the cochlea, the scala vestibule merges into the scala vestibule, which travels parallel to the scala vestibule and the cochlear duct through the cochlear windings and ends at the round window, a connection to the middle ear. The pressure wave is released here.

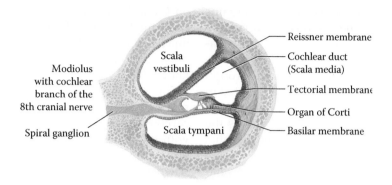

FIGURE 14.4

Schematic cross-sectional view through the cochlea: scala tympani and vestibule both contain perilymph, the cochlear duct contains endolymph.

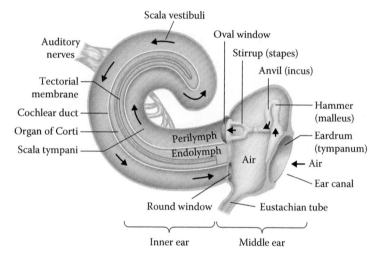

FIGURE 14.5

Schematic drawing of the cochlea. The cochlear aqueduct is a blind sac, located between scala tympany and scala vestibule. Low frequencies are encoded at the basal high frequencies at the apical tune.

The vestibule is located dorsocranial to the cochlea (Figure 14.2). It can easily be recognized on MRI due to its oval shape. The vestibule contains saccule and utricle, which are filled with endolymph. Both are connected with each other via the utriculosaccular duct. The saccule and utricle sense the position of the head within the room and are therefore referred to as the *static labyrinth*. Nerve fibers emerging from the utricle form the superior branch of the vestibular nerve. Nerve fibers arising from the saccule account for the inferior branch of the vestibular nerve. All three semicircular ducts merge into the utricle. The semicircular canals are oriented orthogonal to each other. Each of the semicircular canals forms approximately 240° of a circle. The posterior branch of the superior semicircular canal and the anterior branch of the posterior semicircular canal share one section: the common crus. The hair cells of the semicircular canals are excited by all forms of acceleration and thus constitute the kinetic labyrinth. The hair cells are located within the

ampullae. The ampullae are circumscript broadenings of the semicircular ducts and canals, which are located at the end of the separate branches of the superior and the posterior semicircular canal and in the anterior branch of the lateral semicircular canal, respectively (Figure 14.3). The nerve fibers from the superior and lateral semicircular duct merge into the superior vestibular nerve and the nerve fibers from the posterior semicircular duct into the inferior vestibular nerve. The endolymph-containing semicircular ducts within the semicircular canals measure approximatively one-third of the diameter of the perilymph-conducting outer tube. With the current resolution of heavily T_2-weighted sequences, the two systems cannot be differentiated from each other due to the thin layer separating endolymph and perilymph, regardless of field strength [17]. The cochlear duct also cannot be separated from the perilymphatic system.

The cochlea forms two and a half windings (Figure 14.3). The apical winding is oriented anterior–inferior (Figure 14.2). The cochlear duct contains endolymph. It is placed on the osseous spiral lamina, which merges into the basilar membrane. Both separate the perilymphatic space into the scala vestibule and the scala tympany. The nerve fibers run through the spiral lamina into the modiolus, which is situated in the center of the cochlea (Figure 14.2). The nerve fibers assemble to form the cochlear nerve, which leaves the modiolus at the beginning of the inner auditory canal (Figures 14.6 and 14.7).

The cochlear aqueduct contains perilymph and courses from the scala tympany in the basal winding of the cochlea parallel and inferior to the internal auditory canal to the posterior fossa. The cochlear aqueduct is narrow and is filled with fibrous tissue in a network pattern. Due to this ultrastructure, it is usually only visible on CT in its whole course. In MR imaging, the distal parts might be visible on high-resolution T_2-weighted images (Figure 14.2). In contrast to the vestibular aqueduct, malformations of the cochlear aqueduct are not known [18].

The vestibular aqueduct contains endolymph and plays an important role for equilibrium of the chemical constitution and pressure of the endolymph together with the endolymphatic sac. It courses in the shape of an inverted J from the vestibule to the posterior fossa, where it opens into the endolymphatic sac. The endolymphatic sac is a duplicate of the dura and measures approximately 10–15 mm in wide. The endolymphatic sac is usually not visible. Only in patients with a large vestibular aqueduct/sac syndrome the endolymphatic sac can be visualized using T_2-weighted sequences.

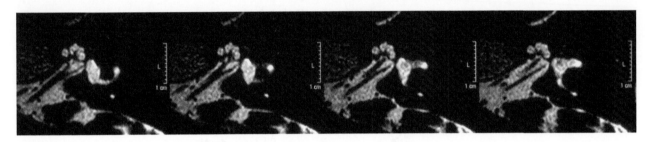

FIGURE 14.6
T_2-TSE, four consecutive slices (left to right) showing the nerve branches within the internal auditory canal. Close to the cochlea, the inferior branch of the vestibular nerve and the cochlear nerve are visible in the inferior part of the internal auditory canal. Since the cochlear is oriented anteriorly, the cochlear nerve is located anterior and the inferior vestibular nerve posterior. In the upper half of the internal auditory canal, the facial nerve (anterior) and the superior vestibular nerve are visible. The superior vestibular nerve is located posterior, according to the position of the vestibule. In the middle third of the internal auditory canal, the inferior and superior vestibular nerve merge to the vestibular nerve and in the proximal part of the internal auditory canal the vestibular nerve and the cochlear nerve merge and form the 8th cranial nerve.

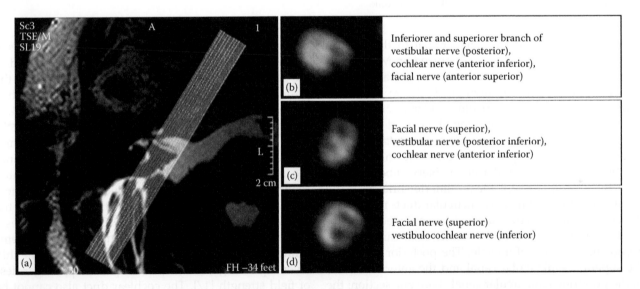

Inferiorer and superiorer branch of vestibular nerve (posterior), cochlear nerve (anterior inferior), facial nerve (anterior superior)

Facial nerve (superior), vestibular nerve (posterior inferior), cochlear nerve (anterior inferior)

Facial nerve (superior) vestibulocochlear nerve (inferior)

FIGURE 14.7
(a) Multiplanar reconstruction (MPR) for delineation of the nerve branches within the internal auditory canal. Slice orientation should be orthogonal to the internal auditory canal and thickness of the reconstructed slices 0.3 mm. (b,c) Three MPR slices through the internal auditory canal: (b) close to the cochlea, (c) middle third of the internal auditory canal, and (d) close to the brain stem.

The internal auditory canal is lined with dura and filled with liquor cerebrospinalis. In the distal third, it contains the superior and inferior branch of the vestibular nerve, the cochlear nerve, and the facial nerve. The two branches of the vestibular nerve merge in the middle third, while the cochlear nerve and the vestibular nerve merge soon thereafter (Figure 14.7). Diameter of the eighth cranial nerve is on average 1.3 mm in adults [19]. The anterior inferior cerebelli artery (AICA) arises from the basilar artery and gives rise to the labyrinthine artery, which courses through the internal auditory canal. The labyrinthine artery supplies cochlea and labyrinth.

14.4 Norm Variations

Norm variants by definition do not cause any symptoms on their own. The main differential diagnoses are malformations, which cause functional impairment of the involved anatomical structures and thus give rise to clinical symptoms. Norm variants must be described in the report, if present, since the deviation from the normal anatomy poses the risk of injury to these structures, if surgery is performed or biopsies are obtained. In the worst case, norm variants can be misinterpreted as pathologies that lead to unnecessary and potentially risk-bearing treatment. One example is diagnosis of a looping artery in the internal auditory canal as vascular conflict (Figure 14.8a). In 25% of people, the AICA forms a loop in the internal auditory canal and is in close contact to the eight cranial nerve [20–24]. In most cases, the vascular loop is symmetric bilateral. In patients with vestibular schwannoma, this common norm variant bears the risk of intraoperative bleeding. Clinically relevant compression of the nerve, on the other hand, most often occur at the entrance or exit points of the nerve (Figure 14.8b) rather than in the middle of the internal auditory canal [25,26]. Circumscript thinning of a nerve at the contact point with an artery is an additional sign for relevant compression. Due to the small caliber of the cranial nerves, which measure on average 1 mm in diameter as well as of the potentially compressing vessels multiplanar reconstruction of isotropic high-resolution T_2-weighted 3D datasets might be required to prove a nerve conflict. In addition, close correlation of imaging findings to the clinical symptoms might allow differentiating between an inconsequential vascular loop and a true compression.

Variable pneumatization of the temporal bone at the petrous apex is a common finding without any pathological relevance [27]. It can be recognized on CT at the first glance. In MRI, the nonpneumatized parts contain fatty marrow and present with high signal intensity on T_1-weighted images as well as on T_2-weighted images, if turbo or fast spin-echo (SE) sequences have been performed (Figure 14.9). The asymmetrical imaging appearance may lead to misinterpretation of the nonpneumatized side as inflammative infiltration or tumor spread [28]. Knowledge of this variant usually prevents such errors.

The jugular bulb is a circumscript dilatation of the jugular vein and is normally located within the jugular foramen. It might be asymmetrically large at one side and extremely small at the other side [29]. In the majority of cases, the right jugular bulb is larger than the left. Similar to asymmetry of the sigmoid sinus, which may be nearly absent at one side and dilated at the contralateral side, this normal variant is insignificant. The highest risks associated with this normal variant are misinterpretation and unnecessary treatment [30]. If blood flow alters the signal at T_1-weighted images, interpretation of jugular bulb asymmetry might be difficult and thrombosis of the jugular vein or schwannoma in the jugular foramen arising from the IX–XII cranial nerves might be considered. In jugular bulb asymmetry, usually the sigmoid sinus and the jugular vein are also asymmetric, a finding that cannot be observed in schwannoma in the jugular foramen. In doubt, time-of-flight sequences or phase contrast angiography might be helpful to demonstrate the vascular nature of the imaging finding and finally exclude thrombus or jugular foramen schwannoma.

FIGURE 14.8
(a) Loop of the AICA within the internal auditory canal (arrow). This finding will be encountered in up to 25% of people and does not have an impact on the nerves. (b,c) Compression of the eighth cranial nerve in a 78-year-old patient. The site of the vascular conflict is close to the brain stem (arrow), a typical finding encountered in patients with compression.

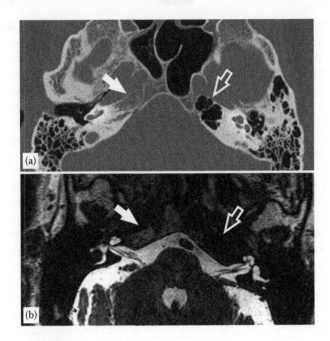

FIGURE 14.9
(a) CT: The right petrous apex contains fat (arrow), while the left petrous apex is pneumonized (open arrow). (b) Accordingly on T_2-TSE the right petrous apex has high signal intensity (arrow), while the left remains signal-free (open arrow).

The jugular bulb usually does not exceed the level of the lateral semicircular canal. A high jugular bulb by definition exceeds the level of the floor of the internal auditory canal. It may or may be not covered by bone (Figure 14.10). The noncovered, dehiscent jugular bulb can be visualized at otoscopy as a *vascular* tumor and the clinical differential diagnosis included paraganglioma of the middle ear. Paraganglioma would be further evaluated by transtympanic biopsy, which would result in bleeding if a dehiscent high jugular bulb would be present instead. Preprocedural MRI needs to differentiate between a real tumor and a vascular variant. Caution must be exercised regarding the important imaging finding of the intact cortical margins of the jugular foramen. In paraganglioma of the jugular foramen, the cortical border is eroded. This finding is not encountered in jugular foramen schwannoma. Both lesions strongly

enhance on T_1-weighted images after intravenous injection of contrast medium. Both lesions demonstrate strong signal on T_2-weighted images.

A jugular bulb diverticulum is a cranially protruding blind sac of the jugular bulb [31]. Cases with pulsatile tinnitus according to this normal variant have been described and are attributed to turbulent flow in the protuberance of the jugular vein [30].

It is important to report all variants of the jugular bulb, since they have to be carefully taken into account during planning of therapy for other entities [32].

The aberrant carotid artery represents another vascular norm variant associated with the appearance of a *highly vascularized tumor* mimicking middle ear paraganglioma at otoscopy [33,34]. This rare variant is due to congenital agenesia of the segment of the internal carotid artery that runs through the vertical portion of the carotid artery canal. The inferior tympanic artery, a branch from the ascendant pharyngeal artery, acts as a collateral so that the internal carotid artery runs through the middle ear (Figure 14.11). The carotid artery enters the middle ear through the inferior tympanic canaliculus, bends anteriorly, and runs over the promontory. It enters the horizontal part of the carotid canal via a dehiscence in the carotid plate [35]. At this reentry point, the vessel might be stenotic, which leads to pulsatile tinnitus. If the transition is not stenotic, this variant remains asymptomatic. At all costs, misinterpretation as a contrast enhancing tumor must be avoided, since biopsy would cause severe bleeding. This variant must be reported, since all invasive therapeutical procedures at the middle or inner ear carry an enormous risk of bleeding. An aberrant internal carotid artery is very difficult to recognize on conventional axial MR images alone, since the bone is not directly visualized as on CT. MR angiography with maximum intensity projection reconstructions shows the unusual course of the vessel, which in most cases is unilateral. In a recent study directly comparing CT and MRI, it was demonstrated that CT had a slightly higher sensitivity for delineation of this variant (100%) than MRA (92%) [36].

The persisting stapedial artery is another rare arterial variant, which is also difficult to assess on conventional

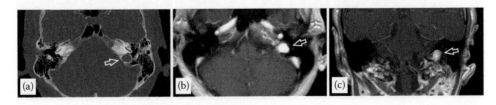

FIGURE 14.10
Unilateral bulbus jugulare diverticulum. (a) axial CT image. The bulbus jugulare diverticulum (arrow) is located behind the cochlear aqueduct. (b) T_1-weighted MRI after injection of contrast medium. (c) Coronal slice.

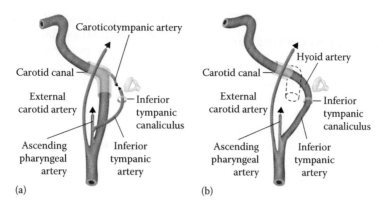

FIGURE 14.11
Schematic drawing of normal carotid artery (a) and aberrant carotid artery, coursing through the middle ear (b). The vertical carotid canal is absent.

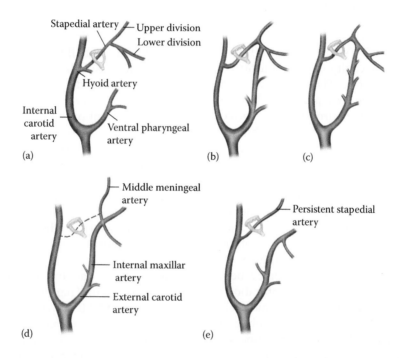

FIGURE 14.12
Persistent stapedial artery, schematic drawing: stages of development. (a) Initially the hyoid artery arises from the internal carotid artery. The stapedial artery is the first branch from the hyoid artery. It passes through the stapes and divides into an upper and lower branch. At this time of development, the ventral pharyngeal artery is present. It later becomes the external carotid artery. (b) Between the ventral pharyngeal artery and the lower branch of the stapedial artery, an anastomosis arises. (c) The stapedial artery shrinks. (d) Normal vascular tree after complete regression of the stapedial artery. (e) Persisting stapedial artery.

MRI [37,38]. Clinical presentation and differential diagnosis resemble those of the aberrant carotid artery. The stapedial artery is usually a transient vessel that persists only up to the third month after gestation. Normally, the development is divided into what later becomes vessels supplying mandibular and maxilla and the middle meningeal artery. If the vessel persists, it enters the middle ear through the stapedial obturator foramen, runs through the middle ear close or through the stapes, and enters the facial nerve canal, which becomes slightly enlarged (Figure 14.12). Then it becomes the middle meningeal artery. Due to its close relation to the stapes, it can cause tinnitus but remains asymptomatic in many cases [39].

Both variants, the aberrant carotid artery and the persistent stapedial artery, have been described in single case reports to block drainage of the middle ear and cause otitis [40,41].

14.5 Congenital Pathologies Malformations

The ear of the adult is a functional unit, composed from highly differentiated and complex mechanical and neuronal structures. The three main components, outer ear, middle ear, and inner ear, develop from three different parts, which arise and mature at different times: the outer ear and the external auditory canal develop from the first branchial cleft and the first and second branchial arches, which belong to the ectoderm. The outer ear develops from 8th week of gestation to the end of the 12th week. At the beginning of the development of the outer ear, a fraction of epithelial cells migrates toward the first pharyngeal pouch and starts to form the outer auditory canal. The outer auditory canal starts to hollow from the side of the middle ear as late as during the sixth and seventh month. Membranous bone growth around the epithelium in this way forms the inner two-third of the external auditory canal. The structures of the middle ear, including the tympanic cavity and the Eustachian tube, develop from endoderm. They arise from an out pocket of the foregut, the first pharyngeal pouch. The inner ear develops from neuroectoderm, located between the first branchial grove and the hindbrain. It develops as the first of the three parts from the fourth to the eighth week. Due to the different origin, and the different times of development, combined malformations of all three structures are rare and occur in less than 10% of patients with malformations. In most of these patients, the mesenchyma is affected by the developmental disorder. Malformations of the outer ear are often combined with malformations of the middle ear, since both develop parallel and formation of the tympanic membrane, at the border of both structures, depends on maturation of the outer auditory canal as well as the tympanic cavity. Malformations of the middle ear are only rarely combined with the developmental disorders of the inner ear.

Malformations of the outer ear are clinically easily assessable [42]. A missing or hypoplastic outer ear can be surgically reconstructed with good functional and cosmetical results. If the outer auditory canal is closed or completely absent, the middle ear might be affected as well [43]. The malformation of the middle ear can range from hypoplasia of the tympanic cavity and ossicular chain to complete absence of the ossicles [44]. Usually it parallels the developmental status of the outer ear: if the outer ear is completely missing, probability is high that the middle ear is hypoplastic and the ossicular chain is missing. The possibility of operative reconstruction depends on the size of the tympanic cavity. CT is the method of choice for delineation of the condition of the middle ear in patients with atresia of the outer auditory canal [44]. Combined malformations of the middle ear and the inner ear rarely occur due to the above-mentioned development from different sources, which takes place at different times. However, MRI should be performed prior to reconstruction of the middle ear, for delineation of the internal auditory canal and assessment of the eighth cranial nerve.

14.6 Inner Ear

Malformations of the inner ear are associated with macroscopic or microscopic changes of morphology and in most cases deterioration or complete loss of function. Severity of morphological change is not necessarily correlated to the severity of hearing loss. The malformations might be prenatally acquired or genetically transmitted. In most cases of acquired congenital hearing deficits, transplacental infections with toxoplasmosis, rubella, cytomegalovirus, or transition of ototoxic substances, such as ototoxic antibiotics or aminoglycoside, are the causing agents. In these cases, the Corti organ is affected while the morphologic development of the inner ear structures remains normal. Consequently, there is no correlate on MRI. If congenital hearing loss is transmitted genetically, all known modes of inheritance are possible. The disease might be either autosomal dominant, autosomal recessive, or X-linked recessive. Genetically transmitted malformations of the inner ear can be syndromal. In Table 14.2, possible syndromes and the probability of a detectable morphologic correlate of the hearing loss are summarized.

In up to 40% of patients, an underlying cause for congenital sensorineural haring loss cannot be detected [45]. In these cases, SNHL as a sequel of unrecognized labyrinthitis represents the main differential diagnosis.

The labyrinth develops from the fourth to the eighth week. As mentioned above, a hill of neuroectoderm, called the otic placode situated between the first branchial groove and the hindbrain, is the origin of the membranous labyrinth. In the first step, the otic placode folds in and builds the otic pit. The otic pit increases in size and forms the otocyst, which is completely separated from the surface. In the next step, the otocyst forms a dorsal and a ventral pouch. The ventral pouch develops into the cochlea and saccule while the dorsal pouch develops into utricle and semicircular canals. A third structure arises dorsomedially at the otocyst as the precursor of the endolymphatic sac. As soon as at the end of the eighth week of gestation, the labyrinth has formed including the 2.5 windings of the cochlea. From the 8th to the 16th week, the membranous labyrinth grows in size and from the 16th to the 24th week, the otic capsule ossifies. The sensory epithelium matures during

TABLE 14.2

Syndromal Sensorineuronal Hearing Loss

Syndrome	Inheritance	Site Affected	Detectable Changes in MRI
Alagile (arteriohepatic dysplasia)	Autosomal dominant	Inner ear	Yes
Alport	X-linked dominant	Inner ear	No
Apert	Autosomal dominant	Inner ear and middle ear	Yes
BOR (branchio-oto-renal syndrome)	Autosomal dominant	External ear, middle ear, inner ear	Yes
CHARGE (retinal coloboma, congenital heart defect, choanal atresia, retarded growth, genital hypoplasia, ear malformation)	Autosomal dominant	Inner ear	Yes
Crouzon (dysostosis craniofacialis)	Autosomal dominant	External ear, middle ear	No (CT: yes)
Franceschetti–Treacher–Collins (dysostosis otomandibularis)	Autosomal dominant	External ear, middle ear, inner ear	Yes
Goldenhar (oculoauriculovertebral dysplasia)	Sporadic	External ear, middle ear, inner ear	Yes
Klippel–Feil	Autosomal recessive	External ear, middle ear, inner ear	Yes
Pendred	Autosomal recessive	inner ear	Yes
Trisomia 13, 18, 21	Abberation of chromosomes	External ear, middle ear, inner ear	Yes
Waaredenburg	Autosomal dominant	Inner ear	Yes
Wildervanck (cervicooculoacustic dysplasia)	In many cases autosomal recessive	Inner ear	Yes

growth and ossification of the otic capsule from week 8th to 24th. After the 24th week, the fetus is able to hear. Any injury that occurs between the 8th and 24th week can cause sensorineural hearing loss while it might not lead to a visible malformation of the inner ear. It is estimated that only 20%–30% of congenital defects of the inner ear are associated with visible deformities of the labyrinth [46].

Malformations of the inner ear are the domain of MRI. In most cases, CT will be performed as the first imaging modality as a complementary technique for delineation of the middle ear structures and the otic capsule. CT and MRI complement each other regarding delineation of the bony details as well as the neural structures [47]. MRI depicts more fine details of the fluid-filled structures of the inner ear and vestibular organ, while CT visualizes the bony structures. Combining CT and high-resolution MRI is of paramount importance if operative correction of a malformation using a cochlear implant or brain-stem implant is planned [48,49].

Approximately, 65% of the malformations of the inner ear occur bilaterally and are symmetrical in extent. This observation has led to the hypothesis that an insult during embryogenesis (between fourth to eighth week) is responsible for the majority of malformations. One general rule, first introduced by Jackler, states that the earlier an insult occurs, the more severe the malformation is, ranging from complete absence of the labyrinth (Michel aplasia) to shortening and widening of the semicircular canals at the other end of the spectrum [50].

Several classifications have been introduced for better understanding the spectrum of malformations

[51]. However, in clinical routine, classifications have not gained importance. For the report description of affected structures and extend of changes are critical for assessing therapeutical options. Since cochlear implants may be considered, the presence of the cochlea and cochlear turns should be reported. The presence of the eighth cranial nerve decides the success of a cochlear implant, and it must be mentioned in the report. Hypoplasia of the cochlear nerve might still be associated with improvement of hearing after insertion of a cochlear implant [52]. Not all malformations are equally associated with absence of the cochlear nerve. As a general rule with increasing severity of cochlear malformation, the probability of cochlear nerve aplasia increases [53].

14.7 Michel Aplasia

The complete absence of the labyrinth and the eighth cranial nerve has first been described by P. Michel in 1863 [51]. Michel aplasia occurs if the development is interrupted in the third week. It is rare and accounts for only 1% of all inner ear malformations. The patients are deaf from birth. Differential diagnosis is complete fibrosis or ossification of the labyrinth as a squeal of labyrinthitis. CT can be applied to differentiate between both the conditions. The bone, overlaying the lateral semicircular, is convex toward the middle ear. If the labyrinth is ossified after labyrinthitis, this shape still remains. In Michel aplasia, the medial wall of the middle ear is flat. In most patients, the volume of the petrous bone usually harvesting the inner ear is relatively small. These findings can more readily be appreciated on CT.

The facial nerve is usually present. Consequently, the inner auditory canal is also present, but it is usually narrow. It contains only one nerve, which is the facial nerve. Clinically, patients with Michel aplasia do not have facial palsy.

14.8 Common Cavity

Common cavity malformation is due to interruption of development in the fourth to fifth week of gestation. The otocyst has already developed from the otic placode, but separation of vestibule, semicircular canals, and cochlea has not yet occurred (Figure 14.13). The organ of Corti, vestibule, and semicircular canals form a cyst. Patients might have residual hearing, since a small number of auditory cells might have been differentiated. Usually the eighth cranial nerve is present and patients with a common cavity malformation can be treated with a cochlear implant.

14.9 Cochlear Aplasia

Aplasia of the cochlea is differentiated from Michel aplasia by the presence of the semicircular canals or remnants of the semicircular canals. Similar to Michel

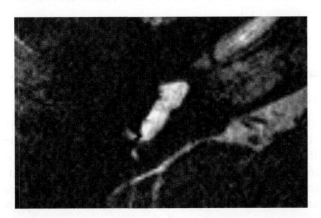

FIGURE 14.13
Common cavity. The high-resolution T_2-weighted TSE image demonstrates the cochlear with no internal structures merged with the vestibule.

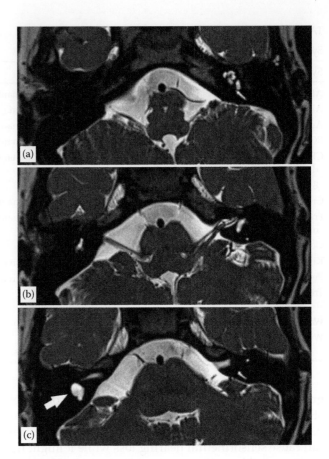

FIGURE 14.14
Aplasia of the cochlea and hypoplasia of the semicircular canals on the right side, the left inner ear is normal. T_2-weighted TSE images. (a) The cochlea is missing, (b,c) the semicircular canals are small and plump (arrow).

aplasia, it is a rare malformation and accounts for maximum 3% of all cochlear malformations [54]. If an insult occurs when the otocyst forms, the ventral and dorsal pouch cochlear aplasia might result. The main differential diagnosis is ossification of the cochlea after labyrinthitis. Usually both the entities can be differentiated since in cochlear aplasia the semicircular canals will be at least slightly affected and might appear shortened and widened (Figure 14.14). Since the vestibule arises from both the pouches, it will be affected in most patients with cochlear aplasia and is in most cases smaller in size.

14.10 Cochlear Hypoplasia

Cochlear hypoplasia accounts for 15% of inner ear malformations. An insult at the sixth week results in the presence of only a single turn of the cochlea. Similar to cochlear aplasia, the semicircular canals and the vestibule are affected in most patients [50]. Cochlear aplasia is often confounded with Mondini malformation, but can clearly be distinguished if the fine details are observed. Cochlear hypoplasia has been described in patients with branchio-oto-renal syndrome.

14.11 Mondini Malformation

Mondini malformation, also termed as *incomplete partition*, is frequently encountered and accounts for 55% of inner ear malformations with a correlate of imaging [55]. It is due to the interruption of the development in the seventh week of gestation. Only 1.5 tunes of the cochlea are present [56]. The basal tune is always present, but the middle and apical tune are not separated. The presence of the basal tune of the cochlea is the key imaging finding in differentiating Mondini malformation from cochlear hypoplasia. The spiral lamina is missing. In one-fifth of patients with Mondini malformation, the vestibule and semicircular canals are affected.

Since the basal tune of the cochlea is present, many of the patients with Mondini malformation initially can hear in the high-frequency spectrum [57]. Mondini malformation is often accompanied by a large vestibular aqueduct. Due to this additional feature, the initial degree of hearing might deteriorate over time (see Section 14.11.1). Moreover, CSF fistula to the middle ear is present in some patients with Mondini malformation. Such patients present with recurrent bacterial meningitis [58–61]. Patients with Mondini malformation have successfully been treated with cochlear implant.

14.11.1 Syndrome of Large Vestibular Aqueduct/ Endolymphatic Sac

The vestibular aqueduct arises from the otocyst at the fourth week of gestation. Initially it is wide and has a straight course. With the development of the posterior fossa, the vestibular aqueduct narrows and obtains the inverted j-shape of the adult. The vestibular aqueduct is considered enlarged if its diameter, measured in the midportion of the canal, is wider than 1.5 mm [62,63]. The endolymphatic sac can only be visualized on MRI and might also be enlarged (Figure 14.15). This malformation is by far the most common of all inner ear malformations with a visible correlate. Patients usually have normal hearing at birth, but stepwise deteriorates during adolescence. In many cases, minor traumata, such as that occurring during sports, cause small but permanent decrease in hearing. This pattern of stepwise deterioration of hearing after minor trauma or pressure changes gave rise to the assumption that internal fistularization in combination with a dysfunctional endolymphatic sac are causing damage to the organ of Corti. Surgical obliteration of the endolymphatic sac has been reported to prevent further deterioration of hearing [64]. In most patients, cochlear implants are applied for restoration of hearing [65].

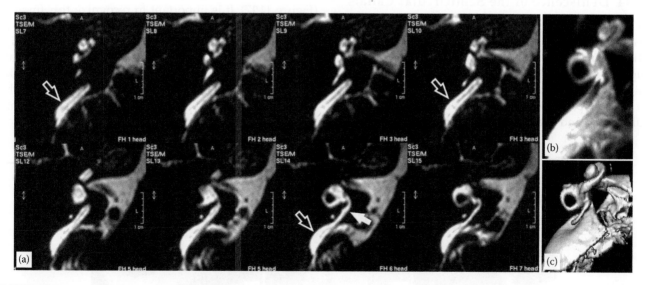

FIGURE 14.15
(a) Large vestibular aqueduct and endolymphatic sac. Eight consecutive slices from a T_2-weighted TSE sequence. The vestibular aqueduct is enlarged (arrow). The large endolymphatic sac can be seen on multiple slices (open arrows). (b) MIP and (c) 3D volume rendering demonstrate the whole extent of the large endolymphatic sac.

14.12 Semicircular Canal Aplasia

Starting at the sixth week of gestation, the superior and posterior semicircular canals develop first, in congruence with the law that embryogenesis resembles phylogenesis: early vertebrates had only a superior and a posterior semicircular canal. The lateral semicircular canal was developed later on in evolution. Embryogenesis imitates phylogenesis in its steps of development. According to the time point, the development is disturbed in an individual, either only the posterior or the lateral semicircular canal might be affected (Figure 14.16). Differential diagnosis is the obliteration after labyrinthitis. Obliteration after labyrinthitis is usually circumscribed and does rarely involve a complete semicircular canal. In addition, in most patients with aplasia of a semicircular canal, the vestibule is slightly deformed.

14.13 Semicircular Canal Hypoplasia

The mildest form of semicircular hypoplasia is shortening and widening of the lateral semicircular canal. The patients are usually asymptomatic, since the sense of balance has been adapted from birth on.

14.14 Dehiscence of the Semicircular Canals

In 1998, the superior canal dehiscence syndrome has been introduced as a morphological cause of vertigo by Minor et al. [66]. In this condition, the bony overly of the superior semicircular canal, which separates it from the middle cranial fossa, is missing. The dura mater serves as the only partition wall between the superior semicircular canal and the CSF in the cranial fossa. The bony

defect acts as a third pathway to the labyrinth in addition to the round and oval window [67]. Changes in pressure in either compartment, namely the external auditory canal, the middle ear, or the intracranial space, can be transmitted through this defect. These pressure transmissions have been shown in experimental animals to increase the firing rate of vestibular nerve afferents [68]. In patients frequently reported symptoms, associated with dehiscence of superior semicircular canal, are recurrent attacks of vertigo following stimuli, which lead to changes in intracranial or inner ear pressure, such as abrupt noise, coughing, or Valsalva maneuver [69,70]. Furthermore, involuntary vertical–torsional eye movements, in which directions can be explained by excitation or inhibition of the superior semicircular canal, have been observed. A few years later, dehiscence was also described for the posterior semicircular canal [71]. The diagnosis of superior or posterior semicircular dehiscence syndrome is based on the typical clinical signs and symptoms, and must be confirmed by cross-sectional imaging, such as CT [66]. However, in some patients, dehiscence of the bone, overlying the superior semicircular canal, remains asymptomatic. In these cases, it is an incidental finding when cross-sectional imaging of the temporal bone is performed or the temporal bone is exposed during surgery [72]. In patients with debilitating symptoms, improvement has been achieved by surgical resurfacing or plugging of the dehiscent semicircular canal [73,74]. If suspected clinically, CT is the modality of choice. Dehiscence of the superior or posterior semicircular canals can easily be assessed on T_2-weighted high-resolution images [75,76]. The high signal intensity of the semicircular canal merges into the high signal of the CSF (Figure 14.17). It is important to recognize this finding, since the clinical tests necessary to reveal dehiscence of semicircular canals are routinely not performed and a clinical question unrelated to the semicircular canals might have been raised. The most important differential diagnosis is a thin bony overlay, which can be below the resolution of imaging. The distinction can be made only based on clinical tests.

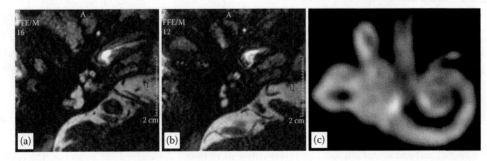

FIGURE 14.16
Aplasia of the posterior semicircular canal, the anterior and lateral semicircular canals are short and plump. (a) Lateral semicircular canal; (b) anterior semicircular canal; (c) MIP.

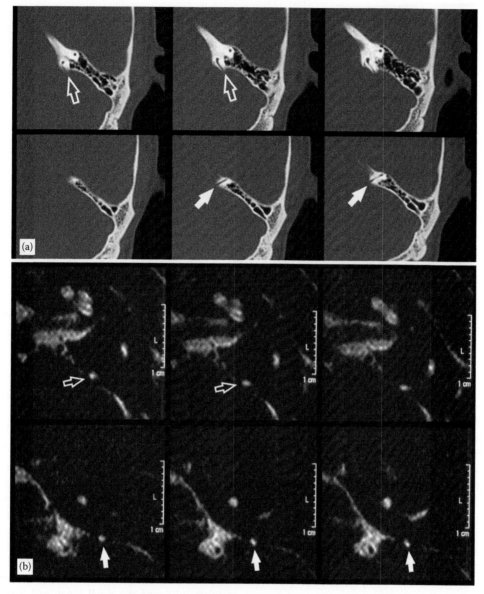

FIGURE 14.17
Dehiscence of the anterior and posterior semicircular canal. (a) Consecutive CT images. The bony overlay of the anterior (arrows) and posterior semicircular canal (open arrows) is not visible. (b) Corresponding T_2-weighted TSE images from the same patient. The anterior (arrows) and posterior semicircular canal (open arrows) are directly adjacent to the liquor of the posterior fossa (the dura is not visible).

14.14.1 Narrow Internal Auditory Canal, Aplasia of the Eighth Cranial Nerve

A narrow internal auditory canal within the range of few millimeters is usually associated with aplasia of the vestibulocochlear nerve. This condition is conspicuous on T_2-weighted high-resolution images (Figure 14.18). If a single nerve is present in the internal auditory canal, it is usually the facial nerve. Clinical examination will confirm that the facial nerve is functioning. Usually aplasia of the vestibulocochlear nerve is combined with severe malformation of the cochlea.

14.15 External Otitis

External otitis is a common condition and does not require imaging. In patients with recurrent or chronic disease not sufficiently responding to therapy, anomalies of the first branchial cleft, first branchial cleft cysts with a sinus opening into the external auditory canal might be the underlying cause [77]. MRI shows these fistulations as a tract of high signal intensity on T_2-weighted images. The walls of the sinus enhance on T_1-weighted images after intravenous administration

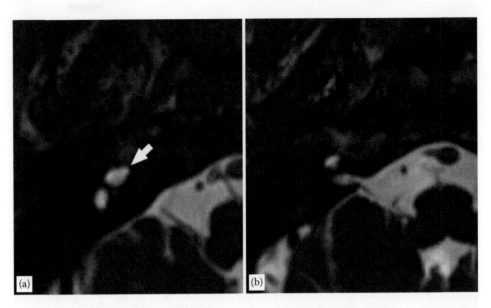

FIGURE 14.18
Aplasia of the vestibulocochlear nerve. (a) The cochlea is hypoplastic (arrow). (b) Only the facial nerve is visible in the narrow internal auditory canal.

of contrast medium. Such sinuses may connect the external auditory canal to a cyst in the parotid gland or retroauricular soft tissue. Surgical resection causes cassation of the inflammation and is the therapy of choice. Several case descriptions reporting on development of carcinoma in first branchial cleft cysts have been published [78,79]. CT is usually performed for exact delineation of the connection to the external auditory canal and planning of the surgical approach. Only few differential diagnoses exist. A foreign body inserted into the wall of the external auditory canal might result in channel-shaped enhancement on T_1-weighted contrast-enhanced images and respective increase in signal intensity on T_2-weighted images. In such cases, the parotid gland remains normal and the apparent *channel* presents with a blind end. Abscesses of the parotid gland may arise as a late sequel of parotitis caused by obstruction due to a concretion in the salivary duct. In these cases, the parotid duct is widened and the parotid gland is edematous and enlarged. Fistulation to the external auditory canal is usually not observed in parotitis or parotic abscesses.

Recurrent exposure to cold water can cause proliferation of fibrous tissue within the external auditory canal, formation of exostosis causing narrowing of the external auditory canal, and chronic inflammation [80,81]. This condition is termed as *surfers* or *swimmers ear*. This differential diagnosis can easily be made from the patients' history. CT is the imaging modality of choice for planning of resection of the exostoses [82].

14.16 Necrotizing/Malignant External Otitis

Elderly patients with poorly controlled type II diabetes are endangered to develop malign external otitis. Less frequently immunosuppressed neutropenic younger patients may be affected. *Pseudomonas aeruginosa* is the causative bacterium in most cases. At first, the soft tissue is involved; later the infection infiltrates the bone and spreads to the intracranial space at end stage [83]. The disease would be lethal without an antibiotic treatment. MRI can show the extent of inflammation, including osteomyelitis of the scull base and intracranial spread [84]. The typical signs in MRI are high signal intensity of the tissue involved on T_2-weighted images due to edema. If the bone of the scull base shows high signal intensity on T_2-weighted images with fat suppression, osteomyelitis is present. On T_1-weighted images with fat suppression, the inflamed tissue enhances. Due to necrosis, the central parts may not enhance, rendering differentiation from squamous cell carcinoma difficult, and malignancies such as mainly squamous cell carcinoma and less likely lymphoma are the major differential diagnoses. In some cases, differentiation of malignant external otitis from malignancy will not be possible based on MRI or other imaging modalities alone. This is especially difficult since both are characterized by spread into neighboring structures and edema of the surrounding tissue. In such cases, biopsy must be obtained. It is also important to recognize that changes of bone marrow signal (increased signal intensity in T_2-weighted images and increased enhancement on T_1-weighted images after IV injection of contrast medium) might persist up to 7 months after successful treatment.

14.16.1 Otitis Media, Granulation Tissue, Cholesterol Granuloma of the Middle Ear, Cholesteatoma

Otitis media does not require imaging. If complications are suspected, CT is the first imaging modality of choice. In patients with onset of neurologic symptoms such as headache or cranial nerve palsy, MRI becomes the modality of choice for assessment of intracranial complications including meningitis, sinus thrombosis, epidural or subdural abscess formation, or cerebritis [85,86]. If unilateral serous effusion in the middle ear is clinically encountered in an adult without a history of attenuated pneumatization of the mastoidal cells, unilateral dysfunction of the Eustachian tube due to tumor growth within the fossa of Rosenmüller (posterolateral recess) might be the cause [10]. Since this region is difficult to assess clinically, MRI can be used to exclude carcinoma (Figure 14.19).

Otitis media is a typical sequel of malfunction of the Eustachian tube. Granulation tissue, cholesterol granuloma, and cholesteatoma are possible complications. Granulation tissue is a reaction to inflammation and owing to reparative processes. It is highly vascularized

and consequently enhances on T_1-weighted images after administration of contrast medium. On precontrast T_1-weighted images, granulation tissue has low signal intensity (Figure 14.20).

Cholesteatoma does not enhance and demonstrates restricted diffusion on diffusion-weighted images (DWIs). Signal is high on T_2-weighted turbo SE (TSE) and does not decrease on fluid-attenuated inversion recovery (FLAIR) (Figure 14.21) [87]. Histologically it is *misplaced keratinizing squamous epithelium*. In most cases (99% of patients), it is acquired, and in most patients secondarily acquired as a consequence of malfunction of the Eustachian tube and otitis media. Cholesteatoma is a benign condition, but erodes the neighboring bone. It can cause fistulation of the endolymph, a complication that in most cases occurs at the lateral semicircular canal after erosion of the promontory. This complication is associated with a higher morbidity. Labyrinthitis is a possible complication (see below).

If not recognized and completely resected, cholesteatoma can expand up to the petrous apex and beyond and infiltrated into the intracranial space. The most important differential diagnoses of such large cholesteatomas are malignancies. On T_1-weighted images, cholesteatoma do not enhance after the administration of contrast medium. Moderate enhancement might only be seen at the border, where neighboring tissue is compressed or altered. In squamous cell carcinoma, typically central

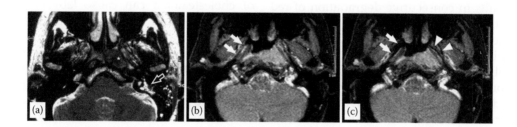

FIGURE 14.19
Squamous cell carcinoma arising from the left Rosenmüller fossa. (a) T_2-TSE. The tumor is hypointense (asterisk), fluid is present in the mastoidal cells (open arrow). (b) Fast T_1-weighted GE image. The contralateral Eustachian tube is closed (arrows). (c) During Valsalva the right Eustachian tube opens (arrows), while the left opens only close to the ostium (arrowheads) due to infiltration from the tumor.

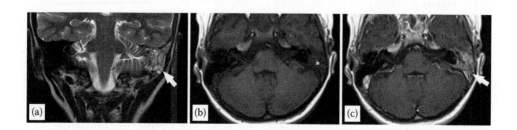

FIGURE 14.20
Chronic otitis media with formation of granulation tissue. (a) The T_2-weighted TSE demonstrates effusion within the middle ear and mastoidal cells (arrow). (b) T_1-weighted image. The granulation tissue appears hypointense. (c) T_1-weighted image obtained after IV administration of contrast medium shows strong enhancement, which is not confined to the mucous membrane (arrow).

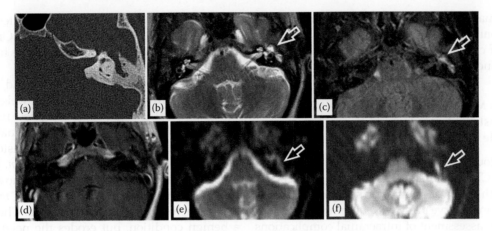

FIGURE 14.21

Cholesteatoma. (a) CT. The bone is destructed. (b) T_2-weighted TSE, The cholesteatoma has high signal intensity (arrow), which does not attenuate at FLAIR B (arrow). (c) T_1-weighted SE after IV injection of contrast medium. The cholesteatoma does not enhance. (d) DWI ($b = 0$), (e) DWI ($b = 1000$). Diffusion is restricted in the cholesteatoma.

necrosis can be found; nevertheless, these tumors have strongly enhancing rims. As mentioned above, a striking finding in cholesteatoma is the restriction of diffusion on DWI (Figure 14.21).

Cholesterol granuloma can come into being after long-standing presence of effusion in the middle ear due to malfunction of the Eustachian tube [88,89]. This condition causes edema and hypervascularization of the mucous membrane, which causes bleeding. Lack of drainage results in consecutive degradation of red blood cells within the middle ear cavity and accumulation of cholesterol crystals, which act as foreign bodies. Accordingly, multinucleated giant cells build up. Cholesterol granuloma remains a highly viscous fluid and in contrast to cholesterol granuloma in the petrous apex does not cause expansive destruction of the bony structures. However, this feature cannot be visualized on MRI. Cholesterol granuloma itself is thought to lead to hypervascularization of the mucous membrane, which triggers recurrent bleeding and sustains the process as a vicious circle. In contrast to cholesteatoma and granulation tissue, cholesterol granuloma has high signal intensity at all SE imaging sequences. This feature allows confident differentiation from cholesteatoma and granulation tissue. Granulation tissue displays low signal intensity on unenhanced T_1-weighted images. On otoscopy cholesterol granuloma of the middle ear resembles a highly vascularized structure such as glomus tympanicum paraganglioma, aberrant carotid artery, or dehiscent jugular bulb. Thus, MRI is obtained in many patients to differentiate between these conditions, as described above. Regarding differential diagnosis it has to be kept in mind that cholesterol granuloma of the middle ear is a rare condition.

14.17 Labyrinthitis

Labyrinthitis might be transmitted per continuitatem from the middle ear after otitis media or after a fracture of the temporal bone, via the meningae or hematogenous. If the infection is transmitted from the middle ear, unilateral involvement is the rule. As described above, labyrinthine fistulation due to erosive expansion of cholesteatoma can cause labyrinthitis. Iatrogenic labyrinthitis may occur after surgical interventions. Meningeal infection causes bilateral labyrinthitis in the vast majority of cases. Typical age of manifestation is early childhood. In many cases permanent hearing loss will result. In adult patients, clear connection to childhood meningitis might be difficult to assess and congenital malformations are the most important differential diagnosis.

In acute labyrinthitis, T_1-weighted images obtained after IV application of contrast medium demonstrated enhancement of the labyrinth, due to the pathologic permeability of vessels [90]. Enhancement might be segmental in the cochlea or semicircular canals, according to segmental inflammation (Figure 14.22) and is usually faint if compared to intralabyrinthine schwannoma [90]. Similar to other inflammations, enhancement might be seen up to 6 months after successful treatment and cessation of clinical symptoms. Permanent hearing loss is dreaded and can be treated once it occurs only by cochlear implants. Fibrosis of the cochlea can be triggered by the inflammation and might precede ossification, which renders insertion of the cochlear implant impossible. Fibrosis is due to differentiation of undifferentiated mesenchymal cells residing in the bony labyrinth into fibroblasts and later osteoblasts and

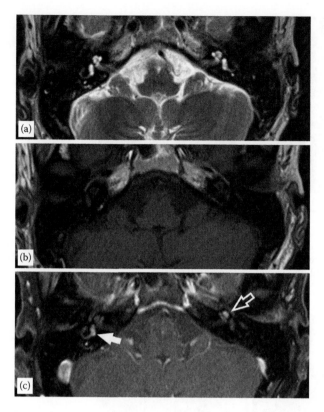

FIGURE 14.22
Acute labyrinthitis. (a) T_2-weighted TSE, the labyrinth is fluid-filled. (b) T_1-TSE prior and (c) after IV injection of contrast medium. The left vestibule (arrow) and right cochlea (open arrow) enhance.

proliferation of these cells. Fibrotic tissue can develop as soon as 12 weeks after the onset of labyrinthitis and in some cases even earlier. Fibrosis can be reliably detected only on T_2-weighted high-resolution images (Figure 14.23) but not be differentiated from ossification [91]. In CT images, the inherent low soft tissue contrast masquerades fibrotic tissue within the cochlea (Figure 14.24). However, CT allows differentiating fibrotic obliterations, which still allows insertion of the electrode from ossification [92].

14.18 Tumors

Location of a tumor in the ear is one of the most important features for differential diagnosis. Tumors of the external ear require imaging only if they have already infiltrated the surrounding soft tissue and infiltration depth can no longer be clinically accessed. Only malignancies demonstrate this growth pattern. Squamous cell carcinomas (Figure 14.25), basalioma, malignant melanoma and lymphoma represent the most common differential diagnosis. Basalioma is considered semimalignant, since it can deeply infiltrate but does not metastasize. In these cases, adequate planning of therapy depends on the exact extent and structures involved. For squamous cell carcinoma of the external ear, histopathological grading of the tumor, size, and infiltration depth have been identified as risk factors for metastases in the cervical lymph nodes [93]. In most cases, MRI cannot replace histologic diagnosis, but histology can easily be obtained.

14.19 Tumors of the Middle Ear

14.19.1 Glomus Tympanicum and Jugulotympanicum Paragangliomas

In the middle ear, congenital cholesteatoma followed by glomus tympanicum paraganglioma are the most common tumors. Schwannoma is the third most common neoplasm of the middle ear. All three are benign. Extremely few cases of metastasized glomus tympanicum tumors have been described as case reports in the literature, but these reports remain anecdotic. Paraganglioma can be multicentric, especially if familiar. In these circumstances multicentricity can be encountered in as much as a quarter of cases.

Congenital cholesteatoma arises from retrotympanic rests of ectodermal tissue and consequently occurs

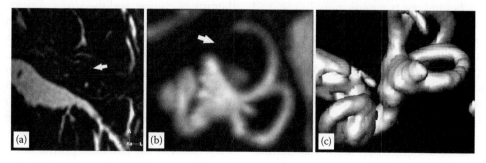

FIGURE 14.23
Partial occlusion of the anterior semicircular canal after labyrinthitis (arrow): (a) axial slice, (b) MIP, (c) 3D volume rendering. MIP and volume rendering provide a comprehensive overview.

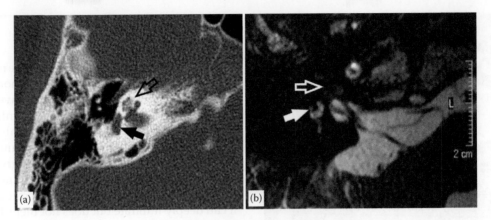

FIGURE 14.24
(a) CT. Ossification within the cochlea after labyrinthitis (open arrow). The basal tune of the cochlea is free from ossification, the presence of fibrotic tissue here and within the vestibule (arrow) cannot be differentiated from fluid. (b) T_2-weighted TSE obtained contemporary with CT. The cochlea (open arrow) is almost completely occluded, parts of the vestibule (arrow) contain fibrotic tissue.

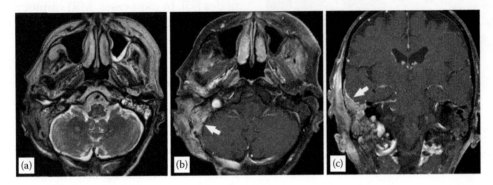

FIGURE 14.25
Squamous cell carcinoma infiltrating the right middle and inner ear. The tumor has reached the dura. (a) T_2-weighted TSE, (b) axial T_1-weighted GE, (c) coronal T_1-weighted GE.

in children or young adults with no history of otitis media or functional deficits of the Eustachian tube. As described above, cholesteatoma erodes the bone and most complications are due to erosion of functionally important structures. Cholesteatoma and glomus tympanicum tumor have different clinical presentations, which renders differential diagnosis between the two straightforward. At otoscopy, cholesteatoma appears as a whitish mass behind the tympanic membrane, and glomus tympanicum tumor is a reddish mass shining through the tympanic membrane in the anteroinferior quadrant. Patients with cholesteatoma suffer from conductive hearing loss, and in many cases have a history of otitis media and a long-standing functional deficit of the Eustachian tube, so that mastoidal pneumatization is not well developed. Clinical symptom in up to 90% of patients suffering from paraganglioma is pulsatile tinnitus, due to the extremely high vascularization of the tumor. Glomus tympanicum and glomus jugulotympanicum meningiomas are both supplied by

branches from the ascendant pharyngeal artery. Facial nerve schwannoma appears as a white mass at the superior half of the tympanic membrane at tympanoscopy. According to this differences, the differential diagnosis of paraganglioma are vascular variants, rather than other tumors. Glomus tympanicum paraganglioma clinically cannot be differentiated from glomus jugulotympanicum paraganglioma. Since complete resection is the therapy of choice, and the surgical approach is different for both entities, preoperative imaging has to differentiate between both and CT and MRI provide complementary information [94,94].

Glomus tympanicum paragangliomas arise from glomus bodies, the so-called paraganglia along the inferior tympanic nerve (the Jacobson nerve, which is the 9th branch of the facial nerve). These cell fractions represent chemoreceptor cells. In the middle ear, they are located at the cochlear promontory so that middle ear glomus tympanicum paragangliomas can only be found at this location. Glomus tumors can arise at all locations, at which chemoreceptor cells occur. Such locations comprise near the middle in addition to the cochlear promontory two additional locations at the jugular bulb: one arises also from Jacobson nerve and causes anterior jugular paraganglioma, while the second is located opposed to the first at the jugular bulb, arises from chemoreceptor cells of the mastoid branch (10th branch) of the facial nerve (also called *Arnolds nerve*). From these two sources, tumor growth can back forward in the direction of the middle ear. In this case, the paraganglioma will be termed *glomus jugulotympanic paraganglioma*. If tumor growth is directed caudally, the tumor is called glomus jugulare paraganglioma. Other locations, distant from the ear, are the carotid bifurcation and the larynx. On MRI, paraganglioma share the signal intensity of muscle on T_1 nonenhanced images and strongly enhance after IV administration of contrast medium [95]. On T_2-weighted images, they usually have high signal intensity (Figure 14.26). In large tumors, with a diameter above 1 cm, flow voids can be seen, resulting from high flow in large vessels (Figure 14.27). The key features for differential diagnosis are the typical pattern of enhancement and the location of the tumor [96]. Main differential diagnoses are a dehiscent jugular bulb or a jugular bulb diverticulum. These entities also demonstrate strong enhancement on T_1-weighted postcontrast images. They can be differentiated from paraganglioma according to their continuity with the jugular vein. Signal intensity at postcontrast images equals that of the jugular vein. Another important differential diagnosis is the aberrant internal carotid artery. This variant can be differentiated from a tumor by the continuity with the carotid artery, the atypical course of the carotid artery, and the location.

Schwannoma of the middle ear arises from the facial nerve including its branches. As mentioned above, on tympanoscopy schwannomas are visible as white masses. Similar to schwannomas of other locations, there is strong enhancement on T_1-weighted images after administration of contrast medium [97]. This feature clearly distinguishes this entity from cholesteatoma. In contrast to paraganglioma, even large schwannomas do not contain signal voids on T_2-weighted images. Small schwannoma might only be differentiated from paraganglioma by consideration of clinical symptoms and the appearance at tympanoscopy.

Other possible entities that typically arise in the middle ear are meningioma or adenoma in adults and rhabdomyosarcoma in children.

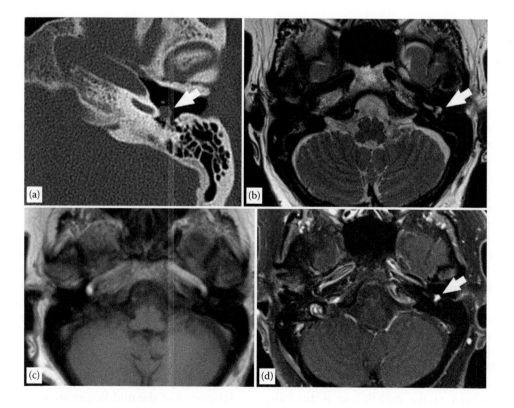

FIGURE 14.26
Glomus tympanicum paraganglioma. (a) CT. The paraganglioma is located at the promontory (arrow). (b) Glomus tympanicum paraganglioma has high signal intensity on T_2-weighted TSE. (c) Prior to administration of contrast medium the tumor is hypointense and strongly enhances (d) after IV injection of contrast medium.

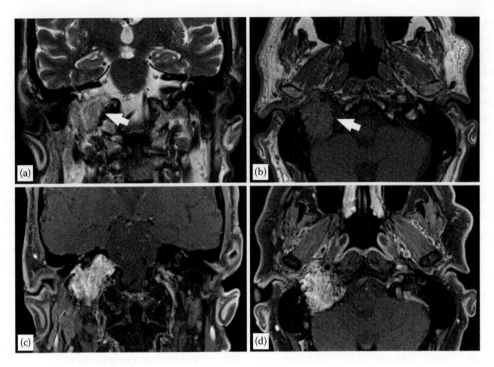

FIGURE 14.27
Glomus jugulotympanicum paraganglioma. (a) T_2-TSE. The tumor has high signal intensity with dark spots (flow voids) and extends from the foramen jugulare to the tympanon and below the internal auditory canal. (b). T_1-weighted TSE prior and after (c,d) IV injection of contrast medium. The tumor enhances strongly.

Meningioma may extend from the jugular foramen or the posterior wall of the temporal bone into the middle ear [98]. In this case, the continuity to the intracranial meningioma that displays the typical imaging appearance of meningiomas (dura tail sign, strong enhancement on T_1-weighted images after contrast injection, in larger tumors cystic degeneration and calcifications, which appear black on MRI) allows for diagnosis. In extreme rare cases, meningioma directly arise within the middle ear, which renders diagnosis difficult [99]. In such cases, glomus tympanicum paraganglioma presents the most important differential diagnosis. Meningioma appears on tympanoscopy as a highly vascularized mass. Compared to paraganglioma, it appears more bluish while the former is more reddish. However, this distinction might be impossible in an individual patient. The role of imaging is the description of the exact extent of the tumor, since aggressive surgical removal is the therapy of choice. The recurrence rate is high due to infiltration of the bone and complete resection is required to avoid growth of remaining cells. Typically, middle-aged women are affected; the most common symptom is conductive hearing loss.

Rhabdomyosarcoma of the middle ear is typically found in children and represents the most common middle ear malignancy in children [100]. Nevertheless, the tumor is extremely rare [101]. Rhabdomyosarcoma of the middle ear arises from the skeletal muscle in the vicinity of the temporal bone, namely the Eustachian tube, or from the mesenchymal stem cells situated in the middle ear [100]. Destructive growth into the bony structures is a typical feature of middle ear rhabdomyosarcoma. Clinical symptoms are similar to those of otitis media with hemorrhagic otorrhea, leading to late diagnosis of the malignancy [102]. Infiltration of the facial canal causes facial nerve palsy, but is a sign of advanced disease. Rhabdomyosarcoma are hypo- to isointense on T_1-weighted images and enhance homogenously after IV injection of contrast media. These tumors are hyperintense on T_2-weighted images. Local margins of the tumor must completely be delineated in order to assess for intracranial extension. This is facilitated by obtaining coronal images. Furthermore, staging for regional lymph node metastasis is important for staging. Histiocytosis X is the major differential diagnosis for a destructive tumor of the middle ear in this age group [103]. Differentiation based on imaging appearance alone can be impossible, since histiocytosis X can demonstrate destructive growth in the middle ear or temporal bone (Figure 14.28) and the tumor shares signal intensity properties [104].

Adenomas of the middle ear are extreme rare lesions, but are mentioned here since they receive increasing attention in the literature and present a differential diagnosis for middle ear tumors [105,106]. They arise from respiratory epithelium and have an age

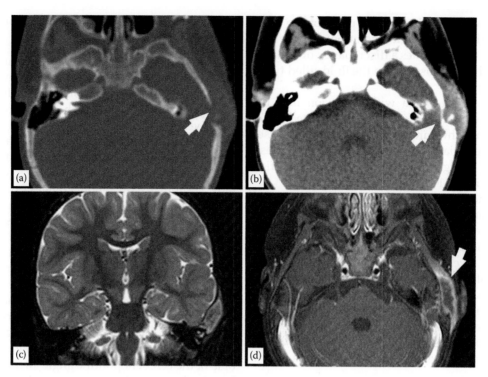

FIGURE 14.28

Histiocytosis X. (a,b) CT shows an osteolysis, caused by a contrast enhancing tumor. (c) T_2-TSE: The tumor is delineated with high signal intensity, (d) T_1-weighted with fat suppression after IV injection of contrast medium. The tumor rim enhances.

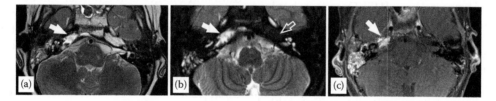

FIGURE 14.29

Ewing sarcoma. (a) T_2-weighted TSE without fat suppression. The sarcoma has high signal intensity (arrow). (b) On T_2-weighted TSE with fat suppression; the sarcoma retains the high signal intensity (arrow), the signal from the contralateral petrous apex is suppressed (open arrow). (c) T_1-weighted images after IV injection of contrast medium. The tumor enhances (arrow).

peak in middle-aged patients. Since they are benign, slow growing tumors that do not infiltrate the bone, conductive hearing loss is the most common clinical symptom. At tympanoscopy, adenomas present as pink to pale masses [107]. Adenomas are hyperintense on T_2-weighted images isointense to muscle at nonenhanced T_1-weighted images and enhance after administration of contrast medium. The last feature allows from differentiation from congenital cholesteatoma.

Other entities that are not specific to the ear but can arise in the temporal bone are metastasis and malignomas related to the resident tissue components such as osteosarcoma, lipoma and liposarcoma, chondrosarcoma, Ewing sarcoma (Figure 14.29), or lymphoma [108]. Characteristic findings on imaging equal those of other body regions [109].

14.20 Tumors of the Inner Ear

The typical tumor of the cochlea or vestibule is schwannoma [110]. In the labyrinth, schwannoma arises from the neural sheath of the distal cochlear and vestibular nerve [111].

Patients present with unilateral sensorineural hearing loss that progresses slowly and develops over decades. On high-resolution T_2-weighted images, a spherical shaped focal mass with low signal intensity can be seen within the fluid-filled labyrinth. On T_1-weighted images, this region enhances intensely [112,113]. The most important differential diagnosis is labyrinthitis with fibrotic tissue formation within the labyrinth. In labyrinthitis, enhancement is not localized but rather involves larger parts of the labyrinth. Correlation with high-resolution T_2-weighted images usually demonstrates a discrepancy of enhancing regions and signal loss within the fluid-filled tubes of the labyrinth [90].

Facial nerve schwannoma with extension into the labyrinth can usually be easily differentiated from intralabyrinthine schwannoma, since the former causes widening of the facial nerve canal by a tubular mass, enhancing on T_1-weighted images after administration of contrast medium [114,115].

Since complete unilateral hearing loss will result from surgically excision of intralabyrinthine schwannoma, no therapy is applied, if the tumor is confined to the cochlea and residual hearing is preserved. Intralabyrinthine schwannoma is a very slowly growing tumor and if left alone usually stops growing after the cochlea is filled. With the approach of *watchful waiting*, the patient can preserve residual hearing on the affected side over many years or decades. If vertigo is the leading symptom, it might be necessary to completely remove the tumor, which will cause postoperative deafness on the affected ear.

Endolymphatic sac tumor is a locally aggressive benign tumor (adenoma) arising from the cells lining the endolymphatic sac [116]. The typical location is at the posterior margin of the temporal bone [117]. Despite its benign nature, it infiltrates the posterior wall of the temporal bone and extends into the labyrinth. Endolymphatic sac tumors have inhomogeneous signal intensity on T_2- and T_1-weighted images with foci of extremely high signal intensity on native T_1-weighted images and correspondingly enhance heterogeneously on T_1-weighted images after the administration of contrast medium. Endolymphatic sac tumor is often associated with von-Hippel–Lindau disease [118]. Differential diagnosis contains meningioma, which might arise at the same location. Meningiomas homogenously enhance. Large tumors may have cystic part and calcifications, which renders contrast enhancement more heterogeneously. Tumor margins of meningioma are smooth even if the tumor itself is large, while endolymphatic sac tumors have irregular borders.

14.20.1 Tumors of the Internal Auditory Canal and Cerebellopontine Angle

Vestibular schwannoma is the most common and prototype tumor of the internal auditory canal. It arises from the nerve sheath of the vestibular nerve. In most patients, it is sporadic. In patients with neurofibromatosis type 2 (NF2) vestibular schwannomas are in 90% of cases bilateral. Due to compression of the cochlear nerve, slowly progressing unilateral sensorineural hearing loss is the most common symptom [119,120]. Age peak of sporadic vestibular schwannoma ranges from 40 to 60 years, without gender preference. In NF2 age peak is 25 years. Vestibular schwannoma is typically accessible as a filling defect within the internal auditory canal on T_2-weighted images (Figure 14.30). On T_1-weighted images, it has intermediate signal intensity. Schwannomas enhance strongly and in 85% homogenously after the administration of contrast medium. Large tumors have a typical ice cone shape, if they fill the internal auditory canal and protrude into the cerebellopontine angle (Figure 14.30) [121].

The most important differential diagnosis is meningioma of the cerebellopontine angle. Meningioma usually has similar signal intensity as vestibular schwannoma. Differential diagnosis can easily be made based on the shape of the tumor. Meningioma has a broad base at the dura and merges at the border smoothly into the dura, which is slightly thickened at the border. This finding is called dura tail sign (Figure 14.31). Vestibular schwannoma are concave to the dura. In patients with NF2, differential diagnosis is different and includes facial schwannoma and metastasis of the internal auditory canal [120].

Neurinomas that arise from other nerves and invade the internal auditory canal are extremely rare. They share signal behavior of vestibular schwannomas (Figure 14.32).

Other lesions at the cerebellopontine angle, which can be easily distinguished, are epidermoid cyst and arachnoid cyst. Epidermoid cysts arise from inclusion of ectodermal cells during closure of the neural tube between the third and fifth week of gestation. Epidermoid cysts are extremely slow-growing lesions, and remain

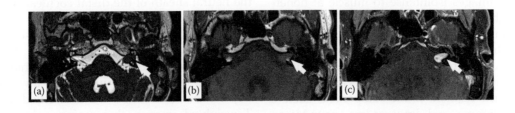

FIGURE 14.30
Vestibular schwannoma. (a) T_2-weighted TSE. The left inner auditory canal is completely filled with a tumor (arrow). (b) T_1-weighted SE. The vestibular schwannoma is isointense to brain tissue. (c) After IV injection of contrast medium vestibular schwannoma strongly enhances.

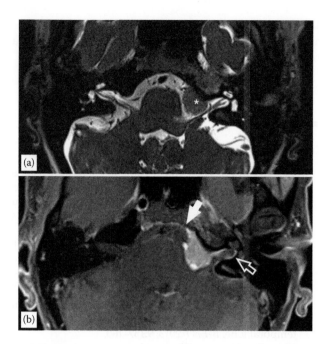

FIGURE 14.31
Meningioma of the cerebellopontine angle. (a) T_2-weighted TSE. The meningioma has a broad base to the dura (asterisk). (b) T_1-SE image after IV injection of contrast medium. The meningioma strongly enhances. At the transition from the tumor to the dura, the dura is thickened and enhances (arrow). This finding is called the *dura tail sign*. The internal auditory canal is lined by dura. Due to the close proximity of the meningioma, the dura of the internal auditory canal enhances in the present case (arrow).

asymptomatic for decades. Age distribution for manifestation of symptoms ranges from 20 to 60 years and the type of symptoms depends on the exact location of the lesion. Possible manifestations are sensorineural hearing loss, headache, or trigeminal or facial neuralgia. Epidermoid cysts do not contain sebaceous glands, fat, or hair follicles, which makes them different from dermoid cysts. The main contents of epidermoid cysts are keratinaceous debris and cholesterol crystals. Due to

their composition, they are isointense or hyperintense to CSF on T_1-weighted images, isointense, or hypointense to CSF on T_2-weighted images. On FLAIR images epidermoid cysts either completely retain their signal or attenuate patchy. Epidermoid tumors do not show enhancement on T_1-weighted images after administration of contrast medium [122,123]. The typical imaging finding is the high signal intensity on DWI, due to restricted diffusion [124]. Another characteristic property of epidermoid cyst is encasement of nerves and vessels, in contrast to replacement, which can be observed with arachnoid cysts [125]. Arachnoid cyst at the cerebellopontine angle as at any other location displays similar signal intensity as CSF at al imaging sequences including DWI. Epidermoid cysts can easily be distinguished from cystic tumors such as cystic (than usually large) meningiomas by the enhancing tissue components of such cystic tumors.

14.21 Intralabyrinthine Hemorrhage

Intralabyrinthine hemorrhage can result from trauma with or without fracture, or as a complication from surgery. It has also been reported in patients with coagulopathy. Since intralabyrinthine hemorrhage results in sudden (sensorineural) hearing loss, MRI is usually obtained early after the event. It can be recognized in the subacute state (degradation to methemoglobin) according to a bright signal within the labyrinth on T_1-weighted nonenhanced images (Figure 14.33) [126]. This imaging feature allows for differentiating hemorrhage from labyrinthitis and intralabyrinthine schwannoma, which are associated with low signal intensity on T_1-weighted images and enhancement after IV injection of contrast medium [127].

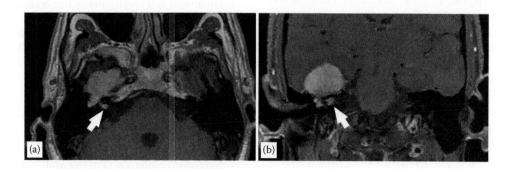

FIGURE 14.32
Neurinoma invading the internal auditory canal via the facial canal. T_1-weighted images after IV injection of contrast medium in the axial (a) and coronal plane (b). In addition to the huge tumor below the temporal lobe, there is enhancement within the facial canal and an enhancing small mass in the internal auditory canal (arrows).

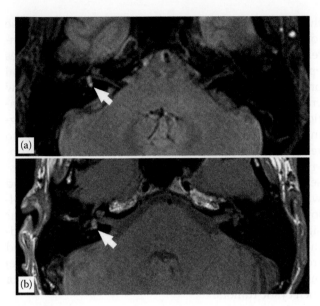

FIGURE 14.33
Intralabyrinthine hemorrhage after stapes surgery. Foci of strong signal within the labyrinth are visible on FLAIR (a) and the T_1-weighted image (b) (arrows).

14.22 Summary

We have witnessed tremendous progress of MR for imaging the ear in the last two decades, due to technical advances. Primarily owing to the increase in signal-to-noise ratio, it was possible to increase the resolution to the submillimeter range. This feature paved the way for sophisticated delineation of the inner ear, which measures together with the vestibular system at maximum 1.5 cm. Malformations and acquired diseases, which affect parts of the labyrinthine structures have become accessible. The range of indications includes congenital and acquired hearing loss, vertigo, or pain. The range of diseases mirroring these symptoms spans from malformations, inflammation, trauma, and associated sequel to tumors and dystrophy of the bone. Depending on symptoms, clinical examination, and laboratory work up, different imaging protocols are applied. The inner ear consists of fluid-filled structures that are ideally suited for delineation using heavily T_2-weighted high-resolution images. Congenital malformations cause deformity of the labyrinthine structures while tumors or fibrosing inflammation cause interruption of the fluid filling within the labyrinth. The middle ear becomes accessible for MRI, if pathologies cause filling of the usually air containing middle ear cavity. Enhancement pattern, exact location, and the margins of lesions allow for differentiation between different entities. The outer ear and external auditory canal is easily assessable for

clinical examination. Indications for MRI are assessment of the extent of malignant external otitis and local staging of carcinoma.

References

1. Gao Z, Chi FL: The clinical value of three-dimensional fluid-attenuated inversion recovery magnetic resonance imaging in patients with idiopathic sudden sensorineural hearing loss: A meta-analysis. *Otol Neurotol* 2014; 35:1730–1735.
2. Maroldi R, Farina D, Palvarini L, Marconi A, Gadola E, Menni K, Battaglia G: Computed tomography and magnetic resonance imaging of pathologic conditions of the middle ear. *Eur J Radiol* 2001; 40:78–93.
3. Bruninx L, Govaere F, Van DJ, Forton GE: Isolated synchronous meningioma of the external ear canal and the temporal lobe. *B-ENT* 2013; 9:157–160.
4. Chin RY, Nguyen TB: Synchronous malignant otitis externa and squamous cell carcinoma of the external auditory canal. *Case Rep Otolaryngol* 2013; 2013:837169.
5. Pietrantonio A, D'Andrea G, Fama I, Volpini L, Raco A, Barbara M: Usefulness of image guidance in the surgical treatment of petrous apex cholesterol granuloma. *Case Rep Otolaryngol* 2013; 2013:257263.
6. Young JY, Ryan ME, Young NM: Preoperative imaging of sensorineural hearing loss in pediatric candidates for cochlear implantation. *RadioGraphics* 2014; 34:E133–E149.
7. Migirov L, Greenberg G, Eyal A, Wolf M: Imaging prior to endoscopic ear surgery: Clinical note. *Isr Med Assoc J* 2014; 16:191–193.
8. Dammann F, Grees H, Kösling S, Kress B, Lell M. Algorithmen für die Durchführung radiologischer Untersuchungen der Kopf-Hals-Region. *AWMF Online.* 2015. http://www.awmf.org/uploads/tx_szleitlinien/039-0931_S1_Radiologische_Diagnostik_Kopf_Hals-Bereich_2015-05.pdf.
9. Vergez S, Morinière S, Dubrulle F, Salaun PY, De Monès E, Bertolus C, Temam S et al. Initial staging of squamous cell carcinoma of the oral cavity, larynx and pharynx (excluding nasopharynx). Part I: Locoregional extension assessment: 2012 SFORL guidelines. *Eur Ann Otorhinolaryngol Head Neck Dis* 2013 Feb;130:39–45.
10. Lukens A, Dimartino E, Gunther RW, Krombach GA: Functional MR imaging of the Eustachian tube in patients with clinically proven dysfunction: Correlation with lesions detected on MR images. *Eur Radiol* 2012; 22:533–538.
11. Purohit B, Hermans R, Op de BK: Imaging in otosclerosis: A pictorial review. *Insights Imaging* 2014; 5:245–252.
12. Stimmer H, Arnold W, Schwaiger M, Laubenbacher C: Magnetic resonance imaging and high-resolution computed tomography in the otospongiotic phase of otosclerosis. *ORL J Otorhinolaryngol Relat Spec* 2002; 64:451–453.
13. Goh JP, Chan LL, Tan TY: MRI of cochlear otosclerosis. *Br J Radiol* 2002; 75:502–505.

14. Czerny C, Gstoettner W, Franz P, Baumgartner WD, Imhof H: CT and MR imaging of acquired abnormalities of the inner ear and cerebellopontine angle. *Eur J Radiol* 2001; 40:105–112.

15. Adachi N, Yoshida T, Nin F, Ogata G, Yamaguchi S, Suzuki T, Komune S, Hisa Y, Hibino H, Kurachi Y: The mechanism underlying maintenance of the endocochlear potential by the K+ transport system in fibrocytes of the inner ear. *J Physiol* 2013; 591:4459–4472.

16. Kim SH, Marcus DC: Regulation of sodium transport in the inner ear. *Hear Res* 2011; 280:21–29.

17. van der Jagt MA, Brink WM, Versluis MJ, Steens SC, Briaire JJ, Webb AG, Frijns JH, Verbist BM: Visualization of Human Inner Ear Anatomy with High-Resolution MR Imaging at 7T: Initial Clinical Assessment. *AJNR Am J Neuroradiol* 2014; 36:378–383.

18. Stimmer H: Enlargement of the cochlear aqueduct: Does it exist? *Eur Arch Otorhinolaryngol* 2011; 268:1655–1661.

19. Nakamichi R, Yamazaki M, Ikeda M, Isoda H, Kawai H, Sone M, Nakashima T, Naganawa S: Establishing normal diameter range of the cochlear and facial nerves with 3D-CISS at 3T. *Magn Reson Med Sci* 2013; 12:241–247.

20. Hoekstra CE, Prijs VF, van Zanten GA: Diagnostic yield of a routine magnetic resonance imaging in tinnitus and clinical relevance of the anterior inferior cerebellar artery loops. *Otol Neurotol* 2015; 36:359–365.

21. Erdogan N, Altay C, Akay E, Karakas L, Uluc E, Mete B, Oygen A et al.: MRI assessment of internal acoustic canal variations using 3D-FIESTA sequences. *Eur Arch Otorhinolaryngol* 2013; 270:469–475.

22. Gorrie A, Warren FM, III, de la Garza AN, Shelton C, Wiggins RH, III: Is there a correlation between vascular loops in the cerebellopontine angle and unexplained unilateral hearing loss? *Otol Neurotol* 2010; 31:48–52.

23. McDermott AL, Dutt SN, Irving RM, Pahor AL, Chavda SV: Anterior inferior cerebellar artery syndrome: Fact or fiction. *Clin Otolaryngol Allied Sci* 2003; 28:75–80.

24. Chadha NK, Weiner GM: Vascular loops causing otological symptoms: A systematic review and meta-analysis. *Clin Otolaryngol* 2008; 33:5–11.

25. van der Steenstraten F, de Ru JA, Witkamp TD: Is microvascular compression of the vestibulocochlear nerve a cause of unilateral hearing loss? *Ann Otol Rhinol Laryngol* 2007; 116:248–252.

26. Herzog JA, Bailey S, Meyer J: Vascular loops of the internal auditory canal: A diagnostic dilemma. *Am J Otol* 1997; 18:26–31.

27. Razek AA, Huang BY: Lesions of the petrous apex: Classification and findings at CT and MR imaging. *RadioGraphics* 2012; 32:151–173.

28. Chapman PR, Shah R, Cure JK, Bag AK: Petrous apex lesions: Pictorial review. *AJR Am J Roentgenol* 2011; 196:WS26–WS37.

29. Friedmann DR, Eubig J, Winata LS, Pramanik BK, Merchant SN, Lalwani AK: A clinical and histopathologic study of jugular bulb abnormalities. *Arch Otolaryngol Head Neck Surg* 2012; 138:66–71.

30. Friedmann DR, Le BT, Pramanik BK, Lalwani AK: Clinical spectrum of patients with erosion of the inner ear by jugular bulb abnormalities. *Laryngoscope* 2010; 120:365–372.

31. Pappas DG, Jr., Hoffman RA, Cohen NL, Holliday RA, Pappas DG, Sr.: Petrous jugular malposition (diverticulum). *Otolaryngol Head Neck Surg* 1993; 109:847–852.

32. Lin YY, Wang CH, Liu SC, Chen HC: Aberrant internal carotid artery in the middle ear with dehiscent high jugular bulb. *J Laryngol Otol* 2012; 126:645–647.

33. McKiever ME, Carlson ML, Neff BA: Aberrant petrous carotid artery masquerading as a glomus tympanicum. *Otol Neurotol* 2014; 35:e228–e230.

34. Nicolay S, de FB, Bernaerts A, Van DJ, Parizel PM: Aberrant internal carotid artery presenting as a retrotympanic vascular mass. *Acta Radiol Short Rep* 2014; 3:1–3.

35. Hitier M, Zhang M, Labrousse M, Barbier C, Patron V, Moreau S: Persistent stapedial arteries in human: From phylogeny to surgical consequences. *Surg Radiol Anat* 2013; 35:883–891.

36. Cappabianca S, Scuotto A, Iaselli F, Pignatelli di SN, Urraro F, Sarti G, Montemarano M, Grassi R, Rotondo A: Computed tomography and magnetic resonance angiography in the evaluation of aberrant origin of the external carotid artery branches. *Surg Radiol Anat* 2012; 34:393–399.

37. Hatipoglu HG, Cetin MA, Yuksel E, Dere H: A case of a coexisting aberrant internal carotid artery and persistent stapedial artery: The role of MR angiography in the diagnosis. *Ear Nose Throat J* 2011; 90:E17–E20.

38. Yilmaz T, Bilgen C, Savas R, Alper H: Persistent stapedial artery: MR angiographic and CT findings. *AJNR Am J Neuroradiol* 2003; 24:1133–1135.

39. Roll JD, Urban MA, Larson TC, III, Gailloud P, Jacob P, Harnsberger HR: Bilateral aberrant internal carotid arteries with bilateral persistent stapedial arteries and bilateral duplicated internal carotid arteries. *AJNR Am J Neuroradiol* 2003; 24:762–765.

40. Stott CE, Kuroiwa MA, Carrasco FJ, Delano PH: Recurrent acute otitis media as the manifestation of an aberrant internal carotid artery. *Otol Neurotol* 2013; 34:e117–e118.

41. Arena P, Portmann D: Persistent stapedial artery and chronic otitis: CT scan aspects, a clinical report. *Rev Laryngol Otol Rhinol (Bord)* 2005; 126:33–36.

42. Bartel-Friedrich S, Wulke C: Classification and diagnosis of ear malformations. *GMS Curr Top Otorhinolaryngol Head Neck Surg* 2007; 6:Doc05.

43. Luquetti DV, Heike CL, Hing AV, Cunningham ML, Cox TC: Microtia: epidemiology and genetics. *Am J Med Genet A* 2012; 158A:124–139.

44. Kosling S, Omenzetter M, Bartel-Friedrich S: Congenital malformations of the external and middle ear. *Eur J Radiol* 2009; 69:269–279.

45. Huang BY, Zdanski C, Castillo M: Pediatric sensorineural hearing loss, part 2: Syndromic and acquired causes. *AJNR Am J Neuroradiol* 2012; 33:399–406.

46. Huang BY, Zdanski C, Castillo M: Pediatric sensorineural hearing loss, part 1: Practical aspects for neuroradiologists. *AJNR Am J Neuroradiol* 2012; 33:211–217.

47. Mukerji SS, Parmar HA, Ibrahim M, Mukherji SK: Congenital malformations of the temporal bone. *Neuroimaging Clin N Am* 2011; 21:603–19, viii.

48. Joshi VM, Navlekar SK, Kishore GR, Reddy KJ, Kumar EC: CT and MR imaging of the inner ear and brain in children with congenital sensorineural hearing loss. *RadioGraphics* 2012; 32:683–698.

49. Kachniarz B, Chen JX, Gilani S, Shin JJ: Diagnostic yield of MRI for pediatric hearing loss: A systematic review. *Otolaryngol Head Neck Surg* 2015; 152:5–22.

50. Jackler RK, Luxford WM, House WF: Congenital malformations of the inner ear: A classification based on embryogenesis. *Laryngoscope* 1987; 97:2–14.

51. Giesemann AM, Goetz F, Neuburger J, Lenarz T, Lanfermann H: Appearance of hypoplastic cochleae in CT and MRI: A new subclassification. *Neuroradiology* 2011; 53:49–61.

52. Wu CM, Lee LA, Chen CK, Chan KC, Tsou YT, Ng SH: Impact of cochlear nerve deficiency determined using 3-dimensional magnetic resonance imaging on hearing outcome in children with cochlear implants. *Otol Neurotol* 2015; 36:14–21.

53. Giesemann AM, Kontorinis G, Jan Z, Lenarz T, Lanfermann H, Goetz F: The vestibulocochlear nerve: Aplasia and hypoplasia in combination with inner ear malformations. *Eur Radiol* 2012; 22:519–524.

54. Kontorinis G, Goetz F, Giourgas A, Lanfermann H, Lenarz T, Giesemann AM: Aplasia of the cochlea: Radiologic assessment and options for hearing rehabilitation. *Otol Neurotol* 2013; 34:1253–1260.

55. Kontorinis G, Goetz F, Giourgas A, Lenarz T, Lanfermann H, Giesemann AM: Radiological diagnosis of incomplete partition type I versus type II: Significance for cochlear implantation. *Eur Radiol* 2012; 22:525–532.

56. Mondini C. Anatomia surdi nati sectio: De Bononiensi Scientiarum et Artium Institute atque Academia commentarii. *Bononiae*. 1791; 7:419–428.

57. Lo WW: What is a "Mondini" and what difference does a name make? *AJNR Am J Neuroradiol* 1999; 20:1442–1444.

58. Iseri M, Ucar S, Derin S, Ustundag E: Cerebrospinal fluid otorrhea and recurrent bacterial meningitis in a pediatric case with Mondini dysplasia. *Kulak Burun Bogaz Ihtis Derg* 2013; 23:57–59.

59. Lien TH, Fu CM, Hsu CJ, Lu L, Peng SS, Chang LY: Recurrent bacterial meningitis associated with Mondini dysplasia. *Pediatr Neonatol* 2011; 52:294–296.

60. Lin CY, Lin HC, Peng CC, Lee KS, Chiu NC: Mondini dysplasia presenting as otorrhea without meningitis. *Pediatr Neonatol* 2012; 53:371–373.

61. Gharib B, Esmaeili S, Shariati G, Mazloomi NN, Mehdizadeh M: Recurrent bacterial meningitis in a child with hearing impairment, Mondini dysplasia: A case report. *Acta Med Iran* 2012; 50:843–845.

62. Spencer CR: The relationship between vestibular aqueduct diameter and sensorineural hearing loss is linear: A review and meta-analysis of large case series. *J Laryngol Otol* 2012; 126:1086–1090.

63. Connor SE, Siddiqui A, O'Gorman R, Tysome JR, Lee A, Jiang D, Fitzgerald-O'Connor A: Magnetic resonance imaging features of large endolymphatic sac compartments: Audiological and clinical correlates. *J Laryngol Otol* 2012; 126:586–593.

64. Wilson DF, Hodgson RS, Talbot JM: Endolymphatic sac obliteration for large vestibular aqueduct syndrome. *Am J Otol* 1997; 18:101–106.

65. Ko HC, Liu TC, Lee LA, Chao WC, Tsou YT, Ng SH, Wu CM: Timing of surgical intervention with cochlear implant in patients with large vestibular aqueduct syndrome. *PLoS One* 2013; 8:e81568.

66. Minor LB, Solomon D, Zinreich JS, Zee DS: Sound- and/or pressure-induced vertigo due to bone dehiscence of the superior semicircular canal. *Arch Otolaryngol Head Neck Surg* 1998; 124:249–258.

67. Merchant SN, Rosowski JJ: Conductive hearing loss caused by third-window lesions of the inner ear. *Otol Neurotol* 2008; 29:282–289.

68. Hirvonen TP, Carey JP, Liang CJ, Minor LB: Superior canal dehiscence: Mechanisms of pressure sensitivity in a chinchilla model. *Arch Otolaryngol Head Neck Surg* 2001; 127:1331–1336.

69. Minor LB, Cremer PD, Carey JP, Della Santina CC, Streubel SO, Weg N: Symptoms and signs in superior canal dehiscence syndrome. *Ann N Y Acad Sci* 2001; 942:259–273.

70. Streubel SO, Cremer PD, Carey JP, Weg N, Minor LB: Vestibular-evoked myogenic potentials in the diagnosis of superior canal dehiscence syndrome. *Acta Otolaryngol Suppl* 2001; 545:41–49.

71. Krombach GA, Dimartino E, Schmitz-Rode T, Prescher A, Haage P, Kinzel S, Gunther RW: Posterior semicircular canal dehiscence: A morphologic cause of vertigo similar to superior semicircular canal dehiscence. *Eur Radiol* 2003; 13:1444–1450.

72. Brantberg K, Bergenius J, Mendel L, Witt H, Tribukait A, Ygge J: Symptoms, findings and treatment in patients with dehiscence of the superior semicircular canal. *Acta Otolaryngol* 2001; 121:68–75.

73. Carter MS, Lookabaugh S, Lee DJ: Endoscopic-assisted repair of superior canal dehiscence syndrome. *Laryngoscope* 2014; 124:1464–1468.

74. Yew A, Zarinkhou G, Spasic M, Trang A, Gopen Q, Yang I: Characteristics and management of superior semicircular canal dehiscence. *J Neurol Surg B Skull Base* 2012; 73:365–370.

75. Krombach GA, Schmitz-Rode T, Haage P, Dimartino E, Prescher A, Kinzel S, Gunther RW: Semicircular canal dehiscence: Comparison of T_2-weighted turbo spin-echo MRI and CT. *Neuroradiology* 2004; 46:326–331.

76. Krombach GA, Di ME, Martiny S, Prescher A, Haage P, Buecker A, Gunther RW: Dehiscence of the superior and/or posterior semicircular canal: Delineation on T_2-weighted axial three-dimensional turbo spin-echo images, maximum intensity projections and volume-rendered images. *Eur Arch Otorhinolaryngol* 2006; 263:111–117.

77. Triglia JM, Nicollas R, Ducroz V, Koltai PJ, Garabedian EN: First branchial cleft anomalies: A study of 39 cases and a review of the literature. *Arch Otolaryngol Head Neck Surg* 1998; 124:291–295.

78. Ida JB, Stark MW, Xiang Z, Fazekas-May MM: Laryngeal cancer involving a branchial cleft cyst. *Head Neck* 2011; 33:1796–1799.

79. Roche JP, Younes MN, Funkhouser WK, Weissler MC: Branchiogenic carcinoma of a first branchial cleft cyst. *Otolaryngol Head Neck Surg* 2010; 143:167–8, 168.

80. Kroon DF, Lawson ML, Derkay CS, Hoffmann K, McCook J: Surfer's ear: external auditory exostoses are more prevalent in cold water surfers. *Otolaryngol Head Neck Surg* 2002; 126:499–504.

81. Kujundzic M, Braut T, Manestar D, Cattunar A, Malvic G, Vukelic J, Puselja Z, Linsak DT: Water related otitis externa. *Coll Antropol* 2012; 36:893–897.

82. Hajioff D, Mackeith S: Otitis externa. *Clin Evid* (Online) 2010; 2010.

83. Lee JE, Song JJ, Oh SH, Chang SO, Kim CH, Lee JH: Prognostic value of extension patterns on follow-up magnetic resonance imaging in patients with necrotizing otitis externa. *Arch Otolaryngol Head Neck Surg* 2011; 137:688–693.

84. Kwon BJ, Han MH, Oh SH, Song JJ, Chang KH: MRI findings and spreading patterns of necrotizing external otitis: Is a poor outcome predictable? *Clin Radiol* 2006; 61:495–504.

85. Minks DP, Porte M, Jenkins N: Acute mastoiditis—The role of radiology. *Clin Radiol* 2013; 68:397–405.

86. Santos VM, Oliveira ER, Barcelos MS, Figueiredo NC, Santos FH, Bergerot PG: Transverse sinus thrombosis associated with otitis media and mastoiditis. *J Coll Physicians Surg Pak* 2012; 22:470–472.

87. Cavaliere M, Di Lullo AM, Caruso A, Caliendo G, Elefante A, Brunetti A, Iengo M: Diffusion-weighted intensity magnetic resonance in the preoperative diagnosis of cholesteatoma. *ORL J Otorhinolaryngol Relat Spec* 2014; 76:212–221.

88. Martin C, Faye MB, Bertholon P, Veyret C, Dumollard JM, Prades JM: Cholesterol granuloma of the middle ear invading the cochlea. *Eur Ann Otorhinolaryngol Head Neck Dis* 2012; 129:104–107.

89. Pisaneschi MJ, Langer B: Congenital cholesteatoma and cholesterol granuloma of the temporal bone: Role of magnetic resonance imaging. *Top Magn Reson Imaging* 2000; 11:87–97.

90. Peng R, Chow D, De SD, Lalwani AK: Intensity of gadolinium enhancement on MRI is useful in differentiation of intracochlear inflammation from tumor. *Otol Neurotol* 2014; 35:905–910.

91. Booth TN, Roland P, Kutz JW, Jr., Lee K, Isaacson B: High-resolution 3-D T_2- weighted imaging in the diagnosis of labyrinthitis ossificans: Emphasis on subtle cochlear involvement. *Pediatr Radiol* 2013; 43:1584–1590.

92. Verbist BM: Imaging of sensorineural hearing loss: A pattern-based approach to diseases of the inner ear and cerebellopontine angle. *Insights Imaging* 2012; 3:139–153.

93. Wermker K, Kluwig J, Schipmann S, Klein M, Schulze HJ, Hallermann C: Prediction score for lymph node metastasis from cutaneous squamous cell carcinoma of the external ear. *Eur J Surg Oncol* 2015; 41:128–135.

94. Amin MF, El Ameen NF: Diagnostic efficiency of multidetector computed tomography versus magnetic resonance imaging in differentiation of head and neck paragangliomas from other mimicking vascular lesions: Comparison with histopathologic examination. *Eur Arch Otorhinolaryngol* 2013; 270:1045–1053.

95. Vogl T, Bruning R, Schedel H, Kang K, Grevers G, Hahn D, Lissner J: Paragangliomas of the jugular bulb and carotid body: MR imaging with short sequences and Gd-DTPA enhancement. *AJR Am J Roentgenol* 1989; 153:583–587.

96. Mafee MF, Raofi B, Kumar A, Muscato C: Glomus faciale, glomus jugulare, glomus tympanicum, glomus vagale, carotid body tumors, and simulating lesions. Role of MR imaging. *Radiol Clin North Am* 2000; 38: 1059–1076.

97. Karandikar A, Tan TY, Ngo RY: Diagnosing features of Jacobson's nerve schwannoma. *Singapore Med J* 2014; 55:e85–e86.

98. Nicolay S, de FB, Bernaerts A, Van DJ, Parizel PM: A case of a temporal bone meningioma presenting as a serous otitis media. *Acta Radiol Short Rep* 2014; 3:2047981614555048.

99. Stevens KL, Carlson ML, Pelosi S, Haynes DS: Middle ear meningiomas: A case series reviewing the clinical presentation, radiologic features, and contemporary management of a rare temporal bone pathology. *Am J Otolaryngol* 2014; 35:384–389.

100. Viswanatha B: Embryonal rhabdomyosarcoma of the temporal bone. *Ear Nose Throat J* 2007; 86:218, 220–218, 222.

101. Vegari S, Hemati A, Baybordi H, Davarimajd L, Chatrbahr G: Embryonal rhabdomyosarcoma in mastoid and middle ear in a 3-year-old girl: A rare case report. *Case Rep Otolaryngol* 2012; 2012:871235.

102. Muranjan M, Karande S, Parikh S, Sankhe S: A mistaken identity: Rhabdomyosarcoma of the middle ear cleft misdiagnosed as chronic suppurative otitis media with temporal lobe abscess. *BMJ Case Rep* 2014; 2014.

103. Marioni G, De FC, Stramare R, Carli M, Staffieri A: Langerhans' cell histiocytosis: Temporal bone involvement. *J Laryngol Otol* 2001; 115:839–841.

104. Angeli SI, Luxford WM, Lo WW: Magnetic resonance imaging in the evaluation of Langerhans' cell histiocytosis of the temporal bone: Case report. *Otolaryngol Head Neck Surg* 1996; 114:120–124.

105. Almuhanna K: Neuroendocrine adenoma of the middle ear with the history of otitis media and carcinoma of the cheek: A case report. *BMC Res Notes* 2014; 7:532.

106. Isenring D, Pezier TF, Vrugt B, Huber AM: Middle ear adenoma: Case report and discussion. *Case Rep Otolaryngol* 2014; 2014:342125.

107. Verhage-Damen GW, Engen-van Grunsven IA, van der Schans EJ, Kunst HP: A white mass behind the tympanic membrane: Adenoma of the middle ear with neuroendocrine differentiation. *Otol Neurotol* 2011; 32:e38–e39.

108. de FB, Kenis C, Vercruysse JP, Somers T, Pouillon M, Offeciers E, Casselman JW: Imaging of temporal bone tumors. *Neuroimaging Clin N Am* 2009; 19:339–366.

109. Pusiol T, Franceschetti I, Bonfioli F, Barberini F, Scalera GB, Piscioli I: Middle ear metastasis from dormant breast cancer as the initial sign of disseminated disease 20 years after quadrantectomy. *Ear Nose Throat J* 2013; 92:121–124.

110. Bouchetemble P, Heathcote K, Tollard E, Choussy O, Dehesdin D, Marie JP: Intralabyrinthine schwannomas: A case series with discussion of the diagnosis and management. *Otol Neurotol* 2013; 34:944–951.

111. Casselman JW: Diagnostic imaging in clinical neuro-otology. *Curr Opin Neurol* 2002; 15:23–30.

112. Donnelly MJ, Daly CA, Briggs RJ: MR imaging features of an intracochlear acoustic schwannoma. *J Laryngol Otol* 1994; 108:1111–1114.

113. Hegarty JL, Patel S, Fischbein N, Jackler RK, Lalwani AK: The value of enhanced magnetic resonance imaging in the evaluation of endocochlear disease. *Laryngoscope* 2002; 112:8–17.

114. Belli E, Rendine G, Mazzone N: Schwannoma of the facial nerve: Indications for surgical treatment. *J Craniofac Surg* 2013; 24:e396–e398.

115. Parnes LS, Lee DH, Peerless SJ: Magnetic resonance imaging of facial nerve neuromas. *Laryngoscope* 1991; 101:31–35.

116. Butman JA, Nduom E, Kim HJ, Lonser RR: Imaging detection of endolymphatic sac tumor-associated hydrops. *J Neurosurg* 2013; 119:406–411.

117. Bastier PL, de ME, Marro M, Elkhatib W, Franco-Vidal V, Liguoro D, Darrouzet V: Endolymphatic sac tumors: Experience of three cases. *Eur Arch Otorhinolaryngol* 2013; 270:1551–1557.

118. Eze N, Huber A, Schuknecht B: De novo development and progression of endolymphatic sac tumour in von Hippel-Lindau disease: An observational study and literature review. *J Neurol Surg B Skull Base* 2013; 74:259–265.

119. Lee SH, Choi SK, Lim YJ, Chung HY, Yeo JH, Na SY, Kim SH, Yeo SG: Otologic manifestations of acoustic neuroma. *Acta Otolaryngol* 2015; 135:140–146.

120. Heier LA, Comunale JP, Jr., Lavyne MH: Sensorineural hearing loss and cerebellopontine angle lesions. Not always an acoustic neuroma—A pictorial essay. *Clin Imaging* 1997; 21:213–223.

121. Mulkens TH, Parizel PM, Martin JJ, Degryse HR, Van de Heyning PH, Forton GE, De Schepper AM: Acoustic schwannoma: MR findings in 84 tumors. *AJR Am J Roentgenol* 1993; 160:395–398.

122. Liu P, Saida Y, Yoshioka H, Itai Y: MR imaging of epidermoids at the cerebellopontine angle. *Magn Reson Med Sci* 2003; 2:109–115.

123. Nguyen JB, Ahktar N, Delgado PN, Lowe LH: Magnetic resonance imaging and proton magnetic resonance spectroscopy of intracranial epidermoid tumors. *Crit Rev Comput Tomogr* 2004; 45:389–427.

124. Dechambre S, Duprez T, Lecouvet F, Raftopoulos C, Gosnard G: Diffusion-weighted MRI postoperative assessment of an epidermoid tumour in the cerebellopontine angle. *Neuroradiology* 1999; 41:829–831.

125. Bonneville F, Savatovsky J, Chiras J: Imaging of cerebellopontine angle lesions: An update. Part 2: Intra-axial lesions, skull base lesions that may invade the CPA region, and non-enhancing extra-axial lesions. *Eur Radiol* 2007; 17:2908–2920.

126. Rosado WM, Jr., Palacios E: Sudden onset of sensorineural hearing loss secondary to intralabyrinthine hemorrhage: MRI findings. *Ear Nose Throat J* 2008; 87:130–131.

127. Dubrulle F, Kohler R, Vincent C, Puech P, Ernst O: Differential diagnosis and prognosis of T_1-weighted post-gadolinium intralabyrinthine hyperintensities. *Eur Radiol* 2010; 20:2628–2636.

15

Magnetic Resonance Imaging of the Salivary Gland

Takashi Nakamura and Misa Sumi

CONTENTS

The salivary gland is a platform for a broad spectrum of benign and malignant tumors that originate from the gland and nongland tissue components, and for inflammatory diseases that occur within and outside of the gland. In addition, as an excretory organ, the salivary gland shows morphological and functional responses to various disorders associated with metabolic and systemic diseases. Although clinical examinations may suggest the presence of a salivary gland disease, these examinations are often insufficient for detecting and characterizing the disease. However, imaging examination is a useful tool for that purpose and magnetic resonance (MR) imaging may be the first choice because of its high tissue contrast and the various aspects of information that can be obtained with different sets of imaging sequences. In this chapter, we review the MR imaging features of the healthy and diseased salivary glands.

15.1 Function and Innervation of the Salivary Glands

The major salivary glands, which include the parotid, submandibular, and sublingual glands, secrete 90% of the saliva in the oral cavity. Additionally, numerous (600–1000) minor salivary glands are present in the submucosa throughout the upper digestive tract, including the oral cavity. The mesenchyme and the nerves in the salivary gland originate from the neural crest, while the salivary gland epithelia arises from multipotent precursors in the embryonic ectoderm [1]. Functional innervation and the factors that exert neuronal-epithelial communication are crucial for maintaining the epithelial progenitor pool and inducing ductal differentiation; thus, these elements are important for morphogenesis and damage repair in the salivary gland [2–4].

The major physiological function of the salivary glands is to produce saliva, which is essential for lubricating the oral cavity and upper digestive tract for eating and vocalization, pH buffering, and immunity [1,2,5]. Saliva secretion is regulated by both parasympathetic and sympathetic autonomic innervation. The salivary glands consist of two types of epithelial parenchyma cells: acinar (80%) and ductal (20%) cells. The acinar cells produce either serous or mucous saliva, which contains 99.3% water, 0.4% proteins, and 0.3% salts. However, the ductal cells modify secretions, primarily by reabsorbing the salts. The innervated stellate myoepithelial cells, which circumscribe the acini and intercalated ducts, are proposed to facilitate secretion by constricting these cells. The ductal cells can be categorized into three types based on their morphological and histological appearances: intercalated, striated, and granular.

The parotid gland is innervated via the parasympathetic fibers of the glossopharyngeal nerve (cranial nerve IX), which originates from the inferior salivatory nucleus (ISN) in the medulla of the brain stem and synapse with the otic ganglion (OG) located just below the foramen ovale on the skull base [2]. Then, the postganglionic fibers emerge from the ganglion and run into the parotid gland via the auriculotemporal nerve of cranial nerve V for the secretion of serous saliva. The submandibular and sublingual glands are innervated by the parasympathetic fibers that originate from the superior salivatory nucleus (SSN) in the pons region of the brain stem and join the facial nerve (cranial nerve VII) in the internal auditory canal. The fibers are conveyed by the chord tympani nerve in the mastoid and enter the infratemporal fossa, where the fibers join the lingual nerve (a branch of the marginal mandibular division of the trigeminal nerve [cranial nerve V]). The fibers subsequently synapse with the submandibular ganglion (SG), whereupon the postganglionic fibers exit the ganglion, enter into the submandibular and sublingual glands, and stimulate serous-mucous and mucous saliva secretion, respectively.

The thoracic ganglion (ThG) sends the ascending sympathetic fibers to the superior cervical ganglion (SCG). The postganglionic sympathetic fibers exit the SCG and join the nerve plexus around the external carotid artery and its branches, including the facial artery [2]. The postganglionic sympathetic fibers from the external carotid plexus extend branches to all three pairs of major salivary glands.

15.2 MR Imaging of Normal Salivary Glands

15.2.1 Parotid Gland

The superficial layer of the deep cervical fascia creates the parotid space in the lateral suprahyoid neck. The parapharyngeal and masticator spaces lie at the medial and anterior aspects of the parotid space, respectively. The parapharyngeal space is filled with fat tissues, and the masticator space contains the ramus and the posterior body of the mandible, four types of masticator muscles (masseter, temporal, and internal and lateral pterygoids), and branches of the trigeminal nerve (V3), and the inferior alveolar artery and vein. The posterior belly of the digastric muscle is between the upper part of the parotid space and the carotid space. The parotid space (and, therefore, the parotid gland) spans from the mastoid tip superiorly to below the angle of the mandible.

The parotid gland is relatively rich in fat tissues compared with the other major salivary glands (submandibular and sublingual glands); thus, the T_1 MR signal levels of the parotid gland are higher than those of the adjacent masticatory muscles and have the signals lower than those of the subcutaneous fat tissues (Figure 15.1a). Accordingly, the T_2 MR signal levels of the parotid glands are greatly diminished on fat suppression (Figure 15.1b).

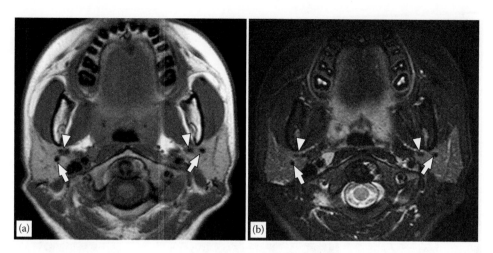

FIGURE 15.1

A 23-year-old woman with healthy parotid glands. Axial T_1-weighted (a) and fat-suppressed T_2-weighted (b) MR images show normal parotid gland architectures. Arrows, retromandibular vein; arrowheads external carotid artery.

The excretory duct system of the parotid gland consists of the main extraglandular duct (Stensen's duct) and the intraglandular main duct and ductules. Stensen's duct penetrates the fascia, runs anteriorly along the anterior margin of the masseter muscle, penetrates the buccal muscles, and finally opens onto the buccal mucosa at the level of the upper molar teeth. The duct system may be detectable on fat-suppressed T_2-weighted images as curvilinear high-intensity signals along the surface of the masseter muscle (Stensen's duct) and radiate in the gland parenchyma (intraglandular main duct and ductules) (Figure 15.1b). However, MR sialography more clearly delineates the entire structure of the parotid duct system (Figure 15.2).

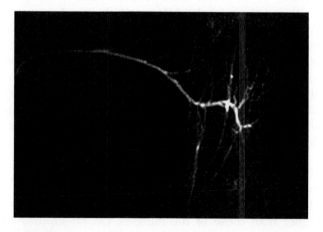

FIGURE 15.2

A 36-year-old woman with healthy parotid glands. MR sialography shows normal duct ramifications (Stensen's duct, intraglandular main duct, and intraglandular ductules) of parotid gland.

The parotid gland has a rich blood supply and shows greater enhancement after contrast medium injection compared with the neighboring masseter muscle. The retromandibular vein, which is a vascular landmark of the parotid gland, runs vertically along the posterior edge of the mandibular ramus (Figure 15.1). The external carotid artery also runs behind the ramus of the mandible medial to the retromandibular vein.

After emerging from the cranial space through the stylomastoid foramen, the facial nerve divides into the temporofacial and cervicofacial branches shortly after entering the gland; in many cases, the nerve further divides into five terminal branches (temporal, zygomatic, buccal, marginal mandibular, and cervical). The facial nerve runs between the superficial and deep lobes of the gland just lateral to the retromandibular vein. Information regarding the course of the intraparotid facial nerve could facilitate surgical planning for a parotid tumor. However, imaging the intraparotid course of the facial nerve is very challenging because of the fine structures and the complex anatomy of the nerve. The intraparotid facial nerve can be differentiated from the small vessels as well as from the intraparotid ducts based on the differences in signal intensity when using balanced steady-state free-precession MR imaging techniques such as fast imaging employing steady-state acquisition (FIESTA) and balanced turbo-field echo (bTFE) sequences [6,7] (Figure 15.3).

The parotid gland contains lymph nodes (~20). Because the volume of the superficial lobe is approximately twice that of the deep lobe, the superficial lobe contains a greater number of nodes than the deep lobe [8]. The parotid

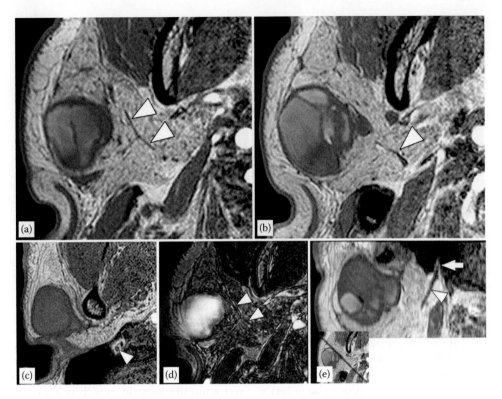

FIGURE 15.3
A 26-year-old man with lymphoepithelial cyst of parotid gland. MR imaging delineates intraparotid facial nerve (arrowheads). (a–c) Axial 3D T_1-weighted fast field echo (FFE) MR images. (d) Axial 3D fat-suppressed balanced turbo field echo (bTFE) MR image. (e) Coronal reformatted image obtained by axial 3D T_1-weighted FFE. Arrow, stylomastoid foramen.

nodes drain the external ear and lateral scalp. Enlarged nodes may be delineated by conventional MR imaging under conditions in which the glands are infected, such as mumps, human immunodeficiency virus (HIV), and juvenile recurrent parotitis (JRP), but this is unusually not possible in healthy glands.

15.2.2 Submandibular Gland

The large superficial lobe of the submandibular gland lies directly beneath the superficial layer of the deep cervical fascia posterior to the facial artery [9]. The facial vein and the cervical branch of the facial nerve cross the superficial lobe. The smaller deep lobe of the gland (referred to as the *lingual extension*) wraps around the posterior border of the mylohyoid muscle, and extends anteriorly, and contacts the posterior aspect of the sublingual gland. The deep lobe runs between the mylohyoid (lateral) and hypoglossal (medial) muscles (Figure 15.4). The submandibular spaces on both sides communicate freely with each other, indicating that fascia do not separate the left and right submandibular glands.

Furthermore, there is no fascial boundary between the posterior part of the sublingual gland and the submandibular gland.

The excretory duct system of the submandibular gland consists of the main extraglandular duct (Wharton's duct) and the intraglandular main duct and ductules. Wharton's duct emerges from the lingual extension of the gland, runs anteriorly in the lateral compartment of the sublingual space medial to the gland in relation to the sublingual artery and lingual nerve, and opens at the apex of the sublingual caruncle. MR sialography can demonstrate the entire course of the extraglandular and intraglandular main ducts and the intraglandular dendrite ramifications (Figure 15.5).

The submandibular gland contains smaller amounts of fat tissue compared with the parotid gland. Accordingly, the T_1 signal intensity levels of the submandibular gland are lower than those of the parotid gland, but are higher than those of the muscles. On fat-suppressed T_2-weighted MR images, the signal intensity levels of the submandibular gland are higher than those of the muscles (Figure 15.6).

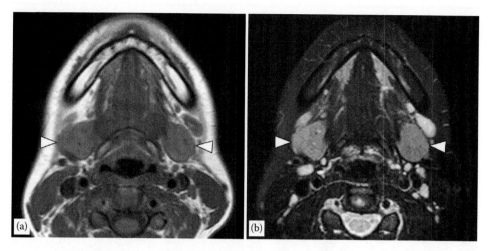

FIGURE 15.4
A 29-year-old woman with healthy submandibular glands. Axial T_1-weighted (a) and fat-suppressed T_2-weighted (b) MR images show normal architectures of submandibular gland. Arrowheads, submandibular glands.

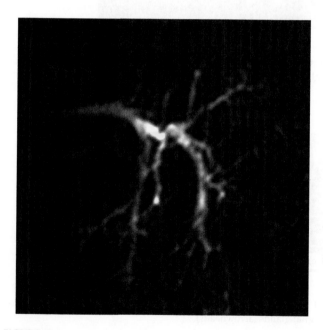

FIGURE 15.5
A 36-year-old man with healthy submandibular glands. MR sialogram shows normal duct ramifications (Wharton's duct, intraglandular main duct, and intraglandular ductules) of submandibular gland.

In contrast to the parotid gland, the submandibular gland does not contain a lymph node. Instead, satellite nodes surround the submandibular gland (anterior, lateral, and posterior nodes) (Figure 15.4); the submandibular nodes (neck level IB) drain the anterior portion of the face, including the oral cavity, anterior paranasal sinuses, and the orbits.

15.2.3 Sublingual Gland

The sublingual gland is a common site for the extension of infection from the oral cavity, and the rate of occurrence of malignant tumors is higher at this site compared to that in the other major salivary glands. The gland is located close to the nerves (the lingual nerve, the sensory branch of the trigeminal nerve combined with the chorda tympani branch of the facial nerve; the glossopharyngeal nerve; and the accessory nerve), veins (lingual artery and vein), the tongue, and the muscles that play important roles in digestion and speech [10]. Therefore, it is critical to know preoperatively the extent of the tumor and infections that occur in the sublingual gland and in the floor of the mouth.

The sublingual gland lies against the sublingual depression on the lingual aspect of the mandible, rests on the mylohyoid muscle, and is separated medially from the genioglossus muscle by Wharton's duct [10] (Figure 15.7). The sublingual spaces, which are not encapsulated by the fascia and sit within the tongue between the mylohyoid muscle inferiorly and the genioglossus and geniohyoid muscles medially, communicate with each other via an isthmus located beneath the frenulum of the tongue; thus, the sublingual glands on both sides also communicate with each other. The absence of fascia between the sublingual and submandibular spaces also allows the sublingual gland to reach posteriorly to the lingual portion of the submandibular gland. On T_1-weighted images, the sublingual gland appears as an area of intermediate signal intensity that is lower than that of the surrounding fat tissues, but is higher

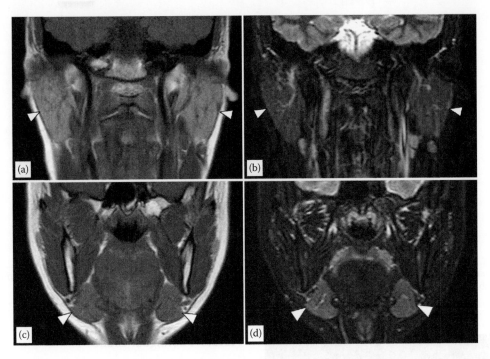

FIGURE 15.6
A 29-year-old woman with healthy parotid and submandibular glands. Coronal T_1-weighted (a, c) and STIR (b, d) MR images of parotid (arrowheads on a and b) and submandibular (arrowheads on c and d) glands show greater fat contents in parotid glands compared with submandibular glands.

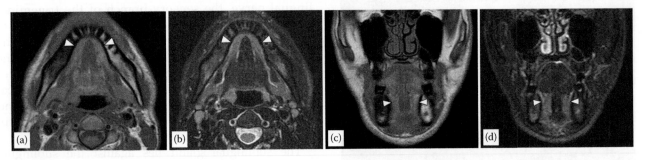

FIGURE 15.7
A 36-year-old man with healthy sublingual glands. Axial T_1-weighted (a) and fat-suppressed T_2-weighted (b) MR images and coronal T_1-weighted (c) and STIR (d) MR images show normal structures of sublingual glands (arrowheads).

than that of the neighboring muscles [10] (Figure 15.7). On T_2-weighted images, carcinomas arising in or extending into the oral floor are isointense or hyperintense compared to the sublingual glands. Sumi et al. showed that 21 carcinomas were hypointense compared to the sublingual gland and 17 carcinomas were isointense or hyperintense compared to the glands after gadolinium enhancement, and this frequently led to a reduction in the contrast between the tumors and the glands [10].

The gland size decreases with age. However, the signal intensity levels of the sublingual gland do not change with age. In contrast, the parotid gland exhibits age-related decreases in density on computed tomography (CT). The signal intensity levels of the submandibular

gland also do not change with age. Age-related variations and reactive changes including oncocyte proliferation, squamous and mucous metaplasia, hyperplasia, atrophy, and degeneration may contribute to changes in MR signal intensity levels of the parotid gland [11].

The sublingual gland has several excretory ducts that drain into the sublingual fold from each of the gland lobules, except for the largest lobule, which drains into the sublingual caruncle lateral to Wharton's duct.

15.2.4 Minor Salivary Glands

The minor salivary glands are distributed in the anterior two-thirds of the tongue, palatal mucosa, buccal mucosa, pharyngeal mucosa, retromolar triangle, and lips [12]. The total number of minor salivary glands is estimated to be within the range of 600–1000. Although the salivary secretion from the minor salivary glands is small in quantity relative to the whole saliva (the minor salivary glands produce 6%–10% of all saliva), the broader distributions of the minor salivary glands play an important role in protecting the oral cavity against pathogens, mediating the sense of taste, moisturizing the oral mucosa, and facilitating smooth pronunciation. In contrast to the major salivary glands, the minor salivary glands lack a branching network of draining system. Instead, each salivary gland lobule has its own simple duct. Most of the minor salivary glands are mucous glands; however, Ebner's glands, which are serous glands located in the circumvallate papillae of the tongue, is an example of an exception of this rule.

The paucity of publication regarding the imaging features of the minor salivary glands may be mainly due to their small size and the underestimation of their role in the maintenance of the oral environment. However, the appropriate use of high-resolution MR imaging techniques with surface coil can successfully delineate the normal and diseased states of the minor salivary glands. For instance, we introduce the MR imaging features of the labial and palate glands in this chapter.

The labial glands are localized in the connective tissue of the lamina propria, extending in depth from the mucous membrane to the orbicularis iris muscle; however, some glands may be found in the muscle layer [12]. The glands are hyperintense on T_1-weighted and fat-suppressed T_2-weighted images (Figure 15.8). The gland parenchyma is enhanced after gadolinium injection, which is very suggestive of rich vascularity, similar to that of the major salivary glands. In both the upper and lower lips, the glands extend throughout the incisor to the premolar regions and are thicker in the posterior part than in the anterior part. Each of the upper and lower labial glands consists of 1–2 and 1–3 layers of multiple gland clusters, respectively; therefore, the maximal thickness of the lower glands (4.1 ± 1.1 mm in women and 4.2 ± 1.1 mm in men) is greater than that of the upper glands (2.5 ± 0.8 mm in women and 2.9 ± 1.2 mm in men) [12].

The palatal glands occupy the posterior two-thirds of the hard palate and the soft palate. The anterior parts of the gland are thinner than the posterior parts of the gland, and are composed of a single layer of gland clusters in the middle third of the hard palate and of multiple cluster layers in the posterior third of the hard palate and in the soft palate (Figure 15.9).

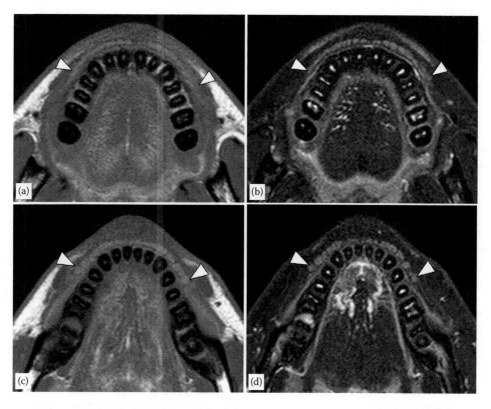

FIGURE 15.8

A 33-year-old man with healthy labial glands. Axial T_1-weighted (a, c) and fat-suppressed T_2-weighted (b, d) MR images show normal structures and distributions of upper (a, b) and lower (c, d) labial glands (arrowheads).

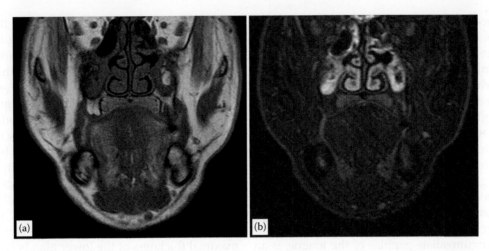

FIGURE 15.9
A 53-year-old man with healthy palatal glands. Coronal T_1-weighted (a) and STIR (b) MR images show normal distributions of palatal glands.

15.3 MR Imaging of Salivary Gland Tumors

Salivary gland tumors represent 3%–5% of head and neck tumors, with 54%–79% of these being benign. The World Health Organization (WHO) classification of salivary gland tumors is listed in Table 15.1 [13]. As shown in the panel, most salivary gland tumors are of epithelial

TABLE 15.1

WHO Histological Classification of Salivary Gland Tumors [13]

Benign epithelial tumors
Pleomorphic adenoma
Basal cell adenoma
Warthin tumor
Oncocytoma
Canalicular adenoma
Sebaceous adenoma
Lymph adenoma
Ductal papilloma
Cystadenoma

Malignant epithelial tumors
Acinic cell carcinoma
Mucoepidermoid carcinoma
Adenoid cystic carcinoma
Polymorphous low-grade adenocarcinoma
Epithelail-myoepithelial carcinoma
Clear cell carcinoma, not otherwise specified
Basal cell adenocarcinoma
Sebaceous carcinoma
Sebaceous lymphadenocarcnoma
Cystadenocarcinoma
Low-grade cribriform cystadenocarcinoma
Mucinous adenocarcinoma

(Continued)

TABLE 15.1 (*Continued*)

WHO Histological Classification of Salivary Gland Tumors [13]

Oncocytic adenocarcinoma
Salivary duct carcinoma
Adenocarcinoma, not otherwise specified
Myoepithelial carcinoma
Carcinoma ex pleomorphic adenoma
Carcinosarcoma
Metastasizing pleomorphic adenoma
Squamous cell carcinoma
Small cell carcinoma
Large cell carcinoma
Lymphoepithelial carcinoma
Sialoblastoma

Soft tissue tumors
Hemangioma

Hematolymphoid tumors
Hodgikin lymphoma
Extranodal marginal zone B-cell lymphoma
Secondary tumors

origin and derived from the epithelial component of the salivary glands. Other salivary gland tumors include lymphomas and hemangiomas. Therefore, the main objective of MR imaging examination is differentiating between benign and malignant tumors. In the parotid gland, the spatial relationship between the tumor and facial nerve would be an important issue in preoperative MR examinations for planning tumor excision (Figure 15.3). An extension of a salivary gland tumor into the deep part of the neck may necessitate differentiating between tumors originating from the salivary gland and those form the adjacent organs and tissues (Figure 15.10).

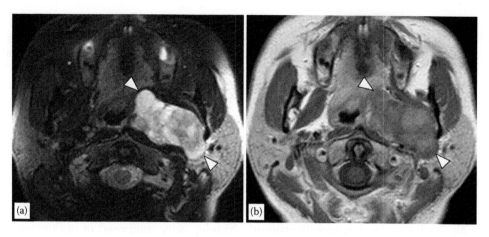

FIGURE 15.10

A 51-year-old man with pleomorphic adenoma of parotid gland. (a) Axial fat-suppressed T_2-weighted MR image shows inhomogeneous tumor extending from deep lobe of parotid gland into neighboring parapharyngeal space (arrowheads). (b) Axial contrast-enhanced T_1-weighted MR image shows inhomogeneous tumor enhancement.

In the parotid gland, which includes lymph nodes and is rich in lymphoid tissues, metastatic nodes, lymphomas, and lymphoepithelial tumors should also be considered in differential diagnosis. However, because salivary gland tumors may have characteristic MR imaging features, this technique is useful for discriminating some types of tumors with distinctive histological features.

15.3.1 Conventional MR Imaging

The marginal morphology of salivary gland tumors on T_1- and T_2-weighted MR images may be useful for differentiating between benign or low-grade malignant tumors (well-defined margins) and intermediate to high-grade malignant tumors (ill-defined or infiltrative margins). Indeed, most benign salivary gland tumors are histologically well encapsulated and, thus, have sharp margins on MR images. However, sharp margins are observed in 58% of primary malignant parotid tumors and in 38% of metastatic tumors, and most lymphomas have sharp margins [14]. Therefore, sharp margins on MR images do not necessarily indicate benign tumors.

Tumor homogeneity may be indicative of the histologic nature and, therefore, the grade of salivary gland tumors. Almost all salivary gland tumors show a homogeneous appearance on T_1-weighted images. The exceptions are tumors that have cystic areas containing mucin or proteinaceous fluids and those with hemorrhagic necrosis; these tumors show large areas of very high signal intensity on T_1-weighted images. An inhomogeneous tumor stroma on T_2-weighted MR images is observed in 54% of benign tumors (pleomorphic adenomas and Warthin tumors) and in 77% of malignant salivary gland tumors [14]. Therefore, inhomogeneity per se

also does not conclusively indicate the benign or malignant nature of salivary gland tumors; furthermore, inhomogeneity is observed in many malignant tumor types with different histological grades (low, intermediate, and high grades), indicating that the inhomogeneity does not necessarily reflect the tumor grades.

Recently, Thoeny's group analyzed the MR imaging findings of parotid tumors to determine effective markers for discriminating malignant lesions from benign lesions and found that the MR imaging findings of T_2 hypointensity and ill-defined tumor margins on contrast-enhanced T_1-weighted MR images are the best indicators of malignant lesions [15]. An ill-defined margin before contrast enhancement is reportedly a less effective marker for malignant lesions. Compared with these findings, cystic/necrotic areas do not help distinguish between benign and malignant tumors. Perineural spread, if present, could be used as a potent indicator for malignant tumors. Below, the clinicopathological and conventional MR imaging properties of the major types of primary benign and malignant salivary gland tumors are summarized.

Pleomorphic adenoma is the most common benign tumor of the salivary gland. Pleomorphic adenomas are well encapsulated, and the major components of this tumor are glandular, ductal, or solid structures of epithelial tumor cells [13]. The tumor also contains mesenchymal tissues associated with cartilaginous or fibromyxomatous tissues. The tumor area that contains abundant fibromyxomatous stroma shows high signal intensity on T_2-weighted images and exhibits gradual enhancement in dynamic study [16–19] (Figure 15.11). However, the tumor area consisting of densely proliferating epithelial tumor cells has low signal intensity on T_2-weighted images with only slightly enhancement.

The lesion may be associated with cysts, hemorrhage, necrosis, or calcification.

Warthin tumor is the second most common tumor of the salivary gland. The tumor is composed of glandular and cystic structures that may exhibit a papillary cystic arrangement and lined by a double layer of epithelium, comprising inner columnar eosinophilic cells lined by smaller basal cells [13]. The stroma contains varying amounts of lymphoid tissues associated with germinal centers. The tumor exclusively occurs in the parotid gland and may occur bilaterally (10%–15%). The solid parts of the tumor show a rapid enhancement with a high washout ratio (WR), similar to other lymphoid tissues [17–20]. On T_1-weighted images, the tumor may show multiple areas of high signal intensity corresponding to cystic areas containing proteinaceous fluids (Figure 15.12).

Other benign salivary gland tumors show less characteristic features than the aforementioned pleomorphic

adenoma and Warthin tumor, and reports regarding the MR imaging features of these begin tumors are very limited. Myoepithelioma and basal cell adenoma have well-defined margins with occasional association of hypointense rim on T_2-weighted images and may contain a central necrotic or cystic area; the periphery of this area may be enhanced, corresponding to proliferating myoepithelial or basal cells [17,18,21–23]. The tumors also have large cystic area(s). The solid area consisting of myoepithelial or basal cells is moderately enhanced, but it does not exhibit the characteristic rapid enhancement with high WR profile that is commonly seen in Warthin tumors (Figure 15.13). Cystadenoma is histologically composed of multiple small cystic spaces and/or large cysts surrounded by lobules of salivary glands or by connective tissues [13]. Accordingly, MR imaging features are characterized by multiple (central and/or peripheral) areas of low signal intensity on T_1-weighted MR images and inhomogeneous signals

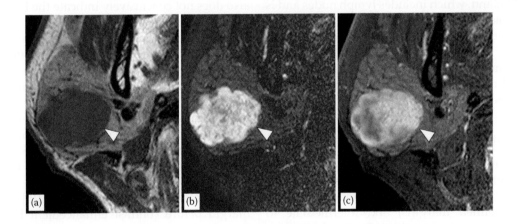

FIGURE 15.11
A 50-year-old woman with pleomorphic adenoma of parotid gland. (a) Axial T_1-weighted MR image shows homogeneous tumor (arrowhead) of right parotid gland. (b) Axial fat-suppressed T_2-weighted MR image shows heterogeneously hyperintense tumor (arrowhead). (c) Axial contrast-enhanced fat-suppressed T_1-weighted MR image shows inhomogeneous tumor enhancement (arrowhead).

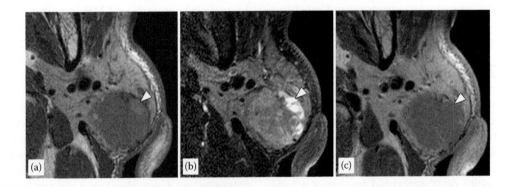

FIGURE 15.12
A 58-year-old man with Warthin tumor. (a) Axial T_1-weighted MR image shows moderately hyperintense foci (arrowhead) in left parotid gland. (b) Axial fat-suppressed T_2-weighted MR image shows hyperintense foci (arrowhead) largely corresponding to hyperintense (cystic) areas on T_1-weighted MR image (a). (c) Axial contrast-enhanced T_1-weighted MR image shows nonenhanced tumor areas (arrowhead) corresponding to hyperintense foci on (a) and (b).

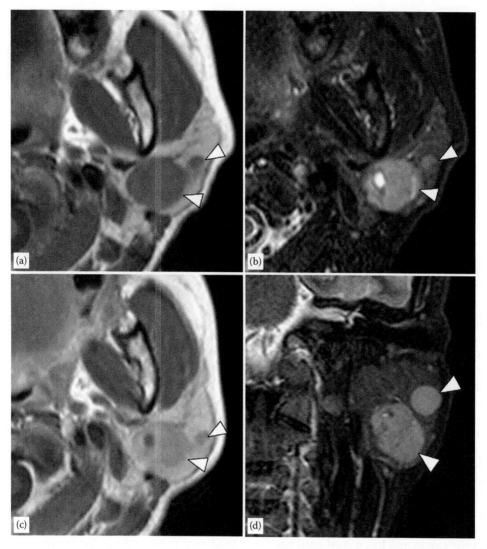

FIGURE 15.13

A 49-year-old woman with basal cell adenoma. (a) Axial T_1-weighted MR image shows multiple homogeneous tumors (arrowheads) of left parotid gland. (b) Axial STIR image shows tumors with intermediate signal intensities (arrowheads). Note that a larger tumor contains hyperintense focus with hypointense demarcations. (c) Axial contrast-enhanced T_1-weighted MR image shows moderate tumor enhancement (arrowheads). (d) Coronal STIR MR image shows multiple tumors with intermediate signal intensities (arrowheads).

on T_2-weighted images. The stromal areas are basically enhanced slowly with small areas of rapid enhancement. Relatively unique among benign salivary gland tumors is oncocytoma, which is composed of mitochondria-rich oncocytes. Oncocytomas are hypointense on T_1-weighted MR images, but are isointense compared to the parotid gland parenchyma on fat-suppressed T_2-weighted and contrast-enhanced T_1-weighted MR images. These MR imaging findings are not observed in any other benign or malignant salivary gland tumors [24].

Mucoepidermoid carcinoma represents ~30% of malignant salivary gland tumors and arises most commonly (~50%) in the parotid gland. The minor salivary glands are the second (~45%) most common sites of occurrence. Low-grade tumors are well circumscribed and usually contain large cystic areas containing mucous components [13], which appear as high-intensity spots within the tumor on both T_1- and T_2-weighted images [19]. The cystic areas are demarcated by abundant mucous cells, which are lined by epidermoid components. The mucous cells are large with a pale cytoplasm and peripherally displaced nuclei. High-grade tumors are poorly circumscribed, infiltrative, and dominantly composed of proliferating epidermoid cells. In contrast to the low-grade counterparts, mucous cells are rarely observed in high-grade mucoepidermoid carcinomas. Generally, the solid tumor areas are hypointense on T_2-weighted MR images owing to the presence of abundant fibrous connective tissues [19] (Figure 15.14).

Adenoid cystic carcinoma constitutes 4%–15% of all salivary gland tumors. The tumor is well circumscribed,

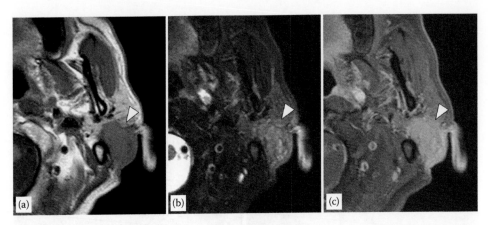

FIGURE 15.14
A 67-year-old man with mucoepidermoid carcinoma. (a) Axial T_1-weighted MR image shows homogeneous tumor (arrowhead) of left parotid gland. Note tumor extension into subcutaneous tissues. (b) Axial fat-suppressed T_2-weighted MR image shows heterogeneous tumor (arrowhead) with speckled distributions of hyperintense spots. (c) Axial contrast-enhanced, fat-suppressed T_1-weighted MR image shows moderately and near-homogeneously enhanced tumor (arrowhead).

but only partially encapsulated, and invariably infiltrative [25]. Cystic changes or hemorrhages are rare in this malignant tumor. Histopathologically, adenoid cystic carcinoma is categorized into tubular, cribriform, or solid type. The tubular type of adenoid cystic carcinoma is composed of tubules lined by inner epithelial and outer myoepithelial cells. The cribriform pattern is most frequently observed and characterized by nests of cells containing microcystic spaces in the cytoplasms. These spaces are filled with hyaline or basophilic mucoid material. The solid type is composed of sheets of uniform basaloid cells lacking tubular structures or microcystic changes in the cytoplasm. Although the degree of cellularity increases from the tubular to solid pattern, every adenoid cystic carcinoma has mixed patterns. On MR images, many adenoid cystic carcinomas are poorly demarcated with infiltrative margins [14] (Figure 15.15). The tumors do not have distinguishing MR signal intensity characteristics. On T_1-weighted images, most tumors exhibit homogeneous signal intensity that is similar to that of the muscles; whereas, the tumors exhibit low-to-high signal intensity with variable homogeneity

on T_2-weighted images [25] (Figure 15.15). Tumor areas with low signal intensity on T_2-weighted images correspond to histological areas with high cellularity (solid type) and areas with high signal intensity on T_2-weighted images correspond to less cellular areas (cribriform or tubular type). However, these MR imaging features may also be observed in benign salivary gland tumors such as pleomorphic adenoma, and differentiating between these tumors based on MR imaging findings alone is difficult. In large adenoid cystic carcinomas, an area with very high signal intensity corresponding to hemorrhagic necrosis may be observed in the solid part of the tumor, providing a clue for discriminating this malignant salivary gland tumor [19]. Adenoid cystic carcinoma has a tendency to spread via the nerves (Figure 15.16). Sigal et al. reported that perineural spread occurred in 6 out of 27 patients with adenoid cystic carcinoma [25]. Although neurons themselves are resistant to tumor, perineural spread occurs by both centripetal and centrifugal tumor growth. Perineural spread of tumor can be observed on MR images as nerve enlargement, or contrast enhancement, or both. Other MR imaging signs of

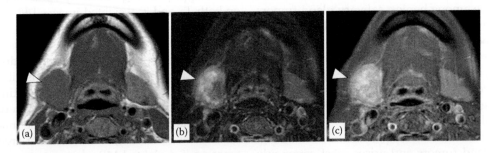

FIGURE 15.15
A 52-year-old woman with adenoid cystic carcinoma. (a) Axial T_1-weighted MR image shows homogeneous tumor (arrowhead) of right submandibular gland. Note irregular tumor margin. (b) Axial fat-suppressed T_2-weighted MR image shows heterogeneous tumor (arrowhead). (c) Axial contrast-enhanced, fat-suppressed T_1-weighted MR image shows heterogeneously enhanced tumor (arrowhead).

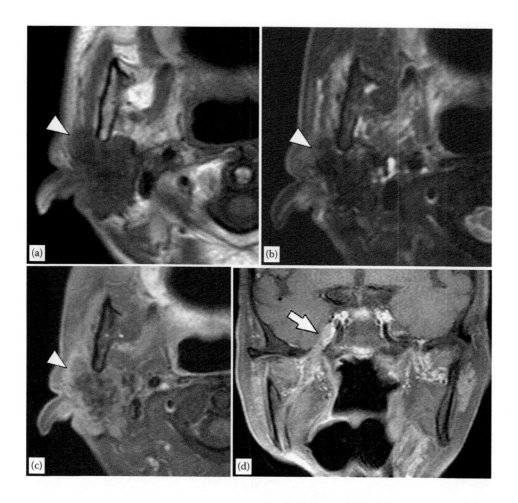

FIGURE 15.16

A 65-year-old woman with adenoid cystic carcinoma. (a) Axial T_1-weighted MR image shows heterogeneously hypointense tumor with irregular margin (arrowhead) of right parotid gland. (b) Axial fat-suppressed T_2-weighted MR image shows heterogeneously hypointense tumor (arrowhead). (c) Axial contrast-enhanced, fat-suppressed T_1-weighted MR image shows heterogeneous tumor enhancement (arrowhead). (d) Coronal contrast-enhanced, fat-suppressed T_1-weighted MR image shows perineural tumor spread into mid-brain via foramen ovale.

perineural spread include an enlargement of the foramen and a replacement of fat within the foramen that the tumor passes.

Acinic cell carcinoma occurs most commonly in the parotid gland (80%–90%) and about 4% and less than 1% of the tumors develop in the submandibular and sublingual glands, respectively [13,26]. Bilateral involvement is observed in 3% of the patients with this tumor [26]. The tumor is solid or partially cystic. Histologically, acinic cell carcinoma is characterized by serous cells with a bubbly basophilic cytoplasm containing zymogen granules [13]. Although the MR imaging features are nonspecific, intratumoral hemorrhage may be observed as a hyperintense area(s) on T_1- and T_2-weighted images [27]. Metastasis to the regional lymph node is not uncommon.

Salivary duct carcinoma is characterized by very aggressive behavior, such as metastasis to the regional lymph nodes, early distant metastasis, locoregional recurrence, and a high rate (>50%) of mortality [28]. Histologically, the tumor is composed of atypical epithelial cells arranged in varying proportions of cribriform, papillary, or solid growth patterns with fibrotic stroma, sometimes displaying a squamoid appearance that is reminiscent of high-grade mucoepidermoid carcinoma or squamous cell carcinoma [13]. Salivary duct carcinoma has ill-defined margins, and the tumors are predominantly heterogeneous on T_2-weighted MR images [28,29] (Figure 15.17). The tumors often contain single or multiple necrotic areas with an irregularly enhancing rim [29]. Intratumoral calcification may be observed, and perineural spread commonly occurs [30]. Lymph node metastasis is not uncommon (Figure 15.17).

Malignant changes in pleomorphic adenoma (carcinoma ex pleomorphic adenoma) occur after a long history of pleomorphic adenoma, a sudden period of rapid growth, and a recurrence. The carcinoma ex

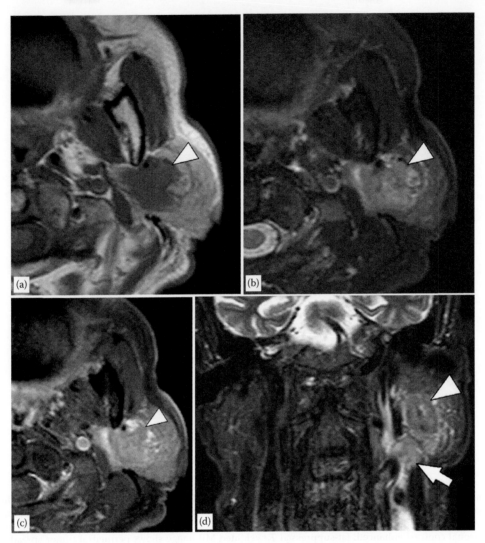

FIGURE 15.17

An 81-year-old man with salivary duct carcinoma. (a) Axial T_1-weighted MR image shows homogeneously hypointense tumor with irregular margin (arrowhead) of left parotid gland. (b) Axial fat-suppressed T_2-weighted MR image shows heterogeneous tumor (arrowhead). (c) Axial contrast-enhanced, fat-suppressed T_1-weighted MR image shows irregularly enhanced tumor periphery (arrowhead). (d) Coronal STIR MR image shows heterogeneous tumor (arrowhead) and regional lymph node metastasis (arrow).

pleomorphic adenoma is typically poorly circumscribed compared with the original tumor. However, pleomorphic adenoma has a broad spectrum of MR imaging features as mentioned above (Figure 15.18). Therefore, differentiation between carcinoma ex pleomorphic adenoma and its benign counterpart is often challenging [31].

15.3.2 Dynamic Contrast-Enhanced MR Imaging

Conventional T_1- and T_2-weighted MR imaging features are characteristic in some types of salivary gland tumors and may be useful for differentiating between benign and malignant tumors. However, conventional MR imaging findings are basically nonspecific, and it

is difficult to discriminate malignant tumors in many cases. Another tool for differentiating between benign and malignant salivary gland tumors is the characterization of tumor vascularity by using a contrast agent. The consensus is that malignant salivary gland tumors are richly vascular to compensate for the increasing demand of the growing tumors.

To quantitate the vascularity of salivary gland tumors, dynamic contrast-enhanced study (DCE MR imaging) has been applied to assess changes in the amounts of contrast agent in the tumor tissues relative to the time after injection of the contrast agent. The most conventional DCE MR imaging assessment of salivary gland tumors involves analysis of the time-signal intensity curve (TIC). The perfusion properties of tumors can be

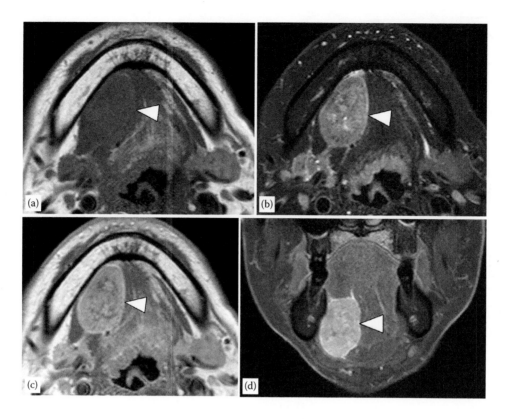

FIGURE 15.18

A 58-year-old man with carcinoma ex pleomorphic adenoma. (a) Axial T_1-weighted MR image shows homogeneously hypointense tumor (arrowhead) of right sublingual gland. (b) Axial fat-suppressed T_2-weighted MR image shows heterogeneous tumor (arrowhead). (c) Axial contrast-enhanced, fat-suppressed T_1-weighted MR image shows heterogeneously enhanced tumor (arrowhead). (d) Coronal contrast-enhanced, fat-suppressed T_1-weighted MR image shows heterogeneously enhanced tumor (arrowhead) displacing but not spreading beyond mylohyoid muscle.

estimated by the TIC patterns (Figure 15.19); the early slope phase corresponds to the contrast agent transfer to the tissues, and it is considered largely dependent upon the blood flow rate, inflow blood volume, and capillary permeability [18]. The magnitude of the peak enhancement is representative of the volume of the extravascular–extracellular space, whereas the time to peak relates is related to the extravascular–extracellular space per contrast agent transfer to the tissues. Although the capillary density and blood volume of the tumors are closely related to the kinetics of contrast enhancement, the cellularity can also contribute to the early phases of tumor enhancement. In cancers, high capillary permeability and poor lymphatic drainage create a condition in which cellularity is very limited [32]. Therefore, low cellularity may lead to slow uptake of the contrast agent. On the other hand, the WR may depend upon the histological type, such as the malignant phenotype and stromal properties; thus, the mechanisms that influence contrast agent WR in these tumors are complex. Several researchers proposed that the contrast-enhancement patterns are based on the vascular permeability of the lesions and the capillary density may be a minor physiologic factor

compared to vascular permeability with regard to the differences in TIC profiles [33,34].

The TIC curve can be characterized by using several parameters. The first is the peak time, which represents the time required to reach the intensity peak (SI_{peak}); the SI_{peak} is the first signal intensity measured within the scan time. The second is the increment ratio (IR), which is calculated by the following formula: ($SI_{peak} - SI_{pre}$)/SI_{pre}, where SI_{pre} represents the signal intensity measured before contrast agent injection. The third is the WR, which is calculated by the following formula: ($SI_{peak} - SI_{180}$) × 100/($SI_{peak} - SI_{pre}$), where SI_{180} represents the signal intensity measured at the last scan time (180 s after the first scan, in this case) [17,18]. Additional TIC parameters can be used such as the slope, which indicates the angle at which the early slope rises. However, two of these parameters, peak time and WR, are essentially sufficient for effectively characterizing the TICs of salivary gland tumors. By using these two parameters, the TICs of salivary gland tumors can be classified into five types, Type 1–Type 5: Type 1 TICs are those with IRs equal to or less than 20% (flat pattern); Type 2 TICs are those with IRs greater than 20% and peak times equal to or

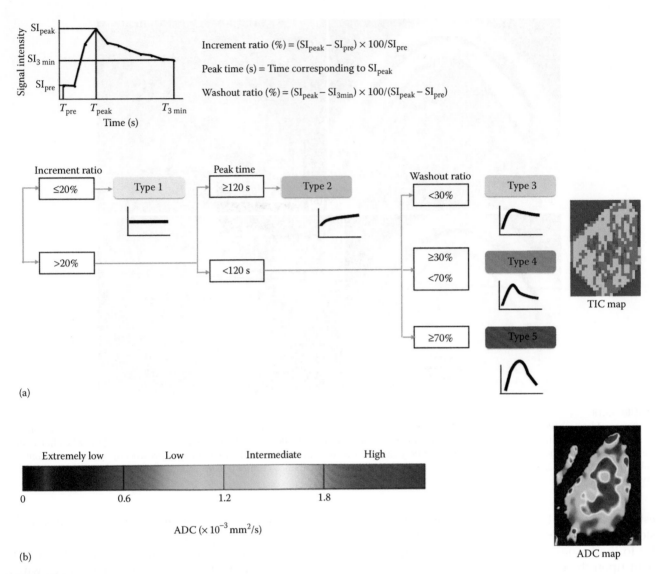

Increment ratio (%) = (SI$_{peak}$ − SI$_{pre}$) × 100/SI$_{pre}$

Peak time (s) = Time corresponding to SI$_{peak}$

Washout ratio (%) = (SI$_{peak}$ − SI$_{3min}$) × 100/(SI$_{peak}$ − SI$_{pre}$)

(a)

(b)

ADC (× 10^{-3} mm^2/s)

FIGURE 15.19

Multiparametric MR imaging of salivary gland tumors. (a) TICs and. Overall TICs and TICs in each pixels of tumor area can be categorized into five types (Type 1–5) based on the increment ratio, peak time, and washout ratio according to formulas in upper panel. SI$_{peak}$, signal intensity at peak; SI$_{pre}$, signal intensity of preenhancement; SI$_{3\,min}$, signal intensity 180 s after the start of contrast medium injection; T_{pre}, time at pre-enhancement; $T_{3\,min}$, 180 s after the start of contrast medium injection; T_{peak}, time required to reach the intensity peak. Color TIC map shows carcinoma ex pleomorphic adenoma in a 58-year-old man. (b) ADC map. Grayscale ADC map images are converted to color ADC map images. ADCs are categorized into extremely low (0 × 10^{-3} mm^2/s–0.6 × 10^{-3} mm^2/s), low (0.6 × 10^{-3} mm^2/s–1.2 × 10^{-3} mm^2/s), intermediate (1.2 × 10^{-3} mm^2/s–1.8 × 10^{-3} mm^2/s), or high (1.8 × 10^{-3} mm^2/s–2.6 × 10^{-3} mm^2/s). Color ADC map shows the same tumor as in (a).

longer than 120 s (slow uptake pattern); Type 3 TICs have IR greater than 20%, with peak times shorter than 120 s, and with WRs smaller than 30% (rapid uptake/low washout pattern); and Type 4 TICs display IRs greater than 20%, with peak times shorter than 120 s and WRs equal to or greater than 30% (rapid uptake/high washout pattern). Additionally, Type 5 TICs can be discriminated from the other TICs as those having IRs greater than 20%, with peak times equal to or shorter than 120 s, and with WRs equal to or greater than 70% [35] (Figure 15.19).

Based on these TIC classification criteria, pleomorphic adenomas characteristically exhibit Type 2 TIC profiles [18,21] (Figure 15.20). In contrast, Warthin tumors are best characterized by Type 4 TIC profiles. The salivary gland tumors showing Type 2 TIC profiles include pleomorphic adenomas, adenoid cystic carcinomas, carcinoma ex pleomorphic adenomas, myoepitheliomas, and adenocarcinomas. The salivary gland tumors displaying Type 3 TIC profiles encompass mucoepidermoid carcinomas, adenocarcinomas, salivary duct carcinomas, squamous cell carcinomas, and epithelial–myoepithelial

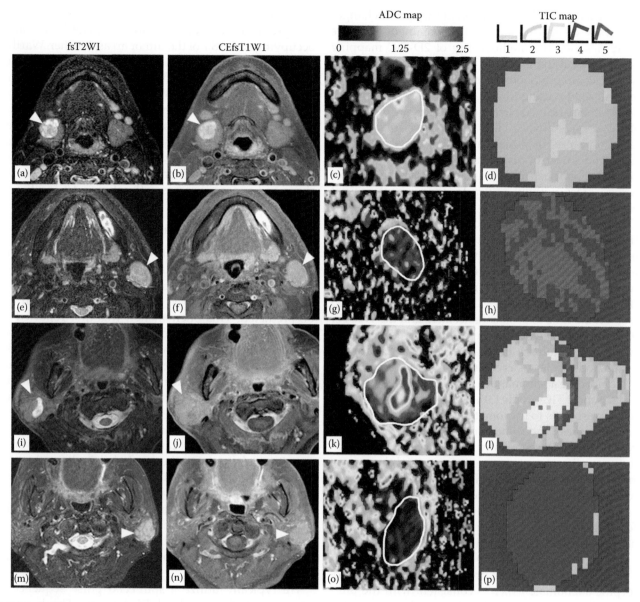

FIGURE 15.20

TIC and ADC maps of salivary gland tumors. (a–d) 58-year-old woman with pleomorphic adenoma. (e–h) 70-year-old man with Warthin tumor. (i–l) 65-year-old man with adenocarcinoma. (m–p) 74-year-old woman with MALT lymphoma.

carcinomas. The salivary gland tumors exhibiting Type 4 TIC profiles include Warthin tumors, oncocytomas, and lymphomas. Therefore, TIC analysis does not effectively differentiate between benign and malignant salivary gland tumors [18]. The overall TIC profiles of benign salivary gland tumors are those that occupy the greatest tumor area, and the minor tissue components are neglected from the overall TIC assessments. On the other hand, large stromal or cystic areas mask the hypercellular tumor areas in salivary gland carcinomas. Consequently, the overall TIC profiles of salivary gland carcinomas may not be differentiated from those of pleomorphic adenomas. In addition, the overall TIC

profiles are not distinctive between Warthin tumors and malignant lymphomas.

Salivary gland tumors comprise distinctive tissues, including proliferating tumor cells, myxomatous tissues, lymphoid tissues, necrotic areas, and cysts [13]. Therefore, TIC analysis using a large region of interest (ROI) in a histologically heterogeneous tumor may result in spurious results with regard to tumor histology. To avoid this potential error, tumor vascularity should be analyzed at a pixel-to-pixel level (Figure 15.20). In 2D TIC analysis, the ultimate ROI is a single pixel. Therefore, by analyzing the elemental TICs, which represent the changes in contrast agent concentration with time in each pixel, the fundamental functional changes in salivary gland tumors could be obtained. Here, we briefly describe the methodology of 2D TIC mapping [18]. Sequential dynamic MR images in a digital imaging and communications in medicine (DICOM) format are analyzed by using ImageJ (http://rsbweb.nih.gov/ij/) and Mathematica (Wolfram Research) or MATLAB® (MathWorks) software. The initial ROI is manually placed on a pre- or postenhancement T_1-weighted MR image and saved by using ImageJ software such that it includes as much tumor area as possible. Cystic or necrotic areas without contrast enhancement are not necessarily omitted from the ROI area. Then, the same ROIs are automatically placed onto the other dynamic MR images by repeatedly copying and pasting the initial ROI by using ImageJ software. TIC analysis is performed by using Mathematica or MATLAB software on a lesion-by-lesion (overall TIC) and a pixel-by-pixel basis (TIC mapping), and each TIC is automatically classified based on the time to peak enhancement and the WR. For TIC mapping analysis, the percentages of tumor areas corresponding to each of the five TIC patterns are calculated and displayed as color TIC maps (Figure 15.20).

Pleomorphic adenomas are largely occupied by areas with a Type 2 TIC profile (31%–97%) (Figure 15.20). This area corresponds to myxomatous or dense fibrous connective tissues. The tumor areas with the Type 3 TIC profile, which are frequently seen in tumor areas containing proliferating tumor cells in stromal tissues, are smaller than those with the Type 2 profile in pleomorphic adenomas. Areas with the type 4 TIC profile, which is consistent with tumor areas containing densely proliferating cells such as lymphoid tissues, clear cells, and epidermoid cells, constitute a minor component in this benign tumor. On the other hand, areas with a Type 1 TIC profile may be predominant in some pleomorphic adenomas, which is consistent with the presence of large cystic areas in pleomorphic adenomas. In sharp contrast to pleomorphic adenomas, Warthin tumors contain predominant areas with a Type 4 TIC profile (41%–93%), but areas with a Type 2 TIC profile are minor components of these tumors (Figure 15.20). The TIC profiles of myoepithelioma are similar to those of pleomorphic adenomas. Papillary cystadenoma contain large areas with a Type 1 TIC profile corresponding to cystic components.

In contrast to benign salivary gland tumors, malignant salivary gland tumors are composed of large areas with a Type 3 TIC profile (Figure 15.20). A tumor area consisting of scattering cancer cell nests displays a Type 2 or Type 3 TIC profile, depending upon the amount of stromal tissue relative to the amount of the cellular component. The tumor areas with a Type 4 TIC profile in malignant tumors occupy greater areas compared to those in pleomorphic adenomas, but much less than those in Warthin tumors. The tumor areas with a Type 4 TIC profile are predominant in malignant lymphomas, occupying 93%–98% of the tumor areas. Further, Warthin tumors contain small but substantial areas displaying Type 1 or Type 2 TIC profiles, but such tumor areas are minimally observed (~0.3%) in malignant lymphomas. In malignant salivary gland tumors, necrotic areas may occupy larger tumor areas compared to benign salivary gland tumors, showing Type 1 TIC profiles. Malignant salivary gland tumors may also contain significant areas of hemoprotein deposition, which results from hemorrhage and appears as areas with a Type 1 TIC profile.

15.3.3 Diffusion-Weighted Imaging

In liquids, water molecules move endlessly, but constantly bump into each other as well as into other molecules. The resulting motion of the water molecules is random, and neither the speed nor the direction can be predicted. This phenomenon was first reported by Brown (although Ingenhousz found the phenomenon earlier than Brown did) and Einstein later provided a sound mathematical description and a physical interpretation [36]. In a homogeneous environment without boundaries, the (water) molecules simply collide with other similar molecules. The probability of finding a specific molecule that was previously identified at the original position ($r = 0$) can be determined by two laws: $\langle r \rangle = 0$ and $\langle r^2 \rangle = 6$ Dt (the symbol $\langle \rangle$ is defined as the average). The first law states that the molecules will spread around the original position but will not move as a whole. The second law states that the probability of finding the molecules at a distance from the original position increases with time. The speed at which this random motion occurs can be characterized by D, the diffusion factor.

We typically utilize a spin-echo pulse sequence to acquire diffusion-weighted (DW) images. This basically consists of first (90°) and second (180°) radiofrequency (RF) pulses and two rectangular gradient pulses with an equal size, which are generated before and after the 180° RF pulse [36,37]. Diffusion-weighted imaging (DWI) is based on the signal attenuation in the moving molecule that is passing through the RF and two additional gradient pulses. The cumulative phase shift in the presence of the magnetic field gradient is proportional to the strength of the field, the duration of the gradient, and the location of the molecule. After the first 90° RF pulse, the water molecule begins to accrue phase. In the case of a static molecule, which does not move significantly between the first and second gradients, the subsequent 180° RF pulse, which is identical to the first pulse, rotates the vectors around the vertical axis. In this case, the molecule will be in phase at the end of the second gradient, but the initial phase shift is reversed by the 180° RF pulse. However, for a moving molecule, the rotation changes between the first and second RF pulses. This dispersion of the phase shift results in the attenuation of signal amplitude, the magnitude of which depends on the gradient intensity, the gradient duration, and the interval between the two gradients. These three parameters can be combined into one parameter, which is called the b-value.

Through DWI, we look at the incoherent motion of water molecules within each voxel [36,37]. The completely free diffusion process is solely dependent on the temperature and viscosity of the fluid. However, the environment of water molecules in the body is not homogeneous because it contains membranes, macromolecules, fibers, and other structures that seriously hamper the free diffusion process. Furthermore, another incoherent motion is present within a voxel that can lead to a reduction in signal intensity. This is particularly true for perfusion. The water molecules in the blood capillaries will be perceived as being in a pseudorandom motion in the tortuous capillaries in tissues. This necessitates the induction of the apparent diffusion coefficient (ADC) instead of the D to describe the diffusion process in biological tissues. The ADC is calculated based on the following equation:

$$S_b = S_0 \cdot \exp(-b \cdot D)$$

indicating that the attenuation of signal intensity varies exponentially with b and D.

The clinical application of the imaging technology has been limited to the brain mainly due to its inherent susceptibility to motion and susceptibility artifacts [38]. However, recent technical innovations (particularly, faster imaging techniques) have enabled the effective application of DWI to the diagnosis of diseases in the extracranial organs [35]. The diffusion process is dominated by the biological environment, which the water molecules encounter. Therefore, DWI allows the assessment of changes in the biological tissue well below the spatial resolution of the MR images in pathological conditions.

The relative contents of the secretory acinus and the relative amounts of interlobular and extralobular fat tissues vary among the different types of the major salivary glands. The parotid gland contains abundant fat tissues, whereas, the other gland types do not. Different gland types consist of various types (serous or mutinous secretory cells) and different densities of excretory cells. Accordingly, the various types of the major salivary glands have different ADC levels; for example, the ADCs of the submandibular glands ($0.87 \pm 0.05 \times 10^{-3}$ mm^2/s) as assessed by using 2 b-values (500 and 1000 s/mm^2) are greater than those of the parotid glands ($0.63 \pm 0.11 \times 10^{-3}$ mm^2/s) and are smaller than those

of the sublingual glands ($0.97 \pm 0.09 \times 10^{-3}$ mm^2/s) [39] (Figures 15.1, 15.4, and 15.7). Notably, considerable differences have been reported in the ADC values of healthy major salivary glands, and these differences may have resulted from the use of different b-values in these studies. Therefore, the results from different studies using various b-values should be carefully interpreted [40].

Salivary gland tumors are heterogeneous in components, consisting of cancer nests, myxomatous tissues, lymphoid tissues, necrosis, and cysts. Therefore, the different tissue and cellular components in salivary gland tumors would greatly affect the ADC levels of the tumors. A pixel-based ADC analysis of salivary gland tumors allows us to know the 2D distributions of the tumor areas having distinctive ADC levels [39] (Figure 15.19). ADC levels can be tentatively classified into four categories: areas with extremely low ADC ($<0.6 \times 10^{-3}$ mm^2/s); low ADC (0.6×10^{-3} mm^2/s $\leq$ ADC $<1.2 \times 10^{-3}$ mm^2/s); intermediate ADC (1.2×10^{-3} mm^2/s $\leq$ ADC $<1.8 \times 10^{-3}$ mm^2/s); or high ADC ($\geq 1.8 \times 10^{-3}$ mm^2/s). The obtained ADC maps can be expressed as percentage areas having particular ADC levels relative to the total tumor area and can further be compared with the histological specimens.

In general, the areas with proliferating and densely packed tumor cells in salivary gland cancers and the areas with lymphoid tissues in malignant lymphomas and Warthin tumors have extremely low ADCs. Further, the areas with cancer cell nests consisting of epidermoid or clear cells scattered in fibrous connective tissues exhibit low ADCs. Finally, the proliferating tumor areas with small cystic or necrotic areas have intermediate ADCs, and fibromyxoid tissues and large cystic or necrotic areas both have high ADCs [36].

In pleomorphic adenomas, the areas with high ADCs corresponding to cystic or myxomatous lesions occupy the predominant (~54%) areas of the tumors (Figure 15.20). The areas with intermediate ADCs, which correspond to the area of proliferating tumor cells, occupy almost all (45%) of the remaining tumor areas. Warthin tumors often exhibit heterogeneous ADC distributions; these tumors are composed of areas with extremely low ADCs, indicating lymphoid tissues, and areas with high ADCs, indicating cystic cavity formation in the lymphoid stroma (Figure 15.20). Other benign salivary gland tumors also contain substantial areas with high ADCs; for example, 28% of the areas of papillary cystadenomas have high ADCs. In contrast, the high ADC areas are very limited in malignant salivary gland tumors, and 0%–3% of the tumor areas of mucoepidermoid carcinomas, salivary duct carcinomas, lymphomas, adenoid cystic carcinomas, and adenocarciomas have high ADCs. Instead, these malignant salivary gland tumors consist of predominant areas of extremely low or low ADCs (Figure 15.20). Accordingly, large areas with high ADCs may be characteristic of benign salivary gland tumors, and the presence of large areas with low or extremely low ADCs could be a hallmark of malignant salivary gland tumors. However, large overlaps exist in the ADCs between different histologic subtypes of salivary gland tumors, and these overlaps are present within the group of benign and malignant tumors as well as between groups [39,41]. For instance, Warthin tumor also contains significant areas with low or extremely low ADCs. Therefore, the use of low or extremely low ADC criterion does not effectively differentiate between benign and malignant salivary gland tumors.

Eida et al. recently introduced the use of the MR imaging system for addressing the issue of effectively differentiating between benign and malignant salivary gland tumors [21]. The system consists of a stepwise approach involving the combined use of DW and DCE MR imaging. An appropriate combination of the tumor-by-tumor (overall ADCs and TICs) and the pixel-by-pixel (ADC and TIC maps) analyses with DW and DCE MR imaging parameters can effectively differentiate between benign and malignant salivary gland tumors, and this approach correctly differentiates

between 52 benign and 18 malignant salivary gland tumors [21] (Figure 15.21). The best criteria indicate that malignant tumors are those with small (<30%) areas of Type 1 TIC profile along with one of the following MR imaging characteristics: the Type 3 overall TIC profile (carcinomas) or the Type 4 overall TIC profile and extremely low (<0.60 × 10^{-3} mm^2/s) overall ADCs (lymphomas), or Type 2 overall TIC profile and large (>40%) areas of low or extremely low ADCs (<1.2 × 10^{-3} mm^2/s) (carcinomas) [21].

15.3.4 Intravoxel Incoherent Motion Imaging

The diffusion and perfusion properties of tumors are physically and biologically distinctive phenomena, and estimating these two distinctive phenomena may be important for the preoperative diagnosis and management of patients. Because the diffusion property of tumor tissues largely depends upon cell-dependent attenuation, the ADC may be useful for predicting malignancy in some tumor types [39,41,42]. On the other hand, perfusion can be used as a biomarker for predicting the responsiveness of some tumor types to chemotherapy [43,44]. However, a combined use of DCE MR

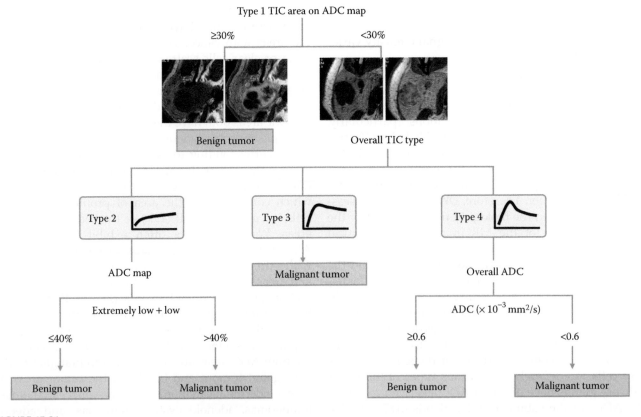

FIGURE 15.21
Multiparametric MR imaging for the differential diagnosis between benign and malignant salivary gland tumors. Schema shows effective differentiation between benign and malignant salivary gland tumors by using apparent diffusion coefficient (ADC) cutoff points and time-signal intensity curve (TIC) profiles.

imaging and DWI may not be clinically feasible because of the extended examination time and postimaging data processing necessary for image analysis.

DWI provides information concerning the tissue diffusion property. However, the diffusion parameter ADC cannot separate pure molecular diffusion from the motion of water molecules in the capillary network [36]. Therefore, the obtained diffusion data are contaminated by blood microcirculation, and this is particularly true at low b-values. To avoid this, we typically use a pair of large (>200 s/mm²) b-values, depending on the tissue of interest. For organs with an intrinsically high ADC (e.g., the prostate), higher b-values are required to provide good contrast of the lesions. However, using higher b-values will result in low signal-to-noise ratios (SNRs). Therefore, the ADC value obtained at a high b-value should be carefully interpreted because the values may have dropped close to or below the noise level and may not be reliable. In such cases, a parallel imaging technique, such as sensitivity encoding (SENSE), may be useful for improving the SNR by reducing the TE as well as the image acquisition time.

Intravoxel incoherent motion imaging (IVIM) imaging can quantitate molecular diffusion and microcirculation in the capillary network in a single-image voxel at MR imaging, provided that multiple b-values encompassing both low b-values (<200 s/mm²) and high b-values (>200 s/mm²) are used [45]. IVIM imaging provides the relationship between signal intensities and b-values by using the following equation:

$$\frac{S_b}{S_0} = (1 - f)\cdot\exp(-bD) + f\cdot\exp[-b(D + D^*)] \quad (15.1)$$

where:

f is a microvascular volume fraction representing the fraction of diffusion linked to microcirculation

D is the diffusion parameter representing pure molecular diffusion (i.e., diffusion coefficient)

D^* is the perfusion-related incoherent microcirculation

S_0 and S_b are signal intensities at $b = 0$ and $b > 0$ s/mm², respectively [46]. Although various sets of b-values can be used, a sufficient number of b-values (<200 s/mm²)

should be used to simulate the biexponential IVIM curve to estimate tumor perfusion. In our institution, we use 11 b-values, including $b = 0$ s/mm² (0, 10, 20, 30, 50, 80, 100, 200, 300, 400, or 800 s/mm²) [47]. Because D^* is much greater than D, the effects of D^* on the signal decay at large b-values (>200 s/mm²) can be neglected. Accordingly, D can be obtained with a linear regression algorithm (i.e., the least-square method by using b-values of 200, 300, 400, and 800 s/mm²). Therefore, Equation 15.1 can be simplified as follows:

$$\frac{S_{b1}}{S_{b2}} = \exp[(b_2 - b_1)\cdot D] \quad (15.2)$$

where S_{b1} and S_{b2} are the signal intensities at two different b-values (>200 s/mm²).

Given an estimated D value with the linear regression algorithm, the corresponding f and D^* values can be calculated using a nonlinear regression algorithm based on Equation 15.1, where the fit f and D^* values can be obtained after substituting the initial f and D^* values into the Levenberg–Marquardt algorithm [47,48].

Differentiation between benign and malignant salivary gland tumors is a successful example of the application of IVIM imaging [47]. A stepwise classification using D and D^* cutoff points effectively differentiated between 20 benign and 11 malignant salivary gland tumors as well as between 12 pleomorphic adenomas and eight Warthin tumors. Accordingly, Warthin tumors have very small D values ($D \leq 0.8 \times 10^{-3}$ mm²/s) and large D^* values ($D^* > 23 \times 10^{-3}$ mm²/s), whereas pleomorphic adenomas have large D values ($D \geq 1.4 \times 10^{-3}$ mm²/s), or have intermediate D values (1.1×10^{-3} mm²/s $\leq D < 1.4 \times 10^{-3}$ mm²/s), and very small D^* values ($D^* < 12 \times 10^{-3}$ mm²/s). Further, malignant salivary gland tumors have very small D values ($D \leq 0.8 \times 10^{-3}$ mm²/s) and very small D^* values ($D^* \leq 12 \times 10^{-3}$ mm²/s), small D values (0.8×10^{-3} mm²/s $<D < 1.1 \times 10^{-3}$ mm²/s), or intermediate D values (1.1×10^{-3} mm²/s $< D < 1.4 \times 10^{-3}$ mm²/s), and intermediate D^* values (12×10^{-3} mm²/s $\leq D^* \leq 23 \times 10^{-3}$ mm²/s) (Figure 15.22). The D values of malignant tumors are smaller than those of pleomorphic adenomas, but those of Warthin tumors are even smaller than those of

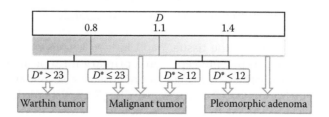

FIGURE 15.22
IVIM parameter-based differentiation of pleomorphic adenomas, Warthin tumors, and malignant salivary gland tumors. The combined use of D ($\times 10^{-3}$ mm²/s) and D^* ($\times 10^{-3}$ mm²/s) cutoff points can achieve effective differentiation of the salivary gland tumors.

malignant salivary gland tumors. The very small D values of Warthin tumors may be attributed to the abundant lymphoid tissues containing densely packed small lymphoid cells in the tumors. Therefore, tumor cellularity may be a dominant factor for the varying levels of D values among the pleomorphic adenomas, Warthin tumors, and malignant salivary gland tumors. The D^* threshold is low for pleomorphic adenomas, but is high for Warthin tumors. D^* is considered proportional to the mean capillary length and average blood velocity [37]. Therefore, the differential criteria of D^* values for the two benign salivary gland tumors may reflect tumor vascularity. The broad range of D^* values among malignant salivary gland tumors may result from variations in tumor vascularity among the different tumor types and among tumors with different sizes.

Equation 15.1 can be transformed as follows:

$$S_b = S_0 \cdot (1 - f) \cdot \exp(-bD) + S_0 \cdot f \cdot \exp[-b(D + D^*)] \quad (15.3)$$

where:

$S_0 \cdot (1 - f)$ indicates the signal intensity of a tumor ($= S_{ti}$)
$S_0 \cdot f$ indicates the signal intensity of blood capillaries ($= S_{ca}$). Therefore, we can obtain the following formula:

$$f = \frac{S_{ca}}{(S_{ca} + S_{ti})} \quad (15.4)$$

Equation 15.4 indicates that the f value is determined as the signal intensity of blood capillaries and tumor tissues [47]. In the IVIM analysis, we ignore the relaxation effects based on the assumption that the relaxation times of tissue and capillary blood are similar. However, the T_2 contributions from tumor tissue and blood capillaries may be greatly different. Therefore, the f value will be wrongly estimated in some types of tumors if the T_2 component for S_{ca} and S_{ti} is ignored. Thus, f value should be interpreted carefully in tumors.

The determination of IVIM parameters (D, f, and D^*) from a multiexponential signal decay curve using the least-squares method is cumbersome, and may not be suitable for routine clinical use. Alternatively, we can estimate the IVIM parameters by using a limited number of b-values (e.g., $b = 0$, 200, and 800 s/mm²) [49]. Similar to the assessment with the least-square method, D can also be estimated as a decline between $b = 200$ and 800 s/mm², calculated as ln $(S_{200}/S_{800})/600$. The estimation of the D value allows us to estimate the tissue perfusion parameter f as $1 - S_{inter}/S_0$ (S_{inter} is the interception of the logarithmic regression line obtained by using b-values of 200 and 800 s/mm² with the y-axis). D^* can be estimated by the formula $D^* = $ ln $(S_0/S_{inter})/200$. By using a limited number of b-values, simplified IVIM imaging has the advantages of achieving DW MR images that have better quality and of examining broader areas of

the head and neck region in a single scan compared with IVIM imaging using more b-values. For example, IVIM imaging using 11 b-values requires 1 min 53 s for obtaining 10 DW image slices per patient; on the other hand, IVIM imaging using three b-values requires 26 s for obtaining the same number of DW images per patient. When compared with authentic IVIM imaging, simplified IVIM imaging provides significantly greater (+2%) D values, smaller (−7%) f values, and smaller (−96%) D^* values for the parotid glands, masseter muscles, and head and neck tumors. However, a stepwise approach using a combination of the D and D^* values obtained by simplified IVIM imaging differentiated between pleomorphic adenomas, Warthin tumors, and malignant salivary gland tumors with the same efficacy (91% accuracy = 21/23). Further, the finding that measurement errors occurred less frequently with the simplified method compared with the least-squares method supports the clinical feasibility of using the simplified method. Therefore, the simplified method appears to be less error-prone than the least-squares method. However, the simplified IVIM technique may lose important information regarding tissue perfusion. For example, the three b-value IVIM imaging technique abandons the idea of using low b-value (<100 s/mm²) DWIs for a fine analysis of the different vascular compartments with different vessel sizes. The DWI at a low b-value has a high SNR and is sensitive to water flows compared with DWIs at high b-values [50]. In addition, low b-value DWIs could also allow visual assessment of diseased areas. Therefore, low b-value DWI may be useful for discriminating between some types of salivary gland tumors, such as those exhibiting low signals at high b-value images.

15.4 MR Imaging of Sjögren's Syndrome and Impaired Salivary Gland Function

Sjögren's syndrome (SS) is a systemic autoimmune disorder primarily affecting the exocrine glands and clinically characterized by impaired function of the salivary and/or lacrimal glands. However, SS may be associated with extraglandular manifestations, such as Raynaud phenomenon, arthritis, and vasculitis, and liver, lung, and kidney lesions may also develop. SS occurs either in a primary form or in association with other autoimmune diseases (secondary form), including rheumatoid arthritis (RA), systemic lupus erythematosus (SLE), and systemic sclerosis (SSc), thus complicating the clinical profile of SS patients. The diagnosis of SS patients is made based on subjective symptoms (dry mouth and dry eyes; sicca symptoms) and objective signs (salivary and lacrimal flow tests, serological tests for anti-SS-A/ Ro and/or anti-SS-B/La antibodies, or rheumatoid factor and antinuclear antibody, and histological assessment of labial gland biopsy specimen) [51].

The American–European Consensus Group (AECG) criteria for SS have been largely accepted in clinical practice and in research (Table 15.2) [51]. Serological tests are negative in many patients with SS, with ant-SS-A and anti-SS-B antibodies being positive in 33%–74% and 23%–52% of the primary SS patients, respectively.

Notably, anti-SS-A antibodies may be present in the absence of anti-SS-B antibodies, but vice versa situations rarely occur [52]. The labial gland of patients with SS shows all the pathologic changes in the major salivary glands of patients with SS, such as infiltrations and aggregations of lymphocytes associated with acinar cell destruction, except for the appearance of epimyoepithelial islands. Therefore, histological examination of the labial gland biopsy specimen has been proposed as the most specific method for diagnosing SS patients.

TABLE 15.2

The American–European Consensus Group Classification Criteria for SS

I. Ocular symptoms: a positive response to at least one of the following questions
1. Have you had daily, persistent, troublesome dry eyes for more than 3 months?
2. Do you have a reorient sensation of sand or gravel in the eyes?
3. Do you use tear substitutions more than three times a day?

II. Oral symptoms: a positive response to at least one of the following questions
1. Have you had a daily feeling of dry mouth for more than 3 months?
2. Have you had recurrently or persistently swollen salivary glands as an adult?
3. Do you frequently drink liquids to aid in swallowing dry food?

III. Ocular signs—that is, objective evidence of ocular involvement defined as a positive results for at least one of the following two tests
1. Schirmer's test, performed without anesthesia ($\leq$5 mm in 5 min)
2. Rose bengal score or other ocular dry eye score ($\geq$4 according to van Bijsterveld's scoring system)

IV. Histopathology: in minor salivary glands focal lymphocytic sialoadenitis, evaluated by an expert histopathologist, with a focus score $\geq$1, defined as a number of lymphocytic foci (which are adjacent to normal-appearing mucous acini and contain more than 50 lymphocytes) per 4 mm^2 of glandular tissue

V. Salivary gland involvement: objective evidence of salivary gland involvement defined by a positive result for at least one of the following diagnostic tests
1. Unstimulated whole salivary flow (<1.5 ml in 15 min)
2. Parotid sialography showing the presence of diffuse sialoectasias without evidence of obstruction in the major ducts
3. Salivary scintigraphy showing delayed uptake, reduced concentration and/or delayed excretion of tracer

VI. Autoantibodies: presence in the serum of the following autoantibodies
1. Antibodies to Ro (SS-A) or La (SS-B) antigens, or both
Rules for classification

For primary SS
In patients without any potentially associated disease, primary SS may be defined as follows:
(A) The presence of any four of the six items is indicative of primary SS, as long as either items IV or VI is positive
(B) The presence of any three of the four objective criteria items (III, IV, V, VI)

For secondary SS
In patients with a potentially associated disease (for instance, another well-defined connective tissue disease), the presence of item I or item II plus any two from among items III, IV, and V may be considered as indicative of secondary SS
Exclusion criteria
  Past head and neck radiation treatment
  Hepatitis C infection
  AIDS
  Pre-existing lymphoma
  Sarcoidosis
  GVHD
  Use of anticholinergic drugs (for a time shorter than 4-fold the half life of the drug)

Source: Vitali C, Bombardieri S, Jonsson R, et al. (2002) Classification criteria for Sjögren's syndrome: A revised version of the European criteria proposed by the American–European Consensus Group. *Ann Rheum Dis* 61:554–558.

However, several researches were against this concept, arguing that the labial gland biopsy is improperly considered more sensitive than sialography, and that both sialography and labial gland biopsy are necessary to confirm the diagnosis of gland diseases.

On the other hand, several researchers have claimed the stringency of the AECG criteria and the leak of prognostic indicator for the disease in the classification items [53–55]. Notably, SS patients must have either anti-SS-S/Ra and anti-SS-B/Lo autoantibodies, or positive results for labial gland biopsy test, or both. Therefore, invasive tests are mandatory even for a substantial number of non-SS patients, who are negative for lip biopsy and autoantibodies but have similar outcomes as the AECG-positive patients.

Imaging techniques are also incorporated as classification items in the AECG criteria [51]. Rubin and Holt first described sialographic changes in the major salivary glands of patients with SS [56]. Indeed, the sialographic findings are well correlated with the severity of the disease [57]. However, the sialography requires radiation exposure, and the technique is very cumbersome because it requires an iodine contrast agent and cannulation of a catheter or insertion of a syringe needle into the extra glandular main duct. Consequently, patients may refuse to undergo sialography and/or clinicians may be reluctant to run the sialographic examination. Given these issues, a noninvasive method for diagnosing SS might benefit patients because treatments using muscarinic antagonists (cevimeline or pilocarpine) are now available and hold promise for improving sicca symptoms if the disease is effectively diagnosed [58].

Recent advances in MR imaging technique allow radiation- and contrast media-free examinations of the major salivary glands of patients with sicca symptoms. The technique enables the assessment of both parenchyma and ductal abnormalities, further characterizing the disease states of affected glands of patents with provisional SS, and thus helps in differentiating glands affected by SS from those with infectious or metabolic diseases. In addition, the technique allows the grading of the gland diseases and thus could be used for predicting the treatment efficacy in the patients with SS [58].

15.4.1 Conventional MR Imaging

The MR imaging features of the parotid gland in patients with SS are characterized by multiple areas of high signal intensity on T_1- and T_2-weighted MR images [57] (Figure 15.23). Fat-suppressed MR imaging with short-tau inversion recovery (STIR) or fat-saturation sequences can decrease the signal intensity levels in areas that correlate with high-intensity signals on T_1-weighted images, indicating that the high-intensity areas on the T_1-weighted images of the parotid glands in patients with SS are due to fat infiltration in the glands [59]. The high-intensity fatty areas on T_1-weighted images appear as small spots in the early stages of the disease, and gradually increase with disease progression (Figure 15.22). Accordingly, when the parotid glands are classified into five grades based on the appearance of high-intensity areas on T_1-weighted images, the grades correlate well with the severity of impaired salivary flow rates, with grading on sialography, and with grading on labial gland biopsy. The use of a small-sized surface coil with resultant high-resolution T_1- and fat-suppressed T_2-weighted images allows quantitative MR imaging for assessing the extent of fat infiltration on T_1-weighted images and gland tissue destruction on fat-suppressed T_2-weighted images [60]. The quantitative MR imaging of fat or the intact gland lobule was significantly and highly correlated with disease severity, and the best cutoff point of $\geq 5\%$ fat area for SS glands yielded 93% sensitivity, 89% specificity, and 92% accuracy. Further, the best cutoff point of $\leq 90\%$ intact lobule area for SS glands yielded 93% sensitivity, 100% specificity, and 95% accuracy for differentiating between xerostomia patients without SS and those with SS.

However, because senile changes in the parotid gland are associated with fat degeneration (Figure 15.24), the discrimination of SS from senile atrophy is important for diagnosing SS patients based on MR imaging examinations. Indeed, the CT values of the parotid gland decrease with age, and the values of healthy parotid glands in subjects aged from 50 to 70 years are similar to those of the glands in SS patients in the early stages of the disease (G1–G2 gland stages) [59]. These results suggest that although fat infiltration in the parotid gland develops with increasing age, the extent is much smaller compared with that of SS patients. In the normal gland of senile subjects, fat deposition occurs in the secretory cells and duct cells, whereas secretory cells are destroyed and replaced by fat tissues in the glands of patients with SS.

15.4.2 MR Sialography

MR sialography is typically based on a heavily T_2-weighted two- or three-dimensional (2- or 3D) spin-echo sequence, thus visualizing the water content of the gland. MR sialography was initially applied by Lomas et al. to a patient with possible SS showing bilateral sialoectasia [61]. Later, MR sialography was evaluated in larger SS patient cohorts [62–64]. MR sialography can overcome the shortcomings of conventional (X-ray) sialography, such as radiation exposure, the cumbersome cannulation technique, and injection of contrast medium. MR sialography is a promising technique for assessing salivary gland diseases, including SS.

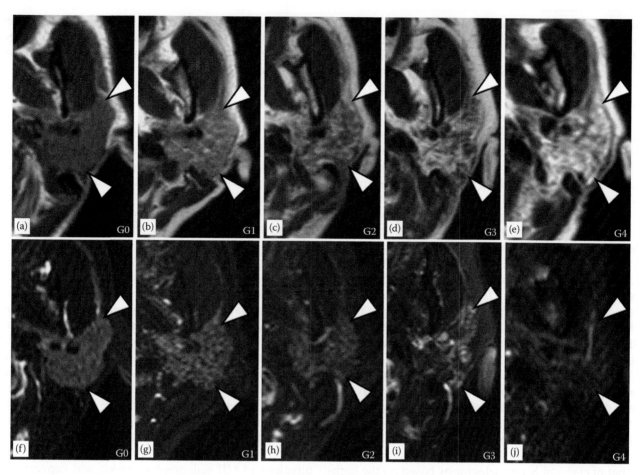

FIGURE 15.23
MR grading of gland disease in patients with Sjögren's syndrome (SS). (a–e) Axial T_1-weighted MR images of parotid glands (arrowheads) in SS patients are categorized into Grade 0 (a), Grade 1 (b), Grade 2 (c), Grade 3 (d), or Grade 4 (e) based on the amounts of fat areas with high signal intensity in glands [60]. (f–j) Axial fat-suppressed T_2-weighted MR images of parotid glands (arrowheads) in SS patients are categorized into Grade 0 (f), Grade 1 (g), Grade 2 (h), Grade 3 (i), or Grade 4 (j) based on the amounts of intact lobe areas with intermediate signal intensity in glands.

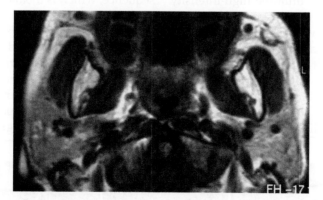

FIGURE 15.24
Axial T_1-weighted MR image of senile parotid glands with irregular fat infiltration in 78-year-old woman.

However, the early applications of MR sialography had several disadvantages; it requires long imaging time and is frequently associated with background noises from the neighboring structures, such as the vessels and the muscles in the neck. These disadvantages limited MR sialography to an adjunctive tool for other imaging techniques such as ultrasonography (US) in diagnosing gland disease in patients with SS. Thereafter, a fast and high-resolution MR sialography technique using a 1.5-T MR imager equipped with a small-sized (= 47 mm) surface coil was introduced [65]. The high-resolution MR sialography is based on a single-shot, single-slice turbo spin-echo (TSE) technique, and we currently use the following sequence: TR/TE/number of signal acquisition = 8000 ms/800 ms/1; imaging time, 8 s; matrix size, 192 × 149; field-of-view, 7 cm; TSE factor, 150; and echo spacing, 10.6 ms. This technique enables the visualization of the second-order branches. Although increasing the number of signal acquisition (NSA) (up to 8) improved the visualization of third-order branches, the background noises were increased and the imaging time was increased (up to 100 s). An increase in TE to 1000 ms also improved the gland image by inhibiting the background signals at the expense of imaging time. The use of a smaller-sized (= 23 mm) coil could improve the sialographic images without increasing the background signals, but the ductules at the gland peripheries are not well visualized due to the smaller coil size. Further, the 3D high-resolution MR sialography is also available (TR/TE/NSA = 6000 ms/1000 ms/4) with a 0.5 mm slice thickness, 256 × 256 matrix size, 7 cm field-of-view, TSE factor of 200, and echo spacing of 9.8 ms. Although the 3D high-resolution MR sialography improves the duct smoothness of the intraglandular branches, the imaging time increases (≥5 min). On the other hand, the use of a larger coil (170-mm superficial coil) results in a poor image with very weak duct signals at any imaging sequence used.

MR sialography of the parotid gland shows the sialoectatic patterns characteristic of SS (Figure 15.25). MR sialographic images can be classified in a manner similar to that of classical X-ray sialography based on the high-intensity spots with varying sizes: Grade 0, no evidence of high-intensity spots; Grade 1, punctate (<1 mm) pattern of multiple high-intensity spots; Grade 2, globular (1–2mm) pattern of multiple high-intensity spots; Grade 3, cavitary (>2 mm) pattern of multiple high-intensity spots; and Grade 4, destructive pattern of larger but fewer high-intensity spots as well as narrowed, dilated, or shortened intraglandular branches [66]. Although MR sialographic gradings are well correlated with the classical X-ray sialographic results, MR sialographic images are not duplicates of conventional sialography; MR sialographic images obtained at the end-stages of the gland disease (Grade 4) are dissimilar to conventional

sialographic images obtained from the same patients. The sizes of the sialoectatic foci in the glands are larger at higher grades on conventional sialographic images, but sialoectatic foci in Grade 4 glands are decreased in number and size compared with those in Grade 3 glands on MR sialographic images. Furthermore, the Grade 0 glands by MR sialography contain significant numbers of Grade 1 glands by conventional sialography. These discrepancies may reflect the fact that the two MR sialographic techniques could depict different pathologic changes in the affected glands. MR sialography is fundamentally a hydrography technique and depicts water-rich foci. Therefore, many large sialoectatic foci in the end-stage glands may not be visualized on MR sialography owing to the absence of saliva or fluid within the abnormal ducts.

Taken together with the abovementioned diagnostic errors in MR sialography, we recommend evaluating the salivary glands of xerostomia patents by using both conventional MR imaging (T_1- and fat-suppressed T_2-weighted MR imaging) and MR sialography [65]. Combinations of MR sialography and conventional T_1- or fat-suppressed T_2-weighted imaging improve the diagnostic abilities for differentiating between the parotid glands of xerostomia patients with SS and those without SS. Compromised MR sialographic criteria (≥6 sialoectatic foci for SS glands) yielded 80% sensitivity and 100% specificity for discriminating SS patients. However, combinations of MR sialographic criteria and conventional T_1- or fat-suppressed T_2-weighted MR imaging criteria resulted in an increased discriminating ability, yielding 96% sensitivity, 100% specificity, and 98% accuracy.

MR sialography can also be achieved by using a bTFE sequence, which is a balanced, steady-state free precession MR technique for achieving rapid and high SNR imaging [66,67] (Figure 15.26). The bTFE sequence yields high signals of the blood vessels and body fluids, and allows effective differentiation between the water contents from fat tissues combined with a fat-suppression prepulse, such as spectral presaturation with inversion recovery. These properties of the bTFE are suitable for visualizing the intraglandular ducts of the parotid

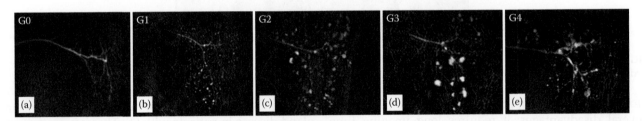

FIGURE 15.25
MR sialographic grading of gland disease in patients with Sjögren's syndrome (SS). MR sialograpms of parotid glands in SS patients are categorized into Grade 0 (a), Grade 1 (b), Grade 2 (c), Grade 3 (d), or Grade 4 (e) according to the grades of conventional sialography [60]. MR sialographic grades are well correlated with those of conventional sialography, except for Grade 4 glands, in which sialoectatic foci are decreased in number and size compared with those in Grade 3 glands.

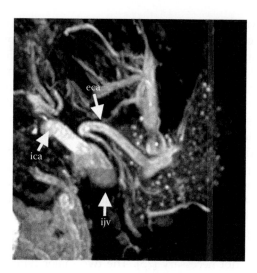

FIGURE 15.26

A 55-year-old woman with Sjögren's syndrome (SS). Axial maximum intensity projection (MIP) bTFE MR image shows parotid gland with multiple sialoectatic foci characteristic of SS. eca, external carotid artery; ica, internal carotid artery; ijv, internal jugular vein.

gland, which contain substantial amounts of fat tissues within the gland parenchyma.

15.4.3 Diffusion-Weighted MR Imaging

Sumi et al. and Regier et al. found that the ADCs of parotid glands with early and mid-stage SS were greater compared to those of the glands in healthy subjects [68,69] (Figure 15.27). However, the ADCs of the glands of patients with later stages of SS were markedly smaller than those of the healthy glands. Similar changes in ADCs occur in the lacrimal glands in patients with SS [70]. More importantly, the ADCs of the parotid glands in patients with SS were correlated with salivary flow rates and with the severity of gland damage as assessed on T_1-weighted MR images, but not with the sialographic grades. Lymphocytic infiltration with subsequent apoptotic cell death occurs in the salivary glands with

the onset of the disease. These events in the affected glands increase the extracellular spaces, leading to an increase in ADCs. However, in the later stages of SS, the effect of fat infiltration gradually surpasses that of the increased extracellular spaces in the damaged glands, thereby markedly limiting the mobility of the water molecules in the glands. Extensive aggregations of lymphocytes and lymphoid follicle formation in the glands may further decrease the ADCs of the glands because the water molecular mobility is restricted by the cell membranes and small intracellular particles. Therefore, DW imaging can reveal the disease states of the salivary glands in patients with SS.

Although the parotid glands produce small amounts of saliva at resting states; however, gustatory stimulation leads to increases in salivary flow rates. Therefore, it is reasonable to consider using DWI to depict changes in the amount of saliva during gustatory stimulation. Thoeny et al. investigated the ADC kinetics of the parotid gland after gustatory stimulation, and reported initial decreases in the ADC during the first 5–7 min of salivary stimulation and latent increases during the following 15–20 min [71]. They postulated that the initial decreases are attributable to the emptying of stored saliva in the glands. However, Kato et al. found that the kinetics of the ADCs after gustatory stimulation in subjects with healthy glands showed initial rapid increases within the first 2 min after stimulation and subsequent gradual decreases [72]. The kinetics can be characterized by calculating the ADC IR and the times to the ADC peak (T_{max}). The IR values of the parotid and submandibular glands are lower in xerostomia patients compared to the values in patients with healthy glands, and the T_{max} values of the glands are higher in xerostomia patients compared to in subjects with healthy glands. The discrepancy between the two reports can be explained by the differences in *b*-values used for calculating the ADCs: Thoeny's group used *b*-values of 400, 600, 800, and 1000 s/mm², while Kato's group used 0 and 1000 s/mm². Therefore, the initial increases in the ADCs may be contaminated by the rapid increases in the

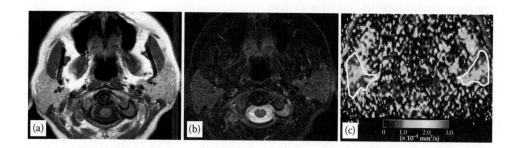

FIGURE 15.27

A 44-year-old woman with Sjögren's syndrome (SS). (a, b) Axial T_1-weighted (a) and fat-suppressed T_2-weighted (b) MR images show parotid glands with diffuse fat infiltration. Note punctate distributions of high-intensity spots in glands on fat-suppressed T_2-weighted image (b), corresponding to sialoectasia. (c) Axial ADC map shows parotid glands with higher ADC levels (left gland, 0.88 × 10⁻³ mm²/s; right gland, 0.93 × 10⁻³ mm²/s) compared with ADC levels of healthy parotid glands (0.63 ± 0.11 × 10⁻³ mm²/s) [39].

blood volume in small vessels and increases in perfusion within the glands in response to the gustatory stimulation. Accordingly, the ADCs determined between small (0–200 s/mm²) and large (400–1000 s/mm²) b-values may differ greatly from those obtained between two different large b-values (i.e., 500 and 1000 s/mm²).

15.4.4 DCE MR Imaging

Another tool for assessing salivary gland function is DCE MR imaging, whereby the vascular nature of the gland can be quantitatively evaluated based on the compartmental tracer kinetic model [73]. The tracer kinetic model provides kinetic parameters in terms of the transfer constant (K^{trans}), the extracellular extravascular space (v_e), and the rate constant between the extravascular space and blood plasma (K_{ep}) according to the following equation:

$$C_t(t) = v_p C_p(t) + K^{trans} \int_0^t C_p(t') \exp\left(\frac{K^{trans}(t-t')}{v_e} \right) dt'$$

where:

$C_t(t)$ is the change in the contrast agent concentration over time
v_p is the vascular plasma volume
t is the time
t' is the time as an integration variable
$C_p(t)$ is the arterial input function
$C_p(t')$ is the concentration of contrast agent in the blood plasma as a function of time [74]

The time-signal intensity curve of the parotid gland after gadolinium enhancement can also be characterized by determining the initial area under the contrast agent concentration-time curve (IAUC). Roberts et al. found that patients with SS have higher levels of the transcapillary contrast agent transfer constant (K^{trans}) and the extracellular extravascular space (v_e) compared with healthy subjects [74]. Further, the glands of patients with SS also have elevated IAUC levels. These results are consistent with the data obtained by the DWI described above, indicating an increased interstitial gland area due to acinar cell apoptosis, destruction of gland lobules, and increased vascular permeability and endothelial-dependent vasodilatation caused by the interaction of endothelial cells with activated lymphocytes in the glands of patients with SS.

15.4.5 Arterial Spin Labeling

Perfusion measurements such as the IAUC and the transfer between the blood plasma and the extracellular extravascular space derived from DCE MR imaging have shown promising initial results in the assessment of the vascular nature of the salivary gland of patients with SS.

However, these estimations of gland perfusion are imperfect because of the contributions of both blood flow and vascular permeability to tissue enhancement [75]. The arterial spin labeling (ASL) technique allows noninvasive quantitative assessment of tissue perfusion without the need for contrast agent administration [76,77]. In ASL, the protons in the blood are magnetically labeled and used as an intrinsic tracer to measure the volume of tissue blood flow. Tissue perfusion can be estimated with the ASL technique by assessing the difference between the images obtained with and without labeling. The major advantages of ASL are that the technique can obtain a direct determination of tissue perfusion that is not affected by vessel permeability and that it allows numerous acquisitions in the same patient because intravenous contrast agent is not necessary [75]. The two major types of the ALS technique are continuous ASL (CASL) and pulsed ASL (PASL) [78]. In the CASL technique, long RF pulses (1–2 s) are applied to a plane proximal to the imaging slices thereby generating a flow-driven adiabatic inversion. In contrast, the PASL technique uses a single inversion pulse. Although the longer steady state tagging in the CASL technique provides a theoretically higher SNR compared to PASL, the tagging efficacy may be hampered by variations in flow velocity. Furthermore, the CASL technique has a hardware demand (continuous-mode RF transmission) that is not required for PASL. The recently developed pseudocontinuous ASL (pCASL) uses a series of discrete RF pulses to mimic the flow-driven adiabatic inversion, which takes advantage of the superior SNR given by CASL and the higher tagging efficacy with PASL [78].

Our preliminary study using pCASL demonstrated that the parotid glands of SS patients exhibit salivary blood flow (SBF) kinetics with significantly higher base SBFs, longer times to SBF peak, and higher SBF IR at peak time compared with healthy glands. These results are consistent with those obtained by DCE MR imaging, implying that the ASL technique is promising for assessing the vascular nature of the impaired function of the salivary glands.

15.5 SS-Mimicking Diseases of the Salivary Glands

15.5.1 IgG4-Related Mikulicz Disease

Mikulicz's disease (MD) is characterized by symmetrical swelling of the salivary and lacrimal glands. The disease was previously regarded as a subtype of SS. However, a separate entity of IgG4-related disease has recently been established to include several diseases that were previously diagnosed and treated separately, including autoimmune pancreatitis and MD [79,80]. These clinical entities share common properties, such as elevated serum IgG4 levels and infiltration of IgG4-positive plasma cells associated with fibrosis in the involved organs and tissues. Acinar cell destruction in the submandibular and lacrimal glands is frequently observed. The formation of lymphoid follicles with a germinal center is also a common histological feature of the disease [81]. A good response to corticosteroid is another characteristic clinical feature of IgG4-related diseases, and is also a major reason for the necessity of differentiating between SS and IgG4-related MD [80].

Recent reports have provided the characteristic ultrasonographic (US) features of the salivary and lacrimal glands of patients with IgG4-MD. The involved glands are enlarged and irregularly hypoechoic, exhibiting multiple hypoechoic areas with varying sizes and shapes. These US features are very similar to those of SS glands. The abnormal US features will rapidly, and in many cases almost completely, disappear in response to oral doses of corticosteroids. However, conventional MR imaging features are less specific, occasionally showing heterogeneous gland signal intensities on T_1 and fat-suppressed T_2-weighted images [83] (Figure 15.28). MR sialography inevitably shows normal sialographic features [84]. Accelerated fat infiltration is never observed in the involved parotid glands. Therefore, conventional MR imaging and/or MR sialography could be useful techniques for distinguishing IgG4-MD from SS.

15.5.2 Hyperlipidemia

Hyperlipidemia is the most common form of dyslipidemia and serologically characterized by elevated levels of plasma triglyceride, total cholesterol, or both. Some patients with hyperlipidemia (hypercholesterolemia and hypertriglyceridemia) are associated with sicca

symptoms reminiscent of SS [85–87]. Several researchers emphasized the correlation between serum lipid profiles and the salivary gland abnormalities. For example, Izumi et al. demonstrated a correlation between plasma triglyceride levels and parotid swellings and a correlation between plasma total cholesterol levels and salivary flow rates [88]. Ramos-Casala et al. found a higher prevalence of associated dyslipidemia, diabetes mellitus, and hyperuricemia in patients with primary SS compared with age- and sex-matched patients without autoimmune diseases [89]. They also demonstrated that hypercholesterolemia is associated with a lower frequency of positive results for anti-SS-A/Ro antibodies and anti-SS-B/La antibodies, and that hypertriglyceridemia is associated with a high prevalence of extraglandular lesions in the kidney, liver, and blood vessels. Izumi et al. reported swelling of the parotid gland, impaired salivary flow, or both were observed in 83% of 24 patients with hyperlipidemia and sicca symptoms [88]. MR imaging revealed that the enlarged parotid glands were infiltrated by varying amounts of fat; however, normal-sized glands in patients with hyperlipidemia showed MR imaging features similar to those of healthy glands. Abnormalities in the submandibular gland were observed in 33% of the 24 patients with hyperlipidemia. However, the submandibular glands were less severely affected by the disease compared with the parotid glands, exhibiting slight increases in signal intensity on T_1-weighted images and mild swelling. These MR imaging features are indistinguishable from those of SS patients. In contrast to the parotid glands involved by SS, the glands in patients with hyperlipidemia and sicca symptoms did not show sialographic features characteristic of SS (Figure 15.29). Furthermore, in 58% of 24 patients with hyperlipidemia and sicca symptoms, maximum area of the parotid gland exceeded the range of the glands in healthy subjects; however, the

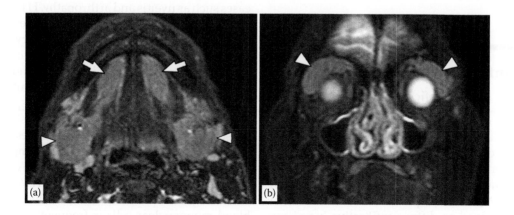

FIGURE 15.28
IgG4-related Mikulicz's disease (IgG4-MD). (a) A 63-year-old man with IgG4-MD. Axial STIR MR image shows enlarged submandibular (arrowheads) and sublingual (arrows) glands. Note that regional lymph nodes are also enlarged. (b) A 60-year-old woman with IgG4-MD. Coronal STIR MR image shows enlarged lacrimal glands (arrowheads).

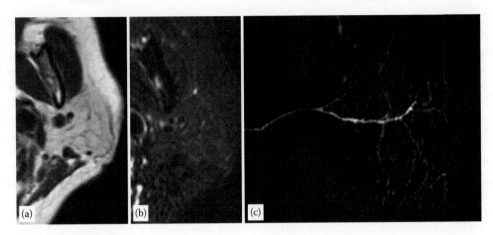

FIGURE 15.29
A 54-year-old man with hyperlipidemia. (a, b) Axial T_1-weighted (a) and STIR (b) MR images show enlarged parotid gland with diffuse fat infiltration. (c) MR sialogram shows normal duct ramifications.

maximum size of the gland in patients with SS did not exceed the normal upper limit. The features that distinguished the patient groups with hyperlipidemia from SS were also evident in the labial glands. The labial glands obtained from patients with hyperglyceridemia showed extensive lipid infiltration in the gland lobes, but fat infiltration was minimally observed in a patient with hypercholesterolemia. The lymphocyte infiltrations and aggregations that are characteristic of SS were minimal in patients with triglyceridemia, but multiple foci of lymphocyte aggregation were observed in the glands of patients with hypercholesterolemia alone. Conversely, lipid infiltration was rarely observed in the labial glands of SS patients. Consistent with the results by Ramos-Casala et al., positive results for serum anti-SS-A/Ro or anti-SS-B/La antibodies were not observed in patients with hyperlipidemia; a single patient with hyperlipidemia had elevated levels of antinuclear antibodies and rheumatoid factors [88,89].

15.6 Viral Infection of the Salivary Glands

15.6.1 Human Immunodeficiency Virus

Several lines of evidence have suggested that retroviral infections play a role in the pathogenesis of SS, including human T-cell leukemia virus (HTLV-I), human immunodeficiency virus (HIV), human intracysternal A-type retroviral particle (HIAP-I), and human retrovirus-5 (HRV-5) [90]. Sicca symptoms associated with HIV infection are defined as diffuse infiltrative lymphocytosis syndrome (DILS) [91]. DILS is clinically indistinguishable from SS, with patients exhibiting bilateral parotid and lacrimal gland swelling associated with

xerostomia and xerophthalmia. The disorder differs from SS in that the lymphocytes that infiltrate in the salivary glands are predominantly $CD8^+$ ($CD4^+$ lymphocytes are predominantly seen in the glands of SS patients) and that the autoantibodies (anti-SS-A/Ro and anti-SS-B/La antibodies, RF, and ANA) are much less frequently positive compared with SS [90]. The recent introduction of highly active antiretroviral treatment (HAART) has reduced the prevalence of DILS in HIV patients [91].

Publications regarding the MR imaging features of the involved parotid glands are scarce; however, some reports described multiple solid or cystic lymphoepithelial lesions in the glands [92–95].

15.6.2 Human T-Cell Leukemia Virus-Associated Myelopathy

HTLV-I infection is associated with a variety of diseases, including adult T-cell leukemia, and inflammatory diseases such as uveitis and arthropathy. HTLV-I-associated myelopathy (HAM) is another example of an HTLV-I-associated nonneoplastic inflammatory disease [96]. HAM is a chronic myelopathy characterized by bilateral involvement of the pyramidal tracts and disturbances of sphincter function. HAM patients exhibit progressive clinical features mainly consisting of spasticity of the lower extremities, urinary bladder disturbances, and muscle weakness of the lower extremities. HAM is occasionally complicated by severe sicca symptoms, exhibiting clinical features reminiscent of SS. Although a high prevalence of SS was reported in HTLV-I-seropositive/HAM-negative patients, the viral load in peripheral blood mononuclear cells is frequently high in patients with HAM [97].

Izumi et al. found that ~40% of 31 HAM patients were diagnosed as positive for SS [98]. The HAM-positive/

SS-positive patients displayed decreased salivary flow rates at levels similar to those of the HAM-negative/SS-positive patients. However, the HAM-positive/SS-positive patients exhibited lower magnitudes of lymphocyte aggregations compared with the HAM-negative/SS-positive patients, and 92% of the HAM-positive/SS-positive patients completely lacked the imaging characteristics of SS, including MR imaging and conventional sialography. Therefore, these findings suggest that SS in patients with HAM may occur via mechanisms that are distinct from those of SS in HAM-negative patients.

15.6.3 Mumps

The mumps virus is an enveloped, single-stranded RNA virus belonging to the family Paramyxoviridae, which causes an acute infectious disease mainly in children and young adults. Mumps is clinically characterized by bilateral swelling of the parotid gland. The incubation period from infection to appearance of the characteristic swelling of the parotid glands is 15–24 days [99]. The infectious period starts several days before the onset of parotitis and continues for several days afterward. The MR imaging features of the affected parotid gland are characterized by bilateral symmetrical swelling of the glands that are homogeneously hyperintense on fat-suppressed T_2-weighted images (Figure 15.30). Extraglandular extensions of inflammation into the neighboring subcutaneous tissues are often observed. Dilatation of the intraglandular ducts, which is a common finding in nonspecific chronic inflammation, is rarely seen.

15.6.4 Hepatitis C Virus

The hepatitis C virus (HCV) is an enveloped, single-stranded RNA virus belonging to the Flaviviridae family.

HCV infections may be associated with sicca complaints [100]. HCV RNA has been found in the saliva and salivary gland tissue (the cytoplasms of salivary epithelial cells) [101]. A chronic focal sialoadenitis associated with infiltration and aggregation of lymphocytes (CD4+ T cells), the histopathological features of which resemble those of SS glands, are observed in ~50% of HCV-infected patients. However, positive results for the serum autoantibodies such as ANA, SS-A/Ro, and SS-B/La are infrequent [102]. These findings evoked a long debate on the possibility of HCV involvement in the pathogenesis of at least some types of SS. However, a recent report showed that the detection of HCV RNA in the saliva or salivary gland tissues is not correlated with the severity of salivary gland disorders such as xerostomia, impaired salivary flow rate, or sialoadenitis [103]. Moreover, Vitali et al. found that the prevalence of HCV infection in patients with primary SS was similar to that observed in the normal population, and that the overestimation previously reported could be due to different classification criteria in different studies, and to the hypergammaglobulinemia frequently present in the sera of patients with SS [104].

An issue that is more important than the clinical and serological similarities between HCV-infected and SS patients is the development of B-cell lymphomas (mainly mucosa-associated lymphoid tissue, MALT; diffuse large B-cell lymphoma, and follicular center cell lymphoma) in these patients [102]. HCV-infected patients with frequent parotid enlargement, vasculitis, and immunologic profiles of positive RA and mixed type II cryoglobulins may develop B-cell lymphomas with a higher prevalence than is observed in the normal population [105]. The salivary glands and the liver are the preferential sites for lymphoma development in HCV-infected patients with sicca symptoms. The MR imaging features of lymphomas developing in the salivary glands of HCV-infected patients would

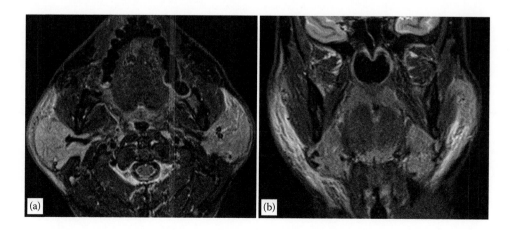

FIGURE 15.30
A 28-year-old man with mumps. (a) Axial STIR MR image shows enlarged and hyperintense parotid glands. (b) Coronal STIR MR image shows extraglandular (subcutaneous) spreads of infection.

be similar to lymphomas occurring in the glands of SS and non-SS patients.

15.7 Inflammation of the Salivary Gland

15.7.1 Sialolithiasis

Although the most common causes of salivary gland inflammation are bacterial and viral infections, obstructive diseases such as tumors and sialolithiasis may cause sialoadenitis. An incomplete obstruction by a sialolith may trigger a secondary infection, and a complete and prolonged obstruction by a large sialolith may lead to glandular atrophy. Sialolithiasis most commonly occurs in the submandibular glands (80%–95%) [106]. The main purpose of plain radiography and CT is to locate and measure a possible sialolith(s) in the duct system. Conventional sialography may be performed to examine the state of the duct; however, a large sialolith would interfere with the visualization of the proximal part of the duct owing to a complete obstruction. A noncalcified or partially calcified sialolith may be identified on sialography. However, these techniques cannot evaluate the diseased states of the gland parenchyma affected by sialolithiasis.

Sumi et al. reported that the submandibular glands could be categorized into three distinctive types based on the clinical symptoms and MR imaging features of the gland parenchyma [107]. Type I glands are hypointense on T_1-weighted MR images and hyperintense on fat-suppressed T_2-weighted images (Figure 15.31). Patients with a Type I gland complain of pain and/or swelling. The increased gland sizes associated with abnormal signal intensities and the symptomatic glands both indicate that Type I glands are in active states of inflammation. Histopathologically, Type I glands show varying amounts of inflammatory cell infiltrations associated with destruction and fibrous tissue replacement of the gland structures. This type was most frequently (56%) observed in sialoadenitis secondary to sialolithiasis. The sublingual glands were also involved in some patients with this type. Follow-up studies of Type I glands showed that the sizes and signal intensity levels had returned to parity compared to those of the glands on the opposite sides after 4–5 months after surgical removal of the sialolith(s), suggesting that the glands are minimally affected by inflammation and that the gland function can be preserved after simple removal of the sialolith(s). Extraglandular extensions of the infection may occur in patients with this gland type, which are discernible as cellulitis and abscess formation in the subcutaneous tissues and extraglandular spaces.

In contrast to patients with Type I glands, patients with Type II glands are free from any clinical symptoms or any history of pain or swelling in the submandibular region. Type II glands are hyperintense on T_1-weighted images (Figure 15.32). The gland intensity levels decreased on fat-suppressed T_2-weighted images, indicating fat infiltration in the affected glands. The gland size was smaller compared with the glands on the opposite side in 75% of Type II glands. An excised submandibular gland from a single patient showed extensive replacement of the gland tissues with fat. The average size of the sialoliths is greater than that of Type I glands (9.2 ± 1.6 mm vs. 5.8 ± 2.9 mm). It may be noteworthy that all of the sialoliths in this gland type were located in the proximal one-third of the main duct, whereas the sialoliths in the Type I glands were also located in the distal- and middle one-third as well as in the proximal one-third of the main ducts.

Type III glands are indistinguishable from the healthy glands on the opposite side. Patients with Type III

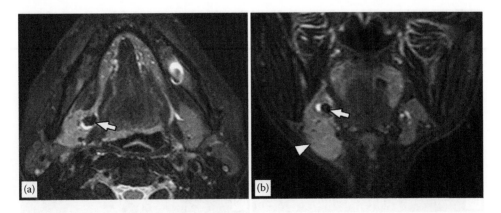

FIGURE 15.31
A 59-year-old woman with submandibular sialolithiasis. (a) Axial fat-suppressed T_2-weighted MR image shows hyperintense submandibular gland with a large sialolith (arrow). Type 1 gland. (b) Coronal STIR MR image shows enlarged submandibular gland (arrowhead). Arrow, sialolith.

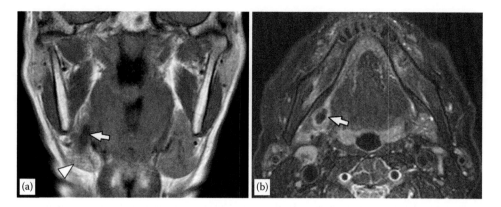

FIGURE 15.32

A 64-year-old woman with submandibular sialolithiasis. (a, b) Coronal T_1-weighted (a) and axial fat-suppressed T_2-weighted (b) MR images show atrophic submandibular gland (arrowhead) with a large sialolith (arrow). Type 2 gland. Note fat degeneration of right submandibular gland.

glands are asymptomatic and have no history of related symptoms. The sialoliths were located in the middle or proximal one-third of the main duct, and the average size was 4.0 ± 1.0 mm.

Therefore, Type I glands are those with secondary infection caused by incomplete obstruction of the main ducts [107]. Type II glands are those with chronic and complete obstruction without secondary infection, and Type III glands are those with a nonobstructing or partially obstructing sialolith that did not cause secondary infection. CT can also depict several features of the active inflammatory changes of the gland affected by sialolithiasis, such as increased density or enhancement after contrast medium injection due to active inflammation in Type I glands and decreased density due to fat infiltration in atrophic Type II glands. Although CT is useful for determining the location and size of sialolith, the diseased states of the affected glands can be more readily evaluated by using MR imaging.

MR sialography could replace conventional sialography for depicting the dilated duct and the sialolith in it. In this context, various studies published by different study groups reported adequate diagnostic ability of MR sialography for detecting a sialolith (or sialolithiasis) [108,109]. Becker et al. showed that MR sialography using a heavily T_2-weighted single-shot fast spin-echo sequence (3D extended-phase conjugate-symmetry rapid spin-echo sequence, 3D-EXPRESS) can detect sialoliths with 91% sensitivity and 94%–97% specificity, and that the MR sialography can detect ductal stenosis with 100% sensitivity and 93%–98% specificity. However, Kalinowski et al. stated that the diagnostic accuracy of successfully completed digital subtraction sialography was superior to that of MR sialography [110].

Therefore, a combination of MR imaging (T_1-weighted and fat-suppressed T_2-weighted or STIR MR imaging)

and CT with or without contrast enhancement may complement each other and may offer a promising strategy for the treatment and follow-up of patients with sialolithiasis.

15.7.2 Juvenile Recurrent Parotitis

JRP is the second most common inflammatory disease in childhood after mumps. The peak incidence is in children between three and seven years of age [111]. JRP is more common in boys than in girls. The number of episodes of parotid swelling per year broadly vary among patients (up to ~100), and each episode lasts 3–7 days. The diagnosis of JRP is usually made based on a clinical history of recurrent unilateral or bilateral parotid swelling associated with tenderness and fever. The affected parotid gland shows characteristic peripheral sialoectasia on sialography and variously sized hypoechoic areas on US [112,113]. The salivary flow rate may be diminished during the active phase of the disease. These sialographic and US findings are not distinguishable from those of the parotid glands of patients with juvenile SS (JSS), in which the symptoms and results of serological tests are more variable compared to those in the adult form of SS [113–115].

Conventional MR imaging of the affected gland shows normal gland architecture, or multiple spots of low (on T_1-weighted images) or high (fat-suppressed T_2-weighted images) signal intensity [112,113] (Figure 15.33). MR sialography characteristically demonstrates multiple punctate to globular sialoectasia distributed throughout the gland parenchyma, which is reminiscent of the MR sialographic findings of the parotid glands in patients with SS (Figure 15.33). However, normal gland architecture is preserved and extensive fat infiltration, which is characteristic of SS glands, is never observed in the glands of patients with JRP.

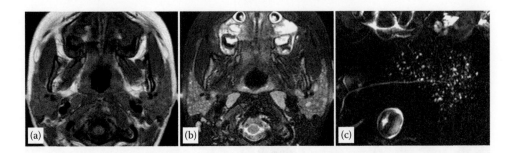

FIGURE 15.33
A 5-year-old girl with juvenile recurrent parotitis. (a) Axial T_1-weighted MR image shows enlarged parotid glands with multiple hypointense foci. (b) Axial fat-suppressed T_2-weighted MR image shows enlarged and diffusely hyperintense parotid glands with multiple hyperintense foci. (c) MR sialogram shows punctate to globular patterns of sialoectasia. Note that near-normal duct ramifications.

15.7.3 Radiation-Induced Injury

Radiation-induced salivary gland injury inevitably occurs when the salivary glands are involved in the radiation field for head and neck malignant tumors and metastatic regional lymph nodes. Salivary gland radiation injury leads to impaired salivary function, which may cause xerostomia and dysphagia. Recently, intensity-modulated radiotherapy (IMRT) has been used in head and neck cancer patients to reduce the dose to the parotid and submandibular glands [116–118]. However, completely sparing these glands;especially, of the submandibular gland from the radiation dose is challenging because of their small volumes and their proximity to the primary lesion and metastatic nodes in the neck. Therefore, MR imaging techniques may be useful for objectively evaluating the salivary gland injury during both radiotherapy and follow-up period.

In patients with oropharyngeal cancer, the parotid and submandibular glands that are involved in the radiation fields reportedly lose their volumes by 17%–26% and 20%–23%, respectively [119,120]. The signal intensity levels of the glands decrease on T_1-weighted images and they increase on T_2-weighted images [120] (Figure 15.34). A significant correlation was found between the changes in the T_2 signal intensity levels and the mean radiation doses, but not between the changes in the T_1 signal intensity levels and the mean doses. Further, the perfusion properties of the salivary glands are changed after irradiation [121,122]. Quantitative assessment of DCE MR imaging revealed increased extracellular extravascular space and decreased vascular permeability in the irradiated parotid and submandibular glands, as evident by higher values for peak enhancement and time to peak in irradiated glands compared with nonirradiated ones. The reduction in gland volume implies the loss of acinar and ductal cells in the irradiated gland and/or reduction in cell size. In this context, Dirix et al. determined the ADC values of the irradiated parotid and submandibular glands by using a set of six high *b*-values (400, 500, 600, 700, 800, and 1000 s/mm^2) to

show that the irradiated glands have elevated ADC levels compared with the contralateral glands, suggesting that the reduced size of the irradiated glands is at least partly attributable to cell loss in the glands [123].

MR sialography can be used to evaluate impairment of salivary gland function after the irradiation. Using the visibility score for the parotid and submandibular ducts at resting states, Astreinidou et al. showed that the visibility of irradiated glands was reduced at six weeks after radiotherapy [124]. A separate study showed that irradiated parotid and submandibular glands exhibit insufficient visualization of the main duct and intraglandular branches, and the responses to secretion stimulation are reduced in the irradiated glands [125].

15.7.4 Salivary Gland Infections Extending from the Neighboring Spaces

Although the layers of the deep cervical fascia act as a barrier to limit the extent of infection, an infection developing in a fascial space may extend beyond the fascial barrier into the neighboring spaces. The main role of imaging is to localize the source of the infection and assess its extent. Infections that spread into the parotid or submandibular spaces most commonly originate from odontogenic infections. A study investigating the pathways for the spread of dental infections revealed that the parotid and submandibular spaces were involved in 45% and 61% of patients with infections in the mandible, respectively, and parotid involvement occurred in 43% of patients with infections in the maxilla [126]. Odontogenic origins of infection can be confirmed by plain radiography of the jaw bones and the patient's clinic course. In some patients, varying degrees of osteitis or osteomyelitis are evident around the roots of the infected tooth on plain radiographs and/or CT. Compared with infectious spread along vertically oriented spaces (parapharyngeal, retropharyngeal, and paravertebral spaces), infectious spread into the nonvertically oriented spaces, such as the parotid and

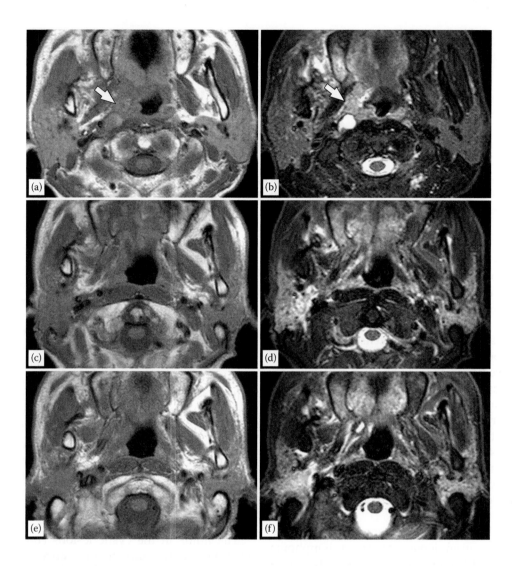

FIGURE 15.34
A 69-year-old man with squamous cell carcinoma (SCC) of oropharynx. (a, b) Axial T_1-weighted (a) and fat-suppressed T_2-weighted (b) MR images show parotid glands (arrowheads) and primary SCC (arrow). (c, d) Axial T_1-weighted (c) and fat-suppressed T_2-weighted (d) MR images show parotid glands (arrowheads) six weeks after completion of radiotherapy (59.4 Gy). Note decreased and increased signal intensities of glands on T_1-weighted and fat-suppressed T_2-weighted MR images, respectively. (e, f) Axial T_1-weighted (e) and fat-suppressed T_2-weighted (f) MR images show parotid glands (arrowheads) 14 weeks after completion of radiotherapy. Note that glands are atrophic and gland signals are still abnormal.

submandibular spaces, have a lower risk of complications that require aggressive treatment [127]. Although CT is a first-choice technique for the assessment of deep neck infections, MR imaging may be useful for monitoring the potential progression of infections from edematous changes to cellulitis and further into the abscesses, which is another important issue for patient management (Figure 15.35).

15.7.5 Kimura's Disease

Kimura's disease (KD) is a rare, chronic inflammatory disease that typically presents with the triad of a firm, painless single/multiple subcutaneous nodules in the head and neck region (especially, in the parotid and submandibular regions), blood and tissue eosinophilia, and markedly elevated levels of serum IgE. Although most patients with KD present with subcutaneous lesions, the disease frequently involves the major salivary glands and regional lymph nodes. Histologically, KD is characterized by dense fibrosis, capillary proliferation, and lymphocyte infiltration, formation of reactive lymphoid follicles, and eosinophil aggregation [128]. The peripheral blood profiles of the eosinophils and IgE levels suggest that KD has an allergic or autoimmune background, which is triggered by an unknown stimulus. Viral infections or toxins have been postulated to be causative events, with subsequent alterations in T-cell immunoregulation or an induction of IgE-mediated Type I hypersensitivity [128,129]. Although the clinical

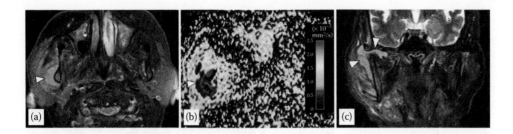

FIGURE 15.35
A 70-year-old man with abscess in masticator space extending from mandibular osteomyelitis. Axial fat-suppressed T2-weighted MR image (a) and coronal STIR image (c) show abscess formation in right masticator space (arrowheads). (b) Axial ADC map shows abscess (arrowhead) with low ADC values due to lesional high viscosity.

course of KD is progressive, the disease is benign and self-limiting. However, KD may be complicated by renal involvement, and the patient may present with proteinuria (12%–16% incidence) and nephrotic syndrome [128,130].

Parotid lesions are usually hypointense on T_1-weighted images and hyperintense on fat-suppressed T_2-weighted images compared with the neighboring muscles, with moderate to marked enhancement on postcontrast T_1-weighted images depending upon the lesional fibrosis and vascularity [131,132] (Figure 15.36). MR imaging may demonstrate serpentine flow-void areas within the parotid lesion, suggestive of abundant vascular proliferation [128]. A regional lymph nodes may be enlarged, but the affected node is usually homogeneous after contrast enhancement without necrotic area within the node [133].

KD in the salivary glands can mimic IgG4-related MD, eosinophilic granuloma, malignant lymphoma, and salivary gland tumors, and patients with KD are often extensively evaluated for these serious disorders. The performance of laboratory examinations to assess periphearl eoshinophilia and elevated serum IgE levels and imaging studies, including MR imaging, could prevent subjecting the patient to potentially harmful and unnecessary invasive diagnostic procedures.

15.8 Cysts of the Salivary Gland

Cystic lesions arising from the major and minor salivary glands include ranulas, mucoceles, and lymphoepithelial lesions. Nonsalivary-gland-derived cystic lesions, such as lymphangiomas, thyroglossal duct cysts, branchial cleft cysts, dermoid cysts, and epidermoid cysts, may occur in the close vicinity of the major salivary glands, and it may be necessary to differentiate these lesions from salivary-gland-derived cystic lesions [134]. MR imaging identifies cystic lesions as nonenhanced areas containing fluids that are hyperintense on fat-suppressed T_2-weighted images. Wall enhancement is very subtle, but the cystic lesion may be enlarged and the wall may be extensively thickened and enhanced if infection is present. However, the MR imaging features of the cystic lesions are often nonspecific.

The fluids in the cystic lesions contain cell debris, macromolecules, and inflammatory cells, and have a broad range of viscosity. Therefore, cystic lesions can be characterized by the fluid areas based on ADCs [135]. For example, ranulas, lymphoepithelial cysts, and mucoceles are among the neck cysts that have high fluid ADC levels (Figures 15.37–15.39): the fluid ADCs of salivary gland-derived cysts, lymphangiomas,

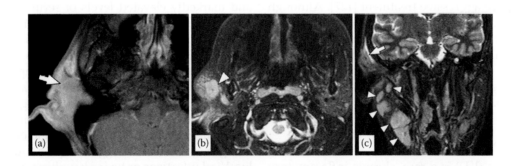

FIGURE 15.36
A 32-year-old man with Kimura's disease (KD). (a) Axial fat-suppressed contrast-enhanced T_1-weighted MR image shows subcutaneous granulation tissues (arrow). (b) Axial fat-suppressed T_2-weighted MR image shows enlarged parotid node (arrowhead). (c) Coronal STIR MR image shows multiple enlarged lymph nodes (arrowheads) in the neck and subcutaneous granulation tissues (arrow).

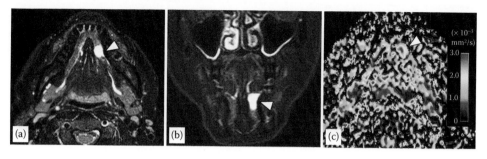

FIGURE 15.37

A 51-year-old woman with simple ranula. (a, b) Axial fat-suppressed T_2-weighted (a) and coronal STIR (b) MR images show simple ranula (arrowheads). (c) Axial ADC map shows simple ranula with high ADC values (2.5×10^{-3} mm^2/s) in sublingual space.

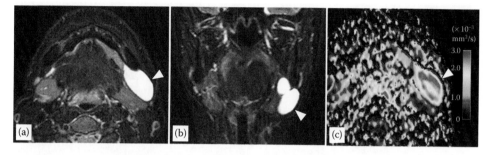

FIGURE 15.38

A 29-year-old woman with plunging ranula. (a, b) Axial fat-suppressed T_2-weighted (a) and corona STIR (b) MR images show plunging ranula extending from sublingual to submandibular spaces (arrowheads). (c) Axial ADC map shows plunging ranula with high ADC values (2.4×10^{-3} mm^2/s).

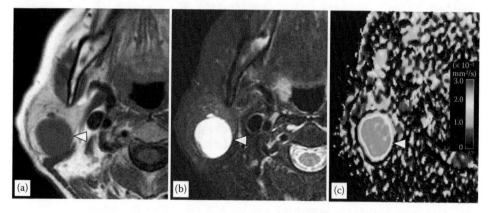

FIGURE 15.39

A 63-year-old woman with lymphoepithelial cyst. (a, b) Axial T_1-weighted (a) and fat-suppressed T_2-weighted (b) MR images show lymphoepithelial cyst of right parotid gland (arrowheads). (c) Axial ADC map shows lymphoepithelial cyst with high ADC values (2.5×10^{-3} mm^2/s).

and thyroglossal duct cysts ($2.38 \pm 0.26 \times 10^{-3}$ mm^2/s; range 1.97–2.93×10^{-3} mm^2/s) are higher than those of branchial cleft cysts, epidermoid cysts, and abscesses ($0.73 \pm 0.37 \times 10^{-3}$ mm^2/s; range 0.11–1.53×10^{-3} mm^2/s) (Figures 15.40 and 15.41). The ADC profiles overlap significantly within the neck cyst group that has high fluid ADCs. However, ADC mapping can effectively discriminate these five different types of neck cysts if appropriate ADC mapping thresholds are used. For example, lymphangiomas can be differentiated from the other types as those containing fluids with >2% areas of extremely low ADCs (<0.8×10^{-3} mm^2/s), lymphoepithelial cysts as those with <40% areas of low ADCs (0.8×10^{-3} mm^2/s) plus >0% areas of intermediate ADCs (1.6×10^{-3} mm^2/s $\leq$ADC <2.4×10^{-3} mm^2/s), thyroglossal duct cysts as those with 40–77% areas of intermediate ADCs, and mucocele as those with $\geq$77% areas of intermediate ADCs. In neck cysts with low ADCs, abscesses are discriminated from the other types as those with fluid areas that have overall ADCs of less

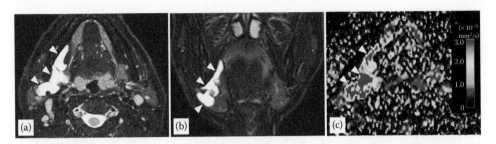

FIGURE 15.40
A 30-year-old man with lymphangioma (cystic hygroma). (a, b) Axial fat-suppressed T_2-weighted (a) and coronal STIR (b) MR images show lymphangioma in submandibular, parotid, and parapharyngeal spaces (arrowheads). (c) Axial ADC map shows lymphangioma with high ADC values (2.9×10^{-3} mm^2/s).

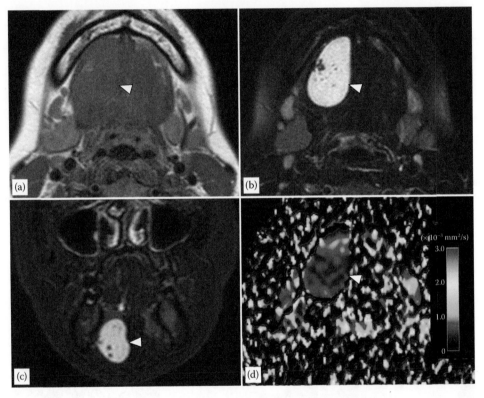

FIGURE 15.41
A 15-year-old woman with dermoid cyst (a–c, arrowheads), Axial T_1-weighted (a) and fat-suppressed T_2-weighted (b), and coronal STIR (c) MR images show dermoid cyst in sublingual space, displacing sublingual gland. Note cystic area contains T_1-hyperintense (a) and T_2-hypointense (b, c) areas. (d) Axial ADC map shows dermoid cyst (arrowhead) with low ADC values (0.6×10^{-3} mm^2/s).

than 0.65×10^{-3} mm^2/s; and branchial cleft cysts as those with <42% areas of extremely low ADCs.

15.8.1 Ranula

Ranula is an extension cyst resulting from inflammation or trauma of the sublingual gland or minor salivary glands in the sublingual space. Ranulas can be classified into simple and plunging types. The simple ranula occurs from inflammation of the sublingual gland or minor salivary glands in the sublingual space, and the cystic cavity is completely lined with epithelium

and is confined within the sublingual space above the mylohyoid muscle (Figure 15.37). On the other hand, the plunging ranula is an extravasation pseudocyst originating from the sublingual glands and extending into the submandibular space and then into the deeper cervical spaces, such as the parapharyngeal space [95,135] (Figure 15.38). Plunging ranulas usually extend through a hiatus or a defect of the mylohyoid muscle into the submandibular space [136–138]. A hiatus or defect is very common, and is observed at the junction of the mylohyoid muscle and the mandible. Therefore, on conventional MR images, plunging ranulas frequently exhibit a comet appearance with its *tail* in the sublingual space and its *head* in the submandibular spaces (Figure 15.38). Some plunging ranulas may wrap around the posterior border of the mylohyoid muscle into the submandibular space. Similar to ranulas, the sublingual gland and the associated tissues may herniate and cause symptomatic submandibular swelling [136,137]. The herniations of plunging ranulas through the mylohyoid muscle can be well delineated on coronal fat-suppressed T_2-weighted images.

Ranulas and lymphangiomas are sometimes clinically indistinguishable. However, lymphangiomas may be multilocular with separate spaces filled with lymphocyte-containing fluids that are lined by a layer of endothelial cells. In addition, lymphangiomas do not involve the sublingual space. Because these imaging and histological features are not observed with ranulas, they may serve as a clue for the differential diagnosis between these two cystic lesions [95]. Although Coit et al. reported that simple ranulas and epidermoid cysts could not be differentiated on CT [134], we now know that the overall ADCs can definitely discriminate ranulas from epidermoid (dermoid) cysts.

15.8.2 Benign Lymphoepithelial Lesion

Benign lymphoepithelial lesions may be associated with SS, HIV infection, and JRP [139,140] (Figure 15.42). Benign lymphoepithelial lesions exhibit a wide spectrum of MR imaging findings, ranging from an almost completely liquid-filled mass to a mixed structure with liquid and solid components [141]. Cervical lymphadenopathy is frequently associated with the lesions. On sonograms, parenchymal changes of the parotid and submandibular glands of patients with SS or HIV may mimic the benign lymphoepithelial lesions [142]. However, MR imaging can readily differentiate between benign lymphoepithelial lesions and the parenchymal changes of the affected glands.

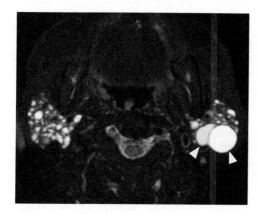

FIGURE 15.42
A 77-year-old woman with SS and lymphoepithelial lesions. Axial fat-suppressed T_2-weighted MR image shows multiple lymphoepithelial lesions of bilateral parotid glands (arrowheads).

References

1. Patel VN, Hoffman MP (2013) Salivary gland development: A template for regeneration. *Semin Cell Dev Biol* 25–26:52–60.
2. Ferreira JN, Hoffman MP (2013) Interactions between developing nerves and salivary glands. *Organogenesis* 9:199–205.
3. Knox SM, Lombaert IMA, Reed X et al. (2010) Parasynpathetic innervation maintains epithelial progenitor cells during salivary organogenesis. *Science* 329: 1645–1647.
4. Knox SM, Lombaert MA, Haddox CL et al. (2013) Parasynpathetic stimulation improved epithelial organ regeneration. *Nat Commun* 4:1494. doi:10.1038/ncommuns2493.
5. Tucker AS (2007) Salivary gland development. *Semin Cell Dev Biol* 18:237–244.
6. Qin Y, Zhang J, Li P et al. (2011) 3D double-ehco steady-state with water excitation MR imaging of the intraparotid facial nerve at 1.5T: A pilot study. *AJNR Am J Neuroradiol* 32:1167–1172.
7. Chu J, Zhou Z, Hong G et al. (2013) High-resolution MRI of the intraparotid facial nerve based on a microsurface coil and a 3D reversed fast imaging with steady-state precession DWI sequence at 3T. *AJNR Am J Neuroradiol* 34:1643–1648.
8. Ergün SS, Gayretli Ö, Büyükpinarbasili et al. (2013) Determining the number of intraparotid lymph nodes: Postmortem examination. *J Craniomaxillofacial Surg* 42:657–460. doi: 10.1016/j.jcms.2013.09.011.
9. Gervasio A, D'Orta G, Mujahed I et al. (2011) Sonographic anatomy of the neck: The supra hyoid region. *J Ultrasound* 14:130–135.
10. Sumi M, Izumi M, Yonetsu K, Nakamura T (1999) Sublingual gland: MR features of normal and diseased states. *AJR Am J Roentgenol* 172:717–722.
11. Martinez-Madrigal F, Micheau C (1989) Histology of the major salivary glands. *Am J Surg Pathol* 13:879–899.
12. Sumi M, Yamada T, Takagi Y, Nakamura T (2007) MR imaging of labial glands. *AJNR Am J Neuroradiol* 28:1552–1556.
13. Barnes L, Eveson JW, Reichart P, Sidransky D (2005) *Pathology and Genetics of Head and Neck Tumors.* IARC Press, Lyon, France.
14. Freling NJM, Molenaar WM, Vermey A et al. (1992) Malignant parotid tumors: Clinical use of MR imaging and histologic correlation. *Radiology* 185:691–696.
15. Christe A, Waldherr C, Hallett R, Zbaeren P, Thoeny H (2011) MR imaging of parotid tumors: Typical lesion characteristics in MR imaging improve discrimination between benign and malignant disease. *AJNR Am J Neuroradiol* 32:1202–1207.

16. Joe VQ, Westesson PL (1994) Tumors of the parotid gland: MR imaging characteristics of various histologic types. *AJR Am J Roentgenol* 163:433–438.

17. Yabuuchi H, Fukuya T, Tajima T, Hachitanda Y, Tomita K, Koga M (2002) Salivary gland tumors: Diagnostic value of gadolinium-enhanced dynamic MR imaging with histologic correlation. *Radiology* 226:345–354.

18. Eida S, Ohki M, Sumi M, Yamada T, Nakamura T (2008) MR factor analysis: Improved technology for the assessment of 2D dynamic structures of benign and malignant salivary gland tumors. *J Magn Reson Imaging* 27:1256–1262.

19. Okahara M, Kiyosue H, Hori Y, Matsumoto A, Mori H, Yokoyama S (2003) Parotid tumors: MR imaging with pathological correlation. *Eur Radiol* 13:L25–L33.

20. Ikeda M, Motoori K, Hanazawa T et al. (2004) Warthin tumor of the parotid gland: Diagnostic value of MR imaging with histopathologic correlation. *AJNR Am J Neuroradiol* 25:1256–1262.

21. Eida S, Sumi M, Nakamura T (2010) Multiparametric magnetic resonance imaging for the differentiation between benign and malignant salivary gland tumors *J Magn Reson Imaging* 31:673–679.

22. Jang M, Park D, Lee SR et al. (2004) Basal cell adenoma in theparotid gland: CT and MR findings. *AJNR Am J Neuroradiol* 25:631–635.

23. Lee DK, Chung KW, Baek CH, Jeong HS, Ko YH, Son YI (2005) Basal cell adenoma of the parotid gland: Characteristics of 2-phase helical computed tomography and magnetic resonance imaging. *J Comput Assist Tomogr* 29:884–888.

24. Patel ND, van Zante A, Eisele DW, Harnsberger HR, Glastonbury CM (2011) Oncocytoma: The vanishing parotid mass. *AJNR Am J Neuroradiol* 32:1703–1706.

25. Sigal R, Monnet O, de Baere T et al. (1992) Adenoid cystic carcinoma of the head and neck: Evaluation with MR imaging and clinical-pathologic correlation in 27 patients. *Radiology* 184:95–101.

26. Lee YYP, Wong KT, King AD, Ahuja AT (2008) Imaging of salivary gland tumours. *Eur J Radiol* 66:419–436.

27. Suh S, Seol HY, Kim TK et al. (2005) Acinic cell carcinoma of the head and neck: Radiologic-pathologic correlation. *J Comput Assist Tomogr* 29:121–126.

28. Weon TC, Park SW, Kim HJ, et al. (2012) Salivary duct carcinomas: Clinical and CT and MR imaging features in 20 patients. *Neuroradiol* 54:631–640.

29. Mooori K, Iida Y, Nagai Y et al. (2005) MR imaging of salivary duct carcinoma. *AJNR Am J Neuroradiol* 26:1201–1206.

30. Nemzek WR, Hecht S, Gandour-Edwards, Donald P, McKennan K (1998) Perineural spread of head and neck tumors: How accurate is MR imaging? *AJNR Am J Neuroradiol* 19:701–706.

31. Kato H, Kanematsu M, Mizuta K, Ito Y, Hirose Y (2008) Carcinoma ex pleomorphic adenoma of the parotid gland: Radiologic-pathologic correlation with MR imaging including diffusion-weighted imaging. *AJNR Am J Neuroradiol* 29:865–867.

32. Donahue KM, Weisskoff RM, Parmelee DJ et al. (1995) Dynamic Gd-DTPA enhanced MRI measurement of tissue cell volume fraction. *Magn Reson Med* 34:423–432.

33. Knopp WV, Weiss E, Sinn HP, et al. (1999) Pathophysiologic basis of contrast enhancement in breast tumors. *J Magn Reson Imaging* 10:260–266.

34. Oshida K, Nagasahima T, Ueda T et al. (2005) Pharmacokinetic analysis of ductal carcinoma in situ of the breast using dynamic MR mammography. *Eur Radiol* 15:1353–1360.

35. Sasaki M, Sumi M, Kaneko K, Ishimaru K, Takahashi H, Nakamura T (2013) Multiparametric MR imaging for differentiating between benign and malignant thyroid nodules: Initial experience in 23 patients. *J Magn Reson Imaging* 38:64–71.

36. Nakamura T, Sumi M, Van Cauteren M (2010) Salivary gland tumors: Preoperative tissue characterization with apparent diffusion coefficient mapping. In: Hayat MA (ed) *Methods of Cancer Diagnosis, Therapy, and Prognosis.* Vol. 7 General overviews, head and neck cancer and thyroid cancer. Springer.

37. Le Bihan D, Breton E, Lallemand D, Aubin ML, Vignaud J, Laval-Jeantet (1988) Separation of diffusion and perfusion in intravoxel incoherent motion MR imaging. *Radiology* 168:497–505.

38. Le Bihan D (2003) Looking into the functional architecture of the brain with diffusion MRI. *Nat Rev Neurosci* 4:469–480.

39. Eida S, Sumi M, Sakihama N, Takahashi H, Nakamura T (2007) Apparent diffusion coefficient mapping of salivary gland tumors: Prediction of the benignancy and malignancy. *Am J Neuroradiol* 28:116–121.

40. Thoeny HC, De Keyzer F, Boesch C, Hermans R (2004) Diffusion-weighted imaging of the parotid gland: Influence of the choice of *b*-values on the apparent diffusion coefficient value. *J Magn Reson Imaging* 20:786–790.

41. Harbermann CR, Arndt C, Graessner J et al. (2009) Diffusion-weighted echo-planar MR imaging of primary parotid gland tumors: Is a prediction of different histologic subtypes possible? *Am J Neuroradiol* 30:591–596.

42. Sumi M, Nakamura T (2009) Diagnostic importance of focal defects in the apparent diffusion coefficient-based differentiation between lymphoma and squamous cell carcinoma nodes in the neck. *Eur Radiol* 19:975–981.

43. Zima A, Carlos R, Gandhi D, Case I, Teknos T, Mukherji SK (2007) Can pretreatment CT perfusion predict response of advanced squamous cell carcinoma of the upper digestive tract treated with induction chemotherapy? *AJNR Am J Neuroradiol* 28:328–334.

44. Lewin M, Fartoux L, Vignaud A, ArrivéL, Menu Y, Rosmorduc O (2011) The diffusion-weighted imaging perfusion fraction f is a potential marker of sorafenib treatment in advanced hepatocellular carcinoma: a pilot study. *Eur Radiol* 21:281–290.

45. Le Bihan D (2008) Intravoxel incoherent motion perfusion MR imaging: A wake-up call. *Radiology* 249:748–752

46. Luciani A, Vignaud A, Cavet M et al. (2008) Liver cirrhosis: Intravoxel incoherent motion MR imaging—Pilot study. *Radiology* 249:891–899.

47. Sumi M, Van Cauteren M, Sumi T, Obara M, Ichikawa Y, Nakamura T (2012) Salivary gland tumors: Use of intravoxel incoherent motion MR imaging for assessment of diffusion and perfusion for the differentiation of benign from malignant tumors. *Radiology* 263:770–777.

48. Gao Q, Srinivasan G, Magin RL, Zhou XJ (2011) Anomalous diffusion measured by a twice-refocused spin echo pulse sequence: Analysis using fractional order calculus. *Magn Reson Imaging* 33:1177–1183.

49. Sumi M, Nakamura T (2013) Head and neck tumors: Assessment of perfusion-related parameters and diffusion coefficients based on the intravoxel incoherent motion model. *Am J Neuroradiol* 34:410–416.

50. Takahara T, Kwee TC (2012) Low *b*-value diffusion-weighted imaging: Emerging applications in the body. *J Magn Reson Imaging* 35:1266–1273.

51. Vitali C, Bombardieri S, Jonsson R, et al. (2002) Classification criteria for Sjögren's syndrome: A revised version of the European criteria proposed by the American–European Consensus Group. *Ann Rheum Dis* 61:554–558.

52. Bournia VK, Vlachoyiannopoulos PG (2012) Subgroups of Sjögren syndrome patients according to serological profiles. *J Autoimmun* 39:15–26.

53. Ramos-Casals M, Brito-Zerón P, Perez-De-Lis M et al. (2010) Sjögren syndrome or Sjögren disease? The histological and immunological bias caused by the 2002 criteria. *Clinc Rev Allerg Immunol* 38:178–185.

54. Baldini C, Talarico R, Tzioufas AG, Bombardieri S (2012) Classification criteria for Sjögren's syndrome: A critical review. *J Autoimmun* 39:9–14.

55. Goules AV, Tzioufas AG, Moutsopoulos M (2014) Classification criteria of Sjögren's syndrome. doi. org/10.1016/j.jaut.2014.01.013.

56. Rubin P, Holt JF (1957) Secretory sialography in disease of the major salivary glands. *AJR Am J Roentgenol* 77:575–598.

57. Izumi M, Eguchi K, Ohki M, Uetani M, Hayashi K, Kita M, Nagataki S, Nakamura T (1996) MR imaging of the parotid gland in Sjögren's syndrome: A proposal for new diagnostic criteria. *AJR Am J Roentgenol* 166:1483–1487.

58. Izumi M, Eguchi K, Nakamura H, Takagi Y, Kawabe Y, Nakamura T (1998) Corticosteroid irrigation of parotid gland for treatment of xerostomia in patients with Sjögren's syndrome. *Ann Rheum Dis* 57:464–469.

59. Izumi M, Eguchi K, Nakamura H, Nagataki S, Nakamura T (1997) Premature fat deposition in the salivary glands associated with Sjögren syndrome: MR and CT evidence. *Am J Neuroradiol* 18:951–958.

60. Takagi Y, Sumi M, Sumi T, Ichikawa Y, Nakamura T (2005) MR microscopy of the parotid glands in patients with Sjögren's syndrome: Quantitative MR diagnostic criteria. *Am J Neuroradiol* 26:1207–1214.

61. Lomas DJ, Carroll NR, Johnson G, Antoun NM, Freer CEL (1996) MR sialography—Work in progress. *Radiology* 200:129–133.

62. Ohbayashi N, Yamada I, Yoshino N, Sasaki T (1998) Sjögren syndrome: Comparison of assessment with MR sialography and conventional sialography. *Radiology* 209:683–688.

63. Tonami H, Ogawa Y, Matoba M et al. (1998) MR sialography in patients with Sjögren syndrome. *AJNR Am J Neuroradiol* 19:1199–1203.

64. Niemelä RK, Pääkkö E, Suramo I, Takalo R, Hakala M (2001) Magnetic resonance imaging and magnetic resonance sialography of parotid glands in primary Sjögren's syndrome. *Arthritis Care Res* 45:512–518.

65. Takagi Y, Sumi M, Van Cauteren M, Nakamura T (2005) Fast and high-resolution MR sialography using a small surface coil. *J Magn Reson Imaging* 22:29–37.

66. Sumi M, Van Cauteren M, Takagi Y, Nakamura T (2007) Balanced turbo field-echo sequence for MRI of parotid gland diseases. *AJR Am J Roentgenol* 188:228–232.

67. Chavhan GB, Babyn PS, Fankharia BG, Cheng HLM, Shroff MM (2008) Steady-state MR imaging sequences: Physics, classification and clinical applications. *RadioGraphics* 28:1147–1160.

68. Sumi M, Takagi Y, Uetani M, Morikawa M, Hayashi K, Kabasawa H, Aikawa K, Nakamura T (2002) Diffusion-weighted echoplanar MR imaging of the salivary glands. *AJR Am J Roentgenol* 178:959–965.

69. Regier M, Ries T, Arndt C et al. (2009) Sjögren's syndrome of the parotid gland: Value of diffusion-weighted echo-planar MRI for diagnosis at an early stage based on MR sialography grading in comparison with healthy volunteers. *Fortschr Röntgenstr* 181:242–248.

70. Kawai Y, Sumi M, Kitamori H, Takagi Y, Nakamura T (2005) Diffusion-weighted MR microimaging of the lacrimal glands in patients with Sjögren's syndrome. *AJR Am J Roentgenol* 184:1320–1325.

71. Thoeny HC, De Keyzer F, Claus FG, Sunaert S, Hermans R (2005) Gustatory stimulation changes the apparent diffusion coefficient of salivary glands: Initial experience. *Radiology* 235:629–634.

72. Kato H, Kanematsu M, Toida M et al. (2011) Salivary gland function evaluated by diffusion-weighted MR imaging with gustatory stimulation: Preliminary results. *J Magn Reson Imaging* 34:904–909.

73. Tofts PS, Brix G, Buckley DL et al. (1999) Estimating kinetic parameters from dynamic contrast-enhanced T_1-weighted MRI of a diffusable tracer: Standardized quantities and symbols. *J Magn Reson Imaging* 10:223–232.

74. Roberts C, Parker GJ, Rose CJ et al. (2008) Glandular function in Sjögren syndrome: Assessment with dynamic contrast-enhanced MR imaging and tracer kinetic modeling—Initial experience. *Radiology* 246:845–853.

75. Lanzman RS, Robson PM, Sun MR et al. (2012) Arterial spin-labeling MR imaging of renal masses: Correlation with histopathologic findings. *Radiology* 265:799–808.

76. Williams DS, Detre JA, Leigh JS, Koretsky AP (1992) Magnetic resonance imaging of perfusion using spin inversion of arterial water. *Proc Natl Acad Sci USA* 89:212–216.

77. Alsop DC, Detre JA (1998) Multisection cerebral blood flow MR imaging with continuous arterial spin labeling. *Radiology* 208:410–416.

78. Wu WC, Fernández-Seara M, Detre JA, Wehrli FW, Wang J (2007) A theoretical and experimental investigation of the tagging efficiency of psuedocontinuous arterial spin labeling. *Magn Reson Med* 58:1020–1027.

79. Masaki Y, Sugai S, Umehara H (2010) IgG4-related diseases including Mikulicz's disease and sclerosing pancreatitis: Diagnostic insights. *J Rheumatol* 37:1380–1385.

80. Masaki Y, Dong L, Kurose N et al. (2009) Proposal for a new clinical entity, IgG4-positive multiorgan lymphoproliferative syndrome: Analysis of 64 cases of IgG4-related disorders. *Ann Rheum Dis* 68:1310–1315.

81. Yamamoto M, Harada S, Ohhara M et al. (2005) Clinical and patological differences between Mikulicz's disease and Sjögren's syndrome. *Rheumatology* 44:227–224.

82. Takagi Y, Nakamura H, Origuchi T, Miyashita T, Kawakami A, Sumi M, Nakamura T (2013) IgG4-related Mikulicz's disease: Ultrasonography of the salivary and lacrimal glands for monitoring the efficacy of corticosteroid therapy. *Clin Exp Rheumatol* 31:773–775.

83. Fujita A, Sakai O, Chapman MN, Sugimoto H (2012) IgG4-related disease of the head and neck: CT and MR imaging manifestations. *RadioGraphics* 32:1945–1958.

84. Sumi T, Takagi Y, Ichikawa Y, Sumi M, Kimura Y, Nakamura T (2012) Imaging features of the lacrimal and salivary glands of patients with IgG4-related Mikulicz's disease: A report of three cases. *Oral Radiol* 28:140–145.

85. Kaltreider BH, Talal N (1969) Bilateral parotid enlargement and hyperlipoproteinemia. *JAMA* 210:2067–2070.

86. Sheikh JS, Sharma M, Kunath A, Frit DA, Glueck CJ, Hess EV (1996) Reversible parotid enlargement and pseudo-Sjögren's syndrome secondary to hypertriglyceridemia. *J Rheumatol* 23:1288–1291.

87. Goldman JA, Julian EH (1977) Pseudo-Sjögren syndrome with hyperlipoproteinemia. *JAMA* 237:1582–1584.

88. Izumi M, Hida A, Takagi Y, Kawabe Y, Eguchi K, Nakamura T (2000) MR imaging of the salivary glands in sicca syndrome: Comparison of lipid profiles and imaging in patients with hyperlipidemia and patients with Sjögren's syndrome. *AJR Am J Roentgenol* 175:829–834.

89. Ramos-Casals M, Brito-Zerón P, SisóA et al. (2007) High prevalence of serum metabolic alterations in primary Sjögren's syndrome: Influence on clinical and immunological expression. *J Rheumatol* 34:754–761.

90. Sipsas NV, Gamaletsou MN, Moutsopoulos HM (2011) Is Sjögren's syndrome a retroviral disease? *Arthritis Res Ther* 13:212–219.

91. Basu D, Williams FM, Ahn CW, Reveille JD (2006) Changing spectrum of the diffuse infiltrative lymphocytosis syndrome. *Arthritis Care Res* 55:466–472.

92. Shugar JM, Som PM, Jacobson AL, Ryan JR, Bernard PJ, Dickman SH (1988) Multicentric parotid cysts ad cervical adenopathy in AIDS patients. A newly recognized entity: CT and MR manifestations. *Laryngoscope* 98:772–775.

93. Kirshenbaum KJ, Nadimpalli SR, Friedman M, Kirshenbaum GL, Cavallino RP (1991) Benign lymphoepithelail parotid tumors in AIDS patients: CT and MR findings in nine cases. *AJNR Am J Neuroradiol* 12:271–274.

94. Gottesman RI, Som PM, Mester J, Silvers A (1996) Observations on two cases of apparent submandibular gland cysts in HIV positive patients: MR and CT findings. *J Comput Assist Tomogr* 20:444–447.

95. Harnsberger HR (2004) Diagnostic imaging. *Head and Neck.* AMIRSYS, Salt Lake City, UT.

96. Osame M, Usuku K, Izumo S et al. (1986) HTLV-1 associated myelopathy, a new clinical entity. *Lancet* 327:1031–1032.

97. Terada K, Katamine S, Eguchi K et al. (1994) Prevalence of serum and salivary antibodies to HTLV-1 in Sjögren's syndrome. *Lancet* 344:1116–1119.

98. Izumi M, Nakamura H, Nakamura T, Eguchi K, Nakamura T (1999) Sjögren's syndrome (SS) in patients with human T cell leukemia virus I associated myelopathy: Paradoxical features of the major salivary glands compared to classical SS. *J Rheumatol* 26:2609–2614.

99. Gupta RK, Best J, MacMahon E (2005) Mumps and the UK epidemic 2005. *BMJ* 330:1132–1135.

100. Jorgensen C, Legouffe MC, Perney P et al. (1996) Sicca syndrome associated with hepatitis C virus infection. *Arthritis Rheum* 39:1166–1171.

101. Arrieta JJ, Rodríguez-Iñigo E, Ortiz-Movilla N et al. (2001) *In situ* detection of hepatitis C virus RNA in salivary glands. *Am J Pathol* 158:259–264.

102. Ramos-Casals M, De Vita S, Tzioufas AG (2005) Hepatitis C vitrus, Sjögren's syndrome and B-cell lymphoma: Linking infection, autoimmunity and cancer. *Autoimmunity Rev* 4:8–15.

103. Grossmann Sde M, Teixeira R, Oliveira GC et al. (2010) Xerostomia, hyposalivation and sialadenitis in patients with chronic hepatitis C are not associated with the detection of HCV RNA in saliva or salivary glands. *J Clin Pathol* 63:1002–1007.

104. Vitali C (2011) Immunopathologic differences of Sjögren's syndrome versus sicca symdrome in HCV and HIV infection. *Arthritis Res Ther* 13:233–239.

105. Ramos-Casals M, La Civita L, De Vita S et al. (2007) Characterization of B cell lymphoma in patients with Sjögren's syndrome and hepatitis C virus infection. *Arthritis Care Res* 57:161–170.

106. Som PM, Brandwein M (1996) Salivary glands. In: Som PM, Curtin HD (eds.) *Head and Neck Imaging*, 3rd edn., Mosby, St. Louis, MO.

107. Sumi M, Izumi M, Yonetsu K, Nakamura T (1999) The MR imaging assessment of submandibular gland sialoadenitis secondary to sialolithiasis: Correlation with CT and histopathologic findings. *AJNR Am J Neuroradiol* 20:1737–1743.

108. Becker M, Marchal F, Becker CD et al. (2000) Sialolithiasis and salivary ductal stenosis: Diagnostic accuracy of MR sialography with a three-dimensional extended-phase conjugate-symmetry rapid spin-echo sequence. *Radiology* 217:347–358.

109. Jäger L, Menauer F, Holzknecht N, Scholz V, Grevers G, Reiser M (2000) Sialolithiasis: MR sialography of the submandibular duct—An alternative to conventional sialography and US? *Radiology* 216:665–671.

110. Kalinowski M, Heverhagen JT, Rehberg E, Klose KJ, Wagner HJ (2002) Comparative study of MR sialography and digital subtraction sialography for benign salivary gland disorders. *Am J Neuroradiol* 23:1485–1492.

111. Leerdam CM, Martin HCO, Isaacs D (2005) Recurrent parotitis of childhood. *J Paediatr Child Health* 41:631–634

112. Gadodia A, Seith A, Sharma R, Thakar A (2010) MRI and MR sialography of juvenile recurrent parotitis. *Pediatr Radiol* 40:1405–1410.

113. Kimura Y, Hotokezaka Y, Sasaki M, Takagi Y, Eida S, Katayama I, Sumi M, Nakamura T (2011) Magnetic resonance imaging-based differentiation between juvenile recurrent parotitis and juvenile Sjögren's syndrome. *Oral Radiol* 27:73–77.

114. Civilibal M, Canpolat N, Yurt A et al. (2007) A child with primary Sjögren syndrome and a review of the literature. *Clin Pediatrics* 46:738–742.

115. Singeer NG, Tomanova-Soltys I, Lowe R (2008) Sjögrens syndrome in childhood. *Cur Rheumatol Rep* 10:147–155.

116. Kam MKM, Leung SF, Zee B et al. (2007) Prospective randomized study of intensity-modulated radiotherapy on salivary gland function in early-stage nasopharyngeal carcinoma patients. *J Clin Oncol* 25:4873–4879.

117. Murdch-Kinch CA, Kim HM, Vineberg KA, Ship JA, Eisbruch A (2008) Dose-effect relationship for the submandibular salivary glands and implications for their sparing by intensity modulated radiotherapy. *Int J Rad Oncol Biol Phys* 72:373–382.

118. Houweling AC, van den Berg CAT, Roesink JM, Terhaard CHJ, Raaijmakers CPJ (2010) Magnetic resonance imaging at 3.0 T for submandibular gland sparing radiotherapy. *Radiother Oncol* 97:239–243.

119. Osorio EMV, Hoogeman MS, Al-Mamgani A, Teguh DN, levendag PC, Heijmen BJM (2008) Local anatomic changes in parotid and submandibular glands during radiotherapy for oropharynx cancer and correlation with dose, studied in detail with nonrigid registration. *Int J Radiation Oncol Biol Phys* 70:875–882.

120. Houweling AC, Schakel T, van den Berg CAT et al. (2011) MRI to quantify early radiation-induced changes in the salivary glands. *Radiother Oncol* 100:386–389.

121. Lee FK, King AD, Kam MK, Ma BB, Yeung DK (2011) Radiation injury of the parotid glands during treatment of head and neck cancer: Assessment using dynamic contrast-enhanced MR imaging. *Radiat Res* 175:291–296.

122. Juan CJ, Chen CY, Jen YM et al. (2009) Perfusion characteristics of late radiation injury of parotid glands: Quantitative evaluation with dynamic contrast-enhanced MRI. *Eur Radiol* 19:94–102.

123. Dirix P, De Keyzer F, Vandecaveye V, Stroobants S, Herman R, Nuyts S (2008) Diffusion-weighted magnetic resonance imaging to evaluate major salivary gland function before and after radiotherapy. *Int J Rad Oncol Biol Phys* 71:1365–1371.

124. Astreinidou E, Roesink JM, Raaijmakers CPJ et al. (2007) 3D MR sialography as a tool to investigate radiation-induced xerostomia: Feasibility study. *Int J Radiation Oncol Biol Phys* 68:1310–1319.

125. Wada A, Uchida N, Yokokawa M, Yoshizako T, Kitagaki H (2009) Radiation-induced xerostomia: Objective evaluation of salivary gland injury using MR sialography. *AJNR Am J Neuroradiol* 30:53–58.

126. Yonetsu K, Izumi M, Nakamura T (1998) Deep facial infections of odontogenic origin: CT assessment of pathways of space involvement. *AJNR Am J Neuroradiol* 19:123–128.

127. Maroldi R, Farina D, Ravanelli M, Lombardi D, Nicolai P (2010) Emergency imaging. Assessment of deep neck space infections. *Semin Ultrasound CT MRI* 33:432–442.

128. Shetty AK, Beaty MW, McGuirt WF, Woods CR, Givner LB (2002) Kimura's disease: A diagnostic challenge. *Pediatrics* 110: e39.

129. Armstrong WB, Allison G, Pena F, Kim JK (1998) Kimura's disease: Two case reports and a literature review. *Ann Oto Rhinol Laryngol* 107:1066–1071.

130. Atar S, Oberman AS, Ben-Izhak O, Flatau E (1994) Recurrent nephritic syndrome associated with Kimura's disease in a young non-oriental male. *Nephron* 68:259–261.

131. Oguz KK, Ozturk A, Cila A (2004) Magnetic resonance imaging findings in Kimura's disease. *Neuroradiology* 46:855–858.

132. Park SW, Kim HJ, Sung KJ, Lee JH, Park IS (2012) Kimura disease: CT and MR imaging findings. *AJNR Am J Neuroradiol* 33:784–788.

133. Takahashi S, Ueda J, Furukawa T et al. (1996) Kimura disease: CT and MR findings. *AJNR Am J Neuroradiol* 17:382–385.

134. Coit W, Harnsberger HR, Osborn AG, Smoker WRK, Stevens MH, Lufkin RB (1987) Ranulas and their mimics: CT evaluation. *Radiology* 163:211–216.

135. Ichikawa Y, Sumi M, Eida S, Takagi Y, Tashiro S, Hotokezaka Y, Katayama I, Nakamura T (2012) Apparent diffusion coefficient characterization of fluid areas in cystic and abscess lesions of the neck. *Oral Radiol* 28:62–69.

136. Keberle M, Eulert S, Relic A, Hahn D (2005) Functional MR imaging of submandibular herniation of sublingual tissues through a gap of the mylohyoid muscle in two cases of submandibular "masses." *Eur Radiol* 15:1326–1328.

137. Kiesler K, Gugatschka M, Friedrich G (2007) Incidence and clinical relevance of herniation of the mylohyoid muscle with penetration of the sublingual gland. *Eur Arch Otorhinolaryngol* 264:1071–1074.

138. Jain P, Jain R, Morton RP, Ahmad Z (2010) Plunging ranulas: High-resolution ultrasound for diagnosis and surgical management. *Eur Radiol* 20:1442–1449.

139. Hamilton BE, Salzman KL, Wiggins RH, Harnsberger HR (2003) Earing lesions of the parotid tail. *AJNR Am J Neuroradiol* 24:1757–1764.

140. Ma Q, Song H (2011) Diagnosis and management of lymphoepithelial lesion of the parotid gland. *Rheumatol Int* 31:959–962.

141. Kirschenbaum KJ, Nadimpalli SR, Friedman M, Kirschenbaum GL, Cavallino RP (1991) Benign lymphoepithelial patorid tumors in AIDS patients: CT and MR findings in nine cases. *AJNR Am J Neuroradiol* 12:271–274.

142. Martinoli C, Pretolesi F, Del Bono V, Derchi LE, Mecca D, Chiaramondia M (1995) Benign lymphoepithelial parotid lesions in HIV-positive patients: Spectrum of findings at gray-scale and Doppler sonography. *AJR Am J Roentgenol* 165:975–979.

16

Infective Pathology of the Neck

Naoko Saito, Joan Cheng, Akifumi Fujita, Hiroyuki Fujii, and Osamu Sakai

CONTENTS

16.1 Introduction

Infective diseases in the head and neck can be life-threatening. Prompt localization of the infection and identification of the pathogen are necessary to direct appropriate antimicrobial treatment as early as possible. Although the diagnosis is often clear by physical examination and clinical history, imaging is crucial in clearly localizing and defining the extent of the infection. Knowing the typical imaging patterns allows rapid and accurate diagnosis and guides subsequent therapeutic decisions.

There are various imaging modalities for the evaluation of infective diseases in the head and neck: ultrasound, radiographs, fluoroscopy, computed tomography (CT), and magnetic resonance imaging (MRI). MRI is a useful tool in the management of head and neck infections. Compared to CT, MRI provides superior soft tissue contrast; better delineating suspected orbital, intracranial, and spinal extension; and soft tissue infection, including salivary gland and thyroid gland. Diffusion-weighted imaging (DWI) is highly sensitive for abscess formation. MR angiography (MRA) and MR venography (MRV) have also been proposed in the evaluation of vascular involvement such as venous or sinus thrombosis. Despite these advantages, MRI may not become a first-line imaging modality in the emergency setting, because it is time consuming and metallic foreign bodies can preclude the patient from safely entering the MR suite.

The focus of this chapter is to demonstrate the typical MRI findings in head and neck infections. Infections discussed include temporal bone (otits media, labyrinthitis), paranasal sinus (rhinosinusitis, fungal infection), orbit (periorbital and orbital infection, dacrocystitis), pharynx and oral cavity (acute tonsillitis, peritonsillar abscess, odontogenic infection, retropharyngeal abscess), salivary gland (sialadeniitis), spinal and prevertebral space (epidural abscess), and cervical lymph node (suppurative lymphadenopathy, tuberculous lymphadenopathy).

16.2 Temporal Bone and the Ear

16.2.1 Acute Otitis Media and Acute Otomastoiditis

16.2.1.1 Definition

Acute otitis media is defined as fluid in the middle ear with signs and symptoms of middle ear inflammation. It is often preceded by a viral upper respiratory infection that causes mucosal edema in the nose and nasopharynx, which leads to Eustachian tube dysfunction. It most commonly occurs in the first 5 years of life [1–3].

16.2.1.2 Pathology

Streptococcus pneumoniae and *Haemophilus influenzae* account for the majority of episodes of acute otitis media [3,4]. Other bacteria include *Moraxella catarrhalis*, *Streptococcus pyogenes*, and *Staphylococcus aureus* [4].

16.2.1.3 Clinical Features

Patients with acute otitis media typically present with otalgia and otorrhea, but nonspecific symptoms of irritability and difficulty with feeding or sleeping are also common in young children. The key diagnostic feature is bulging of the tympanic membrane on otoscopic examination [1–3].

In most patients, the condition resolves after treatment with antibiotics.

16.2.1.4 Imaging Findings

Imaging is not necessary in uncomplicated acute otitis media. If MRI is performed, T_2-weighted images demonstrate hyperintensity occupying the middle ear and mastoid air cells, which may contain air–fluid levels [1–3] (Figure 16.1). MRI is sensitive to identify opacification of the tympanic cavity and mastoid air cells; however, the diagnosis of otitis media and mastoiditis should be made clinically, not by imaging findings. Complication can be intratemporal or intracranial; MRI is more appropriate than CT for diagnosing suspected intracranial complications.

16.2.2 Complications of Otitis Media

Complications of otitis media occur in approximately 18% of cases [1], and MRI of the temporal bone plays a crucial diagnostic role in these cases. There are various complications of acute otitis media, which include mastoiditis, labyrinthitis, petrous apicitis, extracranial abscess (Bezold's abscess and subperiosteal abscess), intracranial abscess, and dural sinus thrombosis.

16.2.2.1 Mastoiditis

Acute mastoiditis (Figure 16.1) results from obstruction of the mastoid antrum. Children with this condition have prolonged symptoms of otitis media with retroauricular pain, erythema, and swelling.

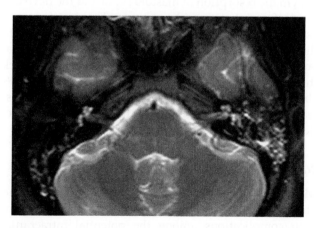

FIGURE 16.1
Opacification of the mastoid air cells in a patient with acute otitis media and mastoiditis. Axial T_2-weighted image demonstrates hyperintensity in the left middle ear and mastoid air cells. Mild right-sided opacification of the mastoid air cells is also seen.

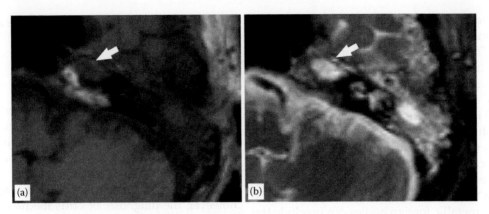

FIGURE 16.2
Petrous apicitis. (a) Axial T_1-weighted image demonstrates a hypointense area, loss of normal T_1 high signal from the fatty marrow within the left petrous apex (arrow). (b) Axial fat-suppressed T_2-weighted image demonstrates hyperintensity (arrow), representing petrous apicitis. Note the associated left-sided tympanic cavity and mastoid opacification consistent with acute otitis media and mastoiditis.

16.2.2.2 Petrous Apicitis

Infection from mastoid air cells and middle ear cavity extends to the petrous apex, which can cause petrous apicitis [1–3]. Clinical symptoms include otalgia and retro-orbital pain. T_2-weighted images demonstrate fluid signal (Figure 16.2) and contrast-enhanced images show peripheral enhancement in the petrous apex.

Gradenigo's syndrome is a rare entity, which is characterized by the triad of the following conditions: suppurative otitis media, pain in the distribution of the first and second divisions of trigeminal nerve due to extension of inflammation into Meckel's cave, and abducens nerve palsy secondary to involvement of Dorello's canal [1–3].

16.2.2.3 Coalescent Mastoiditis

If treatment for acute otomastoiditis fails, there may be enzymatic resorption of mastoid septa and the development of an intramastoid empyema. This is referred to as coalescent mastoiditis [1–3]. Infection may spread to the subcutaneous tissues through a break in the cortical bone or through emissary veins. Spread into the neck through a destroyed mastoid tip is termed a Bezold's abscess [2,3]. The complications also include subperiosteal abscess (Figure 16.3), intracranial extension of infection, and dural sinus thrombosis [1–3].

16.2.2.4 Intracranial Complications

Meningitis, dural sinus thrombosis (Figure 16.4), and otogenic brain abscesses (Figure 16.5) are the most common complications among the potential intracranial complications of otitis media [5].

Otogenic brain abscess is a life-threatening complication of otitis media. Patients with brain abscess may present with headache, fever, nausea, vomiting, seizures, and confusion. Otogenic intracranial abscesses are usually caused by *S. pneumoniae*, Group A *Streptococcus, S. aureus,* or *Proteus* species [6]. These organisms can extend directly into the intracranial cavity, usually through a bony defect in the tegmen tympani (in case of cerebral abscess) or in Trautmann's triangle (in case of cerebellar abscess) [7]. MRI findings of brain abscess include central fluid signal (hyperintensity on T_2-weighted images and hypointensity on T_1-weighted images), with peripheral enhancement following intravenous gadolinium administration and surrounding vasogenic edema. On DWI, restricted diffusion within the brain abscess results in markedly increased signal intensity [8] (Figure 16.5).

Dural sinus thrombosis may occur via direct extension of infection or result from retrograde thrombophlebitis [3]. Perisinus inflammation typically induces formation of mural thrombus, which propagates to obliterate the sinus [3]. The sigmoid and transverse sinuses are most commonly involved because of anatomic proximity [8,9]. Clinical symptoms include headaches, high fevers, sixth nerve palsy, and mental status changes. An absent flow void on spin-echo images and absent flow-related enhancement on gradient-echo sequences are seen in dural sinus thrombosis (Figure 16.4). MR venogram is useful to detect dural sinus thrombosis [9,10].

16.2.3 Labyrinthitis

16.2.3.1 Definition

Labyrinthitis is the inflammation of the inner ear, especially the membranous labyrinth. Labyrinthitis ossificans is considered to be a chronic stage of labyrinthitis, characterized by the pathologic ossification of structures of the membranous labyrinth following an infectious or inflammatory insult to the inner ear [3].

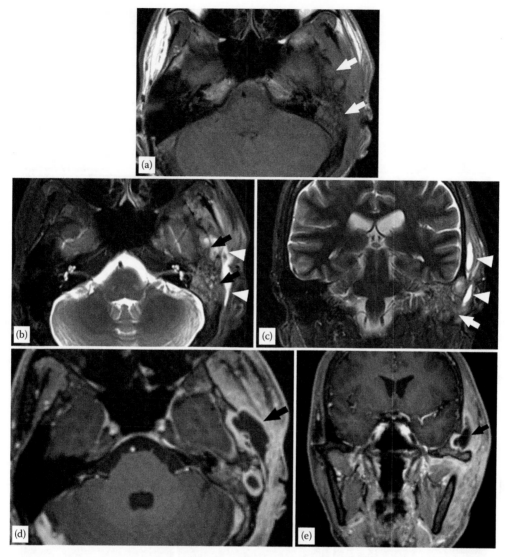

FIGURE 16.3
Coalescent mastoiditis with subperiosteal abscess. (a) Axial T_1-weighted and (b) fat-suppressed T_2-weighted and (c) coronal T_2-weighted images demonstrate complete opacification of the air cells with irregular cortical defects (arrows) in the left mastoid and fluid collection (arrowheads). (d) Axial and (e) coronal contrast-enhanced T_1-weighted images with fat suppression demonstrate a subperiosteal abscess (arrows) with surrounding soft tissue swelling and inflammation.

16.2.3.2 Pathology

The most common cause of labyrinthitis is viral infection [3]. In viral infection, the spread to the inner ear is thought to be hematogenous. Viral labyrinthitis may be the result of a systemic viral illness such as herpes simplex, varicella zoster, cytomegalovirus, influenza, rubeola, rubella, and mumps. Bacterial labyrinthitis is uncommon, but can occur from direct spread of infection from bacterial meningitis, otitis media (Figure 16.6), cholesteatoma (Figure 16.7), trauma, and surgery. *S. pneumoniae* and *H. influenzae* are the most common causes of bacterial labyrinthitis [3].

16.2.3.3 Clinical Features

The major symptoms of labyrinthitis are vertigo and sensorineural hearing loss.

16.2.3.4 Imaging Findings

On high-resolution temporal bone MRI, typical imaging findings are faint and diffuse enhancement of the membranous labyrinth on thin-slice gadolinium T_1-weighted images [3]. It is considered that a breakdown of labyrinthine vasculature permits the accumulation of gadolinium within inflamed labyrinthine membranes, which is the

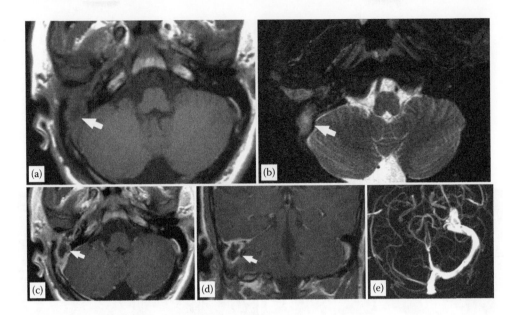

FIGURE 16.4
Acute otitis media with sigmoid sinus thrombosis. (a) Axial T_1-weighted and (b) fat-suppressed T_2-weighted images demonstrate complete opacification of the right mastoid air cells with direct inflammatory extension into the sigmoid sinus (arrows). (c) Axial and (d) coronal contrast-enhanced T_1-weighted images show a filling defect with irregular peripheral enhancement in the right sigmoid and transverse sinuses, which demonstrate increased caliber (arrows), consistent with acute thrombosis and inflammation. (e) MR venogram shows complete absence of flow-related signal in the right sigmoid and transverse sinuses.

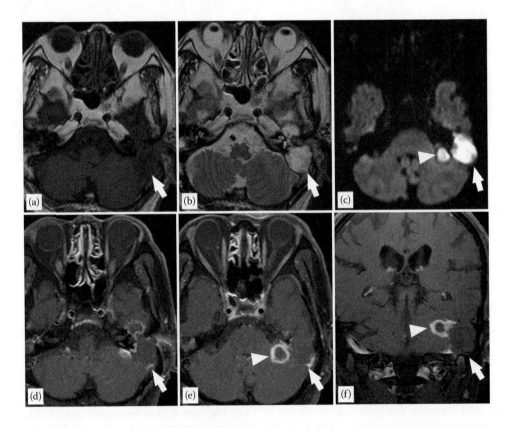

FIGURE 16.5
Choleosteatoma with brain abscess. (a) Axial T_1-weighted and (b) T_2-weighted images demonstrate an area of T_1 hypointensity and T_2 hyperintensity (arrow) in the left middle ear and mastoid air cells, respectively. (c) Axial diffusion-weighted image ($b = 1000$) shows hyperintensity within the mastoid indicative of cholesteatoma (arrow). Note a hyperintense area (arrowheads) adjacent to the cholesteatoma in the left cerebellum, which represents a brain abscess. Axial (d and e) and (f) coronal contrast-enhanced fat-suppressed T_1-weighted images show cholesteatoma (arrow) with thin rim enhancement and cerebellar abscess with thick and irregular rim enhancement (arrowheads).

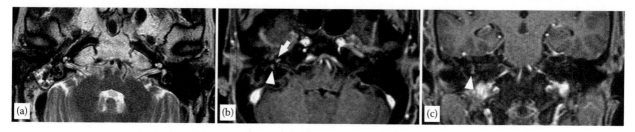

FIGURE 16.6
Labyrinthitis with acute otitis media secondary to bacterial infection. (a) Axial T_2-weighted image demonstrates hyperintense areas in the right middle ear and mastoid air cells. (b) Axial and (c) coronal contrast-enhanced fat-suppressed T_1-weighted images show enhancement in the right basal turn of the cochlea (arrow), vestibule, and lateral semicircular canal (arrowhead).

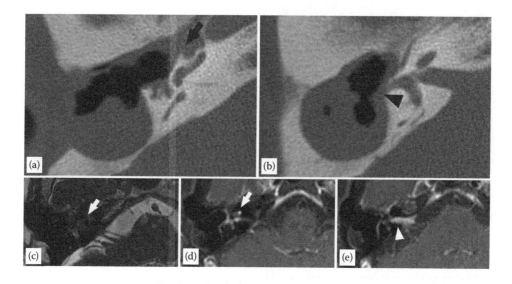

FIGURE 16.7
Labyrinthitis in a patient with cholesteatoma. (a,b) Axial CT demonstrates a residual soft tissue lesion eroding the cochlea (arrow) and lateral semicircular canal (arrowhead). (c) Axial high-resolution T_2-weighted image shows loss of the normal hyperintensity in the membranous labyrinth (arrow). (d,e) Axial contrast-enhanced fat-suppressed T_1-weighted images show enhancement in the cochlea (arrow), vestibule, lateral semicircular canal, and in the internal auditory canal (arrowhead).

cause of the labyrinthine enhancement [11–13]. Imaging alone may not differentiate between the causes; however, meningitis typically affects both ears, whereas otogenic infections typically cause unilateral symptoms [11–13].

Ramsay Hunt syndrome (Figure 16.8), also known as herpes zoster oticus, is viral labyrinthitis due to varicella zoster virus infection [3]. Facial paralysis, hearing loss, and vertigo along with vesicles in the external auditory canal are major symptoms of Ramsay Hunt syndrome. Prominent enhancement in the facial nerve, internal auditory canal, and labyrinth is typical in Ramsay Hunt syndrome [3].

Differential diagnosis for labyrinthine enhancement includes labyrinthine schwannoma, which is segmental and avidly enhancing, and labyrinthine hemorrhage, which is hyperintense on precontrast T_1-weighted images [3,11–13].

Labyrinthitis ossificans is pathologic ossification of the membranous labyrinth, including the cochlea,

vestibule, and semicircular canals [3]. On high-resolution T_2-weighted images such as constructive interference in the steady state and driven equilibrium radiofrequency reset pulse, the normal T_2 hyperintensity in the labyrinth is lost (Figure 16.9). High-resolution temporal bone CT shows calcification or ossification of the membranous labyrinth [3].

16.2.4 Facial Neuritis

16.2.4.1 Definition

Bell's palsy is acute idiopathic peripheral facial paralysis secondary to dysfunction of the facial nerve causing inability to control facial muscles on the affected side.

16.2.4.2 Pathology

By definition, Bell's palsy is idiopathic; however, reactivation of herpes simplex virus within the geniculate ganglion has been advanced as an etiology [3]. Ramsay Hunt syndrome, also known as herpes zoster oticus, is facial neuritis secondary to varicella zoster virus infection [3]. Microbiologic etiologies of facial neuritis include various viral and nonviral infections, such as *Borrelia burgdorferi*, Epstein–Barr virus, and cytomegalovirus [14–16].

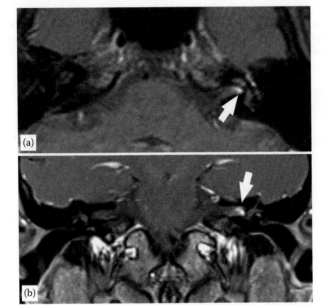

FIGURE 16.8
Ramsay Hunt syndrome. (a) Axial and (b) coronal contrast-enhanced fat-suppressed T_1-weighted images show abnormal enhancement of the intracanalicular segment of the left facial nerve (arrows).

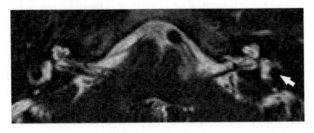

FIGURE 16.9
Labyrinthitis ossificans. Axial high-resolution T_2-weighted image demonstrates loss of the normal high signal intensity in the left lateral semicircular canal (arrow).

16.2.4.3 Clinical Features

Facial paralysis is the typical clinical presentation of facial neuritis, and the diagnosis of facial neuritis is clinically evident and electrophysiologic study confirms the diagnosis. Imaging is usually not indicated except for patients with atypical manifestations and those with intractable palsy despite therapy [17]. In addition to facial paralysis, hearing loss and vertigo along with vesicles in the external auditory canal are seen in patients with Ramsay Hunt syndrome [3].

16.2.4.4 Imaging Findings

Abnormal enhancement of the affected facial nerve is seen, including the meatal and labyrinthine segments in Bell's palsy (Figure 16.10) [18]. In Ramsay Hunt syndrome (Figure 16.8), abnormal enhancement is seen in the internal auditory canal and labyrinth in addition to prominent enhancement of the facial nerve [3].

Differential diagnosis for abnormal enhancement of the facial nerve includes facial schwannoma and perineural tumor spread along the facial nerve [3,19,20].

16.3 Paranasal Sinus

16.3.1 Rhinosinusitis

16.3.1.1 Definition

Sinusitis is inflammation of the mucosa of the paranasal sinuses. It rarely exists without involvement of the nasal mucosa; thus, the term rhinosinusitis may be more accurate. Sinusitis can be subdivided into acute and chronic disease. Chronic sinusitis is defined as inflammation persisting more than 12 weeks [2,21].

16.3.1.2 Pathology

Acute and chronic sinusitis are most commonly caused by bacterial organisms, such as *S. pneumoniae*, *H. influenzae*, *M. catarrhalis*, *S. aureus*, other *Streptococcal* strains, and

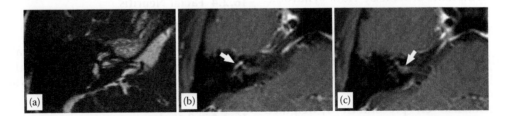

FIGURE 16.10
Bell's palsy. (a) Axial high resolution T_2-weighted image demonstrates no abnormality in the right inner ear. (b,c) Axial postcontrast-enhanced fat-suppressed T_1-weighted images show abnormal enhancement of the genu and the intracanalicular segment of the left facial nerve (arrows).

anaerobic bacteria [22]. Immunosuppressed patients are at risk for other bacterial infections, such as *Pseudomonas* species [22].

16.3.1.3 Clinical Features

Patients with acute and chronic sinusitis have similar symptoms. They typically present with nasal congestion, purulent discharge, headache, facial pain or maxillary tooth discomfort, and reduced sense of smell and taste.

Acute sinusitis is often treated with antibiotics without need for surgery, whereas chronic sinusitis may require surgical intervention to open the sinonasal passageways.

16.3.1.4 Imaging Findings

In acute rhinosinusitis, imaging is usually not necessary as the diagnosis is apparent by the clinical presentation. If MRI is performed, the findings are air–fluid levels or air bubbles in the paranasal sinuses with or without mucosal thickening [2,21] (Figure 16.11). The fluid is hypointense on T_1-weighted images and hyperintense on T_2-weighted images. However, the diagnosis of rhinosinusitis should be made clinically, not by imaging findings alone.

MRI is useful for evaluation of suspected intracranial or orbital complications. T_2-weighted images and short-tau inversion recovery (STIR) images differentiate mucosal thickening from fluid or pus in the sinus lumen [2,21,23]. If there is an obstructing tumor, it is hypointense in comparison with the mucosal thickening [2,21,23].

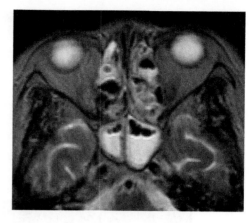

FIGURE 16.11
Acute sinusitis. Axial fat-suppressed T_2-weighted image demonstrates mucosal thickening and opacification of the bilateral ethmoid air cells and air–fluid levels in the sphenoid sinuses.

The imaging findings of chronic rhinosinusitis are similar to acute rhinosinusitis [2,21,23]. In addition, in long-lasting sinusitis, the sinus secretions may become inspissated. Variation in MR signals of the sinonasal secretions has been shown to be related to many properties of the secretions, such as viscosity, protein concentration, fat, temperature, and paramagnetic properties of fungal elements or hemorrhage [2,21,23]. With increasing protein content, T_1-weighted signal intensity increases and T_2-weighted signal intensity initially remains high but then declines progressively. [2,21,23] (Figure 16.12).

16.3.2 Complications of Rhinosinusitis

There are two major complications of rhinosinusitis, which are orbital and intracranial complications. Orbital

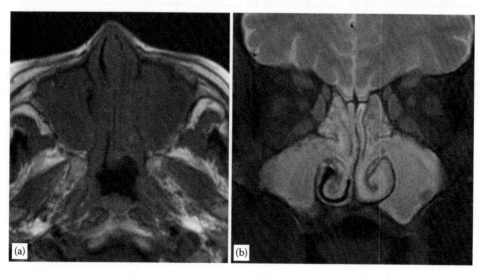

FIGURE 16.12
Chronic sinusitis. (a) Axial T_1-weighted image and (b) coronal fat-suppressed T_2-weighted image demonstrate opacified maxillary and ethmoid sinuses bilaterally.

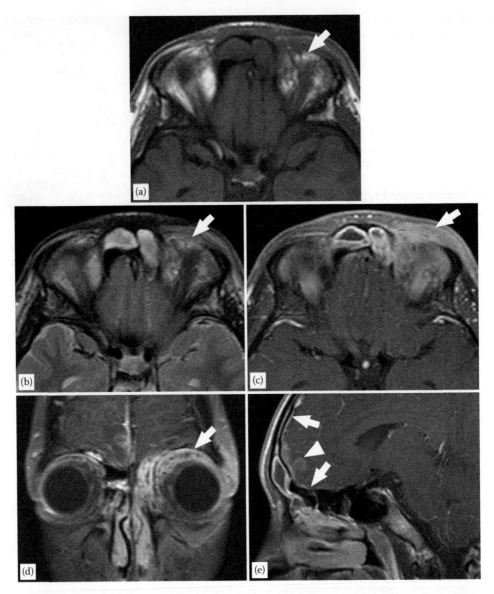

FIGURE 16.13
Sinusitis with orbital and intracranial complications. (a) Axial T_1-weighted and (b) T_2-weighted images demonstrate opacified ethmoid air cells and heterogeneous hypointensity in the left orbital fat, consistent with orbital cellulitis (arrows). (c) Axial and (d) coronal contrast-enhanced fat-suppressed T_1-weighted images show opacified frontal and ethmoid sinuses and heterogeneous enhancement of the left periorbital (arrow) and orbital fat. These findings indicate periorbital and orbital cellulitis secondary to ethmoid and frontal sinusitis. (e) Sagittal contrast-enhanced T_1-weighted image demonstrates meningeal enhancement (arrows) and parenchymal abscess with ring-enhancement (arrowhead).

complications are more common than intracranial complications and are seen in approximately 3% of patients with sinusitis [24]. Orbital complications include periorbital/orbital cellulitis and abscess, subperiosteal abscess, and cavernous sinus thrombosis [2,21]. Ethmoid sinus infection most commonly causes orbital complications. [21]. The thin lamina papyracea and the valveless ethmoid veins cause a predisposition to the spread of infection into the orbit [21].

Intracranial complications of rhinosinusitis are rare. They include meningitis, epidural abscess, subdural abscess, and cerebral abscess. A cerebral abscess may result from the direct spread of infection from the sinuses or hematogenous spread due to septic embolization [21]. Direct intracranial spread is a rare complication of sinusitis, which is most common in frontal sinus infection. Infection can involve the bone (osteitis or osteomyelitis) and spread intracranially to form a cerebral abscess [21] (Figure 16.13). Sinus infection can also form an extracranial subperiosteal phlegmon or abscess (Figure 16.14). Pott's puffy tumor is a complication of acute frontal sinusitis complicated by osteomyelitis and subperiosteal abscess resulting in forehead swelling, "puffy tumor," apparently not a neoplastic condition [2].

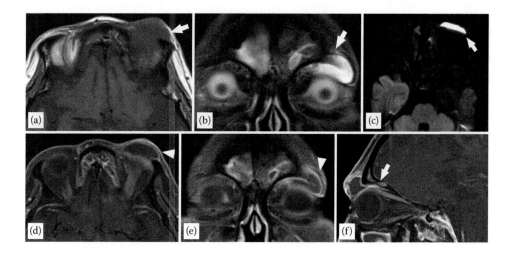

FIGURE 16.14
Periorbital abscess secondary to frontal sinusitis. (a) Axial T_1-weighted and (b) coronal fat-suppressed T_2-weighted images demonstrate opacified left frontal sinus and left periorbital fluid collection (arrow). (c) Axial diffusion-weighted image shows high signal in the lesion (arrow). (d) Axial, (e) coronal, and (f) sagittal contrast-enhanced fat-suppressed T_1-weighted images show periorbital abscess with an enhancing rim (arrowheads). Note cortical defect in the floor of the frontal sinus/roof of the orbit (arrow).

Cavernous sinus thrombosis is a rare intracranial complication of sinusitis, with ethmoid or sphenoid sinusitis as the predisposing factor [21].

16.3.3 Fungal Infection

16.3.3.1 Definition

Fungal infections of the paranasal sinuses are categorized as noninvasive and invasive [25,26]. In noninvasive fungal sinusitis, the fungal elements are limited to the lumen of the involved sinus, whereas in invasive disease, fungal elements cross the mucosa to involve the blood vessels and the bones, and possibly the orbital and intracranial structures [25,26]. Noninvasive fungal sinusitis is subdivided into mycetoma (fungus ball) and allergic fungal sinusitis [25,26]. Invasive fungal sinusitis is subdivided into acute invasive fungal sinusitis, chronic invasive fungal sinusitis, and chronic granulomatous invasive fungal sinusitis [25,26].

16.3.3.2 Pathology

Mycetoma is usually caused by *Aspergillus fumigatus*, although other fungi such as *Pseudallescheria boydii* and *Alternaria* have been implicated [27].

Commonly implicated fungi of allergic fungal sinusitis are the dematiaceous (pigmented) fungi such as *Bipolaris, Curvularia,* and *Alternaria,* and the hyaline molds such as *Aspergillus* and *Fusarium* [25,27].

Up to 80% of invasive fungal infections occurring in patients with diabetes mellitus, especially those with diabetic ketoacidosis, are caused by fungi belonging to the order Zygomycetes, such as *Rhizopus,* *Rhizomucor, Absidia,* and *Mucor* [28]. *Aspergillus* species are responsible for up to 80% of infections in immunocompromised patients with severe neutropenia [28].

Common organisms of chronic invasive fungal sinusitis include *Mucor, Rhizopus, Aspergillus, Bipolaris,* and *Candida* [25,27,28].

16.3.3.3 Clinical Features

Fungal mycetoma tends to occur in older individuals with an apparent female predilection. Afflicted individuals are usually immunocompetent [25,26]. Patients are asymptomatic or have minimal symptoms.

Allergic fungal sinusitis is the most common form of fungal sinusitis [25,26]. The overall prevalence of allergic fungal sinusitis is estimated at 5%–10% of all patients with chronic hypertrophic sinus disease going to surgery [29]. This results from an immunoglobulin E-mediated hypersensitivity reaction in atopic individuals. Allergic fungal sinusitis tends to occur in younger individuals, usually in their third decade [25,26]. Patients usually experience chronic headaches, nasal congestion, and chronic sinusitis for several years.

Acute invasive fungal sinusitis is a rapidly progressing infection seen predominantly in immunocompromised patients and patients with poorly controlled diabetes, and rarely in healthy individuals. It is the most lethal form of fungal sinusitis, with a reported mortality of 50%–80% [30]. Symptoms include fever, facial pain or numbness, nasal congestion, serosanguineous nasal discharge, and epistaxis. Intraorbital, intracranial, or maxillofacial extension is common

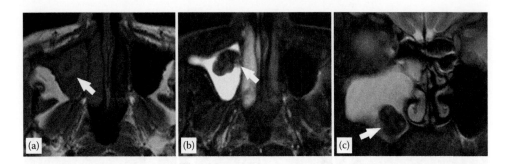

FIGURE 16.15
Fungal sinusitis: Mycetoma. (a) Axial T_1-weighted, (b) axial, and (c) coronal fat-suppressed T_2-weighted images demonstrate mucosal thickening and complete opacification of the right maxillary sinus with a heterogeneous hypointense lesion on both T_1- and T_2-weighted images, representing a fungus ball (arrow).

with resulting proptosis, visual disturbance, headache, mental status changes, seizures, neurologic deficits, coma, and maxillofacial soft tissue swelling. In chronic infection, insidious progression occurs over several months to years.

16.3.3.4 Imaging Findings

Mycetoma appears as a mass within the lumen of a paranasal sinus and is usually limited to one sinus. The maxillary sinus is the most commonly involved sinus, followed by the sphenoid sinus [25]. Mycetoma is hypointense on T_1-weighted and T_2-weighted images due to the absence of free water (Figure 16.15). Calcifications and paramagnetic metals such as iron, magnesium, and manganese also generate areas of signal void on T_2-weighted images [23,25].

Allergic fungal sinusitis tends to be bilateral or involves multiple sinuses, and there is frequent nasal involvement. The majority of the sinuses show near-complete opacification and are expanded. There is characteristic hypointensity or signal void on T_2-weighted images [23,31]. The T_2 signal void is attributed to a high concentration of various metals such as iron, magnesium, and manganese concentrated by the fungal organisms. The T_2 signal void is also attributed to a high protein and low free-water content of the allergic mucin [25] (Figure 16.16).

In invasive fungal sinusitis, there is a predilection for unilateral involvement of the ethmoid and sphenoid sinuses [23,25]. Aggressive bone destruction of the sinus walls (Figure 16.17) occurs rapidly with intracranial and intraorbital extension of the inflammation [23,25]. Intracranial extension of disease from the sphenoid sinus can lead to cavernous sinus thrombosis and even carotid artery invasion, occlusion, or pseudoaneurysm with resulting fatal cerebral infarct or hemorrhage.

16.4 Orbit

16.4.1 Periorbital and Orbital Cellulitis (Abscess)

16.4.1.1 Definition

The location of an orbital infection is described with respect to the orbital septum, as either preseptal (periorbital) or postseptal (orbital) [32]. The orbital septum is a thin band of fibrous tissue that originates in the orbital periosteum and inserts in the palpebral tissues along the tarsal plates. The orbital septum is a barrier to the spread of periorbital infections into the orbit proper [33].

16.4.1.2 Pathology

Periorbital cellulitis arises most commonly from the contiguous spread of infection from adjacent structures such as the face, teeth, and ocular adnexa. It may also arise from local trauma [32–34].

Orbital cellulitis often occurs secondary to sinusitis and, less frequently, foreign bodies. It spreads to the orbit via a perivascular pathway [33], and bone destruction is not usually seen [32].

16.4.1.3 Clinical Features

Patients with periorbital cellulitis present with swelling, erythema, conjunctivitis, fever, and pain. In orbital cellulitis, they may present with similar symptoms as well as proptosis and impaired visual activity.

The distinction between periorbital and orbital processes is clinically important because orbital infections are treated more aggressively to prevent devastating complications such as cavernous sinus thrombosis and meningitis [32,33].

Periorbital cellulitis is often treated with oral antibiotic therapy. However, treatment of orbital cellulitis

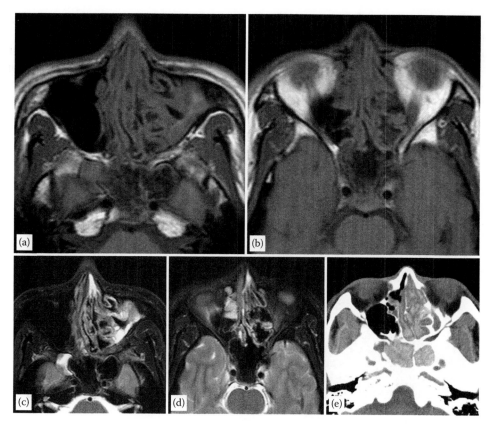

FIGURE 16.16
Fungal sinusitis: Allergic fungal sinusitis. (a,b) Axial T_1-weighted images and (c,d) fat-suppressed T_2-weighted images show marked hypointensity within the ethmoid and sphenoid sinuses. (e) Corresponding unenhanced CT shows increased attenuation in the ethmoid and sphenoid sinuses.

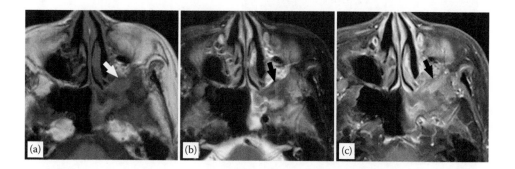

FIGURE 16.17
Fungal sinusitis: Acute invasive fungal sinusitis. (a) Axial T_1-weighted and (b) fat-suppressed T_2-weighted images and (c) contrast-enhanced fat-suppressed T_1-weighted image demonstrate left sphenoid sinusitis with destruction of the wall of the sphenoid sinus. There is invasion of the left pterygopalatine fossa (arrows).

typically requires intravenous antibiotics. In case of abscess formation, surgical drainage may be necessary to avoid rapid increase in intraorbital pressure and resultant visual impairment [32].

16.4.1.4 Imaging Findings

MRI provides excellent contrast resolution in the orbit. Although T_1-weighted contrast-enhanced images with fat suppression are widely held as the gold standard in the detection and characterization of orbital pathology, T_2-weighted fat-suppressed sequences have similar sensitivity for detecting orbital lesions and readily identify postseptal disease [35]. Contrast enhancement, however, is essential for distinguishing abscess from edema and phlegmon [34].

Typical MRI findings of periorbital cellulitis are diffuse soft tissue thickening and enhancement anterior to the orbital septum. The imaging findings of the orbital cellulitis are inflammatory stranding in the intraconal fat, intraconal or extraconal soft tissue mass, edema of the extraocular muscles, intraorbital abscess, or subperiosteal abscess [32–36] (Figures 16.13 and 16.18). Development of a subperiosteal abscess in the orbit is most commonly associated with ethmoid sinusitis [32,33] (Figure 16.19).

Complications of orbital cellulitis include subperiosteal abscess, orbital abscess, superior ophthalmic vein thrombosis, cavernous sinus thrombosis, mycotic aneurysms, meningitis, and intracranial abscess [32,33].

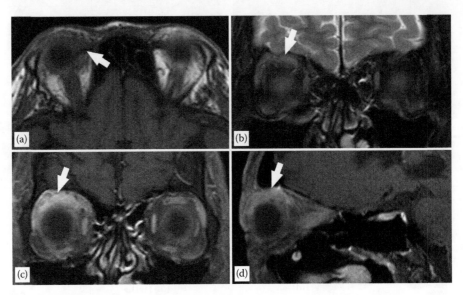

FIGURE 16.18
Orbital cellulitis. (a) Axial T_1-weighted and (b) coronal fat-suppressed T_2-weighted images demonstrate heterogeneous T_1 hypointensity and mild T_2 hyperintensity in the right orbital fat (arrows). (c) Coronal and (d) sagittal contrast-enhanced fat-suppressed T_1-weighted images show heterogeneous enhancement of the right orbital fat (arrows), representing orbital cellulitis.

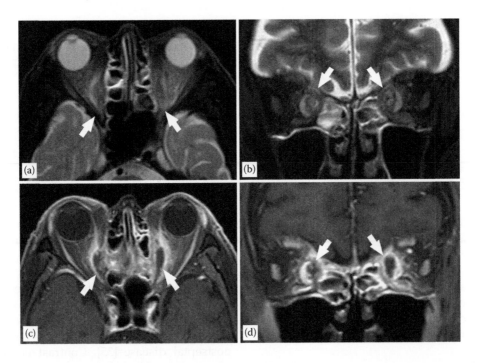

FIGURE 16.19
Orbital subperiosteal abscess secondary to ethmoid sinusitis. (a) Axial and (b) coronal fat-suppressed T_2-weighted images demonstrate heterogeneous signal areas (arrows) along the medial wall of the orbits bilaterally, adjacent to the mucosal thickening of the ethmoid air cells. (c) Axial and (d) coronal fat-suppressed contrast-enhanced T_1-weighted images demonstrate subperiosteal abscesses with enhancing rims (arrows) along the medial walls of the orbits bilaterally with resultant lateral displacement of the medial rectus muscles.

16.4.2 Dacryocystitis

16.4.2.1 Definition

Dacryocystitis results from the obstruction of flow through the lacrimal sac, which is located along the inner canthus [37]. A dacrolith may be present.

16.4.2.2 Pathology

The most common isolates are *S. pneumoniae* (23%), *Streptococcus pyogenes* (14.3%), *S. aureus* (12.1%), *Streptococcus viridans* (9.9%), and *H. influenzae* (9.9%) [38].

16.4.2.3 Clinical Features

Patients with dacryocystitis present as palpable medial canthal mass, conjunctivitis, and purulent drainage [37]. Clinical exam is usually sufficient for diagnosis; however, this may be limited by lid swelling. It can be complicated by periorbital cellulitis and rarely orbital cellulitis. Chronic or recurrent forms can occur.

Treatment options may be medical or surgical, with the approach selected depending on the clinical signs and symptoms [37].

16.4.2.4 Imaging Findings

The typical imaging finding is a well-circumscribed, peripherally enhancing, round mass that is centered at the lacrimal fossa [37] (Figure 16.20).

Differential diagnoses include congenital mucocele, anterior ethmoiditis, and dermoid cyst.

16.5 Pharynx, Oral Cavity, and Larynx

16.5.1 Acute Tonsillitis and Peritonsillar Abscess

16.5.1.1 Definition

Tonsillar infections are the most commonly encountered deep neck infections among adolescents and young adults [2,39]. The complications include cellulitis and abscess formation. A peritonsillar abscess arises between the palatine tonsil and its capsule [2,39].

16.5.1.2 Pathology

Acute tonsillitis is usually due to infection by β-hemolytic *Streptococcus*, *S. aureus*, *Pneumococcus*, and *H. influenzae* [39,40].

16.5.1.3 Clinical Features

Patients with tonsillar infection present with fever, sore throat, and dysphagia.

Because symptoms overlap with viral pharyngitis, clinical presentation, and laboratory testing are used

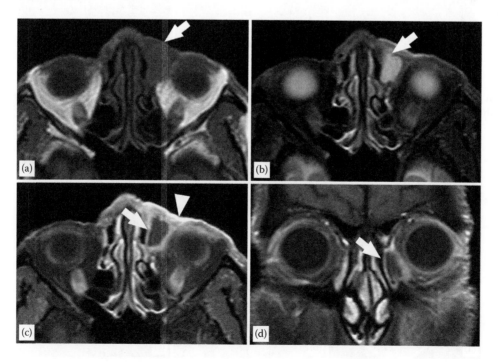

FIGURE 16.20
Dacryocystitis. (a) Axial T_1-weighted and (b) fat-suppressed T_2-weighted images demonstrate dilatation of the left lacrimal sac (arrows). (c) Axial and (d) coronal contrast-enhanced fat-suppressed T_1-weighted images show peripheral enhancement of the left lacrimal sac (arrows). Also note associated preseptal inflammation (arrowhead).

to identify patients who require antibiotic therapy. Peritonsillar abscesses are commonly treated with surgical drainage.

16.5.1.4 Imaging Findings

On MRI, acute tonsillitis appears as diffuse enlargement and enhancement of the tonsil [2,39] (Figure 16.21). MRI findings of peritonsillar abscess include central signal hyperintensity on T_2-weighted images and hypointensity on T_1-weighted images, with peripheral enhancement on contrast-enhanced T_1-weighted images (Figure 16.22). DWI is very sensitive for abscess identification, with pyogenic abscesses demonstrating restricted diffusion on DWI [8].

Differential diagnosis for tonsillar enlargement includes lymphoid hyperplasia, infectious conditions such as mononucleosis, tonsillar neoplasms such as squamous cell carcinoma and lymphoma, and inflammatory conditions such as angioedema and mucositis secondary to Kawasaki disease [2].

16.5.2 Odontogenic Infection

16.5.2.1 Definition

Periodontal abscess and osteomyelitis are common manifestations of odontogenic infection. Bacterial overgrowth and poor hygiene result in chronic inflammation of the gingiva leading to periodontitis, which leads to the destruction of periodontal ligaments, inflammation, and erosion of bone. Bone erosion manifests as a periapical lucency, and penetration through the bone will result in fistulization into adjacent soft tissues and extraosseous abscess formation and cellulitis [2,41].

16.5.2.2 Pathology

The most common species of bacteria isolated in odontogenic infections are the anaerobic gram-positive cocci *Streptococcus milleri* group and *Peptostreptococcus* [42]. Anaerobic gram-negative rods, such as *Bacteroides* (*Prevotella*), also play an important role [42].

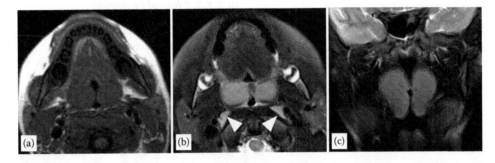

FIGURE 16.21
Acute tonsillitis. (a) Axial T_1-weighted, (b) axial, and (c) coronal fat-suppressed T_2-weighted images show enlargement of the bilateral palatine tonsils with smooth margins and homogeneous internal intensity. Retropharyngeal lymph nodes are enlarged bilaterally (arrowheads).

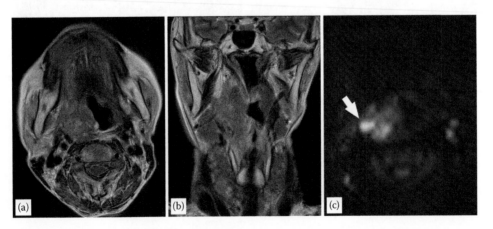

FIGURE 16.22
Peritonsillar abscess. (a) Axial and (b) coronal T_2-weighted images demonstrate enlargement of the right palatine tonsil with irregular margin. (c) Axial diffusion-weighted image shows a hyperintense area (arrow) at the lateral aspect of the tonsil, representing an abscess.

16.5.2.3 Clinical Features

Patients with acute osteomyelitis and periodontal abscess present with fever, tooth pain, facial swelling, dysphagia, trismus, and possibly dyspnea.

Treatment includes antibiotics with possible abscess drainage and definitive management (root canal or extraction) of the offending tooth [2,43]. Acute osteomyelitis usually responds to antibiotics, whereas chronic osteomyelitis usually requires surgical procedure such as curettage of the necrotic bone and granulation tissue, and sequestrectomy.

16.5.2.4 Imaging Findings

Periodontal infection may result in osteomyelitis, intra- or extraosseous abscess, or cellulitis around the jaw (Figures 16.23 and 16.24). Periodontal abscess (Figure 16.25) is a peripherally enhancing collection either along the cortex (subperiosteal) or within adjacent soft tissues [41,43].

It is important to carefully investigate the tooth roots when soft tissue swelling is noted around the jaw. Periapical lucency and periodontal abscess are best evaluated with contrast-enhanced CT [41].

MRI can well demonstrate the bone marrow edema or inflammation due to increase in water content, which often replaces the normal fatty marrow. These signal changes in the bone marrow may be seen in the acute stage before bony changes can be seen on conventional radiography [41,43]. Acute osteomyelitis demonstrates marrow hypointensity on T_1-weighted images and marked hyperintensity on T_2-weighted or STIR images. In the chronic stage, hypointensity areas both on T_1- and T_2-weighted images are seen [41,43].

Ludwig angina is a well-known complication of odontogenic infections. Ludwig angina is a necrotizing infection of the floor of the mouth that involves the submandibular spaces bilaterally [44]. The infection is mostly due to the inflammation of the second and third mandibular molars, because these tooth apices extend inferiorly to the mandibular insertion of the mylohyoid muscle [45], thus allowing direct extension of infection into the submandibular space.

16.5.3 Retropharyngeal Abscess

16.5.3.1 Definition

The retropharyngeal space is a potential space between the prevertebral and visceral spaces, and between the middle and deep layers of the deep cervical fascia.

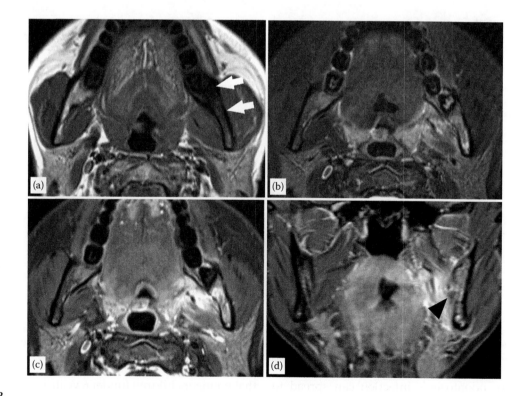

FIGURE 16.23
Osteomyelitis and cellulitis secondary to periodontal infection. (a) Axial T_1-weighted image shows hypointensity, loss of high signal from the normal fatty marrow in the posterior body, and ramus of the left mandible (arrows). (b) Axial fat-suppressed T_2-weighted image shows hyperintensity in the same region and the adjacent medial pterygoid muscle. (c) Axial and (d) coronal contrast-enhanced fat-suppressed T_1-weighted images show enhancement within the left mandible and the adjacent soft tissue and muscles. Cortical bone dehiscence of the mandible is seen (arrowheads).

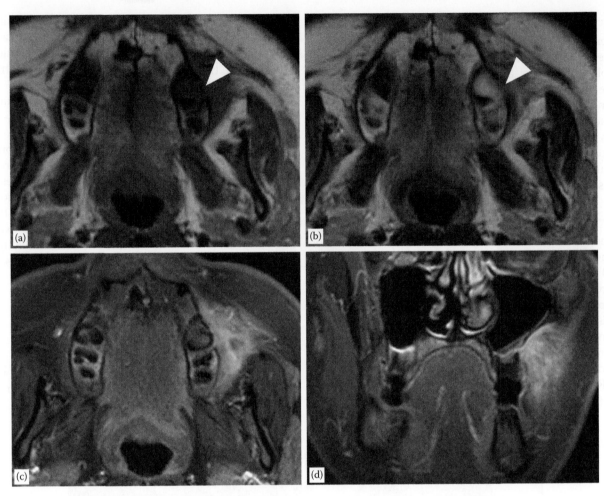

FIGURE 16.24
Facial cellulitis secondary odontogenic infection. (a) Axial T_1-weighted image shows hypointensity in the left upper gingiva. (b) Axial T_2-weighted image shows hyperintensity in the same region. Cortical bone dehiscence of the maxilla is seen (arrowheads). (c) Axial and (d) coronal contrast-enhanced fat-suppressed T_1-weighted images show enhancement of the lesion and the adjacent buccal space.

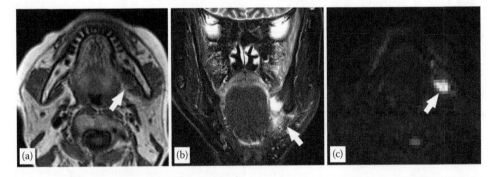

FIGURE 16.25
Periodontal abscess. (a) Axial T_1-weighted image shows thickening of the left medial pterygoid muscle (arrow). (b) Coronal fat-suppressed T_2-weighted image shows a focal area of hyperintensity in the same region and diffuse hyperintensity in the submandibular space (arrow). (c) Axial diffusion-weighted image shows a focal area of hyperintensity (arrow), representing an abscess.

Pharyngeal or odontogenic infection can spread to the retropharyngeal nodes resulting in suppurative adenitis. Suppurated lymph nodes may rupture and result in the formation of a retropharyngeal abscess or retropharyngeal cellulitis [2,39]. It has been postulated that younger children (under 6 years old) [2,46] are more likely to develop such abscesses because retropharyngeal lymph nodes spontaneously regress after 5 years of age [47]. In adults, infection of the retropharyngeal space is usually due to a penetrating injury.

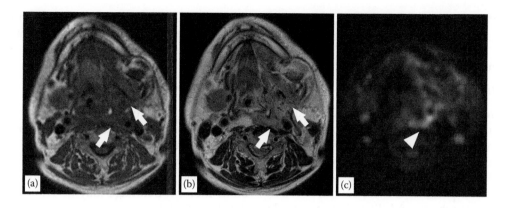

FIGURE 16.26

Retropharyngeal space abscess with mediastinitis. (a) Axial contrast-enhanced fat-suppressed T_1-weighted image shows extensive retropharyngeal inflammation with abscess formation (arrow). (b) Coronal and (c) sagittal contrast-enhanced fat-suppressed T_1-weighted images show extension of the cellulitis and abscess to the mediastinum (arrowheads).

16.5.3.2 Pathology

Retropharyngeal abscess is most commonly due to infection by *S. aureus, Haemophilus parainfluenzae,* and β-hemolytic *Streptococcus* [39].

16.5.3.3 Clinical Features

The clinical symptoms include sore throat, fever, torticollis, dysphagia, and a neck mass. Because of the possibility of life-threatening complications due to airway compromise and spread of infection to the mediastinum (Figure 16.26), the recognition of the retropharyngeal abscess is very important.

Retropharyngeal cellulitis and small retropharyngeal abscesses are typically treated with intravenous antibiotics. Larger abscesses are treated with incision and drainage [2,39].

16.5.3.4 Imaging Findings

In the presence of infection, the retropharyngeal space appears as a bow tie-shaped structure bordered by the pharynx anteriorly, the prevertebral muscles posteriorly, and the internal carotid arteries bilaterally [2,39,44,47].

Retropharyngeal cellulitis is identified by symmetric hypointensity on T_1-weighted images and hyperintensity on T_2-weighted images in the retropharyngeal space. Retropharyngeal abscess manifests as a fluid collection with peripheral enhancement on contrast-enhanced T_1-weighted images [2,39,44,47]. This causes substantial anterior displacement of the posterior wall of the pharynx from the prevertebral muscles. This collection may be asymmetric (Figure 16.27). The differentiation of abscesses from retropharyngeal edema or cellulitis is critical but difficult in the absence of rim enhancement [39].

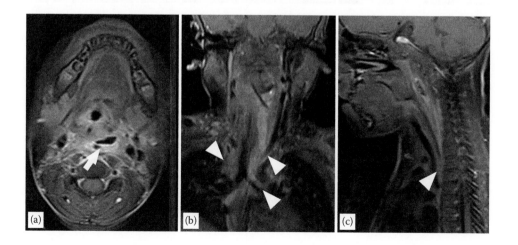

FIGURE 16.27

Retropharyngeal space abscess from submandibular infection. (a) Axial T_1-weighted and (b) T_2-weighted images show heterogeneous signal intensity in the left submandibular and parapharyngeal spaces to the left retropharyngeal space (arrows). (c) Axial diffusion-weighted image shows hyperintensity in the lesion (arrowhead) in the left retropharyngeal space, representing an abscess.

16.6 Salivary Gland

16.6.1 Sialadenitis

16.6.1.1 Definition

Sialadenitis refers to inflammation of the salivary glands. Bacterial infection presents more often as unilateral gland involvement, whereas viral infection is typically bilateral. Calculus-induced inflammation is also common [39,48,49].

In adults, bacterial parotitis is most common among the elderly and those who are debilitated. It can also be seen in postoperative patients with dehydration and history of endotracheal intubation [39,48,49]. Viral parotitis is most common in patients aged 5–9 years [39].

16.6.1.2 Pathology

The most common pathogens in sialadenitis are *S. aureus, S. viridans, S. pneumoniae, H. influenzae, S. pyogenes,* and *Escherichia coli* [39,49,50].

Viral parotitis is associated with systemic viral infection and is bilateral in 75% of cases; the submandibular and sublingual glands may also be involved [39]. The mumps virus (*paramyxovirus*) is the most common

pathogen. Other pathogens include influenza virus, parainfluenza virus, coxsackie virus, cytomegalovirus, and adenovirus [39].

16.6.1.3 Clinical Features

Patients with acute sialadenitis present with painful swelling over the affected salivary gland. The symptoms are exacerbated by eating, which is often referred to as salivary colic [39,49].

Treatment of sialadenitis includes hydration and antibiotic therapy. Drainage may be required if an abscess develops [39,49].

16.6.1.4 Imaging Findings

CT and MRI findings of the sialadenitis include enlargement and enhancement of the affected gland, with inflammatory changes in the surrounding fat [39,49] (Figures 16.28 through 16.30). A dilated duct with an obstructing calculus is often seen (Figure 16.29). Calculi occur most commonly in the submandibular duct because of its large diameter and ascending course, its thicker and more mucinous and alkaline salivary content, and the presence of salivary stasis [39,49]. Abscess formation is sometimes encountered (Figure 16.30).

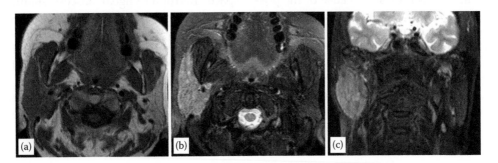

FIGURE 16.28
Acute parotitis. (a) Axial T_1-weighted, (b) axial, and (c) coronal fat-suppressed T_2-weighted images demonstrate enlargement and abnormal signal of the right parotid gland demonstrating T_1 hypointensity and T_2 hyperintensity without focal abnormality.

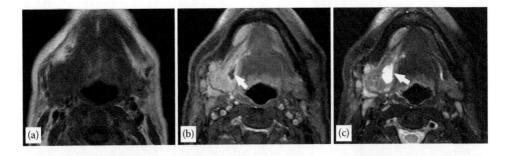

FIGURE 16.29
Sialadenitis: Submandibular gland infection. (a) Axial T_1-weighted and (b) fat-suppressed T_2-weighted images show enlargement and abnormal signal of the right submandibular gland with adjacent inflammation. Note focal dilatation of the submandibular (Wharton's) duct (arrow) (c) Axial contrast-enhanced fat-suppressed T_1-weighted image shows diffuse enhancement of the right submandibular gland and surrounding soft tissues. Note peripheral enhancement of the dilated duct (arrow), which contains a sialolith (not shown).

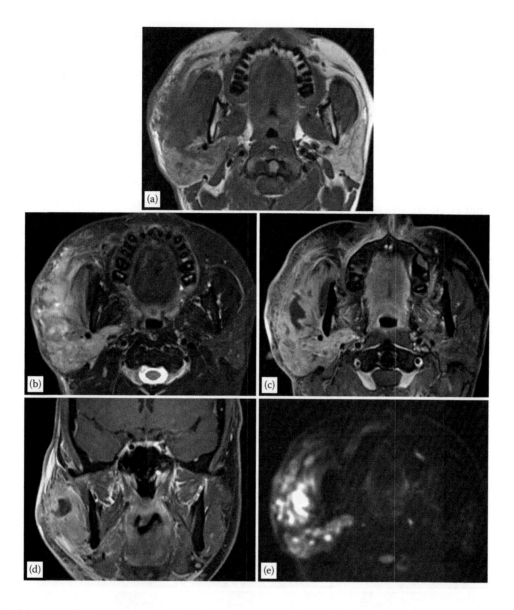

FIGURE 16.30

Acute parotitis with intraparotid abscesses. (a) Axial T_1 and (b) fat-suppressed T_2-weighted images demonstrate marked swelling of the right parotid gland with stranding within the subcutaneous tissues overlying the gland. (c) Axial and (d) coronal contrast-enhanced fat-suppressed T_1-weighted images demonstrate irregular hypointensity areas in the diffusely enhanced gland. (e) Axial diffusion-weighted image shows diffuse hyperintensity of the gland with multiple foci of very high signal consistent with abscesses.

16.7 Spinal and Perivertebral Spaces

16.7.1 Spinal Epidural Abscess

16.7.1.1 Definition

The spinal epidural space is a space between the dura and the ligamentum flavum and the periosteum of the vertebral bodies, pedicles, and laminae.

Most patients with spinal epidural abscess have one or more predisposing conditions: (1) underlying disease (diabetes mellitus, alcoholism, or infection with human immunodeficiency virus), (b) spinal abnormality (degenerative joint disease, trauma, surgery, therapeutic drug injection, or placement of stimulators or catheters), and (c) a potential local or systemic source of infection (skin and soft tissue infections, osteomyelitis, urinary tract infection, sepsis, indwelling vascular access, intravenous drug use, nerve acupuncture, tattoos, epidural analgesia, or nerve block) [51].

16.7.1.2 Pathology

Bacteria gain access to the epidural space through contiguous spread or hematogenous dissemination [51]. Because most predisposing conditions allow for invasion by skin flora, *S. aureus* causes about two-thirds of cases [51–53]. Less common causative pathogens include *Staphylococcus epidermidis* (typically in association with spinal procedures, including placement of catheters for analgesia, glucocorticoid injections, or surgery) and gram-negative bacteria, particularly *E. coli* (usually subsequent to urinary tract infection) and *Pseudomonas aeruginosa* (especially in injection drug users) [51–54].

16.7.1.3 Clinical Features

Patients with spinal epidural abscess most commonly present with neck and back pain at the level of the affected region, fever, and a neurologic deficit [51].

Surgical decompression is indicated if cord compression or spinal instability is present, or if the patient progresses clinically despite appropriate medical therapy [51].

16.7.1.4 Imaging Findings

MRI is the modality of choice for patients with suspected spine and spinal cord lesions. Epidural abscess appears as a fluid collection surrounded by inflammatory tissue, which shows varying degrees of peripheral enhancement [55,56] (Figures 16.31 and 16.32).

Pyogenic osteomyelitis and discitis can spread through the anterior longitudinal ligament, causing a prevertebral abscess. Osteomyelitis and discitis appear as edema signal and enhancement of the involved bone marrow, disk space, and adjacent tissues [55,56] (Figures 16.31 and 16.32).

Epidural infection can injure the spinal cord either directly by mechanical compression or indirectly as a result of vascular occlusion caused by septic thrombophlebitis.

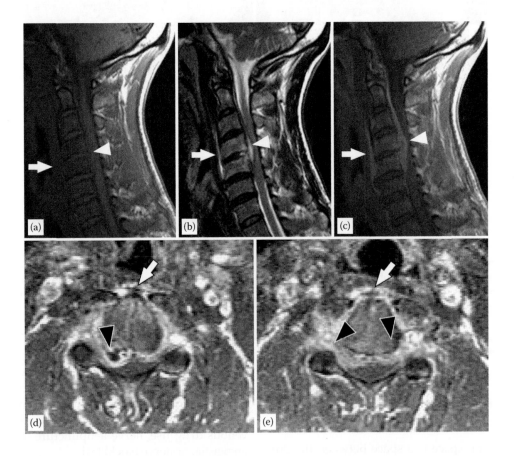

FIGURE 16.31
Epidural abscess secondary to discitis/osteomyelitis. (a) Sagittal T_1-weighted and (b) T_2-weighted images demonstrate an epidural collection (arrowhead). There is prevertebral soft tissue thickening demonstrating hypointensity on (a) T_1-weighted and hyperintensity on (b) T_2-weighted images (arrows) in the prevertebral space. Note a focal T_2 high signal within the C4–5 disc space and mild signal change in C4 and C5 vertebral bodies. (c) Sagittal contrast-enhanced T_1-weighted image and (d,e) axial contrast-enhanced fat-suppressed T_1-weighted images demonstrate peripheral enhancement of epidural collection (arrowheads), representing an abscess. The abscess displaces the spinal cord posteriorly (arrowhead). Diffuse enhancement of the prevertebral space is noted (arrows). C4 and C5 vertebral bodies demonstrate mild enhancement.

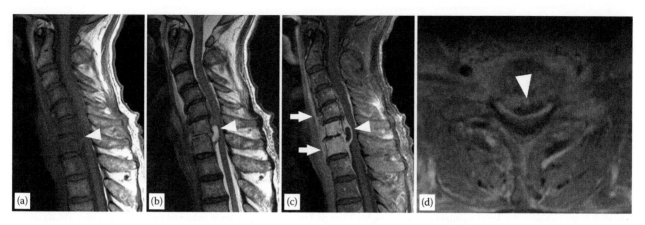

FIGURE 16.32

Epidural abscess secondary to discitis/osteomyelitis. Sagittal (a) T_1-weighted and (b) T_2-weighted images show an epidural collection (arrowhead). The disk space at the C6–7 level demonstrates narrowing and hyperintensity on T_2-weighted image. (c) Sagittal and (d) axial contrast-enhanced fat-suppressed T_1-weighted images demonstrate avid enhancement of C6 and C7 vertebral bodies and peripheral enhancement of the epidural collection (arrowhead), consistent with abscess. Diffuse enhancement of the prevertebral space is also noted (arrows).

16.8 Cervical Lymph Nodes

16.8.1 Suppurative Lymphadenitis

16.8.1.1 Definition

Suppurative lymphadenitis is defined as enlargement of nodes secondary to bacterial infection with pus formation within the nodes. It tends to manifest with inflammatory changes in surrounding tissues [57,58].

16.8.1.2 Pathology

Group A *Streptococcus* and *S. aureus* are the most common pathogens [57–59].

16.8.1.3 Clinical Features

Patients present with fever and warm, tender cervical nodes. In children younger than 6 years of age, retropharyngeal lymph nodes are commonly involved, which can lead to retropharyngeal abscess [2,57,58].

Treatment is antibiotic therapy. Drainage may be required based on size.

16.8.1.4 Imaging Findings

MRI findings include intranodal hyperintensity on T_2-weighted images and hypointensity on T_1-weighted images, with peripheral enhancement on contrast-enhanced T_1-weighted images [2,57,58] (Figure 16.33). Inflammatory changes in adjacent structures are also seen.

Differential diagnoses include metastatic disease (especially squamous cell carcinoma and papillary thyroid carcinoma) and tuberculous lymphadenitis.

16.8.2 Tuberculous Lymphadenitis

16.8.2.1 Definition

Tuberculous infection in the head and neck accounts for 12% of the extrapulmonary tuberculosis [57,60]. Nodal involvement is the most common manifestation of tuberculosis in the head and neck [61]. It is most commonly seen in 20- to 30-year-old patients.

16.8.2.2 Clinical Features

It typically presents as painless enlarged nodes. Complications include sinus tract formation and coalescence to form large abscesses. Treatment is antimycobacterial therapy.

16.8.2.3 Imaging Findings

Tuberculous infection can present with variable imaging findings. Lymphadenopathy may be unilateral or bilateral; the posterior triangle nodes and internal jugular nodes are the most common affected sites [57,58,60,61].

In the acute phase, the nodes are often homogeneous in signal and enhance homogeneously. This reflects tuberculous granuloma with minimal necrosis. In the subsequent subacute phase, intranodal necrosis or abscess can be observed as a central hyperintense area on T_2-weighted images and hypointense area on T_1-weighted images, with peripheral enhancement on contrast-enhanced T_1-weighted images [51,52,54,55] (Figure 16.34). This is the most common imaging finding in tuberculous lymphadenitis. In the chronic phase or after treatment, fibrocalcific changes occur [61].

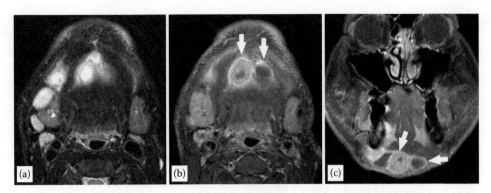

FIGURE 16.33
Suppurative lymphadenitis. (a) Axial fat-suppressed T_2-weighted image demonstrates multiple enlarged level I and II nodes. (b) Axial and (c) coronal contrast-enhanced fat-suppressed T_1-weighted demonstrate enlarged level IA nodes with peripheral enhancement (arrows), consistent with intranodal abscesses.

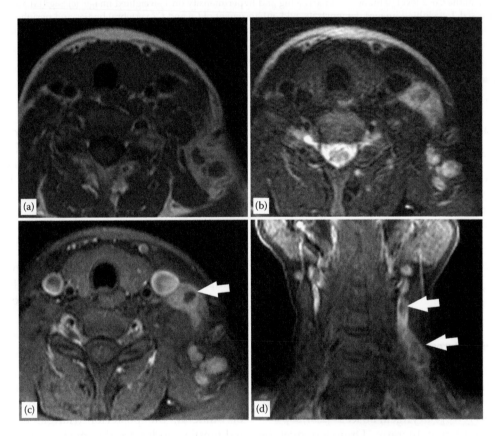

FIGURE 16.34
Tuberculous lymphadenitis. (a) Axial T_1-weighted and (b) fat-suppressed T_2-weighted images show multiple enlarged level IV and V nodes in the left neck. (c) Axial and (d) coronal contrast-enhanced fat-suppressed T_1-weighted demonstrate multiple enlarged left level IV and V nodes with internal necrosis (arrows).

Calcifications are less common than in mediastinal tuberculosis, and may reflect treated disease.

Although both suppurative and tuberculous nodes may be confused with necrotic nodal metastases, the presence of adenopathy involving level V lymph nodes is suggestive of tuberculous lymphadenitis [57,58,60,61].

16.9 Conclusion

Although there are various imaging modalities for the evaluation of head and neck infections, MRI plays an important role. Compared to CT, MRI has the advantages of no ionizing radiation, superior soft tissue contrast, and abscess identification on DWI. Additionally, in patients in whom intravenous contrast is contraindicated, MRA and MRV can be performed to evaluate the vasculature.

Knowledge of various infections involving the head and neck region and their typical imaging findings are helpful for making an accurate diagnosis and directing timely and appropriate patient management.

References

1. Vazquez E, Castellote A, Piqueras J et al. Imaging of complications of acute mastoiditis in children. *RadioGraphics* 2003; 23(2):359–372.

2. Ludwig BJ, Bryan R, Foster BR et al. Diagnostic imaging in nontraumatic pediatric head and neck emergencies. *RadioGraphics* 2010; 30:781–799.

3. Swartz JD, Hagiwara M. Inflammatory disease of the temporal bone. In: Som PM, Curtin HD, eds. *Head and Neck Imaging*, 5th ed. St. Louis, MO: Mosby, 2011; pp. 1183–1230.

4. Heikkinen T, Chonmaitree T. Importance of respiratory viruses in acute otitis media. *Clin Microbiol Rev* 2003; 16(2):230–241.

5. Kangsanarak J, Fooanant S, Ruckphaopunt K et al. Extracranial and intracranial complications of suppurative otitis media. Report of 102 cases. *J Laryngol Otol* 1993; 107:999–1004.

6. Chen PT, Ho KY, Tai CF. Otogenic brain abscess—A case report. *Kaohsiung J Med Sci* 2000; 16:162–165.

7. Sennaroglu L, Sozeri B. Otogenic brain abscess: Review of 41 cases. *Otolaryngol Head Neck Surg* 2000; 123:751–755.

8. Mishra AM, Gupta RK, Saksena S et al. Biological correlates of diffusivity in brain abscess. *Magn Reson Med* 2005; 54:878–885.

9. Leach JL, Fortuna RB, Jones BV, Gaskill-Shipley MF. Imaging of cerebral venous thrombosis: Current techniques, spectrum of findings, and diagnostic pitfalls. *RadioGraphics* 2006; 26(Suppl 1):S19–S41.

10. Linn J, Ertl-Wagner B, Seelos KC, Strupp M, Reiser M, Brückmann H, Brüning R. Diagnostic value of multidetector-row CT angiography in the evaluation of thrombosis of the cerebral venous sinuses. *AJNR Am J Neuroradiol* 2007; 28(5):946–952.

11. Dubrulle F, Kohler R, Vincent C, Puech P, Ernst O. Differential diagnosis and prognosis of T_1-weighted post-gadolinium intralabyrinthine hyperintensities. *Eur Radiol* 2010; 20(11):2628–2636.

12. Mafee MF. MR imaging of intralabyrinthine schwannoma, labyrinthitis and other labyrinthine pathology. *Otolaryngol Clin North Am* 1995; 28(3):407–430.

13. Seltzer S, Mark AS. Contrast enhancement of the labyrinth on MR scans in patients with sudden hearing loss and vertigo: Evidence of labyrinthine disease. *AJNR Am J Neuroradiol* 1991; 12(1):13–16.

14. Lorch M, Teach SJ. Facial nerve palsy: Etiology and approach to diagnosis and treatment. *Pediatr Emerg Care* 2010; 26(10):763–769.

15. Assaf BT, Knight HL, Miller AD. Rhesus cytomegalovirus (Macacine herpesvirus 3)-associated facial neuritis in simian immunodeficiency virus-infected rhesus macaques (*Macaca mulatta*). *Vet Pathol* 2015; 52(1):217–223.

16. Kanerva M, Nissinen J, Moilanen K, Mäki M, Lahdenne P, Pitkäranta A. Microbiologic findings in acute facial palsy in children. *Otol Neurotol* 2013; 34(7):e82–e87.

17. Lim HK, Lee JH, Hyun D, Park JW, Kim JL, Lee Hy, Park S, Ahn JH, Baek JH, Choi CG. MR diagnosis of facial neuritis: Diagnostic performance of contrast-enhanced 3D-FLAIR technique compared with contrast-enhanced 3D-T_1-fast-field echo with fat suppression. *AJNR Am J Neuroradiol* 2012; 33(4):779–783.

18. Engstrom M, Abdsaleh S, Ahlstrom H, Johansson L, Stalberg E, Jonsson L. Serial GD-enhanced MRI and assessment of facial nerve involvement in Bell's palsy. *Otolaryngol Head Neck Surg* 1997; 117:559–566.

19. Thompson AL, Aviv RI, Chen JM, Nedzelski JM, Yuen HW, Fox AJ, Bharatha A, Bartlett ES, Symons SP. Magnetic resonance imaging of facial nerve schwannoma. *Laryngoscope* 2009; 119(12):2428–2436.

20. Caldemeyer KS, Mathews VP, Righi PD, Smith RR. Imaging features and clinical significance of perineural spread or extension of head and neck tumors. *RadioGraphics* 1998; 18(1):97–110.

21. Eggesbø HB. Radiological imaging of inflammatory lesions in the nasal cavity and paranasal sinuses. *Eur Radiol* 2006; 16:872–888.

22. Montone KT. Infectious diseases of the head and neck. *Am J Clin Pathol* 2007; 128:35–67.

23. Som PM, Brandwein MS, Wang BY. Inflammatory diseases of the sinonasal cavities. In: Som PM, Curtin HD, eds. *Head and Neck Imaging*, 5th ed. St. Louis, MO: Mosby, 2011; pp. 167–252.

24. Rak KM, Newell JD 2nd, Yakes WF, Damiano MA, Luethke JM. Paranasal sinuses on MR images of the brain: Significance of mucosal thickening. *AJR Am J Roentgenol* 1991; 156:381–384.

25. Aribandi M, McCoy VA, Bazan C III. Imaging features of invasive and noninvasive fungal sinusitis: A review. *RadioGraphics* 2007; 27:1283–1296.

26. DeShazo RD, Chapin K, Swain RE. Fungal sinusitis. *N Engl J Med* 1997; 337(4):254–259.

27. Ferguson BJ. Fungus balls of the paranasal sinuses. *Otolaryngol Clin North Am* 2000; 33(2):389–398.

28. Gillespie MB, O'Malley BW Jr, Francis HW. An approach to fulminant invasive fungal rhinosinusitis in the immunocompromised host. *Arch Otolaryngol Head Neck Surg* 1998; 124(5):520–526.

29. Schubert MS. Allergic fungal sinusitis. *Otolaryngol Clin North Am* 2004; 37(2):301–326.

30. Waitzman AA, Birt BD. Fungal sinusitis. *J Otolaryngol* 1994; 23(4):244–249.

31. Manning SC, Merkel M, Kriesel K, Vuitch F, Marple B. Computed tomography and magnetic resonance diagnosis of allergic fungal sinusitis. *Laryngoscope* 1997; 107(2):170–176.

32. LeBedis CA, Sakai O. Nontraumatic orbital conditions: Diagnosis with CT and MR imaging in the emergent setting. *RadioGraphics* 2008; 28:1741–1753.

33. Cunnane ME, Sepahadari AR, Gardiner M, Mafee MF. Pathology of the eye and orbit. In: Som PM, Curtin HD, eds. *Head and Neck Imaging*, 5th ed. St. Louis, MO: Mosby, 2011; pp. 591–756.

34. Eustis HS, Mafee MF, Walton C, Mondonca J. MR imaging and CT of orbital infections and complications in acute rhinosinusitis. *Radiol Clin North Am* 1998; 36(6):1165–1183, xi.

35. Jackson A, Sheppard S, Laitt RD, Kassner A, Moriarty D. Optic neuritis: MR imaging with combined fat and water suppression techniques. *Radiology* 1998; 206:57–63.

36. Sepahdari AR, Aakalu VK, Kapur R et al. MRI of orbital cellulitis and orbital abscess: The role of diffusion-weighted imaging. *AJR Am J Roentgenol* 2009; 193:W244–W250.

37. Kassel EE, Schatz CJ. Anatomy, imaging, and pathology of the lacrimal apparatus. In: Som PM, Curtin HD, eds. *Head and Neck Imaging*, 5th ed. St. Louis, MO: Mosby, 2011; pp. 757–854.

38. Kebede A, Adamu Y, Bejiga A. Bacteriological study of dacryocystitis among patients attending in Menelik II Hospital, Addis Ababa, Ethiopia. *Ethiop Med J* 2010; 48(1):29–33.

39. Capps EF, Kinsella JJ, Gupta M, Bhatki AM, Opatowsky MJ. Emergency imaging assessment of acute, nontraumatic conditions of the head and neck. *RadioGraphics* 2010; 30:1335–1352.

40. Tewfik TL, Al Garni M. Tonsillopharyngitis: Clinical highlights. *J Otolaryngol* 2005; 34(Suppl 1):S45–S49.

41. Chapman MN, Nadgir RN, Akman AS et al. Periapical lucency around the tooth: Radiologic evaluation and differential diagnosis. *RadioGraphics* 2013; 33:E15–E32.

42. Hupp JR, Ellis E III, Tucker MR. Contemporary oral and maxillofacial surgery, 5th ed. St. Louis, MO: Mosby, 2008.

43. Kaneda T, Weber AL, Scrivani SJ, Bianchi J, Curtin HD. Cysts, tumors and nontumorous lesions of the jaw. In: Som PM, Curtin HD, eds. *Head and Neck Imaging*, 5th ed. St. Louis, MO: Mosby, 2011; pp. 1469–1546.

44. Vieira F, Allen SM, Stocks RM, Thompson JW. Deep neck infection. *Otolaryngol Clin North Am* 2008; 41(3):459–483.

45. Marcus BJ, Kaplan J, Collins KA. A case of Ludwig angina: A case report and review of the literature. *Am J Forensic Med Pathol* 2008; 29(3):255–259.

46. Daya H, Lo S, Papsin BC et al. Retropharyngeal and parapharyngeal infections in children: The Toronto experience. *Int J Pediatr Otorhinolaryngol* 2005; 69:81–86.

47. Chong VF, Fan YF. Radiology of the retropharyngeal space. *Clin Radiol* 2000; 55(10):740–748.

48. Cohen MA, Docktor JW. Acute suppurative parotitis with spread to the deep neck spaces. *Am J Emerg Med* 1999; 17(1):46–49.

49. Som PM, Brandwein-Gensler MS. Anatomy and pathology of the salivary glands. In: Som PM, Curtin HD, eds. *Head and Neck Imaging*, 5th ed. St. Louis, MO: Mosby, 2011; pp. 2449–2610.

50. Brook I. Acute bacterial suppurative parotitis: Microbiology and management. *J Craniofac Surg* 2003; 14(1):37–40.

51. Darouiche RO. Spinal epidural abscess. *N Engl J Med* 2006; 355:2012–2020.

52. Khan SH, Hussain MS, Griebel RW, Hattingh S. Comparison of primary and secondary spinal epidural abscesses: A retrospective analysis of 29 cases. *Surg Neurol* 2003; 59:28–33.

53. Reihsaus E, Waldbaur H, Seeling W. Spinal epidural abscess: A meta-analysis of 915 patients. *Neurosurg Rev* 2000; 23:175–204.

54. Pereira CE, Lynch JC. Spinal epidural abscess: An analysis of 24 cases. *Surg Neurol* 2005; 63(Suppl 1):S26–S29.

55. Lang IM, Hughes DG, Jenkins JP, St Clair Forbes W, McKenna R. MR imaging appearances of cervical epidural abscess. *Clin Radiol* 1995; 50:466–471.

56. Smith AS, Blaser SI. MR of infectious and inflammatory diseases of the spine. *Crit Rev Diagn Imaging* 1991; 32:165–189.

57. Som PM, Brandwine-Gensler MS. Lymph nodes of the neck. In: Som PM, Curtin HD, eds. *Head and Neck Imaging*, 5th ed. St. Louis, MO: Mosby, 2011; pp. 2287–2384.

58. Ludwig BJ, Wang J, Nadgir RN, Saito N, Castro-Aragon I, Sakai O. Imaging of cervical lymphadenopathy in children and young adults. *AJR Am J Roentgenol* 2012; 199:1105–1113.

59. Chesney PJ. Cervical adenopathy. *Pediatr Rev* 1994; 15:276–284.

60. Moon WK, Han MH, Chang KH et al. CT and MR imaging of head and neck tuberculosis. *RadioGraphics* 1997; 17:391–402.

61. Sakai O, Curtin HD, Romo LV, Som PM. Lymph node pathology: Benign proliferative, lymphoma, and metastatic disease. *Radiol Clin North Am* 2000; 38(5):979–998.

17

Neoplastic Pathology of the Neck

Ahmed Abdel Khalek Abdel Razek

CONTENTS

17.1 Introduction

Head and neck neoplasm is divided into two major groups. The largest group, the epithelial malignancies of the mucosal membranes of the upper aero-digestive tract, is called head and neck squamous cell carcinoma (HNSCC) and accounts for 90% of all head and neck neoplasms. The second important but smaller group are the *glandular neoplasms*, arising in the thyroid and in the salivary glands. Infrequent head and neck tumors include localized lymphoma, soft tissue tumors, peripheral nerve sheath tumors, and neuroectodermal tumors [1].

Head and neck cancer accounts for approximately 3%–5% of all new cancer diagnoses in the United States with 40,000 new cases diagnosed each year. The pattern of tumor staging (T), lymphatic spread (M), and distant spread (M) varies according to the subsite of tumor [2–6]. The American Joint Committee on Cancer (AJCC) and the Union Internationale Contre le Cancer (UICC) published the last seventh edition of tumor–node–metastasis (TNM) for cancer staging in 2010 [7,8]. Table 17.1 shows the nodal staging (N) of head and neck cancer. Tables 17.2 through 17.4 show the tumor staging

(T) of cancer of the pharynx, larynx, and parotid and thyroid glands.

The goal of imaging in head and neck cancer is to differentiate cancer from normal tissue, inflammatory lesions, and benign tumors. In addition, magnetic resonance (MR) imaging is important to determine the extent of the tumor, invasion into the surrounding structure, and nodal and distant spread. Accurate tumor staging is crucial for patient management because it aids in treatment planning and gives some indication of prognosis. Finally, imaging is used for prediction of treatment response, differentiation of recurrence for postradiation changes, and monitoring of patients after chemoradiotherapy [9–15].

When a patient presents with a clinically suspicious head and neck cancer, cross-sectional imaging with computed tomography (CT) scan or MR is commonly used for the assessment of the lesion. Routine and advanced CT techniques such as CT perfusion and dual energy CT can assess head and neck cancer, but they are associated with radiation exposure and of limited value in follow-up after therapy [16–21]. Routine pre- and postcontrast MR imaging enables morphological imaging of head and neck tumors. Addition of advanced MR sequences such as diffusion-weighted MR images, perfusion MR images, and MR spectroscopy (MRS) to routine MR images allows both morphologic and functional imaging of head and neck cancer in a single MR examination [10–15].

17.2 MR Imaging Techniques

17.2.1 Routine MR Imaging

MR imaging of the neck is usually performed with a dedicated neck coil and should include imaging of the entire neck to evaluate nodal metastases. The field of view is usually 20–25 cm^2 and has a slice thickness of 4 mm. The standard MR imaging protocol of head and neck should include a T_2-weighted fast spin-echo sequence, a fat-suppressed T_2-weighted or short-tau inversion recovery (STIR) sequence, and a T_1-weighted spin-echo sequence. The STIR sequence, which is not affected by magnetic field inhomogeneities produced by metallic hardware, may be preferred over the fat-saturated T_2-weighted sequence. The image planes are commonly axial or coronal and may be sagittal. Postcontrast fat-saturated T_1-weighted imaging after gadolinium-DTPA administration in three planes is done. Fat suppression increases conspicuity of the enhancing tumor and enables distinction of tumor from the normal T_1 hyperintense fat present throughout the neck [10–13,22].

17.2.2 Diffusion-Weighted MR Imaging

Diffusion-weighted MR imaging of head and neck can be done using single-shot echo-planar imaging or multishot echo-planar imaging. Automatic shimming and chemical shift selective fat suppression techniques are applied to reduce the artifacts in diffusion-weighted MR images. The applied b-values are 0, 500, and 1000 mm^2/s. Malignant tumors usually show restricted diffusion with low apparent diffusion coefficient (ADC) values and benign tumors tend to show unrestricted diffusion with higher ADC values. However, there is overlap in ADC values of benign and malignant lesions [23–26].

17.2.3 Dynamic Contrast-Enhanced T_1-Weighted MR Imaging

Dynamic contrast-enhanced T_1-weighted MR imaging is done after intravenous injection of gadolinium-DTPA (0.1 mL/kg body weight). MR images can be sequentially obtained every 30 s for 5 min with creation of time-signal intensity curves (TICs). These curves are divided into four types: type A (persistent), curve peaks <120 s after contrast material injection with high washout ratio (>30%); type B (washout), curve peaks <120 s with low washout ratio (<30%); type C (plateau), curve peaks >120 s; and type D (plateau), nonenhanced. Types A, B, and C of the TIC are suggestive of benign lesions, whereas type C exhibited high sensitivity to malignant lesions [27–30].

17.2.4 Dynamic Susceptibility Contrast T_2^*-Weighted Perfusion MR Imaging

Dynamic susceptibilty contrast T_2^*-weighted perfusion MR imaging is performed after bolus injection of gadolinium-DTPA. The injection performed by automatic injector at a rate of 4 mL/s followed by 20 mL saline. The time for data acquisition is 2 min and the time between the data points is 2 s. The TIC is automatically constructed and the dynamic susceptibility contrast percentage (DSC%) is calculated. The threshold of the DSC % is 30.7% and has an accuracy of 84.6% for differentiating malignant from benign head and neck tumors [31,32].

17.2.5 MR Spectroscopy

Proton MRS of the head and neck used with single voxel however; recently multivoxel MRS of the neck has been applied at higher 3 T scanner. Adequate shimming, water suppression, and fat suppression are important for adequate spectrum of neck lesions. MRS can detect metabolic changes within tumors. Proton MRS is used for differentiation of HNSCC from normal tissue and monitoring HNSCC patients undergoing therapy. This technique has not yet found a place in routine clinical use [33,34].

17.2.6 Dynamic Contrast-Enhanced MR Angiography

Dynamic contrast-enhanced MR angiography of the carotid artery is used after the bolus injection of contrast medium with an automatic injector. Contrast-enhanced MR imaging can be used for preoperative mapping of feeding arteries and draining veins of vascular tumors of the head and neck such as paragangliomas and high-flow vascular malformations [35].

17.3 Role of Imaging in Neoplastic Lesions of the Neck

17.3.1 Differentiation of Squamous Cell Carcinoma from Normal Muscles

HNSCCs grow in a submucosal fashion and, therefore, remain undetectable in some patients on clinical and endoscopic examination, and can be easily detected at MR imaging. On T_1-weighted images, tumors demonstrate

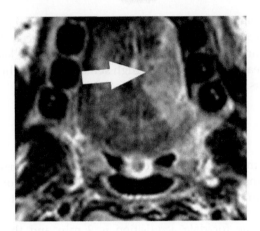

FIGURE 17.1
Oral tongue cancer: Axial T_2-weighted image shows high signal intensity of the mass on the left side of the tongue that is well delineated from the intermediate signal intensity of the adjacent tongue muscles.

the isointense-to-hypointense signal to the muscle. The tumors demonstrate the hyperintense signal on T_2-weighted images and enhance on contrast-enhanced T_1-weighted images (Figure 17.1). Although the conspicuity of lesions is greater with T_2-weighted and contrast-enhanced T_1-weighted images, the extent of tumor may be overestimated on T_2-weighted images because of the similar signal characteristics of the surrounding edema and inflammation. Diffusion-weighted MR imaging can differentiate HNSCC from adjacent muscles as HNSCC shows restricted diffusion with low ADC values compared to unrestricted diffusion of the muscles due to high cellularity of malignant tumors. On proton MRS, the choline/creatine (Cho/Cr) ratio of HNSCC (mean: 5.2) is significantly elevated relative to that in the normal muscle (mean: 0.9) [23,31,33,36,37].

17.3.2 Differentiation of Malignancy from Benignity

Although differentiating malignancy from benignity is often difficult on routine imaging alone, however, certain imaging features such as low signal intensity of

the lesion on T_2-weighted images, ill-defined margin, and inhomogeneous pattern of contrast enhancement are suggestive of malignancy. In addition, benign lesions usually have relatively high signal intensity on T_2-weighted images, well-defined margin, and mild to no enhancement. However, there is overlap in imaging appearance of malignant and benign tumors on routine MR imaging. Advanced MR sequences may differentiate malignant from benign lesions in some cases. On diffusion-weighted MR imaging, most of the benign tumors show unrestricted diffusion with high ADC values (Figure 17.2) apart from Warthin's tumors and lipomas and most of the malignant tumors show restricted diffusion with low ADC values apart from adenoid cystic carcinoma (Figure 17.3). There is difference in the mean and range of ADC values of benign and malignant tumors in the head and neck in different literatures. On dynamic contrast-enhanced MR imaging, the shape of the curves can provide a guide to the type of lesions: benign (ascending plateau) or malignant (descending plateau). In addition, there is a significant difference in the DSC% on dynamic T_2*-weighted perfusion MR imaging with low DSC% of benign tumors (24.3% ± 10.3%) compared to high DSC% of malignant tumors (39.3% ± 9.6%). Finally, there is a significant difference in the Cho/Cr ratio of MRS between benign and malignant tumors; however, its role is limited [31,33,38,39].

17.3.3 Differentiation of Malignancy from Inflammatory Lesions

An early and correct differentiation of malignancy from inflammatory lesions of the head and neck is pivotal for treatment planning. On contrast MR imaging, abscess usually appears as a well-defined marginally enhanced lesion, whereas malignancy shows an inhomogeneous pattern of contrast enhancement with ill-defined margin; however, in some cases, this differentiation is difficult with routine MR imaging. Diffusion-weighted MR

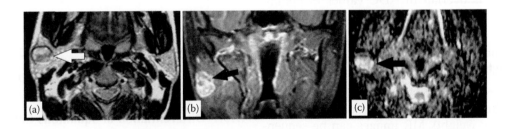

FIGURE 17.2
Pleomorphic adenoma of parotid gland: (a) Axial T_2-weighted image shows a well-defined focal mass (arrow) seen in the right parotid gland. (b) Coronal contrast-enhanced T_1-weighted image shows homogeneous enhancement of the mass (arrow). (c) ADC map shows unrestricted diffusion of the mass (arrow) with high ADC value denoting benign lesion.

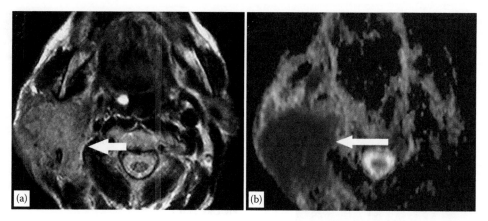

FIGURE 17.3
Cancer parotid: (a) Axial T_2-weighted image shows a large ill-defined infiltrating mass (arrow) of mixed signal intensity seen in the right parotid region. (b) ADC map shows restricted diffusion with low ADC value of the mass (arrow).

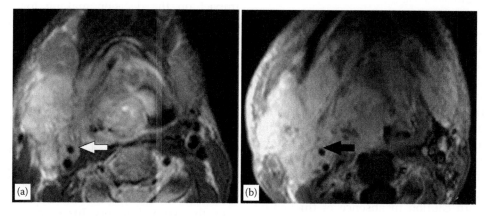

FIGURE 17.4
Carotid artery invasion of right aryepiglottic fold carcinoma: (a) Axial T_2-weighted image shows a large mass arising from the right aryepiglottic fold associated with enlarged metastatic lymph nodes. The mass encases the right carotid artery (arrow). (b) Contrast-enhanced T_1-weighted image shows intense enhancement of the mass that surround and encases the signal void of the right carotid artery (arrow).

imaging helps in this differentiation. Abscess shows restricted diffusion compared to malignancy due to presence of macromolecules in pus that reduce water diffusivity in abscess, whereas bulk necrosis of tumors is less viscous as a serous fluid that increases diffusivity. Proton MRS may demonstrate high acetate peak in abscess [40–43].

17.3.4 Locoregional Tumor Staging of Head and Neck Cancer

17.3.4.1 Vascular Invasion

Carotid encasement is a criterion that upstages HNSCC to stage 4b, and is typically considered unresectable. Circumferential vessel wall involvement of 180° is used as an indicator for vascular invasion on MR imaging. However, this is neither a sensitive nor a specific

indicator. By contrast, 270° circumferential wall involvement is highly specific for the inability to peel the tumor off the affected artery. A combination of imaging findings to include deformation of the carotid artery, encasement of ≥180°, and segmental obliteration of the fat between the tumor and the carotid artery is highly predictive of vascular invasion [15,44–47] (Figure 17.4).

17.3.4.2 Perineural Spread

Perineural spread is a form of metastatic disease in which primary tumors spread along neural pathways. It is confined to the main tumor mass. By contrast, perineural spread is separate from the main bulk of the tumor. Cranial nerves V and VII are most commonly involved because of their extensive distribution. Recognition of this mode of spread is crucial because

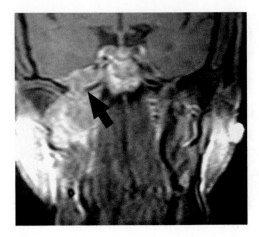

FIGURE 17.5
Perineural spread of nasopharyngeal carcinoma: Coronal fat-suppressed T_1-weighted image shows a large nasopharyngeal mass that extends along the course of V2 nerve through a widened foramen ovale (arrow) into the right cavernous sinus.

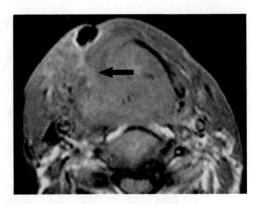

FIGURE 17.6
Cartilage invasion of transglottic cancer: Axial T_2-weighted image shows a large mass that obliterates the laryngeal airway. The mass invades the right thyroid lamina (arrow) with extralaryngeal extension. Note defect in the skin of the right-sided sinus tract.

of its adverse therapeutic and prognostic implications. Many treatment failures are attributed to unrecognized perineural tumor spread. Overall, perineural spread negatively impacts prognosis and jeopardizes long-term survival. The perineural spread can be seen in adenoid cystic carcinoma and squamous cell carcinoma (SCC), and it may be in lymphoma and malignant melanomas. MR imaging findings of perineural tumor spread include thickening and enhancement of the nerve, widening of the neural foramen, and loss of fat surrounding the nerve. Enhancement of the nerve is interpreted with caution in absence of thickening or foraminal changes, because presence of extensive perineural vascular plexus may promote avid contrast uptake [48–52] (Figure 17.5).

17.3.4.3 Cartilaginous Invasion

Cartilaginous invasion by laryngeal and hypopharyngeal cancers is an important imaging finding because it automatically leads to a T4 stage. Cartilage invasion on MR imaging shows an increase in T_2 signal intensity, a low-to-intermediate signal on T_1-weighted images, and enhancement on contrast study. If all of these signs are absent, cartilage infiltration can be ruled out. However, MR imaging findings of cartilage invasion may be false-positive results, because reactive inflammation, edema, and fibrosis may be indistinguishable from neoplastic lesions. Peritumoral inflammatory changes are most commonly observed in the thyroid cartilage, and the specificity of MR imaging to detect invasion of the thyroid is only 56%, as opposed to 87% and 95% in the cricoid and arytenoid cartilages, respectively. Recent study reported that diffusion-weighted MR imaging showed high validity and precision in detecting inner and outer thyroid lamina invasion. This can have an important

impact on the decision making for management of laryngeal carcinoma [53–57] (Figure 17.6).

17.3.4.4 Osseous Invasion

Accurate preoperative determination of mandibular osseous involvement is essential in treatment planning and staging; bone involvement upstages the tumor at least to T4a. Oral cavity carcinomas invade mandible at sites of attachment and can invade the bone through the cortex, or more commonly through the periodontal ligament or at a site of previous tooth extraction. Determining the presence and extent of bone invasion is critical to determine the planned surgery. Tumor invasion of periosteum without gross cortical invasion is managed with marginal mandibulectomy that maintains the integrity of the bone. Tumors with gross cortical erosion or medullary invasion require a segmental resection—resection of the entire involved segment of the mandible [10–15].

MR imaging is excellent in detection of bone marrow invasion. T_1-weighted images best demonstrate the extent of bone marrow invasion, seen as low signal intensity replacing high signal intensity of medullary fat. Bone marrow invasion is depicted as hyperintense signal on T_2-weighted images and enhancement on fat-suppressed contrast images. Replacement of the hypointense signal of the bony cortex on T_1- and T_2-weighted images with the signal intensity of the tumor is a strong indicator for cortical invasion. False-positive cases are attributed to inflammatory change from dental infections or procedures and to chemical shift artifact from bone marrow fat on T_1-weighted images [58–61] (Figure 17.7).

17.3.4.5 Prevertebral Invasion

In patients with hypo- or oropharyngeal cancer, an important staging question is the presence of tumor

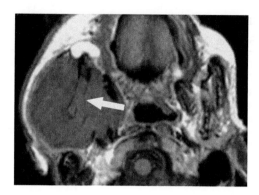

FIGURE 17.7
Bone marrow infiltration of sarcoma: Axial T_1-weighted image shows a large low–intermediate signal intensity mass that extends through the hypointense cortex of the mandible with replacement of the medullary cavity of the mandible (arrow).

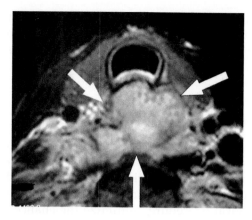

FIGURE 17.9
Esophageal invasion of hypopharyngeal carcinoma: Axial T_2-weighted image shows a large ill-defined mass (arrows) infiltrating the esophagus that extends from the hypopharyngeal mass. The mass extends into the posterior aspect of the trachea and anterior prevertebral space.

fixation to the prevertebral fascia, which typically indicates irresectability. Preservation of the retropharyngeal fat plane between the tumor and the prevertebral compartment on T_1-weighted images appears to predict the absence of prevertebral space fixation. Interruption of high T_1 signal intensity of the retropharyngeal fat stripe is an indicator of invasion of prevertebral fascia or musculature. However, the actual presence of prevertebral invasion cannot accurately ascertain by imaging characteristics and is best determined at surgery [62,63] (Figure 17.8).

17.3.4.6 Esophageal Invasion

MR imaging accurately predicts neoplastic invasion of the cervical esophagus in patients with thyroid and hypopharyngeal cancer. Esophageal wall thickening, effacement of the adjacent fat plane, and any T_2 signal wall abnormality have been reported 100% sensitive, whereas circumferential mass more than 270° is 100% specific for esophageal invasion [64] (Figure 17.9).

17.3.4.7 Tracheal Invasion

MR imaging is the modality of choice for evaluation of tracheal invasion by thyroid cancer. Tracheal invasion diagnosed by the combination of three criteria: tumor abutting circumference of trachea for 180° or more, an intraluminal mass, or a soft-tissue signal within the cartilage. Invasion of the recurrent laryngeal nerve is predicted by the finding of effaced fatty tissue. The accuracy of these criteria is 90% [65].

17.3.4.8 Mediastinal Invasion

Mediastinal lymph node metastasis is thought to spread by lymphatic spread from levels VI and IV, and is significantly correlated with contralateral lateral node metastasis; however, metastasis may occur directly without other lymph node involvement. Metastasis may be obvious on MR imaging but is often occult without imaging evidence. Surgical stress for patients undergoing mediastinal node dissection require median sternotomy [10–15].

17.3.4.9 Brachial Plexus Invasion

Direct invasion of brachial plexus and anterior scalene muscle with head and neck cancer makes the tumor

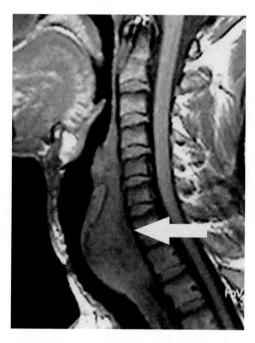

FIGURE 17.8
Prevertebral invasion of postcricoid carcinoma: Sagittal T_1-weighted image shows a large postcricoid mass (arrow) that infiltrates the prevertebral fat plane proved to be prevertebral fat infiltration at surgery.

unresectable. On MR imaging, brachial plexus invasion is identified with high signal intensity on T_2-weighted imaging and enhancement along brachial plexus roots, trunks, divisions, cords, and branches or scalene muscles [10–15].

17.3.4.10 Cutaneous Invasion

It is important to recognize the extension of subcutaneous and cutaneous invasion by cancer for surgical planning for reconstruction of complex skin defects. Accurate surgical planning can reduce the length of the surgical procedure, minimize the chance of failure during reconstruction, and increase the treatment efficacy for the best possible functional and aesthetic results. Direct skin invasion from HNSCC is a prognostically significant factor indicative of a poor prognosis. Presence of subcutaneous and cutaneous fat invasion appears as low signal intensity on T_1-weighted images and high signal on T_2-weighted images [10–13] (Figure 17.10).

17.3.5 Nodal Staging of Head and Neck Cancer

Lymphadenopathy has very important prognostic implications. The presence of a single nodal metastasis reduces the patient's survival rate by 50%. Bilateral lymphadenopathy further reduces the survival rate by another 50%. Extracapsular spread and nodal fixation are also associated with reduced survival. Table 17.1 shows nodal staging of head and neck cancer. Many radiologists use 10 mm as the cutoff for normal lymph node size in the head and neck area, with a greater allowance (15 mm) for jugulodigastric nodes and a smaller cutoff (8 mm) for retropharyngeal nodes. The presence of central necrosis increases specificity. Lymphatic drainage varies according to the location of head and neck cancer [15,66]. On diffusion-weighted MR imaging, metastatic

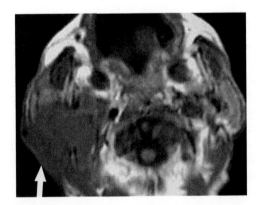

FIGURE 17.10
Subcutaneous invasion of parotid cancer: Axial T_1-weighted image shows a large hypointense mass seen in the right parotid region that infiltrates the subcutaneous soft tissue and the overlying skin (arrow).

TABLE 17.1

Nodal Staging (N) of Head and Neck Cancer

Nasopharynx

- N1—Unilateral metastasis in cervical lymph node(s), <6 cm in greatest dimension, above supraclavicular fossa; and/or unilateral or bilateral metastasis in retropharyngeal lymph nodes, <6 cm in greatest dimension
- N2—Bilateral metastasis in cervical lymph node(s), <6 cm, above supraclavicular fossa
- N3a—Metastasis in lymph node(s), >6 cm in dimension
- N3b—Extension to the supraclavicular fossa

All Other Sites

- N1—Metastasis in a single ipsilateral lymph node, <3 cm in greatest dimension
- N2a—Metastasis in a single ipsilateral lymph node, >3 cm but <6 cm in dimension
- N2b—Metastasis in multiple ipsilateral nodes, none >6 cm in greatest dimension
- N2c—Metastasis in bilateral or contralateral nodes, none >6 cm in greatest dimension
- N3—Metastasis in a lymph node, >6 cm in greatest dimension

Source: Edge, S. et al., *AJCC Cancer Staging Manual*, 7th edn., Springer, Chicago, IL, 2010.

nodes show restricted diffusion with low ADC value compared to reactive lymph nodes due to multitude of enlarged cells and increased mitosis at metastatic nodes. The mean DSC% of metastatic nodes (48.72%) is significantly different ($p = .001$) than that of lymphoma (37.09%). The mean Cho/Cr ratio for metastatic nodes is significantly higher (4.8) than that for reactive nodes (2.2) [66–70] (Figures 17.11 and 17.12).

17.3.6 Distant Metastasis of Head and Neck Cancer

Distant metastasis is reported in about 10% (5%–40%) of patients with HNSCC. The most common sites of metastases include the bone (20%), the lung (13%), and the liver (9%). The risk of distant metastases is dependent upon tumor stage with T4 tumors and advanced regional nodal disease carrying the highest risk. Whole-body MR imaging shows similar diagnostic capacity similar to ^{18}F-fluorodeoxyglucose positron emission tomography (FDG PET)–CT in assessing the distant-site status in patients with HNSCC. Whole-body diffusion-weighted MR imaging and whole-body PET–MR imaging have been used in early detection of distant metastatic deposits in patients with HNSCC [71–73].

17.3.7 Characterization of Neoplastic Lesions

MR imaging may be helpful in characterization of some head and neck neoplasms. Hemangiomas show intense contrast enhancement and are often seen in children. Presence of signal void regions within the mass is highly suggestive of high-flow vascular malformations

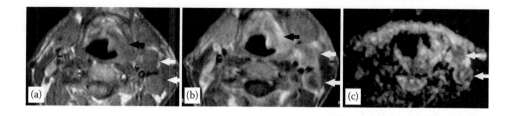

FIGURE 17.11

Nodal metastasis of left aryepiglottic fold carcinoma: (a) Axial T_1-weighted image shows two enlarged left-sided cervical lymph nodes (white arrows) in patients with left aryepiglottic fold carcinoma (black arrow). (b) Axial contrast-enhanced T_1-weighted image demonstrates that the posterior lymph node shows ring enhancement and the anterior one shows an inhomogeneous pattern of enhancement (white arrows). The aryepiglottic fold mass shows the homogenous pattern of enhancement (black arrow). (c) ADC map shows restricted diffusion with low ADC value in the solid part of the metastatic lymph nodes and unrestricted diffusion in the cystic part of the lymph nodes (white arrows).

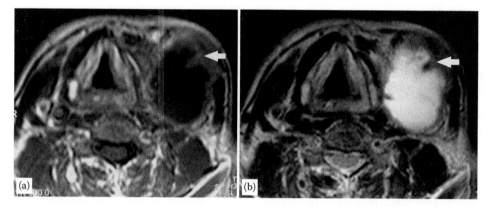

FIGURE 17.12

Nodal metastasis: (a) Axial T_1-weighted image shows the enlarged necrotic left upper deep cervical lymph node that shows a thick wall with small solid part along its anterior aspect (arrow). (b) Axial T_2-weighted image demonstrates that the cystic part of the node shows hyperintensity and the solid part shows hypointensity (arrow).

or paragangliomas. Fatty tumors such as lipomas show high signal intensity on T_1-weighted images that are suppressed on fat suppression sequences. Although malignant tumors show restricted diffusion, some malignant tumors such as adenoid cystic carcinoma and chondrosarcoma show high ADC values because these tumors contain abundance of mucinous, myxoid, or hyalinized material [10–15].

SCC versus lymphoma: The ADC is significantly lower in lymphoma (0.65 and 0.45 × 10⁻³ mm²/s) than in HNSCC (0.96 and 0.86 × 10⁻³ mm²/s) as assessed by diffusion-weighted MR imaging. An ADC value less than 0.56 × 10⁻³ mm²/s predicts lymphoma with an accuracy of 96%. The differences in ADC values attributed to difference in tumor cellularity, with lymphoma having greater cellularity than HNSCC [74–76].

Metastatic versus lymphomatous nodes: The ADC of lymphomatous nodes is significantly lower than that of metastatic nodes from HNSCC. This difference arises because lymphoma has higher

cellularity with restricted diffusion and cancer cells in SCC are surrounded by varying amounts of extracellular matrix and extracellular water, resulting in higher ADC values. Additionally, the ADC values of focal defects in lymphoma nodes are significantly lower than those of focal defects in SCC nodes because focal defects in the SCC nodes contain necrotic material with high amount of extracellular fluid, thereby increasing their ADC values. Conversely, focal defects in lymphoma nodes contain apoptotic tumor cells that decrease the ADC value [77–79].

17.3.8 Correlation with Degree of Differentiation of Malignancy

Differentiation of moderately and well-differentiated HNSCCs from poorly and undifferentiated HNSCCs is important for determining the prognosis and clinical outcome of patients. At diffusion-weighted MR imaging, the ADC values of well- and moderately differentiated tumors are significantly higher ($p = .001$)

than those of poorly differentiated tumors. Poorly differentiated HNSCCs have more cellularity, larger and more angulated nuclei (with more abundant macromolecular proteins), and less extracellular space than well- or moderately-differentiated carcinomas [80,81].

17.3.9 Detection of the Best Site for Biopsy

Detection of the best biopsy site of head and neck neoplasm is important for best results. On routine MR imaging, biopsy is better taken from the areas of lowest signal intensity on T_2-weighted images, and shows contrast enhancement and avoids the areas of necrosis. An ADC map can differentiate viable from necrotic regions of malignant tumors. The viable regions of the tumor show restricted diffusion and the necrotic regions display unrestricted diffusion. A biopsy is better when taken from the region of restricted diffusion with the lowest ADC value on the ADC map [82].

17.3.10 Detection of Carcinoma of Unknown Primary in Metastasis of the Head and Neck

Carcinoma of unknown primary represents 3%–5% of all head and neck cancers. SCC accounts for 70%–90% of these lesions, and most commonly arises from sites in the upper aerodigestive tract, including the tonsils, the base of the tongue, the nasopharynx, and the pyriform sinus. MR imaging and PET–CT play a role in the assessment of patients with cancer of unknown primary. The location of the metastatic lymph node may give an indication of the primary tumor site, and knowledge of lymph node drainage patterns is important for radiologist evaluating these patients. Recently, PET–MR imaging has detected the primary tumor site in these patients [83–87].

17.3.11 Imaging Findings as Prognostic Parameters

Gross target volume (GTV): Measurements of tumor volume on MR imaging referred to as GTV correlate well with local control and outcome for supraglottic, glottis, and pyriform sinus SCCs. In addition, GTV provides a quantitative measure of the response to therapy. Tumor volume appears to be the strongest independent predictor of local failure after radiation therapy for supraglottic tumors. The local control rate is 89% in tumors when the volume is smaller than 6 mL and only 52% when the volume is 6 mL or more. Pyriform sinus tumors with a volume of less than 6.5 mL have an 89% likelihood of local control, whereas those with a volume more than 6.5 mL have a local control rate of 25% [88–93].

Tumor thickness: It is a prognostic factor for oral tongue cancer—tumors smaller than 3 mm have a low incidence of locoregional recurrence and excellent disease-free survival, whereas thickness more than 9 mm carries a 24% probability of local recurrence and a 66% probability of 5-year disease-free survival. A fat-suppressed contrast-enhanced T_1-weighted image is the best MR imaging sequence for measuring the thickness of oral tongue [94,95].

17.3.12 Prediction of Treatment Response

Prediction of response to conservative treatments (radiotherapy, chemotherapy, or combination) is important to determine the optimal treatment on an individual case-by-case basis. If pretreatment probability of success of therapy is predicted, modification of treatment protocol can be made if the chances of success are poor. This approach can minimize the cohort of patients who undergo extensive chemoradiotherapy only to respond poorly but who are thus precluded from salvage surgery because of significant radiation-induced tissue architectural changes that make surgery more complicated and risky. Diffusion-weighted MR imaging provides information about a tumor's composition, particularly about its cellularity, which correlates with response to therapy and survival. There is strong correlation between low ADC (high cellularity) and tumor regression after therapy. Patients with lower ADC values in the baseline scan are more likely to be complete responders to chemoradiotherapy, possibly attributed to better response in hypercellular lesions (the more cellular a lesion, the less hypoxic which is a major impediment to the success of radiotherapy) [25,96–100].

17.3.13 Recurrence versus Postradiation Changes

Patients with unresectable disease have residual or recurrent disease following nonoperative therapy, in 30%–50% of cases. MR imaging can differentiate mature scar tissue, which shows retraction, low T_2 signal, and no contrast enhancement, from tumor, which is expansile and of intermediate T_2 signal with moderate contrast enhancement. However, there may be an overlap between partially treated tumor and immature scar tissue. On diffusion-weighted MR imaging, recurrence with restricted diffusion can differentiate from posttreatment changes with unrested diffusion. Recurrent tumors show hypercellularity that reduces the diffusion space of water protons in extra- and intracellular spaces, with a resultant decrease in the ADC value (Figure 17.13). Posttreatment changes show relatively low cellularity associated with variable degrees of edema and inflammatory reaction that are characterized by an increase

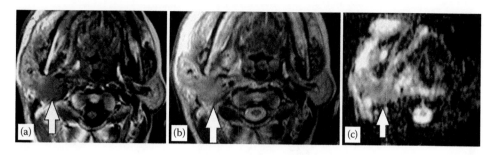

FIGURE 17.13

Recurrence: (a) Axial T_1-weighted image shows hypointense lesion (arrow) seen in the region of the right parotid after surgery. (b) Axial T_2-weighted image shows low signal intensity of the mass (arrow) with ill-defined margin. (c) ADC map shows restricted diffusion with low ADC value of the lesion (arrow) proved to be recurrence malignancy.

in interstitial water content. Percentage changes in the Cho/Cr ratio of MRS after chemoradiotherapy may serve as a marker of residual cancer in a posttreatment mass [101–103].

17.3.14 Monitoring Patients after Radiotherapy

It is advantageous to obtain a posttreatment scan done 3–6 months after radiation therapy that provides a baseline study for any future imaging. Follow-up scans are guided by clinical factors such as suspicion of tumor recurrence or development of a radiation-induced complication. Any enlarging posttreatment soft-tissue mass or any new deep lesion is concerning for recurrent disease. Diffusion-weighted MR imaging can predict the responses to neoadjuvant therapy for HNCCs. An increase in ADC values at 3 weeks after therapy is reported in patients with favorable outcome. Thus, diffusion-weighted MR imaging is an early surrogate biomarker to monitor treatment-induced tissue alterations in patients with HNSCC. The ADC values of complete responders are significantly different ($p < .05$) than those of partial responders. A significant increase in ADC values is observed in complete responders within 1 week of treatment ($p < .01$), and the value remains high until the end of the treatment. Complete responders also show significantly higher ADC values than partial responders by the first week of chemoradiation ($p < .01$) [104–106].

17.4 Imaging Appearance of Head and Neck Tumors

17.4.1 Squamous Cell Carcinoma

HNSCC is the sixth most common cancer with approximately 650,000 incidences and 350,000 deaths worldwide annually. It accounted for approximately

155,400 new cases. HNSCC is the fourth most common cancer in men, whereas it is the ninth most common cancer in women. It develops in the oral cavity, pharynx, larynx, and salivary glands. The most important risk factors for the development of HNSCC are tobacco and alcohol consumption. However, increasing evidence has documented human papillomavirus as a cause of specific subsets of HNSCC. About two-thirds of patients with HNSCC present with advanced stage disease, commonly involving regional lymph nodes. HNSCC demonstrates intermediate-to-high signal intensity on T_2-weighted images, low signal intensity on T_1-weighted images, and moderate enhancement on contrast-enhanced images [2–6,13].

17.4.2 Oral Cavity Carcinoma

The primary role of imaging in oral cavity carcinoma assessing the primary tumor is determination of T4a or T4b disease. The presence or absence of submucosal tumor, extrinsic tongue muscle invasion, bone involvement, perineural extension, or extension to the masticator space is important for all oral cavity carcinomas. Invasion of the extrinsic tongue muscles upstages tumor to T4a and invasion of the masticator space, pterygoid plates, skull base, or encasement of internal carotid artery upstages tumor to T4b. Resection may still be attempted for T4b disease, based upon masticator space invasion, but invasion of the skull base or internal carotid artery often deems a patient unresectable [107–112].

17.4.2.1 Lip Carcinoma

The lip is the most common site of SCC in the oral cavity (40%). Carcinomas of the lip usually arise from the vermillion border and may spread laterally to the adjacent skin or deeply to the orbicularis oris muscle. If the tumor invades the skin or bone, it upstages to T4a [107–110].

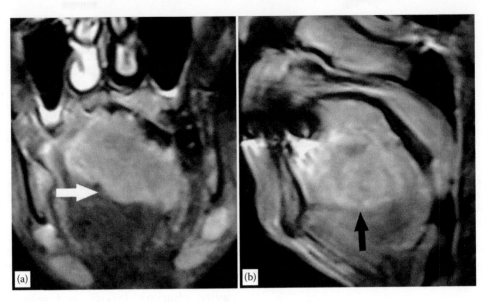

FIGURE 17.14
Oral tongue SCC: (a) Coronal T_2-weighted image shows a large ill-defined mass (arrow) seen infiltrating the oral tongue that crosses the midline. (b) Sagittal contrast-enhanced T_1-weighted image demonstrates that the mass (arrow) shows an intense inhomogeneous pattern of enhancement.

17.4.2.2 Buccal Carcinoma

Buccal carcinomas often originate along the buccal mucosa lining the cheeks. A common route of spread is lateral extension along the buccinators muscle to the retromolar trigone and pterygomandibular raphe. Involvement of the retromolar trigone provides numerous paths of tumor spread and makes surgical management more difficult. More than half of buccal tumors present as deeply invasive tumors that may track along the parotid duct and masseter muscle, or into the palate [108–111].

17.4.2.3 Gingival Carcinoma

Gingival carcinomas along the maxillary or mandibular alveolar ridges account for 10% of oral cavity carcinomas. Because of the proximity to the cortical bone of the mandible, it is critical to assess for bone invasion and perineural extension, particularly along the inferior alveolar nerve for SCCs of the lower alveolar ridge [107,108].

17.4.2.4 Oral Tongue Carcinoma

Nearly all tongue carcinomas occur along the lateral margin or undersurface. Prognosis and treatment depend on the depth of invasion. It is critical that the radiologist assess the extrinsic tongue muscles. Cancer seen adjacent to the neurovascular bundle is highly suggestive of invasion; tumors larger than 2 cm with aggressive margins and deep sublingual extension probably involve the neurovascular bundle. Assessment of

involvement of the midline lingual septum is critical to determine whether the patient requires a hemiglossectomy or total glossectomy (Figure 17.14). Anterior third tumors tend to invade the floor of the mouth. Middle third lesions invade the musculature of the tongue and subsequently the floor of the mouth. Posterior third tumors grow into the anterior tonsillar pillar, tongue base, and glossotonsillar sulcus. Involvement of the tongue base is critical to note because it may necessitate a total laryngectomy [107–112].

17.4.2.5 Retromolar Trigone Cancer

The retromolar trigone refers to the small triangular-shaped mucosa overlying the area just posterior to the last mandibular molar. It is an important region because it represents a crossroads for tumor spread to and from the oral cavity, oropharynx, buccal space, floor of the mouth, and masticator space. In addition, these tumors are in close proximity to the mandible and maxilla, and therefore have a high propensity to invade bone [107–109] (Figure 17.15).

17.4.2.6 Floor of Mouth Carcinoma

Floor of mouth carcinomas tends to arise within 2 cm of the anterior midline and spreads laterally to the mandible or ipsilateral or contralateral neurovascular bundle. The neurovascular bundle, including the lingual artery and hypoglossal nerve, traverses the sublingual space. Ipsilateral tumor involvement necessitates sacrifice,

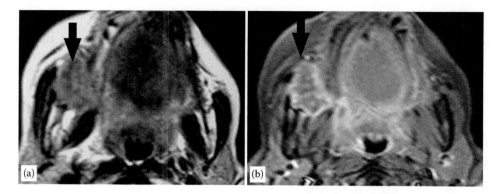

FIGURE 17.15

Retromolar trigone SCC: (a) Axial T_2-weighted image shows a mass (arrow) located in the right retromolar trigone. The mass shows a hypointense appearance. (b) Axial contrast-enhanced T_1-weighted image demonstrates that the mass shows contrast enhancement (arrow).

but the remaining contralateral supply preserves viable tongue function. However, if the tumor extends to the contralateral neurovascular bundle, this necessitates sacrifice of both bundles and total glossectomy [109–112].

17.4.3 Oropharyngeal Carcinoma

Mucosal and smaller lesions of the oropharynx are best assessed clinically. The GTV can be easily determined in discrete and exophytic tumors. However, tumors of the oropharynx are more commonly infiltrative and invasive, and extend along muscle and facial planes, making accurate GTV determination difficult. The base of tongue cancers can be particularly problematic, as dense interdigitation of muscle without intervening fat to define the tissue planes can obscure lesion margins. Attempts made to determine whether the tumor crosses midline, as midline extension increases the likelihood of bilateral/contralateral nodal involvement. At the base of the tongue, cross-midline extension changes the surgical plan, as the contralateral neurovascular bundle is at risk. Approximately 65% of oropharyngeal SCC patients present with metastatic lymphadenopathy. Lesions of the base of the tongue are most likely to present with malignant lymph nodes [111–113]. Table 17.2 shows tumor staging (T) of oropharyngeal cancer.

17.4.3.1 Tonsillar Pillar Carcinoma

The anterior tonsillar pillar and tonsil are the most common locations for tumors of the oropharynx. Cancers arising at the anterior tonsillar pillar can spread superiorly to the lateral soft palate. The tumor may spread to the masticator space, the nasopharynx, and the skull base. Inferior extension along the palatoglossus muscle course results in tumor at the base of the tongue. If the tumor spreads laterally and anteriorly, it can travel along the pharyngeal constrictor muscles and pterygomandibular raphe to the oral cavity at the retromolar

TABLE 17.2

Primary Tumor (T) Staging of the Pharynx

Nasopharynx
- T1—Tumor confined to the nasopharynx or extends to the orophayrnx and/or nasal cavity
- T2—Tumor with parapharyngeal extension
- T3—Tumor involves bony structures of skull base and/or paranasal sinuses
- T4—Tumor with intracranial extension and/or involvement of cranial nerves, hypopharynx, or orbit, or with extension to the infratemporal fossa/masticator space

Oropharynx
- T1—Tumor <2 cm in greatest dimension
- T2—Tumor >2 cm but <4 cm in greatest dimension
- T3—Tumor >4 cm in greatest dimension or extension to lingual surface of epiglottis
- T4a—Tumor invades the larynx, extrinsic tongue muscles, medial pterygoid, hard palate, or mandible
- T4b—Tumor invades lateral pterygoid muscle, pterygoid plates, lateral nasopharynx, or skull base, or encases carotid artery

Hypopharynx
- T1—Tumor limited to one subsite of the hypopharynx and/or <2 cm in greatest dimension
- T2—Tumor invades more than one subsite or adjacent site, or measures >2 cm but <4 cm in greatest dimension, without fixation of hemilarynx
- T3—Tumor >4 cm in largest dimension, fixation of hemilarynx, or esophageal extension
- T4a—Tumor invades the thyroid/cricoid cartilage, hyoid bone, thyroid gland, or central compartment soft tissue
- T4b—Tumor invades prevertebral fascia, encases carotid artery, or involves mediastinal structures

Source: Edge, S. et al., *AJCC Cancer Staging Manual*, 7th edn., Springer, Chicago, IL, 2010.

trigone and into the buccinator muscle. Tumors of the posterior tonsillar spread to the soft palate, the posterior thyroid cartilage, the middle constrictor muscle to the oral cavity, the posterior pharyngeal wall, and the pharyngoepiglottic fold to the top of the pyriform sinus [112,113].

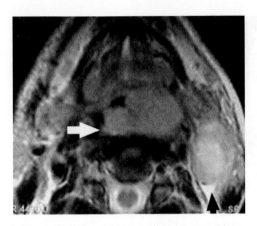

FIGURE 17.16
Left tonsillar SCC: Axial T_2-weighted image shows a large left tonsillar mass (white arrow) exhibiting low signal intensity and abutting the tongue base. There is associated enlarged necrotic left upper deep cervical lymph node that shows high signal intensity (black short arrow).

17.4.3.2 Tonsillar Fossa Carcinoma

Cancers of the tonsillar fossa are often clinically silent and may present as a neck mass from malignant adenopathy. From the tonsillar fossa, a tumor can spread directly into the parapharyngeal space and from there to the carotid space, into the masticator space, and into the mandible. Additionally, a tumor can spread along the anterior and posterior tonsillar pillars with routes of extension as discussed earlier [111–113] (Figure 17.16).

17.4.3.3 Soft Palate Carcinoma

Soft palate carcinoma can extend anteriorly into the hard palate; laterally into the palatine muscles and the parapharyngeal space, and from there to the skull base and nasopharynx; and inferiorly into the tonsillar pillars. Additionally, perineural extension of disease can occur along the palatine nerves and retrograde to pterygopalatine fossa and cavernous sinus along V2 [111].

17.4.3.4 Base of Tongue Carcinoma

Tumors arising in this location spread anteriorly into the root of tongue and extrinsic tongue muscles, and into the sublingual space and neurovascular bundle of the oral cavity. Caudal extension is into the vallecula and potentially the pre-epiglottic fat. Lateral extension is potentially into the lateral wall, pterygomandibular raphe, and mandible. More posteriorly, a tumor can invade parapharyngeal fat and from there, the carotid space. Tumors can extend superiorly along the tonsillar pillars [112,113].

17.4.3.5 Posterior Pharyngeal Wall Carcinoma

Tumors here are often large at the time of diagnosis and can spread superiorly to the nasopharynx, laterally into the parapharyngeal space, inferiorly into the hypopharynx, and anteriorly into the tonsil. If the tumor has deep extension, it invades the prevertebral musculature. Many of these tumors extend past midline. Lymphatic drainage of the posterior pharyngeal wall includes the bilateral jugular chain lymph nodes and the retropharyngeal lymph nodes [111–113].

17.4.3.6 Hypopharyngeal Carcinoma

Hypopharyngeal cancer tends to spread in a submucosal fashion that is often undetectable on clinical and/or endoscopic examination. On imaging, evaluation of fat planes within the postcricoid musculature and around the hypopharynx is essential for mapping of the tumor boundaries. Cartilage invasion, tumor volume, and involvement of the pyriform sinus apex are important variables in stratification of patients into favorable and unfavorable treatment groups. Patients with favorable tumors typically receive radiation therapy, whereas patients with unfavorable tumors usually undergo surgical resection and subsequent reconstructive surgery. The extrahypopharyngeal surrounding structures most commonly invaded by hypopharyngeal SCC are the posterior oropharyngeal wall (superior), the larynx (anterior), and the proximal cervical esophagus (inferior). Posterior extension through the retropharyngeal space occurs less commonly, but it is important because the patient is then truly nonoperable. The exact assessment of the craniocaudal extension of hypopharyngeal tumors into the cervical esophagus is crucial for planning of surgical resection (limited vs. extensive resection) [114–116]. Table 17.2 shows tumor staging (T) of hypopharyngeal cancer.

17.4.3.7 Pyriform Sinus Carcinoma

The growth pattern of pyriform sinus cancers depends on their site of origin. Tumors arising from or infiltrating the lateral wall of the pyriform sinus invade the posterior aspect of the thyroid cartilage, and extend into the soft tissues of the lateral compartment of the neck and the paraglottic space of the true vocal cord. Direct infiltration of the intrinsic laryngeal muscles is rarely seen. By contrast, tumors arising from the medial wall of the pyriform sinuses show early laryngeal invasion and vocal cord fixation in 60% of the patients. They also have a high propensity for contralateral tumor extension with 87%, higher frequency of submucosal tumor spread than the other tumor subtypes with 56%, and perineural tumor invasion in 45% of patients [114–115].

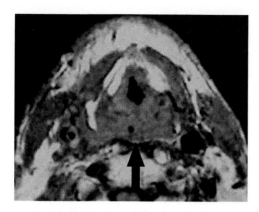

FIGURE 17.17

Postcricoid SCC: Axial T_1-weighted image shows a large hypopharyngeal mass (arrow) that extends posteriorly into the prevertebral fat and anteriorly into the posterior aspect of the larynx.

17.4.3.8 Postcricoid Carcinoma

A rare but well-described syndrome of upper esophageal webs, iron-deficiency anemia, and postcricoid carcinoma is Plummer–Vinson syndrome. Patients are usually Caucasian women between 40 and 70 years of age. Postcricoid carcinoma is relatively uncommon, and is always locally invasive at the time of diagnosis. Intramural fat planes are commonly present in the hypopharyngeal walls, and loss of the planes is seen with tumor. The anteroposterior diameter of the postcricoid region is normally less than 10 mm, and if it is greater than 10 mm, in a patient with hypopharyngeal SCC, it is considered abnormal. Postcricoid region cancers like to invade the posterior aspect of the larynx causing vocal cord paralysis and hoarseness. In addition, these tumors have a high propensity to grow posterior–lateral to involve the pyriform sinuses (100%) and inferiorly to involve the trachea (71%) and/or cervical esophagus (71%) [114,115] (Figure 17.17).

17.4.3.9 Posterior Hypopharyngeal Wall Carcinoma

Posterior hypopharyngeal wall carcinoma extends superiorly to involve the posterior oropharyngeal wall and base of the tongue. When this occurs, the tumor involves a second site: the oropharynx. Inferior extension from the postcricoid hypopharynx to the cervical esophagus is another pattern of spread that by definition involves another subsite: the esophagus. It is critical to identify posterior extension into the retropharyngeal and prevertebral spaces, as invasion of prevertebral fascia is T4b stage [115].

17.4.3.10 Laryngeal Carcinoma

Carcinoma arising in the larynx most often is centered in the glottis (65%), followed by the supraglottis (30%) and the subglottis (5%). The pattern of extension of the

TABLE 17.3

Primary Tumor (T) Staging of the Larynx

Supraglottis
- T1—Tumor limited to one subsite of the supraglottis, with normal vocal cord mobility
- T2—Tumor invades mucosa of more than one subsite of the supraglottis or glottis, without fixation of the larynx
- T3—Tumor limited to the larynx, with vocal cord fixation, and/or invades the postcricoid area, pre-epiglottic space, paraglottic space, and/or inner cortex of thyroid cartilage
- T4a—Tumor invades through the thyroid cartilage and/or tissues beyond the larynx (e.g., trachea, soft tissues of the neck, including deep extrinsic muscle of tongue, strap muscles, thyroid, or esophagus)
- T4b—Tumor invades the prevertebral space, encases the carotid artery, or invades mediastinal structures

Glottis
- T1a—Tumor limited to one vocal cord
- T1b—Tumor involves both cords
- T2—Tumor extends to the supraglottis and/or subglottis, with/without impaired vocal cord mobility
- T3—Tumor limited to the larynx, with cord fixation and/or invasion of paraglottic space and/or thyroid cartilage
- T4a—Tumor invades through the outer cortex of thyroid cartilage and/or invades tissues beyond the larynx
- T4b—Tumor invades the prevertebral space, encases the carotid artery, or invades mediastinal structures

Subglottis
- T1—Tumor limited to the subglottis
- T2—Tumor extends to the vocal cord(s), with normal or impaired mobility
- T3—Tumor limited to the larynx, with vocal cord fixation
- T4a—Tumor invades cricoids or thyroid cartilage and/or invades tissues beyond the larynx (e.g., trachea, soft tissues of the neck, including deep extrinsic muscles of the tongue, strap muscles, thyroid, or esophagus)
- T4b—Tumor invades the prevertebral space, encases the carotid artery, or invades mediastinal structures

Source: Edge, S. et al., *AJCC Cancer Staging Manual*, 7th edn., Springer, Chicago, IL, 2010.

tumor varies according to its subsite within the larynx (Table 17.3) [10,117–119].

17.4.3.11 Supraglottic Carcinoma

Supraglottic tumors often present later and are often large. They may arise from the free edge of epiglottis or from the aryepiglottic fold (Figure 17.18). Imaging is critical for showing the cranial, caudal, and deep extension of supraglottic SCC. Evaluation of the pre-epiglottic (anterior) and para-epiglottic (lateral) fat is key in evaluating supraglottic tumors, as this space is submucosal and not reliably assessed by endoscopy. MR imaging shows enhancing lesion of abnormal signal intensity replacing the fat in these regions. Lesions arising in the epiglottis often invade the pre-epiglottic space and have a propensity to spread to the glottis and subglottis

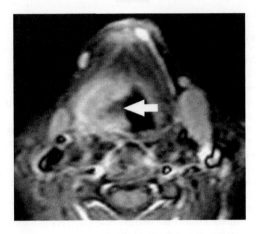

FIGURE 17.18
Aryepiglottic SCC: Axial STIR image shows a hyperintense mass (arrow) arising from the right aryepiglottic fold.

via the anterior commissure (transglottic disease) if the lesion originates in the region of the petiole. Lateral lesions in the false cord, laryngeal ventricle, or aryepiglottic fold tend to infiltrate the paraglottic space in a craniocaudal direction. Superior extension with involvement of the vallecula or base of the tongue is extremely important because it upstages to T_2 and will likely alter management. Supraglottic cancer often presents with nodal metastasis and should prompt a careful search in the regional lymph node areas. Careful inspection of adjacent thyroid cartilage is needed to evaluate cartilaginous involvement [15].

17.4.3.12 Glottic Carcinoma

The key points when staging a glottic primary tumor include arytenoid or thyroid cartilage involvement, transglottic (cranial) or subglottic (caudal) extension, paraglottic and pre-epiglottic extension, and anterior or posterior commissure tumor. Glottic tumors have a propensity to present at an earlier stage, with very small lesions, because they produce hoarseness or airway compromise. Small but symptomatic tumors may be very difficult to detect on imaging and better assessed on laryngoscopy. T1 tumors include those confined to the vocal cord. A tumor is T_2 if there is transglottic or subglottic extension. Once there is hemilarynx fixation and paraglottic or pre-epiglottic space invasion (Figure 17.19), the tumors are T3. In addition, involvement of the anterior commissure is extremely important because these tumors are frequently associated with early cartilage invasion, subglottic extension, and early extralaryngeal extension. Posterior commissure involvement may put the patient at risk for hypopharyngeal extension of tumor. Extension to the anterior or posterior commissure affects survival and choice of treatment [117–119].

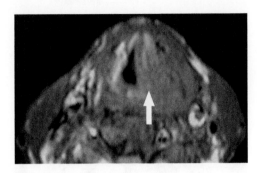

FIGURE 17.19
Glottic SCC: Axial T_1-weighted image shows a well-defined mass (arrow) arising from the left true vocal cord that infiltrates the left paraglottic space.

17.4.3.13 Subglottic Carcinoma

Primary infraglottic tumors are rare but often present late, because they are relatively asymptomatic until large. Staging depends on the extent of involvement of the true vocal cord, fixation of the hemilarynx, whether there is invasion of the cricoid or thyroid cartilage, or extralaryngeal spread below the cricoid cartilage into the cervical trachea [117].

17.4.3.14 Nasopharyngeal Carcinoma

Nasopharyngeal carcinoma is unique in that SCC is a minor subtype, with non-keratinizing undifferentiated carcinoma, followed by keratinizing undifferentiated carcinoma, being more common. Nasopharyngeal carcinoma often begins at the fossa of Rosenmüller. It is important to evaluate this area on MR imaging, especially if there is a mastoid effusion. Subtle abnormalities are brought to the attention of clinicians, as this area is accessible to visual inspection. By contrast, nasopharyngeal carcinoma can often have extensive submucosal extension with minimal mucosal involvement, making imaging critical in the evaluation of this disease. Nasopharyngeal carcinoma has a propensity to spread through the pharyngobasilar fascia into the parapharyngeal space and the retropharyngeal space, which denotes T2 stage. From the parapharyngeal space, it can spread to the masticator and carotid spaces. Involvement of bones at the skull base indicates T3 disease. The most common site of bone involvement is the clivus and petroclival junction, from where tumor can extend intracranially through the foramen lacerum and carotid canal or directly through the bone. Low-resistance paths for intracranial tumor spread at the skull base include the foramen lacerum, jugular, and hypoglossal canals. Intracranial extension indicates T4 disease (Table 17.2). Lymphadenopathy is present in up to 90% of patients, with the retropharyngeal chain often being the first involved nodal site [120–122] (Figure 17.20).

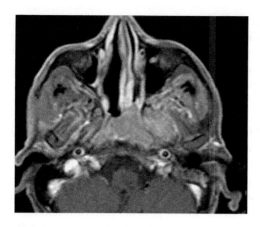

FIGURE 17.20
Nasopharyngeal carcinoma: Axial contrast-enhanced T_1-weighted image shows a large left-sided nasopharyngeal mass with extension into the left parapharyngeal fat. The mass exhibits intense contrast enhancement.

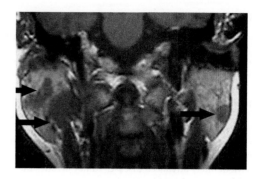

FIGURE 17.21
Warthin's tumor: Coronal T_1-weighted image shows multiple well-defined low signal intensity lesions (arrows) seen in both parotid lobes.

17.4.4 Glandular Tumors

17.4.4.1 Parotid Tumor

The most common benign parotid tumor is pleomorphic adenoma (70%–80%) followed by Warthin's tumors (10%). Malignant tumors form 10%–15% of parotid tumors. In order to determine the most appropriate therapy, routine pre- and postcontrast MR imaging is the best imaging tool to evaluate tumor topography, locoregional extension, and relationship with facial nerve. Diffusion-weighted MR images and dynamic contrast-enhanced sequences are helpful to differentiate benign from malignant parotid tumors [123–125].

Pleomorphic adenoma: It is the most common tumor of the parotid gland, which usually occurs in the middle age (30–50 years), slightly more in females. This tumor is most commonly seen in the superficial lobe of the parotid gland. It shows inhomogeneity of signal with areas of marked hyperintensity on T_2-weighted images. It has a lobulated appearance with a sharply defined fibrous capsule. It exhibits high ADC values on the ADC map, but it may shows areas of low ADC values depending upon the pathological content of the lesions. On dynamic study, it shows that they exhibit low washout of benign tumors [123–125] (Figure 17.2).

Warthin's tumor: It occurs as bilateral masses in 55% and multicentric lesions in 30% in either the ipsilateral or the contralateral gland. It usually occurs in the lower pole or parotid tail. On MR imaging, the lesion is usually sharply defined and heterogeneous low to isointense signal intensity on T_1-weighted images and intermediate to high signal intensity on T_2-weighted images (Figure 17.21). It exhibits low ADC value on the ADC map that is misdiagnosed as a malignant tumor [123,124].

Cancer parotid: Carcinomas of the major salivary glands are composed of a diverse group of histopathologic entities, including at least 20 different histologic subtypes. The most frequent malignancies of the major salivary glands are mucoepidermoid, acinic cell, and adenoid cystic carcinomas. Clinical symptoms suggestive of malignancy are facial nerve paralysis, pain, skin infiltration, a rapidly enlarging lesion, and cervical adenopathy. Prognosis depends on the histologic grade. On MR imaging, cancer parotid shows heterogeneous low signal on T_1- and T_2-weighted images and restricted diffusion on diffusion-weighted MR imaging. High-grade tumors typically are ill-defined and show infiltrative margins (Figure 17.3). Deep extension into the parapharyngeal space and invasion of muscle or bone are most important in prediction of malignancy. MR imaging can delineate the perineural spread of parotid cancer to the skull base via retrograde perineural spread along either cranial nerve VII or the mandibular division of cranial nerve V by way of the auriculotemporal nerve. MR imaging can delineate the intraparotid course of the facial nerve and parotid ducts. Adenoid cystic carcinoma has a particular tendency toward the perineural spread, even with early stage disease. Malignant adenopathy may be present in advanced stages [10–15,123–125]. Table 17.4 shows tumor staging (T) of parotid cancer.

TABLE 17.4

Primary Tumor (T) Staging of Glandular Tumors (Parotid and Thyroid Glands)

Parotid Gland
- T1—Tumor 2 cm or less in greatest dimension without extraparenchymal extension
- T2—Tumor >2 cm but not >4 cm in greatest dimension without extraparenchymal extension
- T3—Tumor >4 cm and/or tumor having extraparenchymal extension
- T4a—Moderately advanced disease. Tumor invades the skin, mandible, ear canal, and/or facial nerve
- T4b—Very advanced disease. Tumor invades the skull base and/or pterygoid plates and/or encases the carotid artery

Thyroid Gland
- T1—Tumor 2 cm or less in greatest dimension limited to the thyroid
- T2—Tumor >2 cm but not >4 cm in greatest dimension limited to the thyroid
- T3—Tumor more than 4 cm in greatest dimension limited to the thyroid or any tumor with minimal extrathyroid extension (e.g., extension to sternothyroid muscle or perithyroid soft tissues)
- T4a—Tumor of any size extending beyond the thyroid capsule to invade the subcutaneous soft tissues, larynx, trachea, esophagus, or recurrent laryngeal nerve
- T4b—Tumor invades the prevertebral fascia or encases the carotid artery or mediastinal vessels
- *All anaplastic carcinomas are considered T4 tumors*
- T4a—Intrathyroidal anaplastic carcinoma
- T4b—Anaplastic carcinoma with extrathyroid extension

Source: Edge, S. et al., *AJCC Cancer Staging Manual*, 7th edn., Springer, Chicago, IL, 2010.

17.4.4.2 Thyroid Tumors

Ultrasound is the most common method for assessment of thyroid gland masses. However, MR imaging is performed for characterization of the thyroid nodule in some atypical cases and extension of thyroid cancer for treatment planning [5–10].

> *Thyroid nodules*: Thyroid nodules occur in up to 50% of adults, whereas *palpable* thyroid nodules occur in only 3%–7%. Malignancy occurs in 5%–7% of all thyroid nodules. Thyroid malignancy has a female predilection. On MR imaging, both benign nodules and malignant tumors of the thyroid gland have a well-defined margin and are often indistinguishable. Most malignant thyroid tumors demonstrate isointense signal to normal thyroid tissue on T_1-weighted images and hyperintense signal on T_2-weighted images, with avid enhancement. Microcalcifications represent psammoma bodies poorly seen on MR imaging. Diffusion-weighted MR imaging shows restricted diffusion with low ADC values of malignant thyroid nodules compared to high ADC values of benign thyroid nodules [126–130] (Figure 17.22).

> *Cancer thyroid*: Papillary carcinoma and follicular carcinoma are tumors of thyroid follicular cells collectively referred to as differentiated thyroid carcinoma. Papillary carcinoma is the most common thyroid malignancy (85%), and often has an indolent clinical course with low mortality and high likelihood for cure. The primary role of imaging in thyroid cancer is to evaluate the extension of tumor beyond the thyroid capsule and to assess nodal metastases. Tumor invasion into the posterior paratracheal tissues and substernal area is better visualized on MR imaging. About 6%–13% of patients with differentiated thyroid cancer have extrathyroid extension, which is associated with an increased incidence of local recurrence, regional and distant metastases, and decreased survival. The most commonly involved structures with invasive thyroid cancer are the strap muscles (53%), the recurrent laryngeal nerve (47%), the trachea (37%), esophagus (21%), and the larynx (12%). Minimal extension of cancer outside the thyroid into the surrounding tissues is reported in about 30% of patients with papillary carcinoma [126–129] (Figure 17.23). Table 17.4 shows tumor staging (T) of thyroid cancer.

17.4.5 Lymphoma

The presence of extranodal involvement of lymphoma in head and neck reflects the increased prevalence of non-Hodgkin lymphoma (25%–40%) over Hodgkin disease (4%–5%). Because both frequently involve widespread, contiguous neck lymphadenopathy, extranodal involvement may be the only clue to suggest non-Hodgkin lymphoma rather than Hodgkin disease. The most common site for extranodal involvement by diffuse large B-cell lymphoma is Waldeyer's ring (i.e., lingual, palatine, and nasopharyngeal tonsils). Despite its prevalence, lymphoma remains second to SSC in head and neck; both may present with widespread nodal and extranodal diseases. The presence of necrosis within nodes and the absence of distant nodal or extranodal disease should suggest SCC rather than lymphoproliferative disease. Infiltrative lesion of lymphoma shows homogeneous low signal on T_1-weighted images and high signal on T_2-weighted images with intense enhancement. The involvement of nodal disease, extranodal lymphatic disease (Waldeyer's ring), and extranodal extralymphatic sites (e.g., sinus, nose, orbit) simultaneously are highly suggestive the diagnosis of lymphoma [131–134] (Figure 17.24).

17.4.6 Soft-Tissue Tumors

Soft-tissue tumors develop from nonepithelial, extraskeletal elements, including adipose tissue, smooth and

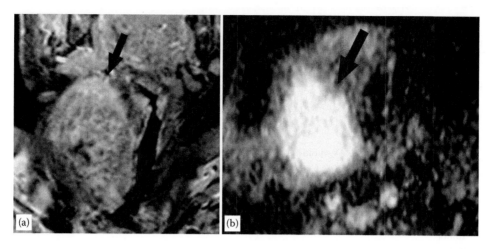

FIGURE 17.22
Thyroid adenoma: (a) Coronal contrast-enhanced T_1-weighted image shows a large nodule (arrow) seen in the right thyroid lobe. (b) ADC map shows unrestricted diffusion of the nodule (arrow) with high ADC value.

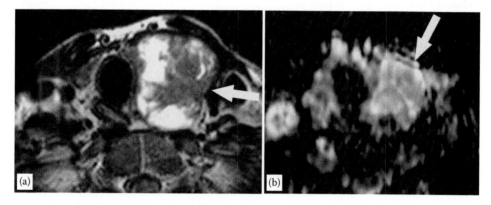

FIGURE 17.23
Cancer thyroid: (a) Axial T_2-weighted image shows a focal nodule (arrow) of mixed signal intensity seen in the left thyroid lobe. (b) ADC map shows restricted diffusion with low ADC value of the nodule (arrow) proved to be cancer thyroid.

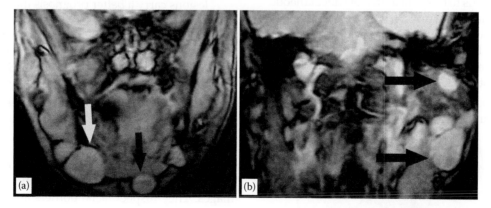

FIGURE 17.24
Lymphoma: Coronal STIR images show bilateral enlarged upper deep (a) and left parotid cervical lymph nodes (b) (arrows). The nodes have well-defined margins with no areas of cystic changes that could be detected.

skeletal muscle, tendon, cartilage, fibrous tissue, blood vessels, and lymphatics. Although soft tissues constitute a large proportion of the human body (12%), soft-tissue tumors account for less than 1% of all tumors. The World Health Organization (WHO) classified tumors of soft tissues into benign, intermediate, and malignant according to their biologic behavior (Table 17.5). The majority of soft-tissue tumors are benign. Routine and advanced MR imaging helps in differentiation of malignant from benign lesions in some cases [135–140].

TABLE 17.5

WHO Classification of Soft-Tissue Tumors of the Neck

Tumor Type	Benign	Intermediate (Locally Aggressive)	Intermediate (Rarely Metastasizing)	Malignant
Adipocytic	Lipoma Lipoblastoma Hibernoma Lipomatosis	Atypical lipomatous tumor/well-differentiated liposarcoma		Liposarcoma
Fibroblastic/ myofibroblastic	Fibromatosis colli Myofibroma Giant cell angiofibroma	Desmoid-type fibromatoses	Solitary fibrous tumor Inflammatory myofibroblastic tumor	Fibrosarcoma
So-called fibrohistiocytic	Benign fibrous histiocytoma Diffuse-type giant cell tumor		Giant cell tumor of soft tissues	Malignant fibrous histiocytoma (undifferentiated pleomorphic sarcoma)
Skeletal muscle	Rhabdomyoma			Rhabdomyosarcoma
Smooth muscle	Leiomyoma Angioleiomyoma			Leiomyosarcoma
Vascular	Hemangioma Lymphangioma	Kaposiform hemangioendothelioma	Kaposi sarcoma	Angiosarcoma
Perivascular	Glomus tumor Myopericytoma			Malignant glomus tumor
Chondro-osseous	Soft-tissue chondroma			Mesenchymal chondrosarcoma Extraskeletal osteosarcoma
Uncertain differentiation	Myxoma		Ossifying fibromyxoid tumor	Synovial sarcoma Alveolar soft part sarcoma primitive neuroectodermal tumor (PNET)/Ewing sarcoma

Source: Razek, A.A. and Huang, B.Y., *RadioGraphics*, 31, 1923–1954, 2011.

17.4.6.1 Benign Soft-Tissue Tumors

Lipomas: On MR imaging, the signal intensity of lipomas tends to be equal to that of subcutaneous fat, including those cases in which MR imaging is obtained with fat suppression. In addition, a surrounding fibrous capsule of low signal intensity on MR imaging similar to that of the muscle may be delineated [135,136] (Figure 17.25).

Lymphangiomas: Lymphatic malformations are most commonly located in the head and neck.

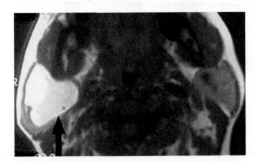

FIGURE 17.25
Lipoma: Axial T_1-weighted image shows a well-defined mass (arrow) located in the right parotid region. The mass has signal intensity similar to subcutaneous fat denoting fatty tissue.

The main division in classifying lymphatic malformations is whether they contain macrocysts (>2 cm), microcysts (<2 cm), or both. Macrocystic lesions carry a better prognosis than their microcystic counterpart. T_1-weighted images show intermediate or hypointense signal, and lymphatic malformations are mostly hyperintense on T_2-weighted images. Lymphatic malformations may be septated. They usually do not show contrast enhancement [136–138] (Figure 17.26).

17.4.6.2 Soft-Tissue Sarcomas

Soft-tissue sarcomas are generally larger than 5 cm and show heterogeneous signal intensity on T_2-weighted images. They respect facial borders and remain within anatomic compartments without infiltrating adjacent structures until late in their development. They show restricted diffusion with low ADC values on diffusion-weighted MR imaging, type C curve on dynamic contrast study, and high Ch/Cr ratio of MRS [135–138].

Rhabdomyosarcomas (RMSs): They are the most common childhood malignancy of soft tissues, accounting for 5%–8% of all childhood cancers.

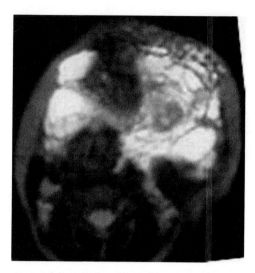

FIGURE 17.26

Lymphangioma: Axial T_2-weighted image shows a multiloculated multicompartmental complex cystic lesion in the anterior part of the neck. The lesion shows hyperintensity with few septations inside.

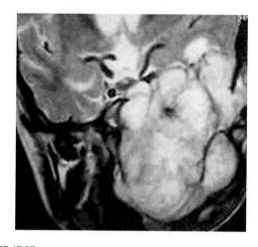

FIGURE 17.27

Rhabdomyosarcoma: Coronal T_2-weighted image shows a large mass in the left masticator space with skull base destruction and intracranial extension. The mass exhibits high signal intensity with hypointense regions of necrotic region.

Approximately 40% of cases of RMSs occur in the head and neck, with locations categorized into parameningeal (50%), non-parameningeal (25%), and orbital (25%) sites. Metastatic cervical adenopathy is seen in 10%–20% of cases, and distant metastases are seen in 15% of cases. The signal intensity of tumors on T_2-weighted images is variable but can occasionally be relatively iso- to hypointense. The embryonal type of RMS may demonstrate hemorrhage, necrosis, and calcifications, resulting in a more heterogeneous enhancement pattern. The botryoid subtype of embryonal RMS may show multiple ring-enhancing areas resembling bunches of grapes. In some alveolar subtypes, serpentine flow voids may be seen [139] (Figure 17.27).

Mesenchymal chondrosarcomas: They are malignant cartilaginous tumors that typically manifest in the fourth and fifth decades of life. The signal intensity of the chondroid matrix is lower than that of the bone matrix on T_1-weighted images. There are hyperintense areas (chondroid tissue) and hypointense areas (calcified regions) on T_2-weighted images. The tumor may show characteristic curvilinear septal and peripheral enhancement of fibrovascular tissue and non-ossified cartilage, a pattern that is described as *ring and arc* [139].

17.4.7 Peripheral Nerve Sheath Tumors

Peripheral nerve sheath tumors include neurofibromas, schwannomas, and malignant peripheral nerve sheath tumors [141–145].

Schwannomas: About 25%–45% of extracranial schwannomas occur in the head and neck region. Schwannomas can arise anywhere from the skull base to the thoracic inlet, but most commonly found in the mid-neck at the carotid space. They appear as well-circumscribed fusiform mass. They demonstrate intermediate signal on T_1-weighted images and hyperintense signal on T_2-weighted images with marked contrast enhancement (Figure 17.28). A large tumor shows a heterogeneous appearance [141,142].

Neurofibromas: They are slowly growing painless tumors and constitute about 5% of all benign soft-tissue tumors. They may be localized, diffuse, or plexiform. Localized form is characterized by heterogeneity on T_2-weighted images, with heterogeneous contrast enhancement. Occasionally, neurofibroma may occasionally exhibit a target pattern of increased peripheral signal intensity and decreased central signal intensity on T_2-weighted images if there is a central fibrous core. Plexiform neurofibromas pass along the nerves in a longitudinal manner and extend along several branches. They most commonly involve branches of the trigeminal nerve and occur in 25%–50% of patients with neurofibromatosis type 1. Malignant transformation occurs in 10% of these patients. On MR imaging, the masses are of low signal intensity on T_1-weighted images and are hyperintense on T_2-weighted images with variable enhancement [142–144] (Figure 17.29).

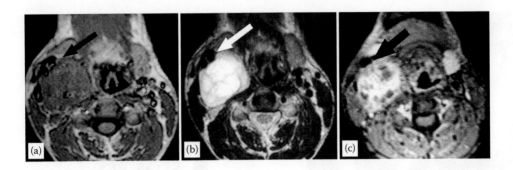

FIGURE 17.28

Schwannoma: (a) Axial T_1-weighted image shows a large inhomogeneous mass seen in the right carotid space that displaces the right carotid artery anteriorly (arrow). (b) Axial T_2-weighted image demonstrates that the mass shows hyperintensity with displaced right carotid artery (arrow). (c) Contrast-enhanced T_1-weighted image demonstrates that the mass shows a heterogonous pattern of contrast enhancement with anterior displaced right carotid artery (arrow).

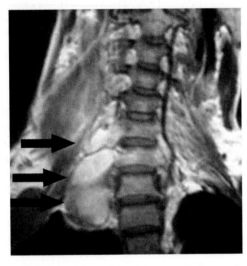

FIGURE 17.29

Neurofibroma: Coronal contrast-enhanced T_1-weighted image shows multiple well-defined fusiform enlarged right spinal nerves (arrows) in neurofibromatosis type I.

Malignant peripheral nerve sheath tumors: Malignant peripheral nerve sheath tumor is a high-grade sarcoma that may arise from the nerves in the head and neck. Hematogenous spread to the lungs, bones, and liver are common in up to a third of patients. MR imaging demonstrates a tubular mass along the course of the nerve. It may be difficult to distinguish malignant from benign counterparts. Large tumor size (≥5 cm), ill-defined infiltrative margins, rapid growth, heterogeneity of signal intensity, and erosion of the skull base foramina out of proportion to tumor size suggest malignancy [141–143].

17.4.8 Neuroendocrine Tumors

Neuroendocrine tumors of the head and neck region are rare tumors divided into two groups: (1) those with epithelial differentiation, including typical carcinoid, atypical carcinoid, and small cell carcinomas, and (2) neutrally derived tumors, including paragangliomas and olfactory neuroblastomas [146].

Paragangliomas: Paragangliomas derived from the extraadrenal paraganglia of the autonomic nervous system are located in the carotid body, vagal nerve, middle ear, and foramen magnum. Glomus tumors are hypervascular benign tumors, which are classified into three types: glomus vagale is derived from the parapharyngeal region, whereas glomus jugulare may extend down from an eroded jugular fossa, and glomus caroticum is typically located in the carotid bifurcation. Glomus tumors show heterogeneous T_1-weighted signal with vascular flow voids, and they are hyperintense on T_2-weighted images. These hypervascular lesions show diffuse intense contrast enhancement. Contrast-enhanced MR angiography is used for preoperative treatment planning [147–150] (Figure 17.30).

Olfactory neuroblastoma: It is a rare malignant neuroectodermal tumor arising from the olfactory epithelium of the olfactory ring of the superior nasal mucosa. A bimodal age distribution in the second and sixth decades of life is reported, although olfactory neuroblastoma may occur at any age. No gender predilection can be found. The tumor is located in the nasal cavity with intracranial extension. It shows an inhomogeneous pattern of enhancement and nonspecific signal intensity on T_1- and T_2-weighted images [146].

17.4.9 Other Neoplasms

Radiation-induced sarcoma: Radiation-induced sarcomas arise 5–10 years in the high-dose field zone after irradiation. They have varied histologies,

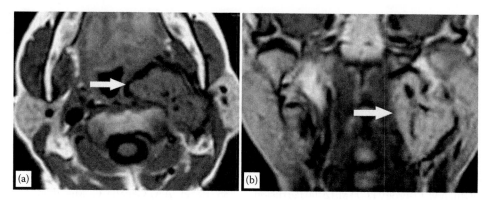

FIGURE 17.30

Paraganglioma: (a) Axial T_1-weighted image shows a well-defined mass (arrow) in the left carotid space. The mass shows multiple signal void regions denoting high vascularity. (b) Coronal contrast-enhanced T_1-weighted image shows an intense contrast enhancement of the mass (arrow) with multiple signal void regions of dilated vessels within the mass.

including osteosarcoma, malignant fibrous histiocytoma, chondrosarcoma, and malignant nerve sheath tumors. The presence of a heterogeneous tumor or rapidly growing large destructive mass that displays different signal intensity from the primary tumor within the radiation field and must occur after a sufficient latency period should suggest the possibility of a radiation-induced sarcoma [151].

Second primary tumor: Patients with head and neck cancer also have increased risk of development of second primary tumors with the majority occurring in the lung or upper aerodigestive tract, including the head and neck area. The risk of developing a second primary tumor in patients with tumors of the upper aerodigestive tract is estimated to be 3%–7% per year [152,153].

17.5 Conclusion

It is concluded that MR imaging is noninvasive imaging modality used for differentiating head and neck cancer from normal tissue, inflammatory lesions, and benign tumors. In addition, MR imaging is important for locoregional tumor, and nodal and distant staging of head and neck cancer. MR imaging is crucial for patient management because it aids in treatment planning, predicts treatment response, differentiates recurrence for postradiation changes, and is used to monitor cancer patients after chemoradiotherapy. In addition, MR imaging has a role in characterization of glandular and soft-tissue tumors of the neck.

References

1. Poorten V (2012) Introduction: Epidemiology, risk factors, pathology, and natural history of head and neck neoplasms. R. Hermans (ed.), *Head and Neck Cancer Imaging*, Medical Radiology: Diagnostic Imaging, 2nd edn., Springer-Verlag, Berlin, Germany.
2. Siegel R, Ma J, Zou Z, Jemal A (2014) Cancer statistics, 2014. *Cancer J Clin* 64:9–29.
3. Vigneswaran N, Williams MD (2014) Epidemiologic trends in head and neck cancer and aids in diagnosis. *Oral Maxillofac Surg Clin North Am* 26:123–1241.
4. Weinstock YE, Alava I 3rd, Dierks EJ (2014) Pitfalls in determining head and neck surgical margins. *Oral Maxillofac Surg Clin North Am* 26:151–162.
5. Shum JW, Dierks EJ (2014) Evaluation and staging of the neck in patients with malignant disease. *Oral Maxillofac Surg Clin North Am* 26:209–221.
6. Argiris A, Karamouzis M, Raben D et al. (2008) Head and neck cancer. *Lancet* 371:1695–1709.
7. Edge S, Byrd D, Compton C et al. (2010) *AJCC Cancer Staging Manual*, 7th edn., Springer, Chicago, IL.
8. Compton CC, Byrd DR, Garcia-Aguilar J et al. (2012) *AJCC Cancer Staging Atlas: A Companion to the Seventh Editions of the AJCC Cancer Staging Manual and Handbook*, 7th edn., Springer-Verlag, Berlin, Germany.
9. Mukherji S, Pillsbury H, Castillo M (1997) Imaging squamous cell carcinomas of the upper aerodigestive tract: What clinicians need to know. *Radiology* 205:629–646.
10. Yousem D, Gad K, Tufano R (2006) Resectability issues with head and neck cancer. *AJNR Am J Neuroradiol* 27:2024–2036.
11. Supsupin EP Jr, Demian NM (2014) Magnetic resonance imaging (MRI) in the diagnosis of head and neck disease. *Oral Maxillofac Surg Clin North Am* 26:253–259.
12. Quon H, Brizel D (2012) Predictive and prognostic role of functional imaging of head and neck squamous cell carcinomas. *Semin Radiat Oncol* 22:220–232.

13. Srinivasan A, Mohan S, Mukherji SK (2012) Biologic imaging of head and neck cancer: The present and the future. *AJNR Am J Neuroradiol* 33:586–594.

14. Shah GV, Wesolowski JR, Ansari SA et al. (2008) New directions in head and neck imaging. *J Surg Oncol* 97:644–648.

15. Walden MJ, Aygun N (2013) Head and neck cancer. *Semin Roentgenol* 48:75–86.

16. Tawfik AM, Razek AA, Elhawary G et al. (2014) Effect of increasing the sampling interval to 2 seconds on the radiation dose and accuracy of CT perfusion of the head and neck. *J Comput Assist Tomogr* 38:469–473.

17. Razek AA, Tawfik AM, Elsorogy LG et al. (2014) Perfusion CT of head and neck cancer. *Eur J Radiol* 83:537–544.

18. Tawfik AM, Razek AA, Elsorogy LG et al. (2011) Perfusion CT of head and neck cancer: Effect of arterial input selection. *AJR Am J Roentgenol* 196:1374–1380.

19. Tawfik AM, Nour-Eldin NE, Naguib NN et al. (2012) CT perfusion measurements of head and neck carcinoma from single section with largest tumor dimensions or average of multiple sections: Agreement between the two methods and effect on intra- and inter-observer agreement. *Eur J Radiol* 81:2692–8196.

20. Tawfik AM, Razek AA, Kerl JM et al. (2014) Comparison of dual-energy CT-derived iodine content and iodine overlay of normal, inflammatory and metastatic squamous cell carcinoma cervical lymph nodes. *Eur Radiol* 24:574–580.

21. Tawfik AM, Kerl JM, Razek AA et al. (2011) Image quality and radiation dose of dual-energy CT of the head and neck compared with a standard 120-kVp acquisition. *AJNR Am J Neuroradiol* 32:1994–1999.

22. Hermans R, De Keyzer F, Vandecaveye V et al. (2012) Imaging techniques. R. Hermans (ed.), *Head and Neck Cancer Imaging*, Medical Radiology: Diagnostic Imaging, 2nd edn., Springer-Verlag, Berlin, Germany.

23. Thoeny HC, De Keyzer F, King AD (2012) Diffusion-weighted MR imaging in the head and neck. *Radiology* 263:19–32.

24. Razek AA (2010) Diffusion-weighted magnetic resonance imaging of head and neck. *J Comput Assist Tomogr* 34:808–815.

25. Schafer J, Srinivasan A, Mukherji S (2011) Diffusion magnetic resonance imaging in the head and neck. *Magn Reson Imaging Clin North Am* 19:55–67.

26. Abdel Razek AA, Gaballa G, Elhawarey G et al. (2009) Characterization of pediatric head and neck masses with diffusion-weighted MR imaging. *Eur Radiol* 19:201–208.

27. Espinoza S, Malinvaud D, Siauve N et al. (2013) Perfusion in ENT imaging. *Diagn Interv Imaging* 94:1225–1240.

28. Furukawa M, Parvathaneni U, Maravilla K et al. (2013) Dynamic contrast-enhanced MR perfusion imaging of head and neck tumors at 3 Tesla. *Head Neck* 35:923–929.

29. Chikui T, Obara M, Simonetti AW et al. (2012) The principal of dynamic contrast enhanced MRI, the method of pharmacokinetic analysis, and its application in the head and neck region. *Int J Dent* 2012:480659.

30. Ai S, Zhu W, Liu Y et al. (2013) Combined DCE- and DW-MRI in diagnosis of benign and malignant tumors of the tongue. *Front Biosci* 18:1098–1111.

31. Razek AA, Elsorogy LG, Soliman NY et al. (2011) Dynamic susceptibility contrast perfusion MR imaging in distinguishing malignant from benign head and neck tumors: A pilot study. *Eur J Radiol* 77:73–79.

32. Abdel Razek AA, Gaballa G (2011) Role of perfusion magnetic resonance imaging in cervical lymphadenopathy. *J Comput Assist Tomogr* 35:21–25.

33. Abdel Razek AA, Poptani H (2013) MR spectroscopy of head and neck cancer. *Eur J Radiol* 82:982–989.

34. Chawla S, Kim S, Loevner LA et al. (2009) Proton and phosphorous MR spectroscopy in squamous cell carcinomas of the head and neck. *Acad Radiol* 16:1366–1372.

35. Razek AA, Gaballa G, Megahed AS et al. (2013) Time resolved imaging of contrast kinetics (TRICKS) MR angiography of arteriovenous malformations of head and neck. *Eur J Radiol* 82:1885–1891.

36. Delaere PR (2012) Clinical and endoscopic examination of the head and neck. R. Hermans (ed.), *Head and Neck Cancer Imaging*, Medical Radiology: Diagnostic Imaging, 2nd edn., Springer-Verlag, Berlin, Germany.

37. Friedrich K, Matzek W, Gentzsch S et al. (2008) Diffusion-weighted magnetic resonance imaging of head and neck squamous cell carcinomas. *Eur J Radiol* 68:493–498.

38. Razek AA, Sieza S, Maha B (2009) Assessment of nasal and paranasal sinus masses by diffusion-weighted MR imaging. *J Neuroradiol* 36:206–211.

39. Sepahdari AR, Politi LS, Aakalu VK et al. (2014) Diffusion-weighted imaging of orbital masses: Multiinstitutional data supports a two ADC threshold model to categorize lesions as benign, malignant, or indeterminate. *AJNR Am J Neuroradiol* 35:170–175.

40. Gonzalez-Beicos A, Nunez D (2012) Imaging of acute head and neck infections. *Radiol Clin North Am* 50:73–83.

41. Rana R, Moonis G (2011) Head and neck infection and inflammation. *Radiol Clin North Am* 49:165–182.

42. Razek AA, Castillo M (2010) Imaging appearance of granulomatous lesions of head and neck. *Eur J Radiol* 76:52–60.

43. Abdel Razek AA, Nada N (2013) Role of diffusion-weighted MRI in differentiation of masticator space malignancy from infection. *Dentomaxillofac Radiol* 42:20120183.

44. Pons Y, Ukkola-Pons E, Clément P et al. (2010) Relevance of 5 different imaging signs in the evaluation of carotid artery invasion by cervical lymphadenopathy in head and neck squamous cell carcinoma. *Oral Surg Oral Med Oral Pathol Oral Radiol Endod* 109:775–778.

45. Rapoport A, Tornin Ode S, Beserra Júnior IM et al. (2008) Assessment of carotid artery invasion by lymph node metastasis from squamous cell carcinoma of aerodigestive tract. *Braz J Otorhinolaryngol* 74:79–84.

46. Cote CR, Goff J, Barry P et al. (2001) The prevalence of occult carotid artery stenosis in patients with head and neck squamous cell carcinoma. *Laryngoscope* 111:2214–2217.

47. Yoo GH, Hocwald E, Korkmaz H et al. (2000) Assessment of carotid artery invasion in patients with head and neck cancer. *Laryngoscope* 110:386–890.

48. Johnston M, Yu E, Kim J (2012) Perineural invasion and spread in head and neck cancer. *Expert Rev Anticancer Ther* 12:359–371.

49. Panizza B, Warren T (2013) Perineural invasion of head and neck skin cancer: Diagnostic and therapeutic implications. *Curr Oncol Rep* 15:128–133.

50. Stambuk HE (2013) Perineural tumor spread involving the central skull base region. *Semin Ultrasound CT MR* 34:445–458.

51. Moonis G, Cunnane MB, Emerick K et al. (2012) Patterns of perineural tumor spread in head and neck cancer. *Magn Reson Imaging Clin North Am* 20:435–446.

52. Razek AA, Castillo M (2009) Imaging lesions of the cavernous sinus. *AJNR Am J Neuroradiol* 30:444–452.

53. Kinshuck AJ, Goodyear PW, Lancaster J et al. (2012). Accuracy of magnetic resonance imaging in diagnosing thyroid cartilage and thyroid gland invasion by squamous cell arcinoma in laryngectomy patients. *J Laryngol Otol* 126:302–306.

54. Becker M, Zbären P, Casselman JW et al. (2008) Neoplastic invasion of laryngeal cartilage: Reassessment of criteria for diagnosis at MR imaging. *Radiology* 249:551–559.

55. Atula T, Markkola A, Leivo I et al. (2001) Cartilage invasion of laryngeal cancer detected by magnetic resonance imaging. *Eur Arch Otorhinolaryngol* 258:272–275.

56. Becker M (2000) Neoplastic invasion of laryngeal cartilage: Radiologic diagnosis and therapeutic implications. *Eur J Radiol* 33:216–229.

57. Taha MS, Hassan O, Amir M et al. (2014) Diffusion-weighted MRI in diagnosing thyroid cartilage invasion in laryngeal carcinoma. *Eur Arch Otorhinolaryngol* 271:2511–2516.

58. Gu DH, Yoon DY, Park CH et al. (2010) CT, MR, (18)F-FDG PET/CT, and their combined use for the assessment of mandibular invasion by squamous cell carcinomas of the oral cavity. *Acta Radiol* 51:1111–1119.

59. Vidiri A, Guerrisi A, Pellini R et al. (2010) Multi-detector row computed tomography (MDCT) and magnetic resonance imaging (MRI) in the evaluation of the mandibular invasion by squamous cell carcinomas (SCC) of the oral cavity. Correlation with pathological data. *J Exp Clin Cancer Res* 29:73.

60. Van Cann EM, Rijpkema M, Heerschap A (2008) Quantitative dynamic contrast-enhanced MRI for the assessment of mandibular invasion by squamous cell carcinoma. *Oral Oncol* 44:1147–1154.

61. Bolzoni A, Cappiello J, Piazza C et al. (2004) Diagnostic accuracy of magnetic resonance imaging in the assessment of mandibular involvement in oral-oropharyngeal squamous cell carcinoma: A prospective study. *Arch Otolaryngol Head Neck Surg* 130:837–843.

62. Hsu WC, Loevner LA, Karpati R et al. (2005) Accuracy of magnetic resonance imaging in predicting absence of fixation of head and neck cancer to the prevertebral space. *Head Neck* 27:95–100.

63. Loevner LA, Ott IL, Yousem DM et al. (1998) Neoplastic fixation to the prevertebral compartment by squamous cell carcinoma of the head and neck. *AJR Am J Roentgenol* 170:1389–1394.

64. Chen B, Yin SK, Zhuang QX et al. (2005) CT and MR imaging for detecting neoplastic invasion of esophageal inlet. *World J Gastroenterol* 11:377–381.

65. Wang JC, Takashima S, Takayama F et al. (2001) Tracheal invasion by thyroid carcinoma: Prediction using MR imaging. *AJR Am J Roentgenol* 177:929–936.

66. Saindane AM (2013) Pitfalls in the staging of cervical lymph node metastasis. *Neuroimaging Clin North Am* 23:147–166.

67. Vandecaveye V, De Keyzer F, Vander Poorten V et al. (2009) Head and neck squamous cell carcinoma: Value of diffusion-weighted MR imaging for nodal staging. *Radiology* 251:134–146.

68. Kimura Y, Sumi M, Sakihama N et al. (2008) MR imaging criteria for the prediction of extranodal spread of metastatic cancer in the neck. *AJNR Am J Neuroradiol* 29:1355–1359.

69. Nakamura T, Sumi M (2007) Nodal imaging in the neck: Recent advances in US, CT and MR imaging of metastatic nodes. *Eur Radiol* 17:1235–1241.

70. Abdel Razek AA, Soliman NY, Elkhamary S et al. (2006) Role of diffusion-weighted MR imaging in cervical lymphadenopathy. *Eur Radiol* 16:1468–1477.

71. Ljumanovic R, Langendijk JA, Hoekstra OS et al. (2006) Distant metastases in head and neck carcinoma: Identification of prognostic groups with MR imaging. *Eur J Radiol* 60:58–66.

72. Noij DP, Boerhout EJ, Pieters-van den Bos IC et al. (2014) Whole-body-MR imaging including DWIBS in the work-up of patients with head and neck squamous cell carcinoma: A feasibility study. *Eur J Radiol* 83:1144–1151.

73. Partovi S, Robbin MR, Steinbach OC et al. (2014) Initial experience of MR/PET in a clinical cancer center. *J Magn Reson Imaging* 39:768–780.

74. Kato H, Kanematsu M, Kawaguchi S et al. (2013) Evaluation of imaging findings differentiating extranodal non-Hodgkin's lymphoma from squamous cell carcinoma in naso- and oropharynx. *Clin Imaging* 37:657–663.

75. Ichikawa Y, Sumi M, Sasaki M et al. (2012) Efficacy of diffusion-weighted imaging for the differentiation between lymphomas and carcinomas of the nasopharynx and oropharynx: Correlations of apparent diffusion coefficients and histologic features. *AJNR Am J Neuroradiol* 33:761–766.

76. Sumi M, Ichikawa Y, Nakamura T (2007) Diagnostic ability of apparent diffusion coefficients for lymphomas and carcinomas in the pharynx. *Eur Radiol* 17:2631–2637.

77. Zhang Y, Chen J, Shen J et al. (2013) Apparent diffusion coefficient values of necrotic and solid portion of lymph nodes: Differential diagnostic value in cervical lymphadenopathy. *Clin Radiol* 68:224–231.

78. Kato H, Kanematsu M, Kato Z et al. (2013) Necrotic cervical nodes: Usefulness of diffusion-weighted MR imaging in the differentiation of suppurative lymphadenitis from malignancy. *Eur J Radiol* 82:e28–e35.

79. Koç O, Paksoy Y, Erayman I et al. (2007) Role of diffusion weighted MR in the discrimination diagnosis of the cystic and/or necrotic head and neck lesions. *Eur J Radiol* 62:205–213.

80. Razek AA, Elkhamary S, Mousa A (2011) Differentiation between benign and malignant orbital tumors at 3-T diffusion MR-imaging. *Neuroradiology* 53:517–522.

81. Abdel Razek A, Mossad A, Ghonim M (2011) Role of diffusion-weighted MR imaging in assessing malignant versus benign skull-base lesions. *Radiol Med* 116:125–132.

82. Razek AA, Megahed AS, Denewer A et al. (2008) Role of diffusion-weighted magnetic resonance imaging in differentiation between the viable and necrotic parts of head and neck tumors. *Acta Radiol* 49:364–370.

83. Strojan P, Ferlito A, Medina JE et al. (2013) Contemporary management of lymph node metastases from an unknown primary to the neck: I. A review of diagnostic approaches. *Head Neck* 35:123–132.

84. Donta TS, Smoker WR (2007) Head and neck cancer: Carcinoma of unknown primary. *Top Magn Reson Imaging* 18:281–292.

85. Park GC, Jung JH, Roh JL et al. (2014) Prognostic value of metastatic nodal volume and lymph node ratio in patients with cervical lymph node metastases from an unknown primary tumor. *Oncology* 86:170–176.

86. Dragan AD, Nixon IJ, Guerrero-Urbano MT et al. (2014) Selective neck dissection as a therapeutic option in management of squamous cell carcinoma of unknown primary. *Eur Arch Otorhinolaryngol* 27:1249–1256.

87. Bree RD, Takes RP, Castelijns JA et al. (2014) Advances in diagnostic modalities to detect occult lymph node metastases in head and neck squamous cell carcinoma. *Head Neck.* doi: 10.1002/hed.23814.

88. Chong VH (2007) Tumour volume measurement in head and neck cancer. *Cancer Imaging* 7:S47–S49.

89. Daisne JF, Duprez T, Weynand B et al. (2004) Tumor volume in pharyngolaryngeal squamous cell carcinoma: Comparison at CT, MR imaging, and FDG PET and validation with surgical specimen. *Radiology* 233:93–100.

90. Gordon A, Loevner L, Shukla-Dave A et al. (2004) Intraobserver variability in the MR determination of tumor volume in squamous cell carcinoma of the pharynx. *AJNR Am J Neuroradiol* 25:1092–1098.

91. Ahmed M, Schmidt M, Sohaib A et al. (2010) The value of magnetic resonance imaging in target volume delineation of base of tongue tumours—A study using flexible surface coils. *Radiotherapy Oncol* 94:161–167.

92. Abdel Razek AA, Kamal E (2013) Nasopharyngeal carcinoma: Correlation of apparent diffusion coefficient value with prognostic parameters. *Radiol Med* 118:534–539.

93. Abdel Razek AA, Elkhamary S, Al-Mesfer S et al. (2012) Correlation of apparent diffusion coefficient at 3T with prognostic parameters of retinoblastoma. *AJNR Am J Neuroradiol* 32:944–948.

94. Okura M, Iida S, Aikawa T et al. (2008) Tumor thickness and paralingual distance of coronal MR imaging predicts cervical node metastases in oral tongue carcinoma. *AJNR Am J Neuroradiol* 29:45–50.

95. Park JO, Jung SL, Joo YH et al. (2011) Diagnostic accuracy of magnetic resonance imaging (MRI) in the assessment of tumor invasion depth in oral/oropharyngeal cancer. *Oral Oncol* 47:381–386.

96. Hermans R (2006) Head and neck cancer: How imaging predicts treatment outcome. *Cancer Imaging* 6:S145–S153.

97. Hong J, Yao Y, Zhang Y et al. (2013) Value of magnetic resonance diffusion-weighted imaging for the prediction of radiosensitivity in nasopharyngeal carcinoma. *Otolaryngol Head Neck Surg* 149:707–713.

98. King AD, Chow KK, Yu KH et al. (2013) Head and neck squamous cell carcinoma: Diagnostic performance of diffusion-weighted MR imaging for the prediction of treatment response. *Radiology* 266:531–538.

99. Chawla S, Kim S, Dougherty L et al. (2013) Pretreatment diffusion-weighted and dynamic contrast-enhanced MRI for prediction of local treatment response in squamous cell carcinomas of the head and neck. *AJR Am J Roentgenol* 200:35–43.

100. Srinivasan A, Chenevert TL, Dwamena BA et al. (2012) Utility of pretreatment mean apparent diffusion coefficient and apparent diffusion coefficient histograms in prediction of outcome to chemoradiation in head and neck squamous cell carcinoma. *J Comput Assist Tomogr* 36:131–137.

101. Bhatnagar P, Subesinghe M, Patel C et al. (2013) Functional imaging for radiation treatment planning, response assessment, and adaptive therapy in head and neck cancer. *RadioGraphics* 33:1909–1929.

102. Lin GW, Wang LX, Ji M et al. (2013) The use of MR imaging to detect residual versus recurrent nasopharyngeal carcinoma following treatment with radiation therapy. *Eur J Radiol* 82:2240–2246.

103. Abdel Razek AA, Kandeel AY, Soliman N et al. (2007) Role of diffusion-weighted echo-planar MR imaging in differentiation of residual or recurrent head and neck tumors and posttreatment changes. *AJNR Am J Neuroradiol* 28:1146–1152.

104. King AD, Mo FK, Yu KH et al. (2010) Squamous cell carcinoma of the head and neck: Diffusion-weighted MR imaging for prediction and monitoring of treatment response. *Eur Radiol* 20:2213–2220.

105. Galbán C, Mukherji S, Chenevert T et al. (2009) A feasibility study of parametric response map analysis of diffusion-weighted magnetic resonance imaging scans of head and neck cancer patients for providing early detection of therapeutic efficacy. *Transl Oncol* 2:184–190.

106. Kim S, Loevner L, Quon H et al. (2009) Diffusion-weighted magnetic resonance imaging for predicting and detecting early response to chemoradiation therapy of squamous cell carcinomas of the head and neck. *Clin Cancer Res* 15:986–994.

107. Aiken A (2013) Pitfalls in the staging of cancer of oral cavity cancer. *Neuroimaging Clin North Am* 23:27–45.

108. Hagiwara M, Nusbaum A, Schmidt BL (2012) MR assessment of oral cavity carcinomas. *Magn Reson Imaging Clin North Am* 20:473–494.

109. Keberle M (2012) Neoplasms of the oral cavity. R. Hermans (ed.), *Head and Neck Cancer Imaging*, Medical Radiology: Diagnostic Imaging, 2nd edn., Springer-Verlag, Berlin, Germany.

110. Kirsch C (2007) Oral cavity cancer. *Top Magn Reson Imaging* 18:269–280.

111. Stambuk HE, Karimi S, Lee N et al. (2007) Oral cavity and oropharynx tumors. *Radiol Clin North Am* 45:1–20.

112. King KG, Kositwattanarerk A, Genden E et al. (2011) Cancers of the oral cavity and oropharynx: FDG PET with contrast-enhanced CT in the posttreatment setting. *RadioGraphics* 31:355–373.

113. Corey A (2013) Pitfalls in the staging of cancer of the oropharyngeal squamous cell carcinoma. *Neuroimaging Clin North Am* 23:47–66.

114. Chen AY, Hudgins PA (2013) Pitfalls in the staging squamous cell carcinoma of the hypopharynx. *Neuroimaging Clin North Am* 23:67–79.

115. Wycliffe ND, Grover RS, Kim PD et al. (2007) Hypopharyngeal cancer. *Top Magn Reson Imaging* 18:243–258.

116. Becker M, Burkhardt K, Dulguerov P et al. (2008) Imaging of the larynx and hypopharynx. *Eur J Radiol* 66:460–479.

117. Baugnon KL, Beitler JJ (2013) Pitfalls in the staging of cancer of the laryngeal squamous cell carcinoma. *Neuroimaging Clin North Am* 23:81–105.

118. Banko B, Dukić V, Milovanović J et al. (2011) Diagnostic significance of magnetic resonance imaging in preoperative evaluation of patients with laryngeal tumors. *Eur Arch Otorhinolaryngol* 268:1617–1623.

119. Blitz AM, Aygun N (2008) Radiologic evaluation of larynx cancer. *Otolaryngol Clin North Am* 41:697–713.

120. Lai V, Khong PL (2014) Updates on MR imaging and (18)F-FDG PET/CT imaging in nasopharyngeal carcinoma. *Oral Oncol* 50:539–548.

121. Abdel Khalek Abdel Razek A, King A (2012) MRI and CT of nasopharyngeal carcinoma. *AJR Am J Roentgenol* 198:11–118.

122. Glastonbury C, Salzman L (2013) Pitfalls in the staging of cancer of nasopharyngeal carcinoma. *Neuroimaging Clin North Am* 23:9–25.

123. Friedman ER, Saindane AM (2013) Pitfalls in the staging of cancer of the major salivary gland neoplasms. *Neuroimag Clin North Am* 23:107–122.

124. Christe A, Waldherr C, Hallett R et al. (2011) MR imaging of parotid tumors: Typical lesion characteristics in MR imaging improve discrimination between benign and malignant disease. *AJNR Am J Neuroradiol* 32:1202–1207.

125. Dubrulle F, Souillard-Scemama R (2012) Parotid gland and other salivary gland tumors. R. Hermans (ed.), *Head and Neck Cancer Imaging*, Medical Radiology: Diagnostic Imaging, 2nd edn., Springer-Verlag, Berlin, Germany.

126. Nachiappan AC, Metwalli ZA, Hailey BS et al. (2014) The thyroid: Review of imaging features and biopsy techniques with radiologic-pathologic correlation. *RadioGraphics* 34:276–293.

127. Abdel Razek AK, Sadek A, Ombar O et al. (2008) Role of apparent diffusion coefficient value in differentiation between malignant and benign solitary thyroid nodule. *AJNR Am J Neuroradiol* 29:563–568.

128. Aiken AH (2012) Imaging of thyroid cancer. *Semin Ultrasound CT MR* 33:138–149.

129. Saindane AM (2013) Pitfalls in the staging of cancer of thyroid. *Neuroimaging Clin North Am* 23:123–145.

130. Takashima S, Takayama F, Wang J et al. (2003) Using MR imaging to predict invasion of the recurrent laryngeal nerve by thyroid carcinoma. *AJR Am J Roentgenol* 180:837–842.

131. Martínez Barbero JP, Rodríquez Jiménez I, Noguerol TM et al. (2013) Utility of MRI diffusion techniques in the evaluation of tumors of the head and neck. *Cancers* 5:875–889.

132. Thomas AG, Vaidhyanath R, Kirke R et al. (2011) Extranodal lymphoma from head to toe: Part 1, the head and spine. *AJR Am J Roentgenol* 197:350–356.

133. Aiken AH, Glastonbury C (2008) Imaging Hodgkin and non-Hodgkin lymphoma in the head and neck. *Radiol Clin North Am* 46:363–378.

134. Chua SC, Rozalli FI, O'Connor SR (2009) Imaging features of primary extranodal lymphomas. *Clin Radiol* 64:574–588.

135. Stramare R, Beltrame V, Gazzola M et al. (2013) Imaging of soft-tissue tumors. *J Magn Reson Imaging* 37:791–804.

136. Razek AA, Huang BY (2011) Soft tissue tumors of the head and neck: Imaging-based review of the WHO classification. *RadioGraphics* 31:1923–1954.

137. Vilanova JC, Woertler K, Narvaez JA et al. (2007) Soft-tissue tumors update: MR imaging features according to the WHO classification. *Eur Radiol* 17:125–138.

138. Wu J, Hochman M (2009) Soft-tissue tumors and tumor-like lesions: A systematic imaging approach. *Radiology* 253:297–316.

139. Abdel Razek AA (2014) Computed tomography and magnetic resonance imaging of lesions at masticator space. *Jpn J Radiol* 32:123–137.

140. Abdel Razek A, Nada N, Ghaniem M et al. (2012) Assessment of soft tissue tumors of the extremities with diffusion echo-planar MR Imaging. *Radiol Med* 117:96–101.

141. Ahlawat S, Chhabra A, Blakely J (2014) Magnetic resonance neurography of peripheral nerve tumors and tumorlike conditions. *Neuroimaging Clin North Am* 24:171–192.

142. Chee DW, Peh WC, Shek TW (2011) Pictorial essay: Imaging of peripheral nerve sheath tumours. *Can Assoc Radiol J* 62:176–182.

143. Abreu E, Aubert S, Wavreille G et al. (2013) Peripheral tumor and tumor-like neurogenic lesions. *Eur J Radiol* 82:38–50.

144. Woertler K (2010) Tumors and tumor-like lesions of peripheral nerves. *Semin Musculoskelet Radiol* 14:547–58.

145. Kamal A, Abd El-Fattah AM, Tawfik A et al. (2007) Cervical sympathetic schwannoma with postoperative first bite syndrome. *Eur Arch Otorhinolaryngol* 264:1109–1111.

146. Subedi N, Prestwich R, Chowdhury F et al. (2013) Neuroendocrine tumours of the head and neck: Anatomical, functional and molecular imaging and contemporary management. *Cancer Imaging* 13:407–422.

147. van den Berg R (2005) Imaging and management of head and neck paragangliomas. *Eur Radiol* 15:1310–1318.

148. Ferré JC, Brunet JF, Carsin-Nicol B et al. (2010) Optimized time-resolved 3D contrast-enhanced MRA at 3T: Appreciating the feasibility of assessing cervical paragangliomas. *J Neuroradiol* 37:104–108.

149. Noujaim SE, Pattekar MA, Cacciarelli A et al. (2002) Paraganglioma of the temporal bone: Role of magnetic resonance imaging versus computed tomography. *Top Magn Reson Imaging* 11:108–122.

150. Rao AB, Koeller KK, Adair CF (2002) Paragangliomas of the head and neck: Radiologic–pathologic correlation. *RadioGraphics* 19:1605–1632.

151. Abrigo J, King A, Leung S et al. (2009) MRI of radiation-induced tumors of the head and neck in post-radiation nasopharyngeal carcinoma. *Eur Radiol* 19:1197–1205.

152. Oeffinger KC, Baxi SS, Novetsky Friedman D et al. (2013) Solid tumor second primary neoplasms: Who is at risk, what can we do? *Semin Oncol* 40:676–689.

153. León X, Pedemonte G, García J et al. (2014) Elective treatment of the neck for second primary tumors of the head and neck. *Eur Arch Otorhinolaryngol* 271:1187–1190.

18

Lymph Nodes of the Neck

Fatih Alper, Irmak Durur-Subasi, and Adem Karaman

CONTENTS

18.1 Introduction

The neck is located between the mandible and the clavicle. Its borders are the mandible and mylohyoid muscles anterosuperiorly, the base of the skull posterosuperiorly, and the scapulae and the thoracic inlet (sternum, first ribs, and first thoracic vertebra) inferiorly [1]. The neck encloses vital structures and a lot of lymph nodes (LNs) within a limited area. Therefore, it is essential that physicians can easily orient themselves via imaging. It also has a complex structural organization, and as such, this region has the potential to mask certain pathological conditions. Still most patients require an early and accurate diagnosis.

There are more than 800 LNs in the human body, and about 40% are located in the head and neck [2]. The neck is also the region that tends to be most responsive to antigen stimulation [3]. Many painfully enlarged LNs do not necessitate advanced imaging because they may be evidence of the typical clinical course of upper respiratory tract infections. However, the main pathway for the metastasis of head and neck carcinomas is lymphatic [2].

Cases of certain pathological conditions demand advanced evaluation or imaging. If a patient presents with a palpable mass, for example, it is essential to establish whether it is an enlarged LN. Magnetic resonance imaging (MRI) serves as a significant tool for the evaluation of diseases related to the neck. Like computed tomography (CT), MRI has the advantage of cross-sectional imaging. In addition, multiplanar MRI capability allows for the simple differentiation of LNs, masses, vessels, and other soft tissue components [4]. However, MRI is superior to CT in that its contrast resolution for soft tissues is high and it does not necessitate iodinated contrast material administration. Thus, another advantage of MRI is that the patient is not exposed to ionizing radiation, which is especially important for patients who may need multiple or follow-up scans. Another advantage of MRI is the protection of the thyroid gland from radiation, especially during childhood. However, one drawback is that smaller children often cannot cooperate by keeping still for the duration of the scan, and general anesthesia or sedation may be necessary to prevent motion artifacts [4].

When a pathologic LN is identified, physicians must establish whether it is a benign/inflamed or a malignant/metastatic node [3]. The number and location are important as well. A single LN metastasis from a squamous cell carcinoma (SCC) of the head and neck has been reported to have a 50% 5-year survival rate, and the involvement of an additional contralateral node reduces the rate to as low as 33% [5]. Imaging serves as an important tool for the evaluation of cervical LNs because the number, level, size, and character of the nodal involvement, as well as extranodal spread and necrosis, are all crucial prognostic features [2]. In addition, the therapeutic response must be evaluated [3] to establish whether there is a need for reevaluation, a different MRI sequence acquisition, or a follow-up on any intervention.

In fact, MRI is usually preferred in complicated conditions or situations requiring differential diagnosis. One of the major objectives of MRI evaluation of the neck is the assessment of a primary tumor, which is most often related to the skin, salivary glands, thyroid, or mucous membranes. The extent and spreading of the disease, which define the appropriate treatment, are important prognostic criteria for head and neck carcinomas. During imaging for the evaluation of known malignancies, other aims include the differential diagnosis of lymphadenopathy and the determination of nodal status. These features can reflect the aggressiveness of the primary tumor.

In such a complex anatomical region and with the difficulties inherent in the evaluation of LNs, understanding their status may be perplexing for the radiologist. The anatomy of the neck can be ominous for both radiologists and clinicians. This chapter on the MRI of neck nodes reviews the normal lymphatic nodal anatomy of the neck, LN levels, pathologic LNs, imaging techniques, and diagnostic issues. Special attention is given to the features of pathologic LNs, MRI protocols, tools, and interpretation.

18.2 Normal Nodal Anatomy of the Neck

18.2.1 Benign LNs

LNs are almost always encountered during routine MRIs of the neck, and it is very rare to see no visible LNs in the neck. Aside from malignancies, many other issues can cause lymphadenopathy in the head and neck region, including infectious, inflammatory, and autoimmune

diseases [6]. For instance, lymphadenopathies due to viral, bacterial, protozoal, or fungal infections; reactive conditions; lymphoproliferative disorders; vascular pathologies; histiocytic disorders; and clinical syndromes such as sarcoidosis, Castleman disease, and Kimura disease may be found and require imaging evaluation. However, the imaging features of all the above diseases are not specific, and their diagnoses rely on the information obtained from the history, clinical and imaging findings, and laboratory data of the patient. Thus, the role of imaging is to locate the node disease and to evaluate any associated findings, such as necrosis and/or abscess formation and soft tissue infiltration [7].

Normal LNs are generally oval- or bean-shaped (Figure 18.1). A slight depression is seen on the side of the hilum, through which arterioles, venules, and efferent lymphatic vessels pass [3]. An LN is surrounded by a fibrous capsule. Fibrous trabeculae or septa spread out from the capsule toward the center of the node and separate the cortical nodules that are located in the sinuses. Sinuses and cords are contained in the medulla. The cortical or medullary sinuses of an LN are furnished with specified reticuloendothelial cells. The lymph fluid and antigens move to a node by means of numerous lymphatic vessels and emerge via one or two efferent vessels. When LNs are stimulated by an immune reaction, they enlarge (reactive hyperplasia); however, a morphological analysis of this reactivity may not be of prognostic value, because each patient and each node can have different response patterns [8–10]. For example, those draining common infection regions (such as the submental and jugulodigastric nodes) tend to enlarge more relative to other neck nodes.

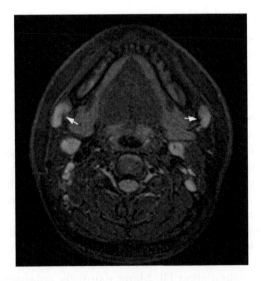

FIGURE 18.1
Transverse STIR image demonstrates benign LNs (arrows) with bean shape.

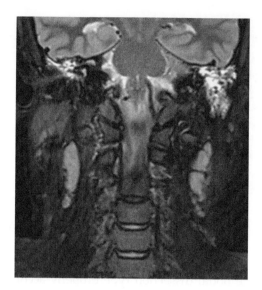

FIGURE 18.2
Coronal T_2-WI demonstrates LNs parallel to skin and vascular structures.

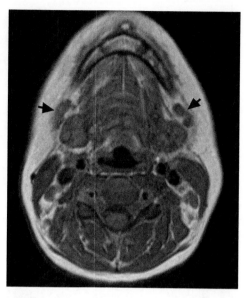

FIGURE 18.3
Transverse T_1-WI demonstrates level Ib LNs (arrows) which are isointense to muscles.

The long axis of normal LNs is directed parallel to the skin and the major blood vessels, along which the nodes are distributed (Figure 18.2). The normal transverse diameter of a node is approximately 10 mm, and the normal maximum longitudinal diameter is 10–15 mm in sagittal or coronal images. The proportion of the maximal diameter to the minimal diameter of an LN is one of the main morphological criteria for metastatic involvement [11]. However, the minimum diameter more closely relates to the volume of the LN, and generally, the scan plane does not affect it. Therefore, the minimum diameter may be the most suitable measurement for use in follow-up examinations [11].

On MRI, normal LNs usually have a similar intensity to muscle on T_1-weighted images (WIs) (Figure 18.3). The hilum may have a small amount of fat. Axillary and inguinal LNs contain more amount of fat than axillary LNs (Figure 18.4). However, a fatty infiltration at the medulla of the LN due to fibrolipomatous degeneration is seen as bright on T_1-WIs. The cortex is hyperintense to the muscle on T_2-WIs and is isointense or hypointense to fat (Figure 18.5). T_2-WIs obtained with fat suppression and short-tau inversion recovery (STIR) show markedly hyperintense LN cortices (Figure 18.6). On enhanced images, the cortex is reniform and homogeneous, and exhibits a slight contrast enhancement (Figure 18.7). The margins of normal LNs are smooth and noninfiltrating. High-resolution MR microimaging allows the detailed structural investigation of the nodes. The medullary sinus and follicles can be evaluated via microscopy coils. This technique, without using contrast material, can effectively characterize benign and malignant

nodes; using this method, the medullary sinuses are seen as hypointense and the follicles are seen as hyperintense [9]. However, microimaging can only be used for superficial nodes and the scan time is longer, as well as requires more time for the evaluation of the imaging data [10].

18.2.2 LN Levels

Among patients with head and neck carcinoma, no less than 80% undergo sectional imaging before treatment. Imaging can reveal clinically silent nodes [12–18]. Smart imaging can provide exact regional or level information. Although the clinical and imaging classifications are designed independently, it is the best to merge the information obtained from both clinical palpation and imaging. LNs that are located at borders between levels may be difficult to classify with imaging. However, the multiplanar imaging capability of MRI can overcome this problem.

For neck LNs, the most used system was developed by Rouvière in 1938 [19]. These and similar classification systems mainly depended on the anatomy of the LNs [20,21]. The landmarks of these anatomic classification systems were identified via palpation. Many of the landmarks were based on classical anatomical triangles of the neck. These were easy to discover and label.

An LN-level-based system, instead of the anatomically based terminology, was first suggested in 1981 [3]. Next, various classifications based on zone, region, or level terminology have been suggested [22–30]. The aim of these more recent classifications was more functional than anatomic. This has allowed the surgeon to

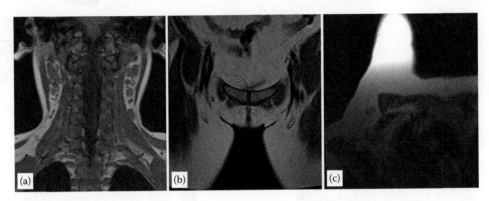

FIGURE 18.4
(a) Coronal T_1-WI of the neck shows LN with lesser hilar fat. However, coronal T_1-WI of the inguinal region (b) and transverse T_1-WI of the axilla (c) demonstrate LNs with more hilar fat tissue.

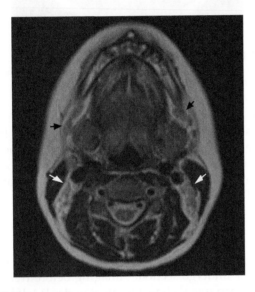

FIGURE 18.5
Transverse T_2-WI demonstrates level Ib (black arrows) and level IIb (white arrows) LNs which are hyperintense to muscles.

select the nodal dissection type properly according to the involved nodes. Cervical node levels are fundamentally based on the pathophysiologic observation of cancer spread [3]. The newest and most acknowledged ones are based on the suggestions of the American Joint Committee on Cancer and the American Academy of Otolaryngology—Head and Neck Surgery [27,28,31].

In sum, level I nodes consist of the submandibular and submental groups, level II includes the upper internal jugular nodes, level III includes the middle internal jugular nodes, and level IV includes the lower internal jugular nodes. Level V nodes are located in the posterior triangle (spinal accessory chain); level VI nodes are the pretracheal, paratracheal, and prelaryngeal nodes; and level VII nodes are the upper mediastinal nodes (Figures 18.8 through 18.17). A comprehensive imaging-based system is shown in Table 18.1.

The nodal groups, including supraclavicular, retropharyngeal, parotid, facial, occipital, postauricular, and the other superficial nodes, are mentioned via their previous anatomic names [3] (Figures 18.18 through 18.23).

18.3 Pathologic LNs

First of all, LNs that are enlarged, round, conglomerating, hyperintense on STIR, diffusion restricting, unenhancing with ultrasmall superparamagnetic iron oxide (USPIO) or necrotic, or have irregular or indistinct margins or focal cortical thickening are considered pathologic (Table 18.2).

A number of cross-sectional imaging criteria have been proposed to determine the nodal involvement in the neck and to differentiate such nodes from reactive nodes [13–18,22]. In addition, in many studies [32–48], the nodal size and the presence of central nodal necrosis (or nodal inhomogeneity) have been found to be the two

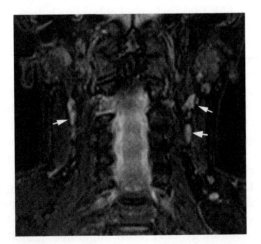

FIGURE 18.6
Coronal STIR image shows very hyperintense LNs (arrows).

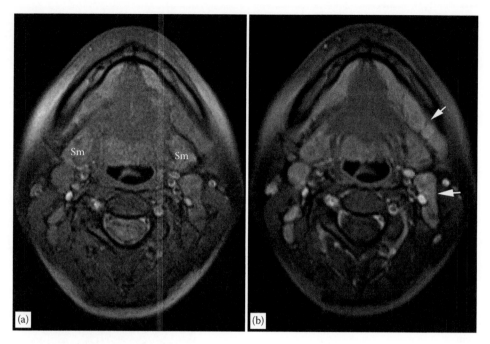

FIGURE 18.7
Precontrast (a) and postcontrast (b) fat-saturated transverse T_1-WIs show level Ib, IIa, IIb, and Va LNs. These reactive appearing LNs have moderate and homogeneous contrast enhancement.

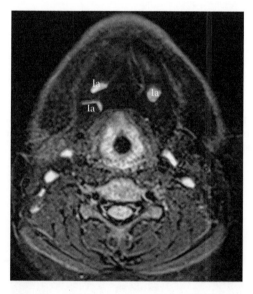

FIGURE 18.8
Transverse STIR image shows level Ia LNs at the blue-colored area.

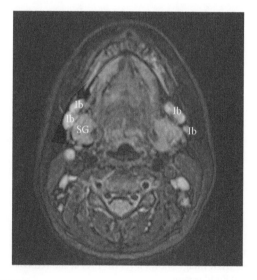

FIGURE 18.9
Transverse STIR image shows level Ib LNs at the blue-colored area. SG, submandibular gland.

major imaging criteria. If there is an LN that is found to be highly suspicious in terms of metastasis, the surgical plan must be modified to include it. In this section, the criteria for a suspicious LN in terms of malignant involvement are discussed.

The involved LNs may have various appearances. They may appear like hyperplastic nodes; show

enhancement; contain small, dystrophic, scattered calcifications; contain complete cavitation, like a benign cyst; or include the areas of hemorrhage [49]. On MRI, both low-to-intermediate T_1-weighted and high T_2-weighted signal intensities can be seen. Enhancement is usually moderate or intense. Some important imaging features of LNs are discussed as well.

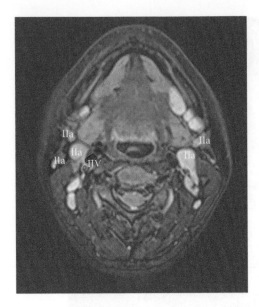

FIGURE 18.10
Transverse STIR image shows level IIa LNs at the red-colored area.
IJV, internal jugular vein.

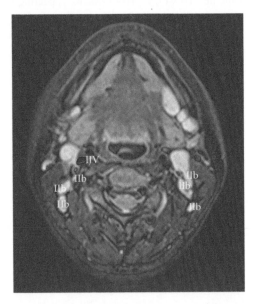

FIGURE 18.11
Transverse STIR image shows level IIb LNs at the red-colored area.
IJV, internal jugular vein.

18.3.1 Size and Shape

Size is the most frequently used criterion for diagnosis. It is based mainly on the measurement of nodal dimensions, such as maximum transverse diameter [11,15–17,50–52] and the ratio of the maximum longitudinal diameter to the maximum transverse diameter [15,53] (Table 18.3). These measures were first described by clinicians. The greatest nodal diameter is typically used by many radiologists. In fact, using the most comfortable

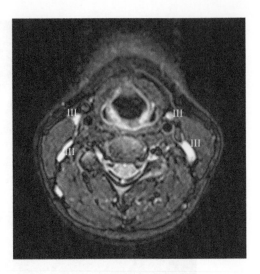

FIGURE 18.12
Transverse STIR image shows level III LNs at the orange-colored area.

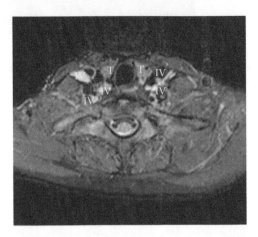

FIGURE 18.13
Transverse STIR image shows level IV LNs at the green-colored area.
T, thyroid gland.

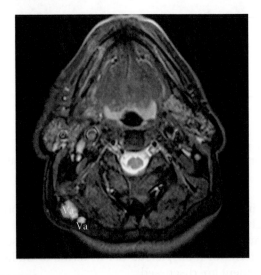

FIGURE 18.14
Transverse STIR image shows right level Va LNs at the yellow-colored area.

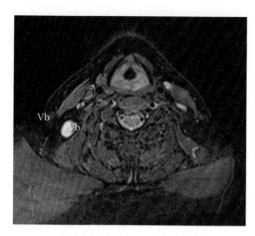

FIGURE 18.15
Transverse STIR image shows right level Vb LNs at the yellow-colored area.

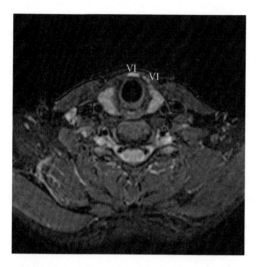

FIGURE 18.16
Transverse STIR image shows pretracheal level VI LNs at the green-colored area.

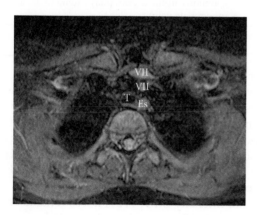

FIGURE 18.17
Transverse STIR image shows level VII LNs at the pink-colored area. Es, esophagus; T, trachea.

method that is also used by the clinician the radiologist works with is also recommended.

In fact, several size criteria for the detection of neck LN disease have been reported, with each cutoff value varying among the studies and constituting a compromise between sensitivity and specificity. The recommendations for a cutoff value between benign and malignant nodes range from 6 to 30 mm. The criterion of a 10 mm short-axis diameter, as proposed by van den Brekel et al. [13], has gained widespread acceptance. Many institutions still use 10 mm for differentiation between benign and malignant nodes. However, when the cutoff value was considered to be 15 mm for levels I–II and 10 mm for others levels, for about one-third cases, over- or underestimates were encountered. Many studies have been conducted to estimate the maximum transverse diameter. It was proposed that an LN exceeding 11 mm in level II or 10 mm in the other regions must be considered suspicious in terms of malignant involvement [12–18,22]. Aside from the retropharyngeal region, 8 mm for the maximum and 5 mm for the minimum diameters were considered as indicating malignant involvement, given sectional imaging criteria [22,44].

The importance of the ratio of the maximum longitudinal diameter to the maximum transverse diameter has also been studied [6]. If this ratio is under 2, the possibility of malignant involvement is increased. Decrease of this ratio leads to the shape of the LN becoming rounder and losing its bean-like appearance. By contrast, normal hyperplastic LNs have a ratio of over 2 (Figure 18.24).

When the size criterion is supported by the number criterion, it is more valuable. When a group of three or more LNs with maximal diameters of 8–15 mm or minimum axial diameters of 9–10 mm in the jugulodigastric region and 8–9 mm in the rest of the neck are seen, metastatic lymphadenopathy must be considered. They may drain the primary tumor site [49].

The size criteria described are valid for homogeneous, sharply outlined LNs. Although they are the most frequently used criteria, sensitivity and specificity vary widely. False-positive and false-negative rates are reported to be between 15% and 20%. van den Brekel and colleagues suggested a minimal axial diameter of 8–9 mm in level II and 7–8 mm for the rest of the neck [54]. To provide a highly negative predictive value, smaller nodal sizes must be used. However, then, the positive predictive value will be decreased. This is why performing sonography-guided aspiration cytology is not recommended [55]. Generally, the accepted criteria could still lead to a high false-negative rate. Nodes less than 1 cm in size can be malignant. Particularly, those located in the primary drainage sites of the tumor should be carefully examined for additional features [56].

TABLE 18.1

Imaging-Based Nodal Level System of the Neck

Level	Sublevel	Region	Location	Sectional Imaging
I	Ia	Submental	Lies between the medial margins of the anterior bellies of the digastric muscles, above the bottom of the body of the hyoid bone, and below the mylohyoid muscle	Lies anteriorly to a transverse line drawn on each axial image through the posterior edge of the submandibular gland (between the bottom of the body of the hyoid bone and the mylohyoid muscles)
	Ib	Submandibular	Lies below the mylohyoid muscle, above the bottom of the body of the hyoid bone, and posterior and lateral to the medial edge of the anterior belly of the digastric muscle	
II	IIa	Upper internal jugular nodes	Lies anterior, lateral, or medial to the internal jugular vein (in close proximity with internal jugular vein)	Lies anterior to the transverse line drawn on each axial image through the posterior edge of the sternocleidomastoid muscle and lies posterior to a transverse line drawn on each axial scan through the posterior edge of the submandibular gland (located between the lower level of the bony margin of the jugular fossa and the level of the lower body of the hyoid bone)
	IIB	Upper spinal accessory nodes	Lies posterior to the internal jugular vein (there is a fat plane separating the nodes and the vein) [3]	
III	–	Midjugular nodes	Lies between the level of the lower body of the hyoid bone and the level of the lower margin of the cricoid cartilage arch	Lies anterior to a transverse line drawn on each axial image through the posterior edge of the sternocleidomastoid muscle (the medial margins of the common or the internal carotid arteries separating level III nodes from level VI nodes) [3]
IV	–	Low jugular nodes	Lies between the level of the lower margin of the cricoid cartilage arch and the level of the clavicle	Lies anterior and medial to an oblique line drawn through the posterior edge of the sternocleidomastoid muscle and the lateral posterior edge of the anterior scalene muscle (the medial aspect of the common carotid artery is the landmark, separating the level IV from level VI nodes) [3]
V	Va	Upper level V (spinal accessory chain)	Lies between the skull base and the level of the lower margin of the cricoid arch	Behind an oblique line through the posterior edge of the sternocleidomastoid muscle and the lateral posterior edge of the anterior scalene muscle
	Vb	Lower level V (spinal accessory chain)	Between the level of the lower margin of the cricoid cartilage arch and the level of the clavicle	
VI	—	Pretracheal, prelaryngeal, and paratracheal nodes	Lies between the lower body of the hyoid bone and the top of the manubrium	Lies between the lower body of the hyoid bone and the top of the manubrium, and also the medial margins of the both common or internal carotid arteries
VII	—	Upper mediastinal nodes	Lies caudal to the top of the manubrium in the superior mediastinum	Lies between the medial margins of the left and right common carotid arteries (may extend to the level of the innominate vein)

18.3.2 Central Necrosis

Heterogeneous and intense enhancement is of great importance for malignant involvement. Rarely tuberculous LNs also show this pattern (Figures 18.25 and 18.26). The morphological features of malignancy, such as necrosis and indistinct margins indicating extranodal spread, are relatively infrequent findings, especially for small (<10 mm) metastatic LNs [30].

In fact, in their necrotic areas, necrotic LNs have those tumor cells, residual LN tissues, and necrosis. To be observed on imaging studies, necrotic areas must be greater than 3 mm [29]. This area generally has a fluid density of 10–25 HU on CT. Classically, nodal necrosis on fat-suppressed T_2-WI images appears as a hyperintense area inside the node. This may correspond to liquefaction necrosis, whereas coagulation necrosis shows a hypointense signal relative to the residual nodal parenchyma [7,48]. MRI shows smaller necrotic areas than CT does for the same node. A nonenhancing area with a low signal intensity on T_1-WI is a less reliable criterion [30] (Table 18.4). Chong et al., in their article on the MRI features of cervical nodal necrosis in metastatic disease, found that the sensitivity and specificity of individual

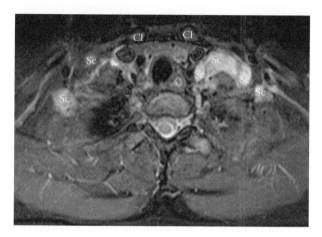

FIGURE 18.18
Transverse STIR image shows supraclavicular LNs (Sc) (Cl defines clavicles).

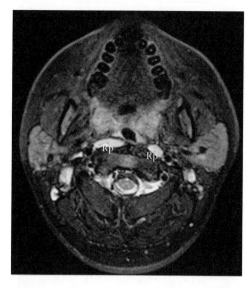

FIGURE 18.20
Transverse STIR image shows retropharyngeal LNs (Rp).

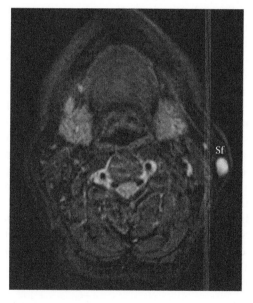

FIGURE 18.19
Transverse STIR image shows superficial LN (Sf).

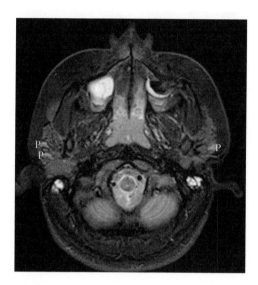

FIGURE 18.21
Transverse STIR image reveals intraparotid LNs (P).

MRI sequences were, respectively, 36% and 100% for T_1-weighted, 47% and 98% for T_2-weighted, and 67% and 100% for enhanced T_1. In combination (for one or more positive sequences), the sensitivity and specificity are as follows: T_1- and T_2-weighted (60%, 99%); T_1-weighted and enhanced scan (67%, 100%); and T_1-weighted, T_2-weighted, and enhanced scan (78%, 99%) when using CT as the standard. They also suggested that the staging of the primary tumor and nodes should be performed using a single modality. MRI was often more advantageous for the evaluation of the tumor. Equivocal MRI findings of nodal involvement required the consideration of CT [57].

The heterogeneous enhancement of nodes with a central necrosis on contrast-enhanced MR images indicates malignant involvement, as mentioned earlier. This may be seen as a rim-like enhancement of the heterogeneous T_1-WI signal of the node, with or without a corresponding hyperintense T_2-WI region [7]. However, a suppurative LN or fatty replacement may be misleading (Figures 18.27 and 18.28). Fatty replacement originates from chronic nodal infection. It occurs almost entirely at the periphery of the node, causing a pronounced lima bean shape. Rarely, this may occur centrally within a node [13]. When a node becomes large enough, it can be determined to have fat attenuation. The use of fat

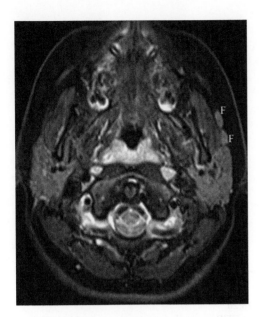

FIGURE 18.22
Transverse STIR image reveals superficially located left facial LNs (F).

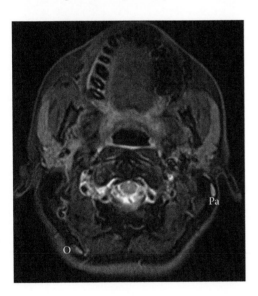

FIGURE 18.23
Transverse STIR image reveals postauricular (Pa) and occipital (O) LNs.

TABLE 18.2

Features of the Pathologic LNs

Enlargement
Round shape
Conglomeration
Irregular margins
Indistinct margins
Focal cortical thickening
Necrosis
Hyperintensity on STIR
Diffusion restriction
Unenhancement with USPIO

TABLE 18.3

Size Assessment on MRI

Short axis >10 mm (15 mm for level 2)
Longitudinal/transverse diameter <2
Supported by number criterion (>3)

saturation and the application of contrast material are essential in differentiating between fatty replacement and necrosis.

King et al. reported, in their study comparing the diagnostic accuracy of CT, MRI, and ultrasonography in the detection of necrosis in metastatic cervical nodes from patients with head and neck SCC, that the sensitivity of MRI and CT is higher than that of ultrasonography. No significant differences were detected between MRI and CT, and MRI was comparable to CT in terms of the detection of necrosis [58].

An additional issue is cystic nodes with purely fluid intensity that have thin walls (Figure 18.29). Papillary thyroid carcinomas and oropharyngeal SCCs may show such involved nodes. Nodal metastases of oropharyngeal squamous cell carcinomas may be purely cystic or necrotic. In such a situation, if the traditional risk factors of smoking and alcohol do not exist, human papillomavirus-associated oropharyngeal SCCs should be considered head and neck SCC [59,60]. Thyroid cystic metastases may have hyperintense signals on T_1-WI MRI because of thyroid protein or blood products. Thus, a cystic neck lesion in an adult must be evaluated carefully [56].

A distorted internal architecture with necrosis appears to be an important characteristic of metastatic nodes of head and neck carcinomas [61]. A distorted internal architecture is seen at 67% of metastatic, 14% of lymphomatous, and 9% of benign nodes [7].

18.3.3 Extracapsular Spread

Tumor cells reach a node via afferent lymphatic vessels. According to pathologists, they may settle down the periphery or center of an LN. Especially, peripherally located metastases tend to show extracapsular spread. When they penetrate the nodal capsule, spread to the adjacent soft tissues will occur. This type of spread can be termed extranodal, extracapsular, or transcapsular tumor spread. In fact, it is a specific feature of a metastatic node. Extranodal spread may result in treatment failure and survival is halved [58–62]. In addition, patients with macroscopic extracapsular spread have an approximately 10-fold greater risk of recurrence compared with patients with either microscopic or no tumor spread [63]. Myers et al. reported that extracapsular spread is the most important predictor of the regional recurrence

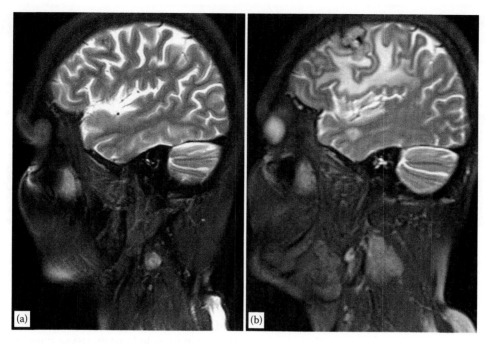

FIGURE 18.24
Breast carcinoma patient. Sagittal fat-saturated T_2-WI (a) shows a round LN. Six months later, sagittal fat-saturated T_2-WI (b) shows the enlargement of the LN and emerging brain metastases.

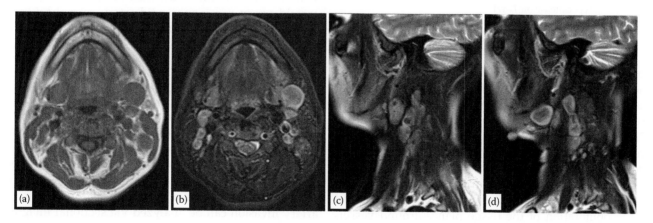

FIGURE 18.25
Transverse T_1-WI (a), transverse STIR image (b), and sagittal T_2-WIs (c,d) show multiple necrotic LNs of a patient with tuberculosis.

and advancement of metastasis. They also concluded that the existence of extracapsular spread is an indication of intensive regional and systemic adjuvant therapy [62]. Ljumanovic et al. reported that extranodal spread with suspicious nodes in the low jugular/posterior triangle (oropharyngeal cancer) or at the paratracheal level (laryngeal cancer) or with contralateral enlarged nodes (laryngeal and oral cavity cancer) on MRI demonstrate a high risk for developing distant metastases. When these features are encountered, supplementary imaging screening (CT-chest, PET-scan) is considered [64].

Lymphoma metastases may show extracapsular spread, especially in the Waldeyer's ring. This can be evaluated via MRI. Indistinct nodal margins, irregular nodal capsular enhancement, or infiltration into the adjacent fat or muscle can be observed [7].

The risk of extracapsular spread increases as a node becomes enlarged; thus, it is easy to identify [65,66]. Histologically, such disease was found in 23% of LNs 1 cm in length at the most and in 40% of LNs 2 cm in length at the most [63,67]. Approximately one-quarter of normal-sized LNs have extracapsular spread. It is reported that 2–3 cm LNs have a 53% rate of extracapsular spread and that LNs larger than 3 cm have a 74% rate. In general, such tumor spread occurs in 60% of nodes less than 3 cm in diameter [68–70].

Macroscopic extracapsular tumor spread is seen on contrast-enhanced CT as an enhancement, often a

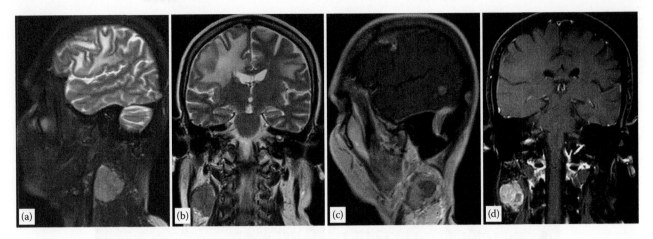

FIGURE 18.26

Sagittal fat-saturated T_2-WI (a) and coronal T_2-WI (b) images show a huge LN involved by breast carcinoma. Right cerebral hemisphere metastases are also seen. Sagittal contrast-enhanced T_1-WI (c) reveals central necrosis of LN and enhancing brain metastases. Coronal fat-saturated contrast-enhanced T_1-WI (d) demonstrates necrosis and irregular borders of the LN.

TABLE 18.4

Central Necrosis on MRI

To see on imaging must be greater than 3 mm
Hypointense on T_1-WI
Hyperintense on T_2-WI
Unenhancing area
No suppression with fat saturation

condensed nodal rim, usually with the infiltration of the adjacent fat planes. MRI reveals indistinct nodal margins, irregular nodal capsular enhancement, fat stranding, or infiltration of the adjacent fat or muscle [71] (Table 18.5). These imaging changes reflect the presence of such macroscopic extranodal tumor spread, and their absence on imaging indicates the lack of this type of macroscopic disease. However, contemporary nodal infection, surgical procedure, and irradiation may lead to the obliteration of the tissue planes around vessels

and LNs, mimicking the extracapsular spread. However, these can be ruled out by using patients' histories.

An additional important issue regarding extracapsular spread interpretation is identifying whether invasion of the adjacent structures, such as the internal carotid artery, common carotid artery, retropharyngeal soft tissues, muscle, neural tissue (brachial plexus), or bone, exists (Figure 18.30).

18.3.4 Carotid Artery Invasion

The extension of a tumor from a node to a nearby carotid artery is a severe prognostic finding. Although the salvage possibility of such a patient is small, it was reported that when the involved artery is resected and grafted, the disease progresses better than with the dissection of the artery and irradiation after surgery [61,72]. Carotid artery invasion can be analyzed via sectional imaging.

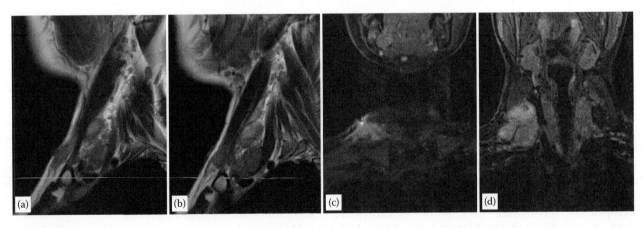

FIGURE 18.27

Sagittal T_2-WIs (a,b) and coronal STIR images (c,d) show suppurative LNs of a patient with neck abscess.

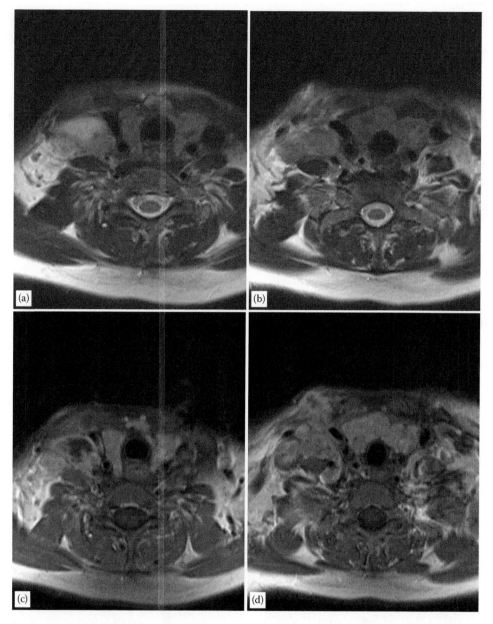

FIGURE 18.28
Transverse STIR images (a,b) and transverse fat-saturated postcontrast T_1-WIs (c,d) reveal suppurative-necrotic LNs on the right side.

The loss of the fatty plane between a metastatic node and the carotid artery and the amount of tumor in the surrounding arterial circumference are indicators of carotid artery invasion. However, a subtle invasion cannot be detected via MRI or CT [6].

Tumor invasions of the arterial adventitia, muscularis, and intima are all of great importance. Involvement of the muscularis and intima can be seen in the form of arterial luminal narrowing. Microscopic adventitial infiltration cannot be detected via imaging [49]. Arterial invasion is evaluated by both observing the tumor and the degree of obliteration of the normal fat plane that surrounds the artery (Table 18.6). When the tumor entirely surrounds the artery, invasion may exist. However, if a small amount of contact is seen between the tumor and the artery, it is possible that there is no invasion. The cutoff value for the degree to which the tumor surrounds the artery is 270°. If more than 270° of the arterial perimeter is enclosed by tumor, it is likely that there is no arterial wall involvement [61]. Yoo et al. reported that more than 180° of artery encasement predicts a poor clinical outcome, but not the extent of tumor invasion into the carotid artery [73]. The demonstration of carotid involvement of 270° or more using MRI, CT, or ultrasonography has a sensitivity of 92%–100% and a specificity of 88%–93% in the prediction of arterial

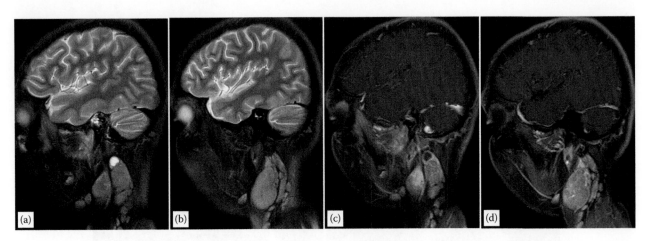

FIGURE 18.29
A patient with nasopharyngeal carcinoma. Sagittal fat-saturated T_2-WIs (a,b) and sagittal fat-saturated postcontrast T_1-WIs (c,d) show multiple lymphadenopathies, one of which has purely cystic change.

TABLE 18.5

Extracapsular Spread on MRI

Indistinct nodal margins
Irregular nodal capsular enhancement
Infiltration of the adjacent fat or muscle

wall invasion [61,74]. Yoo et al. also found that in 83% of cases, carotid encasement greater than 270° indicated direct tumor invasion [73]. In addition, a clinical assessment of fixation may be considered predictive of tumor invasion.

Scar tissue and vascular hyperplasia due to previous radiotherapy or atherosclerosis may lead to false-positive results [75–78].

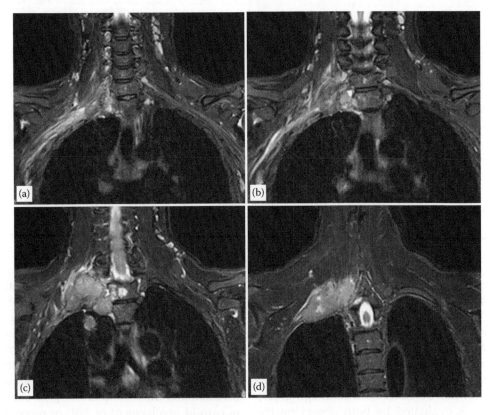

FIGURE 18.30
Coronal STIR images (a–d) show Pancoast tumor on the right side. There are multiple accompanying supraclavicular enlarged and heterogeneous LNs. Right brachial plexus is hyperintense (plexitis). In addition, vertebra and muscle invasions are seen.

TABLE 18.6

Carotid Artery Invasion on MRI

Arterial luminal narrowing

Tumor entirely surrounding the artery

>270° encasement and obliteration of normal fat plane

18.4 Magnetic Resonance Imaging

MRI allows the concurrent evaluation of primary tumor extent and nodal status, and it is often used as a first-line imaging choice to reveal head and neck SCC [79]. However, there is no application technique that is accepted by most head and neck radiologists universally [7]. MRI has a superior contrast resolution compared to CT. However, MRI procedures for the assessment of the LNs of the head and neck are not very standardized. Conventional MR sequences, such as T_2- and T_1-weighted spin-echo (SE) or turbo spin-echo (TSE) sequences, are limited in terms of correctly differentiating between benign and malignant LNs. The creation of new sequences with high resolution in combination with phased-array surface coils has provided an advance in terms of the discovery of and distinction between benign and malignant LNs.

The clinicians must know the indications, advantages, and disadvantages of the various imaging techniques. In addition, the potential risks of MRI, including adverse events related to the contrast material and sedation, must be considered. Before MRI, patients' symptomatology and clinical histories, such as previous imaging studies, have to be reviewed. The radiologist interpreting the MRI should understand the patient's clinical picture in detail and the relevant anatomy–pathophysiology.

In fact, MRI has a wide spectrum of sequences. Their clinical usage and imaging appearances should be known. Also, radiology clinics must create their own case-based imaging protocols. From time to time, these imaging protocols should be checked and renewed. If necessary, each patient could consult with the referring or supervising physician.

Some required information regarding the patient must be supplied for both an appropriate imaging protocol and sufficient and detailed interpretation: signs and symptoms, the history of the patient, previous imaging studies, diagnostic interventions and their results, as well as existing pathology results. Provisional diagnoses must also be reported (Table 18.7).

18.4.1 Examination Technique

MRI of node disease can be performed with conventional MRI surveys. A dedicated head and neck circular polarization surface coil is used.

TABLE 18.7

Some Required Information for an Appropriate Imaging Protocol and Interpretation

Signs and symptoms of the patient

History of the patient

Previous imaging studies

Diagnostic interventions and their results

Pathology results if exist

Provisional diagnoses

The axial plane for the nodal classification is acquired with the patient's hard palate perpendicular to the tabletop and the axial plane aligned along the inferior orbital meatal line [7]. The entire neck must be evaluated for lymphadenopathy, especially in cases of esophageal or thyroid carcinomas or lymphomas that may involve superior mediastinal nodes (Figures 18.31 through 18.33).

Conventional sequences of T_2- and T_1-weighted SE or TSE are inadequate for the correct differentiation between benign and malignant LNs. T_1-WI (short repetition time [TR]/short echo time [TE]) remains the best method for defining adequate anatomic detail for structures surrounded by soft tissues. However, structures surrounded by cerebrospinal fluid can be assessed remarkably well via thin-slab, 3D T_2-WI [80]. Fast spin-echo (FSE) T_2-WI (long TR/long TE FSE) reveals more detail with a shorter scan time than conventional T_2 and is preferred in the head and neck region because physiologic and gross motion artifacts commonly limit the image quality [43,81]. Fat saturation techniques, such as chemical selective partial inversion recovery or STIR, yield better definition of pathology [80]. Fat saturation has come to be a standard for the head and neck region because LNs mostly lie within hyperintense adipose tissue. The two main fat saturation approaches are the frequency-selective fat suppression (chemical shift) and STIR sequences. The chemical shift method has some disadvantages, such as requiring additional scanning time and resulting in nonuniform fat suppression due to magnetic field heterogeneity. Therefore, some investigators prefer STIR because it provides improved fat suppression and augmented contrast between LNs and the surrounding tissue. We find STIR sequences very helpful in detecting even small LNs [2,82]. However, STIR sequences do not permit differentiation between benign and malignant LNs.

Obtaining a high-resolution image of the head and neck region requires thin sections. The image slice thickness should be 3–5 mm with either no interslice gap or an interslice gap of 1–2 mm on a 192 × 256 or 512 × 512 (if possible) matrix. The field of view must be as small as possible. Typically, the imaging method engaged for the primary tumor is used to stage the neck as well.

Contrast administration with fat suppression is essential to unfailingly signify necrosis, and it is often helpful

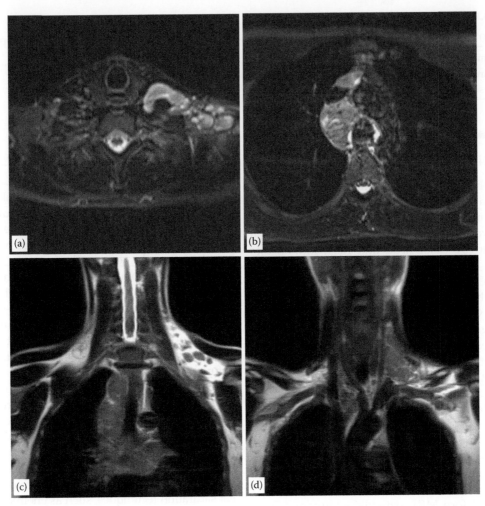

FIGURE 18.31
Transverse STIR image shows multiple left supraclavicular LNs (a). Transverse STIR image shows superior mediastinal LNs (b). Coronal T_2-WI shows multiple mediastinal LNs (c). Coronal T_2-WI shows multiple cervical and superior mediastinal LNs (level IV–VI–VII) (d).

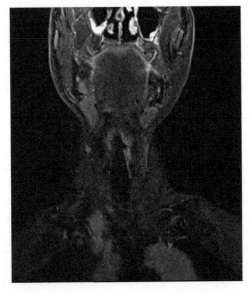

FIGURE 18.32
Coronal STIR image shows superior mediastinal and right supraclavicular LN involvement of large cell lung carcinoma.

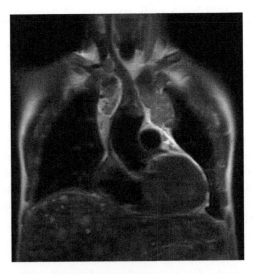

FIGURE 18.33
Coronal T_2-WI reveals supraclavicular and mediastinal LN involvement and diffuse liver metastasis.

in describing the primary tumor as well. T_2-weighted MR images may also be helpful in demonstrating central necrosis. In their study, King et al. showed that MRI can perform as well as CT in the detection of necrosis in metastases in the LNs of the neck. Diagnostic accuracy and sensitivity were similar (91%–99% and 93%, respectively, for MRI, compared with 92%–99% and 91%, respectively, for CT), with no significant difference being observed between the two modalities. The specificities for the total group of 903 benign and malignant nodes were also similar. False-negative results occurred in nodes in which the areas of necrosis were smaller than 3 mm [83].

18.4.2 Conventional MRI Sequences

Metastatic nodes mostly demonstrate low-to-intermediate T_1-weighted and high T_2-weighted signal intensities on MRI. The nodal metastases may show a great variety of imaging findings, including enhancement and distributed calcifications (tuberculosis, previous granulomatous disease, lymphomas before or after irradiation or chemotherapy, metastatic mucinous adenocarcinomas, prostate metastasis and seminomas). Also, a metastatic LN may look like a benign cyst with or without hemorrhagic areas. High T_1-weighted signal intensity may be obtained in nodal metastases of papillary thyroid carcinoma because of the high macroglobulin concentrations. Hemorrhage will also demonstrate high intensity in both T_1-WI and T_2-WI. Malignant melanoma metastases may be seen as bright on T_1-WI as well.

18.4.3 Short-Tau Inversion Recovery

Coronal turbo STIR imaging is a rapid and sensitive MR technique for detecting metastatic nodes in the neck. In addition, the supraclavicular region, which may be difficult to evaluate via cross-sectional imaging, is an easy region to evaluate using coronal STIR images. Kawai et al. proposed this technique for screening before high-resolution MRI. Using a small-sized surface coil for patients with head and neck carcinoma was suggested [84]. However, this technique was not found to be suitable for the assessment of the internal architectures of the nodes by Kawai et al. Especially, smaller nodes (6 mm in short-axis diameter) may often be overlooked by the turbo STIR sequence. Another limitation of coronal turbo STIR imaging is that body background signals from the blood vessels may lead to a failure to distinguish the vessels and LNs [84].

18.4.4 Dynamic Contrast-Enhanced MRI

Dynamic contrast-enhanced MRI (DCE-MRI) has been reported to be useful not only for the evaluation of primary head and neck carcinomas but also in distinguishing involved and benign LNs [85,86]. DCE-MRI contains serial imaging acquired during the passage of a contrast material through the tissue of interest. It is accepted that the parameters obtained via such a study reflect tumor vascularity [87].

Involved LNs behave differently from benign nodes upon kinetic evaluation. Fischbein et al. found, in their study on DCE-MRI of metastatic cervical adenopathy, that metastatic nodes had a longer time to peak, reduced peak enhancement, reduced maximum slope, and slower washout slope [88]. They explained these findings with the fact that lymphoid tissue has higher blood flow than SCCs of the head and neck [89]. This meant that the involved tissue had a lower transfer of the contrast agent than normal tissue (a function of blood flow, blood volume, and vessel permeability, thus lower time to peak and lower maximum up-slope) and that tumor tissue had a decreased volume of extravascular, extracellular space (hence lower peak enhancement) than normal or reactive nodal tissue. In addition, involved tissue does not necessarily have increased blood flow and volume compared with normal lymphoid tissue, especially if the nodal tissue is reactive. They also emphasized that tumor blood flow can be heterogeneous, slow, and even retrograde [89,90].

Altered hemodynamic features can also be recognized via perfusion MRI in terms of blood flow, blood volume, and permeability in the head and neck region. Furukawa et al., in their study on dynamic contrast-enhanced perfusion MRI of head and neck tumors, found that DCE magnetic resonance perfusion imaging could provide pivotal information regarding microcirculation, potentially improving the differentiation of malignant tumors from postradiation changes [91]. Abdel Razek et al. studied the role of perfusion MRI in cervical lymphadenopathy. They found that the mean dynamic susceptibility contrast percentage of malignant nodes was significantly higher than that of benign nodes and lymphoma. They concluded that perfusion MRI is a promising noninvasive method for the characterization of malignant cervical lymphadenopathy [92].

18.4.5 Diffusion-Weighted Imaging

Diffusion-weighted imaging (DWI) is not a routine sequence added to exact protocols. However, we recommend it being included in the head and neck MRI protocols.

DWI has been reported to allow differentiation between malignant and benign LNs, with a reported sensitivity of 52%–98% and a specificity of 88%–97% [79,93,94]. Wang et al. reported that cellularity increases and cellular turnover of poorly differentiated tumors reduces the amount of extracellular matrix, as well as the diffusion space of water protons in the extracellular and intracellular dimensions [95]. Therefore, metastatic nodes show generally high signal intensities in diffusion

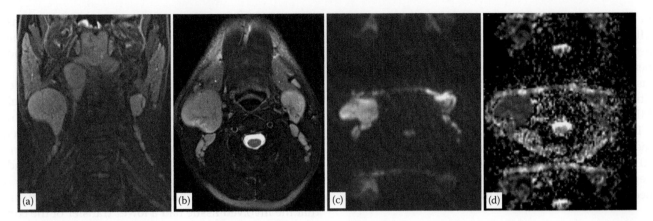

FIGURE 18.34
Coronal (a) and transverse (b) STIR images, DWI (c), and ADC map (d) show multiple retropharyngeal level IIa-b LNs due to nasopharyngeal carcinoma. Size and diffusion restriction (to be hyperintense on DWI and hypointense on ADC map) indicates malignant involvement for this case.

maps (unrelated to the *b*-value) and low ADC values, in other words restricted diffusion (Figure 18.34).

Vandecaveye et al. stated that the mean ADC value of metastatic LNs ($0.85 \pm 0.27 \times 10^{-3}$ mm^2/s) was statistically different from that of benign LNs ($1.19 \pm 0.22 \times 10^{-3}$ mm^2/s) in head and neck SCC and that a mean threshold ADC value of 0.94×10^{-3} mm^2/s has 84% sensitivity, 94% specificity, and 91% accuracy in differentiating between benign and metastatic LNs [79]. Abdel Razek et al. reported a mean threshold ADC value for differentiating benign and metastatic LNs that was 1.38×10^{-3} mm^2/s [94]. King et al. suggested a threshold of metastatic LNs of SCC of the head and neck region that was $1.057 \pm 0.169 \times 10^{-3}$ mm^2/s [96]. Sumi et al. also reported that the ADC value of well-to-moderately differentiated SCC is higher than that of poorly differentiated carcinoma [93].

An ADC map provides helpful information about necrotic areas. An increase in the amount of necrosis will result in increased ADC value [97]. Necrosis is often seen in SCC of the head and neck region. During measurement, necrotic areas should be excluded from region of interests (ROIs) as much as possible. However, it may be difficult to completely exclude micronecrotic areas [98].

Nakamatsu et al. reported that the statistically significant inverse correlation of ADC and standardized uptake values in metastatic neck LNs of SCC supports the idea that DWI and 18F-fluorodeoxyglucose positron emission tomography (FDG PET)/CT could play complementary roles in clinical assessment [98]. Before the surgery, DWI may facilitate the revealing or exclusion of contralateral, ipsilateral, or skip metastases. This situation may alter the extent of neck dissection. Before radiotherapy, the improved detection of small nodal metastases may allow a confined irradiation volume [79].

In sum, the quantification of ADC values is reported to allow a more accurate finding of nodal involvement in SCC than conventional TSE sequences [79]. Although quantitative DWI has a high negative predictive value

in terms of ruling out metastatic disease, it cannot show micro-metastases smaller than 4 mm [79]. Also, because of the associated susceptibility artifacts, DWI may not be able to evaluate the contours and contents of small nodes (usually <0.9 cm) [7].

18.4.6 USPIO-Enhanced MRI

Intravenously injected USPIO particles pass through the vascular endothelium into the interstitial space and are finally engaged by benign LNs. They are phagocytosed by reticuloendothelial system cells, such as macrophages and histiocytes. The magnetic susceptibility and T_2 shortening effects of the iron deposits cause these normal LNs to show a low signal intensity on T_2- and T_2-WI. Conversely, metastatic LNs lose their phagocytotic mechanism and do not show signal intensity. LNs with fractional signal reduction are considered to be partially metastatic involved nodes.

An USPIO-enhanced T_2-weighted two-dimensional gradient echo (GRE 2D) sequence has been found to be superior to USPIO-enhanced T_2-weighted SE because the signal decrease is more pronounced in the former. Newly available high-spatial-resolution T_2-weighted sequences are expected to improve the detection of partial metastatic LN infiltrations in the future.

18.4.7 Gadofluorine M-Enhanced MRI

Gadofluorine M is another LN-specific contrast agent [99]. It is a water-soluble, paramagnetic, gadolinium-based T_1-contrast agent. LN uptakes via the transcapillary passage move through the interendothelial junctions into the medullary sinuses 5–30 min after injection. Only animal experiences exist [99]. Mayer et al. used gadoversetamide and gadofluorine M to evaluate LN enhancement in the head and neck region of healthy dogs. They concluded that either agent could be used to identify those LNs at the highest risk of metastatic disease and guide staging and treatment [100]. Choi et al. compared the accuracy of Gadofluorine M with that of monocrystalline iron oxide nanoparticles (MION)-47 in the depiction of cervical LN metastases with MRI in a rabbit model of head and neck cancer by using histologic analysis as the reference standard. They concluded that gadofluorine-M-enhanced MRI has a higher accuracy in depicting LN metastases than does MION-47-enhanced MRI [101]. Spuentrup et al. investigated the potential of gadofluorine M for targeted LN imaging in a human-sized animal model (swine). They found that gadofluorine M accumulates in the LNs and allows for selectively targeted high-contrast MRI of LN tissue [102].

18.4.8 Proton Magnetic Resonance Spectroscopy

In vivo proton magnetic resonance spectroscopy is a noninvasive procedure that is useful in obtaining information about the cellular chemistry of tissues [103]. A respiratory triggering may be applied to minimize motion artifacts in the laryngeal region. Metastatic nodal disease seems to have characteristic spectra with higher choline concentrations. Elevated levels of choline are believed to reveal membrane synthesis and an elevated cell proliferation rate [104]. This can be used to differentiate metastatic LNs from reactive benign LNs. King et al. reported that both choline and creatine peaks can be recognized in malignant nodes of SCC, undifferentiated carcinoma, and non-Hodgkin's lymphoma, but not in benign tuberculous nodes [103]. Bisdas et al. showed that there is a significant increase in choline/creatine ratios in LN metastases compared with immunostimulated LNs [105]. This method appears to be practical for nodal volumes up to 9 mm^3. This results from the fact that proton magnetic resonance spectroscopy has certain characteristic difficulties, for instance, a strong lipid signal, shimming, inadequate signal reception, and motion artifacts [103,106–112].

In the future, stronger magnetic fields, better shimming processes, and the improved editing of other metabolites (including lactate) may extensively reveal the spectral profile of extracranial head and neck tumors [7].

18.5 Diagnostic Issues

The differential diagnosis of enhancing lymphadenopathy in the neck consists of numerous diseases. The most common cause is infection, and this generally does not require advanced imaging. The list of diseases includes nodal angiomatoses (acute bacterial, viral, and other infections), non-Hodgkin lymphoma, Hodgkin lymphoma, SCC, vascular metastases (thyroid, renal cell carcinoma,

and malignant melanoma), sarcomas (Kaposi's and angiosarcoma), non-neoplastic angiomatoses, angio-immunoblastic lymphadenopathy, angiofollicular LN hyperplasia (Castleman's disease), Kimura's lymphadenopathy, Kikuchi–Fujimoto lymphadenopathy, human immunodeficiency virus lymphadenitis, benign vascular proliferation (distal venous or lymphatic obstruction), and inflammatory pseudotumor of LNs (Figures 18.35 through 18.44).

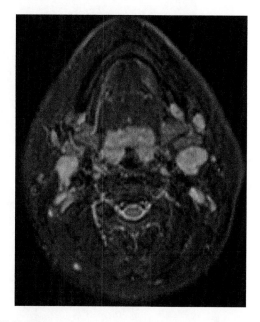

FIGURE 18.35
Transverse STIR image shows hypertrophy of lingual and palatine tonsils. Multiple level Ib, IIa, and IIb LNs are also seen.

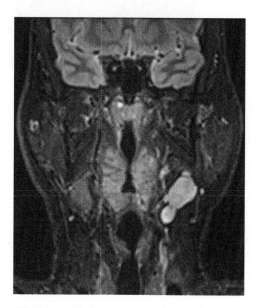

FIGURE 18.36
Coronal STIR image shows hypertrophy of lingual, palatine, and pharyngeal tonsils. Left-sided level II LNs are also seen.

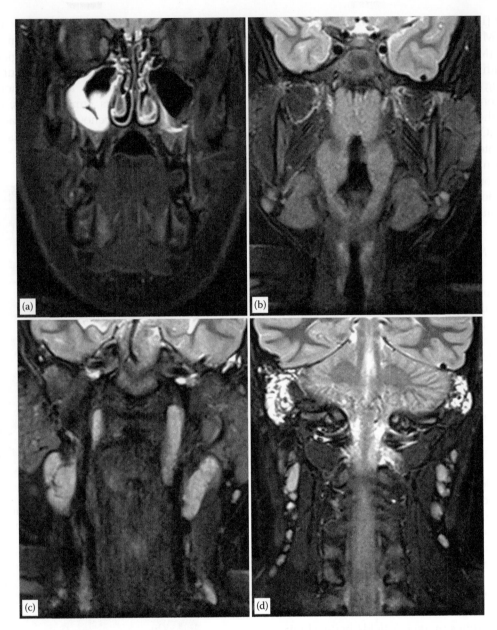

FIGURE 18.37
Coronal fat-saturated T_2-WIs (a–d) show sinusitis; hypertrophic lingual, palatine, and pharyngeal tonsils; otitis; and mastoiditis. Multiple reactive LNs are also seen.

During imaging interpretation, some important diagnostic issues must be kept in mind:

- Small metastases must be searched for (this may change the N status).
- If a single node is determined, another one should be looked for.
- If an ipsilateral disease has been found, the patient must be evaluated for contralateral disease.

- If no abnormality has been detected, the draining region of the primary tumor should be reassessed cautiously before reporting N0 nodal status [56].

In addition, knowledge of the drainage pathways of head and neck tumors is a vital concern. In this way, it is possible to examine the likely sites of metastasis clearly and easily. Furthermore, understanding the nodal drainage pattern permits the examination of probable primary sites for an unknown primary tumor. Asymmetrically

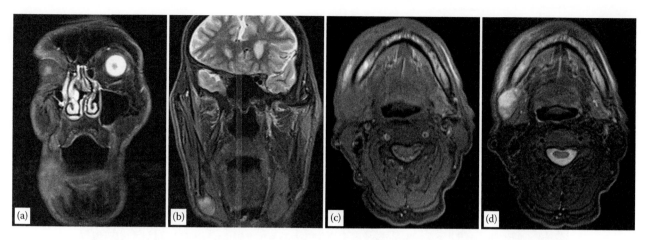

FIGURE 18.38

Coronal T_2-WI (a) shows right buccal tumor with skin, submucosa, and deep tissue involvement. Coronal T_2-WI (b), transverse T_1-WI, and transverse STIR image (c) show level Ib LN. Transverse STIR image (d) demonstrates also central heterogenity (necrosis).

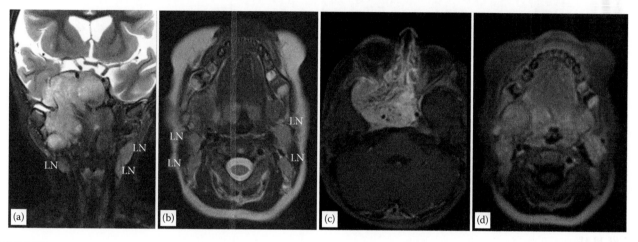

FIGURE 18.39

Coronal T_2-WI (a) reveals pharyngeal huge mass (embryonal rhabdomyosarcoma). Transverse T_2-WI (b) shows level II–III LNs. Postcontrast fat-saturated transverse T_1-WIs show heterogeneous enhancements of primary mass (c) and level IIa–b LNs (d). LN, lymph node.

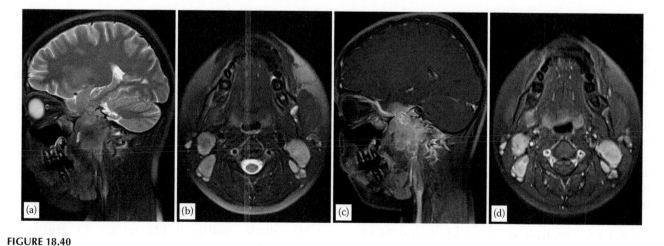

FIGURE 18.40

A patient with nasopharyngeal carcinoma. Sagittal fat-saturated T_2-WIs (a), transverse STIR image (b), and sagittal (c) and transverse (d) fat-saturated postcontrast T_1-WIs show nasopharyngeal carcinoma with intracranial extension. Heterogeneous enhancing necrotic LNs are also seen.

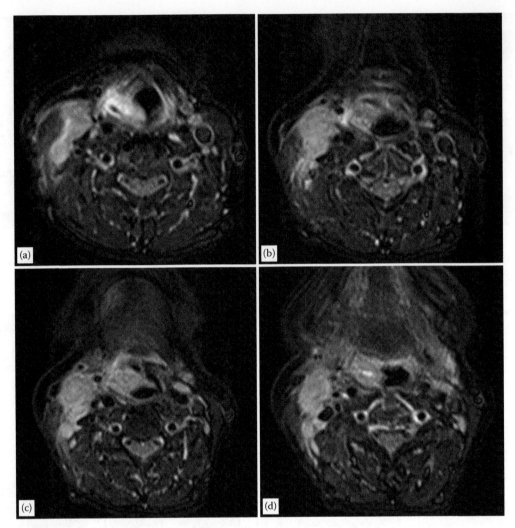

FIGURE 18.41
Transverse STIR images (a–d) show right-sided hypopharynx squamous cell carcinoma and level II and III LNs on the right side.

prominent nodes or three or more neighboring and confluent LNs along a class must be regarded as suspicious [18,56]. Midline malignant tumors, such as nasopharyngeal, epiglottic, and oral cavity tumors, often drain bilaterally [56].

Nodal staging is different from nodal classification. Nodal classification categorizes the involved nodal groups and is supplementary to the determination of the type of surgery. However, nodal staging describes the complete number, size, and location of the affected nodes in order to contribute to creating a prognosis for the disease [3]. A tumor, node, metastasis-based system assesses the size of the primary tumor (T), regional LN involvement (N), and distant metastasis (M). Metastasis to elsewhere than a regional LN reveals distant metastasis. Microscopic involvement of up to 3 mm in the largest dimension is considered

discontinuous tumor extension, and it is classified in a T category [3]. An involvement >3 mm, even though there is no evidence of LN tissue left behind, is classified as a regional LN metastasis. Most nodal masses >3 cm in diameter are not generally single nodes, but they may be confluent nodes or tumors in the soft tissues of the neck [28].

Nodal staging in oropharyngeal, hypopharyngeal, laryngeal, oral cavity, and sinonasal carcinoma is done according to following systems [56]:

N0—No regional LN metastasis

N1—Single, ipsilateral LN <3 cm in greatest dimension

N2a—Single ipsilateral LN 3–6 cm in greatest dimension

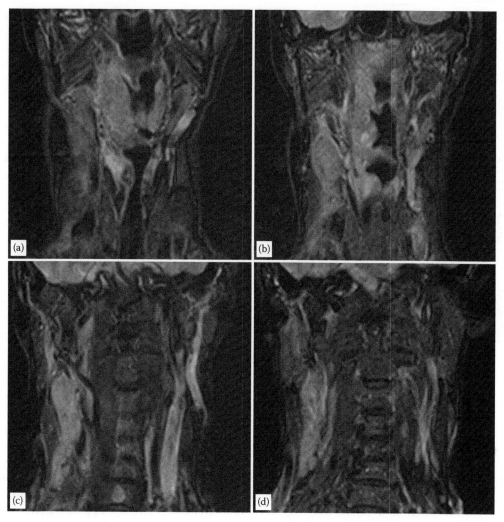

FIGURE 18.42
Coronal STIR images (a–d) show right-sided larynx carcinoma and conglomerating right level II and III LNs.

N2b—Multiple ipsilateral LNs ≤6 cm in greatest dimension

N2c—Bilateral or contralateral LNs ≤6 cm in greatest dimension

N3—LN(s) >6 cm in greatest dimension

Thyroid and nasopharyngeal carcinomas are self-exclusive and require separate nodal (N) classifications [3]. The supraclavicular nodes in nasopharyngeal carcinoma and retropharyngeal nodes in thyroid carcinoma represent more progressive nodal disease. Nasopharyngeal carcinoma may drain to the parotid nodes as well. Retropharyngeal and parotid nodes may not be palpable upon clinical examination; consequently, the radiologist has an important role in the evaluation of both groups [56].

Regional LN involvement of thyroid carcinoma is mutual, but it is of little prognostic importance, especially for papillary, follicular types. The first group that is probably involved is the paralaryngeal, paratracheal, and prelaryngeal (Delphian) nodes adjacent to the gland. However, spread to these nodal stations has no prognostic significance. Hence, they are not part of the staging system. They may spread to level III and IV nodes, supraclavicular nodes, and much less frequently to level IA, IB, and V nodes. Upper mediastinal nodal involvement occurs commonly. Retropharyngeal nodal metastases may be observed in cases of extensive cervical metastases. Bilateral metastases are common. Medullary thyroid carcinoma metastases show a similar pattern [3]. For thyroid carcinomas, N0 indicates no regional nodal metastasis. N1a covers level VI involvement, including

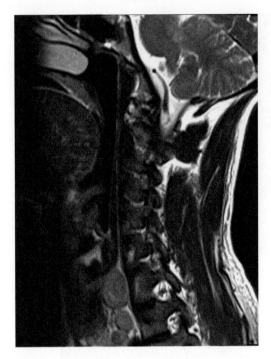

FIGURE 18.43
Sagittal T_2-WI demonstrates multiple paraesophageal LNs of esophageal carcinoma.

pretracheal, paratracheal, and prelaryngeal nodes. N1b consists of unilateral, bilateral, or contralateral nodes (levels I–V) or any retropharyngeal or superior mediastinal nodes (level VII) [56].

For nasopharyngeal carcinoma nodal staging, N0 indicates no regional nodal metastasis. N1 indicates unilateral LNs ≤6 cm in the largest dimension and above the supraclavicular fossa or unilateral or bilateral retropharyngeal nodes measuring ≤6 cm in the largest dimension. N2 indicates bilateral LNs ≤6 cm in the largest dimension and above the supraclavicular fossa. N3a contains unilateral or bilateral LNs measuring >6 cm in the largest dimension. N3b consists of LN extending into the supraclavicular fossa [56].

During imaging evaluation, some solid masses that are not LNs are seen. The wealth of MRI tools can provide a differential diagnosis list to a degree. Therefore, during interpretation, some pathologies have to be kept in mind, such as neural, vascular, and neurovascular tumors (Figures 18.45 and 18.46); mesenchymal tumors; congenital lesions (branchial cysts [Figures 18.47 and 18.48], teratomas, and ectopic thyroid tissues); and infection [7].

In addition, if a skin carcinoma exists, the superficial nodes, such as the parotid, posterior auricular, facial, and occipital nodes, must be examined carefully. Skin carcinoma spreads to level V nodes as well. The Virchow node is located in the left supraclavicular region, near the junction of the thoracic duct and the left subclavian vein. When this is identified as an isolated finding on sectional imaging, the main differential diagnoses are thyroid cancer and thoracic and abdominal malignancy [56].

Lymphomas and leukemia are also kept in mind during the differential diagnosis of the metastatic nodes. The LNs involved by lymphoma have no typical appearance. They generally form a cluster and show variable enhancement and central necrosis. Slightly enlarged, homogeneous, reactive-appearing nodes can also be encountered. In addition, foamy appearing nodes, sometimes with a thin capsule, may be seen [7].

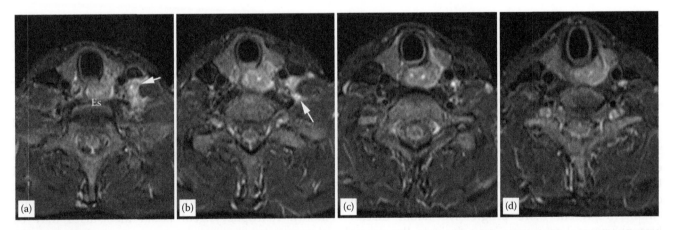

FIGURE 18.44
Transverse STIR images (a–d) demonstrate circular thickening of cervical esophagus (Es) (squamous cell carcinoma) and level III–IV LNs that have irregular borders and heterogeneous intensities (arrows).

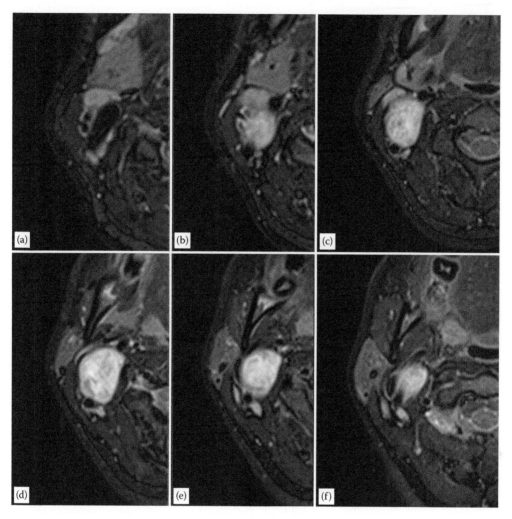

FIGURE 18.45
Transverse STIR images (a–f) reveal a mass located at the carotid bifurcation and splaying the internal and external carotid arteries. It is hyperintense compared to muscle.

When no palpable node is identified in the neck, imaging can assist with identifying occult metastases. Additionally, imaging evaluation may confirm that the neck is really tumor-negative [2]. When all upper aerodigestive tract sites are considered, about 15% of tumor-negative (N0) necks will ultimately progress to metastatic disease [113]. The following risk factors are substantial independent predictors of occult metastases: tumor site, T stage, tumor thickness, histologic grade, vascular embolization, and perineural infiltration [114,115]. The microvessel density of the primary tumor is believed to be closely associated with spread to the regional LNs [3,116].

Approximately 10% of patients with abnormal cervical nodes present without obvious primary tumors. The most common sites for unknown primary tumors are as follows: nasopharynx, pyriform sinus, tongue base, tonsillar crypts, thyroid, and lung. Knowledge of drainage patterns will also help in search of primary tumors.

In children, it is important to remember the retropharyngeal abscessed node. The necrotic retropharyngeal node has an enhancing irregular border around a necrotic nodal center. The remaining margins of the retropharyngeal space are normal. However, a retropharyngeal abscess accompanies the enhancement of the walls of the retropharyngeal space. Mucoid intensity filling the space is seen. A necrotic retropharyngeal node necessitates initial intravenous antibiotic therapy without surgery. However, a retropharyngeal abscess requires surgical drainage.

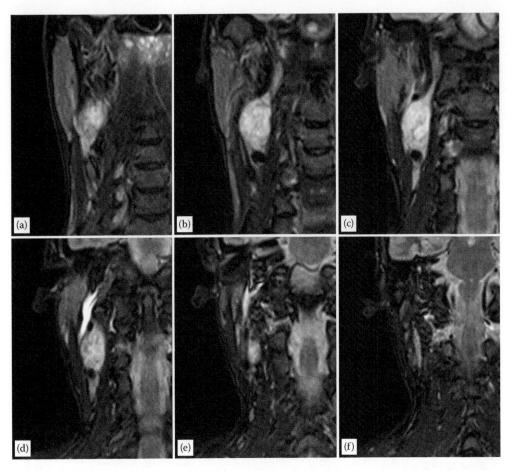

FIGURE 18.46
Coronal fat-saturated T_2-WIs (a–f) reveal a mass located at the carotid bifurcation and splaying the internal and external carotid arteries. It is hyperintense compared to muscle and has a salt-and-pepper appearance.

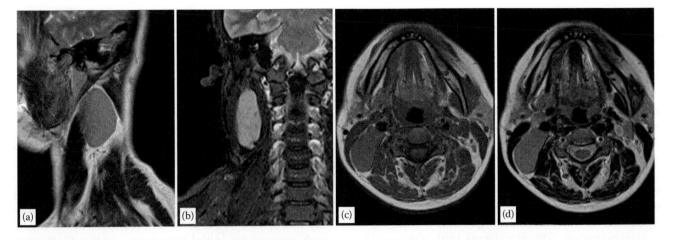

FIGURE 18.47
A huge branchial cyst that is located in the right submandibular region resembles LN intensities on sagittal T_2-WI (a), coronal T_2-WI (b), transverse T_1-WI (c), and transverse T_2-WI (d).

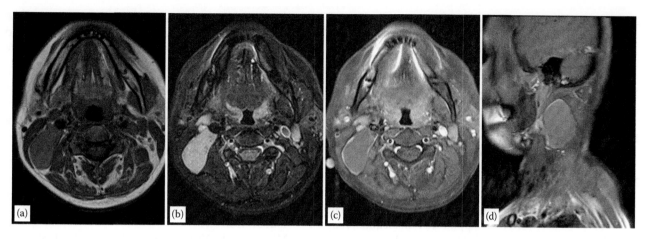

FIGURE 18.48

A case of branchial cyst. The branchial cyst has similar intensities to LN on transverse T_1-WI (a) and transverse STIR image (b). Postcontrast transverse (c) and sagittal (d) T_1-WIs show thin peripheral enhancement.

References

1. Branstetter BF, Weissman JL (2000) Normal anatomy of the neck with CT and MR imaging correlation. *Radiol Clin North Am* 38:925–940.
2. Castelijns JA, van den Brekel MWM (2006) Neck nodal disease. In: Robert H (ed) *Diagnostic Imaging: Head and Neck Imaging*, 1st edn. Springer-Verlag, Berlin, Germany, pp. 568–600.
3. Som PM (2003) Lymph nodes. In: Som PM, Curtin HD (eds) *Head and Neck Imaging*, 4th edn. Mosby, St. Louis, MO, pp. 1865–1934.
4. Wippold FJ (2007) Head and neck imaging: The role of CT and MRI. *J Magn Reson Imaging* 25:453–465.
5. Kao JLA, Teng MS, Huang D, Genden EM (2008) Adjuvant radiotherapy and survival for patients with node-positive head and neck cancer: An analysis by primary site and nodal stage. *Int J Radiat Oncol Biol Phys* 71:362–370.
6. Mack MG, Rieger J, Baghi M, Bisdas S, Vogl TJ (2008) Cervical lymph nodes. *Eur J Radiol* 66:493–500.
7. Vogl T, Bisdas S (2007) Lymph node staging. *Top Magn Reson Imaging* 18:303–316.
8. Okura M, Kagamiuchi H, Tominaga G, Iida S, Fukuda Y, Kogo M (2005) Morphological changes of regional lymph node in squamous cell carcinoma of the oral cavity. *J Oral Pathol Med: Off Publ Int Assoc Oral Pathol Am Acad Oral Pathol* 34:214–219.
9. Sumi M, Van Cauteren M, Nakamura T (2006) MR microimaging of benign and malignant nodes in the neck. *AJR Am J Roentgenol* 186:749–757.
10. Vogl TBS (2009) Cervical adenopathy and neck masses. In: Haaga JR, Dogra VS, Forsting M, Gilkeson RC, Ha HK, Sundaram M (eds) *CT and MRI of the Whole Body*, 5th edn. Mosby, Philadelphia, PA, pp. 639–669.
11. Mancuso AA, Harnsberger HR, Muraki AS, Stevens MH (1983) Computed tomography of cervical and retropharyngeal lymph nodes: Normal anatomy, variants of normal, and applications in staging head and neck cancer. Part I: Normal anatomy. *Radiology* 148:709–714.
12. Mancuso AA, Maceri D, Rice D et al. (1981) CT of cervical lymph node cancer. *AJR Am J Roentgenol* 136:381–385.
13. van den Brekel MW, Stel HV, Castelijns JA et al. (1990) Cervical lymph node metastasis: Assessment of radiologic criteria. *Radiology* 177:379–384.
14. Friedman M, Shelton VK, Mafee M et al. (1984) Metastatic neck disease: Evaluation by computed tomography. *Arch Otolaryngol Head Neck Surg* 110:443–447.
15. Stevens MH, Harnsberger R, Mancuso AA et al. (1985) Computed tomography of cervical lymph nodes: Staging and management of head and neck cancer. *Arch Otolaryngol Head Neck Surg* 111:735–739.
16. Close LG, Merkel M, Vuitch MF et al. (1989) Computed tomographic evaluation of regional lymph node involvement in cancer of the oral cavity and oropharynx. *Head Neck* 11:309–317.
17. Feinmesser R, Freeman JL, Nojek AM et al. (1987) Metastatic neck disease: A clinical/radiographic/pathologic correlative study. *Arch Otolaryngol Head Neck Surg* 113:1307–1310.
18. Som PM (1992) Detection of metastasis in cervical lymph nodes: CT and MR criteria and differential diagnosis. *AJR Am J Roentgenol* 158:961–969.
19. Rouvière H (1938) *Lymphatic System of the Head and Neck*. Edwards Brothers, Ann Arbor, MI.
20. Poirer P, Charpy A (1909) *Traite d'anatomie humaine*, 2nd edn. Masson, Paris, France.
21. Trotter HA (1930) The surgical anatomy of the lymphatics of the head and neck. *Ann Otol Rhinol Laryngol* 39:384–397.

22. Mancuso AA, Harnsberger HR, Muraki AS et al. (1983) Computed tomography of cervical and retropharyngeal lymph nodes: Normal anatomy, variants of normal, and application in staging head and neck cancer. II. Pathology. *Radiology* 148:715–723.

23. Spiro RH (1985) The management of neck nodes in head and neck cancer: A surgeon's view. *Bull N Y Acad Med* 61:629–637.

24. Som PM, Norton KI, Shugar JMA et al. (1987) Metastatic hypernephroma to the head and neck. *AJR Am J Roentgenol* 8:1103–1106.

25. Medina JE (1989) A rational classification of neck dissections. *Otolaryngol Head Neck Surg* 100:169–176.

26. Beahrs OH, Henson DE, Hutter RVP et al. (1988) *Manual for Staging Cancer*, 3rd edn. Lippincott, Philadelphia, PA.

27. Robbins KT (1991) *Pocket Guide to Neck Dissection and TNM Staging of Head and Neck Cancer*. American Academy of Otolaryngology–Head and Neck Surgery Foundation, Alexandria, VA, pp. 1–31.

28. Fleming ID, Cooper JS, Henson DE et al. (1997) *American Joint Committee on Cancer Staging Manual*, 5th edn. Lippincott Raven, Philadelphia, PA.

29. van den Brekel MWM (1992) Assessment of lymph node metastases in the neck: A radiological and histopathological study. University of Amsterdam, Utrecht, the Netherlands, pp. 1–152.

30. Curtin HD, Ishwaran H, Mancuso AA et al. (1998) Comparison of CT and MR imaging in staging of neck metastases. *Radiology* 207:123–130.

31. Robbins KT (1998) Classification of neck dissection: Current concepts and future considerations. *Otolaryngol Clin North Am* 31:639–655.

32. Atula T, Silvoniemi P, Kurki T et al. (1997) The evaluation and treatment of the neck in carcinoma of the oral cavity. *Acta Otolaryngol Suppl* 529:223–225.

33. Atula T, Varpula M, Kurki T et al. (1997) Assessment of cervical lymph node status in head and neck cancer patients: Palpation, computed tomography and low field magnetic resonance imaging compared with ultrasound-guided fine needle aspiration cytology. *Eur J Radiol* 25:152–161.

34. Don D, Anzai Y, Lufkin R et al. (1995) Evaluation of cervical lymph node metastases in squamous cell carcinoma of the head and neck. *Laryngoscope* 105:669–674.

35. Jones A, Roland N, Field J et al. (1994) The level of cervical lymph node metastases: Their prognostic relevance and relationship with head and neck squamous carcinoma primary sites. *Clin Otolaryngol* 19:63–69.

36. Kaji A, Mohuchy T, Swartz J et al. (1997) Imaging of cervical lymphadenopathy. *Semin Ultrasound CT MR* 18:220–249.

37. Maremonti P, Califano L, Longo F et al. (1997) Detection of latero-cervical metastases from oral cancer. *J Craniomaxillofac Surg* 25:149–152.

38. Shah J, Medina J, Shaha A et al. (1993) Cervical lymph node metastasis. *Curr Probl Surg* 30:1–335.

39. Anzai Y, Blackwell K, Hirschowitz S et al. (1994) Initial clinical experience with dextran-coated superparamagnetic iron oxide for detection of lymph node metastases in patients with head and neck cancer. *Radiology* 192:709–715.

40. Anzai Y, Prince M (1997) Iron oxide-enhanced MR lymphography: The evaluation of cervical lymph node metastases in head and neck cancer. *J Magn Reson Imaging* 7:75–81.

41. Dooms G, Hricak H, Crooks L et al. (1984) Magnetic resonance imaging of the lymph nodes: Comparison with CT. *Radiology* 153:719–728.

42. Dooms G, Hricak H, Moseley M et al. (1985) Characterization of lymphadenopathy by magnetic relaxation times: Preliminary results. *Radiology* 155:691–697.

43. Lewin J, Curtin H, Ross J et al. (1994) Fast spin-echo imaging of the neck: Comparison with conventional spin-echo, utility of fat suppression, and evaluation of tissue contrast characteristics. *AJNR Am J Neuroradiol* 15:1351–1357.

44. Mizowaki T, Nishimura Y, Shimada Y et al. (1996) Optimal size criteria of malignant lymph nodes in the treatment planning of radiotherapy for esophageal cancer: Evaluation by computed tomography and magnetic resonance imaging. *Int J Radiat Oncol Biol Phys* 36:1091–1098.

45. Steinkamp H, Hosten N, Richter C et al. (1994) Enlarged cervical lymph nodes at helical CT. *Radiology* 191:795–798.

46. Yousem D, Hurst R (1994) MR of cervical lymph nodes: Comparison of fast spin-echo and conventional spin-echo T_2W scans. *Clin Radiol* 49:670–675.

47. Yousem D, Montone K, Sheppard L et al. (1994) Head and neck neoplasms: Magnetization transfer analysis. *Radiology* 192:703–707.

48. Yousem D, Som P, Hackney D et al. (1992) Central nodal necrosis and extracapsular neoplastic spread in cervical lymph nodes: MR imaging versus CT. *Radiology* 182:753–759.

49. Peter MS, Curtin HD (2011) *Head and Neck Imaging*, 5th edn. Mosby, New York.

50. Bruneton J, Roux P, Caramella E et al. (1984) Ear, nose, and throat cancer: Ultrasound diagnosis of metastasis to cervical lymph nodes. *Radiology* 152:771–773.

51. Friedman M, Roberts N, Kirshenbaum G, Colombo J (1993) Nodal size of metastatic squamous cell carcinoma of the neck. *Laryngoscope* 103:854–856.

52. Som P (1987) Lymph nodes of the neck. *Radiology* 165:593–600.

53. Steinkamp H, Cornhel M, Hosten N et al. (1995) Cervical lymphopathy: Ratio of long- to short-axis diameter as a predictor for malignancy. *Br J Radiol* 68:266–270.

54. van den Brekel MW, Castelijns JA, Snow GB et al. (1998) The size of lymph nodes in the neck on sonograms as a radiologic criterion for metastasis: How reliable is it? *AJNR Am J Neuroradiol* 19:695–700.

55. van den Brekel MW, Castelijns JA (2000) Imaging of lymph nodes in the neck. *Semin Roentgenol* 1:42–53.

56. Hoang JK, Vanka J, Ludwig BJ, Glastonbury CM (2013) Evaluation of cervical lymph nodes in head and neck cancer with CT and MRI: Tips, traps, and a systematic approach. *AJR Am J Roentgenol* 200:17–25.

57. Chong VF, Fan YF, Khoo JB (1996) MRI features of cervical nodal necrosis in metastatic disease. *Clin Radiol* 51:103–109.

58. King AD, Lei KI, Ahuja AT (2004) MRI of neck nodes in non-Hodgkin's lymphoma of the head and neck. *Br J Radiol* 77:111–115.

59. Goldenberg D, Begum S, Westra WH et al. (2008) Cystic lymph node metastasis in patients with head and neck cancer: An HPV-associated phenomenon. *Head Neck* 30:898–903.

60. Hudgins PA, Gillison M (2009) Second branchial cleft cyst: Not!! *AJR Am J Roentgenol* 30:1628–1629.

61. Yousem D, Hatabu H, Hurst R et al. (1995) Carotid artery invasion by head and neck masses: Prediction with MR imaging. *Radiology* 95:715–720.

62. Myers JN, Greenberg JS, Mo V, Roberts D (2001) Extracapsular spread. A significant predictor of treatment failure in patients with squamous cell carcinoma of the tongue. *Cancer* 92:3030–3036.

63. Cummings BJ (1993) Radiation therapy and the treatment of the cervical lymph nodes. In: Cummings CW, Fredrickson JM, Harker LA et al. (eds) *Otolarynogology Head and Neck Surgery*, 2nd edn. Mosby Year Book, St. Louis, MO, pp. 1626–1648.

64. Ljumanovic R, Langendijk JA, Hoekstra OS, Leemans CR, Castelijns JA (2006) Distant metastases in head and neck carcinoma: Identification of prognostic groups with MR imaging. *Eur J Radiol* 60:58–66.

65. King AD, Tse GM, Yuen EH et al. (2004) Comparison of CT and MR imaging for the detection of extranodal neoplastic spread in metastatic neck nodes. *Eur J Radiol* 52:264–270.

66. van den Brekel MW, Castelijns JA (2005) What the clinician wants to know: Surgical perspective and ultrasound for lymph node imaging of the neck. *Cancer Imaging* 5:41–49.

67. Ager A, Humphries M (1990) Use of synthetic peptides to probe lymphocyte high endothelial cell interactions. Lymphocytes recognize a ligand on the endothelial surface which contains the CS1 adhesion motif. *Int Immunol* 2:921–928.

68. Batsakis J (1979) *Tumour of the Head and Neck: Clinical and Pathological Considerations*. Williams & Wilkins, Baltimore, MD, pp. 240–250.

69. Collins S (1987) Controversies in management of cancer of the neck. In: Thawley S, Panje W (eds) *Comprehensive Management of Head and Neck Tumours*. Saunders, Philadelphia, PA, pp. 1336–1443.

70. Snow G, Annyas A, Slooten EV et al. (1982) Prognostic factors of neck node metastasis. *Clin Otolaryngol* 7:185–192.

71. Gor DM, Langer JE, Loevner LA (2006) Imaging of cervical lymph nodes in head and neck cancer: The basics. *Radiol Clin North Am* 44:101–110.

72. Biller H, Urken M, Lawson W et al. (1988) Carotid artery resection and bypass for neck carcinoma. *Laryngoscope* 98:181–183.

73. Yoo GH, Hocwald E, Korkmaz H et al. (2000) Assessment of carotid artery invasion in patients with head and neck cancer. *Laryngoscope* 110:386–390.

74. Gritzmann N, Grasl MC, Helmer M, Steiner E (1990) Invasion of the carotid artery and jugular vein by lymph node metastases: Detection with sonography. *AJR Am J Roentgenol* 154:411–414.

75. Zaragoza L, Sendra F, Solano J, Garrido V, Martinez-Morillo M (1993) Ultrasonography is more effective than computed tomography in excluding invasion of the carotid wall by cervical lymphadenopathies. *Eur J Radiol* 17:191–194.

76. Rothstein SG, Persky MS, Horii S (1988) Evaluation of malignant invasion of the carotid artery by CT scan and ultrasound. *Laryngoscope* 98:321–324.

77. Langman AW, Kaplan MJ, Dillon WP, Gooding GA (1989) Radiologic assessment of tumour and the carotid artery: Correlation of magnetic resonance imaging, ultrasound, and computed tomography with surgical findings. *Head Neck* 11:443–449.

78. Reilly MK, Perry MO, Netterville JL, Meacham PW (1992) Carotid artery replacement in conjunction with resection of squamous cell carcinoma of the neck: Preliminary results. *J Vasc Surg* 15:324–329.

79. Vandecaveye V, De Keyzer F, Vander Poorten V, Dirix P, Verbeken E, Nuyts S, Hermans R (2009) Head and neck squamous cell carcinoma: Value of diffusion-weighted MR imaging for nodal staging. *Radiology* 251:134–146.

80. Maya MM, Lo WM, Kovanlikaya I (2003) Temporal bone tumors and cerebello-pontine angle lesions. In: Som PM, Curtin HD (eds) *Head and Neck Imaging*, 4th edn. Mosby, St. Louis, MO, pp. 1275–1360.

81. Sigal R (1996) Oral cavity, oropharynx, and salivary glands. *Neuroimaging Clin North Am* 6:379–400.

82. Alper F, Turkyilmaz A, Kurtcan S, Aydin Y, Onbas O, Acemoglu H, Eroglu A (2011) Effectiveness of the STIR turbo spin-echo sequence MR imaging in evaluation of lymphadenopathy in esophageal cancer. *Eur J Radiol* 80:625–628.

83. King AD, Tse GM, Ahuja AT, Yuen EH, Vlantis AC, To EW, van Hasselt AC (2004) Necrosis in metastatic neck nodes: Diagnostic accuracy of CT, MR imaging, and US. *Radiology* 230:720–726.

84. Kawai Y, Sumi M, Nakamura T (2006) Turbo short tau inversion recovery imaging for metastatic node screening in patients with head and neck cancer. *AJNR Am J Neuroradiol* 27:1283–1287.

85. Weber AL, Sabates NR (1996) Survey of CT and MR imaging of the orbit. *Eur J Radiol* 22:42–52.

86. Hayes CE, Tsuruda JS, Mathis CM, Maravilla KR, Kliot M, Filler AG (1997) Brachial plexus: MR imaging with a dedicated phased array of surface coils. *Radiology* 203:286–289.

87. Abdel Razek AA, Gaballa G, Elhawarey G, Megahed AS, Hafez M, Nada N (2009) Characterization of pediatric head and neck masses with diffusion-weighted MR imaging. *Eur Radiol* 19:201–208.

88. Fischbein NJ, Noworolski SM, Henry RG, Kaplan MJ, Dillon WP, Nelson SJ (2003) Assessment of metastatic cervical adenopathy using dynamic contrast-enhanced MR imaging. *AJNR Am J Neuroradiol* 24:301–311.

89. Vaupel P (1998) Tumor blood flow. In: Molls M, Vaupel P (eds) *Blood Perfusion and Microenvironment of Human Tumors*. Springer-Verlag, Berlin, Germany, pp. 41–46.

90. Evelhoch JL (1999) Key factors in the acquisition of contrast kinetic data for oncology. *J Magn Reson Imaging* 10:254–259.

91. Furukawa M, Parvathaneni U, Maravilla K, Richards TL, Anzai Y (2013) Dynamic contrast-enhanced MR perfusion imaging of head and neck tumors at 3 Tesla. *Head Neck* 35:923–929.

92. Abdel Razek AA, Gaballa G (2011) Role of perfusion magnetic resonance imaging in cervical lymphadenopathy. *J Comput Assist Tomogr* 35:21–25.

93. Sumi M, Sakihama N, Sumi T, Morikawa M, Uetani M, Kabasawa H et al. (2003) Discrimination of metastatic cervical lymph nodes with diffusion weighted MR imaging in patients with head and neck cancer. *AJNR Am J Neuroradiol* 24:1627–1634.

94. Abdel Razek AA, Soliman NY, Elkhamary S et al. (2006) Role of diffusion-weighted MR imaging in cervical adenopathy. *Eur Radiol* 16:1468–1477.

95. Wang J, Takashima S, Takayama F et al. (2001) Head and neck lesions: Characterization with diffusion weighted echo-planar MR imaging. *Radiology* 220:621–630.

96. King AD, Ahuja AT, Yeung DK et al. (2007) Malignant cervical lymphadenopathy: Diagnostic accuracy of diffusion-weighted MR imaging. *Radiology* 245:806–813.

97. Lang P, Wendland MF, Saeed M et al. (1998) Osteogenic sarcoma: Noninvasive in vivo assessment of tumor necrosis with diffusion-weighted MR imaging. *Radiology* 206:227–235.

98. Nakamatsu S, Matsusue E, Miyoshi H, Kakite S, Kaminou T, Ogawa T (2012) Correlation of apparent diffusion coefficients measured by diffusion-weighted MR imaging and standardized uptake values from FDG PET/CT in metastatic neck lymph nodes of head and neck squamous cell carcinomas. *Clin Imaging* 36:90–97.

99. Misselwitz B, Platzek J, Weinmann HJ (2004) Early MR lymphography with gadoflurine M in rabbits. *Radiology* 231:682–688.

100. Mayer MN, Kraft SL, Bucy DS, Waldner CL, Elliot KM, Wiebe S (2012) Indirect magnetic resonance lymphography of the head and neck of dogs using Gadofluorine M and a conventional gadolinium contrast agent: A pilot study. *Can Vet J* 53:1085–1090.

101. Choi SH, Han MH, Moon WK, Son KR, Won JK, Kim JH, Kwon BJ, Na DG, Weinmann HJ, Chang KH (2006) Cervical lymph node metastases: MR imaging of gadofluorine M and monocrystalline iron oxide nanoparticle-47 in a rabbit model of head and neck cancer. *Radiology* 241:753–762.

102. Spuentrup E, Ruhl K, Weigl S et al (2010) MR imaging of lymph nodes using Gadofluorine M: Feasibility in a swine model at 1.5 and 3T. *Rofo* 182:698–705.

103. King AD, Yeung DK, Ahuja AT, Yuen EH, Ho SF, Tse GM, van Hasselt AC (2005) Human cervical lymphadenopathy: Evaluation with in vivo 1H-MRS at 1.5 T. *Clin Radiol* 60:592–598.

104. Aboagye EO, Bhujwalla ZM (1999) Malignant transformation alters membrane choline phospholipid metabolism of human mammary epithelial cells. *Cancer Res* 59:80–84.

105. Bisdas S, Baghi M, Huebner F, Mueller C, Knecht R, Vorbuchner M, Ruff J, Gstoettner W, Vogl TJ (2007) In vivo proton MR spectroscopy of primary tumours, nodal and recurrent disease of the extracranial head and neck. *Eur Radiol* 17:251–257.

106. Star-Lack JM, Adalsteinsson E, Adam MF et al. (2000) In vivo 1 H MR spectroscopy of head and neck lymph node metastasis and comparison with oxygen tension measurements. *AJNR Am J Neuroradiol* 21:183–193.

107. Mukherji SK, Schiro S, Castillo M, Kwock L, Muller KE, Blackstock W (1997) Proton MR spectroscopy of squamous cell carcinoma of the extracranial head and neck: In vitro and in vivo studies. *AJNR Am J Neuroradiol* 18:1057–1072.

108. Huang W, Roche P, Shindo M, Madoff D, Geronimo C, Button T (2000) Evaluation of head and neck tumour response to therapy using in vivo 1 H MR spectroscopy: Correlation with pathology. *Proc Int Soc Magn Reson Med* 8:552.

109. Maheshwari SR, Mukherji SK, Neelon B et al. (2000) The choline/creatine ratio in five benign neoplasms: Comparison with squamous cell carcinoma by use in vitro spectroscopy. *AJNR Am J Neuroradiol* 21:1930–1935.

110. King AD, Yeung DKW, Ahuja AT, Leung SF, Tse GMK, van Hasselt AC (2004) In vivo proton MR spectroscopy of primary and nodal nasopharyngeal carcinoma. *AJNR Am J Neuroradiol* 25:484–490.

111. Adalsteinsson E, Spielman DM, Pauly JM, Terris DJ, Sommer G, Macovski A (1998) Feasibility study of lactate imaging of head and neck tumours. *NMR Biomed* 11:360–369.

112. Gerstle RJ, Aylward SR, Kromhout Schiro S, Mukherji SK (2000) The role of neural networks in improving the accuracy of MR spectroscopy for the diagnosis of head and neck cancer. *AJNR Am J Neuroradiol* 21:1133–1138.

113. Johnson J (1990) A surgeon looks at cervical lymph nodes. *Radiology* 175:607–610.

114. Kowalski L, Medina J (1998) Nodal metastases: Predictive factors. *Otolaryngol Clin North Am* 31:621–637.

115. Woolgar J, Scott J (1995) Prediction of cervical lymph node metastasis in squamous cell carcinoma of the tongue/floor of mouth. *Head Neck* 17:463–472.

116. Olsen K, Caruso M, Foote R et al. (1994) Primary head and neck cancer. Histopathologic predictors of recurrence after neck dissection in patients with lymph node involvement. *Arch Otolaryngol Head Neck Surg* 120:1370–1374.

19

Magnetic Resonance Imaging of the Eye and Orbit

Pradipta C. Hande

CONTENTS

19.1 Introduction

The globe within the orbit and the visual pathway is a specialized and intricate system of the central nervous system (CNS). Magnetic resonance imaging (MRI) provides an excellent neuroanatomic detail and is therefore the most sensitive tool in evaluation of pathologic processes affecting this system. It is important to be familiar with the detailed anatomy and physiology of these structures to interpret the MRI in a clinical setting.

19.2 Embryology

Knowledge of the embryology of the eye and orbit helps to understand various congenital anomalies and developmental diseases.

Early at the third week of gestation, *optic grooves* (sulci) are formed on either side of the cranial end of the embryo lined by neuroectoderm. The globe develops from the neuroectoderm of the forebrain and neural crest cells, ectodermal cells of the head, and the mesoderm present between these layers.[1] These *diverticulae* extend from the lateral aspects of the forebrain (diencephalon) and hollow out to form *optic vesicles* which are closely apposed to surface ectoderm. The cavity of optic vesicle is continuous with the cavity of diencephalon through the *optic stalk*. The retina, optic nerve (ON) fibers, and smooth muscles are derived from the neuroectoderm.[1] The corneal and conjunctival epithelium, lacrimal and tarsal glands, and lens are formed from the surface ectoderm. The adjacent ectoderm is stimulated to thicken to form the *lens placode* that further invaginates into the double-layered *optic cup* to become the *lens pit* (which hollows out into the *lens vesicle*) below the surface ectoderm attached by the narrow optic stalk that is later absorbed.

At the inferonasal aspect of the optic cup, a groove called *optic (choroidal) fissure* develops and runs along the optic stalk for a distance. The hyaloid artery[1] (from dorsal ophthalmic artery that develops into the definitive ophthalmic artery) and vein run through the fissure up to the posterior part of the lens, which in the third trimester regresses and disappears. By the seventh week, the edges of the fissure close to form a tubelike *optic canal* inside the optic stalk and the failure of closure results in a hole or defect called *coloboma*.

The retina develops from the lining of optic cup with inner neural and outer pigmented layers. The inner neuroblastic layer of the neural part of the retina forms the ganglion cells that develop into nerve fibers passing into the walls of the optic stalk to form the ON. The *primitive epithelial papilla* at the superior end of the embryonic fissure becomes the *ON head* once the axons pass through it. The deepest part of embryonic optic fissure forms the *optic disc* at the center of the floor of optic cup. Lateral to the disc is the macular area where localized increase of nuclei in the ganglion cell layer occurs, and later in the seventh month, displacement of these cells peripherally leaves a central shallow depression, the *fovea centralis*.[1]

The surrounding mesenchyme along with the neural crest cells condenses to form sclera and choroid, corneal stroma, anterior chamber, part of vitreous and vessels, and connective tissue.[1]

19.3 Relevant Imaging Anatomy of the Eye and the Orbit

The orbits are cone-shaped bony cavities with the base (orbital opening) anterolaterally converging posteriorly and medially to the apex, forming a pyramid-like structure.[2] The rigid bony orbital walls are formed by seven bones, namely, frontal, sphenoid, ethmoid, lacrimal,

maxilla, zygomatic, and palatine bones.[2] The orbits are surrounded by air-containing paranasal sinuses and nasal cavity medially. The bony orbital ventral margin (rim) is quadrilateral, and the maximum width of the orbit occurs about 10 mm deep to the anterior orbital opening.[1,2]

The lateral wall is the thickest formed by the greater wing of sphenoid and frontal process of the zygomatic bone. The roof of the orbit is formed by the orbital plate of frontal bone and major part of the lesser wing of the sphenoid bone. The orbital roof forms part of the floor of the anterior cranial fossa. Superiorly and posteriorly, on the medial aspect lies the bony tunnel called *optic canal* through which runs the ON, ophthalmic artery, and sympathetic nerves which exit through the *optic foramen*. The orbit thus communicates with the intracranial compartment, suprasellar cisterns and cavernous sinus via the optic canal. Anterolaterally is the lacrimal fossa for the orbital part of the lacrimal gland. The floor is the roof of the maxillary antrum, which is relatively thin and slopes upward posteromedial aspect being higher than the anteromedial portion. Medially, the orbit is separated from the ethmoidal air cells anteriorly by the very thin orbital plate of ethmoid (lamina papyracea). The anteromedial aspect has the lacrimal groove for the lacrimal sac, which communicates with the nasal cavity through the nasolacrimal duct.

The orbit is a rigid bony box and the walls are prone to fractures due to head injury and facial trauma. The orbital walls are lined by loosely adherent periosteum (periorbita), which thus has potential spaces (extraperiosteal and subperiosteal) and is continuous with the dura of the ON. Thus, posteriorly located trauma or surgery can result in a cerebrospinal fluid (CSF) leak. Anteriorly, the orbital septum is the continuation of the periosteum at the base of the orbit along the anterior margins. The orbicularis oculi is a flat muscle that surrounds the orbital margin and fibers sweep concentrically across the eyelids anteriorly to the orbital septum.

The main contents of the orbit include globe, extraocular muscles (EOMs), blood vessels, cranial nerves (II with meninges, III, IV, V, VI), sympathetic and parasympathetic nerves, lacrimal gland, and apparatus surrounded by fat and connective tissues. At the orbital apex,[2] the optic canal, superior orbital fissure (SOF), and inferior orbital fissure are apertures that transmit vital nerves and vessels to and from the orbit and brain (Table 19.1).

The EOMs include the four recti, namely, the thickest medial rectus (MR), lateral rectus (LR), superior rectus (SR) and inferior rectus (IR) and the superior and inferior obliques, are striated muscles that control ocular movements which can be visualized in the multiplanar

TABLE 19.1

Foramina and Fissures of the Orbit

Anatomic Orbital Fissure/Foramen	Major Contents/Structures Transmitted	Salient Features and Significance on Imaging
Supraorbital foramen/notch	• Supraorbital nerve • Branch of ophthalmic division of V1 CN • Supraorbital blood vessels	Notch along the superior margin of the orbit anteriorly formed by the frontal bone on the medial aspect
Optic canal and optic foramen	• Optic nerve with its meninges (II CN) • ONSC • Ophthalmic artery • Sympathetic nerves	Posterior aspect of the roof of the orbit between the roots of lesser wing of sphenoid and body of sphenoid. Intracanalicular ON well visualized. Check relationship of optic nerve with the sphenoid sinuses, important forewarning for endoscopic sinus surgery
SOF	• Oculomotor (III) • Trochlear (IV), abducens (V1 CN), terminal branches of ophthalmic (V1 CN) • Ophthalmic veins • Lacrimal and frontal nerve • Lacrimal and meningeal arteries	Comma-shaped fissure inferolateral to optic canal separated by bony strut Communicates with middle cranial fossa Gives attachment to the lateral part of common tendinous ring of Zinn for origin of the EOM
IOF	• Maxillary nerve (V2 CN) • Zygomatic nerve • Infraorbital vessels • Inferior ophthalmic vein toward cavernous sinus • Communicating vein from pterygoid plexus	IOF separates the floor from the lateral wall 1–1.5 cm posterior to anterior rim, communicates with pterygopalatine fossa medially and posteriorly, with temporal and retromaxillary fossae anteriorly and laterally
Infraorbital groove/fissure/canal/foramen	• Infraorbital nerve continuation of maxillary (V2 CN)	Notch in the medial lip of the IOF, passes anteriorly along the floor, 1 cm posterior to the inferior orbital margin The IO canal is formed by the groove passing into the floor, opens anteriorly over the maxilla as the IO foramen 1 cm below the inferior rim
Nasolacrimal canal	• Nasolacrimal duct (NLD)	Anterior aspect of the medial wall has the lacrimal groove for lacrimal sac NLD opens in the inferior meatus of nasal cavity
Anterior ethmoidal foramen	• Anterior ethmoidal vessels and nerves	Anterior and posterior aspect of medial wall of orbit, at the level of floor of the anterior cranial fossa Surgical incision for external ethmoidectomy should be inferior to the foramina to prevent entry into anterior cranial fossa
Posterior ethmoidal foramen	• Posterior ethmoidal vessels and nerves	Along the lateral wall
Canals at sphenozygomatic suture	• Zygomaticotemporal nerve and artery • Zygomaticofacial nerve and artery	At the junction of the floor and the lateral wall

MRI images. The recti except inferior rectus have a common origin from the *tendinous ring (annulus) of Zinn*[2] that encloses the optic foramen and medial aspect of SOF and is continuous with the dural sheath of the ON and the *periorbita*. The vertical recti are well seen in coronal (Figure 19.1) and parasagittal planes, and the entire horizontal recti are seen in the axial images (Figure 19.2). The four recti course along the adjacent walls and attach anteriorly onto the globe forming a muscle cone tapering toward the apex.

This muscle cone along with its intermuscular septa divides the orbit into compartments[3,4] or spaces, namely, intraconal (central orbital space), conal, and extraconal (peripheral space), which are well identified on imaging in both axial (Figure 19.3) and sagittal planes and the intraconal space is well seen in coronal sections.[2]

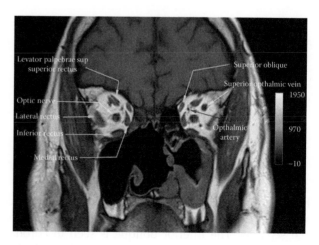

FIGURE 19.1
Coronal T_1WI demonstrating the normal anatomy of the orbit.

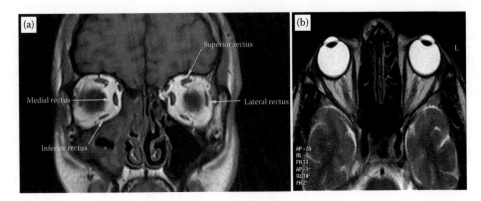

FIGURE 19.2

(a) Coronal T_1WI demonstrating the normal anatomy of the four recti of orbit. (b) Axial T_2WI showing the medial and lateral recti within the orbital fat.

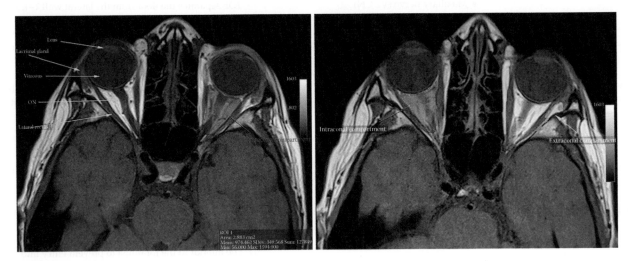

FIGURE 19.3

Axial T_1WI demonstrates the intraconal and extraconal compartments.

FOCUS POINT

This descriptive strategy is very useful for anatomic localization of masses because certain lesions have a predeliction to present in certain compartments.[2] This helps in the radiologic approach to narrow the differential diagnosis of orbital lesions and pathologies which guide management and plan surgery.

The superior oblique muscle is best visualized on coronal images[1] and the trochlea and tendinous portion on axial sections. The inferior oblique is seen in coronal, sagittal, and parasagittal planes. The levator palpebra superioris is the main voluntary muscle that elevates the upper eyelid and lies close to the superior rectus and is seen separately in anterior coronal images and on sagittal and parasagittal MR images.

On MRI, the larger vessels are seen as signal voids and can be identified in the coronal and axial planes (Figure 19.4). Ophthalmic artery can be identified at the apex inferior to the ON and then loops to its superomedial aspect. The superior ophthalmic vein (SOV) originates in the extraconal space in the anteromedial aspect that courses near the trochlea to pass through the muscle cone above the ON and below the superior rectus muscle exiting through the SOF.

Within the orbit the structures are embedded in fat with fibroelastic tissue forming a reticulum which is divided into peripheral orbital fat (outside the muscle cone) and central orbital fat (within muscle cone).

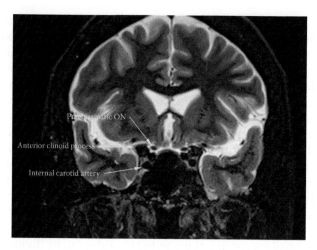

FIGURE 19.4
Coronal T_2WI in the region of the anterior clinoid process in the posterior aspect of the orbit demonstrates various structures in relation to the ON.

FOCUS POINT

It is also important to localize focal abnormalities such as calcifications, cysts, granulomas within the orbit in relation to the globe as *intraocular or extraocular.*[2]

The *globe* is made up of two segments[1] divided by the ciliary body into the smaller, transparent anterior segment, and larger, relatively opaque posterior segment which forms five-sixths of the eyeball. The centers of curvature of each of the segments are the anterior pole at the center of the cornea and the posterior pole at the center of the posterior segment, and are joined by a line called geometric optic axis that does not coincide with the visual axis as the fovea centralis is inferior and temporal to the posterior pole and the equator lies between the two poles. *Tenon's capsule* is a fascial sheath of fibro-elastic membrane that surrounds the globe from the ON to the ciliary body enclosing the posterior four-fifths of the eyeball.

The eyeball contains the most water-laden tissue in the body, namely, vitreous, and the least hydrated tissue, namely, lens which is composed of lens fibers with protein forming liquid crystal covered by lens epithelium and elastic capsule.[1,4] The crescent-shaped anterior chamber containing the fluid aqueous humor and posteriorly the vitreous humor is related to the rigid lens. This difference produces different proton relaxation times for each of these tissues and is the ideal organ for study with MRI. The gel-like vitreous[1] with small protein–water interaction has long T_1 and T_2 relaxation times just shorter than pure water. The normal lens

appears darker than the surrounding fluid-laden tissue on T_2 with ultrashort relaxation time.[1] The ciliary body and lens zonules appear slightly hyperintense to vitreous on T_1-weighted images. The three primary layers consist of outer sclera, middle uvea (choroid, ciliary body, and iris), and the inner retina. The retina is made up of two layers: the inner neural layer and the outer pigment layer, which cannot be resolved by MRI.

Intraocular potential spaces[1,4] are present where fluid can accumulate and separate the layers, which are subhyaloid, subretinal, and suprachoroidal spaces. The hyaloid fossa anterior to the vitreous may be filled with fluid exudates or hemorrhage and is seen on MRI. Tenon's space or episcleral space is another potential space.[1,2]

FOCUS POINT

The orbits are closely related anatomically to the adjacent paranasal sinuses, pterygopalatine fossa,[1] and also the intracranial contents making them vulnerable to extension of disease processes involving these structures. Thus, MR examination should include the imaging of these areas to study the involvement of surrounding anatomic structures in orbital pathologies.

Commonest clinical indications for orbital imaging[2,4] are proptosis, exophthalmos, diminished vision, enophthalmos, diplopia, leukocoria, pain, and epiphora.[5]

19.4 Imaging Modalities

Radiography with standard projections was used earlier, but cross-sectional imaging modalities are more useful[2] and are widely used now.

High-resolution ultrasonography (USG) (B-scan) is used for visualization of the eyeball without harmful effects of ionizing radiation to the lens and is a cost-effective diagnostic tool for ocular abnormalities.[5] Real-time imaging allows dynamic scanning synchronized with ocular movements while performing the scan.[5] The limitation of USG for orbits is its inability to visualize the bony architecture and the orbital apex area, and to assess intracranial extension of various pathologies.[2,5]

Routine computed tomography (CT) exams are obtained parallel to the orbitomeatal line with 3 mm contiguous axial sections, with thinner slices up to 1–2 mm, when required. Multidetector CT (MDCT) has the advantage of rapid acquisition of volumetric dataset (isotropic imaging) allowing excellent multiplanar reconstruction (MPR) images.[5,6]

19.4.1 MRI: Technique and Protocol

The inherent soft tissue contrast of fluid-filled globe, retrobulbar contents within the orbital fat, and bony walls of the orbits delineates the anatomy well.[1,2,5,6]

MR images obtained using head coil on a high-field strength magnet (1.5–3.0 T) enables adequate visualization of the ocular and orbital structures, including the orbital apex and entire visual pathway.[5,6] Short scan times with turbo/fast spin-echo (SE) sequences are preferred to reduce the artifacts due to movements of the eye (Figure 19.5). Routine sequences are T_1-weighted imaging (T_1WI), proton density (PD), T_2-weighted imaging (T_2WI), T_1WI with fat suppression (FS), and inversion recovery techniques such as turbo inversion recovery magnitude (TIRM) with fluid-attenuated inversion recovery (FLAIR) and short-tau inversion recovery (STIR). Thin (3 mm or less) slice thickness with acquisition in at least two scan planes, small field of view (FOV), high-resolution matrix (256 × 256), and T_2WI of the brain is used for evaluation of the entire visual pathways. Three-dimensional sequence acquisition is generally used for MPR. Additional sequences with gradient-echo images can be acquired, depending on the indication. Intravenous (IV) injection of Gadolinium (Gd) chelated contrast agents (upto 0.1 mmol/kg) is useful to study the enhancement patterns of the normal structures and the various lesions of the orbit. Precontrast and postcontrast FS 3D-T_1W images are recommended.

Diffusion-weighted imaging (DWI) with apparent diffusion coefficient (ADC) values are being increasingly used to analyze molecular diffusion at the cellular level.[5,7] This echo-planar imaging technique is being used more these days in evaluation of lesions and tumors of the head and neck. It is likely that diffusion restriction is likely due to intracellular fraction of freely moving water (protons) as well as extracellular molecules and protein, which can alter the viscosity of surrounding tissues.[7] This helps in differentiating similar appearing lesions and pathologic conditions on routine MRI sequences by studying the diffusion restriction and ADC values[5,7] sequences can be acquired with varying *b*-values at least with b = 0, 500, and 1000 s/mm.

Pathologic conditions involving the orbit can be divided into ocular (involving the eyeball or globe) or orbital (excluding the globe).[1]

Proptosis refers to the anterior protrusion of the eyeball and is commonly used synonymously with *exophthalmos* (globe is more prominent) due to increased orbital contents as in thyroid diseases. This can be objectively assessed in relation to globe on axial scans, and at least one-third of the eyeball normally lies posterior to the line joining lateral orbital margins in axial sections[2] (Figure 19.6). The bony interorbital distance (BID)[2] can be measured on axial CT images (Figure 19.6).

Hypertelorism refers to increase in BID and the orbits are farther placed with eyes that are spaced more widely apart than normal compared with *hypotelorism* where BID is reduced.[1,2] The eye and orbits are commonly involved in craniofacial malformations and result in hypotelorism or hypertelorism.[1,2]

Normal adult axial length of the eyeball is 24mm.[1]

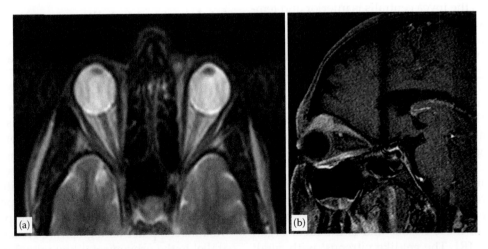

FIGURE 19.5
Artifacts on MRI (a) shows typical ghosting artifacts due to movements of the eyeball during scanning in axial section and (b) thin sagittal section on 1.5 T MRI with head coil shows perpendicular band of loss of signal.

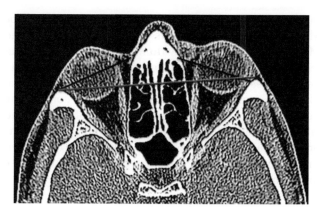

FIGURE 19.6
Axial CT of the orbits at this level shows the globe, the MR, LR muscles within the orbital fat. The BID is drawn joining the zygomatic arch on either side (red line).

19.5 Pathologic Conditions

19.5.1 Pathologies of Globe

19.5.1.1 Congenital Anomalies

Anophthalmia and microphthalmia: Complete absence of the eyeball is rare and may be a result of teratogenic antenatal events with failure of formation of the optic cup.[1,2] Clinical anophthalmos usually reveals a cystic remnant with rudimentary tissue with dystrophic calcification of the orbits. ON along with optic chiasm if present is usually hypoplastic. The EOMs, however, are well developed. MR should image the entire visual pathway, including the lateral geniculate bodies which may be small and rudimentary, and show features of gliosis.[1]

In *microphthalmia*, the globe is underdeveloped and small. The *axial length* is less than 21 mm in an adult or 19 mm in a child at 1 year.[1] This condition may occur as a part of craniofacial syndromes usually as a result of intrauterine infections such as congenital rubella, congenital toxoplasmosis and also seen in persistent hyperplastic primary vitreous (PHPV), and retinopathy of prematurity (ROP). It is due to an insult during the embryologic development *after* the formation of the optic vesicle. However, if the insult is before complete invagination into the optic cup, then there is an associated cyst formation.

Failure of development of eyelids, palpebral fissure, and conjunctiva may cause *cryptophthalmos* with skin-like cornea.[1,2] MRI can demonstrate a rudimentary, cyst-like globe.

At times, microphthalmic eye presents later and can be an acquired abnormality as an outcome of trauma, inflammation, postinfection (herpes, cytomegalovirus), surgery, or radiation, that is, *phthisis bulbi*.

Congenital microphthalmia must be differentiated from phthisis bulbi. There is usually disorganization of the eyeball that is small, and on T_1WI MR, there is hyperintensity peripherally and may show extensive calcification, which is markedly hypointense on all sequences. On CT, the calcification is well seen within a small, irregular shrunken globe (Figure 19.7).

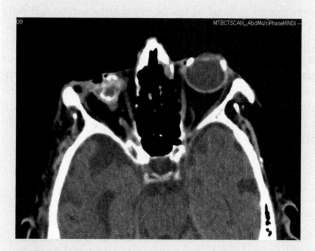

FIGURE 19.7
Phthisis bulbi. Axial CT shows small, disorganized shrunken globe with calcification.

Macrophthalmia: The enlargement of the globe is seen in young patients with congenital glaucoma (buphthalmos) due to increased intraocular pressure stretching of the cornea and sclera or associated with Sturge–Weber syndrome. In neurofibromatosis (NF), there may be facial involvement in 50% patients. Megalophthalmos (axial length >30 mm) is an enlarged eyeball without glaucoma. The commonest cause of increased axial length of the globe is associated with high myopia.[8] This may result in stretching of the scleral–uveal coats due to elongation of eyeball with thinning of the layers causing a *staphyloma*[1] seen posteriorly.

Coloboma: The term means a defect, hole, or fissure where there is absence of tissue and may affect the lens, ciliary body, iris, retina, choroid, and ON,[9] and usually results from the failure of closure of the embryonic optic fissure involving the inferior aspect of the nasal quadrant of the globe.[1,9] Retinal detachments (RDs) are associated with ON colobomas and may be associated with ocular abnormalities such as cyst of the ON or of the hyaloid artery remnant or non-ocular anomalies such as dysplastic ears or encephalocoele.[1]

> **FOCUS POINT**
>
> It is important to exclude orbital encephalocele in suspected colobomatous orbital cyst before surgery and MRI is very useful to identify the ON and the globe unlike the congenital cystic eye.[1,2]

Morning glory anomaly of the disc has been described[10] where the disc is enlarged with a central core of pale glial tissue with annular raised halo of light and may have pigmented subretinal tissue. There may be a colobomatous cyst formation, which at times increases rapidly in size causing progressive proptosis to mimic a neoplasm. MRI reveals a funnel-shaped deformity at the posterior aspect of the globe[1] corresponding to the optic disc associated with a cyst[1,2] and also helps in excluding a mass lesion to differentiate from a tumor.

> **FOCUS POINT**
>
> MRI helps in detecting a typical funnel-shaped deformity at the posterior aspect of the eyeball in congenital coloboma with morning glory anomaly on fundoscopic examination.

19.5.1.2 Ocular Detachments

The separation of the sensory (neural) and the retinal pigment epithelium is termed as *retinal detachment* (RD)[11] and allows fluid to seep into the potential subretinal space, which can be identified on MRI. In rhegmatogenous RD common with degenerative changes, there may be a tear or discontinuity which can be localized on imaging. Non-rhegmatogenous RD is usually seen in children and younger patients due to ocular diseases and can be caused by retraction by a mass or fibroproliferative diseases or secondary to inflammation. It can result from neoplasms such as retinoblastomas (RBs) in children and choroidal tumors such as melanoma and choroidal hemangioma in adults. RD can be primary or secondary, and MRI can help to detect the cause of RD such as tumor or inflammation. There may be an ocular mass with associated subretinal collection which is hyperintense on T_1-weighted images. However, at times, in Coats' disease, subretinal exudates with high protein in inflammatory conditions[11] have similar appearances as opposed to the transudate collection in subretinal space in rhegmatogenous RD, which is hypointense

fluid in T_1WI. As the retinal layers are too thin to be identified due to the limits in resolution in CT and MR, USG is superior in evaluation of RD wherein thin leaves of the detached retina can be visualized and helps in assessing the extent of detachment. In complete RD, membrane-like layers are seen converging posteriorly towards the apex (at the optic disc) and is attached anteriorly towards the ciliary body at the ora serrata. Tractional RD due to scarring of the retina as in proliferative diabetic retinopathy is not so uncommon.

> **FOCUS POINT**
>
> Complete RD has a typical V- or Y-shaped appearance, attached at the ON head on axial views (Figure 19.8) and a characteristic folding membrane on coronal images. Subretinal fluid is well visualized on MRI; transudative collections appear typically hypointense on T_1WI, whereas the exudative fluid is comparatively hyperintense to vitreous due to higher protein content.

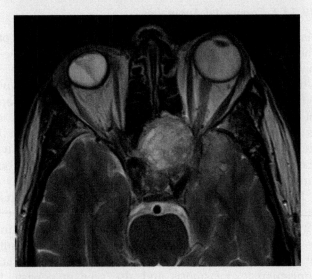

FIGURE 19.8
Axial T_2WI illustrates the right eye with typical Y-shaped complete RD. Also note a large mass lesion at the region of the apex of the left orbit.

Choroidal detachment (CD),[11] however, does not extend up to the optic disc as it is tethered by the vortex veins and has a different appearance. There are two forms: *hemorrhagic* and *serous* type. Hemorrhagic detachment has a C-shaped, arc-like configuration caused by the fixation of the choroid by the posterior ciliary vessels as opposed to the crescent-shaped serous type usually as a consequence of ocular hypotony[1] (Figure 19.9).

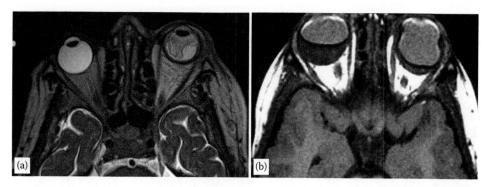

FIGURE 19.9
Axial T_2WI illustrates choroidal detachment: (a) endophthalmitis of the left eye; (b) hypotony of the left eye with arc-like CD.

FOCUS POINT

MRI is excellent and better than CT and USG for choroidal and suprachoroidal fluid especially hematomas. Shifting of subchoroidal fluid in CD detachment is slower than the rapidly shifting fluid in subretinal space in the wall of the eye in RD.

Postsurgical changes: The commonest surgery is cataract and lens extraction with intraocular lens (IOL) implants. On MRI, the biconvex native lens is replaced by a linear hypointense lens implant on T_2WI (Figure 19.10). These are often incidentally visualized during brain imaging for non-orbital problems. MRI is useful for locating displaced or complicated IOL implants and also for detecting associated post-operative complications such as ocular hemorrhage or RD.

19.5.1.3 Intraocular Hemorrhage

Retinal hemorrhage well visualized on MRI in T_1WI may even show layering with fluid–fluid levels depending on the age of the hematoma.

Choroidal hematoma is well seen on MRI as a well-defined, focal biconvex collection appears as a mass-like lesion along the wall of the eyeball which is hyperintense on T_1WI. This needs to be differentiated from choroidal melanoma which is a hyperintense lesion on T_1WI.

FOCUS POINT

Choroidal hematoma resembles choroidal melanoma on T_1WI both appearing hyperintense, but T_2WI helps in differentiating these because the hematomas appear progressively hyperintense as blood ages, whereas the melanomas are hypointense on T_2WI due to its paramagnetic effect.

19.5.1.4 Ocular Inflammation/Infection

Inflammatory disorders affect the globe in local ocular or systemic diseases. Infections may occur usually in immunocompromised patients, and diabetics may cause involvement of the uvea, choroid, and sclera or diffuse endophthalmitis (Figure 19.11).

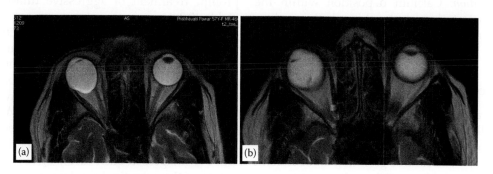

FIGURE 19.10
Post cataract surgery status. (a) Axial T_2WI shows asymmetric elongated right globe with linear hypointense IOL implant with (b) complete RD and associated staphyloma. Compare with normal globe on left..

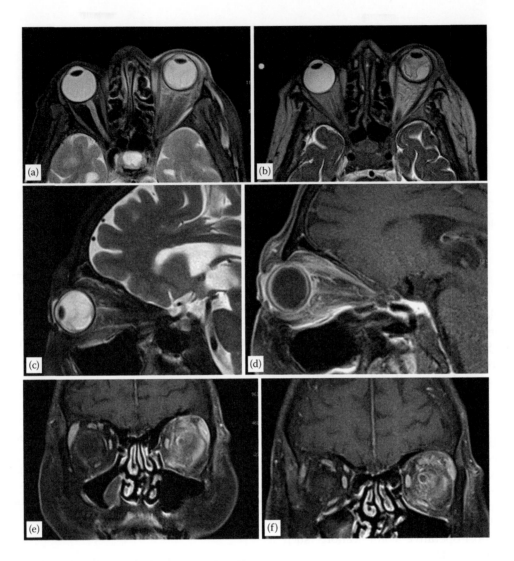

FIGURE 19.11
Panophthalmitis, left. (a) Axial T_2W TIRM illustrate left orbital proptosis with altered signal intensity of the left orbital fat and EOM. Note hyperintensity of eyelid and preseptal space. (b) Axial T_2WI TIRM demonstrates exudative subretinal collection resulting in RD. (c) Sagittal T_2W TIRM shows the exudative RD within the left globe. Postcontrast-enhanced Gd images in (d) sagittal FS T_1WI of left orbit demonstrates thickening of the layers of the left eyeball with blurring of margins with enhancement of the eyelid, orbital septum, globe, orbital fat and EOM which suggests both orbital and ocular inflammation. (e,f) Coronal T_1W FS sections shows the enhancement suggestive of inflammation of the entire left globe and surrounding extraocular orbital contents.

Ocular calcification: Calcium deposition within the globe may be metastatic or dystrophic, which may be a result of inflammation, trauma, or tumors.

19.5.1.5 Leukocoria

This is a clinical sign where there is a white or pink to yellow pupillary reflex due to the reflection of incident light through the pupil.[1] This is due to an opaque intraocular mass, membrane or granulomatous lesions, and even cataractous lens. Other causes are congenital lesions such as Coats' disease, PHPV, or ROP. Identification of the cause is important for early and prompt treatment of aggressive tumors such as RB (Table 19.2).

Coats' disease (primary retinal telengiectasia): This is a retinal disorder where there are idiopathic anomalous telengiectatic retinal vessels with or without aneurysms. Subretinal collection of lipoproteinaceous serum leaking from the telengiectatic vessels may cause exudative RD[1,11] which is hyperintense on T_1WI and it stands out in contrast to the hypointense vitreous. It must be remembered that RB may also be hyperintense on T_1WI and PD images. In such cases, T_2WI helps in differentiation of the subretinal exudates from a mass lesion in patients with RD where the collection is hyperintense compared with an RB that appears hypointense.

TABLE 19.2

Leukokoria

Etiology	Pathologic Condition	MR Imaging Findings: Focus Points
Developmental/congenital	Coats' disease	Subretinal exudates from telengiectatic retinal vessels, protein-rich hyperintense on T_1WI without mass, normal globe and lens morphology
	PHPV	Embryonic hyaloid system persists with fibrovasculature components as enhancing lesion, microphthalmia, Cloquets canal on T_1WI, looks for hemorrhage, excludes mass
	Coloboma	Associated congenital small cystic eye, cystic appearance of ON
	Retinal dysplasis	Abnormal retinal contour, retinal detachment needs to be excluded
	Congenital retinal fold	Undulating retinal layers without associated subretinal fluid exudates
Tumors	Retinoblastomas	Intraocular mass usually with calcification, areas of necrosis, hyperintense to vitreous on T_1WI, markedly hypointense areas due to calcification on T_2WI, various patterns diffuse/focal mass, maybe associated retinal detachments
Infestation	Toxocara	Parasite causes severe endophthalmitis reaction within the globe with inflammatory exudates, intense postcontrast enhancement
Inflammatory	Sclerosing endophthalmitis	Postinflammatory fibrosis with organized exudates, hypointense on T_1WI and T_2WI with no mass lesion
	ROP	Fibrous hyperplasia forming membrane in retrolental region, postcontrast enhancement is variable. History of exposure to prolonged oxygen therapy in premature babies
Degenerative	Cataract	Altered lens signal intensity with shrunken contour, no enhancement on postcontrast T_1WI, looks for associated RD in posterior segment
Trauma	RD (longstanding)	V- or Y-shaped detached membrane, subretinal fluid, looks for associated mass/tumor
	Vitreous hemorrhage (organized)	Longstanding organized blood products identified in T_1WI, fluid–fluid levels, adhesions, vitreous synechiae

---

FOCUS POINT

Coats' disease resembles exophytic RB and has similar signal intensity (SI) on T_1WI appearing hyperintense, but on T_2WI hypointense foci due to the presence of dense calcification in RB are easily seen in contrast to a subretinal collection which is hyperintense. The detached retina may demonstrate postcontrast enhancement.

PHPV: This is a congenital condition where there is failure of complete regression of the embryonic hyaloid vasculature system in the primary vitreous[12] (between the retina and the lens). This remnant of the hyaloid artery is the Cloquet's canal within the secondary vitreous (Figure 19.12), which is seen as a hypointense membrane-like double-layered linear canal within the hyperintense vitreous on T_2-weighted images. This is associated with hyperplasia and proliferation of the embryonic connective tissue.[1] There may be an associated RD also, which may be detected on MRI. The patients usually present with unilateral leukocoria with micro-ophthalmia.

The findings are similar to those seen in ROP in premature infants who are on prolonged oxygen therapy also present with leukocoria. There may be proliferation of peripheral retinal vessels with subsequent hemorrhage. Due to contraction of the organized hematoma and cicatrization, these may be associated with tractional RD and develop a retrolental membrane.

FOCUS POINT

In patients with the white eye reflex, MRI demonstrates the Cloquet's canal in PHPV and more importantly excludes mass lesion in the posterior segment suggestive of RB.

19.5.1.6 Ocular Tumors

RB: It is a primary malignant neoplasm arising from the neuroectodermal cells of the retina and is the commonest intraocular tumor of childhood.[1] Early detection is essential for early management that facilitates better prognosis. It usually affects children below 5 years in up to 90% cases and is thought to be congenital in origin.[26] Clinically, it presents commonly

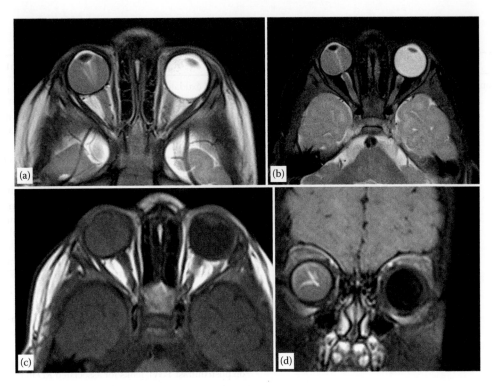

FIGURE 19.12
PHPV right. (a) Axial T_2WI and (b) axial T_2W FS images show smaller right eyeball (microphthalmia) with thin linear hyperintense band from the posterior surface of lens to optic disc within hypointense vitreous (congenital nonattached retina/Cloquet's canal). (c) Axial T_1W reveals mildly hyperintense right globe as compared with left normal globe with hypointense linear band within the vitreous. (d) Coronal T_1W FS image reveals enhancement of the retrolental tissue (hyperplastic tissue) and septum.

with leukocoria[1] associated with proptosis and loss of vision or strabismus may be seen in about 20% cases. In approximately 25%–33% patients, there may be bilateral disease, which represents an autosomal dominant pattern of genetic transmission with a family history.[1,13] The main types of the lesion depending on the growth patterns of the tumor are *exophytic, endophytic,* and *diffuse.*[1,2,5,14] The exophytic type usually grows outward into the subretinal space, resulting in secondary RD which needs to be differentiated from Coats' disease. The endophytic-type lesion grows inward and protrudes into the vitreous (Figure 19.13) and can resemble

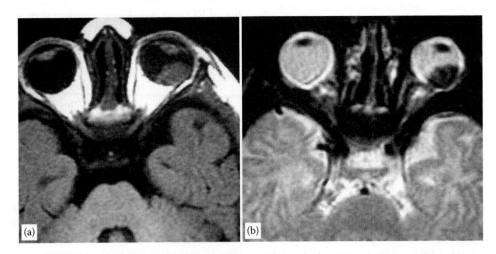

FIGURE 19.13
Retinoblastoma. (a) Axial MR T_1WI shows left intraocular mass which appears as a hyperintense lesion protruding into the hypointense vitreous. (b) Axial T_2WI reveals markedly hypointense lesion to hyperintense vitreous. (From Hande, P.C. and Talwar, I., *Indian J. Radiol. Imaging,* 22, 224, 2012.)

endophthalmitis. A diffuse tumor grows along the retina as a plaque-like mass and simulates inflammatory or hemorrhagic conditions.

On imaging, an intraocular mass with calcification and areas of necrosis, associated RDs, and vitreous involvement can be well visualized. On USG the mass protruding into the globe is well seen. CT can detect the intraocular lesion and orbital extent well and is very sensitive, because >90% of tumors show evidence of calcification[1] (Figure 19.14). However, MRI is very useful in determining the retrobulbar and extraocular spread. When there is bilateral disease, it is important to exclude intracranial midline neuroblastic tumors, involving the pineal gland (trilateral RB), or when there is also an associated suprasellar tumor (tetralateral/quadrilateral RB). On MRI, they are hyperintense to normal vitreous on T_1WI and moderate to markedly hypointense on T_2WI depending on the calcification, which is demonstrated well with contrast administration and thin sections (1.5–3 mm) with FS. It is important to differentiate RB from other benign lesions such as PHPV, ROP, Coats' disease (exophytic), organized subretinal hemorrhage, endophthalmitis (endophytic), and other intraocular masses.[1,4]

FOCUS POINT

USG is useful to visualize the ocular tumor and associated RD. CT is very sensitive for the detection of calcification in the mass, which is an important diagnostic criterion (Figure 19.14). However, MRI is very useful in determining the extraocular spread, including intracranial involvement.[1,2,5]

The staging of RB is essential for the purpose of treatment planning.[14] The newer Murphee classification of groups A–E is more useful for chemotherapy management compared with the older Reese–Ellsworth classification, which was more useful for radiation therapy.[14] It is based on the location and tumor size, extent, invasion of vitreous and adjacent structures, and other prognostic features.

19.5.1.6.1 Malignant Melanomas

These are highly malignant tumors arising from the uvea[1] (choroid, ciliary body, iris). These have a tendency to have early hematogenous dissemination with metastasis in the liver, lung, bone, kidney, and brain.

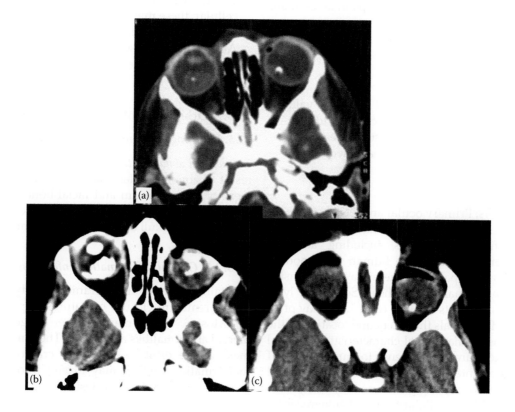

FIGURE 19.14
Bilateral RBs. (a) Axial CT scan shows bilateral intraocular soft tissue density mass lesions arising from the retina with specks of calcifications. (b) Axial CT sections in a case showing dense calcifications within the eyeball bilaterally. (c) Axial CT reveals focal calcification in the retina of left eye. Note the shrunken left globe in cases (a) and (b) suggestive of left phthisis bulbi.

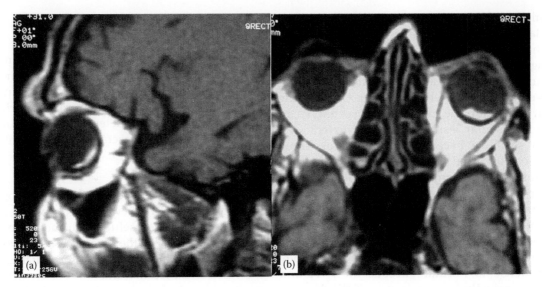

FIGURE 19.15
Ocular melanoma. (a) Sagittal and (b) axial T_1W images show hyperintense ring-shaped posterior intraocular uveal lesion (paramagnetic effect of melanin). (From Hande, P.C. and Talwar, I., *Indian J. Radiol. Imaging*, 22, 224, 2012.)

CT shows a well-marginated, hyperdense, enhancing mass, which must be differentiated from other primary and secondary uveal and choroidal tumors. On MR, the lesions are hyperintense on T_1WI (paramagnetic properties of melanin or hemorrhage) (Figure 19.15) and hypointense on T_2WI. Gd contrast helps in assessment of ON and retrobulbar extension.[1,5,14]

> **FOCUS POINT**
>
> MRI has typical appearances and its role is valuable in evaluation of the extent accurately for surgical planning. Imaging must focus to search for metastatic spread that guides management.

Other ocular tumors: Primary ocular and CNS lymphoma is seen more frequently due to increasing in incidence of immunodeficient conditions, including acquired immune deficiency syndrome in human immunodeficiency virus-positive patients. There may be extensive infiltration of the retinal and ON head. Secondary lymphoma usually manifests as uveal tumor and may present as uveitis. MRI reveals the extent and location of the mass, which has similar signal characteristics as uveal melanomas. These are hyperintense to muscle on T_1WI and mildly hyperintense on T_2WI, and enhance well with contrast. However, on T_2WI these may be hypointense if these are highly cellular tumors. Leukemic infiltration of the eye is known and may involve any layer and may be bilateral; it is usually an indicator of poor prognosis of disease.

> **FOCUS POINT**
>
> MR is useful to image CNS lymphomas which may demonstrate postcontrast enhancement in CT and MR. However, if these tumors are highly cellular, these masses may appear hypointense on T_2WI and resemble uveal melanomas.[1] DWI and ADC values are useful in evaluating these masses distinct from other inflammatory conditions.[7]

19.5.2 Orbital Pathologies

19.5.2.1 Congenital and Development Abnormalities

Craniofacial abnormalities and *craniosynostosis* occur as a result of abnormal blastema stage during the development of the calvarium and facial bones, including the basicranium.[2] This usually cause secondary changes of cranial fossae, skull base, cranial vault, and orbits.

19.5.2.2 Orbital Inflammation/Infections

Orbital malformations in craniofacial dysostosis result from premature closure of sutures, especially coronal synostosis.[2,15] There may be other congenital developmental abnormalities associated. There may be involvement of the orbit in congenital craniofacial vascular malformations, which may extend into the orbit and present as extraconal mass lesions (Figure 19.16).

19.5.2.2.1 Tumorlike Conditions of the Craniofacial Bones

Fibrous dysplasia with extensive involvement of the maxilla, mandible, and skull base may lead to anatomic

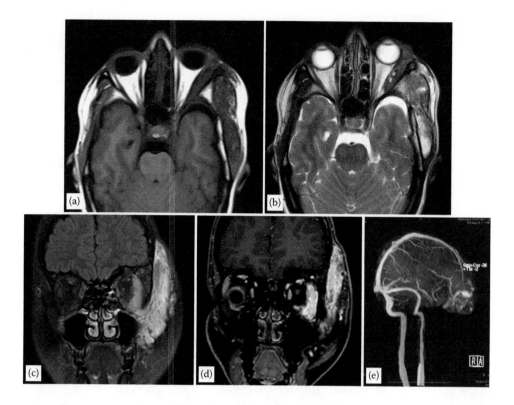

FIGURE 19.16

Facial arteriovenous malformation. (a) Axial T_1W, (b) T_2W, and (c) coronal T_2W FLAIR-TIRM images illustrate a vascular malformation on the left side of the face with extension into the lateral aspect of the orbit in relation to the lateral rectus and inferior rectus muscles. (d) Coronal T_1W FS image reveals intense postcontrast enhancement. (e) MR venogram shows normal intracranial venous sinuses. However, the dilated venous channels are seen within the orbit from the venous malformation.

deformities of the orbit[16] and encroachment of the neurovascular canals.[2] CT is the gold standard for imaging the bony orbit and 3D reformats are very useful (Figure 19.17) in preoperative assessment and postsurgical followup.[17]

> **FOCUS POINT**
>
> MR is the preferred modality for imaging of the orbit and its contents especially the overcrowding at the posterior orbital apex causing compression of the ON and associated brain and spinal abnormalities in evaluation of various craniofacial syndromes.

19.5.2.2.2 Inflammatory Diseases

Infections/orbital cellulitis: Bacterial infections are among the commonest pathologies constituting nearly 60% of primary orbital pathologies,[2,18] majority of which originate from the paranasal sinuses. It is the most common cause of proptosis in children.[2,19] The various stages of orbital cellulitis[2,5] (Figure 19.18) can be well visualized on MRI. Initial inflammatory orbital edema, fat stranding, and phlegmon can progress to the formation of

orbital or subperiosteal abscess (Figure 19.19). Contrast-enhanced CT and MRI are essential to differentiate the different stages which reveal diffuse and intense enhancement of the inflamed orbital contents, including the muscles and also the ON. There can be spread of infection and intraconal and extraconal abscess formation and collections.

> **FOCUS POINT**
>
> MR images of both the orbital and sinonasal inflammatory diseases evaluate the extension of infection from the preseptal compartment into the postseptal compartment. It helps to detect complications such as ophthalmic vein or cavernous sinus thromboses and intracranial extension early[2,19] that helps in the management.

Fungal infections: In patients with diabetes mellitus or immunocompromised states, fungal sinusitis such as mucormycosis or aspergillosis[1] may be aggressive in nature and need to be differentiated from tumors. MRI detects the extent of orbital involvement and extraorbital or intracranial extension (Figure 19.20).

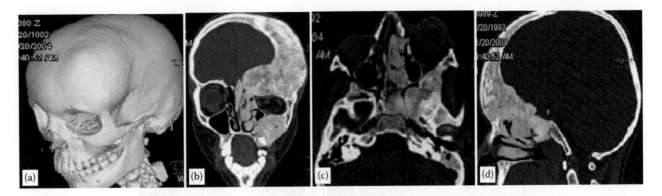

FIGURE 19.17
Craniofacial fibrous dysplasia with left proptosis. CT (a) Surface shaded display (SSD) and MPR (b) coronal, (c) axial, and (d) sagittal images show expansion of diploic space with a ground-glass appearance involving hemicranium and facial bones on the left. Note sparing of the mandible. There is obliteration of sinuses and encroachment of orbit with overcrowding of orbital contents. (From Hande, P.C. and Talwar, I., *Indian J. Radiol. Imaging*, 22, 224, 2012.)

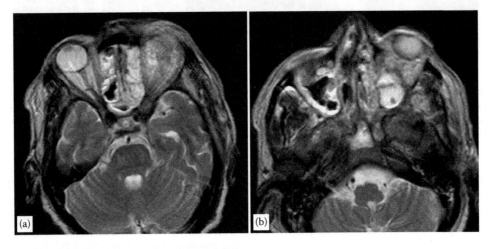

FIGURE 19.18
Orbital cellulitis. (a,b) Axial MR T_2W images show left orbital cellulitis eyelid edema, enhancing thickened sclera, causing proptosis diffusely involving preseptal and postseptal tissues. Retrobulbar fat and EOM inflammation extend up to the orbital apex and associated sinusitis.

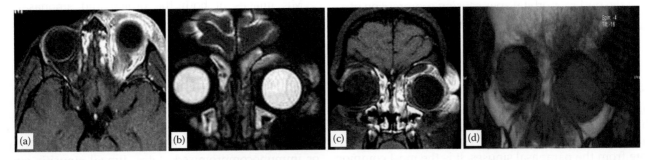

FIGURE 19.19
Orbital cellulitis. (a) Axial MR T_1W FS postcontrast images show left orbital cellulitis with proptosis, enhancing orbital contents and retrobulbar fat with diffuse involvement of preseptal and postseptal compartments and associated subperiosteal abscess which is well localized in coronal images (b,c). (d) VRT images shows boggy collection (phlegmon) with displacement of the globe. (From Hande, P.C. and Talwar, I., *Indian J. Radiol. Imaging*, 22, 224, 2012.)

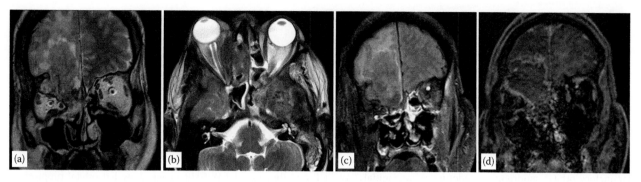

FIGURE 19.20

Fungal sinusitis in a diabetic patient with right proptosis. (a) Coronal T_2WI, (b) axial, and (c) coronal T_2WI TIRM with FLAIR show space-occupying lesion arising from the sinuses with extension into the medial aspect of the right orbit hypointense to orbital fat and intracranial extension with surrounding edema. (d) Postcontrast coronal T_1W FS image shows peripheral enhancement of the lesion. Biopsy from the lesion showed fungal hyphae.

FOCUS POINT

MRI can accurately assess orbital and extraorbital disease; T_2WI demonstrates mycetoma as lesions with hypointensity probably due to the paramagnetic substrates produced by the fungi. However, bony integrity is better demonstrated on CT bone window setting, which is a limitation of MR.

FOCUS POINT

IOS, orbital lymphoid lesions, and edema of orbital cellulitis appear similar in conventional MR sequences and enhance on postcontrast images. DWI can help to differentiate these conditions as the DWI intensity is brighter in lymphoma and lymphoid conditions, and the ADC value is lower than that in IOS and cellulitis.[23]

Orbital inflammatory syndrome[7,20] is simply termed as orbital pseudotumor which refers to nongranulomatous, noninfectious orbital disease wherein no local or systemic cause can be identified.[21,22] It is classified as acute, subacute, or chronic usually presenting with pain, swelling, and restricted eye movements which may be bilateral.[2,21] On T_1WI, it is hypointense to normal muscle and isointense to hyperintense to muscle on T_2WI/STIR. However, sometimes it may be hypointense with intense contrast enhancement.

The common types of pseudotumour are described depending on the part of the orbit involved[2,5]:

- *Anterior orbital group*—It involves the anterior orbit and adjacent globe.[2] MRI can demonstrate uveal–scleral thickening and inflammation at ON junction with postcontrast enhancement on T_1WI FS images. This must be differentiated from orbital cellulitis or leukemic infiltration.[22]

- *Diffuse type*—There is severe inflammation with infiltration of the orbital structures and retrobulbar space extending around the globe, usually multicompartmental and no obvious bony erosion in these cases. On imaging, the appearances mimic lymphoma,[22] which needs to be differentiated.

- *Myositic type*—The inflammatory process involves EOM that may be bilateral. CT and MRI show enlarged muscles with shaggy margins, with infiltration of the surrounding fat. Postcontrast enhancement of the inflamed muscles associated with obliteration of peripheral surgical fat planes is seen. It usually affects the superior group of muscles and medial rectus (Figure 19.21).

FOCUS POINT

This condition should be differentiated from thyroid myopathy[22,24] where muscle bellies are bulky with the tendons characteristically spared compared with IOS where inflammation involves the muscles and the tendons up to their insertions.[2]

- *Lacrimal gland inflammation*[2]: It is a condition in which the lacrimal gland is enlarged with intense postcontrast enhancement appearances similar to granulomatous dacryoadenitis, sarcoid, or lymphoma of the gland.

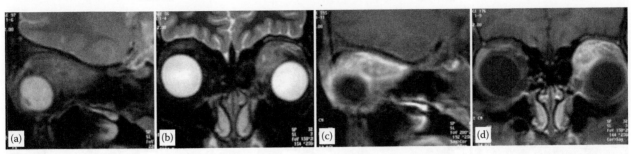

FIGURE 19.21

Inflammatory orbital syndrome (pseudotumor). (a) Sagittal T_2W STIR and (b) coronal T_2W STIR image show left orbital proptosis with hyperintensity of muscle and surrounding orbital fat displacing the globe with uveo-scleral thickening. Postcontrast-enhanced Gd injection in (b) sagittal FS T_1WI and (d) coronal FS T_1WI demonstrate intense enhancement of the involved muscles, fat and uveo-scleral region (Tenon's fasciitis). (From Hande, P.C. and Talwar, I., *Indian J. Radiol. Imaging*, 22, 224, 2012.)

- *Orbital apex inflammation*[2,22] infiltrates posteriorly into *ON sheath complex* (ONSC) or involvement of posterior ends of EOM with intracranial extension. MRI is very useful in tracking the posterior extent of these lesions.

Perineuritis with inflammation of the ONSC may mimic optic neuritis[2] with diminution of visual acuity and mild proptosis. Intense postcontrast enhancement with Gd is noted in involved segments of the ON.

- *Tolosa–Hunt syndrome*[2,5,22]: This variant presents with painful external ophthalmoplegia with visual defects. There is usually unilateral involvement presenting with retro-orbital pain and III, IV, V (i or ii divisions or both), or VI cranial nerve palsies. There may be inflammation along the internal carotid artery (periarteritis) and the cavernous sinus (Figure 19.22). MRI demonstrates the asymmetric inflammation of the orbital apex and enlargement of

the cavernous sinus with intense postcontrast enhancement with involvement of the prepontine cistern. The soft tissue is isointense to muscle on T_1WI and isointense to fat on T_2WI.[24]

Parasitic infestations: These are endemic in regions, commonly seen in the tropics. They may be related to certain occupations and environmental conditions of hygiene and sanitation. Transmission through ingestion of the ova in food or water can result in infestation in humans. Larval forms may be found as cysts within the globe and orbit, which may be detected if they present with orbital signs.[1,2] Cysticercosis may be present in association with neurocysticercosis or disseminated cysticercosis[5] (Figure 19.23).

On MRI, the differential diagnosis includes other granulomatous orbital lesions (Figure 19.24), for example, tuberculomas.

Filarial infestation may occur anywhere in the body and rarely can be seen in the region of the orbit. CT

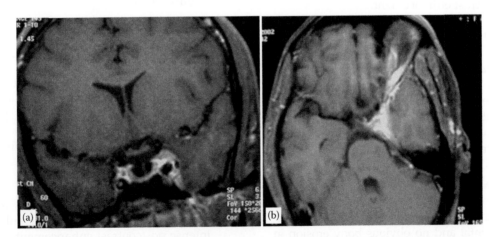

FIGURE 19.22

Tolosa–Hunt variant. (a) Coronal and (b) axial postcontrast-enhanced Gd T_1W FS images demonstrate enhancing inflammatory infiltration in the left orbital apex extending into the cavernous sinus. (From Hande, P.C. and Talwar, I., *Indian J. Radiol. Imaging*, 22, 224, 2012.)

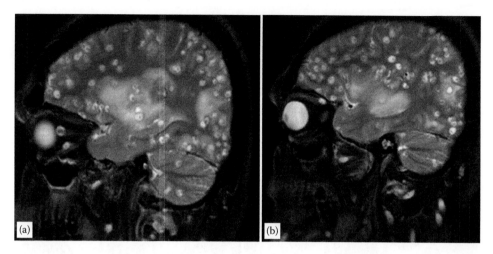

FIGURE 19.23
Cysticercosis. MR T_2W TIRM sagittal sections show multiple hyperintense cysticerci with hypointense eccentric scolices in the orbit within ON and IR muscles (a) and SR muscle (b). Infestation of the brain and other visualized soft tissues is seen. (From Hande, P.C. and Talwar, I., *Indian J. Radiol. Imaging*, 22, 224, 2012.)

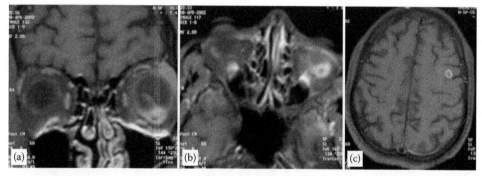

FIGURE 19.24
(a) Coronal and (b) axial postcontrast-enhanced Gd T_1W FS images show enhancing intraocular lesion in left orbit and intracerebral ring-enhancing lesion in left parietal in axial T_1WI of the brain (c) suggestive of typical granulomatous lesions. (From Hande, P.C. and Talwar, I., *Indian J. Radiol. Imaging*, 22, 224, 2012.)

and MR may only reveal nonspecific granulomatous lesion which may enhance on contrast administration (Figure 19.25). However, live filarial worms may show typical movements on real-time USG called *filarial dance* (Figure 19.26).

Toxocara infestation may cause inflammation (endophthalmitis) when they die, which can present with leukocoria in children.

19.5.2.3 Thyroid Ophthalmopathy/Graves' Orbitopathy

The most common cause of exophthalmos in adults is thyroid-related disorders[2] and usually affects females more than males. Bilateral orbital involvement is more common; however, the disease may be asymmetric. It is presumed to be an autoimmune condition consisting of a triad of goiter with hyperthyroidism, infiltrative ophthalmopathy and dermopathy.[2]

FOCUS POINT

Imaging of the CNS can give a clue to the diagnosis, especially in granulomatous diseases where ring-enhancing lesions or cysts may be seen in remote locations.

FOCUS POINT

Most commonly, there is involvement of the inferior rectus followed by medial and superior rectus.[22] Isolated lateral rectus involvement needs to consider a differential other than thyroid orbitopathy.[2]

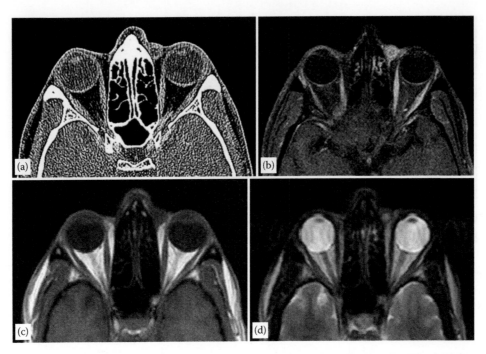

FIGURE 19.25
Swelling medial aspect of the left orbit. (a) Axial CT shows hyperdense lesion medial to the medial canthus, seen as mildly hyperintense on axial MR T_2W image (b) and isointense to muscle on axial T_1W image (c) with postcontrast enhancement of the lesion on axial T_1W FS image (d).

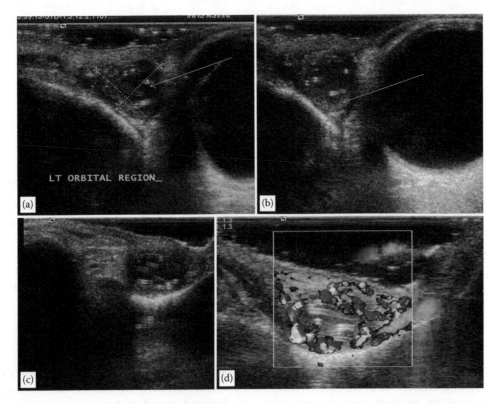

FIGURE 19.26
High-resolution USG left orbit (a–d) shows linear, serpiginious contents within extraocular cystic lesion (arrow) medial to the eyeball. This is an advantage of real-time USG.

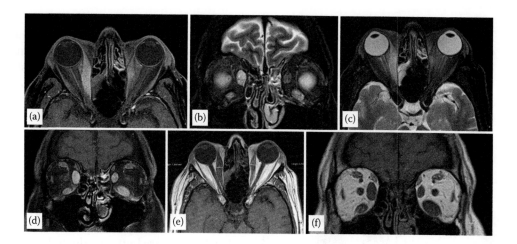

FIGURE 19.27

Thyroid orbitopathy. (a) Axial and (b) coronal MR T_1W images show bilateral exophthalmos with enlargement of EOM, more prominent in medial and inferior rectus muscles, with bulky muscle bellies with sparing of the tendons. (c) Axial and (d) coronal T_2W TIRM FS images show the enlarged MR, IR muscle bellies which are mildly hyperintense to other muscles. (e) Axial and (f) coronal T_1W images demonstrate more intense enhancement of the enlarged muscles.

CT may show low density within the muscles due to deposition of lymphocytes and mucopolysaccharides. MRI reveals hyperintensity on T_2WI due to infiltration and interstitial edema of involved EOM in active disease which can be easily demonstrated in TIRM with FS.[25] MRI demonstrates enlarged EOM with *sparing of tendons*, sharply defined muscle margins with preserved fat planes, increased volume of retro-orbital fat with bulging of orbital septum anteriorly[2] (Figure 19.27). Imaging is used to assess the response to steroid and immunomodulatory therapy and followup. The cross-sectional area of the most inflamed EOM muscle belly can be measured on 3 mm coronal T_2W IR fat-suppressed sequences[10] EOM enlargement can be easily measured at its maximum radial diameter of the belly of the involved muscle in coronal sections.[26] This can be used to assess the response to steroid therapy.

FOCUS POINT

EOM bellies are enlarged with sparing of tendons with preserved fat planes are well visualized on MRI in non-fat-suppressed T_1WI. It is essential to use fat-suppressed sequences to demonstrate active disease. In chronic disease, atrophy of EOM may be seen with fat replacement which can be detected on T_1W images without FS.

Associated compressive ON neuropathy can occur probably as a result of pressure at the orbital apex by increased volume of orbital contents, overcrowding, and stretching of the nerve in exophthalmic Graves'

orbitopathy due to proptosis.[25] High-resolution volume acquisition (T_1W-3D) with curved MPR can be used to measure the ON diameter along its entire length.[26] Decompression surgery may prevent irreversible ON atrophy in the early stage of disease.

FOCUS POINT

MR has an advantage in imaging of the orbital apex, and the curved MPR of the ONSC can detect compression of the ON early in the disease in cases of ON neuropathy.

19.5.2.4 Trauma

CT is the mainstay of imaging in facial and orbital trauma as it demonstrates bony and soft tissue details well. Bony fractures and displacements, involvement of orbital apex, small comminuted fragments, and radio-opaque foreign bodies (FBs) are best seen on CT. *MR is contraindicated if intraocular ferromagnetic FB is suspected.* However, CT may fail to detect non-radio-opaque organic such as wood or plastic FB, which may be intra-ocular and can be localized on MR.[1,27]

Volume rendering techniques (VRTs) with MDCT allow excellent 3D reconstructions that help surgeons to visualize the fractures in relation to other anatomic structures and evaluate treatment plans[1,2,6] (Figure 19.28). Blowout fractures classically involve the floor with sparing of the orbital rim where contents are pushed out by the increased pressure into the roof of the maxillary antrum.[2,6]

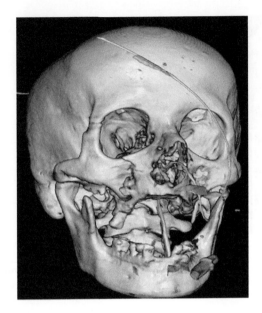

FIGURE 19.28
VRT CT image illustrates fracture in relation to the orbital floor and inferior orbital rim vertically involving the maxilla on the right side.

EOM, especially inferior rectus or inferior oblique muscle, may herniate with orbital fat giving a typical teardrop appearance on MRI (Figure 19.29a).

Disruption and avulsion of EOM or ON in severe trauma are well localized on MRI with its multiplanar capability.

Blunt and penetrating trauma can involve the *globe* and may cause damage due to sudden changes in pressure, distortion with compression, and rebound expansion.[1,2] The role of USG is well established for vitreal, choroidal, and retinal abnormalities,[6] and also for ocular trauma. FB within the globe, RD, CD, and hemorrhage can be identified on real-time USG.[5,6] However, examination in severe injuries is not recommended. MRI on axial and coronal sections can demonstrate ocular injury, air, loss of volume with reduction in vitreous or anterior chamber shrinkage, associated inflammation, and muscle herniation

and entrapment (Figure 19.29b). Ocular and intraorbital hematomas due to acute or delayed hemorrhage are well identified on MRI with SIs depending on the age of the blood[1,2] (Figure 19.30). There may be dislocation of lens with disruption of the zonule and lens capsule in perforating injury.

FOCUS POINT

CT is the modality of choice in imaging of orbital trauma.[2] However, ocular and orbital injuries, vascular complications such as pseudoaneurysms and CCF, nerve injuries, and associated brain and dural spaces are better evaluated with MRI.

19.5.2.5 Orbital Tumors

The orbit may be involved by tumors which are primary masses arising from the various orbital structures.

Lymphoid neoplasms: These range from the benign reactive lymphoid hyperplasias to malignant lymphomas and are more commonly seen in adults and very rare in children.[2] They are seen as the third most common cause of proptosis in adults aged 50–60 years. Proper evaluation is recommended to rule out a systemic lymphoproliferative disorder[2] and may be associated with B-cell non-Hodgkin's lymphoma. Malignant lymphoma may arise as a primary in the orbit and are usually confined within its boundaries. There are four main patterns that may occur in isolation or can be present in combination:

- Anterior: pre/postseptal, superior aspect
- Retrobulbar
- Lacrimal gland: The gland is enlarged with associated displacement of the globe medially[28]

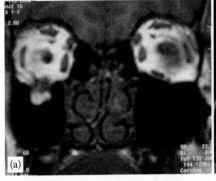
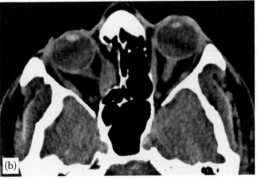

FIGURE 19.29
(a) Coronal T_1W image shows a classic blowout fracture of floor of the right orbit with herniated contents into the roof of the maxillary antrum. (b) Axial CT shows herniation of the medial rectus through the medial wall of the orbit into adjacent ethmoid sinus with fracture of right lamina papyracea. (From Hande, P.C. and Talwar, I., *Indian J. Radiol. Imaging*, 22, 224, 2012.)

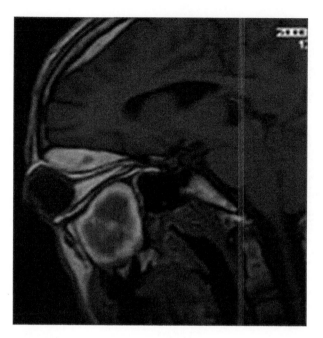

FIGURE 19.30
Subperiosteal hematoma in orbital trauma. Sagittal T_1WI shows subacute hyperintense hematoma right orbit with mass effect. Associated hemosinus within right maxillary sinus. (From Hande, P.C. and Talwar, I., *Indian J. Radiol. Imaging*, 22, 224, 2012.)

and shows postcontrast enhancement on T_1W FS MRI. The imaging features are similar for lymphoid hyperplasia and lymphoma.

- Lymphoma arising primarily in the sinonasal cavities extending into the orbit[6]

There may be bilateral involvement and they may be a diffusely infiltrative lesion or a focal mass causing proptosis. They usually appears as solid, soft tissue masses; these lesions are isointense to muscle on T_1WI, variably hyper- or hypointense on T_2WI depending on

cellularity, and mildly hyperintense to muscle in the retrobulbar region (Figure 19.31) with moderate post-contrast enhancement. Usually, no bony destruction is seen, except rarely in aggressive malignant variants.[29]

> **FOCUS POINT**
>
> MR is more sensitive for imaging the location and extent of disease, and helps detect early infiltration of ocular structures. Imaging findings are not specific and are similar to the findings in inflammatory pseudotumor. However, DWI is helpful in differentiating lymphoma from inflammatory conditions where there is more intense restriction of diffusion with lower ADC values,[23] probably due to increased cellularity with less extracellular space with restriction of random motion of water molecules.[30]

Leukemias: Orbital involvement due to infiltration of soft tissue or orbital bone is seen in acute leukemias which may be seen in upto 75% cases of acute lymphoblastic leukaemia[31] and upto 20% of acute myeloid leukaemia (known as *Chloromas*).[31] CT/MRI reveals subperiosteal lesion (Figure 19.32) that may involve the lateral wall, which may extend into the temporal fossa or a mass on the medial wall involving ethmoid sinuses extending through the cribriform plate into the anterior cranial fossa.[2]

> **FOCUS POINT**
>
> Meningeal infiltration and intracranial extension can be detected early in postcontrast MRI. Bony involvement is well seen on STIR sequences with marrow changes.

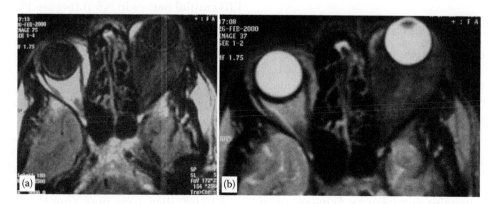

FIGURE 19.31
Lymphoma left orbit in a case of unilateral proptosis. Axial MRI shows diffusely infiltrating lesion involving the left retrobulbar region, appearing hypointense to orbital fat on T_1WI (a) and hypointense on T_2WI (b) with preseptal soft tissue thickening. (From Hande, P.C. and Talwar, I., *Indian J. Radiol. Imaging*, 22, 224, 2012.)

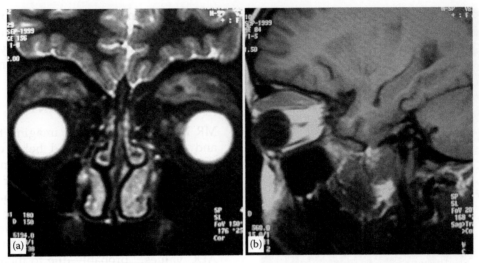

FIGURE 19.32
Orbital leukemic deposits in patient with ALL. Axial T_2WI with FS (a) shows bilateral subperiosteal lesions extending superiorly and laterally up to the lacrimal glands, hypointense to orbital fat extending posteriorly extraconal lesion on sagittal T_1WI (b).

Rhabdomyosarcoma (RMS): About 30% of RMS occurs in the head and neck with orbit and nasopharynx being the most common sites. They usually present in childhood (6–7 years) and may affect adolescents (below 16 years of age). Pathologically, these tumors arise from primitive mesenchymal cells that have the potential to develop into striated muscle. These are one of the commonest primary orbital malignancy of childhood presenting with rapidly developing unilateral painless proptosis, which may be associated with displacement of the eyeball. They are seen usually as retrobulbar intraconal masses or extraconal masses in the superonasal aspect of the orbit with aggressive bony destruction well identified on CT. MRI demonstrates the tumor and assesses the extent of involvement seen as intermediate SI lesions on T_1WI and hyperintense on T_2WI with postcontrast enhancement on T_1W FS images.[2]

malignant transformation.[32] However, only 10% of the patients may have association with NF. They usually present in adults except when associated with NF with progressive proptosis. These are seen commonly originating from the ophthalmic division of the trigeminal nerve as a mass in superior aspect of the orbit. Bone remodeling with expansion is seen with no aggressive erosion or destruction. Schwannomas have a definite capsule and consist of cellular and myxoid components compared with neurofibromas that are nonencapsulated tumors containing all components of nerve origin, including neural and collagen tissue, and may have fatty infiltration with cystic degeneration and calcification.[32] Plexiform neurofibromas may be seen in NF with hypoplasia of the greater wing of sphenoid (bare orbit appearance on radiograph) and macrophthalmia. MRI is the modality of choice in multiple schwannomas more commonly involving the V and VIII cranial nerves in NF II (Figure 19.33).

FOCUS POINT

Because these tumors arise from EOM, they can occupy both intraconal and extraconal compartments.[1,2] However, they can arise primarily in the adjacent structures, for example, paranasal sinuses or nasopharynx.

FOCUS POINT

On MR, neurofibromas may appear nonhomogeneous, usually hypointense or of intermediate SI on T_1WI and hyperintense on T_2WI with characteristic *target sign* seen sometimes, with hypointense central signal and peripheral hyperintensity. The fatty component is seen suppressed on T_1W FS images with postcontrast enhancement. Schwannomas have similar appearances, but if they are highly cellular, then they may appear hypointense on T_2WI and usually show more uniform enhancement[32] (Figure 19.34).

Neurofibroma and schwannoma: These tumors arise from nerves and nerve sheath, are usually extraconal, and have similar imaging appearances. Schwannomas are more common comprising up to 6% of orbital tumors.[32] They are slow growing and usually benign, but when these are associated with NF, they may have

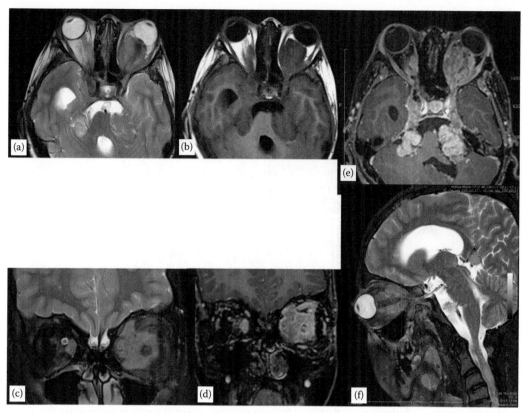

FIGURE 19.33

Neurofibromatosis II. Axial T_2W (a) and T_1W (b) images show mild proptosis on the left side with a large mass lesion in the retrobulbar region surrounding the ON seen well on T_2W TIRM coronal (c) and sagittal (f) images, which is enhancing after contrast administration in coronal T_1W FS image (d) suggestive of left meningioma. Also note a well-defined lesion inferomedial to the ON with enhancement in the intraconal compartment of the right orbit suggestive of a schwannoma. (e) Axial T_1W FS image demonstrates bilateral fifth and acoustic nerve schwannomas.

Vascular lesions represent a large group of vascular malformations involving the orbit.[33] MRI has the advantage over CT as it is very sensitive in detecting blood and blood products and vessels with flow.[2]

Cavernous hemangioma (venous angioma) is the commonest vascular tumor in adults presenting as a progressively increasing mass. These are usually endothelial cell-lined vascular spaces with a fibrous pseudocapsule. It is seen as a well circumscribed, rounded soft tissue density intraconal mass that may have calcifications (phleboliths) in plain CT images (Figure 19.35). There is usually intense enhancement of the lesion in postcontrast CT. T_1WI MR shows an iso- to hypointense mass with a pseudocapsule and hyperintense on T_2WI with septations seen in larger lesions, and demonstrates heterogeneous contrast enhancement that may have central filling in delayed images with slow circulation within the lesion. Differentials of other intraconal masses such as meningiomas, schwannomas, or hemangiopericytoma need to be considered.[2,14,33]

Capillary hemangiomas present in infants within the first year of life and tend to gradually involute and diminish in size.[33] They are made up of thin small vascular spaces

due to capillary proliferation with no distinct capsule and may have definite arterial supply from external or internal carotid artery.[2,14,33] Usually these are extraconal masses with lobulations extending into periocular tissues. They are well localized in axial and coronal MRI sections, usually in the superior aspect of the orbit, or may be retro-orbital in 10% cases. These are hyperintense to muscle and moderate to low SI on T_1WI, and hyperintense on T_2WI with a heterogeneous appearance due to hypointense intratumoral calcifications, blood products, fibrotic components, and flow voids of the vessels within the lesion (Figure 19.36). There may be intracranial extension with diffuse, intense postcontrast enhancement with early wash-in of contrast.[1]

FOCUS POINT

Cavernous hemangiomas present in adults are progressively growing, well-circumscribed intraconal masses compared with capillary hemangiomas that tend to regress in size[2] and may be intraconal and extraconal usually lacking a distinct capsule.

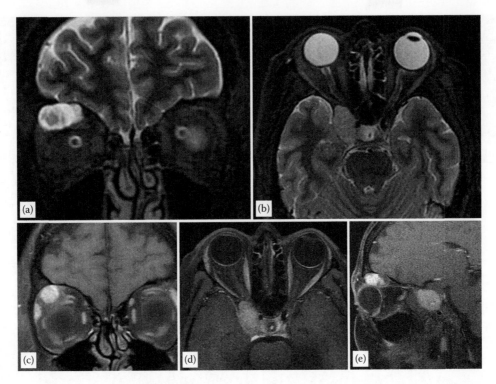

FIGURE 19.34
(a) Coronal T_2W TIRM image shows a hyperintense well-defined lesion in the superlateral aspect of the right orbit. (b) Axial T_2WI TIRM illustrates another lesion isointense to the brain along the visual pathway near the right orbital apex in the region of the optic chiasma. T_1W FS postcontrast images in (c) coronal, (d) axial, and (e) sagittal sections show that both lesions demonstrate intense enhancement.

Congenital hamartomatous lesions are lymphangiomas and lymphatic-venous and arteriovenous malformations.[2,33] These are lobulated, poorly circumscribed without a definite capsule, hypointense on T_1WI, and hyperintense on T_2WI with fluid–fluid levels due to blood and altered blood products depending on the age of the hemorrhage and can be hyperintense due to methemoglobin (subacute stage) (Figure 19.37). No flow voids are seen as in capillary hemangioma.[14]

19.5.2.6 Vascular Lesions

Orbital varix/varices are congenital venous malformations that are usually present with intermittent proptosis more prominent on straining and coughing. This is a common cause of spontaneous orbital hemorrhage.[33] Imaging performed in prone position,[2,5] with or without Valsalva maneuvre, can demonstrate dilated venous channel, or there may be a flow void if there is fast flow within the engorged veins or there may be absent flow with absence of flow void due to thrombosis. Phase-contrast MRI may reveal no flow in the vein suggestive of thrombosis in a dilated SOV.[2]

> **FOCUS POINT**
>
> Enlarged prominent venous channels in orbital varix that may be thrombosed may have variable SI depending on the age of the thrombus and appear hyperintense on both T_1W and T_2W images.

Carotid cavernous fistula (CCF) is an abnormal communication between the carotid artery and the cavernous sinus, which causes pulsating exophthalmos.[2] It may occur spontaneously or as a result of trauma. MRI reveals engorgement of SOV with a signal void with involvement of the cavernous sinus that is enlarged[2,5] (Figure 19.38) and may have associated prominent bulky congested EOM. Digital subtraction angiography (DSA) is very useful for identifying the fistulous communication and demonstrating the vessels involved. Therapeutic interventional procedures can be performed such as coil embolization or occlusion using glue for the fistula.

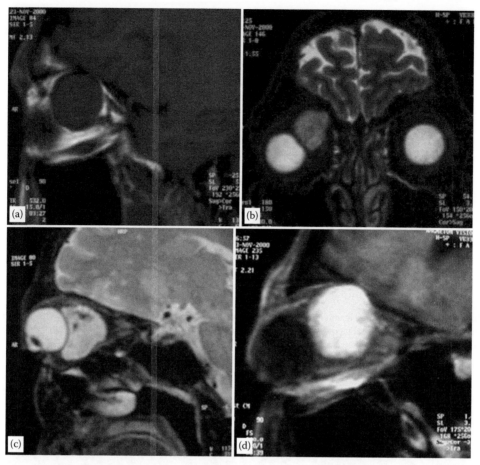

FIGURE 19.35
Cavernous hemangioma right. (a) MRI sagittal T_1WI shows intraconal hypointense, rounded well-circumscribed mass lesion. (b) Coronal T_2WI with TIRM and (c) sagittal T_2WI show the displacement of the globe, hypointense foci (flow voids) within the mass. (d) Sagittal T_1W FS postcontrast-enhanced Gd image reveals intense enhancement of the lesion. (From Hande, P.C. and Talwar, I., *Indian J. Radiol. Imaging*, 22, 224, 2012.)

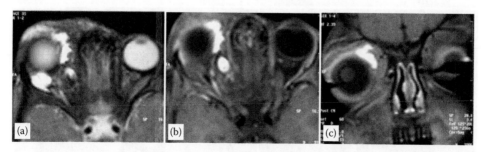

FIGURE 19.36
Capillary hemangioma. (a) Axial T_2WI demonstrates an irregular hyperintense lesion involving the right orbit extending into anterior periocular tissues enhancing well on contrast (Gd) administration seen on axial (b) and coronal (c) T_1W images. (From Hande, P.C. and Talwar, I., *Indian J. Radiol. Imaging*, 22, 224, 2012.)

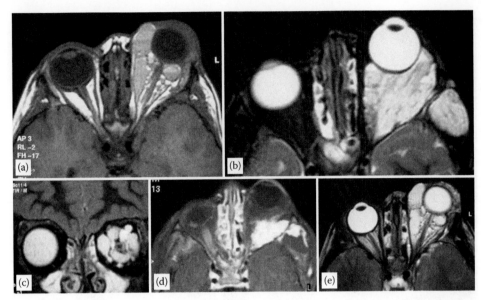

FIGURE 19.37
Lymphangioma. Multiseptate, multilobulated lesion left orbit appears hyperintense on axial T_1W FS image (a), with multiple fluid–fluid levels within multiple locules, hypointense to orbital fat on axial T_1W (d) and coronal STIR images (c), and markedly hyperintense on axial T_2WI (b,e). (From Hande, P.C. and Talwar, I., *Indian J. Radiol. Imaging*, 22, 224, 2012.)

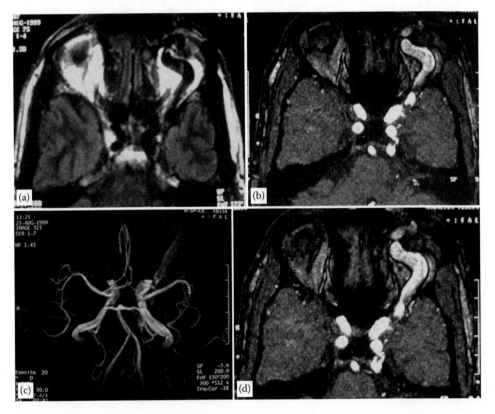

FIGURE 19.38
Carotid cavernous fistula, left. Axial T_1WI (a) shows enlargement of SOV with prominent signal void on the left and distended cavernous sinus on 3D TOF MRA (b–d). (From Hande, P.C. and Talwar, I., *Indian J. Radiol. Imaging*, 22, 224, 2012.)

19.6 ONSC Conditions

MR is the modality of choice for imaging of the entire visual pathway from its origin at the posterior point of the globe to the visual cortex.[1,2,34] Intraorbital ON is seen well on CT due to the attenuation differences with surrounding hypodense retrobulbar fat. Intraorbital part of ON is hypointense on T_1WI within hyperintense fat and isointense on T_2WI with surrounding subarachnoid CSF within the meningeal sheath being hyperintense. The intracanalicular portion is not well visualized on CT due to the dense cortical bone at the apex and is displayed very well on MRI.[34,35]

The intracranial ON, optic chiasma, and tracts are well visualized on coronal and sagittal MR[34,35] images, which appear isointense to cerebral white matter (Figure 19.39). The optic radiation is within the white matter of temporoparietal and occipital lobes extending up to the calcarine cortex (primary visual cortex) flanked by the easily identified calcarine sulcus on the medial aspect of the occipital lobe.[1,34,35]

ON glioma: Intraorbital gliomas appear early within the first decade of life, presenting with painless proptosis, usually with associated loss of vision.[35] There is a high association of gliomas of the visual pathway with NF, especially when there are bilateral lesions.[36] Thus, it is recommended that the entire visual pathway should be imaged. MRI is more sensitive and thin sections (3 mm) can avoid missing small lesions in the intracanalicular portion. Posterior involvement is more commonly seen in children without NF.[2]

CT and MRI show marked fusiform enlargement of ON with kinking or buckling.[2,5] They may be fusiform or nodular and usually iso- to hypointense to orbital fat on T_1WI (isointense to cerebral cortex), and hyperintense to the cortex, white matter, and fat on T_2WI. Peripheral hyperintensity due to arachnoid gliomatosis may be seen in NF type 1, and there may be cystic spaces containing mucinous material. Postcontrast enhancement is common except in areas with cystic changes.[3,36,37] Sometimes, the tumor grows circumferential around the ON into the perineural space to mimic a perioptic meningioma[2,37] (Figure 19.40) or pseudotumor.[37]

ON sheath meningioma: This is usually a benign tumor that arises from the arachnoid cells of the meningeal covering of the ON. Meningiomas are usually more

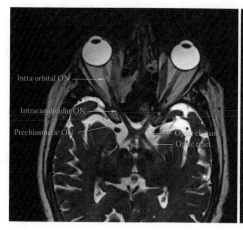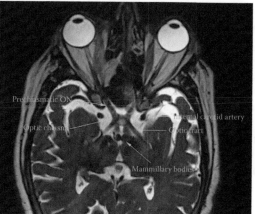

FIGURE 19.39
Axial T_2WI illustrates the ON within the orbit, the apex of the orbit, optic chiasma, and intracerebral visual pathway up to the midbrain.

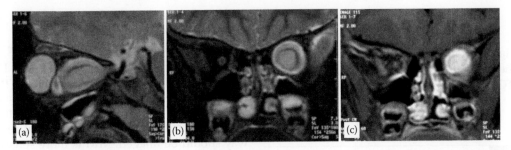

FIGURE 19.40

Optic neuritis glioma. Sagittal T_2WI (a) shows fusiform enlargement ON with hyperintense perineural meningeal circumferential thickening on STIR coronal T_2WI (b) (arachnoid gliomatosis) and significant enhancement on T_1W FS image (c). (From Hande, P.C. and Talwar, I., *Indian J. Radiol. Imaging*, 22, 224, 2012.)

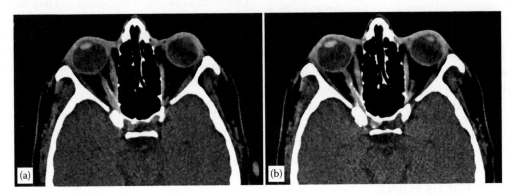

FIGURE 19.41

Perioptic meningioma. Thin-section axial CT image shows (a) tram-track calcification along the right ONSC and hyperostosis at the optic canal. (b) Note postcontrast perineural enhancement. (From Hande, P.C. and Talwar, I., *Indian J. Radiol. Imaging*, 22, 224, 2012.)

common in the fourth to fifth decades, which are commoner in females, but when they present in children, they occur more frequently with NF type 2. On imaging, the usual patterns seen are: i) an eccentric, localized mass on one side of the nerve, ii) a circumferential, tubular type along the length as diffuse thickening, and iii) fusiform enlargement of the ONSC. Intratumoral calcification is common and may cause bony hyperostosis, detected well on CT (Figure 19.41).

This is seen on MRI as decreased SIs on both T_1WI and T_2WI and can result in optic canal enlargement.[3,35] MRI is better than CT for characterizing tumor relative to adjacent orbital structures and defining the extent of disease at the orbital apex, optic canal, and intracranial structures. These are isointense on T_1WI and hyperintense to hypointense on T_2WI (Figure 19.42) with *perioptic cysts*[2] within the ONSC, surrounding the distal ON. These findings are more prominent on STIR images because of suppression of orbital fat signal and have similar appearances as in T_2WI. These exhibit marked postcontrast enhancement (Figure 19.43) and Gd-enhanced MRI with T_1W FS images which help in detection of tiny lesions, especially at the orbital apex[2,34,38] with moderate to

marked enhancement of a *tram-track* pattern in circumferential type of lesions.[2] In children, they tend to be more aggressive and may infiltrate into adjacent other orbital structures.

FOCUS POINT

Presence of calcifications is common in these tumors and they appear hypointense on both T_1W and T_2W images, and if cysts are present, they are well demonstrated on T_2WI with FS or STIR images.

Optic neuritis: Acute inflammation causing enlargement of ON due to infections by microorganisms causing papillitis or following viral infections due to immune mechanisms (parainfections).[1] Noninfective granulomatous diseases such as sarcoidosis and systemic autoimmune diseases such as systemic lupus erythematosus may involve the ON.[38] Fat-suppressed T_1WI pre- and postcontrast images are very useful in optic neuritis and demonstrate localized or diffuse enhancement within the nerve (Figure 19.44). Optic neuritis is an early manifestation of multiple

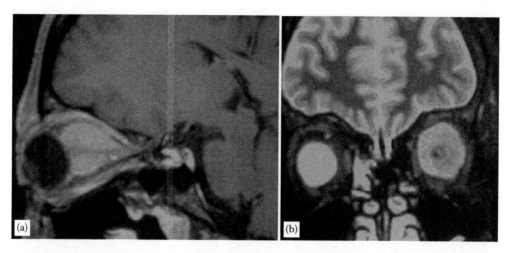

FIGURE 19.42

Left ONSC meningioma. (a) Sagittal postcontrast FST₁WI shows a diffusely enhancing intraconal mass. Note compression of the ON by the fusiform thickening of the perineural portion of the ONSC displacing the IR and SR muscles. (b) Coronal T_2WI demonstrates the compressed ON and splaying of the muscle cone by the mass. (From Hande, P.C. and Talwar, I., *Indian J. Radiol. Imaging*, 22, 224, 2012.)

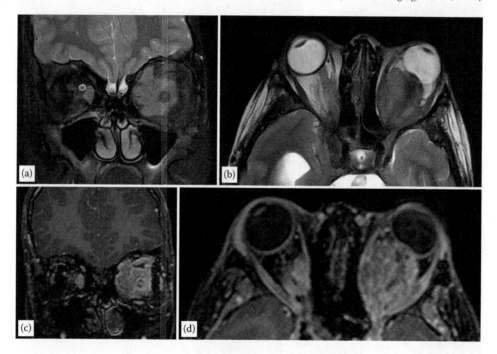

FIGURE 19.43

Left ON meningioma in the case of NF type 2. T_2W coronal and axial images are shown in panels a and b, whereas the T_1W coronal and axial images after the administration of contrast material are shown in panels c and d.

sclerosis (MS) seen well on MRI as thickening with focal hyperintense plaques on T_2WI seen easily with STIR technique.[39]

FOCUS POINT

Brain and spine MRI is usually required while evaluating patients with demyelinating diseases such as MS (Figure 19.45).

19.7 Secondary Tumors of the Orbit and Metastasis

The orbit may be involved secondarily due to spread of malignant tumors from adjacent structures of head and neck by contiguous extension, perineural or perivascular spread and direct invasion. However, metastatic deposits by haematogenous dissemination may occur in the orbit from distant organs.

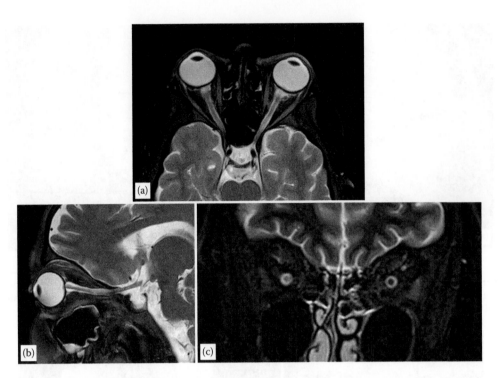

FIGURE 19.44
Optic neuritis. (a) Axial T_2WI and (b) sagittal T_2W shows segmental hyperintensity of ON, and (c) coronal T_2W TIRM images demonstrates hyperintense right ON as compared with that on the left ON.

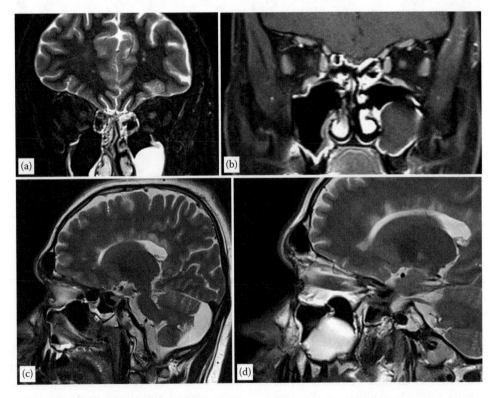

FIGURE 19.45
Multiple sclerosis with optic neuritis, left. (a) Coronal T_2W TIRM shows hyperintense left ON which demonstrates postcontrast enhancement on (b) coronal T_1W FS image. (c,d) Sagittal T_2WI reveals multiple classical demyelinating plaques in brain.

Local spread: There may be involvement of orbital structures due to invasion of malignant tumors from the surrounding anatomical structures, most commonly malignancies from sinonasal cavities[2,5] (Figure 19.46).

Metastatic deposits are from known or unknown primary malignancies.[1,2,3,5] MRI reveals involvement of the globe better and also secondaries to the visual pathway and nerve compressions from mass-like lesions and intracranial metastases (Figure 19.47). The common primary cancers in adults are breast (Figure 19.48), prostate gland, and lung cancers. Chorioretinal metastasis from extraocular melanoma may have similar features as primary ocular melanoma.[1,2] In 19% of the cases, there is no history of cancer when the patient presents with ophthalmic symptoms, and in 10%, the primary site remains obscure despite systemic evaluation. In the pediatric age group, neuroblastoma is one of the common primary tumors that metastasize to the orbits.

> **FOCUS POINT**
>
> MRI with the advantage of multiplanar imaging with STIR and T_1W postcontrast studies helps to assess head and neck tumors and their locoregional extent and invasion into the orbit and surrounding vital structures such as nerves and vessels.

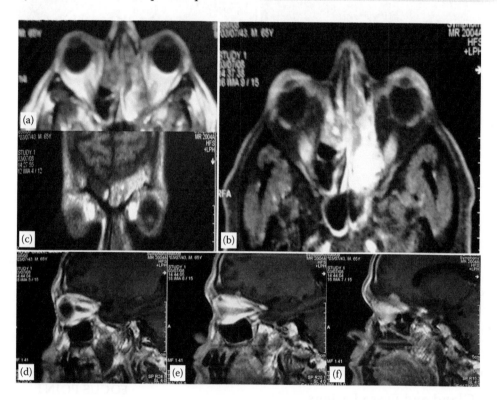

FIGURE 19.46
Orbital involvement in sinonasal malignancy. Lesion arising in the paranasal sinus predominantly on the left side is seen to extend into the left orbit and intracranially on coronal T_1W FS (a), axial postcontrast T_1W FS (b), axial T_1W (c), and T_2W images. Note the lesion is hyperintense on T_1W images suggestive of melanoma arising from the sinus. In panels d, e and f sagittal T_2W images are given.

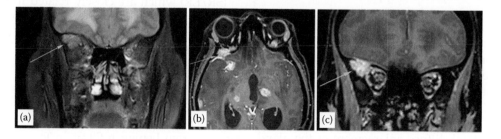

FIGURE 19.47
Metastases in the right orbit with multiple intracerebral secondaries. (a) Coronal T_2W TIRM image shows isointense extraconal lesion in the postero-superior aspect of the right orbit. (b) T_1W FS axial and (c) T_1W FS coronal images demonstrate intense postcontrast enhancement of all the secondary deposits.

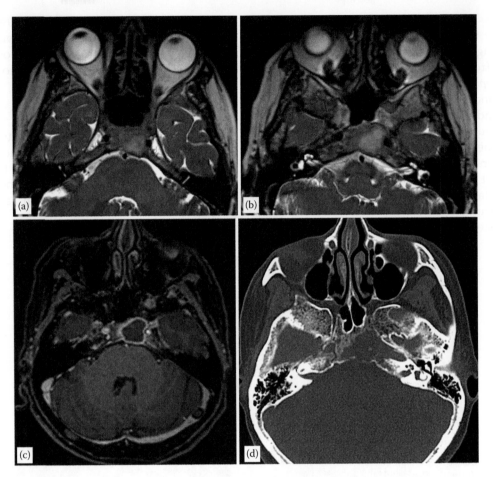

FIGURE 19.48

Metastasis from carcinoma breast involving the skull base seen as hyperintense lesion in the clivus on axial T_2W (a,b) which shows postcontrast peripheral enhancement and causes destruction of the bone seen on CT (d). Also note multiple nodular intracranial lesions in (c) Note evidence of VI nerve involvement on the left side with lateral rectus palsy, medial deviation of the globe.

19.8 Lacrimal Gland and Fossa Lesions

Lacrimal gland is located within the lacrimal fossa in the superolateral aspect of the orbit and lies in the extraconal compartment adjacent to superior and lateral rectus tendons.

Inflammatory diseases are usually acute and may be a part of spectrum of orbital pseudotumor or related to trauma.[40] Chronic dacryoadenitis may be associated with connective tissue disorders, noninfectious granulomatous diseases such as sarcoidosis and Wegener's granulomatosis, Mikulicz syndrome, or other nonspecific infiltrative conditions.[2] CT and MRI demonstrate unilateral or bilateral diffuse enlargement of the gland with marked contrast enhancement (Figure 19.49), with enhancement of adjacent EOMs (SR/LR muscles).

FOCUS POINT

In inflammatory conditions of the orbit as pseudotumor, it is important to assess the involvement of the lacrimal gland. Involvement of the EOMs (SR/LR) which are closely related to the lacrimal gland should be looked for and commented upon, as there may be associated myositis or tendinitis of SR/LR.

Lymphomatous involvement may be unilateral or bilateral, ranging from benign infiltration to malignant lymphomas, which often cause diffuse enlargement of the lacrimal gland (Figure 19.50).

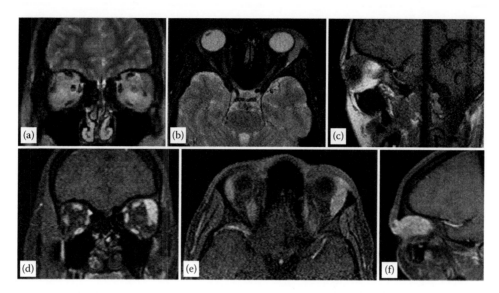

FIGURE 19.49

MR T_1W, coronal (a) T_2W, and axial T_2W (b) images show hypointense enlarged left lacrimal gland with prominent enhancement seen on postcontrast-enhanced Gd T_1W FS (coronal, axial and sagittal, respectively panels d,e and f) images suggestive of dacryoadenitis, which mimics inflammatory pseudotumor. (From Hande, P.C. and Talwar, I., *Indian J. Radiol. Imaging*, 22, 224, 2012.)

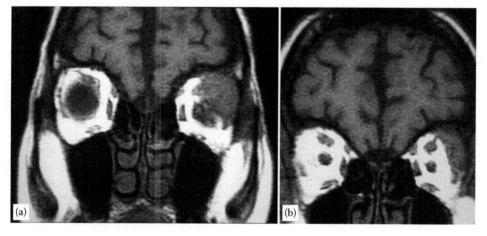

FIGURE 19.50

Lymphoma MR T_1W coronal sections (a,b) show diffuse enlargement of left lacrimal gland.

Lacrimal gland tumors: About 50% the masses are tumors of epithelial origin and half of these are benign pleomorphic adenomas.[40,41] MRI demonstrates the extraconal mass well causing indentation of the globe along with distortion of the muscle cone (Figure 19.51).

FOCUS POINT

CT demonstrates changes in the lacrimal fossa and remodeling of the bone with expansion and enlargement with no erosion destruction suggesting a benign etiology.

19.9 Miscellaneous Conditions

Posttransplantation lymphoproliferative disorder (PTLD): This condition is seen more frequently in the present-day practice as organ transplants are more common.[5] It may be seen in upto 2%–3% patients, within the first year after organ transplantion. This is likely due to uncontrolled posttransplant lymphocytic proliferation response.[42] Orbit is a common site for this condition, and on MRI, it shows a soft tissue mass lesion in lacrimal region. It appears hypointense on T_2WI, with contrast enhancement on T_1WI with FS. On DWI, these masses

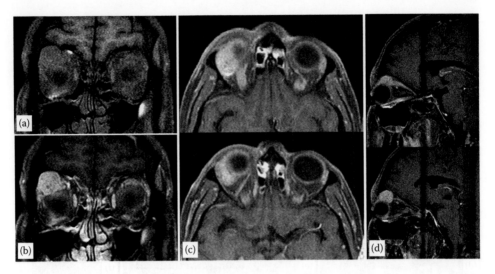

FIGURE 19.51
Lacrimal gland pleomorphic adenoma. MR T_1W FS image (a) shows enlarged globular right lacrimal gland and hypointense with enhancement in postcontrast-enhanced Gd images (b–d) suggestive of benign tumor. (From Hande, P.C. and Talwar, I., *Indian J. Radiol. Imaging*, 22, 224, 2012.)

reveal restriction of diffusion with ADC drop (low values) in both the conditions due to the high cellularity.[23,29] At times, these may become aggressive in appearance as opposed to lymphoma (Figures 19.52 and 19.53).

> **FOCUS POINT**
>
> Follow-up MR studies can help in assessing the response to steroids wherein there is regression noted in PTLD. However, by contrast, in lymphomas there may be progressive enlargement in the gland.

Papilledema: Fundoscopic examination reveals papilledema with elevation of the optic disc[2] suggestive of raised intracranial pressure, which could be due to intracranial tumors and other space-occupying lesions and associated cerebral edema or hydrocephalus.

Idiopathic/benign intracranial hypertension (IIH/BIH) is a condition wherein patients present with throbbing headache, nausea, vomiting with pulsatile tinnitus, and usually affecting obese women in reproductive age group. Occasional diplopia associated with reversible VI nerve palsy may occur, and if untreated, it can lead to visual loss. MR demonstrates flattened posterior margin of the globe with prominent dilated ONSC within the orbits with buckling of ON bilaterally[43] and associated chinked slit-like ventricles in the brain.

> **FOCUS POINT**
>
> Similar findings are seen in raised intracranial pressure due to organic causes. However, when no mass or lesion is visualized with typical findings, IIH/BIH is diagnosed. There is usually reversal of the findings on follow-up imaging. Cerebral MR with TOF venography is recommended to exclude *venous sinus thrombosis* that can have similar clinical presentation and intractable headache. In this condition, smooth compressed intracranial venous sinuses are seen.[44]

Dural ectasia: This is a condition wherein there is prominence of the meningeal sheath around the intraorbital ON. This may be segmental or saccular dilatation well visualized on CT and MRI in axial and coronal sections. It may be primary or secondary type presenting with visual loss or field defects,[2] may be associated with enlarged subarachnoid cisterns and empty sella, or can be due to intracranial tumors such as pilocytic astrocytoma.[2]

Optic atrophy: The ON may be thinned out and atrophic, and can be seen well on MRI, which may be bilateral or unilateral. The condition may be primary (Figure 19.54) or more commonly secondary.

Table 19.3 is the summary of all orbital conditions.

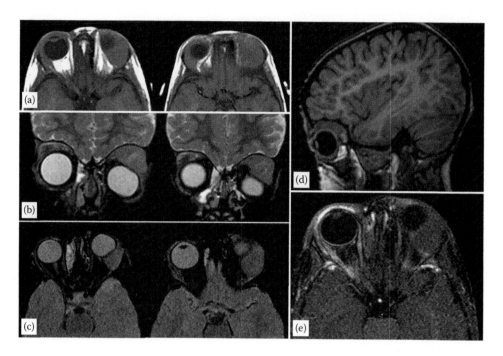

FIGURE 19.52
Post bone marrow transplant in a patient of acute leukemia with left proptosis. (a) Axial and (d) sagittal MR T_1W images show extraconal space infiltration on the left (small on the right) and hypointense to orbital fat. (b) Coronal and (c) axial STIR images show hypointense lesions involving lacrimal glands extending superiorly and displacing the left globe. (e) The T_1W fat sat after contrast material confirm the presence of enhancement.

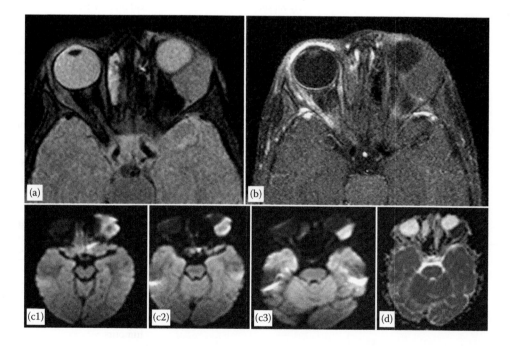

FIGURE 19.53
Lesion on MR axial STIR (a) and postcontrast T_1W FS (b) images shows no enhancement, with true restriction of diffusion appearing bright on DWI axial image (c) with increasing b values (0, 500, 1000) and correspondingly dark on the ADC map (d). (From Hande, P.C. and Talwar, I., *Indian J. Radiol. Imaging*, 22, 224, 2012.)

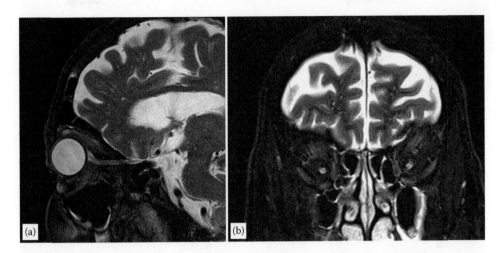

FIGURE 19.54
Bilateral primary optic nerve atrophy. T_2W (a) Sagittal and (b) coronal images reveal pin-head optic nerves with no evidence of altered signal intensity of along the nerve.

TABLE 19.3

Differential Diagnosis of Orbital Lesions

Compartment	Common Differential Diagnosis	Salient Features/Presentation	Focus Points for MR Imaging
Intraconal			
Globe/orbit EOM/ fat	*Infective*: Cellulitis/abscess	Painful proptosis, Oedema, chemosis	T_1W, T_2W, postcontrast enhancement with IV Gd on T_1W FS
	Inflammatory		
	• Pseudotumor	Decreased visual acuity	T_1W, T_2W, postcontrast T_1WI FS, TIRM with FLAIR. Also conal and extraconal
	• Tolosa–Hunt syndrome	Ophthalmoplegia, painful/ reduced eye movements	T_1W, T_2W, postcontrast T_1WI FS, TIRM with FLAIR, DWI, ADC Include orbital apex and cavernous sinus
Optic nerve	*Inflammatory*		
	• Optic neuritis (demyelination)/ neuromyelitis optica (NMO)	Sudden visual loss, usually unilateral Associated neurological signs due spinal/ brain involvement	T_1W, T_2W, Postcontrast-enhanced Gd, 3D T_1W FS, thin sections MR brain/spine for demyelination plaques. Relative brain sparing in NMO
	• Optic neuritis (granulomatous)	Proptosis, loss of vision, painful eye movements	T_1W, T_2W, Gd contrast enhancement T_1W FS, thin sections
	ON tumors		
	• ONSC meningioma	Proptosis, reduced vision or visual loss	T_1W, T_2W, STIR, postcontrast-enhanced Gd T_1W FS images, tram-track enhancement, includes orbital apex, looks for calcification with thin sections
	• ON glioma	Proptosis, reduced vision or visual loss	T_1W, T_2W, STIR, postcontrast-enhanced Gd T_1W FS images, thin sections, apex 3D acquisition recommended Image brain for optic chiasma, pineal gland tumours
	• Schwannoma (II, III, VI CN)	Painless proptosis	Iso- to hypointense on T_1W, hyperintense on T_2W, postcontrast enhancement T_1W FS, may demonstrate hemorrhage extraconal compartment also

(Continued)

TABLE 19.3 (*Continued*)

Differential Diagnosis of Orbital Lesions

Compartment	Common Differential Diagnosis	Salient Features/Presentation	Focus Points for MR Imaging
	Miscellaneous		
	• ON atrophy	Visual loss	Postcontrast-enhanced Gd T_1W FS
	• Colobomatous cyst	Visual loss	Small hypoplastic ON, calcification ON cyst on T_1W
			Looks for hyaloid artery remnant, encephalocoele
Tumor/infiltration	• Lymphoma • Leukemia	Proptosis, reduced vision, may be bilateral	Hyper- to isointense T_1W, hypointense on T_2W, postcontrast enhancement on T_1WI FS, TIRM with FLAIR
	• Metastasis		DWI: restriction, low ADC values involve conal and extraconal also, may have hemorrhage
Vascular	• Cavernous hemangioma	Proptosis in adult age	T_1W low intensity smooth/lobulated, enhancement, well circumscribed, adult age, may have hemorrhage
	• Varix	Postural exophthalmos, prominent in prone position	T_1W, T_2W, PD, STIR usually hyperintense, cystic and solid components, dilated vein with flow void, thrombosis, hemorrhage, phleboliths
	• Vascular malformations/ lymphangioma	Proptosis in child, decreased vision, reduced mobility	T_1W, T_2W cystic spaces, lobulated, altered blood products at different ages may be present, infiltrative, may enhance, involves muscle cone
			Lid, conjunctiva, postseptal compartment, not encapsulated
	• Capillary hemangioma	Proptosis in infant which may reduce as may involute	T_1W hypointense to brain, T_2W, PD hyperintense to brain, infiltrative, paediatric age, usually extraconal
Conal			
Muscle cone	*Infective*		
	• Cellulitis	Chemosis, pain, proptosis, oedema	T_1W, T_2W, postcontrast-enhanced Gd enhancement on T_1W FS
	• Abscess	Eccentric proptosis motility disorder, pain	Postcontrast intense enhancement with nonenhancing collections and rim-enhancing abscess
	Infestation		
	• Cysticercosis	Restriction of eye movements, visual blurring, convulsions if neurocysticercosis	T_2W images classic, cysticerci in EOMs, orbital fat, globe, and soft tissues. Cysts with eccentric scolex+/−, calcification+/−, cysts in brain, with/without edema depending on stage of cysts, other visualized soft tissues/muscles
	Inflammatory		
	• Pseudotumor	Restriction of movements of the eyeball, pain proptosis	T_1W, T_2W, postcontrast T_1W FS images show enhancement of muscle bellies extending to the tendons up to the insertions
	• Myositis/tendinitis	Restricted motility, painful eye movements, isolated/more than one muscle	T_2W images show hyperintensity of the involved EOM with contrast enhancement, may have cystic spaces in the belly of the muscle
	Endocrine		
	• Graves' orbitopathy	Exophthalmos May be bilateral	Enlarged muscle belly, tendons are spared Increased volume of orbital fat

(*Continued*)

TABLE 19.3 (*Continued*)

Differential Diagnosis of Orbital Lesions

Compartment	Common Differential Diagnosis	Salient Features/Presentation	Focus Points for MR Imaging
Orbital Apex			
	Tumors		
	• Metastasis	Proptosis, unilateral/bilateral Loss of vision	T_1W, T_2W, STIR, postcontrast-enhanced Gd T_1W FS images, thin sections at apex 3D acquisition recommended Known distant primary
	• Intracanalicular ON meningioma	Visual loss significant	T_1W FS pre- and postcontrast images show enlarged ON with intense enhancement, T_2W hypointense, calcification, other features of NF type 2
	• ON glioma with chiasmatic glioma	Painless proptosis in childhood, reduced/loss of vision Bilateral in NF type 1	T_1W, T_2W, T_1W FS postcontrast enhancement +, retrobulbar mass with extension into chiasma
	• Inflammatory pseudotumor at apex: Tolosa–Hunt syndrome	Painful ophthalmoplegia	T_1W, T_2W hypointense, T_1W FS postcontrast enhancement orbital inflammation with extension into cavernous sinus which is enlarged, steroid responsive
	• Carotid-cavernous fistula	Congested eye and orbit, pulsatile exopthalmos, edema	Prominent dilated SOV with cavernous sinus-flow void continuous with carotid siphon on MRA DSA confirms fistulous flow with retrograde filling of SOV Coil embolization/occlusion
	Sellar lesions		
	• Empty sella	Visual field loss, headache	T_1W sagittal sections, coronal for pituitary gland, T_1W dynamic postcontrast enhancement Enlarged sella, compressed gland
	• Craniopharyngioma	Presents in childhood	Cystic space occupying mass lesion, hyperintense on T_1W and T_2W, may involve suprasellar cistern
	• Parasellar meningioma		T_1W isointense, T_2W isointense, black rim, dural tail
Extraconal Orbit			
	Infective		
	• Cellulitis/abscess	Fever, malaise, headache, sinusitis	T_1W, T_2W, postcontrast-enhanced Gd enhancement on T_1W FS
	Inflammatory		
	• Pseudotumor	Proptosis	T_1W, T_2W, postcontrast T_1WI FS, TIRM with FLAIR. Also conal and intraconal
	• Granulomatous (sarcoid/ Wegener's)	Systemic disease pulmonary symptoms, involves lacrimal gland, ON(II CN)	T_1W, T_2W, postcontrast T_1WI FS images show enlarged ON with enhancement, lacrimal gland enlargement with enhancement. Associated granulomas in paranasal sinuses CT lungs: cavitating lesions, lymph nodes
	Vascular		
	• Capillary hemangioma	Proptosis that may reduce as may involute	T_1W hypointense to brain, T_2W, PD hyperintense to the brain, infiltrative, paediatric age, may extend intraconal
	• Lymphangioma	Proptosis in child, decreased vision, reduced mobility	T_1W, T_2W cystic spaces, lobulated, altered blood products at different ages may be present, infiltrative, may enhance, involves muscle cone Lid, conjunctiva, postseptal compartment, not encapsulated

(Continued)

TABLE 19.3 (*Continued*)

Differential Diagnosis of Orbital Lesions

Compartment	Common Differential Diagnosis	Salient Features/Presentation	Focus Points for MR Imaging
	Tumors		
	• Neural tumors		
	• Schwannoma (V1 CN)	Proptosis in adults in superior orbit	Iso- to hypointense on T_1W, hyperintense on T_2W, postcontrast enhancement T_1WFS, may demonstrate hemorrhage
	• Neurofibroma (VN)	Proptosis	Circumscribed/plexiform in NF1 Similar to schwannoma, no hemorrhage
	• Dermoid	Proptosis, superolateral	T_1W, T_2W, STIR images, T_1W FS contains fat, fat–fluid levels
	• Rhabdomyosarcoma	Rapid progressive proptosis in child	T_1W isointense to muscle, hyperintense on T_2W, T_1W FS postcontrast enhancement, circumscribed mass may distort globe
	• Epithelial tumors		
	• Lipoma		
	• Metastasis		
Lacrimal gland	*Inflammatory*		
	• Pseudotumor	Enlarged lacrimal gland	T_1W,T_2W, Postcontrast T_1WI FS enlarged ON with enhancement, lacrimal gland enlargement with enhancement. Associated granulomas in paranasal sinuses
	• Dacryoadenitis		
	• Connective tissue disorder		
	• Sarcoid		CT lungs: lymph nodes, cavitating lesions
	• Wegener's		
	Tumors		
	• Lymphoma		
	• Lymphamatoid proliferation		
	• Benign pleomorphic adenoma		
	• Metastasis		
Subperiosteal	*Inflammatory*		
	• Abscess	Orbital cellulitis	T_1W,T_2W, Postcontrast T_1WI FS Rim enhancing collections Orbital/ facial fractures Hyperintense on T_1W,T_2W
	Trauma		
	• Hematoma	History of facialtrauma, subconjunctival hemorrhage	
	Tumors		
	• Metastasis/deposits		
	• Primary bone tumors		
Paranasal sinuses	*Inflammatory*		
	• Sinusitis		T_1W,T_2W, Postcontrast T_1WI FS Mucosal thickening or mass in sinuses, uni/bilateral
	• Mucocoeles		
	Miscellaneous		
	• Osteomas		T_1W,T_2W, Postcontrast T_1WI FS TIRM with FLAIR
	Tumors		
	• Sinus squamous cell carcinomas		Enhancing soft tissue lesions with bony destruction invading the orbit, intracranial extension
Craniofacial	*Dysplasia*		
	• Craniofacial dysplasias	Skull, facial deformity with orbit abnormalities	MRI of brain in detail for congenital abnormalities, NF
	• Fibrous dysplasia		Associated syndromes +/–
	• Neurofibromatosis		
	• Sphenoid dysplasia		
	Tumors		
	• Bone tumors Aneurysmal bone cyst		

19.10 Conclusion

The multimodality approach for imaging of the eye and orbit is useful. However, the strength of each modality needs to be understood by the radiologist to optimize its utility. High-resolution USG has the advantage of real-time imaging and CT is very useful for bony detail, fractures, and calcifications; both modalities are widely used. MDCT is the modality of choice for orbital trauma with the advantage of isotropic imaging that allows 3D VRT and curved MPR images.

MR is definitely superior for evaluation of the visual pathways, the globe, and the soft tissues. It has a vital role in imaging of space occupying lesions of the eye and orbit, and is useful for evaluation in congenital ocular abnormalities and craniofacial developmental anomalies and syndromes. High-field-strength MR along with dynamic contrast-enhanced MRI and DWI with ADC values helps in problem solving and narrowing the differentials leading to a conclusive diagnosis.

The utility of MR spectroscopy, functional MR, and PET-MR as additional adjunctive modalities is being recognized in the recent years to widen the ambit of conventional MRI.

References

1. Mafee MF. The eye. In: Som PM, Curtin HD, eds. *Head and Neck Imaging*. St. Louis, MO: Mosby, 4th edn. Vol. 2. 2003; pp. 441–527.
2. Mafee MF. Orbit: Embryology, anatomy and pathology. In: Som PM, Curtin HD, eds. *Head and Neck Imaging*. St. Louis, MO: Mosby, 4th edn. Vol. 2. 2003; pp. 529–624.
3. Mafee MF, Putterman A, Valvassori GA et al. Orbital space occupying lesions: Role of CT and MRI, an analysis of 145 cases. *Radiol Clin North Am* 1987;25:529–559.
4. Mafee MF. Magnetic resonance imaging: Ocular anatomy and pathology. In: Newton TH, Bilanuik LT, eds. *Modern Neuroradiology*, Vol. 4. *Radiology of the Eye and Orbit*. New York: Calvadel Press, 1990; pp. 2.1–3.45.
5. Hande PC, Talwar I. Multimodality imaging of the orbit. *Indian J Radiol Imaging* 2012;22:224–236.
6. Van Tassel P, Mafee MF, Atlas SW, Galetta SL. Eye, orbit and visual system. In: Atlas SW, ed. *Magnetic Resonance Imaging of the Brain and Spine*. Philadelphia, PA: Lippincott Williams & Wilkins, 4th edn. Vol. 2. 2009; Chapter 23: pp. 1258–1363.
7. Wang J, Takashima S, Takayama T et al. Head and neck lesions: Characterisation with diffusion-weighted echo planar MR imaging. *Radiology* 2001;220:621–630.
8. Char DH, Unsold R. Computed tomography: Ocular and orbital pathology. In: Newton TH, Bilanuik LT, eds. *Modern Neuroradiology*, Vol. 4. *Radiology of the Eye and Orbit*. New York: Calvadel Press, 1990; pp. 9.1–9.65.
9. Kaufman LM, Villabanca PJ, Mafee MF. Diagnostic imaging of cystic lesions in a child's orbit. *Radiol Clin North Am* 1998;36:1149–1163.
10. Kindler P. Morning glory syndrome; unusual congenital optic disc anomaly. *Am J Ophthalmol* 1970;69:376–384.
11. Mafee MF, Peyman GA. Retinal and choroidal detachments: Role of MRI and CT. *Radiol Clin North Am* 1987;25:487–507.
12. Mafee MF, Goldberg MF. CT and MR imaging for diagnosis of persistent hyperplastic primary vitreous (PHPV). *Radiol Clin North Am* 1987;25:683–692.
13. Kaufman LM, Mafee MF, Song CD. Retinoblastoma and simulating lesion: Role of CT, MR imaging and use of GD-DTPA contrast enhancement. *Radiol Clin North Am* 1998;36:1101–1117.
14. Davidson HC. Retinoblastoma. In: Hansberger ER, ed. *Diagnostic Imaging. Head and Neck*. Salt Lake City, UT: Amirsys Inc., Vol. 2. 2004; pp. 52–55.
15. Mafee MF, Valvassori GE. Radiology of craniofacial anomalies. *Otolarynol Clin North Am* 1981;14:929–988.
16. Barnes L, Verbin R, Appel B, Peel R. Tumour and tumour like lesions of the soft tissue. In: Barnes L, ed. *Surgical Pathology of the Head and Neck*. New York: Marcel Dekker, 2nd edn. Vol. 2. 2000; pp. 109–1095.
17. Khanna PC, Thapa MM, Iyer RS, Prasad SS. Pictorial essay: The many faces of craniosynostosis. *Indian J Radiol Imaging* 2011;21:49–56.
18. Rootman J, ed. *Diseases of the Orbit*. Philadelphia, PA: JB Lippincott, 1988.
19. Eustis HS, Mafee MF, Walton C, Mondonca J. MR imaging and CT of orbital infections and complications in acute rhinosinusitis. *Radiol Clin North Am* 1998;36:1165–1183.
20. Karesh JW, Baer JC, Hemady RK. Noninfectious orbital inflammatory disease. In: Tasman W, Jaeger EA, eds. *Duane's Clinical Ophthalmology*. Philadelphia, PA: Lippincott Williams & Wilkins, 2005; pp. 1–45.
21. Weber AL, Vitale Romo L, Sabates NR. Pseudotumour of the orbit. Clinical, pathological and radiologic evaluation. *Radiol Clin North Am* 1999;37:151–168.
22. Flanders AE, Mafee MF, Rao VM et al. CT Characteristics of orbital pseudotumours and other orbital inflammatory processes. *J Comput Assist Tomogr* 1989;13(1):40–47.
23. Kapur R, Sepahdari AR, Mafee MF et al. MR Imaging of orbital inflammatory syndrome, orbital cellulitis, and orbital lymphoid lesions: The role of diffusion-weighted imaging. *AJNR Am J Neuroradiol* 2009;30:64–70.
24. Yousem DM, Atlas SW, Grossman RI et al. MR imaging of Tolosa-Hunt syndrome. *Am J Neuroradiol* 1989;10:1181–1184.
25. Kirsch E, Hammer B, von Arx G. Graves' orbitopathy: Current imaging procedures. *Swiss Med Wkly* 2009;139(43–44):618–623.
26. Dodds NI, Atcha AW, Birchall D, Jackson A. Use of high resolution MRI of the optic nerve in Graves' ophthalmopathy. *BJR Br J Radiol* 2009;82:541–544.
27. Lagouras PA, Langer BG, Peyman GA et al. Magnetic resonance imaging and intraocular foreign bodies. *Arch Ophthalmol* 1987;105:551–553.

28. Flanders AE, Espinosa GA, Markiewicz DA et al. Orbital lymphoma. *Radiol Clin North Am* 1987;25:601–602.

29. Valvassori GE, Sabnis SS, Mafee MF et al. Imaging of orbital lymphoproliferative disorders. *Radiol Clin North Am* 1999;37:135–150.

30. Kind AD, Ahuja AT, Yeung DK et al. Malignant cervical lymphadenopathy: Diagnostic accuracy of diffusion weighted MR imaging. *Radiology* 2007;245:806–813.

31. Shields JA, ed. *Diagnosis and Management of Orbital Tumours*. Philadelphia, PA: WB Saunders, 1989.

32. Aviv RI, Miszkiel K. Orbital imaging: Part 2. Intraorbital pathology. *Clin Radiol* 2005;60:288–307.

33. Bilaniuik LT. Orbital vascular lesions: Role of imaging. *Radiol Clin North Am* 1999;37:169–183.

34. Muller-Forell WS, Pitz S. Orbital pathology. In: Muller-Forell WS (ed). *Imaging of Orbital and Visual Pathway Pathology*. Heidelberg, Germany: Springer, 2002; pp. 147–340.

35. Zimmerman RA, Bilaniuik LT, Savino PJ. Visual pathways. In: Som PM, Curtin HD, eds. *Head and Neck Imaging*. St. Louis, MO: Mosby, 4th edn. Vol. 2. 2003; pp. 735–781.

36. Azar-Kia B, Naheedy MH, Eliad DA et al. Optic nerve tumours: Role of magnetic resonance imaging and computed tomography. *Radiol Clin North Am* 1987; 25:561–581.

37. Davidson HC. Optic pathway glioma orbit. In: Hansberger ER, ed. *Diagnostic Imaging. Head and Neck*. Salt Lake City, UT: Amirsys Inc., Vol. 2. 2004; pp. 60–63.

38. Mafee MF, Goodwin J, Dorodi S. Optic nerve sheath meningiomas: Role of MR imaging. *Radiol Clin North Am* 1999;37:195–202.

39. Weber AL, Mikulis DK. Inflammatory disorders of the paraorbital sinuses and their complications. *Radiol Clin North Am* 1987;25:615–630.

40. Mafee MF, Haik BG. Lacrimal gland and fossa lesions: Role of computed tomography. *Radiol Clin North Am* 1987;25:767–779.

41. Mafee MF, Edward DP, Koeller KK, Dorodi S. Lacrimal gland tumours and simulating lesions. Clinicopathologic and MR imaging features. *Radiol Clin North Am* 1999;37:219–239.

42. Grossman RI, Yousem DM. Orbit. In: *Neuroradiology: The Requisites*. St. Louis, MO: Mosby, 2nd edn. 2003; pp. 469–516.

43. Suzuki H, Takanashi J, Kobayashi K et al. MR imaging of idiopathic intracranial hypertension. *Am J Neuroradiol* 2001;22:196–199.

44. Wall M. Idiopathic intracranial hypertension. *Neurol Clin* 2010;28(3):593–617.

20

Brain Aging and Degenerative Diseases of the Brain

Memi Watanabe, Joshua Thatcher, Yukio Kimura, Ivana Delalle, Samuel Frank, and Osamu Sakai

CONTENTS

20.1 Introduction

Normal aging of the human brain is chiefly characterized by volume decreases and white matter changes on conventional magnetic resonance imaging (MRI) [1]. It is often challenging to differentiate normal brain aging from a pathological process both clinically and radiologically.

This chapter groups different neurodegenerative diseases roughly into categories according to their clinical presentations and symptoms, including normal aging, cognitive disorders, movement disorders, and cerebellar ataxias. In each disease, the characteristic radiological features are reviewed and additional findings in advanced imaging techniques are introduced.

20.2 Normal Brain Aging

As the brain ages, the brain volume decreases and nonspecific white matter changes appear on conventional MRI [1]. As the brain volume decreases, there is compensatory filling of the intracranial space resulting in increased cerebrospinal fluid (CSF) space. Because quantitative assessments were introduced to brain volumetry, a pattern of global and regional brain atrophy in normal brain aging has been investigated. The longitudinal studies revealed that the rates of global brain atrophy gradually increase with age from an annual rate of 0.2% per year at age 30–50 years to 0.3%–0.5% at age 70–80 years [2]. Many morphometry studies of adult brains suggest that aging predominantly and substantially affects gray matter volume, whereas white matter volume is relatively preserved until very

advanced ages [1]. The regional volumetry studies of the normal aging process reveal accelerated volume loss predominantly in the frontal and parietal lobes, specifically in the regions such as the central sulcus, middle frontal gyrus, superior parietal gyrus, and insula [1,3]. Microstructural information obtained by diffusion-weighted imaging (DWI) demonstrates evidence of disrupted integrity of white matter fibers. Many studies have reported a general aging trend of increased apparent diffusion coefficient and mean diffusivity (MD) together with decreased fractional anisotropy (FA), reflecting the degradation of neural connections [4,5].

The presence of multiple nonspecific hyperintensity foci on T_2-weighted imaging is also a common finding in elderly subjects. A few foci can be a nonsignificant finding in most cases. The reported prevalence of white matter hyperintensities ranges from 11% to 21% in adults at age 64 years to 94% at age 82 years in the general population [6]. White matter hyperintensities are more common and extensive in patients with cardiovascular risk factors and symptomatic cerebrovascular disease [6]. Meta-analysis of 46 longitudinal studies reported strong evidence that the presence of white matter hyperintensities is an important indicator of future risks of stroke, cognitive decline, dementia, and mortality [6]. A notable association with cardiovascular mortality was also indicated [6]. Pathological findings in regions of white matter hyperintensity include dilated perivascular spaces, myelin pallor, tissue rarefaction associated with loss of myelin and axons, mild gliosis, and infarction [7]. Enlarged perivascular (Virchow–Robin) spaces have been associated with vascular risk factors, such as hypertension, and are regarded as likely markers of cerebral small vessel disease, possibly due to increased vessel pulsatility (Figure 20.1) [8]. On MRI, nonspecific T_2

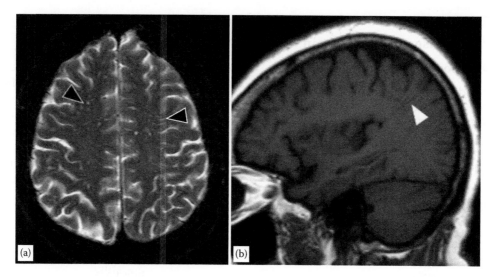

FIGURE 20.1

Perivascular spaces. (a) Axial fat-suppressed gradient and spin-echo T_2-weighted sequence shows multiple hyperintense spots and lines within the cerebral white matter suggestive of perivascular spaces (black arrowheads), whereas they appear as linear hypointense lesions running in a radial direction on sagittal T_1-weighted image (b) (white arrowhead).

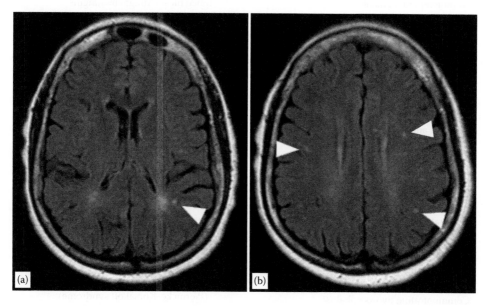

FIGURE 20.2

Nonspecific T_2 hyperintense foci. (a,b) Axial FLAIR images of a 64-year-old subject show scattered small hyperintense foci within cerebral white matter (arrowheads).

hyperintensity foci are seen in the deep white matter, typically sparing subcortical U-fibers, often with magnetic resonance angiography (MRA) evidences of small vessel disease (Figure 20.2) [6]. The age-related white matter changes scale has been proposed for a measure of white matter hyperintensities [7]. In recent years, quantitative assessment of the volume of white matter hyperintensities has been widely used especially in cohort studies. A volumetry study suggested that the volumes of white matter hyperintensities greater than 0.5% of intracranial volume should be considered abnormal [9]. Accelerated increases in volume of white matter hyperintensities have been associated with impairment of multiple areas of cognition, including executive function, processing speed, and memory performance, and have been shown to result in progressive cognitive decline leading to mild cognitive impairment (MCI) [10]. Therefore, increased accumulation of white matter hyperintensities needs to be carefully evaluated.

20.3 Degenerative Diseases of the Brain

20.3.1 Cognitive Disorders

Dementia is defined as disease-related loss of memory and other cognitive abilities of sufficient severity to interfere with activities of daily living. There are a number of possible causes of dementia symptoms (Table 20.1). The most common cause of degenerative dementia in the elderly is Alzheimer's disease (AD) [8,11]. As the aging population continues to grow, AD becomes one of the most threatening problems in public health. Although a diagnosis of dementia is based on clinical evaluation, neuroimaging, such as computed tomography (CT) and MRI, is mandatory to exclude other possible etiologies, including cerebrovascular disease, normal pressure hydrocephalus, and neoplasm. MRI is preferred for the evaluation of dementia. Moreover, together with positron emission tomography (PET) and single-photon emission CT (SPECT), MRI gives more specific information of underlying neuropathology, allowing neuroimaging to play an increasingly important role in the assessment of microscopic changes occurring in neurodegenerative diseases.

TABLE 20.1

Possible Causes of Dementia Symptoms

Degenerative	Infectious
Alzheimer's disease	Human immunodeficiency virus
Dementia with Lewy bodies	Herpes simplex virus
Corticobasal degeneration	Slow virus (PML, SSPE)
Progressive supranuclear palsy	Lime disease
Frontotemporal dementia	Tuberculosis/fungal meningitis
Parkinson's disease	Syphilis
Huntington's disease	Whipple's disease
Argyrophilic grain disease	**Noninfectious Inflammatous**
ALS–dementia complex	Behçet's disease
Wilson's disease	CNS lupus
Cerebrovascular	Neurosarcoidosis
Small vessel disease (Binswanger/lacuna)	**Prion**
Amyloid angiopathy	Creutzfeldt–Jakob disease
Multiple emboli	**Endocrine**
Hypoxic/ischemic injury	Hyper/hypothyroid
Cerebral vasculitis	Hyper parathyroid
Cerebral autosomal dominant arteriopathy with subcortical infarcts and leukoencephalopathy	Cushing syndrome/Addison's disease
Traumatic	**Metabolic**
Chronic subdural hematoma	Uremia
Diffuse axonal injury	Chronic hepatic encephalopathy
Dementia pugilistica (boxer's dementia)	**Malabsorption**
Hydrocephalus	Thiamine (vitamin B_1) deficiency (Wernicke–Korsakoff syndrome)
Communicating hydrocephalus	Other vitamin deficiency (vitamin B_{12}, vitamin E, niacin)
Noncommunicating hydrocephalus	
Normal pressure hydrocephalus	**Toxic**
Neoplastic/Paraneoplastic	Alcohol
Metastatic disease	Anoxia/carbon monoxide poisoning
Primary CNS tumor	Heavy metal poisoning
Paraneoplastic syndrome (limbic encephalitis)	Drugs (benzodiazepines, opiates, tricyclic antidepressants, anticonvulsants, etc.)
Post-radiotherapy	
Demyelinating	**Others**
Multiple sclerosis	Chronic hypercapnea/hypoxemia
Acute disseminated encephalomyelitis	
Psychiatric	
Depression	

ALS, amyotrophic lateral sclerosis; CNS, central nerves system; PML, progressive multifocal leukoencephalopathy; SSPE, subacute sclerosing panencephalitis.

20.3.2 Alzheimer's Disease

20.3.2.1 Clinical Features

AD is the most common cause of degenerative dementia [8,11], accounting for 50%–80% of dementias [12]. Clinically, the cognitive symptoms begin insidiously and, in most cases, progress gradually [11]. The transitional clinical stage of this process is recognized as MCI, which is defined as cognitive decline greater than expected for an individual's age, but does not interfere notably with daily life [12]. The rate of conversion from MCI to AD dementia is approximately 12% per year, which stands in contrast to the 1%–2% per year rate for elderly persons with no cognitive impairment [11]. Within 5 years, approximately 80% of individuals diagnosed with MCI will have progressed to dementia [11]. With advancing age, the majority of dementia is most likely a combination of AD pathology and other types of degenerative pathologies, so-called mixed dementia.

20.3.2.2 Pathological Findings

Pathologically, major distinctive structures seen within brains of patients with Alzheimer-type dementia are senile plaques and neurofibrillary tangles, accompanied by synaptic and neuronal loss [13,14]. Senile plaques are proteinaceous extracellular deposits consisting primarily of amyloid-β (Aβ) peptide fragments [14]. Associated with many of the senile plaques, there is evidence of inflammation and abnormal neuronal processes called dystrophic neurites [14]. Senile plaques occur abundantly in regions of the association areas of the neocortex, posterior cingulate gyrus, and limbic cortex, with only late involvement of the primary sensory and motor areas [14]. Aβ peptide is also deposited in vessel walls, causing cerebral amyloid angiopathy or congophilic angiopathy or because Congo red stain is used in pathology examination to highlight amyloid deposits within senile plaques as well as vessel walls. Neurofibrillary tangles, primarily composed of abnormally phosphorylated tau (τ) protein, are initially formed within neuronal cell bodies and dendrites. With neuronal cell death, the neurofibrillary tangles may remain as extracellular "ghost tangles" or "tombstones" [14]. Normal tau protein promotes the assembly of microtubules, helps the organization of 3D architecture of the neuron, and assists with the transportation of protein and enzyme-containing vesicles [14]. When tau protein is abnormally phosphorylated, it aggregates inside neuronal bodies and processes (neurites), disrupting microtubule function [14]. As a result, some neurites can be highlighted by appropriate tau immunohistochemistry revealing changes in their morphology: these neurites are relatively short and coarse, and thus called "dystrophic neurites" and "neuropil threads." This neurofibrillary pathology may also be observed within senile (neuritic) plaques [15].

Several diagnostic classifications have been proposed to evaluate AD pathology. Braak and Braak staging uses qualitative assessment of the distribution of neurofibrillary tangles and neuropil threads only [13]. The transentorhinal stages (Braak and Braak stages I and II) are characterized by either mild or severe involvement confined to the transentorhinal layer Pre-alpha with only mild involvement of the hippocampus. The limbic stages (Braak and Braak stages III and IV) are marked by severe involvement of layer Pre-alpha in both the transentorhinal region and the proper entorhinal cortex with mild hippocampal and little isocortical involvement. In the isocortical stages (Braak and Braak stages V and VI), neurofibrillary pathology reaches all areas of association cortex and may also involve primary cortices. The Consortium to Establish a Registry of Alzheimer Disease assessment relies on semiquantitative "age-related plaque score" together with the clinical history to establish final diagnosis of definitive AD, neuropathologically probable AD, neuropathologically possible AD, or no AD [16]. The National Institute on Aging-Regan Institute criteria consider the density of all AD lesions (neuritic plaques, neurofibrillary tangles, amyloid deposits, and neuropil threads) to determine the probability that the dementia symptoms are caused by these lesions [17].

20.3.2.3 Genetics

AD is commonly categorized as either early or late onset. Late-onset sporadic disease (after age 65 years) accounts for over 95% of cases [11]. The much less common early onset AD manifests in individuals less than 65 years, some as young as the fourth decade of life [11]. Early work in genetic studies of familial AD identified three autosomal dominant mutations: the amyloid precursor protein (APP), the presenilin-1 (PSEN1), and the presenilin-2 (PSEN2) genes [18]. In the analysis of late-onset AD susceptibility genes, the ε4 allele of the apolipoprotein E (APOE) gene was identified as a major genetic risk factor for brain Aβ deposition [18]. The APOE gene encodes a protein with a key role in cholesterol metabolism and contributes to AD pathogenesis by modulating the metabolism, aggregation, and clearance of Aβ peptide, and also perhaps by directly regulating brain lipid metabolism and synaptic functions [11]. More recently, multiple new genes have been found to be implicated in AD risk [18]. Early onset or familial AD is sometimes referred to as a disease of overproduction of Aβ peptide, whereas late-onset sporadic AD is a disease of inadequate Aβ clearance [11].

20.3.2.4 Clinical Criteria

Brain imaging has been useful to exclude other causes of dementia. It now serves as a biomarker in reformulated diagnostic criteria for AD [8,11,19,20]. The criteria for AD [21], which were based on the clinical status of patients, have been revised, and it is proposed to include more than one supportive neuroimaging or CSF biomarker in addition to a main clinical feature of progressive episodic memory defect for at least 6 months [19]. Biomarkers for Aβ accumulation include abnormal radiotracer retention on amyloid PET images and low levels of CSF $A\beta_{42}$ [11,19,20]. Biomarkers for tau pathology include elevated levels of CSF tau (both total and phosphorylated), decreased ^{18}F-fluorodeoxyglucose (FDG) uptake on PET imaging in the temporoparietal cortex, and atrophy on structural magnetic resonance (MR) images in specific brain regions, typically in medial temporal lobes [11,20]. Decreased FDG uptake is an indicator of impaired synaptic function due to tau pathology, and cerebral atrophy is thought to reflect microscopic neurodegeneration (loss of synapses, dendritic processes, and neurons) as a resultant of tau pathology [11,20].

20.3.2.5 Imaging Features

The major role of MRI is to help excluding other potential causes of dementia, such as cerebrovascular disease, space-occupying lesions, normal pressure hydrocephalus, or subdural hemorrhage.

On MRI, cerebral atrophy beginning in the entorhinal cortex and hippocampus is recognized as a structural imaging biomarker of AD (Figure 20.3) [8]. Accelerated atrophy of these structures is a hallmark of AD and can be seen prior to clinically recognizable stages of AD. Volumetry studies reported a hippocampal atrophy rate of 4%–8% a year in early AD, which is much higher than the reported hippocampal atrophy rate in healthy subjects (0.1%–0.2% at age 30–50 years, 0.8% in the mid-70s, and 1.5%–2% a year at age 80–90 years) [2]. As the disease progresses, diffuse cerebral atrophy is seen, involving the medial, basal, and lateral temporal lobes, as well as the medial and lateral parietal cortices (Figure 20.4) [11,20].

Other imaging for the diagnosis of AD includes amyloid PET imaging and metabolism measurement by FDG PET. Amyloid PET imaging is a molecular imaging technique of specific Aβ ligands, such as Pittsburgh compound B (PIB) (labeled with carbon 11),

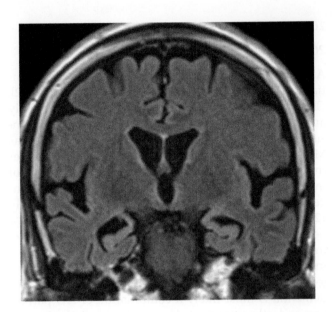

FIGURE 20.3
AD in a 57-year-old female patient with memory loss. Atrophy confined to bilateral medial temporal lobes is demonstrated on coronal FLAIR image.

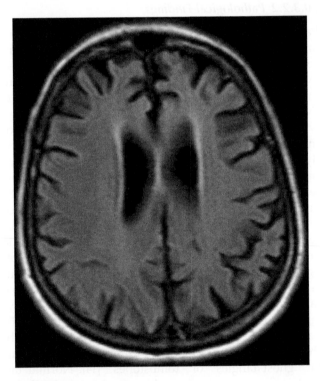

FIGURE 20.4
AD in a 55-year-old female patient with memory loss, *APOE* epsilon4 allele homozygous, and abnormal CSF exam. Axial FLAIR image shows diffuse cortical atrophy.

and flubetapir and flutemetamol (labeled with fluorine 18) [8]. Most subjects who have received a clinical diagnosis of AD have positive amyloid PET imaging studies [11]. Abnormal amyloid PET images are seen in around 30% of cognitively normal elderly subjects.

It matches the proportion of cognitively normal elderly subjects with an autopsy diagnosis of AD [11]; however, the degree of PIB uptake does not distinguish the cognitively normal elderly from patients with MCI or AD [8].

The measures of glucose metabolism on FDG PET imaging reflect net brain metabolism, which predominantly represents synaptic activity [11]. FDG PET studies in AD subjects demonstrate characteristic deficits in posterior temporoparietal, posterior cingulate, and inferior frontal regions (Figure 20.5) [8,11]. The extent of abnormal FDG PET is shown to be greater than the areas of atrophy alone and is thought to be an earlier indicator of tau pathology [8]. Alternative imaging to measure cerebral activity is regional cerebral blood flow (rCBF) SPECT, which is more widely available and cheaper than PET imaging [22]. The rCBF SPECT shows posterior hypoperfusion, specifically in the parietotemporal association cortex in AD (Figure 20.6). Perfusion SPECT has been reported to generally correlate with neurofibrillary pathology [22].

Considering temporal orders of pathophysiological changes, a multiphase model of AD has been proposed [11,20]. The first phase is Aβ accumulation, which is imaged with amyloid biomarkers (amyloid PET findings and CSF $A\beta_{42}$ measures) decades before the first clinical symptoms appear [11,20]. The second phase involves neuronal degeneration or injury, which is measured with biomarkers of tau pathology. CSF tau and FDG PET findings become dynamic later, maybe shortly before clinical symptoms first appear [11,20]. This phase is followed by cell death, which corresponds to characteristic atrophy. Typical atrophy on structural MRI becomes dynamic in the clinically symptomatic phase of the disease, where the MR findings correlate best with clinical symptom severity [11,20]. This multiphase model best explains longitudinal changes in AD biomarkers [8]. Pathological observations also support the position that amyloid develops first, whereas neocortical neurofibrillary tangles appear later [11]. Another possibility is that amyloid and tau pathologies are independent but commonly coexist in the different stages of the disease [8].

In should be noted that, in both MCI and AD dementia criteria, clinical diagnoses are paramount and biomarkers are complementary [20]. These biomarkers have different roles in each clinical phase. In the preclinical phase, biomarkers are used to establish the presence of AD pathology in research subjects with no or very subtle overt symptoms [20]. In the symptomatic predementia, MCI phase, biomarkers are used to establish the underlying etiology responsible for the clinical deficit [20]. In the dementia phase, biomarkers are used to enhance the level of certainty that AD pathology underlies the dementia in an individual [20].

Other imaging techniques include diffusion, perfusion, spectroscopic, and functional MR imaging. However, these techniques have not presently been established as a biomarker. Supported by autopsy evidences of pathologic changes in white matter occurring early in the course of AD (decreased myelin and axonal attenuation, loss of oligodendrocytes, and activation of glial cells) [23], diffusion MR studies revealed the findings compatible with disrupted white matter

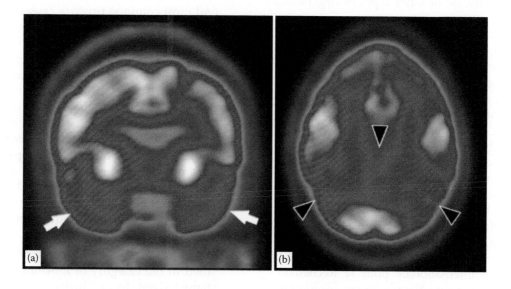

FIGURE 20.5

AD in a 62-year-old female patient with progressive dementia. (a) Coronal FDG-PET image shows marked reduction in metabolic activity in the bilateral temporal lobes (white arrows), whereas FDG uptake is spared in the basal ganglia. (b) Axial image demonstrates additional areas of decreased activity within the bilateral parietal lobes and posterior cingulate gyrus (black arrowheads).

RT.LAT   LT.LAT

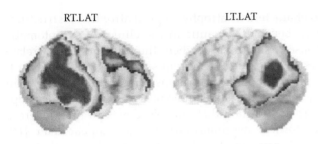

FIGURE 20.6
Alzheimer's disease. Topographic z-score mapping visualizing the degree of perfusion reduction in N-isopropyl-p-[123I]-iodoamphetamine (123I-IMP) SPECT with a color scale transitioning from lower z-score areas (less reduction) in black/purple to higher z-score areas (more reduction) in red. The z-score mapping in a 57-year-old female patient with AD shows asymmetric decreases in rCBF in the temporoparietal lobes, predominantly in the right.

integrity with an increase in diffusivity and a decrease in anisotropy. It is reported that FA and MD correlate with clinical disease severity in AD and can help predict future progression from MCI to AD [11]. Interest in the resting-state functional MRI as a potential noninvasive biomarker has been increasing with the evidence that increased activity in the task-free default mode is seen within the areas that are particularly susceptible to amyloid pathology [11,24].

20.3.2.6 Differential Diagnosis

Although diagnosis of dementia is often complicated by coexisting conditions or overlaps with various other dementias, some specific imaging features can help differentiate AD from other causes of dementia, such as cerebrovascular disease or other types of degenerative diseases.

1. *Cerebrovascular disease*: It is the most common secondary cause of dementia [8]. It can appear with various radiological manifestations depending on the site and size of the vessel involved, ranging from large cortical infarcts, lacunar infarcts, and macro- and microhemorrhages to white matter ischemia [8]. Sudden onset of dysfunction, a stepwise deteriorating course, focal neurologic signs, stroke risk factors, systemic vascular disease, and prior strokes suggest vascular dementia [12]. On MRI, white matter T_2 hyperintensities are an imaging biomarker of vascular disease. The International Workshop of the National Institute of Neurological Disorders and Stroke and the Association Internationale pour la Recherche et l'Enseignement en Neurosciences criteria for vascular dementia include imaging findings of

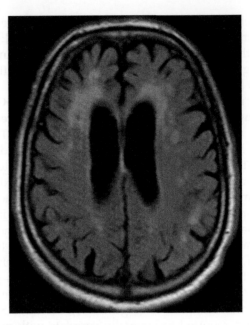

FIGURE 20.7
Cerebrovascular disease. Axial FLAIR image of an 81-year-old male patient with behavioral changes shows diffuse cerebral atrophy with scattered white matter ischemic changes.

multiple large-vessel infarcts, single strategically placed infarcts, multiple basal ganglia and white matter lacunes, or extensive periventricular white matter lesions or combinations thereof [25]. Neuropathology of vascular dementia and AD can coexist; however, vascular lesions on MRI or CT favor vascular dementia over AD (Figure 20.7) [12].

2. *Dementia with Lewy bodies (DLB)*: It is the second most common primary cause of dementia. Clinically, the pattern of cognitive decline can be similar to AD, but in addition, parkinsonism develops within 1 year of onset of cognitive issues. Additionally, fluctuating cognition and recurrent visual hallucinations, and poor response to medications may be seen in DLB. On MRI, characteristic atrophy in the midbrain, hypothalamus, basal forebrain, and thalamus with a relative sparing of the hippocampus and temporoparietal cortex may help differentiate DLB from AD. Dopamine transporter SPECT is usually normal in AD in contrast to significantly reduced tracer uptake observed in DLB [26].

3. *Dementia with Parkinson's disease (PD)*: It is classified in the category of movement disorders, characterized by a cluster of parkinsonism symptoms, such as bradykinesia, resting

tremor, and rigidity. Many individuals with PD develop dementia in later stages of the disease. MRI appears normal in the early stage of the disease with nonspecific general cortical atrophy seen in advanced disease stages. Dopamine transporter SPECT imaging may help differentiate early stages of PD from other disorders such as AD, cerebrovascular disease, essential tremor, or medication-induced parkinsonism.

4. *Frontotemporal dementia (FTD)*: It is rare in general clinical practice. It presents typically as early as the fourth decade of life, and it is the second most common cause of dementia after AD under age 65 years. Radiological differentiation of FTD and early AD may be difficult as both entities show abnormalities of the medial temporal lobe. However, both structural MRI and molecular imaging, such as FDG PET or rCBF SPECT, have good specificity in differentiating FTD from AD based on identification of predominantly anterior-versus-posterior patterns of atrophy or hypometabolism [8]. Amyloid PET would also contribute in detection of amyloid pathology in AD.

5. *Creutzfeldt–Jakob disease (CJD)*: It is a rare, but fatal prion disorder characterized by rapidly progressive dementia. Sporadic CJD typically occurs between ages 50 and 75 years. Characteristic MR imaging findings include hyperintense signal intensity on T_2-weighted images in the basal ganglia, and less often in the cortex, with signal abnormalities on DWI [27,28].

20.3.3 Dementia with Lewy Bodies

20.3.3.1 Clinical Features

DLB is the second most common cause of degenerative dementia [29–31], accounting for approximately 15% of all dementia [8]. DLB diagnosis is useful due to the rapidly progressive course, increased risk of side effects when exposed to neuroleptics, and poor response to cholinesterase inhibitors [12]. Criteria for DLB include, in addition to dementia as a central feature, the development of parkinsonism within 1 year of dementia, fluctuating cognition and visual hallucinations [29,30]. Suggestive features include rapid eye movement sleep behavior disorder and severe neuroleptic sensitivity, as well as imaging findings of low dopamine transmitter uptake in basal ganglia on SPECT or PET [29].

20.3.3.2 Pathological Findings

DLB is one of the neurodegenerative disorders classified as an α-synucleinopathy. α-Synuclein is a normal synaptic protein that has been implicated in vesicle production [30]. In an aggregated and insoluble form, it constitutes the main component of the fibrils that are a major constituent of the Lewy bodies in DLB and other synucleinopathies [30]. Past studies suggest that the number of cortical Lewy bodies does not correlate well with either the severity or the duration of dementia, whereas Lewy neurites and neurotransmitter deficits are more likely associated with clinical symptoms [30]. It is staged into three pathological categories (brain stem-predominant, limbic, or neocortical) depending on the numbers and distribution of Lewy bodies [30]. Pathological features of DLB overlap with those of PD, and dopaminergic cell loss and accumulation of intracellular Lewy bodies in presynaptic terminals are the hallmark of those entities [8]. DLB is probably best thought of as a spectrum with a wide range of Lewy body involvement in the brains of patients with dementia [29].

Pathological, genetic, clinical, and biochemical findings suggest marked overlaps among PD, DLB, and AD [32]. Most patients with DLB also have concurrent AD pathology, including cortical amyloid plaques and neurofibrillary tangles, but only a few meet the tangle-based Braak stages V and VI for AD [29,30]. The presence of AD pathology in DLB modifies the clinical presentation by lowering the rate of visual hallucinations and parkinsonism making such cases harder to differentiate clinically [30].

20.3.3.3 Imaging Features

Brain imaging serves as a supportive feature for DLB diagnosis: occipital hypoperfusion on rCBF SPECT (Figure 20.8) and occipital hypometabolism on FDG PET [12]. Additionally, structural MRI may help to differentiate DLB from AD, showing a pattern of relatively focused atrophy of the midbrain, hypothalamus, basal forebrain, particularly the substantia innominata, and thalamus with a relative sparing of the hippocampus and temporoparietal cortex (Figure 20.9) [8,31]. The specific pattern of volume loss also suggests involvement of multiple neurotransmitter systems, including the cholinergic, serotonergic, and noradrenergic systems, in addition to adrenergic involvement [31]. Quantitative assessment in dopamine transporter SPECT is useful to identify reduced uptake in the striatum (that includes caudate nucleus and putamen), which can discern parkinsonian pathology (including PD and DLB) from AD [8,26].

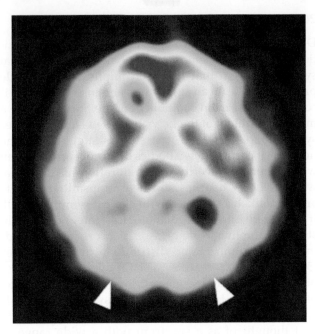

FIGURE 20.8
DLB in a 61-year-old female patient with visual hallucinations and parkinsonism. Cerebral perfusion study with 123I-IMP SPECT shows bilateral occipital deficits (arrowheads).

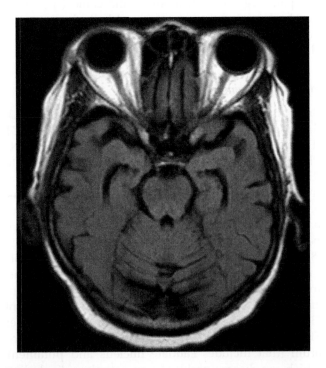

FIGURE 20.9
DLB in a 65-year-old female patient presenting with worsening visual hallucinations. On axial FLAIR image, the hippocampal volume is relatively preserved.

Additionally, reduced cardiac uptake in the cardiac sympathetic imaging by meta-^{123}I-iodobenzylguanidine myocardial scintigraphy has been reported as an early finding in PD and DLB, suggesting a centripetal dieback process before the involvement of the central nervous system [31,33].

20.3.4 Frontotemporal Dementia

20.3.4.1 Clinical and Pathological Features

FTD is a rare clinical disorder, which typically presents earlier than AD. It is the second most common cause of dementia under age 65 years after AD [34,35] and less common in very elderly dementia cases [12].

Clinical FTD syndromes form part of neuropathological syndrome, frontotemporal lobar degeneration (FTLD), which shows common features of selective degeneration in frontotemporal lobes [34,36]. FTLD consists of clinically, genetically, and pathologically nonuniform entities, and its major subdivision is designated by protein abnormalities [36,37]. Approximately half of FTLD cases are associated with accumulation of tau inclusions (tauopathy including FTLD with Pick bodies, FTLD with microtubule-associated protein tau gene [*MAPT*] mutation, corticobasal degeneration [CBD], progressive supranuclear palsy [PS], argyrophilic grain disease, and neurofibrillary tangle dementia), whereas other pathological proteins identified include the transactive response DNA-binding protein 43 and fused in sarcoma protein [35,37].

Clinically, FTD is characterized by progressive alternations in behavior, personality, and/or language with relative preservation of memory [35,37,38]. More than 30% of FTDs are familial [34,35]. The clinical spectrum of FTD consists of distinct syndromes: the behavioral variant of FTD (bvFTD) and language variants, semantic dementia (SD), and progressive nonfluent aphasia (PNFA) [34]. Each clinical syndrome is characterized by topographically distinct cerebral involvement: with bvFTD associated with symmetrical (or right-sided) frontal and anterior temporal dysfunction, PNFA with left frontotemporal dysfunction, and SD with anterior temporal (typically left more than right) deficits [34]. Although all of the subtypes can occur in conjunction with motor neuron disease, it is most commonly seen with bvFTD, occasionally with PNFA, and only very rarely with SD [34].

20.3.4.2 Imaging Features

MRI in patients with FTD typically shows symmetrical atrophy of the frontal and anterior temporal lobes (Figure 20.10) [12,37]. In some cases, asymmetry of atrophy may be seen [37]. In early onset dementia, radiological differentiation of FTD and early AD may be challenging as both entities are characterized with medial temporal

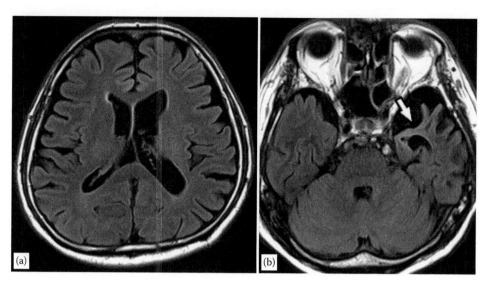

FIGURE 20.10

FTD in a 78-year-old female patient with PNFA. Axial FLAIR images demonstrate diffuse cerebral atrophy predominantly in the left (a) and severe atrophy in the left anterior temporal lobe (b) (arrow).

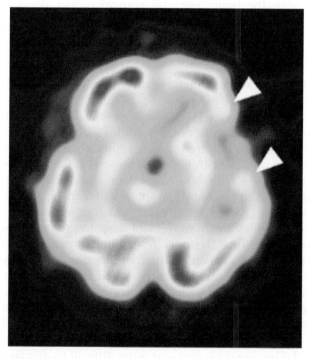

FIGURE 20.11

FTD in a 78-year-old female patient with PNFA (the same patient as in Figure 20.10). 123I-IMP SPECT demonstrates left-sided frontal and temporal hypoperfusion (arrowheads).

lobe atrophy. In established FTD cases, affected regions include the frontal lobes, specifically the ventromedial, orbitofrontal, anterior cingulate, anterior insula, and amygdala [8]. FDG PET or rCBF SPECT shows anterior metabolic or perfusion defects, which are often asymmetric [8,12,34] (Figure 20.11). Frontal lobe hypoperfusion can also be seen in arterial spin labeling MRI [39].

20.3.5 Creutzfeldt–Jakob Disease

20.3.5.1 Clinical Features

CJD is a rare but fatal degenerative prion disease characterized by rapidly progressive dementia with myoclonus, characteristically abnormal EEG, and 14-3-3 protein in the CSF. CJD progress rapidly and most patients die within 1 year [40]. It occurs in four forms: sporadic, familial, iatrogenic, and variant [40]. Sporadic CJD occurs typically between 50 and 70 years of age by unknown causes [40]. Some inherited CJD cases are associated with mutations of the prion protein gene. A few cases per year are iatrogenic, as a result of exposure to contaminated medical instruments or transplant of infected tissues. Variant CJD is linked to the bovine spongiform encephalopathy agent and has different clinical features from other forms of CJD, including a tendency to affect a younger age group, lack of periodic *electroencephalographic* changes [41], and a possibility of transfusion transmission from person to person [42]. Definite diagnosis of CJD is based on histopathological findings, though biopsy is rarely performed [12].

20.3.5.2 Pathological Findings

Definitive diagnosis of CJD requires histopathological examination of the brain and immunostaining for the disease-associated form of prion protein in the brain tissue [40]. The crucial neuropathological features consist of spongiform change accompanied by neuronal loss and gliosis [40]. The amyloid plaques are found in 10% of sporadic CJD [40]. Immunostaining for prion shows synaptic and perivascular staining with striking staining of plaques [40].

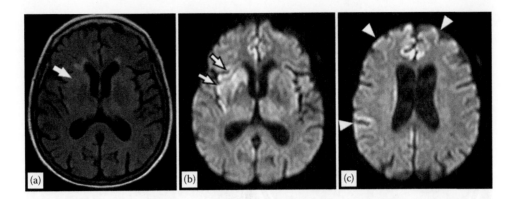

FIGURE 20.12

Sporadic CJD in a 64-year-old female patient with progressive gait instability and dementia. (a) Axial T_2-weighted image demonstrates slight high signals in the basal ganglia (arrow). (b) Axial diffusion-weighted image of the same patient shows signal abnormality more clearly within the basal ganglia predominantly in the right (arrows). (c) Additionally, cortical signal abnormalities are noted in the multiple brain areas (arrowheads) on diffusion-weighted image.

20.3.5.3 Imaging Features

The most common MRI abnormality is hyperintense signal intensity on T_2-weighted or fluid-attenuated inversion recovery (FLAIR) sequence in the basal ganglia, more substantially in the caudate and/or corpus striatum, and less often in the cortex. DWI signal abnormalities are highly sensitive for early clinical diagnosis of CJD (Figure 20.12) [8,27,28]. DWI high signal intensities confined to the cerebral cortex and basal ganglia, particularly the caudate, are specific features in the early stages of CJD; however, differential diagnoses include mitochondrial myopathy, encephalopathy, lactic acidosis, and stroke-like episodes; venous hypertensive encephalopathy; and chronic herpes simplex encephalitis [28].

T_2 high signal intensity in bilateral thalamus, particularly in the pulvinar, is highly suggestive of variant CJD [8,41]. However, pulvinar signs are not specific to variant CJD and can be observed in other disorders, such as benign intracranial hypertension, Alpers' syndrome, postinfectious encephalitis, and cat-scratch disease [41]. Their clinical features are reported to be distinct from variant CJD.

20.3.6 Wilson's Disease

Wilson's disease is an autosomal recessive disorder of hepatic copper deposition caused by various mutations in the gene *ATP7B*, located on chromosome 13 [43,44]. This gene encodes a P-type adenosine triphosphatase (ATPase), known as the Wilson ATPase, which has synthetic and excretory functions to participate in the incorporation of copper into nascent ceruloplasmin and to expedite excretion of copper into the bile [43,44]. Dysfunction of ATPase causes hepatic retention of copper, resulting in liver injury [43]. When hepatocellular storage capacity is exceeded, free copper is slowly liberated into the blood and deposited in various organs, notably the brain, eyes (as corneal copper deposits, known as Kayser–Fleischer rings), and kidneys [43].

20.3.6.1 Clinical Features

The clinical presentations are highly varied, and Wilson's disease can present as hepatic, neurological, or psychiatric disease [43]. Most symptoms first appear in the second and third decades of life [44], but it can present in younger children or older adults. Younger patients tend to present with hepatic disease [43]. Patients who first present with neurological or psychiatric symptoms tend to be older than those with hepatic features alone [44]. Neurological presentation of Wilson's disease is usually with movement disorders (e.g., tremor) or with diffuse dystonia resembling a parkinsonian disorder [43]. Patients with parkinsonian symptoms before age 40 years should be tested for Wilson's disease. Severe depression or various behavioral changes are typical [43], and cognitive dysfunction can accompany neurological deficits [44]. Genetic diagnosis is complex but definitive, and best used for family studies [43]. The disease is rare and often difficult to diagnose; however, treatment is usually effective and lifesaving [43]. Biochemical findings include low concentration of serum ceruloplasmin, elevated basal 24 h urinary excretion of copper, and increased parenchymal copper concentration in the liver.

20.3.6.2 Pathological Findings

Histologically, putamen and caudate nucleus show neuronal loss, pigment- and lipid-laden macrophages, and fibrillary astrocytes. The globus pallidus, subthalamic nucleus, thalamus, and brain stem are less severely

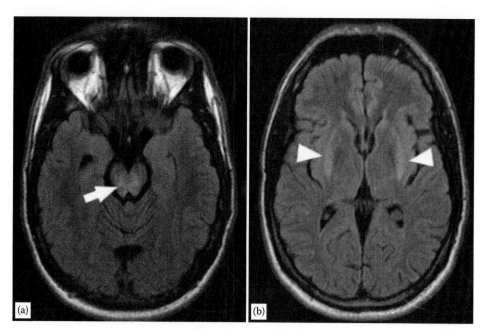

FIGURE 20.13
Wilson's disease in a 24-year-old male patient presenting with neurologic deterioration. Axial FLAIR images show symmetrical hyperintensities within (a) the midbrain (arrow) and (b) the basal ganglia peripherally in the putamen (arrowheads).

involved. A distinctive feature, most readily found in the globus pallidus are Opalski cells that express glial antigens. Spongy degeneration of the cerebral cortex and white matter degenerative changes may be present as well. However, the pathological changes in basal ganglia are thought to be the primary cause of cognitive deficits [44].

20.3.6.3 Imaging Features

Conventional MRI shows widespread lesions, typically T_1-weighted hypointensities and T_2-weighted hyperintensities in the putamen, globus pallidus, caudate nucleus, thalamus, midbrain, pons, and cerebellum, as well as cortical atrophy and white matter changes [44,45]. These changes can be found in presymptomatic patients but tend to be more severe and widespread in patients with neurological Wilson's disease [44]. Lesions in the bilateral basal ganglia are the most common focal changes, including symmetrical concentric-laminar T_2 hyperintensities peripherally in the putamen and T_2 hypointensity in the globus pallidus (Figure 20.13) [45]. Signal abnormality can also be seen in the brain stem and cerebellum. White matter lesions are often asymmetrical with a frontal predilection [45]. Diffuse or focal atrophy can be seen to varying degrees [45].

20.3.6.4 Differential Diagnosis

The symmetrical concentric-laminar T_2 hyperintensities in the peripheral basal ganglia can be seen in patients with other disorders, such as multiple system atrophy with predominant parkinsonian symptoms (MSA-P) or Huntington's disease (HD). Neurometabolic disorders, such as Leigh's disease, may show signal abnormalities in the basal ganglia. Evaluation of the clinical course and laboratory testing can help differentiate those entities.

20.4 Movement Disorders

Parkinsonism refers to a set of neurologic syndromes characterized by variable combinations of tremor, hypokinesia, rigidity, and postural instability. The marked diversity exists in patient age and mode of onset, relative prominence of the cardinal signs, rate of progression, and resultant degree of functional impairment. Although many diseases could be responsible for parkinsonism (Table 20.2), the most common cause is PD.

20.4.1 Parkinson's Disease

20.4.1.1 Clinical Features

PD is a neurodegenerative disorder that is defined clinically by the presence of bradykinesia and at least one additional motor symptom such as resting tremor or rigidity [46]. Diagnosis of idiopathic PD based on history and physical examination alone is frequently confirmed by a dramatic response to dopaminergic therapy [12].

TABLE 20.2

Parkisonian Syndromes

Idiopathic

Parkinson's disease

    Late-onset (>40 years old, generally sporadic)

    Early onset (<40 years old, often familial)

Parkinsonism in Other Neurodegenerative Diseases

Progressive supranuclear palsy

Corticobasal degeneration

Huntington's disease

Multiple system atrophy

Machado–Joseph disease (spinocerebellar ataxia type 3)

Alzheimer's disease

Dementia with Lewy bodies

FTLD: Frontotemporal dementia and parkinsonism linked to
   chromosome 17

Parkinsonism–dementia complex of Guam

Wilson's disease

Secondary

Cerebrovascular (multiple cerebral infarction)

Traumatic (boxer's parkinsonism)

Normal pressure hydrocephalus

Neoplastic (basal ganglia)

Endocrine (hypoparathyroidism)

Metabolic (Wilson's disease, liver cirrhosis, Hallervorden–Spatz
   disease)

Toxic (carbon monoxide, manganese,
   methylphenyltetrahydropyridine)

Drugs (dopamine D2 blockers [neuroleptics], etc.)

Between 2% and 3% of the population are expected to develop PD, with typical onset between 50 and 60 years of age [12]. Subcortical dementia is common and cortical dementia frequently occurs in late stages of PD. When there is prominent dementia seen early in the course of illness, there is a more rapid progression of disability and increased mortality [46].

20.4.1.2 Pathological Findings

The pathology of PD is characterized by the pathological accumulation of abnormal α-synuclein involving only a few predisposed nerve cell types in specific regions of the human nervous system [47]. The primary pathologic changes involve loss of nigrostriatal dopaminergic neurons and accumulation of intraneuronal Lewy bodies. The pathological process is thought to start before clinical symptoms appear and progresses across specific brain regions. Six neuropathological stages have been defined according to distribution and density of Lewy neurites and Lewy bodies [47]. During presymptomatic stages 1–2, inclusion body pathology is confined to the medulla oblongata/pontine tegmentum and olfactory bulb/anterior olfactory nucleus. As the disease progresses

(stages 3–4), the substantia nigra and other nuclear grays of the midbrain and forebrain become the focus of pathological changes affected initially slightly and, then, severely. At this point, most patients probably become symptomatic. In the end stages 5–6, additional lesions arise in non-dopaminergic brain areas, extending to the mature neocortex, and the disease manifests a variety of forms of disabilities. One proposed theory suggests a progression pattern of Lewy body pathology from the brain stem into the midbrain and then to the forebrain before spreading into the cortex [29,47]. Once it reaches to the supratentorial region, the pathological spread ascending within the cerebrum proceeds in the opposite direction of myelination, indicating the early involvement of less densely and relatively late myelinating portions of the neocortex [47]. In contrast to DLB, Aβ pathology is not a typical component in dementia in PD [8].

20.4.1.3 Imaging Features

MRI can aid in the diagnosis to exclude underlying cerebrovascular disease or other structural causes of parkinsonism [46]. It is also important to identify significant causes of secondary parkinsonism such as normal pressure hydrocephalus, frontal neoplasm, or multiple sclerosis [46]. Routine brain MRI assessment, including T_1-weighted, T_2-weighted, and FLAIR sequences, does not show structural abnormalities in early PD (Figure 20.14). Narrowing of the space between the substantia nigra pars compacta and the pars reticulata and nonspecific general cortical atrophy can sometimes be seen, especially in more advanced disease stages [46].

Nuclear medicine imaging (PET and SPECT scans) shows reduced uptake in the striatum in PD (Figure 20.15), as a result of degeneration of the dopaminergic neurons in the substantia nigra. Quantitative assessment in dopamine transporter SPECT is an accurate and reproducible tool to identify the presence of presynaptic dopaminergic denervation [8], and can differentiate the presynaptic parkinsonian syndromes (PD, DLB, PSP, and MSA) from essential tremor or secondary parkinsonism (vascular, drug-induced, and psychogenic parkinsonism) [26]. Striatal tracer binding is usually normal or only slightly decreased in the later group, in contrast to significant reduction in the presynaptic Parkinsonian syndromes [26]. It is, however, more expensive than the clinical examination.

In recent years, DWI has been introduced in PD to identify and quantify microscopic changes. In PD, a decrease of FA in the substantia nigra (particularly caudal side) and along the path of nigrostriatal projection can be seen in early clinical stages of the disease, and the FA values correlate inversely with clinical severity of PD [46,48]. At the time of the clinical motor onset, alternation in FA values spreads throughout the brain, including the motor,

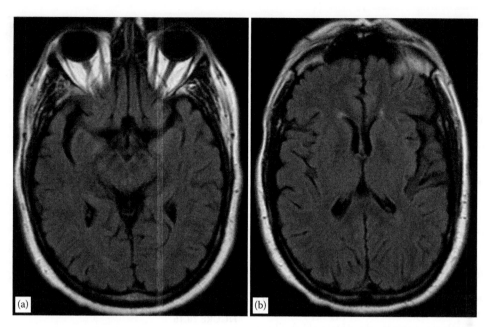

FIGURE 20.14

PD in a 52-year-old male patient with parkinsonism. (a,b) Axial FLAIR images show no structural abnormalities. The major role of conventional MRI is to detect other causes of parkinsonism, such as other neurodegenerative diseases or secondary parkinsonism.

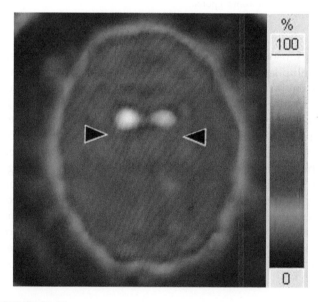

FIGURE 20.15

PD in an 80-year-old female patient with parkinsonism. Dopamine transporter SPECT image shows asymmetric reduction of striatal activity, more markedly in the putamen (black arrowheads) than caudate. Decreased SPECT activity is indicative of presynaptic dopamine transporter deficit, which can be seen in parkinsonian pathology, including PD.

premotor, and supplementary motor cortex, even when no significant atrophy is seen [46]. In advanced stages of PD, changes in FA values are seen in white matter premotor areas, probably a result of extended damage in the extrapyramidal system including the corticostriatal and thalamocortical projections [46].

20.4.1.4 Differential Diagnosis

20.4.1.4.1 Atypical Parkinsonism

Clinical and radiological differentiation of Parkinson's disease from the various forms of neurodegenerative atypical parkinsonism such as MSA, PSP, or cortical basal syndrome can be challenging, especially in early disease stages [46]. The characteristic findings for MSA include hypointensity in the basal ganglia on T_2-weighted images, sometimes in combination with hyperintensity of the lateral putamen. These findings can be observed in healthy elderly persons at a higher magnetic field. Other distinctive imaging features include midbrain atrophy in PSP and asymmetric cortical atrophy in CBD.

20.4.2 Progressive Supranuclear Palsy

20.4.2.1 Clinical Features

PSP is a clinical syndrome comprising supranuclear palsy, postural instability, and dementia [49]. The onset of PSP is usually over 40 years of age, and its symptoms progress gradually [12]. The classic clinical phenotype is referred to as Richardson's syndrome (also known as Steele–Richardson–Olszewski syndrome) (PSP-RS, or classic PSP) and is characterized by lurching gait and unexplained falls backward without loss of consciousness, with personality change or cognitive dysfunction in more than half of patients [49]. Vertical supranuclear gaze palsy is the definitive diagnostic feature but commonly develops in the later stages of the disease. Several clinical variants (atypical PSP) have been identified as associated but distinctive syndromes and can be separated by differences in disease severity, distribution of pathology, and clinical features [49]. The most common variant is PSP-parkinsonism (PSP-P), in which the associated tau pathology is less severe and shows more restricted distribution than that seen in patients with classic PSP [49]. In patients with PSP-P, parkinsonism dominated the early clinical picture, and clinical differentiation from those with idiopathic PD is difficult. PSP-P and CBD may be confused clinically due to a common feature of atypical parkinsonism and limited response to levodopa.

20.4.2.2 Pathological Findings

Neuropathologically, PSP is classified as a primary tauopathy characterized by abnormal aggregates of tau protein in the striatothalamocortical pathway [46]. These aggregates are found in round (globose) neurofibrillary tangles within subcortical neurons, in ("tufted") astrocytes, and as neuropil threads [50]. Similar histological findings can be seen in other tauopathies, thus complicating the pathological diagnosis of PSP [49]. Despite advances in our understanding of the disease process, there are no reliable diagnostic biomarkers for PSP, and accurate diagnosis depends on clinical assessment [49].

20.4.2.3 Imaging Features

Typical MR findings include characteristic rostral brainstem atrophy, more specifically atrophy of the dorsal region of the midbrain and pons. With the volume of the ventral pons relatively preserved, a mid-sagittal MR image shows the hummingbird appearance of the brain stem (hummingbird sign) (Figure 20.16). Atrophy of the superior cerebellar peduncles is also reported [50]. Voxel-based morphometry study also supports greater atrophy in the brain stem in PSP than in CBD, and greater cortical atrophy in CBD, with prominent atrophy in the dorsal frontal and parietal cortical regions [50]. Some cases show slight hyperintense signals on T_2-weighted images in periaqueductal gray matter [12]. Putaminal hypointensity has been described on T_2-weighted images [12].

20.4.3 Corticobasal Degeneration

20.4.3.1 Clinical and Pathological Features

CBD is a rare degenerative disorder characterized by parkinsonism, dystonia, and dementia [46]. Clinical diagnostic criteria are based on motor impairment related to basal ganglia dysfunction (asymmetric

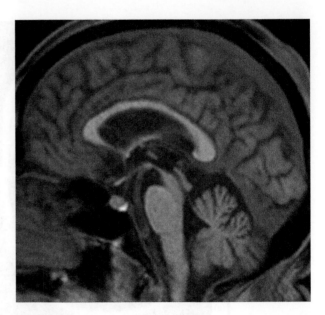

FIGURE 20.16
PSP in a 71-year-old female patient with postural instability. Sagittal T_1-weighted image shows the characteristic hummingbird appearance of the brain stem (hummingbird sign) as a result of atrophy in the dorsal region of the midbrain and pons with relatively preserved volume in the ventral pons.

parkinsonism, rigidity) and cortical dysfunction (alien limb phenomenon, apraxia) [46]. Cognitive decline and behavior abnormalities are typical, and dementia is frequent [46]. CBD shows atypical parkinsonism, which is poorly responsive to levodopa. Patients may also have abnormal eye movements; thus, CBD and PSP may be confused clinically. The pathological similarity between those two entities has been also identified. CBD belongs to the family of tauopathies and is characterized by swollen (ballooned) cortical neurons and various types of neurofibrillary pathology. Tau deposits are found in the cortical ballooned neurons, glial cells, neuropils, and, characteristically, cortical astrocytic plaques that differ from neuritic plaques by lacking amyloid deposition [50,51].

20.4.3.2 Imaging Features

Brain MRI tends to be normal in the early stages of the disease [51]. As the disease progresses, a pattern of asymmetric posterior, frontal, and parietal cortical atrophy becomes evident with dilatation of the lateral ventricle (Figure 20.17). Dopamine transporter SPECT scans are commonly abnormal in CBD in the early stage of the disease [51]. FDG PET and rCBF SPECT may show asymmetric reductions in resting levels of glucose metabolism or blood flow in the posterior frontal, inferior parietal, and superior temporal regions, thalamus, and striatum [51].

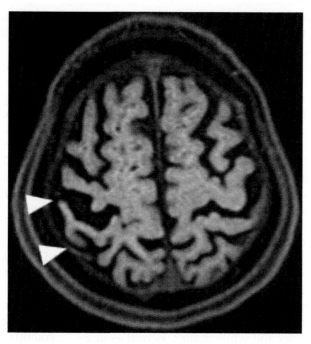

FIGURE 20.17
CBD in a 78-year-old male patient with left-sided motor impairment. Axial T_1-weighted image demonstrates asymmetrical cerebral atrophy with sulcal prominence predominantly in the right (arrowheads).

20.4.4 Huntington's Disease

20.4.4.1 Clinical Features

HD is a rare, autosomal dominant disorder caused by expanded trinucleotide CAG triplet repeat on *HD* chromosome [52]. It is clinically characterized by various degrees of progressive involuntary movement (chorea, dystonia), cognitive impairment, and behavioral disturbance [52–54]. Its symptoms show variability during the course of the disease with chorea early and bradykinesia and dystonia predominating in advanced disease [53,54]. The onset of age and disease progression may vary, but typically presenting during the third and fourth decades. Presymptomatic persons can be identified by genetic testing done by knowledgeable and trained individuals using established protocols [52].

20.4.4.2 Pathological Findings

Pathologically, HD is characterized by striatal degeneration (neuronal loss and astrocytosis), in particular caudate nucleus, with degeneration in related structures, such as globus pallidus, substantia nigra, and thalamus, extending into neocortex and limbic structures (hippocampus and amygdala) in later stages of the disease [55,56]. Considerable variation in the pattern and severity of neurodegeneration has been observed in the striatum and neocortex [53–55]. Increased copper and iron levels in the brain of HD patients have been demonstrated postmortem, and past studies indicate the role of these metals in the pathogenesis of HD [57,58].

20.4.4.3 Imaging Features

Neuroimaging shows characteristic atrophy of the caudate and/or putamen, and consequent dilatation of anterior horn of the lateral ventricles (Figure 20.18) [12,56]. MRI may also show signal intensity changes (low to high T_2 signal) in the striatum (Figure 20.18) [12]. Quantitative MRI studies showed higher levels of iron

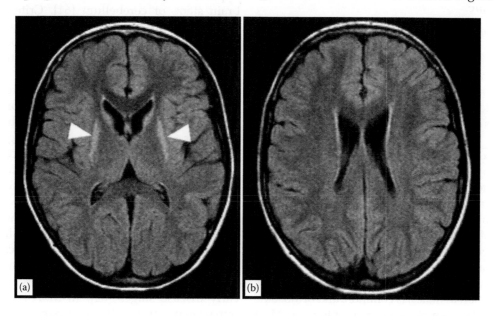

FIGURE 20.18
HD in a 10-year-old female patient with family history of Huntington chorea. Axial FLAIR images demonstrate severe atrophy and increased signal intensities in the bilateral caudate and putamen (a) (arrowheads), accompanied with dilated anterior horns of the lateral ventricles (b).

in the caudate nucleus and putamen in patients with early HD, which is independent of the degree of atrophy [58]. Complicated pathological processes, including neuronal loss accompanied by loss of myelin, gliosis, and copper and iron accumulation, may explain complicated signal intensity abnormalities [12].

20.5 Degenerative Diseases with Cerebellar Atrophy

Cerebellar ataxia originates in progressive degeneration of the cerebellum and is often accompanied by a variety of neurological and other systemic involvement [59]. It is categorized into sporadic and hereditary, and sporadic ataxias may be symptomatic or idiopathic. Cerebellar degeneration can occur as a result of a variety of pathological process, including ischemic or hemorrhagic stroke, inflammation, demyelination, hypothyroidism, and toxic disorders (alcohol, drugs, heavy metals) [59]. Hereditary ataxias can be inherited in an autosomal dominant, autosomal recessive, X-linked, or mitochondrial mode, and the classification is based on an expanding genotype-based testing [59]. Childhood ataxias are more often congenital, metabolic, or associated with a syndrome [59]. The diagnostic process starts with MRI of the brain. Cerebellar atrophy is a nonspecific imaging feature. Additional findings may be more specific and provide clues to aid in making the diagnosis.

20.5.1 Idiopathic

20.5.1.1 Multiple System Atrophy

20.5.1.1.1 Clinical and Pathological Features

MSA is a sporadic and rapidly progressive neurodegenerative disease that manifests with a variable combination of autonomic failure with parkinsonism (that is poorly responsive to levodopa) or cerebellar ataxia or both [46,60–62]. It affects both men and women. It generally starts in the sixth decade of life and progresses rapidly with a mean survival of 6–9 years [59,62]. It is divided by initial symptomatic predominance; patients with predominant parkinsonian symptoms are designated MSA-P and those with predominant cerebellar symptoms are designated MSA-C [46,59–62]. The predominant feature can change over time. Importance of autonomic failure is emphasized as a requiring feature for probable MSA [60]. Genitourinary dysfunction (urinary incontinence, erectile dysfunction in males) is common, and orthostatic hypotension can be asymptomatic or symptomatic. Most

MSA patients develop parkinsonism at some stage. In MSA-P, the nigrostriatal system is the main site of pathology, similar to PD [46,62]. As the disease progresses, the striatal pathology then spreads across the whole cortex [61]. Differential diagnosis of MSA-P and PD may be very difficult in the early stages owing to several overlapping features [62]. However, MSA-P usually progresses substantially faster than PD [61], and 90% of the MSA-P patients are unresponsive to levodopa in the long term, allowing for a clinical diagnosis during follow-up [62]. MSA-C, previously recognized as olivopontocerebellar atrophy, comprises gait ataxia, limb kinetic ataxia, scanning dysarthria, and cerebellar oculomotor disturbances [62], having the brunt of pathological changes in the oligopontocerebellar system [46,62].

Pathologically, MSA has been classified as an α-synucleinopathy, together with PD and DLB [60,61]. The histological hallmark of MSA is the glial cytoplasmic inclusion (GCI). These oligodendroglial inclusions are present in the supplementary and primary motor cortex and white matter, as well as in the "pyramidal," "extrapyramidal," corticocerebellar, and supraspinal autonomic systems and their targets. The distribution and the density of GCIs correlate with the severity of MSA [63]. Both GCIs and neuronal inclusions contain α-synuclein [46,61].

20.5.1.1.2 Imaging Features

The revised diagnostic criteria include neuroimaging features [61]. Neuroimaging features for possible MSA-P include atrophy of the putamen, middle cerebellar peduncle, pons, or cerebellum on MRI, and hypometabolism on FDG-PET within the putamen, brain stem, or cerebellum [61]. Criteria for possible MSA-C include atrophy of the putamen, middle cerebellar peduncle, pons, or hypometabolism on FDG-PET in the putamen, and presynaptic nigrostriatal dopaminergic denervation on SPECT or PET [61]. In addition to atrophy in a specific pattern, MR signal changes can be seen. As the disease progresses, pontocerebellar atrophy is seen with T_2 hyperintensities as a characteristic "hot cross bun" sign in the pons, reflecting degeneration of the pontocerebellar fibers (Figure 20.19) [12,62,64]. Profound putaminal atrophy and T_2 hypointensity, equal to or greater than pallidal hypointensity, is visible correlating with severity of rigidity, and more common in MSA than in PD [12,59]. This finding may be combined with a slit-like hyperintense band lateral to the putamen, which is specific for MSA-P (Figure 20.20) [59,62]. DWI may detect signal abnormality in the brain stem or basal ganglia during early disease stages [62]. In MSA-P, motor function impairment can be observed particularly in the early

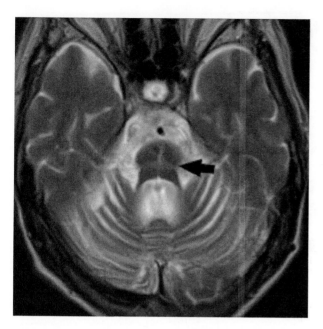

FIGURE 20.19

MSA-C in a 62-year-old female patient with dysarthria and gait ataxia. Axial T_2-weighted image shows prominent atrophy of pons and cerebellum with T_2 hyperintensities as a characteristic "hot cross bun" sign in the pons (arrow), which reflects degeneration of the pontocerebellar fibers.

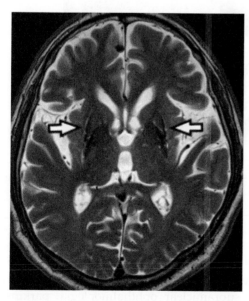

FIGURE 20.20

MSA-P in a 63-year-old male patient with parkinsonism. Axial T_2-weighted image exhibits putaminal atrophy and T_2 hypointensity in the putamen with a slit-like hyperintense band lateral to the putamen (arrows). These findings are characteristic of MSA-P.

stages, accompanied by increased diffusivity in the putamen on DWI [61,62].

20.5.1.2 Idiopathic and Symptomatic Cerebellar Ataxia

20.5.1.2.1 Clinical Features

Idiopathic cerebellar ataxia (IDCA), or idiopathic late-onset cerebellar ataxia (ILOCA), is a group of sporadically occurring degenerative diseases of the cerebellum and the brain stem with unknown origin [65]. IDCA is distinguished from symptomatic cerebellar ataxia which is caused by alcoholism, hypothyroidism, inflammation, immune disorders (gluten-sensitivity ataxia, antiglutamic acid decarboxylase antibody-related ataxia), malignancy (paraneoplastic phenomenon), demyelinating diseases, and other causes [59,65]. When all diagnostic tests are negative, the descriptive acronym ILOCA can be used [59].

20.5.1.2.2 Imaging Features

In many cases of IDCA, atrophy is most severe in the upper vermis and anterior part of the cerebellum. Cerebellar cortical atrophy appears in the late stages with or without brain-stem atrophy (Figure 20.21) [65].

20.5.2 Hereditary

20.5.2.1 Hereditary Spinocerebellar Ataxia

Hereditary spinocerebellar ataxia (SCA) comprises a genetically and clinically heterogeneous group of autosomal dominant disorders, characterized by progressive ataxia caused by degeneration of the cerebellum [66]. It is genetically categorized into more than 30 different subtypes [59,66–68]. Most common subtypes are SCA types 1, 2, 3, 6, and 7, with SCA3 being the most common subtype [59]. Most of the known SCA mutations link to the CAG triplet repeat expansions [59,66–68]. The diagnosis of SCA rests on molecular genetic testing. The mean age of onset in SCA is in the third or fourth decade, but the onset varies considerably between the SCA genotypes and also between patients of the same SCA subtype [59]. There are many reports describing that the length of CAG repeats accumulates in later generations, and that the number of CAG repeats correlates with disease severity and the patient's age at onset, known as the phenomenon of "anticipation" [59,69]. The different mutations in each subtype result in different patterns of neurodegeneration [67] and may associate with characteristic atrophy patterns on MRI. Brain MRI generally shows progressive cerebellar atrophy in the course of the disease, sometimes combined with atrophy of the brain stem or spinal cord [59].

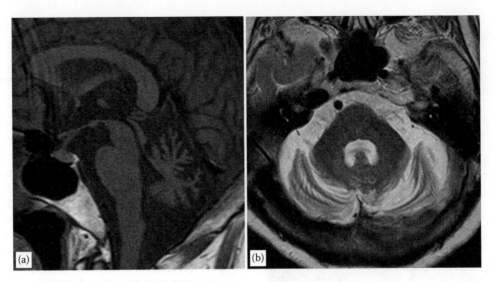

FIGURE 20.21
ILOCA in a 58-year-old female patient with dysarthria and gait ataxia. (a) Sagittal T_1-weighted and (b) axial T_2-weighted images show profound cerebellar atrophy with relatively preserved volume in the brain stem.

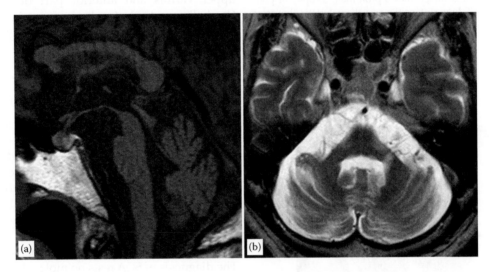

FIGURE 20.22
Spinocerebellar ataxia type 2 in a 70-year-old male patient with gait ataxia. (a) Sagittal T_1-weighted and (b) axial T_2-weighted images show severe atrophy in the cerebellum and brain stem with T_2 hyperintensities within the pons.

20.5.2.1.1 SCA Type 1

SCA type 1 is a multisystem neurodegenerative disease characterized by progressive ataxia, dysarthria, and dysmetria [69]. Pathological examination shows loss of neurons in the cerebellar cortex, dentate nuclei, and brain stem [69].

MRI shows severe atrophy in the cerebellum, brain stem, and basal ganglia with longitudinal atrophy progression most significantly in the mesencephalon, bilateral putamen, and caudate nuclei [68].

20.5.2.1.2 SCA Type 2

SCA type 2 is characterized by progressive cerebellar ataxia, supranuclear ophthalmoplegia, parkinsonism, and pyramidal symptoms (depressed or absent deep tendon reflexes) [70].

MRI demonstrates early atrophy in the brain stem (particularly in cerebellar peduncles), sometimes with T_2 hyperintensities being the "hot cross bun" sign in the pons (Figure 20.22), which is accompanied with cerebellar atrophy and later with frontotemporal atrophy and

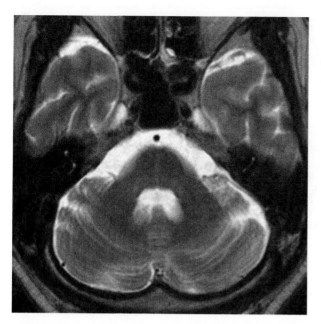

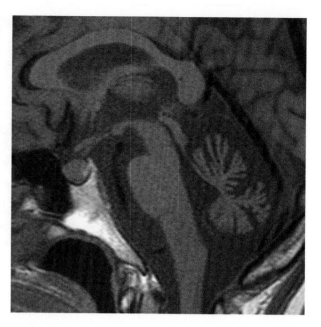

FIGURE 20.23

Spinocerebellar ataxia type 3/Machado–Joseph disease in a 39-year-old male patient with gait ataxia. Axial T_2-weighted image demonstrates atrophy predominantly in the pons and cerebellar peduncles with T_2 hyperintensities within the pons showing the degeneration of the pontine fibers.

FIGURE 20.24

Spinocerebellar ataxia type 6 in a 53-year-old female patient with gait unsteadiness and dysarthria. Sagittal T_1-weighted image demonstrates pronounced, disproportionate atrophy of the cerebellum.

general ventricular enlargement [71,72]. The imaging features may resemble sporadic MSA.

20.5.2.1.3 SCA Type 3/Machado–Joseph Disease

SCA type 3 (SCA3)/Machado–Joseph disease (MJD) predominantly involves cerebellar, pyramidal, extrapyramidal, motor neuron, and oculomotor systems [73], presenting in young to older adulthood. The SCA3/MJD (allelic mutation) is localized to the gene *MJD1* on chromosome 14 [67,73]. Degenerative process involves wide parts of the nervous system, including the cerebral cortex, basal ganglia, pontomedullary systems, and peripheral nerves in addition to cerebellar structures [67].

MRI shows pontocerebellar atrophy predominantly affecting the pontine nuclei and fiber tracts [67], specifically in the superior and middle cerebellar peduncles, cerebellar white matter, pons, and globus pallidus (Figure 20.23) [67,74]. High T_2 signal intensity in the transverse pontine fibers can be observed in almost half of the patients [74].

20.5.2.1.4 SCA Type 6

SCA type 6 (SCA6) is a slowly progressive degenerative disease with dominant involvement of the cerebellar cortex, clinically characterized by relatively late-onset and predominant findings of gait and limb ataxia and dysarthria [74]. Neuropathologically, degeneration is

confined to the cerebellar cortex, particularly to Purkinje cells [67].

MRI in patients with SCA6 typically shows the pronounced, disproportionate atrophy of the cerebellar cortex and the superior vermis, compared to the relatively preserved brain stem (Figure 20.24) [66,67,74]. The brain stem may show mild atrophy.

20.5.2.2 Dentatorubral-Pallidoluysian Atrophy

20.5.2.2.1 Clinical Features

Dentatorubral-pallidoluysian atrophy (DRPLA) is an autosomal dominant neurodegenerative disorder, characterized by a combination of progressive myoclonus, epilepsy, ataxia, choreoathetosis, and dementia [75]. Although rare, DRPLA is most common in the Japanese population [75]. The average age of onset of DRPLA is 30 years, but this condition can appear anytime from the first to the seventh decade of life [76]. DRPLA is one of the CAG triplet repeat diseases, caused by unstable expansion of CAG repeats in the *ATN1* gene on chromosome 12 [75]. Like other CAG repeat diseases, the greater number of CAG repeats leads to earlier onset and a more severe clinical condition [75,76].

20.5.2.2.2 Imaging Features

MRI shows severe atrophy in the brain stem (especially in the midbrain and pontine tegmentum), cerebellum, and cerebrum (predominantly in the frontotemporal region)

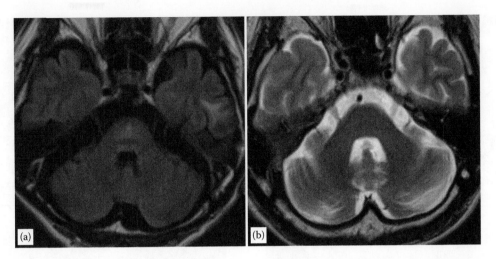

FIGURE 20.25
DRPLA in a 52-year-old male patient with gait ataxia, limb kinetic ataxia, and dysarthria. (a) Axial FLAIR image demonstrates pontocerebellar atrophy with signal abnormality in the pons. (b) Axial T_2-weighted image of his 24-year-old son with epilepsy shows more pronounced pontine atrophy.

(Figure 20.25) [76]. In the advanced stage, T_2-weighted images demonstrate symmetric patchy or diffuse hyperintensity within the brain stem, thalamus, and periventricular and/or deep cerebral white matter [76].

20.5.2.3 Marinesco–Sjögren Syndrome

20.5.2.3.1 Clinical Features

Marinesco–Sjögren syndrome (MSS) is an autosomal recessive disorder characterized by cerebellar ataxia with cerebellar atrophy, early onset (not necessarily congenital) cataracts, mild-to-severe intellectual disability, hypotonia, and muscle weakness [77]. Additional features are short stature and various skeletal abnormalities, including scoliosis [77].

20.5.2.3.2 Imaging Features

MRI shows cerebellar atrophy, usually more pronounced in the vermis than in the hemispheres [78]. T_2 hyperintensity within cerebellar cortex has been reported in individuals with MSS who have *SIL1* mutations [77,78].

20.5.2.4 Hereditary Spastic Paraplegia

20.5.2.4.1 Clinical Features

The hereditary spastic paraplegias (HSPs) comprise a large group of inherited neurologic disorders in which the predominant symptom is lower extremity spastic weakness [79]. They are classified according to the mode of inheritance (autosomal dominant, autosomal recessive, and X-linked HSP) and HSP locus (designated spastic gait loci 1 through 30) [79]. The spastic paraplegia syndrome can occur alone (uncomplicated HSP) or with additional neurologic or systematic abnormalities (complicated HSP), such as intellectual disability, ataxia, peripheral neuropathy, deafness, cataracts, or muscle atrophy [79]. Subjects who experience HSP symptom onset in the first several years of life often show very little worsening through the first two decades [79].

20.5.2.4.2 Pathological Findings

In uncomplicated HSP, axon degeneration is limited to the central nervous system affecting primarily the distal ends of the longest descending motor fibers (corticospinal tracts, with maximal involvement in the thoracic spinal cord) and the distal ends of the longest ascending fibers (fasciculus gracilis fibers, with maximal involvement in the cervicomedullary region) [79]. In complicated HSP, the distal degeneration of long sensory and motor axons within the central nervous system is seen, associated with neuropathology of Charcot–Marie–Tooth type II, in which distal motor and sensory axon degeneration is limited to the peripheral nervous system. The genetic studies of HSP indicate that multiple biochemical abnormalities are responsible for axon degeneration in HSP pathogenesis, including cytoskeletal and axonal transport abnormalities, mitochondrial disturbance, altered Golgi function, primary myelin disturbance, and corticospinal tract developmental abnormality [79].

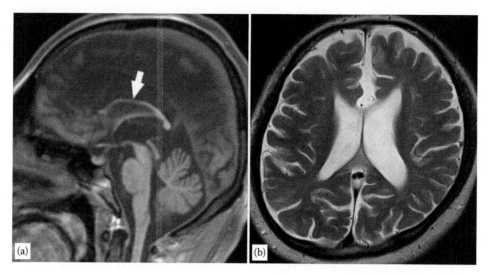

FIGURE 20.26

HSPs in a 31-year-old male patient with lower extremity spastic paralysis. (a) Sagittal T_1-weighted image reveals thin corpus callosum (arrow). (b) Axial T_2-weighted image shows severe atrophy in the corpus callosum and enlarged lateral ventricles.

20.5.2.4.3 Imaging Features

MRI of the brain and spinal cord are important to exclude alternative disorders, including multiple sclerosis, transverse myelitis, leukodystrophies, and structural abnormalities affecting the brain or spinal cord. Conventional brain MRI is normal in uncomplicated HSP. Brain MRI in several forms of complicated HSP reveals syndrome-specific abnormalities such as thin corpus callosum (Figure 20.26) in SPG11, cerebral or cerebellar abnormalities in SPG7, and hydrocephalus in SPG1. MRI of the spinal cord in uncomplicated HSP may appear normal or show atrophy, particularly involving the thoracic spinal cord [79].

20.6 Summary

The diagnosis of brain degenerative disease is primarily based on clinical evaluation. Clinical diagnostic criteria of many of the classical diseases have been reevaluated following recent achievements in genetic and pathophysiological understanding. Although conventionally the role of neuroimaging has been to exclude potential secondary causes of neurodegenerative diseases, recent advances in neuroimaging with improved specificity for detecting these diseases have made neuroimaging play a more central role in diagnosis of the brain degenerative diseases.

References

1. Watanabe M, Liao JH, Jara H et al. (2013) Multispectral quantitative MR imaging of the human brain: Lifetime age-related effects. *RadioGraphics* 33(5):1305–1319.
2. Fox NC, Schott JM (2004) Imaging cerebral atrophy: Normal ageing to Alzheimer's disease. *Lancet* 363(9406):392–394.
3. Taki Y, Thyreau B, Kinomura S et al. (2011) Correlations among brain gray matter volumes, age, gender, and hemisphere in healthy individuals. *PLoS One* 6(7):e22734.
4. Watanabe M, Sakai O, Ozonoff A (2013) Age-related apparent diffusion coefficient changes in the normal brain. *Radiology* 266(2):575–582.
5. Sullivan EV, Pfefferbaum A (2006) Diffusion tensor imaging and aging. *Neurosci Biobehav Rev* 30(6):749–761.
6. Debette S, Markus HS (2010) The clinical importance of white matter hyperintensities on brain magnetic resonance imaging: Systematic review and meta-analysis. *BMJ* 341:c3666.
7. Wahlund LO, Barkhof F, Fazekas F et al. (2001) A new rating scale for age-related white matter changes applicable to MRI and CT. *Stroke* 32(6):1318–1322.
8. Murray AD (2012) Imaging approaches for dementia. *AJNR Am J Neuroradiol* 33(10):1836–1844.
9. DeCarli C, Murphy DG, Tranh M et al. (1995) The effect of white matter hyperintensity volume on brain structure, cognitive performance, and cerebral metabolism of glucose in 51 healthy adults. *Neurology* 45(11):2077–2084.
10. Silbert LC, Dodge HH, Perkins LG et al. (2012) Trajectory of white matter hyperintensity burden preceding mild cognitive impairment. *Neurology* 79(8):741–747.

11. Jack CR Jr (2012) Alzheimer disease: New concepts on its neurobiology and the clinical role imaging will play. *Radiology* 263(2):344–361.

12. Dormont D, Seidenwurm DJ (2008) Dementia and movement disorders. *AJNR Am J Neuroradiol* 29(1):204–206.

13. Braak H, Braak E (1991) Neuropathological staging of Alzheimer-related changes. *Acta Neuropathol* 82(4):239–259.

14. Wippold FJ 2nd, Cairns N, Vo K et al. (2008) Neuropathology for the neuroradiologist: Plaques and tangles. *AJNR Am J Neuroradiol* 29(1):18–22.

15. Braak H, Braak E, Grundke-Iqbal I et al. (1986) Occurrence of neuropil threads in the senile human brain and in Alzheimer's disease: A third location of paired helical filaments outside of neurofibrillary tangles and neuritic plaques. *Neurosci Lett* 65:351–355.

16. Mirra SS, Heyman A, McKeel D et al. (1991) The Consortium to Establish a Registry for Alzheimer's Disease (CERAD). Part II. Standardization of the neuropathologic assessment of Alzheimer's disease. *Neurology* 41:479–486.

17. The National Institute on Aging, and Reagan Institute Working Group on Diagnostic Criteria for the Neuropathological Assessment of Alzheimer's Disease (1997) Consensus recommendations for the postmortem diagnosis of Alzheimer's disease. *Neurobiol Aging* 18:S1–S2.

18. Schellenberg GD, Montine TJ (2012) The genetics and neuropathology of Alzheimer's disease. *Acta Neuropathol* 124(3):305–323.

19. Dubois B, Feldman HH, Jacova C et al. (2007) Research criteria for the diagnosis of Alzheimer's disease: Revising the NINCDS-ADRDA criteria. *Lancet Neurol* 6(8):734–746.

20. Jack CR Jr, Albert MS, Knopman DS et al. (2011) Introduction to the recommendations from the National Institute on Aging-Alzheimer's Association workgroups on diagnostic guidelines for Alzheimer's disease. *Alzheimers Dement* 7(3):257–262.

21. McKhann G, Drachman D, Folstein M et al. (1984) Clinical diagnosis of Alzheimer's disease: Report of the NINCDS-ADRDA Work Group under the auspices of Department of Health and Human Services Task Force on Alzheimer's Disease. *Neurology* 34(7):939–944.

22. Matsuda H (2007) Role of neuroimaging in Alzheimer's disease with emphasis on brain perfusion SPECT. *J Nucl Med* 48(8):1289–1300.

23. Fieremans E, Benitez A, Jensen JH et al. (2013) Novel white matter tract integrity metrics sensitive to Alzheimer disease progression. *AJNR Am J Neuroradiol* 34(11):2105–2112.

24. Koch W, Teipel S, Mueller S et al. (2012) Diagnostic power of default mode network resting state fMRI in the detection of Alzheimer's disease. *Neurobiol Aging* 33(3):466–478.

25. Román GC, Tatemichi TK, Erkinjuntti T et al. (1993) Vascular dementia: Diagnostic criteria for research studies: Report of the NINDS-AIREN International Workshop. *Neurology* 43(2):250–260.

26. Djang DS, Janssen MJ, Bohnen N et al. (2012) SNM practice guideline for dopamine transporter imaging with 123I-ioflupane SPECT 1.0. *J Nucl Med* 53(1):154–163.

27. Shiga Y, Miyazawa K, Sato S et al. (2004) Diffusion-weighted MRI abnormalities as an early diagnostic marker for Creutzfeldt-Jakob disease. *Neurology* 63(3):443–449.

28. Ukisu R, Kushihashi T, Tanaka E et al. (2006) Diffusion-weighted MR imaging of early-stage Creutzfeldt-Jakob disease: Typical and atypical manifestations. *RadioGraphics* Suppl 1:S191–S204.

29. Weisman D, McKeith I (2007) Dementia with Lewy bodies. *Semin Neurol* 27(1):42–47.

30. McKeith I, Mintzer J, Aarsland D et al. (2004) Dementia with Lewy bodies. *Lancet Neurol* 3(1):19–28.

31. Whitwell JL, Weigand SD, Shiung MM et al. (2007) Focal atrophy in dementia with Lewy bodies on MRI: A distinct pattern from Alzheimer's disease. *Brain* 130(Pt 3):708–719.

32. Jellinger KA (2004) Lewy body-related alpha-synucleinopathy in the aged human brain. *J Neural Transm* 111(10–11):1219–1235.

33. Orimo S, Amino T, Itoh Y (2005) Cardiac sympathetic denervation precedes neuronal loss in the sympathetic ganglia in Lewy body disease. *Acta Neuropathol* 109(6):583–588.

34. Seelaar H, Rohrer JD, Pijnenburg YA et al. (2011) Clinical, genetic and pathological heterogeneity of frontotemporal dementia: A review. *J Neurol Neurosurg Psychiatry* 82(5):476–486.

35. Neumann M, Sampathu DM, Kwong LK et al. (2006) Ubiquitinated TDP-43 in frontotemporal lobar degeneration and amyotrophic lateral sclerosis. *Science* 314(5796):130–133.

36. Mackenzie IR, Neumann M, Bigio EH et al. (2009) Nomenclature for neuropathologic subtypes of frontotemporal lobar degeneration: Consensus recommendations. *Acta Neuropathol* 117(1):15–18.

37. Cairns NJ, Bigio EH, Mackenzie IR et al. (2007) Neuropathologic diagnostic and nosologic criteria for frontotemporal lobar degeneration: Consensus of the Consortium for Frontotemporal Lobar Degeneration. *Acta Neuropathol* 114(1):5–22.

38. Neumann M, Rademakers R, Roeber S et al. (2009) A new subtype of frontotemporal lobar degeneration with FUS pathology. *Brain* 132(Pt 11):2922–2931.

39. Du AT, Jahng GH, Hayasaka S et al. (2006) Hypoperfusion in frontotemporal dementia and Alzheimer disease by arterial spin labeling MRI. *Neurology* 67(7):1215–1220.

40. Johnson RT, Gibbs CJ Jr (1998) Creutzfeldt-Jakob disease and related transmissible spongiform encephalopathies. *N Engl J Med* 339(27):1994–2004.

41. Zeidler M, Sellar RJ, Collie DA et al. (2000) The pulvinar sign on magnetic resonance imaging in variant Creutzfeldt-Jakob disease. *Lancet* 355(9213):1412–1418.

42. Llewelyn CA, Hewitt PE, Knight RS et al. (2004) Possible transmission of variant Creutzfeldt-Jakob disease by blood transfusion. *Lancet* 363(9407):417–421.

43. Roberts EA (2011) Wilson's disease. *Medicine* 39(10):602–604.

44. Ala A, Walker AP, Ashkan K et al. (2007) Wilson's disease. *Lancet* 369(9559):397–408.

45. Sinha S, Taly AB, Ravishankar S et al. (2006) Wilson's disease: Cranial MRI observations and clinical correlation. *Neuroradiology* 48(9):613–621.

46. Meijer FJ, Bloem BR, Mahlknecht P et al. (2013) Update on diffusion MRI in Parkinson's disease and atypical parkinsonism. *J Neurol Sci* 332(1–2):21–29.

47. Braak H, Ghebremedhin E, Rüb U et al. (2004) Stages in the development of Parkinson's disease-related pathology. *Cell Tissue Res* 318(1):121–134.

48. Vaillancourt DE, Spraker MB, Prodoehl J et al. (2009) High-resolution diffusion tensor imaging in the substantia nigra of de novo Parkinson disease. *Neurology* 72(16):1378–1384.

49. Williams DR, Lees AJ (2009) Progressive supranuclear palsy: Clinicopathological concepts and diagnostic challenges. *Lancet Neurol* 8(3):270–279.

50. Boxer AL, Geschwind MD, Belfor N et al. (2006) Patterns of brain atrophy that differentiate corticobasal degeneration syndrome from progressive supranuclear palsy. *Arch Neurol* 63(1):81–86.

51. Mahapatra RK, Edwards MJ, Schott JM et al. (2004) Corticobasal degeneration. *Lancet Neurol* 3(12):736–743.

52. The Huntington's Disease Collaborative Research Group (1993) A novel gene containing a trinucleotide repeat that is expanded and unstable on Huntington's disease chromosomes. *Cell* 72(6):971–983.

53. Tippett LJ, Waldvogel HJ, Thomas SJ et al. (2007) Striosomes and mood dysfunction in Huntington's disease. *Brain* 130(Pt 1):206–221.

54. Thu DC, Oorschot DE, Tippett LJ et al. (2010) Cell loss in the motor and cingulate cortex correlates with symptomatology in Huntington's disease. *Brain* 133(Pt 4): 1094–1110.

55. Vonsattel JP, Myers RH, Stevens TJ et al. (1985) Neuropathological classification of Huntington's disease. *J Neuropathol Exp Neurol* 44(6):559–577.

56. Kassubek J, Juengling FD, Kioschies T et al. (2004) Topography of cerebral atrophy in early Huntington's disease: A voxel based morphometric MRI study. *J Neurol Neurosurg Psychiatry* 75(2):213–220.

57. Fox JH, Kama JA, Lieberman G et al. (2007) Mechanisms of copper ion mediated Huntington's disease progression. *PLoS One* 2(3):e334.

58. Dumas EM, Versluis MJ, van den Bogaard SJ et al. (2012) Elevated brain iron is independent from atrophy in Huntington's disease. *Neuroimage* 61(3):558–564.

59. Brusse E, Maat-Kievit JA, van Swieten JC (2007) Diagnosis and management of early- and late-onset cerebellar ataxia. *Clin Genet* 71(1):12–24.

60. Gilman S, Wenning GK, Low PA et al. (2008) Second consensus statement on the diagnosis of multiple system atrophy. *Neurology* 71(9):670–676.

61. Stefanova N, Bücke P, Duerr S et al. (2009) Multiple system atrophy: An update. *Lancet Neurol* 8(12):1172–1178.

62. Wenning GK, Colosimo C, Geser F et al. (2004) Multiple system atrophy. *Lancet Neurol* 3(2):93–103.

63. Inoue M, Yagishita S, Ryo M et al. (1997) The distribution and dynamic density of oligodendroglial cytoplasmic inclusions (GCIs) in multiple system atrophy: A correlation between the density of GCIs and the degree of involvement of striatonigral and olivopontocerebellar systems. *Acta Neuropathol* 93:585–591.

64. Schrag A, Kingsley D, Phatouros C et al. (1998) Clinical usefulness of magnetic resonance imaging in multiple system atrophy. *J Neurol Neurosurg Psychiatry* 65:65–71.

65. Klockgether T, Schroth G, Diener HC et al. (1990) Idiopathic cerebellar ataxia of late onset: Natural history and MRI morphology. *J Neurol Neurosurg Psychiatry* 53(4):297–305.

66. Eichler L, Bellenberg B, Hahn HK et al. (2011) Quantitative assessment of brain stem and cerebellar atrophy in spinocerebellar ataxia types 3 and 6: Impact on clinical status. *AJNR Am J Neuroradiol* 32(5):890–897.

67. Lukas C, Schöls L, Bellenberg B et al. (2006) Dissociation of grey and white matter reduction in spinocerebellar ataxia type 3 and 6: A voxel-based morphometry study. *Neurosci Lett* 408(3):230–235.

68. Reetz K, Costa AS, Mirzazade S et al. (2013) Genotype-specific patterns of atrophy progression are more sensitive than clinical decline in SCA1, SCA3 and SCA6. *Brain* 136(Pt 3):905–917.

69. Chung MY, Ranum LP, Duvick LA et al. (1993) Evidence for a mechanism predisposing to intergenerational CAG repeat instability in spinocerebellar ataxia type I. *Nat Genet* 5(3):254–258.

70. Inagaki A, Iida A, Matsubara M et al. (2005) Positron emission tomography and magnetic resonance imaging in spinocerebellar ataxia type 2: A study of symptomatic and asymptomatic individuals. *Eur J Neurol* 12(9):725–728.

71. Lastres-Becker I, Rüb U, Auburger G (2008) Spinocerebellar ataxia 2 (SCA2). *Cerebellum* 7(2):115–124.

72. Brenneis C, Bösch SM, Schocke M et al. (2003) Atrophy pattern in SCA2 determined by voxel-based morphometry. *Neuroreport* 14(14):1799–1802.

73. Dürr A, Stevanin G, Cancel G et al. (1996) Spinocerebellar ataxia 3 and Machado-Joseph disease: Clinical, molecular, and neuropathological features. *Ann Neurol* 39(4):490–499.

74. Murata Y, Yamaguchi S, Kawakami H et al. (1998) Characteristic magnetic resonance imaging findings in Machado-Joseph disease. *Arch Neurol* 55(1):33–37.

75. Koide R, Ikeuchi T, Onodera O et al. (1994) Unstable expansion of CAG repeat in hereditary dentatorubral-pallidoluysian atrophy (DRPLA). *Nat Genet* 6(1):9–13.

76. Miyazaki M, Kato T, Hashimoto T et al. (1995) MR of childhood-onset dentatorubral-pallidoluysian atrophy. *AJNR Am J Neuroradiol* 16(9):1834–1836.

77. Anttonen AK, Lehesjoki AE (2006) Marinesco-Sjögren syndrome. In: Pagon RA, Adam MP, Bird TD et al. (eds.) *GeneReviews*™ [Internet]. University of Washington, Seattle, WA.

78. Harting I, Blaschek A, Wolf NI et al. (2004) T2-hyperintense cerebellar cortex in Marinesco-Sjögren syndrome. *Neurology* 63(12):2448–2449.

79. Fink JK (2006) Hereditary spastic paraplegia. *Curr Neurol Neurosci Rep* 6(1):65–76.

21

Congenital Brain Malformations

Elzbieta Jurkiewicz and Katarzyna Nowak

CONTENTS

21.1 Cerebral Commissure Anomalies

The cerebral commissures consist of corpus callosum, anterior commissure, and hippocampal commissure. Agenesis of the corpus callosum is one of the most common malformations of the brain. Its etiology is heterogeneous, including cytogenetic abnormalities, genetic syndromes, chromosomal errors, metabolic disorders, prenatal infections or injuries, or environmental features.

Congenital anomalies of commissural malformations can be isolated or more often associated with other brain malformations such as cephalocele, polymicrogyria (PMG), schizencephaly, holoprosencephaly (HPE), or genetic syndromes, and other cerebral anomalies such as Aicardi syndrome, Apert syndrome, fetal alcohol syndrome, Chiari II malformation, and trisomy 18 and 13.

There are different forms of cerebral commissure malformations, depending on abnormalities of developmental processes.

The anomalies of cerebral commissures are as follows:

1. Complete agenesis of all cerebral commissures (Figure 21.1a and b)

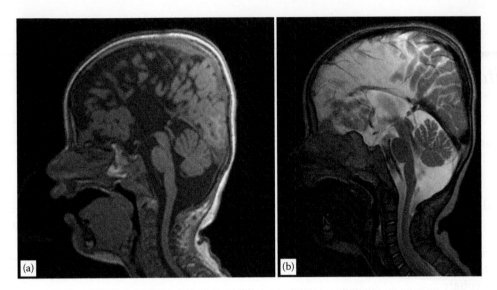

FIGURE 21.1
Complete commissural agenesis. (a) Sagittal T_1-weighted and (b) sagittal T_2-weighted images show a lack of all three commissures and cingulate gyrus. Radially oriented sulci of the hemispheres are seen.

2. Classic callosal agenesis—agenesis of corpus callosum and hippocampal commissures with visible anterior commissure (Figures 21.2a, 21.3a–c, and 21.4)

3. Partial agenesis of corpus callosum—agenesis of the posterior part of corpus callosum and hippocampal commissures with anterior commissure being visible or not (Figure 21.5)

4. Posterior callosal hypogenesis with callosal tapering. Anterior and hippocampal commissures may be normal (Figure 21.6a–c)

The common imaging characteristics associated with agenesis of the corpus callosum are as follows: absent cingulate gyrus, radially oriented medial hemispheric sulci that reach the third ventricle, high-riding third ventricle open superiorly to the interhemispheric fissure, lateral ventricles parallel to each other with dilatation of the atrium and occipital horn (colpocephaly), dilated temporal horns with incomplete rotation and abnormal shape of the hippocampi, a *bull's horn* appearance of the lateral ventricles on coronal views (Figure 21.2b), Probst bundles in most cases, and sometimes sigmoid bundle can be seen (Figure 21.2c and d).

Longitudinal callosal fascicles (Probst bundles) (Figure 21.2c and d)—These are the neurons that fail to cross the midline. They run along the medial wall of the lateral ventricle, parallel to the interhemispheric fissure. They can be seen in complete or partial corpus callosum agenesis and connect ipsilateral regions of the hemispheres. They form

intrahemispheric connection between different areas within the same hemisphere.

Sigmoid bundle—These are heterotopic tracts that are sometimes seen in cases of partial agenesis of the corpus callosum (Figure 21.7a and b). They connect the frontal lobe (forceps minor) with the contralateral occipitoparietal lobe (forceps major).

In the classic commissural agenesis, in about half of the cases, the anterior commissure is either absent or too thin to be recognized; in some cases it may be enlarged, as if it compensates for the missing corpus callosum.

A hippocampal commissure may mimic the splenium of the corpus callosum, but a precise visual inspection will show that this structure connects the fornices, and not the hemispheres.

21.1.1 Commissural Abnormalities/Dysgenesis and Meningeal Dysplasia (Formerly: Agenesis of the Corpus Callosum with Interhemispheric Cyst)

Agenesis of the commissures with interhemispheric cysts is possibly related to a meningeal rather than neural disorder. In this anomaly two categories were recognized:

Type 1, when cysts communicate with the ventricular system (Figures 21.5, 21.8, and 21.9)

Type 2, when they are multiloculated and independent from ventricles and do not communicate with ventricles. Signal intensity of the cysts in type 2 is different, mostly hyperintense to cerebrospinal fluid on T_1-weighted and fluid-attenuated inversion recovery (FLAIR) images (protein

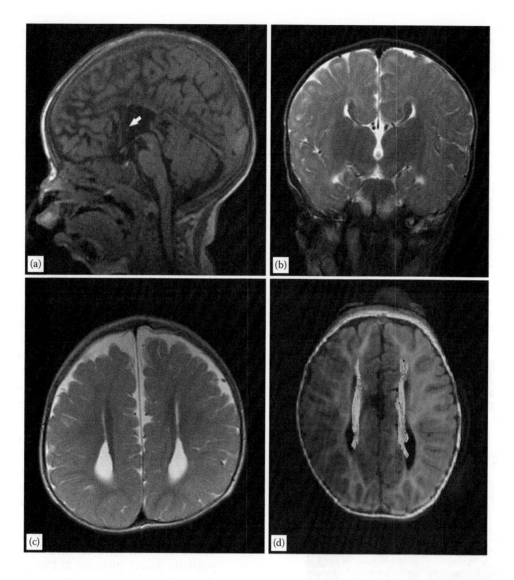

FIGURE 21.2

Agenesis of the corpus callosum. (a) Sagittal T_1-weighted image shows agenesis of the corpus callosum, anterior commissure is present but very thin (white arrow), and hippocampal commissures are not seen. The cingulate gyrus is absent. (b) Coronal T_2-weighted image shows a *bull's horn* appearance of the lateral ventricle. Probst bundles are seen medially to ventricles. Note incomplete rotation and abnormal shape of the hippocampi. (c) Axial T_2-weighted image shows widening of the ventricular atria, and widely separated and parallel lateral ventricles. Medially to the lateral ventricles, myelinated Probst bundles are seen. (d) Diffusion tensor imaging–tractography shows parallel Probst bundles.

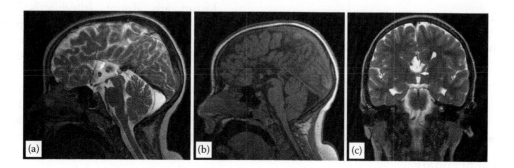

FIGURE 21.3

Agenesis of the corpus callosum. (a) Sagittal T_1-weighted and (b) sagittal T_2-weighted images show a very thick anterior commissure (white arrow), but no callosal or hippocampal commissure. The sulci of the hemispheres are oriented radially. No cingulate gyrus. (c) Coronal T_2-weighted image shows a thick anterior commissure (white arrow) and abnormal hippocampi with extension of the temporal horns.

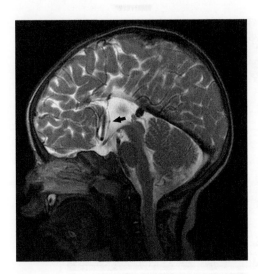

FIGURE 21.4
Agenesis of the corpus callosum and hippocampal commissure.
Sagittal T_2-weighted image shows a very thin anterior commissure
(black arrow). The posterior fossa is normal.

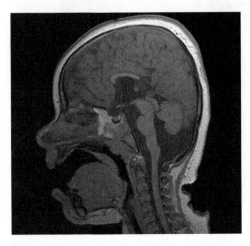

FIGURE 21.5
Corpus callosum hypoplasia. (a) Sagittal T_1-weighted image shows a
thin and short corpus callosum. Very thin anterior and hippocampal
commissures are seen. The quadrigeminal plate is enlarged. Mega
cisterna magna is present.

content) (Figure 21.10). The cysts are frequently
multiple (Figure 21.11). Enhancement of the
walls of the cysts in type 2 may be seen (Figure
21.10c). Type 2 is often associated with heterotopia, PMG, or Aicardi syndrome (Figure 21.12).

21.2 Holoprosencephaly

HPE is a complex congenital malformation of the brain
characterized by a lack or an incomplete separation of
the cerebral hemispheres. According to severity, HPE

is classified as alobar, semilobar, lobar, and middle
interhemispheric variants—syntelencephaly. Currently,
an additional variant of HPE, septopreoptic HPE is
described.

The most severe form of HPE is the alobar form, but
HPE is a continuum of malformations so it may be difficult to precisely recognize the type of anomaly.

HPE is caused by teratogens or different gene mutations: *SIX3*, *SHH*, *TGIF*, *ZIC2*, *PTCH1*, and *GLI12*. It
can be a part of different congenital syndromes such
as Smith–Lemli–Opitz, Pallister–Hall, and Edwards
syndrome.

Alobar HPE: There is complete or nearly complete
lack of separation of the cerebral hemispheres
with complete absence of the interhemispheric
fissure and falx. A single monoventricle communicating with a dorsal cyst is present.
Corpus callosum, septum pellucidum, sylvian fissure, and the third ventricle are absent.
Thalami, basal ganglia, and hypothalami are
fused. Neuronal migration anomalies may be
noted. In angiography, only the internal carotid
and basilar arteries with small vessels are seen
in most cases of this type of HPE.

Semilobar HPE: Frontal lobes are not separated
(Figures 21.13a and c, 21.14 through 21.16a and e).
Typically, the splenium of the corpus callosum is
seen (Figures 21.13d and 21.16b). Interhemispheric
fissure and falx are present posteriorly. Posterior
horns of the lateral ventricles and a small III ventricle are seen. Dorsal cyst may be present or not.
Thalami, basal ganglia, and hypothalami are
partially fused (Figures 21.13b and 21.16c and
d). In arteriography, the azygos anterior cerebral
artery is seen (Figure 21.13e).

Lobar HPE: The base of frontal lobes is not divided
(Figure 21.17c and d). Splenium and a part of
the body of corpus callosum are noted (Figure
21.17a and b). Rostrum and genu of corpus callosum are absent. Interhemispheric fissure
and falx are present posteriorly, partially, and
also anteriorly. Anterior horns of the lateral
ventricles are seen, and the third ventricle is
present. Dorsal cyst is absent. Thalami are separated, but basal ganglia are often not divided,
like hypothalami (Figure 21.17e). Subcortical
heterotopia may be seen. In angiography, the
azygos anterior cerebral artery may be seen.

Middle interhemispheric variant of HPE: Posterior
frontal and parietal lobes are not separated
(Figure 21.18a and b). Genu and splenium
of the corpus callosum are present, but the
body is absent (Figure 21.18c). Lateral and third

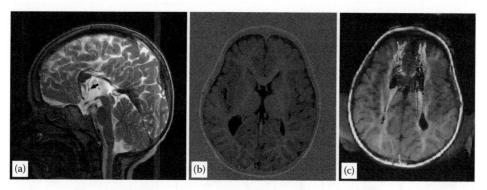

FIGURE 21.6
Partial agenesis of the corpus callosum. (a) Sagittal T_2-weighted image shows a short anterior segment of the corpus callosum with thin anterior (black arrow) and hippocampal commissures. (b) Axial T_1-weighted image shows a myelinated anterior part of the corpus callosum. (c) Diffusion tensor imaging–tractography shows callosal fibers connecting contralateral hemispheres (red color) and longitudinal Probst fibers connecting ipsilateral regions of hemispheres (green color).

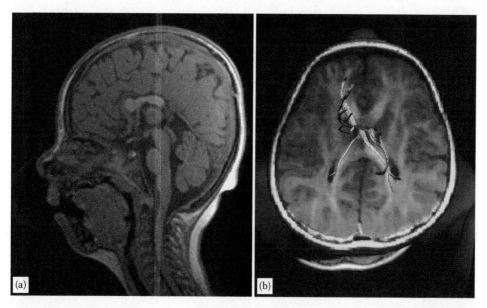

FIGURE 21.7
Partial agenesis—posterior callosal hypogenesis. (a) Sagittal T_1-weighted image shows shortening and abnormal shape of the corpus callosum. Anterior commissure is seen. There is no hippocampal commissure. (b) Diffusion tensor imaging–tractography shows sigmoid bundle connecting the right frontal lobe with the left occipital lobe (green color). Crossing fibers of the callosal axons are also seen (red color).

ventricles are normal. Dorsal cyst can be seen. Septum pellucidum is absent. Basal ganglia and hypothalami are separated, and thalami may be fused. In angiography, the azygos anterior cerebral artery may be seen.

Septopreoptic HPE: It is a very mild form of HPE, characterized by fusion of septal and preoptic regions. Rostrum is absent or hypoplastic and genu is hypoplastic. Anterior hypothalamus is often fused. In angiography, the azygos anterior cerebral artery may be seen.

21.3 Septo-Optic Dysplasia

Septo-optic dysplasia (SOD) encompasses a heterogeneous spectrum of malformations. Typically, it consists of hypoplasia or absent septum pellucidum (Figure 21.19a) (sometimes corpus callosum), hypoplasia of the optic nerves (Figures 21.19c and 21.20a) (bilateral or more rarely unilateral, Figure 21.21), and chiasm (Figure 21.19b); sometimes pituitary gland and hypothalamus are hypoplastic.

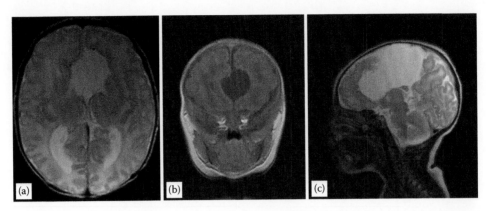

FIGURE 21.8
Corpus callosum partial agenesis with interhemispheric cyst—type 1. (a) Axial T_2-weighted and (b) coronal T_1-weighted images show interhemispheric cyst connecting with the third ventricle. (c) Sagittal T_2-weighted image shows a large midline cyst. A small portion of the splenium of the corpus callosum is present.

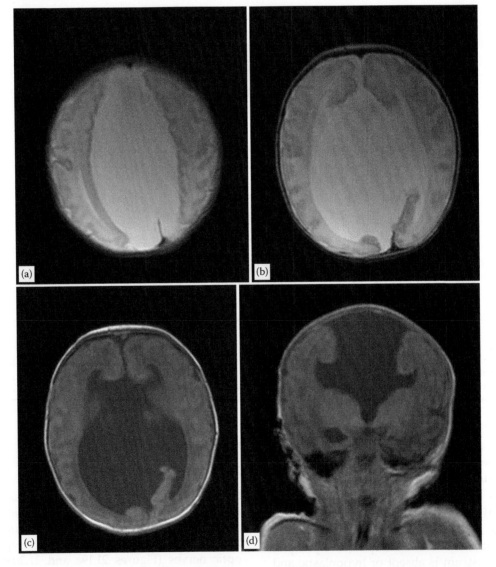

FIGURE 21.9
Corpus callosum agenesis with interhemispheric cyst—type 1. (a,b) Axial T_2-weighted and (c) axial T_1-weighted images show a very large interhemispheric cyst connecting with the ventricular system. (d) Coronal T_1-weighted image shows interhemispheric cyst connecting with the lateral and third ventricle. It shows a *bull's horn* appearance of the lateral ventricle.

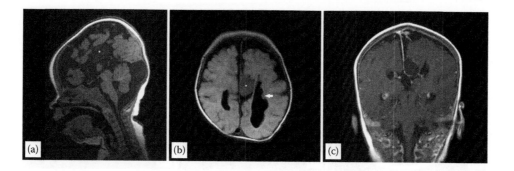

FIGURE 21.10

Corpus callosum agenesis with interhemispheric cyst—type 2. (a) Sagittal T_1-weighted image shows agenesis of the corpus callosum and hyperintense (compared to CSF) midline cyst (white star). Quadrigeminal plate is enlarged. In the posterior fossa, mega cisterna magna is seen. (b) Axial T_1-weighted image shows hyperintense midline cyst (white star) and periventricular, nodular heterotopias (white arrow). (c) Postcontrast coronal T_1-weighted image shows enhancement of the cystic wall.

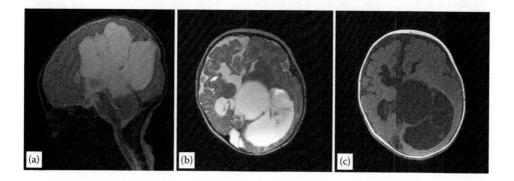

FIGURE 21.11

Corpus callosum agenesis with interhemispheric cyst—type 2. (a) Sagittal T_2-weighted image shows multiple cysts, displacing the vermis inferiorly and the occipital lobe posteriorly. (b) Axial T_2-weighted and (c) axial T_1-weighted images show asymmetry of the cerebral hemispheres: the right hemisphere is smaller with nodular, periventricular heterotopias (white arrow).

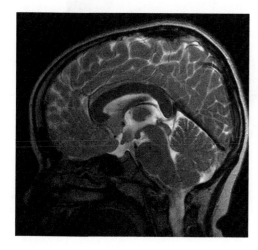

FIGURE 21.12

Sagittal T_2-weighted image shows hyperplasia of the corpus callosum in a patient with neurofibromatosis type 1.

The wide spectrum of features in SOD includes also ectopic posterior lobe of the pituitary, defects of eyes (e.g., coloboma), schizencephaly, PMG, and heterotopia. Genetic abnormalities causing SOD have been identified: *HESX1*, *SOX2*, *SOX3*, and *OTX2*. The presence of normal septum pellucidum does not rule out SOD. The absence of septum pellucidum changes the shape of the frontal horns of the lateral ventricles which are pointed down. The anterior recess of the third ventricle and the suprasellar cistern are widened.

Recently, SOD has been divided into two groups: one with partial absence of septum pellucidum with coexisting schizencephaly and/or heterotopia, and the second one with complete absence of septum pellucidum and ventriculomegaly without abnormalities of cortical malformations. The second group of patients are believed to overlap with lobar HPE group.

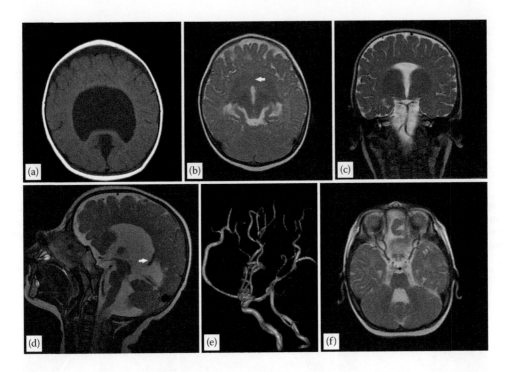

FIGURE 21.13
Semilobar HPE. (a) Axial T_1-weighted image shows fusion of the frontal lobes, monoventricle, and absence of the anterior interhemispheric fissure. Interhemispheric fissure is seen posteriorly. (b) Axial T_2-weighted image shows fusion of the basal ganglia (white arrow), no anterior interhemispheric fissure, and small third ventricle and temporal horns. (c) Coronal T_2-weighted image shows nonseparated frontal lobes and monoventricle continuous with the third ventricle. (d) Sagittal T_2-weighted image shows hypogenetic, partially formed splenium (white arrow). (e) Magnetic resonance angiography shows a single azygos artery located anteriorly. (f) Axial T_2-weighted image shows both middle cerebral arteries and no anterior cerebral arteries.

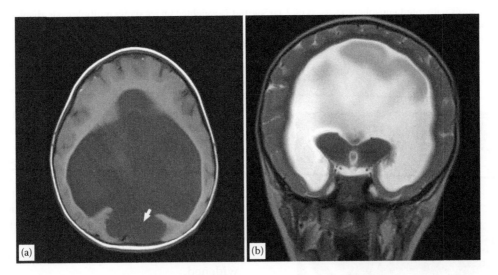

FIGURE 21.14
Semilobar HPE. (a) Axial T_1-weighted image shows nonseparated frontal lobes without interhemispheric fissure anteriorly. There is a single ventricular cavity connecting with a small dorsal cyst (white arrow). (b) Coronal T_2-weighted image shows no interhemispheric fissure and a large monoventricle.

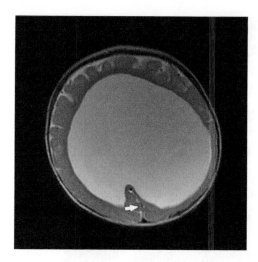

FIGURE 21.15
Semilobar HPE. Axial T_2-weighted image shows a large monoventricle without dorsal cyst and no anterior interhemispheric fissure; posteriorly fissure is seen (white arrow).

21.4 Schizencephaly

Schizencephaly is a rare congenital brain anomaly characterized by a cleft of the cerebral hemisphere, extending from the cortical surface of the brain (pia mater) to the lateral ventricle (ependyma) (Figure 21.22a and b). This cleft is lined by the dysplastic gray matter (mostly polymicrogyric) and filled with cerebrospinal fluid.

According to Barkovich, schizencephaly is considered a malformation due to abnormal postmigrational development (formerly malformations of cortical organization). The etiology of schizencephaly is probably caused by prenatal vascular disruption and ischemia, teratogens, *in utero* infections (cytomegalovirus and herpes virus), or maternal alcohol and drug abuse. Association with mutations in homeobox of the *EMX2* gene is suspected.

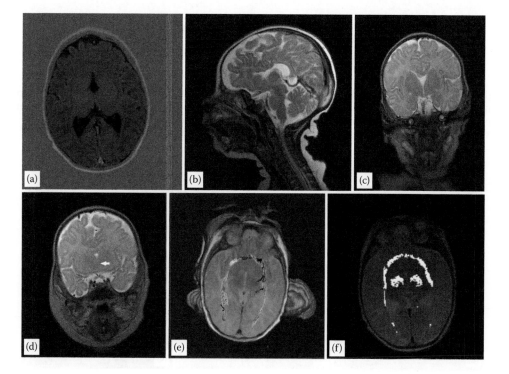

FIGURE 21.16
Semilobar HPE. (a) Axial T_1-weighted image shows fusion of the anterior portion of the frontal lobes without anterior interhemispheric fissure. The third ventricle, occipital horns, and posterior interhemispheric fissure are present. (b) Sagittal T_2-weighted image shows unmyelinated splenium and posterior part of the body of the corpus callosum. (c,d) Coronal T_2-weighted images show fusion of the basal ganglia (white arrow), and narrow and connected lateral ventricles with rudimentary temporal horns. (e,f) Diffusion tensor imaging–tractography shows transversally located white matter fibers within the fused regions.

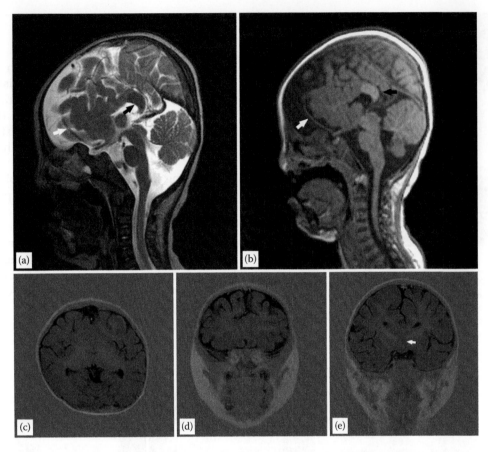

FIGURE 21.17
Lobar HPE. (a) Sagittal T_2-weighted and (b) sagittal T_1-weighted images show a partially formed splenium of the corpus callosum (black arrow). The single azygos artery is seen on the anterior surface of the brain (white arrow). (c) Axial T_1- and (d) coronal T_1-weighted images show a nondivided base of the frontal lobes. The anterior interhemispheric fissure is seen. (e) Coronal T_1-weighted image shows fusion of the basal ganglia (white arrow) and rudimentary frontal horns of the lateral ventricles. Interhemispheric fissure is seen.

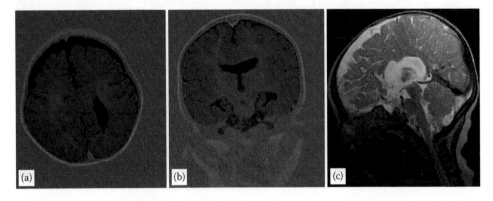

FIGURE 21.18
Middle interhemispheric variant of HPE. (a) Axial T_1-weighted image shows fusion of the posterior frontal lobes without interhemispheric fissure. Interhemispheric fissure is seen anteriorly and posteriorly to the fusion. (b) Coronal T_1-weighted image shows nonseparated lateral ventricles, small third ventricle, and fusion of the lobes. (c) Sagittal T_2-weighted image shows the absence of the body of the corpus callosum. Genu and splenium are present but hypoplastic.

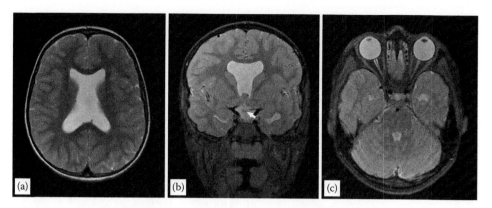

FIGURE 21.19
Septo-optic dysplasia. (a) Axial T_2-weighted image shows the absence of the septum pellucidum. (b) Coronal PD-weighted image shows a typical for this anomaly configuration of the lateral ventricle. The frontal horns of the lateral ventricles are pointed down. The optic chiasm is hypoplastic (white arrow). Note that temporal horns are asymmetrically enlarged. (c) Axial T_2-weighted image shows hypoplasia of both optic nerves.

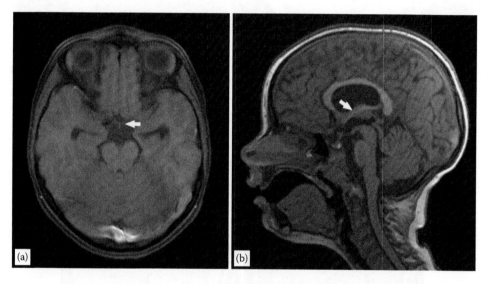

FIGURE 21.20
Septo-optic dysplasia. (a) Axial and (b) sagittal T_1-weighted images show hypoplastic chiasm (white arrow). Low position of the fornices is visible on a sagittal view (white arrow).

Insula and frontal lobe are mostly affected. The clefts may be unilateral (more common) (Figure 21.22) or bilateral (Figure 21.23). Depending on the width of the cleft, two types of schizencephaly are described:

Schizencephaly type I—Closed lip cleft, walls of the cleft opposed to each other, cerebrospinal fluid not seen. Sometimes, a visible deformity of the lateral ventricle helps in diagnosis (Figures 21.23 and 21.24).

Schizencephaly type II—Open lip cleft, walls of the cleft being wide, separated with cerebrospinal fluid, communicating with hydrocephalus (Figures 21.23 and 21.25). Type II is more common than type I.

Schizencephaly is associated with cortical malformation (e.g., heterotopia, Figure 21.24), optic nerve hypoplasia, SOD, and hippocampal malformation. The absence of the septum pellucidum is often seen (Figure 21.26), mostly in bilateral frontal schizencephaly.

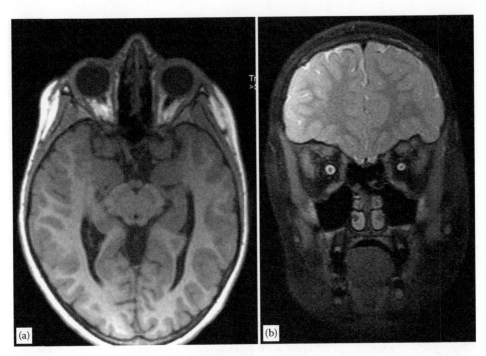

FIGURE 21.21
Septo-optic dysplasia. (a) Axial T_1-weighted image shows hypoplasia of the left optic nerve. (b) Coronal PD-weighted image shows asymmetry of the optic nerves.

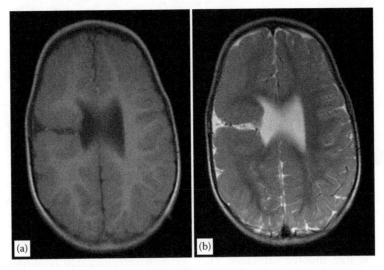

FIGURE 21.22
Schizencephaly. (a) Axial T_1-weighted and (b) T_2-weighted images show unilateral open-lip schizencephaly. The dysplastic, irregular, thick, dysplastic gray matter lines a small cleft. The cerebrospinal fluid connects the ventricle and the subarachnoid space. Abnormal, pachygyric cortex in the right frontal lobe is also seen.

21.5 Lissencephaly

Lissencephaly is a congenital malformation due to abnormal neuronal migration. Different groups of lissencephaly are recognized, according to associated malformations and genetic alterations. Several genes such as *LIS1, YWHAE, DCX, ARX, TUBA1 A*, and *RELN* are responsible for lissencephaly. The acquired causes of lissencephaly include *in utero* CMV infection, radiation injury, and fetal alcohol syndrome.

Lissencephaly is characterized by a complete or partial lack of gyri and thick cortex (Figures 21.27 through 21.29). A cell sparse zone separates the outer cortex from a deeper cortical layer (Figures 21.27a and b and 21.28a–c), and the three-layer appearance on T_2-weighted imaging

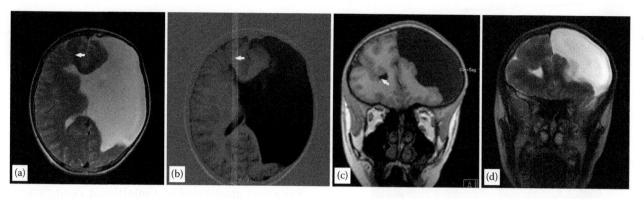

FIGURE 21.23

Schizencephaly. (a) Axial T_2-weighted and (b) T_1-weighted images show large, left-sided open-lip schizencephaly. The calvarium expands over the malformation. There is also closed-lip schizencephaly (white arrow) in the right frontal lobe which is better visible on (c) coronal T_1- and (d) T_2-weighted images. A typical dimple of the lateral wall confirms the presence of the cleft (white arrow).

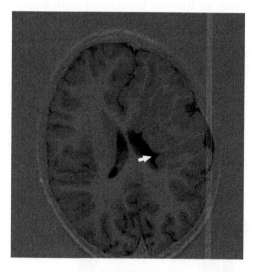

FIGURE 21.24

Schizencephaly. Axial T_1-weighted image shows unilateral, left-sided closed-lip schizencephaly (white arrow), the lips of the cleft are fused, and the cerebrospinal fluid is not seen. Linear gray matter extends from the cortex to the ventricular surface. Large transmantle heterotopia is present in the left frontal lobe, which is asymmetrically smaller.

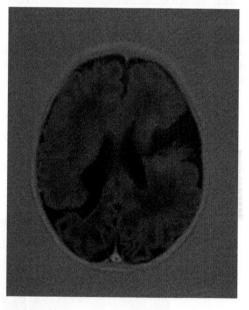

FIGURE 21.25

Schizencephaly. Axial T_1-weighted image shows bilateral open-lip schizencephaly with large clefts.

is characteristic of LIS1 and DCX lissencephaly. The cerebrum usually has a *figure-of-eight* shape on axial imaging, with the white matter volume being reduced (Figure 21.27a, d, and e). The cerebral cortex is thick, measuring up to 15 mm.

The spectrum of anomalies includes: complete absence of gyri in agyria and reduction of normal gyri in pachygyria. The mildest form classical lissencephaly is subcortical band heterotopia.

Patients with *LIS1* mutations have more posterior malformations (parieto-occipital regions), while the frontal lobes are most severely affected in patients with *DCX* mutation. Mutations in *TUBA1A* were connected with a more severe cortical anomaly posteriorly (with posterior-to-anterior gradient) and often associated

with corpus callosum dysgenesis (Figures 21.27f and 21.28d), brain stem and cerebellar hypoplasia, and band heterotopia.

21.5.1 Pachygyria or Incomplete Lissencephaly

The name *pachygyria* means thickened cortex with broad gyri and shallow sulci. Pachygyria can be focal or diffuse so areas of abnormal cortex may be seen together with areas of normal cortex. Usually, it is located bilaterally and posteriorly (with *LIS1* mutation) (Figure 21.29) or anteriorly (when *DCX* mutation is revealed) (Figure 21.30). The cerebral cortex is thick but thinner than in lissencephaly and measures about 6 mm.

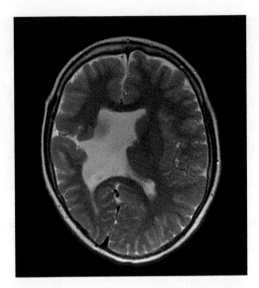

FIGURE 21.26
Schizencephaly. Axial T_2-weighted image shows small open-lip schizencephaly with gray matter along the cleft. Of note is the asymmetry of the cerebral hemispheres and absence of the septum pellucidum.

Band heterotopia (double cortex) is a band of gray matter located between two layers of normal-appearing white matter (Figures 21.31 through 21.33). It can be complete or partial with preference to frontal in *DCX*

or posterior in *LIS1* mutation. Band heterotopia mostly affects females. The brain cortex has normal thickness, but the sulci may be shallow.

21.6 Microcephaly

Congenital microcephaly belongs to malformations related to reduced proliferation or excess apoptosis and clinically characterized by a small head, head circumference <3 SD below normal for age. This primary microcephaly is caused by mutations in several genes such as *MCPH1, CENPJ, CDK5RAP2, ASPM, STIL,* and *WDR62*. By contrast, acquired microcephaly results from brain injury, such as the one associated with hypoxic–ischemic injury, intracranial infection, fetal alcohol syndrome, or metabolic disease. Microcephaly is also a part of different syndromes: Down, Edwards, and Cornelia de Lange.

Microcephaly with a simplified gyral pattern is characterized by microcephaly with too few gyri and abnormally shallow sulci (an abnormally simplified gyral pattern), but cortical thickness may be normal. Corpus callosum may be normal, thinner, or absent (Figures 21.34 and 21.35).

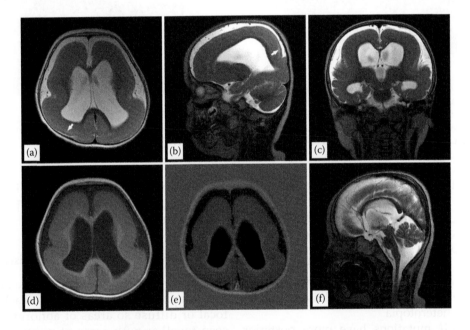

FIGURE 21.27
Lissencephaly. (a) Axial T_2-weighted and (b) sagittal T_2-weighted images show a completely smooth surface of the brain. The cortex is very thick and Sylvian fissures are shallow and vertically oriented. The brain has a characteristic *figure of eight* shape (on axial image). The cell-sparse zone is visible between the outer and inner cortical layers as hyperintense band (white arrow). (c) Coronal T_2-weighted image shows that some of the irregular gyri are seen in the temporal lobes. Note the enlargement of the ventricular system. (d,e) Axial T_1-weighted images show extensively diminished volume of the white matter. There is a low signal intensity of the thin cell-sparse zone. (f) Sagittal T_2-weighted image shows an abnormal shape of the corpus callosum: The body is arched, the splenium and genu are thin, and the rostrum is absent.

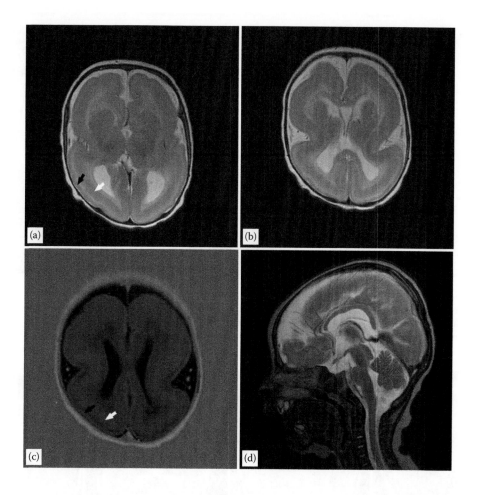

FIGURE 21.28

Lissencephaly in a neonate. (a) Axial T_1-weighted, (b) axial T_2-weighted, and (c) axial T_1-weighted images show smooth brain with the *figure of eight* appearance. A three-layer cortex, more pronounced posteriorly, is present. Thick inner (white arrow) and outer (black arrow) layers with a low signal intensity are seen. Between them, there is a high signal of the cell-sparse zone. (d) Sagittal T_2-weighted image shows arched corpus callosum with its hypoplasia, more pronounced anteriorly.

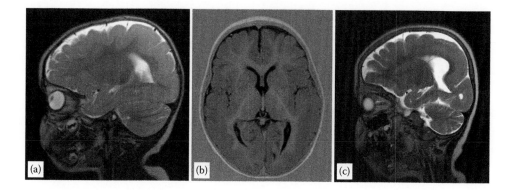

FIGURE 21.29

Pachygyria: From posterior to anterior gradient. (a) Parasagittal T_2-weighted and (b) axial T_1-weighted images show diffuse, bilateral, symmetric pachygyria with thickened cortex in the parietal and occipital lobes—posterior–anterior gradient. Note the diminished volume of the white matter in the affected regions. (c) Parasagittal T_2-weighted   *(Continued)*

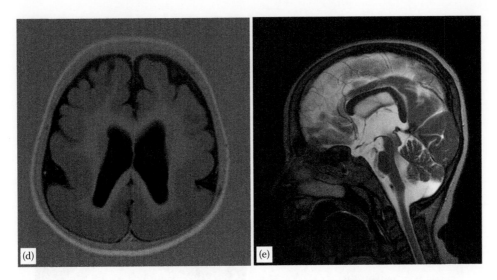

FIGURE 21.29 (Continued)
Pachygyria: From posterior to anterior gradient. (d) Axial T_2-weighted images show diffuse, bilateral, symmetric pachygyria with thickened cortex in the parietal and temporal lobes—posterior–anterior gradient. The white matter thickness is narrowing. (e) Sagittal T_2-weighted image shows arched corpus callosum.

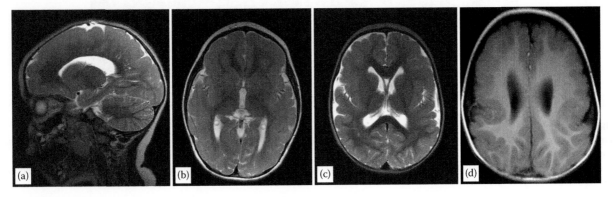

FIGURE 21.30
Pachygyria: From anterior to posterior gradient. (a) Parasagittal T_2-weighted and (b) axial T_2-weighted images show diffuse, bilateral, symmetric pachygyria with thickened cortex in the frontal and parietal lobes—anterior–posterior gradient. (c) Axial T_2-weighted and (d) axial T_1-weighted images show bilateral, asymmetric pachygyria with thickened cortex in the temporal and frontal lobes. The lesions are more pronounced in the right hemisphere.

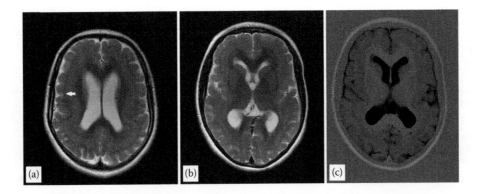

FIGURE 21.31
Band heterotopia. (a) Axial T_2-weighted image shows band heterotopia with a typical thick layer of neurons (gray matter) located between ventricles and the cortex (white arrow). The white matter is normally myelinated. The thickness of the cortex is in normal limits, but the sulci are shallow. (b) Axial T_2-weighted and (c) T_1-weighted images show that the abnormal layer of neurons is slightly thicker in the anterior brain.

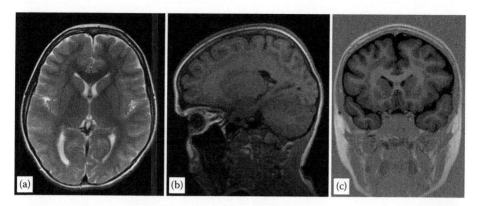

FIGURE 21.32
Band heterotopia. (a) Axial T_2-weighted, (b) parasagittal T_1-weighted, and (c) coronal T_1-weighted images show slightly asymmetric band heterotopia. The layer of heterotopic neurons is more pronounced in the right hemisphere.

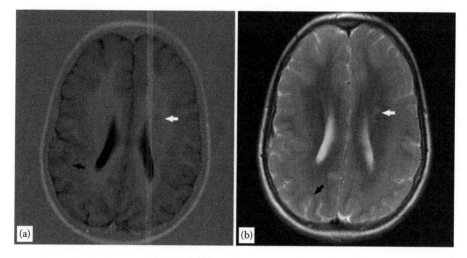

FIGURE 21.33
Band heterotopia. (a) Axial T_1-weighted and (b) axial T_2-weighted images show band heterotopia in the posterior regions of the brain (black arrow). Pachygyria is seen anteriorly: anterior pachygyria–posterior band heterotopia pattern. Note periventricular heterotopic nodules on the left side (white arrow).

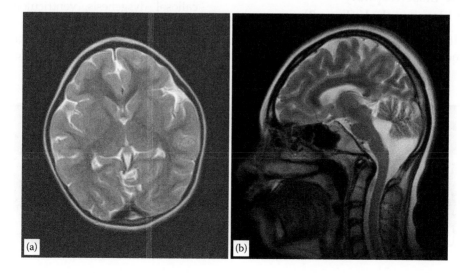

FIGURE 21.34
Microcephaly. (a) Axial T_2-weighted and (b) sagittal T_2-weighted images show a small volume of the cerebral hemispheres compared with the size of the posterior fossa. The thickness of the cortex looks normal. The corpus callosum is fully formed but thin.

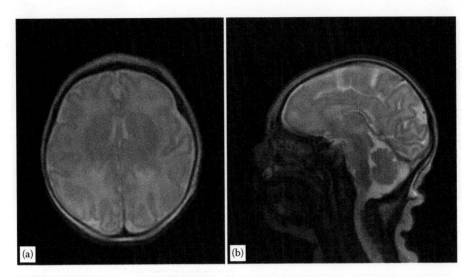

FIGURE 21.35
Microcephaly in a neonate. (a) Axial T_2-weighted image shows microcephaly with a simplified gyral pattern in the anterior cerebral hemispheres. The thickness of the cortex looks normal. (b) Sagittal T_2-weighted image shows normal corpus callosum.

A more severe form is microlissencephaly, characterized by a severe microcephaly, thickened cortex, and smooth brain (Figure 21.36). At least two main types are known: one without infratentorial anomalies (Norman–Robert syndrome) and the second one with hypoplasia of the corpus callosum and cerebellum (Barth syndrome). Additionally, congenital anomalies such as corpus callosum malformations, periventricular heterotopia, and delayed myelination may be present.

21.7 The Cobblestone Malformation—Cobblestone Lissencephaly (Formerly: Lissencephaly Type 2)

This group of malformations is due to abnormal neuronal migration and overmigration of neurons into the pial layer associated with anomalies of the brain, eyes, and muscles, and usually seen in children with congenital muscular dystrophies (but it can be isolated). There is a wide spectrum of abnormalities caused by mutations in several genes such as *POMT1*, *POMT2*, *POMGnT1*, *FCMD*, *FKRP*, *FKTN*, *LAMC3*, *LAMA1A*, and *LARGE*.

Congenital muscular dystrophies are divided into Walker–Warburg syndrome, Fukuyama congenital muscular dystrophy, and muscle–eye–brain disease.

Walker–Warburg syndrome is the most severe form. Imaging features include supra- and infratentorial abnormalities: cobblestone lissencephaly, prominent hypomyelination of the white matter, callosal dysgenesis, ventriculomegaly, pontine hypoplasia with a kink at the mesencephalic–pontine junction, collicular fusion, and cerebellar hypoplasia. Occipital cephalocele may be seen.

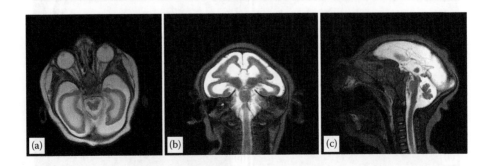

FIGURE 21.36
Microlissencephaly. (a) Axial and (b) coronal T_2-weighted images show a very small, smooth brain without gyri or sulci. There are only Sylvian fissures visible. (c) Sagittal T_2-weighted image shows a very thin, hypogenetic corpus callosum, seen only anteriorly and a very thin anterior cerebral commissure. Hypogenetic vermis is seen.

The second kink may be seen at the ventral cervico-medullary junction, and hypoplasia/agenesis of the corpus callosum has also been reported (Figures 21.37 through 21.41).

> *Fukuyama congenital muscular dystrophy* includes different types of cortex malformation: PMG predominantly in frontal lobes and cobblestone cortex in occipito-parietal lobes. Multiple cerebellar cortical cysts are present. Delayed myelination is also seen.
>
> *Muscle–eye–brain disease* presents a diffuse abnormal cortex, a hypoplastic brain stem, and a large tectum.

21.8 Polymicrogyria

PMG is a malformation of cortical development. According to Barkovich, this is a malformation of abnormal postmigrational development (formerly known as malformation of cortical organization).

The etiology of PMG is diverse. PMG can be caused by *in utero* infection (cytomegalovirus), vascular factors (associated with schizencephaly), inborn errors of metabolism, genetically defined congenital anomalies, or genetic mutations of *WDR62*, *SRPX2*, *PAX6*, *TBR2*, and *TUBB2*.

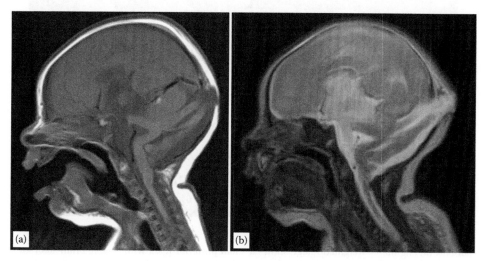

FIGURE 21.37

Cobblestone malformations. (a) Sagittal T_1-weighted and (b) T_2-weighted images show a large pontine kink. The hypoplastic cerebellum, pons, and midbrain are stretched posteriorly toward a small occipital cephalocele. Hypoplasia of the corpus callosum is seen.

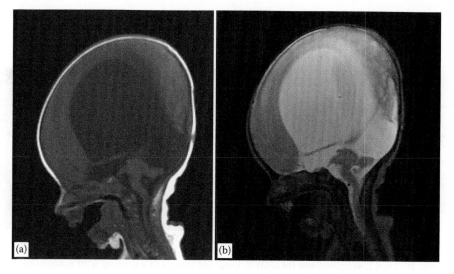

FIGURE 21.38

Cobblestone malformations. (a) Sagittal T_1-weighted and (b) T_2-weighted images show macrocephaly and hydrocephalus. The pontine, cervicomedullary junction kinks, and enlarged quadrigeminal plate are seen. Hypoplasia of the cerebellum is also seen.

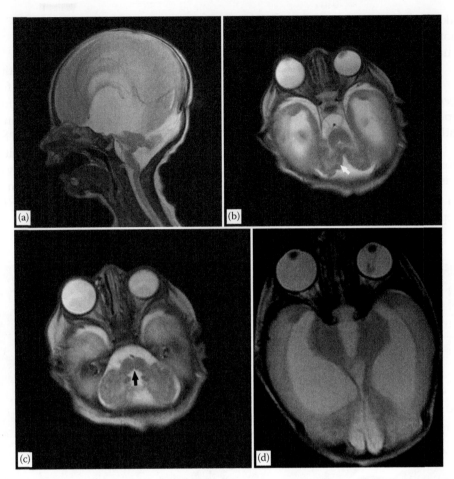

FIGURE 21.39
Cobblestone malformations. (a) Sagittal T_2-weighted image shows a small posterior fossa with double brain-stem kinks, small vermis, and enlarged quadrigeminal plate. (b,c) Axial T_2-weighted images show small cerebellar hemispheres with abnormal cortex with small cysts (white arrow). A cleft of the pons is also visible (black arrow). (d) Axial T_2-weighted images shows nonattachment of the retina of the left globe. Note the hydrocephalus.

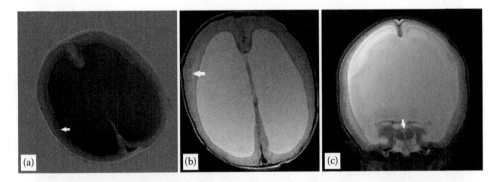

FIGURE 21.40
Cobblestone malformations. (a) Axial T_1-weighted and (b) axial T_2-weighted images in a neonate with massive hydrocephalus. A typical for the cobblestone malformation appearance of the cortex: irregular inner layer of the cortex (white arrow). (c) Coronal T_2-weighted image shows a midline cleft of the pons (white arrow) and a characteristic nodular inner layer of the cortex.

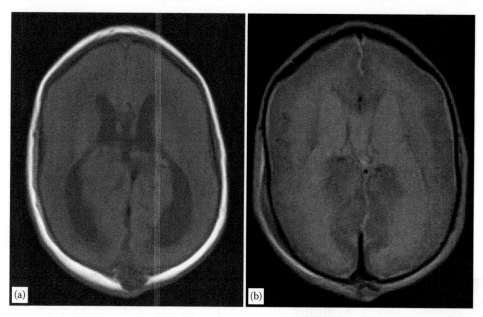

FIGURE 21.41
Cobblestone malformations. (a) Axial T_1-weighted and (b) axial T_2-weighted images in a neonate with slightly enlarged ventricles. Dysgenesis of the posterior component of the corpus callosum is not seen on images. Cavum septi pellucidi and a small occipital cephalocele are seen. Smooth cerebral hemispheres with irregular nodular cortex are seen.

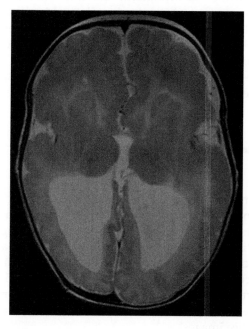

FIGURE 21.42
Polymicrogyria. Axial T_2-weighted image shows bilateral PMG with predominance of the left hemisphere.

PMG can be isolated or may be associated with different anomalies such as abnormal Sylvian fissures, agenesis/hypogenesis of the corpus callosum, schizencephaly, and cerebellar hypoplasia. It can also be associated with several syndromes: bilateral perisylvian PMG syndrome, megalencephaly syndrome with PMG,

Aicardi syndrome, Zellweger syndrome, DiGeorge syndrome (deletions of 22q 11.2), and Dellemann syndrome.

PMG is characterized by overfolded cortex with irregular surface of the cerebral hemispheres (it can be paradoxically smooth) and irregular gray–white matter interface. In areas of PMG, multiple small gyri are seen and no normal sulci. PMG can be diffuse, multifocal or focal, bilateral or unilateral, symmetric or asymmetric (Figures 21.42 through 21.45). Predilection for perisylvian regions is noticed (Figure 21.46).

The most common features are generalized PMG with periventricular gray matter heterotopia; anomalous venous drainage in malformed cortex is also common. PMG may simulate pachygyria, and correct diagnosis depends on the thickness of slices (volumetric imaging sequences are needed) and the degree of myelination. In unmyelinated regions, the polymicrogyric cortex looks thin, whereas in myelinated areas, it looks thicker, so the degree of myelination affects the appearance.

21.9 Hemimegalencephaly

Hemimegalencephaly is a rare brain malformation characterized by hypertrophy of the whole cerebral hemisphere with disturbed proliferation, and migration and apoptosis

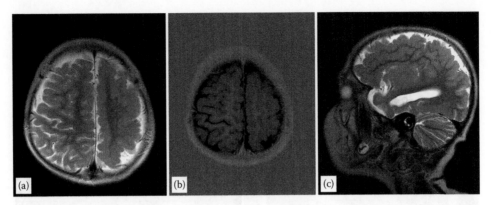

FIGURE 21.43
Polymicrogyria. (a) Axial T_2-weighted and (b) axial T_1-weighted images show asymmetry of the hemispheres, with polymicrogyric cortex on the left side. Cortical thickening and shallow sulci are present. (c) Sagittal T_2-weighted image shows undulation and irregularity of the cortex.

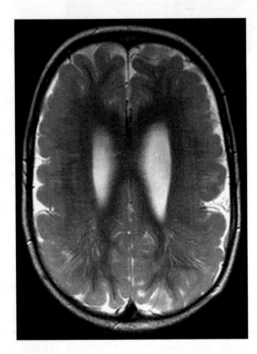

FIGURE 21.44
Polymicrogyria. Axial T_2-weighted image shows bilateral PMG with irregularity of the gray–white matter interface.

of neurons. It is rare that the hypertrophy concerns only a part of the hemisphere (Figures 21.47 and 21.48).

The etiology remains uncertain. It can be isolated or associated with neurocutaneous syndromes such as epidermal nevus syndrome, Klippel–Trenaunay syndrome, Proteus syndrome, hypomelanosis of Ito, and NF1. The affected hemisphere or its part is enlarged with increased volume of the white matter and with thickened abnormal cortex (Figures 21.49a and b through 21.51). The interface between the white and the gray matter is flattened and blurred

with abnormal signal of the white matter in affected regions. The sulci are shallow and the brain surface is smooth. The asymmetry of the lateral ventricles is seen with dilation of the ipsilateral lateral ventricle. If the whole cerebral hemisphere is affected, its frontal horn is elongated, pointed anteriorly and superiorly. The enlarged hemisphere or its part can bulge across the midline. Contralateral cortical malformations can also occur and involved ipsilateral brain stem or cerebellum can also be seen (total hemimegalencephaly). Hemimegalencephaly may sometimes change over time; in such cases, serial brain imaging shows progressive atrophy of the affected hemispheres (Figure 21.49c).

21.10 Heterotopia

Gray matter heterotopia are the malformations resulting from abnormal neuron migration. These are collections of heterotopic neurons arrested during their migration route from the periventricular germinal zone to the cortex. There are three types of heterotopia: periventricular nodular (Figures 21.52 through 21.55), focal subcortical (Figures 21.56 and 21.57), and leptomeningeal heterotopia (the last one is not seen on magnetic resonance images).

They are caused by mutations of *FLNA* gene or *ARFGEF2* gene. They can be isolated or may be associated with different brain malformations such as pachygyria, agenesis of the corpus callosum, and schizencephaly (Figure 21.57). The signal is isointense to the gray matter and lesions do not enhance after contrast injection.

Periventricular heterotopia are located close to the ventricular wall, commonly in the trigone and occipital horns; they can be focal or diffuse, uni- or bilateral (Figures 21.52 through 21.55 and 21.58).

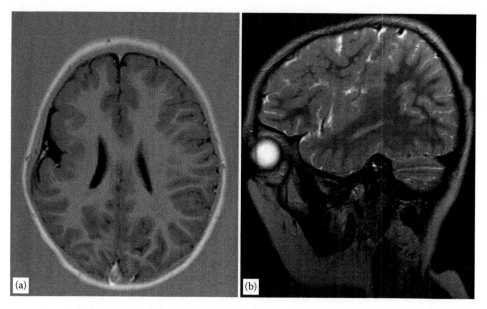

FIGURE 21.45

Polymicrogyria. (a) Axial T_1-weighted and (b) sagittal T_2-weighted images show polymicrogyric cortex in the frontal and temporal regions of the right hemisphere.

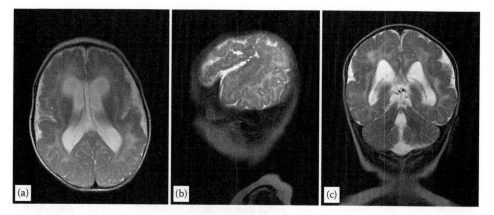

FIGURE 21.46

Polymicrogyria. (a) Axial T_2-weighted, (b) sagittal T_2-weighted, and (c) coronal T_2-weighted images show bilateral frontal and perisylvian PMG. Irregular pattern of the cortex with shallow sulci surrounding the Sylvian fissure is better demonstrated on the sagittal image (white arrow).

Subcortical heterotopia are located in the subcortical area or within deep white matter. They can be nodular or curvilinear; sometimes they are a large mass. The ipsilateral hemisphere or a part of it can be smaller (Figures 21.56 and 21.57).

21.11 Focal Cortical Dysplasia

Focal cortical dysplasia (FCD) is a heterogeneous group of lesions. Both genetic and acquired factors are involved in the pathogenesis of cortical dysplasia. There are three types of FCD: type I and III belong to malformations secondary to abnormal postmigrational development. FCD I occurs in children with pre- and perinatal asphyxia or bleeding also in patients with severe prematurity, whereas FCD III is associated with hippocampal sclerosis, tumors, and vascular malformations (Figure 21.59).

FCD II belongs to cortical dysgenesis with abnormal proliferation but without neoplasia, and is characterized by the presence of dysmorphic neurons in type IIa or dysmorphic neurons and balloon cells in type IIb. It is characterized by cortical thickening; gyri may be expanded and the gray–white matter junction may be

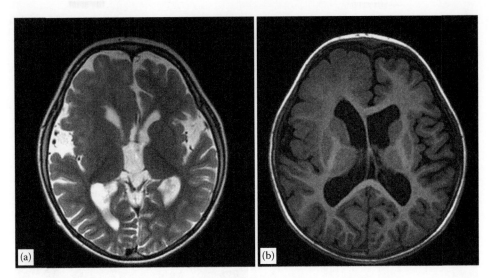

FIGURE 21.47
Localized hemimegalencephaly. (a) Axial T_2-weighted and (b) axial T_1-weighted images show hemimegalencephaly localized to the right frontotemporal lobes. The gray and white matter interface is blurred, gyri are broad, and the cortex is thick in the affected region. The frontal horn of the right ventricle is pointed anteriorly.

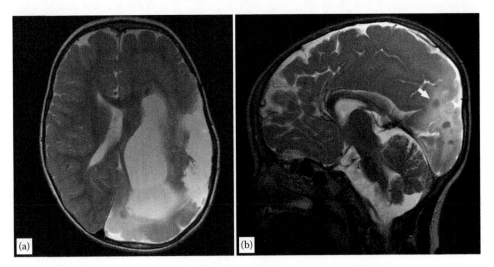

FIGURE 21.48
Localized hemimegalencephaly. (a) Axial T_2-weighted image shows enlargement of the left hemisphere with posterior predominance. The white matter shows abnormal signal intensity. The left ventricle is enlarged. (b) Sagittal T_2-weighted image shows nodular myelin heterotopias (white arrow). Abnormal shape of the posteriorly hypoplastic corpus callosum is seen.

blurred. Signal intensity of the white matter is increased on T_2-weighted and FLAIR images and decreased on T_1-weighted images, but it depends on the age of patients (stage of white matter myelination) (Figures 21.60 through 21.63).

In transmantle dysplasia (which is seen in FCD II only), altered white matter signal is often extended from the gyri toward the ventricle (Figure 21.64).

21.12 Cerebellar Hypoplasia

It belongs to midbrain–hindbrain malformations. It can occur in isolation or be accompanied by supratentorial defects as complex malformations. Knowledge of etiology is limited, and gene mutations, teratogens, and metabolic derangements are taken into account.

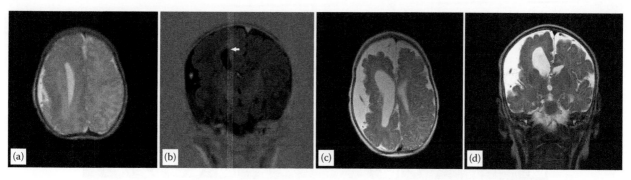

FIGURE 21.49
Hemimegalencephaly. (a) Axial T_2-weighted and (b) coronal T_1-weighted images in a neonate show enlarged right hemisphere with abnormal cortex in the frontal and parietal lobes. The ipsilateral ventricle is also enlarged and the frontal horn is elongated and pointed superiorly (white arrow). (c) Axial T_2-weighted and (d) coronal T_2-weighted images performed 2 years later show atrophy of the affected regions of the right hemisphere and atrophic enlargement of the right ventricle.

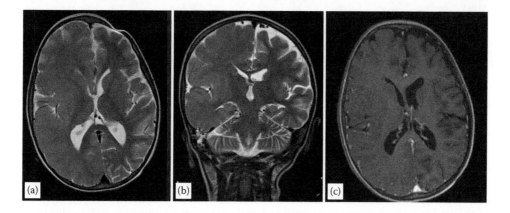

FIGURE 21.50
Hemimegalencephaly. (a) Axial T_2-weighted, (b) coronal T_2-weighted, and (c) axial T_1-weighted images show hemimegalencephaly with enlargement of the whole right hemisphere. Some gyri, especially in the area of the sylvian fissure, are pachygyric. The signal intensity of the white matter is abnormal and the interface between the gray and the white matter is blurred. The anterior horn of the lateral ventricle is pointed anteriorly. The left hemisphere is normal.

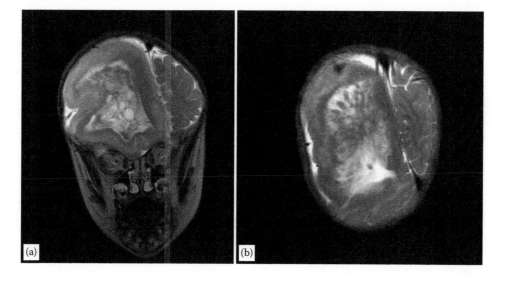

FIGURE 21.51
Hemimegalencephaly. (a,b) Coronal T_2-weighted images show a severe form of hemimegalencephaly with massive, hamartomatous enlargement of the right hemisphere. The brain surface is smooth with diffuse gyral thickening and diminished sulcation. The white matter signal is completely abnormal with patchy regions of increased signal intensity.

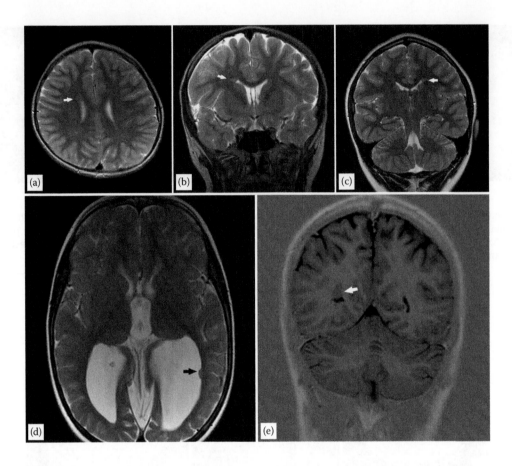

FIGURE 21.52
Heterotopia—Examples of small, periventricular nodular heterotopia. (a) Axial T_2-weighted and (b,c) coronal T_2-weighted images show single heterotopic nodules with signal intensity isointense in comparison with the gray matter. Heterotopic nodules slightly protrude into the ventricles (white arrows). (d) Axial T_2-weighted image shows a small heterotopic nodule protruding into the left ventricle (black arrow). Note the enlargement of the occipital horns of both ventricles and the third ventricle. (e) Coronal T_1-weighted image shows nodular heterotopia in the occipital horn of the right ventricle (white arrow).

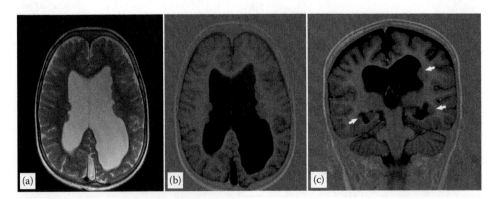

FIGURE 21.53
Multiple periventricular nodular heterotopia. (a) Axial T_2-weighted, (b) axial T_1-weighted, and (c) coronal T_1-weighted images show several, bilateral nodules which are isointense with gray matter (white arrows). They protrude into the lateral ventricles.

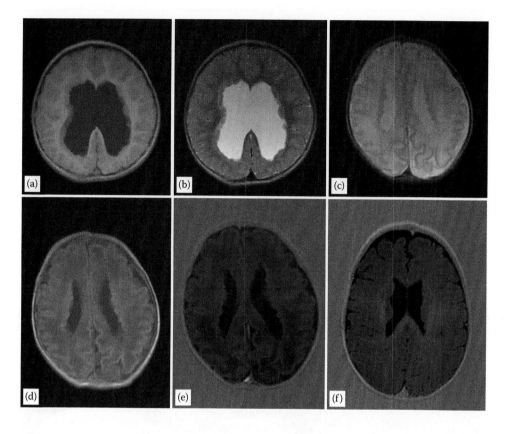

FIGURE 21.54

Diffuse periventricular heterotopia. (a) Axial T_1-weighted and (b) axial T_2-weighted images show multiple heterotopic nodules connected with each other (linear heterotopia), lining the walls of both lateral ventricles. Lateral ventricles are enlarged. (c) Axial T_2-weighted and (d,e) axial T_1-weighted images in this neonate show multiple nodules protruding into the ventricles. (f) Axial T_1-weighted image shows wide, linear periventricular heterotopia seen in the margin of both bodies of the lateral ventricles.

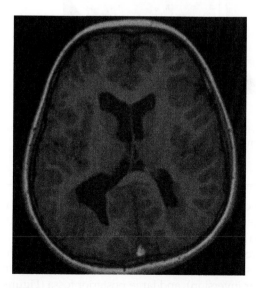

FIGURE 21.55

Bilateral subcortical heterotopia. Axial T_1-weighted image shows a large concentration of gray matter heterotopia localized near the trigonum of the ventricles bilaterally. Heterotopia on the left side protrudes into the ventricle and extends from the wall of the ventricle to the cortex. The right lateral ventricle is slightly enlarged.

A wide spectrum of disease severity can be noticed. In cerebellar hypoplasia, normal fissures are seen in small hemispheres, unlike in cerebellar dysgenesis, where abnormal fissures are present. The vermis and pons are usually small as well (Figures 21.65 through 21.68).

21.13 Pontocerebellar Hypoplasia

Pontocerebellar hypoplasia (PCH) is a heterogeneous group of disorders which belong to midbrain–hindbrain malformations. At least seven subtypes of PCH have been described, mostly caused by mutations of a complex of *TSEN* genes, *VRK1* and *RARS2*.

PCH is characterized by atrophy of the brain stem (mostly ventral pons) and cerebellar hemispheres with variable involvement. The hemispheres are affected more severely than the vermis, which can be relatively spared in some subtypes. PCH shows progressive atrophy seen on serial brain examinations (Figures 21.69 through 21.71).

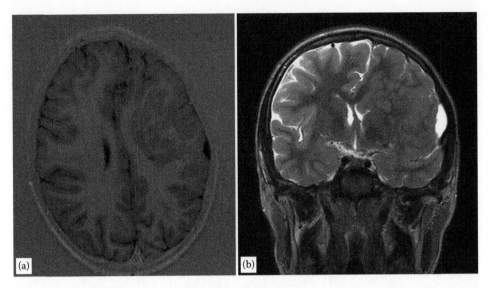

FIGURE 21.56
Transmantle heterotopia. (a) Axial T_2-weighted and (b) coronal T_2-weighted images show a large mass of heterotopic gray matter extending from the left ventricle to the cortex. The affected part of the left cerebral hemisphere is small. A heterotopic nodule of the gray matter protrudes into the ventricle on (b).

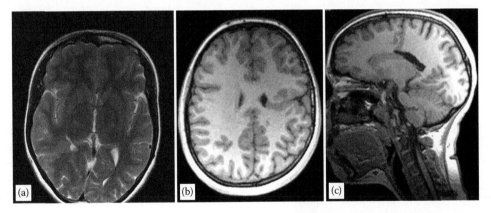

FIGURE 21.57
Transmantle heterotopia. (a) Axial T_2-weighted, (b) axial T_1-weighted, and (c) sagittal T_1-weighted images show a thin layer of gray matter extending from the ventricle to the cortex. Note that there are no ventricular dimples.

21.14 Dandy–Walker Malformation

Dandy–Walker malformation (DWM) belongs to midbrain–hindbrain malformations and disorders with cerebellar hypoplasia. It is usually a sporadic disorder and its genetic background remains poorly understood. Changes in *FOXC1* and *ZIC1* and *ZIC4* genes were reported as causative of malformations. Dandy–Walker syndrome (DWM) can also be caused by environmental factors, teratogens, and infections such as rubella or toxoplasmosis as postulated by some authors.

The classic DWM is characterized by complete or partial agenesis of the vermis (if vermis is seen, it is upward rotated), large dilatation of the fourth ventricle, elevation of the torcula herophili, elevation of the tentorium, elevation of transverse sinuses above the lambdoid suture (lambdoid torcular inversion), and large posterior fossa (Figures 21.72 through 21.75). Cerebellar hemispheres are hypoplastic, and the brain stem is typically compressed. Hydrocephalus is frequently seen. Additionally, malformations such as agenesis of the corpus callosum, heterotopia, schizencephaly, and parietal cephalocele (Figure 21.75) can be found. There is a wide spectrum of disease severity.

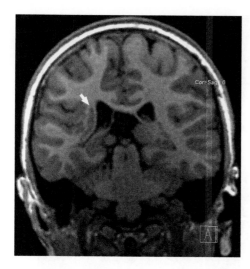

FIGURE 21.58
Heterotopia. Coronal T_1-weighted image shows a linear cluster of gray matter heterotopias (white arrow).

Disorders related to DWMs are mega cisterna magna, cerebellar vermian hypoplasia, and Blake's pouch cyst.

Mega cisterna magna is a common posterior fossa anomaly characterized by a dilatation of the cisterna magna (Figure 21.76). The fourth ventricle, vermis, and cerebellar hemispheres are mostly normal. There is no hydrocephalus.

Cerebellar vermian hypoplasia is characterized by normal size or slightly enlarged posterior fossa as well as hypoplastic and hypogenetic vermis which is in normal position. Retrocerebellar fluid communicates with the fourth ventricle. The tentorium cerebelli is not elevated; it is in normal position.

Blake's pouch cyst is a persisting Blake's pouch characterized by a herniated collection of the cerebrospinal fluid, which communicates with

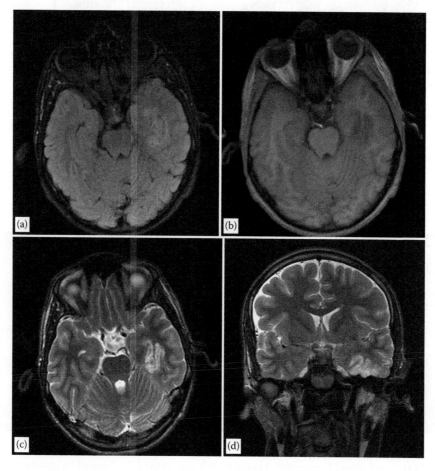

FIGURE 21.59
Focal cortical dysplasia—FCD III with ganglioglioma. (a) Axial FLAIR, (b) axial T_1-weighted, (c) axial T_2-weighted, and (d) coronal T_2-weighted images show abnormal signal intensity of the cortex of the left inferior temporal lobe with many small cysts.

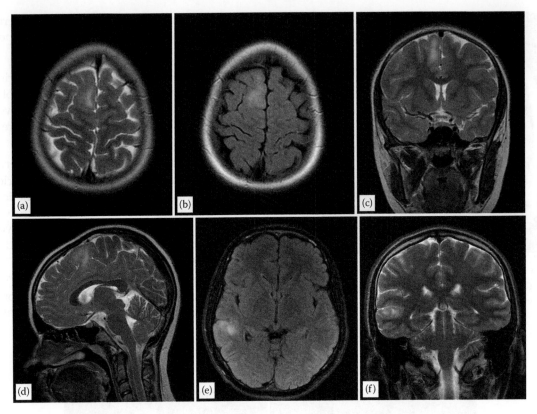

FIGURE 21.60
Focal cortical dysplasia—FCD II. (a) Axial T_2, (b) axial FLAIR, (c) coronal T_2, and (d) sagittal T_2-weighted images show abnormal signal of the superior frontal gyrus. The volume of the gyrus is slightly enlarged without mass effect. (e) Axial FLAIR and (f) coronal T_2-weighted images show a focal hyperintense lesion located in the middle temporal gyrus.

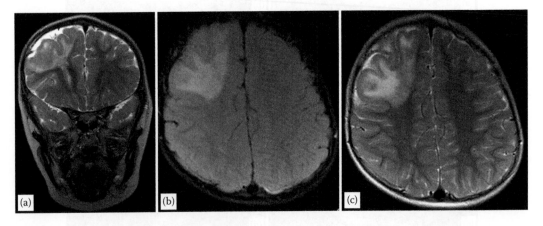

FIGURE 21.61
Focal cortical dysplasia—FCD II. (a) Coronal T_2-weighted, (b) axial FLAIR, and (c) axial T_2-weighted images show a large lesion in the right frontal lobe. A hyperintense signal in the cortex and the subcortical white matter is seen.

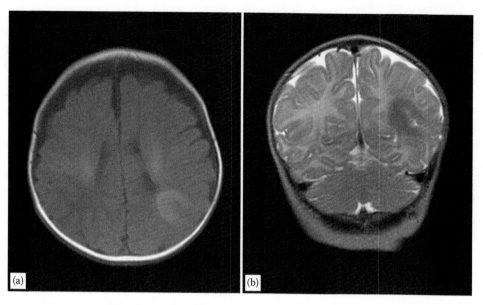

FIGURE 21.62
Focal cortical dysplasia—FCD II in a neonate. (a) Axial T_1-weighted and (b) coronal T_2-weighted images show focal lesions in the left hemisphere. The dysplastic region is hyperintense on T_1-weighted and hypointense on T_2-weighted images in the unmyelinated brain.

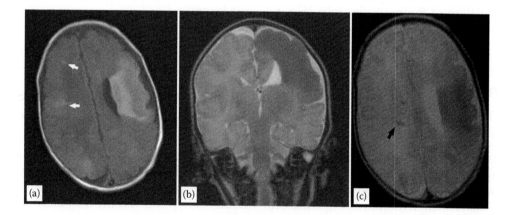

FIGURE 21.63
Focal cortical dysplasia—FCD II in a neonate with tuberous sclerosis complex (TSC). (a) Axial T_1-weighted, (b) coronal T_2-weighted, and (c) axial T_2-weighted images show very large dysplasia in the left hemisphere with widening of the affecting gyri. Enlargement of the left lateral ventricle is seen. Subependymal nodules in the right ventricle (white arrow) and a few cortical tubers in the right hemisphere (white arrows)—changes related to TSC.

a dilated and deformed fourth ventricle but does not communicate with the cisterna magna posteriorly. The cyst is located infra- or retrocerebellarly; the vermis is normal and hydrocephalus is present.

Sometimes, displacement of the choroid plexus of the fourth ventricle is visible on a sagittal view (Figure 21.77).

21.15 Cerebellar Cortical Dysgenesis— Focal or Diffuse

Malformation of the cerebellar cortex may be isolated, without supratentorial anomalies, or it may be associated with malformations of the cerebral hemispheres. It is caused by genetic mutations, fetal infection, or fetal injury. Cortical dysgenesis may also be a result of toxins,

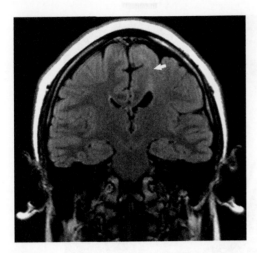

FIGURE 21.64
Transmantle cortical dysplasia. Coronal FLAIR image shows hyperintense signal extending from the ventricle to the frontal gyrus (white arrow).

drug, and alcohol abuse. Disorganized development of the cerebellum causes cerebellar dysplasia with abnormal shape and pattern of the cerebellar folia (Figures 21.78 and 21.79).

21.16 Molar Tooth Syndrome (Joubert Syndrome and Related Disorders)

According to Barkovich, Joubert syndrome and related disorders are included in the group of malformations due to abnormal postmigrational development (formerly known as malformations of cortical organization).

They result from a midbrain–hindbrain malformation and belong to disorders with cerebellar dysgenesis.

The molar tooth sign is caused by cerebellar vermian hypoplasia. This sign is a radiological hallmark of a group of syndromes termed Joubert syndrome and related disorders: Dekaban–Arima syndrome, Senior-Loken syndrome, Cerebellar vermis hypoplasia/aplasia, Oligophrenia, Ataxia, Coloboma, and Hepatic fibrosis (COACH) syndrome, Malta syndrome, and Varadi–Papp syndrome. Changes in different genes are recognized: *AHI1*, *ARL13B*, *CC2D2A*, *CEP290*, *INPP5E*, *NPHP1*, *OFD1*, *RPGRIP1L*, *TMEM216*, and *TMEM67*.

The molar tooth syndrome is characterized by thickened and elongated superior cerebellar peduncles without their normal decussation (Figures 21.80 through 21.82). Superior cerebellar peduncles are perpendicular to the dorsal pons. Dysplasia of the superior vermis, agenesis of mild and inferior vermian lobules and an abnormally deep interpeduncular fossa, and narrow isthmus are also noticed (Figures 21.80 through 21.83). The fourth ventricle is dilated with a *bat-wing* or *umbrella* shape (Figure 21.80a, b, and d). It may be accompanied by brain-stem hypoplasia (Figure 21.83). Additionally, anomalies of hippocampi, corpus callosum, and cortex can be noticed.

21.17 Brain-Stem Disconnection Syndrome (Pontomedullary Disconnection)

Brain-stem disconnection syndrome (BDS) belongs to midbrain–hindbrain malformations (Figure 21.84). The etiology of BDS remains uncertain. Dysgenesis of the vertebral arteries and posterior cerebral circulation results in hypoxia, causing anomalies of brain stem and

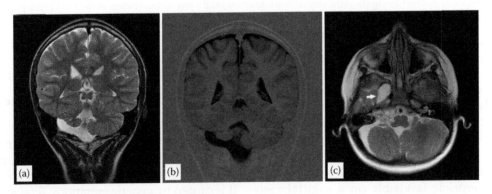

FIGURE 21.65
Cerebellar hypoplasia. (a) Coronal T_2-weighted, (b) coronal T_1-weighted, and (c) axial T_2-weighted images show hypoplasia of the right cerebellar hemisphere. Note a small temporal arachnoid cyst on the right side (white arrow).

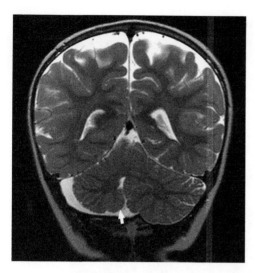

FIGURE 21.66
Cerebellar hypoplasia. Coronal T_2-weighted image shows small hypoplasia of the right cerebellar hemisphere with hypoplasia of the right cerebellar tonsil (white arrow).

cerebellar development. No genetic mutations have been detected in humans. The lack of *EN2* gene is suspected by some authors.

The disconnection of brain stem can be located between the midbrain and the lower pons or between the pons and the medulla oblongata. Additionally, hypoplasia of cerebellar hemispheres and vermis as well as absent cerebellar peduncles are seen. The basilar artery may also be absent.

21.18 Rhombencephalosynapsis

Rhombencephalosynapsis (RES) is a rare congenital defect that belongs to midbrain–hindbrain malformations. The cause of RES is unknown; RES is sporadic or occurs as a part of different syndromes such as VACTERL-H and Gómez–López–Hernandez. It can also be associated with HPE.

According to variable severity of presented features, RES can be complete (the most severe form) or partial with a different severity range of features.

Complete RES is characterized by a complete fusion of cerebellar hemispheres, no anterior or posterior vermis. The nodulus may be absent. Fusion of dentate nuclei and superior cerebellar peduncles is also seen. Closely apposed or fused dentate nuclei may be present as well (Figure 21.85).

The posterior fossa is typically small in patients with complete RES and the fourth ventricle has a keyhole shape. Anomalies of mesencephalic structures may be seen. Additional abnormalities include fusion of the thalami, inferior colliculi, and cerebral peduncles. Septum pellucidum may be absent.

Partial RES includes partial fusion of the hemispheres and deficient posterior vermis with normal anterior vermis and nodulus (Figures 21.86 and 21.87).

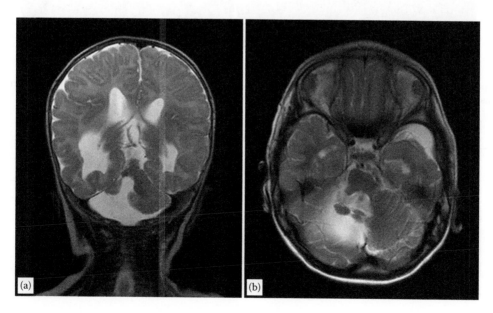

FIGURE 21.67
Cerebellar hypoplasia. (a) Coronal T_2-weighted and (b) axial T_2-weighted images show severe hypoplasia of the right cerebellar hemisphere. Note PMG in the right occipital lobe and enlargement of both ventricles. A temporal arachnoid cyst is also present.

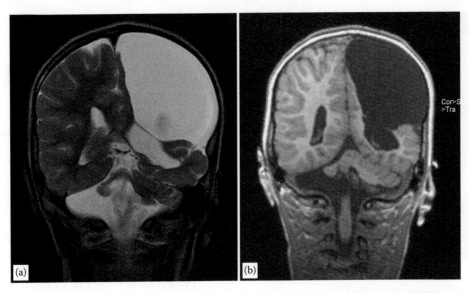

FIGURE 21.68
Cerebellar hypoplasia. (a) Coronal T_2-weighted and (b) coronal T_1-weighted images show severe hypoplasia of both cerebellar hemispheres; the right hemisphere is significantly smaller. A huge arachnoid cyst of the right cerebral hemisphere is seen.

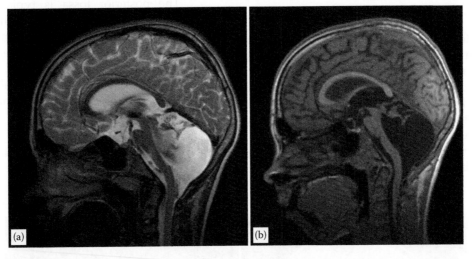

FIGURE 21.69
Pontocerebellar hypoplasia. (a) Sagittal T_2-weighted and (b) sagittal T_1-weighted images show an extremely hypoplastic vermis and hypoplasia of the brain stem (especially medulla oblongata). The corpus callosum is developed, but the splenium is slightly thin.

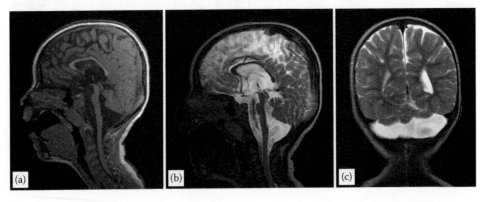

FIGURE 21.70
Pontocerebellar hypoplasia. (a) Sagittal T_2-weighted and (b) sagittal T_1-weighted images show profound hypoplasia of the brain stem and hypoplasia of the vermis. The corpus callosum is thin. (c) Coronal T_2-weighted image shows also hypoplasia of both cerebellar hemispheres.

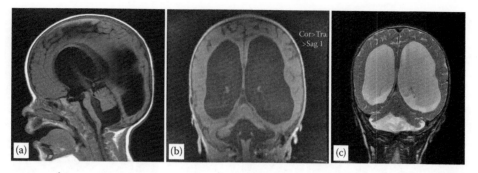

FIGURE 21.71

Pontocerebellar hypoplasia. (a) Sagittal T_1-weighted image shows hypoplasia of the brain stem and slight hypoplasia of the vermis. Note the hydrocephalus. (b) Coronal T_1-weighted and (c) coronal T_2-weighted images show hypoplasia of both hemispheres of the cerebellum and significant enlargement of the lateral ventricles.

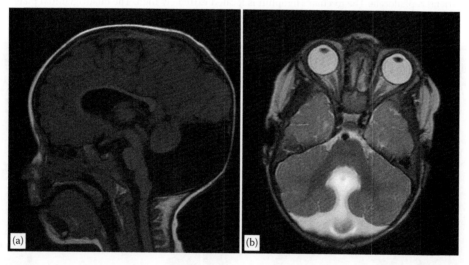

FIGURE 21.72

Dandy–Walker malformation. (a) Sagittal T_1-weighted and (b) axial T_2-weighted images show partial agenesis and an upwardly rotated vermis. The fourth ventricle is continuous with a large cisterna magna. The torcular herophili is elevated. Enlargement of the posterior fossa is seen.

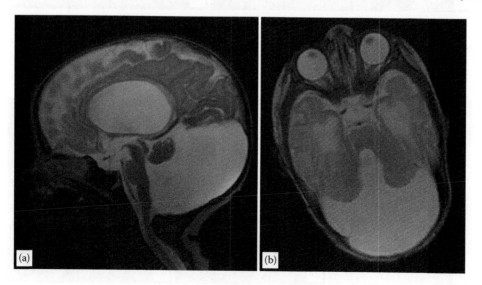

FIGURE 21.73

Dandy–Walker malformation. (a) Sagittal T_2-weighted and (b) axial T_2-weighted images show a large posterior fossa with severe partial agenesis of a rotated vermis. A large fourth ventricle is continuous with a very large cisterna magna. The torcular herophili is elevated. Enlargement of the temporal horns due to hydrocephalus is present.

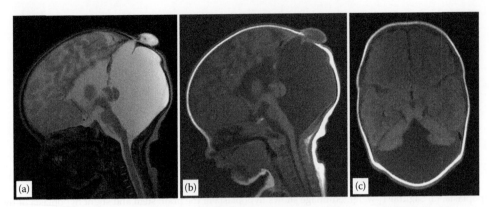

FIGURE 21.74
DWM with parietal cephalocele. (a) Sagittal T_2-weighted and (b) sagittal T_1-weighted images show an extremely small vermis, enormous posterior fossa, and parietal cephalocele. The torcular Herophili is located extremely high. The quadrigeminal plate is enlarged. Complete agenesis of the corpus callosum is seen. (c) Axial T_1-weighted image shows hypoplasia of the cerebellar hemispheres.

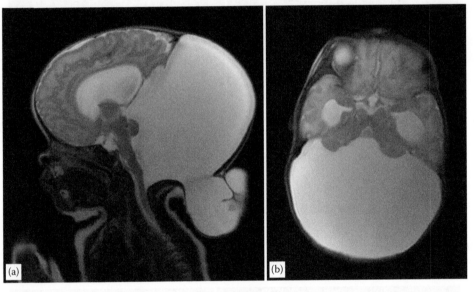

FIGURE 21.75
DWM with a large occipital cephalocele. (a) Sagittal T_2-weighted and (b) axial T_2-weighted images show a huge posterior fossa with a wide occipital bone defect and cephalocele. Severe hypoplasia of the vermis and cerebellar hemispheres is visible. The brain stem is also hypoplastic, anteriorly displaced, and slightly compressed.

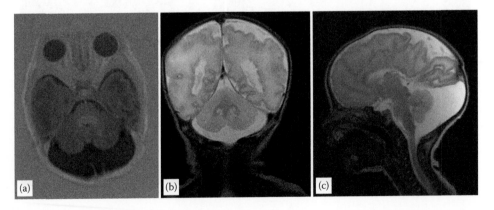

FIGURE 21.76
Mega cisterna magna. (a) Axial T_1-weighted and (b) coronal T_2-weighted images show a large cisterna magna with normal cerebellum. Bilateral periventricular heterotopia is present. (c) Sagittal T_2-weighted image shows enlargement of the posterior fossa and mega cisterna magna.

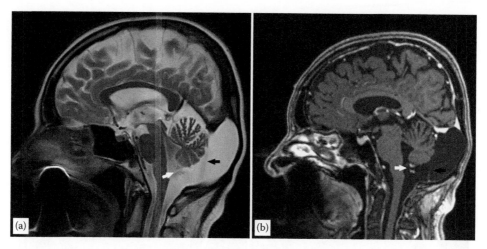

FIGURE 21.77

Blake's pouch cyst. (a) Sagittal T_2-weighted and (b) sagittal contrast-enhanced T_1-weighted images show the continuation of the plexus (white arrows) of the fourth ventricle into the Blake's pouch cyst. Note a slightly enlarged posterior fossa with a thin septum dividing (black arrows) the cyst and an enlarged retrocerebellar cistern.

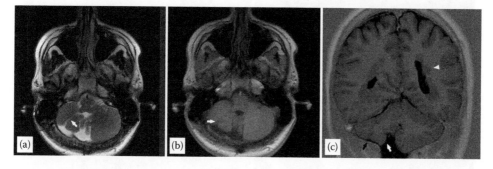

FIGURE 21.78

Cerebellar cortical dysgenesis. (a) Axial T_2-weighted and (b) axial T_1-weighted images show a cleft of the right cerebellar hemisphere (white arrows). Asymmetry of the cerebellar hemispheres: the right one is smaller. (c) Coronal T_1-weighted image shows a vertical cleft of the dysgenetic cerebellar hemisphere (black arrow). Note the lack of the right tonsil (white arrow) and the bilateral periventricular nodular heterotopias (white arrowhead).

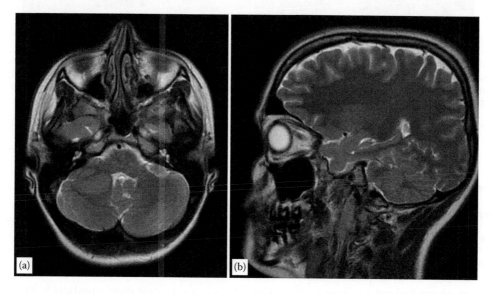

FIGURE 21.79

Cerebellar cortical dysgenesis. (a) Axial T_2-weighted and (b) parasagittal T_2-weighted images show abnormal cerebellar cortex of the right cerebellar hemisphere with abnormally extending fissures.

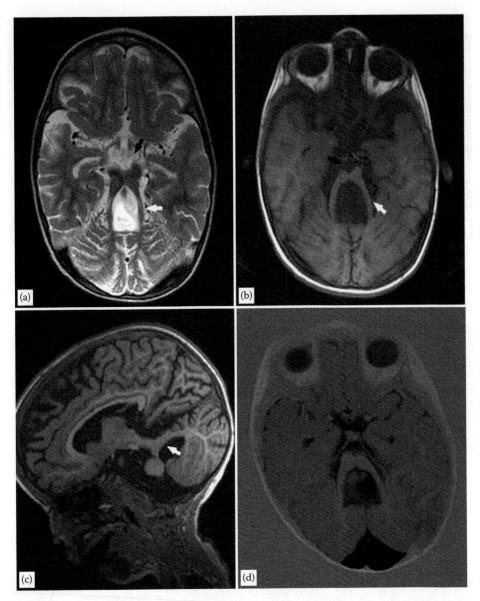

FIGURE 21.80
Molar tooth syndrome. (a) Axial T_2-weighted and (b) axial T_1-weighted images show large superior cerebellar peduncles (white arrows) and a deep interpeduncular fossa (black arrows). The midbrain has a *molar tooth* appearance. The fourth ventricle is typically deformed. (c) Parasagittal T_1-weighted image shows a large superior cerebellar peduncle perpendicular to the pons (white arrow). (d) Axial T_1-weighted image shows large superior cerebellar peduncles and a typical *bat-wing* appearance of the fourth ventricle.

21.19 Malformations of the Pons and Medulla

21.19.1 Pontine Tegmental Cap Dysplasia

Pontine tegmental cap dysplasia (PTCD) belongs to hindbrain malformations and its pathogenesis is still unknown. PTCD is characterized by flattening of the ventral pons, beak-shaped (tegmental cap) of the dorsal pons, hypoplasia of the middle cerebellar peduncles, and abnormal shape and orientation of the superior cerebellar peduncles. Cerebellar vermis hypoplasia and alteration of the inferior olivary nucleus can been seen as well (Figure 21.88).

21.19.2 Medullary Cap Dysplasia

This is a very rare malformation, with hypoplastic pons and thickened medulla oblongata. Anomalies of the corpus callosum are noted as well (Figure 21.89).

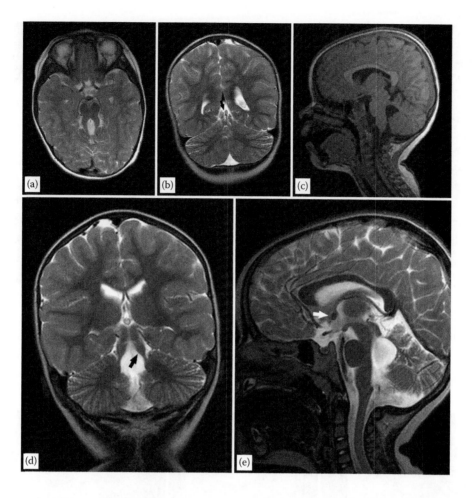

FIGURE 20.81
Molar tooth syndrome. (a) Axial T_2-weighted image shows a *molar tooth* appearance of the midbrain with a deep interpeduncular fossa. (b) Coronal T_2-weighted image shows a cleft of the dysplastic vermis (black arrow). (c) Parasagittal T_1-weighted image shows a large superior cerebellar peduncle perpendicular to the brain stem. (d) Coronal T_2-weighted image shows large superior cerebellar peduncles (black arrow). (e) Sagittal T_2-weighted image shows a narrow isthmus. Note the thick anterior commissure (white arrow).

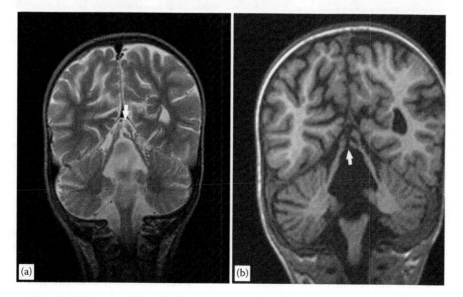

FIGURE 21.82
Molar tooth syndrome. (a) Coronal T_2-weighted and (b) coronal T_1-weighted images show the vermian midline cleft (white arrow) and large superior cerebellar peduncles.

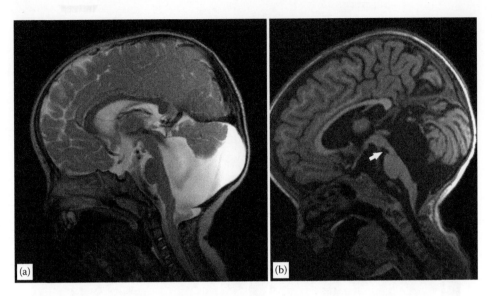

FIGURE 21.83
Molar tooth syndrome. (a) Sagittal T_2-weighted and (b) sagittal T_1-weighted images show a very narrow isthmus (white arrow) and severe dysplasia of the cerebellum.

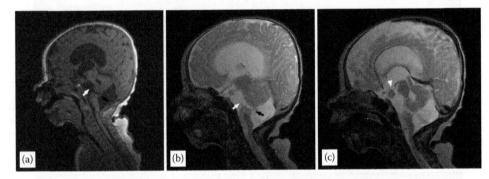

FIGURE 21.84
Brain stem disconnection. (a) Sagittal T_1-weighted and (b,c) sagittal T_2-weighted images demonstrated absent pons, hypoplastic cerebellum and vermis, and a malpositioned fourth ventricle. Thin cords connecting the midbrain and the medulla (white arrow) oblongata as well as the midbrain and the cerebellar hemisphere (black arrow). A large hamartoma is also present (arrowhead).

21.20　Chiari Malformations

Chiari malformations refer to a spectrum of congenital hindbrain abnormalities.

> *Chiari I* is a malformation characterized by herniation of the cerebellar tonsils through the foramen magnum into the cervical spinal canal (Figure 21.90). Asymptomatic tonsillar ectopia is of 5 mm in adults and 6 mm in children between 5 and 15 years of age. The posterior fossa may be small.

Syringohydromyelia of the spinal cord is often associated with Chiari I malformation. Additional abnormalities include basilar invagination, short and horizontal clivus, and vertical straight sinus.

> *Chiari II* is a complex malformation including infra- and supratentorial anomalies with myelomeningocele (Figures 21.91 and 21.92). The posterior fossa is small with low position of the torcula; narrow and elongated fourth ventricle can be moved inferiorly into the spinal canal; the brain stem is elongated and displaced caudally, like the vermis; the tectum is deformed

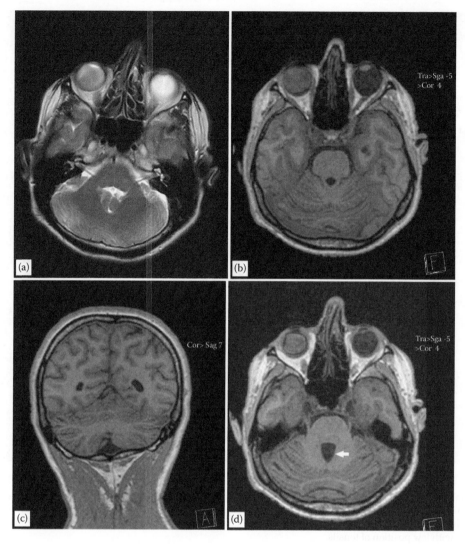

FIGURE 21.85
Rhombencephalosynapsis. (a) Axial T_2-weighted, (b) T_1-weighted, and (c) coronal T_1-weighted images show fusion of the cerebellar hemispheres with continuity of the cerebellar folia and the white matter. The vermis is absent. Note bilateral ocular abnormalities. (d) Axial T_1-weighted image shows a typical *teardrop* shape of the fourth ventricle (white arrow) and fusion of the dentate nuclei and cerebellar peduncles.

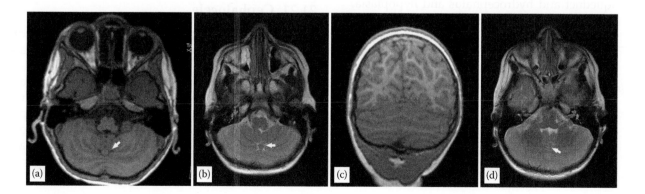

FIGURE 21.86
Partial rhombencephalosynapsis. (a) Axial T_1-weighted, (b) axial T_2-weighted, and (c) coronal T_1-weighted images show continuity of the white matter of the cerebellar hemispheres. The anterior vermis is absent. A residual uvula is present (white arrow). (d) Axial T_2-weighted image shows a small uvula in another patient (white arrow).

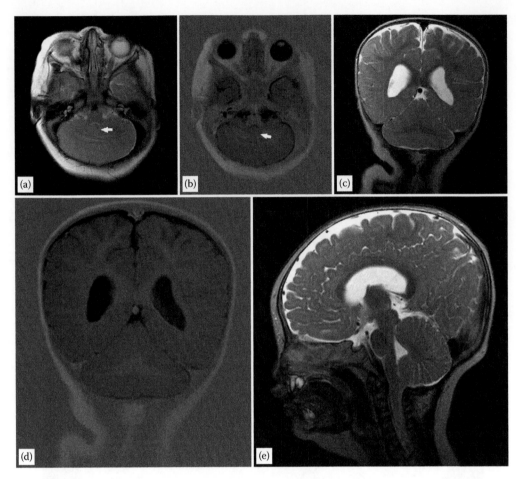

FIGURE 21.87
Partial rhombencephalosynapsis. (a) Axial T_2-weighted, (b) axial T_1-weighted, (c) coronal T_2-weighted, and (d) coronal T_1-weighted images show fusion of the cerebellar hemispheres with continuity of the cerebellar folia and the white matter. The vermis is absent, only a residual nodulus is present (white arrows). Note a typical shape of the cerebellar hemispheres. (e) Parasagittal T_2-weighted image shows an abnormal cerebellar hemisphere with low position of tonsils.

and stretched posteriorly; and stenosis of the aqueduct and hydrocephalus and hypoplasia/dysgenesis of the corpus callosum are very often seen. The severity of these malformations may be variable. In the spinal cord, syringohydromyelia can be seen.

Chiari III is a very uncommon malformation, accompanied by a cephalocele (which contains the whole or a part of the dislocated cerebellum or the occipital lobe or both mentioned structures of the brain) herniating through a C1–C2 defect and signs typical for Chiari II.

21.21 Cephalocele

Cephalocele is a congenital defect in the skull and dura resulting in herniation of intracranial contents extracranially. Based on their contents, cephaloceles are classified into the following types: meningoencephalocele (contains the brain, meninges, and cerebrospinal fluid), meningocele (contains the meninges and cerebrospinal fluid), atretic cephalocele (forme fruste of cephalocele containing a small nodule of meninges, degenerative brain tissue, and fibrous–fatty tissue), and gliocele

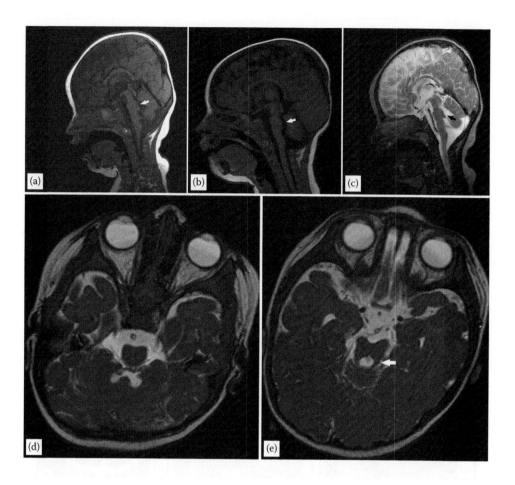

FIGURE 21.88
Pontine tegmental cap dysplasia. (a,b) Sagittal T_1-weighted and (c) sagittal T_2-weighted images show flattening of the ventral pons and an abnormal cap protruding into the fourth ventricle (white arrows). Hypoplasia of the vermis is also visible. (d) Axial T_1-weighted image shows the absence of the middle cerebellar peduncles. (e) Axial T_2-weighted image shows slightly widened superior cerebellar peduncles (white arrow).

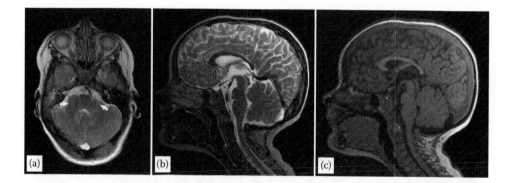

FIGURE 21.89
Medullary cap dysplasia. (a) Axial T_2-weighted image shows an abnormal rounded cap over the medulla oblongata (white arrows). (b) Sagittal T_2-weighted and (c) sagittal T_1-weighted images show a small pons with an abnormal anterior outline and a thickened medulla oblongata. An oversized and dysplastic cerebellar vermis is visible. Hypogenesis of the splenium of the corpus callosum is also seen.

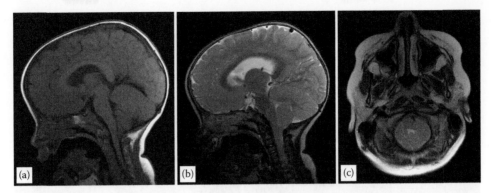

FIGURE 21.90
Chiari I malformation. (a) Sagittal T_1-weighted and (b) parasagittal T_2-weighted images show a low position of the cerebellar tonsils displaced inferiorly through the foramen magnum into the upper cervical spinal canal. The corpus callosum is normal. A small and slightly elongated fourth ventricle is seen. (c) Axial T_2-weighted image shows tonsils engulfing the medulla oblongata.

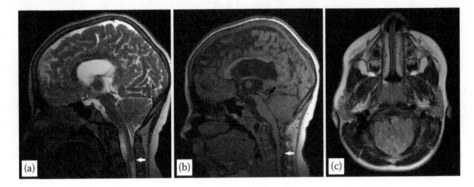

FIGURE 21.91
Chiari II malformation. (a) Sagittal T_2-weighted and (b) sagittal T_1-weighted images show a small posterior fossa and elongated cerebellar tonsils moving into the cervical canal up to the C3 level (white arrows). The fourth ventricle is narrowed and elongated. The brain stem is compressed. An abnormally shaped tectum is elongated posteriorly. Hypoplasia of the corpus callosum is seen. (c) Axial T_2-weighted image shows the cerebellum engulfing the medulla oblongata and not the craniocervical subarachnoid space at this level.

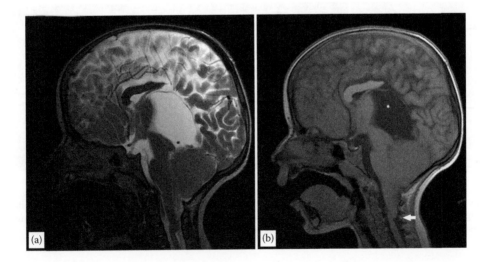

FIGURE 21.92
Chiari II malformation. (a) Sagittal T_2-weighted and (b) sagittal T_1-weighted images show a small posterior fossa. The cerebellar tonsils are moved down into the cervical canal to the C3 level (white arrows). The brain stem is compressed. The posterior portion of the corpus callosum is hypoplastic (not the splenium). The arachnoid cyst of the quadrigeminal cistern is visible (white and black stars).

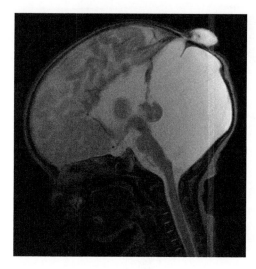

FIGURE 21.93
Parietal cephalocele in a patient with DWM. Sagittal T_2-weighted image shows a small meningocele. The posterior fossa and the fourth ventricle are enlarged, and the vermis is hypoplastic—features typical for DWM. The tentorium is elevated above the lambdoid suture.

(contains the cerebrospinal fluid and is lined by glial tissue). Additionally, meningoencephalocystocele is included—it contains the cerebrospinal fluid, brain, and ventricles.

Based on bone defects through which the structures herniate, they are classified as occipitocervical, occipital, parietal (Figure 21.93), frontal (Figure 21.94), temporal, frontoethmoidal (Figure 21.95), sphenomaxillar, spheno-orbital, nasopharyngeal, and lateral. The occipital cephalocele is the most common one.

The causes are unknown. They can be a result of several developmental disorders; environmental factors are reported as well. They may be isolated anomalies or may constitute a part of genetic syndromes (Figure 21.96).

Occipital cephalocele: It is located between the foramen magnum and the lambda (Figure 21.97). Sometimes supra- and infratentorial

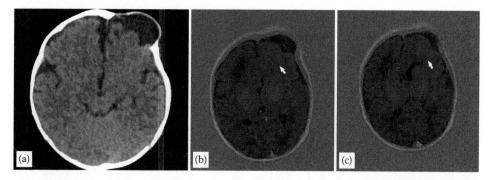

FIGURE 21.94
Frontal cephalocele. (a) Axial computed tomography image shows a large left frontal bone defect and meningocele. (b,c) Axial T_1-weighted images show the meningocele and dysplasia of the gray matter in the left frontal lobe (white arrows).

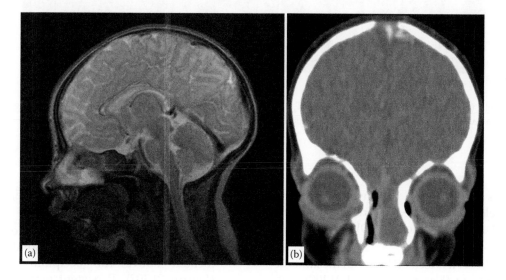

FIGURE 21.95
Frontonasal cephalocele. (a) Sagittal T_2-weighted image shows a large mass containing meninges and cerebrospinal fluid, herniated through a skull base defect. (b) Coronal reformatted CT image shows ethmoid bone cephalocele.

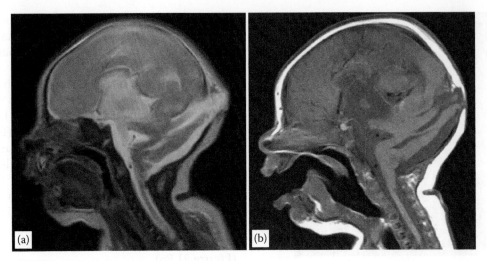

FIGURE 21.96
Occipital cephalocele in a patient with the Walker–Warburg syndrome. (a) Sagittal T_2-weighted and (b) sagittal T_1-weighted images show a small defect in the occipital bone with posterior displacement of infratentorial brain tissue, stretched toward the bone defect. Features typical for the Walker–Warburg syndrome are present: callosal dysgenesis, pontine hypoplasia with a kink of the mesencephalic–pontine junction, and cerebellar hypoplasia.

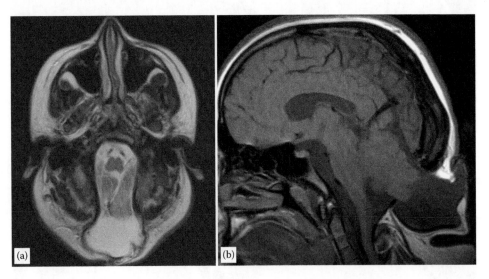

FIGURE 21.97
Occipital cephalocele. (a) Axial T_2-weighted and (b) sagittal T_1-weighted images show the occipital cephalocele containing the meninges, the cerebral fluid, and a part of cerebellar hemispheres. A large occipital bone defect is seen.

structures are involved. It can be associated with myelomeningocele, Chiari II, Chiari III, Klippel–Feil syndrome, or DWM (Figures 21.98 and 21.99).

Parietal cephalocele: It is located between the lambda and the bregma (Figures 21.93 and 21.100). It is often associated with different anomalies such as DWM, Walker–Warburg syndrome (Figure 21.101) or lissencephaly, lobar HPE, or agenesis of the corpus callosum.

Atretic cephalocele: It is a small lesion that contains mostly meninges and glial rests. It is located above the occipital protuberance (Figure 21.102). Persistent embryonic falcine sinus is often seen without a normal straight sinus. It can be isolated or associated with

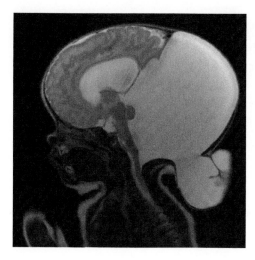

FIGURE 21.98
Occipital cephalocele in a patient with DWM. Sagittal T_2-weighted image shows a large, multinodular meningocele with a large occipital bone defect.

other brain malformations such as Walker–Warburg syndrome or lissencephaly, lobar HPE, corpus callosum hypogenesis, and gray matter heterotopia.

Transalar sphenoidal cephalocele: It belongs to basal cephaloceles (such as transphneoid, spheno-maxillar, spheno-orbital, transethmoidal, and sphenoethmoidal) (Figure 21.103).

21.22 Intracranial Lipoma

Intracranial lipoma is a congenital malformation which is a result of abnormal differentiation of embryologic meninx primitiva. The majority of intracranial lipomas are located at or near the midline (Figures 21.104 and 21.105), mostly as interhemispheric lipomas in the

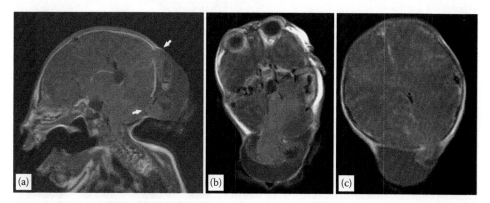

FIGURE 21.99
Occipital and parietal cephaloceles. (a) Sagittal T_1-weighted and (b,c) axial T_1-weighted images show double enecephaloceles, both containing brain parenchyma (white arrows).

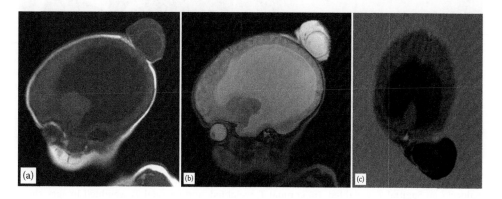

FIGURE 21.100
Parietal cephalocele. (a) Parasagittal T_1-weighted and (b) T_2-weighted images show a multinodular, cystic mass without brain tissue. (c) Axial T_1-weighted image shows ventriculomegaly and no septum pellucidum.

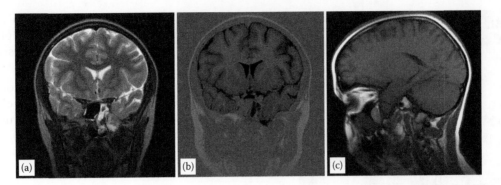

FIGURE 21.101
Transalar sphenoidal cephalocele. (a) Coronal T_2-weighted, (b) coronal T_1-weighted, and (c) sagittal T_2-weighted images show herniation of the left temporal lobe through a defect of the left greater wing of the sphenoid bone.

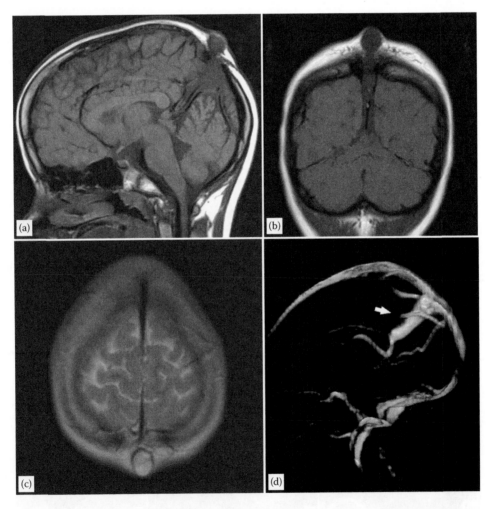

FIGURE 21.102
Atretic parietal cephalocele. (a) Sagittal T_1-weighted, (b) coronal T_1-weighted, and (c) axial T_2-weighted images show atretic cephalocele communicating with the interhemispheric fissure. (d) Magnetic resonance venogram shows persistent embryonic falcine sinus (white arrow). A normal straight sinus is absent.

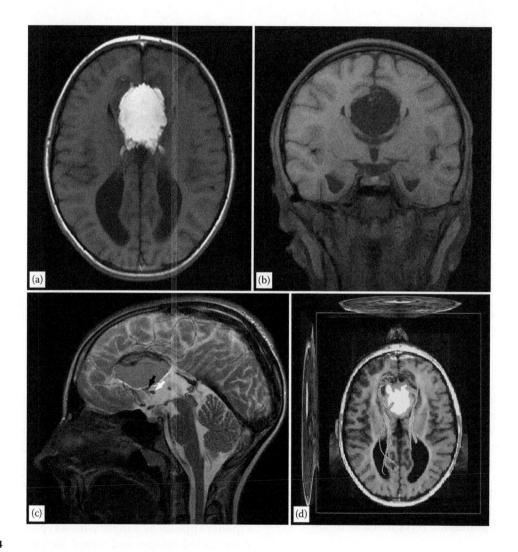

FIGURE 21.103
Transsphenoidal cephalocele. (a) Sagittal T_2-weighted image shows an abnormal shape and enlarged empty sella turcica filled with the cerebrospinal fluid. The stalk is seen in the midline, extending for a long distance, that is, from the hypothalamus to the sella turcica, but the pituitary gland is not visible (white arrow). Note the agenesis of the corpus callosum. (b) coronal T_2-weighted and (c) coronal reformatted CT images show a defect of the sphenoid bone.

FIGURE 21.104
Lipoma—tubulonodular type. (a) Axial T_1-weighted image shows a large, midline, interhemispheric lipoma, located anteriorly between the frontal horns of the lateral ventricle. A typical agenesis of the corpus callosum is seen with enlargement of the atria of the lateral ventricles—colpocephaly. (b) Coronal T_1-weighted image with fat saturation. Fat signal is suppressed. (c) Sagittal T_2-weigthed image shows a large anterior lipoma with linear low signals of the branches of the anterior arteries inside the lipoma. The anterior commissure (white arrow) and fornix (black arrow) are seen. The corpus callosum is not present. (d) Diffusion tensor imaging–tractography shows the Probst bundles.

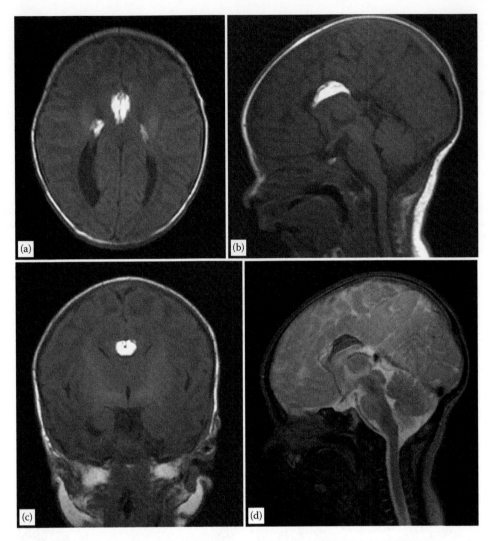

FIGURE 21.105
Lipoma—tubulonodular type. (a) Axial T_1-weighted, (b) sagittal T_1-weighted, (c) coronal T_1-weighted, and (d) sagittal T_2-weighted images show an anterior, interhemispheric lipoma. Bilateral small lipomas are also seen in the lateral ventricles. A hypogenetic and unmyelinated corpus callosum is poorly visible.

pericallosal cistern of different size. Other locations include quadrigeminal (Figure 21.106), suprasellar (Figure 21.107), cerebellopontine angle, and sylvian fissure cisterns (Figure 21.108).

They may be sporadic or associated with brain anomalies, mostly with defects of the corpus callosum (Figures 21.104 and 21.105). Rarely, they can be associated with congenital neurocutaneous syndromes.

Two types of lipomas are recognized as follows:

Tubulonodular type—A more severe form, located predominantly anteriorly. It is a bulky nodular mass, larger than 2 cm in size and often associated with corpus callosum and/or other encephalic anomalies (Figures 21.104 and 21.105).

Curvilinear type—Mostly located posteriorly, around the splenium, typically asymptomatic. It is thin and rarely associated with corpus callosum and/or parenchymal anomalies (Figure 21.109).

Lipoma appears as a mass with a typical high signal on T_1-weighted images and signal suppression on fat-saturated images. The chemical shift artifact may be seen around the mass. Lipoma does not enhance. Vascular structures and flow void may be occasionally seen inside it.

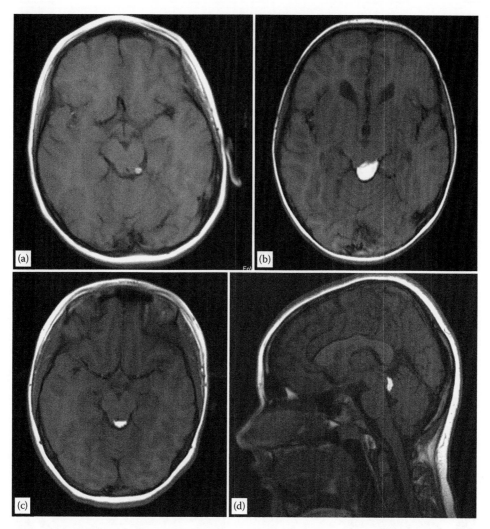

FIGURE 21.106
Lipoma of the quadrigeminal plate. (a–c) Axial and (d) sagittal T_1-weighted images show examples of quadrigeminal plate lipomas.

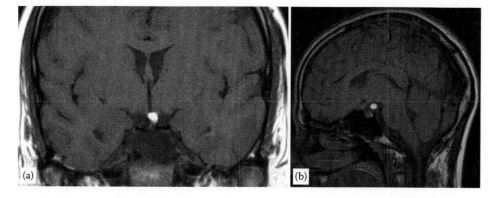

FIGURE 21.107
Lipoma. (a) Coronal T_1-weighted image shows a small hypothalamic lipoma. (b) Sagittal T_1-weighted image shows an interpeduncular cistern lipoma.

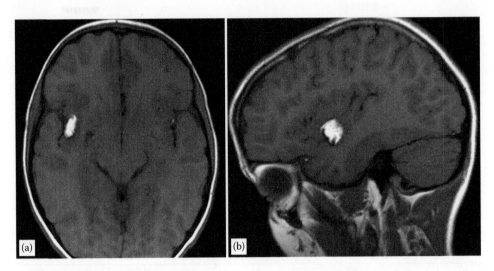

FIGURE 21.108
Lipoma. (a) Axial T_1-weighted and (b) parasagittal T_1-weighted images show lateral fissure lipomas.

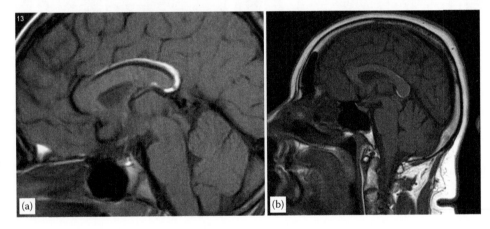

FIGURE 21.109
Lipoma—curvilinear type. (a,b) Sagittal T_1-weighted images show two examples of posterior interhemispheric lipoma. The splenium of the corpus callosum is hypoplastic on (a). On image (b), the corpus callosum is fully formed.

21.23 Arachnoid Cysts

Intracranial arachnoid cysts are congenital developmental anomalies, located within the intraarachnoid space (Figures 21.110 through 21.118). They can occur anywhere, mostly supratentorially (most frequently in the middle cranial fossa—within the sylvian fissure; Figures 21.110 through 21.112). When located infratentorially, the cerebellopontine angle is preferred (Figure 21.118).

Other locations are cisterna magna (Figure 21.117), quadrigeminal cistern, and choroidal fissure.

Arachnoid cysts do not communicate with the intraventricular system and they usually contain fluid resembling the cerebrospinal fluid.

On magnetic resonance images, they reveal the same intensity with the cerebrospinal fluid in all sequences, also in diffusion-weighted images (differentiation with epidermoid cyst). Sometimes scalloping of the adjacent calvarium may be seen (Figure 21.115).

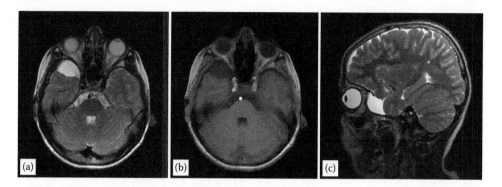

FIGURE 21.110

Temporal arachnoid cyst. (a) Axial T_2-weighted, (b) axial T_1-weighted, and (c) parasagittal T_2-weighted images show a small, middle cranial fossa arachnoid cyst (galassi type 1).

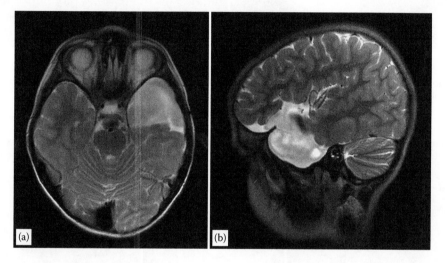

FIGURE 21.111

Temporal arachnoid cyst. (a) Axial and (b) parasagittal T_2-weighted images show a left-sided middle cranial fossa large cyst with displacement of the anterior portion of the temporal lobe (galassi type 2).

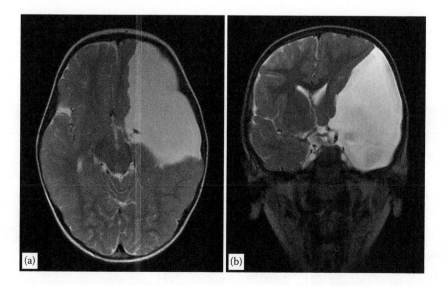

FIGURE 21.112

Frontotemporal arachnoid cyst. (a) Axial T_2-weighted and (b) coronal T_2-weighted images show a very large arachnoid cyst located in the middle and anterior cranial fossas (galassi type 3). The left hemisphere is compressed with partial hypoplasia of the frontal and temporal lobes. Mass effects with midline shift are also present.

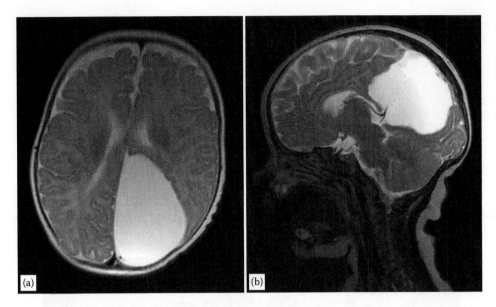

FIGURE 21.113
Parieto-occipital arachnoid cyst. (a) Axial T_2-weighted and (b) parasagittal T_2-weighted images show a parieto-occipital arachnoid cyst with displacement and hypoplasia of a part of parietal and occipital lobes.

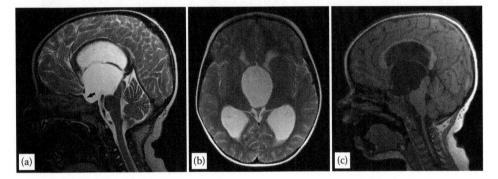

FIGURE 21.114
Suprasellar arachnoid cyst. (a) Sagittal T_2-weighted, (b) axial T_2-weighted, and (c) sagittal T_1-weighted images show a large suprasellar cyst displaced posteriorly the midbrain and anteriorly the pituitary stalk (black arrow). Enlargement of the lateral ventricles, and especially the occipital horns, is seen.

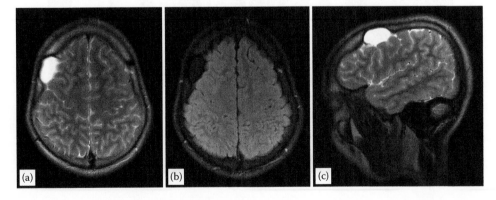

FIGURE 21.115
Frontal arachnoid cyst. (a) Axial T_2-weighted, (b) axial FLAIR, and (c) parasagittal T_2-weighted images show a small right frontal arachnoid cyst causing scalloping of the inner table of the frontal bone.

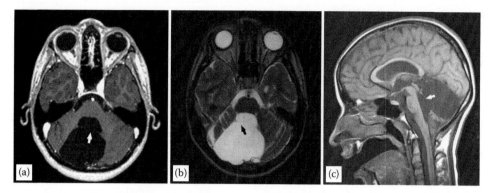

FIGURE 21.116
Posterior fossa arachnoid cyst. (a) Axial T_1-weighted, (b) axial T_2-weighted, and (c) sagittal T_1-weighted images show a large posterior fossa cyst without communication with the fourth ventricle; a thin membrane is present (white and black arrows). Hypoplasia of the vermis is seen.

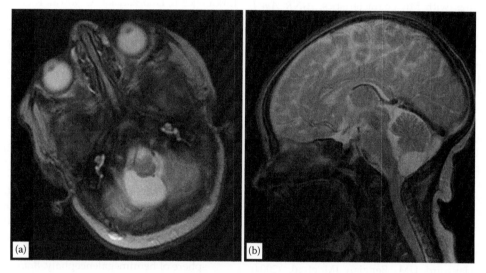

FIGURE 21.117
Posterior fossa, infracerebellar arachnoid cyst. (a) Axial T_2-weighted and (b) sagittal T_2-weighted images show a small infravermian arachnoid cyst with small hypoplasia, laterally displacing the inferomedial portion of the right cerebellar hemisphere.

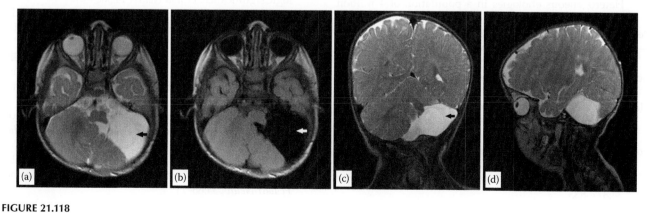

FIGURE 21.118
Posterior fossa arachnoid cyst. (a) Axial T_2-weighted, (b) axial FLAIR, (c) coronal T_2-weighted, and (d) parasagittal T_2-weighted images show a cerebellopontine angle arachnoid cyst (white and black arrows) located in the left widening of cerebrospinal fluid collection connected with the fourth ventricle. The left cerebellar hemisphere and vermis are hypoplastic.

Bibliography

Abdel Razek AA, Kandell AY, Elsorogy LG et al. (2009) Disorders of cortical formation: MR imaging features. *AJNR Am J Neuroradiol* 30(1):4–11.

Andrade CS, da Costa Leite C (2011) Malformations of cortical development. Current concepts and advanced neuroimaging review. *Arq Neuropsiquaiatr* 69(1):130–138.

Bakshi R, Shaikh ZA, Kamran S et al. (1999) MRI findings in 32 consecutive lipomas using conventional and advanced sequences. *J Neuroimaging* 9:134–140.

Barkovich AJ (2010) Current concepts of polymicrogyria. *Neuroradiology* 52:479–487.

Barkovich AJ, Guerrini R, Kuzniecky RI et al. (2012) A developmental and genetic classification for malformations of cortical development: Update 2012. *Brain* 135:1348–1369.

Barkovich AJ, Kjos BO (1992) Schizencephaly: Correlation of clinical findings with MR characteristics. *AJNR Am J Neuroradiol* 13:85–94.

Barkovich AJ, Kuzniecky RI, Bollen AW et al. (1997) Focal transmantle dysplasia: A specific malformation of cortical development. *Neurology* 49:1148–1152.

Barkovich AJ, Millen KJ, Dobyns WB (2009) A developmental and genetic classification for midbrain-hindbrain malformations. *Brain* 132:3199–3230.

Barth PG (2012) Rhombencephalosynapsis: New findings in a larger study. *Brain* 135:1346–1347.

Barth PG, Majoie CB, Caan MWA et al. (2007) Pontine tegmental cap dysplasia: A novel brain malformation with a defect in axonal guidance. *Brain* 130(9):2258–2266.

Bonneville F, Savatovsky J, Chiras J (2007) Imaging of cerebellopontine angle lesions: An update. Part 2: Intraaxial lesions, skull base lesions that may invade the CPA region, and non-enhancing extra-axial lesions. *Eur Radiol* 17:2908–2920.

Broumandi DD, Hayward UM, Benzian JM et al. (2004) Hemimegalencephaly 1. *RadioGraphics* 24:843–848.

Brunelli S, Faiella A, Capra V et al. (1996) Germline mutations in the homebox gene EMX2 in patients with severe schizencephaly. *Nat Genet* 12:94–96.

Curry CJ, Lammer EJ, Nelson V et al. (2005) Schizencephaly: Heterogeneous etiologies in a population of 4 million California births. *Am J Med Genet A* 137:181–189.

Denis D, Chateil JF, Brun M et al. (2000) Schizencephaly: Clinical and imaging features in 30 infantile cases. *Brain Dev* 22:475–483.

Flores-Sarnat L (2002) Hemimegalencephaly. I. Genetic, clinical, and imaging aspects. *J Child Neurol* 17:373–384.

Gleeson JG, Keler LC, Parisi MA et al. (2004) Molar tooth sign of the midbrain-hindbrain junction: Occurrence in multiple distinct syndromes. *Am J Med Genet* 125:125–134.

Gonzales G, Vedolin L, Barry B et al. (2013) Location of periventricular nodular heterotopia is related to the malformation phenotype on MRI. *AJNR Am J Neuroradiol* 34:877–883.

Grinberg I, Northrup H, Ardinger H et al. (2004) Heterozygous deletion of the linked genes ZIC1 and ZIC4 is involved in Dandy-Walker malformation. *Nat Genet* 36:1053–1055.

Hahn JS, Barnes PD (2010) Neuroimaging advances in holoprosencephaly: Refining the spectrum of midline malformation. *Am J Med Genet Part C Semin Med Genet* 154C:120–132.

Hahn JS, Barnes PD, Clegg NJ et al. (2010) Septopreoptic holoprosencephaly: A mild subtype associated with midline craniofacial anomalies. *AJNR Am J Neuroradiol* 31(9):1596–1601.

Ishak GE, Dempsey JC, Shaw DWW et al. (2012) Rhombencephalosynapsis: A hindbrain malformation associated with incomplete separation of midbrain and forebrain, hydrocephalus, and a broad spectrum of severity. *Brain* 135:1370–1386.

Jansen A, Andermann E (2005) Genetics of the polymicrogyria syndromes. *J Med Genet* 42:369–378.

Jurkiewicz E, Dobrzanska A, Nowak K et al. (2010) MRI findings in the young infant with brainstem disconnection and extracerebral features. Report of one case and review of the literature. *Brain Dev* 32(6):495–498.

Jurkiewicz E, Nowak K (2015) Medullary cap dysplasia. MR and diffusion tensor imaging of a hindbrain malformation. *Neurology* 84:102–103.

JurkiewiczE, Pakula-Kosciesza I, Walecki J (2007) Transalar sphenoidal encephalocele. A case report. *The Neuroradiology J* 20:200–202.

Jha VC, Kumar R, Srivastav AK et al. (2012) A case series of 12 patients with incidental asymptomatic Dandy-Walker syndrome and management. *Childs Nerv Syst* 28:861–867.

Kara S, Jissendi-Tchofo P, Barkovich AJ (2010) Developmental differences of the major forebrain commisures in lissencephalies. *AJNR Am J Neuroradiol* 31:1602–1607.

Kometani H, Sugai K, Saito Y et al. (2010) Postnatal evolution of cortical malformation in the "non-affected" hemisphere of hemimegalencephaly. *Brain Dev* 32:412–416.

Leventer RJ, Jansen A, Pilz DT et al. (2010) Clinical and imaging heterogeneity of polymicrogyria: A study of 328 patients. *Brain* 133:1415–1427.

Maria BL, Quisling RG, Rosainz LC et al. (1999) Molar tooth sign in Joubert syndrome: Clinical, radiologic and pathologic significance. *J Child Neurol* 14:368–376.

Parisi MA, Dobyns WB (2003) Human malformations of the midbrain and hindbrain: Review and proposed classification scheme. *Mol Gen Metab* 80:36–53.

Pasquier L, Marcorelles P, Loget P et al. (2008) Rhombencephalosynapsis and related anomalies: A neuropathological study of 40 fetal cases. *Acta Neuropathologica* 117:185–200.

Patel S, Barkovich AJ (2002) Analysis and classification of cerebellar malformations. *AJNR Am J Neuroradiol* 23:1074–1087.

Polizzi A, Pavone P, Ianetti P et al. (2006) Septo-optic dysplasia complex: A heterogeneous malformation syndrome. *Pediatr Neurol* 34:66–71.

Romano S, Boddaert N, Desguerre I et al. (2006) Molar tooth sign and superior vermian dysplasia: A radiological, clinical and genetic study. *Neuropediatrics* 37:42–45.

Sato N, Yagishita A, Oba H et al. (2007) Hemimegalencephaly: A study of abnormalities occurring outside the involved hemisphere. *J Neuroradiol* 28:678–682.

Szczaluba K, Szymanska K, Bekiesinska-Figatowska M et al. (2010) Pontine tegmental cap dysplasia A hindbrain malformation caused by defective neuronal migration. *Neurology* 74:1835.

Takanashi J, Barkovich A (2003) The changing MR imaging appearance of polymicrogyria: A consequence of myelination. *AJNR Am J Neuroradiol* 24:788–793.

Truwit CL, Barkovich AJ (1990) Pathogenesis of intracranial lipoma: An MR study in 42 patients. *AJR Am J Roentgenol* 155:855–864.

Verity C, Firth H, Ffrench-Constant C (2003) Congenital abnormalities of the central nervous system. *J Neurol Neurosurg Psychiatry* 74(Suppl I):i3–i8.

Verrotti A, Spalice A, Yrsitti F et al. (2010) New trends in neuronal migration disorders. *Eur J Paediatr Neurol* 14(1):1–12.

Yildiz H, Hakyemez B, Koroglu M et al. (2006) Intracranial lipomas: Importance of localization. *Neuroradiology* 48:1–7.

Zaki MS, Saleem SN, Dobyns WB et al. (2012) Diencephalic-mesencephalic junction dysplasia a novel recessive brain malformation. *Brain* 135:2416–2427.

22

Infectious Diseases of the Brain

John H. Rees and James G. Smirniotopoulos

CONTENTS

22.1 Introduction

Since the dawn of time, different life forms have competed for the resources they need to sustain and perpetuate their existence. The struggle began when unicellular organisms developed or evolved from complex molecular networks. As one-celled organisms became more complex and more specialized, multicellular organisms arose [1]. This process eventually led to the array of life forms that we see on the planet today. The human organism is a complex multicellular structure with many strengths and potentials; however, with great strength and greater complexity also comes many vulnerabilities.

We live a peaceful symbiosis with a vast array of micro- and some macroorganisms which populate our skin, alimentary tract, and respiratory tract, and are present in the air we breathe and in our food and water.

In order for these symbiotic relationships to turn pathologic, special circumstances or preconditions must exist (see Table 22.1). There are a wide variety of infectious organisms that can invade the body and more specifically the central nervous system (CNS) (see Table 22.2). Infections of the CNS are unique in several ways. The brain and spine are protected from exposure and invasion by a number of anatomic factors,

including the bony calvarium, the vertebral column, and also the leptomeningeal membranes, including the dura mater. On a microscopic level, the CNS is immunologically cloistered and distinct from all other tissue due to the presence of the blood–brain barrier, and this results in both protections and limitations in preventing and fighting CNS infection [2]. Regardless of the infectious agent, brain infections display a number of typical anatomic and pathogenetic patterns and common pathways (see Table 22.3). Similarly there are recurrent patterns of infection in the spine (see Table 22.4).

To complicate matters further, infectious disease is dynamic and continually reinvents itself by mutation, evolution, geographic migration, host translation, and ever-changing patterns of resistance to our attempts at therapy.

In addition, more recent study suggests that infectious agents may use molecular signals to control the behavior of their hosts to their own benefit [3]. Iridovirus make crickets increase sexual activity resulting in increased transmission and host availability [4]. The parasite *Toxoplasma gondii* (Toxo) is believed to influence host behavior in several ways, including making rats less afraid of cats, possibly creating a fatal attraction, and increasing their transmission (manipulation hypothesis) [5–7]. There is a fungus that induces carpenter ants to leave their colony

TABLE 22.1

Preconditions for Infection

1—Breach of barrier: surgical, medical, or traumatic
2—Immune dysfunction: AIDS, steroids, chemotherapy, and genetic
3—Change in micro- or macroenvironment
4—Loss of natural competitive regulatory factors
5—Evolution or mutation in the organism resulting in increased propensity for pathogenicity: HIV, other viruses, and bacterial resistance
6—Weaponization

TABLE 22.2

Vectors of CNS Infection

I—Bacterial
II—Viral
III—Fungal
IV—Parasitic
V—Transmissible protein or prion

TABLE 22.3

Patterns of Brain Infection: From Peripheral to Central

Neuritis—peripheral nerve
Meningitis—coverings of the brain
Empyema—surface collection
Ventriculitis—inner surface of the brain
Cerebritis—infection within brain parenchyma
Abscess—infected collection within brain parenchyma

TABLE 22.4

Patterns of Infection: Spine

Discitis/osteomyelitis—infection of disc/endplate
Epidural abscess—within the canal
Paraspinal abscess, phlegmon, or cellulitis
Meningitis
Myelitis—infection of cord parenchyma
Radiculitis/neuritis—infection of nerve roots

FIGURE 22.1
Parasitized ants hang upside down as they die in order to spread fungal spore in a host manipulation first postulated in the late 1800s but proved in 2009.

and hang upside down and die spreading fungal spores all over the ground. This last example was observed by A. R. Wallace, a contemporary of Darwin, in the late 1800s, but not till 2009 did D. Hughes prove the connection [8] (Figure 22.1). Other examples of pathogen host behavioral influence are being actively studied.

In sum, CNS infections involve many different infectious agents, some working solo and some working together, which display nonspecific but occasionally typical behaviors. We as physicians must continually educate ourselves and our patients, and use every tool at our disposal in order to detect, characterize, and treat these conditions.

In the pages that follow, we will consider imaging and pathologic examples of all the major forms of CNS infections, as well as the major patterns of infection.

22.2 Bacterial Infections

Bacterial CNS Infections may arise by several pathways, including local spread from sinus, facial, ear, or scalp infection, or hematogenously through either systemic bacteremia or septic emboli. There are a wide variety of organisms, including *Streptococcus pneumoniae* and *Staphylococcus aureus* such as methicillin-resistant *S. aureus* (MRSA) and other staphylococci, *Haemophilus influenzae*, *Neisseria meningitidis*, *Listeria monocytogenes*, and *Mycobacterium* such as tuberculosis (TB). Some of these infections have been significantly decreased due to usage of childhood vaccinations, although vaccination programs remain politically controversial for a variety of different reasons in different places.

22.2.1 Bacterial Meningitis

Case history: A 14-month-old boy with history of upper respiratory infection (URI), followed by otitis media, followed by severe febrile illness. Complete workup reveals blood culture and cerebrospinal fluid (CSF) positive for *Streptococcus pneumoniae*. Initial computed tomography (CT) before lumbar puncture (LP) showed subtle low-density areas, probably surface infarcts related to thrombosis of superficial veins. Post-gadolinium magnetic resonance imaging (MRI) shows enhancement in deep sulcal spaces and superficial cortex (Figure 22.2).

More subtle MRI findings of meningitis, seen in a different case (Figure 22.3), include indistinct or barely effaced sulci, or slightly hyperintense subarachnoid space, especially on fluid-attenuated inversion recovery (FLAIR) [9–15].

22.2.2 Bacterial Abscess: Bacteremia

Staphylococcus and *Streptococcus* species may invade the brain via bacteremia resulting in solitary or multiple brain abscesses. Contributing factors include cardiac, open fractures, dental procedures, osteomyelitis, and other conditions leading to bacteremia.

Aortic valvular vegetations may be a source of septic cerebral emboli (Figure 22.4). Incidence of rheumatic heart disease significantly decreased possibly due to aggressive treatment of streptococcal infections in the

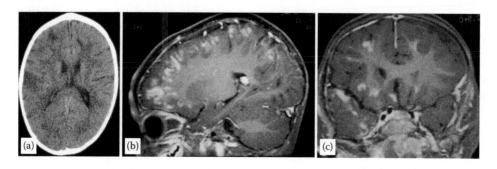

FIGURE 22.2
Bacterial meningitis. (a) Noncontrast CT image shows peripheral low-density likely surface infarcts from occlusion superficial veins. (b) Sagittal and (c) coronal T_1 post-gadolinium on the same case shows diffuse enhancement of brain surfaces, sulcal clefts.

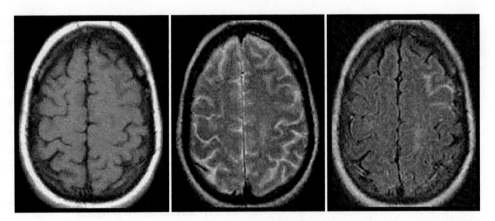

FIGURE 22.3
Bacterial meningitis. Subtle MRI findings in a different case left frontal T_1, T_2, and FLAIR.

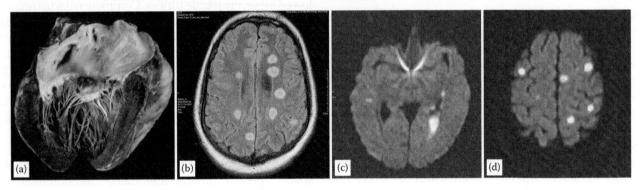

FIGURE 22.4
Septic embolic shower. (a) Cardiac valve vegetations. (b) Axial T_1 post-gadolinium shows multiple small enhancing ring lesions. (c) DWI shows multiple rounded foci of diffusion restriction. (d) DWI shows restriction indicating left choroid plexitis.

United States, and this was followed by decrease in cardiac embolic disease [16].

22.2.3 Bacterial Mycotic Aneurysm

Mycotic aneurysms are uncommon but occur when bacteremia seeds a vessel wall leading to a breakdown of normal vessel lamina and aneurysm. This case was associated with intravenous drug abuse, but it may also be seen in bacteremia of other etiologies (Figure 22.5) [17–21].

22.2.4 Bacterial Abscess: Sinusitis

Acute bacterial sinusitis may be transmitted to the brain via the valveless veins of the face resulting in meningitis, cerebritis, and ultimately abscess (Figure 22.6).

22.2.5 Bacterial Abscess: Imaging Features with Pathologic Correlation

In general, bacterial brain abscesses have a nonspecific appearance regardless of organism. They begin as a focal cerebritis that either resolves or evolves into a focal

fluid collection surrounded by a thin enhancing rim with exuberant edema (Figure 22.7). There is typically a trilaminar wall that consists of granulation tissue, collagen capsule, and then gliosis surrounded by edema. The fluid center is bright on T_2WI and also shows restricted proton diffusivity on diffusion-weighted imaging (DWI) (Figure 22.8) [22–26].

On MRI, the dark, hypointense signal intensity (SI) of the ring was previously attributed to the production of free radical molecular species and/or elemental O_2 by superoxide dismutase, an enzyme in white blood cells which is used to destroy bacteria. However, free radicals are very transient molecular species, and the hypointense wall likely represents the breakdown products of red blood cells (RBCs) or other paramagnetic materials within the inner granulation ring (Figure 22.9).

22.2.6 Bacterial Cerebritis and Ventriculitis Due to MRSA

Case history: A 73-year-old man suffered aneurysm rupture and underwent subsequent clipping, with complicated course, including hydrocephalus and subsequent shunt.

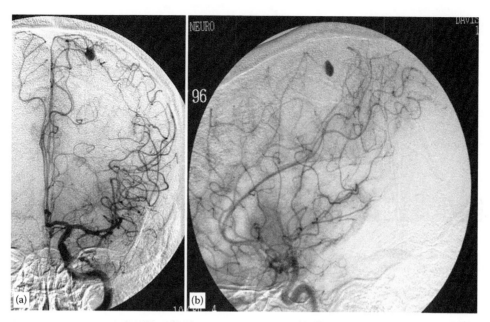

FIGURE 22.5
Bacterial mycotic aneurysm in a patient with history of intravenous drug abuse, previous bacteremia, and sepsis treated with occlusion: (a) anterior-posterior (AP); (b) lateral projections of a left internal carotid artery (ICA) injection.

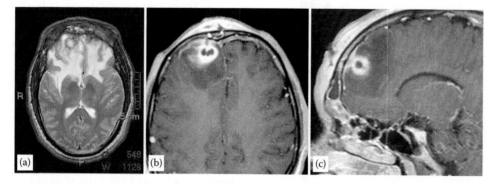

FIGURE 22.6
Brain abscess from adjacent sinus disease. (a) Axial T_2 small encapsulated fluid collection with surrounding exuberant edema. (b) Axial T_1 post-gadolinium enhancing wall. (c) Sagittal T_1 post-gadolinium with adjacent sinusitis.

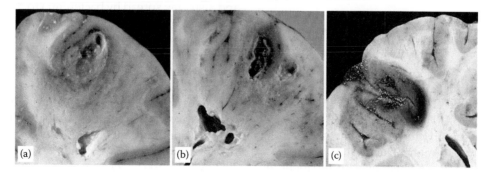

FIGURE 22.7
Gross pathologic stages of brain abscess. (a) Focal cerebritis; (b) early cavity with phlegmon; (c) mature abscess cavity.

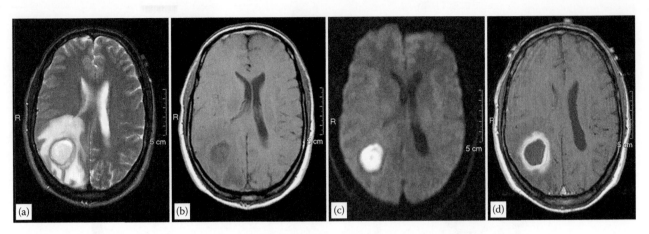

FIGURE 22.8
Bacterial brain abscess, hematogeneous. (a) Axial T_2, thin-walled fluid collection with exuberant edema. (b) Axial T_1 without gad. (c) Axial T_1 with gad, thin, smooth, densely enhancing rim. (d) DWI robust diffusion restriction in abscess contents.

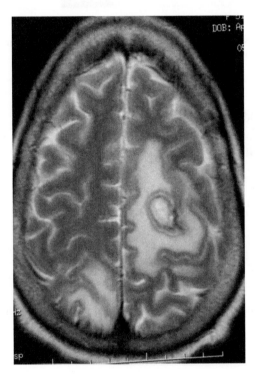

FIGURE 22.9
Trilaminar abscess wall: Inner dark layer—breakdown products; middle—granulaton tissue; outer—collagen, scar tissue.

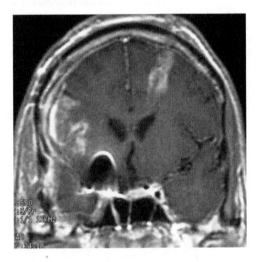

FIGURE 22.10
Coronal T_1 postcontrast MR image of an intensive care unit patient following aneurysm rupture and subsequent clipping with focal right frontal meningitis and also shunt tract infection culture-proven to be MRSA.

Subsequent clinical deterioration and imaging revealed infected shunt tract, cerebritis, and ventriculitis with positive cultures for MRSA (Figure 22.10).

MRSA is a common bacterium (*S. aureus*) which has acquired resistance to many frontline antibiotics and has become an important cause of both inpatient and community-acquired infections. It is estimated that 33% of all people carry *S. aureus* on their nasal mucosa and 2%

carry MRSA. MRSA may cause infections of the skin and also may spread by bacteremia to deeper structures, including the lung, bone, and brain. In addition, it may also colonize surgical equipment giving rise to infection in surgical equipment and catheters [27–30].

Over the past 20 years, knowledge and awareness about MRSA have led to greatly improved infection rates, particularly inpatient and postsurgical, but MRSA remains a very important pathogen. The Centers for Disease Control (CDC) documents that invasive MRSA infections in hospitals declined 54% between 2005 and 2011, with 30,800 fewer severe MRSA infections. In addition, the study showed 9000 fewer deaths in hospital patients in 2011 versus 2005 [31].

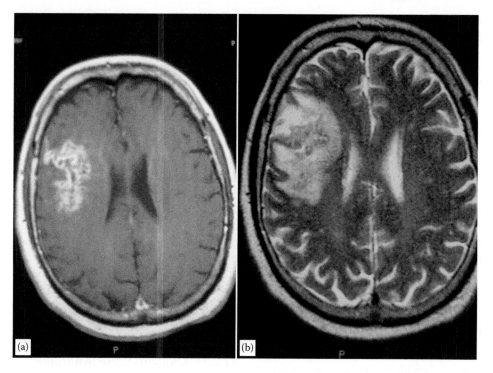

FIGURE 22.11

Listeria cerebritis. (a) Axial T_2-weighted image shows areas of mixed high signal intensity with exuberant perilesional edema. (b) Postcontrast T_1 images show irregular areas of enhancement without discrete abscess.

22.2.7 Bacterial Cerebritis Due to *Listeria*

Named after the great pioneer of antisepsis, Joseph Lister, *Listeria monocytogenes* is an uncommon but important CNS pathogen which is most commonly associated with food-borne infections of many types, including dairy products and produce. It is most dangerous to the very young, the elderly, pregnant women, and those patients whose immune systems are compromised. CNS manifestations of listeriosis include meningitis as well as focal cerebritis, and mortality rates may be as high as 20%. Although the imaging findings are nonspecific, focal cerebritis in the clinical setting of known exposure to contaminated food should increase suspicion for this condition (Figure 22.11) [32–36].

22.2.8 Bacterial Empyema and Epidural Abscess Due to Brucellosis

Case history: A 28-year-old Floridian with severe headache and fever. MRI shows subdural empyema. Neurosurgical drainage obtains purulent fluid, but no known organism is cultured at regional medical center laboratory. Three months later, patient returns with a recurrent fluid collection; repeat neurosurgical drainage was sent to the CDC for culture and found to be positive for *Brucella melitensis* (Figure 22.12).

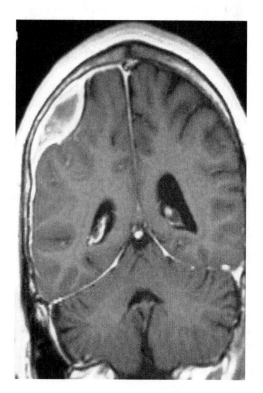

FIGURE 22.12

Brucella melitensis empyema: 1, 2—coronal T_1 post-gadolinium shows peripherally enhancing extra-axial fluid collection; 3—feral hog aka Hogzilla (endemic brucellosis host).

Additional history was obtained revealing that the patient lives in swampy central Florida and hunts and butchers feral pigs aka razorback hogs. These animals are believed to be descendants of animals brought by Spanish colonists several centuries earlier and are now indigenous from Florida to Texas and Arizona in different varieties.

Brucellosis infection is also known as undulant fever due to the pattern of fever and Malta fever due to endemic conditions on the island of the same name. It is a fastidious organism difficult to culture. Exposure is usually associated with goats and animal husbandry. Neurobrucellosis usually presents as ameningoencephalitis. Brucellosis cerebral empyema is rare in world literature, never previously reported in the United States. However, the imaging is relatively nonspecific and cerebral empyema by other organisms could look identical [37–40].

22.2.9 Tuberculosis

Mycobacterial infections are endemic throughout the world, worsened by the incidence of human immunodeficiency virus (HIV) and other forms of immune compromise, as well as by the evolution of drug-resistant strains in the United States and elsewhere, exacerbated by noncompliance and incomplete treatment. TB has been called *a forgotten epidemic* and *humanity's greatest killer.* It is a major global and U.S. health problem complicated by the financial complexities of developing, producing, administering, and paying for new drug regimens [41].

Imaging of CNS TB is complicated by the protean manifestations of acute and chronic infection in bone and soft tissue (Table 22.5) [42–45].

22.2.10 Chédiak–Higashi Syndrome

Chédiak–Higashi syndrome (CHS) is a rare autosomal recessive condition, one of whose characteristic features

TABLE 22.5

Imaging Manifestations of CNS TB

1—Tuberculoma: solitary lesion, possibly calcified, possibly dormant focus (Figure 22.13)

2—Tubercular abscess, either solitary or multiple and miliary (as in millet seed) (Figure 22.13)

3—Meningitis: diffuse enhancing basilar meningitis with purulent exudate and chronic thickening of the meninges, dense enhancement, leading to vasculitis, white matter disease, and superficial infarcts (Figure 22.14)

4—Diffuse spinal meningitis

5—Vertebral osteomyelitis and collapse (Pott's disease) leading to chronic deformity (Figure 22.15)

6—Paraspinal abscess, frequently anterior (Figure 22.15)

is impaired immune function resulting in increased number of bacterial infections. The specific defect is mutation of the lysosomal trafficking regulator in chromosome 1q42–44 [46]. Lysosome function is critical in cellular protection against and destruction of bacterial infectious agents. Other features include neurologic dysfunction and oculocutaneous albinism. Different phenotypes have been reported, but in the most severe cases death occurs during childhood.

CHS is included here to represent the category of immunodeficiency syndromes, which specifically leave the individual susceptible to bacterial infections. Other examples include B-cell defects such as selective immunoglobulin A deficiency, phagocytic cell defects other than CHS, such as chronic granulomatous disease, and many other rare conditions.

In the case presented, a child with known CHS presented with neurologic deterioration and fever, and imaging reveals multiple areas of cerebritis and abscess formation (Figure 22.16).

22.2.11 Guillain–Barré Syndrome

Guillain–Barré syndrome is a rare post infectious syndrome involving ascending motor paralysis, which may involve life-threatening respiratory failure in up to 25% of patients as well as autonomic dysfunction. Although the exact mechanism is still a topic of ongoing study, there is evidence that immune system activation by an infectious agent, most commonly *Campylobacter jejuni,* results in autoimmune injury to neural cells through the process of molecular mimicry (Figure 22.17) [47–50].

22.3 Viral CNS Infection

Viruses are critically important infectious agents which are both ubiquitous and incompletely understood. Composed primarily of genetic material, either RNA or DNA, with varying degrees of surrounding protein, and in some cases lipids, but without cell membrane or cell wall, they only partially meet the criteria to be called a living organism. They do reproduce but only within living cells pirating their host's reproductive apparatus. They do not actually possess their own metabolic apparatus but rather rely on the energy supplies of their host cells.

The role of viruses in evolution and life in general is a subject of extensive active research and speculation and can be studied elsewhere. We will present representative viruses of different types and display their typical patterns of CNS infection and in some cases their secondary effects [51].

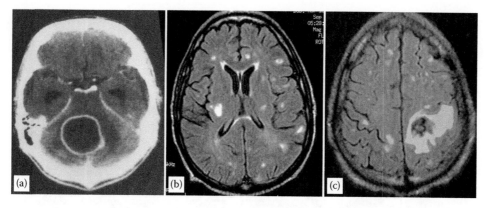

FIGURE 22.13
Parenchymal TB. (a) Axial postcontrast CT image shows large solitary posterior fossa abscess. (b,c) Axial FLAIR images show multiple small foci of miliary TB, with one area of hemorrhage left parietal lobe.

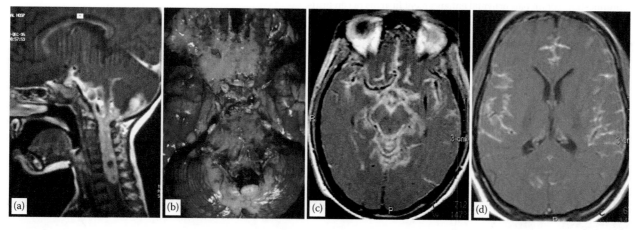

FIGURE 22.14
Tuberculous meningitis. (a) Sagittal T_1 postcontrast image shows diffuse thick enhancing basilar meningitis. (b,c) Axial T_1 postcontrast images show diffuse meningeal enhancement in bilateral cerebral hemispheres. (d) Brain autopsy with thick exudative covering of the basilar surfaces of the brain.

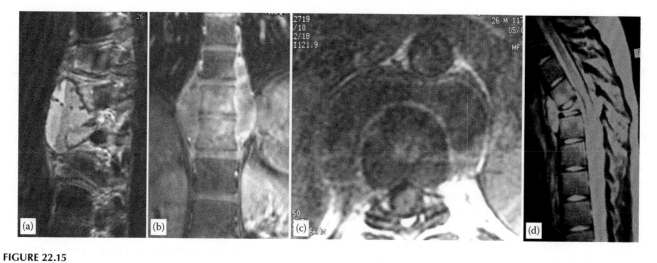

FIGURE 22.15
Spinal TB. (a) Sagittal T_2, (b) coronal T_1, and (c) axial T_1 postcontrast images demonstrate a large pre- and paravertebral tuberculous abscess encircling the upper lumbar spine. (d) Sagittal T_2 image in a different patient demonstrates acute vertebral collapse, vertebra plana, due to TB osteomyelitis.

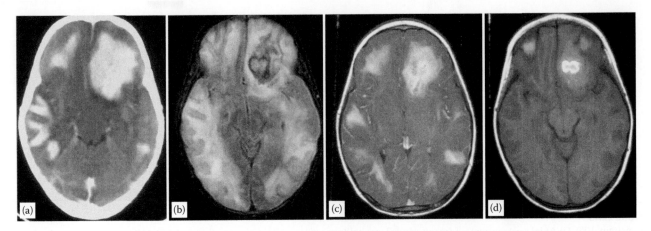

FIGURE 22.16
Encephalitis in patient with Chédiak–Higashi syndrome. (a) Contrast CT image shows extensive enhancing meningoencephalitis and cerebritis. (b) FLAIR shows hemorrhage left frontal with extensive cerebritis. (c) T_1 pre-gadolinium shows subacute hemorrhage. (d) T_1 post-gadolinium shows brain parenchymal enhancement in areas of cerebritis.

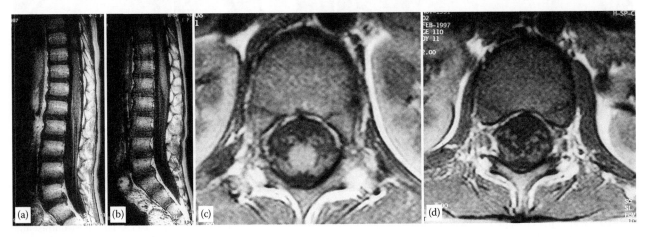

FIGURE 22.17
Guillain–Barré syndrome. (a,b) Sagittal T_1 pre- and postcontrast and (c,d) axial T_1 postcontrast images show diffusely enhancing conus and cauda equine.

22.3.1 Human Immunodeficiency Virus

HIV-1 is a very important pathogen with primary and secondary pathologic effects as well as more complex societal and cultural implications. This virus was first isolated in the early 1980s after clinical observations of fatal immune impairment, primarily in gay men, appeared to exhibit an infectious pattern of spread. Subsequent study, including sequencing and research into its mode of effect, has yielded an effective treatment strategy for controlling the virus as well as a large body of information that contains many lessons important in understanding other viruses as well. The HIV-1 virus replicates using the reverse transcriptase enzyme system, and current therapy focuses on this key step with antiretroviral therapy to control its reproduction. Ongoing research focuses on more effective ways to prevent or eradicate this globally toxic pathogen [52–57].

All living organisms undergo mutation over time, some more than others, and viruses are no exception. Study of the mutations of the HIV-1 virus has led to specific information about its geographic pattern of spread (Figure 22.18) and has contributed to our understanding of the difficulties in development of specific viral vaccines.

The pathologic effects of the HIV-1 virus include primary acute leukoencephalitis (Figure 22.19), which is often not clinically detected or imaged but leads over time to chronic encephalitis and atrophy. As the name implies, secondary infections due to immune compromise are a critical feature giving rise to the acquired immunodeficiency syndrome (AIDS) when full blown but able now to be controlled with the use of reverse transcriptase inhibitors. Many secondary pathogens exist, including Toxo and progressive multifocal leuko-encephalopathy (PML) (Figure 22.20) [58].

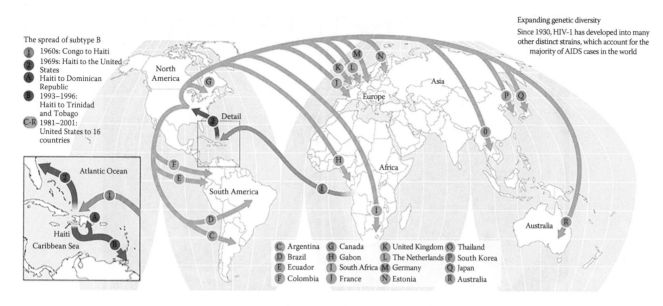

FIGURE 22.18

HIV geography. This chart documents the geographic spread of the HIV using data based on the progressive sequential mutations, from its origin in Africa, from a simian immune deficiency virus, to its clinical detection in the Caribbean, to the continental United States, Europe, and Asia.

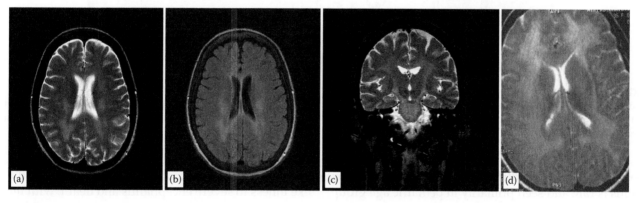

FIGURE 22.19

HIV leukoencephalitis. (a) Axial FLAIR, (b) axial T_2, and (c) coronal T_2 images show extensive white matter signal abnormality. (d) In a different patient prominent bifrontal white matter signal abnormality was biopsy proven, to be primary HIV leukoencephalitis.

22.3.2 Progressive Multifocal Leukoencephalopathy

PML is caused by a specific strain of papova virus which is a nonpathogenic pulmonary commensal but becomes pathogenic in immunocompromised conditions, including AIDS, chemotherapy for neoplasms, and other immunosuppressive therapy for immune-related conditions, including multiple sclerosis (MS).

Imaging features include nonenhancing areas of white matter edema primarily peripheral involving subcortical U-fiber bundles. Enhancement actually does occur but is transient and often not seen.

Toxoplasmosis is caused by the parasite Toxo that lives in the intestinal systems of domestic cats and their relatives. Unsporulated oocysts are shed in cat feces and are further transmitted by intermediate hosts, including birds and rodents. Humans may become infected by eating meat containing tissue cysts, consuming other food or water contaminated with oocysts, blood transfusion, or organ transplantation or transplacentally.

Imaging features of CNS toxoplasmosis are large ring-enhancing lesions in the basal ganglia or rarely in the spinal cord. See Section 22.6.3.

22.3.3 HIV Vasculitis

Less commonly, and particularly in patients who acquired the virus *in utero*, HIV-1 has led to CNS HIV vasculitis. This is a primary cerebral vasculitis,

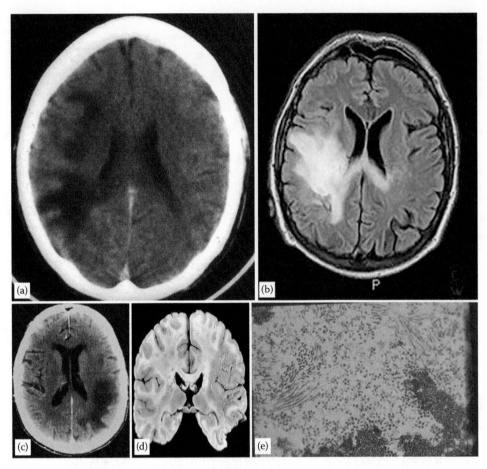

FIGURE 22.20
Progressive multifocal leukoencephalopathy (PML): (a) Contrast CT, (b) FLAIR MRI show typical nonenhancing subcortical white matter pattern. In a different patient (c) postcontrast CT and (d) brain autopsy again show typical involvement of subcortical U-fibers. (e) Scanning electron micrograph shows dark intracellular viral particles.

possibly involving multiple infectious agents. HIV vasculitis is rare and seen primarily in pediatric age group. Pathologic findings include infected mononuclear cell invasion of vessel wall with destruction of elastic lamina and subintimal fibrosis. This condition may lead to recurrent cerebrovascular accidents, or hemorrhage.

Case history: A 9-year-old child with history of congenital HIV and recurrent strokes. MRI imaging reveals multiple areas of ischemic injury, diffuse arteriomegaly, and ectasia of the entire circle of Willis and proximal branches, as well as enhancing vessel walls (Figure 22.21) [59,60].

22.3.4 West Nile Virus

West Nile virus (WNV) is an RNA flavivirus related to St. Louis encephalitis and Japanese encephalitis viruses. This virus was first isolated following the death of a washerwoman on the banks of the West Nile in Uganda in 1937, hence the name. It is spread by mosquitoes that have fed on infected birds, often

crows. Clinical symptoms vary widely from a mild flu-like illness to severe encephalitis and death, with more severe symptoms usually in older patients with less robust health.

WNV has become the most common arboviral infection affecting the CNS in the United States and increasingly in Europe as well. The exact reason and manner of its spread is uncertain but is likely related to either wild or domestic bird population movements.

Severe headache and disorientation may be early indicators of infection with a neuroinvasive subtype, and symptoms may progress to flaccid paralysis or poliomyelitis-like syndrome. The exact percentage of cases that involve the neuroinvasive subtype is difficult to calculate because so many patients may have mild to no symptoms but is likely in the range of 1% or possibly much less [61–63].

Case history: A 58-year-old diabetic woman with 10-day course beginning with upper respiratory flu-like symptoms progressing to cognitive decline and altered level of consciousness, coma, and death (Figure 22.22).

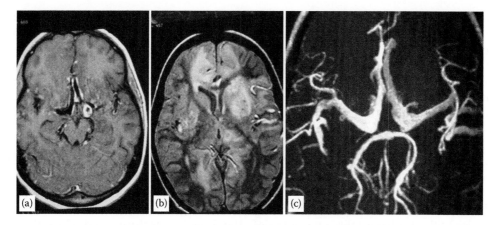

FIGURE 22.21

HIV vasculitis. (a) Axial FLAIR image shows multiple areas of ischemic injury. (b) Axial T_1 postcontrast image shows enlarged enhancing vessel walls, especially left ICA. (c) MR angiography (MRA) shows marked diffuse vascular ectasia.

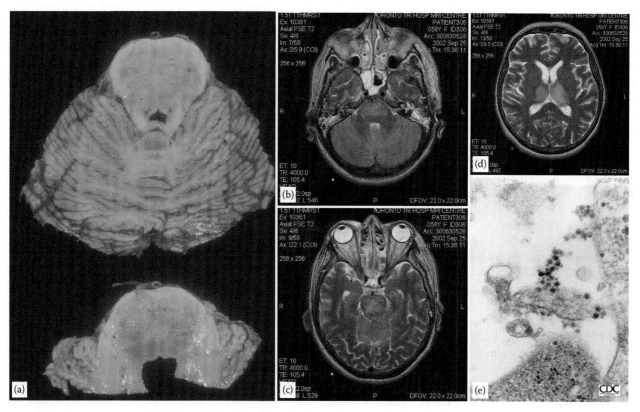

FIGURE 22.22

West Nile virus. (a–c) Axial T_2 images show sinusitis and abnormal signal in the brain stem and thalami. (d) Autopsy shows edema and necrosis in the brain stem. (e) Electron micrograph shows WNV viral particles in tissue.

22.3.5 Avian Flu—H5N1 and H7N9

Avian flu, H5N1, is a very dangerous virus which, as the name suggests, may be endemic in bird populations, both wild and domestic. Nonepidemic infection (epizootic) has been documented in non-avian animal hosts, including pigs, cats, and dogs, but luckily it does not have a mutation that readily allows human-to-human transmission with only sporadic human cases reported. This virus evolved, mutated, and traveled geographically from the Far East over 10 years to Russia and Europe in 2005. The CDC has confirmed a strain in wild ducks in Pennsylvania in 2006 and most recently in geese 2007 in Canada [64–68]. The mortality for humans infected with H5N1 may be greater than 50% based on limited data, but one key mutation could change this. Human-to-human transmission has not evolved—outside of the laboratory—but Ron Fouchier at Erasmus Medical Center developed a mutant that is contagious among mammals, specifically ferrets, and could possibly become contagious among humans. After his original work was publicized, there was discussion and great concern. Publication and research were halted temporarily but now have begun again under carefully controlled conditions [69,70]. Dr. Fouchier described the strain he developed as "probably one of the most dangerous viruses you can make."

A similar and likely related virus H7N9 was isolated in China in 2013, also with a high mortality rate, but this appears to be geographically limited.

22.3.6 Herpes Simplex Virus Type 1

Herpes simplex virus type 1 (HSV-1) encephalitis (HSE) is the most common cause of nonepidemic encephalitis in the United States in nonimmunocompromised patients and is a true neurologic and neuroradiologic emergency. The accurate and timely diagnosis of herpes encephalitis is of critical importance because if detected and treated early in its course by antiviral agents such as acyclovir, permanent neurologic damage can be minimized, and if not, severe brain injury and death can occur with an estimated mortality estimated between 30% and 70% in untreated patients. This early diagnosis begins with clinical suspicion and recognition followed by early MRI, including postcontrast scans, and is made definitive by specific identification of the virus in the CSF using polymerase chain reaction or other antibody-based testing protocols. Often, critical therapy with acyclovir will be instituted prior to specific diagnosis because the combination of clinical and MRI findings may be so characteristic and delay may be deleterious. CT imaging is significantly less sensitive and should not be relied on if MRI is available.

The HSV virion is a neurotropic DNA virus of family Herpesviridae (the word *herpes* derived from the Greek, meaning to creep or crawl), which may live as a latent infection in dorsal root ganglia of the peripheral nerves, or in the trigeminal ganglion, becoming active in times of stress and spreading by axonal transmission to the face resulting in *fever blisters*. It is believed that the virion may enter a phase of greatly increased reproduction resulting in either axonal spread or shedding of active virus into the CSF surrounding the trigeminal ganglion, in Meckel's caves and the base of the brain, then spreading along CSF flow pathways upward and around the brain surfaces. This results in a superficial meningoencephalitis most commonly affecting the anterior temporal lobes in a bilateral but asymmetric pattern and but also frequently spreading via the interhemispheric fissure to the cingulate lobe in the anterior midline (Figure 22.23). These patterns although more common are not absolute, and HSV-1 encephalitis has less commonly been initially seen as a superficial meningoencephalitis in more distal brain surfaces such as the occipital and parietal lobes [71,72].

It should be noted that the exact mechanisms of initiation and spread of HSE remain a focus of research and speculation, and uncertainty remains as to whether the spread of HSV-1 is primarily via the CSF or via axonal transport and even whether HSE may in fact result from a *de novo* infection rather than reactivation as described above. Regardless of the exact underlying mechanism, the imaging features and clinical management are unchanged [73,74].

22.3.7 Herpes Simplex Virus Type 2

Herpes simplex virus type 2 (HSV-2) is a double-stranded DNA virus closely related to HSV-1 with over 70% genetic homology, which is clinically most commonly associated with neonatal encephalitis. In this condition, HSV-2 is acquired by the infant in the birth canal from an active maternal genital herpetic infection. This condition has become less common with active and aggressive prenatal care and screening but is still an important cause of neonatal brain injury. Most typically, HSV-2 affects the paracentral lobule in the frontoparietal lobes first and may subsequently cause diffuse encephalomyelitis with near total brain necrosis (Figure 22.24). Physical exam may reveal transillumination of the skull in advanced cases [75,76].

22.3.8 Herpes Zoster Virus

Herpes zoster virus (HZV) causes chickenpox in childhood and then may remain dormant in ganglion cells reactivating and spreading via axonal transport resulting in the painful skin condition known as shingles. Arising by the same mechanism, Ramsay Hunt syndrome is a varicella zoster neuritis of the seventh and eighth nerves in the internal auditory canal, which may be isolated or may be accompanied by typical skin lesions (Figure 22.25) [77–80].

22.3.9 Rabies

Rabies infection is a significant worldwide problem that has been controlled but never completely eliminated in a number of countries and is responsible for approximately 50,000 deaths annually worldwide according to the WHO. Rabies is caused by one of several RNA viruses—genus *Lyssavirus* in the Rhabdoviridae family. *Lyssa* comes from a Greek root meaning rage or wildness,

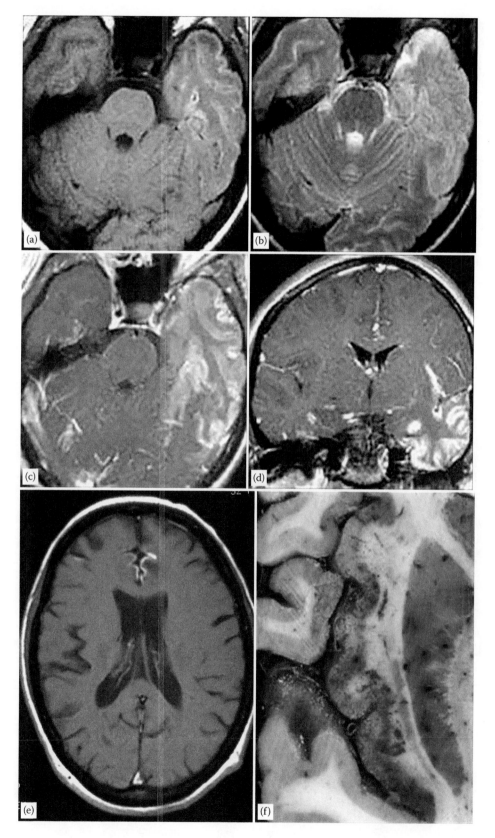

FIGURE 22.23

Herpes 1: (a,b,c,d) Axial FLAIR, T_2, and postcontrast axial and coronal T_1 images show asymmetric bitemporal hyperintensity and enhancement. (e) Axial postcontrast T_1 image shows focal enhancement of the cingulate lobe. (f) Autopsy specimen shows hemorrhagic meningoencephalitis.

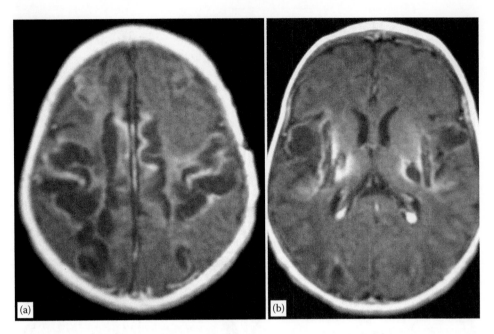

FIGURE 22.24
HSV simplex virus type 2. (a,b) Axial T_1 postcontrast images show encephalomalacia and cortical enhancement in the bilateral paracentral lobules and adjacent regions.

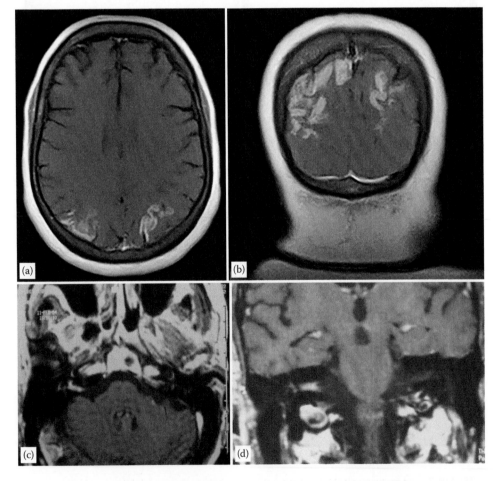

FIGURE 22.25
Herpes zoster. (a) Axial and (b) coronal T_1 postcontrast images show enhancement of the right internal auditory canal in Ramsay Hunt syndrome. (c) Axial and (d) coronal T_1 postcontrast images show enhancing cortical meningoencephalitis due to herpes zoster.

and animals and humans with rabies may exhibit wild or violent behavior. Worldwide rabies is most commonly caused by dogs. In the United States where animal control and screening has been rigorously pursued, the most common sources of human infection have been bats and fox, and more recently domestic cats and raccoons. In Italy and Greece, rabies was felt to have been eliminated but reemerged in fox populations.

Following a bite by an infected animal, the virus travels via axonal streaming to the brain. The infection may be treated in its early stages, but by the time the virion reaches the CNS, treatment is ineffective with death caused by diffuse hemorrhagic encephalitis (Figure 22.26) [81–83].

The first rabies vaccine was developed by Louis Pasteur in the 1880s and was used up until the 1970s. This treatment used a series of subcutaneous injections of vaccine, but more current treatment involves monoclonal antibodies (human rabies immuno globulin [HRIG]) coupled with a rabies vaccine series [84].

Diagnosis of rabies is accomplished by fluorescent antibody testing using a rabies-specific antibody. Pathologically infected tissue is identified due to the presence of the characteristic Negri bodies which are intracellular inclusions seen on electron microscopy (Figure 22.26).

22.3.10 Poliomyelitis

Poliomyelitis is caused by the polio virus, subgroup enterovirus, of the family Picornidae and is transmitted by oral fecal contamination. Polio is highly contagious but in 95% or more of cases causes only minor flu-like symptoms or is asymptomatic. In a small number of cases, often in patients who are immunocompromised or the elderly, the virus invades the spinal cord and produces a paralytic syndrome or in the case of bulbar poliomyelitis invades the brain stem leading to respiratory failure and death.

Initial polio vaccine was an attenuated strain developed by Hilary Koprowski, then improved by Jonas Salk; however, greater success was achieved with live but altered viral vaccine developed by Albert Sabin (oral polio vaccine [OPV]). OPV was replaced in 2000 by an improved inactivated polio vaccine (IPV) which is used today.

Epidemic poliomyelitis peaked in the early 1950s in the United States and led to the development of modern respiratory support beginning with external device (iron lung) and then endotracheal intubation and mechanical ventilation.

Postvaccinal poliomyelitis occurred prior to 2000 when attenuated live vaccine was given inadvertently to an immunocompromised child. This was found rarely to result in active poliomyelitis syndrome with motor paralysis and brain-stem dysfunction resulting in aspiration. This syndrome could vary in severity but was potentially fatal, and this led to the use of the killed vaccine IPV. Current immunizations utilize killed virus [85].

22.3.10.1 Postvaccinal Poliomyelitis—Case History

A 2-year-old child was given normal childhood immunizations but in retrospect was noted to have displayed an unusual propensity toward infections. The child subsequently developed lower extremity paralysis and loss of control of the muscles of deglutition resulting in aspiration and death (Figure 22.27).

22.3.10.2 STOP Program by the WHO

Polio incidence has dropped more than 99% since the launch of global polio eradication efforts in 1988. According to global polio surveillance data from January 14, 2014, 385 polio cases have been reported from the following countries: Afghanistan, Cameroon, Ethiopia, Kenya, Nigeria, Pakistan, Somalia, and Syrian Arab Republic. In 2012, a total of 223 polio cases were reported from five countries: Afghanistan, Chad, Niger, Nigeria, and Pakistan. Of the 2012 polio cases, 97% (217 of the 223 cases) were reported from the three remaining endemic countries: Afghanistan, Pakistan, and Nigeria. Ongoing vaccination and eradication efforts are hampered by civil unrest and misguided beliefs, which have led to fatal attacks on vaccination workers in Pakistan and Nigeria [86–88].

22.3.11 Cytomegalovirus

Cytomegalovirus (CMV) is a common virus that can cause a wide range of clinical symptoms from being asymptomatic to severe and permanent neurologic dysfunction (Figure 22.28). Substantially greater morbidity is seen in patients who are immunocompromised or are exposed to the virus *in utero*. CMV can cause diffuse cerebritis, rhombencephalitis, or ventriculitis. Congenital CMV infection may result in neuronal migration abnormalities, microcephaly, and other developmental disorders. Nonspecific white matter disease, coarse calcifications, atrophy, and periventricular cysts may all be seen [89–91].

22.4 Secondary or Late Effects of Viral Infections

22.4.1 TORCH Syndrome

TORCH is an acronym describing the late sequelae of a number of childhood viral and other infections either alone or in combination. The infections are

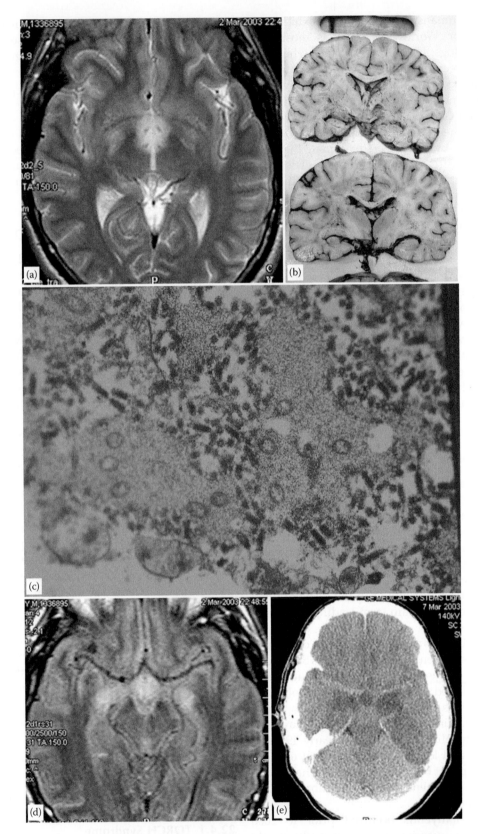

FIGURE 22.26

Rabies encephalitis. (a) Postcontrast CT image shows low density in medial temporal lobes bilaterally. (b,c) Axial T_2 MRI images show high signal intensity in bilateral medial temporal lobes. (d) Brain autopsy shows diffuse petechial hemorrhage encephalitis. (e) Electron microscopy shows characteristic Negri inclusion bodies.

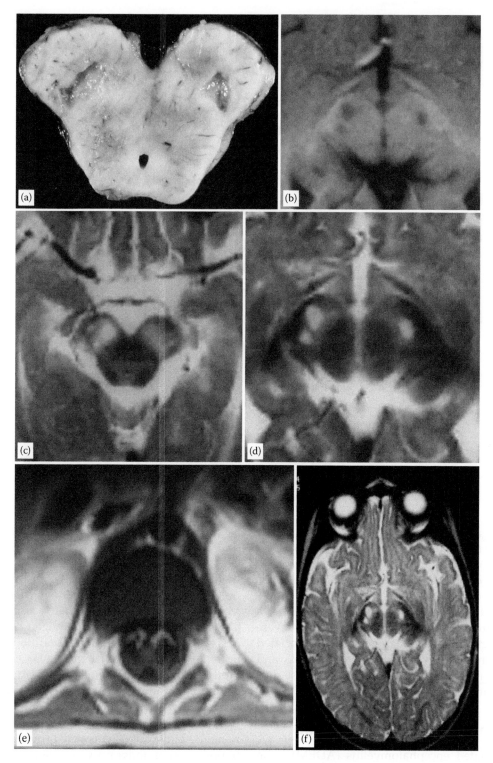

FIGURE 22.27
Bulbar poliomyelitis (postvaccinal). (a) Axial T_1 image shows low signal lesion in the cerebral peduncle. (b–d) Axial T_2 images show high signal lesions in the brain stem and cerebral peduncles. (e) Axial T_1 postcontrast image shows enhancing anterior motor radicles in the lower cord. (f) Autopsy section shows areas of encephalomalacia in cerebral peduncles.

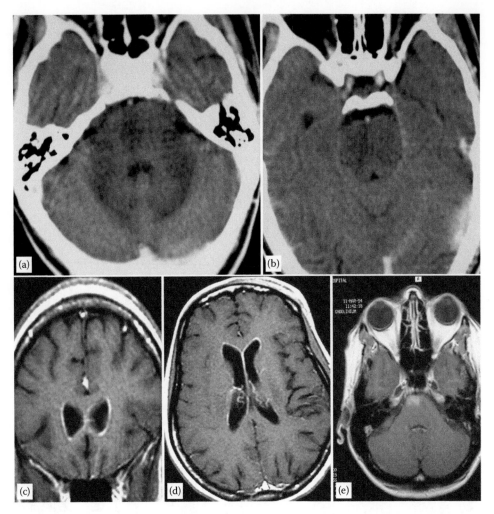

FIGURE 22.28
Cytomegalovirus. (a) Axial and (b) coronal T_1 postcontrast images show CMV ventriculitis with enhancing ependymal lining. (c) Axial T_1 postcontrast shows focal CMV rhombencephalitis. (d,e) Axial noncontrast CT images of a diffuse, ultimately fatal rhombencephalitis.

toxoplasmosis, other, rubella, CMV, and herpes. It is syndrome less commonly seen and an acronym less commonly used as greater progress has been made in diagnosing more specifically and in eradicating severe childhood infections. The imaging features associated with TORCH include severe diffuse volume loss and chronic encephalomalacia as well as coarse diffuse dystrophic calcifications (Figure 22.29) [92–95].

22.4.2 Acute Disseminated Encephalomyelitis

Acute disseminated encephalomyelitis (ADEM) is an immune-mediated inflammatory and demyelinating disease of the CNS typically occurring within several weeks of a viral infection. In its clinical and initial imaging appearance, ADEM bears many similarities to MS, with the primary difference being that the white matter changes seen in ADEM resolve over time and do not recur. This monophasic time course is in distinction to the typical relapsing and regressing course seen in MS.

In addition, ADEM is primarily a disease of childhood, occurring in a significantly younger demographic than MS which is most common in early adulthood.

One current theory of the pathogenesis of ADEM suggests that it is an autoimmune condition in which T cells become activated against molecular components of a viral pathogen and then mistakenly attack molecularily similar component of central myelin. This invokes the concept of *molecular mimicry*. Although this hypothesis explains many of the known features of ADEM, there remain significant unanswered questions and other hypotheses are under active investigation.

The severity of cases of ADEM occur across a continuous spectrum but for the purposes of illustration can be roughly divided into mild, medium, and severe, both clinically and from an imaging standpoint. Mild cases involve some encephalopathy and may exhibit more focal neurologic deficits. From an imaging standpoint, mild lesions may most resemble small ovoid white matter lesions in MS (Figure 22.30).

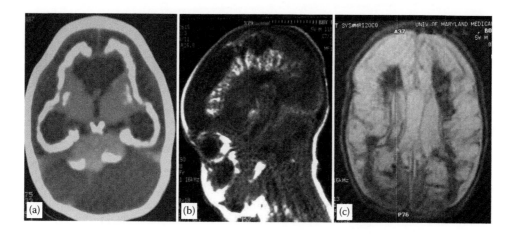

FIGURE 22.29

Torch syndrome: (a) Axial noncontrast CT shows chronic cortical volume loss and encephalomalacia with dense calcifications. (b) Sagittal noncontrast T_1 MRI in a different patient shows diffuse cortical volume loss and bright T_1 signal likely refelecting areas of some petechial hemorrhage and calcification. (c) Axial proton density image from same patient as in (b) shows diffuse cortical encephalomalacia and low signal reflecting areas of calcification and chronic petechial hemorrhage.

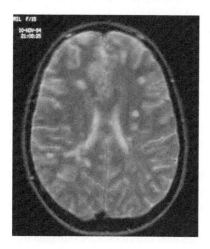

FIGURE 22.30

ADEM, Mild: axial T_2 images show ovoid periventricular white matter lesions similar to typical MS plaques.

Moderate ADEM may clinically exhibit more profound encephalopathy, including coma, and may follow a more prolonged course. On imaging, this may correspond to larger and or more numerous and confluent areas of white matter abnormality (Figure 22.31).

Severe ADEM may exhibit a number of atypical features, including greater persistence and even recurrence, making differentiation from MS more difficult. It may in some rare instances become hemorrhagic and fatal, and this has been referred to as Weston-Hurst syndrome or hemorrhagic leukoencephalitis (Figure 22.32) [96–99].

22.4.3 Subacute Sclerosing Panencephalitis

Subacute sclerosing panencephalitis (SSPE) is a post viral white matter disease related to an initial measles infection and is relatively uncommon, particularly since measles itself has been so well controlled by

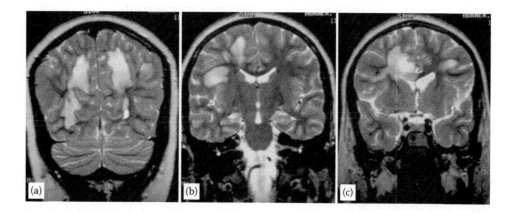

FIGURE 22.31

Acute disseminated encephalomyelitis (ADEM) (a), (b), (c)—coronal T_2 images show extensive increased T_2 signal in a patient who developed coma but recovered; (d)—axial T_2 images show scattered T_2 hyperintensities somewhat suggestive of MS in a patient with mild transient CNS changes.

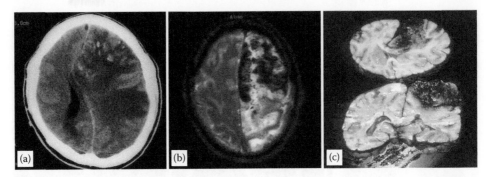

FIGURE 22.32
Acute hemorrhagic leukoencephalitis, Hurst syndrome, and severe variant of ADEM: (a)—noncontrast CT shows hemorrhagic, edema, and mass effect with midline shift; (b)—T_2 MRI shows extensive hemorrhage, edema, and mass effect; and (c)—autopsy shows large area of hemispheric hemorrhage.

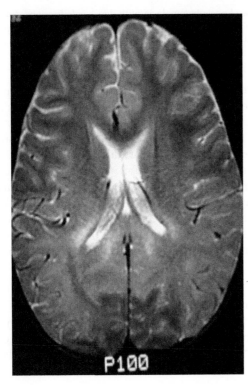

FIGURE 22.33
Subacute sclerosing panencephalitis (post measles). Axial T_2-weighted image shows diffuse periventricular white matter hyperintensity.

vaccination. The pathogenesis of SSPE has been linked to persistent defective measles virus in intracellular inclusions, which cause a slow persistent damage to both neuronal and glial cells (Figure 22.33) [100,101,102].

22.5 Fungal Infections of the CNS

Fungi are a diverse kingdom of organisms numbering between 1.5 and 5 million species with only an estimated 5% which have been fully classified. They possess characteristics of both plants and animals and had been previously considered to fall under the province of botany; however, genetic studies have shown that they are more closely related to animals but yet distinct from both [103–106].

Metabolically, fungi lack chloroplasts or a photosynthetic pathway, and therefore lack direct access to the energy of the sun. In order to survive, fungi exist in a variety of symbiotic relationships that range from beneficial to neutral to parasitic with a wide variety of other life forms. Fungi possess certain other unique characteristics, including the presence of glucans (found in plants) and chitin (found in the exoskeleton or arthropods) rather than cellulose in their cell walls.

Human fungal infections are primarily and initially respiratory, due to inhalation, as well as cutaneous with secondary infections arising due to hematogenous spread as well as colonization in sinuses and other bodily cavities and crevices. CNS fungal infections may arise via hematogenous dissemination, as an invasive vasculitis, or by direct spread from adjacent structures, including sinuses and the skin. Less commonly fungal infections may arise via open or surgical wounds that become contaminated.

As with other vectors, CNS fungal infections may manifest themselves as a surface infection or meningitis, single or multiple areas of fungal cerebritis, single or multiple abscesses, or as a vasculitis. Less commonly fungal infections may be seen as a solid appearing extradural mass or as a focal choroid plexitis [107].

22.5.1 Aspergillosis

Aspergillus fumigatus may be indolent in sinuses or lung cavity for many years and subsequently become invasive, possibly due to altered host immune status, spreading via hematogenous spread to the brain. Diagnostic tools are evolving but may involve some form of direct biopsy and culture, although antigen-antibody testing and other more sophisticated diagnostic modalities are becoming increasingly important. Risk factors include bone marrow transplantation, steroid use, advancing age, or other conditions leading to immune compromise. It may cause single or multiple areas of focal cerebritis and abscess (Figure 22.34). It may manifest as an invasive vasculitis leading to cavernous sinus thrombosis (Figure 22.35) or other vascular complications. In this uncommon case, aspergillosis has spread to brain parenchyma forming and abscess then colonized within the cavity to form an intracranial aspergilloma, mimicking the more common pulmonary aspergilloma (Figure 22.36) [108–110].

22.5.2 Cryptococcosis

Cryptococcus neoformans (and other species; *C. gatti*, etc.) is a relatively indolent fungus seen as a pathogen primarily in immunocompromised patients. It displays imaging features similar to TB and other granulomatous disease, presenting most commonly as meningitis. Although it may begin as a surface infection, it tends to spread along brain surfaces leading to cryptococcomas within deep sulcal spaces and crevices, and may appear to invade the brain parenchyma traveling via the Virchow–Robin spaces. Frank cerebritis is extremely uncommon but has been reported. Diagnosis traditionally has involved India Ink test staining of CSF following lumbar punctures, specific antigen-antibody tests, and other more advanced methods that are increasingly used (Figure 22.37) [111–117].

22.5.3 Mucormycosis

Mucormycosis is an uncommon but highly virulent fungal infection caused by organisms in the subphylum Mucormycotinia, order Mucorales, including *Rhizopus oryzae* and *Mucor circinelloides*. The exact genetic traits that contribute to the clinical features are under active study; however, it is known that these organisms flourish in environments of high glucose and iron concentrations, and are prevalent in immunocompromised populations, including patients who have undergone

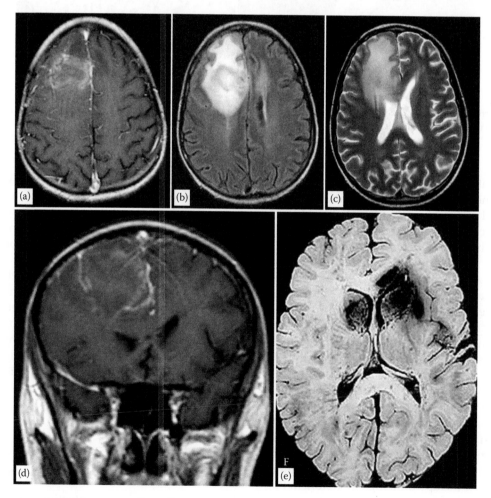

FIGURE 22.34

Aspergillosis. (a) Axial T_1 postcontrast, (b) FLAIR, (c) T_2, and (d) coronal T_1 postcontrast and (e) autopsy correlate of fatal cerebral aspergillosis images show large enhancing lesion with extensive perilesional edema right frontal lobe.

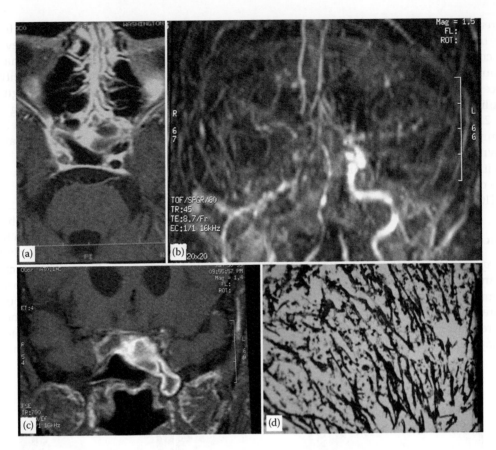

FIGURE 22.35
Aspergillosis cavernous sinus thrombosis. (a) Axial and (b) coronal postcontast images show sinusitis and thrombosis of cavernous sinus. (c) MRA image shows marked loss of flow distal right ICA. (d) Micrograph shows typical *Aspergillus* hyphae.

organ transplantation, AIDS, diabetics, and other conditions of immune impairment. In diabetic ketoacidosis, the lowered pH in the bloodstream is believed to contribute to dissociation of iron from protein resulting in increased free iron. In patients treated with the iron-chelating agent deferoxamine, the fungal organism is believed to utilize the chelator as an iron transport molecule for itself, and thus, there is an increase in mucormycosis in these patients as well.

The clinical features of mucormycosis include a blackish residue on the surface of the infected mucosa related to iron metabolism, a highly invasive growth pattern that does not respect tissue planes, and a predilection for the endothelial lining of blood vessel walls leading to a fulminant invasive vasculitis, which in turn leads to thrombotic occlusions and ischemic tissue damage. Once active mucor infection is detected, either in sinus or in orbit, aggressive surgical treatment, including maxillectomy and orbital exenteration, may be indicated, in combination with antibiotic therapy utilizing amphotericin B.

Rhinocerebral mucormycosis is the direct spread of infection from an infected sinus or orbit to the brain, typically with a concomitant vasculitis, leading to

thrombosis and infarction (Figure 22.38). Thrombosis may involve the cavernous sinus and may involve both arterial and venous structures. Less commonly mucor may spread to the brain via hematogenous spread from elsewhere in the body or by injection of contaminated materials (Figure 22.39).

Despite vigilant surveillance of patients at risk, and aggressive medical and surgical treatment following diagnosis, the mortality for rhinocerebral mucormucosis remains approximately 40%, with a mortality of approximately 10% if the infection is confined to a sinus cavity. Hopefully ongoing research will lead to further advances in both early diagnosis and treatment of this deadly disease [118–120].

22.6 Parasitic Infections of the CNS

The category of parasitic infections of the human CNS is quite large, encompassing a broad array of vastly different organisms, routes of infection, and clinical

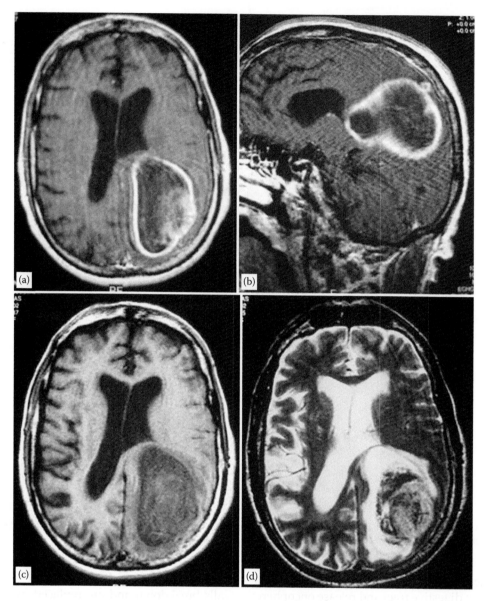

FIGURE 22.36
Aspergilloma. T_1 (a) pre- and (b) postcontrast images show cavity with complex signal contents and enhancing rim. (c) Sagittal postcontrast image shows enhancing rim with complex nonenhancing contents. (d) Axial T_2 image shows mixed high and low signal of fungus ball or aspergilloma.

circumstances. According to the CDC, "a parasite is an organism that lives on or in a host organism and gets its food from or at the expense of its host" [121]. There are over 370 documented human parasitic infections, many quite rare but at least 90 that are relatively common, and of these, many may have CNS manifestations. In general, these diseases fall into one of two large categories: parasitic worms that are often large and visible to the naked eye and the much smaller, microscopic protozoa. I will present specific details of a selected group of these infections based upon prevalence and specific interest to the neuroscience and imaging community [122].

22.6.1 Cysticercosis

Cysticercosis is the most common CNS parasitic infection worldwide and the most common etiology of acquired epilepsy. It is caused by the encysted larva of pork tapeworm *Taenia solium* (Figure 22.40) and displays a variety of clinical and imaging features. The parasite is spread by ingestion of live larval eggs secondary to fecal oral contamination and is seen in situations of poor public sanitation systems in combination with domestic pig farming, but is also not uncommonly seen throughout the developed world due to the complexities of human travel and migration.

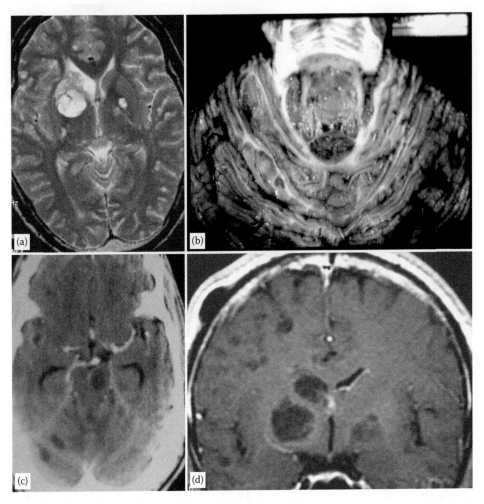

FIGURE 22.37
Cryptococcus. (a) Axial T_2, (b) axial and (c) coronal T_1 postcontrast images, and (d) autopsy show nonenhancing cyst-like lesions in locations typical for prominent Virchow–Robin spaces. On autopsy, the contents are glistening and gelatinous.

Following ingestion of contaminated meat, the parasite eggs enter the digestive tract and release oncopheres which migrate through the intestinal walls into the blood stream where they may spread throughout the body primarily into tissues with rich blood flow such as the muscle, eye, and brain. Brain involvement is estimated at 60%–90% (Figure 22.41).

Once in their final tissue destination within the host, the larva goes through four stages: The first is *the vesicular stage* which consists of fluid-filled cysts with thin transparent walls (Figure 22.42). During this stage, the parasite is invisible to our immune system because it secretes a protease inhibitor, taeniaestatin, which inhibits complement activation, and paramyosin, thereby suppressing the cellular immune response. The larvae are alive during this stage and can live for years while remaining asymptomatic to the host. Eventually, as the cysts start to degenerate, an inflammatory response occurs leading to the second *colloidal stage* with gelatinous material in the cyst, followed by the third *granular*

nodular stage in which the cyst is contracted and the walls breakdown and are replaced by granulation tissue and lymphoid aggregates, and the fourth and final *nodular calcified stage* with granulation tissue replaced by collagen and calcification (Figure 22.43).

During each of these phases, the larva may be solitary, presenting as small regular inclusions (cysticercus cellulosae), or they may be found in larger clustered inclusions (cysticercus racemosus) (Figure 22.44). In the brain and less commonly in the spinal cord, the encysted larva may be found in cisternal spaces, in the subarachnoid spaces, or within the brain or spinal cord parenchyma.

Clinically, the cysticercal larva may be symptomatic, particularly during second and third stages, colloidal leading to granular nodular, resulting in seizures, headache, and other focal and diffuse neurologic symptoms.

Imaging appearances range from the common incidental calcified lesion, to a wide variety of appearances

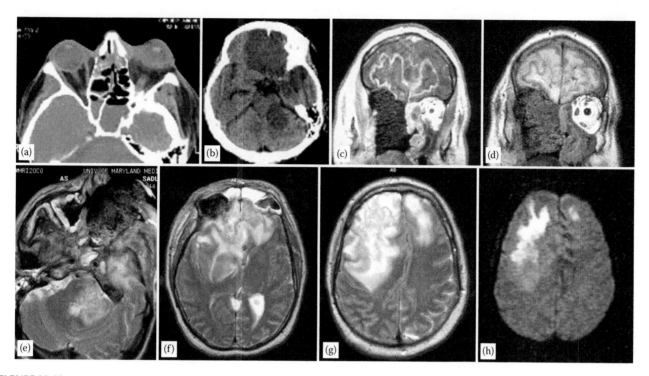

FIGURE 22.38

Rhinocerebral mucormycosis. (a,b) Axial CT images show sinusitis and cerebritis. (c) Coronal T_2 and (d) coronal T_1 postcontrast images show ring enhancement around cerebritis even after aggressive maxillectomy and orbital exenteration. Axial T_2 (e,f,g) and DWI (h) show posterior spread of cerebritis and restricted diffusion in evolving abscess.

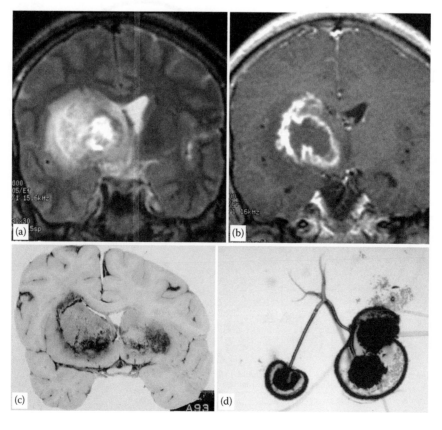

FIGURE 22.39

Hematogenous mucormycosis. (a) Coronal T_2, (b) coronal T_1 postcontrast, and (c) brain autopsy show central areas of cerebritis not contiguous with extracalvarial source of infection. (d) Photomicrograph of mucor species.

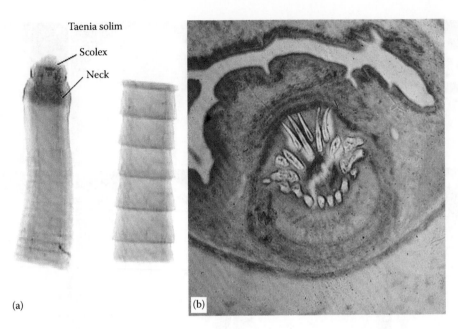

FIGURE 22.40
Cysticercosis, organism *Taenia solium*: (a)—longitudinal microscopy of *Taenia solium* larva and (b)—*Taenia solium* cross-sectional head.

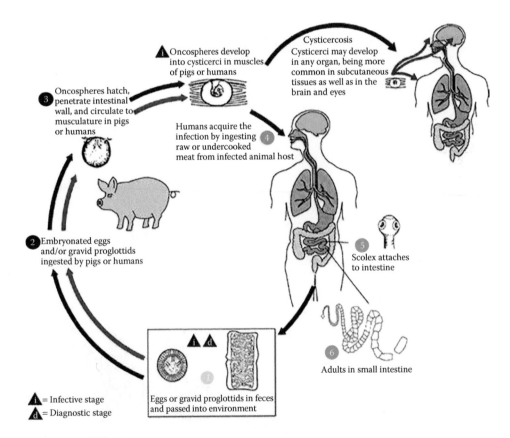

FIGURE 22.41
Cysticercosis life cycle: *Taenia solium* organism is ingested through fecal contaminated water and food, and attaches to intestinal wall. Larval form enters the bloodstream and migrates throughout the body coming to rest in skin, muscles, brain, and other tissues of human and pigs. (From CDC, Atlanta, GA.)

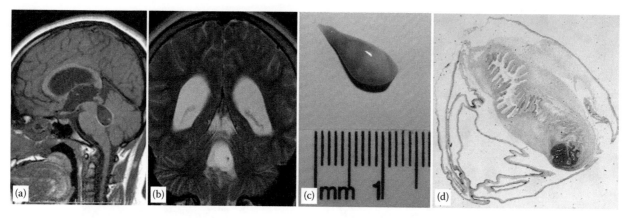

FIGURE 22.42
Cysticercosis, solitary intraventricular: (a)—sagittal T_1 post-gadolinium shows thin rim enhancement in cystic lesion in the forth ventricle, (b)—coronal T_2-weighted images shows cystic thin-walled structure in the forth ventricle, (c)—gross path of intact scolex, and (d)—microscopic section solitary scolex.

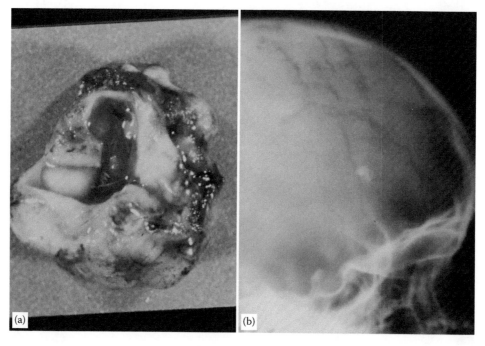

FIGURE 22.43
(a) Lateral skull X-ray shows solitary calcified lesion nonspecific but typical and common for chronic cysticercosis. (b) Gross pathologic specimen of chronic cystercercosis lesion.

and locations through the brain and spinal cord, including vesicular lesion in the fourth ventricle, multiple clustered lesions in cisternal spaces, and multiple lesions scattered throughout the brain parenchyma (Figure 22.45) [123–125].

22.6.2 Cerebral Malaria

Malaria is thought by some to be the most extensive and important infectious disease in the world. It is caused by a parasitic protozoa in the *Plasmodium* genus, including species *falciparum, vivax, ovale*, and *knowlesii*. The infection is transmitted by mosquito bite, primarily anopheles species, and occurs when the mosquito regurgitates a small amount of infected saliva into the blood of its victim. Although eliminated or substantially limited in much of the developed world, malaria remains a serious and extensive public health problem in tropical regions of Sub-Saharan Africa, South Asia, and South and Central America, and is not infrequently reintroduced or *imported* by returning travelers. An estimated 3.4 billion people live in areas of endemic malaria, and the World Health Organization (WHO) estimates that in

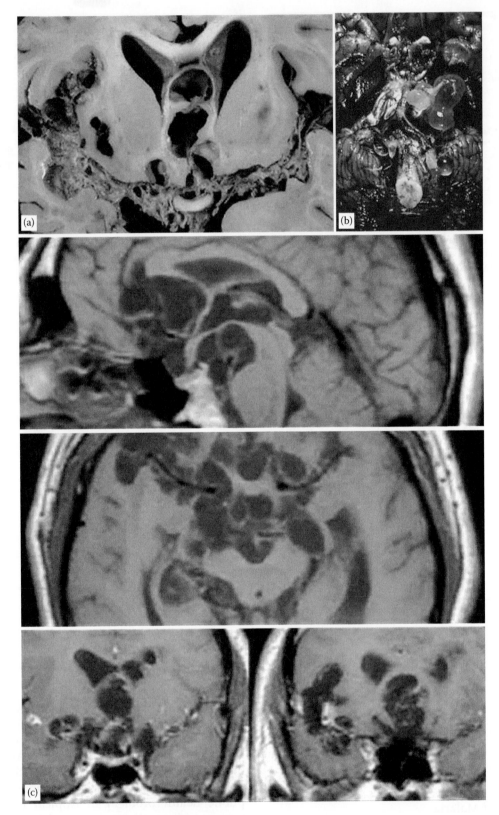

FIGURE 22.44
Cysticercosis racemose: (a) Coronal brain autopsy shows multiple clustered cysts throughout subarachnoid and ventriculr spaces. (b) Gross path surface photograph shows clustered cysts on the surface of the base of the brain. (c) Triplanar MRI T_1-weighted images show multiple clustered cysticercal cysts scattered through the fluid spaces of the brain.

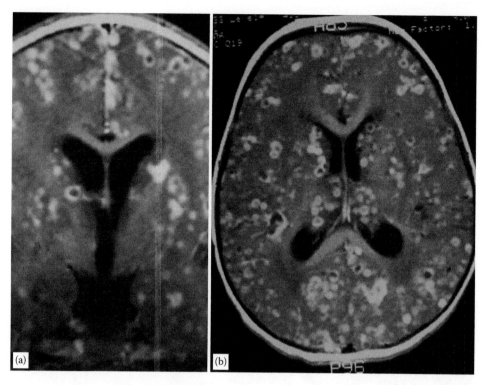

FIGURE 22.45

Lyme's disease: (a) Axial T_2 image of the brain shows ovoid white matter lesions of Lyme's disease which may mimic demyelinating disease such as multiple sclerosis (MS). (b) Sagittal and axial MRI of the lower spine show enhancing nerve roots of the cauda equina due to Lyme's disease.

2012 malaria resulted in 207 million clinical episodes resulting in 627,000 deaths.

Malaria is a complex clinical, genetic, environmental, and epidemiologic problem with ramifications and research beyond the scope of this work. The basic pathology of the disease is that parasitized erythrocytes (RBCs) clump together and adhere to vascular endothelial surfaces leading to sludging and ischemia. In addition, circulating toxins released by the parasite are believed to play a role in altering membrane permeability and other nonvascular effects. Studies into pathogenesis, mosquito eradication, vaccine development, antibiotic therapies, and many other related areas are all in active ongoing investigation and research.

Clinically, malarial infection may present with a wide range of symptoms of varying severity and time course, but an acute infection is typically characterized by severe fever, chills, and flu-like symptoms. Cerebral malaria represents a small subset of clinically active patients and may often be seen in children resulting in headache, convulsions, focal neurologic deficits, coma, and death. Imaging of cerebral malaria is nonspecific but may include areas of focal cerebritis and/or subacute ischemic injury and/or hemorrhage (Figure 22.46) [126–134].

22.6.3 Toxoplasmosis

Toxo is an intracellular protozoal infection whose primary host is *felinus domesticus* (domestic cats) but which has been found to infect, often in a asymptomatic manner, a wide variety of mammals, including humans, dogs, rodents, deer, cattle goats, sea otters as well as geese, chicken, and others. Many animals and humans may be infected by Toxo with either no or very subtle symptoms for long periods of time perhaps as long as they live. Because of this ability to persist and parasitize within an unsuspecting host for long periods of time, it is considered by some to be not only the most prevalent but the most successful parasite on earth. It continues to be widely studied as an example of pathogen host behavioral manipulation, including inducing changes in rat behavior which makes them more vulnerable to cats and also possibly inducing subtle long-term behavioral changes in otherwise asymptomatic human hosts.

Toxo exhibits a complex life cycle, as mentioned, involving numerous hosts, although sexual reproduction can only occur in the domestic cat making it the definitive or primary host (Figure 22.47). Humans may come to harbor Toxo by three main routes: direct ingestion of oocytes by contaminated soil or water, ingestion

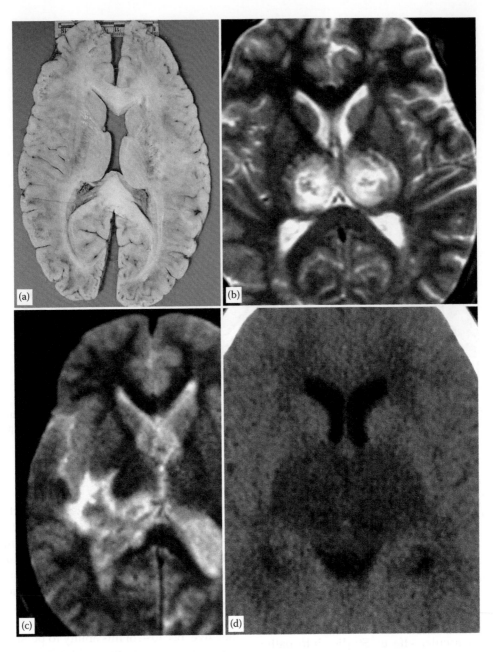

FIGURE 22.46
Cerebral malaria. (a) Noncontrast CT shows edema and mass effect in bilateral thalami. (b) Axial T_2 MRI in a different patient also shows marked edema and swelling in bilateral thalami possibly with some hemorrhage. (c) Axial T_2 image shows hemorrhagic infarction adjacent to ventricle. (d) Brain autopsy in a different patient shows diffuse areas of hemorrhage in the brain bilaterally.

of poorly cooked meat containing encysted organisms (trophozoites), or via transplacental migration from an infected maternal unit (congenital Toxo).

Congenital Toxo may produce a diffuse encephalitis either alone or in conjunction with viral pathogens resulting in diffuse encephalomalacia and coarse basal ganglia calcifications referred to as the TORCH syndrome. Toxo may also produce chronic chorioretinitis and is an important cause of congenital blindness.

In adults, Toxo is commonly seen as an opportunistic infection in immunocompromised hosts, including

patients on chemotherapy, patients with congenital immune deficiencies, and patients with AIDS who are not on antiretroviral therapy. Typical manifestations include meningoencephalitis and multiple cerebral abscesses frequently in the basal ganglia (Figure 22.48). The appearance of CNS Toxo is nonspecific and there may be considerable overlap with other potential pathogens and with neoplasms, particularly in AIDS patients. Often antibiotic therapy is initially used empirically, with positive response as evidence of Toxo. A more specific diagnostic tool involves the use

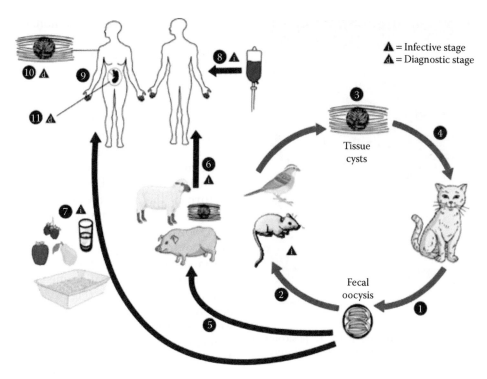

FIGURE 22.47
Toxoplasmosis gondii life cycle showing the definitive host felinus domesticus, as well as patterns of dissemination to other animals, soil, and water to humans. (From CDC, Atlanta, GA.)

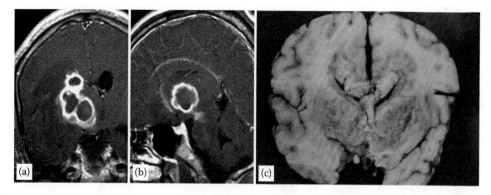

FIGURE 22.48
Toxoplasmosis brain. (a) Axial and (b) coronal T_1 postcontrast images show clustered ring enhancing lesions in the basal ganglia and (c) autopsy specimen demonstrates necrotic abscess in the basal ganglia.

of proton diffusivity with increased ADC ratio greater than 1.6 correlating with likely Toxo and ratios of less than 1 being more suggestive of a neoplastic etiology such as CNS lymphoma where a diagnostic biopsy may be indicated.

Although uncommon, Toxo may also present as a ring-enhancing lesion in the cord, and given that delay of treatment can significantly increase morbidity, this possibility should be considered in the appropriate clinical setting (Figure 22.49) [135–140].

22.6.4 Neurosyphilis

Syphilis is a sexually transmitted disease that was first described after European explorers returned from The New World in the 1600s. The causative organism is the spirochete, *Treponema pallidum*, and for 400–500 years, prior to the discovery and routine use of penicillin, it was the cause of widespread morbidity and mortality. Since the use of penicillin, syphilis has remained a persistent and elusive, but still widespread pathogen perhaps in

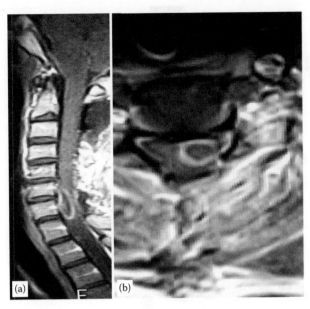

FIGURE 22.49
Toxoplasmosis spine. (a) Sagittal and (b) axial T_1 postcontrast images show ring-enhancing lesion cervical cord biopsy-proven toxoplasmosis.

part due to the convoluted and secretive nature of human sexual psychology as well as the occurrence of immunodeficiency syndromes, including AIDS [141–153].

Syphilitic infection occurs in a series of overlapping stages beginning with a painless skin ulcer usually on the genitalia or perioral region with later stages, which may result in permanent and even life-threatening damage to the cardiovascular system and CNS.

Neurosyphilis may be divided into two stages: early and late. Early neurosyphilis occurs less than 1 year from initial exposure and is general characterized by meningitis, cranial nerve dysfunction, and/or cerebrovascular

disease. Late neurosyphilis is more diffuse in manifestation but may result in generalized or focal cognitive deficits, weakness, and gait disturbance due to dorsal column injury in the spinal cord (tabes dorsalis).

For all our knowledge and experience preventing and treating syphilis and neurosyphilis, one description remains apt, so-called The Great Imitator, because the manifestations of the disease are so protean and the etiology is sometimes so unsuspected that it has been seen to masquerade and fool the earnest clinician repeatedly. The incidence of syphilis was markedly diminished by the turn of the millennium but has increased due to among other things, namely, the HIV epidemic, human psychology, and the variable availability of effective antiretroviral therapy.

From an imaging standpoint, neurosyphilis may take many forms, from a coating meningoencephalitis, to a focal area of brain stem (Figure 22.50) or supratentorialcerebritis, to extraaxial masses or *gummas* that may mimic other more common dural-based processes such as meningioma, to nonspecific white matter changes that may not enhance, to vasculitis with or without areas of ischemic injury, to diffuse atrophic changes in either the brain or the spinal cord. Diagnosis must be based on serum and CSF antibody levels coupled with other clinical information and history. Treatment by penicillin, as well as other agents, including tetracycline and erythromycin, when penicillin allergic must be of appropriate length and cure is demonstrated by decreased titers and clinical remission.

22.6.5 Echinococcosis

Echinococcosis or hydatid disease is caused by the larval stage of tiny tapeworms of the genus *Echinococcus*. There are two main causes of human disease: cystic and alveolar.

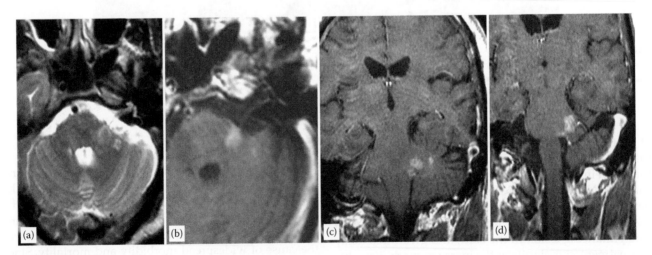

FIGURE 22.50
Neurosyphilis. (a) Axial T_1 postcontrast and (b,c) coronal T_1 postcontrast images show enhancing cluster of lesions in the left pons near the cerebellopontine angle. (d) Axial T_2 image shows T_2 hyperintensity of same lesions.

Echinococcus granulosus causes cystic hydatid disease and is found in fox, dog, and other canids, and humans and sheep are intermediate hosts. It is endemic in the Mediterranean, the Middle East, and the Latin America. Cystic hydatid disease is relatively benign becoming symptomatic primarily when the cysts exert sufficient mass effect on adjacent structures. In general, cystic hydatid disease is most commonly found in areas where sheepherding is an important source of food.

Echinococcus multilocularis causes alveolar hydatid disease and is endemic in Alaska, central Europe, Turkey, and China. The definitive hosts are arctic and red fox, but rodents, dogs, other canids, and cats may also become infected. Alveolar hydatid disease is potentially more significant clinically, consisting of parasitic tumors that may invade locally and spread to other organs, including lung and brain with a reported mortality rate of greater than 50% due in part to the generally remote conditions in which it is found.

Craniospinal hydatid disease is rare, even in endemic regions, and as in other forms is found more commonly in children possibly because of greater likelihood of physical proximity. Intracranial disease is usually due to *E. granulosis* and is generally seen as a large single or several cystic structures which are well circumscribed without adjacent brain parenchymal changes (Figure 22.51). Similarly spinal lesions are nonspecific simple appearing cysts [154–162].

Case history: A 9-year-old Bosnian girl refugee airevaced by relief agencies to Canada due to unilateral proptosis. Imaging reveals unilocular cyst in the upper outer orbit with an enhancing rim or capsule, which is later determined to be within the lacrimal gland. Intraorbital hydatid disease is quite rare but when present is most commonly found in the lacrimal gland for reasons that are not clear (Figure 22.52).

22.6.6 *Naegleria fowleri*—Primary Amoebic Encephalitis

Naegleri fowleri is a thermophillic amoebae that is relatively common in the environment worldwide, but only rarely pathogenic in humans. Found in bodies of freshwater, including lakes, rivers, ponds, industrial drainages, and hot springs, it causes disease when inhaled or splashed into the nasal cavity. From

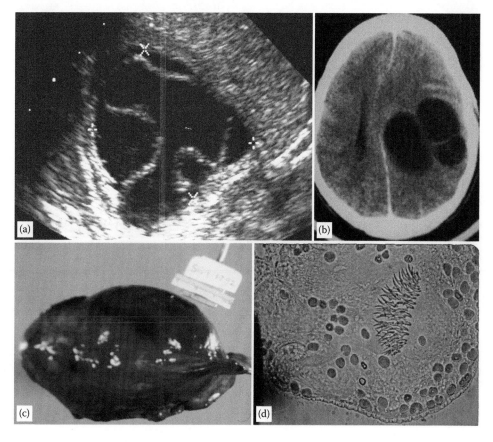

FIGURE 22.51
Echinococcosis. (a) Large clustered cystic lesions in brain. (b) Cystic lesion in the liver with collapsed capsule forming *sign of the camelotte* or floating lily. (c) Microscopic view of *spiny coccus*. (d) Path specimen of intact echinococcoma.

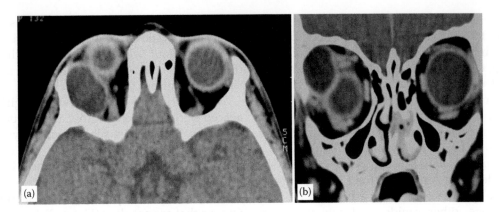

FIGURE 22.52
Echinococcoma, orbit, lacrimal gland. (a) Axial and (b) coronal postcontrast CT images demonstrate large cystic mass in the right lacrimal gland.

the nasal mucosa, it extends through the cribriform plate into the brain causing hemorrhagic encephalitis (Figure 22.53) which has a fatality rate of roughly 98%, giving rise to the term *brain-eating amoeba* in the popular press.

The term *primary amoebic encephalitis* was coined to describe amoebic encephalitis in which the protozoans' initial and primary pathologic focus is the brain, in distinction from other amoebic encephalitides which are metastatic or blood borne from primary infections

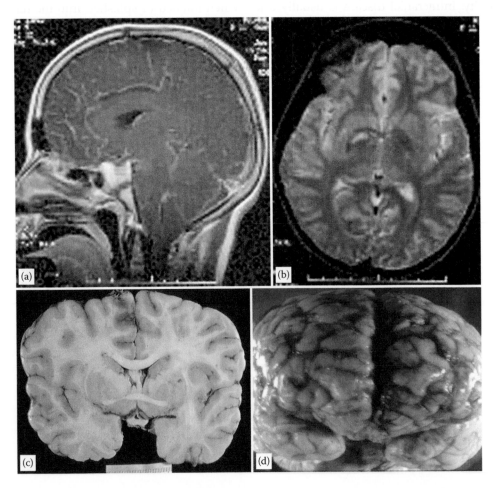

FIGURE 22.53
Naegleria fowleri encephalitis. (a) Sagittal postcontrast T_1 MRI shows mild diffuse surface enhancement. (b) Axial T_2 image shows abnormal globi pallidus and subtle cortical edema. (c,d) Brain autopsy shows surface hemorrhage and basal ganglia petechiae.

elsewhere in the body such as the gut, liver, or spleen. This category of infection, including its transnasal mode of spread, and its main human pathogen, *Naegleri fowleri*, were first fully studied and understood in the 1960s, although in retrospect, isolated examples had been reported earlier.

Reports of this very rare disease are increased during summer months when the bacteria itself may multiply and be found in greater concentration, and children may be more likely to swim in whatever body of water is available. In rare instances, it has also been reported to have spread by nasal inhalation of tap water for nasal cleansing and in Australia was discovered to have been caused when freshwater conduits were exposed to higher than typical temperatures in transit from reservoir to people's homes. Microbiologic studies of *N. fowleri* versus other *Naegleri* species show differences in characteristics related to protein adhesion and invasion, which may account for its particular pathogenicity and virulence [162–165].

22.6.7 Granulomatous Amoebic Encephalitis

The more general term *amoebic encephalitis* refers to cerebral infestation by a wider category of parasitic amoebae, including *Acanthamoeba* species, and the closely related organism from the order Leptomyxida, *Balamuthia mandrillaris*. This infection may be acquired due to ingestion of contaminated local water supplies, resulting in lower respiratory tract and genitourinary and skin infection, which then spread hematogenously to the brain. Increased susceptibility to these infections is seen in the setting of immune compromise such as AIDS. In at least one instance, this infection was also believed to have been acquired due to the use of tap water to prepare intravenous drug injection (Figure 22.54). Imaging

features are nonspecific with scattered foci of localized cerebritis with enhancement and edema [166–170].

22.6.8 Schistosomiasis

Schistosomiasis, also referred to as bilharziasis, is caused by infection with the trematode flatworms of the genus *Schistosoma*, most commonly *S. haematobium* and *S. mansoni*. These organisms hatch from eggs in freshwater giving rise to miracidia, which enter freshwater snails and reproduce asexually forming cercariae. When humans come into contact with infested freshwater, motile cercariae penetrate the human skin and transform into schistosomulae, which enter the bloodstream and migrate to the heart, lung, and liver, then to more permanent homes in small venules in the bladder, rectum, and small bowel. In these more permanent homes, pairs of organisms engage in sexual reproduction producing a multitude of eggs, which subsequently migrate throughout the body occasionally into spinal and cerebral venous spaces of the CNS (Figure 22.55). The WHO estimates that 200–300 million people worldwide suffer from bilharziasis, second only to malaria in worldwide prevalence, and many more are at risk. The clinical manifestations of this disease are primarily due to the host immune response to these organisms, and therefore diagnosis may be difficult. Although uncommon in nontropical and developed countries, it is important to maintain awareness of this condition.

Neuroschistosomiasis is most commonly caused by *Schistosoma japonicum*, and also occasionally by *S. mansoni*. Clinical manifestations vary widely from asymptomatic to an acute febrile illness with headache and behavioral changes, to seizures to a variety of neurologic deficits depending on where the organism is

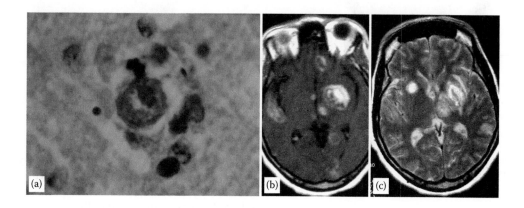

FIGURE 22.54
Granulomatous amoebic abscess. (a) Axial T_1 post-gadolinium shows multiple irregularly enhancing ring lesions in brain parenchyma. (b) Axial T_2 image shows multiple T_2 hyperintense lesions of varying fluid content. (c) Hematoxylin and eosin slide shows live amoebae, *Balamuthia mandrillaris*, recovered from brain biopsy.

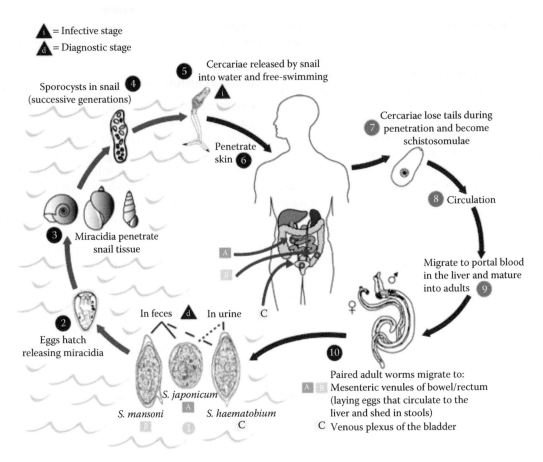

FIGURE 22.55
Schistosomiasis life cycle : As Illustrated, schistosomasis speciues move from freshwater to snails to humans where they migrate and mature in te liver with subsequent dissemination throughout the body. (From CDC, Atlanta, GA.)

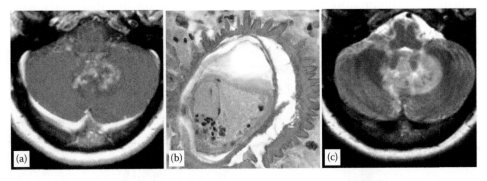

FIGURE 22.56
Schistosomiasis. (a) Axial T_2 image shows complex lesion with edema centered in the deep cerebellar nuclei. (b) T_1 postcontrast image shows irregular areas of enhancement surrounded by edema. (c) Schistosomula.

located and the robustness of the host immune response (Figure 22.56) [172–176].

22.6.9 Paragonimiasis

Paragonimus is a trematode or flatform referred to as the lung fluke that causes disease in humans following ingestion of undercooked crab or crayfish contaminated by the organism. Although primarily seen as a pulmonary infection, more serious infection may occur when the worm, most typically *Paragonimus westermani*, spreads hematogenously to the brain resulting in focal abscess, cerebritis, and/or hemorrhage (Figure 22.57). Due to the uncommon occurrence of this method of

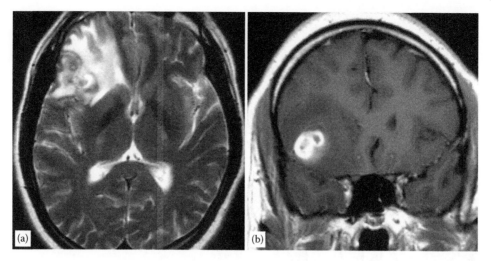

FIGURE 22.57
Paragonimiasis. (a) Axial T_2 image shows focal ring lesion with exuberant perilesional edema. (b) Coronal T_1 postcontrast image shows paired ring-enhancing lesions with surrounding edema.

spread, and the nonspecific clinical and imaging findings, the etiology may not be suspected resulting in delayed treatment with increased morbidity. Outbreaks of this condition may occur in clusters due to a batch of contaminated crabs [177–181].

22.6.10 Lyme Disease

Lyme disease is a parasitic infection, which unlike many we have discussed, is not tropical, but in fact it is the most common parasitic disease in temperate latitudes, including the United States, Canada, Europe, and portions of Asia. Named after the town of Lyme, Connecticut, where it was initially studied, it is caused by the spirochete *Borrelia burgdorferi* in the United States and other *Borrelia* species in Europe and Asia, and it is transmitted by the bite of an infected tick, most typically the deer tick, *Ixodes* species *scapularis* and *pacificus*, in the United States, but actually carried by many tick species. *Borrelia* species live in small mammals such as rodents and birds, as well as deer. The tick becomes infected when feeding on a host animal, then transmits the parasite in a subsequent prolonged blood meal on a noninfected host believed to require 36 h or more for transmission.

The initial clinical symptoms of Lyme disease include an enlarging erythematous macule at the site of the original bite, which sometimes but not always develops a slightly raised border as it expands creating a target-like appearance and may contain some central vesicles. This cutaneous lesion is called erythema chronicum migrans. This skin lesion may be single and occasionally multiple, and in appearance may overlap with a number of other skin lesions and may in fact be complicated by other

infections acquired concomitantly from the original tick bite. Local symptoms may include mild urticarial and irritation. Systemic symptoms may occur at the time of the original infection, including fever and headache; however, the initial infection may be asymptomatic and unrecognized. If recognized and treated correctly with a course of doxycycline or other appropriate therapy (doxycycline is contraindicated in pregnant women and children less than 12 years old), primary Lyme disease has a 100% cure rate.

Secondary Lyme disease occurs after an unrecognized or incompletely treated initial infection and may result in cardiac, rheumatologic as well as neurologic symptoms, and is effectively treated with appropriate antibiotic therapy. In addition, there are reports of other chronic, vague, and/or systemic symptoms, which have been referred to as *chronic Lyme disease* about which there is significant controversy and which is not definitively caused by borreliosis.

CNS Lyme disease is generally detected within weeks to months following infection, and may exhibit numerous manifestations, including cranial nerve dysfunction such as CN7 resulting in facial palsy and CN8 resulting in hearing loss, or may be asymptomatic. Neuroimaging findings may include focal areas of meningeal and/or radicular enhancement, and may also result in white matter lesions, which in some cases have been noted to be similar to those seen in demyelinating disease (Figure 22.58). Immunologic studies have raised the possibility that molecular mimicry may trigger T-cell activation resulting in MS-like lesions in neuroborreliosis. Another area of ongoing research suggests that different subtypes of borrelia have different proclivities toward CNS involvement [182–199].

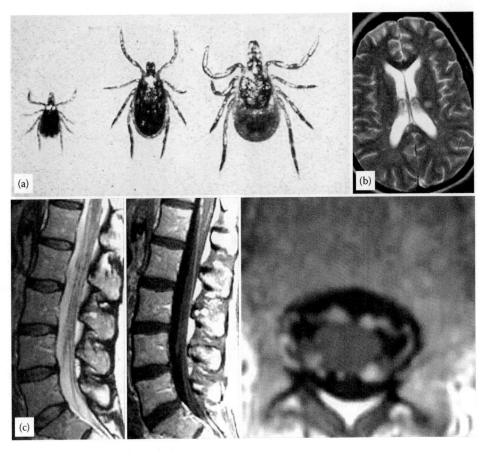

FIGURE 22.58

Lymes *Disease Host*: (a) Causative organism *Borrelia burgdorferi lives* multiple tick species most predominantly the smaller wood tick on the left, but may also be found in black legged ticks middle, and deer ticks right. Cysticercosis intraparenchymal: (b) coronal and (c) axial T_1 post-gadolinium images show multiple cysticercal cysts throughout brain parenchyma.

22.7 Prion Disease

Prion diseases are progressive fatal neurodegenerative disorders affecting humans, livestock, and other primates that are included in this discussion because the pathogenesis of the disease invokes the concept of an *infectious* protein or prion, which is able to induce abnormal protein folding in otherwise normal proteins. This transmissible protein was named and first postulated by Stanley Prusiner who eventually won a Nobel Prize for his work following a period of substantial controversy regarding this concept.

The most common human prion disease is Creutzfeldt–Jakob disease (CJD), which most commonly occurs due to a sporadic mutation (85%–90%), with a small number of cases genetically inherited (5%–15%) and a small number of cases acquired through ingestion of beef products from cows with bovine spongiform encephalopathy (BSE) or *mad cow disease*. This last category has been specifically termed variant CJD (vCJD) and has been substantially minimized by more careful surveillance to eliminate sick

cows from entering the human food chain. vCJD tends to occur in a younger demographic and appears to have a more rapid progressive course than typical sporadic CJD but was otherwise very similar.

The pathogenesis of CJD involves a normal cell wall glycoprotein, PrP, whose exact function is uncertain, which normally exists in a certain conformation or folding pattern within the neuron. Upon exposure to abnormally folded PrP molecules, either endogenous (sporadic inherited mutation) or exogenous (ingestion/exposure), the normal proteins are recruited and change into the pathologic conformation (Figure 22.59). These abnormal or *bent* proteins are nonfunctional and essentially build up within the cell, eventually leading to neuronal dysfunction and cell death.

The exact mechanism of prion neurotoxicity is uncertain and may be partially due to protein crowding; however, recent research suggests that a more complex dysregulation of cellular protein control processes also plays a role.

The neuropathological hall mark of CJD is neuronal vacuolization, atrophy, and cell death (Figure 22.60). On autopsy or biopsy, definitive diagnosis is provided

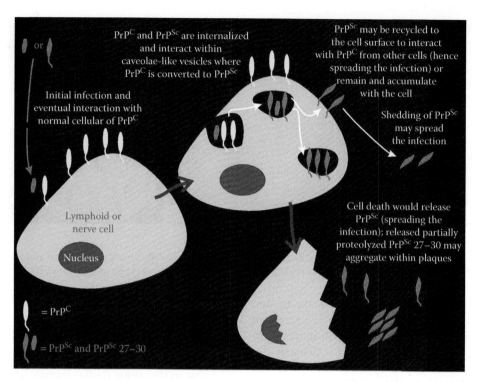

FIGURE 22.59
Prion PrP pathogenesis: Abnormal conformation of PrP protein is introduced into neuronal cells and then *induces* conformational change in normal PrP, ultimately leading to neuronal dysfunction and death.

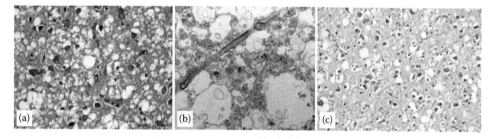

FIGURE 22.60
Creutzfeldt–Jakob disease neuropathology: (a) Electron micrograph. (b and c) Low and high power H&E stain micrographs show neuronal vacuolization and degeneration.

by detection of protease-resistant PrP by Western blot. Clinically patients exhibit myoclonus and progressive cognitive decline. On imaging studies, FLAIR and DWI sequences are most sensitive and are considered critical in the diagnosis. Various patterns have been described, including cortical, basal ganglia, cortical and basal ganglia, and the more specific *hockey stick* or pulvinar sign involving the dorsomedial and posterior regions of the thalamus, which at one time was thought to be more specific for vCJD but can actually be seen in sporadic and familial CJD as well (Figure 22.61) [200–207].

One uncommon but notable subtype of CJD is the Heidenhain variant, first described clinically in 1929. Clinically these patients exhibit rapidly progressive homonymous field defects, disturbed perception of colors or structures, optical hallucinations, cortical blindness, and optical anosognosia in the setting of cognitive decline progressing to severe dementia and death. Imaging in these patients reveals bi-occipital hypometabolism on positron emission tomography, and abnormal visual pathway metabolism, and on MRI subtly poorly defined areas of signal in the bilateral occipital cortices on FLAIR and DWI may be seen [208–214].

Other known human prion diseases include Gerstmann–Sträussler–Scheinker syndrome (rare), fatal familial insomnia (rarer), and kuru, first reported by Gajdusek as a wasting disease among Fore Islanders who practiced ritual cannibalism, including consumption of human brain, but which has since been eradicated. Known animal prion diseases include BSE,

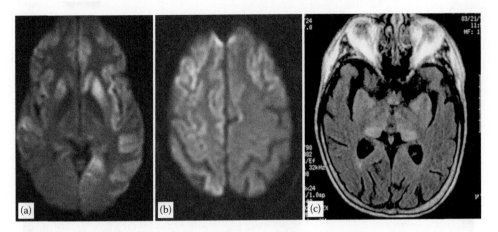

FIGURE 22.61

Creutzfeldt–Jakob disease. (a) Axial DWI shows cortical restricted diffusion. (b) Axial DWI shows both cortical and basal ganglia restricted diffusion. (c) Axial FLAIR image shows a partial hockey stick or pulvinar sign, once thought to be specific for vCJD.

scrapie in sheep, and chronic wasting disease in deer and elk. Although controversial, there is mounting evidence for prion involvement in other human degenerative diseases, including Alzheimer's disease (tau protein), parkinsonism (alpha synnuclein), frontotemporal dementia, chronic posttraumatic encephalopathy, amyotrophic lateral sclerosis, and Huntington's disease.

22.8 Conclusion

There is an immense array of human infectious disease that may affect the CNS. In this chapter, we have described and outlined the major categories with a few examples of each and hopefully have provided some framework in which to consider these potentially life-threatening conditions. Correct diagnosis leading to appropriate therapy will continue to depend on the efforts of conscientious clinicians using all tools at their disposal, including careful history, physical examination, laboratory studies, imaging studies, and when needed invasive tests and biopsies to obtain direct tissue confirmation.

References

1. Willensdorfer M. (2009). On the evolution of differentiated multicellularity. *Evolution.* 63(2):306–323. doi: 10.1111/j.1558-5646.2008.00541.x.

2. Muldoon LL; Alvarez JI; Begley DJ; Boado RJ; Del Zoppo GJ; Doolittle ND; Engelhardt B et al. Immunologic privilege in the central nervous system and the blood-brain barrier. *Journal of Cerebral Blood Flow & Metabolism.* 33(1):13–21, Jan 2013.

3. Biron DG; Loxdale HD. Host-parasite molecular crosstalk during the manipulative process of a host by its parasite. *Journal of Experimental Biology.* 216(Pt 1):148–160, Jan 1, 2013.

4. Adamo SA; Kovalko I; Easy RH; Stoltz D. A viral aphrodisiac in the cricket *Gryllus texensis. Journal of Experimental Biology.* 217(Pt 11):1970–1976, Jun 1, 2014.

5. Flegr J; Markos A. Masterpiece of epigenetic engineering—How *Toxoplasma gondii* reprogrammes host brains to change fear to sexual attraction. *Molecular Ecology.* 23(24):5934–5936, Dec 2014.

6. Carruthers VB; Suzuki Y. Effects of *Toxoplasma gondii* infection on the brain. *Schizophrenia Bulletin.* 33(3):745–751, May 2007.

7. Webster JP. The effect of *Toxoplasma gondii* on animal behavior: Playing cat and mouse. *Schizophrenia Bulletin.* 33(3):752–756, May 2007.

8. Hughes DP; Andersen SB; Hywel-Jones NL; Himaman W; Billen J; Boomsma JJ. Behavioral mechanisms and morphological symptoms of zombie ants dying from fungal infection. *BMC Ecology.* 11:13, 2011.

9. Upadhyayula S. Question 2 * Is there a role for MRI as an adjunct for diagnosing bacterial meningitis? *Archives of Disease in Childhood.* 98(5):388–390, May 2013.

10. Kasanmoentalib ES; Brouwer MC; van de Beek D. Update on bacterial meningitis: Epidemiology, trials and genetic association studies. *Current Opinion in Neurology.* 26(3):282–288, Jun 2013.

11. Kowalsky RH; Jaffe DM. Bacterial meningitis post-PCV7: Declining incidence and treatment. *Pediatric Emergency Care.* 29(6):758–766; quiz 767–768, Jun 2013.

12. Barichello T; Fagundes GD; Generoso JS; Elias SG; Simoes LR; Teixeira AL. Pathophysiology of neonatal acute bacterial meningitis. *Journal of Medical Microbiology.* 62(Pt 12):1781–1789, Dec 2013.

13. Radetsky M. Fulminant bacterial meningitis. *Pediatric Infectious Disease Journal.* 33(2):204–207, Feb 2014.

14. Bleck TP. Bacterial meningitis and other nonviral infections of the nervous system. *Critical Care Clinics.* 29(4):975–987, Oct 2013.

15. Kasanmoentalib ES; Brouwer MC; van de Beek D. Update on bacterial meningitis: Epidemiology, trials and genetic association studies. *Current Opinion in Neurology.* 26(3):282–288, Jun 2013.

16. Ferreira NP; Otta GM; do Amaral LL; da Rocha AJ. Imaging aspects of pyogenic infections of the central nervous system. *Topics in Magnetic Resonance Imaging.* 16(2):145–154, Apr 2005.

17. Lee KS; Liu SS; Spetzler RF; Rekate HL. Intracranial mycotic aneurysm in an infant: Report of a case. *Neurosurgery.* 26(1):129–133, Jan 1990.

18. Williams MT; Jiang H. Diffuse cerebral petechial hemorrhage in an 8-year-old girl with MRSA pneumonia and sepsis. *Neurology.* 82(3):282, Jan 21, 2014.

19. Gupta V; Jain V; Mathuria SN; Khandelwal N. Endovascular treatment of a mycotic intracavernous carotid artery aneurysm using a stent graft. *Interventional Neuroradiology.* 19(3):313–319, Sep 2013.

20. Sonneville R; Mirabel M; Hajage D; Tubach F; Vignon P; Perez P; Lavoue S et al.; ENDOcardite en REAnimation Study Group. Neurologic complications and outcomes of infective endocarditis in critically ill patients: The ENDOcardite en REAnimation prospective multicenter study. *Critical Care Medicine.* 39(6):1474–1481, Jun 2011.

21. Minnerup J; Schilling M; Wersching H; Olschlager C; Schabitz WR; Niederstadt T; Dziewas R. Development of a mycotic aneurysm within 4 days. *Neurology.* 71(21):1745, Nov 18, 2008.

22. Mishra AM; Gupta RK; Jaggi RS; Reddy JS; Jha DK; Husain N; Prasad KN; Behari S; Husain M. Role of diffusion-weighted imaging and in vivo proton magnetic resonance spectroscopy in the differential diagnosis of ring-enhancing intracranial cystic mass lesions. *Journal of Computer Assisted Tomography.* 28(4):540–547, Jul–Aug 2004.

23. Lai PH; Ho JT; Chen WL; Hsu SS; Wang JS; Pan HB; Yang CF. Brain abscess and necrotic brain tumor: Discrimination with proton MR spectroscopy and diffusion-weighted imaging. *AJNR American Journal of Neuroradiology.* 23(8):1369–1377, Sep 2002.

24. Cartes-Zumelzu FW; Stavrou I; Castillo M; Eisenhuber E; Knosp E; Thurnher MM. Diffusion-weighted imaging in the assessment of brain abscesses therapy. *AJNR American Journal of Neuroradiology.* 25(8):1310–1317, Sep 2004.

25. Luthra G; Parihar A; Nath K; Jaiswal S; Prasad KN; Husain N; Husain M; Singh S; Behari S; Gupta RK. Comparative evaluation of fungal, tubercular, and pyogenic brain abscesses with conventional and diffusion MR imaging and proton MR spectroscopy. *AJNR American Journal of Neuroradiology.* 28(7):1332–1338, Aug 2007.

26. Soares-Fernandes JP; Valle-Folgueral JM; Morais N; Ribeiro M; Moreira-da-Costa JA. Diffusion-weighted MR imaging findings in an isolated abscess of the clivus. *AJNR American Journal of Neuroradiology.* 29(1):51–52, Jan 2008.

27. Aguilar J; Urday-Cornejo V; Donabedian S; Perri M; Tibbetts R; Zervos M. *Staphylococcus aureus* meningitis: Case series and literature review [44 refs]. *Medicine.* 89(2):117–125, Mar 2010.

28. Landrum ML; Neumann C; Cook C; Chukwuma U; Ellis MW; Hospenthal DR; Murray CK. Epidemiology of Staphylococcus aureus blood and skin and soft tissue infections in the US military health system, 2005-2010. *JAMA.* 308(1):50–59, Jul 4, 2012.

29. Holland TL; Arnold C; Fowler VG Jr. Clinical management of *Staphylococcus aureus* bacteremia: A review. *JAMA.* 312(13):1330–1341, Oct 1, 2014.

30. Cosgrove SE; Sakoulas G; Perencevich EN; Schwaber MJ; Karchmer AW; Carmeli Y. Comparison of mortality associated with methicillin-resistant and methicillin-susceptible *Staphylococcus aureus* bacteremia: A meta-analysis. *Clinical Infectious Diseases.* 36(1):53–59, Jan 1, 2003.

31. Centers for Disease Control and Prevention. 2012. Active Bacterial Core Surveillance Report, Emerging Infections Program Network, Methicillin-Resistant *Staphylococcus aureus*, 41–42, 2012.

32. Hooper DC; Pruitt AA; Rubin RH. Central nervous system infection in the chronically immunosuppressed [196 refs]. *Medicine.* 61(3):166–188, May 1982.

33. Bowie D; Marrie TJ; Haldane EV; Noble MA. Ataxia in Listeria monocytogenes infections of the central nervous system. *Southern Medical Journal.* 76(5):567–570, May 1983.

34. Katz RI; McGlamery ME; Levy R. CNS listeriosis: Rhomboencephalitis in a healthy, immunocompetent person. *Archives of Neurology.* 36(8):513–514, Aug 1979.

35. Rocha PR; Lomonaco S; Bottero MT; Dalmasso A; Dondo A; Grattarola C; Zuccon F et al. Ruminant rhombencephalitis-associated *Listeria monocytogenes* strains constitute a genetically homogeneous group related to human outbreak strains. (Erratum appears in *Applied & Environmental Microbiology.* 79(22):7114, Nov 2013.) *Applied & Environmental Microbiology.* 79(9):3059–3066, May 2013.

36. Choi MJ; Jackson KA; Medus C; Beal J; Rigdon CE; Cloyd TC; Forstner MJ et al.; Centers for Disease Control and Prevention (CDC). Notes from the field: Multistate outbreak of listeriosis linked to soft-ripened cheese—United States, 2013. *MMWR—Morbidity & Mortality Weekly Report.* 63(13):294–295, Apr 4, 2014.

37. Yetkin MA; Bulut C; Erdinc FS; Oral B; Tulek N. Evaluation of the clinical presentations in neurobrucellosis. *International Journal of Infectious Diseases.* 10(6):446–452, Nov 2006.

38. Drutz JE. Brucellosis of the central nervous system. A case report of an infected infant. *Clinical Pediatrics.* 28(10):476–478, Oct 1989.

39. Strannegard IL; Araj GF; Fattah HA. Neurobrucellosis in an eight-year-old child. *Annals of Tropical Paediatrics.* 5(4):191–194, Dec 1985.

40. Sanchez-Sousa A; Torres C; Campello MG; Garcia C; Parras F; Cercenado E; Baquero F. Serological diagnosis of neurobrucellosis. *Journal of Clinical Pathology.* 43(1):79–81, Jan 1990.

41. Herbert N; George A; Baroness Masham of Ilton; Sharma V; Oliver M; Oxley A; Raviglione M; Zumla AI. World TB Day 2014: Finding the missing 3 million. *Lancet.* 383(9922):1016–1018, Mar 22, 2014.

42. Chou PS, Liu CK, Lin RT, Lai CL, Chao AC. Central nervous system tuberculosis: a forgotten diagnosis. *Neurologist.* 18(4):219–222, Jul 2012. doi:10.1097/NRL.0b013e3182610347.

43. Thwaites GE; Schoeman JF. Update on tuberculosis of the central nervous system: Pathogenesis, diagnosis, and treatment [65 refs]. *Clinics in Chest Medicine.* 30(4):745–754, ix, Dec 2009.

44. Bernaerts A; Vanhoenacker FM; Parizel PM; Van Goethem JW; Van Altena R; Laridon A; De Roeck J; Coeman V; De Schepper AM. Tuberculosis of the central nervous system: Overview of neuroradiological findings [46 refs]. *European Radiology.* 13(8):1876–1890, Aug 2003.

45. Garg RK; Sharma R; Kar AM; Kushwaha RA; Singh MK; Shukla R; Agarwal A; Verma R. Neurological complications of miliary tuberculosis. *Clinical Neurology & Neurosurgery.* 112(3):188–192, Apr 2010.

46. Antunes H; Pereira A; Cunha I. Chediak-Higashi syndrome: pathognomonic feature. *Lancet.* 382(9903):1514, Nov 2, 2013.

47. van den Berg B; Walgaard C; Drenthen J; Fokke C; Jacobs BC; van Doorn PA. Guillain-Barre syndrome: Pathogenesis, diagnosis, treatment and prognosis. *Nature Reviews Neurology.* 10(8):469–482, Aug 2014.

48. Honavar M; Tharakan JK; Hughes RA; Leibowitz S; Winer JB. A clinicopathological study of the Guillain-Barre syndrome. Nine cases and literature review. *Brain.* 114 (Pt 3):1245–1269, Jun 1991.

49. Wakerley BR; Uncini A; Yuki N; GBS Classification Group; GBS Classification Group. Guillain-Barre and Miller Fisher syndromes—New diagnostic classification. *Nature Reviews Neurology.* 10(9):537–544, Sep 2014.

50. Taboada EN; van Belkum A; Yuki N; Acedillo RR; Godschalk PC; Koga M; Endtz HP; Gilbert M; Nash JH. Comparative genomic analysis of *Campylobacter jejuni* associated with Guillain-Barre and Miller Fisher syndromes: Neuropathogenic and enteritis-associated isolates can share high levels of genomic similarity. *BMC Genomics.* 8:359, 2007.

51. Leite C; Barbosa A Jr; Lucato LT. Viral diseases of the central nervous system [287 refs]. *Topics in Magnetic Resonance Imaging.* 16(2):189–212, Apr 2005.

52. Ciuffi A; Telenti A. State of genomics and epigenomics research in the perspective of HIV cure. *Current Opinion in HIV & AIDS.* 8(3):176–181, May 2013.

53. Dennis AM; Herbeck JT; Brown AL; Kellam P; de Oliveira T; Pillay D; Fraser C; Cohen MS. Phylogenetic studies of transmission dynamics in generalized HIV epidemics: An essential tool where the burden is greatest? *Journal of Acquired Immune Deficiency Syndromes: JAIDS.* 67(2):181–195, Oct 1, 2014.

54. Lenjisa JL; Woldu MA; Satessa GD. New hope for eradication of HIV from the body: The role of polymeric nanomedicines in HIV/AIDS pharmacotherapy. *Journal of Nanobiotechnology.* 12:9, 2014.

55. Kumarasamy N; Krishnan S. Beyond first-line HIV treatment regimens: The current state of antiretroviral regimens, viral load monitoring, and resistance testing in resource-limited settings. *Current Opinion in HIV & AIDS.* 8(6):586–590, Nov 2013.

56. Smith MK; Rutstein SE; Powers KA; Fidler S; Miller WC; Eron JJ Jr; Cohen MS. The detection and management of early HIV infection: A clinical and public health emergency. *Journal of Acquired Immune Deficiency Syndromes: JAIDS.* 63(Suppl 2):S187–S199, Jul 2013.

57. Baeten J; Celum C. Systemic and topical drugs for the prevention of HIV infection: Antiretroviral pre-exposure prophylaxis. *Annual Review of Medicine.* 64:219–232, 2013.

58. Smith AB; Smirniotopoulos JG; Rushing EJ. From the archives of the AFIP: Central nervous system infections associated with human immunodeficiency virus infection: Radiologic-pathologic correlation. (Erratum appears in *RadioGraphics.* 29(2):638, Mar–Apr 2009.) *RadioGraphics.* 28(7):2033–2058, Nov–Dec 2008.

59. Dubrovsky T; Curless R; Scott G; Chaneles M; Post MJ; Altman N; Petito CK; Start D; Wood C. Cerebral aneurysmal arteriopathy in childhood AIDS. *Neurology.* 51(2):560–565, Aug 1998.

60. Rhodes RH; Ward JM; Cowan RP; Moore PT. Immunohistochemical localization of human immunodeficiency viral antigens in formalin-fixed spinal cords with AIDS myelopathy. *Clinical Neuropathology.* 8(1):22–27, Jan–Feb 1989.

61. Tyler KL. Current developments in understanding of West Nile virus central nervous system disease. *Current Opinion in Neurology.* 27(3):342–348, Jun 2014.

62. Gaensbauer JT; Lindsey NP; Messacar K; Staples JE; Fischer M. Neuroinvasive arboviral disease in the United States: 2003 to 2012. *Pediatrics.* 134(3):e642–e650, Sep 2014.

63. Asadi L; Bunce PE. West Nile virus infection. *CMAJ Canadian Medical Association Journal.* 185(18):E846, Dec 10, 2013.

64. Mertz D; Kim TH; Johnstone J; Lam PP; Science M; Kuster SP; Fadel SA et al. Populations at risk for severe or complicated Avian Influenza H5N1: A systematic review and meta-analysis. *PLoS One [Electronic Resource].* 9(3):e89697, 2014.

65. Kidd M. Influenza viruses: Update on epidemiology, clinical features, treatment and vaccination. *Current Opinion in Pulmonary Medicine.* 20(3):242–246, May 2014.

66. Van Kerkhove MD. Brief literature review for the WHO global influenza research agenda—Highly pathogenic avian influenza H5N1 risk in humans. *Influenza & Other Respiratory Viruses.* 7(Suppl 2):26–33, Sep 2013.

67. Liu Q; Liu DY; Yang ZQ. Characteristics of human infection with avian influenza viruses and development of new antiviral agents. *Zhongguo Yao Li Xue Bao/Acta Pharmacologica Sinica.* 34(10):1257–1269, Oct 2013.

68. Sambhara S; Poland GA. H5N1 Avian influenza: Preventive and therapeutic strategies against a pandemic [71 refs]. *Annual Review of Medicine.* 61:187–198, 2010.

69. Herfst S; Schrauwen EJ; Linster M; Chutinimitkul S; de Wit E; Munster VJ; Sorrell EM et al. Airborne transmission of influenza A/H5N1 virus between ferrets. *Science.* 336(6088):1534–1541, Jun 22, 2012.

70. Russell CA; Fonville JM; Brown AE; Burke DF; Smith DL; James SL; Herfst S et al. The potential for respiratory droplet-transmissible A/H5N1 influenza virus to evolve in a mammalian host. *Science*. 336(6088):1541–1547, Jun 22, 2012.

71. Provenzale JM. Centennial dissertation. Honoring Arthur W. Goodspeed, MD and James B. Bullitt, MD. CT and MR imaging and nontraumatic neurologic emergencies [64 refs]. *AJR American Journal of Roentgenology*. 174(2):289–299, Feb 2000.

72. Maschke M; Kastrup O; Forsting M; Diener HC. Update on neuroimaging in infectious central nervous system disease. *Current Opinion in Neurology*. 17(4):475–480, Aug 2004.

73. Steiner I. Herpes simplex virus encephalitis: New infection or reactivation? *Current Opinion in Neurology*. 24(3):268–274, Jun 2011.

74. Ward KN; Ohrling A; Bryant NJ; Bowley JS; Ross EM; Verity CM. Herpes simplex serious neurological disease in young children: Incidence and long-term outcome. *Archives of Disease in Childhood*. 97(2):162–165, Feb 2012.

75. Berger JR; Houff S. Neurological complications of herpes simplex virus type 2 infection. *Archives of Neurology*. 65(5):596–600, May 2008.

76. Bajaj M; Mody S; Natarajan G. Clinical and neuroimaging findings in neonatal herpes simplex virus infection. *Journal of Pediatrics*. 165(2):404–407.e1, Aug 2014.

77. Gershon AA; Gershon MD. Pathogenesis and current approaches to control of varicella-zoster virus infections. *Clinical Microbiology Reviews*. 26(4):728–743, Oct 2013.

78. Johnson RW; Rice AS. Clinical practice. Postherpetic neuralgia. *New England Journal of Medicine*. 371(16):1526–1533, Oct 16, 2014.

79. Nagel MA; Gilden D. Neurological complications of varicella zoster virus reactivation. *Current Opinion in Neurology*. 27(3):356–360, Jun 2014.

80. Yawn BP; Gilden D. The global epidemiology of herpes zoster. *Neurology*. 81(10):928–930, Sep 3, 2013.

81. Stahl JP; Mailles A. What is new about epidemiology of acute infectious encephalitis? *Current Opinion in Neurology*. 27(3):337–341, Jun 2014.

82. Crowcroft NS; Thampi N. The prevention and management of rabies. *BMJ*. 350:g7827, 2015.

83. Tyler KL. Emerging viral infections of the central nervous system: Part 1. *Archives of Neurology*. 66(8):939–948, Aug 2009.

84. Mittal MK. Revised 4-dose vaccine schedule as part of postexposure prophylaxis to prevent human rabies. *Pediatric Emergency Care*. 29(10):1119–1121; quiz 1122–1124, Oct 2013.

85. Mateen FJ; Shinohara RT; Sutter RW. Oral and inactivated poliovirus vaccines in the newborn: A review. *Vaccine*. 31(21):2517–2524, May 17, 2013.

86. Nathanson N; Kew OM. From emergence to eradication: The epidemiology of poliomyelitis deconstructed. *American Journal of Epidemiology*. 172(11):1213–1229, Dec 1, 2010.

87. Khan T; Qazi J. Hurdles to the global antipolio campaign in Pakistan: An outline of the current status and future prospects to achieve a polio free world. *Journal of Epidemiology & Community Health*. 67(8):696–702, Aug 2013.

88. Grassly NC. The final stages of the global eradication of poliomyelitis. *Philosophical Transactions of the Royal Society of London—Series B: Biological Sciences*. 368(1623):20120140, Aug 5, 2013.

89. Fink KR; Thapa MM; Ishak GE; Pruthi S. Neuroimaging of pediatric central nervous system cytomegalovirus infection. *RadioGraphics*. 30(7):1779–1796, Nov 2010.

90. Barkovich AJ; Moore KR; Jones BV; Vezina GK; Bernadette L; Raybaud C; Grant PE et al. *Diagnostic Imaging: Pediatric Neuroradiology*. Salt Lake City, UT: Amirsys, 2007.

91. Goderis J; De Leenheer E; Smets K; Van Hoecke H; Keymeulen A; Dhooge I. Hearing loss and congenital CMV infection: A systematic review. *Pediatrics*. 134(5):972–982, Nov 2014.

92. Sherman RA. Charts: The TORCH syndrome revisited. *Pediatric Infectious Disease Journal*. 8(1):62–63, Jan 1989.

93. Del Pizzo J. Focus on diagnosis: Congenital infections (TORCH). (Erratum appears in *Pediatr. Rev.* 33(3):109, Mar 2012.) *Pediatrics in Review*. 32(12):537–542, Dec 2011.

94. Altman NR. Intracranial infection in children. *Topics in Magnetic Resonance Imaging*. 5(3):143–160, 1993.

95. Greenough A. The TORCH screen and intrauterine infections. *Archives of Disease in Childhood Fetal & Neonatal Edition*. 70(3):F163–F165, May 1994.

96. Tenembaum S; Chitnis T; Ness J; Hahn JS; International Pediatric MS Study Group. Acute disseminated encephalomyelitis. *Neurology*. 68(16 Suppl. 2):S23–S36, Apr 17, 2007.

97. Alper G. Acute disseminated encephalomyelitis. *Journal of Child Neurology*. 27(11):1408–1425, Nov 2012.

98. Elias MD; Narula S; Chu AS. Acute disseminated encephalomyelitis following meningoencephalitis: Case report and literature review. *Pediatric Emergency Care*. 30(4):254–256, Apr 2014.

99. Wingerchuk DM; Lucchinetti CF. Comparative immunopathogenesis of acute disseminated encephalomyelitis, neuromyelitis optica, and multiple sclerosis. *Current Opinion in Neurology*. 20(3):343–350, Jun 2007.

100. Yuksel D; Diren B; Ulubay H; Altunbasak S; Anlar B. Neuronal loss is an early component of subacute sclerosing panencephalitis. *Neurology*. 83(10):938–944, Sep 2, 2014.

101. Colpak AI; Erdener SE; Ozgen B; Anlar B; Kansu T. Neuro-ophthalmology of subacute sclerosing panencephalitis: Two cases and a review of the literature. *Current Opinion in Ophthalmology*. 23(6):466–471, Nov 2012.

102. Garg RK. Subacute sclerosing panencephalitis. *Postgraduate Medical Journal*. 78(916):63–70, Feb 2002.

103. Baldauf SL; Palmer JD. Animals and fungi are each other's closest relatives: Congruent evidence from multiple proteins. *Proceedings of the National Academy of Sciences of the United States of America*. 90(24):11558–11562, Dec 15, 1993.

104. Moran GP; Coleman DC; Sullivan DJ. Comparative genomics and the evolution of pathogenicity in human pathogenic fungi. *Eukaryotic Cell.* 10(1):34–42, Jan 2011.

105. Mendoza L; Taylor JW; Ajello L. The class mesomycetozoea: A heterogeneous group of microorganisms at the animal-fungal boundary. *Annual Review of Microbiology.* 56:315–344, 2002.

106. May GS; Adams TH. The importance of fungi to man. *Genome Research.* 7(11):1041–1044, Nov 1997.

107. Singh H; Irwin S; Falowski S; Rosen M; Kenyon L; Jungkind D; Evans J. Curvularia fungi presenting as a large cranial base meningioma: Case report. *Neurosurgery.* 63(1):E177; discussion E177, Jul 2008.

108. Kourkoumpetis TK; Desalermos A; Muhammed M; Mylonakis E. Central nervous system aspergillosis: A series of 14 cases from a general hospital and review of 123 cases from the literature. *Medicine.* 91(6):328–336, Nov 2012.

109. Almutairi BM; Nguyen TB; Jansen GH; Asseri AH. Invasive aspergillosis of the brain: Radiologic-pathologic correlation. *RadioGraphics.* 29(2):375–379, Mar-Apr 2009.

110. Robinson MR; Fine HF; Ross ML; Mont EK; Bryant-Greenwood PK; Hertle RW; Tisdale JF et al. Sino-orbital-cerebral aspergillosis in immunocompromised pediatric patients. *Pediatric Infectious Disease Journal.* 19(12):1197–1203, Dec 2000.

111. Gupta AO; Singh N. Immune reconstitution syndrome and fungal infections. *Current Opinion in Infectious Diseases.* 24(6):527–533, Dec 2011.

112. O'Meara TR; Alspaugh JA. The *Cryptococcus neoformans* capsule: A sword and a shield. *Clinical Microbiology Reviews.* 25(3):387–408, Jul 2012.

113. Voelz K; May RC. Cryptococcal interactions with the host immune system. *Eukaryotic Cell.* 9(6):835–846, Jun 2010.

114. Kwee RM; Kwee TC. Virchow-Robin spaces at MR imaging. *RadioGraphics.* 27(4):1071–1086, Jul–Aug 2007.

115. Bicanic T; Harrison TS. *Cryptococcal meningitis. British Medical Bulletin.* 72:99–118, 2004.

116. Scozzafava J; Block H; Asdaghi N; Siddiqi ZA. Teaching NeuroImage: Cryptococcal brain pseudocysts in an immunocompetent patient. *Neurology.* 69(9):E6–E7, Aug 28, 2007.

117. Kumari R; Raval M; Dhun A. Cryptococcal choroid plexitis: Rare imaging findings of central nervous system cryptococcal infection in an immunocompetent individual. *British Journal of Radiology.* 83(985):e14–e17, Jan 2010.

118. Spellberg B; Edwards J Jr; Ibrahim A. Novel perspectives on mucormycosis: Pathophysiology, presentation, and management. *Clinical Microbiology Reviews.* 18(3):556–569, Jul 2005.

119. Ibrahim AS; Kontoyiannis DP. Update on mucormycosis pathogenesis. *Current Opinion in Infectious Diseases.* 26(6):508–515, Dec 2013.

120. Ibrahim AS; Spellberg B; Edwards J Jr. Iron acquisition: A novel perspective on mucormycosis pathogenesis and treatment. *Current Opinion in Infectious Diseases.* 21(6):620–625, Dec 2008.

121. Loke P; Lim YA. Helminths and the microbiota: parts of the hygiene hypothesis. *Parasite Immunology.* 37(6):314–323, Jun 2015.

122. Walker MD; Zunt JR. Neuroparasitic infections: Cestodes, trematodes, and protozoans. *Seminars in Neurology.* 25(3):262–277, Sep 2005.

123. Kimura-Hayama ET; Higuera JA; Corona-Cedillo R; Chavez-Macias L; Perochena A; Quiroz-Rojas LY; Rodriguez-Carbajal J; Criales JL. Neurocysticercosis: Radiologic-pathologic correlation. *RadioGraphics.* 30(6):1705–1719, Oct 2010.

124. Osborn AG; Preece MT. Intracranial cysts: Radiologic-pathologic correlation and imaging approach. *Radiology.* 239(3):650–664, Jun 2006.

125. Singhi P; Singhi S. Neurocysticercosis in children. *Journal of Child Neurology.* 19(7):482–492, Jul 2004.

126. MacCormick IJ; Beare NA; Taylor TE; Barrera V; White VA; Hiscott P; Molyneux ME; Dhillon B; Harding SP. Cerebral malaria in children: Using the retina to study the brain. *Brain.* 137(Pt 8):2119–2142, Aug 2014.

127. Crompton PD; Moebius J; Portugal S; Waisberg M; Hart G; Garver LS; Miller LH; Barillas-Mury C; Pierce SK. Malaria immunity in man and mosquito: Insights into unsolved mysteries of a deadly infectious disease. *Annual Review of Immunology.* 32:157–187, 2014.

128. Riedel B; Browne K; Silbert B. Cerebral protection: Inflammation, endothelial dysfunction, and post-operative cognitive dysfunction. *Current Opinion in Anaesthesiology.* 27(1):89–97, Feb 2014.

129. Epstein JE; Richie TL. The whole parasite, pre-erythrocytic stage approach to malaria vaccine development: A review. *Current Opinion in Infectious Diseases.* 26(5):420–428, Oct 2013.

130. Potchen MJ; Kampondeni SD; Ibrahim K; Bonner J; Seydel KB; Taylor TE; Birbeck GL. NeuroInterp: A method for facilitating neuroimaging research on cerebral malaria. *Neurology.* 81(6):585–588, Aug 6, 2013.

Baird JK. Evidence and implications of mortality associated with acute *Plasmodium vivax* malaria. *Clinical Microbiology Reviews.* 26(1):36–57, Jan 2013.

131. Laishram DD; Sutton PL; Nanda N; Sharma VL; Sobti RC; Carlton JM; Joshi H. The complexities of malaria disease manifestations with a focus on asymptomatic malaria. *Malaria Journal.* 11:29, 2012.

132. Landfear SM. Nutrient transport and pathogenesis in selected parasitic protozoa. *Eukaryotic Cell.* 10(4):483–493, Apr 2011.

133. Buffet PA; Safeukui I; Deplaine G; Brousse V; Prendki V; Thellier M; Turner GD; Mercereau-Puijalon O. The pathogenesis of *Plasmodium falciparum* malaria in humans: Insights from splenic physiology. *Blood.* 117(2):381–392, Jan 13, 2011.

134. Sadanand S. Malaria: An evaluation of the current state of research on pathogenesis and antimalarial drugs. *Yale Journal of Biology & Medicine.* 83(4):185–191, Dec 2010.

135. Lee GT; Antelo F; Mlikotic AA. Best cases from the AFIP: Cerebral toxoplasmosis. *RadioGraphics.* 29(4):1200–1205, Jul–Aug 2009.

136. Caselli D; Andreoli E; Paolicchi O; Savelli S; Guidi S; Pecile P; Arico M. Acute encephalopathy in the immune-compromised child: Never forget toxoplasmosis. *Journal of Pediatric Hematology/Oncology.* 34(5):383–386, Jul 2012.

137. Popli MB; Popli V. Congenital toxoplasmosis infection. *Neurology India.* 51(1):125, Mar 2003.

138. Camacho DL; Smith JK; Castillo M. Differentiation of toxoplasmosis and lymphoma in AIDS patients by using apparent diffusion coefficients. *AJNR American Journal of Neuroradiology.* 24(4):633–637, Apr 2003.

139. Pietrucha-Dilanchian P; Chan JC; Castellano-Sanchez A; Hirzel A; Laowansiri P; Tuda C; Visvesvara GS; Qvarnstrom Y; Ratzan KR. *Balamuthia mandrillaris* and *Acanthamoeba* amebic encephalitis with neurotoxoplasmosis coinfection in a patient with advanced HIV infection. *Journal of Clinical Microbiology.* 50(3):1128–1131, Mar 2012.

140. Boothroyd JC; Grigg ME. Population biology of *Toxoplasma gondii* and its relevance to human infection: Do different strains cause different disease? *Current Opinion in Microbiology.* 5:438–442, 2002.

141. Fadil H; Gonzalez-Toledo E; Kelley BJ; Kelley RE. Neuroimaging findings in neurosyphilis. *Journal of Neuroimaging.* 16(3):286–289, Jul 2006.

142. Primary and secondary syphilis—United States, 2003–2004. *MMWR—Morbidity & Mortality Weekly Report.* 55(10):269–273, 2006.

     Fraser CM; Norris SJ; Weinstock GM; White O; Sutton GG; Dodson R; Gwinn M et al. Complete genome sequence of *Treponema pallidum*, the syphilis spirochete. *Science.* 281:375, 1998.

143. Golden MR; Marra CM; Holmes KK. Update on syphilis: Resurgence of an old problem. *JAMA.* 290:1510–1514, 2003.

144. Karsan N; Barker R; O'Dwyer JP. Clinical reasoning: The "great imitator." *Neurology.* 83(22):e188–e196, Nov 25, 2014.

145. Clement ME; Okeke NL; Hicks CB. Treatment of syphilis: A systematic review. *JAMA.* 312(18):1905–1917, Nov 12, 2014.

146. Sakai K; Fukuda T; Iwadate K; Maruyama-Maebashi K; Asakura K; Ozawa M; Matsumoto S. A fatal fall associated with undiagnosed parenchymatous neurosyphilis. *American Journal of Forensic Medicine & Pathology.* 35(1):4–7, Mar 2014.

147. Land AM; Nelson GA; Bell SG; Denby KJ; Estrada CA; Willett LL. Widening the differential for brain masses in human immunodeficiency virus-positive patients: Syphilitic cerebral gummata. *American Journal of the Medical Sciences.* 346(3):253–255, Sep 2013.

148. Brightbill TC; Ihmeidan IH; Post MJ; Berger JR; Katz DA. Neurosyphilis in HIV-positive and HIV-negative patients: Neuroimaging findings. *AJNR American Journal of Neuroradiology.* 16(4):703–711, Apr 1995.

149. Dhasmana D; Joshi J; Manavi K. Intracerebral and spinal cord syphilitic gummata in an HIV-negative man: A case report. *Sexually Transmitted Diseases.* 40(8):629–631, Aug 2013.

150. Proudfoot M; McLean B. Old adversaries, modern mistakes: Neurosyphilis. *Practical Neurology.* 13(3):174–177, Jun 2013.

151. Agayeva N; Karli-Oguz K; Saka E. Teaching NeuroImages: A neurosyphilis case presenting with atypical neuroradiologic findings. *Neurology.* 80(11):e119, Mar 12, 2013.

152. Crenner C. The Tuskegee Syphilis Study and the scientific concept of racial nervous resistance. *Journal of the History of Medicine & Allied Sciences.* 67(2):244–280, Apr 2012.

153. Ventura N; Cannelas R; Bizzo B; Gasparetto EL. Intracranial syphilitic gumma mimicking a brain stem glioma. *AJNR American Journal of Neuroradiology.* 33(7):E110–E111, Aug 2012.

154. McManus DP; Gray DJ; Zhang W; Yang Y. Diagnosis, treatment, and management of echinococcosis. *BMJ.* 344:e3866, Jun 11, 2012.

155. Bakoyiannis A; Delis S; Triantopoulou C; Dervenis C. Rare cystic liver lesions: A diagnostic and managing challenge. *World Journal of Gastroenterology.* 19(43):7603–7619, Nov 21, 2013.

156. Atanasov G; Benckert C; Thelen A; Tappe D; Frosch M; Teichmann D; Barth TF et al. Alveolar echinococcosis-spreading disease challenging clinicians: A case report and literature review. *World Journal of Gastroenterology.* 19(26):4257–4261, Jul 14, 2013.

157. Kantarci M; Bayraktutan U; Karabulut N; Aydinli B; Ogul H; Yuce I; Calik M et al. Alveolar echinococcosis: Spectrum of findings at cross-sectional imaging. *RadioGraphics.* 32(7):2053–2070, Nov–Dec 2012.

158. Brunetti E; Junghanss T. Update on cystic hydatid disease. *Current Opinion in Infectious Diseases.* 22(5):497–502, Oct 2009.

159. Guzel A; Tatli M; Maciaczyk J; Altinors N. Primary cerebral intraventricular hydatid cyst: A case report and review of the literature. *Journal of Child Neurology.* 23(5):585–588, May 2008.

160. Gunecs M; Akdemir H; Tugcu B; Gunaldi O; Gumucs E; Akpinar A. Multiple intradural spinal hydatid disease: A case report and review of literature. *Spine.* 34(9):E346–E350, Apr 20, 2009.

161. Eckert J; Deplazes P. Biological, epidemiological, and clinical aspects of echinococcosis, a zoonosis of increasing concern. *Clinical Microbiology Reviews.* 17(1):107–135, Jan 2004.

162. Polat P; Kantarci M; Alper F; Suma S; Koruyucu MB; Okur A. Hydatid disease from head to toe. *RadioGraphics.* 23(2):475–494; quiz 536–537, Mar–Apr 2003.

     Lockey MW. Primary amoebic meningoencephalitis. *Laryngoscope.* 88(3):484–503, Mar 1978.

163. Visvesvara GS; Moura H; Schuster FL. Pathogenic and opportunistic free-living amoebae: *Acanthamoeba* spp., *Balamuthia mandrillaris*, *Naegleria fowleri*, and *Sappinia diploidea*. *FEMS Immunology and Medical Microbiology.* 50(1):1–26, Jun 2007.

164. Jamerson M; da Rocha-Azevedo B; Cabral GA; Marciano-Cabral F. Pathogenic *Naegleria fowleri* and non-pathogenic *Naegleria lovaniensis* exhibit differential adhesion to, and invasion of, extracellular matrix proteins. *Microbiology.* 158(Pt 3):791–803, Mar 2012.

165. Cooter R. The history of the discovery of primary amoebic meningoencephalitis. *Australian Family Physician.* 31(4):399–400, Apr 2002.

166. Silva RA; Araujo Sde A; Avellar IF; Pittella JE; Oliveira JT; Christo PP. Granulomatous amoebic meningoencephalitis in an immunocompetent patient. *Archives of Neurology.* 67(12):1516–1520, Dec 2010.

167. Cox FE. History of human parasitology. (Erratum appears in *Clinical Microbiology Reviews.* 16(1):174, Jan 2003.) *Clinical Microbiology Reviews.* 15(4):595–612, Oct 2002.

168. Siddiqui R; Khan NA. *Balamuthia* amoebic encephalitis: An emerging disease with fatal consequences. *Microbial Pathogenesis.* 44(2):89–97, Feb 2008.

169. Perez MT; Bush LM. *Balamuthia mandrillaris* amebic encephalitis. *Current Infectious Disease Reports.* 9(4):323–328, Jul 2007.

170. Perez MT; Bush LM. Fatal amebic encephalitis caused by *Balamuthia mandrillaris* in an immunocompetent host: A clinicopathological review of pathogenic free-living amebae in human hosts. *Annals of Diagnostic Pathology.* 11(6):440–447, Dec 2007.

171. Vale TC; de Sousa-Pereira SR; Ribas JG; Lambertucci JR. *Neuroschistosomiasis mansoni*: Literature review and guidelines. *Neurologist.* 18(6):333–342, Nov 2012.

172. Morgan OW; Brunette G; Kapella BK; McAuliffe I; Katongole-Mbidde E; Li W et al. Schistosomiasis among recreational users of upper Nile River, Uganda, 2007. *Emerging Infectious Diseases.* 16(5):866–868, 2010 May.

173. Hotez PJ; Fenwick A. Schistosomiasis in Africa: An emerging tragedy in our new global health decade. *PLoS Neglected Tropical Diseases.* 2(9):e485, Sep 29, 2009. doi:10.1371/journal.pntd.0000485.

174. Bierman WF; Wetsteyn JC; van Gool T. Presentation and diagnosis of imported schistosomiasis: Relevance of eosinophilia, microscopy for ova, and serology. *Journal of Travel Medicine.* 12(1):9–13, Jan–Feb 2005.

175. Ross AG; Bartley PB; Sleigh AC; Olds GR; Li Y; Williams GM; McManus DP. Schistosomiasis. *New England Journal of Medicine.* 346:1212–1220, Apr 18, 2002.

176. Rollinson D; Simpson AJG (eds.). *The Biology of Schistosomes from Genes to Latrines.* London: Academic Press, 1987.

177. Xia Y; Ju Y; Chen J; You C. Hemorrhagic stroke and cerebral paragonimiasis. *Stroke.* 45(11):3420–3422, Nov 2014.

178. Diaz JH. Paragonimiasis acquired in the United States: Native and nonnative species. *Clinical Microbiology Reviews.* 26(3):493–504, Jul 2013.

179. Boland JM; Vaszar LT; Jones JL; Mathison BA; Rovzar MA; Colby TV; Leslie KO; Tazelaar HD. Pleuropulmonary infection by *Paragonimus westermani* in the United States: A rare cause of eosinophilic pneumonia after ingestion of live crabs. *American Journal of Surgical Pathology.* 35:707–713, 2011.

180. Human paragonimiasis after eating raw or undercooked crayfish—Missouri, July 2006–September 2010. *MMWR—Morbidity & Mortality Weekly Report.* 59(48):1573–1576, 2010.

181. Lane MA; Barsanti MC; Santos CA; Yeung M; Lubner SJ; Weil GJ. Human paragonimiasis in North America following ingestion of raw crayfish. *Clinical Infectious Diseases.* 49:e55–e61, 2009.

182. Peeters N; van der Kolk BY; Thijsen SF; Colnot DR. Lyme disease associated with sudden sensorineural hearing loss: Case report and literature review. *Otology & Neurotology.* 34(5):832–837, Jul 2013.

183. Traisk F; Lindquist L. Optic nerve involvement in Lyme disease. *Current Opinion in Ophthalmology.* 23(6):485–490, Nov 2012.

184. Samuels DS. Gene regulation in *Borrelia burgdorferi*. *Annual Review of Microbiology.* 65:479–499, 2011.

185. Stanek G; Strle F. Lyme borreliosis: A European perspective on diagnosis and clinical management. *Current Opinion in Infectious Diseases.* 22(5):450–454, Oct 2009.

186. Ogden NH; Lindsay LR; Morshed M; Sockett PN; Artsob H. The emergence of Lyme disease in Canada. (Erratum appears in *CMAJ.* 181(5):291, Sep 1, 2009.) *CMAJ Canadian Medical Association Journal.* 180(12):1221–1224, Jun 9, 2009.

187. Bratton RL; Whiteside JW; Hovan MJ; Engle RL; Edwards FD. Diagnosis and treatment of Lyme disease. *Mayo Clinic Proceedings.* 83(5):566–571, May 2008.

188. Feder HM Jr; Johnson BJ; O'Connell S; Shapiro ED; Steere AC; Wormser GP; Ad Hoc International Lyme Disease Group et al. A critical appraisal of "chronic Lyme disease". (Erratum appears in *New England Journal of Medicine.* 358(10):1084, Mar 6, 2008.) *New England Journal of Medicine.* 357(14):1422–1430, Oct 4, 2007.

189. Steere AC. Lyme disease. *New England Journal of Medicine.* 345(2):115–125, Jul 12, 2001.

190. Donta ST; Noto RB; Vento JA. SPECT brain imaging in chronic Lyme disease. *Clinical Nuclear Medicine.* 37(9):e219–e222, Sep 2012.

191. Agarwal R; Sze G. Neuro-lyme disease: MR imaging findings. *Radiology.* 253(1):167–173, Oct 2009.

192. Centers for Disease Control and Prevention (CDC). Lyme disease—United States, 2003–2005. *MMWR—Morbidity & Mortality Weekly Report.* 56(23):573–576, Jun 15, 2007.

193. Haass A. Lyme neuroborreliosis. *Current Opinion in Neurology.* 11(3):253–258, Jun 1998.

194. Kacinski M; Zajac A; Skowronek-Bala B; Kroczka S; Gergont A; Kubik A. CNS Lyme disease manifestation in children. *Przeglad Lekarski.* 64(Suppl. 3):38–40, 2007.

195. Martin R; Gran B; Zhao Y; Markovic-Plese S; Bielekova B; Marques A; Sung MH et al. Molecular mimicry and antigen-specific T cell responses in multiple sclerosis and chronic CNS Lyme disease. *Journal of Autoimmunity.* 16(3):187–192, May 2001.

196. Campbell J; McNamee J; Flynn P; McDonnell G. Teaching NeuroImages: Facial diplegia due to neuroborreliosis. *Neurology.* 82(2):e16–e17, Jan 14, 2014.

197. Dubrey SW; Bhatia A; Woodham S; Rakowicz W. Lyme disease in the United Kingdom. *Postgraduate Medical Journal.* 90(1059):33–42, Jan 2014.

198. Hildenbrand P; Craven DE; Jones R; Nemeskal P. Lyme neuroborreliosis: Manifestations of a rapidly emerging zoonosis. *AJNR American Journal of Neuroradiology.* 30(6):1079–1087, Jun 2009.

199. Rupprecht TA; Koedel U; Fingerle V; Pfister HW. The pathogenesis of lyme neuroborreliosis: From infection to inflammation. *Molecular Medicine.* 14(3–4):205–212, Mar–Apr 2008.

200. Jeong BH; Kim YS. Genetic studies in human prion diseases. *Journal of Korean Medical Science.* 29(5):623–632, May 2014.

201. Degnan AJ; Levy LM. Neuroimaging of rapidly progressive dementias, part 2: Prion, inflammatory, neoplastic, and other etiologies. *AJNR American Journal of Neuroradiology.* 35(3):424–431, Mar 2014.

202. Halliday M; Radford H; Mallucci GR. Prions: Generation and spread versus neurotoxicity. *Journal of Biological Chemistry.* 289(29):19862–19868, Jul 18, 2014.

203. Fraser PE. Prions and prion-like proteins. *Journal of Biological Chemistry.* 289(29):19839–19840, Jul 18, 2014.

204. Ma Q; Hu JY; Chen J; Liang Y. The role of crowded physiological environments in prion and prion-like protein aggregation. *International Journal of Molecular Sciences.* 14(11):21339–21352, 2013.

205. Prusiner SB. Biology and genetics of prions causing neurodegeneration. *Annual Review of Genetics.* 47:601–623, 2013.

206. Prusiner SB. Novel proteinaceous infectious particles cause scrapie. *Science.* 216(4542):136–144, Apr 9, 1982.

207. Prusiner SB. Creutzfeldt-Jakob disease and scrapie prions. *Alzheimer Disease & Associated Disorders.* 3(1–2):52–78, Spring–Summer 1989.

208. Kalp M; Gottschalk CH. Mystery case: Heidenhain variant of Creutzfeldt-Jakob disease. *Neurology.* 83(22):e187, Nov 25, 2014.

209. Vachalova I; Gindl V; Heckmann JG. Acute inferior homonymous quandrantanopia in a 71-year-old woman. *Journal of Clinical Neuroscience.* 21(4):683–685, Apr 2014.

210. Parker SE; Gujrati M; Pula JH; Zallek SN; Kattah JC. The Heidenhain variant of Creutzfeldt-Jakob disease—A case series. *Journal of Neuro-Ophthalmology.* 34(1):4–9, Mar 2014.

211. Prasad S; Lee EB; Woo JH; Alavi A; Galetta SL. Photo essay. MRI and positron emission tomography findings in Heidenhain variant Creutzfeldt-Jakob disease. *Journal of Neuro-Ophthalmology.* 30(3):260–262, Sep 2010.

212. Cornelius JR; Boes CJ; Ghearing G; Leavitt JA; Kumar N. Visual symptoms in the Heidenhain variant of Creutzfeldt-Jakob Disease. *Journal of Neuroimaging.* 19(3):283–287, Jul 2009.

213. Appleby BS; Appleby KK; Crain BJ; Onyike CU; Wallin MT; Rabins PV. Characteristics of established and proposed sporadic Creutzfeldt-Jakob disease variants. *Archives of Neurology.* 66(2):208–215, Feb 2009.

214. Tsuji Y; Kanamori H; Murakami G; Yokode M; Mezaki T; Doh-ura K; Taniguchi K et al. Heidenhain variant of Creutzfeldt-Jakob disease: Diffusion-weighted MRI and PET characteristics. *Journal of Neuroimaging.* 14(1):63–66, Jan 2004.

23

Ischemic and Hemorrhagic Stroke

Shahmir Kamalian, Supada Prakkamakul, and Albert J. Yoo

CONTENTS

23.1 Introduction

Stroke is the third leading cause of death and the leading cause of severe disability in adults. In total, 795,000 strokes occur annually in the United States, of which 692,000 (87%) are ischemic and 103,000 (13%) are hemorrhagic.[1] The diagnostic evaluation and treatment approach vary between these two major classes of stroke. Current treatment approaches for acute ischemic stroke include intravenous tissue plasminogen activator (IV tPA), thrombolytic and/or mechanical intra-arterial therapies (IATs), and combined IV and intra-arterial (IA) therapy. Current therapeutic options for hemorrhagic stroke vary by the location and etiology.

Neuroimaging is central to stroke management because it is the primary method for differentiating between ischemic and hemorrhagic diseases. Moreover, by identifying treatable etiologies, imaging can guide therapy and predict treatment response. This chapter outlines magnetic resonance imaging (MRI) findings in ischemic and hemorrhagic strokes, and the role of imaging in the proper triage of patients.

23.2 Ischemic Stroke

Ischemic stroke may be classified as hyperacute, acute, subacute, and chronic based on the timing from the symptom onset. Generally, hyperacute stroke is within 6 hours of stroke onset, when patients are potentially eligible for reperfusion therapies. Acute stroke is defined as a stroke of less than 24-h duration. Subacute and chronic stages range from 24 h to 4–8 weeks and more than 4–8 weeks old, respectively.

Stroke-related edema and mass effect peak at 3–4 days after symptom onset and begin to subside after 7 days. Hemorrhagic transformation occurs in 10%–20% of ischemic infarctions, usually 2–7 days after ictus. The chronic phase is characterized by volume loss, cavitation, and gliosis. Large infarcts may be associated with Wallerian degeneration and crossed cerebellar diaschisis as evidenced by decreased volume of the ipsilateral cerebral peduncle and atrophy of the contralateral cerebellum, respectively.

23.2.1 MRI Sequences and Time Courses

MRI, particularly diffusion-weighted imaging (DWI), provides the most accurate evaluation of ischemic stroke within the hyperacute setting. The pattern of ischemia on DWI may reveal important information regarding stroke etiology, and hence appropriate and timely management of potential underlying causes. In addition, MR angiography (MRA) can identify vascular occlusions that are amenable to IAT. Other early vessel signs include gradient-recalled echo (GRE) clot-related susceptibility and fluid-attenuated inversion recovery (FLAIR) hyperintense vessels.[2] The major disadvantages of MRI for emergency stroke evaluation are inability to image patients with pacemakers or other ferromagnetic implants, and lower availability compared to CT in the emergency departments (Table 23.1).

23.2.1.1 Diffusion-Weighted Imaging

DWI is the most sensitive and specific MRI sequence to identify early infarction.[3] It is a technique designed to depict the diffusion of water molecules within the tissue, by the application of a pair of diffusion-sensitizing gradients to a spin-echo pulse sequence.[4,5] Image acquisition is rapid due to the use of a single-shot echo-planar technique as a readout module, and can usually be obtained in about 2 min. Faster acquisition and better image quality may be obtained using other types of fast imaging techniques such as parallel imaging. Recently, there has been increased interest in fast spin-echo DWI with radial *k*-space sampling (periodically rotated overlapping parallel lines with enhanced reconstruction [PROPELLER]) as this method is less sensitive to susceptibility artifact.[6–10] This technique is especially useful in the presence of metallic implants such as dental hardware. It can also be used to correct for patient motion, although it takes longer to acquire.

Because the incorporation of a pair of diffusion-sensitizing gradients adds time, longer echo time (*TE*) is

TABLE 23.1

Pearls and Pitfalls of MRI Sequences for the Evaluation of Acute Ischemic Stroke

MRI Sequences	Technical Pearls	Pitfalls
DWI	DWI offers the greatest sensitivity and specificity for infarct detection and volume assessment of the infarct core in the hyperacute setting.	T_2 shine-through T_2 blackout DWI false negative (3%–9%) depending on imaging time point; infarct size and location; and severity of hypoperfusion
FLAIR	FHVS is an indicator of severe stenosis or occlusion of the large arteries in patients with hyperacute stroke. Parenchymal FLAIR hyperintensity may be helpful in estimating the time of stroke when the onset is unknown. HARM is a potential marker of BBB disruption and risk of hemorrhagic transformation.	FHVS may not be depicted on 3D FLAIR sequences.
Susceptibility-sensitive sequences (GRE T_2*WI, SWI, SWAN)	Susceptibility-sensitive sequences are highly sensitive and specific for the detection of hemorrhages and microbleeds. SVS may enhance clot detection within blood vessels, particularly in distal segments. Prominent hypointense veins on SWI may represent tissue at risk around the infarct core.	Susceptibility artifact precludes evaluation in the skull base region.
MRA	Brain MRA in combination with MRI is useful for patient selection for IA therapy. Neck MRA is useful for the determination of stroke etiology and for IAT planning. Commonly used techniques are TOF (3D, MOTSA) and static CE MRA.	MRA is prone to overestimation of arterial stenosis.
MR perfusion	DWI–PWI mismatch aims to define ischemic penumbra and the benefit of reperfusion therapy. DSC is the most commonly used PWI technique in acute ischemic stroke.	MRP is not validated for the detection of tissue at risk or decision making regarding reperfusion therapy.

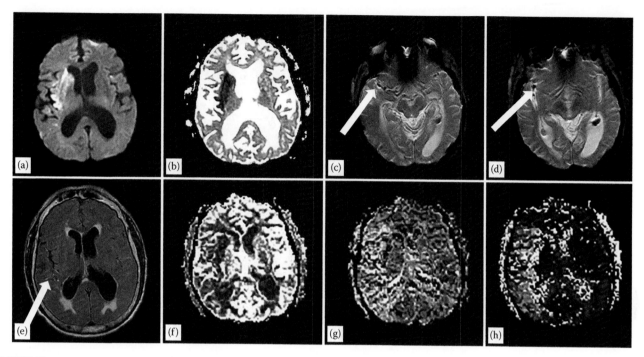

FIGURE 23.1

An 86-year-old man with atrial fibrillation presented with sudden fall and unresponsiveness. He was treated with IV tPA. CTA (not shown) demonstrated clot in the right MCA distal M1 and M2 segments, and bilateral vertebral and proximal basilar arteries. The DWI trace image (a) and ADC map (b) show an area of restricted diffusion in the right basal ganglia and insula. The gradient-echo T_2^*WI (c,d) shows thrombus-related susceptibility blooming artifact in the distal M1 and M2 segments (arrows). The FLAIR sequence (e) reveals hyperintense vessels (arrow) in the Sylvian fissure consistent with slow flow within collateral vessels. The postprocessed perfusion images (f, CBF; g, MTT; and h: TTP) demonstrate a large perfusion–diffusion mismatch in the right MCA territory.

required in the diffusion-weighted sequence. Therefore, trace diffusion-weighted images represent a mixture of diffusion-weighted and T_2-weighted (T_2W) contrasts. In order to determine whether hyperintense signal on the trace image is related to restricted diffusion or underlying T_2 hyperintensity, the trace images should always be assessed in conjunction with the apparent diffusion coefficient (ADC) map, which is free of T_2 weighting effect and where restricted diffusion appears hypointense (Figures 23.1a and b and 23.2a and b). An exponential map also eliminates T_2 weighting information and portrays restricted diffusion as hyperintense signal, but is less commonly used due to diminished image contrast. A lesion demonstrating high signal intensity (SI) on both trace and ADC images does not have true restricted diffusion but appears as such due to "T_2 shine-through" effect. Sole interpretation of the ADC map is not recommended, as a lesion showing intrinsically low SI on T_2WI may appear dark on trace and ADC images, such as in an ischemic infarct with hemorrhagic transformation. This condition is called "T_2 blackout" effect.

In ischemic stroke, when inadequate cerebral perfusion leads to failure of adenosine triphosphate pumps, shifting of extracellular ions into the intracellular space and accompanying fluid shifts result in cell swelling (i.e., cytotoxic edema). The mechanism of diffusion restriction in cerebral infarction is not fully elucidated, but hypothesized to result from a reduction of extracellular volume in concert with hindered water movement in the intracellular compartment.[11,12]

Sensitivities and specificities of DWI for detecting acute ischemic stroke are high and reported to be 90%–97% and 75%–97%, respectively.[13–15] DWI can detect hyperacute ischemic infarction as early as 15 min after onset of symptoms,[16,17] seen as high SI on trace and low SI on the corresponding ADC images. ADC signal continues to decrease until 3–4 days after onset, then gradually increases due to cell lysis and vasogenic edema resulting in ADC pseudonormalization at up to 7–10 days (Figure 23.3).[18,19] Subsequently, as the infarct evolves and ages, the trace images show variable signal and the ADC map can appear unremarkable or show increased diffusion (i.e., hyperintense ADC signal). In rare cases, restricted diffusion can persist up to 4.5 months after the event, especially in small infarcts.[20]

In most cases, the restricted diffusion associated with acute ischemia represents irreversible tissue injury. In a minority of cases, there may be reversal of diffusion restriction.[21,22] However, the mean volume of tissue reversal is usually relatively small (2–15 mL), and delayed regrowth into the previously seen diffusion abnormality is frequent.[23–25] The clinical significance

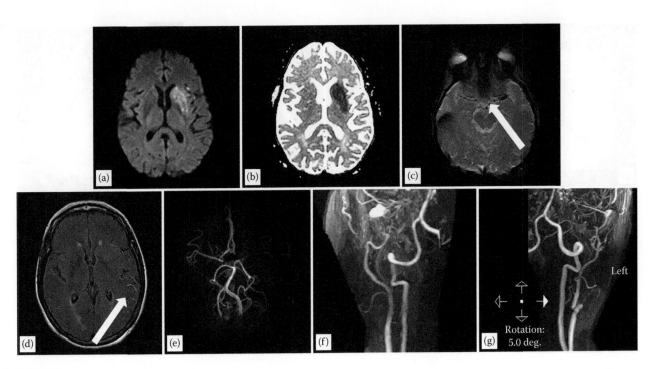

FIGURE 23.2
A 57-year-old woman presented with right-sided weakness and aphasia. She was treated with IV tPA. CTA (not shown) demonstrated occlusion of bilateral intracranial internal carotid arteries extending into the A1 segment of the left ACA and into the M1 segment of the left MCA. The DWI trace image (a) and ADC map (b) reveal an area of restricted diffusion in the basal ganglia and insula. The gradient-echo T_2*WI (c) shows susceptibility blooming artifact from the M1 segment thrombus (arrows). The FLAIR sequence (d) shows hyperintense vessels (arrow) in the Sylvian fissure consistent with slow flow within collateral vessels. The 3D TOF MRA (e) demonstrates the lack of flow-related enhancement in bilateral intracranial ICAs, left ICA terminus, left proximal A1 segment, and left MCA M1 segment and distal branches corresponding to the occluded vessels. The post-gadolinium MRA of the neck reveals occlusion of the bilateral cervical ICAs (f, right; g, left).

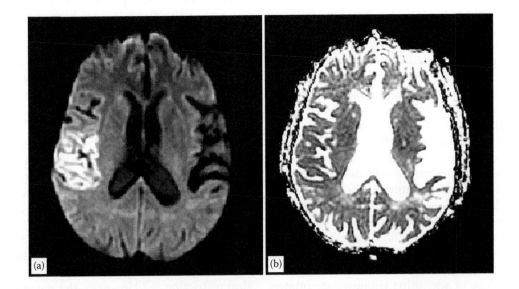

FIGURE 23.3
An 82-year-old woman presented with temporary worsening of her symptoms (aphasia and left-sided weakness) 5 days following her initial right MCA territory infarction. The diffusion-weighted trace map (a) demonstrates hyperintense signal in the posterior right insula and inferior aspect of the right pre- and postcentral gyri with near pseudonormalization of the ADC map (b).

of DWI reversal appears minimal, as reperfusion is a prerequisite condition and any associated clinical improvement is more likely related to the salvaged penumbral tissue rather than the negligible region of DWI lesion reversal.[24] Taken together, DWI is considered the most accurate method for imaging the infarction core in the hyperacute setting (class I—level of evidence [LOE] A).[26]

Despite the high sensitivity of DWI, false negatives can be found in up to 3%–9% of cases.[13,15,27,28] Various explanations have been proposed, including image acquisition that is too early or too late (e.g., pseudonormalization), lesion size that is too small,[28] brain stem or posterior fossa location, proximity to the skull base and associated susceptibility signal loss, or perfusion reduction not severe enough to manifest cytotoxic edema.[27] Stroke mimics may produce false-positive cases and will be discussed in a Section 23.2.5.

23.2.1.2 Fluid-Attenuated Inversion Recovery

During the hyperacute stage of ischemic stroke, cytotoxic edema is followed by the blood–brain barrier (BBB) breakdown.[29,30] Between 1 and 4 h after stroke onset, there is a net increase of tissue water, termed vasogenic edema. Fluid in tissue has prolonged T_2 relaxation time, and therefore vasogenic edema appears bright on both T_2 and FLAIR images. The inversion recovery pulse in FLAIR nulls cerebrospinal fluid (CSF) signal, resulting in higher sensitivity for distinguishing the bright SI of edema from dark CSF signal.[31] The development of parenchymal FLAIR hyperintensity is variable and may take about 6 h to become apparent.[31,32] Hyperintensity on T_2WI becomes conspicuous later, at about 8 h after onset.[33] The edema and mass effect peak at 3–4 days, with T_2 hyperintensity, gyral swelling, and sulcal effacement. Although T_2 hyperintensity persists, the edema and mass effect begin to decrease after 7 days. In the chronic phase, the infarct demonstrates volume loss, gliosis, and variable degrees of cavitation (i.e., cystic encephalomalacia). Cystic encephalomalacia mirrors CSF signal (dark on FLAIR and hyperintense on T_2WI). Gliosis is hyperintense on both T_2W and FLAIR sequences.

Normally, a flow void is visualized in rapidly flowing arteries due to signal loss on spin-echo T_2W and FLAIR images.[34,35] In the presence of severe stenosis or occlusion of the large arteries, loss of the normal flow voids can readily be seen in the affected vessel segment or in the vascular bed distal to the lesion. Increased SI in the vessels is more robustly seen on FLAIR images due to background CSF suppression, which is called the FLAIR hyperintense vessel sign (FHVS). FHVS represents anterograde flow through a stenotic segment, or more commonly, sluggish retrograde flow in leptomeningeal collaterals distal to the occlusion site[36–38] (Figures 23.1e

and 23.2d). It is a reflection of altered hemodynamics and can be seen prior to the appearance of diffusion restriction. Beyond its association with vessel occlusion, the utility of FHVS in acute ischemic stroke is controversial.[2,39,40] FHVS can also be seen in the setting of transient ischemic attack (TIA), chronic arterial steno-occlusive disease, or other vasculopathies such as moyamoya disease.[36,41–43]

An important pitfall is that FLAIR vascular hyperintensities can be absent on 3D FLAIR imaging due to black blood effect resulting from dephasing of spins induced by the long echo train.[44] Therefore, nonvisualization of FHVS on 3D FLAIR images does not exclude sluggish flow in leptomeningeal collaterals.

FLAIR sulcal hyperintensity can be seen in acute ischemic stroke patients that have received gadolinium-based contrast agent. This phenomenon has been termed the hyperintense acute reperfusion marker (HARM) and represents leakage of contrast agent into the CSF through a disrupted BBB.[45,46] It can also be seen on contrast-enhanced (CE) T_1WI. Whether this sign is a predictor of hemorrhagic transformation is still debated.[47–49]

23.2.1.3 Susceptibility-Sensitive Sequences

GRE T_2*WI, susceptibility-weighted imaging (SWI), and susceptibility-weighted angiography (SWAN) are highly sensitive to paramagnetic substances such as certain blood breakdown products (deoxyhemoglobin, ferritin, and hemosiderin). On these sequences, paramagnetic substances result in signal loss and appear as regions of dark SI.

In acute stroke, GRE T_2*WI is important for the detection of intracranial hemorrhage (Figure 23.12) and may identify clot within the occluded vessel segment. It is as accurate as noncontrast CT (NCCT) for detection of acute intracranial hemorrhage and is superior to CT in the detection of chronic hemorrhages.[50]

The susceptibility vessel sign (SVS) is a highly specific indicator of clot within the vessel[51] and is defined as a prominent hypointense signal within a vessel segment on gradient T_2*WI (Figures 23.1c and d and 23.2c). The signal loss is attributable to deoxyhemoglobin within the thrombus.[52] SVS visualization depends on many factors, including clot composition, age, and size, as well as scanning parameters. Fibrin-rich thrombi (white thrombi) are less likely to cause susceptibility-related signal loss due to a relative paucity of red blood cells. SVS has been associated with red blood cell-rich thrombi from a cardiac source, older clot, and larger clot size.[53] Alteration of scan parameters, including higher field strength, type of readout module, and longer TE, may result in greater susceptibility-related signal loss. A recent study has shown that SVS may provide a reliable estimate of clot length.[51] This may have therapeutic implications as larger clot burden, longer clot length

(>20 mm), and irregular clot shape appear to decrease the potential for recanalization.[54,55] In addition, these sequences sometimes depict distally located clots, which may be missed by routine MRA.[56]

SWI is an MRI technique that utilizes both magnitude and phase information to create a novel contrast based on the magnetic susceptibility of tissue. This sequence is very sensitive to the paramagnetic effects of deoxyhemoglobin in both intravascular and extravascular compartments, which again appear as dark SI. Besides its high sensitivity to acute hemorrhage, chronic microbleeds, and early hemorrhagic transformation of ischemic stroke, blood oxygen level-dependent effects may indicate the presence of a large ischemic region and potential tissue at risk around the infarct core.[57,58] Compensatory vasodilatation due to hypoperfusion and ischemia causes relative slowing of the blood flow in the capillary bed and increased oxygen extraction from the blood in the ischemic tissue. Hence, the concentration of deoxyhemoglobin increases in the regional capillaries and draining veins. Due to the paramagnetic effect of deoxyhemoglobin, prominent hypointense veins can be detected by susceptibility-sensitive sequences in the vicinity of the ischemic region.

SWAN allows high-resolution visualization of both cerebral arteries and veins from the same acquisition by utilizing different reconstruction techniques: maximum intensity projection (MIP) technique for MRA and minimum intensity projection for MRV.[59] On SWI, arteries are hyperintense due to time-of-flight (TOF) effects and lack of T_2^* effects, whereas veins are hypointense due to deoxyhemoglobin.[60,61] This contrast can be used to detect arteriovenous shunting, as will be discussed later (Figure 23.13).

23.2.1.4 MR Angiography

Vascular imaging is a vital part of the stroke workup. In the hyperacute setting, evaluation of the intracranial vasculature to identify major arterial occlusion is mandatory for patient selection for IAT. Although digital subtraction angiography (DSA) is considered the reference standard for both anatomic and physiologic evaluation of the intracranial circulation, noninvasive vascular studies are usually performed initially. CT angiography (CTA) or MRA can be chosen as a diagnostic tool, but should not delay the administration of IV tPA if indicated.[62]

MRA techniques can be divided into two main categories: noncontrast and CE techniques. Two noncontrast MRA approaches are TOF and phase contrast sequences. However, phase contrast is not commonly used in acute stroke because it is prone to artifactual signal loss when arterial flow does not fall within the prespecified velocity encoding range, as well as lengthy scan time. CE MRA can be acquired in one phase (static CE MRA) or multiple phases (time-resolved MRA). Although CTA offers a fast and sensitive evaluation of intracranial arteries, MRA is

particularly useful when the patient requires MRI scanning for other essential information such as infarct volume assessment. Also, MRA avoids radiation exposure. However, MRA is very sensitive to flow effect and tends to overestimate the degree of vascular stenosis.[63,64] This being said, MRA is sufficiently accurate for identifying proximal artery occlusion amenable to IAT.[65]

TOF technique is a noncontrast MRA technique that utilizes multiple repetitive radiofrequency pulses to suppress signal in the stationary tissue, enabling depiction of flow-related signal in fresh blood entering the imaged slice. 3D or multiple overlapping thin-slab acquisition (MOTSA) are commonly used TOF techniques (Figure 23.2e). The thicker the 3D slab, the longer the flowing blood is exposed to the repetitive RF pulses, causing blood saturation and signal loss near the exiting edge of the slab. MOTSA solves this problem by dividing a large 3D volume into multiple thinner overlapping slabs to preserve signal. 3D TOF has sensitivities of 60%–85% for the detection of intracranial stenoses and 80%–90% for occlusions.[62,66,67] The advantages of this technique are high spatial resolution and no requirement for contrast administration. The major drawbacks include in-plane signal loss, particularly in horizontally oriented arteries around the circle of Willis, and lengthy scan time. Intracranial 3D TOF MRA takes approximately 5–10 min to acquire, making it prone to motion artifact.[64]

CE MRA offers faster acquisition time, wider coverage, and less flow-related signal loss.[68] The extended coverage enables both extracranial and intracranial arteries to be imaged in the same setting (Figure 23.2f and g).[69,70] However, CE MRA suffers from lower spatial resolution.[68] With the advent of high-field and parallel imaging, spatial resolution of CE MRA has improved, resulting in diagnostic performance comparable to TOF MRA and CTA.[64,68,71] Another problem that plagues static CE MRA is venous contamination, which can be alleviated by time-resolved MRA.

Time-resolved MRA offers dynamic flow information within the extracranial and intracranial arteries, with high temporal resolution but limited spatial resolution.[72] The sensitivity and negative predictive value for high-grade stenoses and occlusions are reported to be high, despite limited specificity and positive predictive value.[73] Time-resolved MRA can potentially give information about collateral status, analogous to multiphasic CTA,[74] but this technique has not yet been popularly implemented for this purpose.

Evaluation of the extracranial vasculature may be performed in the same setting or later, and is helpful to determine the stroke mechanism and provides valuable information regarding aortic arch anatomy and cervical steno-occlusive disease for IA treatment planning.[75] Common causes of ischemic stroke include atherothrombosis or dissection. A common pitfall is the focal signal loss that is often observed on 3D TOF at the carotid bifurcation and in the skull base region due to turbulent flow and spin dephasing. CE MRA is less prone to this problem. Nevertheless, both techniques still overestimate the degree of vascular stenoses compared to angiography.[76] If there is clinical suspicion for dissection (e.g., young patient), a noncontrast T_1 sequence with fat saturation should be performed through the neck, which may reveal T_1 hyperintensity related to methemoglobin within the false lumen (Figure 23.4).

23.2.1.5 MR Perfusion

MR perfusion-weighted imaging (MR PWI) is a technique that depicts hemodynamic changes at the tissue microvascular level. The most common MR PWI technique used in acute ischemic stroke is dynamic susceptibility contrast (DSC) perfusion.[77] The brain is imaged multiple times in rapid succession after administration of a bolus of gadolinium-based contrast agent to measure its first-pass transit through the brain tissue. Single-shot

T_2^*W echo-planar imaging is commonly used to detect susceptibility signal loss generated by the contrast bolus. A time–SI curve is generated and later transformed into a concentration–time curve. Several perfusion parameters are calculated based on the concentration–time curve and the central volume principle [cerebral blood flow (CBF) = cerebral blood volume (CBV)/mean transit time (MTT)]. The commonly used parameters are listed in Table 23.2.

DSC MR PWI technique may be confounded by contrast recirculation, leakage, delay, and dispersion. However, the first two are more problematic in brain tumors than in acute stroke. Delay and dispersion are mitigated by the application of an arterial input function and deconvolution.[77] A disadvantage of MR PWI is the nonlinear relationship between MR SI and paramagnetic contrast concentration, preventing truly quantitative measurements of MR PWI parameters.

In acute ischemic stroke, occlusion of a major artery typically causes regional tissue hypoperfusion and progressive neuronal damage. Affected tissue can be

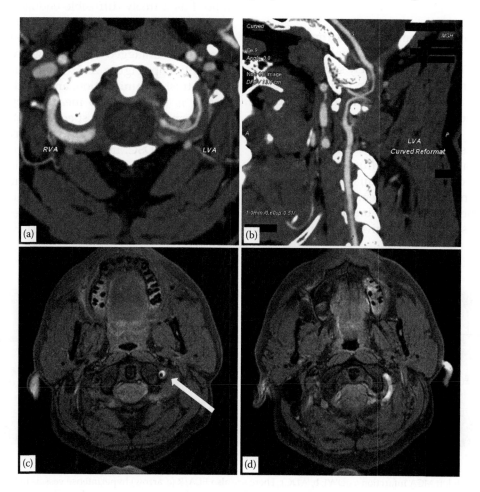

FIGURE 23.4

A 29-year-old man with right neck pain and occipital headache for 10 days following lifting weights. The CTA shows abrupt change in caliber of the distal cervical segment of the left vertebral artery (a, axial MIP; b, curved reformat). T_1WI with fat saturation through the neck (c,d) depicts intrinsic T_1 hyperintensity along the dissected segment (arrow).

TABLE 23.2

MRI Perfusion Parameters

Parameters	Definition	Unit of Measurement
MTT	Average time taken by a tracer to pass through a tissue	Seconds
TTP	A non-deconvolution-based metric, representing time between the first arrival of contrast agent within major arterial vessels included in the section and the local bolus peak in the brain tissue	Seconds
T_{max}	A deconvolution-based metric, containing similar measures as TTP, representing the time at which the tissue residue function reaches maximum intensity	Seconds
CBF	Flow through a given region of interest	mL/100 g of tissue per minute
CBV	Volume occupied by the blood in a region of interest	mL/100 g of tissue

divided into three concentric compartments.[78] The innermost region is irreversibly damaged tissue, called the infarct core. The area immediately surrounding the infarct core is termed the ischemic penumbra, representing tissue at risk of infarction if early revascularization is not achieved. The outermost region represents benign oligemia, where hypoperfused tissue will survive despite persistent occlusion owing to sufficient leptomeningeal collateral supply. Diffusion-weighted MRI performs well in delineating the infarct core. Perfusion imaging, including MR PWI, attempts to differentiate between the other two regions, but the reliability of this assessment is questionable given the limitations of quantifying perfusion. A more qualitative approach has been employed in the clinical setting, where a perfusion lesion that is larger than the diffusion lesion (i.e., DWI–PWI mismatch) is taken as a marker of salvageable tissue and potential treatment response (Figure 23.1f–h).[79–83] However, DWI–PWI mismatch size may be affected by several variables, including (1) postprocessing software platform, (2) assessment approach (visual or quantitative), (3) parameter map [time parameters are commonly used, but discrepancies exist between T_{max}, time to peak (TTP), and MTT], and (4) cutoff value for each parameter.[84–88] The lack of consensus on how to best depict the DWI–PWI mismatch is a major obstacle to guiding decision making regarding thrombolytic therapy.[89] Also, as previously mentioned, there is no perfusion imaging approach that can accurately differentiate between ischemic penumbra and benign oligemia in the hyperacute setting.

Arterial spin labeling (ASL) is an MR PWI technique that does not require administration of exogenous contrast agent. Instead, magnetically labeled blood water is used as a freely diffusible endogenous tracer.[90] The blood water is first tagged with an inversion recovery pulse prior to entering the brain. Subtraction of the tagged and untagged brain images yields a quantifiable CBF map (Figure 23.5). In addition to not requiring contrast administration, another distinct advantage of ASL is the ability to measure regional perfusion, which can be done by specifically tagging one vessel at a time.[90] There are several shortcomings of ASL compared to DSC, including low signal-to-noise ratio, lengthy scan time, and delayed arterial arrival. Hence, it is less commonly used in the setting of acute ischemic stroke. Nevertheless, several studies have shown that prolonged arterial arrival time, seen as bright SI in large brain arteries, and ASL perfusion defect can reflect

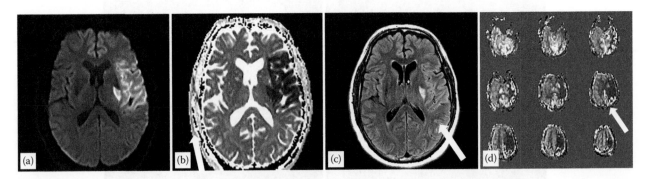

FIGURE 23.5

A 61-year-old man with left MCA infarction (a, DWI; b, ADC). There are also FLAIR (c, arrow) hyperintense vessels in the left Sylvian fissure and over the left parietal and temporal lobes, likely reflecting slow flow within the collateral vessels. The ASL perfusion imaging (d) shows a large area of perfusion abnormality involving the entire left middle cerebral artery territory with a large area of perfusion–diffusion mismatch within the left temporal, anterior frontal, and parietal lobes. In addition, there are curvilinear high signal intense foci within the left temporal, anterior frontal, and parietal lobes (arrow), which are due to delay in arrival of the labeled blood via the collateral flow.

hemodynamic abnormality and delayed arterial flow in acute stroke patients.[91–94] Accurate quantification of CBF is limited by labeling efficiency and arterial transit delay.[90,95,96]

Dynamic CE (DCE) MR PWI is not commonly used in the evaluation of acute stroke patients. It has been shown, along with DSC and ASL, to predict postischemic hemorrhagic transformation.[97–99]

23.2.2 Role of MRI in Patient Selection

Imaging plays a critical role in patient selection for stroke reperfusion therapy. The imaging biomarkers that identify patients who are likely to benefit from treatment vary based on the treatment approach. The two main approaches are intravenous therapy and IAT (i.e., catheter-based).

23.2.2.1 IV Thrombolysis

The efficacy of IV tPA is supported by multiple randomized controlled trials,[100,101] and has a class I recommendation from the American Heart Association for use within the first 3–4.5 h after stroke onset.[62] There is overwhelming evidence that treatment benefit erodes with time, underscoring the importance of rapid door-to-needle times.[102] For this reason, imaging should be performed and interpreted rapidly, preferably within 45 min of patient arrival to the hospital.[103]

The trials that demonstrated the efficacy of IV tPA utilized NCCT imaging, and this approach remains the dominant imaging method for patient evaluation given its widespread availability and speed. The primary question is the presence of intracranial hemorrhage, which is an absolute contraindication to treatment.[62] Additionally, within 3 h of onset, parenchymal evaluation of the NCCT images is used to exclude patients with well-established, clearly hypodense strokes larger than one-third of the MCA territory (class III—LOE A).[62] Early ischemic changes such as loss of gray–white matter distinction are used to exclude patients in the 3–4.5 h window, if these changes are extensive (i.e., greater than one-third of the MCA territory).[62]

Although less common, the use of MRI for IV tPA decision making within the 4.5 h window has been used in centers that have access to this modality in the hyperacute stroke setting.[104–107] Similar to NCCT, MRI should be examined for the presence of acute intracranial hemorrhage as an exclusion criterion. The presence of microbleeds should not deter treatment. Despite the conflicting findings within the literature, there is no compelling evidence that microbleeds increase the risk of symptomatic hemorrhage after thrombolysis, although there are few data in the setting of numerous

(≫5) microbleeds.[108–114] Additionally, there are no data regarding whether patients with large diffusion lesions benefit or not from IV thrombolysis, leading some centers to treat regardless of DWI lesion size.[107,111] Finally, MRI may allow better identification and exclusion of stroke mimics.[104]

23.2.2.2 IA Therapy

More than one-third of patients with acute ischemic stroke have a proximal occlusion of a major intracranial artery such as the distal internal carotid or proximal middle cerebral artery.[115] In the absence of reperfusion, the prognosis is very poor in this subgroup of patients. Unfortunately, some of these patients present outside of the time window or are otherwise ineligible for intravenous thrombolysis. Moreover, IV tPA has limited efficacy for these major occlusions. Less than 10% of patients with distal ICA and approximately 30% of patients with proximal MCA occlusion achieve early recanalization.[116,117] IAT is a promising option for these patients (Figure 23.6).

Recently, the phase 3, Multicenter Randomized Clinical Trial of Endovascular Treatment for Acute Ischemic Stroke in the Netherlands (MR CLEAN), which enrolled 500 patients, demonstrated that IA treatment in addition to current standard of care therapies, including IV tPA, was more effective than standard of care treatment alone in patients with a proven occlusion of a proximal anterior circulation artery and presenting within 6 h after symptom onset.[118] Other recent trials have confirmed these results.[119–122]

A major lesson learned from MR CLEAN is the importance of noninvasive vessel imaging to identify a proximal intracranial occlusion. Without such an occlusion, IATs cannot impact the patient's natural history. Although this idea seems intuitive, two previous IAT trials did not mandate vascular imaging prior to enrollment, and both failed to demonstrate the efficacy of the IA approach.[123,124] Therefore, MRA or CTA is a key component of the workup of the stroke patient who is eligible for IAT (Figure 23.6a).

The other critical imaging question is the size of the pretreatment infarct (Figure 23.6b). Numerous studies have shown that smaller core infarcts are associated with better functional outcomes, less mortality, and lower rates of reperfusion hemorrhage after IAT.[80,125–128] In particular, patients who present with extensive infarcts (e.g., DWI lesion volume >70–100 mL or NCCT Alberta stroke program early CT score [ASPECTS] <5) are highly unlikely to achieve functional independence after catheter-based treatment. This idea is supported by MR CLEAN, where a subgroup analysis demonstrated a neutral treatment effect among patients with a baseline NCCT ASPECTS <5.[118]

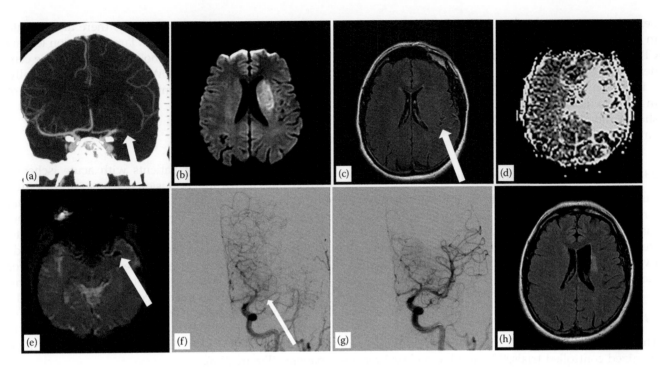

FIGURE 23.6
A 48-year-old man with new onset atrial fibrillation presented with sudden right hemiplegia and expressive aphasia (NIHSS of 10). The CTA (a, coronal reformat) shows occlusion of the left MCA M1 segment (arrow). DWI (b) shows restricted diffusion in the basal ganglia and left corona radiata. FLAIR (c) shows hyperintense vessels in the Sylvian fissure consistent with slow flow in the collaterals (arrow). The perfusion imaging shows an area of large perfusion–diffusion mismatch (d, MTT). The perfusion source image (e) shows susceptibility blooming artifact related to the clot (arrow). The left common carotid injection (f) confirms an occlusive thrombus in the left M1 segment (arrow). Reinjection after endovascular clot retrieval demonstrates successful recanalization (g). The follow-up FLAIR sequence (h) after 3 months shows a small area of gliosis corresponding to the focus of restricted diffusion on the initial MRI. The patient had subtle residual facial asymmetry and slight difficulty in naming.

There appears to be a limited role for perfusion imaging and mismatch-based selection in patients being considered for IAT. MR and CT perfusion are very sensitive for identifying hypoperfused tissue, but are not specific for differentiating true tissue at risk from benign oligemia. Because large parenchymal territories are at risk following proximal arterial occlusion, there is virtually always a large perfusion deficit and mismatch in the presence of a small infarct core (Figure 23.6d). An analysis of 116 patients with distal ICA and proximal MCA occlusions showed that all patients with a DWI lesion <70 mL had at least 100% mismatch.[129] These findings highlight the more important role of DWI over MR or CT perfusion in patient selection for endovascular therapy. Rather than perfusion imaging, the clinical exam can be used to infer the presence of a large amount of tissue at risk, specifically a moderate–severe neurological deficit (i.e., National Institute of Health Stroke Scale [NIHSS] score ≥10) in the setting of a proximal vessel occlusion and a small DWI lesion. Among patients with proximal MCA occlusions in the Prolyse in acute cerebral thromboembolism (PROACT) II trial, IA pro-urokinase resulted in better outcomes among patients with NIHSS scores 11 or higher.[130] There was no difference

in outcomes between the treatment and control arms when the NIHSS score was 10 or lower. This is owing to the relatively good natural history in the milder stroke group.

In short, patients with a significant clinical deficit, a proximal vessel occlusion confirmed on MRA or CTA, and a small DWI lesion are most likely to benefit from IAT, and patients with large DWI lesions are very likely to have a poor outcome and elevated risk with treatment. It should be kept in mind that a single infarct volume threshold for treatment decision making may not be optimal for all patients, and that the threshold is likely to depend on patient age. A recent study found that the target infarct volume cutoff to predict a good outcome decreased with increasing age, such that among octogenarians the threshold was as low as 15 mL.[131]

23.2.2.3 Hemorrhagic Conversion

Hemorrhagic transformation is a potentially life-threatening complication after ischemic stroke, and its incidence increases with revascularization therapy.[132,133] The risk of hemorrhagic transformation is associated with BBB breakdown and loss of CBF autoregulation.[134]

The severity can range from asymptomatic petechial hemorrhage to large IPH leading to significant mass effect, clinical deterioration, and death.

In the European Cooperative Acute Stroke Study (ECASS) trials, hemorrhagic transformations after ischemic stroke were classified as hemorrhagic infarction types 1 and 2 (HI1 and HI2) and parenchymal hematoma types 1 and 2 (PH1 and PH2).[101,135,136] HI1 and HI2 indicate isolated and confluent petechial hemorrhage in the infarcted tissue, respectively, without any associated mass effect.[101] PH1 indicates hematomas with mild mass effect occupying less than 30% of the infarcted tissue.[101] PH2 refers to large hematomas occupying more than 30% of the infarcted area with significant mass effect and possible extension into ventricles and outside the infarcted tissue.[101]

Greater degrees of hemorrhagic transformation are associated with worse outcomes.[137] Parenchymal hematomas, particularly PH2, have shown a clear association with early neurological deterioration and 3-month mortality.[132,138] The association between petechial hemorrhage and outcome is less clear.[137,139]

The ECASS criteria are based on NCCT evaluation. Because MRI is more sensitive for detecting hemorrhagic transformation than CT, it may lead to discrepancies in grading the severity of bleeding.[140] Studies that have evaluated various MRI sequences with respect to the ECASS classification system have found that MRI, particularly using T_2*WI alone, may cause a shift from HI1 to HI2 and misclassification of HI2 as PH1.[140–143] Unfortunately, no standardized imaging criteria have been developed that equate MRI findings to the ECASS classification on CT.

There is strong evidence that a larger DWI lesion at baseline is associated with a statistically significant increase in risk of hemorrhagic transformation after reperfusion therapy.[79,144–146] Other MRI findings have been identified as potential markers of increased risk, including extensive leukoaraiosis,[147] very low ADC values,[148,149] early BBB breakdown as evidenced by delayed gadolinium enhancement of CSF on FLAIR (i.e., HARM),[47] and marked reductions in various perfusion parameters[79,150]; however, the lack of validation studies and the methodologic limitations of perfusion imaging limit the clinical utility of these potential imaging biomarkers.

time windows. The mismatch hypothesis has been the primary focus of the majority of these studies. Patients with a significant mismatch (i.e., much larger region of hypoperfusion) are considered ideal candidates for reperfusion therapy. The mismatch can be identified on MRI using DWI and PWI or with CT perfusion imaging using various summary maps.

Mismatch imaging-based selection has been studied with the novel thrombolytic desmoteplase in the 3–9 h window and with alteplase in the 3–6 h window. In both settings, a significant mismatch was defined as a PWI lesion at least 20% larger than the DWI lesion (i.e., 20% mismatch). These studies provide conflicting data regarding the utility of this selection paradigm.[87,151–158] A meta-analysis of mismatch-based extended window thrombolytic trials found that there was no significant increased odds of a favorable outcome with thrombolysis in patients with a mismatch.[158]

A major drawback of the mismatch approach is that it is a relative measure. Therefore, the same percentage mismatch may correspond to vastly different volumes of tissue. This results in a nonspecific characterization of stroke physiology. Despite using the same mismatch criteria, the Desmoteplase in Acute Ischemic Stroke-2 (DIAS-2) trial population had lower NIHSS scores (9 vs. 12), smaller DWI and absolute mismatch volumes, and lower rates of vessel occlusion compared to DIAS and Dose Escalation of Desmoteplase in Acute Ischemic Stroke (DEDAS), which likely explains the higher than the expected rate of good outcome in the control arm of DIAS-2, and in turn, the failure to demonstrate the benefit of desmoteplase.[24] Furthermore, as previously mentioned, there are several limitations of perfusion imaging for identifying tissue at risk (penumbra), which include the lack of consensus about the optimal perfusion parameter,[86,159] inter- and intraobserver variability in visual lesion assessment, and the lack of a validated quantitative threshold for discriminating threatened tissue from benign oligemia.[160,161]

Instead of mismatch imaging, posthoc data from the Diffusion and Perfusion Imaging Evaluation for Understanding Stroke Evolution (DEFUSE) study support a selection approach for extended window thrombolysis that combines a proven vessel occlusion with a small core infarct, analogous to selection for IAT.[162] This approach, as well as the mismatch hypothesis, is being evaluated in ongoing trials.

23.2.3 Future Directions in Reperfusion Therapy

23.2.3.1 Extended Window Reperfusion Therapy Based on Penumbral Imaging

There have been numerous studies examining the use of advanced imaging to treat patients with intravenous thrombolysis beyond the currently approved

23.2.3.2 Wakeup Strokes and Strokes with Unknown Onset

Up to 25% of stroke patients, including patients with wakeup strokes, present with an unknown time of symptom onset.[163] Given the strict time windows for treatment, these patients are excluded from reperfusion therapy.

However, similar to coronary artery disease, most of these strokes occur in the morning,[164] and studies suggest that a large number of patients with wakeup stroke may be within the thrombolysis time window when they present to the hospital.[165] Recently, studies have investigated whether MRI can identify these patients.

During the first few hours, an acute infarct will be visible on DWI but not on FLAIR.[166] Therefore, a DWI–FLAIR mismatch (positive DWI with negative FLAIR) may be useful for identifying patients with strokes of less than 3–6 h duration.[167] However, the use of FLAIR signal change is controversial as the reported sensitivity and specificity of DWI–FLAIR mismatch are variable between studies, which may be related to differences in magnetic field strength and acquisition technique, definition of the DWI–FLAIR mismatch, infarct size, and the extent of leukoaraiosis.[31,32,163,167–170] The FLAIR–DWI mismatch is less sensitive and specific using 3 T MRI than 1.5 T.[171] With every 10 mL increase in size of the diffusion abnormality, the odds of a FLAIR signal abnormality increase by 7%.[169] Finally, approximately half of patients with strokes less than 4.5 h old may have FLAIR signal changes, and conversely, up to a third of patients beyond 4.5 h may not show any FLAIR signal change.[171]

Currently, there are two ongoing multicenter trials, MR WITNESS and WAKE-UP, that aim to test the DWI–FLAIR mismatch as a selection tool for treating patients with unknown-onset strokes with IV tPA. In the phase IIa MR WITNESS trial, patients with no FLAIR abnormality or minimal FLAIR abnormality (normalized ratio <1.15) in the region of diffusion abnormality are treated with IV tPA. In the phase III WAKE-UP trial, the mismatch will be determined based on visual assessment alone.

23.2.4 Imaging Workup of TIA

The clinical definition of TIA is a focal neurologic deficit that lasts less than 24 h. Approximately 15% of all ischemic strokes are heralded by a TIA.[1] The estimated prevalence of physician-diagnosed TIA in the United States is about 2.3% (approximately five million people).[172] The true prevalence of TIA is likely higher as many patients with a TIA may fail to report it to their physician.[172]

TIAs represent a critical opportunity to identify patients at high risk for early stroke, so that treatments aimed at stroke prevention can be instituted. The ABCD[2] scoring system incorporates clinical variables to reliably predict stroke risk after TIA.[173,174] Importantly, recent studies have shown that infarction on DWI and a carotid stenosis of at least 50% can further improve prognostication over ABCD[2] (age, blood pressure, clinical features of TIA, duration of symptoms, history of diabetes) alone.[175,176] Therefore, early neuroimaging (i.e., within

24 h) is an integral part of the TIA workup. In the most recent American Heart Association guidelines, cervical and intracranial vascular imaging and brain MRI (or CT if MRI is not available) are class I recommendations for patients with TIA.[62]

23.2.5 Stroke Mimics

It is estimated that 9%–30% of patients with suspected stroke[177–182] and 2.8%–17% of patients treated with IV tPA have stroke mimics.[183–185] There are numerous vascular and nonvascular diseases that can present with signs and symptoms similar to ischemic stroke (Tables 23.3 and 23.4). The majority of stroke mimics are due to seizures, migraines, tumors, and toxic–metabolic disturbances.[180,186]

Multiple recent studies have demonstrated that thrombolysis is probably safe in most cases of stroke mimics.[183,185,187–189] Nevertheless, thrombolysis is not safe in cases of specific mimics such as intracranial neoplasm, brain abscess, or intracranial hemorrhage.

Neuroimaging studies, particularly DWI, facilitate diagnosis, but do not always provide the answer. Stroke mimics may produce similar MR findings to ischemic stroke, including decreased diffusion (Table 23.3). Conversely, more than 50% of patients with stroke-like symptoms and normal DWI scan have true ischemia.[196] Finally, the imaging appearance of ischemic stroke changes with time.

Despite these issues, the topographic pattern of the abnormalities in combination with vascular imaging and MR perfusion findings can help narrow the differential diagnosis (Table 23.4). These topographic patterns include (1) regional gray and white matter (Figure 23.7), (2) cortical and deep gray matter (Figure 23.8), (3) deep gray matter, (4) white matter, (5) multiple scattered foci (Figure 23.9), (6) borderzone pattern, and (7) splenial involvement (Figure 23.10).

TABLE 23.3

Causes of Decreased Diffusion

Failure of Na+/K+-ATPase (ischemic and/or excitotoxic injury)	Hypoglycemia[190]; hyperglycemia[191]; ketosis[191]; seizures; transient global amnesia[192]; drug induced encephalopathies including metronidazole, methotrexate, or vigabatrin; necrotizing infections such as herpes simplex virus (HSV); Wernicke encephalopathy[193]
Tissue vacuolization or spongiform changes	Creutzfeldt–Jakob disease; Heroin leukoencephalopathy; demyelination and dysmyelination; diffuse axonal injury[194,195]
High protein concentration or increased viscosity	Pyogenic infection; hemorrhage; extracellular methemoglobin or oxyhemoglobin
Dense cell packing	High grade glioma; lymphoma; small cell lung cancer

TABLE 23.4

Topographic Patterns of Stroke Mimics

1. Regional gray and white matter	Seizures
	Migraine
	Primary brain tumors
	Hypoglycemia
	Transient global amnesia
	HSV encephalitis
	MELAS
	Venous thrombosis
2. Cortical and deep gray matter	Hypoxic–ischemic encephalopathy (HIE)
	Wernicke encephalopathy
	Hepatic encephalopathy
	Creutzfeldt–Jakob disease
	Eastern equine encephalitis
3. Deep gray matter	HIE
	CO poisoning
	Osmotic myelinolysis
	Wernicke encephalopathy
	Creutzfeldt–Jakob disease
	Vigabatrin toxicity
	Nonketotic hyperglycemia
	Deep venous thrombosis
4. White matter	Cerebral autosomal dominant arteriopathy with subcortical infarcts and leukoencephalopathy
	Susac syndrome
	HIE
	Metronidazole toxicity
	Methotrexate toxicity
	Heroin leukoencephalopathy
	Demyelinating diseases
	Cerebritis/abscess
5. Multiple scattered foci	Diffuse axonal injury
	Fat emboli
	Metastases
6. Border zone pattern	Reversible cerebral vasoconstriction syndrome
	Moyamoya disease
	HIE
	Posterior reversible encephalopathy syndrome
	Hyperperfusion syndrome
7. Splenial lesions	Antiepileptic medications
	Seizures
	Hypoglycemia
	Trauma
	Wernicke encephalopathy
	Osmotic myelinolysis
	Marchiafava–Bignami disease
	Radiation therapy
	Altitude sickness

23.3 Hemorrhagic Stroke

23.3.1 Imaging Evaluation of Acute Nontraumatic Intracranial Hemorrhage

Approximately 10%–15% of all strokes are hemorrhagic. The diagnosis and imaging workup of hemorrhagic stroke is largely performed using NCCT and CTA. However, MRI can provide additional information that may aid in the diagnosis of the underlying etiology.

Acute hemorrhage is most easily identified on NCCT, where it appears hyperdense (60–80 HU) to brain parenchyma. On MRI, hemorrhage has a variable appearance depending on the age and breakdown products of the blood. The imaging workup of nontraumatic intracranial hemorrhage seeks to identify treatable vascular etiologies. The most common causes include cerebral aneurysms and pial or dural arteriovenous malformations (AVMs). The location of the hemorrhage (extra- vs. intraparenchymal) provides clues to the likely cause and will influence the subsequent imaging workup.

23.3.1.1 Subarachnoid Hemorrhage

Acute subarachnoid hemorrhage (SAH) is best identified on NCCT as hyperdensity in the CSF spaces surrounding the brain, and may often be seen layering in the ventricles. On MRI, SAH may be identified using FLAIR imaging, where it appears as hyperintense signal within the CSF spaces. FLAIR may be helpful in identifying SAH in cases where NCCT is negative, but the clinical suspicion is high.[197] However, it should not replace lumbar puncture.[198] SWI may further increase the sensitivity for SAH detection.[199]

The primary cause of nontraumatic or spontaneous SAH is a ruptured cerebral aneurysm. Other vascular causes include intracranial dissection, vasculitis, mycotic aneurysm, dural AVM, and cervical arteriovenous fistulas. For diagnostic purposes, spontaneous SAH may be divided into perimesencephalic or diffuse patterns. This distinction has important bearing on the yield of the aneurysm workup, as well as the clinical course. A perimesencephalic pattern accounts for approximately 10% of SAH cases, is centered anterior to the brain stem, and can extend to involve the ambient and suprasellar cisterns. There can be involvement of the proximal anterior interhemispheric and proximal Sylvian cisterns, but the remainder of these cisterns should not be filled with blood and there should be no intraventricular extension. Patients with this SAH pattern are highly unlikely to have an intracranial aneurysm (~5%), and in the vast majority of cases have an excellent clinical course.

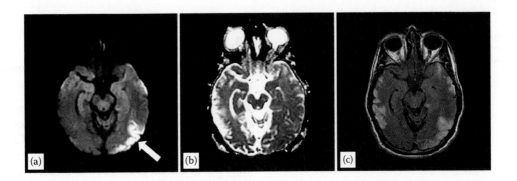

FIGURE 23.7
A 49-year-old man who was diagnosed with four episodes of stroke-like events during the past 2 years. The most recent MRI shows an area of hyperintense signal on trace DWI (a) with cortical swelling on FLAIR sequence (c) in the left posterior temporal and occipital lobes (arrow). The ADC map (b) shows areas with and without restricted diffusion. There are additional areas of cortical T_2 hyperintensity on the FLAIR sequence involving the bilateral temporal lobes with increased diffusivity on the ADC map consistent with expected evolution of older lesions. The patient was finally diagnosed with mitochondrial encephalomyopathy, lactic acidosis, stroke-like episodes (MELAS).

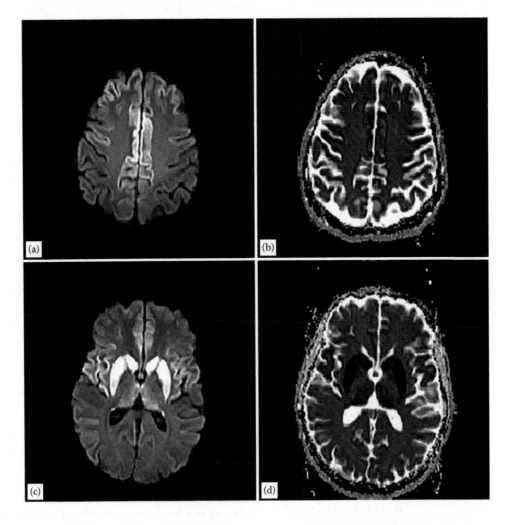

FIGURE 23.8
A 41-year-old woman with a 1 year history of decline in memory with gait disturbance and more recently, changes in personality. The diffusion-weighted trace images (a,c) show cortical hyperintense signal in bilateral frontal lobes, cingulate gyri, and insulae, as well as bilateral basal ganglia and thalami. The ADC maps (b,d) confirm diminished diffusion in these regions. She was diagnosed with Creutzfeldt–Jakob disease.

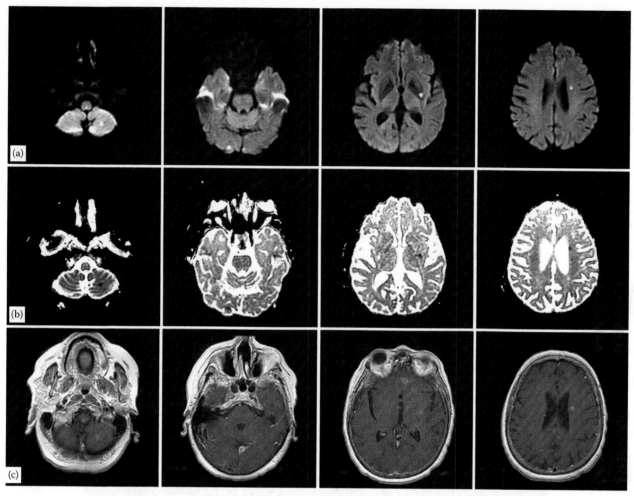

FIGURE 23.9

A 71-year-old woman with a history of small cell lung cancer and brain metastases presented with transient visual changes and gait imbalance. MRI demonstrates multiple foci of restricted diffusion in the cerebellum, right occipital lobe, left lentiform nuclei, and left corona radiata (a, DWI; b, ADC). All of the lesions demonstrated contrast enhancement on the postcontrast T_1WI (c).

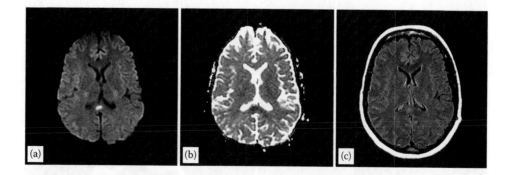

FIGURE 23.10

A 26-year-old woman with brain-stem astrocytoma presented with 3 days of weakness, altered mental status, and difficulty speaking after finishing a cycle of radiation therapy 5 days before presentation. MRI shows a focus of restricted diffusion (a,b) with T_2 FLAIR hyperintensity (c) in the splenium of the corpus callosum. Focal T_2 FLAIR hyperintense signal in the splenium of the corpus callosum is a common finding after brain radiation therapy and can be associated with restricted diffusion.

Noninvasive vascular imaging has become the first-line test for aneurysm detection and treatment planning. With its high spatial resolution, CTA has excellent accuracy for identifying aneurysms. A large comparative study of 64-detector CTA against reference standard DSA revealed 100% sensitivity and specificity for aneurysms sized 3 mm or greater.[200] However, the sensitivity for tiny aneurysms (<3 mm) is lower, with variable rates reported in the literature (60%–95%).[200,201] Several authors have recommended that a negative CTA in the setting of perimesencephalic SAH is sufficient and obviates the need for DSA.[202] By contrast, in patients with a diffuse SAH pattern, DSA is warranted even if initial CTA is negative, and a second DSA at 5–7 days should be strongly considered if the first DSA is negative as the yield for detection of a causative vascular lesion may be 5%–7% in this setting.[203]

Although MRA is less frequently used for this workup, a systematic review found that the overall accuracy of 3D TOF MRA for detection of intracranial aneurysms was similar to CTA and approximately 90%[201] (Figure 23.11). Similar findings were reported with MRA at 3 T.[204,205] The most recent systematic review examining the accuracy of MRA against the reference standard DSA reported a pooled sensitivity of 95% and a pooled specificity of 89%, with a trend toward better performance at 3 T.[206] The authors further noted that false diagnoses were more frequent at the skull base and middle cerebral artery.

For proximal aneurysms arising from or near the circle of Willis, CTA provides suitable anatomic information for treatment planning in the majority of cases. Important information for deciding between surgical clipping and endovascular coiling includes aneurysm size and location, dome-to-neck ratio, location of adjacent branch vessels, and presence of intra-aneurysmal thrombus (Figure 23.11c and d). Postprocessed images such as multiplanar reformats and volume-rendered images are often helpful for depicting these aneurysm characteristics. A recent study found a greater accuracy for identifying branch vessels arising from the aneurysm sac with 3 T TOF MRA compared to CTA.[207] Imaging assessment and notification of findings to the neurovascular team should be performed urgently as early aneurysm closure (e.g., within 24 h) is associated with more favorable outcomes.[208]

If SAH is predominantly in a peripheral sulcal distribution, there should be a high clinical suspicion for vasculitis/vasculopathy or mycotic aneurysm. In these cases, MRI may be helpful for depicting the additional findings associated with these entities, which include focal infarcts, parenchymal hematomas, and microbleeds.[209] DSA is superior to both CTA and MRA for evaluation of the small caliber vessels that are commonly affected by these processes. In an older patient,

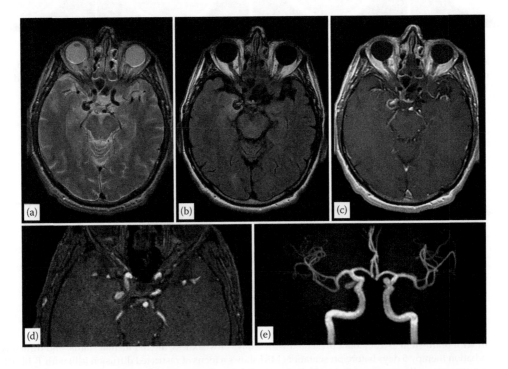

FIGURE 23.11
A 70-year-old man with a 1 cm aneurysm at the origin of the right posterior communicating artery found incidentally during evaluation of intermittent headache and lightheadedness. MRI shows an aneurysm with mass effect on the adjacent right mesial temporal lobe (a, T_2WI) with associated parenchymal T_2 FLAIR hyperintensity (b, FLAIR). There is a filling defect in the aneurysm on the postcontrast T_1WI (c) indicating partial thrombosis. The MRA source image (d) and 3D reconstruction (e) depict the aneurysm.

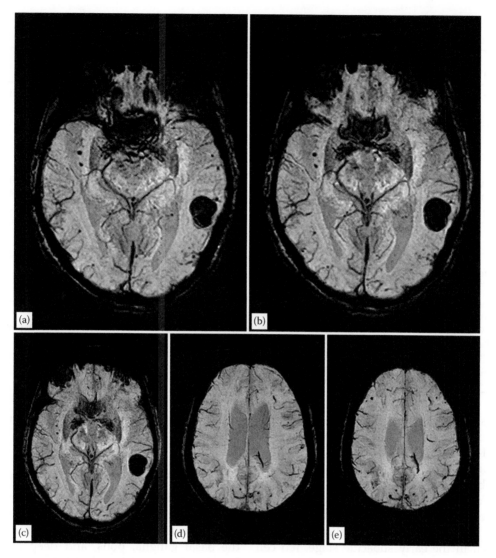

FIGURE 23.12
A 86-year-old woman with an IPH in the posterior left temporal lobe in the presence of multiple lobar microhemorrhages, suggesting cerebral amyloid angiopathy at her age.

peripheral SAH raises the suspicion for cerebral amyloid angiopathy, which may be further confirmed by the presence of multiple microbleeds[210] (Figure 23.12).

23.3.1.2 Intraparenchymal Hemorrhage: Temporal Evolution of MRI Findings

The MRI appearance of an intraparenchymal hematoma (IPH) is influenced by two major factors: its intrinsic molecular structure and how it is imaged (e.g., MRI field strength and scanning parameters).[211–213]

The temporal evolution of IPHs can be divided into hyperacute, acute, early subacute, late subacute, and chronic stages[214] (Table 23.5). During these stages, there are alterations in hemoglobin oxygenation status, integrity of the red cell membrane, and protein concentration.[215] Imaging findings depend primarily on the oxygen tension around and inside the hematoma. Importantly, these stages are superimposed, with the changes starting in the periphery and progressing toward the center.[216]

Intracellular oxyhemoglobin predominates in the hyperacute stage. Oxyhemoglobin is a diamagnetic substance (i.e., no unpaired electrons) and demonstrates intermediate to slightly low SI on T_1WI and high SI on T_2WI compared to the brain parenchyma.[217] After a few hours, the acute stage is characterized by transformation of intracellular oxyhemoglobin into deoxyhemoglobin. Deoxyhemoglobin contains four unpaired electrons, making it paramagnetic and resulting in signal loss on T_2WI and T_2^*WI. The molecular configuration of deoxyhemoglobin prevents its interaction with

TABLE 23.5

Evolution of IPH on MRI

Stage of Hematoma	Time Course	Molecular Component	T_1WI	T_2WI	Blooming (Low SI) on T_2*WI	DWI
Hyperacute	<6 h	Oxyhemoglobin	Intermediate to slightly low SI	High SI	No	High DWI, low ADC
Acute	6–72 h	Deoxyhemoglobin	Intermediate to slightly low SI	Low SI	Yes	—
Early subacute	3–7 days	Intracellular methemoglobin	High SI	Low SI	Yes	—
Late subacute	1 week– months	Extracellular methemoglobin	High SI	High SI	No	High DWI, variable ADC
Chronic	Months–years	Hemosiderin, ferritin	Low SI	Low SI	Yes	—

TABLE 23.6

Technical Pearls for Hemorrhagic Stroke Evaluation

Technical Pearls

- Vascular imaging is critical to identify treatable causes of spontaneous intracranial hemorrhage.
- CTA is highly accurate for identifying aneurysms and AVMs, and is a first-line test for both diagnosis and treatment planning.
- MRA with TOF technique has similar accuracy to CTA for aneurysm and AVM detection, and also allows visualization of arteriovenous shunting. MRI is a complementary test in the evaluation of causes of nontraumatic intracranial hemorrhage.
- Catheter angiography should be performed in most cases when noninvasive imaging does not reveal a source of hemorrhage.
- Contrast extravasation on CTA is a strong predictor of early and significant hematoma growth.

water molecules, thus causing no T_1 shortening effect.[218] Therefore, acute hematomas show an intermediate to low SI on T_1WI and low SI on T_2WI and T_2*WI. After about a week, the hematoma enters the subacute stage, in which hemoglobin is oxidized to methemoglobin whose five unpaired electrons confer paramagnetic properties. The molecular configuration of methemoglobin allows access of water molecules, causing T_1 shortening and high SI on T_1WI. In the early subacute stage, red cell membrane integrity is preserved. The intracellular methemoglobin causes magnetic field inhomogeneity seen as low SI on T_2WI and T_2*WI. In the late subacute stage, the red cell membrane breaks down, causing homogenous dilution of the paramagnetic substance and thus loss of susceptibility effect. Regaining of bright SI on T_2WI is subsequently seen at this stage. Methemoglobin is resorbed over time, leaving ferritin and hemosiderin which are highly paramagnetic and demonstrate persistent dark SI on T_1WI and T_2WI along the rim of the chronic hematoma.[214,219,220]

The appearance of hematomas on diffusion imaging also varies by stage. Hyperacute and late subacute hematomas appear bright on DWI.[221] In the hyperacute stage, the high SI on DWI and low SI on ADC are believed to result from clot retraction and high viscosity. In late subacute hematomas, high SI on DWI is accompanied by variable ADC findings.[221,222] High viscosity or infiltration of inflammatory cells probably causes the late subacute hematoma to restrict diffusion. In acute, early subacute, and chronic stages, susceptibility effect precludes accurate evaluation of DWI in the hematoma.

23.3.1.3 IPH: Imaging Workup and Diagnosis

Approximately 85% of IPHs are primary (i.e., no causative anatomic lesion) and most often associated with hypertension. Hypertensive bleeds typically affect the basal ganglia, pons, and deep cerebellar nuclei. Imaging is critical for identifying potential causes of secondary IPH, which include AVM, aneurysm, venous sinus thrombosis, tumor, and vasculitis (Table 23.6). Vascular lesions have a high risk of recurrent hemorrhage, and should be treated when discovered. Young (e.g., <45 years) normotensive patients with IPH have a particularly high incidence (50%–65%) of underlying vascular abnormality, the majority of which are AVMs and aneurysms.[223]

On both CTA and MRA, anatomic clues to the diagnosis of AVM include numerous and enlarged vessels corresponding to feeding arteries, the vascular nidus or draining veins as well as associated calcifications (Figure 23.13). On MRI, the abnormal vessels appear as enlarged flow voids on T_2WI. However, small AVMs may be easily overlooked on anatomic imaging. Unlike standard CTA images that are static, MRA techniques offer a noninvasive means of imaging arteriovenous shunting, and thus it is a complementary diagnostic test for the noninvasive evaluation of brain AVMs and dural fistulas. Specifically, 3D TOF MRA may reveal

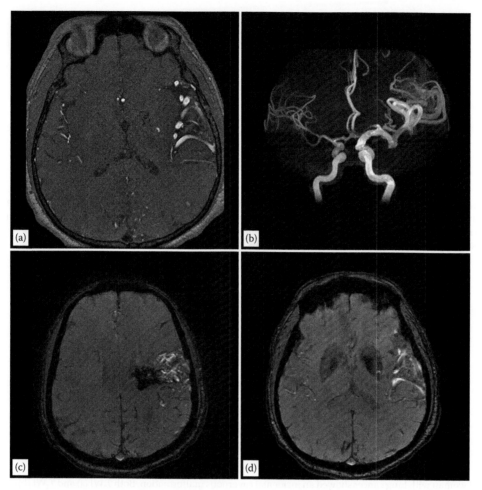

FIGURE 23.13
A 63-year-old man with longstanding seizure due to a large left frontal AVM. The MRA source images (a) and 3D reconstruction (b) demonstrate enlarged left intracranial ICA and left MCA supplying the AVM with flow-related enhancement in numerous enlarged draining veins. The SWAN images demonstrate areas of susceptibility effect in the left centrum semiovale due to parenchymal hemorrhage (c) and high signal in the draining veins that are similar to the arteries due to rapid shunting and decreased venous deoxyhemoglobin content (d).

arterial flow-related enhancement within venous structures secondary to shunting (Figure 23.13a and b). Flow dynamics may also be depicted using time-resolved MRA or CTA, but these techniques are more technically demanding. A recent Cochrane systematic review evaluated the accuracy of CTA and MRA compared to reference standard DSA for detecting vascular causes of IPH (including aneurysms, brain AVMs, and dural arteriovenous fistulae), and found excellent accuracy for both techniques (sensitivity 95%–98%, specificity 99%) and no evidence that one was superior.[224]

More recently, studies have demonstrated the utility of SWI for identifying arteriovenous shunting (Figure 23.13c and d). In the normal patient, arteries are bright and veins are dark on SWAN. The presence of bright signal in the veins is a reliable marker for arteriovenous shunting, with one study reporting 93% sensitivity and 99% specificity among 60 patients with known or suspected brain AVMs.[225] The authors noted that

false-positive cases may be seen in patients who are on supplemental oxygen, which can result in diffuse hyperintense signal in the major veins. In a follow-up study of 17 patients, the same group found that post-gadolinium SWI may be superior to SWI alone for detecting shunting.[226] This is thought to be related to the enhanced contrast between arteries and veins after gadolinium administration.

Dural sinus or cortical vein thrombosis is a less common cause of IPH, but should always be considered, particularly in young to middle-aged females who are postpartum or on oral contraceptives (Figure 23.14). Intracranial veins are often opacified on CTA, and if so, should be evaluated for filling defects or occlusion. Important mimics within the dural sinuses include arachnoid granulations, which are often seen on CTA as lobulated filling defects in the lateral aspects of the transverse sinuses, as well as hypoplasia of a transverse sinus. Suspected thrombosis should be confirmed

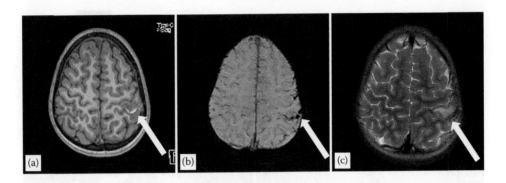

FIGURE 23.14

A 7-year-old boy presented with persistent headache, right facial droop, and right upper extremity weakness 2 weeks after a febrile illness. MRI shows serpiginious T_1 hyperintensity (a, T_1WI) with associated susceptibility blooming effect (b, SWI) within a cortical vein in the left parietal lobe (arrows). Similar findings were seen in other cortical veins and the superior sagittal sinus (not shown). The T_2WI (c) demonstrates parenchymal edema in the postcentral gyrus consistent with venous infarct.

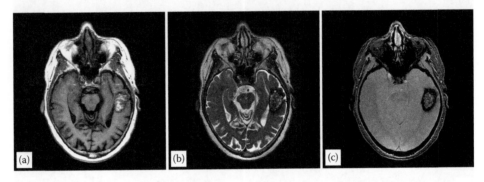

FIGURE 23.15

MRI in a 89-year-old woman with lung cancer showed an incidental cavernous malformation. Both T_1WI (a) and T_2WI (b) demonstrate a typical lesion with a "popcorn" morphology and a rim of signal loss especially on T_2WI and gradient-echo T_2*WI (c) due to hemosiderin. The lesion demonstrates heterogeneous signal on both T_1WI and T_2WI due to presence of blood products of different ages.

on NCCT as hyperdense clot within the vein or sinus. Dedicated CT or MR venography may be performed. Gradient-echo imaging is often helpful for cortical vein thrombosis, which appears as a serpentine area of signal loss (blooming artifact) (Figure 23.14b). DWI may reveal restricted diffusion (cytotoxic edema), elevated diffusion (vasogenic edema from venous congestion), or both within venous infarcts.

As mentioned earlier, vasculopathy or mycotic aneurysm is another uncommon cause of IPH. These entities typically involve the medium and small vessels, and should be considered when parenchymal hemorrhage is seen in combination with peripheral SAH and scattered focal infarcts. Vasculitic changes include beaded irregularity and narrowing, which may be difficult to detect in smaller vessels on noninvasive imaging. Collapsed MIP images are often helpful to evaluate these vessels, and may also help to detect distal saccular outpouchings consistent with mycotic aneurysms, which are often seen in the distal MCA branches given their higher flow. MRI may

suggest the diagnosis by revealing associated infarcts on DWI. Moreover, gradient-echo images may help to detect mycotic aneurysms, which often appear as focal areas of signal loss.

In addition to 3D TOF MRA, routine MRI is important for identifying other causes of IPH that would obviate the need for further vascular imaging. DWI may reveal restricted diffusion to suggest hemorrhagic transformation of an ischemic stroke. Hemorrhagic tumors such as metastases may be evident on post-gadolinium imaging as multiple abnormally enhancing lesions. Cavernous malformations have a classic appearance on T_2W images as a focal area of central T_2 bright signal surrounded by a dark rim of hemosiderin (Figure 23.15). Moreover, post-gadolinium T_1W images may reveal an associated developmental venous anomaly, which has a characteristic caput medusa appearance. Amyloid angiopathy results in numerous small foci of dark signal (microbleeds) distributed throughout the brain on gradient-echo imaging or SWI (Figure 23.12).

Despite the high accuracy of CTA and MRA for identifying vascular causes for IPH, small lesions such as AVMs can be obscured by associated mass effect from the hematoma, and therefore, DSA should be strongly considered in the setting of a negative noninvasive test. Deciding which patients should go on to DSA is a matter of debate.[227] In older patients (e.g., >45 years) with a history of hypertension who present with IPH in the basal ganglia or thalamus, the yield of DSA for identifying an underlying vascular abnormality has been reported to be exceedingly low.[228] Although the posterior fossa is another common location for hypertensive bleeds, one study could not identify an independent predictor of underlying vascular lesion among 68 patients with posterior fossa IPH,[229] suggesting that DSA should be performed in this setting when noninvasive imaging is negative. In all other patients, DSA is the appropriate next step in evaluation. If the DSA is negative, repeat angiography should be considered after the resolution of the hematoma and associated mass effect (e.g., after 2–4 weeks).

In cases of primary IPH (i.e., no cause is identified), imaging may be helpful for prognostication and to identify patients at high risk for hematoma expansion. In addition to a poor clinical exam at presentation (e.g., Glasgow Coma Scale score <9), imaging findings that are highly predictive of mortality and poor outcomes include a large hematoma volume (e.g., >60 mL) and the presence of intraventricular hemorrhage. Approximately one-third of patients found to have IPH within 3–6 h of onset demonstrate further significant (>33% or 6 mL) hematoma growth, which is another independent predictor of worse outcomes.

Although treatments to prevent hematoma expansion require further clinical validation, identifying high-risk populations for hematoma growth may provide important prognostic information, guide decisions regarding early surgery, and allow more targeted patient selection in future trials of hemostatic therapy. Recent studies have supported the ability of CTA to predict hematoma growth by revealing contrast extravasation within the hemorrhage (i.e., the spot sign). There are various working definitions of the spot sign. In general, a spot must be within the hematoma, demonstrate density much higher than the surrounding blood (e.g., ≥120 HU), and be discontinuous from vessels. It can take any morphology, and more than one may be identified within the hematoma. Additional spot sign characteristics that may further stratify rebleeding risk include the presence of ≥3 spots, maximum diameter of the largest spot ≥5 mm, and maximum attenuation of the largest spot ≥180 HU. These findings have been demonstrated to independently predict significant hematoma growth, higher in-hospital mortality, and poor outcomes among survivors.[230] Similar contrast extravasation may be seen within hematomas on MRI, but much less is known

regarding this phenomenon and its relationship to hematoma growth and clinical outcome.

References

1. Mozaffarian D, Benjamin EJ, Go AS, Arnett DK, Blaha MJ, Cushman M et al. Heart disease and stroke statistics—2015 update: A report from the American Heart Association. *Circulation*. 2015;131:e29–e322.
2. Schellinger PD, Chalela JA, Kang DW, Latour LL, Warach S. Diagnostic and prognostic value of early MR imaging vessel signs in hyperacute stroke patients imaged <3 hours and treated with recombinant tissue plasminogen activator. *AJNR American Journal of Neuroradiology*. 2005;26:618–624.
3. Brazzelli M, Sandercock PA, Chappell FM, Celani MG, Righetti E, Arestis N et al. Magnetic resonance imaging versus computed tomography for detection of acute vascular lesions in patients presenting with stroke symptoms. *The Cochrane Database of Systematic Reviews*. 2009:CD007424.
4. de Figueiredo EH, Borgonovi AF, Doring TM. Basic concepts of MR imaging, diffusion MR imaging, and diffusion tensor imaging. *Magnetic Resonance Imaging Clinics of North America*. 2011;19:1–22.
5. de Carvalho Rangel C, Hygino Cruz LC, Jr., Takayassu TC, Gasparetto EL, Domingues RC. Diffusion MR imaging in central nervous system. *Magnetic Resonance Imaging Clinics of North America*. 2011;19:23–53.
6. Fries P, Runge VM, Kirchin MA, Stemmer A, Naul LG, Wiliams KD et al. Diffusion-weighted imaging in patients with acute brain ischemia at 3 T: Current possibilities and future perspectives comparing conventional echoplanar diffusion-weighted imaging and fast spin echo diffusion-weighted imaging sequences using blade (propeller). *Investigative Radiology*. 2009;44:351–359.
7. Yiping L, Hui L, Kun Z, Daoying G, Bo Y. Diffusion-weighted imaging of the sellar region: A comparison study of blade and single-shot echo planar imaging sequences. *European Journal of Radiology*. 2014;83:1239–1244.
8. Lee CY, Li Z, Pipe JG, Debbins JP. Turboprop+: Enhanced turboprop diffusion-weighted imaging with a new phase correction. *Magnetic Resonance in Medicine: Official Journal of the Society of Magnetic Resonance in Medicine/Society of Magnetic Resonance in Medicine*. 2013;70:497–503.
9. Li Z, Pipe JG, Lee CY, Debbins JP, Karis JP, Huo D. X-prop: A fast and robust diffusion-weighted propeller technique. *Magnetic Resonance in Medicine: Official Journal of the Society of Magnetic Resonance in Medicine/Society of Magnetic Resonance in Medicine*. 2011;66:341–347.
10. Attenberger UI, Runge VM, Stemmer A, Williams KD, Naul LG, Michaely HJ et al. Diffusion weighted imaging: A comprehensive evaluation of a fast spin echo DWI sequence with blade (propeller) k-space sampling at 3 T, using a 32-channel head coil in acute brain ischemia. *Investigative Radiology*. 2009;44:656–661.

11. Mintorovitch J, Yang GY, Shimizu H, Kucharczyk J, Chan PH, Weinstein PR. Diffusion-weighted magnetic resonance imaging of acute focal cerebral ischemia: Comparison of signal intensity with changes in brain water and Na+,K(+)-atpase activity. *Journal of Cerebral Blood Flow and Metabolism: Official Journal of the International Society of Cerebral Blood Flow and Metabolism.* 1994;14:332–336.

12. Benveniste H, Hedlund LW, Johnson GA. Mechanism of detection of acute cerebral ischemia in rats by diffusion-weighted magnetic resonance microscopy. *Stroke: A Journal of Cerebral Circulation.* 1992;23:746–754.

13. Simonsen CZ, Madsen MH, Schmitz ML, Mikkelsen IK, Fisher M, Andersen G. Sensitivity of diffusion- and perfusion-weighted imaging for diagnosing acute ischemic stroke is 97.5%. *Stroke: A Journal of Cerebral Circulation.* 2015;46:98–101.

14. Brunser AM, Hoppe A, Illanes S, Diaz V, Munoz P, Carcamo D et al. Accuracy of diffusion-weighted imaging in the diagnosis of stroke in patients with suspected cerebral infarct. *Stroke: A Journal of Cerebral Circulation.* 2013;44:1169–1171.

15. Rosso C, Drier A, Lacroix D, Mutlu G, Pires C, Lehericy S et al. Diffusion-weighted MRI in acute stroke within the first 6 hours: 1.5 or 3.0 tesla? *Neurology.* 2010;74:1946–1953.

16. Hjort N, Christensen S, Solling C, Ashkanian M, Wu O, Rohl L et al. Ischemic injury detected by diffusion imaging 11 minutes after stroke. *Annals of Neurology.* 2005;58:462–465.

17. Mullins ME, Schaefer PW, Sorensen AG, Halpern EF, Ay H, He J et al. CT and conventional and diffusion-weighted MR imaging in acute stroke: Study in 691 patients at presentation to the emergency department. *Radiology.* 2002;224:353–360.

18. Schlaug G, Siewert B, Benfield A, Edelman RR, Warach S. Time course of the apparent diffusion coefficient (ADC) abnormality in human stroke. *Neurology.* 1997;49:113–119.

19. Fiebach JB, Jansen O, Schellinger PD, Heiland S, Hacke W, Sartor K. Serial analysis of the apparent diffusion coefficient time course in human stroke. *Neuroradiology.* 2002;44:294–298.

20. Geijer B, Lindgren A, Brockstedt S, Stahlberg F, Holtas S. Persistent high signal on diffusion-weighted MRI in the late stages of small cortical and lacunar ischaemic lesions. *Neuroradiology.* 2001;43:115–122.

21. Fiehler J, Knudsen K, Kucinski T, Kidwell CS, Alger JR, Thomalla G et al. Predictors of apparent diffusion coefficient normalization in stroke patients. *Stroke: A Journal of Cerebral Circulation.* 2004;35:514–519.

22. Schaefer PW, Hassankhani A, Putman C, Sorensen AG, Schwamm L, Koroshetz W et al. Characterization and evolution of diffusion MR imaging abnormalities in stroke patients undergoing intra-arterial thrombolysis. *AJNR American Journal of Neuroradiology.* 2004;25:951–957.

23. Kidwell CS, Saver JL, Mattiello J, Starkman S, Vinuela F, Duckwiler G et al. Thrombolytic reversal of acute human cerebral ischemic injury shown by diffusion/perfusion magnetic resonance imaging. *Annals of Neurology.* 2000;47:462–469.

24. Yoo AJ, Pulli B, Gonzalez RG. Imaging-based treatment selection for intravenous and intra-arterial stroke therapies: A comprehensive review. *Expert Review of Cardiovascular Therapy.* 2011;9:857–876.

25. Chemmanam T, Campbell BC, Christensen S, Nagakane Y, Desmond PM, Bladin CF et al. Ischemic diffusion lesion reversal is uncommon and rarely alters perfusion-diffusion mismatch. *Neurology.* 2010;75:1040–1047.

26. Latchaw RE, Alberts MJ, Lev MH, Connors JJ, Harbaugh RE, Higashida RT et al. Recommendations for imaging of acute ischemic stroke: A scientific statement from the american heart association. *Stroke: A Journal of Cerebral Circulation.* 2009;40:3646–3678.

27. Bulut HT, Yildirim A, Ekmekci B, Eskut N, Gunbey HP. False-negative diffusion-weighted imaging in acute stroke and its frequency in anterior and posterior circulation ischemia. *Journal of Computer Assisted Tomography.* 2014;38:627–633.

28. Oppenheim C, Stanescu R, Dormont D, Crozier S, Marro B, Samson Y et al. False-negative diffusion-weighted MR findings in acute ischemic stroke. *AJNR American Journal of Neuroradiology.* 2000;21:1434–1440.

29. Schuier FJ, Hossmann KA. Experimental brain infarcts in cats. II. Ischemic brain edema. *Stroke: A Journal of Cerebral Circulation.* 1980;11:593–601.

30. Watanabe O, West CR, Bremer A. Experimental regional cerebral ischemia in the middle cerebral artery territory in primates. Part 2: Effects on brain water and electrolytes in the early phase of mca stroke. *Stroke: A Journal of Cerebral Circulation.* 1977;8:71–76.

31. Thomalla G, Rossbach P, Rosenkranz M, Siemonsen S, Krutzelmann A, Fiehler J et al. Negative fluid-attenuated inversion recovery imaging identifies acute ischemic stroke at 3 hours or less. *Annals of Neurology.* 2009;65:724–732.

32. Ebinger M, Galinovic I, Rozanski M, Brunecker P, Endres M, Fiebach JB. Fluid-attenuated inversion recovery evolution within 12 hours from stroke onset: A reliable tissue clock? *Stroke: A Journal of Cerebral Circulation.* 2010;41:250–255.

33. Allen LM, Hasso AN, Handwerker J, Farid H. Sequence-specific MR imaging findings that are useful in dating ischemic stroke. *RadioGraphics: A Review Publication of the Radiological Society of North America, Inc.* 2012;32:1285–1297; discussion 1297–1289.

34. Axel L. Blood flow effects in magnetic resonance imaging. *AJR American Journal of Roentgenology.* 1984;143:1157–1166.

35. Bradley WG, Jr., Waluch V, Lai KS, Fernandez EJ, Spalter C. The appearance of rapidly flowing blood on magnetic resonance images. *AJR American Journal of Roentgenology.* 1984;143:1167–1174.

36. Kamran S, Bates V, Bakshi R, Wright P, Kinkel W, Miletich R. Significance of hyperintense vessels on flair MRI in acute stroke. *Neurology.* 2000;55:265–269.

37. Cosnard G, Duprez T, Grandin C, Smith AM, Munier T, Peeters A. Fast flair sequence for detecting major vascular abnormalities during the hyperacute phase of stroke:

A comparison with MR angiography. *Neuroradiology.* 1999;41:342–346.

38. Cheng B, Ebinger M, Kufner A, Kohrmann M, Wu O, Kang DW et al. Hyperintense vessels on acute stroke fluid-attenuated inversion recovery imaging: Associations with clinical and other MRI findings. *Stroke: A Journal of Cerebral Circulation.* 2012;43: 2957–2961.

39. Legrand L, Tisserand M, Turc G, Naggara O, Edjlali M, Mellerio C et al. Do flair vascular hyperintensities beyond the DWI lesion represent the ischemic penumbra? *AJNR American Journal of Neuroradiology.* 2014.

40. Liu W, Xu G, Yue X, Wang X, Ma M, Zhang R et al. Hyperintense vessels on flair: A useful non-invasive method for assessing intracerebral collaterals. *European Journal of Radiology.* 2011;80:786–791.

41. Kawashima M, Noguchi T, Takase Y, Ootsuka T, Kido N, Matsushima T. Unilateral hemispheric proliferation of ivy sign on fluid-attenuated inversion recovery images in moyamoya disease correlates highly with ipsilateral hemispheric decrease of cerebrovascular reserve. *AJN. American Journal of Neuroradiology.* 2009;30:1709–1716.

42. Iancu-Gontard D, Oppenheim C, Touze E, Meary E, Zuber M, Mas JL et al. Evaluation of hyperintense vessels on flair MRI for the diagnosis of multiple intracerebral arterial stenoses. *Stroke: A Journal of Cerebral Circulation.* 2003;34:1886–1891.

43. Sanossian N, Ances BM, Shah SH, Kim D, Saver JL, Liebeskind DS. Flair vascular hyperintensity may predict stroke after TIA. *Clinical Neurology and Neurosurgery.* 2007;109:617–619.

44. Hodel J, Leclerc X, Rodallec M, Gerber S, Blanc R, Outteryck O et al. Fluid-attenuated inversion recovery vascular hyperintensities are not visible using 3D cube flair sequence. *European Radiology.* 2013;23:1963–1969.

45. Kohrmann M, Struffert T, Frenzel T, Schwab S, Doerfler A. The hyperintense acute reperfusion marker on fluid-attenuated inversion recovery magnetic resonance imaging is caused by gadolinium in the cerebrospinal fluid. *Stroke: A Journal of Cerebral Circulation.* 2012;43:259–261.

46. Ostwaldt AC, Rozanski M, Schmidt WU, Nolte CH, Hotter B, Jungehuelsing GJ et al. Early time course of flair signal intensity differs between acute ischemic stroke patients with and without hyperintense acute reperfusion marker. *Cerebrovascular Diseases.* 2014;37:141–146.

47. Warach S, Latour LL. Evidence of reperfusion injury, exacerbated by thrombolytic therapy, in human focal brain ischemia using a novel imaging marker of early blood-brain barrier disruption. *Stroke: A Journal of Cerebral Circulation.* 2004;35:2659–2661.

48. Rozanski M, Ebinger M, Schmidt WU, Hotter B, Pittl S, Heuschmann PU et al. Hyperintense acute reperfusion marker on flair is not associated with early haemorrhagic transformation in the elderly. *European Radiology.* 2010;20:2990–2996.

49. Merino JG, Warach S. Imaging of acute stroke. *Nature Reviews. Neurology.* 2010;6:560–571.

50. Kidwell CS, Chalela JA, Saver JL, Starkman S, Hill MD, Demchuk AM et al. Comparison of MRI and CT for detection of acute intracerebral hemorrhage. *JAMA.* 2004;292:1823–1830.

51. Naggara O, Raymond J, Domingo Ayllon M, Al-Shareef F, Touze E, Chenoufi M et al. T2* "susceptibility vessel sign" demonstrates clot location and length in acute ischemic stroke. *PLoS One.* 2013;8:e76727.

52. Liebeskind DS, Sanossian N, Yong WH, Starkman S, Tsang MP, Moya AL et al. CT and MRI early vessel signs reflect clot composition in acute stroke. *Stroke: A Journal of Cerebral Circulation.* 2011;42:1237–1243.

53. Cho KH, Kim JS, Kwon SU, Cho AH, Kang DW. Significance of susceptibility vessel sign on T2*-weighted gradient echo imaging for identification of stroke subtypes. *Stroke: A Journal of Cerebral Circulation.* 2005;36:2379–2383.

54. Legrand L, Naggara O, Turc G, Mellerio C, Roca P, Calvet D et al. Clot burden score on admission T2*-MRI predicts recanalization in acute stroke. *Stroke: A Journal of Cerebral Circulation.* 2013;44:1878–1884.

55. Yan S, Hu H, Shi Z, Zhang X, Zhang S, Liebeskind DS et al. Morphology of susceptibility vessel sign predicts middle cerebral artery recanalization after intravenous thrombolysis. *Stroke: A Journal of Cerebral Circulation.* 2014;45:2795–2797.

56. Chalela JA, Haymore JB, Ezzeddine MA, Davis LA, Warach S. The hypointense MCA sign. *Neurology.* 2002;58:1470.

57. Mittal S, Wu Z, Neelavalli J, Haacke EM. Susceptibility-weighted imaging: Technical aspects and clinical applications, part 2. *AJNR American Journal of Neuroradiology.* 2009;30:232–252.

58. Hermier M, Nighoghossian N. Contribution of susceptibility-weighted imaging to acute stroke assessment. *Stroke: A Journal of Cerebral Circulation.* 2004;35: 1989–1994.

59. Boeckh-Behrens T, Lutz J, Lummel N, Burke M, Wesemann T, Schopf V et al. Susceptibility-weighted angiography (swan) of cerebral veins and arteries compared to TOF-MRA. *European Journal of Radiology.* 2012;81:1238–1245.

60. Haacke EM, Mittal S, Wu Z, Neelavalli J, Cheng YC. Susceptibility-weighted imaging: Technical aspects and clinical applications, part 1. *AJNR American Journal of Neuroradiology.* 2009;30:19–30.

61. Barnes SR, Haacke EM. Susceptibility-weighted imaging: Clinical angiographic applications. *Magnetic Resonance Imaging Clinics of North America.* 2009;17:47–61.

62. Jauch EC, Saver JL, Adams HP, Jr., Bruno A, Connors JJ, Demaerschalk BM et al. Guidelines for the early management of patients with acute ischemic stroke: A guideline for healthcare professionals from the American Heart Association/American Stroke Association. *Stroke: A Journal of Cerebral Circulation.* 2013;44:870–947.

63. Heiserman JE, Drayer BP, Keller PJ, Fram EK. Intracranial vascular stenosis and occlusion: Evaluation with three-dimensional time-of-flight MR angiography. *Radiology.* 1992;185:667–673.

64. Huang BY, Castillo M. Neurovascular imaging at 1.5 tesla versus 3.0 tesla. *Magnetic Resonance Imaging Clinics of North America*. 2009;17:29–46.

65. Tomanek AI, Coutts SB, Demchuk AM, Hudon ME, Morrish WE, Sevick RJ et al. MR angiography compared to conventional selective angiography in acute stroke. *The Canadian Journal of Neurological Sciences. Le Journal Canadien des Sciences Neurologiques*. 2006;33:58–62.

66. Bash S, Villablanca JP, Jahan R, Duckwiler G, Tillis M, Kidwell C et al. Intracranial vascular stenosis and occlusive disease: Evaluation with CT angiography, MR angiography, and digital subtraction angiography. *AJNR American Journal of Neuroradiology*. 2005;26:1012–1021.

67. Hirai T, Korogi Y, Ono K, Nagano M, Maruoka K, Uemura S et al. Prospective evaluation of suspected stenoocclusive disease of the intracranial artery: Combined MR angiography and CT angiography compared with digital subtraction angiography. *AJNR American Journal of Neuroradiology*. 2002;23:93–101.

68. Alfke K, Jensen U, Pool C, Rohr A, Bruning R, Weber J et al. Contrast-enhanced magnetic resonance angiography in stroke diagnostics: Additional information compared with time-of-flight magnetic resonance angiography? *Clinical Neuroradiology*. 2011;21:5–10.

69. Nael K, Khan R, Choudhary G, Meshksar A, Villablanca P, Tay J et al. Six-minute magnetic resonance imaging protocol for evaluation of acute ischemic stroke: Pushing the boundaries. *Stroke: A Journal of Cerebral Circulation*. 2014;45:1985–1991.

70. Nael K, Meshksar A, Ellingson B, Pirastehfar M, Salamon N, Finn P et al. Combined low-dose contrast-enhanced MR angiography and perfusion for acute ischemic stroke at 3T: A more efficient stroke protocol. *AJNR American Journal of Neuroradiology*. 2014;35:1078–1084.

71. Villablanca JP, Nael K, Habibi R, Nael A, Laub G, Finn JP. 3 T contrast-enhanced magnetic resonance angiography for evaluation of the intracranial arteries: Comparison with time-of-flight magnetic resonance angiography and multislice computed tomography angiography. *Investigative Radiology*. 2006;41:799–805.

72. Parmar H, Ivancevic MK, Dudek N, Gandhi D, Geerts L, Hoogeveen R et al. Neuroradiologic applications of dynamic MR angiography at 3 T. *Magnetic Resonance Imaging Clinics of North America*. 2009;17:63–75.

73. Seeger A, Klose U, Poli S, Kramer U, Ernemann U, Hauser TK. Acute stroke imaging: Feasibility and value of MR angiography with high spatial and temporal resolution for vessel assessment and perfusion analysis in patients with wake-up stroke. *Academic Radiology*. 2015.

74. Sheth SA, Liebeskind DS. Collaterals in endovascular therapy for stroke. *Current Opinion in Neurology*. 2015;28:10–15.

75. Lee LJ, Kidwell CS, Alger J, Starkman S, Saver JL. Impact on stroke subtype diagnosis of early diffusion-weighted magnetic resonance imaging and magnetic resonance angiography. *Stroke: A Journal of Cerebral Circulation*. 2000;31:1081–1089.

76. Riles TS, Eidelman EM, Litt AW, Pinto RS, Oldford F, Schwartzenberg GW. Comparison of magnetic resonance angiography, conventional angiography, and duplex scanning. *Stroke: A Journal of Cerebral Circulation*. 1992;23:341–346.

77. Petrella JR, Provenzale JM. MR perfusion imaging of the brain: Techniques and applications. *AJR American Journal of Roentgenology*. 2000;175:207–219.

78. Harris AD, Coutts SB, Frayne R. Diffusion and perfusion MR imaging of acute ischemic stroke. *Magnetic Resonance Imaging Clinics of North America*. 2009;17:291–313.

79. Albers GW, Thijs VN, Wechsler L, Kemp S, Schlaug G, Skalabrin E et al. Magnetic resonance imaging profiles predict clinical response to early reperfusion: The diffusion and perfusion imaging evaluation for understanding stroke evolution (defuse) study. *Annals of Neurology*. 2006;60:508–517.

80. Lansberg MG, Straka M, Kemp S, Mlynash M, Wechsler LR, Jovin TG et al. MRI profile and response to endovascular reperfusion after stroke (defuse 2): A prospective cohort study. *The Lancet. Neurology*. 2012;11:860–867.

81. De Silva DA, Brekenfeld C, Ebinger M, Christensen S, Barber PA, Butcher KS et al. The benefits of intravenous thrombolysis relate to the site of baseline arterial occlusion in the Echoplanar Imaging Thrombolytic Evaluation Trial (EPITHET). *Stroke: A Journal of Cerebral Circulation*. 2010;41:295–299.

82. Neumann-Haefelin T, Wittsack HJ, Wenserski F, Siebler M, Seitz RJ, Modder U et al. Diffusion- and perfusion-weighted MRI. The DWI/PWI mismatch region in acute stroke. *Stroke: A Journal of Cerebral Circulation*. 1999;30:1591–1597.

83. Jovin TG, Liebeskind DS, Gupta R, Rymer M, Rai A, Zaidat OO et al. Imaging-based endovascular therapy for acute ischemic stroke due to proximal intracranial anterior circulation occlusion treated beyond 8 hours from time last seen well: Retrospective multicenter analysis of 237 consecutive patients. *Stroke: A Journal of Cerebral Circulation*. 2011;42:2206–2211.

84. Kane I, Carpenter T, Chappell F, Rivers C, Armitage P, Sandercock P et al. Comparison of 10 different magnetic resonance perfusion imaging processing methods in acute ischemic stroke: Effect on lesion size, proportion of patients with diffusion/perfusion mismatch, clinical scores, and radiologic outcomes. *Stroke: A Journal of Cerebral Circulation*. 2007;38:3158–3164.

85. Christensen S, Mouridsen K, Wu O, Hjort N, Karstoft H, Thomalla G et al. Comparison of 10 perfusion MRI parameters in 97 sub-6-hour stroke patients using voxel-based receiver operating characteristics analysis. *Stroke: A Journal of Cerebral Circulation*. 2009;40:2055–2061.

86. Coutts SB, Simon JE, Tomanek AI, Barber PA, Chan J, Hudon ME et al. Reliability of assessing percentage of diffusion-perfusion mismatch. *Stroke: A Journal of Cerebral Circulation*. 2003;34:1681–1683.

87. Rivers CS, Wardlaw JM, Armitage PA, Bastin ME, Carpenter TK, Cvoro V et al. Do acute diffusion- and perfusion-weighted MRI lesions identify final infarct volume in ischemic stroke? *Stroke: A Journal of Cerebral Circulation*. 2006;37:98–104.

88. Takasawa M, Jones PS, Guadagno JV, Christensen S, Fryer TD, Harding S et al. How reliable is perfusion MR in acute stroke? Validation and determination of the penumbra threshold against quantitative pet. *Stroke: A Journal of Cerebral Circulation.* 2008;39:870–877.

89. Dani KA, Thomas RG, Chappell FM, Shuler K, MacLeod MJ, Muir KW et al. Computed tomography and magnetic resonance perfusion imaging in ischemic stroke: Definitions and thresholds. *Annals of Neurology.* 2011;70:384–401.

90. Petersen ET, Zimine I, Ho YC, Golay X. Non-invasive measurement of perfusion: A critical review of arterial spin labelling techniques. *The British Journal of Radiology.* 2006;79:688–701.

91. Wang DJ, Alger JR, Qiao JX, Hao Q, Hou S, Fiaz R et al. The value of arterial spin-labeled perfusion imaging in acute ischemic stroke: Comparison with dynamic susceptibility contrast-enhanced MRI. *Stroke: A Journal of Cerebral Circulation.* 2012;43:1018–1024.

92. MacIntosh BJ, Lindsay AC, Kylintireas I, Kuker W, Gunther M, Robson MD et al. Multiple inflow pulsed arterial spin-labeling reveals delays in the arterial arrival time in minor stroke and transient ischemic attack. *AJNR American Journal of Neuroradiology.* 2010;31:1892–1894.

93. Chalela JA, Alsop DC, Gonzalez-Atavales JB, Maldjian JA, Kasner SE, Detre JA. Magnetic resonance perfusion imaging in acute ischemic stroke using continuous arterial spin labeling. *Stroke: A Journal of Cerebral Circulation.* 2000;31:680–687.

94. Zaharchuk G, Bammer R, Straka M, Shankaranarayan A, Alsop DC, Fischbein NJ et al. Arterial spin-label imaging in patients with normal bolus perfusion-weighted MR imaging findings: Pilot identification of the borderzone sign. *Radiology.* 2009;252:797–807.

95. Buxton RB. Quantifying CBF with arterial spin labeling. *Journal of Magnetic Resonance Imaging: JMRI.* 2005;22:723–726.

96. Wong EC. Quantifying CBF with pulsed ASL: Technical and pulse sequence factors. *Journal of Magnetic Resonance Imaging: JMRI.* 2005;22:727–731.

97. Kassner A, Roberts T, Taylor K, Silver F, Mikulis D. Prediction of hemorrhage in acute ischemic stroke using permeability MR imaging. *AJNR American Journal of Neuroradiology.* 2005;26:2213–2217.

98. Scalzo F, Alger JR, Hu X, Saver JL, Dani KA, Muir KW et al. Multi-center prediction of hemorrhagic transformation in acute ischemic stroke using permeability imaging features. *Magnetic Resonance Imaging.* 2013;31:961–969.

99. Yu S, Liebeskind DS, Dua S, Wilhalme H, Elashoff D, Qiao XJ et al. Postischemic hyperperfusion on arterial spin labeled perfusion MRI is linked to hemorrhagic transformation in stroke. *Journal of Cerebral Blood Flow and Metabolism: Official Journal of the International Society of Cerebral Blood Flow and Metabolism.* 2015;35:630–637.

100. Brott T, Broderick J, Kothari R, O'Donoghue M, Barsan W, Tomsick T et al. Tissue plasminogen activator for acute ischemic stroke. The National Institute of Neurological Disorders and Stroke rt-PA Stroke Study Group. *The New England Journal of Medicine.* 1995;333:1581–1587.

101. Hacke W, Kaste M, Bluhmki E, Brozman M, Davalos A, Guidetti D et al. Thrombolysis with alteplase 3 to 4.5 hours after acute ischemic stroke. *The New England Journal of Medicine.* 2008;359:1317–1329.

102. Lees KR, Bluhmki E, von Kummer R, Brott TG, Toni D, Grotta JC et al. Time to treatment with intravenous alteplase and outcome in stroke: An updated pooled analysis of ECASS, ATLANTIS, NINDS, and EPITHET trials. *Lancet.* 2010;375:1695–1703.

103. Fonarow GC, Smith EE, Saver JL, Reeves MJ, Hernandez AF, Peterson ED et al. Improving door-to-needle times in acute ischemic stroke: The design and rationale for the American Heart Association/American Stroke Association's target: Stroke initiative. *Stroke: A Journal of Cerebral Circulation.* 2011;42:2983–2989.

104. Paolini S, Burdine J, Verenes M, Webster J, Faber T, Graham CB et al. Rapid short MRI sequence useful in eliminating stroke mimics among acute stroke patients considered for intravenous thrombolysis. *Journal of Neurological Disorders.* 2013;1:137.

105. Schmitz ML, Simonsen CZ, Hundborg H, Christensen H, Ellemann K, Geisler K et al. Acute ischemic stroke and long-term outcome after thrombolysis: Nationwide propensity score-matched follow-up study. *Stroke: A Journal of Cerebral Circulation.* 2014;45:3070–3072.

106. Turc G, Aguettaz P, Ponchelle-Dequatre N, Henon H, Naggara O, Leclerc X et al. External validation of the MRI-dragon score: Early prediction of stroke outcome after intravenous thrombolysis. *PLoS One.* 2014;9:e99164.

107. Apoil M, Turc G, Tisserand M, Calvet D, Naggara O, Domigo V et al. Clinical and magnetic resonance imaging predictors of very early neurological response to intravenous thrombolysis in patients with middle cerebral artery occlusion. *Journal of the American Heart Association.* 2013;2:e000511.

108. Charidimou A, Kakar P, Fox Z, Werring DJ. Cerebral microbleeds and the risk of intracerebral haemorrhage after thrombolysis for acute ischaemic stroke: Systematic review and meta-analysis. *Journal of Neurology, Neurosurgery, and Psychiatry.* 2013;84:277–280.

109. Koennecke HC. Cerebral microbleeds on MRI: Prevalence, associations, and potential clinical implications. *Neurology.* 2006;66:165–171.

110. Kakuda W, Thijs VN, Lansberg MG, Bammer R, Wechsler L, Kemp S et al. Clinical importance of microbleeds in patients receiving IV thrombolysis. *Neurology.* 2005;65:1175–1178.

111. Fiehler J, Albers GW, Boulanger JM, Derex L, Gass A, Hjort N et al. Bleeding risk analysis in stroke imaging before thrombolysis (BRASIL): Pooled analysis of T2*-weighted magnetic resonance imaging data from 570 patients. *Stroke: A Journal of Cerebral Circulation.* 2007;38:2738–2744.

112. Derex L, Nighoghossian N, Hermier M, Adeleine P, Philippeau F, Honnorat J et al. Thrombolysis for ischemic stroke in patients with old microbleeds on pretreatment MRI. *Cerebrovascular Diseases.* 2004;17:238–241.

113. Kidwell CS, Saver JL, Villablanca JP, Duckwiler G, Fredieu A, Gough K et al. Magnetic resonance imaging detection of microbleeds before thrombolysis: An emerging application. *Stroke: A Journal of Cerebral Circulation.* 2002;33:95–98.

114. Huang P, Chen CH, Lin WC, Lin RT, Khor GT, Liu CK. Clinical applications of susceptibility weighted imaging in patients with major stroke. *Journal of Neurology.* 2012;259:1426–1432.

115. Heldner MR, Zubler C, Mattle HP, Schroth G, Weck A, Mono ML et al. National Institutes of Health stroke scale score and vessel occlusion in 2152 patients with acute ischemic stroke. *Stroke: A Journal of Cerebral Circulation.* 2013;44:1153–1157.

116. Wolpert SM, Bruckmann H, Greenlee R, Wechsler L, Pessin MS, del Zoppo GJ. Neuroradiologic evaluation of patients with acute stroke treated with recombinant tissue plasminogen activator. The rt-PA Acute Stroke Study Group. *AJNR American Journal of Neuroradiology.* 1993;14:3–13.

117. Lee KY, Han SW, Kim SH, Nam HS, Ahn SW, Kim DJ et al. Early recanalization after intravenous administration of recombinant tissue plasminogen activator as assessed by pre- and post-thrombolytic angiography in acute ischemic stroke patients. *Stroke: A Journal of Cerebral Circulation.* 2007;38:192–193.

118. Berkhemer OA, Fransen PS, Beumer D, van den Berg LA, Lingsma HF, Yoo AJ et al. A randomized trial of intra-arterial treatment for acute ischemic stroke. *The New England Journal of Medicine.* 2015;372:11–20.

119. Goyal M, Demchuk AM, Menon BK, Eesa M, Rempel JL, Thornton J et al. Investigators, Escape Trial. Randomized assessment of rapid endovascular treatment of ischemic stroke. *The New England Journal of Medicine.* 2015;372:1019–1030.

120. Jovin TG, Chamorro A, Cobo E, de Miquel MA, Molina CA, Rovira A et al. Investigators, Revascat Trial. Thrombectomy within 8 hours after symptom onset in ischemic stroke. *The New England Journal of Medicine.* 2015;372:2296–2306.

121. Saver JL, Goyal M, Bonafe A, Diener HC, Levy EI, Pereira VM, … Investigators, Swift Prime. Stent-retriever thrombectomy after intravenous t-PA vs. t-PA alone in stroke. *The New England Journal of Medicine.* 2015;372:2285–2295.

122. Campbell BC, Mitchell PJ, Kleinig TJ, Dewey H M, Churilov L, Yassi N et al. Investigators, Extend-Ia. Endovascular therapy for ischemic stroke with perfusion-imaging selection. *The New England Journal of Medicine.* 2015;372:1009–1018.

123. Broderick JP, Palesch YY, Demchuk AM, Yeatts SD, Khatri P, Hill MD et al. Endovascular therapy after intravenous t-PA versus t-PA alone for stroke. *The New England Journal of Medicine.* 2013;368:893–903.

124. Ciccone A, Valvassori L, Nichelatti M, Sgoifo A, Ponzio M, Sterzi R et al. Endovascular treatment for acute ischemic stroke. *The New England Journal of Medicine.* 2013;368:904–913.

125. Jovin TG, Yonas H, Gebel JM, Kanal E, Chang YF, Grahovac SZ et al. The cortical ischemic core and not the consistently present penumbra is a determinant of clinical outcome in acute middle cerebral artery occlusion. *Stroke: A Journal of Cerebral Circulation.* 2003;34:2426–2433.

126. Yoo AJ, Verduzco LA, Schaefer PW, Hirsch JA, Rabinov JD, Gonzalez RG. MRI-based selection for intra-arterial stroke therapy: Value of pretreatment diffusion-weighted imaging lesion volume in selecting patients with acute stroke who will benefit from early recanalization. *Stroke: A Journal of Cerebral Circulation.* 2009;40:2046–2054.

127. Olivot JM, Mosimann PJ, Labreuche J, Inoue M, Meseguer E, Desilles JP et al. Impact of diffusion-weighted imaging lesion volume on the success of endovascular reperfusion therapy. *Stroke: A Journal of Cerebral Circulation.* 2013;44:2205–2211.

128. Yoo AJ, Zaidat OO, Chaudhry ZA, Berkhemer OA, Gonzalez RG, Goyal M et al. Impact of pretreatment noncontrast CT alberta stroke program early CT score on clinical outcome after intra-arterial stroke therapy. *Stroke: A Journal of Cerebral Circulation.* 2014;45:746–751.

129. Hakimelahi R, Yoo AJ, He J, Schwamm LH, Lev MH, Schaefer PW et al. Rapid identification of a major diffusion/perfusion mismatch in distal internal carotid artery or middle cerebral artery ischemic stroke. *BMC Neurology.* 2012;12:132.

130. Furlan A, Higashida R, Wechsler L, Gent M, Rowley H, Kase C et al. Intra-arterial prourokinase for acute ischemic stroke. The PROACT II study: A randomized controlled trial. Prolyse in acute cerebral thromboembolism. *JAMA.* 1999;282:2003–2011.

131. Ribo M, Flores A, Mansilla E, Rubiera M, Tomasello A, Coscojuela P et al. Age-adjusted infarct volume threshold for good outcome after endovascular treatment. *Journal of Neurointerventional Surgery.* 2014;6:418–422.

132. Fiorelli M, Bastianello S, von Kummer R, del Zoppo GJ, Larrue V, Lesaffre E et al. Hemorrhagic transformation within 36 hours of a cerebral infarct: Relationships with early clinical deterioration and 3-month outcome in the European Cooperative Acute Stroke Study I (ECASS I) cohort. *Stroke: A Journal of Cerebral Circulation.* 1999;30:2280–2284.

133. Wardlaw JM, Murray V, Berge E, del Zoppo GJ. Thrombolysis for acute ischaemic stroke. *The Cochrane Database of Systematic Reviews.* 2014;7:CD000213.

134. Shen Q, Du F, Huang S, Duong TQ. Spatiotemporal characteristics of postischemic hyperperfusion with respect to changes in T1, T2, diffusion, angiography, and blood-brain barrier permeability. *Journal of Cerebral Blood Flow and Metabolism: Official Journal of the International Society of Cerebral Blood Flow and Metabolism.* 2011;31:2076–2085.

135. Hacke W, Kaste M, Fieschi C, Toni D, Lesaffre E, von Kummer R et al. Intravenous thrombolysis with recombinant tissue plasminogen activator for acute hemispheric stroke. The European Cooperative Acute Stroke Study (ECASS). *JAMA.* 1995;274:1017–1025.

136. Hacke W, Kaste M, Fieschi C, von Kummer R, Davalos A, Meier D et al. Randomised double-blind placebo-controlled trial of thrombolytic therapy with intravenous alteplase in acute ischaemic stroke (ECASS II). Second European-Australasian Acute Stroke Study Investigators. *Lancet.* 1998;352:1245–1251.

137. Dzialowski I, Pexman JH, Barber PA, Demchuk AM, Buchan AM, Hill MD et al. Asymptomatic hemorrhage after thrombolysis may not be benign: Prognosis by hemorrhage type in the Canadian alteplase for stroke effectiveness study registry. *Stroke: A Journal of Cerebral Circulation.* 2007;38:75–79.

138. Berger C, Fiorelli M, Steiner T, Schabitz WR, Bozzao L, Bluhmki E et al. Hemorrhagic transformation of ischemic brain tissue: Asymptomatic or symptomatic? *Stroke: A Journal of Cerebral Circulation.* 2001;32: 1330–1335.

139. Paciaroni M, Agnelli G, Corea F, Ageno W, Alberti A, Lanari A et al. Early hemorrhagic transformation of brain infarction: Rate, predictive factors, and influence on clinical outcome: Results of a prospective multicenter study. *Stroke: A Journal of Cerebral Circulation.* 2008;39:2249–2256.

140. Arnould MC, Grandin CB, Peeters A, Cosnard G, Duprez TP. Comparison of CT and three MR sequences for detecting and categorizing early (48 hours) hemorrhagic transformation in hyperacute ischemic stroke. *AJNR American Journal of Neuroradiology.* 2004;25:939–944.

141. Renou P, Sibon I, Tourdias T, Rouanet F, Rosso C, Galanaud D et al. Reliability of the ECASS radiological classification of postthrombolysis brain haemorrhage: A comparison of CT and three MRI sequences. *Cerebrovascular Diseases.* 2010;29:597–604.

142. Fiebach JB, Bohner G. T2*-weighted imaging enables excellent interobserver concordance but should not be considered as sole gold standard imaging for hemorrhagic transformation classification after thrombolysis. *Cerebrovascular Diseases.* 2010;29:605–606.

143. Neeb L, Villringer K, Galinovic I, Grosse-Dresselhaus F, Ganeshan R, Gierhake D et al. Adapting the computed tomography criteria of hemorrhagic transformation to stroke magnetic resonance imaging. *Cerebrovascular Diseases Extra.* 2013;3:103–110.

144. Lansberg MG, Thijs VN, Bammer R, Kemp S, Wijman CA, Marks MP et al. Risk factors of symptomatic intracerebral hemorrhage after tPA therapy for acute stroke. *Stroke: A Journal of Cerebral Circulation.* 2007;38:2275–2278.

145. Singer OC, Humpich MC, Fiehler J, Albers GW, Lansberg MG, Kastrup A et al. Risk for symptomatic intracerebral hemorrhage after thrombolysis assessed by diffusion-weighted magnetic resonance imaging. *Annals of Neurology.* 2008;63:52–60.

146. Edgell RC, Vora NA. Neuroimaging markers of hemorrhagic risk with stroke reperfusion therapy. *Neurology.* 2012;79:S100–S104.

147. Neumann-Haefelin T, Hoelig S, Berkefeld J, Fiehler J, Gass A, Humpich M et al. Leukoaraiosis is a risk factor for symptomatic intracerebral hemorrhage after thrombolysis for acute stroke. *Stroke: A Journal of Cerebral Circulation.* 2006;37:2463–2466.

148. Tong DC, Adami A, Moseley ME, Marks MP. Relationship between apparent diffusion coefficient and subsequent hemorrhagic transformation following acute ischemic stroke. *Stroke: A Journal of Cerebral Circulation.* 2000;31:2378–2384.

149. Selim M, Fink JN, Kumar S, Caplan LR, Horkan C, Chen Y et al. Predictors of hemorrhagic transformation after intravenous recombinant tissue plasminogen activator: Prognostic value of the initial apparent diffusion coefficient and diffusion-weighted lesion volume. *Stroke: A Journal of Cerebral Circulation.* 2002;33:2047–2052.

150. Yassi N, Parsons MW, Christensen S, Sharma G, Bivard A, Donnan GA et al. Prediction of poststroke hemorrhagic transformation using computed tomography perfusion. *Stroke: A Journal of Cerebral Circulation.* 2013;44:3039–3043.

151. Thomalla G, Schwark C, Sobesky J, Bluhmki E, Fiebach JB, Fiehler J et al. Outcome and symptomatic bleeding complications of intravenous thrombolysis within 6 hours in MRI-selected stroke patients: Comparison of a german multicenter study with the pooled data of ATLANTIS, ECASS, and NINDS tPA trials. *Stroke: A Journal of Cerebral Circulation.* 2006;37:852–858.

152. Schellinger PD, Thomalla G, Fiehler J, Kohrmann M, Molina CA, Neumann-Haefelin T et al. MRI-based and CT-based thrombolytic therapy in acute stroke within and beyond established time windows: An analysis of 1210 patients. *Stroke: A Journal of Cerebral Circulation.* 2007;38:2640–2645.

153. Ribo M, Molina CA, Rovira A, Quintana M, Delgado P, Montaner J et al. Safety and efficacy of intravenous tissue plasminogen activator stroke treatment in the 3- to 6-hour window using multimodal transcranial doppler/MRI selection protocol. *Stroke: A Journal of Cerebral Circulation.* 2005;36:602–606.

154. Hacke W, Albers G, Al-Rawi Y, Bogousslavsky J, Davalos A, Eliasziw M et al. The Desmoteplase in Acute Ischemic Stroke Trial (DIAS): A phase II MRI-based 9-hour window acute stroke thrombolysis trial with intravenous desmoteplase. *Stroke: A Journal of Cerebral Circulation.* 2005;36:66–73.

155. Furlan AJ, Eyding D, Albers GW, Al-Rawi Y, Lees KR, Rowley HA et al. Dose escalation of desmoteplase for acute ischemic stroke (DEDAS): Evidence of safety and efficacy 3 to 9 hours after stroke onset. *Stroke: A Journal of Cerebral Circulation.* 2006;37:1227–1231.

156. Hacke W, Furlan AJ, Al-Rawi Y, Davalos A, Fiebach JB, Gruber F et al. Intravenous desmoteplase in patients with acute ischaemic stroke selected by MRI perfusion-diffusion weighted imaging or perfusion CT (DIAS-2): A prospective, randomised, double-blind, placebo-controlled study. *The Lancet. Neurology.* 2009;8:141–150.

157. Davis SM, Donnan GA, Parsons MW, Levi C, Butcher KS, Peeters A et al. Effects of alteplase beyond 3 h after stroke in the Echoplanar Imaging Thrombolytic Evaluation Trial (EPITHET): A placebo-controlled randomised trial. *The Lancet. Neurology.* 2008;7:299–309.

158. Mishra NK, Albers GW, Davis SM, Donnan GA, Furlan AJ, Hacke W et al. Mismatch-based delayed thrombolysis: A meta-analysis. *Stroke: A Journal of Cerebral Circulation.* 2010;41:e25–33.

159. Kane I, Sandercock P, Wardlaw J. Magnetic resonance perfusion diffusion mismatch and thrombolysis in acute ischaemic stroke: A systematic review of the evidence to date. *Journal of Neurology, Neurosurgery, and Psychiatry.* 2007;78:485–491.

160. Bandera E, Botteri M, Minelli C, Sutton A, Abrams KR, Latronico N. Cerebral blood flow threshold of ischemic penumbra and infarct core in acute ischemic stroke: A systematic review. *Stroke: A Journal of Cerebral Circulation.* 2006;37:1334–1339.

161. Kamalian S, Kamalian S, Konstas AA, Maas MB, Payabvash S, Pomerantz SR et al. CT perfusion mean transit time maps optimally distinguish benign oligemia from true "at-risk" ischemic penumbra, but thresholds vary by postprocessing technique. *AJNR American Journal of Neuroradiology.* 2012;33:545–549.

162. Lansberg MG, Thijs VN, Bammer R, Olivot JM, Marks MP, Wechsler LR et al. The MRA-DWI mismatch identifies patients with stroke who are likely to benefit from reperfusion. *Stroke: A Journal of Cerebral Circulation.* 2008;39:2491–2496.

163. Wouters A, Lemmens R, Dupont P, Thijs V. Wake-up stroke and stroke of unknown onset: A critical review. *Frontiers in Neurology.* 2014;5:153.

164. Elliott WJ. Circadian variation in the timing of stroke onset: A meta-analysis. *Stroke: A Journal of Cerebral Circulation.* 1998;29:992–996.

165. Rimmele DL, Thomalla G. Wake-up stroke: Clinical characteristics, imaging findings, and treatment option—An update. *Frontiers in Neurology.* 2014;5:35.

166. Moseley ME, Kucharczyk J, Mintorovitch J, Cohen Y, Kurhanewicz J, Derugin N et al. Diffusion-weighted MR imaging of acute stroke: Correlation with T2-weighted and magnetic susceptibility-enhanced MR imaging in cats. *AJNR American Journal of Neuroradiology.* 1990;11:423–429.

167. Aoki J, Kimura K, Iguchi Y, Shibazaki K, Sakai K, Iwanaga T. Flair can estimate the onset time in acute ischemic stroke patients. *Journal of the Neurological Sciences.* 2010;293:39–44.

168. Cho AH, Sohn SI, Han MK, Lee DH, Kim JS, Choi CG et al. Safety and efficacy of MRI-based thrombolysis in unclear-onset stroke. A preliminary report. *Cerebrovascular Diseases.* 2008;25:572–579.

169. Thomalla G, Cheng B, Ebinger M, Hao Q, Tourdias T, Wu O et al. DWI-flair mismatch for the identification of patients with acute ischaemic stroke within 4.5 h of symptom onset (pre-flair): A multicentre observational study. *The Lancet. Neurology.* 2011;10:978–986.

170. Petkova M, Rodrigo S, Lamy C, Oppenheim G, Touze E, Mas JL et al. MR imaging helps predict time from symptom onset in patients with acute stroke: Implications for patients with unknown onset time. *Radiology.* 2010;257:782–792.

171. Emeriau S, Serre I, Toubas O, Pombourcq F, Oppenheim C, Pierot L. Can diffusion-weighted imaging-fluid-attenuated inversion recovery mismatch (positive diffusion-weighted imaging/negative fluid-attenuated inversion recovery) at 3 tesla identify patients with stroke at <4.5 hours? *Stroke: A Journal of Cerebral Circulation.* 2013;44:1647–1651.

172. Johnston SC, Fayad PB, Gorelick PB, Hanley DF, Shwayder P, van Husen D et al. Prevalence and knowledge of transient ischemic attack among us adults. *Neurology.* 2003;60:1429–1434.

173. Giles MF, Rothwell PM. Risk prediction after TIA: The ABCD system and other methods. *Geriatrics.* 2008;63: 10–13, 16.

174. Giles MF, Albers GW, Amarenco P, Arsava EM, Asimos AW, Ay H et al. Early stroke risk and ABCD2 score performance in tissue- vs time-defined tia: A multicenter study. *Neurology.* 2011;77:1222–1228.

175. Giles MF, Albers GW, Amarenco P, Arsava MM, Asimos A, Ay H et al. Addition of brain infarction to the ABCD2 score (ABCD2I): A collaborative analysis of unpublished data on 4574 patients. *Stroke: A Journal of Cerebral Circulation.* 2010;41:1907–1913.

176. Merwick A, Albers GW, Amarenco P, Arsava EM, Ay H, Calvet D et al. Addition of brain and carotid imaging to the ABCD² score to identify patients at early risk of stroke after transient ischaemic attack: A multicentre observational study. *The Lancet. Neurology.* 2010;9:1060–1069.

177. Moritani T, Smoker WR, Sato Y, Numaguchi Y, Westesson PL. Diffusion-weighted imaging of acute excitotoxic brain injury. *AJNR American Journal of Neuroradiology.* 2005;26:216–228.

178. Hand PJ, Kwan J, Lindley RI, Dennis MS, Wardlaw JM. Distinguishing between stroke and mimic at the bedside: The brain attack study. *Stroke: A Journal of Cerebral Circulation.* 2006;37:769–775.

179. Hemmen TM, Meyer BC, McClean TL, Lyden PD. Identification of nonischemic stroke mimics among 411 code strokes at the University of California, San Diego, stroke center. *Journal of Stroke and Cerebrovascular Diseases: The Official Journal of National Stroke Association.* 2008;17:23–25.

180. Libman RB, Wirkowski E, Alvir J, Rao TH. Conditions that mimic stroke in the emergency department. Implications for acute stroke trials. *Archives of Neurology.* 1995;52:1119–1122.

181. Allder SJ, Moody AR, Martel AL, Morgan PS, Delay GS, Gladman JR et al. Limitations of clinical diagnosis in acute stroke. *Lancet.* 1999;354:1523.

182. Merino JG, Luby M, Benson RT, Davis LA, Hsia AW, Latour LL et al. Predictors of acute stroke mimics in 8187 patients referred to a stroke service. *Journal of Stroke and Cerebrovascular Diseases: The Official Journal of National Stroke Association.* 2013;22:e397–e403.

183. Winkler DT, Fluri F, Fuhr P, Wetzel SG, Lyrer PA, Ruegg S et al. Thrombolysis in stroke mimics: Frequency, clinical characteristics, and outcome. *Stroke: A Journal of Cerebral Circulation.* 2009;40:1522–1525.

184. Thrombolytic therapy with streptokinase in acute ischemic stroke. The Multicenter Acute Stroke Trial—Europe Study Group. *The New England Journal of Medicine.* 1996;335:145–150.

185. Chernyshev OY, Martin-Schild S, Albright KC, Barreto A, Misra V, Acosta I et al. Safety of tPA in stroke mimics and neuroimaging-negative cerebral ischemia. *Neurology.* 2010;74:1340–1345.

186. Forster A, Griebe M, Wolf ME, Szabo K, Hennerici MG, Kern R. How to identify stroke mimics in patients eligible for intravenous thrombolysis? *Journal of Neurology.* 2012;259:1347–1353.

187. Zinkstok SM, Engelter ST, Gensicke H, Lyrer PA, Ringleb PA, Artto V et al. Safety of thrombolysis in stroke mimics: Results from a multicenter cohort study. *Stroke: A Journal of Cerebral Circulation.* 2013;44:1080–1084.

188. Guillan M, Alonso-Canovas A, Gonzalez-Valcarcel J, Garcia Barragan N, Garcia Caldentey J, Hernandez-Medrano I et al. Stroke mimics treated with thrombolysis: Further evidence on safety and distinctive clinical features. *Cerebrovascular Diseases.* 2012;34:115–120.

189. Spokoyny I, Raman R, Ernstrom K, Meyer BC, Hemmen TM. Imaging negative stroke: Diagnoses and outcomes in intravenous tissue plasminogen activator-treated patients. *Journal of Stroke and Cerebrovascular Diseases: The Official Journal of National Stroke Association.* 2014;23:1046–1050.

190. Kang EG, Jeon SJ, Choi SS, Song CJ, Yu IK. Diffusion MR imaging of hypoglycemic encephalopathy. *AJNR Am J Neuroradiol.* 2010;31:559–564.

191. Glaser N, Ngo C, Anderson S, Yuen N, Trifu A, O'Donnell M. Effects of hyperglycemia and effects of ketosis on cerebral perfusion, cerebral water distribution, and cerebral metabolism. *Diabetes.* 2012;61:1831–1837.

192. Winbeck K, Etgen T, von Einsiedel HG, Rottinger M, Sander D. DWI in transient global amnesia and TIA: proposal for an ischaemic origin of TGA. *J Neurol Neurosurg Psychiatry.* 2005;76:438–441.

193. Moritani T, Smoker WR, Sato Y, Numaguchi Y, Westesson PL. Diffusion-weighted imaging of acute excitotoxic brain injury. *AJNR Am J Neuroradiol.* 2005;26:216–228.

194. Wolters EC, van Wijngaarden GK, Stam FC, Rengelink H, Lousberg RJ, Schipper ME, Verbeeten B. Leucoencephalopathy after inhaling "heroin" pyrolysate. *Lancet.* 1982;2:1233–1237.

195. Muccio CF, De Simone M, Esposito G, De Blasio E, Vittori C, Cerase A. Reversible post-traumatic bilateral extensive restricted diffusion of the brain. A case study and review of the literature. *Brain Inj.* 2009;23:466–472.

196. Ay H, Buonanno FS, Rordorf G, Schaefer PW, Schwamm LH, Wu O et al. Normal diffusion-weighted MRI during stroke-like deficits. *Neurology.* 1999;52:1784–1792.

197. da Rocha AJ, da Silva CJ, Gama HP, Baccin CE, Braga FT, Cesare Fde A et al. Comparison of magnetic resonance imaging sequences with computed tomography to detect low-grade subarachnoid hemorrhage: Role of fluid-attenuated inversion recovery sequence. *Journal of Computer Assisted Tomography.* 2006;30:295–303.

198. Mohamed M, Heasly DC, Yagmurlu B, Yousem DM. Fluid-attenuated inversion recovery MR imaging and subarachnoid hemorrhage: Not a panacea. *AJNR American Journal of Neuroradiology.* 2004;25:545–550.

199. Verma RK, Kottke R, Andereggen L, Weisstanner C, Zubler C, Gralla J et al. Detecting subarachnoid hemorrhage: Comparison of combined FLAIR/SWI versus CT. *European Journal of Radiology.* 2013;82:1539–1545.

200. Li Q, Lv F, Li Y, Luo T, Li K, Xie P. Evaluation of 64-section CT angiography for detection and treatment planning of intracranial aneurysms by using dsa and surgical findings. *Radiology.* 2009;252:808–815.

201. White PM, Wardlaw JM, Easton V. Can noninvasive imaging accurately depict intracranial aneurysms? A systematic review. *Radiology.* 2000;217:361–370.

202. Agid R, Andersson T, Almqvist H, Willinsky RA, Lee SK, terBrugge KG et al. Negative CT angiography findings in patients with spontaneous subarachnoid hemorrhage: When is digital subtraction angiography still needed? *AJNR American Journal of Neuroradiology.* 2010;31:696–705.

203. Delgado Almandoz JE, Jagadeesan BD, Refai D, Moran CJ, Cross DT, 3rd, Chicoine MR et al. Diagnostic yield of repeat catheter angiography in patients with catheter and computed tomography angiography negative subarachnoid hemorrhage. *Neurosurgery.* 2012;70:1135–1142.

204. Hiratsuka Y, Miki H, Kiriyama I, Kikuchi K, Takahashi S, Matsubara I et al. Diagnosis of unruptured intracranial aneurysms: 3T MR angiography versus 64-channel multi-detector row CT angiography. *Magnetic Resonance in Medical Sciences: MRMS: An Official Journal of Japan Society of Magnetic Resonance in Medicine.* 2008;7:169–178.

205. Numminen J, Tarkiainen A, Niemela M, Porras M, Hernesniemi J, Kangasniemi M. Detection of unruptured cerebral artery aneurysms by MRA at 3.0 tesla: Comparison with multislice helical computed tomographic angiography. *Acta Radiologica.* 2011;52:670–674.

206. Sailer AM, Wagemans BA, Nelemans PJ, de Graaf R, van Zwam WH. Diagnosing intracranial aneurysms with MR angiography: Systematic review and meta-analysis. *Stroke: A Journal of Cerebral Circulation.* 2014;45:119–126.

207. Goto M, Kunimatsu A, Shojima M, Mori H, Abe O, Aoki S et al. Depiction of branch vessels arising from intracranial aneurysm sacs: Time-of-flight MR angiography versus CT angiography. *Clinical Neurology and Neurosurgery.* 2014;126:177–184.

208. Phillips TJ, Dowling RJ, Yan B, Laidlaw JD, Mitchell PJ. Does treatment of ruptured intracranial aneurysms within 24 hours improve clinical outcome? *Stroke: A Journal of Cerebral Circulation.* 2011;42:1936–1945.

209. Abdel Razek AA, Alvarez H, Bagg S, Refaat S, Castillo M. Imaging spectrum of CNS vasculitis. *RadioGraphics: A Review Publication of the Radiological Society of North America, Inc.* 2014;34:873–894.

210. Khurram A, Kleinig T, Leyden J. Clinical associations and causes of convexity subarachnoid hemorrhage. *Stroke: A Journal of Cerebral Circulation.* 2014;45:1151–1153.

211. Zyed A, Hayman LA, Bryan RN. MR imaging of intracerebral blood: Diversity in the temporal pattern at 0.5 and 1.0 T. *AJNR American Journal of Neuroradiology.* 1991;12:469–474.

212. Gomori JM, Grossman RI, Yu-Ip C, Asakura T. NMR relaxation times of blood: Dependence on field strength, oxidation state, and cell integrity. *Journal of Computer Assisted Tomography.* 1987;11:684–690.

213. Clark RA, Watanabe AT, Bradley WG, Jr., Roberts JD. Acute hematomas: Effects of deoxygenation, hematocrit, and fibrin-clot formation and retraction on T2 shortening. *Radiology.* 1990;175:201–206.

214. Bradley WG, Jr. MR appearance of hemorrhage in the brain. *Radiology.* 1993;189:15–26.

215. Hayman LA, Taber KH, Ford JJ, Bryan RN. Mechanisms of MR signal alteration by acute intracerebral blood: Old concepts and new theories. *AJNR American Journal of Neuroradiology.* 1991;12:899–907.

216. Linfante I, Llinas RH, Caplan LR, Warach S. MRI features of intracerebral hemorrhage within 2 hours from symptom onset. *Stroke: A Journal of Cerebral Circulation.* 1999;30:2263–2267.

217. Pauling L, Coryell CD. The magnetic properties and structure of hemoglobin, oxyhemoglobin and carbonmonoxyhemoglobin. *Proceedings of the National Academy of Sciences of the United States of America.* 1936;22:210–216.

218. Fabry TL, Reich HA. The role of water in deoxygenated hemoglobin solutions. *Biochemical and Biophysical Research Communications.* 1966;22:700–703.

219. Naidich TP, Castillo M, Cha S, Smirniotopoulos J. *Imaging of the brain.* 2013. Philadelphia, PA: Saunders/Elsevier.

220. Schelhorn J, Gramsch C, Deuschl C, Quick HH, Nensa F, Moenninghoff C et al. Intracranial hemorrhage detection over time using susceptibility-weighted magnetic resonance imaging. *Acta Radiologica.* 2014;doi:10.1177/0284185114559958.

221. Kang BK, Na DG, Ryoo JW, Byun HS, Roh HG, Pyeun YS. Diffusion-weighted MR imaging of intracerebral hemorrhage. *Korean Journal of Radiology: Official Journal of the Korean Radiological Society.* 2001;2:183–191.

222. Atlas SW, DuBois P, Singer MB, Lu D. Diffusion measurements in intracranial hematomas: Implications for MR imaging of acute stroke. *AJNR American Journal of Neuroradiology.* 2000;21:1190–1194.

223. Romero JM, Artunduaga M, Forero NP, Delgado J, Sarfaraz K, Goldstein JN et al. Accuracy of CT angiography for the diagnosis of vascular abnormalities causing intraparenchymal hemorrhage in young patients. *Emergency Radiology.* 2009;16:195–201.

224. Josephson CB, White PM, Krishan A, Al-Shahi Salman R. Computed tomography angiography or magnetic resonance angiography for detection of intracranial vascular malformations in patients with intracerebral haemorrhage. *The Cochrane Database of Systematic Reviews.* 2014;9:CD009372.

225. Jagadeesan BD, Delgado Almandoz JE, Benzinger TL, Moran CJ. Postcontrast susceptibility-weighted imaging: a novel technique for the detection of arteriovenous shunting in vascular malformations of the brain. *Stroke.* 2011;42:3127–3131.

226. Jagadeesan BD, Delgado Almandoz JE, Moran CJ, Benzinger TL. Accuracy of susceptibility-weighted imaging for the detection of arteriovenous shunting in vascular malformations of the brain. *Stroke.* 2011;42:87–92.

227. Cordonnier C, Klijn CJ, van Beijnum J, Al-Shahi Salman R. Radiological investigation of spontaneous intracerebral hemorrhage: Systematic review and trinational survey. *Stroke: A Journal of Cerebral Circulation.* 2010;41:685–690.

228. Zhu XL, Chan MS, Poon WS. Spontaneous intracranial hemorrhage: Which patients need diagnostic cerebral angiography? A prospective study of 206 cases and review of the literature. *Stroke: A Journal of Cerebral Circulation* 1997;28:1406–1409.

229. Delgado Almandoz JE, Schaefer PW, Forero NP, Falla JR, Gonzalez RG, Romero JM. Diagnostic accuracy and yield of multidetector CT angiography in the evaluation of spontaneous intraparenchymal cerebral hemorrhage. *AJNR American Journal of Neuroradiology.* 2009;30:1213–1221.

230. Delgado Almandoz JE, Yoo AJ, Stone MJ, Schaefer PW, Goldstein JN, Rosand J et al. Systematic characterization of the computed tomography angiography spot sign in primary intracerebral hemorrhage identifies patients at highest risk for hematoma expansion: The spot sign score. *Stroke: A Journal of Cerebral Circulation.* 2009;40:2994–3000.

24

Vascular Pathologies of the Brain (Vasculitis–Arteriovenous Malformation/Arteriovenous Fistula–Aneurysm)

Karl-Olof Lövblad, Sven Haller, Vitor Mendes Pereira, and Maria Isabel Vargas

CONTENTS

24.1 Introduction

Vascular etiologies have always played an important role in pathologies of the central nervous system, but there has been an increased awareness of this due to important progresses in both diagnostic and interventional therapies [1]. The 1990s saw the parallel development of both diagnostic and therapeutic means that has led to our current situation. Although the most obvious impact of technology and therapy has been in the domain of brain ischemia [2], all other pathologies have gained from the advances made over the past few decades. The aim of this chapter is to discuss the technical modalities that have evolved and that have found a large application in diseases of the vasculature of the central nervous system. Vascular diseases of the nervous system, in general, and of the brain especially, will cause morbidity and mortality mainly through hemorrhage and stroke. For stroke, magnetic resonance imaging (MRI) is a superior technique, thanks to the capacity of diffusion-weighted techniques to demonstrate acute ischemia [2]. Hemorrhage remains slightly more controversial, because despite well-established theoretical advantages to the use of magnetic resonance (MR) in hemorrhage, it has not clearly established itself as a replacement for the acute detection of hemorrhage [3,4]. Indeed, very often, although visible on most images, the hyperacute and acute stages of hemorrhage are somewhat confusing to the untrained eye; this is due to the fact that sometimes its visibility on T_1-weighted images (T_1WIs) is suboptimal; with time however hemorrhage becomes more visible on both T_1- and T_2-weighted images (T_2WIs), and acquires more hyperintense aspect on both sequences in the subacute stage, whereas later the image will show liquid intensity that is bright on T_2WI but dark on T_1WI. Although stroke due to an embolic event remains the most frequent cerebrovascular disorder we are confronted with, pathologies such as cerebral hemorrhage (SAH) and/or subarachnoid hemorrhage are associated with severe outcomes if left untreated. Until the development of computed tomography (CT) angiography and MR angiography (MRA) [5], very little was possible to do in order to exactly determine the cause of hemorrhage or even to help planning treatment [1]. Nowadays, due to the technological advances, it is very often in one session to determine the nature of the disease, to delineate the extent of the damage caused to the brain, and then to plan treatment that can be done under more optimal circumstances [5]. Traditionally, brain vascular malformations were classified into brain arteriovenous malformations, arteriovenous fistulas, aneurysms, capillary telangiectasias, and cavernous angiomas. Conventional angiography techniques were used for the diagnosis of these diseases, but techniques such as digital subtraction angiography (DSA) are now more reserved for instances when an intervention is being planned. In this chapter, we will consider these entities together with vascular inflammatory diseases and discuss their MR appearance and what sequences to perform in order to display them in an optimal way for diagnostic and therapeutic purposes.

24.2 Technical Considerations

The diagnosis of vascular diseases of the brain, besides in some cases well-established clinical signs, has largely relied on the use of conventional angiography. Neuroimaging was profoundly revolutionized by the development of CT, then by the advent of fast MRI [6]. Indeed, although most techniques until then had mostly provided indirect signs of pathological conditions, or very semi-invasive like angiography, these two techniques allowed to visualize the brain *in situ* and *in vivo* like never before. Although CT remains the cornerstone for most neuroradiological workups, MRI has proven to be a much superior tool for the assessment of diseases of the nervous system. Indeed, due to the absence of X-rays, the capacity to obtain directly multiplanar images and to perform sequences that provide different contrasts of the tissues visualized, the technique has shown itself superior. However, for a rather long time after its introduction, MRI was a rather slow technique and complete examinations of the neuraxis could take up to an hour or even more. It was with the clinical implementation of at first echo-planar techniques, then of parallel imaging schemes that imaging times could be reduced with even an increase in signal and resolution. It is also becoming more clear that higher magnetic fields such as 3 T are the standard requirements for centers dealing with advanced treatment of cerebrovascular disorders. MR, due to its inherent sensitivity to motion, and especially to vascular motion, was predestined for the development of MRA techniques [5]: this happened early on in the course of the history of MR; however, a lot of improvements were necessary because early images took a long time to acquire, required a lot of postprocessing, and did not provide full coverage of the brain. With the development of fast sequences, the development of stronger gradients and improved coils, it was possible to improve resolution to an important degree. Initially, T_1 and T_2 images were acquired with spin-echo sequences that required long acquisition times; time was an even bigger issue if one was to consider multiplanar imaging. Indeed MRI has the distinct advantage of being able to obtain directly images in multiple planes in any contrast, but this is of course at the cost of additional imaging time; thus, at times in cases where an extensive imaging protocol had to be performed, it was possible for MR examinations to run well beyond an hour. This, together with the very motion-sensitive nature of the technique, would make the examination only accessible to patients who were able to cooperate and remain still for extensive periods of time. Indeed, it was the advent of echo-planar-imaging that was a revolution [6] that allowed to speed up imaging and to be able to perform complex examinations within an acceptable period of time. Not only did conventional imaging become faster, all functional techniques that had remained somewhat dormant would develop quickly and become established as clinical tools. Diffusion imaging that was devised by Le Bihan consists of a simple modification of a spin-echo sequence, which is sensitized to movement by the application of two gradient pulses [7]: this makes the image extremely sensitive to changes in water motility in the tissue that is being investigated. The first real application is to ischemic stroke. It has then been applied to a multitude of diseases, but its capacity to detect with certainty acute ischemia is unparalleled: indeed because of this it is very often used also to monitor neurovascular interventions such as thrombectomy [8], coiling, or embolization. In additional to diffusion, perfusion techniques were improved to a new point [9]: these could now be performed using either contrast material (with mostly T_2* images) or without contrast (with arterial spin-labeling techniques): one could now obtain reliable maps of relative cerebral blood flow and volume. These techniques, in addition to diffusion-weighted imaging (DWI). allowed clinicians to establish an initial, slightly rudimentary, but working model of an ischemic penumbra in human stroke [10]. The advantages of MRI over CT were multiple from the start; besides being nonirradiating, it provided multiplanar images that would cover the whole brain. One technique to further profit from fast imaging has been functional activation imaging of the brain: here T_2* images are very sensitive to changes in oxygenation that occur with activation. During the performance of a task (or paradigm), there is an increase in blood flow, whereas extraction remains stable: this will cause a decrease in deoxygenated blood and thus a signal increase on the images—this is the so-called BOLD effect [11]. Functional imaging with MR can now be adapted for any task involving any region of the brain. However, for most therapeutic purposes, motor or language paradigms will be mostly used. Additionally, noncontrast perfusion techniques can allow to improve the visualization of brain hemo-dynamics in function and dysfunction [11,12]. The use of conventional gadolinium chelates or of compounds with higher relaxativity may also impact the way we see both the vascular system and the blood–brain barrier [13]. A further refinement of diffusion imaging, diffusion tensor imaging (DTI), allows to reconstruct the direction and strength of water motion along the cerebral white fibers [14]: this allows to perform tractography that brings another information in addition to anatomy and function that we have just discussed. Further refinements in T_2^* images such as susceptibility-weighted imaging (SWI) have become a great help in the diagnosis of diseases of the brain [15] where blood degradation products are involved [16]. Taken together, all these techniques allow us to acquire information about anatomy, pathology, and function of the brain.

24.3 Pathological Conditions

24.3.1 Aneurysms

FOCUS POINT

- Cerebral aneurysms are the main cause of non-traumatic SAH.
- They may be seen as flow voids on T_2WI and T_1WI.
- Angio-MRI of the time-of-flight (TOF) type with and without contrast has become the standard for follow-up.
- Protocol contains whole-brain T_1WI, T_2WI, T_2^*WI, MRA, and DWI.

Intracranial aneurysms [17], which are the most frequent cause of non-traumatic SAH, contribute to some degree to morbidity and mortality due to diseases of the central nervous system [18–20]. The treatment today consists of either clipping or endovascular treatment [21], depending on a variety of factors, among which are the location of the aneurysm, clinical experience of the neurosurgeon or neurointerventional-ist, and the availability of techniques. Previously, mortality was high due to a combination of early and late morbidity and mortality [22]: initially the first

hemorrhage is at risk of killing the patient as is a subsequent early re-bleed; however, after 1 week approximatively, due to the presence of extravasated blood, the possibility of delayed ischemic injury due to vaso-spasm will increase dramatically. Also, the question of an early small bleeding, called the warning leak, is being debated, which could herald an impending bleeding. Thus, the roles of imaging are multiple: first to permit screening in patients with a familial risk or those who present with a possible warning leak [23]. Then, once an aneurysm has been found, imaging should help in establishing its risk of rebleeding. Here, MR techniques will have a more important role: indeed, in addition to demonstrating the size and location of the aneurysm, MR techniques with velo-cimetry should allow to obtain curves of flow in order to determine the potential dangerosity of the lesion. MRI techniques are used more for the diagnosis and follow-up of unruptured aneurysms: indeed due to an increase of the use of neuroimaging in the past few years, more patients are found to have these aneu-rysms. The risk for rupture seems to be related to size and location. Sometimes, the aneurysm itself will be seen on the basic T_1WI and T_2WI (Figure 24.1); in cases of a simple aneurysm, one will see an area of round flow void that is associated with a close vessel. If the aneurysm is thrombosed, one will see areas of hyper- and hypointensity on both T_1WI and T_2WI. MR is also the chosen technique when follow-up is necessary in the case of treated or untreated cases. One typical case of SAH is the so-called perimesencephali, where often no aneurysm is seen initially but every follow-up studies are mandatory (Figure 24.2). When considering imaging with MRA techniques, it is also very important to not just look at the reconstructed vessels but look at the basal axial slices from the source images. Indeed, if the signal is low or lower due to slow, perpendicular, or turbulent flow, the signal may be suppressed by such reconstruction algorithms such as maximum intensity projection. Whenever treatment has been performed, it is important to dispose of a method that can perform post-therapeutic imaging. First of all, it is very important to ensure that the clip or material used is compatible with MRI (Figures 24.3 through 24.5): this information is usually provided by the manufacturer or is available in publications and web sites dedicated to MR safety. Here it may be of interest to obtain additionally images with contrast in order to obtain better vessel filling and to assess possible aneurysm regrowth. CT has very strong beam-hardening artifacts that very often render the images useless. Sometimes, on basic MR images, that is, T_1WI and T_2WI, the aneurysm may be seen as an area of

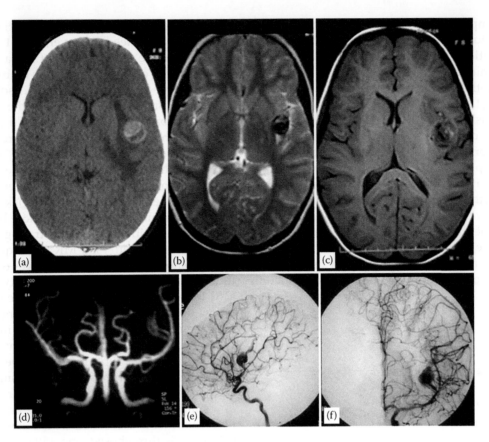

FIGURE 24.1
Left-sided MCA aneurysm. On CT, partially hyperdense (a) round lesion is seen in the sylvian fissure. T_2WI (b) and T_1WI (c) show an inhomogeneous lesion with flow void. The aneurysm is seen on the 3D TOF MRA (d) and the DSA images (e,f).

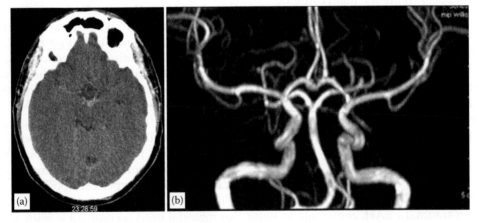

FIGURE 24.2
Perimesencephalic subarachnoid hemorrhage. On the acute CT, there is blood in front of the mesencephalon (a). On the MRA, there is no vascular lesion to be found (b).

enlarged flow void (Figures 24.6 through 24.8). MRA, however, has established itself as a cornerstone for the detection of aneurysms [24]. However, MRA, using a combination of shorter echo times, can allow to reduce the presence of susceptibility artifacts [25,26]. Also, the addition of contrast will help to demonstrate if there is any aneurismal remnant. These techniques are valid for patients treated by surgical clipping or by endovascular techniques such as coiling and/or stenting. One problem is that regular TOF techniques are sensitive to the strength and directionality of the flow, so sometimes aneurysms may be less well seen. Some authors

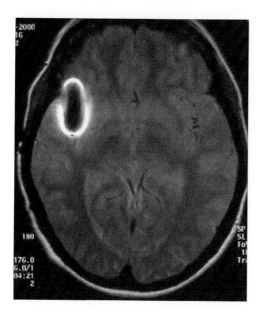

FIGURE 24.3
Clip artifact on proton-density image in the right sylvian fissure.

have thus advocated phase contrast techniques. Also, it is always necessary to look at the source images when looking for aneurysms because sometimes while doing maximum intensity reconstructions, it is possible to subtract the flow inside the aneurysm. It may also be important to assess the possibility of the occurrence of post-SAH vasospasm (Figure 24.9): it

is well known that the extravasation of blood and its concomitant degradation products will induce a vasoconstriction that may cause delayed neurological deficits after a week which are a high case of morbidity and mortality in these patients. MR techniques such as diffusion and perfusion may demonstrate these changes early on [27]. Arterial spin labeling can show sometimes the presence of collateral flow in these cases [28]. However, as discussed earlier, when considering treatment and cases that have been treated, it is important to choose correctly the patients and the sequences according to the clip [29] or stent that has been implanted [30]. The use of contrast at 1.5 or 3 T will help to enhance the needed luminographic effect [31].

Giant aneurysms (Figures 24.7 and 24.8) are another entity altogether because the symptomatology they cause is very seldom due to hemorrhage but will be caused by the increase in size with nerve palsies or headache [32]. Due to their size, it is also very often possible on T_2WI to see a flow void (absence of signal) in the lumen and to have hyperintensities in the wall due to thrombosis.

24.3.2 Cerebral Vascular Malformations

Cerebral vascular malformations are classified into cerebral arteriovenous malformations (AVMs), dural arteriovenous fistulas, cavernomas, and telangiectasias.

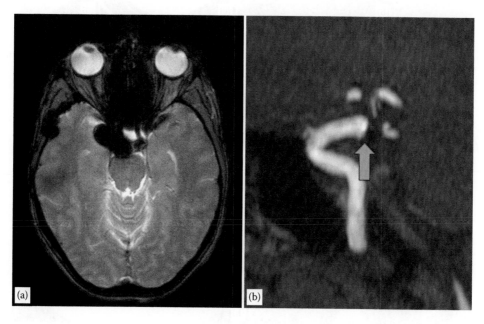

FIGURE 24.4
Combination of clip and coil artifacts on posterior communicating artery. Artifacts are prominent on T_2^* image (a) and there is interruption of vessel on MRA (b).

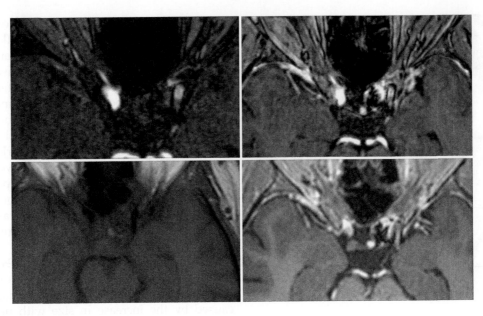

FIGURE 24.5
Stent with residual flow inside.

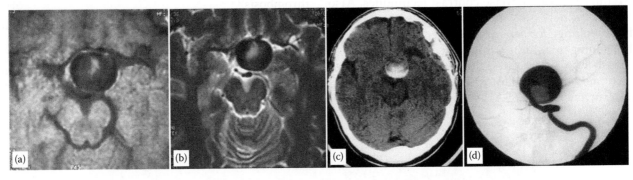

FIGURE 24.6
Giant aneurysm. Turbulence and flow void seen on T_1WI (a) and T_2WI (b). CT shows partly hyperdense aneurysm (c) which is confirmed by DSA (d).

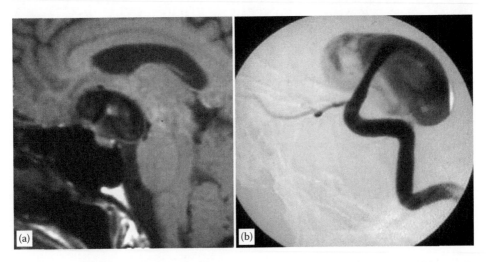

FIGURE 24.7
Giant aneurysm of the carotid artery (a). There is signal inhomogeneity on the T_1WI with both flow voids and absence of flow voids, which can be a sign of turbulence or thrombosis. Digital subtraction image confirms this (b).

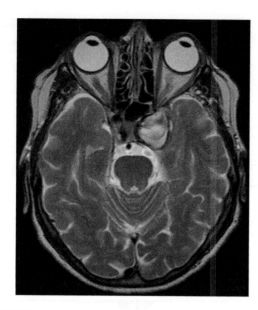

FIGURE 24.8
Giant thrombosed carotid artery aneurysm. There is a T_2 hyperintensity in the vessel.

24.3.2.1 Arteriovenous Malformations

FOCUS POINT

- They consist of conglomerates of abnormal veins and arteries.
- They are often wedge-shaped going from the cortex to the ventricle.

- They may be seen as multiple flow voids on T_2WI.
- They require functional MRI (fMRI) and DTI to plan treatment.
- Protocol contains T_1WI, multiplanar T_2WI, MRA, DWI, sometimes fMRI and DTI.

AVMs are a group of rare congenital vascular lesions of the brain that cause stroke and death. The most common symptoms are seizures and headaches. They consist of anomalous vessels, arteries, and veins that coalesce into a nidus where there may be aneurysms associated. On acute imaging, they should be suspected when a younger person is affected or when there is no clear underlying cause for hemorrhage (such as hypertension) (Figure 24.10), or of course in the case that one sees the presence of contrast pooling in pathological vessels in patients with acute hemorrhage. They are often organized in a special wedge-like shape, with multiple vessels being present at the cortex and the whole structure being oriented with the tip of the triangle into the depth of the brain toward the ventricular system (Figure 24.11). Besides abnormal arteries that may feed from multiple sources, it is possible to have aneurysms inside.

These vascular lesions are usually classified according to the criteria devised by Spetzler [33–35]: account size, eloquence, and the pattern of venous drainage. It is important to use the tools provided by MRI to help in assessing these criteria: MRI can determine the size by using conventional T_1WI and T_2WI (Figure 24.12)

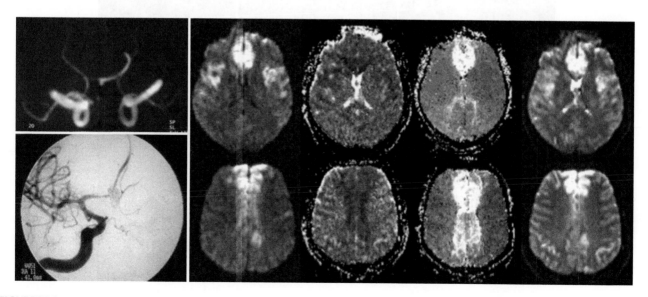

FIGURE 24.9
Vasospasm post-subarachnoid hemorrhage in a patient with an anterior communicating aneurysm. On the perfusion images, one sees hypoperfusion in the territory of both anterior cerebral arteries and ischemia in these regions on the diffusion-weighted image.

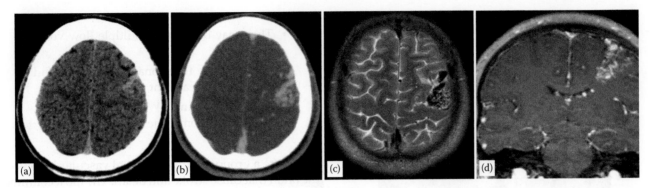

FIGURE 24.10
Left frontal AVM. On nonenhanced CT (a), there is slight hyperintensity of the cortical frontal area; on contrast-enhanced image (b), there is strong enhancement, signaling vessels. This is confirmed on the T_2 images that show flow voids (c). On the coronal T_1-weighted image (d), there is the typical wedge-shaped AVM running from the cortex to the ventricle.

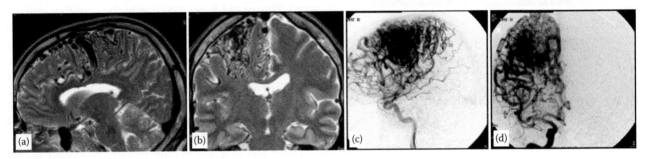

FIGURE 24.11
Right frontal AVM. On T_2WI (a), there are multiple small flow voids of different diameters running from the cortex to the ventricle again in a triangular way (b). This is confirmed on DSA (c,d).

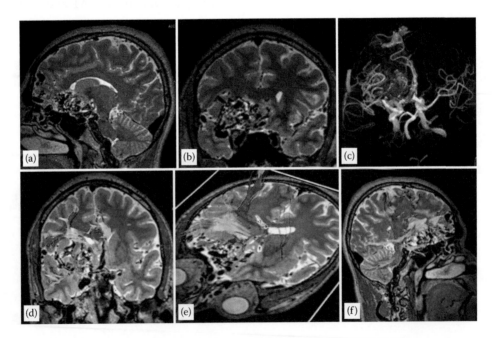

FIGURE 24.12
Frontobasal AVM with DTI. On T_2WI (a,b), there are multiple flow voids. The AVN is seen on 3D TOF angio-MR (c). DTI helps show the relationship between the malformation and the underlying white fibers (d–f).

or angio-MR techniques, whereas techniques such as fMRI can help to determine whether they are close to so-called eloquent areas of the brain by pinpointing the presence of activation due to a paradigm. On conventional MRI, multiple flow voids will be seen especially on T_2WI (Figures 24.13 through 24.15).

Treatment, which is still controversial, may require a combination of therapies [36,37]: operation, radio surgery [38,39], embolization, or a combination of these techniques. Although in the emergency setting, treatment is often necessary in order to evacuate an acute hematoma (Figures 24.15 and 24.16), there have been

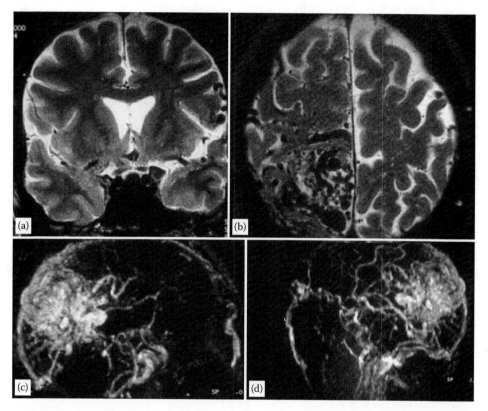

FIGURE 24.13
Posterior AVM. On the coronal T_2 image (a), one can see dilated flow voids on the cortex above the left frontal lobe. The AVM extends medially and posteriorly (b). The extension is better on the 3D sagittal reconstructions of the contrast-enhanced MRA of the head (c,d).

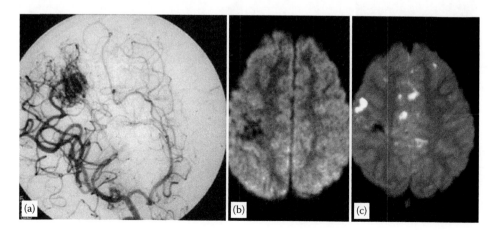

FIGURE 24.14
Deep AVM seen on DSA (a). On DWI (b), the flow voids give no signal. fMRI shows cortical activation to be located more frontally to it, signaling AVM to be far from the activated area (c).

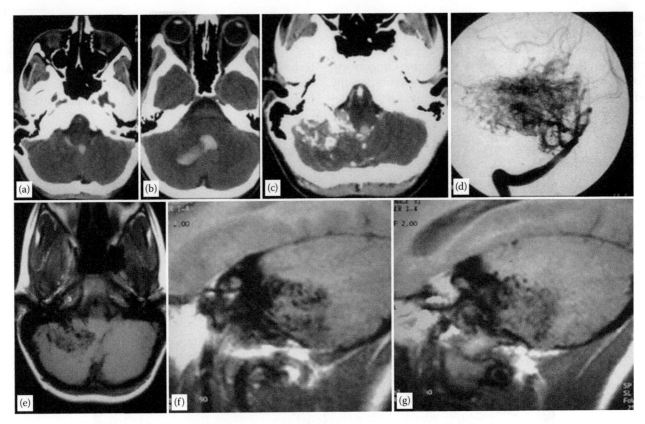

FIGURE 24.15
Young female patient with hemorrhage into the fourth ventricle due to a cerebellar AVM. On the CT, there is blood in the fourth ventricle (a,b) and enhancement (c), signaling an AVM. The AVM is seen on the DSA (d) but also on the axial (e) and sagittal T_1W images where one sees the multiple flow voids (f,g).

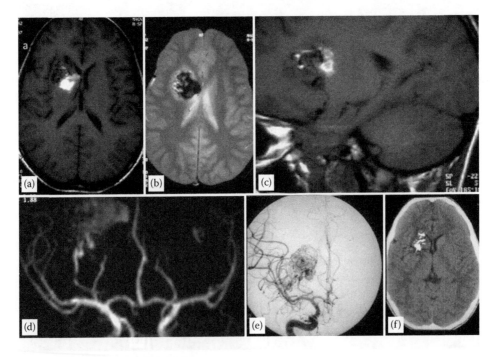

FIGURE 24.16
Deep AVM. There is visible hemorrhage on T_1, with subacute blood being both hyperintense and some flow voids (a). On the T_2* image, there is a signal loss due to hemorrhage (b). The coexistence of hemorrhage and flow voids is better seen on the sagittal T_1WI (c). The MRA shows the proximal part of the AVM (d), better seen on DSA (e). Post-interventional CT shows the embolization material in the AVM (f).

concerns regarding the long-term outcome of patients with yet unruptured AVMs. Thus, a randomized trial of unruptured brain AVMs (ARUBA) study was performed [40,41], whose aim was to determine long-term efficacy on outcomes; however, the trial failed to demonstrate any positive effects in the short term. This trial is at the moment very controversial and long-term effects need to be assessed before any final conclusions are drawn.

For both diagnostic and clinical purposes, imaging is based on a combination of MRA techniques and multiplanar images (Figure 24.17). When considering "traditional" MRA techniques (Figures 24.18 and 24.19), the so-called TOF techniques have a stronger signal for normal vessels, but in cases of turbulence and flow in other directions than straight out of the plane, there may be decreases or even abolition of luminographic signal. This led very often people to prefer phase contrast-based MRA techniques even if the anatomical resolution is lower. This paradigm has been somewhat changed with the arrival of 3D and even 4D (i.e., with time resolution) MRA techniques: here it is now possible to fill the vasculature completely with an almost perfect luminographic effect on the one hand, on the other hand one can also obtain time-resolved images. It is possible to visualize early and late vascular phases from the arterial to the venous side and also to demonstrate the nidus which was only possible until now

with DSA. The use of multiplanar images, especially T_2W ones, will allow additionally to monitor the presence of the flow voids and their eventual disappearance and replacement by hyperintensities due to embolization material and/or thrombosis.

If hemorrhage occurs, MRI can very often demonstrate this well especially in the subacute or chronic stages when the bleeding is small; if there is massive bleeding, this will of course be seen on MRI as a mass that is sometimes not well defined on T_1W images and somewhat inhomogeneous or dark on T_2WI. MRI is the method of choice for the follow-up of course. In case that there is an acute bleeding that is suspect but where all acutely performed imaging modalities are negative: CT, MRI, and DSA show no clear sign of an AVM. It is recommended to re-perform an MRI for at least 3 months, if not earlier. Indeed, the early hematoma and the early edema may render visualization of any small AVM remnant impossible. Of course, additional techniques such as fMRI will help in planning [42] and diffusion techniques will help in demonstrating pre- or even post-therapeutic ischemic changes.

Vein of Galen malformations are special malformations of the pediatric age where a dilatation of the vein of Galen is present that progresses probably due to a fistulous process. The clinical signs will be those associated with hydrocephalus and it may be discovered by

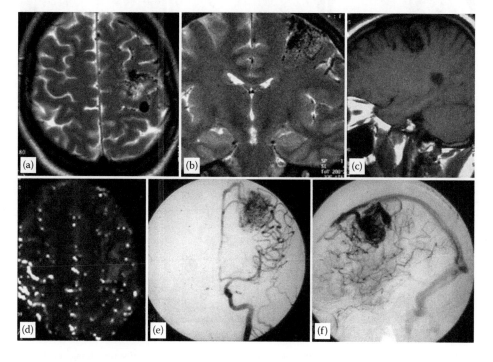

FIGURE 24.17
Left-sided cortical AVM. Flow voids are visible on the axial T_2WI (a) as well as on the coronal T_2WI, where there is the typical wedge-shaped AVM (b). On sagittal T_1WI, one can see the flow voids (c). Cortical activation map obtained with fMRI shows activation on the opposite side in relation to the AVM (d). Angio-MRI confirms the AVM (e,f).

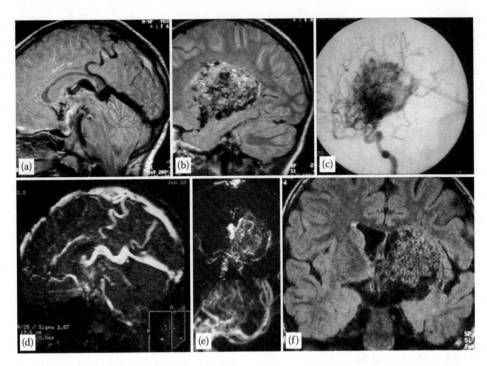

FIGURE 24.18
Big AVM in the basal ganglia on the left. The flow voids are seen on the sagittal proton-density images (a,b) as well as on DSA (c). Angio-MRI confirms this (d,e), as does the coronal FLAIR image that displays multiple flow voids in the basal ganglia (f).

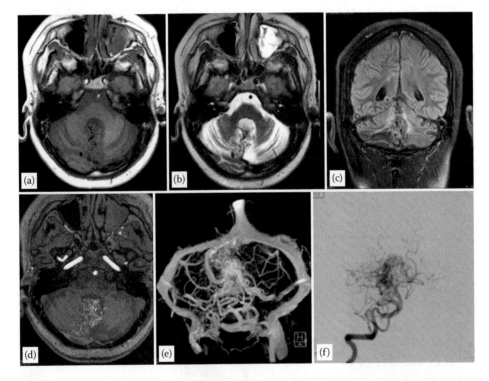

FIGURE 24.19
AVM of the cerebellum. Cerebellar atrophy as well as flow voids are present (a–c). On the axial source image of the TOF MRA, there is flow inside the vessels (d). 3D MRA confirms this (e) as well as DSA (f).

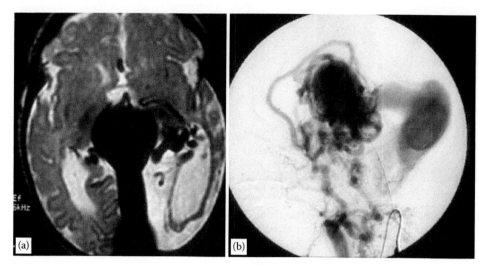

FIGURE 24.20
Vein of Galen aneurysm. Giant venous pouch visible on T_2WI (a) and on DSA (b).

ultrasound where dilated vessels are seen converging into a venous pouch. A technique such as MR venography or phlebography demonstrates this well [43]. On classical MR images, one can observe dilated vessels with abnormal areas of low voids on both T_1WI and T_2WI (Figure 24.20). An endovascular approach may then be discussed in order to better delineate the malformation and to discuss treatment. Follow-up imaging is then done by MR and phlebo-MR.

24.3.2.2 Dural Arteriovenous Fistulas

> **FOCUS POINT**
>
> - They have a direct communication between the artery and the vein.
> - They may be due to sinus thrombosis.
> - They may be seen directly on TOF angiography: direct visualization of the vein on arterial images.
> - Protocol contains whole-brain T_1WI, T_2WI, T_2*WI, MRA, and DWI.

There are mainly two types of fistulas: carotid cavernous fistulas and dural arteriovenous fistulas, both of which are acquired. Here we deal with the so-called dural AVM.

Dural arteriovenous fistulas consist of a pathological direct connection between an artery and a dural vein or sinus. In adulthood, it has been postulated that they are the consequence of a sinus thrombosis [44], whereas other etiologies such as trauma are possible [45]. Usually the clinical sign will be a bruit, such as tinnitus for which the patients will come to consult.

These vascular lesions are often classified according to Cognard et al. into five types depending on the pattern of venous drainage [46]: type I, which is located in the main sinus, has antegrade flow; type II, also in the main sinus, has reflux into the sinus (IIa), cortical veins (IIb), or both (IIa and b); type III has a direct cortical venous drainage without venous ectasia; type IV, however, has direct cortical venous drainage with venous ectasia; and type V has a spinal venous drainage leading to myelopathy in half of the cases. These criteria determine the degree of dangerosity of the fistula, with type 1 being considered entirely benign and type 2 being associated with some intracranial hypertension and hemorrhage in 10%. However, in the higher grades, hemorrhage increased dramatically: 40% in type 3 and 65% in type 4. This will be very well examined and explored by using the so-called time-resolved contrast-enhanced MRA techniques.

Treatment will be either surgical or endovascular. Very often the fistulas are discovered on MRI or CT with dilated vessels, especially seen on postcontrast images (CT), or with increased flow voids (seen on T_2W MR images). MRA helps to demonstrate the presence of direct communication by showing even venous opacification on noncontrast TOF MRA images. Also increased vascularization may be seen on postcontrast MR images (Figure 24.21).

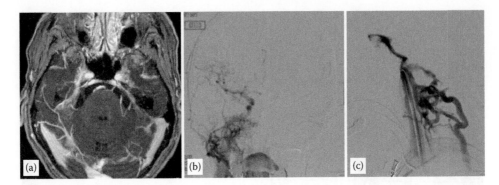

FIGURE 24.21
Arteriovenous fistula: On the fat-saturated T_1WI at the level of the tentorium, there is increased tentorial enhancement as well as some hypervascularity (a). The fistula is conformed by the early appearance of vessels on DSA (b,c).

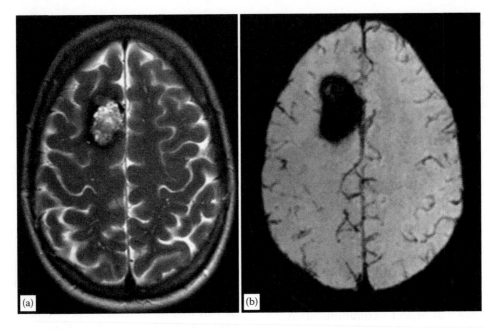

FIGURE 24.22
Right frontal cavernoma. A popcorn pattern is seen on T_2WI (a). SWI shows signal loss (b).

Other types of vascular malformations are capillary telangiectasias, cavernomas, and developmental venous anomalies (DVA). Cavernomas, or cavernous malformations, are often not visible on CT but are apparent on MRI as areas where there has been often recurrent bleeding. They consist of malformations with "caverns" of blood vessels that may bleed, inducing symptomatology such as headache and epilepsy. These caverns may induce a so-called popcorn pattern on MRI (Figure 24.22). For cavernomas, one will have to look for signal drops on the T_2* images (Figure 24.23) or even better nowadays on the SWI (Figure 24.24) [47]: there is a signal drop with prominent blooming artifact. Cavernomas may have to be treated surgically

sometimes due to the secondary epilepsy they induce by irritating the cortex.

For capillary telangiectasias, where there are multiple capillaries, one must use T_1W images before and after gadolinium. However, SWI is also becoming a possibility [48] (Figures 24.25 and 24.26)

Sometimes, a DVA will be found alone or in association with cavernoma, which is visible on contrast-enhanced T_1W images. A DVA is a vascular venous malformation that sometimes will have the typical "medusa head" appearance on contrast images (Figure 24.27). SWI seems to be finding a new role here as well because it is especially suited for venous imaging (Figure 24.28).

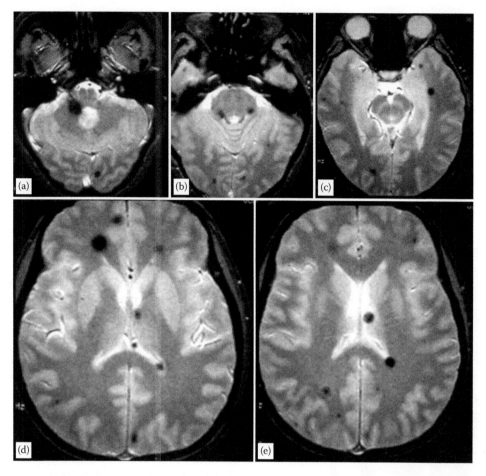

FIGURE 24.23
Cavernomatosis on T_2*: multiple punctate signal losses. Hypointensities are seen in the brainstem (a,b), the temporal lobes (c), and the frontal and parietal lobes (d,e).

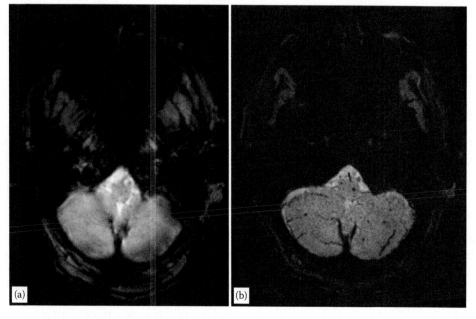

FIGURE 24.24
Cavernomatosis: T_2* (a) vs. SWI (b). Lesions are much more visible on T_2*.

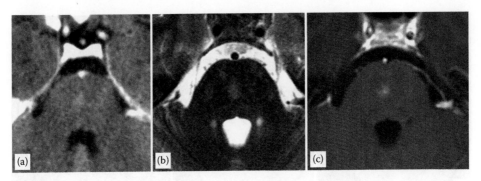

FIGURE 24.25
Pontine telangiectasia: slight midline enhancement on CT (a), hyperintensity on T_2WI (b) with enhancement on T_1WI (c).

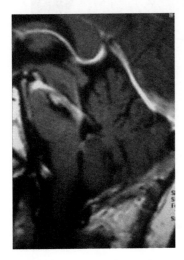

FIGURE 24.26
Giant pontine telangiectasia visible on sagittal T_1W postcontrast image.

24.3.2.3 Vasculitis

FOCUS POINT

- The diagnosis of cerebral vasculitis is often difficult.
- Typical images show multiple ischemic lesions
- Protocol contains whole-brain T_1WI, T_2WI, T_2*WI, MRA, DWI, postcontrast images (T_1WI), as well as SWI.

Vasculitides, that is, inflammatory and infectious diseases of the brain, are very often clinical challenge and very often difficult to diagnose. Clinically, the findings consist of headache, stroke, and seizures associated with a systemic inflammation. The correct approach requires the combination of imaging and laboratory results and even vascular biopsy [49,50]. Very often the radiologist is going to be confronted with queries for imaging for vasculitis with very little information and thus has to often ask for clarification. Often imaging studies will not reveal the typical stigmata of vasculitis [51–55] (Figure 24.29). It is often necessary to know if a known vasculitis is present and/or if there are any biological findings. Unfortunately, very often this is not the case and the radiologist has to exclude or demonstrate vasculitis based on very thin data. A variety of biomarkers may be looked at, both selective and nonselective ones. There are mainly two types of vasculitis: primary vasculitis of the central nervous system and vasculitis of the brain within the context of systemic vasculitis. The most common of the systemic vasculitis will be lupus. Any kind of vasculitis may cause an involvement of the central nervous system. The most common type is primary angiitis of the central nervous system [56].

Basically, the nervous system can be affected by any kind of vasculitis [57]. Imaging is thus going to be based on a mixture of axial and angiographic imaging: the aim is to look for secondary stigmata (ischemia, bleeds) (Figure 24.30) as well as vascular anomalies that are a direct expression of the disease. Typically, one is to look for multiple ischemic events of different ages (on T_2WI and DWI) (Figure 24.31), with maybe contrast-enhancement as well. The presence of irregularities of the vessels will help in clarifying the diagnosis.

The overall workup is going to necessitate anamnesis and physical examination findings together with laboratory results from the blood as well as the cerebrospinal fluid as well as imaging (mostly MRI) and eventually brain and vascular biopsy [58].

The involvement of the central nervous system by lupus is very challenging to diagnose and is finally based on the combination of imaging and biological criteria.

MRI has the distinct advantage also that once diagnosis is made on biological and imaging grounds, one can monitor the disease: it is possible with MRA to see if the treatment makes the vascular irregularities go away (Figure 24.32).

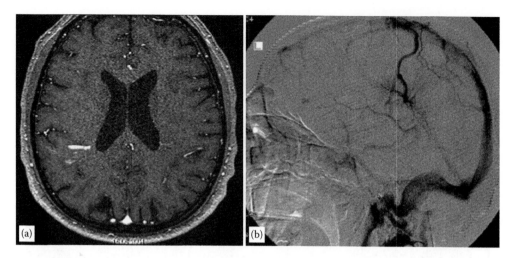

FIGURE 24.27
Developmental venous anomaly. The venous anomaly is visible on the postcontrast axial TOF source image (a) and DSA (b).

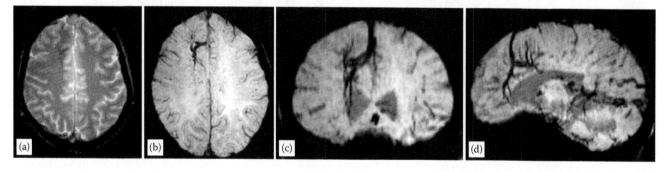

FIGURE 24.28
Developmental venous anomaly. The venous anomaly is well visible on SWI. These SWI datasets can then be easily reconstructed in any plane. Discrete frontal hypointensity visible on the conventional T2* image (a), while the DAVM is more visible on SWI in the axial (b) coronal (c) and sagittal (d) projections.

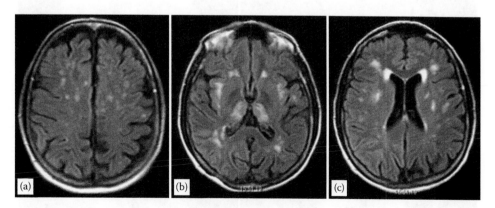

FIGURE 24.29
Vasculitis. There are multiple hyperintensities on fluid-attenuated inversion recovery (FLAIR): subcortically (a), in the basal ganglia (b), and in the white matter (c).

(*Continued*)

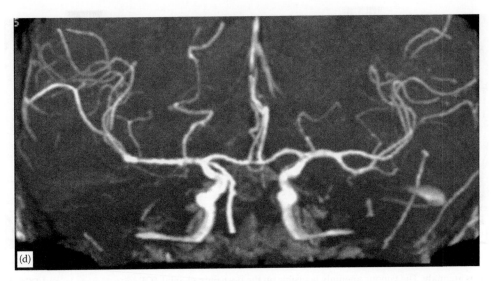

FIGURE 24.29   (Continued)
Vasculitis. There are multiple hyperintensities on fluid-attenuated inversion recovery (FLAIR): narrowing of the vessels on MRA (d) as well.

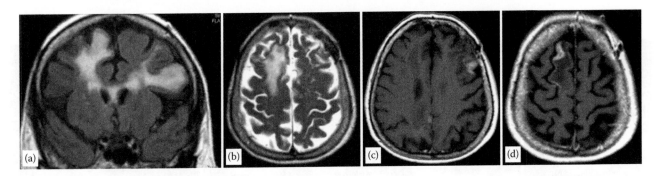

FIGURE 24.30
Active vasculitis with multiple areas of ischemia with old and subacute infarcts. Old infarcts are visible on FLAIR (a) and T_2 (b) as constituted lesions, but subacute ones will enhance (c,d).

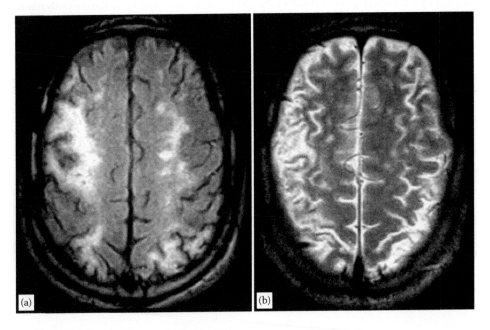

FIGURE 24.31
Vasculitis with lesions of many stages seen on FLAIR (a), T_2 (b).

(Continued)

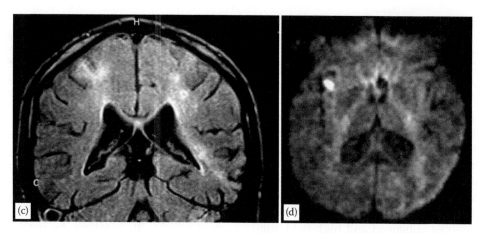

FIGURE 24.31 (Continued)
Vasculitis with lesions of many stages seen on new lesions on coronal FLAIR (c) and on DWI (d) as well.

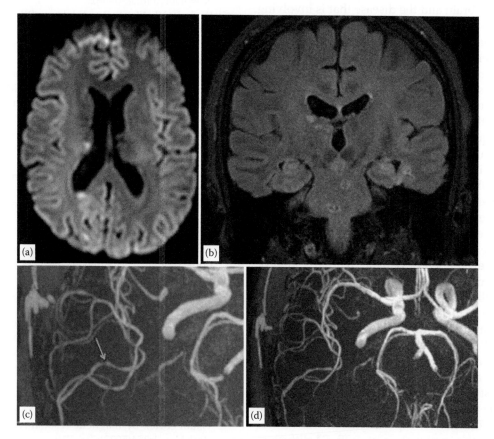

FIGURE 24.32
Vasculitis on DWI. There are multiple lesions in the brain (a) and brain stem (FLAIR) (b). The vessels show narrowing in the distal MCA branches (c) (arrow) that go away with treatment (d).

24.4 Conclusions

MRI techniques for the detection of diseases of the central nervous system have evolved immensely over the past few decades. Imaging of vascular diseases of the brain relied for a very long time on cerebral angiography, which is still the gold standard but which is being reserved more for cases where an intervention is needed. Over the past few decades, a series of evolutions and revolutions have happened in the so-called non-invasive imaging techniques which enabled us to have techniques with a very high spatial and temporal resolution. Indeed, we are now able to not only have high-resolution anatomic imaging in three dimensions but also obtain images that allow to quantify physiological parameters that are of prime importance when dealing with patients who have cerebrovascular disorders. The combination of high-resolution MRA techniques, especially those are possible now with 3 T scanners together with imaging such as diffusion perfusion and SWI, allows to image vascular irregularities, ischemic changes, alterations in hemodynamics, and the presence of hemorrhage. Vascular imaging with MRA has progressed greatly in the development of techniques such as angio-MR with techniques such as TOF or phase-contrast images, before being overtaken by the combination of contrast-enhanced and high-field angiography techniques. These techniques very often have to be used in combination in order to obtain a more complete view of the brain and the disease that is involving it. Diffusion imaging, which was once seen as purely a research tool, has now fully entered the clinical arena and belongs even to the standard workhorse techniques for brain imaging as does angio-MR. At first MR was rather time consuming, and it still partly is, but there has been an increase in speed of acquisition and image reconstruction that make an unparalleled number of images and sequences possible: although examination time may not have decreased much, the number of possible sequences in the same time has increased, and this has been possible by going from spin echo to fast spin echo to echo planar and finally to parallel imaging techniques that have revolutionized the way we acquire the images and have also entered quickly the clinical arena. Although CT very often is still the method of choice for the emergency evaluation of patients with any kind of acute disease, mainly due to practical reasons, MRI is the tool that will allow radiologists and clinicians to do the full workup of these patients. While imaging aneurysms, it is recommended nowadays to use 3 T units and to perform TOF before and after contrast. While imaging AVMs, one must use a combination of multiplanar T_2 images with angio-MR, which may be TOF or even dynamic; for fistulas, dynamic MRA is recommended in addition to TOF techniques. For lesions such as cavernomas, T_2* imaging, and now especially SWI, will be useful. For angiitis, the combination of TOF, T_2* diffusion, and contrast-enhanced T_1WI will be necessary.

References

1. Pereira VM, Vargas MI, Marcos A, Bijlenga P, Narata AP, Haller S, Lövblad KO. Diagnostic neuroradiology for the interventional neuroradiologist. *World J Radiol.* 2013 Nov 28;5(11):386–397.

2. Lövblad KO, Laubach HJ, Baird AE, Curtin F, Schlaug G, Edelman RR, Warach S. Clinical experience with diffusion-weighted MR in patients with acute stroke. *AJNR Am J Neuroradiol.* 1998 Jun–Jul;19(6):1061–1066.

   Mohamed M, Heasly DC, Yagmurlu B, Yousem DM. Fluid-attenuated inversion recovery MR imaging and subarachnoid hemorrhage: Not a panacea. *AJNR Am J Neuroradiol.* 2004 Apr;25(4):545–550.

3. Shimoda M, Hoshikawa K, Shiramizu H, Oda S, Matsumae M. Problems with diagnosis by fluid-attenuated inversion recovery magnetic resonance imaging in patients with acute aneurysmal subarachnoid hemorrhage. *Neurol Med Chir (Tokyo).* 2010;50(7): 530–537.

4. El-Koussy M, Guzman R, Bassetti C, Stepper F, Barth A, Lövblad KO, Schroth G. CT and MRI in acute hemorrhagic stroke. *Cerebrovasc Dis.* 2000 Nov–Dec;10(6):480–482.

5. Wedeen VJ, Meuli RA, Edelman RR, Geller SC, Frank LR, Brady TJ, Rosen BR. Projective imaging of pulsatile flow with magnetic resonance. *Science.* 1985 Nov 22;230(4728):946–948.

   Lövblad KO, Altrichter S, Viallon M, Sztajzel R, Delavelle J, Vargas MI, El-Koussy M, Federspiel A, Sekoranja L. Neuro-imaging of cerebral ischemic stroke. *J Neuroradiol.* 2008 Oct;35(4):197–209.

6. Edelman RR, Wielopolski P, Schmitt F. Echo-planar MR imaging. *Radiology.* 1994 Sep;192(3):600–612.

7. Le Bihan D, Breton E, Lallemand D, Grenier P, Cabanis E, La-val-Jeantet M. MR Imaging of intravoxel incoherent motions: Application to diffusion and perfusion in neurologic disorders. *Radiology.* 1986;161:401–407.

8. Lövblad KO, Plüschke W, Remonda L, Gruber-Wiest D, Do DD, Barth A, Kniemeyer HW et al. Diffusion-weighted MRI for monitoring neurovascular interventions. *Neuroradiology.* 2000 Feb;42(2):134–138.

9. Rosen BR, Belliveau JW, Vevea JM, Brady TJ. Perfusion imaging with NMR contrast agents. *Magn Reson Med.* 1990 May;14(2):249–265.

10. Schlaug G, Benfield A, Baird AE, Siewert B, Lövblad KO, Parker RA, Edelman RR, Warach S. The ischemic penumbra: Operationally defined by diffusion and perfusion MRI. *Neurology.* 1999 Oct 22;53(7):1528–1537.

    Belliveau JW, Kennedy DN Jr, McKinstry RC, Buchbinder BR, Weisskoff RM, Cohen MS, Vevea JM, Brady TJ, Rosen BR. Functional mapping of the human visual cortex by magnetic resonance imaging. *Science.* 1991 Nov 1;254(5032):716–719.

11. Wong EC. Quantifying CBF with pulsed ASL: Technical and pulse sequence factors. *J Magn Reson Imaging.* 2005 Dec;22(6):727–731.

12. Edelman RR, Siewert B, Darby DG, Thangaraj V, Nobre AC, Mesulam MM, Warach S. Qualitative mapping of cerebral blood flow and functional localization with echo-planar MR imaging and signal targeting with alternating radio frequency. *Radiology.* 1994 Aug;192(2):513–520.

13. Lovblad KO. Impact of contrast-enhanced CT and MRI on the management of patients with neurological diseases. *Neuroradiology* 2007;49(1 Suppl):S1–S2.

14. Le Bihan D, Mangin JF, Poupon C, Clark CA, Pappata S, Molko N, Chabriat H. Diffusion tensor imaging: Concepts and applications. *J Magn Reson Imaging*. 2001 Apr;13(4):534–546.

15. Haacke EM, Mittal S, Wu Z, Neelavalli J, Cheng YC. Susceptibility-weighted imaging: Technical aspects and clinical applications, part 1. *AJNR Am J Neuroradiol*. 2009 Jan;30(1):19–30.

16. Thulborn KR, Brady TJ. Iron in magnetic resonance imaging of cerebral haemorrhage. *Magn Reson Q*. 1989;5(1):23–38.

17. Krayenbühl HA, Yaşargil MG, Flamm ES, Tew JM Jr. Microsurgical treatment of intracranial saccular aneurysms. *J Neurosurg*. 1972 Dec;37(6):678–686.

18. Wiebers DO, Whisnant JP, Huston J 3rd, Meissner I, Brown RD Jr, Piepgras DG, Forbes GS; International Study of Unruptured Intracranial Aneurysms Investigators et al. Unruptured intracranial aneurysms: Natural history, clinical outcome, and risks of surgical and endovascular treatment. *Lancet*. 2003 Jul 12;362(9378):103–110.

19. Molyneux AJ, Kerr RS, Birks J, Ramzi N, Yarnold J, Sneade M, Rischmiller J; ISAT Collaborators. Risk of recurrent subarachnoid haemorrhage, death, or dependence and standardised mortality ratios after clipping or coiling of an intracranial aneurysm in the International Subarachnoid Aneurysm Trial (ISAT): Long-term follow-up. *Lancet Neurol*. 2009 May;8(5):427–433.

20. Bijlenga P, Ebeling C, Jaegersberg M, Summers P, Rogers A, Waterworth A, Iavindrasana J et al.; @neurIST Investigators. Risk of rupture of small anterior communicating artery aneurysms is similar to posterior circulation aneurysms. *Stroke*. 2013 Nov;44(11):3018–3026.

21. Pereira VM, Bijlenga P, Marcos A, Schaller K, Lovblad KO. Diagnostic approach to cerebral aneurysms. *Eur J Radiol*. 2013 Oct;82(10):1623–1632.

22. Ljunggren B, Säveland H, Brandt L, Zygmunt S. Early operation and overall outcome in aneurysmal subarachnoid hemorrhage. *J Neurosurg*. 1985 Apr;62(4):547–551.

23. Bassi P, Bandera R, Loiero M, Tognoni G, Mangoni A. Warning signs in subarachnoid hemorrhage: A cooperative study. *Acta Neurol Scand*. 1991 Oct;84(4):277–281.

24. Maeder PP, Meuli RA, de Tribolet N. Three-dimensional volume rendering for magnetic resonance angiography in the screening and preoperative workup of intracranial aneurysms. *J Neurosurg*. 1996 Dec;85(6):1050–1055.

25. Gönner F, Heid O, Remonda L, Nicoli G, Baumgartner RW, Godoy N, Schroth G. MR angiography with ultrashort echo time in cerebral aneurysms treated with Guglielmi detachable coils. *AJNR Am J Neuroradiol*. 1998 Aug;19(7):1324–1328.

26. Gönner F, Lövblad KO, Heid O, Remonda L, Guzman R, Barth A, Schroth G. Magnetic resonance angiography with ultrashort echo times reduces the artefact of aneurysm clips. *Neuroradiology*. 2002 Sep;44(9):755–758.

27. Lövblad KO, el-Koussy M, Guzman R, Kiefer C, Remonda L, Taleb M, Reinert M et al. Diffusion-weighted and perfusion-weighted MR of cerebral vasospasm. *Acta Neurochir Suppl*. 2001;77:121–126.

28. Altrichter S, Kulcsar Z, Jägersberg M, Federspiel A, Viallon M, Schaller K, Rüfenacht DA, Lövblad KO. Arterial spin labeling shows cortical collateral flow in the endovascular treatment of vasospasm after posttraumatic subarachnoid hemorrhage. *J Neuroradiol*. 2009 Jun;36(3):158–161.

29. Wichmann W, Von Ammon K, Fink U, Weik T, Yasargil GM. Aneurysm clips made of titanium: Magnetic characteristics and artifacts in MR. *AJNR Am J Neuroradiol*. 1997 May;18(5):939–944.

30. Lövblad KO, Yilmaz H, Chouiter A, San Millan Ruiz D, Abdo G, Bijlenga P, de Tribolet N, Ruefenacht DA. Intracranial aneurysm stenting: Follow-up with MR angiography. *J Magn Reson Imaging*. 2006 Aug;24(2):418–422.

31. Pierot L, Portefaix C, Boulin A, Gauvrit JY. Follow-up of coiled intracranial aneurysms: Comparison of 3D time-of-flight and contrast-enhanced magnetic resonance angiography at 3T in a large, prospective series. *Eur Radiol*. 2012 Oct;22(10):2255–2263.

32. Schubiger O, Valavanis A, Wichmann W. Growth-mechanism of giant intracranial aneurysms; demonstration by CT and MR imaging. *Neuroradiology*. 1987;29(3):266–271.

33. Spetzler RF, Martin NA. A proposed grading system for arteriovenous malformations. *J Neurosurg*. 1986 Oct;65(4):476–483.

34. Spetzler RF, Zabramski JM. Grading and staged resection of cerebral arteriovenous malformations. *Clin Neurosurg*. 1990;36:318–337.

35. Ponce FA, Spetzler RF. Arteriovenous malformations: Classification to cure. *Clin Neurosurg*. 2011;58:10–12.

36. Wilson DA, Abla AA, Uschold TD, McDougall CG, Albuquerque FC, Spetzler RF. Multimodality treatment of conus medullaris arteriovenous malformations: 2 decades of experience with combined endovascular and microsurgical treatments. *Neurosurgery*. 2012 Jul;71(1):100–108.

37. Zaidi HA, Abla AA, Nakaji P, Spetzler RF. Prospective evaluation of preoperative stereotactic radiosurgery followed by delayed resection of a high grade arteriovenous malformation. *J Clin Neurosci*. 2014 Jun;21(6):1077–1080.

38. Leksell L. The stereotaxic method and radiosurgery of the brain. *Acta Chir Scand*. 1951 Dec 13;102(4):316–319.

39. Steiner L, Leksell L, Forster DM, Greitz T, Backlund EO. Stereotactic radiosurgery in intracranial arterio-venous malformations. *Acta Neurochir (Wien)*. 1974;(Suppl 21): 195–209.

40. Stapf C. The rationale behind "A Randomized Trial of Unruptured Brain AVMs" (ARUBA). *Acta Neurochir Suppl*. 2010;107:83–85.

41. Bambakidis NC, Cockroft K, Connolly ES, Amin-Hanjani S, Morcos J, Meyers PM, Alexander MJ, Friedlander RM. Preliminary results of the ARUBA study. *Neurosurgery*. 2013 Aug;73(2):E379–E381.

42. Ozdoba C, Nirkko AC, Remonda L, Lövblad KO, Schroth G. Whole-brain functional magnetic resonance imaging of cerebral arteriovenous malformations involving the motor pathways. *Neuroradiology*. 2002 Jan;44(1):1–10.

43. Yasargil MG, Antic J, Laciga R, Jain KK, Boone SC. Arteriovenous malformations of vein of Galen: Microsurgical treatment. *Surg Neurol*. 1976 Sep;(3):195–200.

44. Kutluk K, Schumacher M, Mironov A. The role of sinus thrombosis in occipital dural arteriovenous malformations—Development and spontaneous closure. *Neurochirurgia (Stuttg)*. 1991 Sep;34(5):144–147.

45. Mironov A. Classification of spontaneous dural arteriovenous fistulas with regard to their pathogenesis. *Acta Radiol.* 1995 Nov;36(6):582–592.

46. Cognard C, Gobin YP, Pierot L, Bailly AL, Houdart E, Casasco A, Chiras J, Merland JJ. Cerebral dural arteriovenous fistulas: Clinical and angiographic correlation with a revised classification of venous drainage. *Radiology.* 1995 Mar;194(3):671–680.

47. Kivelev J, Niemelä M, Hernesniemi J. Characteristics of cavernomas of the brain and spine. *J Clin Neurosci.* 2012 May;19(5):643–648.

48. El-Koussy M, Schroth G, Gralla J, Brekenfeld C, Andres RH, Jung S, Shahin MA, Lovblad KO, Kiefer C, Kottke R. Susceptibility-weighted MR imaging for diagnosis of capillary telangiectasia of the brain. *AJNR Am J Neuroradiol.* 2012 Apr;33(4):715–720.

49. Monach PA. Biomarkers in vasculitis. *Curr Opin Rheumatol.* 2014 Jan;26(1):24–30.

50. Berlit P, Kraemer M. Cerebral vasculitis in adults: What are the steps in order to establish the diagnosis? Red flags and pitfalls. *Clin Exp Immunol.* 2014 Mar;175(3): 419–424.

51. Pistracher K, Gellner V, Riegler S, Schökler B, Scarpatetti M, Kurschel S. Cerebral haemorrhage in the presence of

52. Sciascia S, Bertolaccini ML, Baldovino S, Roccatello D, Khamashta MA, Sanna G. Central nervous system involvement in systemic lupus erythematosus: Overview on classification criteria. *Autoimmun Rev.* 2013 Jan;12(3):426–429.

53. Salvarani C, Brown RD Jr, Hunder GG. Adult primary central nervous system vasculitis. *Lancet.* 2012 Aug 25;380(9843):767–777.

54. Zuccoli G, Pipitone N, Haldipur A, Brown RD Jr, Hunder G, Salvarani C. Imaging findings in primary central nervous system vasculitis. *Clin Exp Rheumatol.* 2011 Jan–Feb;29(1 Suppl 64):S104–S109.

55. Weiss PF. Pediatric vasculitis. *Pediatr Clin North Am.* 2012 Apr;59(2):407–423.

56. Twilt M, Benseler SM. The spectrum of CNS vasculitis in children and adults. *Nat Rev Rheumatol.* 2011 Dec 20;8(2):97–107.

57. Garg A. Vascular brain pathologies. *Neuroimaging Clin N Am.* 2011 Nov;21(4):897–926.

58. Villa I, Agudo Bilbao M, Martínez-Taboada VM. Advances in the diagnosis of large vessel vasculitis: Identification of biomarkers and imaging studies. *Reumatol Clin.* 2011 Dec;7(Suppl 3):S22–S27.

primary childhood central nervous system vasculitis—A review. *Childs Nerv Syst.* 2012 Aug;28(8):1141–1148.

25

Neoplasms of the Brain and Pituitary Gland

Sara E. Kingston, Daniel S. Treister, Willa Jin, Megha Nayyar, Benita Tamrazi, Francesco D'Amore, Bavrina Bigjahan, Alexander Lerner, Bruno A. Telles, Chia-Shang J. Liu, and Mark S. Shiroishi

CONTENTS

25.1 Tumors of Neuroepithelial Tissue

25.1.1 Astrocytic Tumors

25.1.1.1 Pilocytic Astrocytoma

25.1.1.1.1 General Features

Pilocytic astrocytoma (PA) is the most common pediatric central nervous system (CNS) glial tumor and the most common tumor of the cerebellum in children. The age of presentation is typically in the first two decades of life (peak incidence 5–13 years of age), and clinical signs and symptoms depend on the location of the mass. Patients with cerebellar PA often present with headache, blurred vision, diplopia, neck pain, and gait disturbance. PA in adults occurs more often in the cerebral hemispheres. No gender predilection is known. Other than the cerebellum, PA can also occur in the optic nerve and chiasm, hypothalamic region, cerebral hemispheres, ventricles, velum interpositum, and spinal cord. A well-known association with neurofibromatosis type 1 exists with PAs seen in the optic pathway. They are World Health Organization (WHO) grade I tumors and generally have

an excellent prognosis with up to 20-year survival of 79%. Hypothalamic PAs carry a less favorable prognosis. Rare instances of metastatic disease have been reported. Surgical resection is the treatment of choice and is often curative. Recurrences are rare. Therapeutic options for PA of the optic nerve/chiasm or hypothalamic region range from conservative to chemotherapy and radiation therapy. Rare instances of spontaneous regression have also been reported.

25.1.1.1.2 Imaging Features

The classic appearance for PA is that of a well-circumscribed cyst with mural nodule arising from the cerebellar hemisphere (Figure 25.1). On computed tomography (CT), the cystic component appears hypodense and the mural nodule appears isodense. Calcifications can occasionally be present. On magnetic resonance imaging (MRI), the cyst appears iso- to slightly hyperintense on T_1-weighted imaging (T_1WI) and iso- to hyperintense on T_2-weighted imaging (T_2WI). The solid portion appears iso- to hypointense on T_1WI. On T_2WI, the solid portion shows hyperintensity compared to gray matter that is slightly hypo- to isointense to the cerebrospinal

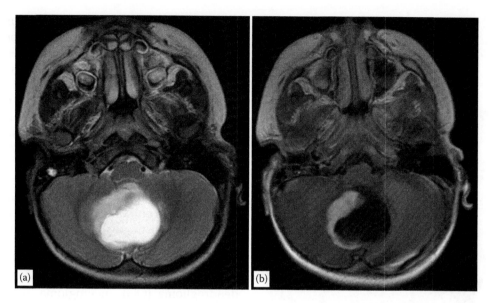

FIGURE 25.1
Pilocytic astrocytoma. Axial T_2WI (a) and axial contrast-enhanced T_1WI (b) demonstrate a well-circumscribed cystic mass of the cerebellum with an enhancing mural nodule.

fluid (CSF). Lower signal intensity on T_2WI is present in medulloblastoma (MB) or high-grade astrocytoma. It is typical to have little to no edema and no diffusion abnormalities. Hemorrhage is uncommon. The mural nodule demonstrates avid contrast enhancement. Other imaging appearances include a mass that demonstrates solid contrast enhancement, ringlike enhancement of the cysts, or a solid-enhancing mass with necrotic center. Variable amounts of contrast enhancement have been noted in optic pathway PAs. Despite its relatively benign course, the solid portions of PAs can demonstrate aggressive appearing elevation of choline: N-acetylaspartate (Cho:NAA) ratios as well as the presence of lactate on magnetic resonance (MR) spectroscopy.

Imaging differential diagnostic considerations include MB, ependymoma, high-grade astroctyoma, hemangioblastoma (HGBL), ganglioglioma, and pilomyxoid astrocytoma.

25.1.1.2 Subependymal Giant Cell Astrocytoma

25.1.1.2.1 General Features

Subependymal giant cell astrocytoma (SEGA) is an indolent low-grade glioneuronal, classified as a WHO grade I tumor. The age range of presentation is wide, but it occurs most commonly in the first and second decades of life with a mean age of 11 years. There is no gender predilection. The most common location of the tumor is near the foramen of Monro, likely arising from a subependymal nodule in the ventricular wall. Almost all cases are associated with the neurocutaneous phakomatosis tuberous sclerosis (TS), and SEGA is the most common brain tumor occurring in TS patients. Common

presenting neurologic symptoms include seizures or signs of obstructive hydrocephalus. Surgical resection is indicated for symptomatic SEGAs or those demonstrating interval growth. Gamma knife stereotactic radiosurgery as well as medical management with rapamycin or everolimus are other treatment options.

25.1.1.2.2 Imaging Features

On CT, SEGAs typically appear as an intraventricular mass with calcifications at the foramen of Monro (Figure 25.2). They are iso- to hypoattenuating on CT and may have hyperattenuating foci indicating internal calcifications or hemorrhage. They demonstrate contrast enhancement and this, along with interval growth, is more suggestive of SEGA rather than a subependymal nodule. On MRI, these lesions are hypointense on T_1WI and heterogenously hyperintense on T_2WI. These lesions demonstrate marked enhancement with contrast. Calcifications can be noted as blooming on T_2*W gradient echo (GRE) imaging. After surgical resection of the tumor, an annual MRI should be performed for active surveillance.

Imaging differential diagnostic considerations include subependymal nodule of TS, central neurocytoma (CN), subependymoma (SE), ependymoma, choroid plexus papilloma (CPP)/choroid plexus carcinoma (CPCA), and astrocytoma.

25.1.1.3 Pleomorphic Xanthoastrocytoma

25.1.1.3.1 General Features

Pleomorphic xanthoastrocytoma (PXA) is a rare WHO grade II astrocytic tumor. It has a relatively favorable prognosis with a 10-year overall survival of 70%. The typical

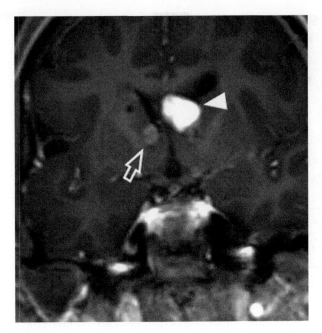

FIGURE 25.2
Subependymal giant cell astrocytoma. Coronal contrast-enhanced T_1WI demonstrates a left-sided enhancing intraventricular mass at the foramen of Monro (white arrowhead) consistent with a SEGA in a tuberous sclerosis patient. A smaller mildly enhancing subependymal nodule is seen on the right side (white hollow arrow).

clinical presentation of PXA is a child or young adult with a long history of seizures. Surgical resection may be curative; however, these tumors exhibit patterns of recurrence and/or malignant transformation in up to 20% of cases.

25.1.1.3.2 Imaging Features

The classic imaging features are a peripheral, cortically based cystic neoplasm with an enhancing mural nodule (Figure 25.3). However, a solid appearance is a common variant. It can be found throughout the brain, spinal cord, and globe, but the temporal lobe is the most commonly involved site. On CT, the tumor most often demonstrates a hypodense cystic component and a nodular, superficial mixed density component. Hemorrhage and calcification are uncommon. On conventional MRI, the cystic portion has a signal that is isointense to the CSF in T_1WI, T_2WI, and fluid-attenuated inversion recovery (FLAIR) sequences. The solid nodular component shows a mixed hypointense-to-isointense signal on T_1WI and a mixed hyperintense signal on T_2WI. Surrounding edema is uncommon. On postcontrast T_1WI, the nodular component usually shows avid contrast enhancement. Because of the superficial cortical location, a *dural tail* is commonly present. Cortical dysplasia is not usually seen with PXAs, distinguishing it from dysembryoplastic neuroepithelial tumor (DNET).

Imaging differential diagnostic considerations include DNET, ganglioglioma, glioblastoma (GBM), PA, oligodendroglioma (OG), meningioma, metastatic disease, and infection.

25.1.1.4 Low-Grade Diffuse Astrocytoma

25.1.1.4.1 General Features

Low-grade diffuse astrocytoma (DA) is a WHO grade II diffusely infiltrating tumor, which is usually centered in the supratentorial white matter. Other less commonly involved areas include the brain stem as well as the deep gray matter structures. Extension into the cortex is also common. DAs make up less than 10% of all gliomas. Three histological subtypes of DA exist: fibrillary (most common) followed by gemistocytic and protoplasmic. Though they are low-grade tumors, they are infiltrative like high-grade gliomas, and this makes treatment success difficult. The mean age of presentation is 34 years of age, but they can also occur in children. A slight male predilection has been noted. Clinical signs and symptoms depend on the location involved, but commonly

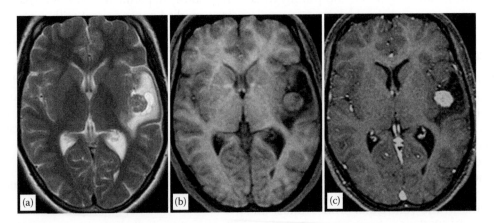

FIGURE 25.3
Pleomorphic xanthoastrocytoma. Axial T_2WI (a), T_1WI (b), and fat-saturated contrast-enhanced T_1WI (c) demonstrate a cortically based cystic mass with enhancing nodule of the left insular cortex.

headache, seizures, increased intracranial pressure, progressive neurologic deficit, and cognitive/behavior decline are seen. The median survival is 5–7 years during which malignant degeneration into GBM is common at the terminal phase. Currently, no standard form of DA treatment exists and various options ranging from observation to surgical resection, chemotherapy, and radiation therapy are used.

25.1.1.4.2 Imaging Features

On CT, DA classically appears as an ill-defined iso- to hypodense mass within the white matter (Figure 25.4). Calcifications are present in about 20% of cases. MRI is the optimal method for evaluation. DA appears hypointense on T_1WI and hyperintense on T_2WI and FLAIR. It can appear well circumscribed; however, tumor infiltration extends beyond the visible signal abnormalities. Contrast enhancement is usually not present. It should be noted, however, that although contrast enhancement is typically a sign of higher grade tumors, up to one-third of high-grade gliomas do not show contrast enhancement and some types of low-grade gliomas (PAs, gangliogliomas, OGs, and DAs) can show contrast enhancement. Decreased diffusion is typically not seen on DWI. There is contradictory evidence supporting the use of apparent diffusion coefficient (ADC) to determine glioma grade. Generally, lower relative cerebral blood volume (rCBV) from T_2*W dynamic susceptibility contrast (DSC) MRI is seen in DAs compared with high-grade gliomas. MR spectroscopy (MRS) can show nonspecific mild decrease

in NAA and mild increase in Cho; however, increased myo-inositol (mI) is thought to be a characteristic finding.

Imaging differential diagnostic considerations include anaplastic astrocytoma (AA), OG, infarct, cerebritis, encephalitis, and cortical dysplasia.

25.1.1.5 Brain-Stem Gliomas/Diffuse Intrinsic Pontine Glioma

25.1.1.5.1 General Features

As a group, brain-stem gliomas (BSGs) account for 5%–11% of all intracranial tumors and 15%–30% of infratentorial tumors in children. These tumors are divided into three main groups based on location: diffuse intrinsic pontine glioma (DIPG), exophytic gliomas of the lower brain stem/medulla oblongata, and tectal gliomas. DIPGs are the most common type of BSG and represent 60%–75% of all posterior fossa gliomas. They possess the worst prognosis of any pediatric brain tumor with a median survival of 9–12 months after therapy. However, exophytic gliomas of the lower brain stem/medulla oblongata and tectal gliomas usually have very good prognoses. Children with DIPG usually present between 5 and 10 years of age, and there is no gender predilection. Clinical presentation usually includes long tract signs, ataxia, and multiple cranial nerve palsies, particularly of the sixth and seventh nerves. Malignant fibrillary astrocytoma (WHO grade III or IV) features are usually found in histopathology; however, biopsy is usually not performed because it will

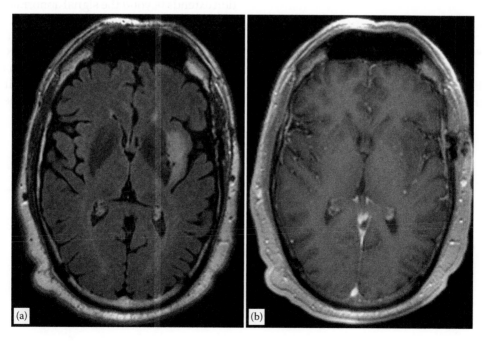

FIGURE 25.4
Low-grade DA. Axial FLAIR (a) and an axial contrast-enhanced T_1WI (b) demonstrate an ill-defined FLAIR hyperintense mass without contrast enhancement in the left subinsular region.

not affect management. Treatment options are primarily radiation therapy (over the age of 3), although conventional chemotherapy appears to be ineffective. Their infiltrative nature precludes surgical resection.

25.1.1.5.2 Imaging Features

On CT, DIPGs appear expansile and hypodense (Figure 25.5). At presentation, DIPGs usually are larger than 2 cm and occupy most of the pons. Extension into the adjacent fourth ventricle is not common; however, caudal and rostral extension is common. They are hypointense on T_1WI and heterogeneously hyperintense on T_2WI. Due to their infiltrative nature, the margins appear indistinct. FLAIR may better demonstrate the extent of the tumor rather than T_2WI. Basilar and vertebral artery encasement by the tumor is common. Areas of contrast enhancement can be seen; however, this does not appear to have significant prognostic utility. As the tumors progress, more focal areas of contrast enhancement and necrosis will become evident. CSF dissemination can occur in up to one-third of cases. Elevated Cho/creatine (Cr) and Cho/NAA ratios suggest a worse prognosis. Areas of decreased diffusion on DWI may imply higher grade of tumor or necrosis. DTI can be used to visualize white matter tracts that are usually preserved. The imaging findings are usually pathognomonic and biopsy is rarely needed.

Imaging differential diagnostic considerations include infection, demyelinating disease, cavernous malformation, Langerhans cell histiocytosis, and neurofibromatosis type 1.

25.1.1.6 Anaplastic Astrocytoma

25.1.1.6.1 General Features

AA is a WHO grade III infiltrative astrocytic tumor. Along with WHO grade IV GBM, it is considered a *high-grade glioma*. However, AAs are much less common compared with GBM. They are classically seen within the supratentorial hemispheric white matter; however, less common locations include the brain stem, thalamus, and spinal cord. AAs can occur in adults and children, but the median age at diagnosis is 43–52 years with a median survival of 2–3 years. They are more common in males, and they often develop from a preexisting low-grade DA. They may represent a molecular precursor of GBM, and they tend to recur or degenerate into GBM IV tumor regardless of treatment. Clinical signs and symptoms depend on the location involved, but commonly headache, seizures, increased intracranial pressure, progressive neurologic deficit, and cognitive/behavioral decline are seen. Like GBM, standard therapy usually involves maximal safe resection with temozolomide chemoradiation.

25.1.1.6.2 Imaging Features

CT shows a poorly defined inhomogenous hypo- to isodense mass (Figure 25.6). Hemorrhage, cysts, or calcifications are uncommon. On MRI, T_1WI displays an inhomogeneous hypointense signal depending on the extent of necrotic and cystic changes inside the tumor mass, although these changes are more suggestive of GBM. Contrast enhancement is usually not seen; however, when present, it may be nodular, patchy, or less frequently ringlike. In the latter case, malignant transformation to GBM should be considered. T_2WI and FLAIR images show a heterogeneous hyperintense mass. Although a well-circumscribed appearance might be seen, like low-grade DAs and GBMs, tumor infiltration extends beyond the signal abnormalities. Restricted diffusion on DWI is usually not present. There is contradictory evidence supporting the use of ADC to determine glioma grade. Generally higher rCBV from T_2^*W DSC MRI is seen in AAs compared with low-grade DAs. MRS can show decrease in NAA and increase in Cho, with lower mI compared with low-grade DAs.

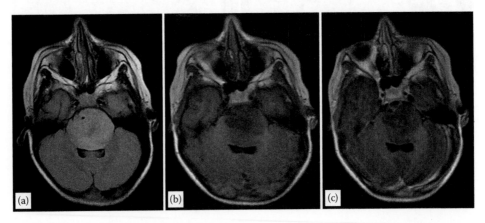

FIGURE 25.5
Diffuse intrinsic pontine glioma. Axial FLAIR (a), T_1WI (b), and contrast-enhanced T_1WI (c) demonstrate an expansile hyperintense FLAIR mass of the pons. Mild contrast enhancement and necrosis are evident as well as encasement of the basilar artery.

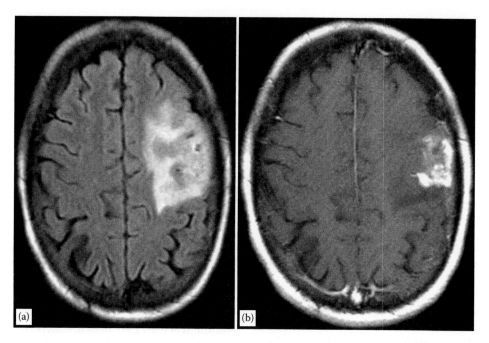

FIGURE 25.6
Anaplastic astrocytoma. Axial FLAIR (a) and contrast-enhanced T_1WI (b) demonstrate an ill-defined FLAIR hyperintense left frontal lobe mass. Peripheral areas of patchy and nodular contrast enhancement are noted.

Imaging differential diagnostic considerations include low-grade DA, glioblastoma multiforme, OG, infarct, cerebritis, encephalitis, and cortical dysplasia.

25.1.1.7 Glioblastoma

25.1.1.7.1 General Features

GBM is an aggressive WHO grade IV malignant tumor of neuroepithelial tissue arising from astrocytes. GBMs account for 54.4% of all gliomas and are the most common type of primary brain malignancy. Most GBMs arise *de novo* and are referred to as primary GBMs, whereas secondary GBMs arise from an existing astrocytoma. Primary GBMs present at a mean age of 55 years, whereas secondary GBMs present earlier (mean age 40 years). They are more common in males. Clinical presentation depends on the brain structures involved by the tumor mass and by the amount of surrounding edema. Common presentation includes headache, seizures, and focal neurologic deficits. Regardless of the standard of care therapy with surgery and temozolomide chemoradiation, GBMs generally demonstrate relentless progression with a median overall survival of 15–18 months.

25.1.1.7.2 Imaging Feature

CT imaging typically shows a hypodense to isodense mass centered within white matter with irregular, ill-defined margins (Figure 25.7). The tumor core is often hypoattenuating, representing the areas of necrosis.

Diffuse peritumoral hypodensity is also commonly present and should be interpreted as a combination of edema and tumor infiltration. CT might demonstrate the areas of hemorrhage or calcifications, although the latter is a rare finding, particularly in primary GBM. Contrast enhancement is typically heterogeneous and appears as thick, irregular rim enhancement with central areas of nonenhancing necrosis.

Conventional MRI adds more valuable findings compared to CT. T_1WI of GBMs show various signal intensities that reflect necrosis, degrees of cellularity, and hemorrhage at different stages. Contrast-enhanced T_1WI demonstrates a wide variety of heterogeneous patterns of enhancement, frequently thick irregular enhancement with areas of central necrosis. Although less common, these tumors may also have a solid or nodular appearance. Enhancement may also be seen along the meninges, ependymal, and ventricules. GBMs can be solitary lesions involving any lobe, most commonly the frontal lobe. However, they may also infiltrate to the contralateral hemisphere through the anterior commissure or corpus callosum, and GBMs that assume a bilateral configuration are commonly described as having a *butterfly* appearance. If multiple enhancing masses are detected, the term *multicentric* is applied to GBMs that have arisen independently at different locations, whereas the term *multifocal* implies tumor spreading to other locations from a primary focus. T_2WI of the tumor core is characterized by heterogeneous signal dependent on the extent of necrosis, presence of fluid-debris

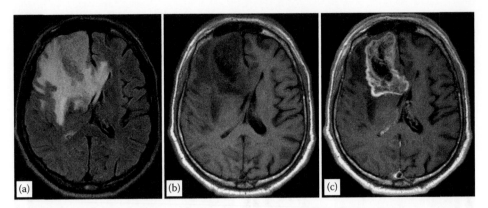

FIGURE 25.7

Glioblastoma. Axial FLAIR (a), T_1WI (b), and contrast-enhanced T_1WI (c) show a large right frontal lobe mass with thick, irregular enhancement and central necrosis. Surrounding areas of hyperintense FLAIR signal represent a combination of vasogenic edema as well as tumor infiltration. Furthermore, like with low-grade DAs and AAs, tumor infiltration extends beyond MRI signal abnormalities.

levels, blood products, and flow-related signal void due to tumor vessels. The area surrounding the tumor mass on T_2W and FLAIR images typically is of increased signal intensity representing a combination of surrounding vasogenic edema and tumor infiltration. T_2^*W images may display foci of signal loss related to blood products. Because of their infiltrative nature, GBM tumor involvement is present even in normal appearing brain parenchyma at imaging.

The T_2^*W DSC image-derived parameter rCBV is higher in GBM than in low-grade astrocytoma. MRS usually shows high Cho/Cr and Cho/NAA ratios with a decreased mI.

Imaging differential diagnostic considerations include abscess, lymphoma, necrotic metastasis, lower grade or cystic astrocytoma, resolving hematoma, subacute infarct, and tumefactive demyelinating lesion.

25.1.1.8 Gliomatosis Cerebri

25.1.1.8.1 General Features

Gliomatosis cerebri (GC) is a rare WHO grade III CNS malignancy that is a diffuse, extensively infiltrating glioma. It is often bilateral, involves at least two contiguous cerebral lobes, and has the potential to extend even into the spinal cord. Diagnosis can be difficult because the clinical manifestations and imaging appearances appear nonspecific. It can occur in all ages and has a peak incidence between 40 and 50 years with no gender predilection. Clinical presentation depends on the involved structures and encompasses focal neurological deficits as well as symptoms related to an increased intracranial pressure. Evolution of GC is unpredictable and may be slow or rapid. Overall prognosis is generally poor. No standard therapies exist for GC and options include surgery, chemotherapy, and radiation therapy.

25.1.1.8.2 Imaging Features

Although at least two lobes are involved in GC, there may be a relative lack of architectural distortion or mass effect (Figure 25.8). Extension from the white matter into the cortex is relatively common. Basal ganglia and thalamic involvement can also be seen. The tumor may show a wide variety of features on CT, including presenting as a diffuse, poorly defined hypodensity. In some cases, the CT examination may appear seemingly normal. On MRI T_1WI, GC appears as a hypointense to isointense homogeneous infiltrating mass. Contrast-enhanced T_1WI may rarely demonstrate patchy areas of enhancement and, if present, raises the possibility of a focus of GBM. T_2WI and FLAIR images demonstrate high signal areas with possible loss of gray–white interface. On DWI, GC usually does not show restricted diffusion. rCBV from perfusion-weighted imaging is not elevated. On MRS, normal Cho, low NAA, and increased mI peak can be seen.

Imaging differential diagnostic considerations include low-grade DA, AA, lymphoma, encephalitis, demyelinating disease, progressive multifocal leukoencephalopathy, and metabolic disease.

25.1.2 Oligodendroglial Tumors

25.1.2.1 Oligodendroglioma

25.1.2.1.1 General Features

OG is a WHO grade II tumor and the third most common glial neoplasm. WHO grade III anaplastic OGs can arise *de novo* or from a preexisting, well-differentiated WHO grade II OG. OGs are more common in adults with an incidence peak between the fourth and fifth decades. Loss of heterozygosity on the short arm of chromosome 1 (1p) and the long arm of chromosome 19 (19q) is associated with better prognosis. Males are more commonly affected. Seizure is a frequent symptom at

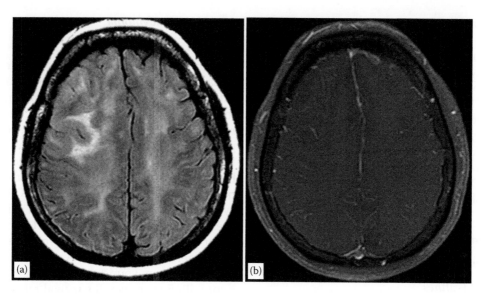

FIGURE 25.8
Gliomatosis cerebri. Axial FLAIR (a) and axial fat-saturated contrast-enhanced T_1WI (b) demonstrate areas of diffuse hypintense FLAIR signal in both frontal and parietal lobes without significant architectural distortion, mass effect, or contrast enhancement.

presentation. An unusual variant that contains both oligodendroglial and astrocytic components is known as an oligoastrocytoma. The median survival for low-grade OG is 10 years, whereas for anaplastic OG, it is 4 years. Treatment options include surgery, chemotherapy, and radiation therapy.

25.1.2.1.2 Imaging Features

Classically, OG is a cortically/subcortically based partially calcified mass most often seen in the frontal lobe (Figure 25.9). Other cerebral lobes can be affected as well as less commonly in the brain stem, intraventricular

regions, and presentation as leptomeningeal disease. The margins of the tumor are usually well marginated but can appear ill-defined. Differentiation between a low-grade OG and anaplastic OG may be difficult with imaging alone. On CT, the mass is usually hypodense and less commonly iso- or hyperdense. Most calcifications appear coarse. Hemorrhage and cystic degeneration can sometimes be seen. About 50%–60% of low-grade OGs demonstrate contrast enhancement, and this is more often seen in anaplastic OG. Erosion and remodeling of adjacent calvarium have also been reported. On T_1WI, OGs are hypo- to isointense to gray matter. On T_2WI and

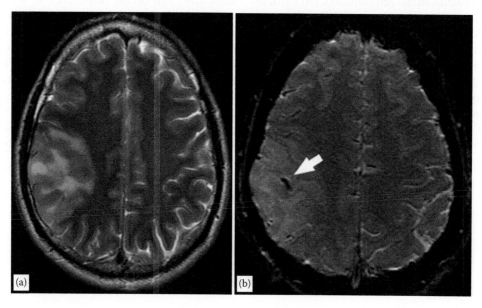

FIGURE 25.9
Oligodendroglioma. Axial T_2WI (a) and susceptibility-weighted imaging (SWI) (b) demonstrate a cortical/subcortical right frontoparietal mass that is hyperintense on T_2WI and demonstrates blooming from calcification on SWI (white arrow, b).

FLAIR images, a heterogeneous hyperintensity signal is often present. Edema is often minimal. Necrosis and hemorrhage are more commonly seen in anaplastic OG. T_2*W images may show areas of *blooming*, usually due to calcification blood products. T_2*W DSC MRI may demonstrate high rCBV value independent of tumor grade. OGs with 1p and 19q deletions are more likely to contain calcifications and ill-defined margins, extend across midline, and involve the frontal lobe.

Imaging differential diagnostic considerations include ganglioglioma, DNET, PXA, infarct, low-grade AA, cerebritis, and encephalitis.

25.1.3 Ependymal Tumors

25.1.3.1 Subependymoma

25.1.3.1.1 General Features

SE is an uncommon intraventricular WHO grade I tumor with ependymal differentiation. SEs are most frequently seen in 40–60-year-old males. Most patients are asymptomatic at presentation, but in selected cases, the tumor may cause hydrocephalus. There is an overall good prognosis with surgical resection. Recurrences are very uncommon.

25.1.3.1.2 Imaging Features

SE is classically a well-circumscribed intraventricular mass within the inferior fourth ventricle in a middle-aged/elderly male (Figure 25.10). They can also be seen within the lateral ventricles and rarely in the third ventricle or in the central canal of the spinal cord. On CT, SE appears as a lobulated hypo- to isodense intraventricular mass. Internal calcifications are common. On MRI, SE demonstrates hypo- to isointensity on T_1WI and hyperintensity on T_2WI. Slow progression and minimal to no enhancement may help differentiate SE from more aggressive intraventricular tumors. Brain invasion or edema is not seen, unlike the case for ependymomas. Calcifications can result in blooming on T_2*W GRE imaging.

Imaging differential diagnostic considerations include ependymoma, CN, HGBL, SEGA, and choroid plexus tumor.

25.1.3.2 Ependymoma

25.1.3.2.1 General Features

Ependymoma is the third most common posterior fossa tumor in the pediatric population after PA and MB. About two-thirds of ependymomas are infratentorial with one-third occurring in the supratentorial region. Low-grade ependymomas are considered WHO grade II and anaplastic ependymomas are WHO grade III. WHO grade II tumors are divided into four major histologic subtypes with varying prognosis: cellular, papillary, clear cell, and tanycytic. Clinical signs and symptoms include a long history of raised intracranial pressure, ataxia, and lower cranial nerve palsies. They occur most often in the 3–5-year-old age group and have a male gender predilection. A smaller peak of incidence is also seen in young adults. Prognosis is generally poor with 5-year overall survival rates of 50%–84% and recurrence rates from 24% to 54%. Treatment consists of surgical resection, radiotherapy, and chemotherapy. Surgical resection can be challenging, given their location, involvement of blood vessels and cranial nerves, and adherence to the brain stem.

25.1.3.2.2 Imaging Features

Infratentorial ependymomas arise from the floor of the fourth ventricle (MB, the main differential diagnostic consideration, typically arises from the roof of the fourth ventricle), whereas supratentorial tumors develop from the outer walls of the lateral ventricles (Figure 25.11). Intracranial ependymomas are more common than spinal ependymomas in children. Although they are usually well circumscribed, infiltration into adjacent brain structures can occur. Adherence to the brain stem and upper cervical cord is suggestive, but not pathognomonic for ependymoma. After expanding within the fourth ventricle, they typically extend out through the foramina of Luschka and Magendie. This growth pattern has also been described for MB, but it is much more typical for ependymoma. Furthermore, this extension through the foramina may appear more bulbous in MB and more thin in ependymomas. Posterior fossa ependymomas can also arise within the cerebellopontine

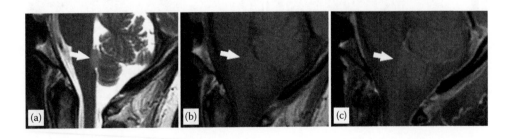

FIGURE 25.10
Subependymoma. Sagittal T_2WI (a), T_1WI (b), and fat-saturated contrast-enhanced T_1WI (c) demonstrate a mass in the inferior fourth ventricle (white arrows) that shows hyperintensity on T_2WI, isointensity on T_1WI, and no contrast enhancement.

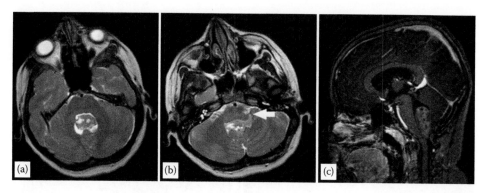

FIGURE 25.11
Ependymoma. Axial T_2WI (a) and (b) and sagittal fat-saturated contrast-enhanced T_1WI (c) demonstrate an enhancing fourth ventricular mass with extension through the left foramen of Luschka (white arrow, b).

angle from rests of ependymal cells. On CT, infratentorial ependymomas are iso- to hyperdense masses within the fourth ventricle. Calcification and hemorrhage are common and appear as hyperdense foci. On MRI, they demonstrate iso- to hypointense signal intensity on T_1WI and range from hypo- to hyperintense on T_2WI depending on the amount of cystic formation and hemorrhage/calcification. Calcification and hemorrhage can result in hypointensity on T_2*W GRE sequences. Solid portions of the tumor may demonstrate decreased diffusion on DWI. Areas of cystic formation and necrosis may be seen, and heterogeneous enhancement is usually noted following contrast agent administration. CSF dissemination can be seen throughout the brain and spine in 10%–12% of cases, and thus spinal MRI should be performed as well.

Imaging differential diagnostic considerations include MB, PA, atypical teratoid rhabdoid tumor (ATRT), BSG, HGBL, metastasis, and lymphoma.

25.1.4 Choroid Plexus Tumors

25.1.4.1 CPP/CPCA

25.1.4.1.1 General Features

Choroid plexus tumors are uncommon neoplasms arising from the choroid plexus, most often in the lateral ventricle, followed by the fourth and third ventricles. Most (80%) are CPPs and usually present during the first decade of life. CPPs are benign WHO grade I tumors and have a favorable prognosis. They are the most common lateral ventricle trigonal mass in young children, usually under the age of 5. CPCA is seen in about 5%–20% of cases and is considered an aggressive WHO grade III lesion. It is more commonly seen in children compared to adults. Tumors within the lateral ventricles show no gender predilection and present in the first decade of life, whereas tumors of the fourth ventricle are more common in males and can be seen in both children and adults. Choroid plexus tumors produce CSF, and

consequently patients present with signs and symptoms of hydrocephalus and raised intracranial pressure. Treatment for CPP is surgical resection with excellent long-term outcomes, whereas for CPCA, treatment is surgical resection and chemoradiation with 5-year survival of 26%–50%.

25.1.4.1.2 Imaging Features

Choroid plexus tumors are lobulated tumors that are usually attached to the choroid plexus with a vascular pedicle (Figure 25.12). Imaging may not reliably differentiate CPP from CPCA and the diagnosis frequently relies on histology. Brain invasion, heterogeneous appearance, vasogenic edema, and necrosis suggest CPCA over CPP. Hydrocephalus is commonly seen. On CT, choroid plexus tumors appear as iso- to hyperdense intraventricular masses with calcifications occurring in a quarter of cases. Some studies have suggested that it is more common for tumors to arise in the left lateral ventricle trigone than in the right side. On MRI, they appear iso- to hypointense on T_1WI and variable signal intensity on T_2WI. Flow voids may be evident. Tumors in the lateral ventricle are usually supplied by the choroidal arteries, whereas fourth ventricles are supplied by choroidal branches from the posterior inferior cerebellar artery. Contrast-enhanced spinal MRI should be performed to evaluate for the presence of CSF dissemination.

Imaging differential diagnostic considerations include meningioma, ependymoma, metastasis, MB, and lymphoma.

25.1.5 Neuronal and Mixed Neuronal–Glial Tumors

25.1.5.1 Dysplastic Cerebellar Gangliocytoma

25.1.5.1.1 General Features

Dysplastic cerebellar gangliocytoma (DCG), also known as Lhermitte–Duclos disease, consists of thickening of cerebellar folia with both features of a benign

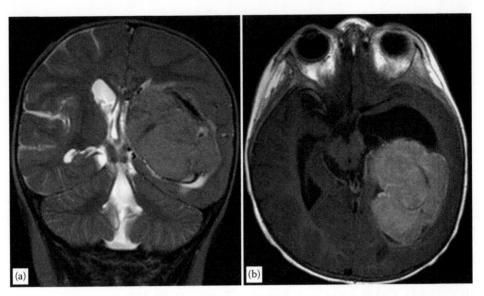

FIGURE 25.12
Choroid plexus carcinoma. Coronal T_2WI (a) and axial contrast-enhanced T_1WI (b) demonstrate a large intraventricular mass centered in the trigone of the left lateral ventricle. There is enlargement of the temporal horn of the left lateral ventricle with T_2 hyperintensity in the surrounding brain parenchyma suggestive of transependymal flow and/or vasogenic edema. At surgery, this was found to be a CPCA.

neoplasm and malformation. However, its exact pathogenesis remains unclear. These WHO grade I tumors may involve a large portion of cerebellar parenchyma and extend from a cerebellar hemisphere to the vermis, and occasionally contralaterally. An association with Cowden's syndrome has also been described. DCG is found in pediatric and adult patients, but it is most often seen in the third and fourth decades. This lesion may cause hydrocephalus, and consequently patients may present with signs and symptoms of raised intracranial pressure. Treatment is surgical

resection. Progression or recurrence may occur after surgery.

25.1.5.1.2 Imaging Features

DCG has a very characteristic striated or *tiger-striped* appearance of the cerebellum (Figure 25.13). On CT, the tumor is hypo- or isodense. Skull remodeling from mass effect can occur. MR images frequently demonstrate an alternating striated appearance of isointensity alternating with hypointensity on T_1WI and alternating iso- and hyperintensity on T_2WI. Calcifications and contrast

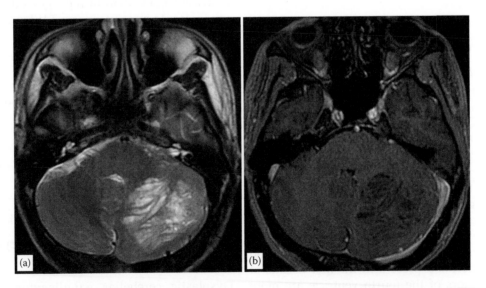

FIGURE 25.13
Dysplastic cerebellar gangliocytoma. Axial T_2WI (a) and axial fat-saturated contrast-enhanced T_1WI (b) demonstrate a left cerebellar mass with a striated or *tiger-striped* appearance. No contrast enhancement is noted.

enhancement are rare. Both T_2-shine through and diffusion restriction on DWI as well as decreased choline, NAA, and mI on MRS have been reported. Mass effect resulting in deformation of the fourth ventricle and hydrocephalus is frequently seen. There can also be cerebellar tonsillar herniation with syrinx formation in the spinal cord.

Imaging differential diagnostic considerations include infarct, cerebellitis, MB, ependymoma, astrocytoma, HGBL, and metastasis.

25.1.5.2 Desmoplastic Infantile Ganglioglioma

25.1.5.2.1 General Features

Desmoplastic infantile ganglioglioma (DIG) is a rare WHO grade I supratentorial tumor seen most often in the first 2 years of life. There is a slight male predominance. Rare occurrences of adolescent and young adult presentation have been reported. It is an uncommon variant of ganglioglioma, and it is generally very large in size and appear partially cystic. Clinical presentation includes enlarging head circumference, lethargy, hemiparesis, seizures, and bulging fontanelle. Prognosis is good with long-term survival and recurrence-free intervals of up to 14 years. Surgery is the primary method of treatment. If gross total resection is not possible, close interval follow-up imaging is important.

25.1.5.2.2 Imaging Features

The classic appearance of a DIG is a very large supratentorial cystic tumor with a cortically based enhancing mural nodule (Figure 25.14). Its imaging features are very similar to PXA or ganglioglioma, only much larger. The frontal, parietal, and temporal lobes are most commonly affected. On CT, it appears as a well-defined very large isodense solid and hypodense cystic mass. Calcifications are not usually seen. On MRI, the solid portion appears cortically based, isointense on T_1WI and heterogeneous or isointense on T_2WI, and shows prominent contrast enhancement. An adjacent enhancing dural tail can also be seen, very similar to that of a PXA. The cystic component appears hypointense on T_1WI and hyperintense on T_2WI.

Imaging differential diagnostic considerations include ganglioglioma, PXA, high-grade astrocytoma, supratentorial primitive neuroectodermal tumor (PNET), supratentorial ependymoma, and PA.

25.1.5.3 Dysembryoplastic Neuroepithelial Tumors

25.1.5.3.1 General Features

DNETs are benign and slow-growing tumors of mixed glioneuronal origin carrying a WHO grade I classification. They are more commonly seen in individuals less than 20 years of age. They are cortically based tumors, most commonly from the temporal lobe, and are usually associated with cortical dysplasia. Patients may present with drug-resistant partial seizures. Although these tumors carry an excellent prognosis, treatment is often surgical in order to treat medically refractive partial seizures and to prevent tumor progression. Malignant transformation of DNETs is rare.

25.1.5.3.2 Imaging Features

On CT, these tumors typically appear as cortically based hypodense, nonenhancing masses with or without calcifications (Figure 25.15). They may cause remodeling of

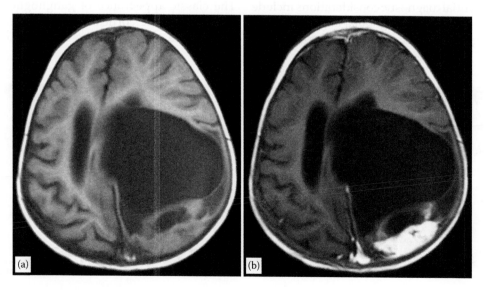

FIGURE 25.14
Desmoplastic infantile ganglioglioma. Axial T_1WI (a) and axial contrast-enhanced T_1WI (b) demonstrate a very large cystic mass with a cortically based enhancing mural nodule.

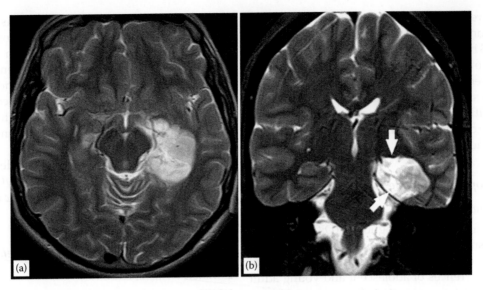

FIGURE 25.15
Dysembryoplastic neuroepithelial tumor. Coronal (a) and axial (b) T_2WI demonstrate a cortically based T_2 hyperintense mass of the left medial temporal lobe. Areas of *soap bubble* appearance from multicystic degeneration are seen (white arrows, b).

the overlying skull. Calcifications can be seen in up to one-third of cases. On MRI, these lesions are hypointense on T_1WI and hyperintense on T_2WI. There is no associated mass effect or edema seen unless hemorrhage occurs. Some of the DNET tumors may have multicystic degeneration, classically referred to as the *soap bubble* appearance, which is hyperintense on T_2WI. FLAIR imaging may demonstrate complete or incomplete hyperintense rings surrounding them. These lesions generally do not enhance, and the appearance of enhancement may raise the possibility of a more aggressive neoplasm.

Imaging differential diagnostic considerations include low-grade astrocytoma, ganglioglioma, oligodendroglioma, other low-grade tumors, and Taylor dysplasia.

25.1.5.4 Ganglioglioma

25.1.5.4.1 General Features

Ganglioglioma is a mixed neural/glial WHO grade I or II cortically based tumor. It is the most common neoplastic cause of chronic temporal lobe epilepsy. It occurs in children and young adults, and the peak incidence is between 10 and 30 years of age. Males are slightly more commonly affected. Excellent outcomes can be obtained with complete surgical resection. Malignant degeneration is rare.

25.1.5.4.2 Imaging Features

The classic appearance of ganglioglioma is that of a solid and cystic cortically based mass with calcification in the temporal lobe (Figure 25.16). However,

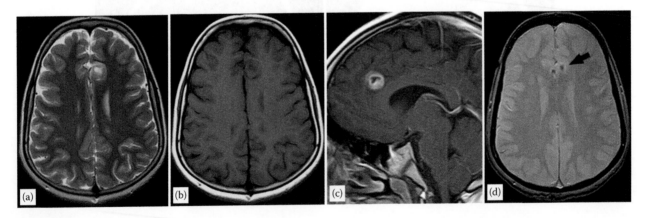

FIGURE 25.16
Ganglioglioma. Axial T_2WI (a), T_1WI (b), contrast-enhanced T_1WI (c) and T_2*W GRE (d) demonstrate a solid and cystic cortically based mass of the medial left frontal lobe. Contrast enhancement is seen (c) and blooming from calcification is present on the GRE sequence (black arrow, d).

other regions of the brain, optic nerve/chiasm, brain stem, and spinal cord can be involved. A solid appearance is also common, whereas a completely cystic mass is a rare presentation. Little mass effect or vasogenic edema is typical. Superficial lesions may remodel the skull. On CT, ganglioglioma typically has both a hypoattenuating cystic component and a mixed density solid portion. Calcifications are common. Variable and nonspecific appearances can be seen on MRI. The mass may appear hypo- to isointense on T_1WI compared to the cortex and may be accompanied by cortical dysplasia. On T_2WI/FLAIR, the solid portion is hyperintense compared to gray matter. Intratumoral calcifications may alter MRI signals and T_2*W GRE images may demonstrate *blooming* from calcification. Contrast enhancement is variable. Leptomeningeal spread is rare.

Imaging differential diagnostic considerations include PXA, DNET, PA, oligodendrobglioma, low-grade DA, metastatic disease, and infection.

25.1.5.5 Central Neurocytoma

25.1.5.5.1 General Features

CN is a rare WHO grade II intraventricular tumor of both neuronal and glial differentiation. It represents half of intraventricular tumors in the 20–40-year age range. Rarely, it may occur outside the ventricles in the brain parenchyma, cerebellum, and spinal cord. It is seen in both children and adults, but its mean age of presentation is 29 years. It may cause symptoms related to hydrocephalus. Treatment is usually complete surgical resection; however, CSF dissemination and recurrence can occur.

25.1.5.5.2 Imaging Features

The classical appearance of CN is that of a well-circumscribed, multicystic intraventricular mass in the lateral ventricle attached to the septum pellucidum (Figure 25.17). Less common intraventricular locations include the third and fourth ventricles. CT demonstrates a well-demarcated iso- to hyperdense lesion, reflecting cystic and solid portions, respectively. Associated findings include ventricular enlargement. Calcifications are present in more than half of cases. Hemorrhage is uncommon. On MRI, T_1WI shows a signal intensity similar to that of gray matter, although heterogeneous signal can still be observed. Following contrast administration, moderate to avid contrast enhancement is common; however, this is variable. On T_2WI, the signal is iso- to hyperintense to gray matter. Cystic components are usually hyperintense. The presence of flow voids has also been reported. On T_2*W images, intratumoral calcification produces *blooming* foci.

Imaging differential diagnostic considerations include ependymoma, SE, SEGA, choroid plexus tumor, metastasis, OG, and intraventricular meningioma.

25.1.6 Tumors of the Pineal Region

25.1.6.1 Overview

For tumors in the pineal region, neuroimaging findings cannot distinguish germ cell tumors (GCTs) from pineal parenchymal tumors, and laboratory and histological

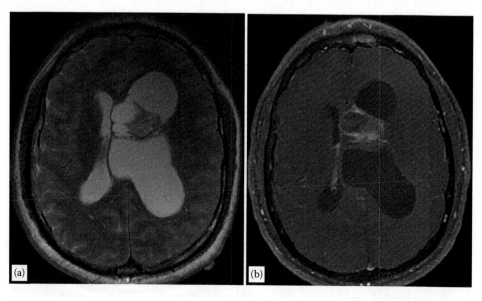

FIGURE 25.17
Central neurocytoma. Axial T_2WI (a) and axial fat-saturated contrast-enhanced T_1WI (b) demonstrate a multicystic intraventricular mass in the left lateral ventricle attached to the septum pellucidum. Heterogeneous contrast enhancement is noted.

findings are needed to establish a diagnosis. General imaging differential diagnostic considerations include GCTs (germinoma, etc.), pineocytoma, pineoblastoma (also known as pinealoblastoma [PB]), pineal parenchymal tumor of intermediate differentiation, astrocytoma, meningioma, and pineal cyst.

25.1.6.2 Pineocytoma

25.1.6.2.1 General Features

Pineocytomas are grade I pineal tumors according to the 2007 WHO classification. They account for 0.4%–1% of intracranial lesions and are more common in adults (mean age 38 years) than children. Very rarely, they can show aggressive behavior in the form of metastatic seeding throughout CSF spaces. No gender predilection is known. Clinical presentation includes signs and symptoms of raised intracranial pressure and Parinaud syndrome (paralysis of upward gaze, light-near dissociation of the pupils, convergence retraction nystagmus, and lid retraction). The outcome for pineocytomas is good with 86%–100% 5-year survival and there are no reports of relapses following gross total surgical resection.

25.1.6.2.2 Imaging Features

On CT, pineocytomas appear as round, well-defined masses that are iso- to hypodense and are often smaller 3 cm (Figure 25.18). If sufficiently large, they may cause obstruction to CSF flow and secondary hydrocephalus. Intratumoral calcifications may also be present, classically peripheral (*exploded*). This is because tumors arising from the pineal parenchyma result in peripheral displacement of the normal parenchymal calcification. MRI demonstrates iso/hypointensity on T_1WI and hyperintensity on T_2WI. Pineocytomas often show intense enhancement following contrast administration.

T_2*W GRE images show blooming due to calcifications. Cystic formation of pineocytomas can simulate the appearance of a pineal cyst; however, cystic-appearing pineocytomas will usually demonstrate nodular wall or internal contrast enhancement.

Imaging differential diagnostic considerations include germinomas (classically central *engulfed* calcifications—but not always seen), PB, pineal parenchymal tumor of intermediate differentiation, astrocytoma, meningioma, and pineal cyst.

25.1.6.3 Pineoblastomas

25.1.6.3.1 General Features

PB is a rare, aggressive WHO grade IV pineal parenchymal tumor. It is considered a PNET and PBs can have histologic features that resemble retinoblastoma. Some cases of inherited retinoblastoma (*trilateral retinoblastoma*) involve tumors of both the eyes and the pineal gland. Pineal parenchymal tumors account for 15% of pineal region tumors, and PBs represent 30%–45% of pineal parenchymal tumors. PB can occur in young adults, but it is more commonly a pediatric tumor that is seen in the first two decades of life. No gender predilection is known. Clinical presentation may include signs and symptoms related to raised intracranial pressure. Tumors in the pineal gland are also associated with Parinaud syndrome, which consists of the following: paralysis of upward gaze, light-near dissociation of the pupils, convergence retraction nystagmus, and lid retraction. However, pineal region astrocytomas commonly do not present with Parinaud syndrome. Imaging of the neuroaxis is recommended as these highly malignant tumors are prone to CSF seeding. Despite treatment with surgery, radiation, and chemotherapy options, these tumors show very poor survival.

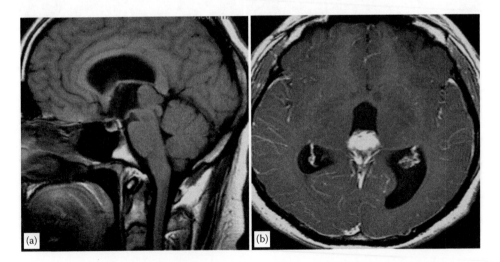

FIGURE 25.18
Pineocytoma. Sagittal T_1WI (a) and axial contrast-enhanced T_1WI (b) demonstrate an intensely contrast enhancing mass in the pineal region.

25.1.7.3.2 Imaging Features

The imaging findings of PB appear similar to other pineal region tumors, especially germinoma (Figure 25.19). On CT, the solid component of this tumor is hyperdense with calcifications. The classic descriptions of peripheral *exploded* calcifications of PB and more central *engulfed* calcifications of germinoma are not always seen. On both T_1WI and T_2WI MRI, PBs appear heterogeneously isointense and show heterogeneous enhancement. Areas of necrosis may be evident. Calcifications and hemorrhage may be seen on T_2*W GRE imaging and restricted diffusion is noted on DWI.

Imaging differential diagnostic considerations include GCTs, pineocytoma, pineal parenchymal tumor of intermediate differentiation, astrocytoma, meningioma, and pineal cyst.

25.1.7 Embryonal Tumors

25.1.7.1 Medulloblastoma

25.1.7.1.1 General Features

MB is a highly aggressive neoplasm that is the most frequent malignant childhood brain tumor and the most common posterior fossa primary tumor in children. Overall, it is the second most common pediatric brain tumor following astrocytoma. There is a bimodal peak of incidence between 3–4 and 8–9 years of age, younger than that is seen in PA. It is twice as common in boys than in girls. Its occurrence in adults is uncommon and usually appears between 20 and 40 years of age. It is a highly cellular, WHO grade IV PNET. The classic histological subtype is most often seen in children. In adults, the desmoplastic subtype is common. The *MB with extensive nodularity and advanced neuronal differentiation* is mostly seen in children under the age of 3 and is associated with *grapelike* nodularity on imaging. The worst prognosis is seen in the large cell/anaplastic subtype. Prognosis is generally poor with 50%–80% survival at 5 years. Clinical presentation includes signs of raised intracranial pressure and ataxia, spasticity, and sixth nerve palsy. The clinical symptomatology is typically brief, reflective of the aggressive biology of MB. Treatment is surgical resection with chemotherapy and radiation therapy. Recurrence is common following surgical resection and there is an overall poor prognosis.

25.1.7.1.2 Imaging Features

MB most frequently arises from the midline cerebellar vermis in the region of the roof of the fourth ventricle (Figure 25.20). However, a lateral cerebellar hemispheric location is more frequently seen in older children and adults. MB can have a variable appearance, but it typically appears as a round, lobulated midline mass. Involvement of the fourth ventricle, other parts of the brain, and the spinal cord can also occur. Extension through the foramina of Magendie and Luschka, commonly seen in ependymoma, is less common. Imaging evaluation of the entire neuraxis is recommended because of its propensity to have early subarachnoid spread of tumor. Within the spine, these drop metastases have been described as *sugar coating*. On CT, the increased cellularity of MB results in hyperdensity, distinguishing it from PA. Calcifications can be seen in 20% of cases and nonenhancing, cystic or necrotic regions are present in about 50%. Hemorrhage is uncommon. On MRI, MB can have variable appearances. It is iso- to hypointense on T_1WI and iso- to hypointense on T_2WI, often lower in signal than PA. The lower signal intensity on T_2WI is thought to be secondary to high cellularity, which also results in decreased diffusion on DWI. Postcontrast T_1WI demonstrates enhancement of the mass in the vast majority of cases as well as possible subarachnoid spread of disease in the brain and spine. Elevation of taurine has been reported on short T_e MRS.

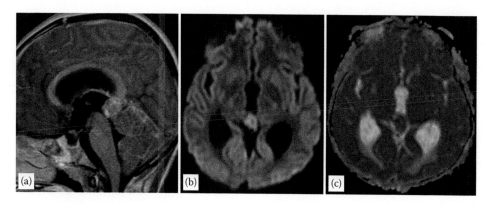

FIGURE 25.19

Pineoblastoma. Sagittal contrast-enhanced T_1WI (a), axial DWI (b), and ADC map (c) demonstrate a heterogeneous enhancing mass with diffusion restriction in the pineal region.

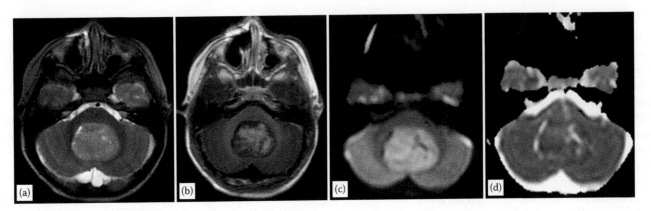

FIGURE 25.20

Medulloblastoma. Axial T_2WI (a), contrast-enhanced T_1WI (b), DWI (c), and ADC map (d) demonstrate a fourth ventricular mass that shows iso- to hypointense T_2 signal (a) and heterogeneous contrast enhancement (b). Diffusion restriction is evident in (c) and (d).

Imaging differential diagnostic considerations of include ependymoma, PA, ATRT, BSG, HGBL, metastasis, lymphoma, and DCG (Lhermitte–Duclos disease).

25.1.7.2 Metastatic Neuroblastoma

25.1.7.2.1 General Features

Neuroblastoma (NB) is a malignant embryonal tumor derived from primordial neural crest cell derivatives found in the adrenal glands or along the sympathetic chain. It is the most common solid pediatric extracranial tumor in children, with widespread metastases present in more than 50% of patients at the time of diagnosis. It is the most common malignant tumor to metastasize to the skull in children. Metastatic spread to the bones and liver are also common and carry a poor prognosis. Calvarial involvement indicates stage IV disease. Metastatic disease involving the CNS is rare, but improved survival from better therapies may underlie an increasing incidence of CNS parenchymal and leptomeningeal disease. CNS metastatic disease is most often detected at the time of recurrence and carries a dismal prognosis, particularly with leptomeningeal disease. Classically, involvement of the bony orbit manifests as *raccoon eyes*. Symptoms of CNS metastases are location dependent and can be clinically occult. Headache, nausea, and emesis are associated with parenchymal involvement; pain, fever, and motor deficits are associated with meningeal involvement. Treatment options include surgical resection, bone marrow transplant, chemotherapy, and radiation therapy.

25.1.7.2.2 Imaging Features

Craniocerebral metastatic disease is most commonly extradural and calvarial-based around the orbit and sphenoid wings and the skull base (Figure 25.21). Metastatic calvarial involvement can be seen on CT as thickened bone, *hair-on-end* periosteal changes, lytic lesions, and sutural separation. Dural deposits can be seen as thick and irregular nodular enhancement on contrast-enhanced T_1WI. It often occurs with bony involvement and can contain hemorrhage.

CNS intra-axial lesions may occur anywhere in the parenchyma, ventricles, or spinal cord. Neuroparenchymal

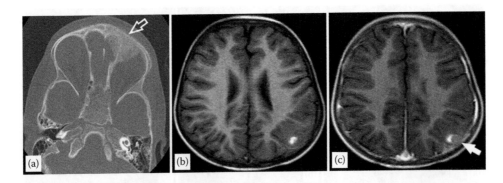

FIGURE 25.21

Metastatic neuroblastoma. Axial CT (a) demonstrates *hair-on-end* periosteal changes of the bone (white hollow arrow, a) of the left fronto-orbital region. Axial T_1WI (b) and contrast-enhanced T_1WI (c) demonstrate a hemorrhagic lesion as demonstrated by T_1 hyperintensity in the left parietal lobe with a posterolateral focus of mild contrast enhancement (white arrow, c).

metastases are more commonly found in supratentorial regions and may appear as cystic or hemorrhagic lesions. On CT, they present as cystic lesions with dense mural nodules with or without calcific foci or as intraparenchymal hemorrhagic lesions. Contrast-enhanced CT shows peripheral enhancement and intense enhancement of mural nodules. On MRI, T_1WI reveals heterogenous signal intensity with hyperintense foci likely secondary to intratumoral hemorrhage, whereas T_2WI and FLAIR are often hyperintense with minimal surrounding edema. Susceptibility-weighted imaging (SWI) may show hypointense signal intensity if hemorrhage is present. Leptomeningeal involvement can be seen as nodular leptomeningeal enhancement on contrast-enhanced T_1WI as well as FLAIR hyperintensity in the subarachnoid spaces. Nuclear medicine techniques such as meta-iodobenzylguanidine and Tc-99m-methylene diphosphonate (MDP) bone scans can be helpful to further evaluate disease status.

Imaging differential diagnostic considerations include the following: for skull and dural involvement—leukemia, lymphoma, sarcoma, metastases, Langerhans cell histiocytosis, and subdural/epidural hematoma; for brain parenchymal involvement—leukemia, primary brain tumors such as astrocytoma, OG, ependymoma, and PNET; and for leptomeningeal involvement—meningitis and leukemia.

25.1.7.3 Atypical Teratoid Rhabdoid Tumor

25.1.7.3.1 General Features

ATRT is a rare, aggressive WHO grade IV pediatric brain tumor. It is most often seen in the posterior fossa involving the cerebellopontine angle, cerebellum, and brain stem. Supratentoroial brain involvement is also a common location, with much less frequent manifestations involving the pineal and spinal regions as well as multifocal disease. A wide age range has been reported in the literature; however, it most commonly presents in children under the age of 2 years. No gender predilection is known. Common signs and symptoms relate to raised

intracranial pressure, whereas cerebellar involvement can present as gait disturbance and ataxia. Prognosis is very poor with most affected children dying within months of diagnosis. Optimal therapy remains undefined and options include surgery, radiation, and chemotherapy.

25.1.7.3.2 Imaging Features

The imaging and histologic features of ATRT overlap considerably with MB, and it is thought that many cases of ATRT have been misdiagnosed as MB (Figure 25.22). However, ATRT is thought to be relatively more heterogenous appearing than MB because of areas of hemorrhage, cysts, and necrosis. Also, it tends to be more off-midline as opposed to the more central location MB. On CT, ATRT appears as a heterogeneous hyperdense mass. Calcifications can be seen. On MRI, hypointensity and heterogeneous signal can be present in the solid portions on T_1WI, whereas T_2WI exhibits heterogeneous hypo- to hyperintense regions. Variable surrounding edema may be seen. Contrast enhancement is usually avid and heterogeneous, but mild or no enhancement can also be seen. Restricted diffusion is present on DWI. CSF dissemination is common even at presentation.

Imaging differential diagnostic considerations of include MB/PNET, ependymoma, PA, teratoma, GCT, DIG, and CPCA.

25.2 Tumors of the Meninges

25.2.1 Tumors of Meningothelial Cells

25.2.1.1 Meningioma

25.2.1.1.1 General Features

Meningioma is the most common primary intracranial neoplasm in adults. It is also the most common extra-axial tumor. It is generally a benign WHO grade I tumor;

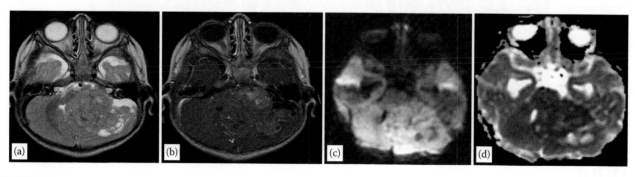

FIGURE 25.22
Atypical teratoid rhabdoid tumor. Axial T_2WI (a), contrast-enhanced T_1WI (b), DWI (c), and ADC map (d) demonstrate a heterogeneously enhancing large cerebellar mass with diffusion restriction.

however, more aggressive atypical WHO grade II and anaplastic WHO grade III tumors are less common variants. Its peak incidence is the fifth and sixth decades and it is more common in females. Meningiomas are dural-based tumors that arise from arachnoid cap cells. Ninety percent of tumors occur in the supratentorial compartment (25% parasagittal, 20% convexity, 40% anterior skull base). The cerebellopontine angle, foramen magnum, clivus, cerebral ventricles, and pineal region are less common locations. Ten percent of meningiomas arise in the posterior fossa or spinal canal. A wide variety of histologic subtypes exist, including meningothelial, fibroblastic, and transitional subtypes. The clinical presentation depends on the tumor location, but fewer than 10% of meningiomas result in clinical symptoms. Grade I meningiomas are generally slow growing and are amenable to surgical resection with a recurrence rate of 9%.

25.2.1.1.2 Imaging Features

The general appearance of a meningioma is that of an extra-axial dural-based mass that results in CSF clefts and underlying cortical buckling (Figure 25.23). Multiple tumors are not uncommon. On CT, meningiomas most often appear hyperdense, with a smaller fraction appearing isodense and rarely hypodense. Calcifications are present in up to 20% of cases. Bony changes, including hyperostosis and bone destruction, can be seen. On MRI, meningiomas appear classically isointense to gray matter on both T_1WI and T_2WI. Lower signal intensities can be seen on T_2WI if there are significant amounts of fibrous changes and calcification. Areas of signal heterogeneity can also result from necrosis, cystic degeneration, pseudocysts, and bony metaplasia. Peritumoral edema can be seen in the adjacent brain parenchyma. An adjacent enhancing *dural tail* is a characteristic, though not pathognomonic, feature of meningiomas. Variable appearances can be seen on

DWI. MRS can show evidence of alanine (1.3–1.5 ppm). Intense, homogeneous enhancement is usually present. Diffusion restriction, marked vasogenic edema, local invasion, and more necrosis are imaging features that have been noted in the more aggressive grade II and III meningioma variants.

Imaging differential diagnostic considerations include atypical/malignant meningioma, granulomatous disease (tuberculosis/sarcoidosis), lymphoma, hemangiopericytoma, and sarcoma.

25.2.2 Other Neoplasms Related to the Meninges

25.2.2.1 Hemangioblastoma

25.2.2.1.1 General Features

HGBL is a highly vascular, benign tumor of the CNS. It comprises approximately 2% of all intracranial tumors. It is also the most common *primary* tumor of the cerebellum in adults. It is more common among men and occurs sporadically around the fifth or sixth decade of life. Nonetheless, when associated with the von Hippel–Lindau syndrome, it tends to present in younger adults between the third and fourth decades. HGBL is classified as a WHO grade I tumor, characterized by a rich blood supply that is usually derived from the intracranial vertebrobasilar circulation. Symptoms include headache, disequilibrium, and mental confusion. Surgical resection of the tumor can result in good long-term survival.

25.2.2.1.2 Imaging Features

HGBL most commonly manifests as a cystic mass with a mural nodule, but can appear as a solid mass or as a purely cystic or solid mass with internal cysts (Figure 25.24). It is most commonly located in the cerebellar hemisphere, but it may also occur in the spine and in other places in the CNS. On CT, the cystic portion appears hypodense.

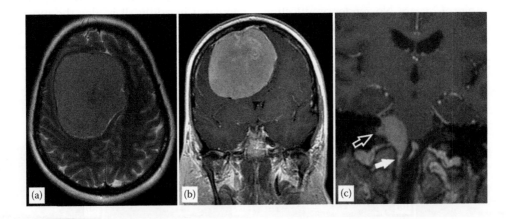

FIGURE 25.23

Meningioma. Axial T_2WI (a) and coronal contrast-enhanced T_1WI (b) show an interhemispheric meningioma which demonstrates CSF cleft and underlying cortical buckling (a). Intense, homogeneous enhancement is noted (b). Coronal contrast-enhanced T_1WI (c) shows a cerebellopontine angle meningioma extending into the internal auditory canal (white hollow arrow). A dural tail is also evident (white arrow).

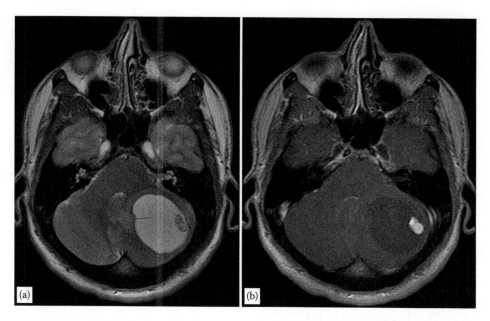

FIGURE 25.24

Hemangioblastoma. Axial T_2WI (a) and axial contrast-enhanced T_1WI (b) demonstrate cystic cerebellar mass with an enhancing mural nodule.

The mural nodule tends to be isodense and shows avid homogeneous enhancement after contrast injection. When a solid mass is present, it typically exhibits iso- or hyperdensity and enhances homogeneously. Associated findings of hydrocephalus and edema may also be seen. MRI is the best imaging modality for the characterization of these lesions. On T_1WI, the cystic component tends to be slightly hyperintense on T_1WI and more hyperintense on T_2WI due to elevated protein content. The solid portion of the tumor enhances, whereas the cystic component usually does not unless its wall is lined by tumor. Blood products may be evident as blooming on T_2* GRE sequences. Flow voids are sometimes seen as well. Contrast-enhanced MRI of the spine should also be obtained to search for spinal HGBLs.

Imaging differential diagnostic considerations include metastasis (the most common intra-axial posterior fossa tumor in adults), astrocytoma, ependymoma, and cavernous malformation.

25.3 Lymphomas and Hematopoietic Neoplasms

25.3.1 Primary CNS Lymphoma

25.3.1.1 General Features

Primary CNS lymphoma (PCNSL) is a rare and malignant form of non-Hodgkin lymphoma that may affect the brain, spinal cord, eyes, and leptomeninges. It constitutes around 3% of all primary brain tumors and is often associated with immunocompromised states, such as acquired immune deficiency syndrome. During the past few decades, an increased incidence of PCNSL among immunocompetent patients has also been noted. The mean age of presentation of PCNSL is 53–57 years old for immunocompetent patients, with a slight male predominance. Immunocompromised patients tend to present at earlier ages. Lesions are solitary in about 65% and the vast majority are diffuse large B-cell lymphomas with a small portion being composed of Burkitt, lymphoblastic, marginal zone, and T-cell lymphoma types. Associated symptoms can include headache, fatigue, mental confusion, focal neurologic deficits, and seizures. PCNSLs are aggressive tumors that can penetrate through the subependymal tissues and CSF to the meninges. Prognosis is poor and treatment typically involves chemotherapy, radiation therapy, and corticosteroids.

25.3.1.2 Imaging Features

The supratentorial compartment along the periventricular and superficial regions of the brain is most often involved (Figure 25.25). Corpus callosum involvement is common. However, the imaging features of PCNSL are often difficult to discern from other intracranial processes such as toxoplasmosis, malignant glioma, bacterial abscess, metastases, and multiple sclerosis. In some cases, enhancement along the Virchow–Robin spaces is seen. PCNSL appears iso- to hyperdense on non-contrast-enhanced CT (NECT) and enhances on postcontrast imaging. On MRI, PCNSL appears as a hypo- to isointense lesion on T_1WI that typically enhances homogeneously

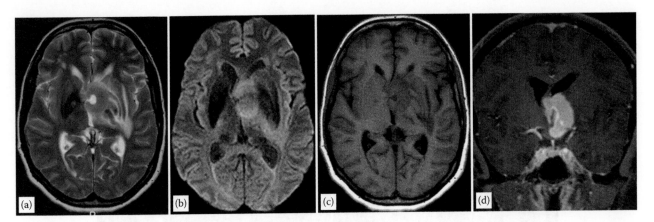

FIGURE 25.25
Primary CNS lymphoma. Axial T_2WI (a), axial DWI (b), axial T_1WI (c), and coronal contrast-enhanced T_1WI (d) demonstrate a left periventricular enhancing mass with diffusion restriction.

in immunocompetent patients. On T_2WI, the signal intensity can vary, but it is often hypointense to gray matter. Rarely, it may show isolated high T_2W signal intensity in white matter without contrast enhancement. In immunocompromised patients, PCNSL may also present as a necrotic, rim-enhancing rather than a homogeneously enhancing mass. A rare variant of PCNSL can present as a dural-based mass that can appear similar to a meningioma. PCNSL often demonstrates reduced diffusion on DWI due to its hypercellularity. On proton MRS, PCNSL generally shows higher lipid peaks and Cho/Cr ratios. On SWI, PCNSL generally does not cause microhemorrages, a fact that can be helpful in the differentiation from high-grade gliomas. F-18 FDG PET may differentiate lymphoma from infectious/inflammatory lesions as well as monitor treatment response. Other nuclear medicine techniques such as Tl-201 and Ga-67 SPECT may also be helpful for diagnosis.

Imaging differential diagnostic considerations include toxoplasmosis, GBM, bacterial abscess, metastases, progressive multifocal leukoencephalopathy, and multiple sclerosis.

25.3.2 Intravascular (Angiocentric) Lymphoma

25.3.2.1 General Features

Intravascular (angiocentric) lymphoma (IVL) is a very rare form of non-Hodgkin lymphoma, constituting less than 1% of all lymphomas. It is characterized by proliferation of B lymphocytes exclusively within blood vessel lumina without affecting lymphoid tissue or blood. The median age of presentation is 70 years of age and it does not show gender bias. IVL can occur in the small vessels of nearly any organ, but most commonly those supplying the CNS and skin. Symptoms include nonspecific neurologic and cognitive symptoms as well as seizures. The average survival time is typically under a year.

25.3.2.2 Imaging Features

In the brain, IVL is commonly located in the supratentorial region in the deep white matter (Figure 25.26). On CT, it generally appears as focal, bilateral, low-density lesions in the white matter, cortex, or brain stem. MRI

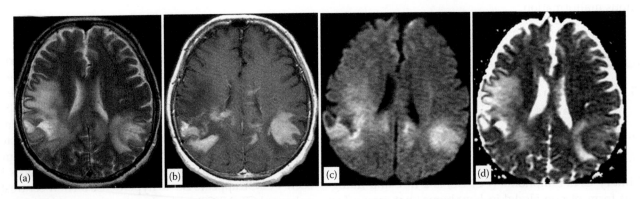

FIGURE 25.26
Intravascular (angiocentric) lymphoma. Axial T_2WI (a), axial contrast-enhanced T_1WI (b), axial DWI (c), and axial ADC map (d) demonstrate enhancing bilateral T_2 hyperintense lesions that demonstrate areas of diffusion restriction.

shows multifocal abnormalities resulting from vessel occlusion, ranging from nonspecific white matter changes to infarct-like abnormalities. IVLs appear as multifocal hypointense on T_1WI and hyperintense on T_2W/FLAIR imaging lesions; blood products may be seen on both on T_2*W GRE images. Diffusion restriction can also be seen on DWI, as well as variable enhancement of the lesion. Enhancing intra-axial mass lesions as well as leptomeningeal enhancement can also be seen.

Imaging differential diagnostic considerations include PCNSL, neurosarcoidosis, vasculitis, and vascular dementia.

25.4 Germ Cell Tumors

25.4.1 Overview

GCTs are a group of tumors that develop from primordial ectoderm, mesoderm, or endoderm. They are the most common tumors of the pineal region, accounting for more than 50% of pineal region tumors. They are classified by the WHO into two types: germinomas and nongerminomatous GCTs (NGGCTs) (teratomas, embryonal carcinoma, yolk sac tumor, choriocarcinoma, and mixed GCTs). Other than the pineal region, intracranial GCTs are also found in the suprasellar region. GCTs can be hormonally active, and increased serum and CSF levels of different oncoproteins can be detected (ion, intracranial human chorionic gonadotropin (hCG), and placental alkaline phosphatase). A higher prevalence of GCTs is seen in Asian countries compared to Western nations. Germinomas are the most common GCT and teratomas are the second most common. The age of presentation is between 10 and 30 years of age, and they occur more often in males. In general, germinomas have a much more favorable prognosis compared to NGGCTs. It is difficult to differentiate pineal parenchymal tumors from GCTs using imaging findings alone, and laboratory findings are needed to determine the diagnosis.

25.4.2 Germinoma

25.4.2.1 General Features

Germinomas typically involve midline structures such as the pineal (50%–65%) and suprasellar regions (25%–35%). Off-midline localization occurs 5% to 10% of the time, with involvement of the basal ganglia, thalamus, and internal capsule. Pineal region germinomas occur 10 times more commonly in males, whereas those in the suprasellar region show no gender predilection. Pure germinomas are considered WHO grade II, whereas the syncytiotrophoblastic germina is considered WHO grade II–III. Symptoms depend on the size and location of the tumor. Suprasellar germinomas may present with hypothalamic–pituitary dysfunction, most commonly diabetes insipidus, whereas pineal gland involvement may manifest as Parinaud syndrome and headache from hydrocephalus. Due to their sensitivity to chemoradiotherapy, germinomas have a favorable prognosis with a 5 year-survival rate estimated to be greater than 90%.

25.4.2.2 Imaging Features

On CT, germinomas are classically well-circumscribed and iso- to hyperdense masses with *engulfed* calcifications (as opposed to *exploded* peripheral calcifications classically seen in pineal parenchymal tumors (Figure 25.27). Hydrocephalus may be present. Suprasellar involvement is typically retrochiasmatic in location, and no cystic degeneration or calcifications are seen. On MRI, germinomas appear isointense or hyperintense on T_1WI and hyperintense on T_2WI. They enhance homogeneously with gadolinium, though heterogenous enhancement can be seen if cysts are present. T_2*W GRE images can detect the presence of calcification as areas of blooming. Cystic regions and hemorrhage are more common in large rapidly growing germinomas, as found in the thalamus and basal ganglia. When these regions are involved, the tumors tend to be more infiltrative and may invade the internal capsule, causing cerebral hemiatrophy. On DWI, germinomas are associated with low ADC values due to the high cellularity of the tumor. CSF dissemination is common, and thus, contrast-enhanced MRI of the entire neuraxis should be considered.

Imaging differential diagnostic considerations include the following: neuroimaging findings cannot distinguish GCTs from pineal parenchymal tumors, and laboratory and histological findings are needed to help establish a diagnosis. Other tumors/lesions to consider are other GCTs, pineocytoma, PB, pineal parenchymal tumor of intermediate differentiation, astrocytoma, meningioma, pineal cyst, metastases, Langerhans cell histiocytosis, sarcoid, and craniopharyngioma (CP).

25.4.3 Teratoma and Other GCTs

25.4.3.1 General Features

Though teratomas are rare intracranial tumors, they are also the most common type of congenital brain tumor as well as the second most common pineal tumor in children. Their size is variable and they can be massive, appearing as holocranial tumors in the newborn. Intracranial teratomas are most often midline and supratentorial. They can also arise in the suprasellar

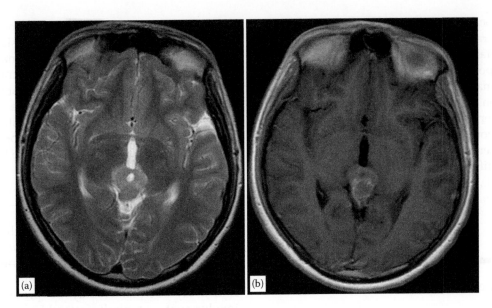

FIGURE 25.27
Germinoma. Axial T_2WI (a) and contrast-enhanced T_1WI (b) demonstrate a heterogeneously enhancing mass in the pineal region with a small central cystic area.

region and the cerebral hemispheres. However, it may be impossible with imaging or even autopsy to determine the exact point of origin when they are very large tumors. Compared to the intracranial compartment, extracranial teratomas are much more common and are seen most often in the sacrococcygeal region, followed by the head and neck, chest, and retroperitoneum. Three general types of teratomas are recognized: mature, immature, and teratoma with malignant transformation. They are derived from totipotent cells and are therefore characterized by a heterogenous composition

of cells derived from all three germinal layers (ectoderm, mesoderm, and endoderm).

25.4.3.2 Imaging Features

Teratomas have a heterogenous appearance on imaging due to the variable contents within them (Figure 25.28). They are generally midline masses that contain fat, calcifications, soft tissue, and cysts. On prenatal ultrasound, they can appear as a multiloculated solid–cystic lesion with internal shadowing from

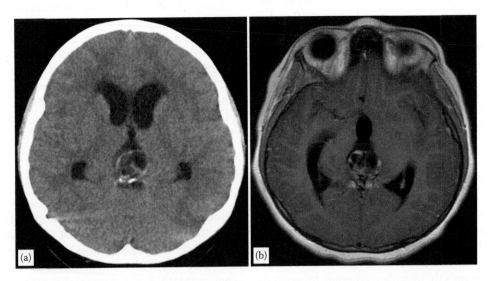

FIGURE 25.28
Teratoma. Axial CT (a) and axial contrast-enhanced T_1WI demonstrate a partially calcified, solid–cystic lesion (a) that demonstrates heterogeneous contrast enhancement (b).

calcifications. CT may show fatty elements, areas of contrast enhancement, intratumoral hyperdense regions consistent with calcification or hemorrhage, and areas of cystic attenuation. On MRI, the appearance of teratomas also appears heterogeneous. They are frequently hyperintense on T_1WI due to fatty components with calcifications resulting in heterogeneous signal intensity. On T_2WI, the soft tissue portions can appear iso- to hyperintense. Multiple cystic components may manifest with a *honeycomb-like appearance*. Contrast enhancement is heterogeneous and MRS can demonstrate the presence of lipids.

Other GCTs such as embryonal carcinoma, yolk sac tumor, and choriocarcinoma are much more rare. Their imaging appearance can appear similar to other pineal region GCTs or pineal parenchymal tumors. Correlation with the serum and CSF oncoproteins such as such asimaging appea, and placental alkaline phosphatase can be helpful to improve diagnostic accuracy.

25.5 Tumors of the Sellar Region

25.5.1 Pituitary Adenomas

25.5.1.1 General Features

Pituitary adenomas are the most common lesions found in the region of the sella turcica and represent 10%–15% of all intracranial tumors. These tumors are classified by both their size and their ability to secrete hormones. Pituitary adenomas smaller than 10 mm are termed *microadenomas*, whereas their larger counterparts are referred to as *macroadenomas*. Functional adenomas make up to 75% of pituitary adenomas and may secrete any of the hormones produced by the pituitary, with the most common being the prolactinoma, accounting for up to 30% of cases. Prolactinomas are diagnosed earlier in women because their symptoms of amenorrhea, infertility, and galactorrhea are more clearly recognized, whereas the symptoms in men, namely, impotence and libido, are less clinically recognizable. In the case of nonfunctioning adenomas, the lesion may be clinically silent until reaching a large enough size to cause symptoms related to mass effect. Patients with functional adenomas typically present with symptoms of hormone excess, depending on the cell-type origin. Treatment for microadenomas ranges from conservative (for *incidentalomas*) to surgical and medical management. For macroadenomas, treatment options include surgical resection, radiation therapy, and medical management.

25.5.1.2 Imaging Features

Pituitary adenomas are best evaluated with contrast-enhanced MRI using a dedicated sella protocol of thin-section coronal and sagittal images (Figures 25.29 and 25.30). Microadenomas can be of variable signal intensity on MRI, typically iso/hypointense on T_1WI and iso/hyperintense on T_2WI. Contrast-enhanced MRI demonstrates enhancement, but it tends to be slower compared to the normal pituitary gland. Up to 30% of microadenomas may be missed with traditional MRI. Detection may be improved with the use of dynamic contrast-enhanced MRI scans, which demonstrate a slower rate of contrast enhancement in the microadenoma. Macroadenomas are typically easier to identify on imaging due to their extensive growth and extension outside of the sella, including invasion of the cavernous sinuses, clivus, and sphenoid sinus. These tumors have a propensity to extend superiorly to the suprasellar region, where they assume an *hour-glass* appearance caused by compression of the tumor at the region of the diaphragm sellae. Macroadenomas typically appear isointense to gray matter on T_1WI and T_2WI on MRI, and are usually strongly enhancing on postcontrast images. Heterogeneous areas of signal intensity can be seen in the presence of hemorrhage or necrosis within macroadenomas.

Imaging differential diagnostic considerations include the following: for microadenoma—Rathke cleft cyst (RCC), and pituitary hyperplasia; and for macroadenoma—meningioma, metastases, lymphocytic

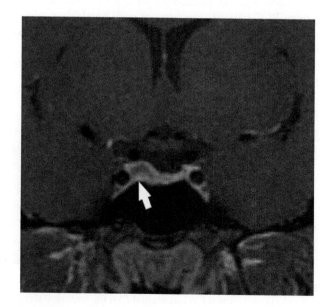

FIGURE 25.29

Microadenoma. Coronal contrast-enhanced MRI demonstrates a small hypoenhancing mass (white arrow) in the right posterior pituitary gland.

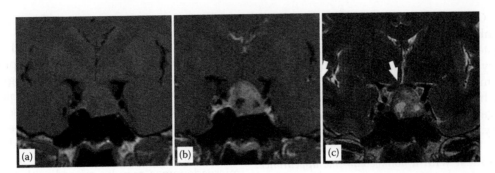

FIGURE 25.30
Pituitary macroadenoma. Coronal T_1WI (a), coronal contrast-enhanced T_1WI (b), and coronal T_2WI demonstrates a large sellar/suprasellar mass with central areas of necrosis and mild mass effect on the optic chiasm (white arrow, c).

hypophysitis, RCC, pituitary hyperplasia, CP, and pituicytoma, aneurysm.

25.5.2 Rathke Cleft Cyst

25.5.2.1 General Features

RCCs are non-neoplastic lesions resulting from remnants of embryonic ectoderm known as Rathke's pouch. There is a histologic continuum between RCC and CP. RCCs are the most common incidental sellar lesion followed by pituitary adenomas. They have been reported in up to 22% of normal pituitary glands during routine autopsies. The majority are asymptomatic and are often discovered incidentally on imaging. Asymptomatic RCCs are conservatively managed, whereas symptomatic RCCs require surgery. Symptoms occur when the cyst reaches a large enough size to cause compression of surrounding structures. Recurrence or persistence of RCCs is common after surgery. RCC patients can present with headaches, visual disturbances, and pituitary hormone deficits. These lesions have been described in all age groups, but have a peak frequency between the fourth and sixth decades of life, with a possible female predominance.

25.5.2.2 Imaging Features

RCCs are typically intrasellar, with or without suprasellar extension, and arise in the pars intermedia between the anterior and posterior pituitary lobes (Figure 25.31). On imaging, RCCs are well-circumscribed masses with smooth edges that compress adjacent pituitary. On CT, they typically appear as homogenously nonenhancing hypodense or isodense masses. The appearance of RCCs on MRI is dependent on the fluid contents of the cyst where hypointensity on T_1WI imaging is suggestive of serous fluid, whereas hyperintensity suggests mucinous content. The appearance on T_2WI varies from hypo- to hyperintensity. Intracystic nodules, which appear hyperintense on T_1WI and hypointense on T_2WI, corresponding to proteinaceous material, are seen in up to 77% of cases. Increased signal on both T_1WI and T_2WI suggests the presence of blood products. RCCs rarely enhance after the administration of contrast agent, and enhancing RCCs demonstrate ringlike enhancement of only the cyst wall. This is seen in cases of aggressive RCCs with inflammation or squamous metaplasia or recurrent RCCs. Presence of an enhancing nodule associated with a cystic lesion is suggestive for the diagnosis of CP.

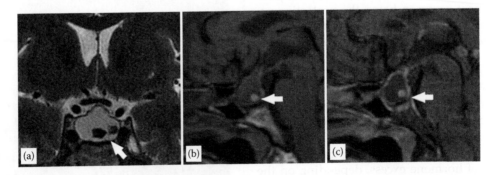

FIGURE 25.31
Rathke cleft cyst. Coronal T_2WI (a), sagittal T_1WI (b), and sagittal contrast-enhanced T_1WI demonstrate a sellar/suprasellar mass coronal cystic-appearing mass. Characteristic T_1 hyperintense/T_2 hypointense nodules are seen within the mass (white arrow). Mild enhancement of the cyst wall is seen in (c).

25.5.3 Hypothalamic Hamartoma

25.5.3.1 General Features

Hypothalamic hamartoma (HH) is a benign, non-neoplastic, congenital malformation found in the hypothalamus made up of heterotopic nodules resembling normal gray matter. HHs arise from the tuber cinereum, mammillary bodies, or floor of the third ventricle. Patients with HHs usually present within the first 2–3 years of life with symptoms of neurologic and/or endocrine dysfunction. Neurologic manifestations of the disease are most commonly seizures of the gelastic type characterized by episodes of laughing or crying. However, many types of seizures have been reported in patients with HH, and patients often develop debilitating epilepsy. Other neurological manifestations of HH include cognitive, behavioral, and psychiatric decline and dysfunction. HH is also strongly associated with the development of central precocious puberty, usually before the age of 2. The morphology of these masses, which may be pedunculated or sessile, is reported to be predictive of clinical symptomatology. Sessile lesions are more likely to exhibit seizures, whereas the pedunculated types are more commonly present with precocious puberty. There is no gender or racial preponderance in the development of HH.

25.5.3.2 Imaging Features

On CT, HH appears as a nonenhancing mass in the region of the interpeduncular and suprasellar cisterns that is isodense to brain parenchyma (Figure 25.32). The mass may cause obliteration of the anterior third ventricle or suprasellar cistern. MRI imaging of HH demonstrates a solid, nonenhancing mass arising from the region of the tuber cinereum that is iso- or slightly hypointense to gray matter on T_1WI and may be isointense or hyperintense on T_2WI. HHs do not have significant growth potential, so serial imaging should show consistency in size. The presence of enhancement, calcifications, or interval growth of a lesion should exclude HH from the differential diagnosis, although in rare cases, calcifications and cystic components can be found in very large HHs.

Imaging differential diagnostic considerations include astrocytoma, CP, germinoma, and Langerhans cell histiocytosis.

25.5.4 Craniopharyngioma

25.5.4.1 General Features

CP is a congenital benign sellar/parasellar tumor derived from remnants of squamous epithelium of Rathke's Pouch. CPs are WHO grade I tumors. Most involve the suprasellar region with other cases involving the intrasellar as well as other intracranial regions. CP exhibits a bimodal age distribution with a larger peak in childhood (adamantinomatous histological subtype) and a second smaller peak in the sixth and seventh decades (papillary histological subtype). There is no

Imaging differential diagnostic considerations include pituitary adenoma, CP, arachoid cyst, and epidermoid cyst.

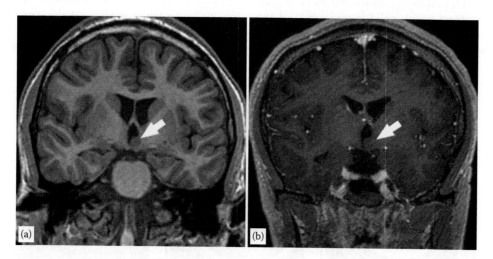

FIGURE 25.32
Hypothalamic hamartoma. Coronal T_1WI (a) and contrast-enhnanced T_1WI (b) demonstrate a hypothalamic mass (white arrows) that is isointense to gray matter and demonstrates no contrast enhancement.

gender predisposition for these lesions. CPs account for 10% of all intracranial pediatric tumors. They are also the most common suprasellar tumor as well as the most common nonglial primary intracranial tumors in children. Despite their designation as a benign lesion, CPs have a propensity to recur following surgical intervention and may also exhibit local invasion and adherence to surrounding structures such as the hypothalamus, optic chiasm, and third ventricle. Symptomology of CPs in childhood include growth disturbance, symptoms of raised intracranial pressure, and visual deficits, whereas those in adults often include endocrine abnormalities. Treatment options include surgery and radiation therapy.

25.5.4.2 Imaging Features

CPs typically display a combination of cystic and solid components, which are visible as regions of hypoattenuation and hyperattenuation on CT, respectively (Figure 25.33). The adamantinomatous subtype of CP tends to calcify, demonstrate contrast enhancement, and have both cystic and solid components, whereas the papillary subtype of CPs tends to be more solid and infrequently calcifies. CT is also particularly sensitive to the presence of intratumoral calcifications, which are found in 60%–93% of CPs, within solid areas or cyst walls. The presence of calcifications strongly favors the diagnosis of CP over RCC. MRI of CPs usually demonstrates a well-demarcated and lobular lesion in the suprasellar region between 2 and 4 cm in size. Larger tumors that invade multiple intracranial regions can occasionally be seen. Solid components are of variable intensity on T_1WI and T_2WI. Cystic parts are usually hyperintense on both T_1WI and T_2WI. Following administration of gadolinium-based contrast agent, solid components of the tumor usually enhance homogeneously, whereas cystic parts exhibit rim enhancement of the cyst wall.

Imaging differential diagnostic considerations include RCC, pituitary adenoma, astrocytoma, arachnoid/epidermoid cyst, and aneurysm.

25.6 Metastatic Tumors

25.6.1 Metastases

25.6.1.1 General Features

Brain metastases are the most common intracranial tumors in adults, outnumbering primary brain tumors and occurring in approximately 15%–40% of all cancer patients. Following treatment, they can recur at the original location in the brain or metastasize to other parts of the brain. In adults, brain metastases commonly originate from lung cancer, breast cancer, melanoma, and colon cancer, and present as single or multiple lesions. They cause severe neurological complications that present progressively or abruptly depending on the type and location of the metastasis. Signs and symptoms include seizures, headaches, hemiparesis, aphasia, and visual field defects. Prognosis of a patient with brain metastases is generally poor. Treatment options are variable and can include surgery, radiation therapy, and chemotherapy.

25.6.1.2 Imaging Features

Brain metastases are typically located at the gray–white matter junction, affecting the cerebral hemispheres (80%), cerebellum (15%) as well as the brain stem (5%). On CT images, metastatic tumors present as iso- to hypodense masses at the gray–white interface (Figure 25.34). Contrast-enhanced MRI is the preferred modality for imaging metastases due to its sensitivity

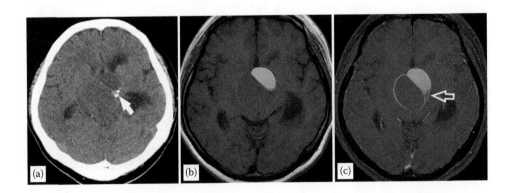

FIGURE 25.33
Craniopharyngioma. Axial CT (a), axial T_1WI (b), and axial fat-saturated contrast-enhanced T_1WI (c) demonstrate a large suprasellar mass that extends up into the third ventricle. Areas of calcifications are seen on the CT (white arrow, a). The cystic components demonstrate both iso- and hyperintensity on T_1WI (b). Contrast enhancement is noted in the cyst wall as well as in the solid portion of the tumor (white hollow arrow, c).

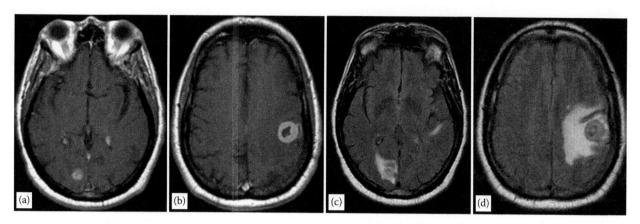

FIGURE 25.34
Metastases. Axial contrast-enhanced T_1WI (a,b) and axial FLAIR (c,d) images demonstrate contrast-enhancing masses at the gray–white junction of the cerebral hemispheres (a,b) with surrounding vasogenic edema (c,d) in this patient with metastatic lung cancer.

in detecting smaller lesions. Metastatic tumors appear iso- to hypointense on T_1WI; however, they may appear hyperintense in some cancers (e.g., melanoma). They are typically hyperintense on T_2WI, though this feature may vary. On FLAIR, metastases are likewise inconsistent, but are usually hyperintense with significant vasogenic edema. Nearly all metastatic tumors demonstrate contrast enhancement. Hemorrhage can be evident on T_2^*W GRE sequences. Based on the lack of infiltration of surrounding brain parenchyma and complete absence of a blood-brain barrier (BBB) in metastases relative to high-grade gliomas, some studies have reported that T_2^*W DSC MRI-derived rCBV, peak height, and percentage of signal intensity recovery as well as ^{1}H MRS may potentially differentiate solitary metastases from high-grade glioma.

Imaging differential diagnostic considerations include glioblastoma, infection (fungal, tuberculosis, parasites, septic emboli, toxoplasmosis), small vessel disease, and demeylinating disease.

Bibliography

Abrey LE, Batchelor TT, Ferreri AJ et al. (2005) Report of an international workshop to standardize baseline evaluation and response criteria for primary CNS lymphoma. *J Clin Oncol.* Aug 1;23(22):5034–5043.

Ajlan A, Recht L (2014) Supratentorial low-grade diffuse astrocytoma: Medical management. *Semin Oncol.* Aug;41(4):446–457.

Alexiou GA, Stefanaki K, Sfakianos G et al. (2008) Desmoplastic infantile ganglioglioma: A report of 2 cases and a review of the literature. *Pediatr Neurosurg.* 44(5):422–425.

Altman DA, Atkinson DS Jr, Brat DJ (2007) Best cases from the AFIP: Glioblastoma multiforme. *RadioGraphics* May–Jun;27(3):883–888.

Arita K, Kurisu K, Kiura Y et al. (2005) Hypothalamic hamartoma. *Neurol Med Chir (Tokyo).* May;45(5):221–231.

Baehring JM, Henchcliffe C, Ledezma CJ et al. (2005) Intravascular lymphoma: Magnetic resonance imaging correlates of disease dynamics within the central nervous system. *J Neurol Neurosurg Psychiatry.* 76:540–544.

Balaji R, Ramachandran K, Kusumakumari P (2009) Neuroimaging patterns of central nervous system metastases in neuroblastoma: Report of 2 recent cases and literature review. *J Child Neurol.* Oct;24(10):1290–1293.

Barkovich AJ (2007) Brainstem tumors. In: Barkovich AJ et al. (eds.) *Diagnostic Imaging Pediatric Neuroradiology,* pp. I-4-2–I-4-5. Salt Lake City, UT: Amirsys.

Batchelor T, Leoffler JS (2006) Primary CNS lymphoma. *J Clin Oncol.* 24:1281–1288.

Biswas A, Goyal S, Puri T et al. (2009) Atypical teratoid rhabdoid tumor of the brain: Case series and review of literature. *Childs Nerv Syst.* Nov;25(11):1495–1500.

Blaser SI, Harwood-Nash DC (1996) Neuroradiology of pediatric posterior fossa medulloblastoma. *J Neurooncol.* Jul;29(1):23–34.

Brandão LA, Shiroishi MS, Law M (2013) Brain tumors: A multimodality approach with diffusion-weighted imaging, diffusion tensor imaging, magnetic resonance spectroscopy, dynamic susceptibility contrast and dynamic contrast-enhanced magnetic resonance imaging. *Magn Reson Imaging Clin N Am.* May;21(2):199–239.

Buetow MP, Buetow PC, Smirniotopoulos JG (1991) Typical, atypical, and misleading features in meningioma. *RadioGraphics* Nov;11(6):1087–1106.

Campen CJ, Porter BE. (2011) Subependymal giant cell astrocytoma (SEGA) treatment update. *Curr Treat Options Neurol.* 13(4):380–385.

Carlson JJ, Milburn JM, Barré GM (2006) Lhermitte-Duclos disease: Case report. *J Neuroimaging.* Apr;16(2):157–162.

Cha S, Lupo JM, Chen M-H et al. (2007) Differentiation of glioblastoma multiforme and single brain metastasis by peak height and percentage of signal intensity recovery derived from dynamic susceptibility-weighted contrast-enhanced perfusion MR imaging. *AJNR Am J Neuroradiol.* 28:1078–1084.

Chiechi MV, Smirniotopoulos JG, Mena H (1995) Pineal parenchymal tumors: CT and MR features. *J Comput Assist Tomogr*. Jul–Aug;19(4):509–517.

Choi JY, Chang KH, Yu IK et al. (2002) Intracranial and spinal ependymomas: Review of MR images in 61 patients. *Korean J Radiol*. Oct–Dec;3(4):219–228.

Coates TL, Hinshaw DB Jr, Peckman N et al. (1989) Pediatric choroid plexus neoplasms: MR, CT, and pathologic correlation. *Radiology* Oct;173(1):81–88.

D'Ambrosio N, Lyo JK, Young RJ et al. (2010) Imaging of metastatic CNS neuroblastoma. *AJR Am J Roentgenol*. May;194(5):1223–1229.

del Carpio-O'Donovan R, Korah I, Salazar A et al. (1996) Gliomatosis cerebri. *Radiology*. Mar;198(3):831–835.

Desclée P, Rommel D, Hernalsteen D et al. (2010) Gliomatosis cerebri, imaging findings of 12 cases. *J Neuroradiol*. Jul;37(3):148–158.

Echevarría ME, Fangusaro J, Goldman S (2008) Pediatric central nervous system germ cell tumors: A review. *Oncologist* Jun;13(6):690–699. doi:10.1634/theoncologist.2008-0037.

Eran A, Ozturk A, Aygun N et al. (2010) Medulloblastoma: Atypical CT and MRI findings in children. *Pediatr Radiol*. 40:1254–1262.

Farrukh HM (1996) Cerebellar hemangioblastoma presenting as secondary erythrocytosis and aspiration pneumonia. *West J Med*. 164(2):169–171.

Ferreri AJM, Marturano E (2012) Primary CNS lymphoma. *Best Pract Res Clin Haematol*. 25:119–130.

Fiegl M, Greil R, Pechlaner C et al. (2002) Intravascular large B-cell lymphoma with a fulminant clinical course: A case report with definite diagnosis post mortem. *Ann Oncol*. 13:1503–1506.

Fischbein NJ, Prados MD, Wara W et al. (1996) Radiologic classification of brain stem tumors: Correlation of magnetic resonance imaging appearance with clinical outcome. *Pediatr Neurosurg*. 24(1):9–23.

FitzPatrick M, Tartaglino LM, Hollander MD et al. (1999) Imaging of sellar and parasellar pathology. *Radiol Clin North Am*. 37(1):101–121.

Fouladi M, Jenkins J, Burger P et al. (2001) Pleomorphic xanthoastrocytoma: Favorable outcome after complete surgical resection. *Neuro Oncol*. Jul;3(3):184–192.

Furie DM, Provenzale JM (1995) Supratentorial ependymomas and subependymomas: CT and MR appearance. *J Comput Assist Tomogr*. Jul–Aug;19(4):518–526.

Gelabert-Gonzalez M, Serramito-García R, Arcos-Algaba A (2010) Desmoplastic infantile and non-infantile ganglioglioma. Review of the literature. *Neurosurg Rev*. Apr;34(2):151–158.

Ginsberg LE (1996) Radiology of meningiomas. *J Neurooncol*. Sep;29(3):229–238.

Goh S, Butler W, Thiele EA (2004) Subependymal giant cell tumors in tuberous sclerosis complex. *Neurology* 63(8):1457–1461.

Grossman RI, Yousem DM (2003). Neoplasms of the brain. In: Grossman RI, Yousem DM (eds.). *Neuroradiology: The Requisites*, pp. 97–172. Philadelphia, PA: Mosby.

Haldorsen IS, Espeland A, Larsson EM (2011) Central nervous system lymphoma: Characteristic findings on traditional and advanced imaging. *Am J Neuroradiol*. 32:984–992.

Hedlund G (2007) Teratoma. In: Barkovich AJ et al. (eds.) *Diagnostic Imaging Pediatric Neuroradiology*, pp. I-3-8–I-3-11. Salt Lake City, UT: Amirsys.

Ho VB, Smirniotopoulos JG, Murphy FM et al. (1992) Radiologic-pathologic correlation: Hemangioblastoma. *Am J Neuroradiol*. 13:1343–1352.

Hoeffel C, Boukobza M, Polivka M et al. (1995) MR manifestations of subependymomas. *AJNR Am J Neuroradiol*. Nov–Dec;16(10):2121–2129.

Illner A (2007) Pineoblastoma. In: Barkovich AJ et al. (eds.) *Diagnostic Imaging Pediatric Neuroradiology*, pp. I-3-12–I-3-15. Salt Lake City, UT: Amirsys.

Jallo GI, Biser-Rohrbaugh A, Freed D (2004) Brainstem gliomas. *Childs Nerv Syst*. Mar;20(3):143–153.

Jane JA Jr, Laws ER (2006) Craniopharyngioma. *Pituitary* 9(4):323–326.

Katzman GL (2007a). Atypical and malignant meningioma. In: Osborn AG et al. (eds.) *Diagnostic Imaging Brain*, pp. II-4-56–II-4-59. Salt Lake City, UT: Amirsys.

Katzman GL (2007b). Meningioma. In: Osborn AG et al. (eds.) *Diagnostic Imaging Brain*. Salt Lake City, UT: Amirsys.

Khalid L, Carone M, Dumrongpisutikul N et al. (2012) Imaging characteristics of oligodendrogliomas that predict grade. *AJNR Am J Neuroradiol*. May;33(5):852–857.

Klisch J, Juengling F, Spreer J et al. (2001) Lhermitte-Duclos disease: Assessment with MR imaging, positron emission tomography, single-photon emission CT, and MR spectroscopy. *AJNR Am J Neuroradiol*. May;22(5):824–830.

Koeller K, Rushing EF (2004) From the archives of the AFIP. Pilocytic astrocytoma: Radiologic-pathologic correlation. *RadioGraphics* 24:1693–1708.

Koeller KK, Dillon WP (1992) MR appearance of dysembryoplastic neuroepithelial tumors (DNT). *AJNR Am J Neuroradiol*. 13:1319–1325.

Koeller KK, Henry JM (2001) From the archives of the AFIP: Superficial gliomas: Radiologic-pathologic correlation. Armed Forces Institute of Pathology. *RadioGraphics* Nov–Dec;21(6):1533–1556.

Koeller KK, Rushing EJ (2003) From the archives of the AFIP: Medulloblastoma: A comprehensive review with radiologic-pathologic correlation. *RadioGraphics* 23:1613–1637.

Koeller KK, Rushing EJ (2005) From the archives of the AFIP: Oligodendroglioma and its variants: Radiologic-pathologic correlation. *RadioGraphics* Nov–Dec;25(6):1669–1688.

Koeller KK, Sandberg GD (2002) From the archives of the AFIP: Cerebral intraventricular neoplasms: Radiologic-pathologic correlation. *RadioGraphics* 22(6):1473–1505.

Kuchelmeister K, Demirel T, Schlorer E et al. (1995) Dysembryoplastic neuroepithelial tumour of the cerebellum. *Acta Neuropathol*. (Berl.) 89:385–390.

Kuroiwa T, Kishikawa T, Kato A et al. (1994) Dysembryoplastic neuroepithelial tumors: MR findings. *J Comput Assist Tomogr*. 18:352–356.

Lafitte F, Morel-Precetti S, Martin-Duverneuil N et al. (2001) Multiple glioblastomas: CT and MR features. *Eur Radiol*. 11(1):131–136.

Lassman AB, DeAngelis LM (2003) Brain metastases. *Neurologic Clin N Am*. 21:1–23.

Lee Y, Van Tassel P, Bruno JM et al. (1989) Juvenile pilocytic astrocytomas: CT and MR characteristics. *AJR Am J Roentgenol*. 152:1263–1270.

Lee YY, Van Tassel P (1989) Intracranial oligodendrogliomas: Imaging findings in 35 untreated cases. *AJR Am J Roentgenol*. Feb;152(2):361–369.

Lefton DR, Pinto RS, Martin SW (1998) MRI features of intracranial and spinal ependymomas. *Pediatr Neurosurg*. 28:97–105.

Leonardi MA, Lumenta CB (2001) Oligodendrogliomas in the CT/MR-era. *Acta Neurochir (Wien)*. Dec;143(12):1195–1203.

Lipper MH, Eberhard DA, Phillips CD et al. (1993) Pleomorphic xanthoastrocytoma, a distinctive astroglial tumor: Neuroradiologic and pathologic features. *AJNR Am J Neuroradiol*. Nov–Dec;14(6):1397–1404.

Loto MG, Danilowicz K, González Abbati S et al. (2014) Germinoma with involvement of midline and off-midline intracranial structures. *Case Rep Endocrinol*. Vol. 2014, Article ID 936937, 5 pages.

Louis DN, Ohgaki H, Wiestler OD et al. (2007) The 2007 WHO classification of tumours of the central nervous system. *Acta Neuropathol*. Aug;114(2):97–109.

Lucas JW, Zada G (2012) Imaging of the pituitary and parasellar region. *Semin Neurol*. 32(4):320–331.

Lui PCW, Wong GKC, Poon WS et al. (2003) Intravascular lymphomatosis. *J Clin Pathol*. 56:468–470.

Matthay KK, Brisse H, Couanet D et al. (2003) Central nervous system metastases in neuroblastoma: Radiologic, clinical, and biologic features in 23 patients. *Cancer* Jul 1;98(1):155–165.

Meyers SP, Kemp SS, Tarr RW (1992) MR imaging features of medulloblastomas. *AJR Am J Roentgenol*. 158:859–865.

Mittal S, Mittal M, Montes JL et al. (2013) Hypothalamic hamartomas. Part 1. Clinical, neuroimaging, and neurophysiological characteristics. *Neurosurg Focus*. Jun;34(6):E6.

Moazzam AA, Wagle N, Shiroishi MS (2014) Malignant transformation of DNETs: A case report and literature review. *Neuroreport* Aug 20;25(12):894–899.

Moonis G, Ibrahim M, Melhem ER (2004) Diffusion-weighted MRI in Lhermitte-Duclos disease: Report of two cases. *Neuroradiology* May;46(5):351–354.

Naeini RM, Yoo JH, Hunter JV (2009) Spectrum of choroid plexus lesions in children. *AJR Am J Roentgenol*. Jan;192(1):32–40.

Nakamura M, Saeki N, Iwadate Y et al. (2000) Neuroradiological characteristics of pineocytoma and pineoblastoma. *Neuroradiology* 42:509–514.

Norden AD, Wen PY, Kesari S (2005) Brain metastases. *Curr Opin Neurol*. 18:654–661.

Osborn AG (2007) Parenchymal metastases. In: Osborn AG et al. (eds.) *Diagnostic Imaging Brain*, pp. I-6-140–I-6-143. Salt Lake City, UT: Amirsys.

Plaza MJ, Borja MJ, Altman N et al. (2013) Conventional and advanced MRI features of pediatric intracranial tumors: Posterior fossa and suprasellar tumors. *AJR Am J Roentgenol*. May;200(5):1115–1124.

Ponzoni M, Ferreri AJ, Campo E et al. (2007) Definition, diagnosis, and management of intravascular large B-cell lymphoma: Proposals and perspectives from an international consensus meeting. *J Clin Oncol*. 25:3168–3173.

Poretti A, Meoded A, Huisman TA (2012) Neuroimaging of pediatric posterior fossa tumors including review of the literature. *J Magn Reson Imaging*. Jan;35(1):32–47.

Ramos A, Hilario A, Lagares A et al. (2013) Brainstem gliomas. *Semin Ultrasound CT MR*. Apr;34(2):104–112.

Raz E, Zagzag D, Saba L, Mannelli L et al. (2012) Cyst with a mural nodule tumor of the brain. *Cancer Imaging*. Aug 10;12:237–244.

Rees JH (2011) Diagnosis and treatment in neuro-oncology: An oncological perspective. *Br J Radiol*. Dec;84(Spec No 2):S82–S89.

Rees JH, Smirniotopoulos JG, Jones RV et al. (1996) Glioblastoma multiforme: Radiologic-pathologic correlation. *RadioGraphics* Nov;16(6):1413–1438.

Rennert J, Doerfler A (2007) Imaging of sellar and parasellar lesions. *Clin Neurol Neurosurg*. 109(2):111–124.

Roman-Goldstein SM, Goldman DL (1992) MR of Primary CNS Lymphoma in immunologically normal patients. *Am J Neuroradiol*. 13:1207–1213.

Ruda R, Bertero L, Sanson M (2014) Gliomatosis cerebri: A review. *Curr Treat Options Neurol*. 16:273.

Salzman KL (2007a) Anaplastic astrocytoma. In: Osborn AG et al. (eds.) *Diagnostic Imaging Brain*, pp. I-6-16–I-6-19. Salt Lake City, UT: Amirsys.

Salzman KL (2007b) Central neurocytoma. In: Osborn AG et al. (eds.) *Diagnostic Imaging Brain*, pp. I-6-80–I-6-83. Salt Lake City, UT: Amirsys.

Salzman KL (2007c) Gliomatosis cerebri. In: Osborn AG et al. (eds.) *Diagnostic Imaging Brain*, pp. I-6-26–I-6-29. Salt Lake City, UT: Amirsys.

Salzman KL (2007d). Intravascular (angiocentric) lymphoma. In: Osborn AG et al. (eds.) *Diagnostic Imaging Brain*, pp. I-6-126–I-6-127. Salt Lake City, UT: Amirsys.

Salzman KL (2007e) Low grade diffuse astrocytoma. In: Osborn AG et al. (eds.) *Diagnostic Imaging Brain*, pp. I-6-8–I-6-11. Salt Lake City, UT: Amirsys.

Salzman KL (2007f) Pineocytoma. In: Osborn AG et al. (eds.) *Diagnostic Imaging Brain*, pp. I-6-88–I-6-91. Salt Lake City, UT: Amirsys.

Sandow BA, Dory CE, Aguiar MA et al. (2004) Best cases from the AFIP: Congenital intracranial teratoma. *RadioGraphics* Jul–Aug;24(4):1165–1170.

Sheporaitis LA, Osborn AG, Smirniotopoulos JG et al. (1992) Intracranial meningioma. *AJNR Am J Neuroradiol*. Jan–Feb;13(1):29–37.

Shin JH, Lee HK, Khang SK et al. (2002) Neuronal tumors of the central nervous system: Radiologic findings and pathologic correlation. *RadioGraphics* Sep–Oct; 22(5):1177–1189.

Sinson G, Sutton L, Yachnis A et al. (1994) Subependymal giant cell astrocytomas in children. *Pediatr Neurosurg*. 20:233–239.

Slater A, Moore NR, Huson SM (2003) The natural history of cerebellar hemangioblastomas in von Hippel-Lindau Disease. *Am J Neuroradiol*. 24:1570–1574.

Smirniotopoulos JG, Rushing EJ, Mena H (1992a) From the archives of the AFIP. Pineal region masses: Differential diagnosis. *RadioGraphics* 12:577–596.

Smirniotopoulos JG, Rushing EJ, Mena H (1992b) Pineal region masses: Differential diagnosis. *RadioGraphics* 12(3):577–596.

Smith AB, Rushing EJ, Smirniotopoulos JG (2010) From the archives of the AFIP: Lesions of the pineal region: Radiologic-pathologic correlation. *RadioGraphics* Nov;30(7):2001–2020.

Smith AB, Smirniotopoulos JG, Horkanyne-Szakaly I (2013) From the radiologic pathology archives: Intraventricular neoplasms: Radiologic-pathologic correlation. *Radiographics* Jan–Feb;33(1):21–43.

Strother D (2005) Atypical teratoid rhabdoid tumors of childhood: Diagnosis, treatment and challenges. *Expert Rev Anticancer Ther*. Oct;5(5):907–915.

Taillibert S, Chodkiewicz C, Laigle-Donadey F et al. (2006) Gliomatosis cerebri: A review of 296 cases from the ANOCEF database and the literature. *J Neurooncol*. Jan;76(2):201–205.

Tamburrini G, Colosimo C Jr, Giangaspero F et al. (2003) Desmoplastic infantile ganglioglioma. *Childs Nerv Syst*. Jun;19(5–6):292–297.

Tien RD, Cardenas CA, Rajagopalan S (1992) Pleomorphic xanthoastrocytoma of the brain: MR findings in six patients. *AJR Am J Roentgenol*. Dec;159(6):1287–1290.

Tien RD, Tuori SL, Pulkingham N et al. (1992) Ganglioglioma with leptomeningeal and subarachnoid spread: Results of CT, MR, and PET imaging. *AJR Am J Roentgenol*. Aug;159(2):391–393.

Treister D, Kingston S, Hoque KE et al. (2014) Multimodal magnetic resonance imaging evaluation of primary brain tumors. *Semin Oncol*. Aug;41(4):478–495.

Trifanescu R, Ansorge O, Wass JA et al. (2012) Rathke's cleft cysts. *Clin Endocrinol (Oxf)*. Feb;76(2):151–160.

U-King-Im JM, Taylor MD, Raybaud C (2010) Posterior fossa ependymomas: New radiological classification with surgical correlation. *Childs Nerv Syst*. 26:1765–1772.

Upadhyay N, Waldman AD (2011) Conventional MRI evaluation of gliomas. *Br J Radiol*. Dec;84(Spec No 2): S107–S111.

Vieco PT, del Carpio-O'Donovan R, Melanson D et al. (1992) Dysplastic gangliocytoma (Lhermitte-Duclos disease): CT and MR imaging. *Pediatr Radiol*. 22(5):366–369.

Weller M, Cloughesy T, Perry JR et al. (2013) Standards of care for treatment of recurrent glioblastoma—Are we there yet? *Neuro Oncol*. Jan;15(1):4–27.

Wiestler O, Lopes B, Green A et al. (2000) Tuberous sclerosis complex and subependymal giant cell astrocytoma. In: Kleihues P, Cavenee W (eds.) *Pathology and Genetics of Tumours of the Nervous System*, pp. 72–76. Lyon, France: IARC.

Williams RL, Meltzer CC, Smirniotopoulos JG et al. (1998) Cerebral MR imaging in intravascular lymphomatosis. *Am J Neuroradiol*. 19:427–431.

Woodward PJ, Sohaey R, Kennedy A et al. (2005) From the archives of the AFIP: A comprehensive review of fetal tumors with pathologic correlation. *RadioGraphics* Jan–Feb;25(1):215–242.

Yuh EL, Barkovich AJ, Gupta N (2009) Imaging of ependymomas: MRI and CT. *Childs Nerv Syst*. 25:1203–1213.

Zada G, Lin N, Ojerholm E, Ramkissoon S et al. (2010) Craniopharyngioma and other cystic epithelial lesions of the sellar region: A review of clinical, imaging, and histopathological relationships. *Neurosurg Focus*. Apr;28(4):E4.

Zee CS, Go JL, Kim PE et al. (2003) Imaging of the pituitary and parasellar region. *Neurosurg Clin N Am*. 14(1):55–80.

Zentner J, Wolf HK, Ostertun B et al. (1994) Gangliogliomas: Clinical, radiological, and histopathological findings in 51 patients. *J Neurol Neurosurg Psychiatry*. Dec;57(12): 1497–1502.

Zhang D, Henning TD, Zou LG et al. (2008) Intracranial ganglioglioma: Clinicopathological and MRI findings in 16 patients. *Clin Radiol*. Jan;63(1):80–91.

Zuckerman D, Seliem R, Hochberg E (2006) Intravascular lymphoma: The oncologist's 'great Imitator.' *The Oncologist* 11:496–502.

26

Demyelinating and Metabolic Diseases of the Brain

Antonia Ceccarelli, Eytan Raz, and Matilde Inglese

CONTENTS

26.1 Introduction

Demyelinating and metabolic diseases of the brain include a heterogeneous and large group of acquired and inherited disorders affecting primarily the white matter (WM) of the brain whose classification and diagnostic workup is complex. Traditionally, demyelinating diseases are characterized by an acquired damage/loss of properly formed myelin, whereas metabolic disorders of the brain include diseases that present an inherited dysmyelination/hypomyelination resulting from an altered synthesis, developmental, or maintenance disorder of myelin sheets. A comprehensive classification of WM diseases of the central nervous system (CNS) is presented in Table 26.1.

In the last decades, our knowledge regarding this heterogeneous group of brain diseases has increased tremendously due to the advent of sophisticated diagnostic tools such as magnetic resonance imaging (MRI). MRI has become the most important tool for the investigation of brain WM diseases, due to its high sensitivity to WM abnormalities (Guleria and Kelly, 2014). In particular, conventional MRI has improved visualization, recognition, diagnostic, prognostic, and treatment response monitoring ability. Indeed, normal myelinated WM have short longitudinal (T_1) and transverse (T_2) relaxation times of the mobile protons resulting in signal loss or hypointensity (dark appearance) on T_2-weighted images and hyperintensity on T_1-weighted images (bright appearance) compared to gray matter (GM) tissue. Building on these basic principles, different degrees of myelination due to aging (Figure 26.1) or pathological processes (Guleria and Kelly, 2014) can be visualized by conventional MRI sequences ranging from mild to prominent hyperintensity on T_2-weighted images and from absent to prominent hypointensity on T_1-weighted images of the brain WM compared to GM tissue. On the other hand, advanced quantitative MRI techniques, such as diffusion-weighted images (DWI), diffusion-tensor imaging (DTI), magnetization transfer imaging (MTI), and magnetic resonance spectroscopy (MRS), coupled with high and ultrahigh field strength

TABLE 26.1

Classification of WM Demyelinating and Metabolic Diseases of the CNS

Dysmyelinating Disorders (Inherited and Metabolic)

1. *Hypomyelinating leukodistrophies*
   a. PMD
   b. Pelizaeus–Merzbacher like disease
   c. Pol III-related leukodystrophies/4H
   d. 18q-syndrome
   e. Cockayne syndrome
   f. Hypomyelination with atrophy of the basal ganglia and cerebellum
   g. Hypomyelination with congenital cataracts
   h. Hypomyelination of early myelinated structures
   i. Hypomyelination with brain stem and spinal cord involvement and leg spasticity
   j. Free sialic acid storage disease
   k. Fucosidosis
   l. Oculodentodigital dysplasia
   m. RARS-associated hypomyelination, SOX10-associated disorders
   n. Trichothiodystrophy with hypersensitivity to sunlight
2. *Dysmyelinating leukodistrophies*
   a. Lysosomial storage disorders:
      Metachromatic leukodystrophy
      Globoid cell leukodystrophy (Krabbe disease)
      Fabry's disease
      GM1 gangliosidosis
      GM2 gangliosidosis
      Gaucher's disease
      Niemann–Pick disease, types A and B
      Farber disease
      Sialic acid storage disorders
      Mucopolysaccharidosis
      Multiple sulfatase deficiency
   b. Peroxisomal disorders (Poll et al., 2012)
      X-linked adrenoleukodystrophy and adrenomyeloneuropathy
      Peroxisome biogenesis defects (Zellweger spectrum disorders, Zellweger syndrome, Neonatal adrenoleukodystrophy, infantile refsum disease)
      Refsum disease
      Bifunctional protein deficiency
      Acyl-CoA oxidase deficiency
      Rhizomelic chondrodysplasia punctata (type 1, 2, 3)
      2-Methylacyl-CoA racemase deficiency
   c. Leukodystrophies with astrocytic dysfunction (Rodriguez et al., 2013)
      Alexander's disease
      Childhood ataxia with hypomyelination/leukoenchephalopathies with vanishing white matter (WM)
      Megalenchephalic leucoenchephalopathy with subcortical cysts
      Canavan disease
   d. Mitochondrial dysfunction with leukoencephalopathy
      Mitochondrial myopathy encephalopathy, lactic acidosis, and stroke-like episodes (MELAS)
      Leber hereditary optic neuropathy
      Kearns–Sayre syndrome
      Mitochondrial neurogastrointestinal encephalomyopathy (MNGIE)
      Leigh disease and mitochondrial leukoencephalopathies
      Pyruvate carboxylase deficiency
      Multiple carboxylase deficiency
      Cerebrotendinous xanthomatosis

(Continued)

TABLE 26.1 (*Continued*)

Classification of WM Demyelinating and Metabolic Diseases of the CNS

    e. Disorders of amino acid and organic acid metabolism

       Phenylketonuria

       Glutaricaciduria type 1

       Propionic acidemia

       Nonketotic hyperglycinemia

       Maple syrup urine disease

       L-2-Hydroxyglutaric aciduria

       D-2-Hydroxyglutaric aciduria

       Hyperhomocysteinemias

       Urea cycle defects

       Serine synthesis defects

Demyelinating Disorders (Acquired)

1. *Idiopathic*

    a. Multiple sclerosis

    b. Multiple sclerosis variants

       Radiologically isolated syndrome (RIS)

       Pediatric MS

       Balo's concentric sclerosis

       Marburg variant

       Schilder

       Tumefactive demyelination

       Neuromyelitis optica (NMO)

2. *Secondary*

    a. Allergic/infections/vaccinations/immunological

       Acute disseminated encephalomyelitis (ADEM) and acute hemorrhagic encephalomyelitis

       Progressive multifocal leukoencephalopathy (PML) and immune reconstitution inflammatory syndrome (IRIS)

       Lyme disease

       Susac syndrome

       Sarcoidosis

       Subacute human immunodeficiency virus (HIV) enchephalities

       Subacute sclerosing panencephalitis (SSPE)

       Congenital cytomegalovirus infection

       Whipple disease

       Other infections

    b. Vascular

       Cerebral autosomal-dominant arteriopathy with subcortical infarcts and leukoencephalopathy (CADASIL)

       Cerebral autosomal-recessive arteriopathy with subcortical infarcts and leukoencephalopathy (CARASIL)

       Cerebral amyloid angiopathy (CAA)

       Posterior reversible encephalopathy syndrome (PRES)

       Subcortical arteriosclerotic encephalopathy (SAE) or Binswanger disease

       Vasculitis

       Vasculopathy of other origin

    c. Toxic leukoencephalopathies (endogenous and exogenous toxins for nutritional/vitamin deficiency or physical/chemical agents and medical therapy)

       Central pontine and extrapontine myelinolysis

       Vitamin B12 deficiency (subacute combined degeneration)

       Wernicke encephalopathy

       Marchiafava–Bignami disease

       Malnutrition

       Paraneoplastic syndromes

       Toxic drug exposure

       Radiation necrosis

(Continued)

TABLE 26.1 (Continued)

Classification of WM Demyelinating and Metabolic Diseases of the CNS

Diffuse radiation-induced leukoencephalopathy
Disseminated necrotizing leukoencephalopathy or chemiotheraphic leukoencephalopathy
Mineralizing microangiopathy (from radiation and chemiotherapic treatment)
d. Hypoxic-ischemic disorders
Posthypoxic-ischemic leukoencephalopathy of neonates
Delayed posthypoxic-ischemic leukoencephalopathy
Global hypoperfusion syndromes
Hypertensive encephalopathy
Eclampsia and PRES
e. Traumatic disorders
Diffuse axonal injury
Compression-induced demyelination

Source: Adapted from Kanekar, S. and Gustas, C., *Semin. Ultrasound CT MR*, 32, 590, 2011.

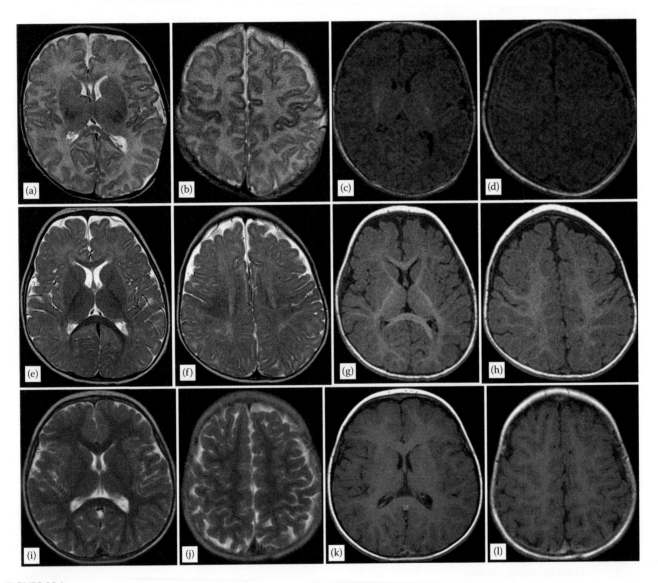

FIGURE 26.1

Normal myelination according to age. T_2- and T_1-weighted images at 3 weeks old (a–d), 6 months old (e–h), and 2 years old (i–l). Myelination of fibers increases in the first 2 years of life and white matter myelinated fibers appear progressively more hypointense on T_2 images (i, j) and more hyperintense on T_1 images (k, l) compared to GM intensity.

MRI, have further improved and facilitated early and differential diagnosis of these diseases, prognostication, and monitoring ability after treatment, revealing the underlying brain tissue pathology (Ceccarelli et al., 2012). Therefore, the MRI has provided unique insights into our understanding of brain WM diseases and has offered significant promise toward the past knowledge on brain WM diseases.

Taking this in mind, this chapter will provide an overview of the brain WM diseases, highlighting for each disease the importance of the MRI in the diagnosis, prognosis, treatment monitoring, and its role in revealing the underlying pathology. We will discuss the diseases with respect to the involvement of the brain WM.

26.2 Inherited Metabolic Disorders of the Brain or Leukodystrophies

Leukodystrophies (LDs) are usually genetic disorders. They are mostly progressive and the treatment is often not available. Defects in CNS myelin, synthesis, development, maintenance and catabolism can be the cause of LDs. (Aicardi, 1993; Kanekar and Gustas, 2011; Kanekar and Verbrugge, 2011, Perlman and Mar, 2012). Classification of LDs is challenging (Kanekar and Gustas, 2011; Kanekar and Verbrugge, 2011; Vanderver et al., 2014; Yang and Prabhu, 2014). First, it is continuously revised since new research advancements (Köhler, 2010). Second, classification can change according to their genetic, biochemical, clinical, pathological, and MRI findings. Although the genetic target of the majority of LDs is still unknown, in some of them a biochemical or an enzymatic abnormality has been found. Of note, recently, emerging evidences have shown that astrocytic dysfunction plays an important role in myelin damage in some LDs such as Alexander disease (AxD), megalencephalic leukoencephalopathy with subcortical cysts (MLC), and vanishing WM (VWM) disease (Lanciotti et al., 2013; Rodriguez et al., 2013). Clinically, LDs are progressive diseases affecting mostly children and manifests with a prevalence of symmetrical and bilateral motor and cerebellar symptoms, including spasticity, weakness, and ataxia. Seizures, myoclonus, and cognitive deterioration could appear later. However, particular clinical profiles diverse from this general one could be observed (Aicardi, 1993; Vanderver et al., 2014). The presence of peripheral nervous system and other organs' involvement is frequent and could help in the diagnosis of myelination (Yang and Prabhu, 2014). Pathologically, there is solely or primarily WM damage with different grades: delayed myelination, hypomyelination, and dysmyelination. Hypomyelinating LDs are due to a defect of myelin synthesis and production and refer to low or absence of myelin (Pouwels et al., 2014). Dysmyelination is due to a defect of myelin biogenesis and maintenance or myelin degeneration for astrocytic dysfunction and refers to the presence of abnormal myelin with or without demyelination, whereas demyelination refers to loss of properly formed myelin (Hatten, 1991). While traditionally, the diagnosis was relying mostly on clinical findings and detection on brain computerized tomography (CT) scan of WM hypodensity (Harwood-Nash et al., 1975), recently a complex diagnostic work-up has been employed including evaluation of family history, neurological examination, CNS imaging, specialized laboratory testing, and genetic counseling (Vanderver et al., 2014). Among other diagnostic tools, MRI has shown a pivotal role and dramatically revolutionized the diagnostic workup of these diseases (Osterman et al., 2012; Ratai et al., 2012; Schiffmann and van der Knaap, 2009; Yang and Prabhu, 2014). Using conventional MRI, delayed myelination, hypomyelination, and dysmyelination can be identified. Usually, hyperintensity on T_2-weighted images relative to GM contrast is milder in hypomyelination compared to demyelination, and it is associated with a variable signal on T_1-weighted images ranging from hyperintensity (normal amount of myelin), to isointensity (low amount of myelin) or hypointensity (absent or poor myelination). On the other hand, prominent hyperintensity on T_2-weighted images and hypointensity on T_1-weighted images is usually found in dysmyelination or demyelination (Schiffmann and van der Knaap, 2009) (Key Points 26.1). Once brain abnormal WM signal has been identified, shape (isolated, confluent, or multifocal lesions), extension (diffuse, bilateral, or focal), and localization of WM damage (regional distribution) can be extremely helpful in differentiating these diseases (Schiffmann and van der Knaap, 2009). A 6-month follow-up scan is needed to distinguish delayed myelination from permanent hypomyelination, where permanent hypomyelination is suggested by unchanged MRI pattern in subjects older than 1 year. Distinct MRI recognition patterns are typical of some WM diseases (Osterman et al., 2012; Schiffmann and van der Knaap, 2009; Yang and Prabhu, 2014). Classification of LDs according to their major radiological pattern is provided in Figures 26.2 and 26.3 (Osterman et al., 2012).

Improved diagnostic conventional MRI sensitivity and specificity can be further facilitated by the detection of other peculiar features, such as enhancement, presence of cysts in the WM or in the anterior part of the temporal lobe, involvement of deep GM areas, megalocephaly, enlarged perivascular spaces,

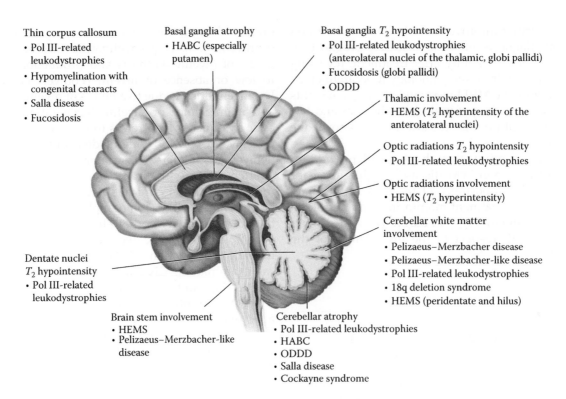

Thin corpus callosum
• Pol III-related
  leukodystrophies
• Hypomyelination with
  congenital cataracts
• Salla disease
• Fucosidosis

Basal ganglia atrophy
• HABC (especially
  putamen)

Basal ganglia T_2 hypointensity
• Pol III-related leukodystrophies
  (anterolateral nuclei of the thalamic, globi pallidi)
• Fucosidosis (globi pallidi)
• ODDD

Thalamic involvement
• HEMS (T_2 hyperintensity of the
  anterolateral nuclei)

Optic radiations T_2 hypointensity
• Pol III-related leukodystrophies

Optic radiations involvement
• HEMS (T_2 hyperintensity)

Cerebellar white matter
involvement
• Pelizaeus–Merzbacher disease
• Pelizaeus–Merzbacher-like disease
• Pol III-related leukodystrophies
• 18q deletion syndrome
• HEMS (peridentate and hilus)

Dentate nuclei
T_2 hypointensity
• Pol III-related
  leukodystrophies

Brain stem involvement
• HEMS
• Pelizaeus–Merzbacher-like
  disease

Cerebellar atrophy
• Pol III-related leukodystrophies
• HABC
• ODDD
• Salla disease
• Cockayne syndrome

FIGURE 26.2
Structures specifically involved in hypomyelinating leukodystrophies. Graphical representation of different brain structures with the corresponding hypomyelinating leukodystrophies presenting with specific involvement or preservation of each structure. HABC, hypomyelination with atrophy of the basal ganglia and cerebellum; HEMS, hypomyelination of early myelinating structures; ODDD, oculodentodigital dysplasia; Pol III, polymerase III. (Reproduced from Osterman, B. et al., *Future Neurol.*, 7, 595, 2012. With permission from the publisher.)

cortical dysplasia, cortical lesions, basal calcium deposits, microbleeds, spinal cord involvement, and evolution over time (Yang and Prabhu, 2014). For example, today the use of CT in these diseases is mostly confined to the search for calcifications, which are peculiar of some LDs (Yang and Prabhu, 2014). Saying that, at the first observation, the minimal MRI protocol suggested to make a diagnosis of LDs consisted of a T_2, fluid-attenuated inversion recovery (FLAIR), and T_1-weighted images with and without contrast enhancement plus a spinal cord imaging (Schiffmann and van Der Knaap, 2009) (Key Points 26.2). Advanced MRI techniques could further enrich the MRI diagnostic algorithm (Mar and Noetzel 2010; Patay, 2005; Pouwels et al., 2014; Ratai et al., 2012; Rossi and Biancheri, 2013). MRS, DWI, DTI, and magnetization transfer ratio (MTR) coupled with the use of high and ultrahigh field strength MRI are becoming increasingly important in the early diagnosis, in the differential diagnosis, and in monitoring progression and treatment response. For example, in normal-appearing WM (NAWM), early demyelination can be detected by using MRS, DWI, and MTI (as suggested respectively from an increased choline peak,

a decreased fractional anisotropy (FA), an increased radial diffusivity [RD] coupled with an increased mean diffusivity, and a decreased MTR). In addition, in some LDs, MRS can show specific patterns of metabolites changes (Rossi and Biancheri, 2013). Extremely increased peak of N-acetyl aspartate (NAA) is typical of Canavan disease, while myo-inositol peak has been detected as a marker of glial cells dysfunction, for example, in AxD. Myelin edema (due to vacuolating or spongiform alterations of myelin) can be differentiated by DWI (restricted diffusion) from vasogenic edema (Patay, 2005). RD has been proposed as a more specific marker of hypomyelination (Pouwels et al., 2014) and more sensitive than other markers of the degree of severity of myelin damage. Moreover, ongoing investigations in the last years are pointing out on newer qualitative and quantitative advanced MRI techniques more pathologically specific to myelin (Ceccarelli et al., 2012). Among these, myelin water fraction (MWF) imaging has emerged as a specific biomarker of myelin damage and repair. However, due mostly to technical limitations, its use is still confined to the research settings. Similarly, other several new imaging sequences

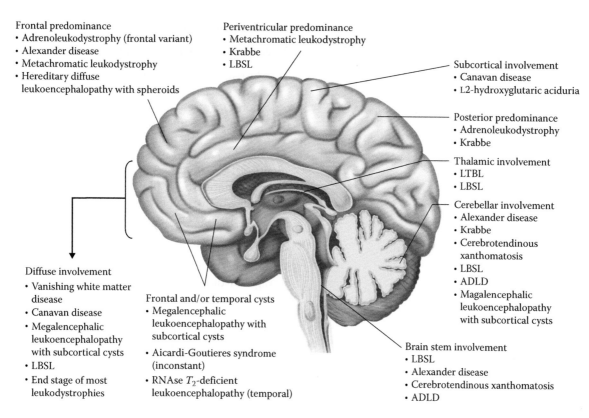

Frontal predominance
• Adrenoleukodystrophy (frontal variant)
• Alexander disease
• Metachromatic leukodystrophy
• Hereditary diffuse leukoencephalopathy with spheroids

Periventricular predominance
• Metachromatic leukodystrophy
• Krabbe
• LBSL

Subcortical involvement
• Canavan disease
• L2-hydroxyglutaric aciduria

Posterior predominance
• Adrenoleukodystrophy
• Krabbe

Thalamic involvement
• LTBL
• LBSL

Cerebellar involvement
• Alexander disease
• Krabbe
• Cerebrotendinous xanthomatosis
• LBSL
• ADLD
• Magalencephalic leukoencephalopathy with subcortical cysts

Diffuse involvement
• Vanishing white matter disease
• Canavan disease
• Megalencephalic leukoencephalopathy with subcortical cysts
• LBSL
• End stage of most leukodystrophies

Frontal and/or temporal cysts
• Megalencephalic leukoencephalopathy with subcortical cysts
• Aicardi-Goutieres syndrome (inconstant)
• RNAse T_2-deficient leukoencephalopathy (temporal)

Brain stem involvement
• LBSL
• Alexander disease
• Cerebrotendinous xanthomatosis
• ADLD

FIGURE 26.3
Regional distribution of white matter abnormalities in the different types of demyelinating leukodystrophies. Graphical representation of the different brain regions with the corresponding demyelinating leukodystrophies presenting a prominent involvement of each specific region. ADLD, adult-onset autosomal dominant leukodystrophy; LBSL, leukoencephalopathy with brain stem and spinal cord involvement and lactic acidosis; LTBL, leukoencephalopathy with thalamus and brain stem involvement and high lactate. (Reproduced from Osterman, B. et al., *Future Neurol.*, 7, 595, 2012. With permission from the publisher.)

and refinements are expected to notably impact the clinical imaging of LDs in the coming years. Together with the advent of powerful diagnostic and research tools, last decades have seen also an increased array of diverse novel therapeutics options for metabolic WM diseases, such as hematologic stem cell transplantation (HCT), enzyme replacement therapy, gene therapy, specific cell therapy, and small molecules therapy (Köhler, 2010). After their development, some of them are recently translated to clinical trials. Further development of these treatment approaches and possible novel therapeutics will be closely linked in turn with a better understanding of the disease pathogenesis and disease course. In this section we will describe the classical LDs that occur with major involvement of the CNS in combination with the peculiar MRI findings used for their characterization. A review of LDs with substantial involvement of other organs and systems and of neonatal onset WM diseases is beyond the scope of the present chapter.

KEY POINTS 26.1 ABNORMAL WM SIGNAL ON CONVENTIONAL MRI

• On T_2-weighted images, hypomyelination and dysmyelination/demyelination appeared as mild or clear WM hyperintensity compared to GM signal intensity.

• On T_1-weighted images, hypomyelination and dysmyelination/demyelination appeared as WM isointensity (low amount of myelin) or hypointensity (absent or poor myelination).

• Different shape (isolated, confluent, or multifocal lesions), extension (diffuse, bilateral, or focal), and localization (regional distribution) of abnormal WM signal on conventional MRI are observed and provided clues for diagnosis of brain WM diseases.

**KEY POINTS 26.2   MRI PROTOCOL IN
SUBJECTS WITH SUSPECTED LD**

- T_2-weighted images of the brain and spinal cord.
- T_1-weighted images with and without contrast enhancement of the brain and spinal cord.
- MRS of the brain.
- DWI of the brain.
- Susceptibility-weighted images and MTR of the brain at least once.
- Follow up MRI is recommended.

26.2.1 Hypomyelinating LDs

This is a group of LDs characterized by arrested or absent development of myelin. The prototypical disease of this group is Pelizaeus–Merzbacher disease (PMD).

26.2.1.1 Pelizaeus–Merzbacher Disease

PMD is caused by a defect in the *PLP*1 gene localized on the X chromosome and coding the proteolipid protein (PLP) and its isoform DM20. The *PLP* gene is predominantly expressed in the oligodendrocytes. Several mutations of this gene have been described (Hobson and Garbern, 2012). Clinically, it appears at birth with nystagmus, ataxia, cognitive as well as psychomotor delay, and hypotonia that progresses to spasticity. Usually, it is a fatal disease in the infancy. So far, no treatments are available, but pharmacological and cell-based therapies are being explored (Pouwels et al., 2014).

> *Conventional MRI findings.* It is used to observe loss of myelin maturation with mild diffuse and homogeneous hyperintensity on T_2-weighted images and variable signal at T_1-weighted images, compared to GM.

> *Advanced MRI findings.* Reduction of NAA can be frequently detected by MRS. However, while MRS may not be useful in quantifying hypomyelination, DTI and MT have recently shown sensitivity and specificity to hypomyelination and should be included in the study of hypomyelinating diseases (Pouwels et al., 2014). Specifically reduced MTR, increased MT saturation, and increased RD on DTI images are found to be associated strongly with hypomyelination (Pouwels et al., 2014).

26.2.2 Dysmyelinating LDs

26.2.2.1 Lysosomal Storage Disorders

The lysosomal storage disorders represent a mixed group of genetic diseases characterized by the accumulation of non-metabolized macromolecules for dysfunction of lysosomes in different tissues and organs. More than 50 disorders in this category have been described (Kohlschütter, 2013). Among all, only two, such as metachromatic LD (MLD) and Krabbe disease, present only noteworthy involvement of the CNS. We will focus on them. The others are mostly multi-organs disorders.

26.2.2.1.1 Metachromatic Leukodystrophy

MLD is a rare inherited autosomal recessive lysosomal storage disorder caused mainly by the deficiency of enzyme arylsulfatase A (ARSA) due to mutations on gene located on chromosome 22q13, responsible for the accumulation of sulfatide sphingolipids in the brain and peripheral nerves that brings to fragile myelin sheath (Gieselmann, 2008). It is probably the most common LD and it has worldwide distribution, encompassing different ages according to the severity of the gene damage. Late infantile form (onset around 2 years) is the most severe variant due to complete deficiency of the enzymatic activity. Incomplete deficiency can onset as juvenile (onset around 4–16 years) or adult forms (after 16 years old) (Aicardi, 1993).

Typical clinical symptoms are rapid motor disorders, gait disorders, spasticity coupled with loss of peripheral reflexes, and cerebellar involvement. Cognitive deterioration appears progressively. However, in the late juvenile and in the adult form neuropsychiatric symptoms schizophrenia-like or dementia-like including behavioral abnormalities, social and language problems, cognitive impairment, mood disorders, and hallucinations can manifest at onset followed only later by central and peripheral motor symptoms. Furthermore, peripheral nerve involvement, always present, is more pronounced in younger age, whereas cognitive abnormalities are more typical of adult forms (Gieselmann and Krägeloh-Mann, 2010). The pathological stamps of MLD are damage of myelin and metachromatic granules present on the inside (free, glial cells, and macrophages) and outside of the nervous system (renal tubules, bile duct, epithelium, gallbladder, islet cell and ductal epithelium of pancreas, reticular zone of adrenal cortex, and liver). However, the highest accumulation and impaired function is a prerogative of the nervous system (Kohlschütter, 2013). While cerebrospinal fluid (CSF) proteins are usually increased, oligoclonal bands are absent. Serum and urine ARSA levels are usually reduced. Some therapeutic options have been explored such as gene therapy and hematopoietic stem cells in this disease (Patil and Maegawa, 2013). Mainly experimental, these treatments

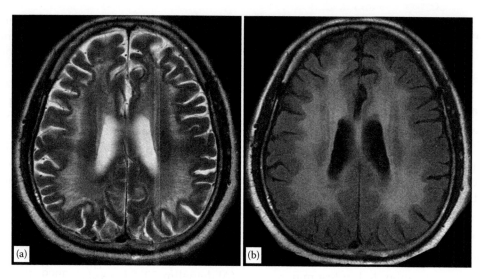

FIGURE 26.4
Metachromatic leukodystrophy. A 43-year-old patient presenting with early dementia. Axial T_2 (a) and FLAIR (b) demonstrate confluent, symmetric, hyperintense signal in the white matter with multiple hypointense lines within the white matter, which create the characteristic *tigroid* pattern of metachromatic leukodystrophy.

have shown variable efficacy linked to the disease stages and are currently being under clinical trials. Recently, gene therapy approach showed an extraordinary prevention of MLD progression (Biffi et al., 2013).

26.2.2.1.1.1 Conventional MRI Findings In the late stage of the disease, conventional MRI shows a diffuse and symmetrical WM dysmyelination of the brain including cerebellum. At the beginning, WM dysmyelination follows a centrifugal pattern starting from periventricular areas. Symmetric and bilaterally periventricular damage with occipital prevalence is typical of early onset form, while frontal involvement is frequent in the juvenile and adult forms. Corpus callosum could be damaged, while U fibers are usually spared. A *tigroid* feature (stripes in the demyelinating WM) (Yang and Prabhu, 2014) of the centrum semiovale is observed and it is due to the relative sparing of the myelin around the transmedullary vessels (Figure 26.4).

Involvement of cerebellum and basal ganglia are usually present in the severe forms of the disease. Based on these MRI features, Eichler et al. (2009) have elaborated an MR severity scoring method to classify the MLD in mild (involvement of occipital, frontal, and parietal WM), moderate (involvement of subcortical WM and presence of tigroid pattern), and severe disease (involvement of projection fibers, cerebellar WM, and basal ganglia/thalamic involvement). Contrast enhancement has typically never been present in the WM, while can be present at the level of cranial nerves and cauda equina (Maia et al., 2007) (Key Points 26.3).

26.2.2.1.1.2 Advanced MRI Findings Advanced MRI has a pivotal role in identifying early myelin damage, myelin

edema, and in monitoring progression and treatment response. Using MRS, decreased NAA and elevated lactate and myo-inositol peaks may be detected before any damage of the WM is evident on conventional MRI sequences (Yang and Prabhu, 2014) and as the disease progresses motor and cognitive functions worsen (í Dali et al., 2010). Recent studies have also shown atrophy of cerebral and cerebellar cortex as well as thalamic changes suggesting an early neuronal involvement in MLD (Groeschel et al., 2012; Martin et al., 2012) (Key Points 26.4).

KEY POINTS 26.3   CONVENTIONAL MRI FEATURES IN MLD

- Symmetrical and confluent WM hyperintensity begin periventricularly and progress with a centrifugal gradient toward subcortical WM with sparing of the U fibers.

- At the onset of infantile forms, symmetrical and confluent WM hyperintensity begin periventricularly in occipital areas.

- At the onset of juvenile and adult forms, symmetrical and confluent WM hyperintensity begin periventricularly in frontal areas.

- In moderate disease, tigroid pattern appeared.

- In severe disease, involvement of cerebellar WM and deep GM structures is presented.

- Contrast enhancement of cranial nerves and cauda equine are frequent.

26.2.2.1.2 Krabbe Disease or Globoid Cell Leukodystrophy

Krabbe disease is a rare inherited autosomal recessive lysosomal storage disease caused by deficiency of the enzyme β-galactocerebrosidase, which results in abnormal accumulation of β-galactocerebroside and galactosylsphingosine in WM, in both the central and peripheral nervous systems (Kohlschütter, 2013). It has worldwide distribution and encompass different ages according to the severity of the enzyme damage (Duffner et al., 2011, 2012; Lyon et al., 1991). Clinical symptoms vary according to the age of onset. Early onset (before the age of 2 years) Krabbe disease is a devastating disease. Infantile or late infantile forms start with fever, spasticity, and hyperirritability followed by fast mental and motor decline. A seizure-like syndrome could manifest with tonic–clonic spasms unresponsive to epileptic drugs. Before 3 years of age, death occurs. The late-onset Krabbe disease (after the age of 2) could appear as juvenile, late juvenile, or adults. These forms are milder in symptoms and progression than the early onset forms. Motor and mental deterioration could be accompanied by seizure-like

syndrome and optic nerve damage. Pathologically the hallmark of Krabbe disease is the accumulation around blood vessels in both GM and WM of globoid cells, which are multinucleated giant cells with ballooned cytoplasm containing fibrillary to granular material (Kohlschütter, 2013). As in MLD, peripheral neuropathy and absence of oligoclonal bands are observed. Although, at present only HCT, offered before onset, has shown some efficacy, several experimental therapy are under investigation and some of them have shown promising results in animal models, such as gene therapy (Lattanzi et al., 2014).

26.2.2.1.2.1 Conventional MRI Findings

Different patterns of conventional MRI abnormalities are shown by several researchers according to the disease onset (Abdelhalim, 2014; Loes et al., 1999). Early infantile forms present mainly a characteristic involvement of cerebellar WM and dentate as well as deep cerebral WM and basal ganglia and thalami. Late infantile forms present involvement of the parieto-occipital WM, periventricular regions, posterior corpus callosum, and posterior internal capsule with a relative sparing of the U fibers (Figure 26.5).

On the other hand, in the adult onset, selective involvements of the cortical spinal tract has been observed with extension toward upper rolandic area, posterior internal capsule, cerebral peduncles, and the corpus callosum but sparing of the periventricular WM. In addition, involvement of basal ganglia is rare in late-onset Krabbe disease. On the other hand, even if it is rare, predominant involvement of bilateral cortical spinal tract in a child should raise the suspicion of Krabbe disease (Kamate and Hattiholi, 2011).

The use of CT could be complementary to the MRI showing presence of bilateral calcification particularly in the early stages of the disease at the level of deep GM,

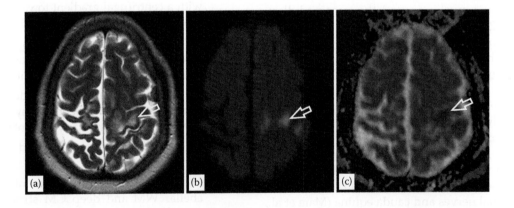

FIGURE 26.5

Krabbe disease. A 36-year-old female with adult-onset Krabbe disease. (a) T_2 signal hyperintensity within the white matter posteriorly predominant, corpus callosum, and centrum semiovale, right greater than left, with associated volume loss on the right in the region of the pre- and postcentral gyri. (b—DWI, c—apparent diffusion coefficient) Punctate focus of diffusion restriction within the left perirolandic subcortical white matter.

cortical spinal tract, and WM (Livingston et al., 2012). As for MLD, the *tigroid* pattern could be observed. Contrast enhancement has typically never been present in the WM, while can be present at the level of cranial nerves and cauda equina (Yang and Prabhu, 2014) (Key Points 26.5).

26.2.2.1.2.2 Advanced MRI Findings As for MLD, advanced MRI techniques are useful for detecting the early demyelination, degree of myelination, progression of the disease, and treatment response. MRS may show elevated choline and lactate and decreased NAA levels. A recent report has shown a higher sensitivity of RD over the FA for the myelin damage in Krabbe disease (Poretti et al., 2014). Restricted diffusion can be also observed (Figure 26.5) (Key Points 26.6).

KEY POINTS 26.5   CONVENTIONAL MRI FEATURES IN KRABBE DISEASE

- Symmetrical and confluent WM hyperintensity begin periventricularly in occipital areas and progress with a centrifugal gradient toward subcortical WM with sparing of the U fibers.
- Combination of GM and WM signal abnormalities with preferential thalamic involvement is observed.
- In early infantile form, cerebellar WM and dentate involvement were observed.
- In adult form, selective cortical spinal tract involvement that progress to involvement of rolandic area, posterior internal capsule, cerebral peduncles, and the corpus callosum but sparing of the periventricular WM were observed.
- Contrast enhancement of cranial nerves and cauda equine, and mainly on the optic nerve is frequent.
- Parenchymal calcifications, better visualized with CT, are found.

KEY POINTS 26.6   ADVANCED MRI FEATURES IN KRABBE DISEASE

- MRS reveals decreased NAA and elevated choline and lactate peaks at the beginning of the disease.
- Restricted diffusion can be observed.
- RD is a sensitive marker of myelin damage in Krabbe disease.

26.2.2.2 Peroxisomal Disorders

Peroxisomal disorders are a group of genetic diseases in which there is impairment in one or more peroxisomal functions. As lysosomes, peroxisomes are organelles involved in metabolism, but mainly for very long chain fatty acids (VLCFAs), branched chain fatty acids, D-amino acids, and other molecules important for brain and other organs. According to their type of deficiency, peroxisomal disorders can be differentiated in peroxisome biogenesis deficiency; (2) the single peroxisomal (enzyme) protein deficiencies; and (3) the single peroxisomal substrate transport deficiencies (Aubourg and Wanders, 2013; Poll-The and Gärtner, 2012). Different diseases are grouped in this classification as you can see from Table 26.1. As for other LDs, the diagnosis is made by combination of clinical findings, age of onset, battery of biochemical assays in blood and/or urine, genetic testing, and imaging (Loes et al., 1994). X-linked adrenoleukodystrophy (X-ALD) is the only disease included in the group (3) and it is the most common peroxisomal disorders.

26.2.2.2.1 X-Linked Adrenoleukodystrophy

X-ALD is caused by mutations in the *ABCD*1 gene that encodes the peroxisomal membrane protein ALDP that is involved in the trans-membrane transport of VLCFA. A defect in ALDP results in elevated levels of VLCFA in plasma and tissues (Moser et al., 2007). It has worldwide distribution and is frequently a disease affecting children, especially males, damaging the nervous system, adrenal glands, and testicles (Engelen et al., 2012). While in males, X-ALD ranges from isolated adrenocortical insufficiency and slowly progressive myelopathy to devastating cerebral demyelination, usually women, when affected, have a milder form of the disease, which is adrenomyeloneuropathy (AMN) phenotype. The X-ALD clinical symptoms are strongly associated with age: the earliest disease onset is the worst (Engelen et al., 2014). Indeed, there are three forms of the disease—(1) the rapidly progressive form that usually affects children, cerebral demyelinating form of X-ALD, but in milder form can affect adolescents and adults; (2) the AMN; and (3) the isolated adrenocortical insufficiency form. Clinically, a fast cognitive decline and behavioral disturbance, respectively, more frequent in infantile and in the adults form, are described (Aubourg and Wanders, 2013; Poll-The and Gärtner, 2012). AMN, manifests most commonly in the late twenties as progressive paraparesis, sphincter disturbances, sexual dysfunction, and often, impaired adrenocortical function and peripheral neuropathy; all symptoms are progressive over decades (Aubourg and Wanders, 2013; Poll-The and Gärtner, 2012). Pathologically, while X-ALD is characterized by inflammation leading to a demyelination,

AMN is a noninflammatory axonopathy (Engelen et al., 2012). Although MRI is an important tool to identify these demyelinating diseases, however the diagnosis of such diseases relies on the demonstration of increased/decreased levels of plasma enzymatic activity. Clinical symptoms and in particular associated clinical features such as age of onset, adrenal dysfunction symptoms (abnormalities in skin pigmentations), and genetic counseling could help in differentiating these disease from acquired demyelinating diseases such as MS. The most effective treatment for X-ALD is allogeneic HCT. HCT can arrest or even reverse cerebral demyelination whether it is provided in a very early stage or endocrine replacement therapy (Engelen et al., 2012). Obviously, the use of advanced MRI is fundamental to monitor treatment response.

26.2.2.2.1.1 Conventional MRI Findings Conventional MRI shows symmetrical hyperintensity on T_2-weighted images starting from the splenium of the corpus callosum and extending toward parieto-occipital WM. Different grades of hypointensity on T_1-weighted images can been seen according to the disease severity as well as cavitations on T_2-weihghted images. Contrast enhancement is usually confined to the border of the areas of abnormal WM and appears when the disease progresses rapidly and it is positively associated with the disease severity (Kim and Kim, 2005; Loes et al., 1994; Patel et al., 1995; Yang and Prabhu, 2014) (Figure 26.6). As the disease advances, demyelination progress in a rostro-caudal direction. Frontal lobe involvement is typically asymmetric and occurs in the later stage of the disease. Loes et al. (1994) have elaborated an MRI scoring system that correlates well with clinical symptoms at the beginning of the disease. AMN, while, as well as X-ALD, it is more frequent in men than women, being an X-linked disease, it is characterized only by

damage of corticospinal tracts, spinal cord, and peripheral neuropathy. Brain conventional MRI could be normal or shows pontomedullary corticospinal tract and cerebellar involvement and no contrast enhancement is observed (Elenein et al., 2013; Kumar et al., 1995). In adolescent forms a reversal pattern with the first demyelinating lesion in the genu of the corpus callosum and bilateral involvement of frontal lobes could be observed and the involvement of the parieto-occipital areas of the brain is only secondary (Loes et al., 2003; Yang and Prabhu, 2014) (Key Points 26.7).

26.2.2.2.1.2 Advanced MRI Findings Advanced MRI such as MRS, MTR, and DTI are useful for early detection of WM damage, monitoring progression, and treatment response (Rossi and Biancheri, 2013; van der Voorn et al., 2011). Metabolites at MRS well correlated with clinical status in child and in distinguishing between inflammatory demyelination and axonal loss in AMN (Rossi and Biancheri, 2013). NAA is usually decreased and lactate is increased. Recently, using 7-T MRS, Ratai and collaborators (2008) have shown that global elevated myo-inositol-to-creatine ratios correlate with the severity of clinical phenotypes. On the other hand, 7-T qMTR has shown potential to detect spinal cord and long tracts damage in the AMN (Smith et al., 2009). Furthermore, decreased brain MRI perfusion is a sign of early lesion development in X-ALD, while no changes in cerebral perfusion were found in AMN and females with X-ALD (Musolino et al., 2012). In all forms of X-ALD there is no documented damage of the GM (Yang and Prabhu, 2014). However, a very recent study (Salsano et al., 2014) has shown that using brain FDG-PET MRI the presence of functional abnormalities in the GM of all forms of ALD, which are mainly located at the level of frontal/cingulate (hypermetabolism) and cerebellar cortex (hypometabolism) and are completely

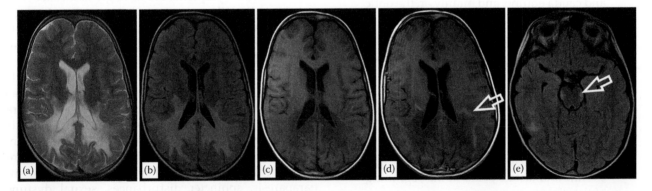

FIGURE 26.6
X-linked adrenoleukodystrophy. Preteen male with typical childhood cerebral X-adrenoleukodystrophy. On T_2-weighted image (a) and FLAIR (b), T_1-weighted image without (c) and with contrast enhancement (d), MRI reveals peritrigonal white matter lesions with involvement of the splenium of the corpus callosum. Note the multiple layers of demyelination with the outer enhancing layer consisting of active destruction surrounding the most severely damaged parietal white matter (d). Note also the extension of the white matter signal abnormality along the corticospinal tracts inferiorly in the midbrain (e).

independent of abnormalities detected on conventional MRI. These functional abnormalities seem to be also related to the severity of the disease (Key Points 26.8).

KEY POINTS 26.7   CONVENTIONAL MRI FEATURES IN X-ALD

- Symmetrical and confluent WM hyperintensiy begin periventricularly in parieto-occipital areas with characteristic involvement of the splenium of the corpus callosum and progress with a centrifugal gradient toward subcortical WM.

- Reversal pattern of WM abnormalities (fronto-occipital) could be observed in some adolescent forms.

- Contrast enhancement at the borders of WM abnormalities is typical.

- In adult form, involvement of brainstem long tracts (AMN) and cerebellar WM without contrast enhancement is observed.

KEY POINTS 26.8   ADVANCED MRI FEATURES IN X-ALD DISEASE

- MRS reveals decreased NAA, elevated Cho and lactate peaks at the beginning of the disease.

- Restricted water diffusion in diffuse WM demyelination.

- MRS metabolite levels correlated with clinical status and phenotypes.

- Functional abnormalities in the metabolism of glucose have been revealed by PET imaging in GM of patients with all forms of ALD.

26.2.2.3 LDs Due to Astrocytic Dysfunction

Although astrocytic dysfunction plays an important role in many neurological diseases (De Keyser et al., 2008; Lundgaard et al., 2014), a primary astrocytic dysfunction can lead to myelin defects (Lanciotti et al., 2013; Rodriguez, 2013). LDs with astrocytic dysfunctions include AxD, which is the prototype; MLC, childhood ataxia with CNS hypomyelination/VWM disease (CACH/VWM); and Canavan disease (Lanciotti et al., 2013; Rodriguez, 2013). Shared features of this group are cystic or spongiform degeneration of myelin, while the presence of Rosenthal fibers is pathognomonic of AxD;

megalocephaly is present in all these LDs. Although specific biochemical test is present only for Canavan disease and genetic and pathological confirmations are available for most of these diseases, the role of MRI is very important in the diagnosis of these WM diseases.

26.2.2.3.1 Alexander Disease

AxD or fibrinoid leukodystrophy is a progressive, fatal, genetic neurodegenerative disorder, usually affecting infants (Messing et al., 2012). Several mutations (Prust et al., 2011), that can be sporadic or autosomal dominant, have been found in the gene for glial fibrillary acidic protein (GFAP) on the chromosome 17q21 and genetic testing is now available (Quinlan et al., 2007). It is considered a *toxic gain of function disorder* (Messing et al., 2012), characterized by intracytoplasmatic accumulation in astrocytes of GFAP (Messing et al., 2012), resulting in the pathological manifestation of Rosenthal fibers. Simplifying, the disease could be compared to an *exaggerate form of gliosis in the brain* (Messing et al., 2012). An extensive description of possible pathogenetic mechanisms responsible for astrocytic dysfunction involved in this disease could be found in the review of Messing et al. (2012). Traditionally, a clinical classification based on the age of onset was used (Messing et al., 2012). However, a recent classification based on clinical and CNS involvement has been formulated (Prust et al., 2011). According to this new classification, AxD could be divided into types I and II. Characteristics of type I and type II AxD are the following according to Prust et al. (2011):

1. *Type I AxD*: It is characterized by the onset often before age of 4 years, usually with symptoms such as seizures, macrocephaly, encephalopathy, paroxysmal deterioration, failure to thrive, developmental delay, and classic radiologic features (van der Knaap et al., 2001). Possible development of hydrocephalus has been observed in some cases (Vanderver et al., 2014, *Gene Reviews*) coupled with other atypical radiological findings (Barreau et al., 2011).

2. *Type II AxD*: It manifests across the lifespan, and presents autonomic dysfunction; bulbar symptoms; ocular movement abnormalities; recurrent vomiting, often negative for neurocognitive or developmental deficits; and atypical radiologic features.

Several treatments for AxD are under investigation and they can be classified into treatment that reduce the expression and accumulation of GFAP and treatment that reduce the toxic effect of GFAP accumulation (Messing et al., 2010). In type II AxD, the intravenous administration of the antibiotic ceftriaxone has recently shown the potential to slow the disease progression and improve quality of life (Sechi et al., 2013).

26.2.2.3.1.1 Conventional MRI Findings Classical radiological features of this disease have been provided by van der Knaap et al. (2001). *Type I AxD* is characterized by[*]

1. Symmetrical cerebral WM abnormalities mainly in frontal lobe. Usually the initial involvement of the frontal lobe is centripetal starting from subcortical WM (Yang and Prabhu, 2014).
2. Presence of periventricular garlands and rim of decreased signal intensity on T_2-weighted images and elevated signal intensity on T_1-weighted images.
3. Abnormalities of the basal ganglia and thalami, either in the form of elevated signal intensity and some swelling or atrophy and elevated or decreased signal intensity on T_2-weighted images.
4. Brain-stem abnormalities, in particular involving the midbrain and medulla.
5. Contrast enhancement involving one or more of the following structures: ventricular lining, periventricular rim of tissue, WM of the frontal lobes, optic chiasm, fornix, basal ganglia, thalamus, dentate nucleus, cerebellar cortex, and brain-stem structures. To diagnose AxD, four of these five criteria have to be fulfilled (van der Knaap et al., 2001).

A recent study (Graff-Radford et al., 2014) has expanded our knowledge regarding the type II AxD focusing on clinical and radiological findings, previously defined as atypical. *Type II AxD form* is characterized by symptoms and MRI damage based of the major involvement of medulla and spinal cord. Involvement of dentate nucleus and middle cerebral peduncle enhancement are frequent while periventricular signal change is rare. Garland-like feature along the ventricular wall (ependymal nodularity) are less frequent than in child form. Myelopathy is the most common clinical feature at presentation, while palatal tremor is rare (Key Points 26.9).

26.2.2.3.1.2 Advanced MRI Findings At MRS in type I AxD, the decreased NAA and elevated lactate and myo-inositol peaks may be useful in grading the severity of the disease (Nelson et al., 2013; Ratai et al., 2012;

Rossi and Biancheri, 2013). No other specific findings are provided by advanced MRI techniques (Key Points 26.10).

KEY POINTS 26.9 CONVENTIONAL MRI FEATURES IN AxD

- Symmetrical and confluent WM hyperintensity involved frontal subcortical areas and progress centripetally.
- There is involvement of brain stem and deep GM structures.
- Contrast enhancement is present.
- In adult forms, involvement of brain stem and spinal cord is typical.

KEY POINTS 26.10 ADVANCED MRI FEATURES IN AxD

- MRS reveals decreased NAA and elevated Cho and lactate peaks at the beginning of the disease. Myo-inositol peak is typically elevated in type I forms.

26.2.2.3.2 CACH or VWM Disease

Being a recently added entity, VWM (van der Knaap et al., 1997) disease or CACH (Schiffmann et al., 1994) is an autosomal recessive leukoencephalopathy caused by a mutation in any of the genes that encodes the five subunits of eIF2B, the eukaryotic translation initiation factor (van der Knaap et al., 2002). Clinically, common symptoms are slowly progressive cerebellar ataxia, spasticity, optic atrophy, and cognitive deterioration usually precipitated by minimal stress such as fever leading to coma (Bugiani et al., 2010). Epilepsy and hypotonia appear as the disease progresses (Bugiani et al., 2010). Contrarily to the others LDs characterized by astrocytes dysfunction, involvement of other organs and systems beyond the CNS is commonly present (hepatosplenomegaly, kidney hypoplasia, pancreas involvement, and mainly ovarian dysgenesis). Macrocephaly and ovarian failure are typically present (Phelan et al., 2008). Typically children are affected, but increasing evidences suggest involvement of all ages. Clinically, four variants have been described ranging from antenatal onset (van der Knaap et al., 2006) to adult onset. The adult form, characterized by a greater variability of clinical features, has also a milder and subtler disease course compared to the child onset, with motor and behavioral symptoms, dementia, and seizures. Diagnosis is done

---

[*] Adapted from van der Knaap, M.S. et al., *AJNR Am. J. Neuroradiol.*, 22, 541, 2001.

by genetic testing but MRI is extremely helpful in suggesting the disease. Treatment is not available and the disease is mostly fatal.

26.2.2.3.2.1 Conventional MRI Findings Conventional MRI shows symmetrical and diffuse WM damage starting from periventricular areas toward subcortical WM with evolution over time and appearance of confluent cystic formation until WM rarefaction fulfilled with CSF (van der Knaap et al., 2006). Cerebellar atrophy is variably present. In the adult form, the progression of the disease could be extremely slow and cystic and a rarefaction of the WM could really be observed only in the very later stages. While the MRI diagnosis is simple in the well-defined form of the disease, the diagnosis is very difficult in the early phases. van der Lei et al. (2012a) have suggested some clues to identify at the very early stage the disease with conventional MRI. At the very early stage, looking for the involvement of the inner rim of the corpus callosum is very helpful (van der Lei et al.). Furthermore, even in the earlier stage, brain conventional MRI is never completely normal and lesions in tegmental tracts could be found (van der Lei et al., 2012a). Cortex and basal ganglia are usually spared (Rossi and Biancheri, 2013) (Key Points 26.11).

26.2.2.3.2.2 Advanced MRI Findings MRS pattern is not specific with reduction of all metabolites peaks with only lactate and glucose observed. Myo-inositol peak, considered the hallmark of astrocytes dysfunction, is not observed in VWM disease (Ding et al., 2012). In cystic WM complete absence of all metabolites is observed (Ratai et al., 2012), but Lactate peak can be encountered (Yang and Prabhu, 2014). Restricted diffusion is found even when the WM is not damaged at conventional MRI (van der Lei et al., 2012b). The above mentioned are usually early occurring features and typical in younger patients (van der Lei et al., 2012b) (Key Points 26.12).

KEY POINTS 26.11 CONVENTIONAL MRI FEATURES IN VWM DISEASE

- Symmetrical and diffuse WM hyperintensiy begin periventricularly and progress centrifugally.
- At initial stage, the involvement of the inner rim of the corpus callosum is typical.
- Central tegmental tracts abnormalities are present.
- Cerebellar atrophy is variably present.
- At later stages, confluent cystic formation and rarefaction of WM are observed.

KEY POINTS 26.12 ADVANCED MRI FEATURES IN VWM DISEASE

- MRS reveals reduction of all metabolites peaks, with only lactate and glucose peak revealed. Myo-inositol peak is not observed.
- Restricted water diffusion in diffuse WM demyelination.

26.2.2.3.3 Canavan Disease

Canavan disease or spongy deterioration of the CNS is a rapid progressive autosomal recessive disease due to abnormalities in the encoding of the enzyme aspartoacyclase (Hoshino and Kubota, 2014). This enzyme is responsible for the catabolism of NAA; thus, its deficiency leads to NAA accumulation. The hallmark of this disease is the presence of NAA in all body fluids (Hoshino and Kubota, 2014) and in typical WM markedly increased peak of NAA at MRS is observed (Rossi and Biancheri, 2013; Yang and Prabhu, 2014). Urinary NAA concentrations are useful for diagnosing (Hoshino and Kubota, 2014). Several pathogenetic hypotheses have been formulated and the most acclaimed is that there is a dysfunction of NAA catabolism, which serves as water pump, thus leading to astrocytic edema (Lanciotti et al., 2013). Clinically, it is mostly frequent in infants; although three forms of the disease are recognized such as neonatal, infantile, and juvenile. Normally it is a rare disease with disproportionally high frequency in the Ashkenazi Jewish population and the juvenile form is extremely rare. Typical symptoms are hypotonia and visual-cognitive dysfunction, and psychomotor delay, followed by seizures, fevers of unknown origin, difficulties in feeding, and spasticity that rapidly progress to death. Macrocephaly is always present (Hoshino and Kubota, 2014). Pathologically, spongy deterioration and abnormal astrocytic mitochondria are observed (Matalon et al., 1988). The diagnosis is strongly suggested by the presence of macrocephaly and spectroscopy findings. Although the cause is well known, so far there is no available cure for this disease. Dietary supplements, antioxidants and gene therapies are under investigations (Hoshino and Kubota, 2014; Rodriguez, 2013).

26.2.2.3.3.1 Conventional MRI Findings Conventional MRI shows subcortical cerebral U fibers and cerebellar WM involvement with centripetal gradient and a diffuse deep WM involvement in the later stages (Michel and Given, 2006; Sreenivasan and Purushothaman, 2013). The involvement of globus pallidum and thalami with sparing of other deep GM structures is pathognomonic

(Sreenivasan and Purushothaman, 2013; Yang and Prabhu, 2014). Occasionally, the brain stem and dentate nucleus may be involved (Yang and Prabhu, 2014) (Key Points 26.13).

26.2.2.3.3.2 Advanced MRI Findings Markedly increased NAA peak at MRS (Key Points 26.14).

KEY POINTS 26.13   CONVENTIONAL MRI FEATURES IN CANAVAN DISEASE

- Symmetrical and confluent WM hyperintensiy involved subcortical areas of the brain and cerebellum and progress centripetally.
- Typical involvement of globus pallidus with relative sparing of other deep GM structure.

KEY POINTS 26.14   ADVANCED MRI FEATURES IN CANAVAN DISEASE

- At MRS, a marked increase of NAA is typically present.
- Restricted water diffusion in diffuse WM demyelination.

26.2.2.3.4 MLC or van der Knaap–Singhal Disease

MLC is a very rare LD associated with the mutation of two genes, *MLC1* mapped on chromosome 22 and *HEPACAM*. Several mutations have been described and the most frequent is the autosomal recessive. Interestingly, the *MLC1* is an astrocytic-specific protein, not present on oligodendrocytes. Although its function and role is still not clearly understood, it seems to be involved together with the *HEPACAM* traffic protein in regulating brain homeostasis through transport of water and/or ions between astrocytes and the blood or CSF (van der Knaap et al., 2012). Thus, MLC is a disease related to abnormalities in the regulation of ions and water content (Lanciotti et al., 2013). However, the cases without mutations of *MLC1* are also described (van der Knaap et al., 2010), usually associated with improvement or disappearance of MRI findings over time. Clinically, two forms have been described. The classical form is characterized by macrocephaly within the first year of life and later appearance of ataxia spasticity and cognitive decline. The second form is the remitting form with a very mild course starting with macrocephaly but developing very mild motor disturbance and a complete remission with lack of mental deterioration. Usually this form is associated with autosomal dominant mutation of the *HEPACAM* gene (van der Knaap et al., 2012). Pathologically, the hallmark of this disease is the myelin vacuolation and gliosis (van der Knaap et al., 1996). Laboratory and neurophysiological test are usually normal. Essentially, MRI is used for diagnosis. Treatment is under investigation (Perlman and Mar, 2012).

26.2.2.3.4.1 Conventional MRI Findings At conventional MRI, a macrocephalic brain presents abnormal subcortical WM with marked swelling and cysts. Cysts are usually bilateral, they increase in size overtime, and they are mainly located in the anterior temporal lobe and less frequently in the frontal lobe. Typically, the relative sparing of the central WM areas such as corpus callosum, internal capsule, and brain stem may be seen. Atrophy is a common feature and the GM is always spared (van der Knaap et al., 2012; Yang and Prabhu, 2014) (Key Points 26.15).

26.2.2.3.4.2 Advanced MRI Findings At MRS, a marked decrease of all metabolites is observed associated with an increase of the myo-inositol peak in severely affected WM. At diffusion, there is an increase of apparent diffusion coefficient (ADC) maps and diffusivity with reduced anisotropy (van der Voorn et al., 2006) (Key Points 26.16).

KEY POINTS 26.15   CONVENTIONAL MRI FEATURES IN MLC

- Macrocephaly with subcortical diffuse WM hyperintensity is characterized by edema and bilateral anterior temporal cysts.
- Relative sparing of central WM structures such as corpus callosum, internal capsule, and brain stem.
- GM is never involved.

KEY POINTS 26.16   ADVANCED MRI FEATURES IN MLC

- At MRS, in severely affected WM, reduction of all metabolites peaks with the exception of myo-inositol peak that is increased.

26.3 Demyelinating Diseases

Demyelinating diseases are mostly acquired CNS disorders due to the damage of properly formed myelin in adults or children (Hatten, 1991). This heterogeneous group of CNS diseases can be classified according to their etiopathogenesis (see Table 26.1). The first group will include idiopathic demyelinating diseases that are demyelinating diseases of unknown etiopathogenesis. However, an autoimmune pathogenesis is suspected for the majority of them. Multiple sclerosis (MS) is the most common idiopathic disease. Last decades have seen rising and promising progress in the diagnosis, pathological specificity, monitoring, and treatment of MS and its variants. The second group of diseases includes diseases due to allergic or immunological reactions, infections, vaccinations, nutritional/vitamin deficiency, physical/chemical agents, medical therapy, and vascular (Bester et al., 2013). Thus, as for LDs, although, classifying these diseases is a basic way to approach them, however, the classification is challenging, in part for the overlapping etiopathogenesis, and also because it is continuously updated (Table 26.1). Pathologically, they were classically considered diseases in which only myelin is damaged with sparing of axons, however, recent advancement in research have highlighted that in some of them damage of axons is present as well as damage of the GM. A good example of this is MS. In MS, both damage of axons and GM has been lately discovered. Clinically, all symptoms related to CNS involvement can be found, while the involvement of other organs or systems is generally rare compared to LDs. Although the diagnosis is still driven by clinical observations, in the last decades MRI has played a pivotal role in helping with differential diagnosis, in monitoring progression and treatment response, and dramatically changed the diagnostic workup of these diseases improving our related knowledge. For example, so far, diagnostic criteria for MS, neuromyelitis optica (NMO), acute disseminated encephalomyelitis (ADEM), and pediatric MS have assimilated conventional MRI findings. Furthermore, as for LDs, advanced MRI techniques coupled with high and ultrahigh field strength have shed light on the pathological substrates of the occult tissue damage and can help in the differential diagnosis. For example, in MS, using ultrahigh field strength, it has been possible to visualize perivenular localization of WM lesions. Perivenular localization is very helpful in differentiating between MS and lesions that mimic MS. Furthermore, newer techniques, such as double inversions recovery (DIR) and T_1-weighted phase-sensitive inversion recovery (PSIR), have allowed the detection of cortical lesions. Thus, ongoing investigations in the last years are pointing out to newer qualitative and quantitative advanced MRI techniques and

new post-processing software that will dramatically impact this field (Ceccarelli et al., 2012). Together with the advent of powerful diagnostic and research tools, the last decades have also seen an increased array of diverse novel therapeutics options, in particular for MS. In this section, we will describe the classical and most common demyelinating diseases in which peculiar MRI findings can be observed and are able to help in disease characterization and diagnostic criteria. A full review of all demyelinating disease is beyond the scope of this chapter.

26.3.1 Idiopathic Demyelinating Diseases

26.3.1.1 Multiple Sclerosis

MS is a chronic progressive disabling disease of the CNS affecting mostly young adults (between 20 and 40 years old) and mainly women (female/male ratio 3:1) (Bove and Chitnis, 2013; Compston, 2008). The etiology of MS is unknown, while, although the exact mechanism is still obscure, its complex pathogenesis and pathophysiology include immune system dysregulation, with inflammation and axonal damage (Lassmann et al., 2007; Wu and Alvarez, 2011). Furthermore, genetic, behavioral, and environmental risk factors modulate MS disease susceptibility and progression (Ascherio, 2013; Sawcer et al., 2014). Indeed, MS incidence and prevalence are strictly associated with latitude (Kingwell et al., 2013; Simpson et al., 2011), with exception of some population (Rosati, 2001) and ethnicities (Aguirre-Cruz et al., 2011). Pathologically, the hallmark of MS consists of brain and spinal cord focal demyelinated lesions, with variable degrees of inflammation, gliosis, and neurodegeneration (Popescu et al., 2013). Furthermore, emerging evidences have suggested presence of meningeal and cortical inflammation as an early event (Popescu et al., 2013). Clinically, MS usually starts with a clinically isolated syndrome (CIS) consisting of optic neuritis, brain stem, or spinal cord event that last more than 24 h (Brownlee and Miller, 2014). Multifocal presentation is also possible in 15% of the cases (Brownlee and Miller, 2014) and is a sign of dissemination in space of the disease. The majority of these patients have already brain lesions at MRI. Within 2 years, 80% of CIS patients will develop MS (Brownlee and Miller, 2014). Risk factors for conversion to MS have been extensively studied and several of them have been found such as younger age, clinical dissemination in space at onset, cognitive impairment at onset, the presence of the allele HLA-DRB1/1501, and the presence of oligoclonal bands. However, the stronger risk factor for conversions is the presence of conventional MRI abnormalities (Brownlee and Miller, 2014). Once MS is developed, the most common disease course is the relapsing remitting (RR) course characterized by clinical attack (relapse) followed by periods of remission. Every relapse can resolve spontaneously or after treatment or leave clinical sequelae (Compston, 2008). Approximately 60% of patients will develop secondary progressive (SP) MS (SPMS) with the development of severe and irreversible disability (Vukusic and Confavreux, 2003). A 10% of patients will start with a progressive onset affecting mostly spinal cord characterized by constant progression of disability from onset without improvement (Rice et al., 2013) called primary progressive (PP) form. Patients with PPMS manifest some differences in gender ratio, age of onset, and CNS involvement compared to RRMS suggesting a prevalence of neurodegenerative mechanisms (Antel et al., 2012). On the contrary, MS patients with long disease duration and minimal disability are called benign MS (Rovaris et al., 2009). Around 45%–60% of MS patients present cognitive impairment together with other neurological symptoms (Amato et al., 2010), due to the involvement of motor, sensory, visual, and autonomic systems. Some features are clinically specific of MS such as Lhermitte and Uhthoff signs (Compston and Coles, 2008). There are currently various numbers of Food and Drug Administration (FDA)-approved treatments for MS, ranging from immunomodulatory drugs to immunosuppressive drugs (Inglese and Petracca, 2014). They target mostly RRMS patients. There are no treatments for PPMS and only one drug is targeting SPMS (mitoxantrone). High doses of corticosteroids are used for the acute treatment of relapses. First-line long-term treatments for MS are injectable drugs such as interferons (beta-1a and -1b), glatimer acetate, but also new oral drugs such as fingolimod, teriflunomide, and dimethyl fumarate. If first-line treatment fails, switching to second-line treatment is recommended. Second-line therapeutic options are natalizumab, fingolimod, or cyclofosfamide. Other approaches such as hematopoietic and mesenchymal stem cell transplantation are under investigation. Neurorehabilitation, commonly administered to MS patients, is also being improved. The diagnosis of MS has been historically a clinical diagnosis (Poser and Brinar, 2001); however, in the last few decades, the use of MRI has revolutionized and played a pivotal role in the diagnosis of MS. In CIS patients, an MRI protocol has been suggested (Filippi et al., 2013; Simon et al., 2006). This MRI protocol includes brain and spinal MRI sequences. Mandatory brain MRI sequences in suspected MS or in the follow up study of established MS are axial T_2-weighted sequence, proton density, and T_1-weighted sequence pre and postcontrast. Sagittal FLAIR sequences are also needed. Spinal cord MRI will include a sagittal T_2 FSE and STIR-weighted sequence, a sagittal T_1-weighted SE, axial T_2-weighted, axial T_1-weighted SE post-Gd, and sagittal T_1-weighted SE post-Gd. All these sequences should have a slice thickness of 3 mm. Optional sequence for the brain should be an axial DIR (Filippi et al., 2013) to study cortical lesions. Once MS diagnosis is made, follow-up conventional MRI is useful for detecting disease activity (new T_2 and active lesions, formation of black holes and atrophy) and monitoring treatment response with MRI findings.

26.3.1.1.1 Conventional MRI Findings

At conventional MRI, MS is characterized by the presence of focal hyperintense on T_2-weighted sequence lesions in the deep WM of the brain and in the spinal cord. Based on morphology and distribution in space and time of WM lesions in MS, diagnostic criteria to improve the earlier diagnosis have been proposed and revised several times. The new diagnostic criteria have been summarized in Table 26.2 (Polman et al., 2011) (Key Points 26.17).

Brain focal WM T_2-hyperintense lesions. MS focal WM lesions appear bright or hyperintense on T_2- and FLAIR-weighted sequences. Typical WM lesions morphology and localization help to distinguish MS lesions from those of other WM diseases. Hyperintense WM lesions are characteristically focal, oval to ovoid in shape, and greater than 5 mm in size. They are preferentially located periventricular, on the surface of the corpus callosum, the juxtacortical GM/WM junction, in the infratentorial brain regions, and in the spinal cord (Filippi et al., 2013; Geurts et al., 2005). Their distribution is always asymmetric (Figure 26.7). Periventricular lesions are

TABLE 26.2

Current MRI Diagnostic Criteria for Multiple Sclerosis

For Dissemination in Space
Dissemination in space is fulfilled by >1 T_2 lesion in at least two of four CNS areas of the following:
• Periventricular
• Juxtacortical
• Infratentorial
• Spinal cord

For Dissemination in Time
Dissemination in time is fulfilled by
• A new T_2 or gadolinium-enhancing lesions on follow-up MR scan irrespective of the timing of the baseline MRI
• Simultaneous presence of enhancing and non-enhancing lesions at any time.

Source: Adapted from Polman, C.H. et al., *Ann. Neurol.*, 69, 292, 2011.
Note: Gadolinium-enhancing lesions are not more needed for defining dissemination in space and symptomatic lesions are excluded from the criteria and do not contribute to lesion count in brain stem and spinal cord syndrome.

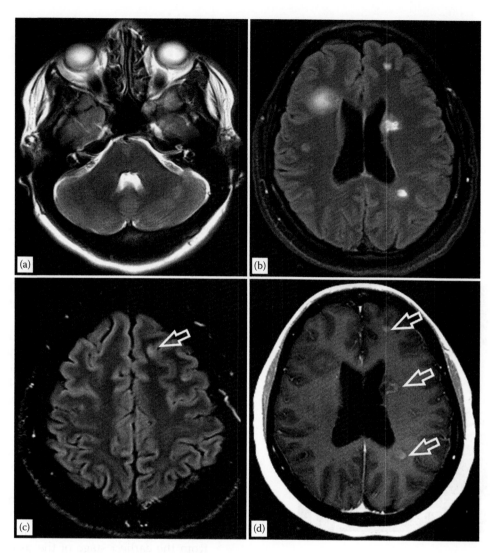

FIGURE 26.7
Focal white matter lesions in multiple sclerosis. Focal white matter lesions are noticed in the brain stem and cerebellum (a), periventricular white matter, adjacent to the corpus callosum (b), in the juxtacortical white matter (c) (arrow). Multiple enhancing lesions with different enhancing pattern are also noticed bilaterally, in the periventricular region (d) (arrows).

defined as WM lesions that are attached to the wall of lateral ventricles, while juxtacortical lesions are lesions attached to the cortex and with the extension in the WM. Infratentorial lesions are mainly located in pons, cerebellum, and cerebellar peduncles. Using ultrahigh field scanner, MS lesions have shown a pathognomonic perivenular location (Tallantyre et al., 2011) (Figure 26.8). A typical perivenular location directly adjacent to the corpus callosum as well as perpendicular to the major axis of the lateral ventricles defines the *Dawson finger* pattern (Figure 26.9). Over time, WM T_2-hyperintense lesions can remain stable, disappear or increase in size (Meier et al., 2007). FLAIR sequence is superior compared to T_2-weighted sequence in

detecting periventricular and cortical–juxtacortical lesions, on the other hand T_2-weighted sequence is more sensitive compared to FLAIR sequence in detecting MS lesions at the level of posterior fossa and spinal cord. WM T_2-hyperintense lesions are not pathologically specific. Although they show weak correlation with clinical measures, they have a high predictive value for conversion to MS and have been used as surrogate markers of disease activity in clinical trials for MS.

Brain gadolinium-enhancing lesions. A subset of WM T_2-hyperintense lesions could enhance or appear bright on T_1-weighted images after infusion of an intravenous Gadolinium (Gd) contrast agent. Usually Gd enhancement precedes

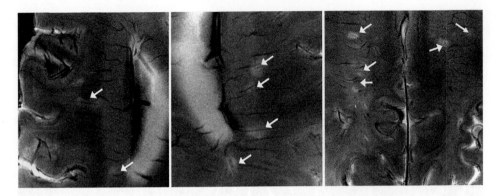

FIGURE 26.8
7 Tesla T_2^*-weighted sequence with in-plane resolution of 200 μm × 200 μm in a patients with multiple sclerosis. Several white matter lesions in a patient with multiple sclerosis with perivenuos appeerence (arrows). (Courtesy of Dr. Yulin Ge.)

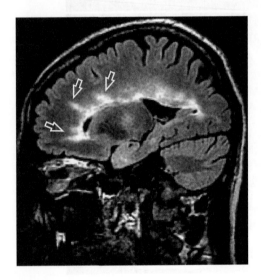

FIGURE 26.9
Dawson fingers pattern. Sagittal FLAIR MR image of the brain in a patient with relapsing remitting multiple sclerosis demonstrating white matter lesions oriented perpendicularly to corpus callosum, along medullary veins (Dawson fingers, arrows).

or accompanies new T_2-hyperintense lesion formation and lasts for an average of 3 weeks, pathologically reflecting blood–brain barrier inflammatory disruption and indicating acute phase of MS disease. Enhancement pattern may be concentric ring, open ring, tumor-like, and heterogeneous or homogeneous (Filippi et al., 2013; Minneboo et al., 2005; Rovira et al., 1999). The ring-enhancement pattern is indicative of a more severe damage (Figure 26.7). Gd lesions are more frequent in patients with RRMS.

Brain T_1-hypointense lesions. Another subset of T_2-hyperintense WM lesions could appear dark or hypointense on the corresponding T_1-weighted sequence. T_1-hypointense lesions over time can be transient due to remyelination

or remain stable and permanent. The latter are called *black holes* and are sign of irreversible severe tissue damage (Sahraian et al., 2010). Black holes are seen mostly in progressive forms and better correlate with clinical measures than WM T_2-hyperintense lesions. They are more pathologically specific than T_2-hyperintense lesions reflecting more neurodegeneration and gliosis and better correlated with clinical measures (Figure 26.10).

Brain T_2 hypointensity. Compared to controls, MS patients show dark appearance or hypointensity on T_2-weighted sequence at the level of subcortical GM (mainly in thalami and basal ganglia) (Figure 26.11). Deep GM T_2 hypointensity have been described in all MS phenotypes from the earliest stage of the disease and correlated with clinical measures and progression. Pathologically T_2 hypointensity is considered a putative marker of iron deposition (Ceccarelli et al., 2009; Stankiewicz et al., 2014).

Cortical lesions. While FLAIR sequence is able to detect mostly juxtacortical lesions, the introduction of the DIR sequence, which suppresses the signal from CSF and WM, in the standard MRI protocol for MS (Filippi et al., 2013) has enabled the detection of cortical lesions in MS, which are not visible on T_2-weighted sequences. Cortical lesions appear as bright lesions in the cortex on DIR sequence and dark lesions on T_1-weighted sequence. The combined use of DIR and a PSIR can aid the differentiation between purely cortical and mixed GM–WM lesions, which is challenging using DIR or FLAIR images alone (Neema et al., 2012). However, so far, classification of cortical lesions into juxtacortical (type I), intracortical (type II), and subpial (type III) remains difficult with MRI, predominantly for

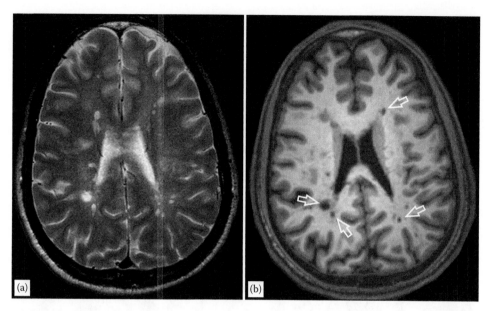

FIGURE 26.10
Black holes. Axial T_2 (a) and axial MPRAGE (b) MR images of the brain in a patient with primary progressive multiple sclerosis demonstrating white matter lesions and corresponding black holes (arrows).

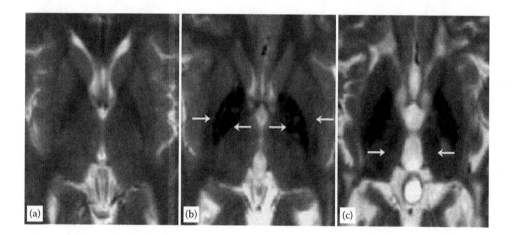

FIGURE 26.11
Deep GM T_2 hypointensity in multiple sclerosis. Axial T_2-weighted MRI scans of an age-matched healthy control subject (a—51-year-old woman), a patient with benign MS (b—47-year-old woman), and SPMS (c—43-year-old man). Note that both MS patients have more prominent T_2 hypointensity in the deep GM nuclei, particularly in the bilateral putamen and globus pallidus (b [arrows], c) versus the healthy control. However, note that the patient with SPMS has brain atrophy (widening of the ventricles and sulci) as compared with the two other subjects. In addition, the patient with SPMS has T_2 hypointensity of the bilateral thalamus (c, arrows). (Reproduced from Ceccarelli, A. et al., *Mult. Scler.*, 15, 678, 2009. With permission from the publisher.)

the subpial cortical lesions. Recent MRI studies have shown that cortical lesions that appear early in the disease are present in all MS phenotypes (Ceccarelli et al., 2012; Neema et al., 2012) and in CIS their detection increase the risk of conversion, tends to accumulate over time, and better correlate with clinical status than WM lesions (Neema et al., 2012). They are an expression of cortical demyelination that is widespread in MS, although less inflammatory than WM demyelination (Figure 26.12).

Brain atrophy. Global brain atrophy is a typical feature of MS. Brain atrophy could be subtle or macroscopically clearly visible at conventional MRI images, in particular in the advance disease stages. If visible at conventional MRI, brain atrophy appear with enlargement of CSF spaces, including sulci, cisterns, fissures, and the ventricular system, combined with reduced size of parenchymal structures, in comparison with age-matched healthy controls (Neema et al., 2012) (Figure 26.13). In addition

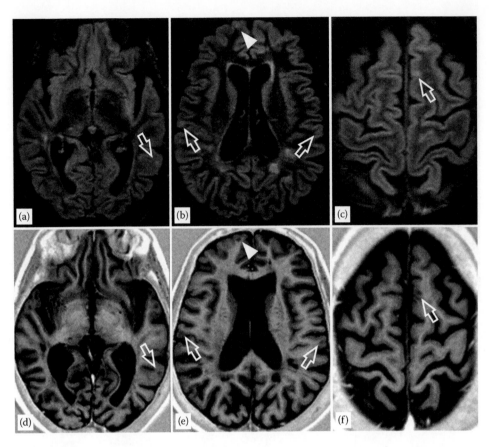

FIGURE 26.12
Cortical GM lesions. Axial DIR (a–c) and axial PSIR (d–f) MR images of the brain in a patient with primary progressive MS demonstrating focal lesions in the cortical GM (arrows). Head arrow shows change of classification of cortical GM lesions. DIR leuco-cortical lesion (head arrow, b), appear as juxta-cortical white matter lesion on PSIR (head arrow, e).

to visual inspection, several methods have been employed to uncover subtle brain atrophy. In particular, in the earlier disease stages the methods range from measure of the ventricular width (lateral and third), or intercaudate distance, or area of the corpus callosum on a sagittal section, to more complex post-processing software, such as voxel-based morphometry (Bermel and Bakshi, 2009; Ceccarelli et al., 2012; Neema et al., 2012). By using automated image segmentation tools on T_1-weighted sequences, it is possible to quantify global brain atrophy but also GM, WM atrophy, and atrophy of specific regions separately. These analyses have shown in MS that although MS patients present both GM and WM atrophy from the onset of the disease, GM atrophy is a better sensitive marker of neurodegeneration. It starts earlier than WM atrophy in the disease course, better correlates with clinical status, and accumulates to a greater degree than global brain and WM atrophy over time (Ceccarelli et al., 2012; Neema et al., 2012). Furthermore, distinct topographical

distribution of regional GM atrophy has been shown to differentiate MS patients from controls, according to their phenotypes (Figure 26.14) (Ceccarelli et al., 2008) and clinical status (cognitive dysfunction and fatigue) (Ceccarelli et al., 2012). GM atrophy of thalamus and basal ganglia seems to occur early while progressive patients manifest greater amount of cortical atrophy (Ceccarelli et al., 2008).

Spinal cord lesions and atrophy. Lesions in the spinal cord are also typical of MS. Hyperintense T_2-weighted lesions are mostly located in the cervical spinal cord, involving no more than two segments in lengths and strongly correlated with MS patients' disability. They are located more peripherally and in less than half of the cross-sectional area of the cord and usually not visible on T_1-weighted sequence (Bot and Barkhof, 2009) (Figure 26.15). Greater lesion sensitivity can be achieved by the addition of STIR (short-tau inversion recovery) or PSIR sequences (Neema et al., 2012). Gd lesions are less frequent than in the brain. Atrophy of the spinal cord is present

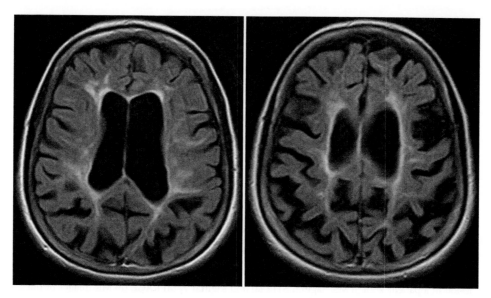

FIGURE 26.13
Patient with multiple sclerosis and brain atrophy. FLAIR images of a patient with 19 years of disease duration demonstrate multiple confluent areas of hyperintensity, associate with dilation of the ventricular system and of the cortical sulci, suggesting severe brain atrophy.

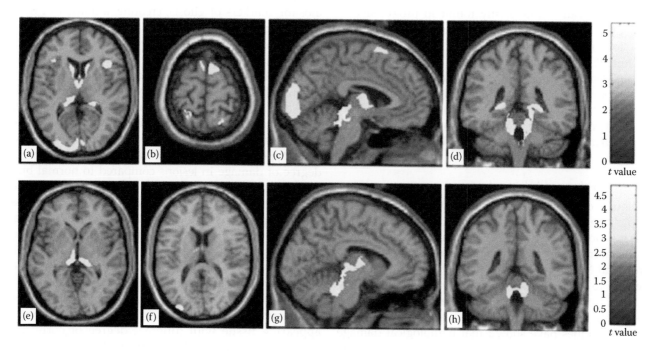

FIGURE 26.14
Patterns of regional GM atrophy between different multiple sclerosis phenotypes. Statistical parametric mapping (SPM) regions with decreased GM concentration, overlaid on a high-resolution T_1-weighted image, contrasting patients with different multiple sclerosis (MS) phenotypes ($p < .001$, corrected for multiple comparisons at a voxel level). a–d: SPM regions with significant GM loss in SPMS patients compared with relapsing-remitting MS patients: (a) bilateral insula, bilateral thalamus, bilateral caudate nucleus, bilateral cuneus, and left middle occipital gyrus; (b) bilateral postcentral gyrus and bilateral superior frontal gyrus; (c) right superior and inferior colliculus, right thalamus, right cuneus, and right superior frontal gyrus; (d) bilateral parahippocampal gyrus and bilateral anterior lobe of the cerebellum. e–h: SPM regions with significant GM loss in SPMS patients compared with primary progressive (PP) MS patient: (e) bilateral thalamus; (f) left thalamus and left middle occipital gyrus; (g) left thalamus, left anterior lobe of the cerebellum, and left superior and inferior colliculus; (h) bilateral anterior lobe of the cerebellum. Images are in neurological convention. See text for further details. (Reproduced from Ceccarelli, A. et al., *Neuroimage*, 42, 315, 2008. With permission from the publisher.)

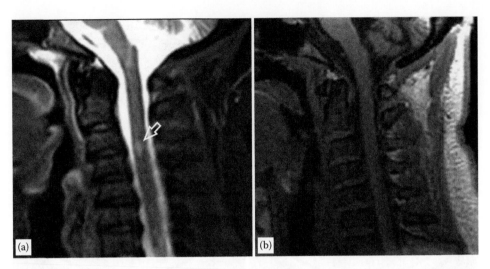

FIGURE 26.15
Cervical spinal cord lesion in a patient with SPMS. Sagittal STIR (a) and sagittal T_1-MPRAGE (b) MR images of the spinal cord in a SPMS patient demonstrating focal lesion in upper cervical cord (arrow).

in MS and it mostly affects the cervical cord (Bot and Barkhof, 2009). Although quantification of cord atrophy is challenging (Neema et al., 2012), in MS, it is strongly correlated with disability measures and mostly found in progressive phases. Interestingly, cord abnormalities appear to be independent of brain abnormalities (Neema et al., 2012). A very recent study has shown than GM spinal cord atrophy is present in RRMS with lack of WM cord atrophy, it is higher in progressive MS than in RRMS, and contributes more to patient disability than spinal cord WM or brain GM atrophy (Schlaeger et al., 2014).

26.3.1.1.2 Advanced MRI Findings

Although conventional MRI is sensitive to MS macroscopic abnormalities, it is not specific pathologically and it is not able to detect occult tissue damage present in the normal-appearing brain tissue. Thus, advanced structural MRI techniques such as DTI, MTI, and MRS are able to examine more pathologically specific injury in both lesions and normal-appearing brain tissue (Ceccarelli et al., 2012). In general, all these advanced MRI techniques have shown that pathological damage is not confined within the macroscopic lesions visible on conventional MRI, but it is present in the NAWM and normal-appearing GM in all MS phenotypes, from the earliest stages of the disease and increase notably in the progressive phases. Interestingly, the damage in the normal-appearing brain tissue may precede lesions formation and it is more pronounced in black holes than in active and stable WM lesions. Furthermore, it is better correlated with clinical measures than conventional MRI features and the degree of damage may vary in different

clinically eloquent brain regions. Specifically, at MRS, NAA levels, as marker of axonal damage, are reduced in active lesion and in normal-appearing brain tissue, while lactate, a marker of energy failure; choline and myo-inositol, markers of gliosis; creatine, a marker of membrane turnover in demyelination and remyelination; and lipids levels, marker of myelin breakdown, appear normal in most lesions after an initial surge, suggestive of remyelination and the resolution of edema. In chronic lesions, NAA is markedly reduced while myo-inositol is markedly increased (Neema et al., 2012) (Figure 26.16).

MTR reduction and DTI metrics abnormalities, reflecting myelin and axonal damage, are altered in every MS phenotypes starting from the earliest stage with higher degree of damage in lesions compared to normal brain tissue. Regional analysis of DTI and MTR has provided further insight into occult tissue damage in normal-appearing brain tissue and correlations with clinical measures, showing that the degree of damage vary according to regions and tracts (Neema et al., 2012; Raz et al., 2010) (Figure 26.17). In addition, a recovery of MTR is supposed to reveal remyelination and repair (Neema et al., 2012).

As for the brain, advanced MRI analysis of the cord has shown abnormalities in both lesions and normal-appearing tissue with different degree of correlation with clinical status (Ceccarelli et al., 2012). Recently regional analysis of the cord also has been employed showing ability in detecting clinically relevant information (Schlaeger et al., 2014; Toosy et al., 2014).

Furthermore, functional MRI studies (including connectivity and resting-state studies) in both brain and spinal cord of patients with MS have shown that cortical reorganization varies according to the disease phase and phenotypes, having mainly an adaptive role, consisting of an increased recruitment of brain regions inside and

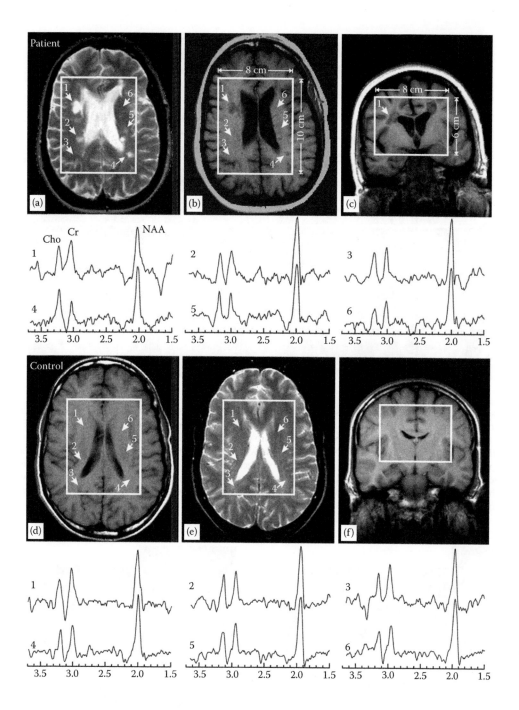

FIGURE 26.16

MRS in multiple sclerosis. (a) Transverse T_2-weighted (2500/90) MR image in a 26-year-old woman with MS. (b) Transverse T_1-weighted (450/14) MR image. (c) Coronal T_1-weighted (450/14) MR image. MR spectroscopic volume of interest has been superimposed on the images. Spectra from hypo- (arrows *1* and *4*) and isointense (arrows *2* and *5*) lesions, as well as two NAWM regions (arrows *3* and *6*), are shown below the images on common intensity and chemical shift (in parts per million) scales. (d–f) Corresponding sections from a matched control subject. Numbered arrows indicate equivalent regions to a–c for metabolite levels and spectra comparisons. (Reproduced from He, J. et al., *Radiology*, 234, 211, 2005. With permission from the publisher.)

outside the classical task-related network, to offset the functional impact of the diffuse irreversible occult tissue damage in MS (Filippi and Rocca, 2013) (Figure 26.18).

Other advanced MRI techniques, more confined to the research setting, such as MWF, a specific marker of myelin damage, or molecular imaging, new contrast agents, sodium imaging (Figure 26.19), multimodal MRI in conjunction with histopathological assessments, and novel post-processing techniques, are further providing new insights into the underlying disease processes and improvements are expected in the coming years (Ceccarelli et al., 2012) (Key Points 26.18).

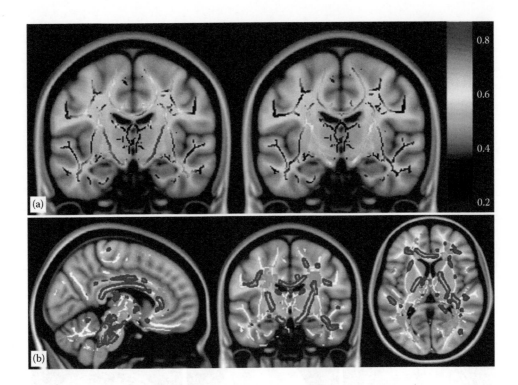

FIGURE 26.17
Regional DTI analysis in clinically isolated syndromes: comparison with controls. (a) Fractional anysotrophy (FA) values colorimetric map on brain coronal section. On the left, is the average FA values in the healthy volunteers group are shown, while on the right FA values in the CIS patients are shown. Notice how the FA values in the patients group are grossly reduced, above all in the corticospinal tracts and corpus callosum, even where there are no lesions, that is, in the NAWM. (b) Voxelwise analysis of the FA in CIS patients ($n = 34$) versus healthy controls ($n = 16$). Clusters of reduced FA in patients compared with healthy subjects are shown in red-orange; shown together are the average lesion mask (dark blue) and the mean skeleton mask (light blue). There is a widespread bilateral decrease in regional FA in many white matter fiber tracts of the whole brain of MS patients if compared with healthy controls subjects. (Reproduced from Raz, E. et al., *Radiology*, 254, 227, Jan 2010. With permission from the publisher.)

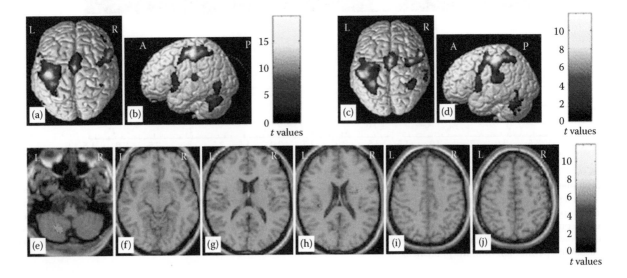

FIGURE 26.18
Functional MRI in patients with primary progressive multiple sclerosis. Cortical activations on a rendered brain from right-handed healthy controls (a, b) and patients with primary progressive multiple sclerosis (PPMS) (c, d) during the performance of a simple motor task with their clinically unimpaired and fully normal functioning right hands (within-group analysis, one-sample *t*-tests, $p < .05$, corrected for multiple comparisons). The between-group differences (areas with more significant activations in PPMS patients than in controls) are shown in the bottom row (e–j) (two-sample *t*-test, $p < .001$). Images are in neurological convention. See text for further details. L, left; R, right; A, anterior; P, posterior. (Reproduced from Ceccarelli, A. et al., *Eur. J. Neurosci.*, 31, 1273, 2010. With permission from the publisher.)

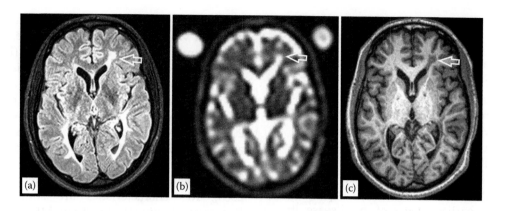

FIGURE 26.19

7 T sodium imaging: selected brain axial FLAIR (a), ^{23}Na images (b), and T_1-weighted (c) from a patient with multiple sclerosis. Two calibration tubes with different concentrations of sodium (100 and 50 mM) in 4% agar gel were placed in the field of view as references and allowed quantification of the sodium concentration. Note the higher TSC value in the periventricular lesion that appears hyperintense on FLAIR (a) and hypointense (arrow) on the T_1-weighted image (c).

KEY POINTS 26.17   CONVENTIONAL MRI IN MS

- Shape, size, and location of focal WM T_2-hyperintense lesions: ovoid, bigger than 5 mm, located mainly in periventricular, subcortical, infratentorial, and in the spinal cord. They are not pathologically specific.

- Dawson finger lesion defines the perivenular location adjacent to corpus callosum and perpendicular to the major axis of the lateral ventricles.

- Gadolinium-enhancing lesions are indicative of active inflammation that usually corresponds to new T_2 lesions. Enhancement pattern is heterogeneous. Gadolinium-enhancing lesions last usually 3 weeks.

- Focal WM T_1-hypointense lesions can be transient or permanent. Permanent T_1-hypointense lesions are called black holes and are a sign of irreversible tissue damage.

- T_2 hypointensity in thalami and basal ganglia are present in all MS phenotypes and are thought to be indicative of iron deposition.

- Brain atrophy is visible in progressive phases of the disease on T_1-weighted sequence appearing as enlargement of ventricles, perivascular spaces, sulci, and reduction of parenchymal structures volume. More sophisticated measures of brain volume other visualization are needed in early MS phases to detect brain volume reduction.

- Brain atrophy is a progressive feature in MS, mostly due to GM atrophy. GM atrophy affects early stages of the disease and progress over time and it is better correlated with clinical status than global brain atrophy and WM atrophy.

- Distinct topographical distribution of regional GM atrophy varies according to clinical phenotypes and clinical disability and could explain different clinical symptoms in MS.

- Cortical lesions better visualized with DIR sequence, or PSIR are found in all MS phenotypes from the earliest stages of the disease and becoming more frequent in the progressive phases and correlated with GM, cognitive impairment, and disability. They are due to cortical demyelination.

- Spinal cord lesions: T_2-hyperintense lesions, better visualized by STIR sequence, are usually involving no more than two segments and mainly located at the level of the cervical cord. Enhancement is less frequent than in the brain.

- Cord atrophy is mostly found at the level of the cervical cord.

- Both cord lesions and atrophy strongly correlated with disability and are independent from brain abnormalities.

KEY POINTS 26.18   ADVANCED MRI IN MS

- Using advanced MRI techniques, more pathological specificity can be achieved.

- Advanced MRI has shown that MS tissue damage is present outside conventional MRI-visible WM lesions. Although the tissue damage in the normal-appearing brain and spinal cord tissue has less degree of damage compared to damage inside WM lesions, however it is present from the earliest stages of the disease, it may precede the lesions formation, worsen in progressive phase, and accumulate over time. Furthermore, the degree of damage can vary according to the region or tracts involved.

- Functional MRI studies have shown that cortical reorganization has an adaptive role in limiting the clinical impact of occult tissue damage in brain and spinal cord at least in some MS phases.

- Newer advanced MRI techniques will further improve our knowledge regarding MS pathogenetic mechanisms.

26.3.1.2 MS Variants

26.3.1.2.1 Radiologically Isolated Syndrome

Radiologically isolated syndrome (RIS), a term coined by Okuda and colleagues, is defined as the detection of conventional MRI abnormalities in brain WM highly suggestive of MS according to location, size, and shape in individuals with normal neurological examination and no history of MS (Okuda et al., 2009). While previous postmortem studies have already shown pathologically MS-related incidental lesions in the brain WM of neurologically asymptomatic subjects with a prevalence of 0.7% (Granberg et al., 2013), the advent of high and ultrahigh field MRI has further facilitated the detection of brain WM lesions in neurologically asymptomatic subjects. Consequently, diagnostic MRI criteria to uncover this new complex entity have been recently proposed (Table 26.3) (Okuda et al., 2009) and involved the fulfillment of Barkhof criteria for dissemination in space for MS (Barkhof et al., 1997). However, although fulfillment of Barkhof criteria has undoubtedly helped in the diagnosis of RIS, these criteria are not specific for differentiating MS lesions from brain lesions related to disorders other than MS (Bourdette and Yadav 2012; Charil et al., 2006; De Stefano 2013; Liu et al., 2013; Miller et al., 2008). Indeed, MRI studies have shown that brain

WM T_2 hyperintensities, due to aging, migraine (Figure 26.20), or vascular damage, for example, can mimic MS lesions and even meet Barkhof criteria (Liu et al., 2013). In this context, the use of Fazekas criteria (Fazekas et al., 1998; Hachinski et al., 1987) to validate and distinguish MS lesions from those of small vessel disease and the capacity to define a series of MRI red flags (Charil et al., 2011; Miller et al., 2008) should be exercised in reviewing and reconsidering other possible differential diagnosis (Table 26.4).

At present, based on Okuda criteria (Okuda et al., 2009), RIS characteristics are summarized below. Reasons for having the first brain MRI in clinically silent subjects are headache (50%), trauma, endocrinological and psychiatric disorders, and less commonly research control, epilepsy, pain, tinnitus, and medical screening of follow-up (Granberg et al., 2013). However, RIS prevalence and incidence are still unknown (Granberg et al., 2013) and it is not clear whether RIS is an independent

TABLE 26.3

Diagnostic Criteria for Radiologically Isolated Syndromes

- The presence of incidentally identified CNS WM anomalies meeting the following MRI criteria:
  1. Ovoid, well-circumscribed, and homogeneous foci with or without involvement of the corpus callosum
  2. T_2 hyperintensities measuring >3 mm and fulfilling Barkhof criteria (1997) (at least three out of four) for dissemination in space
  3. CNS WM anomalies not consistent with a vascular pattern
- No historical accounts of remitting clinical symptoms consistent with neurologic dysfunction
- The MRI anomalies do not account for clinically apparent impairments in social, occupational, or generalized areas of functioning
- The MRI anomalies are not due to the direct physiologic effects of substances (recreational drug abuse, toxic exposure) or a medical condition
- Exclusion of individuals with MRI phenotype suggestive of leukoaraiosis or extensive WM pathology lacking involvement of the corpus callosum
- The CNS MRI anomalies are not better accounted for by another disease process

Source: Okuda DT et al. Incidental MRI anomalies suggestive of multiple sclerosis: The radiologically isolated syndrome. *Neurology* 2009;72:800–805. Erratum in: *Neurology* 2009;72:1284.

Table 26.4

Non Demyelinating WM Lesions Identification Criteria

- Smaller than 3 mm
- Mainly located in subcortical WM away from ventricles
- If lesions are periventricular, they are more symmetric
- Involvement of basal ganglia and sparing of corpus callosum, spinal cord, and U fibers.

Sources: Fazekas, F. et al., *J. Neurol. Neurosurg. Psychiatry,* 64, S2, 1998; Hachinski, V.C. et al., *Arch. Neurol.,* 44, 21, 1987.

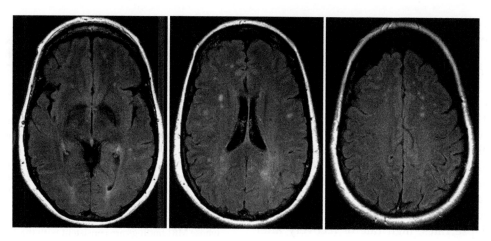

FIGURE 26.20
Migraine. A 42-year-old woman with history of migraine and MRI demonstrating multiple white matter hyperintensity lesions mainly subcortical on FLAIR sequence.

entity or is a subclinical MS (Bourdette and Simon, 2009; De Stefano and Siva, 2013; Lana-Peixoto, 2012). Indeed, precise MS conversion risk stratification for RIS subjects has become increasingly important in order to understand this entity, improve clinical decision-making, and to determine whether treatment is needed. Around 30% of subjects with RIS convert to a first MS clinical event within 5 years (Okuda et al., 2014; Siva et al., 2009) and 66% have a radiological progression in time (Granberg et al., 2013). Risk factors for conversion to CIS are cervical spinal cord lesions, Gd-enhancing lesions, abnormal visual-evoked potentials, younger age, male sex (Okuda et al., 2011, 2014), pregnancy (Lebrun et al., 2012), familiarity (Gabelić et al., 2014), and the presence of sorcin antibody (Sehitoğlu et al., 2014), whereas number of T_2 lesions, presence of oligoclonal bands, and IgG index were not predictive (Lebrun et al., 2009). However, in searching for biomarkers of RIS, another study has shown that 14% of RIS patients have abnormal visual-evoked potentials while 44% have oligoclonal bands (Gabelić et al., 2013). Furthermore, recent studies have suggested that subjects with RIS manifest the same cognitive profile as patients with definite MS and CIS and that cognitive impairment is a pejorative marker in converting to MS (Amato et al., 2012; Lebrun et al., 2010). So far, treatment management in RIS is highly controversial (Brassat et al., 2012; Hutchinson, 2012; Sellner et al., 2010; Spain and Bourdette, 2011). Three possible treatment decision choices have been explored: wait, follow, or treat (Granberg et al., 2013). A suggested rationale for treating RIS (Weiner and Stankiewicz, 2012) may be the combined presence of brain and spinal cord lesions.

26.3.1.2.1.1 Conventional MRI Findings WM T_2-hyperintense lesions have typical MS lesions features (shape, morphology, and location, such as the

predominant perivenular distribution) (Barkhof et al., 1997). Furthermore, WM lesion volumes, distribution, and brain atrophy are similar between RIS and RRMS (Amato et al., 2012; De Stefano et al., 2011). Cortical lesions are also present in 40% of RIS, mainly in frontotemporal lobes and are more frequent in subjects with oligoclonal bands, radiological progression in time, and cervical cord lesions (Giorgio et al., 2011).

26.3.1.2.1.2 Advanced MRI Findings Beyond macroscopic damage detected by conventional MRI, advanced MRI such as MTR and MRS (decreased level of NAA/Cr) has shown that submicroscopic tissue damage in NAWM and normal-appearing GM is present in patients with RIS, similarly to CIS, but milder than RRMS (De Stefano et al., 2011; Stromillo et al., 2013).

26.3.1.2.2 Pediatric MS

Pediatric MS (onset of the disease before 18 years of age [Krupp et al., 2007; 2013] occurs in up to 5%–10% of all MS cases) (Banwell et al., 2007a,b). Although it has been traditionally considered a rare disorder, in the latest decades, the advent of new clinical, laboratory, and MRI diagnostic criteria has significantly increased the detection of pediatric cases and provided clues for the differential diagnosis with several other conditions that affects children (Banwell et al., 2014). An unknown etiology, an autoimmune pathophysiology, and both genetic and environmental risk factors are similar between pediatric and adult MS (Banwell et al., 2007a,b, 2014; Di Santo et al., 2011; Kennedy et al., 2006; van Pelt et al., 2013; Vargas-Lowy and Chitnis, 2012). However, compared to the adult MS, some differences are found in ethnicity characteristics, gender predominance, clinical course and presentation, MRI findings, and treatment (Banwell, 2014; Banwell et al., 2007c;

Bigi and Banwell, 2014; Callen et al., 2009; Chitnis et al., 2009; Suppiej and Cainelli 2014; Van Haren and Waubant, 2013; Verhey et al., 2013a,b,c). Pediatric MS is more frequent and aggressive in Afro-Americans, while in terms of gender, the female predominance, typical of adult onset MS, is only observed in children with adolescent onset MS (age at onset between 12 and 17 years) (Banwell, 2014; Chitnis et al., 2009; Vargas-Lowy and Chitnis, 2012). In terms of clinical course, pediatric MS patients usually manifest a RR course, while a progressive course is very rare and indicative of other child diseases such as LDs (Banwell, 2014). Furthermore, compared to adults, pediatric patients have higher relapse rates, high rate of recovery after relapse, and longer time to reach a sustained physical disability, although this is then reached at younger age (Boiko et al., 2002; Gorman et al., 2009; Simone et al., 2002). Disease onset can be both polyfocal and monofocal. Polyfocal clinical presentation is more common and usually includes optic neuritis, sensory, brain stem, and gait disorders (Banwell et al., 2014). On the other hand, common monofocal presentations in children are optic neuritis and transverse myelitis that can progress to MS or NMO (Banwell et al., 2014). Furthermore, pediatric MS patients also have a relatively high rate of cognitive impairment, with 30% of children showing some deficits within the first few years from the disease onset mainly involving language, visuomotor skills, attention, and processing speed (Suppiji and Cainelli, 2014). Fatigue and depression are also common (Suppiji and Cainelli, 2014). Unlike adult onset MS, ADEM is a common feature at presentation in children who will develop MS, mainly in child onset MS (<12 years old) and, has been recently incorporated in the new criteria for defining pediatric MS (Krupp et al., 2013) (Table 26.5). Compared to old criteria (Krupp et al., 2007), the new definition offers several novel points such as the inclusion of an ADEM event (see Section 26.3.2.1.1) as a possible first demyelinating event (more in child onset than in adolescent onset) and the introduction of the 2010 MRI revised McDonald criteria (Polman et al., 2011). Although the use of DIS and DIT from the revised 2010 McDonald criteria on a baseline MRI (Polman et al., 2011) have extremely increased the likelihood of diagnosis in pediatric MS, still specific criteria for the pediatric form have to be developed. Beyond this, conventional MRI in pediatric MS is extremely important for the diagnosis (Callen et al., 2009; Krupp et al., 2013). Like in adult MS, therapeutic options for pediatric patients consist of first-line treatments and, when possible, second-line treatment choices, such as rituximab (Beres et al., 2014) or natalizumab (Ghezzi et al., 2013) are under investigation. Although there is a lack of randomized clinical trials in children, the first-line treatment choices

TABLE 26.5

Revised Definition of MS and Clinical Isolated Syndromes in Children by the Pediatric MS International Study Group

Pediatric MS is defined as:

1. Two or more clinical events separated by more than 30 days and involving more than one area of the CNS.
2. A single clinical event plus a baseline MRI evidence for DIS and DIT that meets the recent 2010 revised McDonald criteria (Polman et al., 2011).
3. ADEM followed more than 3 months later by a nonencephalopathic clinical event with new lesions on brain MRI consistent with MS according to the revised diagnosis criteria (Polman et al., 2011).
4. A first, single, acute event that does not meet ADEM criteria and whose MRI findings are consistent with the 2010 Revised McDonald criteria for DIS and DIT (Polman et al., 2011) (applies only to children ≥12 years old).

A single clinical event is defined as:

1. A monofocal or polyfocal clinical neurological event with presumed inflammatory demyelinating cause.
2. Absence of encephalopathy that cannot be explained by fever.
3. Absence of previous clinical history of CNS demyelinating disease.
4. Other etiologies have been excluded.
5. The most recent 2010 revised MS McDonald criteria on a baseline MRI are not met.

Source: Adapted from Krupp, L.B. et al., International Pediatric Multiple Sclerosis Study Group criteria for pediatric multiple sclerosis and immune-mediated central nervous system demyelinating disorders: Revision to the 2007 definitions, *Mult. Scler.*, 19, 1261, 2013.

for pediatric MS involve the use of interferon and glatiramer acetate (van Haren and Waubant, 2013) that can be administered with a full dose or with a dose that is adjusted for age and body weight.

26.3.1.2.2.1 Conventional MRI Findings Several studies have reported key conventional MRI features in pediatric MS and have compared these features with those of the adult forms (Banwell et al., 2007c; Ghassemi et al., 2014; Verhey et al., 2010, 2013b; Waubant et al., 2009; Yeah et al., 2009). Supratentorial WM-hyperintense T_2 lesions are mostly located in the occipital and frontal periventricular WM and in the juxacortical WM. The lesion burdens and volumes are similar or higher than in adult forms (Ghassemi et al., 2014; Waubant et al., 2009; Yeah et al., 2009), while infratentorial T_2 and mainly T_1 lesion volumes and number have been found higher than in adult forms (Ghassemi et al., 2014). As an exception, younger children (child onset MS) could have large confluent T_2 lesions, with poorly defined borders that can also be transient over time (Chabas et al., 2008), which resemble more the lesions of ADEM. Gadolinium-enhancing lesions are present in 22% of children who will develop MS (Verhey et al., 2013b). On the other hand, conventional spinal cord imaging features are similar to those found in adults (Verhey et al., 2010). A standardized MRI protocol has been proposed in pediatric MS (Verhay et al., 2013b), including brain T_2-weighted sequences, T_1-weighted sequences with and without gadolinium for studying lesions characteristics, and evaluation of 2010 McDonald criteria (Polman et al., 2011). DWI, optical, and spinal cord imaging are added to increase the differential diagnosis of stroke, NMO, and myelitis (Key Points 26.19).

26.3.1.2.2.2 Advanced MRI Findings

While conventional MRI has played a key role in the diagnostic work-up of patients with pediatric MS, advanced MRI techniques have contributed to gradation of the extent of occult brain damage in these patients. Overall, studies using MRS, MTR and DTI, and tract-based analysis have suggested that, compared with healthy children, children with MS have diffuse submicroscopic damage of the brain NAWM and cord normal-appearing tissues (Absinta et al., 2010; Bauer, 1992; Bethune et al., 2011; Mezzapesa et al., 2004; Oguz et al., 2009; Tortorella et al., 2006; Verhey et al., 2013c; Vishwas et al., 2010). On the other hand, a selective sparing of the brain GM has been shown in these patients (Absinta et al., 2011; Ceccarelli et al., 2011; Kerbrat et al., 2012; Mesaros et al., 2008). Indeed, cortical and deep GM lesions (Absinta et al., 2011) are very unusual in pediatric MS as well as iron deposition measured by T_2 hypointensity is limited to head of caudate nucleus (Ceccarelli et al., 2011). While global brain volume and head size are reduced in MS children, regional atrophy is confined to thalami (Kerbrat et al., 2012; Mesaros et al., 2008). Of note, overall, cognitive impaired pediatric MS patients showed higher submicroscopic damage in both NAWM and normal-appearing GM than cognitive preserved pediatric patients, as measured by atrophy and diffusivity abnormalities (Rocca et al., 2014a; Till et al., 2011). In particular, cognitive functions were inversely correlated with volume of thalami (Till et al., 2011). Using MTR (Yeh et al., 2009) and tract-based spatial statistics analysis of DTI (Aliotta et al., 2014), compared to adult onset MS, adults with pediatric onset MS showed higher diffuse damage in NAWM and normal-appearing GM, while no regional GM atrophy differences were found (Donohue et al., 2014). According to structural MRI studies, functional MRI studies have provided evidence of a relative preservation of brain functional integrity in pediatric onset MS (Rocca et al., 2009, 2010, 2014b) compared to MS adults. Interestingly, a selective structural and functional damage of posterior brain regions, compensated by preservation of neuronal activity in frontal

regions (Rocca et al., 2014b), has been recently showed in pediatric patients with cognitive impairment (Key Points 26.20).

KEY POINTS 26.19 CONVENTIONAL MRI FEATURES IN PEDIATRIC MS

- Demyelinating lesions are preferentially located in frontal and occipital periventricular and in juxtacortical WM.

- Morphology and shape of demyelinating lesions in children with MS are similar to those of adults MS, with the exception of younger patients who could show large confluent T_2-hypeintense lesions, with poorly defined borders that can also be transient over time.

- Enhancement is observed in 22% of patients who will develop MS.

- Spinal cord lesions are similar to those found in adults with MS.

- While burden of supratentorial T_2 and T_1 lesions in children with MS is similar to adults with MS, burden of infratentorial T_2 and T_1 lesions is higher in children with MS.

26.3.1.2.3 Marburg Variant

Marburg variant is a rare disease, usually rapidly progressive, relapsing, classically considered as a fatal MS variant with death within 1 year. Clinically, it manifests with rapid deterioration of the patient's clinical condition, including bulbar symptoms and epileptic paroxysms and ending with persistent coma and tetra paresis. Pathologically, Marburg variant lesions are characterized by important demyelination, axonal damage, and necrosis. First-line treatment is high dose of corticosteroids, while second-line treatment is plasma exchange and mitoxantrone (Rahmlow and Kantarci, 2013).

26.3.1.2.3.1 Conventional MRI Findings

T_2-weighted sequence shows multifocal-hyperintense WM lesions of different size that tend to become confluent characterized by mass effect and edema. Lesions can be located everywhere in the deep WM, but also in the brain stem and cortex, without a clear perivenular location. These lesions are also typically hypointense on T_1-weighted sequence. Enhancement can be present (Simon and Kleinschmidt-DeMasters, 2008).

> **KEY POINTS 26.20   ADVANCED MRI FEATURES IN PEDIATRIC MS**
>
> - Advanced MRI techniques have shown, beyond visible demyelinating lesions, subtle microscopic tissue damage in NAWM of the brain and spinal cord in children with MS.
> - Brain GM is relatively spared in children with MS compared to adults with MS as advocated by structural and functional MRI studies.
> - Cognitive impaired children show a higher brain damage.

26.3.1.2.4 Balo's Concentric Sclerosis

Balo's concentric sclerosis is a rare disease, usually rapidly progressive, classically considered as a fatal MS variant. However, so far, the link with MS is still not clear. As MS, age of onset is set around 30 years old, and females are more affected than males (female/male ratio = 2:1). However, Balo's concentric sclerosis mostly affects Asian ethnicity; manifests with symptoms of a mass, such as headache, cognitive disturbance, behavioral changes, and hemiparesis; its clinical course is mainly monophasic; and exhibits distinct radiological and pathological features compared to MS. The advent of MRI has improved notably the prognosis of this entity. Pathologically, Balo's concentric sclerosis lesions are characterized by alternate ring of myelinated and demyelinating areas with perivenular location. First-line therapy is represented by corticosteroid, while a second-line treatment consisted of plasma exchange. MS treatment for Balo's concentric sclerosis has been proposed in the chance that the disease fulfill MS diagnostic criteria for DIS and DIT (Table 26.2) (Hardy and Miller, 2014).

26.3.1.2.4.1 Conventional MRI Findings

Balo's concentric sclerosis has typical MRI lesions appearance. WM lesions (often one) appear concentrically layered of preserved and damaged myelin, forming concentric rings or irregular stripes (Figure 26.21). Lesions usually spare U fibers and, beyond the deep WM, can be located also in basal ganglia, pons, cerebellum, spinal cord, and optic nerves.

T_2-weighted images of Balo's lesions show hyperintense and hypo/isointense rings neighboring a lesions center that could show diverse patterns (storm like, mosaic like, rosette like, or bars like). T_1-weighted images show alternating hypointense and isointense rings. Mass effect is absent and gadolinium enhancement could be present, mainly in the peripheral rings.

26.3.1.2.4.2 Advanced MRI Findings

At MRS, reduction in NAA/creatine and the increase in choline/creatine are more evident in the central lesion than in the internal and outermost ring. Reduction of MTR is also markedly evident in the central lesion than in the surrounding rings. Restricted diffusion is shown as high ring DWI, which is typically associated with ADC reduction (Figure 26.21). Advanced MRI abnormalities develop gradually and centrifugally (Chen et al., 2014).

26.3.1.2.5 Tumefactive Demyelinating Disease

Tumefactive demyelinating disease is characterized by the occurrence of large lesion (>2 cm) as a solitary entity or in the context of MS. In both cases, differential diagnosis with tumor lesions or abscess is very challenging. Epidemiologically, tumefactive demyelinating disease is a rare entity, occurs mainly in females in their 30s

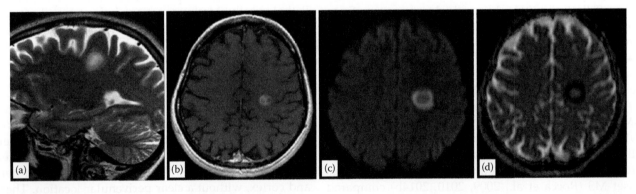

FIGURE 26.21
Balo's concentric sclerosis. Patient presenting with right hemiparesis. An MRI—sagittal T_2 (a) and axial T_1 postcontrast (b)—demonstrated a solitary white matter lesion with a concentric appearance, corresponding to the pathology finding of rings of demyelination alternating with rings of intact myelin. DWI (c) shows a hyperintense ring associated with corresponding ADC reduction (d), a pathognomonic finding of Balo. (This image first appeared online in the *American Journal of Neuroradiology*; Case of the Week from August 18, 2008.)

and manifests with symptoms that are uncommon in MS such as focal neurological signs, seizures, and aphasia. Etiology and pathogenesis are unknown as well as its relationship with MS. The prognosis varies; it can be benign or evolve to MS or Marburg variant of MS. For acute lesions, first-line treatment is corticosteroids, and second-line treatment is plasma exchange. If MS diagnosis is met, treatment for MS should be started (Hardy and Chataway, 2013).

26.3.1.2.5.1 Conventional MRI Findings WM T_2-hyperintense demyelinating lesion has size bigger than 2 cm, is located preferentially in frontal and parietal lobes, and involves U fibers. Mass effect is always observed and enhancement can be present. The most frequent enhancement pattern is the open ring enhancement with the incomplete portion of the ring on the GM side (Hardy and Chataway, 2013). Additional typical findings of demyelination include T_2-hypointense rim, peripheral restriction on DWI, and venular enhancement (Figure 26.22).

26.3.1.2.5.2 Advanced MRI Findings ADC is mildly increased and can help to differentiate these lesions from abscess (that have restricted diffusion) but not

from tumors. Compared to tumors, PET scanning has showed that mean relative cerebral blood volume is lower than in tumors (Hardy and Chataway, 2013).

26.3.1.2.6 Schilder Variant

Schilder variant of MS is extremely rare. It can affect both children and adults. Typical symptoms are psychiatric, progressive deterioration, and intracranial hypertension. Pathologically, gliosis and demyelination are observed. Differential diagnosis is with X-ALD when symmetrical lesions are present or with tumor and abscess (Rovira Cañellas et al., 2007). Diagnostic criteria proposed by Poser (Poser et al., 1992) are aimed to facilitate the differential diagnosis between Schilder, MS, and X-ALD, and they include (1) clinical symptoms and signs often atypical for the early course of MS; (2) CSF normal or atypical for MS; (3) bilateral large areas of demyelination of cerebral WM; (4) no fever, viral or mycoplasma infection, or vaccination preceding the neurological symptoms; (5) normal serum concentrations of VLCFAs; and (6) adrenal normal function. Treatment is by high dose of corticosteroids and prognosis is poor.

26.3.1.2.6.1 Conventional MRI Findings T_2-weighted images show one or two symmetric, bilateral large (>2 cm)

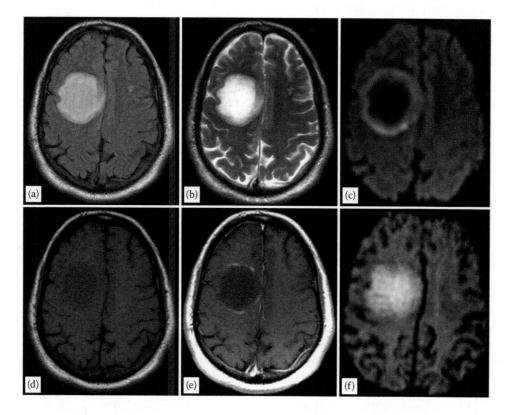

FIGURE 26.22
Tumefactive multiple sclerosis. FLAIR (a), T_2 (b), diffusion (c), pre- (d) and post-contrast (e) T_1, perfusion (f) demonstrates a single large peripherally enhancing lesion associated with mass effect. Interestingly the perfusion image demonstrates venous susceptibility through the lesion consistent with the typical vein passing through the large demyelinating plaque.

lesions, mainly enhanced and preferentially located in the centrum semiovale. Open ring enhancement is pathognomonic.

26.3.1.2.7 NMO and NMO Spectrum Disorders

NMO, also known as Devic's disease, is an idiopathic, severe, demyelinating disease of the CNS characterized by involvement of the optic nerves and the spinal cord with a monophasic (10%–20%) or multiphasic (80%–90%) course (Wingerchuk and Weinshenker, 2014). In the past considered as a severe variant of MS, emerging evidences have shown that NMO presents unique clinical, biochemical, radiologic, and pathologic features compared to MS. Clinically, optic neuritis and extensive transverse myelitis, presenting with pain and visual loss and tetraplegia or paraplegia, a well-defined sensory level and bladder dysfunction, respectively, are pathognomonic of NMO. The optic neuritis can be monolateral or bilateral, usually more severe than in MS and associated with a poor recovery (Levin et al., 2013; Wingerchuk and Weinshenker, 2014). Symptoms due to the involvement of the brain stem can be also observed (Wingerchuk and Weinshenker, 2014), involving hiccups, nausea, and fatal damage of the respiratory centers. NMO is more common in non-Caucasian ethnicity and affects mostly women (female/male ratio of 9:1) in an older age than MS (around 40 years old) (Asgari, 2013). However, NMO can affect also children (Absoud et al., 2014) and old subjects (Quek et al., 2012). Studies of prevalence and incidence have been confounded by the presence of several NMO variants among various age groups (Marrie and Gryba, 2013). In 2004 (Lennon et al., 2004), antibody IgG targeting aquaporin 4 (AQP4), the most abundant astrocytes water channel in the brain, have been associated to NMO, becoming a highly sensitive and specific biomarker for NMO (99%) and suggesting that NMO may be a novel autoimmune channelopathy (Wingerchuk and Weinshenker, 2014). The advent of AQP4 antibody has incredibly facilitated the differential diagnosis between MS and NMO and has further improved the characterization of NMO spectrum disorders. NMO spectrum disorders include recurrent myelitis associated with longitudinally extensive spinal cord lesions, Asian optic-spinal MS, and recurrent isolated optic neuritis (Flanagan and Weinshenker, 2014). Often autoimmune diseases such as Sjogren, lupus, and myasthenia gravis are associated with NMO. NMO spectrum disorders positive for anti-AQP4 antibody may also present cognitive impairment from the early stages associated with cortical neurodegeneration (Saji et al., 2013). Based on the discovery of NMO-IgG anti-AQP4, new diagnostic criteria for NMO (Wingerchuk et al., 2006) have been elaborated (with sensitivity of 99% and specificity of 90%). These criteria required the recurrence of one optic neuritis and one transverse myelitis coupled with two of three of the following conditions:

- Spinal MRI with spinal cord lesion extending over three or more spinal segments.
- Brain MRI that does not fulfill Paty's diagnostic criteria for MS (Paty et al., 1998).
- Evidence of NMO-IgG in the serum.

However, 10%–30% of NMO are AQP4 seronegative and it is still not clear if seropositive and seronegative NMO are pathologically distinct (Sato et al., 2013). Thus, new biomarkers are under investigation. Recently, distinct clinical and MRI features and prognosis have been shown to differentiate between AQP4 antibody positive relapsing NMO and anti-MOG positive NMO (Kitley et al., 2014; Sato et al., 2014). Indeed, further studies will be needed to better define seronegative NMO. While corticosteroids can be used for acute symptoms management, some treatments for MS are ineffective and can even aggravate NMO (Bienia and Balabanov, 2013). Although only recently controlled trials in NMO are beginning, several therapeutics options are being currently explored, beyond the azathioprine that was used in the past. New line treatments are classified as (1) targeting B cells line such as rituximab; (2) targeting antibodies such as aquaporumab; (3) targeting IL-6 such as tocilizumab (Bienia and Balabanov, 2013).

26.3.1.2.7.1 Conventional MRI Findings
As suggested from the 2006 diagnostic criteria (Wingerchuk et al., 2006), conventional MRI is very important for the diagnosis of NMO and its differentiation from MS. The MRI protocol for NMO has to include both brain and spinal cord MRI sequences.

Spinal cord MRI: A longitudinally extensive transverse myelitis, shown as hyperintense lesions on T_2-weighted sequence, extending for more than three segments, usually in both cervical and thoracic level and involving central spinal cord with damage of the GM, beyond WM, is specific for NMO. Thoracic level is more affected than cervical level as opposed to MS. Very hyperintense spotty lesions on axial T_2-weighted imaging, named *bright spotty lesions* (BSLs), have recently been defined (Yonezu et al., 2014). Combined with longitudinally extensive transverse myelitis, BSLs are claimed to distinguish NMO from MS with 88% sensitivity and >97% specificity (Tackley et al., 2014; Yonezu et al., 2014). The enhancement is not so common even during acute relapse. Diffuse atrophy is also associated with spinal

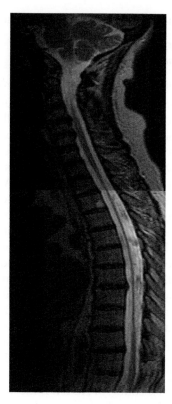

FIGURE 26.23
Neuromyelitis optica. Patient with longstanding NMO in which a sagittal T_2 of the spinal cord demonstrates extensive severe spinal cord atrophy involving most of the thoracic segment.

cord lesions (Figure 26.23). A corresponding T_1-hypointensity lesion in the spinal cord is a frequent feature of NMO, while is rare in MS. Some lesions could enhance with patchy cloud pattern.

Brain MRI: Conventional MRI of the brain is generally normal. However, nonspecific T_2-hyperintense lesions or even meeting the Barkhof criteria for MS can be observed and are common in advanced disease. A recent work has shown that patients with NMO and a concomitant autoimmune disease have higher frequency of brain abnormalities (Zhang et al., 2014) than patients with only NMO. MRI criteria for distinguishing between NMO spectrum disorders and MS lesions have been proposed (Matthews et al., 2013), such as paucity of lateral ventricle, inferior temporal lobe, Dawson finger, and S-shaped U-fiber lesions in NMOSD seropositive patients help differentiate NMO from MS. In NMO, T_2-hyperintense lesions are more located in infratentorial regions and in particular in enriched AQP4 channels (periaqueductal GM lesions or medullary lesions, hypothalamus, and area postrema). A recent study has shown that differences in border's shape in periventricular

WM lesions between MS and NMO lesions are present (Raz et al., 2014). Furthermore, NMO lesions do not show perivenular localization that instead is typical of MS (Sinnecker et al., 2012). Although AQP4 channels are also widely distributed in the cortex (Papadopoulos and Verkman, 2012), cortical lesions and cortical demyelination have not been so far observed neither in post-mortem studies (Popescu et al., 2010; Saji et al., 2013) nor in MRI DIR studies (Calabrese et al., 2012). Furthermore, on T_2-weighted sequences, optic nerves appear hyperintense, swollen, and enhanced (Key Points 26.21).

26.3.1.2.7.2 Advanced MRI Findings Compared to MS, advanced MRI techniques have shown more severe damage in brain and spinal cord lesions, while in brain and spinal cord NAWM the damage is present but less severe than in MS (Tackley et al., 2014). Using DTI, several recent studies have shown a diffuse widespread brain and spinal cord damage outside T_2-weighted lesions compared to healthy subjects (Kimura et al., 2014; Klawiter et al., 2012; Liu et al., 2012; Pichiecchio et al., 2012; Qian et al., 2011; Rivero et al., 2014; von Glehn et al., 2014; Zhao et al., 2012). Using MTR, a diffuse damage of the cerebral GM in NMO compared to controls has been shown (Rocca et al., 2004). Using MRS, while no abnormalities were found in NAWM and normal-appearing GM in brain and spinal cord of NMO patients (de Seze et al., 2010), a selective decrease of myo-inositol, a marker of astrocyte dysfunction, has been observed in NMO spinal cord lesions, improving ability to differentiate between NMO and MS (Ciccarelli et al., 2013). While spinal cord atrophy is clearly more severe in NMO than MS patients, global brain atrophy is less severe in NMO and it is mostly caused by global WM atrophy (Chanson et al., 2013). Using advanced post-processing techniques, such as voxel-based morphometry, regional WM and GM atrophy have been found widespread in NMO, compared to controls, but usually less diffuse than in MS patients (Duan et al., 2012, 2013; von Glehn et al., 2014). Although regional GM atrophy affected several areas of NMO patient's brain, including visual cortex, frontal, temporal, parietal, and insula cortex, it seems to relatively spare deep GM compared to MS patients (Duan et al., 2012). Furthermore in NMO, extension of regional GM atrophy seems to be correlated with disease duration (Duan et al., 2013b), while extension of regional WM atrophy correlated with severity of cognitive impairment (Blanc et al., 2012). Optical coherence tomography studies have demonstrated a severe reduction in the thickness of the retinal nerve fiber layer in NMO as a consequence of Wallerian degeneration following optic neuritis, more severe than in MS (Naismith et al., 2009) (Key Points 26.22).

KEY POINTS 26.21 CONVENTIONAL MRI FEATURES IN NMO

- Differential diagnosis between MS and NMO is improved by the analysis of brain and spinal MRI.

- Brain MRI is usually normal or can present aspecific or sometimes MS like lesions. Frequently, lesion distribution (AQP4-enriched areas) and shape can help distinguish between MS and NMO. Perivenular location of brain lesions is not observed in NMO.

- Cortical lesions are absent in NMO.

- Optic nerves appear hyperintense on T_2-weighted sequences, swollen and enhanced.

- Spinal cord MRI with transverse long lesions hyperintense on T_2 and hypointense on T_1-weighted images, mostly at the thoracic or cervical levels are pathognomonic of NMO.

- Very hyperintense spotty lesions on spinal axial T_2-weighted images, called *bright spotty lesions* (BSLs) are prototypical.

- Some lesions could enhance with patchy cloud pattern.

- Atrophy of the spinal cord is more severe than in MS.

KEY POINTS 26.22 ADVANCED MRI FEATURES IN NMO

- Advanced MRI techniques have shown more severe damage in brain and spinal cord lesions compared to MS.

- Occult normal-appearing GM and NAWM damage is widespread in NMO, but it is less severe than in MS.

- Regional GM and WM atrophy are widespread in NMO but showed different patterns of regional distribution compared to MS.

- Selective myo-inositol peak decrease in spinal cord lesions of NMO patients could differentiate NMO from MS.

- NMO is associated with more widespread axonal injury in the affected optic nerves compared to MS as revealed by optical coherence tomography.

26.3.2 Secondary Demyelinating Diseases

26.3.2.1 *Allergic/Infections/Vaccinations/Immunological*

26.3.2.1.1 *Acute Disseminated Encephalomyelitis*

ADEM is an acute or subacute inflammatory immune-mediated syndrome of the CNS that usually affects children (under the age of 10) within 4 weeks from an infection or a vaccination (Krupp et al., 2013). Revised diagnostic criteria for ADEM are shown in Key Points 26.23 (Krupp et al., 2013). The clinical presentation of ADEM in children manifests with encephalopathy (consciousness and behavioral changes), usually not explained by the previous fever associated with the infection or vaccination, complemented by multifocal neurological symptoms related to the location of the CNS inflammation (Fernández Carbonell and Citnis, 2013; Tenembaum, 2013). The most common neurological symptoms are acute transverse myelitis, cranial nerve palsy, acute cerebellar ataxia, and optic neuritis (Fernández Carbonell and Citnis, 2013; Javed and Khan, 2014; Tenembaum, 2013). In adult forms (average age = 33 years old), encephalopathy is rare, while more common are focal neurological symptoms (Javed and Khan, 2014). The disease course of ADEM is mostly monophasic (90%), so that the neurological symptoms vary in severity and evolve within 3 months. A multiphasic course is also possible and its definition has been recently revised (Krupp et al., 2013) and consists of only two episodes of ADEM separated by 3 months, with either new or a return of prior neurologic symptoms, signs, and MRI findings. When a third event of CNS inflammation occurs over time, alternative diagnosis such as MS or NMO should be considered (Krupp et al., 2013). Indeed, ADEM could represent the first demyelinating event of pediatric MS or NMO, respectively, if the subsequent event appears after 3 months from the ADEM onset and meets the criteria for dissemination in space or a positive tier for NMO-IgG is found (Krupp et al., 2013). Pathologically, the hallmark of ADEM is the presence of perivascular inflammatory infiltrates of lymphocytes and macrophages in the WM and GM with resulting myelin loss (Javed and Khan, 2014). The diagnosis of ADEM is made by combination of clinical symptoms, CSF evaluation, and MRI. CSF proteins and cell counts with lymphocytes predominance are elevated while oligoclonal bands are usually rare (Dale et al., 2000). Specific treatment involves high doses of steroids, plasmapheresis, and intravenous immunoglobulin therapy. The prognosis is usually excellent but some neurological signs could persist due to a not complete resolution of the syndrome (Tenembaum, 2014), such as focal neurological deficits as motor or visual impairment but also behavioral and neurocognitive deficits.

26.3.2.1.1.1 *Conventional MRI Findings* Typical conventional MRI findings are widespread, bilateral, asymmetric,

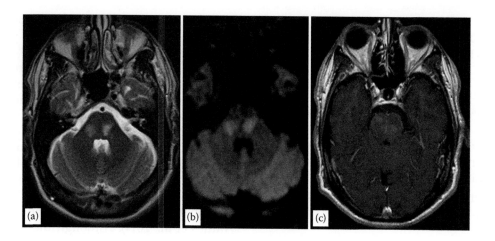

FIGURE 26.24

ADEM. A 6-year-old patient with headache, fever 4 weeks after a flu. Axial T_2 (a), diffusion (b), and postcontrast T_1-weighted sequence (c) demonstrate hyperintense lesions in the pons bilaterally, typical for demyelination. Enhancement of several of the lesions was present on postcontrast T_1 images.

patchy, poorly marginated (due to edema) lesions which are bright or hyperintense on T_2-weighted imaging within the cerebral subcortical WM, deep GM (usually symmetrical and mostly involving the thalamus), and spinal cord (Marin and Callen, 2013; pp. 245–266; Rossi, 2008). Juxtacortical, deep WM, and infratentorial lesions are common while periventricular WM, corpus callosum, and cortical lesions are rare. Lesion size varies from small to tumefactive appearance with modest mass effect (Bester et al., 2013). T_1 hypointensity is rare and if present is a predictor of conversion to MS (Krupp et al., 2013). Contrast enhancement could be present (Tenebaum et al., 2014) on several lesions contemporaneously, but the pattern is not specific (Figure 26.24). ADEM should be suspected if enhancement is not present. MRI follow-up examinations play an important role for the diagnosis of ADEM. In monophasic ADEM, complete regression of MRI lesions is described in 37%–75% of patients, with a partial remission still occurring in 25%–53% of cases (Bester et al., 2013; Rossi, 2008) (Key Points 26.23).

26.3.2.1.1.2 Advanced MRI Findings   In acute phase, WM T_2-hyperintense lesions are characterized by restricted diffusion and in the subacute stage by free diffusion and a decrease in NAA/choline ratio (Balasubramanya et al., 2007). With the normalization of MRI findings after the acute phase, an increase of NAA is observed (Rossi, 2008). During the early stage of ADEM acute phase, a decrease of myo-inositol levels and its increase in the chronic phase has been suggested as a marker of this entity, improving differential diagnosis with MS (Ben Sira et al., 2010). So far, advanced MRI techniques such as DTI, MTR, and MRS have not shown any abnormalities in normal-appearing bran tissue, in contrast to MS (Bester et al., 2013; Inglese et al., 2002) (Key Points 26.24).

ADEM variants. Of note, several variants of ADEM have been described. The most common is acute hemorrhagic

leukoencephalopathy (AHL), also variably called AHEM or acute necrotizing hemorrhagic leukoencephalitis (ANHLE). Similar to ADEM regarding onset and presentation, it is worse than ADEM in severity and MRI findings, showing larger and more diffuse lesions with mass effect, edema, and hemorrhage pathologically due to the destruction of the small blood vessels. Rapid intervention with high dose of steroids, plasma exchange, and immunosuppressive is needed (Javed and Khan, 2014).

KEY POINTS 26.23   DIAGNOSTIC CRITERIA FOR ADEM

The following criteria need to be fulfilled:

- A first polyfocal, clinical CNS event with presumed inflammatory demyelinating cause.
- Encephalopathy that cannot be explained by fever.
- No new clinical and MRI findings emerge 3 months or more after the onset.
- Brain MRI is abnormal during the acute (3-month) phase.
- Typically on brain MRI: (a) diffuse, poorly demarcated, large (>1–2 cm) lesions involving predominantly the cerebral WM; (b) T_1-hypointense lesions in the WM are rare.
- Deep GM lesions (e.g., thalamus or basal ganglia) can be present.

Source: Adapted from Krupp, L.B. et al., International Pediatric Multiple Sclerosis Study Group criteria for pediatric multiple sclerosis and immune-mediated central nervous system demyelinating disorders: Revision to the 2007 definitions, *Mult. Scler.*, 19, 1261, 2013.

KEY POINTS 26.24 ADVANCED MRI FEATURES IN ADEM

- In acute phase, restricted diffusion, decrease of NAA and myo-inositol/creatine ratio is observed in WM T_2-hyperintense lesions.
- At follow up MRI, with the resolution of T_2-hyperintense lesions, increased NAA and myo-inositol, and free diffusion are found.
- No occult tissue damage is detected outside WM-hyperintense T_2 lesions.

26.3.2.1.2 Progressive Multifocal Leukoencephalopathy

PML is an acquired slow progressive demyelinating disorder caused by the reactivation of a ubiquitous polyomavirus named John Cunningham virus (JCV) (Bellizzi et al., 2013; Berger 2014). It generally manifests in immunosuppressed such as patients with human immunodeficiency virus (HIV) patients or in patients under immunosuppressive therapy (Major, 2010). Although it is considered a rare disorder, PML has got a lot of attention in the last years in MS patients since it represent the worse adverse event of the treatment with natalizumab (Tysabri®, Biogen-Idec Inc, Cambridge, MA), a monoclonal antibody directed against α4β1 and α4β7 integrins (Kleinschmidt-DeMasters and Tyler, 2005). Risk factors for developing PML in MS patients treated with natalizumab are (1) positive anti-JCV serum antibodies level, (2) prior use of immuno-suppressant medication, and (3) prolonged treatment (more than 2 years) (Baldwin and Hogg, 2013). Based on these risk factors, incidence of PML in MS patients with prior use of immunosuppressant medication has been estimated of 11.1 or 4.6 without prior immunosuppression cases per 1000 patients treated with natalizumab and treatment duration longer than 2 years (Tur and Montalban, 2014). Pathologically, the key feature of PML is an extensive multifocal demyelination characterized mainly by enlarged oligodendrocytes and enlarged reactive astrocytes and macrophages (Gheuens et al., 2013). Emerging evidences have suggested that PML can also cause demyelinating lesions within the GM infecting neurons (Gheuens et al., 2013). Clinically, depending on the demyelination site in the brain, behavioral, cognitive (48%), motor (37%), language (31%), and visual deficit (26%) and seizure can be present. While in the past years the diagnosis was only made pathologically, today the diagnosis of definitive PML relies on clinical symptoms, coupled with detection of JCV in the CSF by PCR (virologically confirmed PML) or tissue by biopsy (histologically confirmed PML) and MRI distinctive findings (Berger et al., 2013; Yoursy et al., 2012). The diagnosis of probable PML is made if lack of MRI key features or absence of JCV in the CSF is observed. However, probable PML has to be treated as a definitive PML (Berget et al., 2013). MRI has a pivotal role in PML diagnosis, and further improving the diagnostic challenge of asymptomatic PML (Wattjes et al., 2014). Red flag for asymptomatic PML would be the advent of small new focal lesions with PML characteristics (Berger, 2013) supported by JCV CSF testing. New diagnostic criteria though will be necessary for asymptomatic PML coupled with additional diagnostic tests and consensus guidelines (Wattjes et al., 2014). Although the prognosis of PML associated to natalizumab is usually better than that of HIV-associated PML, there is no secure treatment for PML. So far, natalizumab-associated PML is managed by discontinuation of natalizumab and plasmapheresis/immunoadsorption (Baldwin and Hogg, 2013; Berger, 2014). However, although this treatment has notably increased survivors, it could lead to immune reconstitution inflammatory syndrome (IRIS) caused by sudden renovation of cellular immunity (Tan et al., 2011). IRIS-associated natalizumab is defined by the development of neurological deterioration after natalizumab interruption supported by inflammatory changes on MRI (Tan et al., 2011, *Neurology*). Although no controlled studies have been performed yet, intravenous methylprednisolone is the usual treatment for IRIS related to PML.

26.3.2.1.2.1 Conventional MRI Findings

It is very difficult to distinguish between MS and PML lesions with conventional MRI. However, some MRI features can help distinguish between these two diseases (Yoursy et al., 2012). PML WM lesions are usually new demyelinating lesions mostly monofocal, subcortical, and juxtacortical (also extending to the cortex) (Khoury et al., 2014) involving preferentially the frontal lobe and parieto-occipital regions (Yousry et al., 2012). Dawson finger pattern is uncommon in PML associated to natalizumab. Isolated or associated involvement of the basal ganglia, external capsule, and posterior fossa structures (cerebellum and brain stem) may be present in PML associated to natalizumab (Tortorella et al., 2013). WM lesions are usually hyperintense on T_2-weighted images and hypointense on T_1-weighted images. Furthermore, in contrast to AIDS-associated PML, 30%–40% of patients suffering from natalizumab-associated PML display Gd-enhancing lesions on MRI at the time of diagnosis (Berger et al., 2013; Yoursy et al., 2012). No mass effect is usually present. As disease progress, lesions can become multifocal, diffuse, and confluent. Of note, in PML associated to natalizumab, FLAIR sequences are more sensitive of WM lesions even in the posterior fossa

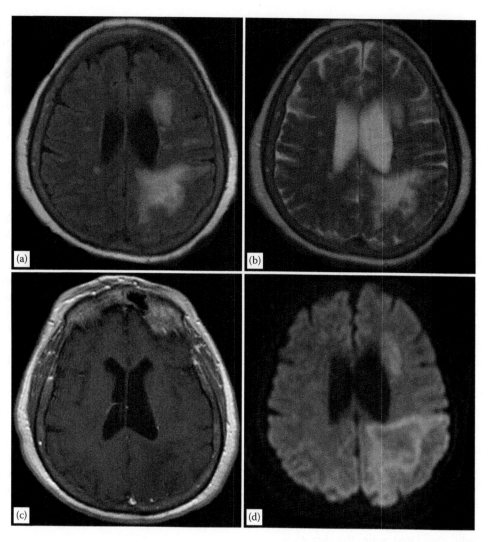

FIGURE 26.25

PML in a patient with multiple sclerosis on natalizumab. Association of typical findings of PML: the FLAIR (a), T_2 (b), and T_1 postcontrast (c) signals are respectively hyperintense and hypointense as most of the brain lesions. Another feature of PML is that usually there is no mass effect associated with it (a–c) and the peripheral diffusion hyperintensity (d). The PML findings are superimposed on multiple lesions typical of MS (visible on FLAIR and T_2), and this association should raise the thought of natalizumab-related PML.

(Berger et al., 2013; Yoursy et al., 2012). GM lesions can be also present and strongly correlated with seizures development (Khoury et al., 2014). Lesions that appeared as hyperintense on T_1-weighted images are strongly correlated with developing IRIS. A follow-up conventional MRI study, performed in any case of clinical relapse, can help discriminate between PML and MS according to the specific radiological aspects seen above (Figure 26.25 and Key Points 26.25).

26.3.2.1.2.2 Advanced MRI Findings The role of advanced MRI features in PML associated to natalizumab is limited so far. DWI show peripheral hyperintensity due to a restricted diffusion (Yoursy et al., 2012). This high peripheral signal has been correlated with active infection (da Pozzo et al., 2006). A central hypointense core

on DWI is strongly associated with progression and long disease duration (Cosottin et al., 2008). Thus, old lesions exhibit higher signal on ADC maps (Bergui et al., 2004). MTR may be useful in further differentiating between MS and PML. NAWM MTR is shown higher in PML compared to that of MS patients, while MTR in WM lesions is lower in PML and stable over time, suggesting more focal pathological changes in PML compared to MS (Boster et al., 2009). Using MRS (Gheuens et al., 2014), major differences in the metabolic profile of the brain lesions of patients with PML with IRIS compared to patients with PML without IRIS were found. Decrease in NAA and increase in choline/creatine, myo-inositol/creatine, Lip1/creatine, and Lip2/creatine ratios is present in lesions of patients with PML-IRIS. While Lip1/creatine ratio and Gd enhancement in PML lesions are diagnostic indicators of IRIS, and myo-inositol/creatine remains elevated in lesions of patients with PML-IRIS over time. However, further advanced MRI studies are needed (Key Points 26.26).

26.3.2.1.3 *Lyme Disease or Neuroborreliosis*

Lyme disease is a multisystemic inflammatory infection due to the tick-borne spirochete *Borrelia burgdoferi*. It has a worldwide distribution and is the most common vectorborne disease in the United States. Fifteen percent of patients can have neurological symptoms (Hildenbrand et al., 2009). Sometimes there is a long latency between the infection and the neurological manifestation. The exact pathogenesis is unknown, suspected mechanisms are vasculitis, cytotoxicity, neurotoxic mediators, or autoimmune reaction through molecular mimicry (Hildenbrand et al., 2009). Clinically, it is characterized by erythema migrant and neurological triad: meningitis (or meningoencephalitis), radiculoneuritis, and cranial neuritis (commonly facial palsy less commonly optic neuritis); spinal cord involvement is rare. Diagnosis is built using the clinical history, blood test, CSF samples, MRI findings, and response to antibiotics. Treatment consisted of antibiotics and usually, unlike the MRI; the clinical recovery is rapid after treatment.

26.3.2.1.3.1 Conventional MRI Findings   Brain MRI could be normal but commonly hyperintense WM lesions on T_2-FLAIR weighted sequences are present and they are similar to those found in MS, mostly found periventricularly and in deep WM. Some lesions may enhance after contrast administration (Foerster et al., 2007). Cranial nerve enhancement is also common (Figure 26.26). Spinal cord could be rarely affected. When affected, diffuse or multifocal T_2-weighted cord lesions can be observed, coupled with nerve root enhancement on postcontrast T_1-weighted sequences (Hattingen et al., 2004).

26.3.2.1.3.2 Advanced MRI Imaging   Unlike MS, using MTR and DTI, no occult tissue damage has been found

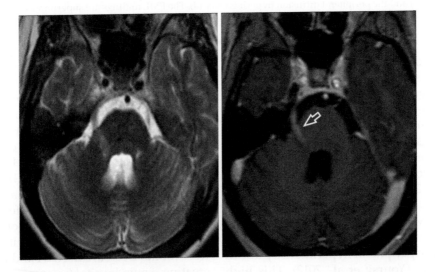

FIGURE 26.26
Lyme disease. Patient with lyme disease and trigeminal nerve lesion. MRI demonstrates an enhancing lesion along the pontine and cisternal portion of the left trigeminal nerve (arrow). The patient was subsequently found to have lyme disease (serologically documented).

in NAWM and normal-appearing GM cerebral tissue and in spinal cord (Agosta et al., 2006) in patients with Lyme disease.

KEY POINTS 26.26   ADVANCED MRI FEATURES IN PML

- Restricted diffusion and peripheral diffusion hyperintense rim of lesions.
- Low and stable over time MTR in PML lesions.
- MRS may be useful for differential diagnosis between PML with and without IRIS.

26.3.2.1.4 Susac Syndrome

Susac syndrome is an autoimmune endoteliopathy affecting the brain, retina, and inner ear (cochlea and semicircular canals), leading to encephalopathy, branch retinal artery occlusions (BRAOs), and asymmetric neurosensory hearing loss (García-Carrasco et al., 2014; Susac, 1994). Discovered in 1979 (Susac et al., 1979), its prevalence and incidence are unknown (Dörr et al., 2013). It is more common in young women (female/male ratio 3:1) around 40 years old and no racial preferences (García-Carrasco et al., 2014). The pathogenesis is not clear, but it is considered as an autoimmune microangiopathy (García-Carrasco et al., 2014). Antibodies and anti-endothelial cells have been found in patients affected by Susac syndrome (Jarius et al., 2009). It can be monophasic, relapsing, or chronic (García-Carrasco et al., 2014). Usually self-limited in time and lifelong monitoring is needed. High dose corticosteroids are the first therapeutic approach, followed by immunosuppressive treatment such as intravenous immunoglobulin, or plasma exchange (García-Carrasco et al., 2014). The diagnosis is made using the typical clinical symptoms, the documentation of branch retinal artery occlusion by fluorescence angiography, and characteristic findings on cerebral MRI. Although the clinical symptoms are very pathognomonic, they can occur separately so the role of MRI is pivotal in differentiating Susac syndrome from MS and ADEM.

26.3.2.1.4.1 Conventional MRI Findings   WM hyperintense roundly shaped lesions involving preferentially the corpus callosum are pathognomonic of Susac syndrome and they are visible on T_2-weighted images but improved visualization is provided by FLAIR sequence. These lesions are called *snow-ball* lesions and appeared as holes in the corpus callosum on T_1-weighted images (García-Carrasco et al., 2014; Rennebohm et al., 2010; Susac et al., 2003). Callosal holes are typical of Susac syndrome and help to differentiate this entity from MS and ADEM. In corpus callosum, atrophy on conventional

T_1-weighted sequence and liner lesions called *spoke* for their configuration due to micro-infarcts are frequently visible. When several micro-infarcts are found at the level of internal capsule, the configuration is called *string of pearls* (García-Carrasco et al., 2014; Rennebohm et al., 2010; Susac et al., 2003). Deep GM and cortex involvement is frequent up to 70% and, as opposite, to MS is visible on conventional MRI. In ADEM, although present in deep GM involvement, cortical involvement is never found. Up to 30% of patients show also leptomenigeal enhancement, mainly in the cerebellum, never seen in MS or ADEM (García-Carrasco et al., 2014; Rennebohm et al., 2010; Susac et al., 2003). Contrast enhancement can be seen also for WM lesions of the corpus callosum (García-Carrasco et al., 2014; Rennebohm et al., 2010; Susac et al., 2003) (Key Points 26.27).

26.3.2.1.4.2 Advanced MRI Findings   Diffusion studies have shown reduced ADC in WM-hyperintense lesions, *string of pearls* in the internal capsule and in multiple lesions in the genu and splenium of corpus callosum. DTI studies have shown occult microstructural damage of the corpus callosum, mainly in the genus (García-Carrasco et al., 2014; Kleffner et al., 2010). Although microstructural damage of the genu of corpus callosum is considered prototypical of Susac syndrome, other areas were found microstructurally damaged such as brain stem and prefrontal areas (Kleffner et al., 2010). Using a 7 T scanner, recently differences in lesion morphology have been shown between Susac syndrome and MS. As opposed to MS, WM lesions in Susac syndrome do not show any perivenular location (Wuerfel et al., 2012) (Key Points 26.28).

KEY POINTS 26.27   CONVENTIONAL MRI FEATURES IN SUSAC SYNDROME

- Multiple hyperintense lesions visible on FLAIR sequence in the corpus callosum *snow-balls* like. They can sometimes enhance.
- Corresponding hypointense *holes like* lesions on T_1-weighted images in corpus callosum.
- Hypointense on T_1-weighted images of linear lesions called *spoke*.
- Multiple hyperintense on T_2-weighted and diffusion images of lesions in the internal capsulae *string of pearls* like.
- Leptomeningeal enhancement of the cerebellum.
- Cortex and deep GM involvement visible on conventional MRI up to 70% of cases.
- Visible atrophy of the corpus callosum.

26.3.2.2 Vascular

26.3.2.2.1 *Cerebral Autosomal-Dominant Arteriopathy with Subcortical Infarcts and Leukoencephalopathy*

Cerebral autosomal dominant arteriopathy with subcortical infarcts and leukoencephalopathy (CADASIL) is an inherited disorder due to the mutations of the *NOTCH3* gene on chromosome 19 (Joutel et al., 1996). CADASIL affects adults in their middle age and symptoms are migraine, transient ischemic attack, strokes, psychiatric disorders, and cognitive impairment (Federico et al., 2012). It is a progressive disease with death occurring during the sixth decade of life (Chabriat et al., 1995). The exact pathogenetic mechanism is poorly understood (Chabriat et al., 2009). Pathologically, deposition of granular osmophilic agglomerates in the basal membrane of arteries determines ischemic brain injury. There is no treatment for CADASIL so far. Therapeutic studies involving medications and cognitive interventions are strongly needed in CADASIL (Andrè, 2010). Recently therapeutic modulation of *NOTCH3* signal is under investigation (Andersson and Lendahl, 2014).

26.3.2.2.1.1 *Conventional MRI Findings*

Conventional MRI is important for the diagnosis of CADASIL (Ryan et al., 2014). The presence at the same time of several features at conventional MRI could facilitate the diagnosis of CADASIL. First, WM hyperintense T_2-weighted lesions are present, and they can be multiple lesions or large confluent lesions located periventricularly but preferentially affecting subcortical WM of the temporal and frontal lobe, insula, and external capsule (Skehan et al., 1995; Yousry et al., 1999). Lesions of basal ganglia and thalamus are observed (Chabriat et al., 2009). A recent study has shown that T_2 hyperintensities in the temporal pole of CADASIL patients are explained by enlarged perivascular spaces and degeneration of myelin (Yamamoto et al., 2009). These are subcortical lacunar lesions, which are linear groups of round lesions, seen with signal intensity matching CSF on all sequences. Cerebral microbleeds are also commonly seen in CADASIL (Lesnik Oberstein et al., 2001; Ryan et al., 2014). Microbleeds are defined as focal areas of signal loss on T_2-weighted sequences that increase in size, so-called blooming effect (Ryan et al., 2014; van Den Boom et al., 2002). Lacunar infarcts are also observed. A new sequence has improved detection of subcortical infarcts in CADASIL (Mendes Coelho et al., 2014) (Figure 26.27 and Key Points 26.29).

26.3.2.2.1.2 *Advanced MRI Findings*

Restricted diffusion corresponds to WM hyperintensity. Using MTR and DTI, a submicroscopic tissue damage beyond WM lesions and subcortical infarcts have been described in CADASIL in both brain NAWM, deep GM (Iannucci et al., 2001; Molko et al., 2001), and spinal cord (Rocca et al., 2001). Occult damage of the cord was strongly correlated with the extension of brain lesions, suggesting Wallerian degeneration as the substrate of the cord MTR changes (Rocca et al., 2001). Using ultrahigh field MRI, early occult damage of cortex (reduction of thickness and

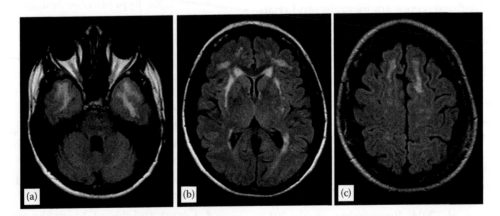

FIGURE 26.27
CADASIL. Typical CADASIL hyperintense confluent white matter lesions in the anterior temporal lobe (a), external capsule (b) and superior frontal gyri (c). Notice the presence of the subcortical lacunar lesions, affecting the anterior part of the temporal lobe (a) and subinsular region (b) at the junction of gray and white matter, very specific for CADASIL.

volume) that progress over time has been recently shown in CADASIL (De Guio et al., 2014; Jouvent et al., 2012), supporting previous pathological reports (Viswanathan et al., 2006) and explaining the cognitive deficits in these patients (Key Points 26.30).

KEY POINTS 26.29   CONVENTIONAL MRI FEATURES IN CADASIL

- WM hyperintense T_2-weighted lesions preferentially located juxacortical in anterior temporal lobe, external capsule, frontal lobe, and insula.

- Small subcortical lacunar infarcts with CSF-like intensity, thus they appear dark on T_1-weighted images or in FLAIR and bright on T_2-weighted sequence. Typically they are present in the anterior temporal lobe.

- Micro-bleeds have CSF-like intensity in all sequences, thus they appear dark on T_2-weighted sequences too.

- Deep GM and thalamus lesions.

KEY POINTS 26.30   ADVANCED MRI FEATURES IN CADASIL

- Restricted diffusion in WM-hyperintense lesions.

- Occult tissue damage in brain WM and GM including cortex and spinal cord beyond conventional MRI visible lesions is present.

26.3.2.2.2 Posterior Reversible Encephalopathy Syndrome

The posterior reversible encephalopathy syndrome (PRES) is an entity that can follow several clinical conditions such as hypertension, eclampsia, autoimmune diseases, immunosuppressive therapy, chemotherapy, infections, hyperperfusion, and miscellaneous conditions. It is defined by a posterior brain involvement that can revert. While the pathogenesis is incompletely understood, it is probably linked with endothelial dysfunction. Clinically, it manifests with headache, confusion, visual disturbance, and seizures. Pathologically, subcortical infarction and edema are characteristic (Lamy et al., 2014). The diagnosis is made by consulting the conventional MRI findings. The treatment is based on recognition of the primary disease that is responsible for the development of the PRES (Lamy et al., 2014).

26.3.2.2.2.1 Conventional MRI Findings   WM-hyperintense lesions on T_2-weighted sequences and FLAIR, mainly in the posterior cerebral WM, and in the adjacent GM. Frontal lobe involvement is possible, but as an extension of the posterior damage. On T_1-postcontrast images, when present, enhancement is sparse. Within a few weeks, WM lesions can revert and clinical recovery can be obtained if the cause is treated (Lamy et al., 2014) (Figure 26.28).

26.3.2.2.2.2 Advanced MRI Findings   DWI shows increased ADC. Higher ADC is associated with a better prognosis with recovery (Lamy et al., 2014). At MRS, decreased NAA and increased creatine and choline in lesions have been found.

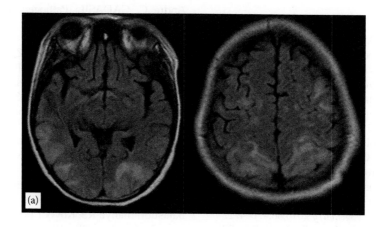

FIGURE 26.28
PRES. (a): MRI of a patient presenting with headache, nausea, vomiting, altered mental status and severe hypertension; the MRI demonstrated multiple areas of symmetric hemispheric edema involving mainly the parietal and occipital lobes, but also the frontal lobes, the temporal-occipital junction.

(Continued)

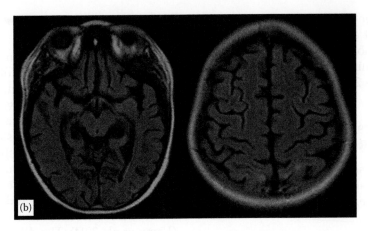

FIGURE 26.28 (Continued)
PRES. Follow-up MRI (b) performed after one month demonstrated the resolution of these areas of edema.

26.4 Conclusions

Using conventional and nonconventional MRI techniques, significant advancements have been made in the diagnostic work-up and in the understanding of pathological mechanisms of both brain demyelinating and metabolic disorders. Future progresses in MRI technology, incorporating ultrahigh field MRI, multimodal approaches, and newer techniques targeting myelin damage, molecular imaging, and histopathological observations are expected to further expand the current knowledge of pathologic disease mechanisms in order to improve diagnostic, prognostic, monitoring ability and to promote new treatment developments.

References

Abdelhalim AN, Alberico RA, Barczykowski AL, Duffner PK. Patterns of magnetic resonance imaging abnormalities in symptomatic patients with Krabbe disease correspond to phenotype. *Pediatr Neurol* 2014;50:127–134.

Absinta M, Rocca MA, Moiola L et al. Brain macro and microscopic damage in patients with paediatric MS. *J Neurol Neurosurg Psychiatry* 2010;81:1357–1362.

Absinta M, Rocca MA, Moiola L, Copetti M, Milani N, Falini A, Comi G, Filippi M. Cortical lesions in children with multiple sclerosis. *Neurology* 2011;76:910–913.

Absoud M, Lim MJ, Appleton R et al. Paediatric neuromyelitis optica: Clinical, MRI of the brain and prognostic features. *J Neurol Neurosurg Psychiatry* 2015;86:470–472.

Agosta F, Rocca MA, Benedetti B, Capra R, Cordioli C, Filippi M. MR imaging assessment of brain and cervical cord damage in patients with neuroborreliosis. *AJNR Am J Neuroradiol* 2006;27:892–894.

Aguirre-Cruz L, Flores-Rivera J, De La Cruz-Aguilera DL, Rangel-López E, Corona T. Multiple sclerosis in Caucasians and Latino Americans. *Autoimmunity* 2011;44:571–575.

Aicardi J. The inherited leukodystrophies: A clinical overview. *J Inherit Metab Dis* 1993;16:733–743. Review.

Aliotta R, Cox JL, Donohue K, Weinstock-Guttman B, Yeh EA, Polak P, Dwyer MG, Zivadinov R. Tract-based spatial statistics analysis of diffusion-tensor imaging data in pediatric- and adult-onset multiple sclerosis. *Hum Brain Mapp.* 2014;35:53–60.

Amato MP, Hakiki B, Goretti B et al.; Italian RIS/MS Study Group. Association of MRI metrics and cognitive impairment in radiologically isolated syndromes. *Neurology* 2012;78:309–314.

Amato MP, Portaccio E, Goretti B, Zipoli V, Hakiki B, Giannini M, Pastò L, Razzolini L. Cognitive impairment in early stages of multiple sclerosis. *Neurol Sci* 2010;31:S211–S214.

Andersson ER, Lendahl U. Therapeutic modulation of Notch signalling—Are we there yet? *Nat Rev Drug Discov* 2014;13:357–378.

André C. CADASIL: Pathogenesis, clinical and radiological findings and treatment. *Arq Neuropsiquiatr* 2010;68: 287–299. Review.

Antel J, Antel S, Caramanos Z, Arnold DL, Kuhlmann T. Primary progressive multiple sclerosis: Part of the MS disease spectrum or separate disease entity? *Acta Neuropathol* 2012;123:627–638.

Ascherio A. Environmental factors in multiple sclerosis. *Expert Rev Neurother* 2013;13:3–9.

Asgari N. Epidemiological, clinical and immunological aspects of neuromyelitis optica (NMO). *Dan Med J* 2013;60:B4730. Review.

Aubourg P, Wanders R. Peroxisomal disorders. *Handb Clin Neurol* 2013;113:1593–1609.

Balasubramanya KS, Kovoor JM, Jayakumar PN, Ravishankar S, Kamble RB, Panicker J, Nagaraja D. Diffusion-weighted imaging and proton MR spectroscopy in the characterization of acute disseminated encephalomyelitis. *Neuroradiology* 2007;49:177–183.

Baldwin KJ, Hogg JP. Progressive multifocal leukoencephalopathy in patients with multiple sclerosis. *Curr Opin Neurol* 2013;26:318–323.

Banwell B, Ghezzi A, Bar-Or A, Mikaeloff Y, Tardieu M. Multiple sclerosis in children: Clinical diagnosis, therapeutic strategies, and future directions. *Lancet Neurol* 2007a;6:887–902.

Banwell B, Krupp L, Kennedy J et al. Clinical features and viral serologies in children with multiple sclerosis: A multinational observational study. *Lancet Neurol* 2007b;6:773–781.

Banwell B, Shroff M, Ness JM, Jeffery D, Schwid S, Weinstock-Guttman B. MRI features of paediatric multiple sclerosis. *Neurology* 2007c;68:S46–S53.

Banwell BL. Multiple sclerosis in children. *Handb Clin Neurol* 2014;122:427–441.

Banwell B, Bar-Or A, Arnold DL et al. Clinical, environmental, and genetic determinants of multiple sclerosis in children with acute demyelination: A prospective national cohort study. *Lancet Neurol* 2014;10:436–445.

Barkhof F, Filippi M, Miller DH, Scheltens P, Campi A, Polman CH, Comi G, AdèrHJ, Losseff N, Valk J. Comparison of MRI criteria at first presentation to predict conversion to clinically definite multiple sclerosis. *Brain* 1997;120:2059–2069.

Barreau P, Prust MJ, Crane J, Loewenstein J, Kadom N, Vanderver A. Focal central white matter lesions in Alexander disease. *J Child Neurol* 2011;26:1422–1424.

Bellizzi A, Anzivino E, Rodio DM, Palamara AT, Nencioni L, Pietropaolo V. New insights on human polyomavirus JC and pathogenesis of progressive multifocal leukoencephalopathy. *Clin Dev Immunol* 2013;2013:839719. Review.

Ben Sira L, Miller E, Artzi M, Fattal-Valevski A, Constantini S, Ben Bashat D. 1H-MRS for the diagnosis of acute disseminated encephalomyelitis: Insight into the acute-disease stage. *Pediatr Radiol* 2010;40:106–113.

Beres SJ, Graves J, Waubant E. Rituximab use in pediatric central demyelinating disease. *Pediatr Neurol* 2014;51:114–118.

Berger JR. Progressive multifocal leukoencephalopathy. *Handb Clin Neurol* 2014;123:357–376.

Berger JR, Aksamit AJ, Clifford DB, Davis L, Koralnik IJ, Sejvar JJ, Bartt R, Major EO, Nath A. PML diagnostic criteria: Consensus statement from the AAN Neuroinfectious Disease Section. *Neurology* 2013;80:1430–1438.

Bergui M, Bradac GB, Oguz KK et al. Progressive multifocal leukoencephalopathy: Diffusion-weighted imaging and pathological correlations. *Neuroradiology* 2004;46:22–25.

Bermel RA, Bakshi R. The measurement and clinical relevance of brain atrophy in multiple sclerosis. *Lancet Neurol* 2006;5:158–170.

Bester M, Petracca M, Inglese M. Neuroimaging of multiple sclerosis, acute disseminated encephalomyelitis, and other demyelinating diseases. *Semin Roentgenol* 2014;49:76–85.

Bethune A, Tipu V, Sled JG, Narayanan S, Arnold DL, Mabbott D, Rockel C, Ghassemi R, Till C, Banwell B. Diffusion tensor imaging and cognitive speed in children with multiple sclerosis. *J Neurol Sci* 2011;309:68–74.

Bienia B, Balabanov R. Immunotherapy of neuromyelitis optica. *Autoimmune Dis* 2013;2013:741490.

Biffi A, Montini E, Lorioli L et al. Lentiviral hematopoietic stem cell gene therapy benefits metachromatic leukodystrophy. *Science* 2013;341(6148):1233158.

Bigi S, Banwell B. Pediatric multiple sclerosis. *J Child Neurol* 2012;27:1378–1383. Review.

Blanc F, Noblet V, Jung B et al. White matter atrophy and cognitive dysfunctions in neuromyelitis optica. *PLoS One* 2012;7:e33878.

Boiko A, Vorobeychik G, Paty D, Devonshire V, Sadovnick D; University of British Columbia MS Clinic Neurologists. Early onset multiple sclerosis: A longitudinal study. *Neurology* 2002;57:1006–1010.

Boster A, Hreha S, Berger JR et al. Progressive multifocal leukoencephalopathy and relapsing-remitting multiple sclerosis: A comparative study. *Arch Neurol* 2009;66:593–599.

Bot JC, Barkhof F. Spinal-cord MRI in multiple sclerosis: Conventional and nonconventional MR techniques. *Neuroimaging Clin N Am* 2009;19:81–99.

Bourdette D, Simon J. The radiologically isolated syndrome: Is it very early multiple sclerosis? *Neurology* 2009;72:780–781.

Bourdette D, Yadav V. Treat patients with radiologically isolated syndrome when the MRI brain scan shows dissemination in time: No. *Mult Scler* 2012;18:1529–1530.

Bove R, Chitnis T. Sexual disparities in the incidence and course of MS. *Clin Immunol* 2013;149:201–210.

Brassat D, Lebrun-Frenay C; Club Francophone de la SEP. Treat patients with radiologically isolated syndrome when the MRI brain scan show dissemination in time: Yes. *Mult Scler* 2012;18:1531–1532.

Brownlee WJ, Miller DH. Clinically isolated syndromes and the relationship to multiple sclerosis. *J Clin Neurosci.* 2014;21:2065–2071.

Bruhn H, Frahm J, Merboldt KD, Hänicke W, Hanefeld F, Christen HJ, Kruse B, Bauer HJ. Multiple sclerosis in children: Cerebral metabolic alterations monitored by localized proton magnetic resonance spectroscopy in vivo. *Ann Neurol* 1992;32:140–150.

Bugiani M, Boor I, Powers JM, Scheper GC, van der Knaap MS. Leukoencephalopathy with vanishing white matter: A review. *J Neuropathol Exp Neurol* 2010;69:987–989.

Calabrese M, Oh MS, Favaretto A et al. No MRI evidence of cortical lesions in neuromyelitis optica. *Neurology* 2012;79:1671–1676.

Callen DJ, Shroff MM, Branson HM et al. MRI in the diagnosis of paediatric multiple sclerosis. *Neurology* 2009;72:961–967.

Ceccarelli A, Bakshi R, Neema M. MRI in multiple sclerosis: A review of the current literature. *Curr Opin Neurol* 2012;25:402–409.

Ceccarelli A, Filippi M, Neema M, Arora A, Valsasina P, Rocca MA, Healy BC, Bakshi R. T2 hypointensity in the deep gray matter of patients with benign multiple sclerosis. *Mult Scler* 2009;15:678–686.

Ceccarelli A, Rocca MA, Pagani E, Colombo B, Martinelli V, Comi G, Filippi M. A voxel-based morphometry study of grey matter loss in MS patients with different clinical phenotypes. *Neuroimage* 2008;42:315–322.

Ceccarelli A, Rocca MA, Perego E, Moiola L, Ghezzi A, Martinelli V, Comi G, Filippi M. Deep grey matter T2 hypo-intensity in patients with paediatric multiple sclerosis. *Mult Scler* 2011;17:702–707.

Ceccarelli A, Rocca MA, Valsasina P, Rodegher M, Falini A, Comi G, Filippi M. Structural and functional magnetic resonance imaging correlates of motor network dysfunction in primary progressive multiple sclerosis. *Eur J Neurosci* 2010;31:1273–1280.

Chabas D, Castillo-Trivino T, Mowry EM, Strober JB, Glenn OA, Waubant E. Vanishing MS. T2-bright lesions before puberty: A distinct MRI phenotype? *Neurology* 2008;71:1090–1093.

Chabriat H, Joutel A, Dichgans M, Tournier-Lasserve E, Bousser MG. Cadasil. *Lancet Neurol* 2009;8:643–653.

Chabriat H, Vahedi K, Iba-Zizen MT et al. Clinical spectrum of CADASIL: A study of 7 families. Cerebral autosomal dominant arteriopathy with subcortical infarcts and leukoencephalopathy. *Lancet* 1995;346:934–939.

Chanson JB, Lamy J, Rousseau F, Blanc F, Collongues N, Fleury M, Armspach JP, Kremer S, de Seze J. White matter volume is decreased in the brain of patients with neuromyelitis optica. *Eur J Neurol* 2013;20:361–367.

Charil A, Yousry TA, Rovaris M et al. MRI and the diagnosis of multiple sclerosis: Expanding the concept of "no better explanation." *Lancet Neurol* 2006;5:841–852. Review.

Chen F, Liu T, Li J, Xing Z, Huang S, Wen G, Lu G. Eccentric development of Balo's concentric sclerosis: Detected by magnetic resonance diffusion-weighted imaging and magnetic resonance spectroscopy. *Int J Neurosci* 2015;125:433–440.

Chitnis T, Glanz B, Jaffin S, Healy B. Demographics of pediatric-onset multiple sclerosis in an MS center population from the Northeastern United States. *Mult Scler* 2009;15:627–631.

Ciccarelli O, Thomas DL, De Vita E et al. Low myo-inositol indicating astrocytic damage in a case series of neuromyelitis optica. *Ann Neurol* 2013;74:301–305.

Compston A, Coles A. Multiple sclerosis. *Lancet* 2008;372:1502–1517.

Correale J, Ysrraelit MC, Fiol MP. Benign multiple sclerosis: Does it exist? *Curr Neurol Neurosci Rep* 2012;12:601–609.

Cosottini M, Tavarelli C, Del Bono L, Doria G, Giannelli M, De Cori S, Michelassi MC, Bartolozzi C, Murri L. Diffusion-weighted imaging in patients with progressive multifocal leukoencephalopathy. *Eur Radiol* 2008;18:1024–1030.

da Pozzo S, Manara R, Tonello S, Carollo C. 2006. Conventional and diffusion-weighted MRI in progressive multifocal leukoencephalopathy: New elements for identification and follow-up. *Radiol Med* 111:971–977.

Dale RC, de Sousa C, Chong WK, Cox TC, Harding B, Neville BG. Acute disseminated encephalomyelitis, multiphasic disseminated encephalomyelitis and multiple sclerosis in children. *Brain* 2000;123:2407–2422.

De Guio F, Reyes S, Vignaud A, Duering M, Ropele S, Duchesnay E, Chabriat H, Jouvent E. In vivo high-resolution 7 Tesla MRI shows early and diffuse cortical alterations in CADASIL. *PLoS One* 2014;9:e106311.

De Keyser J, Mostert JP, Koch MW. Dysfunctional astrocytes as key players in the pathogenesis of central nervous system disorders. *J Neurol Sci* 2008;267:3–16.

de Seze J, Blanc F, Kremer S, Collongues N, Fleury M, Marcel C, Namer IJ. Magnetic resonance spectroscopy evaluation in patients with neuromyelitis optica. *J Neurol Neurosurg Psychiatry* 2010;81:409–411.

De Stefano N, Siva A. The radiologically isolated syndrome dilemma: Just an incidental radiological finding or pre-symptomatic multiple sclerosis? *Mult Scler* 2013;19:257–258.

De Stefano N, Stromillo ML, Rossi F et al. Improving the characterization of radiologically isolated syndrome suggestive of multiple sclerosis. *PLoS One* 2011;6:e19452.

Ding XQ, Bley A, Ohlenbusch A, Kohlschütter A, Fiehler J, Zhu W, Lanfermann H. Imaging evidence of early brain tissue degeneration in patients with vanishing white matter disease: A multimodal MR study. *J Magn Reson Imaging* 2012;35:926–932.

Disanto G, Magalhaes S, Handel AE, Morrison KM, Sadovnick AD, Ebers GC, Banwell B, Bar-Or A; Canadian Pediatric Demyelinating Disease Network. HLA-DRB1 confers increased risk of pediatric-onset MS in children with acquired demyelination. *Neurology* Mar 1, 2011;76(9):781–786.

Donohue K, Cox JL, Dwyer MG, Aliotta R, Corwin M, Weinstock-Guttman B, Ann Yeh E, Zivadinov R. No regional gray matter atrophy differences between pediatric and adult-onset relapsing-remitting multiple sclerosis. *J Neuroimaging* 2014;24:63–67.

Dörr J, Krautwald S, Wildemann B, Jarius S, Ringelstein M, Duning T, Aktas O, Ringelstein EB, Paul F, Kleffner I. Characteristics of Susac syndrome: A review of all reported cases. *Nat Rev Neurol* 2013;9:307–316.

Duan Y, Liu Y, Liang P, Jia X, Ye J, Dong H, Li K. White matter atrophy in brain of neuromyelitis optica: A voxel-based morphometry study. *Acta Radiol* 2013;55:589–593.

Duan Y, Liu Y, Liang P, Jia X, Yu C, Qin W, Sun H, Liao Z, Ye J, Li K. Comparison of grey matter atrophy between patients with neuromyelitis optica and multiple sclerosis: A voxel-based morphometry study. *Eur J Radiol* 2012;81:e110–e114.

Duffner PK, Barczykowski A, Jalal K et al. Early infantile Krabbe disease: Results of the world-wide Krabbe registry. *Pediatr Neurol* 2011;45:141–148.

Duffner PK, Barczykowski A, Kay DM et al. Later onset phenotypes of Krabbe disease: Results of the world-wide registry. *Pediatr Neurol* 2012;46:298–306.

Eichler F, Grodd W, Grant E, Sessa M, Biffi A, Bley A, Kohlschuetter A, Loes DJ, Kraegeloh-Mann I. Metachromatic leukodystrophy: A scoring system for brain MR imaging observations. *AJNR Am J Neuroradiol* 2009;30:1893–1897.

Elenein RA, Naik S, Kim S, Punia V, Jin K. Teaching neuroimages: Cerebral adrenoleukodystrophy: A rare adult form. *Neurology* 2013;80:e69–e70.

Engelen M, Barbier M, Dijkstra IM et al. X-linked adrenoleukodystrophy in women: A cross-sectional cohort study. *Brain* 2014;137(Pt 3):693–706.

Engelen M, Kemp S, de Visser M, van Geel BM, Wanders RJ, Aubourg P, Poll-The BT. X-linked adrenoleukodystrophy (X-ALD): Clinical presentation and guidelines for diagnosis, follow-up and management. *Orphanet J Rare Dis* 2012;7:51.

Fazekas F, Barkhof F, Filippi M. Unenhanced and enhanced magnetic resonance imaging in the diagnosis of multiple sclerosis. *J Neurol Neurosurg Psychiatry* 1998;64:S2–S5.

Federico A, Di Donato I, Bianchi S, Di Palma C, Taglia I, Dotti MT. Hereditary cerebral small vessel diseases: A review. *J Neurol Sci* 2012;322:25–30.

Fernández Carbonell C, Chitnis T. Inflammatory demyelinating diseases in children: An update. *Minerva Pediatr* 2013;65:307–323. Review.

Filippi M, Rocca MA. Present and future of fMRI in multiple sclerosis. *Expert Rev Neurother* 2013;13:27–31.

Filippi M, Rocca MA, Bastianello S et al.; Neuroimaging and MS Study Groups of the Italian Society of Neurology; Functional Neuroradiology Section of the Italian Association of Neuroradiology. Guidelines from The Italian Neurological and Neuroradiological Societies for the use of magnetic resonance imaging in daily life clinical practice of multiple sclerosis patients. *Neurol Sci* 2013;34:2085–2093.

Flanagan EP, Weinshenker BG. Neuromyelitis optica spectrum disorders. *Curr Neurol Neurosci Rep* 2014;14:483.

Foerster BR, Thurnher MM, Malani PN, Petrou M, Carets-Zumelzu F, Sundgren PC. Intracranial infections: Clinical and imaging characteristics. *Acta Radiol* 2007; 48:875–893.

Gabelić T, Radmilović M, Posavec V, Skvorc A, Bošković M, Adamec I, Milivojević I, Barun B, Habek M. Differences in oligoclonal bands and visual evoked potentials in patients with radiologically and clinically isolated syndrome. *Acta Neurol Belg* 2013;113:13–17.

Gabelić T, Ramasamy DP, Weinstock-Guttman B, Hagemeier J, Kennedy C, Melia R, Hojnacki D, Ramanathan M, Zivadinov R. Prevalence of radiologically isolated syndrome and white matter signal abnormalities in healthy relatives of patients with multiple sclerosis. *AJNR Am J Neuroradiol* 2014;35:106–112.

García-Carrasco M, Mendoza-Pinto C, Cervera R. Diagnosis and classification of Susac syndrome. *Autoimmun Rev* 2014;13:347–350.

Geurts JJ, Pouwels PJ, Uitdehaag BM, Polman CH, Barkhof F, Castelijns JA. Intracortical lesions in multiple sclerosis: Improved detection with 3D double inversion-recovery MR imaging. *Radiology* 2005;236:254–260.

Ghassemi R, Narayanan S, Banwell B, Sled JG, Shroff M, Arnold DL; Canadian Pediatric Demyelinating Disease Network. Quantitative determination of regional lesion volume and distribution in children and adults with relapsing-remitting multiple sclerosis. *PLoS One* 2014;9:e85741.

Gheuens S, Ngo L, Wang X, Alsop DC, Lenkinski RE, Koralnik IJ. Metabolic profile of PML lesions in patients with and without IRIS: An observational study. *Neurology* 2012;79:1041–1048.

Gheuens S, Wüthrich C, Koralnik IJ. Progressive multifocal leukoencephalopathy: Why gray and white matter. *Annu Rev Pathol* 2013;8:189–215.

Ghezzi A, Pozzilli C, Grimaldi LM et al.; Italian MS Study Group. Natalizumab in pediatric multiple sclerosis: Results of a cohort of 55 cases. *Mult Scler* 2013;19:1106–1112.

Gieselmann V. Metachromatic leukodystrophy: Genetics, pathogenesis and therapeutic options. *Acta Paediatr Suppl* 2008;97:15–21.

Gieselmann V, Krägeloh-Mann I. Metachromatic leukodystrophy—An update. *Neuropediatrics* 2010;41:1–6.

Giorgio A, Stromillo ML, Rossi F, Battaglini M, Hakiki B, Portaccio E, Federico A, Amato MP, De Stefano N. Cortical lesions in radiologically isolated syndrome. *Neurology* 2011;77:1896–1899.

Gorman MP, Healy BC, Polgar-Turcsanyi M, Chitnis T. Increased relapse rate in pediatric-onset compared with adult-onset multiple sclerosis. *Arch Neurol* 2009;66:54–59.

Graff-Radford J, Schwartz K, Gavrilova RH, Lachance DH, Kumar N. Neuroimaging and clinical features in type II (late-onset) Alexander disease. *Neurology* 2014;82:49–56.

Granberg T, Martola J, Kristoffersen-Wiberg M, Aspelin P, Fredrikson S. Radiologically isolated syndrome—Incidental magnetic resonance imaging findings suggestive of multiple sclerosis, a systematic review. *Mult Scler* 2013;19:271–280.

Groeschel S, í Dali C, Clas P, Böhringer J, Duno M, Krarup C, Kehrer C, Wilke M, Krägeloh-Mann I. Cerebral gray and white matter changes and clinical course in metachromatic leukodystrophy. *Neurology* 2012;79:1662–1670.

Guleria S, Kelly TG. Myelin, myelination, and corresponding magnetic resonance imaging changes. *Radiol Clin North Am* 2014;52:227–239.

Hachinski VC, Potter P, Merskey H. Leuko-araiosis. *Arch Neurol* 1987;44:21–23.

Hardy TA, Chataway J. Tumefactive demyelination: An approach to diagnosis and management. *J Neurol Neurosurg Psychiatry* 2013;84:1047–1053.

Hardy TA, Miller DH. Baló's concentric sclerosis. *Lancet Neurol* 2014;13:740–746.

Harwood-Nash DC, Fitz CR, Reilly BJ. Cranial computed tomography in infants and children. *Can Med Assoc J* 1975;113:546–549.

Hatten HP Jr. Dysmyelinating leukodystrophies: "LACK Proper Myelin." *Pediatr Radiol* 1991;21:477–482.

Hattingen E, Weidauer S, Kieslich M, Boda V, Zanella FE. MR imaging in neuroborreliosis of the cervical spinal cord. *Eur Radiol* 2004;14:2072–2075.

He J, Inglese M, Li BSY, Babb JS, Grossman RI, Gonen O. Relapsing-remitting multiple sclerosis: metabolic abnormality in nonenhancing lesions and normal-appearing white matter at MR imaging: initial experience. *Radiology* 2005;234:211–217.

Hildenbrand P, Craven DE, Jones R, Nemeskal P. Lyme neuroborreliosis: Manifestations of a rapidly emerging zoonosis. *AJNR Am J Neuroradiol* 2009;30:1079–1087.

Hobson GM1, Garbern JY. Pelizaeus-Merzbacher disease, Pelizaeus-Merzbacher-like disease 1, and related hypomyelinating disorders. *Semin Neurol* 2012;32:62–67.

Hoshino H, Kubota M. Canavan disease: Clinical features and recent advances in research. *Pediatr Int* 2014;56:477–483.

Hutchinson M. Treat patients with radiologically isolated syndrome when the MRI brain scan shows dissemination in time: Commentary. *Mult Scler* 2012;18:1533.

í Dali C, Hanson LG, Barton NW, Fogh J, Nair N, Lund AM. Brain N-acetylaspartate levels correlate with motor function in metachromatic leukodystrophy. *Neurology* 2010;75:1896–1903.

Iannucci G, Dichgans M, Rovaris M, Brüning R, Gasser T, Giacomotti L, Yousry TA, Filippi M. Correlations between clinical findings and magnetization transfer imaging metrics of tissue damage in individuals with cerebral autosomal dominant arteriopathy with subcortical infarcts and leukoencephalopathy. *Stroke* 2001;32:643–648.

Inglese M, Petracca M. Therapeutic strategies in multiple sclerosis: A focus on neuroprotection and repair and relevance to schizophrenia. *Schizophr Res* 2015;161:94–101.

Inglese M, Salvi F, Iannucci G, Mancardi GL, Mascalchi M, Filippi M. Magnetization transfer and diffusion tensor MR imaging of acute disseminated encephalomyelitis. *AJNR Am J Neuroradiol* 2002;23:267–272.

Jarius S, Neumayer B, Wandinger KP, Hartmann M, Wildemann B. Anti-endothelial serum antibodies in a patient with Susac's syndrome. *J Neurol Sci* 2009;285:259–261.

Javed A, Khan O. Acute disseminated encephalomyelitis. *Handb Clin Neurol* 2014;123:705–717.

Joutel A, Corpechot C, Ducros A et al. Notch3 mutations in CADASIL, a hereditary adult-onset condition causing stroke and dementia. *Nature* 1996;383:707–710.

Jouvent E, Mangin JF, Duchesnay E et al. Longitudinal changes of cortical morphology in CADASIL. *Neurobiol Aging* 2012;33:1002.e29–e36.

Kamate M, Hattiholi V. Predominant corticospinal tract involvement in early-onset Krabbe disease. *Pediatr Neurol* 2011;44:155–156.

Kanekar S, Gustas C. Metabolic disorders of the brain: Part I. *Semin Ultrasound CT MR* 2011;32:590–614.

Kanekar S, Verbrugge J. Metabolic disorders of the brain: Part II. *Semin Ultrasound CT MR* 2011;32(6):615–636.

Kennedy J, O'Connor P, Sadovnick AD, Perara M, Yee I, Banwell B. Age at onset of multiple sclerosis may be influenced by place of residence during childhood rather than ancestry. *Neuroepidemiology* 2006;26:162–167.

Kerbrat A, Aubert-Broche B, Fonov V, Narayanan S, Sled JG, Arnold DA, Banwell B, Collins DL. Reduced head and brain size for age and disproportionately smaller thalami in child-onset MS. *Neurology* 2012;78:194–201.

Khoury MN, Alsop DC, Agnihotri SP, Pfannl R, Wuthrich C, Ho ML, Hackney D, Ngo L, Anderson MP, Koralnik IJ. Hyperintense cortical signal on magnetic resonance imaging reflects focal leukocortical encephalitis and seizure risk in progressive multifocal leukoencephalopathy. *Ann Neurol* 2014;75(5):659–669.

Kim JH, Kim HJ. Childhood X-linked adrenoleukodystrophy: Clinical-pathologic overview and MR imaging manifestations at initial evaluation and follow-up. *RadioGraphics* 2005;25:619–631.

Kimura MC, Doring TM, Rueda FC, Tukamoto G, Gasparetto EL. In vivo assessment of white matter damage in neuromyelitis optica: A diffusion tensor and diffusion kurtosis MR imaging study. *J Neurol Sci* 2014;345:172–175.

Kingwell E, Marriott JJ, Jetté N et al. Incidence and prevalence of multiple sclerosis in Europe: A systematic review. *BMC Neurol* 2013;13:128.

Kitley J, Waters P, Woodhall M, Leite MI, Murchison A, George J, Küker W, Chandratre S, Vincent A, Palace J. Neuromyelitis optica spectrum disorders with aquaporin-4 and myelin-oligodendrocyte glycoprotein antibodies: A comparative study. *JAMA Neurol* 2014;71:276–283.

Klawiter EC, Xu J, Naismith RT, Benzinger TL, Shimony JS, Lancia S, Snyder AZ, Trinkaus K, Song SK, Cross AH. Increased radial diffusivity in spinal cord lesions in neuromyelitis optica compared with multiple sclerosis. *Mult Scler* 2012;18:1259–1268.

Kleffner I, Deppe M, Mohammadi S, Schwindt W, Sommer J, Young P, Ringelstein EB. Neuroimaging in Susac's syndrome: Focus on DTI. *J Neurol Sci* 2010;299:92–96.

Kleinschmidt-DeMasters BK, Tyler KL. Progressive multifocal leukoencephalopathy complicating treatment with natalizumab and interferon beta-1a for multiple sclerosis. *N Engl J Med* 2005;353:369–374.

Köhler W. Leukodystrophies with late disease onset: An update. *Curr Opin Neurol* 2010;23:234–241.

Kohlschütter A. Lysosomal leukodystrophies: Krabbe disease and metachromatic leukodystrophy. *Handb Clin Neurol* 2013;113:1611–1618.

Krupp LB, Banwell B, Tenembaum S; Consensus definitions proposed for pediatric multiple sclerosis and related disorders. *Neurology* 2007;68:S7–S12. Review.

Krupp LB, Tardieu M, Amato MP et al.; International Pediatric Multiple Sclerosis Study Group. International Pediatric Multiple Sclerosis Study Group criteria for pediatric multiple sclerosis and immune-mediated central nervous system demyelinating disorders: Revisions to the 2007 definitions. *Mult Scler* 2013;19:1261–1267.

Kumar AJ, Kohler W, Kruse B et al. MR findings in adult-onset adrenoleukodystrophy. *AJNR Am J Neuroradiol* 1995;16(6):1227–1237.

Lamy C, Oppenheim C, Mas JL. Posterior reversible encephalopathy syndrome. *Handb Clin Neurol* 2014; 121:1687–1701.

Lana-Peixoto MA. How much radiologically isolated syndrome suggestive of multiple sclerosis is multiple sclerosis? *Arq Neuropsiquiatr* 2012;70:2–4.

Lanciotti A, Brignone MS, Bertini E, Petrucci TC, Aloisi F, Ambrosini E. Astrocytes: Emerging stars in leukodystrophie pathogenesis. *Transl Neurosci* 2013;4. doi: 10.2478/s13380-013-0118-1.

Lassmann H, Brück W, Lucchinetti CF. The immunopathology of multiple sclerosis: An overview. *Brain Pathol* 2007;17:210–218.

Lattanzi A, Salvagno C, Maderna C, Benedicenti F, Morena F, Kulik W, Naldini L, Montini E, Martino S, Gritti A. Therapeutic benefit of lentiviral-mediated neonatal intracerebral gene therapy in a mouse model of globoid cell leukodystrophy. *Hum Mol Genet* 2014;23:3250–3268.

Lebrun C, Bensa C, Debouverie M et al.; Club Francophone de la Sclérose en Plaques. Association between clinical conversion to multiple sclerosis in radiologically isolated syndrome and magnetic resonance imaging, cerebrospinal fluid, and visual evoked potential: Follow-up of 70 patients. *Arch Neurol* 2009;66:841–846.

Lebrun C, Blanc F, Brassat D, Zephir H, de Seze J; CFSEP. Cognitive function in radiologically isolated syndrome. *Mult Scler* 2010;16:919–925.

Lebrun C, Le Page E, Kantarci O, Siva A, Pelletier D, Okuda DT; Club Francophone de Sclerose en Plaques (CFSEP); Radiologically Isolated Syndrome Consortium (RISC) Group. Impact of pregnancy on conversion to clinically isolated syndrome in a radiologically isolated syndrome cohort. *Mult Scler* 2012;18:1297–1302.

Lennon VA, Wingerchuk DM, Kryzer TJ, Pittock SJ, Lucchinetti CF, Fujihara K, Nakashima I, Weinshenker BG. A serum autoantibody marker of neuromyelitis optica: Distinction from multiple sclerosis. *Lancet* 2004;364:2106–2112.

Lesnik Oberstein SA, van den Boom R, van Buchem MA et al. Cerebral microbleeds in CADASIL. *Neurology* 2001;57:1066–1070.

Levin MH, Bennett JL, Verkman AS. Optic neuritis in neuromyelitis optica. *Prog Retin Eye Res* 2013;36:159–171.

Liu S, Kullnat J, Bourdette D, Simon J, Kraemer DF, Murchison C, Hamilton BE. Prevalence of brain magnetic resonance imaging meeting Barkhof and McDonald criteria for dissemination in space among headache patients. *Mult Scler* 2013;19:1101–1105.

Liu Y, Duan Y, He Y, Yu C, Wang J, Huang J, Ye J, Butzkueven H, Li K, Shu N. A tract-based diffusion study of cerebral white matter in neuromyelitis optica reveals widespread pathological alterations. *Mult Scler* 2012;18: 1013–1021.

Livingston JH, Graziano C, Pysden K, Crow YJ, Mordekar SR, Moroni I, Uziel G. Intracranial calcification in early infantile Krabbe disease: Nothing new under the sun. *Dev Med Child Neurol* 2012;54:376–379.

Loes DJ, Fatemi A, Melhem ER et al. Analysis of MRI patterns aids prediction of progression in X-linked adrenoleukodystrophy. *Neurology* 2003;61:369–374.

Loes DL, Hite S, Moser H, Stillman AE, Shapiro E, Lockman L, Latchaw RE, Krivit W. Adrenoleukodystrophy: A scoring method for brain MR observations. *Am J Neuroradiol* 1994;15:1761–1766.

Loes DJ, Peters C, Krivit W. Globoid cell leukodystrophy: Distinguishing early-onset from late-onset disease using a brain MR imaging scoring method. *AJNR Am J Neuroradiol* 1999;20:316–323.

Lundgaard I, Osório MJ, Kress BT, Sanggaard S, Nedergaard M. White matter astrocytes in health and disease. *Neuroscience* 2014;276C:161–173. Review.

Lyon G, Hagberg B, Evrard PH et al. Symptomatology of late-onset Krabbe's leukodystrophy: The European experience. *Dev Neurosci* 1991;13:240–244.

Maia AC Jr, da Rocha AJ, da Silva CJ, Rosemberg S. Multiple cranial nerve enhancement: A new MR imaging finding in metachromatic leukodystrophy. *AJNR Am J Neuroradiol* 2007;28:999.

Major EO. Progressive multifocal leukoencephalopathy in patients on immunomodulatory therapies. *Annu Rev Med* 2010;61:35–47.

Mar S, Noetzel M. Axonal damage in leukodystrophies. *Pediatr Neurol* 2010;42:239–242.

Marin SE, Callen DJ. The magnetic resonance imaging appearance of monophasic acute disseminated encephalomyelitis: An update post application of the 2007 consensus criteria. *Neuroimaging Clin N Am* May 2013;23:245–266.

Marrie RA, Gryba C. The incidence and prevalence of neuromyelitis optica: A systematic review. *Int J MS Care* 2013;15:113–118.

Martin A, Sevin C, Lazarus C, Bellesme C, Aubourg P, Adamsbaum C. Toward a better understanding of brain lesions during metachromatic leukodystrophy evolution. *AJNR Am J Neuroradiol* 2012;33:1731–1739.

Matalon R, Michals K, Sebesta D, Deanching M, Gashkoff P, Casanova J. Aspartoacylase deficiency and N-acetylaspartic aciduria in patients with Canavan disease. *Am J Med Genet* 1988: 29:463–471.

Matthews L, Marasco R, Jenkinson M et al. Distinction of seropositive NMO spectrum disorder and MS brain lesion distribution. *Neurology* 2013;80:1330–1337.

Meier DS, Weiner HL, Guttmann CR. Time-series modeling of multiple sclerosis disease activity: A promising window on disease progression and repair potential? *Neurotherapeutics* 2007;4:485–498.

Mendes Coelho VC, Bertholdo D, Ono SE, de Carvalho Neto A. MRI hydrographic 3D sequences in CADASIL. *Neurology* 2014;82:371.

Mesaros S, Rocca MA, Absinta M, Ghezzi A, Milani N, Moiola L, Veggiotti P, Comi G, Filippi M. Evidence of thalamic gray matter loss in pediatric multiple sclerosis. *Neurology* 2008;70:1107–1112.

Messing A, Brenner M, Feany MB, Nedergaard M, Goldman JE. Alexander disease. *J Neurosci* 2012;32:5017–5023.

Messing A, LaPash Daniels CM, Hagemann TL. Strategies for treatment in Alexander disease. *Neurotherapeutics* 2010;7:507–515. Review.

Mezzapesa DM, Rocca MA, Falini A et al. A preliminary diffusion tensor and magnetization transfer magnetic resonance imaging study of early-onset multiple sclerosis. *Arch Neurol* 2004;61:366–368.

Michel SJ, Given CA 2nd. Case 99: Canavan disease. *Radiology* 2006;241:310–314.

Miller DH, Weinshenker BG, Filippi M et al. Differential diagnosis of suspected multiple sclerosis: A consensus approach. *Mult Scler* 2008;14:1157–1174.

Minneboo A, Uitdehaag BM, Ader HJ, Barkhof F, Polman CH, Castelijns JA. Patterns of enhancing lesion evolution in multiple sclerosis are uniform within patients. *Neurology* 2005;65:56–61.

Molko N, Pappata S, Mangin JF, Poupon C, Vahedi K, Jobert A, LeBihan D, Bousser MG, Chabriat H. Diffusion tensor imaging study of subcortical gray matter in cadasil. *Stroke* 2001;32:2049–2054.

Moser HW, Mahmood A, Raymond GV (2007). X-linked adrenoleukodystrophy. *Nat Clin Pract Neurol* 3:140–151.

Musolino PL, Rapalino O, Caruso P, Caviness VS, Eichler FS. Hypoperfusion predicts lesion progression in cerebral X-linked adrenoleukodystrophy. *Brain* 2012;135(Pt 9): 2676–2683.

Naismith RT, Tutlam NT, Xu J, Klawiter EC, Shepherd J, Trinkaus K, Song SK, Cross AH. Optical coherence tomography differs in neuromyelitis optica compared with multiple sclerosis. *Neurology* 2009;72:1077–1082.

Neema M, Ceccarelli A, Jackson J, Bakshi R. Magnetic resonance imaging in multiple sclerosis. In: H. Weiner, J. Stankiewicz (Eds.), *Multiple Sclerosis: Diagnosis and Therapy*, Wiley-Blackwell, West Sussex, England (2012), pp. 136–162.

Nelson A, Kelley RE, Nguyen J, Palacios E, Neitzschman HR. MRS findings in a patient with juvenile-onset Alexander's leukodystrophy. *J La State Med Soc* 2013;165:14–17.

Oguz KK, Kurne A, Aksu AO, Karabulut E, Serdaroglu A, Teber S, Haspolat S, Senbil N, Kurul S, Anlar B.

Assessment of citrullinated myelin by 1H-MR spectroscopy in early-onset multiple sclerosis. *AJNR Am J Neuroradiol* 2009;30:716–721.

Okuda DT, Mowry EM, Beheshtian A, Waubant E, Baranzini SE, Goodin DS, Hauser SL, Pelletier D. Incidental MRI anomalies suggestive of multiple sclerosis: The radiologically isolated syndrome. *Neurology* 2009;72:800–805. Erratum in: *Neurology* 2009;72:1284.

Okuda DT, Mowry EM, Cree BA, Crabtree EC, Goodin DS, Waubant E, Pelletier D. Asymptomatic spinal cord lesions predict disease progression in radiologically isolated syndrome. *Neurology* 2011;76:686–692.

Okuda DT, Siva A, Kantarci O et al.; Radiologically Isolated Syndrome Consortium (RISC); Club Francophone de la Sclérose en Plaques (CFSEP). Radiologically isolated syndrome: 5-year risk for an initial clinical event. *PLoS One* 2014;9:e90509.

Osterman B, La Piana R, Bernard G. Advances in the diagnosis of leukodystrophies: Neuroradiology: The importance of MRI pattern recognition. *Future Neurol* 2012;7:595–612.

Papadopoulos MC, Verkman AS. Aquaporin 4 and neuromyelitis optica. *Lancet Neurol* 2012;11:535–544.

Patay Z. Diffusion-weighted MR imaging in leukodystrophies. *Eur Radiol* 2005;15:2284–2303.

Patel PJ, Kolawole TM, Malabarey TM, al-Herbish AS, al-Jurrayan NA, Saleh M. Adrenoleukodystrophy: CT and MRI findings. *Pediatr Radiol* 1995;25:256–258.

Patil SA, Maegawa GH. Developing therapeutic approaches for metachromatic leukodystrophy. *Drug Des Devel Ther* 2013;7:729–745.

Paty DW, Oger JJ, Kastrukoff LF et al. MRI in the diagnosis of MS: A prospective study with comparison of clinical evaluation, evoked potentials, oligoclonal banding and CT. *Neurology* 1998;38:180–185.

Perlman SJ, Mar S. Leukodystrophies. *Adv Exp Med Biol* 2012;724:154–171.

Phelan JA, Lowe LH, Glasier CM. Pediatric neurodegenerative white matter processes: Leukodystrophies and beyond. *Pediatr Radiol* 2008;38:729–749.

Pichiecchio A, Tavazzi E, Poloni G et al. Advanced magnetic resonance imaging of neuromyelitis optica: A multiparametric approach. *Mult Scler* 2012;18:817–824.

Poll-The BT, Gärtner J. Clinical diagnosis, biochemical findings and MRI spectrum of peroxisomal disorders. *Biochim Biophys Acta* 2012;1822:1421–1429.

Polman CH, Reingold SC, Banwell B et al. Diagnostic criteria for multiple sclerosis: 2010 revisions to the McDonald criteria. *Ann Neurol* 2011;69:292–302.

Popescu BF, Parisi JE, Cabrera-Gómez JA, Newell K, Mandler RN, Pittock SJ, Lennon VA, Weinshenker BG, Lucchinetti CF. Absence of cortical demyelination in neuromyelitis optica. *Neurology* 2010;75:2103–2109.

Popescu BF, Pirko I, Lucchinetti CF. Pathology of multiple sclerosis: Where do we stand? *Continuum (Minneap Minn)* 2013;19:901–921.

Poretti A, Meoded A, Bunge M, Fatemi A, Barrette P, Huisman TA, Salman MS. Novel diffusion tensor imaging findings in Krabbe disease. *Eur J Paediatr Neurol* 2014;18:150–156.

Poser CM, Brinar VV. Diagnostic criteria for multiple sclerosis. *Clin Neurol Neurosurg* 2001;103:1–11.

Poser S, Luer W, Bruhn H, Frahm J, Bruck Y, Felgenhauer K. Acute demyelinating disease. Classification and noninvasive diagnosis. *Acta Neurol Scand* 1992;86:579–585.

Pouwels PJ, Vanderver A, Bernard G et al. Hypomyelinating leukodystrophies: Translational research progress and prospects. *Ann Neurol* 2014;76:5–19.

Prust M, Wang J, Morizono H et al. GFAP mutations, age at onset, and clinical subtypes in Alexander disease. *Neurology* 2011;77:1287–1294.

Qian W, Chan Q, Mak H, Zhang Z, Anthony MP, Yau KK, Khong PL, Chan KH, Kim M. Quantitative assessment of the cervical spinal cord damage in neuromyelitis optica using diffusion tensor imaging at 3 Tesla. *J Magn Reson Imaging* 2011;33:1312–1320. Erratum in: *J Magn Reson Imaging* 2011;34:727.

Quek AM, McKeon A, Lennon VA et al. Effects of age and sex on aquaporin-4 autoimmunity. *Arch Neurol* 2012;69:1039–1043.

Quinlan RA, Brenner M, Goldman JE, Messing A. GFAP and its role in Alexander disease. *Exp Cell Res* 2007;313:2077–2087.

Rahmlow MR, Kantarci O. Fulminant demyelinating diseases. *Neurohospitalist* 2013;3:81–91.

Ratai E, Kok T, Wiggins C, Wiggins G, Grant E, Gagoski B, O'Neill G, Adalsteinsson E, Eichler F. Seven-Tesla proton magnetic resonance spectroscopic imaging in adult X-linked adrenoleukodystrophy. *Arch Neurol* 2008;65: 1488–1494.

Ratai EM, Caruso P, Eichler F. Advances in MR imaging of leukodystrophies. In: Prof. P. Bright (Ed.), *Neuroimaging—Clinical Applications*, ISBN: 978-953-51-0200-7, InTech (2012), pp. 559–576. Available from: http://www.intechopen.com/books/neuroimaging-clinical-applications/advances-in-mr-imaging-of-leukodystrophies and from http://cdn.intechopen.com/pdfs-wm/31427.pdf.

Raz E, Cercignani M, Sbardella E, Totaro P, Pozzilli C, Bozzali M, Pantano P. Clinically isolated syndrome suggestive of multiple sclerosis: Voxelwise regional investigation of white and gray matter. *Radiology* Jan 2010;254(1):227–234.

Raz E, Loh JP, Saba L, Omari M, Herbert J, Lui Y, Kister I. Periventricular lesions help differentiate neuromyelitis optica spectrum disorders from multiple sclerosis. *Mult Scler Int* 2014;2014:986923.

Rennebohm R, Susac JO, Egan RA, Daroff RB. Susac's syndrome—Update. *J Neurol Sci* 2010;299:86–91.

Rice CM, Cottrell D, Wilkins A, Scolding NJ. Primary progressive multiple sclerosis: Progress and challenges. *J Neurol Neurosurg Psychiatry* 2013;84:1100–1106.

Rivero RL, Oliveira EM, Bichuetti DB, Gabbai AA, Nogueira RG, Abdala N. Diffusion tensor imaging of the cervical spinal cord of patients with Neuromyelitis Optica. *Magn Reson Imaging* 2014;32:457–463.

Rocca MA, Absinta M, Amato MP et al. Posterior brain damage and cognitive impairment in pediatric multiple sclerosis. *Neurology* 2014a;82:1314–1321.

Rocca MA, Absinta M, Ghezzi A, Moiola L, Comi G, Filippi M. Is a preserved functional reserve a mechanism limiting clinical impairment in pediatric MS patients? *Hum Brain Mapp* 2009;30:2844–2851.

Rocca MA, Absinta M, Moiola L, Ghezzi A, Colombo B, Martinelli V, Comi G, Filippi M. Functional and structural connectivity of the motor network in pediatric and adult-onset relapsing-remitting multiple sclerosis. *Radiology* 2010;254:541–550.

Rocca MA, Agosta F, Mezzapesa DM, Martinelli V, Salvi F, Ghezzi A, Bergamaschi R, Comi G, Filippi M. Magnetization transfer and diffusion tensor MRI show gray matter damage in neuromyelitis optica. *Neurology* 2004;62:476–478.

Rocca MA, Filippi M, Herzog J, Sormani MP, Dichgans M, Yousry TA. A magnetic resonance imaging study of the cervical cord of patients with CADASIL. *Neurology* 2001;56:1392–1394.

Rocca MA, Valsasina P, Absinta M et al. Intranetwork and internetwork functional connectivity abnormalities in pediatric multiple sclerosis. *Hum Brain Mapp* 2014b;35:4180–4192.

Rodriguez D. Leukodystrophies with astrocytic dysfunction. *Handb Clin Neurol* 2013;113:1619–1628.

Rosati G. The prevalence of multiple sclerosis in the world: An update. *Neurol Sci* 2001;22:117–139. Review.

Rossi A. Imaging of acute disseminated encephalomyelitis. *Neuroimaging Clin N Am* 2008;18:149–161.

Rossi A, Biancheri R. Magnetic resonance spectroscopy in metabolic disorders. *Neuroimaging Clin N Am* 2013;23: 425–448.

Rovaris M, Barkhof F, Calabrese M et al. MRI features of benign multiple sclerosis: Toward a new definition of this disease phenotype. *Neurology* 2009;72:1693–1701.

Rovira A, Alonso J, Cucurella G et al. Evolution of multiple sclerosis lesions on serial contrast-enhanced T1-weighted and magnetization-transfer MR images. *AJNR Am J Neuroradiol* 1999;20:1939–1945.

Rovira Cañellas A, Rovira Gols A, Río Izquierdo J, Tintoré Subirana M, Montalban Gairin X. Idiopathic inflammatory-demyelinating diseases of the central nervous system. *Neuroradiology* 2007;49:393–409.

Ryan M, Ibrahim M, Parmar HA. Secondary demyelination disorders and destruction of white matter. *Radiol Clin North Am* 2014;52:337–354.

Sahraian MA, Radue EW, Haller S, Kappos L. Black holes in multiple sclerosis: Definition, evolution, and clinical correlations. *Acta Neurol Scand* 2010;122:1–8.

Saji E, Arakawa M, Yanagawa K et al. Cognitive impairment and cortical degeneration in neuromyelitis optica. *Ann Neurol* 2013;73:65–76.

Salsano E, Marotta G, Manfredi V, Giovagnoli AR, Farina L, Savoiardo M, Pareyson D, Benti R, Uziel G. Brain fluorodeoxyglucose PET in adrenoleukodystrophy. *Neurology* 2014;83:981–989.

Sato DK, Callegaro D, Lana-Peixoto MA et al. Distinction between MOG antibody-positive and AQP4 antibody-positive NMO spectrum disorders. *Neurology* 2014;82: 474–481.

Sato DK, Lana-Peixoto MA, Fujihara K, de Seze J. Clinical spectrum and treatment of neuromyelitis optica spectrum disorders: Evolution and current status. *Brain Pathol* 2013;23:647–660.

Sawcer S, Franklin RJ, Ban M. Multiple sclerosis genetics. *Lancet Neurol* 2014;13:700–709.

Schiffmann R, van der Knaap MS. Invited article: An MRI-based approach to the diagnosis of white matter disorders. *Neurology* 2009;72(8):750–759.

Schiffmann R, Moller JR, Trapp BD, Shih HH, Farrer RG, Katz DA, Alger JR, Parker CC, Hauer PE, Kaneski CR. Childhood ataxia with diffuse central nervous system hypomyelination. *Ann Neurol* 1994;35:331–340.

Schlaeger R, Papinutto N, Panara V et al. Spinal cord gray matter atrophy correlates with multiple sclerosis disability. *Ann Neurol* 2014;76:568–580.

Sechi G, Ceccherini I, Bachetti T, Deiana GA, Sechi E, Balbi P. Ceftriaxone for Alexander's disease: A four-year follow-up. *IMD Rep.* 2013;9:67–71.

Sehitoğlu E, Cavuş F, Ulusoy C, Küçükerden M, Orçen A, Akbaş-Demir D, Coban A, Vural B, Tüzün E, Türkoğlu R. Sorcin antibody as a possible predictive factor in conversion from radiologically isolated syndrome to multiple sclerosis: A preliminary study. *Inflamm Res* 2014;63:799–801.

Sellner J, Schirmer L, Hemmer B, Mühlau M. The radiologically isolated syndrome: Take action when the unexpected is uncovered? *J Neurol* 2010;257:1602–1611.

Simon JH, Kleinschmidt-DeMasters BK. Variants of multiple sclerosis. *Neuroimaging Clin N Am* 2008;18:703–716.

Simon JH, Li D, Traboulsee A et al. Standardized MR imaging protocol for multiple sclerosis: Consortium of MS Centers consensus guidelines. *AJNR Am J Neuroradiol* 2006;27: 455–461.

Simone IL, Carrara D, Tortorella C, Liguori M, Lepore V, Pellegrini F, Bellacosa A, Ceccarelli A, Pavone I, Livrea P. Course and prognosis in early-onset MS: Comparison with adult-onset forms. *Neurology* 2002;59:1922–1928.

Simpson S Jr, Blizzard L, Otahal P, Van der Mei I, Taylor B. Latitude is significantly associated with the prevalence of multiple sclerosis: A meta-analysis. *J Neurol Neurosurg Psychiatry* 2011;82:1132–1141.

Sinnecker T, Dörr J, Pfueller CF, Harms L, Ruprecht K, Jarius S, Brück W, Niendorf T, Wuerfel J, Paul F. Distinct lesion morphology at 7-T MRI differentiates neuromyelitis optica from multiple sclerosis. *Neurology* Aug 14, 2012;79:708–714.

Siva A, Saip S, Altintas A, Jacob A, Keegan BM, Kantarci OH. Multiple sclerosis risk in radiologically uncovered asymptomatic possible inflammatory-demyelinating disease. *Mult Scler* 2009;15:918–927.

Skehan SJ, Hutchinson M, MacErlaine DP. Cerebral autosomal dominant arteriopathy with subcortical infarcts and leukoencephalopathy: MR findings. *AJNR Am J Neuroradiol* 1995;16:2115–2119.

Smith SA, Golay X, Fatemi A, Mahmood A, Raymond GV, Moser HW, van Zijl PC, Stanisz GJ. Quantitative magnetization transfer characteristics of the human cervical spinal cord in vivo: Application to adrenomyeloneuropathy. *Magn Reson Med* 2009;61:22–27.

Spain R, Bourdette D. The radiologically isolated syndrome: Look (again) before you treat. *Curr Neurol Neurosci Rep* 2011;11:498–506.

Sreenivasan P, Purushothaman KK. Radiological clue to diagnosis of Canavan disease. *Indian J Pediatr* 2013;80:75–77.

Stankiewicz JM, Neema M, Ceccarelli A. Iron and multiple sclerosis. *Neurobiol Aging* 2014;35(Suppl 2):S51–S58.

Stromillo ML, Giorgio A, Rossi F et al. Brain metabolic changes suggestive of axonal damage in radiologically isolated syndrome. *Neurology* 2013;80:2090–2094.

Suppiej A, Cainelli E. Cognitive dysfunction in pediatric multiple sclerosis. *Neuropsychiatr Dis Treat* 2014;10:1385–1392.

Susac JO. Susac's syndrome: The triad of microangiopathy of the brain and retina with hearing loss in young women. *Neurology* 1994;44:591–593.

Susac JO, Hardman JM, Selhorst JB. Microangiopathy of the brain and retina. *Neurology* 1979;29:313–316.

Susac JO, Murtagh FR, Egan RA et al. MRI findings in Susac's syndrome. *Neurology* 2003;61:1783–1787.

Tackley G, Kuker W, Palace J. Magnetic resonance imaging in neuromyelitis optica. *Mult Scler* 2014.

Tallantyre EC, Dixon JE, Donaldson I, Owens T, Morgan PS, Morris PG, Evangelou N. Ultra-high-field imaging distinguishes MS lesions from asymptomatic white matter lesions. *Neurology* 2011;76:534–539.

Tan IL, McArthur JC, Clifford DB, Major EO, Nath A. Immune reconstitution inflammatory syndrome in natalizumab-associated PML. *Neurology* 2011;77:1061–1067.

Tenembaum SN. Acute disseminated encephalomyelitis. *Handb Clin Neurol* 2013;112:1253–1262.

Till C, Ghassemi R, Aubert-Broche B, Kerbrat A, Collins DL, Narayanan S, Arnold DL, Desrocher M, Sled JG, Banwell BL. MRI correlates of cognitive impairment in childhood-onset multiple sclerosis. *Neuropsychology* 2011;25:319–332.

Toosy AT, Kou N, Altmann D, Wheeler-Kingshott CA, Thompson AJ, Ciccarelli O. Voxel-based cervical spinal cord mapping of diffusion abnormalities in MS-related myelitis. *Neurology* 2014;83:1321–1325.

Tortorella C, Direnzo V, D'Onghia M, Trojano M. Brainstem PML lesion mimicking MS plaque in a natalizumab-treated MS patient. *Neurology* 2013;81:1470–1471.

Tortorella P, Rocca MA, Mezzapesa DM et al. MRI quantification of grey and white matter damage in patients with early-onset multiple sclerosis. *J Neurol* 2006;253:903–907.

Tur C, Montalban X. Natalizumab: Risk stratification of individual patients with multiple sclerosis. *CNS Drugs* 2014;28:641–648.

van Den Boom R, Lesnik Oberstein SA, van Duinen SG, Bornebroek M, Ferrari MD, Haan J, van Buchem MA. Subcortical lacunar lesions: An MR imaging finding in patients with cerebral autosomal dominant arteriopathy with subcortical infarcts and leukoencephalopathy. *Radiology* 2002;224:791–796.

van der Knaap MS, Barth PG, Gabreels FJ, Franzoni E, Begeer JH, Stroink H, Rotteveel JJ, Valk J. A new leukoencephalopathy with vanishing white matter. *Neurology* 1997;48:845–855.

van der Knaap MS, Barth PG, Vrensen GF et al. Histopathology of an infantile-onset spongiform leukoencephalopathy with a discrepantly mild clinical course. *Acta Neuropathol* 1996;96:206–212.

van der Knaap MS, Boor I, Estévez R. Megalencephalic leukoencephalopathy with subcortical cysts: Chronic white matter oedema due to a defect in brain ion and water homoeostasis. *Lancet Neurol* 2012;11:973–985.

van der Knaap MS, Lai V, Köhler W et al. Megalencephalic leukoencephalopathy with cysts without MLC1 defect. *Ann Neurol* 2010;67:834–837.

van der Knaap MS, Leegwater PA, Könst AA, Visser A, Naidu S, Oudejans CB, Schutgens RB, Pronk JC. Mutations in each of the five subunits of translation initiation factor eIF2B can cause leukoencephalopathy with vanishing white matter. *Ann Neurol* 2002;51:264–270.

van der Knaap MS, Naidu S, Breiter SN et al. Alexander disease: Diagnosis with MR imaging. *AJNR Am J Neuroradiol* 2001;22:541–552.

van der Knaap MS, Pronk JC, Scheper GC. Vanishing white matter disease. *Lancet Neurol* 2006;5:413–423.

van der Lei HD, Steenweg ME, Barkhof F, de Grauw T, d'Hooghe M, Morton R, Shah S, Wolf N, van der Knaap MS. Characteristics of early MRI in children and adolescents with vanishing white matter. *Neuropediatrics* 2012a;43:22–26.

van der Lei HD, Steenweg ME, Bugiani M, Pouwels PJ, Vent IM, Barkhof F, van Wieringen WN, van der Knaap MS. Restricted diffusion in vanishing white matter. *Arch Neurol* 2012b;69:723–727.

van der Voorn JP, Pouwels PJ, Powers JM, Kamphorst W, Martin JJ, Troost D, Spreeuwenberg MD, Barkhof F, van der Knaap MS. Correlating quantitative MR imaging with histopathology in X-linked adrenoleukodystrophy. *AJNR Am J Neuroradiol* 2011;32:481–489.

Van der Voorn JP, Pouwels PJW, Hart AA et al. Childhood white matter disorders: Quantitative MR imaging and spectroscopy. *Radiology* 2006;241:510–517.

Van Haren K, Waubant E. Therapeutic advances in pediatric multiple sclerosis. *J Pediatr* 2013;163:631–637. Review.

van Pelt ED, Mescheriakova JY, Makhani N et al. Risk genes associated with pediatric-onset MS but not with monophasic acquired CNS demyelination. *Neurology* 2013;81:1996–2001.

Vanderver A, Tonduti D, Schiffmann R, Schmidt J, Van der Knaap MS. *Leukodystrophy Overview 2014*. In: R.A. Pagon, M.P. Adam, H.H. Ardinger, T.D. Bird, C.R. Dolan, C.T. Fong, R.J.H. Smith, K. Stephens (Eds.), *GeneReviews®* [Internet], University of Washington, Seattle, WA (1993–2014).

Vargas-Lowy D, Chitnis T. Pathogenesis of pediatric multiple sclerosis. *J Child Neurol* 2012;27:1394–1407. Review.

Verhey LH, Branson HM, Makhija M, Shroff M, Banwell B. Magnetic resonance imaging features of the spinal cord in pediatric multiple sclerosis: A preliminary study. *Neuroradiology* 2010;52:1153–1162.

Verhey LH, Narayanan S, Banwell B. Standardized magnetic resonance imaging acquisition and reporting in pediatric multiple sclerosis. *Neuroimaging Clin N Am* 2013a;23:217–226.e1–e7. Review.

Verhey LH, Shroff M, Banwell B. Pediatric multiple sclerosis: Pathobiological, clinical, and magnetic resonance imaging features. *Neuroimaging Clin N Am* 2013b;23:227–243. Review.

Verhey LH, Sled JG. Advanced magnetic resonance imaging in pediatric multiple sclerosis. *Neuroimaging Clin N Am* 2013c;23:337–354. Review.

Vishwas MS, Chitnis T, Pienaar R, Healy BC, Grant PE. Tract-based analysis of callosal, projection, and association pathways in pediatric patients with multiple sclerosis: A preliminary study. *AJNR Am J Neuroradiol* 2010;31:121–128.

Viswanathan A, Gray F, Bousser MG, Baudrimont M, Chabriat H. Cortical neuronal apoptosis in CADASIL. *Stroke* 2006;37:2690–2695.

von Glehn F, Jarius S, Cavalcanti Lira RP et al. Structural brain abnormalities are related to retinal nerve fiber layer thinning and disease duration in neuromyelitis optica spectrum disorders. *Mult Scler* 2014. [Epub ahead of print].

Vukusic S, Confavreux C. Prognostic factors for progression of disability in the secondary progressive phase of multiple sclerosis. *J Neurol Sci* 2003;206:135–137.

Wattjes MP, Vennegoor A, Mostert J, van Oosten BW, Barkhof F, Killestein J. Diagnosis of asymptomatic natalizumab-associated PML: Are we between a rock and a hard place? *J Neurol* 2014;261:1139–1143.

Waubant E, Chabas D, Okuda DT, Glenn O, Mowry E, Henry RG, Strober JB, Soares B, Wintermark M, Pelletier D. Difference in disease burden and activity in pediatric patients on brain magnetic resonance imaging at time of multiple sclerosis onset vs adults. *Arch Neurol* 2009;66:967–971.

Wingerchuk DM, Lennon VA, Pittock SJ, Lucchinetti CF, Weinshenker BG. Revised diagnostic criteria for neuromyelitis optica. *Neurology* 2006;66:1485–1489.

Wingerchuk DM, Weinshenker BG. Neuromyelitis optica (Devic's syndrome). *Handb Clin Neurol* 2014;122:581–599.

Wu GF, Alvarez E. The immunopathophysiology of multiple sclerosis. *Neurol Clin* 2011;29:257–278.

Wuerfel J, Sinnecker T, Ringelstein EB, Jarius S, Schwindt W, Niendorf T, Paul F, Kleffner I, Dörr J. Lesion morphology at 7 Tesla MRI differentiates Susac syndrome from multiple sclerosis. *Mult Scler* 2012;18:1592–1599.

Yamamoto Y, Ihara M, Tham C, Low RW, Slade JY, Moss T, Oakley AE, Polvikoski T, Kalaria RN. Neuropathological correlates of temporal pole white matter hyperintensities in CADASIL. *Stroke* 2009;40:2004–2011.

Yang E, Prabhu SP. Imaging manifestations of the leukodystrophies, inherited disorders of white matter. *Radiol Clin North Am* 2014;52:279–319.

Yeh EA, Weinstock-Guttman B, Ramanathan M, Ramasamy DP, Willis L, Cox JL, Zivadinov R. Magnetic resonance imaging characteristics of children and adults with paediatric-onset multiple sclerosis. *Brain* 2009;132(Pt 12):3392–3400.

Yonezu T, Ito S, Mori M, Ogawa Y, Makino T, Uzawa A, Kuwabara S. "Bright spotty lesions" on spinal magnetic resonance imaging differentiate neuromyelitis optica from multiple sclerosis. *Mult Scler* 2014;20:331–337.

Yousry TA, Pelletier D, Cadavid D, Gass A, Richert ND, Radue EW, Filippi M. Magnetic resonance imaging pattern in natalizumab-associated progressive multifocal leukoencephalopathy. *Ann Neurol* 2012;72:779–787.

Yousry TA, Seelos K, Mayer M, Brüning R, Uttner I, Dichgans M, Mammi S, Straube A, Mai N, Filippi M. Characteristic MR lesion pattern and correlation of T1 and T2 lesion volume with neurologic and neuropsychological findings in cerebral autosomal dominant arteriopathy with subcortical infarcts and leukoencephalopathy (CADASIL). *AJNR Am J Neuroradiol* 1999;20:91–100.

Zhang B, Zhong Y, Wang Y, Dai Y, Qiu W, Zhang L, Li H, Lu Z. Neuromyelitis optica spectrum disorders without and with autoimmune diseases. *BMC Neurol* 2014;14:162.

Zhao DD, Zhou HY, Wu QZ, Liu J, Chen XY, He D, He XF, Han WJ, Gong QY. Diffusion tensor imaging characterization of occult brain damage in relapsing neuromyelitis optica using 3.0T magnetic resonance imaging techniques. *Neuroimage* 2012;59:3173–3177.

27

Traumatic Disease of the Brain and Skull

Eytan Raz

CONTENTS

This chapter reviews the findings of intracranial and skull base trauma. Even though computed tomography (CT) is the most efficient method of triage for acute head trauma because it is very fast and very accurate for the diagnosis of life-threatening conditions, magnetic resonance imaging (MRI) has a few indications both hyperacutely and in the subacute and chronic evaluation. After a few epidemiological and clinical notes, we will review the different kinds of intracranial injury.

27.1 Introduction

Traumatic brain injury (TBI) is one of the most important causes of morbidity and mortality in the Western world. Each year in the United States alone, more than two million people sustain head trauma, and 10% of these injuries are fatal. Ten percent of survivors experience neurological deficits of varying degrees (Gentry, 1994). It is estimated that as many as 5.3 million people are living in the United States with disability related to TBI, approximately 2% of the population. The leading cause of TBI is injury related to falls, followed by motor

vehicle or traffic collisions, and external cause of being *struck by or against* (Brown, Elovic, Kothari, Flanagan, & Kwasnica, 2008).

The classification of the clinical severity of TBI is based on the Glasgow Coma Scale (GCS) (Table 27.1) (Teasdale & Jennett, 1974). The GCS is a neurological scale that allows the recording of the level of consciousness through the assessment of eye, motor, and verbal responses. The severity distribution is approximately 80% mild (GCS score of 13–15), 10% moderate (GCS score of 12–9), and 10% severe (GCS scores of 8 or less). It is known that patients with the same GCS score can have extremely different outcomes. Duration of posttraumatic amnesia and loss of consciousness are other important clinical factors to establish the severity of TBI (Table 27.2).

The role of neuroimaging in head trauma is well established. According to the American College of Radiology (ACR) appropriateness criteria, there is a general consensus that patients with moderate or high risk for intracranial injury should have a head CT done early (Davis et al., 2000). CT is the best modality because it is widely available, and it is fast and highly accurate in the detection of skull fractures and intracranial hemorrhage; moreover, life support and monitoring equipment can

TABLE 27.1

Glasgow Coma Scale

Eye Opening	Verbal Response	Motor Response
Opens spontaneously—4	Normal conversation—5	Normal—6
Opens to voice—3	Disoriented conversation—4	Localizes pain—5
Opens to pain—2	Words, incoherent—3	Withdraws from pain—4
None—1	Incomprehensible sounds—2	Decorticate posturing—3
	None—1	Decerebrate posturing—2
		None—1

TABLE 27.2

Severity of Traumatic Brain Injury

	Mild	Moderate	Severe
GCS	13–15	<1 day	0–30 min
PTA	9–12	>1 to <7 days	>30 min to <24 h
LOC	3–8	>7 days	>24 h

be fitted in the CT room without restrictions and generally, there are no contraindications to emergency patient scanning. There has been more discordance among patients with GCS > 13, and head CT has been proposed to this population as a screening tool, due to the low cost and the wide availability of CT.

27.2 Role of MRI

In the imaging of head trauma, MRI is hampered by its limited availability in the acute trauma setting, by the long imaging times, by the sensitivity to patient motion, incompatibility with multiple medical devices; moreover, MRI is less accurate than CT for the diagnosis of subarachnoid hemorrhage (SAH). MRI is more sensitive than CT to identify diffuse axonal injury (DAI) and to evaluate the presence of ischemia, which can be a secondary injury related to trauma.

Among conventional MR sequences, gradient-recalled-echo (GRE) T_2 weighted is particularly sensitive to the presence of breakdown products of blood, because hemosiderin and ferritin alter the local magnetic susceptibility of tissue, resulting in areas of signal loss on GRE-T_2 weighted images (Gomori, Grossman, Goldberg, Zimmerman, & Bilaniuk, 1985). Gradient-echo sequence has some problems in the evaluation of the basal frontal and temporal lobes because of the susceptibility artifact

given by the adjacent air in the frontal sinus and in the temporal bone (Gentry, 1994; Gomori et al., 1985).

In this view, the development of susceptibility-weighted imaging (SWI) technique has improved the ability to detect hemorrhage, as compared to conventional GRE-T_2, given its major sensitivity and specificity in identifying presence of blood breakdown products (Tong et al., 2003).

Despite the improvement determined by MRI, there are some issues in which doubts persist: (1) conventional MR sequences underestimate the extent of injury; (2) the correlation between MR and functional deficits is often poor; and (3) MR does not provide quantitative pathophysiological hints to determine prognosis and to monitor the therapy.

To determine which injury is treatable, to prevent secondary damage, and to provide useful prognostic information, other neuroimaging functional tools have been used, such as diffusion-weighted imaging (DWI)/diffusion tensor imaging (DTI), Magnetic Resonance Spectroscopy (MRS), and, secondarily, BOLD imaging.

While severe trauma is usually evident on conventional imaging as visible hemorrhagic lesions or contusions, mild and moderate trauma is often associated with undetectable damage. These subgroups are hence those that can mostly benefit from unconventional imaging tools.

27.3 Classification of Injury

TBI may be divided into two types: primary injuries, the direct result of trauma to the head, and secondary injuries, which arise as complications of primary lesions. Secondary injuries are potentially preventable, whereas primary injuries, by definition, have already occurred by the time the patient first presents (Table 27.3). TBI can be further divided according to location (intra-axial or extra-axial) and mechanism (blunt/closed and penetrating/open).

27.3.1 Skull Fractures

For the evaluation of skull fractures, a CT is necessary. The role of MRI is mainly for the evaluation of secondary consequences to the brain parenchyma that may arise from a skull base or a calvarial fracture. For example, depressed fractures arising from blunt force trauma are typically comminuted with inwardly displaced broken bone and commonly associated with underlying brain injury. Associate infection occurs in up to 10% of cases, with high morbidity (high frequency of epilepsy)

TABLE 27.3

Imaging Classification of Traumatic Brain Injury

Primary Injury	Secondary Injury
Extra-axial injury	*Acute*
Skull fracture	Diffuse cerebral swelling/ dysautoregulation
Epidural hematoma	Brain herniation
Subdural hematoma	Infarction
Subarachnoid hemorrhage	Infection
Intraventricular hemorrhage	*Chronic*
Intra-axial injury	Hydrocephalus
Diffuse axonal injury	Encephalomalacia
Cerebral contusion	Cerebrospinal fluid leak
Intraparenchymal hematoma	Leptomeningeal cyst
Vascular injury	
Dissection	
Carotid-cavernous fistula	
Arteriovenous dural fistula	
Pseudoaneurysm	

and high mortality. Depressed skull fracture may require neurosurgical evaluation to elevate the fracture and remove bone fragments if the depressed fracture is greater than the thickness of the cranium or if there is evidence of dural penetration, associated intracranial hematoma, pneumocephalus, or gross cosmetic deformity (Bullock et al., 2006).

Basilar fractures occur through the skull base and are very rare and sometimes difficult to detect. When a basilar fracture is seen, a vascular study such as an MR angiography (MRA) or a CT angiography (CTA) is recommended to evaluate internal carotid arteries.

Growing skull fracture is a kind of fracture seen most commonly in children and occurs when there is dural tear with development of a cerebrospinal fluid (CSF)-filled collection in the fracture and progressive enlargement of the fracture margins caused by CSF pulsations (Ciurea, Gorgan, Tascu, Sandu, & Rizea, 2011). This is also termed *leptomeningeal cyst* and typically occurs in the base of the skull.

27.3.2 Epidural Hematoma

The epidural space is the potential space located between the inner table of the skull and the dura, which represents the functional periosteum of the inner table of the skull. The epidural hemorrhage often occurs in the temporo-parietal region in the so-called *Marchant zone* (90% of the cases), typically occurring when a fracture crosses the vascular territory of the middle meningeal artery or vein. When there is tearing of these vessels, the blood moves in the epidural space resulting in an epidural hematoma. The dura is tightly adherent to the inner table of the skull at cranial sutures, and this is the reason

why the extension of the epidural hematoma tend to be limited across the sutures, although exceptions are common (Kubal, 2012). Clinically, after the injury, there may be a lucid interval, defined as a temporary improvement in the patient's condition after the injury, followed by deterioration; 50% of patients with epidural hematoma experience the lucid interval. Epidural hematomas are usually biconvex collections: the shape is related to the firm attachment of the dura to the skull. An epidural hematoma can sometimes contain different components, showing the *swirl sign*, which is a bad prognostic sign, representing an active component of bleeding within a chronic hematoma (Al-Nakshabandi, 2001). The active component is isodense to the brain and represents actively extravasating unclotted blood, while the chronic component is the hyperdense epidural hematoma (Kubal, 2012; Provenzale, 2007). Chronic epidural hematomas are low density with peripheral enhancement and may often lose the biconvex shape. They have to be differentiated from other epidural lesions such as tumors. The venous epidural hematoma occurs most frequently in the middle cranial fossa in the pediatric population and carries a lower morbidity than arterial epidural hematomas (Gean et al., 2010). MRI can help to differentiate some mimickers of an epidural hematoma, or to evaluate the coexistence with other lesions such as in the case shown in Figure 27.1 of a hemorrhagic meningioma.

27.3.3 Subdural Hematoma

Between the dura and the underlying arachnoid, there is a potential space named *subdural space*, which usually contains a minimal amount of fluid similar to CSF and is traversed by bridging cortical veins; hemorrhage can insinuate in this subdural space, constituting the subdural hematoma (SDH). SDHs are caused by rupture of bridging veins, which, from the cerebral convexities, traverse the subarachnoid space to reach the dural sinuses. The entry point of these veins into the dural sinus are fixed: with trauma, the rotational motion of the brain causes the shearing of the bridging veins, which eventually tear in the subdural portion (which is the weakest part) due to the lack of arachnoid trabeculae sheating (Provenzale, 2007).

In elderly patients, cerebral atrophy allows an increased movement between the brain parenchyma and the calvarium, resulting in an increased incidence of SDHs. An acute SDH has a high mortality rate up to 40%. Also with SDH, like with the epidural hematoma, a lucid clinical interval may be seen, which can last up to few days.

The acute SDH appears as a homogeneous, crescent-shaped extra-axial collection. The hematoma can be variably thick. In patients with anemia, the hematoma can have a different appearance (Provenzale, 2007).

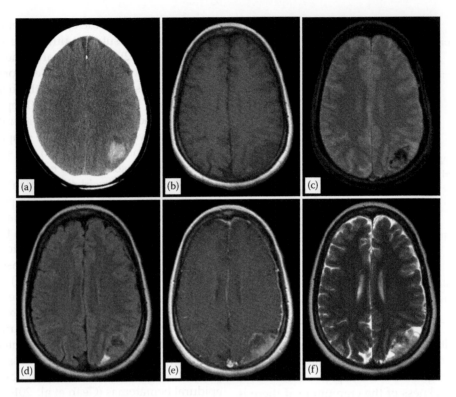

FIGURE 27.1
Hemorrhagic meningioma. A 31-year-old female undergoes CT for trauma (a), which demonstrates an intracranial parietal extra-axial hemorrhage with atypical features, but possibly related to an epidural hematoma. The atypical features and the lack of an overlying calvarial fractures prompted the further evaluation with an MRI: unenhanced T_1 (b), gradient-echo (c), FLAIR (d), and postcontrast T_1 (e) demonstrate an enhancing hemorrhagic extra-axial mass, which was removed and was an hemorrhagic meningioma. In panel (f) a T_2W image is given.

Most SDHs are supratentorial and are located along the convexity. SDHs are frequently seen along the whole hemispheric convexity from the anterior falx to the posterior falx cerebri. They are also frequently seen along the falx and tentorium, contrary to the epidural hematoma, which, by definition, cannot be located along these meningeal structures. Subacute and chronic SDHs may be concave, simulating epidural hematomas and are associated with peripheral enhancement caused by the vascularization of the subdural membrane. This vascularization is formed by vessels without tight junctions, which can easily leak resulting in repeated bleeding within the hematoma. SDHs can hence be very heterogeneous because of these repeated episodic bleeding that result in compartments separated by septations (Hellwig, Kuhn, Bauer, & List-Hellwig, 1996). Chronic SDH can also be seen without history of head trauma, in elderly patient with coagulopathy. SDHs have to be differentiated from other lesions such as empyema or tumors. In addition, in children, the presence of different aged SDHs should raise suspicion for child abuse (see Section 27.3.8).

Bilateral SDHs can sometimes be difficult to call because they can be very thin and symmetric. In order to not to be misleaded, the sulcal size has to be carefully evaluated and the gray–white matter interface should be scrutinized in order to determine if it is buckled inward (Penchet, Loiseau, & Castel, 1998).

Rarely, SDH can be the presentation of an aneurysm rupture, even without an associated SAH (Gilad, Fatterpekar, Johnson, & Patel, 2007).

27.3.4 Subarachnoid and Intraventricular Hemorrhages

Often after trauma, a small amount of subarachnoid and intraventricular hemorrhages is present and can be very subtle. SAH in trauma results from an injury to small cortical veins traversing the subarachnoid space or the subdural or intraparenchymal hematomas, which extend through the subarachnoid space. In neither case, the volume of hemorrhage is comparable to that faced in aneurysmal SAH. Another difference is that typically aneurysmal SAH is located in the suprasellar, ambiens, middle cerebral artery cistern, and interhemispheric fissure, while traumatic SAH is located peripherally. SAH is associated with secondary complications such as vasospasm, which brings to ischemic infarction; communicating hydrocephalus, which is related to the engorgement of the arachnoid villi with phagocytized blood cells; and impaired absorption of cerebrospinal fluid.

27.3.5 Contusion

Cortical contusion is the most common parenchymal injury. Contusions occur on the gyral surface with possible extension to the subjacent white matter. Pathologically, they are characterized by petechial hemorrhage, due to the high vascularity of the gray matter, with surrounding edema. Rarely the hemorrhage can extend also superficially toward the subarachnoid space (Provenzale, 2007).

The contusion occurring at the site of the impact is called *coup contusion*, whereas a contrecoup injury is located 180° from the impact site. Contusions are more probable to occur where the bone is irregular and present roughened edges, such as in the anterior cranial fossa, where the frontal lobe may be contused by sliding on the cribriform plate, along the petrous ridges, and in the greater sphenoid wings, where the inferior temporal lobe is prone to contusions. A particular kind of contusion is the gliding contusion, typically occurring in the superior parasagittal frontal lobes. Also, a contusion can occur if the brain hits against the falx or the tentorium (Figure 27.2). Contusions are more conspicuous on MRI than on CT, especially when they are not hemorrhagic; also, MRI has a higher sensitivity for the detection of contusions in the basal frontal and temporal lobes, very common sites as previously noted and very well seen using coronal sequences (Hahnel et al., 2008). The edema may increase for several days after trauma, contributing to an increase in the intracranial pressure. After recovery, the lesion shrinks, leading to the formation of encephalomalacia with gliosis (Figure 27.3).

27.3.6 Diffuse Axonal Injury

DAI is a type of TBI commonly associated with loss of consciousness and vegetative state. DAI is a shearing injury that develops when the skull undergoes a rapid

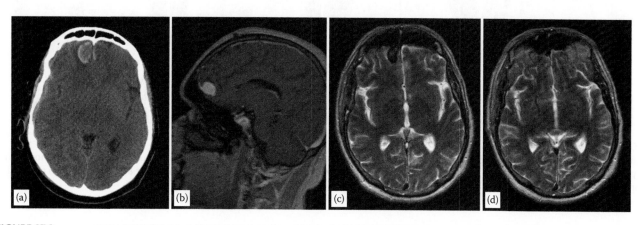

FIGURE 27.2
Typical appearance of an intraparenchymal hematoma related to contusion against the falx. Notice the hematoma on the parasagittal frontal lobe seen as hyperdensity on CT (a) and as a *hyperintense on T_1* (b) *hypointense on T_2* lesion (axial contiguous slices c and d).

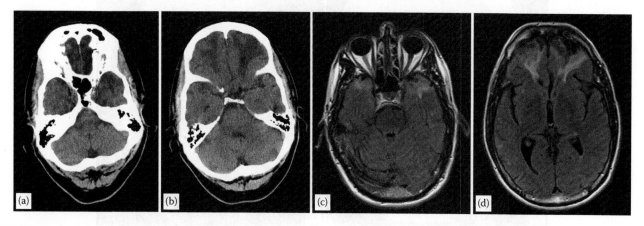

FIGURE 27.3
Chronic traumatic brain injury. A patient with a history of brain trauma 8 years before this scan was acquired. CT scan (a and b) demonstrates extensive low attenuation changes involving the frontal basal cortex and subcortical white matter with associated volume loss. Similar changes are noted in the temporal poles (a). The findings are compatible with posttraumatic encephalomalacia. FLAIR images from an MRI (c and d) confirm the above-mentioned findings.

rotation, causing some parts of the brain to accelerate or decelerate faster than other regions; this results in axonal stretching, edema, and DAI (Li & Feng, 2009). Some anatomical structures are electively involved in the case of DAI: the gray–white junction, because the fibers have different environment and hence different inertial force; in the body and in the splenium of the corpus callosum, which hit the falx cerebri during the movement of the brain; in the dorsolateral midbrain, where the superior cerebellar peduncles are located, in relation to the contact with the tentorial notch. CT is not sensitive to DAI and CT scan is often normal in these patients (Figure 27.4); CT is positive only when hemorrhages are present together with the DAI. In these cases, CT shows small regions of hyperattenuation surrounded by a small amount of edema in the above-mentioned locations. But, since DAI is often (80% of the times) non-hemorrhagic, DAI is one of the few brain traumatic conditions for which MRI is recommended over CT (Li & Feng, 2009) (Figure 27.5); gradient-echo sequence is the most sensitive to evaluate hemorrhagic foci (Figure 27.6).

27.3.7 Vascular Traumatic Injury

Arterial dissection in the setting of trauma is usually extracranial and caused by a fracture through the carotid canal at the base of the skull (internal carotid artery)

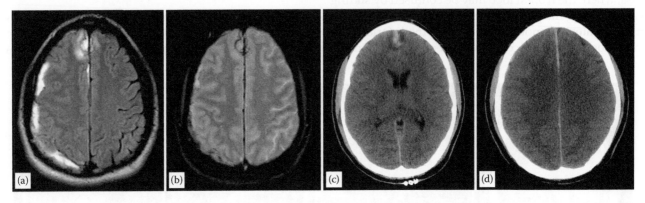

FIGURE 27.4
Contusion along the falx cerebri. A patient with rotational trauma demonstrates a hyperdense contusion focus in the bilateral anterior medial aspect of the frontal lobes, adjacent to the falx cerebri as seen on the CT (a and b) and on the MRI (FLAIR on c and gradient-echo on d). This happens when the brain hits against the falx.

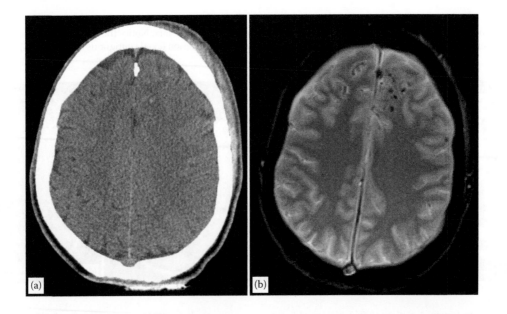

FIGURE 27.5
Low sensitivity of CT for DAI. A patient falls into coma after a major trauma. CT (a) demonstrates soft tissue swelling of the left frontal region and some small foci of hyperdensity within the left frontal lobe. MRI (gradient-echo, b) demonstrates multiple additional foci of susceptibility, more conspicuous compared to the CT.

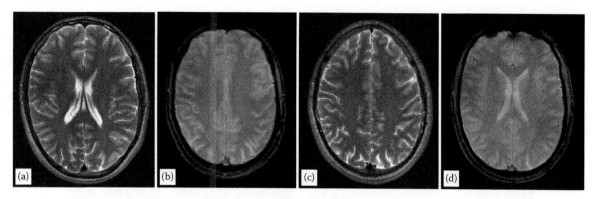

FIGURE 27.6
Gradient-echo to visualize hemorrhagic foci. A patient with DAI and multiple hemorrhagic foci within the brain parenchyma not visible on T_2 (a and b) but visualized on the gradient-echo sequence as multiple small black dots (c and d).

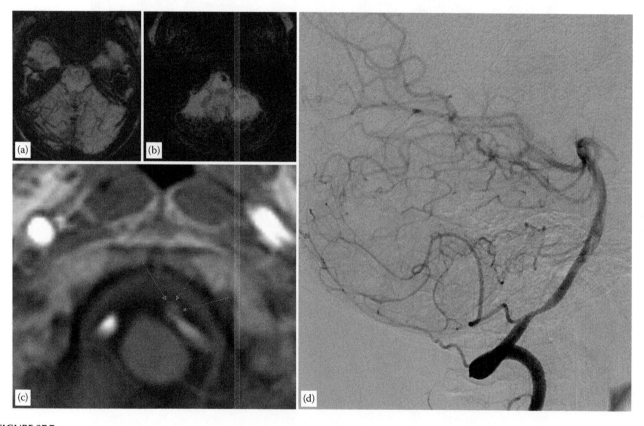

FIGURE 27.7
Patient complaining of headache for a few weeks after a traumatic fall. SWI images demonstrate the presence of susceptibility in the folia of the cerebellum (a) consistent with superficial siderosis, expression of prior subarachnoid bleeding. SWI axial image at the level of the medulla (b) demonstrate a focal area of susceptibility in the wall of the left vertebral artery. The T_1 without contrast (c) demonstrates T_1 bright signal at the same region within the wall of this vessel (arrows). The findings are consistent with a vascular mural hematoma related to vessel dissection. Angiography (d) was done demonstrating stenosis of the vessel without evidence of flap or extravasation of contrast.

and a cervical fracture through the transverse foramina (vertebral artery). Sometimes, the dissection can be intracranial, in this case most commonly involving the supraclinoid internal carotid artery. The bifurcation of the anterior cerebral artery in pericallosal and callosomarginal arteries is a highly vulnerable site to vascular injury because of the anatomic relationship with the anterior aspect of the falx cerebri (Soria, Paroski, & Schamann, 1988).

Traumatic intracranial aneurysms are usually pseudo-aneurysm (false aneurysms), in which there is a rupture of the artery wall layers, with the surrounding hematoma preventing blood extravasation (Figure 27.7) (Acosta, Williams, & Clark, 1972). The intracranial compartment into which the hemorrhage occurs depends on the location of the segment of the vessel involved, and can hence be either extradural (middle meningeal artery) or subarachnoid. The risk of hemorrhage in traumatic pseudoaneurysms is around 20%, and the peak incidence of rupture is 2 weeks after trauma (Cohen et al., 2008).

Another potential vascular complication of head trauma is the carotid cavernous fistula (CCF), which is a pathologic communication between the internal carotid artery and the cavernous sinus (Figure 27.8). The trauma-related CCF is usually a direct communication, while the indirect type, in which dural branches communicate with the cavernous sinus, is related to other etiologies, such as hypertension, collagen vascular disease, and atherosclerosis. Clinically, a CCF manifests with pulsating exophthalmos, orbital bruit, motility disturbance,

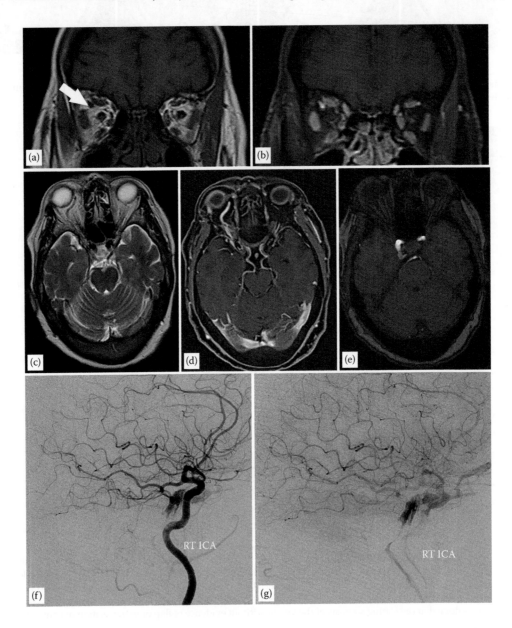

FIGURE 27.8
A patient with history of right orbital pain. The right superior ophthalmic vein (a) is prominent (arrow). The right extraocular muscles are slightly enlarged (b). The right cavernous sinus is slightly prominent (b). The T_2 images (c,d) demonstrate focal complex flow void in the region of the junction of the precavernous and cavernous segments of the right internal carotid artery, within the posterior aspect of the right cavernous sinus. Corresponding flow-related enhancement is noted on the time-of-flight MRA (e). The findings are consistent with a carotid cavernous fistula. Digital subtraction angiography of the intracranial right internal carotid circulation in the lateral projection (f), arterial phase images (g), are notable for a focal region of arteriovenous shunting localized to the right cavernous sinus consistent with an intracranial dural arteriovenous fistula. The lesion is supplied from the right internal carotid artery through the medial division of the meningohypophyseal trunk arising from the cavernous segment of the right internal carotid artery and drains from the posterior aspect of the right cavernous sinus, contralaterally through the left inferior petrosal vein and anteriorly through an enlarged right superior ophthalmic vein.

chemosis, glaucoma, and ultimately vision loss. Notably, CCF is a treatable entity, with endovascular embolization or surgical therapy. CT/CTA findings include proptosis with enlargement of the extraocular muscle and enlargement of the superior ophthalmic vein. The enlargement of the cavernous sinus is sometimes seen. With the use of CTA, the early opacification of the cavernous sinus can be appreciated during the arterial phase. CTA is particularly useful in treatment planning by precisely identifying the location of the fistula relative to surrounding anatomical structures.

27.3.8 Nonaccidental Trauma

Nonaccidental trauma is the radiology term for child abuse. Attention should be drawn to skull fractures, which, in this setting, result from contact injury. Fractures are usually depressed, cross the midline, bilateral, and involve the occiput. Shaken baby syndrome is a triad of SDH, retinal hemorrhage, and brain edema caused by intentional shaking, which is often fatal and can cause severe brain damage (Figure 27.9). There is often little evidence of external injury. Retinal hemorrhages are intraretinal or preretinal, whereas the SDHs are often interhemispheric and of different ages. Interhemispheric SDH is an early and a specific finding in shaken baby syndrome (Oehmichen, Meissner, & Saternus, 2005). Strangulation can cause hypoxia, producing distinctive CT findings (Figure 27.9): loss of gray–white matter differentiation and diffuse hypodensity of the cortex with frequent preservation of the basal ganglia and cerebellum (white cerebellum sign). Other than being helpful for the diagnosis, imaging provides evidence for potential forensic investigation (Hoskote, Richards, Anslow, & McShane, 2002; Jaspan, Griffiths, McConachie, & Punt, 2003).

27.4 Functional Neuroimaging in the Setting of TBI

Multiple conventional and unconventional MRI sequences have been exploited for the evaluation of brain trauma. Among these, SWI and DWI/DTI have the bigger potential to give extra information.

27.4.1 Susceptibility-Weighted Imaging

SWI is a new neuroimaging technique, based on a three-dimensional gradient-echo sequence, that creates a contrast different from T_1, T_2, and GRE-T_2* (Haacke, Xu, Cheng, & Reichenbach, 2004); using magnitude and phase information, different maps are created, obtaining complementary information to conventional MRI; this sequence is strongly sensitive to magnetic susceptibility effects and, for this purpose, is significantly more sensitive than GRE-T_2* gradient-echo sequences (Haacke et al., 2004). The neuroimaging evaluation of DAI with CT or conventional MRI, is often unsatisfying, due to the lack of sensibility of these techniques in the detection of the deep white matter punctate hemorrhages. SWI, as opposed, maximizes the magnetic susceptibility differences determined by the presence of punctate hemorrhages, as observable in the phase maps. In the setting of DAI, this is often gained by the GRE-T_2 sequence, but neuropathological studies demonstrated that DAI lesions are much more widespread as seen with microscopic analysis (Adams et al., 1989). Babikian et al. (2005) put together SWI data at baseline with long-term neuropsychological outcome results of various parameters, for example, intelligence, attention, and memory academic achievement. To obtain these results, they studied a cohort of 18 patients

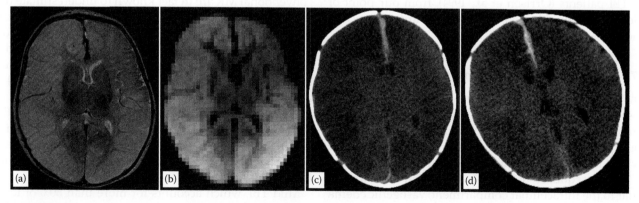

FIGURE 27.9
Child abuse, CT, and MRI. A 7-week-old infant status post strangulation. Unenhanced CT (a and b) demonstrates interhemishperic subdural hematoma (early and specific finding of nonaccidental trauma) extensive bilateral loss of the gray–white matter differentiation and diffuse sulcal effacement indicating diffuse hypoxic/ischemic brain injury, indicating strangulation. There is relative sparing of the basal ganglia and a few portions of both cerebral hemispheres. Subdural hemorrhage is noted in the inter-hemispheric fissure and subarachnoid hemorrhage in the medial right frontal sulcus. MRI with T_2 (c) and DWI (d) sequences demonstrates the cortical swelling with diffusion restriction, thus confirming the hypoxic-ischemic injury.

after accidental trauma; significant correlation was found between number of lesions on SWI, total volume of lesions on SWI, and neuropsychological test results; for example, lesion burden was found to be inversely correlated with intelligent quotient. As expected, a strong correlation was found between the amount of lesions in specific brain regions and the specific neuropsychological domains deficits. Chastain et al. (2009) compared CT, T_2WI, FLAIR, and SWI to identify which modality best predicted clinical outcome after trauma; hence, they selected 38 adult patients and by using Glasgow Outcome Score as clinical tool, they compared lesion number and lesion volume. According to their data, FSE-T_2 and T_2-FLAIR best-predicted poor and good outcomes; SWI was more sensitive in detecting lesions, but had a lower efficacy in predicting the outcome. As known, the involvement of brain stem is an important predictor of TBI sequelae (Mannion et al., 2007); in this context, it is worth considering that, although there are no studies demonstrating the difference in sensitivity of GRE and SWI in the brain stem, the capability of this sequence is shown in some SWI case series.

Putting these data together, SWI seems to have an important role in providing a prognostic marker in patients after trauma in the pediatric population. In adults, the limited utility in predicting prognosis of TBI can be due to the fact that edema and non-hemorrhagic contusion may have greater impact on prognosis than they have on pediatric patients.

SWI detects significantly more lesions, but the possibility to give useful prognostic information in adult population after trauma is still a matter of debate. Given these limitations in the adult population, the presence of SWI visible lesions and quantification of SWI lesion burden in TBI can be a valuable tool for clinician in giving the degree of injury; when an MR has to be performed in TBI patients (Orrison et al., 1994), from our point of view, SWI should be considered a standard sequence in the protocol.

27.4.2 DWI and Diffusion Tensor Imaging

DWI is a noninvasive functional MRI sequence that represents the random movement of water molecules in the brain. It allows differentiation between cytotoxic and vasogenic edema (Schaefer, Grant, & Gonzalez, 2000): cytotoxic edema is classically present in ischemia and is the consequence of shift of water from the extracellular to the intracellular compartment; vasogenic edema is determined by the shift of water from the vessels toward the extracellular space (Schaefer et al., 2000).

In DAI, both cytotoxic and vasogenic edema have been described; hence DWI may enable differentiating those two types of edema in trauma patients (Hergan, Schaefer, Sorensen, Gonzalez, & Huisman, 2002). This can be important because, while cytotoxic is described as irreversible, vasogenic edema is typically considered a reversible damage. The cause of restriction of water associated with TBI is thought to be related to (a) a failure of energy metabolism and the consequent membrane pump failure and (b) a reduction in the extracellular space volume caused by cell swelling (Hergan et al., 2002). Other than revealing the different nature of the injury-related lesions, DWI reveals more DAI lesions than fast spin-echo T_2-weighted or GRE T_2*-weighted images in patients after trauma; DWI is however less sensitive than GRE T_2-weighted in detecting microhemorrhagic lesions (Huisman, Sorensen, Hergan, Gonzalez, & Schaefer, 2003). Because of this higher sensitivity to the foci of acute shearing injury, DWI is particularly useful for the detection of DAI. When trying to correlate DWI lesion burden with clinical outcome, Schaefer et al. (2000) revealed that the total volume of DWI hyperintense lesions has the strongest correlation with the subacute Rankin scale at discharge and, among the MR sequences, has the strongest correlation with acute GCS score.

The integrity of white matter tracts can be evaluated with DTI. DTI is a three-dimensional technique and is considered an evolution of DWI (Mori & Zhang, 2006). The technical aspects of DTI are beyond the scope of this chapter. DTI can be used to demonstrate, in vivo, the diffusion properties of tissues; using DTI obtained maps of fractional anisotropy (FA), it is also possible to track fibers and obtain the representation of white matter bundles (Mori & Zhang, 2006). Earlier, DTI was considered a valuable tool in evaluating white matter; in particular, due to the frequent lack of correlation between clinical status and lesion burden, DTI and its derived parameters, overall FA, and mean diffusivity were found to be suitable for studying the white matter in DAI; CT and conventional MR are in fact known to underestimate the extent of DAI and to poorly correlate with final clinical outcome. Histopathological correlation of FA alteration and DAI damage was demonstrated in a mouse model of TBI. Many studies focused on determining the value of the FA measured with an MR performed early—24 h after trauma—as a predictor of outcome after trauma at 6 or 12 months. Arfanakis et al. (2002) compared the FA values in a cohort of trauma patients first demonstrating that FA values were significantly abnormal immediately after trauma, understanding the potential role of this tool.

Mild TBI, which accounts for more than 80% of total amount of head trauma, demonstrates a faint correlation between post-concussional syndrome and MR findings; this has led our group to try to find correlations by using DTI (Inglese et al., 2005). By using the whole-brain DTI histogram-derived measures, no differences in any of the histogram-derived measures were found between patients and controls. A significant reduction in FA was found in mTBI corpus callosum, internal capsule, and centrum semiovale, and there were significant increases of mean diffusivity in the corpus callosum and internal capsule. While no abnormalities could be detected in the whole-brain analysis, some brain areas, namely those most susceptible to DAI lesions, showed abnormalities in FA and mean diffusivity.

The two most common dysfunctions after mild TBI are attention and memory, and Niogi et al. (2008) demonstrate a clinico-anatomical correlation in a cohort of patients: ROI drawn on DTI maps in the anterior corona radiata and the uncinate fasciculus correlated with neuropsychological tests.

Voxel-based approach, only recently implemented, allowed to repeat measures of DTI parameters without a priori hypothesis; with that in mind, the brain region damaged in a subgroup of TBI patients with unfavorable outcome was studied and the FA in inferior longitudinal fasciculus, cerebral peduncle, posterior limb of internal capsule, and splenium of corpus callosum were the regions involved (Perlbarg et al., 2009).

These reported studies though were cross-sectional, without identifying the long-term evolution of the FA abnormalities; Sidaros et al. (2008), in a prospective DTI-based longitudinal study, examined 30 patients with severe TBI in the late subacute phase and after 12 months on average; other than the alteration of baseline FA values, in line with other studies, FA of patients with unfavorable outcome deviated more from controls than of patients with better outcome.

27.5 Summary

This chapter described the different types of intracranial injuries that can be seen in patients after head trauma. Discussion of the different entities associated with brain trauma was followed by the description of how MRI can help in the diagnosis and the follow-up of patients after TBI.

References

Acosta, C., Williams, P. E., Jr., & Clark, K. (1972). Traumatic aneurysms of the cerebral vessels. *J Neurosurg*, 36(5), 531–536. doi: 10.3171/jns.1972.36.5.0531.

Adams, J. H., Doyle, D., Ford, I., Gennarelli, T. A., Graham, D. I., & McLellan, D. R. (1989). Diffuse axonal injury in head injury: Definition, diagnosis and grading. *Histopathology*, 15(1), 49–59.

Al-Nakshabandi, N. A. (2001). The swirl sign. *Radiology*, 218(2), 433. doi: 10.1148/radiology.218.2.r01fe09433.

Arfanakis, K., Haughton, V. M., Carew, J. D., Rogers, B. P., Dempsey, R. J., & Meyerand, M. E. (2002). Diffusion tensor MR imaging in diffuse axonal injury. *AJNR Am J Neuroradiol*, 23(5), 794–802.

Babikian, T., Freier, M. C., Tong, K. A., Nickerson, J. P., Wall, C. J., Holshouser, B. A., ... Ashwal, S. (2005). Susceptibility weighted imaging: Neuropsychologic outcome and pediatric head injury. *Pediatr Neurol*, 33(3), 184–194. doi: 10.1016/j.pediatrneurol.2005.03.015.

Brown, A. W., Elovic, E. P., Kothari, S., Flanagan, S. R., & Kwasnica, C. (2008). Congenital and acquired brain injury. 1. Epidemiology, pathophysiology, prognostication, innovative treatments, and prevention. *Arch Phys Med Rehabil*, 89(3 Suppl 1), S3–S8. doi: 10.1016/j.apmr.2007.12.001.

Bullock, M. R., Chesnut, R., Ghajar, J., Gordon, D., Hartl, R., Newell, D. W., ... Surgical Management of Traumatic Brain Injury Author Group. (2006). Surgical management of depressed cranial fractures. *Neurosurgery*, 58(3 Suppl), S56–S60; discussion Si–Siv. doi: 10.1227/01.NEU.0000210367.14043.0E.

Chastain, C. A., Oyoyo, U. E., Zipperman, M., Joo, E., Ashwal, S., Shutter, L. A., & Tong, K. A. (2009). Predicting outcomes of traumatic brain injury by imaging modality and injury distribution. *J Neurotrauma*, 26(8), 1183–1196. doi: 10.1089/neu.2008.0650.

Ciurea, A. V., Gorgan, M. R., Tascu, A., Sandu, A. M., & Rizea, R. E. (2011). Traumatic brain injury in infants and toddlers, 0-3 years old. *J Med Life*, 4(3), 234–243.

Cohen, J. E., Gomori, J. M., Segal, R., Spivak, A., Margolin, E., Sviri, G., ... Spektor, S. (2008). Results of endovascular treatment of traumatic intracranial aneurysms. *Neurosurgery*, 63(3), 476–485; discussion 485–476. doi: 10.1227/01.NEU.0000324995.57376.79.

Davis, P. C., Drayer, B. P., Anderson, R. E., Braffman, B., Deck, M. D., Hasso, A. N., ... Masdeu, J. C. (2000). Head trauma. American College of Radiology. ACR Appropriateness Criteria. *Radiology*, 215(Suppl), 507–524.

Gean, A. D., Fischbein, N. J., Purcell, D. D., Aiken, A. H., Manley, G. T., & Stiver, S. I. (2010). Benign anterior temporal epidural hematoma: Indolent lesion with a characteristic CT imaging appearance after blunt head trauma. *Radiology*, 257(1), 212–218. doi: 10.1148/radiol.10092075.

Gentry, L. R. (1994). Imaging of closed head injury. *Radiology*, 191(1), 1–17. doi: 10.1148/radiology.191.1.8134551.

Gilad, R., Fatterpekar, G. M., Johnson, D. M., & Patel, A. B. (2007). Migrating subdural hematoma without subarachnoid hemorrhage in the case of a patient with a ruptured aneurysm in the intrasellar anterior communicating artery. *AJNR Am J Neuroradiol*, 28(10), 2014–2016. doi: 10.3174/ajnr.A0726.

Gomori, J. M., Grossman, R. I., Goldberg, H. I., Zimmerman, R. A., & Bilaniuk, L. T. (1985). Intracranial hematomas: Imaging by high-field MR. *Radiology*, 157(1), 87–93. doi: 10.1148/radiology.157.1.4034983.

Haacke, E. M., Xu, Y., Cheng, Y. C., & Reichenbach, J. R. (2004). Susceptibility weighted imaging (SWI). *Magn Reson Med*, 52(3), 612–618. doi: 10.1002/mrm.20198.

Hahnel, S., Stippich, C., Weber, I., Darm, H., Schill, T., Jost, J., … Meyding-Lamade, U. (2008). Prevalence of cerebral microhemorrhages in amateur boxers as detected by 3T MR imaging. *AJNR Am J Neuroradiol*, 29(2), 388–391. doi: 10.3174/ajnr.A0799.

Hellwig, D., Kuhn, T. J., Bauer, B. L., & List-Hellwig, E. (1996). Endoscopic treatment of septated chronic subdural hematoma. *Surg Neurol*, 45(3), 272–277.

Hergan, K., Schaefer, P. W., Sorensen, A. G., Gonzalez, R. G., & Huisman, T. A. (2002). Diffusion-weighted MRI in diffuse axonal injury of the brain. *Eur Radiol*, 12(10), 2536–2541. doi: 10.1007/s00330-002-1333-2.

Hoskote, A., Richards, P., Anslow, P., & McShane, T. (2002). Subdural haematoma and non-accidental head injury in children. *Childs Nerv Syst*, 18(6-7), 311–317. doi: 10.1007/s00381-002-0616-x.

Huisman, T. A., Sorensen, A. G., Hergan, K., Gonzalez, R. G., & Schaefer, P. W. (2003). Diffusion-weighted imaging for the evaluation of diffuse axonal injury in closed head injury. *J Comput Assist Tomogr*, 27(1), 5–11.

Inglese, M., Makani, S., Johnson, G., Cohen, B. A., Silver, J. A., Gonen, O., & Grossman, R. I. (2005). Diffuse axonal injury in mild traumatic brain injury: A diffusion tensor imaging study. *J Neurosurg*, 103(2), 298–303. doi: 10.3171/jns.2005.103.2.0298.

Jaspan, T., Griffiths, P. D., McConachie, N. S., & Punt, J. A. (2003). Neuroimaging for non-accidental head injury in childhood: A proposed protocol. *Clin Radiol*, 58(1), 44–53.

Kubal, W. S. (2012). Updated imaging of traumatic brain injury. *Radiol Clin North Am*, 50(1), 15–41. doi: 10.1016/j.rcl.2011.08.010.

Li, X. Y., & Feng, D. F. (2009). Diffuse axonal injury: Novel insights into detection and treatment. *J Clin Neurosci*, 16(5), 614–619. doi: 10.1016/j.jocn.2008.08.005.

Mannion, R. J., Cross, J., Bradley, P., Coles, J. P., Chatfield, D., Carpenter, A., … Hutchinson, P. J. (2007). Mechanism-based MRI classification of traumatic brainstem injury and its relationship to outcome. *J Neurotrauma*, 24(1), 128–135. doi: 10.1089/neu.2006.0127.

Mori, S., & Zhang, J. (2006). Principles of diffusion tensor imaging and its applications to basic neuroscience research. *Neuron*, 51(5), 527–539. doi: 10.1016/j.neuron.2006.08.012.

Niogi, S. N., Mukherjee, P., Ghajar, J., Johnson, C. E., Kolster, R., Lee, H., … McCandliss, B. D. (2008). Structural dissociation of attentional control and memory in adults with and without mild traumatic brain injury. *Brain*, 131(Pt 12), 3209–3221. doi: 10.1093/brain/awn247.

Oehmichen, M., Meissner, C., & Saternus, K. S. (2005). Fall or shaken: Traumatic brain injury in children caused by falls or abuse at home—A review on biomechanics and diagnosis. *Neuropediatrics*, 36(4), 240–245. doi: 10.1055/s-2005-872812.

Orrison, W. W., Gentry, L. R., Stimac, G. K., Tarrel, R. M., Espinosa, M. C., & Cobb, L. C. (1994). Blinded comparison of cranial CT and MR in closed head injury evaluation. *AJNR Am J Neuroradiol*, 15(2), 351–356.

Penchet, G., Loiseau, H., & Castel, J. P. (1998). [Chronic bilateral subdural hematomas]. *Neurochirurgie*, 44(4), 247–252.

Perlbarg, V., Puybasset, L., Tollard, E., Lehericy, S., Benali, H., & Galanaud, D. (2009). Relation between brain lesion location and clinical outcome in patients with severe traumatic brain injury: A diffusion tensor imaging study using voxel-based approaches. *Hum Brain Mapp*, 30(12), 3924–3933. doi: 10.1002/hbm.20817.

Provenzale, J. (2007). CT and MR imaging of acute cranial trauma. *Emerg Radiol*, 14(1), 1–12. doi: 10.1007/s10140-007-0587-z.

Schaefer, P. W., Grant, P. E., & Gonzalez, R. G. (2000). Diffusion-weighted MR imaging of the brain. *Radiology*, 217(2), 331–345. doi: 10.1148/radiology.217.2.r00nv24331.

Sidaros, A., Engberg, A. W., Sidaros, K., Liptrot, M. G., Herning, M., Petersen, P., … Rostrup, E. (2008). Diffusion tensor imaging during recovery from severe traumatic brain injury and relation to clinical outcome: A longitudinal study. *Brain*, 131(Pt 2), 559–572. doi: 10.1093/brain/awm294.

Soria, E. D., Paroski, M. W., & Schamann, M. E. (1988). Traumatic aneurysms of cerebral vessels: A case study and review of the literature. *Angiology*, 39(7 Pt 1), 609–615.

Teasdale, G., & Jennett, B. (1974). Assessment of coma and impaired consciousness. A practical scale. *Lancet*, 2(7872), 81–84.

Tong, K. A., Ashwal, S., Holshouser, B. A., Shutter, L. A., Herigault, G., Haacke, E. M., & Kido, D. K. (2003). Hemorrhagic shearing lesions in children and adolescents with posttraumatic diffuse axonal injury: Improved detection and initial results. *Radiology*, 227(2), 332–339. doi: 10.1148/radiol.2272020176.

Index

Note: Locators followed by '*f*' and '*t*' refer to figures and tables, respectively

For Product Safety Concerns and Information please contact our EU
representative GPSR@taylorandfrancis.com Taylor & Francis Verlag GmbH,
Kaufingerstraße 24, 80331 München, Germany

Printed and bound by CPI Group (UK) Ltd, Croydon, CR0 4YY

02/05/2025

01859401-0001